Handbook of Plastics, Elastomers, and Composites

Other McGraw-Hill Books of Interest

Harper ELECTRONIC MATERIALS AND PROCESSES HANDBOOK

Petrie HANDBOOK OF ADHESIVES AND SEALANTS

Harper MODERN PLASTICS HANDBOOK

Harper HANDBOOK OF CERAMICS, GLASSES, AND DIAMONDS

Harper HANDBOOK OF MATERIALS FOR PRODUCT DESIGN

Harper HANDBOOK OF MATERIALS IN FIRE PROTECTION

Harper, Petrie PLASTIC MATERIALS AND PROCESSES HANDBOOK

Martin ELECTRONIC FAILURE ANALYSIS HANDBOOK: TECHNIQUES AND APPLICATIONS FOR ELECTRONIC AND ELECTRICAL PACKAGES, COMPONENTS, AND ASSEMBLIES

CONTENTS

Cataloging-in-Publication Data is on File with the Library of Congress.

McGraw-Hill

A Division of The McGraw-Hill Companies

1 2 3 4 5 6 7 8 9 0 DOC/DOC 0 8 7 6 5 4 3 2

ISBN 0-07-138476-6

The sponsoring editor for this book was Kenneth P. McCombs and the production supervisor was Pamela Pelton. It was set in Times by J. K. Eckert & Company, Inc.

Printed and bound by R. R. Donnelley & Sons Company.

This book is printed on recycled acid-free paper containing a minimum of 50% recycled, de-inked fiber.

ac

Handbook of Plastics, Elastomers, and Composites

Charles A. Harper Editor-in-Chief
Technology Seminars, Inc.
Lutherville, Maryland

Fourth Edition

McGRAW-HILL
New York Chicago San Francisco
Lisbon London Madrid Mexico City Milan
Montreal New Delhi San Juan Seoul Singapore
Sydney Toronto

PREFACE

Welcome to this new, heavily revised and updated Fourth Edition of *Handbook of Plastics, Elastomers, and Composites.* The continued development of new and improved polymers, and their application in new and improved products, have led to almost unlimited product opportunities. In fact, there are probably few who would not rate this area of product growth as one of the most important in industry growth areas. The impact of polymers-plastics, elastomers, and composites—in all of their material forms—has been little short of phenomenal. New polymers and improvements in established polymer groups regularly extend the performance limits of plastics, elastomers, and composites. These achievements in polymer and plastic technology offer major benefits and opportunities for the myriad of products in which they can be used.

With all these achievements, however, a major impediment exists to the successful use of plastics, elastomers, and composites in products. This impediment is the lack of fundamental understanding of plastics, elastomers, and composites by product designers. Along with this lack of understanding is the absence of a useful consolidated source of information, data, and guidelines that can be practically used by product designers, most of whom do not "speak plastics." The usual practice is to use random supplier data sheets and data tables for guidance. It is, therefore, the object of this handbook to present, in a single source, all of the fundamental information required to understand the large number of materials and material forms, and to provide the necessary data and guidelines for optimal use of these materials and forms in the broad range of industry products. At the same time, this handbook will be invaluable to the plastics industry in acquainting its specialists with product requirements for which they must develop, manufacture, and fabricate plastics materials and forms.

This new Fourth Edition of *Handbook of Plastics, Elastomers, and Composites* has been prepared as a thorough sourcebook of practical data for all ranges of interests. It contains an extensive array of property and performance data, presented as a function of the most important product variables. Further, it presents all important aspects of application guidelines, fabrication-method trade-offs, design, finishing, performance limits, and other important application considerations. It also fully covers chemical, structural, and other basic polymer properties. The handbook's other major features include thorough lists of standards and specifications sources, a completely cross-referenced easy-to-use index, a comprehensive glossary, useful end-of-chapter reference lists, and several appendices containing individual data and information for product engineers.

The chapter organization and coverage of the handbook is equally well suited for reader convenience. The first three chapters present the fundamentals and the important information, data, and guidelines for the three basic material categories of thermoplastics, thermosets, and elastomers, thus enabling readers to more fully understand the presentation in the following chapters. The next four chapters are devoted to major plastic product forms that are so important to product design. The first two of these chapters cover composites, one chapter covering basic and structural composites and one chapter covering the increasingly growing area of composites in electronics. These two chapters are followed by one chapter each on plastics in coatings and finishes, and plastics in adhesives. After this, one chapter very thoroughly covers the critical and important subject of area of join-

ing of plastic parts. The understanding of this design area, almost always a major factor in the quality of plastic products, is most expertly covered in this chapter, which provides excellent guidelines for designers of plastic products.

Next, appropriately following the above listed chapters on basic plastics and plastic forms, a special chapter is devoted to a clearly illustrated presentation of all the important considerations for the design and fabrication of molded plastic products. The following two chapters thoroughly cover the use of plastics and elastomers in two of the largest application fields, namely, automotive and packaging.

The final chapter is an excellent presentation on a subject of increasingly vital importance to all of those in all areas of plastics and elastomers—the recycling of waste products.

The result of these presentations is an extremely comprehensive and complete single reference and text—a must for the desk of anyone involved in any aspect of product design, development, or application of plastics, elastomers, and composites. This handbook will be invaluable for every reference library.

Charles A. Harper
Technology Seminars, Inc.
Lutherville, Maryland

CONTRIBUTORS

Anne-Marie Baker, University of Massachusetts, Lowell, Massachusetts (Chap. 1)

Ralph E. Wright, R. E. Wright Associates, Yarmouth, Maine (Chap. 2)

Ruben Hernandez, University of Michigan, School of Packaging, East Lansing, Michigan (Chap. 11)

John L. Hull, Hull/Finmac, Inc., Warminster, Pennsylvania (Chap. 9)

Carl P. Izzo, Industry Consultant, Export, Pennsylvania (Chap. 6)

James Margolis, Industry Consultant, Montreal, Quebec, Canada (Chaps. 3, 10)

Joey Mead, University of Massachusetts, Lowell, Massachusetts (Chap. 1)

Stanley T. Peters, Process Research, Mountain View, California (Chap. 4)

Edward M. Petrie, ABB Transmission Technology Institute, Raleigh, North Carolina (Chaps. 7, 8)

Susan E. Selke, University of Michigan, School of Packaging, East Lansing, Michigan (Chap. 11)

Carl H. Zweben, Industry Consultant, Devon, Pennsylvania (Chap. 5)

ABOUT THE EDITOR

Charles A. Harper is President of Technology Seminars, Inc., an organization located in Lutherville, MD, devoted to providing educational training courses on an industry-wide basis. He has had an esteemed career both in industry and teaching. Mr.Harper is Series Editor for both the Materials Science and Engineering Series, and the Electronic Packaging and Interconnecting Series, published by McGraw-Hill. He serves on the Advisory Board for several professional and business organizations. He is a graduate of The Johns Hopkins University, where he has also served as Adjunct Professor. Mr. Harper has had active and leadership roles in several plastics, materials, and electronic packaging professional societies, and holds the honorary level of Fellow in both the Society for the Advancement of Materials and Process Engineering (SAMPE) and the International Microelectronics and Packaging Society (IMAPS), for which he is also a Past President.

Chapter

1

Thermoplastics

Anne-Marie M. Baker

Joey Mead

Plastics Engineering Department
University of Massachusetts Lowell
Lowell, Massachusetts

1.1 Introduction

Plastics are an important part of everyday life; products made from plastics range from sophisticated articles, such as prosthetic hip and knee joints, to disposable food utensils. One of the reasons for the great popularity of plastics in a wide variety of industrial applications is the tremendous range of properties exhibited by plastics and their ease of processing. Plastic properties can be tailored to meet specific needs by varying the atomic composition of the repeat structure; and by varying molecular weight and molecular weight distribution. The flexibility can also be varied through the presence of side chain branching and according to the lengths and polarities of the side chains. The degree of crystallinity can be controlled through the amount of orientation imparted to the plastic during processing, through copolymerization, by blending with other plastics, and via the incorporation of an enormous range of additives (fillers, fibers, plasticizers, stabilizers). Given all of the avenues available to pursue in tailoring any given polymer, it is not surprising that the variety of choices available to us today exists.

Polymeric materials have been used since early times, even though their exact nature was unknown. In the 1400s, Christopher Columbus found natives of Haiti playing with balls made from material obtained from a tree. This was natural rubber, which became an important product after Charles Goodyear discovered that the addition of sulfur dramatically improved the properties; however, the use of polymeric materials was still limited to natural-based materials. The first true synthetic polymers were prepared in the early 1900s using phenol and formaldehyde to form resins—Baekeland's Bakelite. Even with the development of synthetic polymers, scientists were still unaware of the true nature of the materials they had prepared. For many years, scientists believed they were colloids—a substance that is an aggregate of molecules. It was not until the 1920s that Herman

Staudinger showed that polymers were giant molecules or *macromolecules*. In 1928, Carothers developed linear polyesters and then polyamides, now known as nylon. In the 1950s, Ziegler and Natta's work on anionic coordination catalysts led to the development of polypropylene; high-density, linear polyethylene; and other stereospecific polymers.

Materials are often classified as metals, ceramics, or polymers. Polymers differ from the other materials in a variety of ways but generally exhibit lower densities, thermal conductivities, and moduli. Table 1.1 compares the properties of polymers to some representative ceramic and metallic materials. The lower densities of polymeric materials offer an advantage in applications where lighter weight is desired. The addition of thermally and/or electrically conducting fillers allows the polymer compounder the opportunity to develop materials from insulating to conducting. As a result, polymers may find application in electromagnetic interference (EMI) shielding and antistatic protection.

TABLE 1.1 Properties of Selected Materials[451]

Material	Specific gravity	Thermal conductivity, (Joule-cm/°C cm^2 s)	Electrical resistivity, μΩ-cm	Modulus MPa
Aluminum	2.7	2.2	2.9	70,000
Brass	8.5	1.2	6.2	110,000
Copper	8.9	4.0	1.7	110,000
Steel (1040)	7.85	0.48	17.1	205,000
Al_2O_3	3.8	0.29	$>10^{14}$	350,000
Concrete	2.4	0.01	–	14,000
Bororsilicate glass	2.4	0.01	$>10^{17}$	70,000
MgO	3.6	–	10^5 (2000°F)	205,000
Polyethylene (H.D.)	0.96	0.0052	10^{14}–10^{18}	350–1,250
Polystyrene	1.05	0.0008	10^{18}	2,800
Polymethyl methacrylate	1.2	0.002	10^{16}	3,500
Nylon	1.15	0.0025	10^{14}	2,800

Polymeric materials are used in a vast array of products. In the automotive area, they are used for interior parts and in under-the-hood applications. Packaging applications are a large area for thermoplastics, from carbonated beverage bottles to plastic wrap. Application requirements vary widely, but, luckily, plastic materials can be synthesized to meet these varied service conditions. It remains the job of the part designer to select from the array of thermoplastic materials available to meet the required demands.

1.2 Polymer Structure and Synthesis

A polymer is prepared by stringing together a series of low-molecular-weight species (such as ethylene) into an extremely long chain (polyethylene), much as one would string together a series of bead to make a necklace (see Fig. 1.1). The chemical characteristics of

the starting low-molecular-weight species will determine the properties of the final polymer. When two different low-molecular-weight species are polymerized the resulting polymer is termed a copolymer such as ethylene vinylacetate. This is depicted in Fig. 1.2. Plastics can also be separated into thermoplastics and thermosets. A thermoplastic material is a high-molecular-weight polymer that is not cross-linked. It can exist in either a linear or a branched structure. Upon heating, thermoplastics soften and melt, which allows them to be shaped using plastics processing equipment. A thermoset has all of the chains tied together with covalent bonds in a three dimensional network (cross-linked). Thermoset materials will not flow once cross-linked, but a thermoplastic material can be reprocessed simply by heating it to the appropriate temperature. The different types of structures are shown in Fig. 1.3. The properties of different polymers can vary widely; for example, the modulus can vary from 1 MPa to 50 GPa. Properties can be varied for each individual plastic material as well, simply by varying the microstructure of the material.

There are two primary polymerization approaches: step-reaction polymerization and chain-reaction polymerization.[1] In step-reaction (also referred to as *condensation polymerization*), reaction occurs between two polyfunctional monomers, often liberating a small molecule such as water. As the reaction proceeds, higher-molecular-weight species

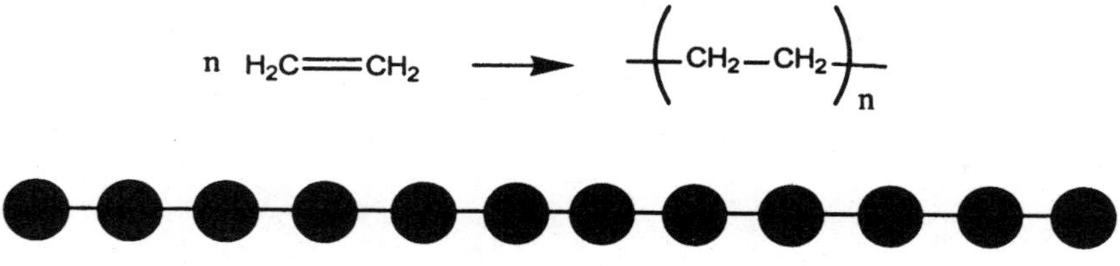

Figure 1.1 Polymerization.

Figure 1.2 Copolymer structure.

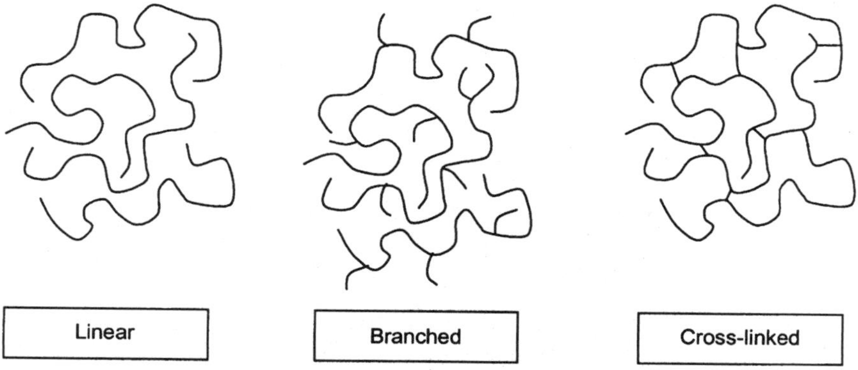

Figure 1.3 Linear, branched, and cross-linked polymer structures.

are produced as longer and longer groups react together. For example, two monomers can react to form a dimer, then react with another monomer to form a trimer. The reaction can be described as *n*-mer + *m*-mer → (*n* + *m*)mer, where *n* and *m* refer to the number of monomer units for each reactant. Molecular weight of the polymer builds up gradually with time, and high conversions are usually required to produce high-molecular-weight polymers. Polymers synthesized by this method typically have atoms other than carbon in the backbone. Examples include polyesters and polyamides.

Chain-reaction polymerizations (also referred to as *addition polymerizations*) require an initiator for polymerization to occur. Initiation can occur by a free radical or an anionic or cationic species, which opens the double bond of a vinyl monomer and the reaction proceeds as shown above in Fig. 1.1. Chain-reaction polymers typically contain only carbon in their backbone and include such polymers as polystyrene and polyvinyl chloride.

Unlike low-molecular-weight species, polymeric materials do not possess one unique molecular weight but rather a distribution of weights as depicted in Fig. 1.4. Molecular weights for polymers are usually described by two different average molecular weights, the number average molecular weight, $\overline{M_n}$, and the weight average molecular weight, $\overline{M_w}$. These averages are calculated using the equations below:

$$\overline{M_n} = \sum_{i=1}^{\infty} \frac{n_i M_i}{n_i}$$

$$\overline{M_w} = \sum_{i=1}^{\infty} \frac{n_i M_i^2}{n_i M_i}$$

where n_i is the number of moles of species *i*, and M_i is the molecular weight of species *i*. The processing and properties of polymeric materials are dependent on the molecular weights of the polymer.

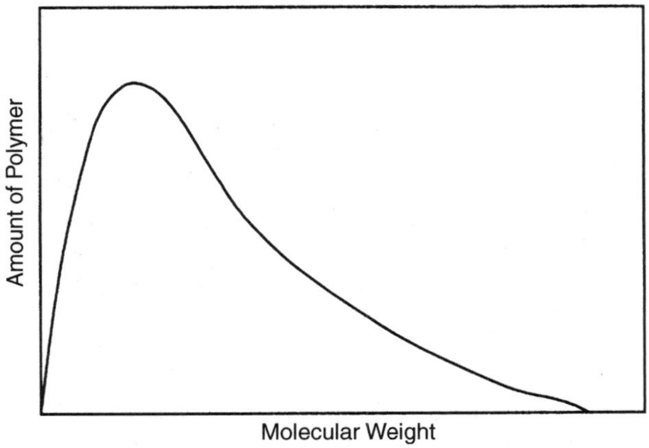

Figure 1.4 Molecular weight distribution.

1.3 Solid Properties of Polymers

1.3.1 Glass Transition Temperature (T_g)

Polymers come in many forms, including plastics, rubber, and fibers. Plastics are stiffer than rubber yet have reduced low-temperature properties. Generally, a plastic differs from a rubbery material due to the location of its glass transition temperature (T_g), which is the temperature at which the polymer behavior changes from glassy to leathery. A plastic has a T_g above room temperature, whereas a rubber has a T_g below room temperature. T_g is most clearly defined by evaluating the classic relationship of elastic modulus to temperature for polymers as presented in Fig. 1.5.

At low temperatures, the material can best be described as a glassy solid. It has a high modulus, and behavior in this state is characterized ideally as a purely elastic solid. In this temperature regime, materials most closely obey Hooke's law:

$$\sigma = E\varepsilon$$

where σ is the stress being applied, and ε is the strain. Young's modulus, E, is the proportionality constant relating stress and strain.

In the leathery region, the modulus is reduced by up to three orders of magnitude from the glassy modulus for amorphous polymers. The rubbery plateau has a relatively stable modulus until further temperature increases induce rubbery flow. Motion at this point does not involve entire molecules, but, in this region, deformations begin to become nonrecoverable as permanent set takes place. As temperature is further increased, eventually the onset of liquid flow takes place. There is little elastic recovery in this region, and the flow involves entire molecules slipping past each other. This region models ideal viscous materials, which obey Newton's law as follows:

$$\sigma = \eta\dot{\varepsilon}$$

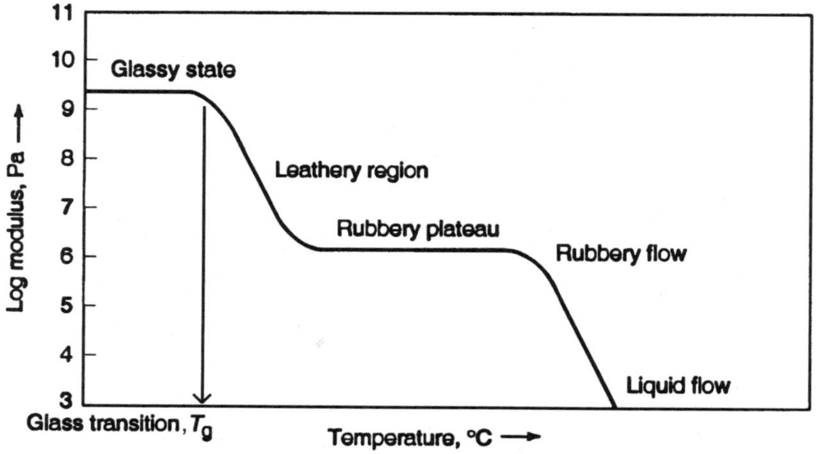

Figure 1.5 Relationship between elastic modulus and temperature.

1.3.2 Crystallization and Melting Behavior, T_m

In its solid form, a polymer can exhibit different morphologies, depending on the structure of the polymer chain as well as the processing conditions. The polymer may exist in a random unordered structure termed *amorphous*. An example of an amorphous polymer is polystyrene. If the structure of the polymer backbone is a regular, ordered structure, then the polymer can tightly pack into an ordered crystalline structure, although the material will generally be only semicrystalline. Examples are polyethylene and polypropylene. The exact makeup and architecture of the polymer backbone will determine whether the polymer is capable of crystallizing. This microstructure can be controlled by different synthetic methods. As mentioned above, the Ziegler-Natta catalysts are capable of controlling the microstructure to produce stereospecific polymers. The types of microstructure that can be obtained for a vinyl polymer are shown in Fig. 1.6. The isotactic and syndiotactic structures are capable of crystallizing because of their highly regular backbone. The atactic form is amorphous.

1.4 Mechanical Properties

The mechanical behavior of polymers is dependent on many factors, including polymer type, molecular weight, and test procedure. Modulus values are obtained from a standard tensile test with a given rate of crosshead separation. In the linear region, the slope of a stress-strain curve will give the elastic or Young's modulus, E. Typical values for Young's modulus are given in Table 1.2. Polymeric material behavior may be affected by other factors such as test temperature and rates. This can be especially important to the designer when the product is used or tested at temperatures near the glass transition temperature

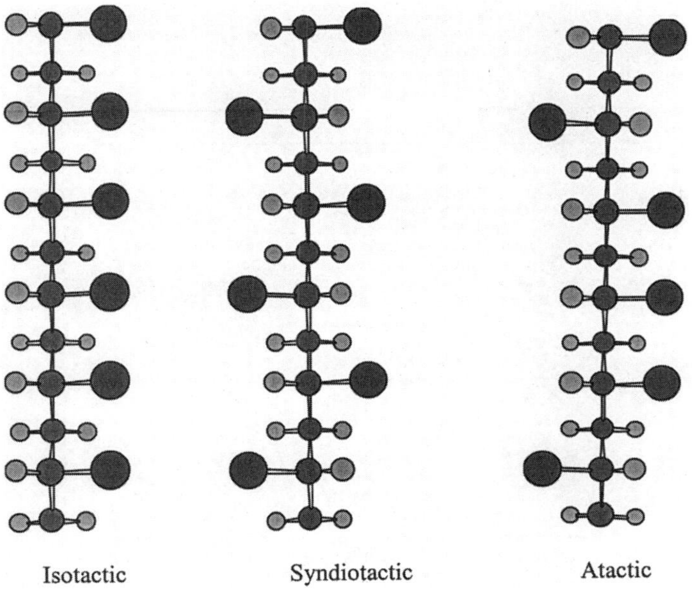

Figure 1.6 Isotactic, syndiotactic, and atactic polymer chains.

TABLE 1.2 Comparative Properties of Thermoplastics[452,453]

Material	Heat deflection temperature @1.82 MPa (°C)	Tensile strength MPa	Tensile modulus GPa	Impact strength J/m	Density g/cm^3	Dielectric strength MV/m	Dielectric constant @ 60 Hz
ABS	99	41	2.3	347	1.18	15.7	3.0
CA	68	37.6	1.26	210	1.30	16.7	5.5
CAB	69	34	.88	346	1.19	12.8	4.8
PTFE		17.1	.36	173	2.2	17.7	2.1
PCTFE		50.9	1.3	187	2.12	22.2	2.6
PVDF	90	49.2	2.5	202	1.77	10.2	10.0
PB	102	25.9	.18	NB*	0.91		2.25
LDPE	43	11.6	.17	NB*	0.92	18.9	2.3
HDPE	74	38.2		373	0.95	18.9	2.3
PMP		23.6	1.10	128	0.83	27.6	
PI		42.7	3.7	320	1.43	12.2	4.1
PP	102	35.8	1.6	43	0.90	25.6	2.2
PUR	68	59.4	1.24	346	1.18	18.1	6.5
PS	93	45.1	3.1	59	1.05	19.7	2.5
PVC–rigid	68	44.4	2.75	181	1.4	34.0	3.4
PVC–flexible		9.6		293	1.4	25.6	5.5
POM	136	69	3.2	133	1.42	19.7	3.7
PMMA	92	72.4	3	21	1.19	19.7	3.7
Polyarylate	155	68	2.1	288	1.19	15.2	3.1
LCP	311	110	11	101	1.70	20.1	4.6
Nylon 6	65	81.4	2.76	59	1.13	16.5	3.8
Nylon 6/6	90	82.7	2.83	53	1.14	23.6	4.0
PBT	54	52	2.3	53	1.31	15.7	3.3
PC	129	69	2.3	694	1.20	15	3.2
PEEK	160	93.8	3.5	59	1.32		
PEI	210	105	3	53	1.27	28	3.2
PES	203	84.1	2.6	75	1.37	16.1	3.5
PET	224	159	9.96	101	1.56	21.3	3.6
PPO (modified)	100	54	2.5	267	1.09	15.7	3.9
PPS	260	138	11.7	69	1.67	17.7	3.1
PSU	174	73.8	2.5	64	1.24	16.7	3.5

*NB = no break.

where dramatic changes in properties occur as depicted in Fig. 1.5. The time-dependent behavior of these materials is discussed below.

1.4.1 Viscoelasticity

Polymer properties exhibit time-dependent behavior, which is dependent on the test conditions and polymer type. Figure 1.7 shows a typical viscoelastic response of a polymer to changes in testing rate or temperature. Increases in testing rate or decreases in temperature cause the material to appear more rigid, while an increase in temperature or decrease in rate will cause the material to appear softer. This time-dependent behavior can also result in long-term effects such as stress relaxation or creep.[2] These two time-dependent behaviors are shown in Fig. 1.8. Under a fixed displacement, the stress on the material will decrease over time, and this is called *stress relaxation.* This behavior can be modeled using a

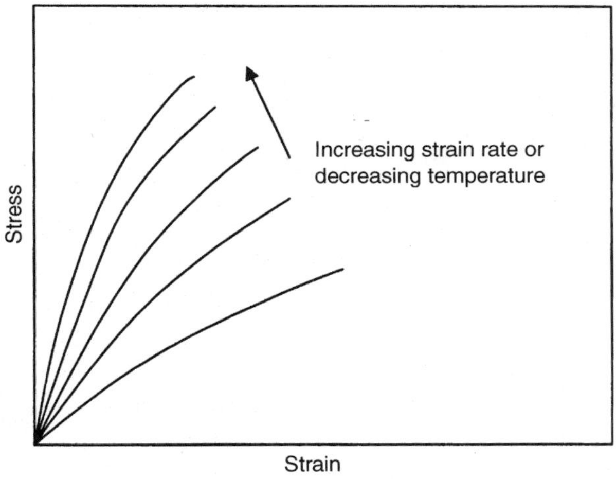

Figure 1.7 Effect of strain rate or temperature on mechanical behavior.

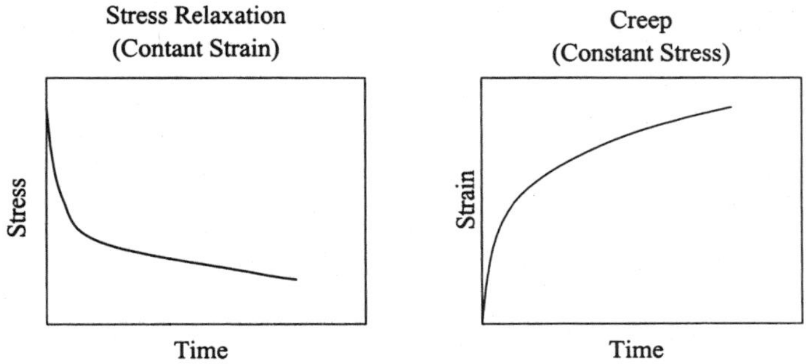

Figure 1.8 Creep and stress relaxation behavior.

spring and dashpot in series as depicted in Fig. 1.9. The equation for the time dependent stress using this model is

$$\sigma(t) = \sigma_o e^{\frac{-t}{\tau}}$$

where τ is the characteristic relaxation time *(η/k)*. Under a fixed load, the specimen will continue to elongate with time, a phenomenon termed *creep,* which can be modeled using a spring and dashpot in parallel as seen in Fig. 1.9. This model predicts the time-dependent strain as

$$\varepsilon(t) = \varepsilon_o e^{\frac{-t}{\tau}}$$

For more accurate prediction of the time-dependent behavior, other models with more elements are often employed. In the design of polymeric products for long-term applications, the designer must consider the time-dependent behavior of the material.

If a series of stress relaxation curves is obtained at varying temperatures, it is found that these curves can be superimposed by horizontal shifts to produce a master curve[3]. This demonstrates an important feature in polymer behavior: the concept of time-temperature equivalence. In essence, a polymer at temperatures below room temperature will behave in a manner as if it were tested at a higher rate at room temperature. This principle can be applied to predict material behavior under testing rates or times that are not experimentally accessible through the use of shift factors *(a_T)* and the equation below:

$$\ln a_T = \ln\left(\frac{t}{t_o}\right) = -\frac{17.44(T - T_g)}{51.6 + T - T_g}$$

where T_g is the glass transition temperature of the polymer, T is the temperature of interest, t_o is the relaxation time at T_g, and t is the relaxation time.

1.4.2 Failure Behavior

Design of plastic parts requires the avoidance of failure without overdesign of the part, which leads to increased part weight. The type of failure can depend on temperatures,

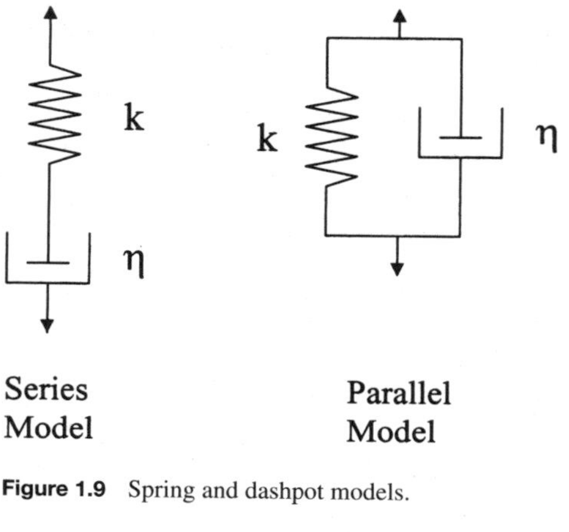

Figure 1.9 Spring and dashpot models.

rates, and materials. Some information on material strength can be obtained from simple tensile stress-strain behavior. Materials that fail at rather low elongations (1% strain or less) can be considered to have undergone brittle failure.[4] Polymers that produce this type of failure include general-purpose polystyrene and acrylics. Failure typically starts at a defect where stresses are concentrated. Once a crack is formed, it will grow as a result of stress concentrations at the crack tip. Many amorphous polymers will also exhibit what are called *crazes*. Crazes look like cracks, but they are load bearing, with fibrils of material bridging the two surfaces as shown in Fig. 1.10. Crazing is a form of yielding that, when present, can enhance the toughness of a material.

Ductile failure of polymers is exhibited by yielding of the polymer or slip of the molecular chains past one another. This is most often indicated by a maximum in the tensile stress-strain test or what is termed the *yield point*. Above this point, the material may exhibit lateral contraction upon further extension, termed *necking*.[5] Molecules in the necked region become oriented and result in increased local stiffness. Material in regions adjacent to the neck are thus preferentially deformed and the neck region propagates. This process is known as *cold drawing* (see Fig. 1.11). Cold drawing results in elongations of several hundred percent.

Under repeated cyclic loading, a material may fail at stresses well below the single-cycle failure stress found in a typical tensile test.[6] This process is called *fatigue* and is usually depicted by plotting the maximum stress versus the number of cycles to failure.

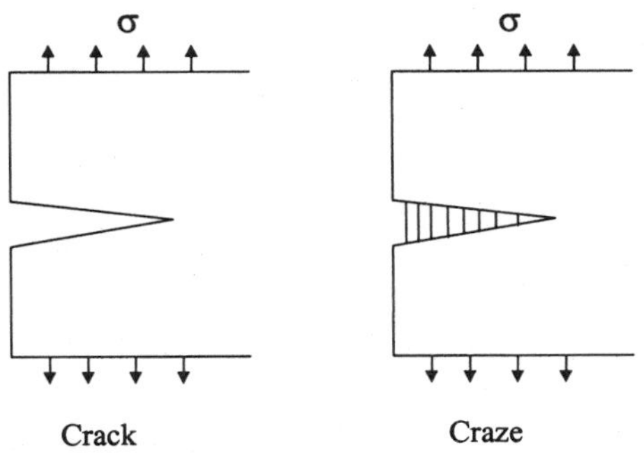

Figure 1.10 Cracks and crazes.

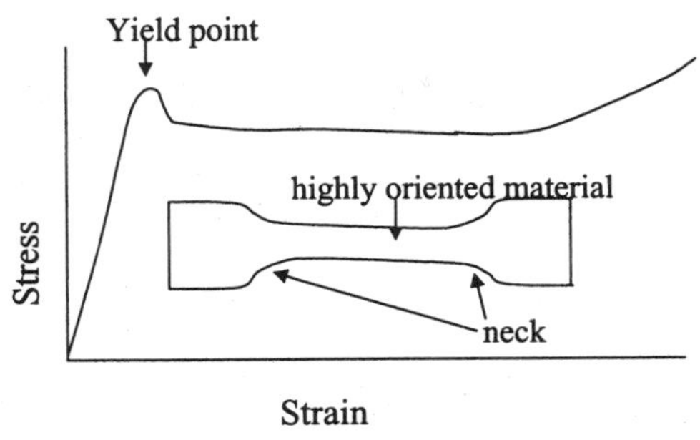

Figure 1.11 Ductile behavior.

Fatigue tests can be performed under a variety of loading conditions as specified by the service requirements. Thermal effects and the presence or absence of cracks are other variables to be considered when the fatigue life of a material is to be evaluated.

1.4.3 Effect of Fillers

The term *fillers* refers to solid additives that are incorporated into the plastic matrix.[7] They are generally inorganic materials and can be classified according to their effect on the mechanical properties of the resulting mixture. Inert or extender fillers are added mainly to reduce the cost of the compound, while reinforcing fillers are added to improve certain mechanical properties such as modulus or tensile strength. Although termed *inert,* inert fillers can nonetheless affect other properties of the compound besides cost. In particular, they may increase the density of the compound, reduce the shrinkage, increase the hardness, and increase the heat deflection temperature. Reinforcing fillers typically will increase the tensile, compressive, and shear strengths; increase the heat deflection temperature; reduce shrinkage; increase the modulus; and improve the creep behavior. Reinforcing fillers improve the properties via several mechanisms. In some cases, a chemical bond is formed between the filler and the polymer; in other cases, the volume occupied by the filler affects the properties of the thermoplastic. As a result, the surface properties and interaction between the filler and the thermoplastic are of great importance. A number of filler properties govern their behavior. These include the particle shape, the particle size, and distribution of sizes, and the surface chemistry of the particle. In general, the smaller the particle, the greater the improvement of the mechanical property of interest (such as tensile strength).[8] Larger particles may give reduced properties compared to the pure thermoplastic. Particle shape can also influence the properties. For example, plate-like particles or fibrous particles may be oriented during processing. This may result in properties that are *anisotropic*. The surface chemistry of the particle is important to promote interaction with the polymer and to allow for good interfacial adhesion. It is important that the polymer wet the particle surface and have good interfacial bonding so as to obtain the best property enhancement.

Examples of inert or extender fillers include china clay (kaolin), talc, and calcium carbonate. Calcium carbonate is an important filler with a particle size of about one micron.[9] It is a natural product from sedimentary rocks and is separated into chalk, limestone, and marble. In some cases, the calcium carbonate may be treated to improve interaction with the thermoplastic. Glass spheres are also used as thermoplastic fillers. They may be either solid or hollow, depending on the particular application. Talc is a filler with a lamellar particle shape.[10] It is a natural, hydrated magnesium silicate with good slip properties. Kaolin and mica are also natural materials with lamellar structures. Other fillers include wollastonite, silica, barium sulfate, and metal powders. Carbon black is used as a filler primarily in the rubber industry, but it also finds application in thermoplastics for conductivity, UV protection, and as a pigment. Fillers in fiber form are often used in thermoplastics. Types of fibers include cotton, wood flour, fiberglass, and carbon. Table 1.3 shows the fillers and their forms. An overview of some typical fillers and their effect on properties is shown in Table 1.4.

1.5 General Classes of Polymers

1.5.1 Acetal (POM)

Acetal polymers are formed from the polymerization of formaldehyde. They are also given the name *polyoxymethylenes (POMs)*. Polymers prepared from formaldehyde were

TABLE 1.3 Forms of Various Fillers

Spherical	Lamellar	Fibrous
Sand/quartz powder	Mica	Glass fibers
Silica	Talc	Asbestos
Glass spheres	Graphite	Wollastonite
Calcium carbonate	Kaolin	Carbon fibers
Carbon black		Whiskers
Metallic oxides		Cellulose
		Synthetic fibers

studied by Staudinger in the 1920s, but thermally stable materials were not introduced until the 1950s when DuPont developed Delrin.[11] Hompolymers are prepared from very pure formaldehyde by anionic polymerization as shown in Fig. 1.12. Amines and the soluble salts of alkali metals catalyze the reaction.[12] The polymer formed is insoluble and is removed as the reaction proceeds. Thermal degradation of the acetal resin occurs by unzipping with the release of formaldehyde. The thermal stability of the polymer can be increased by esterification of the hydroxyl ends with acetic anhydride. An alternative method to improve the thermal stability is copolymerization with a second monomer such as ethylene oxide. The copolymer is prepared by cationic methods.[13] This was developed by Celanese and marketed under the trade name Celcon. Hostaform is another copolymer marketed by Hoescht. The presence of the second monomer reduces the tendency for the polymer to degrade by unzipping.[14]

There are four processes for the thermal degradation of acetal resins. The first is thermal or base-catalyzed depolymerization from the chain, resulting in the release of formaldehyde. End capping the polymer chain will reduce this tendency. The second is oxidative attack at random positions, again leading to depolymerization. The use of antioxidants will reduce this degradation mechanism. Copolymerization is also helpful. The third mechanism is cleavage of the acetal linkage by acids. It is therefore important not to process acetals in equipment used for PVC, unless it has been cleaned, due to the possible presence of traces of HCl. The fourth degradation mechanism is thermal depolymerization at temperatures above 270°C. It is important that processing temperatures remain below this temperature to avoid degradation of the polymer.[15]

Acetals are highly crystalline, typically 75 percent crystalline, with a melting point of 180°C.[16] Compared with polyethylene (PE), the chains pack closer together because of the shorter C–O bond. As a result, the polymer has a higher melting point. It is also harder than PE. The high degree of crystallinity imparts good solvent resistance to acetal polymers. The polymer is essentially linear with molecular weights (M_n) in the range of 20,000 to 110,000.[17]

Acetal resins are strong, stiff thermoplastics with good fatigue properties and dimensional stability. They also have a low coefficient of friction and good heat resistance.[22] Acetal resins are considered similar to nylons but are better in fatigue, creep, stiffness, and water resistance.[18] Acetal resins do not, however, have the creep resistance of polycarbon-

Figure 1.12 Polymerization of formaldehyde to polyoxymethylene.

TABLE 1.4 Effect of Filler Type on Properties[454]

	Glass fiber	Asbestos	Wollastonite	Carbon fiber	Whiskers	Synthetic fibers	Cellulose	Mica	Talc	Graphite	Sand/quartz powder	Silica	Kaolin	Glass spheres	Calcium carbonate	Metallic oxides	Carbon black
Tensile strength	++	+		+	−+			+	0					+			
Compressive strength	+								+		+			+	+		
Modulus of elasticity	++	++	++	++	+			++	+		+	+		+	+	+	+
Impact strength	−+	−	−	−	−	++	+	−+	−		−	−	−	−	−+	−	+
Reduced thermal expansion	+	+			+			+	+		+	+	+			+	
Reduced shrinkage	+	+	+	+				+	+	+	+	+	+	+	+	+	+
Better thermal conductivity		+	+	+					+	+	+	+			+		+
Higher heat deflection temperature	++	+	+	++				+	+				+		+	+	
Electrical conductivity				+						+							+
Electrical resistance			+					++	+			+	++			+	
Thermal stability			+					+	+		+	+	+			+	+
Chemical resistance		+	+					+	0	+			+	+			
Better abrasion behavior				+				+	+	+			+				
Extrusion rate	−+	+						+					+		+		
Machine abrasion	−	0			0	0	0		0	0	−			0	0		0
Price reduction	+	+	+				+	+	+	+	++	+	+	+	++		

++ large influence, + influence, 0 no influence, – negative influence

ate. As mentioned previously, acetal resins have excellent solvent resistance with no organic solvents found below 70°C; however, swelling may occur in some solvents. Acetal resins are susceptible to strong acids and alkalis as well as to oxidizing agents. Although the C–O bond is polar, it is balanced and much less polar than the carbonyl group present in nylon. As a result, acetal resins have relatively low water absorption. The small amount of moisture absorbed may cause swelling and dimensional changes but will not degrade the polymer by hydrolysis.[12] The effects of moisture are considerably less dramatic than for nylon polymers. Ultraviolet light may cause degradation, which can be reduced by the addition of carbon black. The copolymers have generally similar properties, but the homopolymer may have slightly better mechanical properties and higher melting point but poorer thermal stability and poorer alkali resistance.[21] Along with both homopolymers and copolymers, there are also filled materials (glass, fluoropolymer, aramid fiber, and other fillers), toughened grades, and UV stabilized grades.[22] Blends of acetal with polyurethane elastomers show improved toughness and are available commercially.

Acetal resins are available for injection molding, blow molding, and extrusion. During processing, it is important to avoid overheating, or the production of formaldehyde may cause serious pressure buildup. The polymer should be purged from the machine before shut-down to avoid excessive heating during startup.[23] Acetal resins should be stored in a dry place. The apparent viscosity of acetal resins is less dependent on shear stress and temperature than polyolefins, but the melt has low elasticity and melt strength. The low melt strength is a problem for blow molding applications, and copolymers with branched structures are available for this application. Crystallization occurs rapidly with post mold shrinkage complete within 48 hr of molding. Because of the rapid crystallization, it is difficult to obtain clear films.[24]

The market demand for acetal resins in the United States and Canada was 368 million pounds in 1997.[25] Applications for acetal resins include gears, rollers, plumbing components, pump parts, fan blades, blow molded aerosol containers, and molded sprockets and chains. They are often used as direct replacements for metal. Most of the acetal resins are processed by injection molding, with the remainder used in extruded sheet and rod. Their low coefficient of friction makes acetal resins good for bearings.[26]

1.5.2 Biodegradable Polymers

Disposal of solid waste is a challenging problem. The United States consumes over 53 billion pounds of polymers a year for a variety of applications.[27] When the life cycle of these polymeric parts is completed, they may end up in a landfill. Plastics are often selected for applications based of their stability to degradation; however, this means that degradation will be very slow, adding to the solid waste problem. Methods to reduce the amount of solid waste include recycling and biodegradation.[28] Considerable work has been done to recycle plastics, both in the manufacturing and consumer area. Biodegradable materials offer another way to reduce the solid waste problem. Most waste is disposed of by burial in a landfill. Under these conditions, oxygen is depleted, and biodegradation must proceed without the presence of oxygen.[29] An alternative is aerobic composting. In selecting a polymer that will undergo biodegradation, it is important to ascertain the method of disposal. Will the polymer be degraded in the presence of oxygen and water, and what will be the pH level? Biodegradation can be separated into two types—chemical and microbial degradation. Chemical degradation includes degradation by oxidation, photodegradation, thermal degradation, and hydrolysis. Microbial degradation can include both fungi and bacteria. The susceptibility of a polymer to biodegradation depends on the structure of the backbone.[30] For example, polymers with hydrolyzable backbones can be attacked by acids or bases, breaking down the molecular weight. They are therefore more likely to be degraded. Polymers that fit into this category include most natural-based polymers, such as

polysaccharides, and synthetic materials, such as polyurethanes, polyamides, polyesters, and polyethers. Polymers that contain only carbon groups in the backbone are more resistant to biodegradation.

Photodegradation can be accomplished by using polymers that are unstable to light sources or by the used of additives that undergo photodegradation. Copolymers of divinyl ketone with styrene, ethylene, or polypropylene (Eco Atlantic) are examples of materials that are susceptible to photodegradation.[31] The addition of a UV absorbing material will also act to enhance photodegradation of a polymer. An example is the addition of iron dithiocarbamate.[32] The degradation must be controlled to ensure that the polymer does not degrade prematurely.

Many polymers described elsewhere in this book can be considered for biodegradable applications. Polyvinyl alcohol has been considered in applications requiring biodegradation because of its water solubility; however, the actual degradation of the polymer chain may be slow.[33] Polyvinyl alcohol is a semicrystalline polymer synthesized from polyvinyl acetate. The properties are governed by the molecular weight and by the amount of hydrolysis. Water soluble polyvinyl alcohol has a degree of hydrolysis near 88 percent. Water insoluble polymers are formed if the degree of hydrolysis is less than 85 percent.[34]

Cellulose based polymers are some of the more widely available naturally-based polymers. They can therefore be used in applications requiring biodegradation. For example, regenerated cellulose is used in packaging applications.[35] A biodegradable grade of cellulose acetate is available from Rhone-Poulenc (Bioceta and Biocellat), where an additive acts to enhance the biodegradation.[36] This material finds application in blister packaging, transparent window envelopes, and other packaging applications.

Starch-based products are also available for applications requiring biodegradability. The starch is often blended with polymers for better properties. For example, polyethylene films containing between 5 and 10 percent cornstarch have been used in biodegradable applications. Blends of starch with vinyl alcohol are produced by Fertec (Italy) and used in both film and solid product applications.[37] The content of starch in these blends can range up to 50 percent by weight, and the materials can be processed on conventional processing equipment. A product developed by Warner-Lambert call Novon is also a blend of polymer and starch, but the starch contents in Novon are higher than in the material by Fertec. In some cases, the content can be over 80 percent starch.[38]

Polylactides (PLA) and copolymers are also of interest in biodegradable applications. This material is a thermoplastic polyester synthesized from ring opening of lactides. Lactides are cyclic diesters of lactic acid.[39] A similar material to polylactide is polyglycolide (PGA). PGA is also thermoplastic polyester but formed from glycolic acids. Both PLA and PGA are highly crystalline materials. These materials find application in surgical sutures, resorbable plates and screws for fractures, and new applications in food packaging are also being investigated.

Polycaprolactones are also considered in biodegradable applications such as films and slow-release matrices for pharmaceuticals and fertilizers.[40] Polycaprolactone is produced through ring opening polymerization of lactone rings with a typical molecular weight in the range of 15,000 to 40,000.[41] It is a linear, semicrystalline polymer with a melting point near 62°C and a glass transition temperature about –60°C.[42]

A more recent biodegradable polymer is polyhydroxybutyrate-valerate copolymer (PHBV). These copolymers differ from many of the typical plastic materials in that they are produced through biochemical means. It is produced commercially by ICI using the bacteria *Alcaligenes eutrophus*, which is fed a carbohydrate. The bacteria produce polyesters, which are harvested at the end of the process.[43] When the bacteria are fed glucose, the pure polyhydroxybutyrate polymer is formed, while a mixed feed of glucose and propionic acid will produce the copolymers.[44] Different grades are commercially available that vary in the amount of hydroxyvalerate units and the presence of plasticizers. The pure hy-

droxybutyrate polymer has a melting point between 173 and 180°C and a T_g near 5°C.[45] Copolymers with hydroxyvalerate have reduced melting points, greater flexibility, and impact strength, but lower modulus and tensile strength. The level of hydroxyvalerate is 5 to 12 percent. These copolymers are fully degradable in many microbial environments. Processing of PHBV copolymers requires careful control of the process temperatures. The material will degrade above 195°C, so processing temperatures should be kept below 180°C and the processing time kept to a minimum. It is more difficult to process unplasticized copolymers with lower hydroxyvalerate content because of the higher processing temperatures required. Applications for PHBV copolymers include shampoo bottles, cosmetic packaging, and as a laminating coating for paper products.[46]

Other biodegradable polymers include Konjac, a water soluble natural polysaccharide produced by FMC, Chitin, another polysaccharide that is insoluble in water, and Chitosan, which is soluble in water.[47] Chitin is found in insects and in shellfish. Chitosan can be formed from chitin and is also found in fungal cell walls.[48] Chitin is used in many biomedical applications, including dialysis membranes, bacteriostatic agents, and wound dressings. Other applications include cosmetics, water treatment, adhesives, and fungicides.[49]

1.5.3 Cellulose

Cellulosic polymers are the most abundant organic polymers in the world, making up the principal polysaccharide in the walls of almost all of the cells of green plants and many fungi species.[50] Plants produce cellulose through photosynthesis. Pure cellulose decomposes before it melts and must be chemically modified to yield a thermoplastic. The chemical structure of cellulose is a heterochain linkage of different anhydrogluclose units into high-molecular-weight polymer, regardless of plant source. The plant source, however, does affect molecular weight, molecular weight distribution, degrees of orientation, and morphological structure. Material described commonly as "cellulose" can actually contain hemicelluloses and lignin.[51] Wood is the largest source of cellulose and is processed as fibers to supply the paper industry and is widely used in housing and industrial buildings. Cotton-derived cellulose is the largest source of textile and industrial fibers, with the combined result being that cellulose is the primary polymer serving the housing and clothing industries. Crystalline modifications result in celluloses of differing mechanical properties, and Table 1.5 compares the tensile strengths and ultimate elongations of some common celluloses.[52]

TABLE 1.5 Selected Mechanical Properties of Common Celluloses

	Tensile strength, MPa		Ultimate elongation, %	
Form	Dry	Wet	Dry	Wet
Ramie	900	1060	2.3	2.4
Cotton	200–800	200–800	12–16	6–13
Flax	824	863	1.8	2.2
Viscose Rayon	200–400	100–200	8–26	13–43
Cellulose Acetate	150–200	100–120	21–30	29–30

Cellulose, whose repeat structure features three hydroxyl groups, reacts with organic acids, anhydrides, and acid chlorides to form esters. Plastics from these cellulose esters are

extruded into film and sheet and are injection molded to form a wide variety of parts. Cellulose esters can also be compression molded and cast from solution to form a coating. The three most industrially important cellulose ester plastics are cellulose acetate (CA), cellulose acetate butyrate (CAB), and cellulose acetate propionate (CAP), with structures as shown below in Fig. 1.13.

These cellulose acetates are noted for their toughness, gloss, and transparency. CA is well suited for applications requiring hardness and stiffness, as long as the temperature and humidity conditions don't cause the CA to be too dimensionally unstable. CAB has the best environmental stress cracking resistance, low-temperature impact strength, and dimensional stability. CAP has the highest tensile strength and hardness. Comparison of typical compositions and properties for a range of formulations are given in Table 1.6.[53] Properties can be tailored by formulating with different types and loadings of plasticizers.

TABLE 1.6 Selected Mechanical Properties of Cellulose Esters

Composition, %	Cellulose acetate	Cellulose acetate butyrate	Cellulose acetate propionate
Acetyl	38–40	13–15	1.5–3.5
Butyrl	–	36–38	–
Propionyl	–	–	43–47
Hydroxyl	3.5–4.5	1–2	2–3
Tensile strength at fracture, 23°C, MPa	13.1–58.6	13.8–51.7	13.8–51.7
Ultimate elongation, %	6–50	38–74	35–60
Izod impact strength, J/m			
notched, 23°C	6.6–132.7	9.9–149.3	13.3–182.5
notched, –40°C	1.9–14.3	6.6–23.8	1.9–19.0
Rockwell hardness, R scale	39–120	29–117	20–120
% moisture absorption at 24 hr	2.0–6.5	1.0–4.0	1.0–3.0

Formulation of cellulose esters is required to reduce charring and thermal discoloration, and it typically includes the addition of heat stabilizers, antioxidants, plasticizers, UV stabilizers, and coloring agents.[54] Cellulose molecules are rigid due to the strong intermolecular hydrogen bonding that occurs. Cellulose itself is insoluble and reaches its decomposition temperature prior to melting. The acetylation of the hydroxyl groups re-

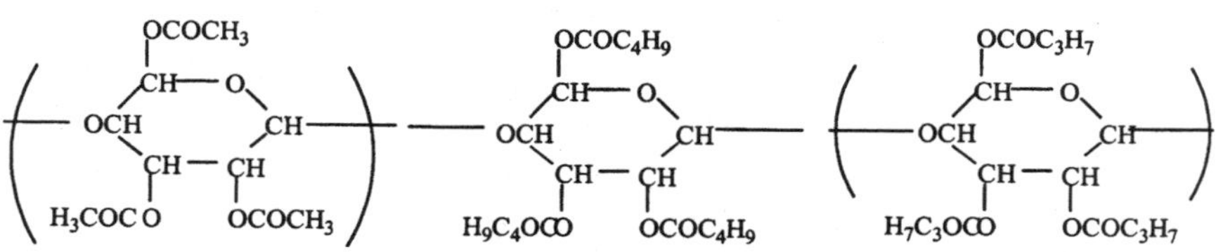

Figure 1.13 Structures of cellulose acetate, cellulose acetate butyrate, and cellulose acetate propionate.

duces intermolecular bonding, and increases free volume, depending upon the level and chemical nature of the alkylation.[55] CAs are thus soluble in specific solvents but still require plasticization for rheological properties appropriate to molding and extrusion processing conditions. Blends of ethylene vinyl acetate (EVA) copolymers and CAB are available. Cellulose acetates have also been graft-copolymerized with alkyl esters of acrylic and methacrylic acid and then blended with EVA to form a clear, readily processable thermoplastic.

CA is cast into sheet form for blister packaging, window envelopes, and file tab applications. CA is injection molded into tool handles, tooth brushes, ophthalmic frames, and appliance housings and is extruded into pens, pencils, knobs, packaging films, and industrial pressure-sensitive tapes. CAB is molded into steering wheels, tool handles, camera parts, safety goggles, and football nose guards. CAP is injection molded into steering wheels, telephones, appliance housings, flashlight cases, and screw and bolt anchors, and it is extruded into pens, pencils, toothbrushes, packaging film, and pipe.[56] Cellulose acetates are well suited for applications that require machining and then solvent vapor polishing, such as in the case of tool handles, where the consumer market values the clarity, toughness, and smooth finish. CA and CAP are likewise suitable for ophthalmic sheeting and injection molding applications, which require many post-finishing steps.[57]

Cellulose acetates are also commercially important in the coatings arena. In this synthetic modification, cellulose is reacted with an alkyl halide, primarily methylchloride to yield methylcellulose or sodium chloroacetate to yield sodium cellulose methylcellulose (CMC). The structure of CMC is shown below in Fig. 1.14. CMC gums are water soluble and are used in food contact and packaging applications. CMC's outstanding film forming properties are used in paper sizings and textiles, and its thickening properties are used in starch adhesive formulations, paper coatings, toothpaste, and shampoo. Other cellulose esters, including cellulosehydroxyethyl, hydroxypropylcellulose, and ethylcellulose, are used in film and coating applications, adhesives, and inks.

1.5.4 Fluoropolymers

Fluoropolymers are noted for their heat resistance properties. This is due to the strength and stability of the carbon-fluorine bond.[58] The first patent was awarded in 1934 to IG Farben for a fluorine containing polymer, polychlorotrifluoroethylene (PCTFE). This polymer had limited application, and fluoropolymers did not have wide application until the discovery of polytetrafluoroethylene (PTFE) in 1938.[59] In addition to their high-temperature properties, fluoropolymers are known for their chemical resistance, very low coefficient of friction, and good dielectric properties. Their mechanical properties are not high unless reinforcing fillers, such as glass fibers, are added.[60] The compressive properties of fluoropolymers are generally superior to their tensile properties. In addition to their high temperature resistance, these materials have very good toughness and flexibility at low temperatures.[61]

A wide variety of fluoropolymers are available, including polytetrafluoroethylene (PTFE), polychlorotrifluoroethylene (PCTFE), fluorinated ethylene propylene (FEP), eth-

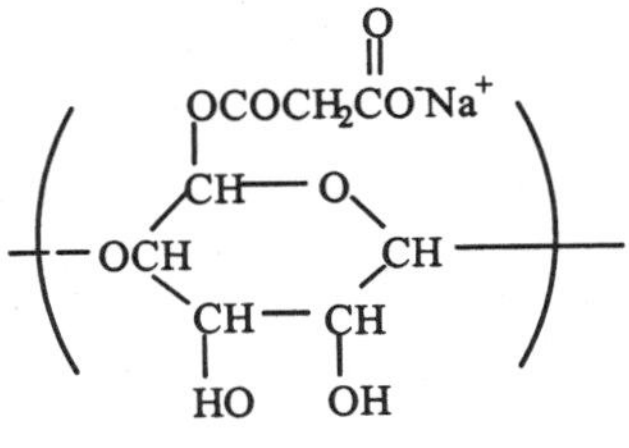

Figure 1.14 Sodium cellulose methylcellulose structure.

ylene chlorotrifluoroethylene (ECTFE), ethylene tetrafluoroethylene (ETFE), polyvinylidene fluoride (PVDF), and polyvinyl fluoride (PVF).

1.5.4.1 Copolymers. Fluorinated ethylene propylene (FEP) is a copolymer of tetrafluoroethylene and hexafluoropropylene. It has properties similar to PTFE, but with a melt viscosity suitable for molding with conventional thermoplastic processing techniques.[62] The improved processability is obtained by replacing one of the fluorine groups on PTFE with a trifluoromethyl group as shown in Fig. 1.15.[63]

FEP polymers were developed by DuPont, but other commercial sources are available, such as Neoflon (Daikin Kogyo) and Teflex (Niitechem, USSR).[64] FEP is a crystalline polymer with a melting point of 290°C, and it can be used for long periods at 200°C with good retention of properties.[65] FEP has good chemical resistance, a low dielectric constant, low friction properties, and low gas permeability. Its impact strength is better than PTFE, but the other mechanical properties are similar to those of PTFE.[66] FEP may be processed by injection, compression, or blow molding. FEP may be extruded into sheets, films, rods, or other shapes. Typical processing temperatures for injection molding and extrusion are in the range of 300 to 380°C.[67] Extrusion should be done at low shear rates because of the polymer's high melt viscosity and melt fracture at low shear rates. Applications for FEP include chemical process pipe linings, wire and cable, and solar collector glazing.[68] A material similar to FEP, Hostaflon TFB (Hoechst), is a terpolymer of tetrafluoroethylene, hexafluoropropylene, and vinylidene fluoride.

Ethylene chlorotrifluoroethylene (ECTFE) is an alternating copolymer of chlorotrifluoroethylene and ethylene. It has better wear properties than PTFE along with good flame resistance. Applications include wire and cable jackets, tank linings, chemical process valve and pump components, and corrosion-resistant coatings.[69]

Ethylene tetrafluoroethylene (ETFE) is a copolymer of ethylene and tetrafluoroethylene similar to ECTFE but with a higher use temperature. It does not have the flame resistance of ECTFE, however, and will decompose and melt when exposed to a flame.[70] The polymer has good abrasion resistance for a fluorine containing polymer, along with good impact strength. The polymer is used for wire and cable insulation where its high-temperature properties are important. ETFE finds application in electrical systems for computers, aircraft, and heating systems.[71]

1.5.4.2 Polychlorotrifluoroethylene. Polychlorotrifluoroethylene (PCTFE) is made by the polymerization of chlorotrifluoroethylene, which is prepared by the dechlorination of trichlorotrifluoroethane. The polymerization is initiated with redox initiators.[72] The replacement of one fluorine atom with a chlorine atom as shown in Fig. 1.16 breaks up the symmetry of the PTFE molecule, resulting in a lower melting point and allowing PCTFE to be processed more easily than PTFE. The crystalline melting point of PCTFE at 218°C is lower than that of PTFE. Clear sheets of PCTFE with no crystallinity may also be prepared.

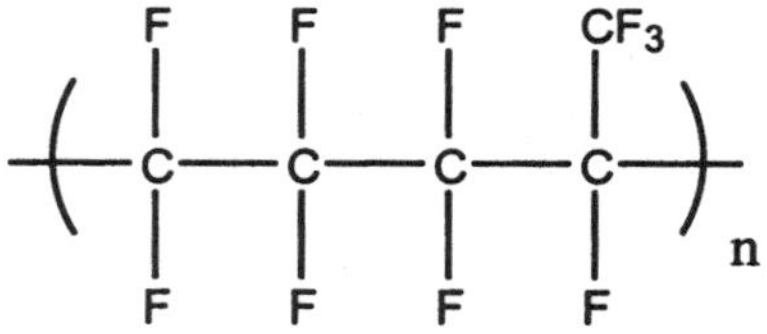

Figure 1.15 Structure of FEP.

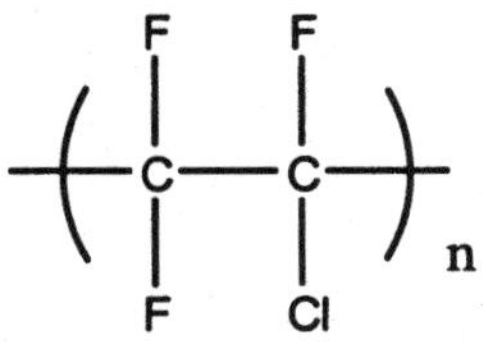

Figure 1.16 Structure of PCTFE.

PCTFE is resistant to temperatures up to 200°C and has excellent solvent resistance, with the exception of halogenated solvents or oxygen containing materials, which may swell the polymer.[73] The electrical properties of PCTFE are inferior to those of PTFE, but PCTFE is harder and has higher tensile strength. The melt viscosity of PCTFE is low enough that it may be processed using most thermoplastic processing techniques.[74] Typical processing temperatures are in the range of 230 to 290°C.[75]

PCTFE is higher in cost than PTFE, somewhat limiting its use. Applications include gaskets, tubing, and wire and cable insulation. Very low vapor transmission films and sheets may also be prepared.[76]

1.5.4.3 Polytetrafluoroethylene (PTFE). Polytetrafluoroethylene (PTFE) is polymerized from tetrafluoroethylene by free radical methods.[77] The reaction is shown below in Fig. 1.17. Commercially, there are two major processes for the polymerization of PTFE, one yielding a finer particle size dispersion polymer with lower molecular weight than the second method, which yields a "granular" polymer. The weight average molecular weights of commercial materials range from 400,000 to 9,000,000.[78] PTFE is a linear crystalline polymer with a melting point of 327°C.[79] Because of the larger fluorine atoms, PTFE takes up a twisted zigzag in the crystalline state, while polyethylene takes up the planar zigzag form.[80] There are several crystal forms for PTFE, and some of the transitions from one crystal form to another occur near room temperature. As a result of these transitions, volume changes of about 1.3 percent may occur.

PTFE has excellent chemical resistance but may go into solution near its crystalline melting point. PTFE is resistant to most chemicals. Only alkali metals (molten) may attack the polymer.[81] The polymer does not absorb significant quantities of water and has low permeability to gases and moisture vapor.[82] PTFE is a tough polymer with good insulating properties. It is also known for its low coefficient of friction, with values in the range of 0.02 to 0.10.[83] PTFE, like other fluoropolymers, has excellent heat resistance and can withstand temperatures up to 260°C. Because of the high thermal stability, the mechanical and electrical properties of PTFE remain stable for long times at temperatures up to 250°C. However, PTFE can be degraded by high energy radiation.

One disadvantage of PTFE is that it is extremely difficult to process by either molding or extrusion. PFTE is processed in powder form by either sintering or compression mold-

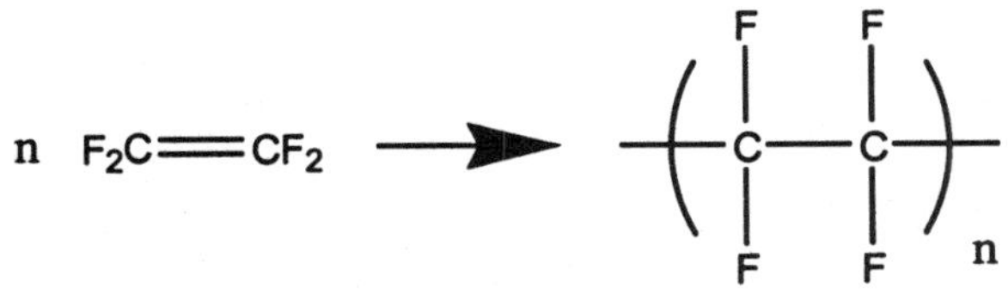

Figure 1.17 Preparation of PTFE.

ing. It is also available as a dispersion for coating or impregnating porous materials.[84] PTFE has very high viscosity, prohibiting the use of many conventional processing techniques. For this reason, techniques developed for the processing of ceramics are often used. These techniques involve preforming the powder, followed by sintering above the melting point of the polymer. For granular polymers, the preforming is carried out with the powder compressed into a mold. Pressures should be controlled, as too low a pressure may cause voids, while too high a pressure may result in cleavage planes. After sintering, thick parts should be cooled in an oven at a controlled cooling rate, often under pressure. Thin parts may be cooled at room temperature. Simple shapes may be made by this technique, but more detailed parts should be machined.[85]

Extrusion methods may be used on the granular polymer at very low rates. In this case, the polymer is fed into a sintering die that is heated. A typical sintering die has a length about 90 times the internal diameter. Dispersion polymers are more difficult to process by the techniques previously mentioned. The addition of a lubricant (15 to 25 percent) allows the manufacture of preforms by extrusion. The lubricant is then removed and the part sintered. Thick parts are not made by this process, because the lubricant must be removed. PTFE tapes are made by this process; however, the polymer is not sintered, and a nonvolatile oil is used.[86] Dispersions of PTFE are used to impregnate glass fabrics and to coat metal surfaces. Laminates of the impregnated glass cloth may be prepared by stacking the layers of fabric, followed by pressing at high temperatures.

Processing of PTFE requires adequate ventilation for the toxic gases that may be produced. In addition, PTFE should be processed under high cleanliness standards, because the presence of any organic matter during the sintering process will result in poor properties as a result of the thermal decomposition of the organic matter. This includes both poor visual qualities and poor electrical properties.[87] The final properties of PTFE are dependent on the processing methods and the type of polymer. Both particle size and molecular weight should be considered. The particle size will affect the amount of voids and the processing ease, while crystallinity will be influenced by the molecular weight.

Additives for PTFE must be able to undergo the high processing temperatures required. This limits the range of additives available. Glass fiber is added to improve some mechanical properties. Graphite or molybdenum disulphide may be added to retain the low coefficient of friction while improving the dimensional stability. Only a few pigments are available that can withstand the processing conditions. These are mainly inorganic pigments such as iron oxides and cadmium compounds.[88]

Because of the excellent electrical properties, PTFE is used in a variety of electrical applications such as wire and cable insulation and insulation for motors, capacitors, coils, and transformers. PTFE is also used for chemical equipment such as valve parts and gaskets. The low friction characteristics make PTFE suitable for use in bearings, mold release devices, and anti-stick cookware. Low-molecular-weight polymers may be used in aerosols for dry lubrication.[89]

1.5.4.4 Polyvinylindene fluoride (PVDF). Polyvinylindene fluoride (PVDF) is crystalline with a melting point near 170°C.[90] The structure of PVDF is shown in Fig. 1.18. PVDF has good chemical and weather resistance, along with good resistance to distortion and creep at low and high temperatures. Although the chemical resistance is good, the polymer can be affected by very polar solvents, primary amines, and concentrated acids. PVDF has limited use as an insulator, because the dielectric properties are frequency dependent. The polymer is important because of its relatively low cost compared with other fluorinated polymers.[91] PVDF is unique in that the material has piezoelectric properties, meaning that it will generate electric current when compressed.[92] This unique feature has been utilized for the generation of ultrasonic waves.

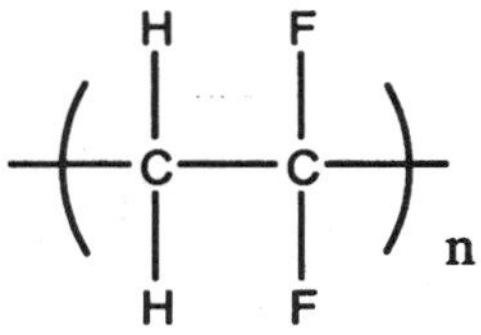

Figure 1.18 Structure of PVDF.

PVDF can be melt processed by most conventional processing techniques. The polymer has a wide range between the decomposition temperature and the melting point. Melt temperatures are usually 240 to 260°C.[93] Processing equipment should be extremely clean, as any contaminants may affect the thermal stability. As with other fluorinated polymers, the generation of HF is a concern. PVDF is used for applications in gaskets, coatings, wire and cable jackets, chemical process piping, and seals.[94]

1.5.4.5 Polyvinyl fluoride (PVF). Polyvinyl fluoride (PVF) is a crystalline polymer available in film form and used as a lamination on plywood and other panels.[95] The film is impermeable to many gases. PVF is structurally similar to polyvinyl chloride (PVC) except for the replacement of a chlorine atom with a fluorine atom. PVF exhibits low moisture absorption, good weatherability, and good thermal stability. Similar to PVC, PVF may give off hydrogen halides at elevated temperatures. However, PVF has a greater tendency to crystallize and better heat resistance than PVC.[96]

1.5.5 Polyamides

Nylons were one of the early polymers developed by Carothers.[97] Today, nylons are an important thermoplastic, with consumption in the United States of about 1.2 billion lb in 1997.[98] Nylons, also known as polyamides, are synthesized by condensation polymerization methods, often an aliphatic diamine and a diacid. Nylon is a crystalline polymer with high modulus, strength, and impact properties; low coefficient of friction; and resistance to abrasion.[99] Although the materials possess a wide range of properties, they all contain the amide (–CONH–) linkage in their backbone. Their general structure is shown in Fig. 1.19.

There are five main methods to polymerize nylon.

1. Reaction of a diamine with a dicarboxylic acid
2. Condensation of the appropriate amino acid
3. Ring opening of a lactam
4. Reaction of a diamine with a dicarboxylic acid
5. Reaction of a diisocyanate with a dicarboxylic acid[100]

The type of nylon (nylon 6, nylon 10, etc.) is indicative of the number of carbon atoms. The are many different types of nylons that can be prepared, depending on the starting

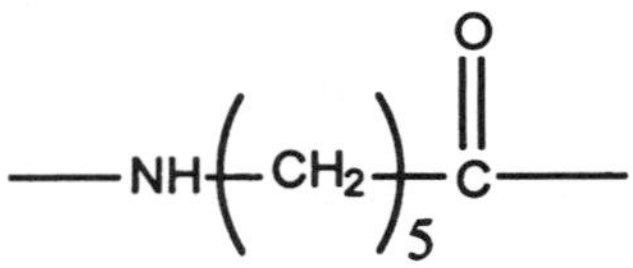

Figure 1.19 Structure of nylon.

monomers used. The type of nylon is determined by the number of carbon atoms in the monomers used in the polymerization. The number of carbon atoms between the amide linkages also controls the properties of the polymer. When only one monomer is used (lactam or amino acid), the nylon is identified with only one number (nylon 6, nylon 12).

When two monomers are used in the preparation, the nylon will be identified using two numbers (nylon 6/6, nylon 6/12).[101] This is shown in Fig. 1.20. The first number refers to the number of carbon atoms in the diamine used (a) and the second number refers to the number of carbon atoms in the diacid monomer (b + 2), due to the two carbons in the carbonyl group.[102]

The amide groups are polar groups and significantly affect the polymer properties. The presence of these groups allows for hydrogen bonding between chains, improving the interchain attraction. This gives nylon polymers good mechanical properties. The polar nature of nylons also improves the bondability of the materials, while the flexible aliphatic carbon groups give nylons low melt viscosity for easy processing.[103] This structure also yields polymers that are tough above their glass transition temperature.[104]

Nylons are relatively insensitive to nonpolar solvents; however, because of the presence of the polar groups, nylons can be affected by polar solvents, particularly water.[105] The presence of moisture must be considered in any nylon application. Moisture can cause changes in part dimensions and reduce the properties, particularly at elevated temperatures.[106] As a result, the material should be dried before any processing operations. In the absence of moisture, nylons are fairly good insulators, but, as the level of moisture or the temperature increases, the nylons are less insulating.[107]

The strength and stiffness will be increased as the number of carbon atoms between amide linkages is decreased, because there are more polar groups per unit length along the polymer backbone.[108] The degree of moisture absorption is also strongly influenced by the number of polar groups along the backbone of the chain. Nylon grades with fewer carbon atoms between the amide linkages will absorb more moisture than grades with more carbon atoms between the amide linkages (nylon 6 will absorb more moisture than nylon 12). Furthermore, nylon types with an even number of carbon atoms between the amide groups have higher melting points than those with an odd number of carbon atoms. For example, the melting point of nylon 6/6 is greater than that of either nylon 5/6 or nylon 7/6.[109] Ring opened nylons behave similarly. This is due to the ability of the nylons with the even number of carbon atoms to pack better in the crystalline state.[110]

Nylon properties are affected by the amount of crystallinity. This can be controlled to a great extent in nylon polymers by the processing conditions. A slowly cooled part will have significantly greater crystallinity(50 to 60 percent) than a rapidly cooled, thin part (perhaps as low as 10 percent).[111] Not only can the degree of crystallinity be controlled, but also the size of the crystallites. In a slowly cooled material, the crystal size will be larger than for a rapidly cooled material. In injection molded parts where the surface is

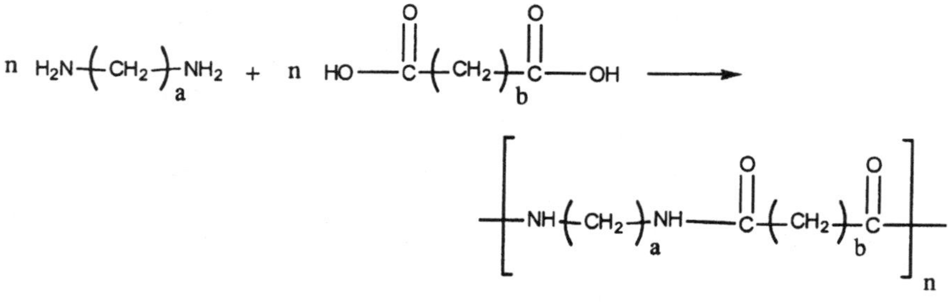

Figure 1.20 Synthesis of nylon.

rapidly cooled, the crystal size may vary from the surface to internal sections.[112] Nucleating agents can be utilized to create smaller spherulites in some applications. This creates materials with higher tensile yield strength and hardness, but lower elongation and impact.[113] The degree of crystallinity will also affect the moisture absorption, with less crystalline polyamides being more prone to moisture pick-up.[114]

The glass transition temperature of aliphatic polyamides is of secondary importance to the crystalline melting behavior. Dried polymers have T_g values near 50°C, while those with absorbed moisture may have T_g values in the neighborhood of 0°C.[115] The glass transition temperature can influence the crystallization behavior of nylons; for example, nylon 6/6 may be above its T_g at room temperature, causing crystallization at room temperature to occur slowly, leading to post mold shrinkage. This is less significant for nylon 6.[116]

Nylons are processed by extrusion, injection molding, blow molding, and rotational molding, among other methods. Nylon has a very sharp melting point and low melt viscosity, which is advantageous in injection molding but causes difficulty in extrusion and blow molding. In extrusion applications, a wide molecular weight distribution (MWD) is preferred, along with a reduced temperature at the exit to increase melt viscosity.[117]

When used in injection molding applications, nylons have a tendency to drool due to their low melt viscosity. Special nozzles have been designed for use with nylons to reduce this problem.[118] Nylons show high mold shrinkage as a result of their crystallinity. Average values are about 0.018 cm/cm for nylon 6/6. Water absorption should also be considered for parts with tight dimensional tolerances. Water will act to plasticize the nylon, relieving some of the molding stresses and causing dimensional changes. In extrusion, a screw with a short compression zone is used, with cooling initiated as soon as the extrudate exits the die.[119]

A variety of commercial nylons are available, including nylon 6, nylon 11, nylon 12, nylon 6/6, nylon 6/10, and nylon 6/12. The most widely used nylons are nylon 6/6 and nylon 6.[120] Specialty grades with improved impact resistance, improved wear, or other properties are also available. Polyamides are used most often in the form of fibers, primarily nylon 6/6 and nylon 6, although engineering applications are also of importance.[121]

Nylon 6/6 is prepared from the polymerization of adipic acid and hexamethylenediamine. The need to control a 1:1 stoichiometric balance between the two monomers can be ameliorated by the fact that adipic acid and hexamethylenediamine form a 1:1 salt that can be isolated. Nylon 6/6 is known for high strength, toughness, and abrasion resistance. It has a melting point of 265°C and can maintain properties up to 150°C.[122] Nylon 6/6 is used extensively in nylon fibers that are used in carpets, hose and belt reinforcements, and tire cord. Nylon 6/6 is used as an engineering resin in a variety of molding applications such as gears, bearings, rollers, and door latches because of its good abrasion resistance and self-lubricating tendencies.[123]

Nylon 6 is prepared from caprolactam. It has properties similar to those of nylon 6/6 but has a lower melting point (255°C). One of the major applications is in tire cord. Nylon 6/10 has a melting point of 215°C and lower moisture absorption than nylon 6/6.[124] Nylon 11 and nylon 12 have lower moisture absorption and also lower melting points than nylon 6/6. Nylon 11 has found applications in packaging films. Nylon 4/6 has found applications in a variety of automotive products due to its ability to withstand high mechanical and thermal stresses. It is used in gears, gearboxes, and clutch areas.[125] Other applications for nylons include brush bristles, fishing line, and packaging films.

Additives such as glass or carbon fibers can be incorporated to improve the strength and stiffness of the nylon. Mineral fillers are also used. A variety of stabilizers can be added to nylon to improve the heat and hydrolysis resistance. Light stabilizers are often added as well. Some common heat stabilizers include copper salts, phosphoric acid esters, and phenyl-β-naphthylamine. In bearing applications, self-lubricating grades are available, which may incorporate graphite fillers. Although nylons are generally impact resistant, rubber is

sometimes incorporated to improve the failure properties.[126] Nylon fibers do have a tendency to pick up a static charge, so antistatic agents are often added for carpeting and other applications.[127]

1.5.5.1 Aromatic polyamides. A related polyamide is prepared when aromatic groups are present along the backbone. This imparts a great deal of stiffness to the polymer chain. One difficulty encountered in this class of materials is their tendency to decompose before melting.[128] However, certain aromatic polyamides have gained commercial importance. The aromatic polyamides can be classified into three groups.

1. Amorphous copolymers with a high T_g
2. Crystalline polymers that can be used as a thermoplastic
3. Crystalline polymers used as fibers

The copolymers are noncrystalline and clear. The rigid aromatic chain structure gives the materials a high T_g. One of the oldest types is poly (trimethylhexamethylene terephthalamide) (Trogamid T®). This material has an irregular chain structure, restricting the material from crystallizing, but a T_g near 150°C.[129] Other glass-clear polyamides include Hostamid®, with a T_g also near 150°C but with better tensile strength than Trogamid T®. Grilamid TR55® is a third polyamide copolymer, with a T_g about 160°C and the lowest water absorption and density of the three.[130] The aromatic polyamides are tough materials and compete with polycarbonate, poly(methyl methacrylate), and polysulfone. These materials are used in applications requiring transparency. They have been used for solvent containers, flow meter parts, and clear housings for electrical equipment.[131]

An example of a crystallizable aromatic polyamide is poly-*m*-xylylene adipamide. It has a T_g near 85 to 100°C and a T_m of 235 to 240°C.[132] To obtain high heat deflection temperature, the filled grades are normally sold. Applications include gears, electrical plugs, and mowing machine components.[133]

Crystalline aromatic polyamides are also used in fiber applications. An example of this type of material is Kevlar®, a high-strength fiber used in bulletproof vests and in composite structures. A similar material, which can be processed more easily, is Nomex®. It can be used to give flame retardance to cloth when used as a coating.[134]

1.5.6 Polyacrylonitrile

Polyacrylonitrile is prepared by the polymerization of acrylonitrile monomer using either free radical or anionic initiators. Bulk, emulsion, suspension, solution, or slurry methods may be used for the polymerization. The reaction is shown in Fig. 1.21.

Polyacrylonitrile will decompose before reaching its melting point, making the materials difficult to form. The decomposition temperature is near 300°C.[135] Suitable solvents,

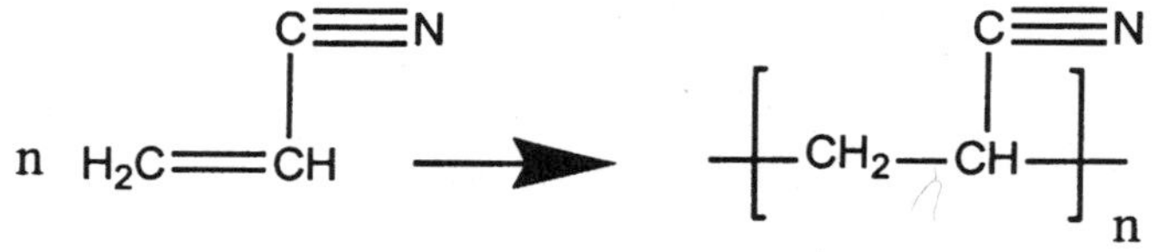

Figure 1.21 Preparation of polyacrylonitrile.

such as dimethylformamide and tetramethylenesulfone, have been found for polyacrylonitrile, allowing the polymer to be formed into fibers by dry and wet spinning techniques.[136]

Polyacrylonitrile is a polar material, giving the polymer good resistance to solvents, high rigidity, and low gas permeability.[137] Although the polymer degrades before melting, special techniques allowed a melting point of 317°C to be measured. The pure polymer is difficult to dissolve, but the copolymers can be dissolved in solvents such as methyl ethyl ketone, dioxane, acetone, dimethyl formamide, and tetrahydrofuran. Polyacrylonitrile exhibits exceptional barrier properties to oxygen and carbon dioxide.[138]

Copolymers of acrylonitrile with other monomers are widely used. Copolymers of vinylidene chloride and acrylonitrile find application in low-gas-permeability films. Styrene-acrylonitrile (SAN polymers) copolymers have also been used in packaging applications. Although the gas permeability of the copolymers is higher than for pure polyacrylonitrile, the acrylonitrile copolymers have lower gas permeability than many other packaging films. A number of acrylonitrile copolymers were developed for beverage containers, but the requirement for very low levels of residual acrylonitrile monomer in this application led to many products being removed from the market.[139] One copolymer currently available is Barex (BP Chemicals). The copolymer has better barrier properties than both polypropylene and polyethylene terephthalate.[140] Acrylonitrile is also used with butadiene and styrene to form ABS polymers. Unlike the homopolymer, copolymers can be processed by many methods including extrusion, blow molding, and injection molding.[141]

Acrylonitrile is often copolymerized with other monomers to form fibers. Copolymerization with monomers such as vinyl acetate, vinyl pyrrolidone, and vinyl esters gives the fibers the ability to be dyed using normal textile dyes. The copolymer generally contains at least 85 percent acrylonitrile.[142] Acrylic fibers have good abrasion resistance, flex life, toughness, and high strength. They have good resistance to stains and moisture. Modacrylic fibers contain between 35 percent and 85 percent acrylonitrile.[143]

Most of the acrylonitrile consumed goes into the production of fibers. Copolymers also consume large amounts of acrylonitrile. In addition to their use as fibers, polyacrylonitrile polymers can be used as precursors to carbon fibers.

1.5.7 Polyamide-imide (PAI)

Polyamide-imide (PAI) is a high-temperature amorphous thermoplastic that has been available since the 1970s under the trade name of Torlon.[144] PAI can be produced from the reaction of trimellitic trichloride with methylenedianiline as shown in Fig. 1.22.

Polyamide-imides can be used from cryogenic temperatures to nearly 260°C. They have the temperature resistance of the polyimides but better mechanical properties, including good stiffness and creep resistance. PAI polymers are inherently flame retardant, with little smoke produced when they are burned. The polymer has good chemical resistance, but, at high temperatures, it can be affected by strong acids and bases and steam.[145] PAI has a heat deflection temperature of 280°C along with good wear and friction properties.[146] Polyamide-imides also have good radiation resistance and are more stable than standard nylons under different humidity conditions. The polymer has one of the highest glass transition temperatures in the range of 270 to 285°C.[147]

Polyamide-imide can be processed by injection molding, but special screws are needed due to the reactivity of the polymer under molding conditions. Low-compression-ratio screws are recommended.[148] The parts should be annealed after molding at gradually increased temperatures.[149] For injection molding, the melt temperature should be near 355°C with mold temperatures of 230°C. PAI can also be processed by compression molding or used in solution form. For compression molding, preheating at 280°C, followed by molding between 330 to 340°C with a pressure of 30 MPa, is generally used.[150]

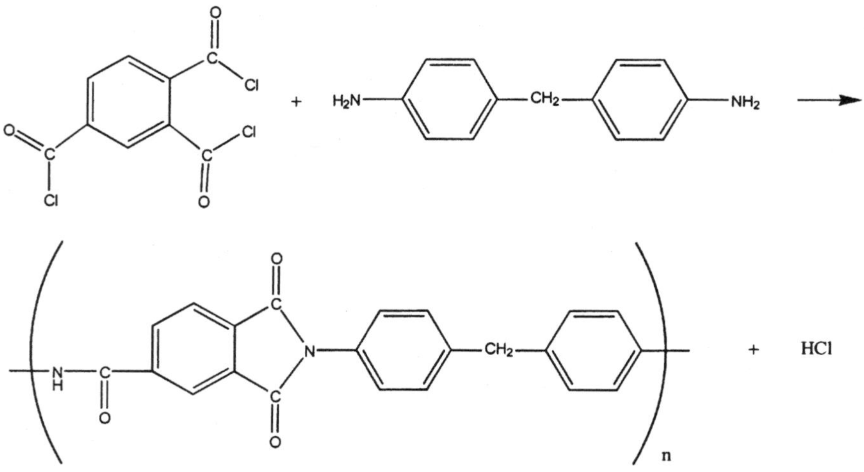

Figure 1.22 Preparation of polyamide-imide.

Polyamide-imide polymers find application in hydraulic bushings and seals, mechanical parts for electronics, and engine components.[151] The polymer in solution has application as a laminating resin for spacecraft, a decorative finish for kitchen equipment, and as wire enamel.[152] Low coefficient of friction materials may be prepared by blending PAI with polytetrafluoroethylene and graphite.[153]

1.5.8 Polyarylate

Polyarylates are amorphous, aromatic polyesters. Polyarylates are polyesters prepared from dicarboxylic acids and bis-phenols.[154] Bis-phenol A is commonly used along with aromatic dicarboxylic acids, such as mixtures of isophthalic acid and terephthalic acid. The use of two different acids results in an amorphous polymer; however, the presence of the aromatic rings gives the polymer a high T_g and good temperature resistance. The temperature resistance of polyarylates lies between polysulfone and polycarbonate. The polymer is flame retardant and shows good toughness and UV resistance.[155] Polyarylates are transparent and have good electrical properties. The abrasion resistance of polyarylates is superior to polycarbonate. In addition, the polymers show very high recovery from deformation.

Polarylates are processed by most of the conventional methods. Injection molding should be performed with a melt temperature of 260 to 382°C with mold temperatures of 65 to 150°C. Extrusion and blow molding grades are also available. Polyarylates can react with water at processing temperatures, and they should be dried prior to use.[156]

Polyarylates are used in automotive applications such as door handles, brackets, and headlamp and mirror housings. Polyarylates are also used in electrical applications for connectors and fuses. The polymer can be used in circuit board applications, because its high-temperature resistance allows the part to survive exposure to the temperatures generated during soldering.[157] The excellent UV resistance of these polymers allows them to be used as a coating for other thermoplastics for improved UV resistance of the part. The good heat resistance of polyarylates allows them to be used in applications such as fire helmets and shields.[158]

1.5.9 Polybenzimidazole (PBI)

Polybenzimidazoles (PBI) are high-temperature-resistant polymers. They are prepared from aromatic tetramines (for example tetra amino-biphenol) and aromatic dicarboxylic acids (diphenylisophthalate).[159] The reactants are heated to form a soluble prepolymer that is converted to the insoluble polymer by heating at temperatures above 300°C.[160] The general structure of PBI is shown below in Fig. 1.23.

The resulting polymer has high temperature stability, good chemical resistance, and nonflammability. The polymer releases very little toxic gas and does not melt when exposed to pyrolysis conditions. The polymer can be formed into fibers by dry-spinning processes. Polybenzimidazole is usually amorphous with a T_g near 430°C.[161] Under certain conditions, crystallinity may be obtained. The lack of many single bonds and the high glass transition temperature give this polymer its superior high-temperature resistance. In addition to the high-temperature resistance, the polymer exhibits good low-temperature toughness. PBI polymers show good wear and frictional properties along with excellent compressive strength and high surface hardness.[162] The properties of PBI at elevated temperatures are among the highest of the thermoplastics. In hot, aqueous solutions, the polymer may absorb water with a resulting loss in mechanical properties. Removal of moisture will restore the mechanical properties. The heat deflection temperature of PBI is higher than most thermoplastics, and this is coupled with a low coefficient of thermal expansion. PBI can withstand temperatures up to 760°C for short durations and exposure to 425°C for longer durations.

The polymer is not available as a resin and is generally not processed by conventional thermoplastic processing techniques, but rather by a high-temperature, high-pressure sintering process.[163] The polymer is available in fiber form, certain shaped forms, finished parts, and solutions for composite impregnation. PBI is often used in fiber form for a variety of applications such as protective clothing and aircraft furnishings.[164] Parts made from PBI are used as thermal insulators, electrical connectors, and seals.[165]

1.5.10 Polybutylene (PB)

Polybutylene polymers are prepared by the polymerization of 1-butene using Ziegler-Natta catalysts The molecular weights range from 770,000 to 3,000,000.[166] Copolymers with ethylene are often prepared as well. The chain structure is mainly isotactic and is shown in Fig. 1.24.[167]

The glass transition temperature for this polymer ranges from –17 to –25°C. Polybutylene resins are linear polymers exhibiting good resistance to creep at elevated temperatures and good resistance to environmental stress cracking.[168] They also show high impact strength, tear resistance, and puncture resistance. As with other polyolefins, polybutylene

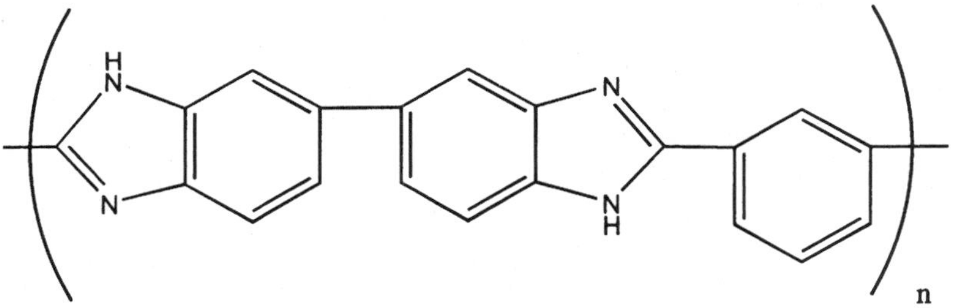

Figure 1.23 General structure of polybenzimidazoles.

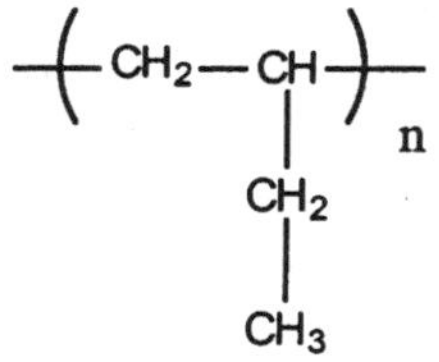

Figure 1.24 General structure for polybutylene.

shows good resistance to chemicals, good moisture barrier properties, and good electrical insulation properties. Pipes prepared from polybutylene can be solvent welded, yet the polymer still exhibits good environmental stress cracking resistance.[169] The chemical resistance is quite good below 90°C, but at elevated temperatures the polymer may dissolve in solvents such as toluene, decalin, chloroform, and strong oxidizing acids.[170]

Polybutylene is a crystalline polymer with three crystalline forms. The first crystalline form is obtained when the polymer is cooled from the melt. The first crystalline form is unstable and will change to a second crystalline form upon standing over a period of 3 to 10 days. The third crystalline form is obtained when polybutylene is crystallized from solution. The melting point and density of the first crystalline form are 124°C and 0.89 g/cm^3, respectively.[171] On transformation to the second crystalline form, the melting point increases to 135°C, and the density is increased to 0.95 g/cm^3. The transformation to the second crystalline form increases the polymer's hardness, stiffness, and yield strength.

Polybutylene can be processed on equipment similar to that used for low-density polyethylene. Polybutylene can be extruded and injection molded. Film samples can be blown or cast. The slow transformation from one crystalline form to another allows polybutylene to undergo post forming techniques such as cold forming of molded parts or sheeting.[172] A range of 160 to 240°C is typically used to process polybutylene.[173] The die swell and shrinkage are generally greater for polybutylene than for polyethylene. Because of the crystalline transformation, initially molded samples should be handled with care.

An important application for polybutylene is plumbing pipe for both commercial and residential use. The excellent creep resistance of polybutylene allows for the manufacture of thinner wall pipes as compared with pipes made from polyethylene or polypropylene. Polybutylene pipe can also be used for the transport of abrasive fluids. Other applications for polybutylene include hot-melt adhesives and additives for other plastics. The addition of polybutylene improves the environmental stress cracking resistance of polyethylene and the impact and weld line strength of polypropylene.[174] Polybutylene is also used in packaging applications.[175]

1.5.11 Polycarbonate

Polycarbonate (PC) is often viewed as the quintessential engineering thermoplastic, due to its combination of toughness, high strength, high heat-deflection temperatures, and transparency. The worldwide growth rate, predicted in 1999 to be between eight and ten percent, is hampered only by the resin cost and is paced by applications where PC can replace ferrous or glass products. Global consumption is anticipated to be more than 1.4 billion kilograms (3 billion pounds) by the year 2000.[176] The polymer was discovered in 1898 and by the year 1958 both Bayer in Germany and General Electric in the United States had commenced production. Two current synthesis processes are commercialized, with the economically most successful one said to be the "interface" process, which involves the dissolution of bisphenol A in aqueous caustic soda and the introduction of phosgene in the presence of an inert solvent such as pyridine. The bisphenol A monomer is dissolved in the aqueous caustic soda, then stirred with the solvent for phosgene. The water and solvent remain in separate phases. Upon phosgene introduction, the reaction occurs at the interface

with the ionic ends of the growing molecule being soluble in the catalytic caustic soda solution and the remainder of the molecule soluble in the organic solvent.[177] An alternative method involves transesterification of bisphenol A with diphenyl carbonate at elevated temperatures.[178] Both reactions are shown in Fig. 1.25.

Molecular weights of between 30,000 and 50,000 g/mol can be obtained by the second route, while the phosgenation route results in higher-molecular-weight product.

The structure of PC with its carbonate and bisphenolic structures has many characteristics that promote its distinguished properties. The para-substitution on the phenyl rings results in a symmetry and lack of stereospecificity. The phenyl and methyl groups on the quartenary carbon promote a stiff structure. The ester-ether carbonate groups –OCOO– are polar, but their degree of intermolecular polar bond formation is minimized due to the steric hindrance posed by the benzene rings. The high level of aromaticity on the backbone, and the large size of the repeat structure, yield a molecule of very limited mobility. The ether linkage on the backbone permits some rotation and flexibility, producing high impact strength. Its amorphous nature, with long, entangled chains, contributes to the unusually high toughness. Upon crystallization, however, PC is brittle. PC is so reluctant to crystallize that films must be held at 180°C for several days to impart enough flexibility and thermal mobility to conform to a structured three-dimensional crystalline lattice.[179] The rigidity of the molecule accounts for strong mechanical properties, elevated heat deflection temperatures, and high dimensional stability at elevated temperatures. The relative high free volume results in a low-density polymer, with unfilled PC having a 1.22 g/cm^3 density.

A disadvantage includes the need for drying and elevated temperature processing. PC has limited chemical resistance to numerous aromatic solvents, including benzene, toluene, and xylene, and has a weakness to notches. Selected mechanical and thermal properties are given in Table 1.7.[180]

When PC is blended with ABS, this increases the heat-distortion temperature of the ABS and improves the low-temperature impact strength of the PC. The favorable ease of processing and improved economics make PC/ABS blends well suited for thin-walled electronic housing applications such as laptop computers. Blends with PBT are useful for improving the chemical resistance of PC to petroleum products and its low-temperature impact strength. PC alone is widely used as vacuum cleaner housings, household appliance housings, and power tools. These are arenas in which PC's high impact strength, heat resistance, durability, and high-quality finish justify its expense. It is also used in safety helmets, riot shields, aircraft canopies, traffic light lens housings, and automotive battery cases. Design engineers take care not to design with tight radii where PC's tendency to

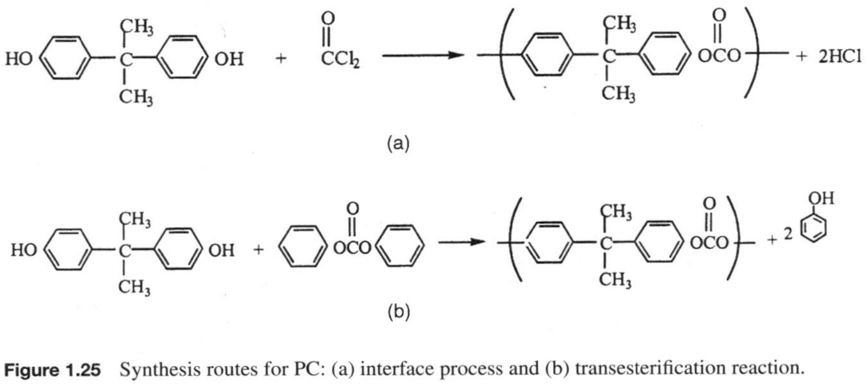

Figure 1.25 Synthesis routes for PC: (a) interface process and (b) transesterification reaction.

TABLE 1.7 PC Thermal and Mechanical Properties

	Polycarbonate	30 Percent glass-filled polycarbonate	Makroblend PR51, Bayer	Xenoy, CL101 GE
Heat-deflection temperature, °C Method A	138	280	90	95
Heat-deflection temperature, °C Method B	12	287	105	105
Ultimate tensile strength, N/mm^2	>65	70	56	>100
Ultimate elongation, %	110	3.5	120	>100
Tensile modulus, N/mm^2	2300	5500	2200	1900

stress crack could be a hindrance. PC cannot withstand constant exposure to hot water and can absorb 0.2% of its weight in water at 33°C and 65 percent RH. This does not impair its mechanical properties but, at levels greater than 0.01 percent, processing results in streaks and blistering.

1.5.12 Polyester Thermoplastic

The broad class of organic chemicals called *polyesters* are characterized by the fact that they contain an ester linkage,

$$-(\overset{\overset{\displaystyle O}{\|}}{C}-O)-$$

and may have either aliphatic or aromatic hydrocarbon units. As an introduction, Table 1.8 offers some selected thermal and mechanical properties as a means of comparing polybutylene terephthalate (PBT), polycyclohexylenedimethylene terephthalate (PCT), and poly(ethylene terephthalate) (PET).

1.5.12.1 Liquid crystal polymers (LCPs). Liquid crystal polyesters, known as *liquid crystal polymers,* are aromatic copolyesters. The presence of phenyl rings in the backbone of the polymer gives the chain rigidity, forming a rod-like chain structure. Generally, the phenyl rings are arranged in para linkages to give good rod-like structures.[181] This chain structure orients itself in an ordered fashion both in the melt and in the solid state as shown in Fig. 1.26. The materials are self-reinforcing, with high mechanical properties, but, as a result of the oriented liquid crystal behavior, the properties will be anisotropic. The designer must be aware of this so as to properly design the part and gate the molds.[182] The phenyl ring also helps increase the heat distortion temperature.[183]

The basic building blocks for liquid crystal polyesters are *p*-hydroxybenzoic acid, terephthalic acid, and hydroquinone. Unfortunately, the use of these monomers alone gives materials that are difficult to process, with very high melting points. The polymers often degrade before melting.[184] Various techniques have been developed to give materials with lower melting points and better processing behavior. Some methods include the incorporation of flexible units in the chain (copolymerizing with ethylene glycol), the addi-

TABLE 1.8 Comparison of Thermal and Mechanical Properties of PBT, PCT, PCTA, PET, PETG, and PCTG

	PBT unfilled	30% glass filled PBT	30% glass filled PCT	30% glass filled PCTA	PET unfilled	30% glass filled PET	PETG unfilled	PCTG unfilled
T_m, °C	220–267	220–267	–	285	212–265	245–265	–	–
Tensile modulus, MPa	1930–3000	8,960–10,000	–	–	2760–4140	8,960–9,930	–	–
Ultimate tensile strength, MPa	56–60	96–134	124–134	97	48–72	138–165	28	52
Ultimate elongation, %	50–300	2–4	1.9–2.3	3.1	30–300	2–7	110	330
Specific gravity	1.30–1.38	1.48–1.54	1.45	1.41	1.29–1.40	1.55–1.70	1.27	1.23
HDT, °C								
264 lb/in^2	50–85	196–225	260	221	21–65	210–227	64	65
66 lb/in^2	115–190	216–260	>260	268	75	243–249	70	74

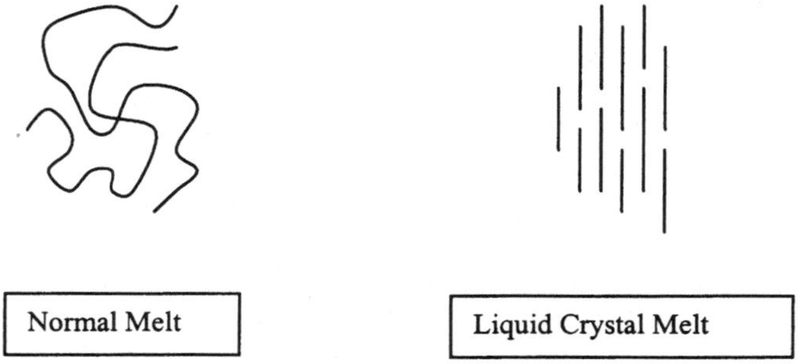

Figure 1.26 Melt configurations.

tion of nonlinear rigid structures, and the addition of aromatic groups to the side of the chain.[185]

Liquid crystal polymers based on these techniques include Victrex (ICI), Vectra (Hoescht Celanese), and Xydar (Amoco). Xydar is based on terephthalic acid, *p*-hydroxybenzoic acid, and *p,p′*-dihydroxybiphenyl, while Vectra is based on *p*-hydroxybenzoic acid and hydroxynaphthoic acid.[186] These materials are known for their high-temperature resistance, particularly heat distortion temperature. The heat distortion temperature can vary from 170 to 350°C. They also have excellent mechanical properties, especially in the flow direction. For example, the tensile strength varies from 165 to 230 MPa, the flexural strength varies from 169 to 256 MPa, and the flexural modulus varies from 9 to 12.5 GPa.[187] Filled materials exhibit even higher values. LCPs are also known for good solvent resistance and low water absorption as compared with other heat-resistant polymers. They have good electrical insulation properties, low flammability, with a limiting oxygen index in the range of 35 to 40, but a high specific gravity (about 1.40).[188] LCPs show little dimensional change when exposed to high temperatures and a low coefficient of thermal expansion.[189]

These materials can be high priced and often exhibit poor abrasion resistance, due to the oriented nature of the polymer chains.[190] Surface fibrillation may occur quite easily.[191] The materials are processable on a variety of conventional equipment. Process temperatures are normally below 350°C, although some materials may need to be processed higher. They generally have low melt viscosity as a result of their ordered melt and should be dried before use to avoid degradation.[192] LCPs can be injection molded on conventional equipment, and regrind may be used. Mold release is generally not required.[193] Part design for LCPs requires careful consideration of the anisotropic nature of the polymer. Weld lines can be very weak if the melt meets in a "butt" type of weld line. Other types of weld lines show better strength.[194]

Liquid crystal polymers are used in automotive, electrical, chemical processing, and household applications. One application is for oven and microwave cookware.[195] Because of their higher costs, the material will be used in applications only where their superior performance justifies the additional expense.

1.5.12.2 Polybutylene terephthalate (PBT). With the expiry of the original PET patents, manufacturers pursued the polymerization of other polyalkene terephthalates, particularly polybutylene terephthalate (PBT). The polymer is synthesized by reacting terephthalic acid with butane 1,4-diol to yield the structure shown below in Fig. 1.27.

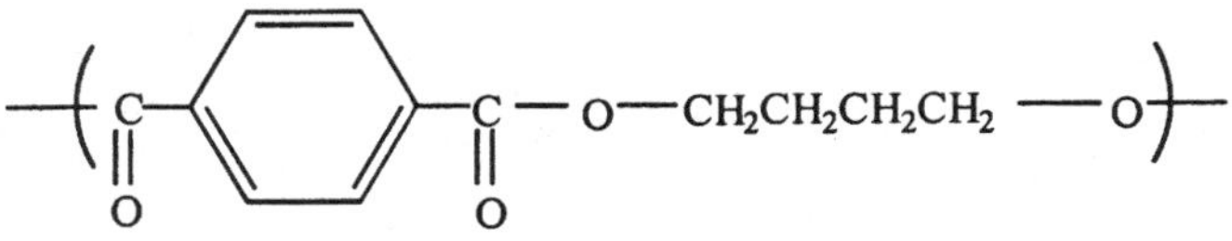

Figure 1.27 Repeat structure of PBT.

The only structural difference between PBT and PET is the substitution in PBT of four methylene repeat units rather than two present in PET. This feature imparts additional flexibility to the backbone and reduces the polarity of the molecule resulting in similar mechanical properties to PET (high strength, stiffness, and hardness). PBT growth is at least ten percent annually, in large part due to automotive exterior and under-hood applications such as electronic stability control and housings, which are made out of a PBT/ASA (acrylonitrile/styrene/acrylic ester) blend. PBT/ASA blends are sold by BASF and GE Plastics Europe. Another development involving the use of PBT is coextrusion of PBT and a copolyester thermoplastic elastomer. This can then be blow molded into under-hood applications to minimize noise vibration. Highly filled PBTs are also making inroads into the kitchen and bathroom tile industries.[196] As with PET, PBT is also often glass fiber filled so as to increase its flexural modulus, creep resistance, and impact strength. PBT is suitable for applications requiring dimensional stability, particularly in water, and resistance to hydrocarbon oils without stress cracking.[197] Hence, PBT is used in pump housings, distributors, impellers, bearing bushings, and gear wheels.

To improve PBT's poor notched impact strength, copolymerization with five percent ethylene and vinyl acetate onto the polyester backbone improves its toughness. PBT is also blended with PMMA, PET, PC, and polybutadiene to provide enhanced properties tailored to specific applications. Table 1.9 shows a breakdown of the U.S. market use for PBT.[198]

TABLE 1.9 U.S. Markets for PBT Use in l997 and l 998

Market	Millions of pounds, 1997	Millions of pounds, 1998
Appliances	29	31
Consumer/recreational	13	14
Electrical/electronic	58	65
Industrial	36	40
Transportation, including PC/PBT blends	136	154
Other	13	14
Export	28	28
Total	313	346

1.5.12.3 Polycyclohexylenedimethylene terephthalate (PCT). Another polyalkylene terephthalate polyester of significant commercial importance is PCT—a condensation product of the reaction between dimethyl terephthalate and 1,4-cyclohexylene glycol as shown below in Fig. 1.28.

This material is biaxially oriented into films and, while it is mechanically weaker than PET, it offers superior water resistance and weather resistance.[199] As seen in the introductory Table 1.8, PCT differentiates itself from PET and PBT with its high heat distortion

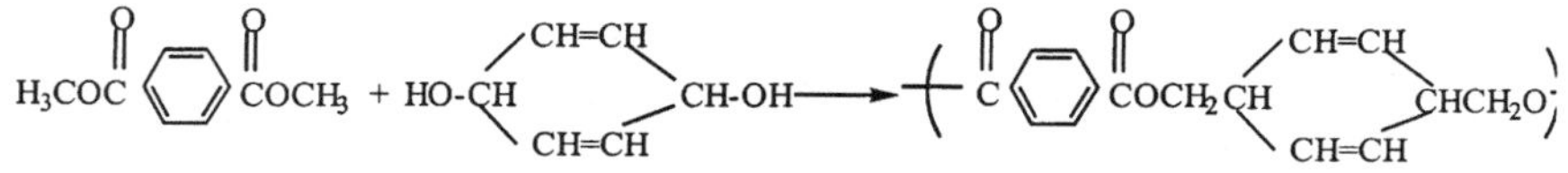

Figure 1.28 Synthesis route of PCT.

temperature. As with PET and PBT, PCT has low moisture absorption, and its good chemical resistance to engine fluids and organic solvents lend it to under-hood applications such as alternator armatures and pressure sensors.[200]

Copolymers of PCT include PCTA, an acid-modified polyester, and PCTG, a glycol-modified polyester. PCTA is used primarily for extruded film and sheet for packaging applications. PCTA has high clarity, tear strength, and chemical resistance, and when PCTA is filled, it is used for dual ovenable cookware. PCTG is primarily injection molded, and PCTG parts have notched Izod impact strengths similar to polycarbonate, against which it often competes. It also competes with ABS, another clear polymer. It finds use in medical and optical applications.[201]

1.5.12.4 Poly(ethylene terephthalate) (PET). There are tremendous commercial applications for PET, as an injection molding grade material, for blow molded bottles, and for oriented films. In 1998, the U.S. consumption of PET was 4,330 million pounds, while domestic consumption of PBT was 346 million pounds.[202] PET, also known as poly(oxyethylene oxyterephthaloyl), can be synthesized from dimethyl terephthalate and ethylene glycol by a two-step ester interchange process, as shown in Fig. 1.29.[203] The first stage involves a solution polymerization of one mole of dimethyl terephthalate with 2.1 to 2.2 moles of ethylene glycol.[204] The excess ethylene glycol increases the rate of formation of bis(2-hydroxyethyl) terephthalate. Small amounts of trimer, tetramer, and other oligomers are formed. A metal alkanoate, such as manganese acetate, is often added as a catalyst; this is later deactivated by the addition of a phosphorous compound such as phosphoric acid. The antioxidant phosphate improves the thermal and color stability of the polymer during the higher-temperature second-stage process.[205] The first stage of the reaction is run at 150 to 200°C with continuous methanol distillation and removal.[206]

The second step of the polymerization, shown in Fig. 1.30, is a melt polymerization as the reaction temperature is raised to 260 to 290°C. This second stage is carried out under either partial vacuum (0.13 kPa)[207] to facilitate the removal of ethylene glycol or with an inert gas being forced through the reaction mixture. Antimony trioxide is often used as a polymerization catalyst for this stage.[208] It is critical that excess ethylene glycol be completely removed during this alcoholysis stage of the reaction to proceed to high-molecu-

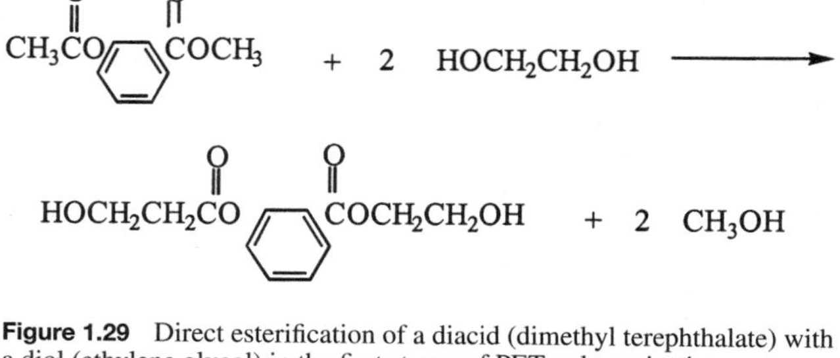

Figure 1.29 Direct esterification of a diacid (dimethyl terephthalate) with a diol (ethylene glycol) in the first stages of PET polymerization.

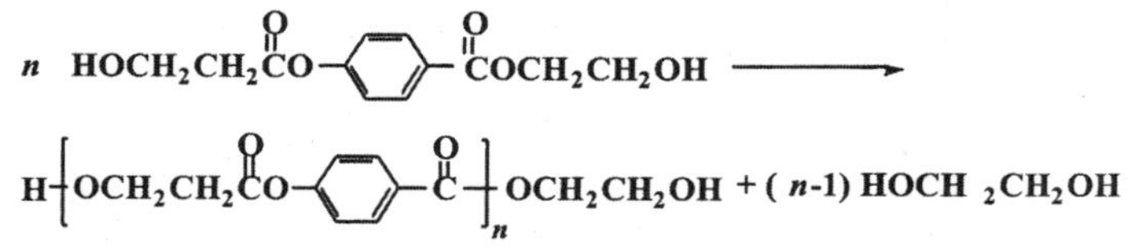

Figure 1.30 Polymerization of bis(2-hydroxyethyl) terephthalate to PET.

lar- weight products; otherwise, equilibrium is established at an extent of reaction of less than 0.7. This second stage of the reaction proceeds until a number-average molecular weight, M_n, of about 20,000 g/mol is obtained. The very high temperatures at the end of this reaction cause thermal decomposition of the end groups to yield acetaldehyde. Thermal ester scission also occurs, which competes with the polymer step-growth reactions. It is this competition that limits the ultimate M_n that can be achieved through this melt condensation reaction.[209] Weight-average molecular weights of oriented films are around 35,000 g/mol.

Other commercial manufacturing methods have evolved to a direct esterification of acid and glycol in place of the ester-exchange process. In direct esterification, terephthalic acid and ethylene glycol are reacted, rather than esterifying terephthalic acid with methanol to produce the dimethyl terephthalate intermediate. The ester is easier to purify than the acid, which sublimes at 300°C and is insoluble; however, better catalysts and purer terephthalic acid offer the elimination of the intermediate use of methanol.[210] Generally, PET resins made by direct esterification of terephthalic acid contain more diethylene glycol, which is generated by an intermolecular ether-forming reaction between ß-hydroxyethyl ester end groups. Oriented films produced from these resins have reduced mechanical strength and melting points as well as decreased thermo-oxidative resistance and poorer UV stability.[211]

The degree of crystallization and direction of the crystallite axis govern all of the resin's physical properties. The percentage of structure existing in crystalline domains is primarily determined through density measurements or by thermal means using a differential scanning calorimeter (DSC). The density of amorphous PET is 1.333 g/cm^3, while the density of a PET crystal is 1.455 g/cm^3.[212] Once the density is known, the fraction of crystalline material can be determined.

An alternate means of measuring crystallinity involves comparing the ratio of the heat of cold crystallization, ΔH_{cc}, of amorphous polymer to the heat of fusion, ΔH_f, of crystalline polymer. This ratio is 0.61 for an amorphous PET and a fully crystalline PET sample should yield a value close to zero.[213] After the sample with its initial morphology has been run once in the DSC, the heat of fusion determined in the next run can be considered as ΔH_{cc}. The lower the $\Delta H_{cc}/\Delta H_f$ ratio, the more crystalline the original sample was.

In the absence of nucleating agents and plasticizers, PET crystallizes slowly, which is a hindrance in injection molding applications, as either hot molds or costly extended cooling times are required. In the case of films, however, where crystallinity can be mechanically induced, PET resins combine rheological properties that lend themselves to melt extrusion with a well defined melting point, making them ideally suited for biaxially-oriented film applications. The attachment of the ester linkage directly to the aromatic component of the backbone means that these linear, regular PET chains have enough flexibility to form stress-induced crystals and achieve enough molecular orientation to form strong, thermally stable films.[214]

Methods for producing oriented PET films have been well documented and will be only briefly discussed here. The process as described in the *Encyclopedia of Polymer Science and Engineering* usually involves a sequence of five steps that include melt extrusion and slot casting, quenching, drawing in the longitudinal machine direction (MD), drawing in

the transverse direction (TD), and annealing.[215] Dried, highly viscous polymer melt is extruded through a slot die with an adjustable gap width onto a highly polished quenching drum. If very high output rates are required, a cascade system of extruders can be set up to first melt and homogenize the PET granules, then to use the next in-line extruder to meter the melt to the die. Molten resin is passed through filter packs with average pore sizes of 5 to 30 μm. Quenching to nearly 100 percent amorphous morphology is critical to avoid embrittlement; films that have been allowed to form spherulites are brittle, translucent, and unable to be further processed.

The sheet is then heated to about 95°C (above the glass transition point of approximately 70°C), where thermal mobility allows the material to be stretched to three or four times its original dimension in the MD. This uniaxially oriented film has stress-induced crystals whose main axes are aligned in the machine direction. The benzene rings, however, are aligned parallel to the surface of the film in the <1,0,0> crystal plane. The film is then again heated, generally to above 100°C, and stretched to three to four times its initial dimension in the TD. This induces further crystallization bringing the degree of crystallinity to 25 to 40 percent and creates a film that has isotropic tensile strength and elongation properties in the machine and transverse directions. The film at this point is thermally unstable above 100°C and must be annealed in the tenter frame to partially relieve the stress.

The annealing involves heating to 180 to 220°C for several seconds to allow amorphous chain relaxation, partial melting, recrystallization, and crystal growth to occur.[216] The resultant film is approximately 50 percent crystalline and possesses good mechanical strength and a smooth surface that readily accepts a wide variety of coatings, and it has good winding and handling characteristics. PET films are produced from 1.5 μm thick as capacitor films to 350 μm thick for use as electrical insulation in motors and generators.[217]

Due to the chemically inert nature of PET, films that are used in coatings applications are often treated with a variety of surface modifiers. Organic and inorganic fillers are often incorporated in relatively thick films to improve handling characteristics by roughening the surface slightly. For thin films, however, many applications require transparency that would be marred by the incorporation of fillers. Therefore an in-line coating step of either aqueous or solvent-based coatings is set up between the MD and TD drawing stations. The drawing of the film after the coating has been applied helps to achieve very thin coatings.

1.5.13 Polyetherimide (PEI)

Polyetherimides (PEIs) are a newer class of amorphous thermoplastics with high temperature resistance, impact strength, creep resistance, and rigidity. They are transparent with an amber color.[218] The polymer is sold under the trade name of Ultem (General Electric) and has the structure shown in Fig. 1.31. It is prepared from the condensation polymerization of diamines and dianhydrides.[219]

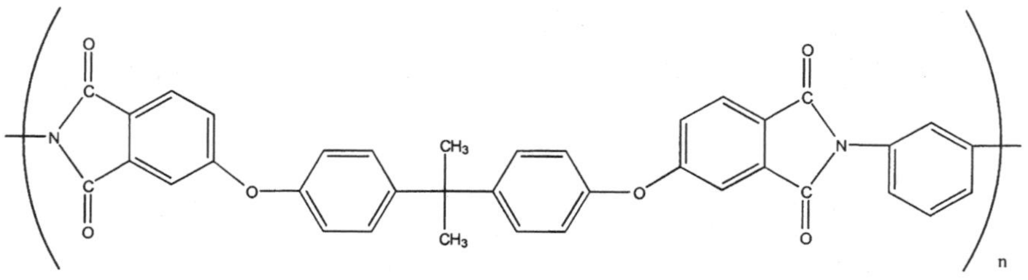

Figure 1.31 General structure of polyetherimide.

The material can be melt processed because of the ether linkages present in the backbone of the polymer, but it still maintains properties similar to those of the polyimides.[220] The high-temperature resistance of the polymer allows it to compete with the polyketones, polysulfones, and poly(phenylene sulfides). The glass transition temperature of PEI is 215°C. The polymer has very high tensile strength, a UL temperature index of 170°C, flame resistance, and low smoke emission.[221] The polymer is resistant to alcohols, acids, and hydrocarbosolvents but will dissolve in partially halogenated solvents.[222] Both glass and carbon fiber reinforced grades are available.[223]

The polymer should be dried before processing, and typical melt temperatures are 340 to 425°C.[224] Polyetherimides can be processed by injection molding and extrusion. In addition, the high melt strength of the polymer allows it to be thermoformed and blow molded. Annealing of the parts is not required.

Polyetherimide is used in a variety of applications. Electrical applications include printed circuit substrates and burn-in sockets. In the automotive industry, PEI is used for under-hood temperature sensors and lamp sockets. PEI sheet has also been used to form an aircraft cargo vent.[225] The dimensional stability of this polymer allows its use for large flat parts such in hard disks for computers.

1.5.14 Polyethylene (PE)

Polyethylene (PE) is the highest-volume polymer in the world. Its high toughness, ductility, excellent chemical resistance, low water vapor permeability, and very low water absorption, combined the ease with which it can be processed, make PE of all different density grades an attractive choice for a variety of goods. PE is limited by its relatively low modulus, yield stress, and melting point. PE is used to make containers, bottles, film, and pipes, among other things. It is an incredibly versatile polymer with almost limitless variety due to copolymerization potential, a wide density range, a molecular weight (MW) that ranges from very low (waxes have an MW of a few hundred) to very high (6×10^6), and the ability to vary molecular weight distribution (MWD).

Its repeat structure is $(–CH_2CH_2–)_x$, which is written as polyethylene rather than polymethylene $(–CH_2)_x$, in deference to the various ethylene polymerization mechanisms. PE has a deceptive simplicity. PE homopolymers are made up exclusively of carbon and hydrogen atoms, and, just as the properties of diamond and graphite (which are also materials made up entirely of carbon and hydrogen atoms) vary tremendously, different grades of PE have markedly different thermal and mechanical properties. While PE is generally a whitish, translucent polymer, it is available in grades of density that range from 0.91 to 0.97 g/cm^3. The density of a particular grade is governed by the morphology of the backbone; long, linear chains with very few side branches can assume a much more three-dimensionally compact, regular, crystalline structure. Commercially available grades are

- Low-density PE (LDPE)
- Linear low-density PE (LLDPE)
- High-density PE (HDPE)
- Ultrahigh-molecular-weight PE (UHMWPE)

Figure 1.32 demonstrates figurative differences in chain configuration that govern the degree of crystallinity, which, along with MW, determines final thermomechanical properties.

Four established production methods are (1) a gas phase method known as the Unipol process, practiced by Union Carbide, (2) a solution method used by Dow and DuPont, (3) a slurry emulsion method practiced by Phillips, and (4) a high-pressure method.[226] Generally, yield strength and the melt temperature increase with density, while elongation decreases with increased density.

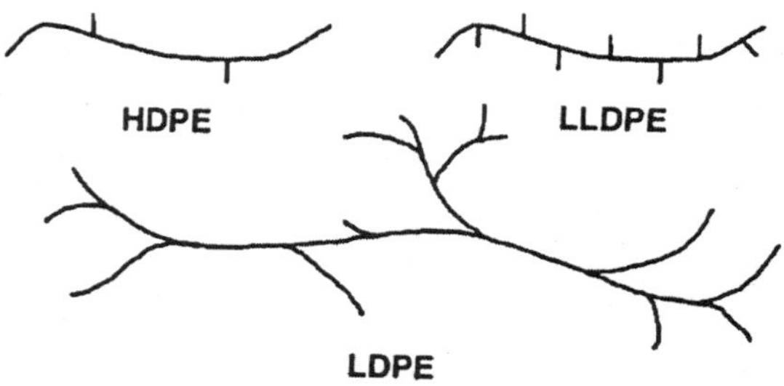

Figure 1.32 Chain configurations of polyethylene.

1.5.14.1 Very low-density polyethylene (VLDPE). This material was introduced in 1985 by Union Carbide, is very similar to LLDPE, and is principally used in film applications. VLDPE grades vary in density from 0.880 to 0.912 g/cm^3.[227] Its properties are marked by high elongation, good environmental stress cracking resistance, and excellent low-temperature properties, and it competes most frequently as an alternative to plasticized polyvinyl chloride (PVC) or ethylene-vinyl acetate (EVA). The inherent flexibility in the backbone of VLDPE circumvents plasticizer stability problems that can plague PVC, and it avoids odor and stability problems that are often associated with molding EVAs.[228]

1.5.14.2 Low-density polyethylene (LDPE). LDPE combines high impact strength, toughness, and ductility to make it the material of choice for packaging films, which is one of its largest applications. Films range from shrink film, thin film for automatic packaging, heavy sacking, and multilayer films (both laminated and coextruded) where LDPE acts as a seal layer or a water vapor barrier.[229] It has found stiff competition from LLDPE in these film applications due to LLDPE's higher melt strength. LDPE is still very widely used, however, and is formed via free radical polymerization, with alkyl branch groups [given by the structure $-(CH_2)_xCH_3$] of two to eight carbon atom lengths. The most common branch length is four carbons long. High reaction pressures encourage crystalline regions. The reaction to form LDPE is shown in Fig. 1.33, where "n" approximately varies in commercial grades between 400 and 50,000.[230]

Medium-density PE is produced via the reaction above, carried out at lower polymerization temperatures.[231] The reduced temperatures are postulated to reduce the randomizing Brownian motion of the molecules, and this reduced thermal energy allows crystalline formation more readily at these lowered temperatures.

1.5.14.3 Linear low-density polyethylene (LLDPE). This product revolutionized the plastics industry with its enhanced tensile strength for the same density, as compared

$$n\ CH_2{=}CH_2 \xrightarrow[\text{small amounts of } O_2 \text{ or organic peroxide present}]{200\text{ C},\ 20{,}000-35{,}000\text{ psi}} -(CH_2CH_2)_n-$$

Figure 1.33 Polymerization of PE.

with LDPE. Table 1.10 compares mechanical properties of LLDPE to LDPE. As is the case with LDPE, film accounts for approximately three-quarters of the consumption of LLDPE. As the name implies, it is a long linear chain without long side chains or branches. The short chains that are present disrupt the polymer chain uniformity enough to prevent crystalline formation and hence prevent the polymer from achieving high densities. Developments of the past decade have enabled production economies compared with LDPE due to lower polymerization pressures and temperatures. A typical LDPE process requires 35,000 lb/in^2, which is reduced to 300 lb/in^2 in the case of LLDPE, and reaction temperatures as low as 100°C rather than 200 to 300°C are used. LLDPE is actually a copolymer containing side branches of 1-butene most commonly, with 1-hexene or 1-octene also present. Density ranges of 0.915 to 0.940 g/cm^3 are polymerized with Ziegler catalysts, which orient the polymer chain and govern the tacticity of the pendant side groups.[232]

TABLE 1.10 Comparison of Blown Film Properties of LLDPE and LDPE[455]

	LLDPE	LDPE
Density, g/cm^3	0.918	0.918
Melt index, g/10 min	2.0	2.0
Dart impact, g	110	110
Puncture energy, J/mm	60	25
Machine-direction tensile strength, MPa	33	20
Cross-direction tensile strength, MPa	25	18
Machine-direction tensile elongation, %	690	300
Cross-direction tensile elongation, %	740	500
Machine-direction modulus, MPa	210	145
Cross-direction modulus, MPa	350	175

1.5.14.4 High-density polyethylene (HDPE). HDPE is one of the highest-volume commodity chemicals produced in the world. In 1998, the worldwide demand was 1.8×10^{10} kg.[233] The most common method of processing HDPE is blow molding, where resin is turned into bottles (especially for milk and juice), housewares, toys, pails, drums, and automotive gas tanks. It is also commonly injection molded into housewares, toys, food containers, garbage pails, milk crates, and cases. HDPE films are commonly found as bags in supermarkets and department stores, and as garbage bags.[234] Two commercial polymerization methods are most commonly practiced; one involves Phillips catalysts (chromium oxide) and the other involves Ziegler-Natta catalyst systems (supported heterogeneous catalysts such as titanium halides, titanium esters, and aluminum alkyls on a chemically inert support such as PE or PP). Molecular weight is governed primarily through temperature control, with elevated temperatures resulting in reduced molecular weights. The catalyst support and chemistry also play an important factor in controlling molecular weight and molecular weight distribution.

1.5.14.5 Ultrahigh-molecular-weight polyethylene (UHMWPE). UHMWPE is identical to HDPE but, rather than having a MW of 50,000 g/mol, it typically has a MW of between 3×10^6 and 6×10^6. The high MW imparts outstanding abrasion resistance, high toughness (even at cryogenic temperatures), and excellent stress cracking resistance, but it

does not generally allow the material to be processed conventionally. The polymer chains are so entangled due to their considerable length that the conventionally considered melt point doesn't exist practically, as it is too close to the degradation temperature, although an injection molding grade is marketed by Hoechst. Hence, UHMWPE is often processed as a fine powder that can be ram extruded or compression molded. Its properties are taken advantage of in uses that include liners for chemical processing equipment, lubrication coatings in railcar applications to protect metal surfaces, recreational equipment such as ski bases, and medical devices.[235] A recent product has been developed by Allied Chemical that involves gel-spinning UHMWPE into light weight, very strong fibers that compete with Kevlar in applications for protective clothing.

1.5.15 Polyethylene Copolymers

Ethylene is copolymerized with many non-olefinic monomers, particularly acrylic acid variants and vinyl acetate, with EVA polymers being the most commercially significant. All of the copolymers discussed in this section necessarily involve disruption of the regular, crystallizable PE homopolymer and as such feature reduced yield stresses and moduli, with improved low-temperature flexibility.

1.5.15.1 Ethylene-acrylic acid (EAA) copolymers. EAA copolymers, first identified in the 1950s, experienced renewed interest when, in 1974, Dow introduced new grades characterized by outstanding adhesion to metallic and nonmetallic substrates.[236] The presence of the carboxyl and hydroxyl functionalities promotes hydrogen bonding, and these strong intermolecular interactions are taken advantage of to bond aluminum foil to polyethylene in multilayer extrusion-laminated toothpaste tubes and as tough coatings for aluminum foil pouches.

1.5.15.2 Ethylene-ethyl acrylate (EEA) copolymers. EEA copolymers typically contain 15 to 30 percent by weight of ethyl acrylate (EA) and are flexible polymers of relatively high molecular weight suitable for extrusion, injection molding, and blow molding. Products made of EEA have high environmental stress cracking resistance, excellent resistance to flexural fatigue, and low-temperature properties down to as low as –65°C. Applications include molded rubber-like parts, flexible film for disposable gloves and hospital sheeting, extruded hoses, gaskets, and bumpers.[237] Typical applications include polymer modifications where EEA is blended with olefin polymers (since it is compatible with VLDPE, LLDPE, LDPE, HDPE, and PP[238]) to yield a blend with a specific modulus, yet with the advantages inherent in EEA's polarity. The EA presence promotes toughness, flexibility, and greater adhesive properties. EEA blending can cost-effectively improve the impact resistance of polyamides and polyesters.[239]

The similarity of ethyl acrylate monomer to vinyl acetate predicates that these copolymers have very similar properties, although EEA is considered to have higher abrasion and heat resistance, while EVA tends to be tougher and of greater clarity.[240] EEA copolymers are FDA approved up to 8 percent EA content in food contact applications.[241]

1.5.15.3 Ethylene-methyl acrylate (EMA) copolymers. EMA copolymers are often blown into film with very rubbery mechanical properties and outstanding dart-drop impact strength. The latex rubber-like properties of EMA film lend to its use in disposable glove and medical devices without the associated hazards to people with allergies to latex

rubber. Due to their adhesive properties, EMA copolymers, like their EAA and EEA counterparts, are used in extrusion coating, coextrusions, and laminating applications as heat-seal layers. EMA is one of the most thermally stable of this group, and as such it is commonly used to form heat and RF seals as well in multiextrusion tie-layer applications. This copolymer is also widely used as a blending compound with olefin homopolymers (VLDPE, LLDPE, LDPE, and PP) as well as with polyamides, polyesters, and polycarbonate to improve impact strength and toughness and to increase either heat seal response or to promote adhesion.[242] EMA is also used in soft blow-molded articles such as squeeze toys, tubing, disposable medical gloves, and foamed sheet. EMA copolymers and EEA copolymers containing up to 8 percent ethyl acrylate are approved by the FDA for food packaging.[243]

1.5.15.4 Ethylene-n-butyl acrylate (EBA) copolymers. EBA copolymers are also widely blended with olefin homopolymers to improve impact strength, toughness, and heat sealability, and to promote adhesion. The polymerization process and resultant repeat unit of EBA are shown in Fig. 1.34.

1.5.15.5 Ethylene-vinyl acetate (EVA) copolymers. EVA copolymers are given by the structure shown in Fig. 1.35 and find commercial importance in the coating, laminating, and film industries. EVA copolymers typically contain between 10 and 15 mole percent vinyl acetate, which provides a bulky, polar pendant group to the ethylene and provides an opportunity to tailor the end properties by optimizing the vinyl acetate content. Very low vinyl-acetate content (approximately 3 mole percent) results in a copolymer that is essentially a modified low density polyethylene,[244] with an even further reduced regular structure. The resultant copolymer is used as a film due to its flexibility and surface gloss.

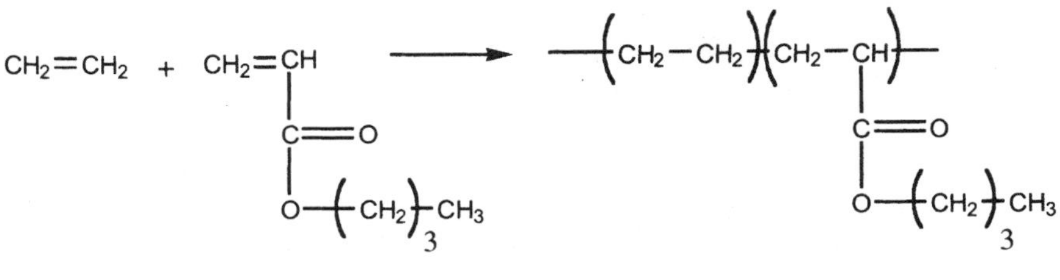

Figure 1.34 Polymerization and structure of EBA.

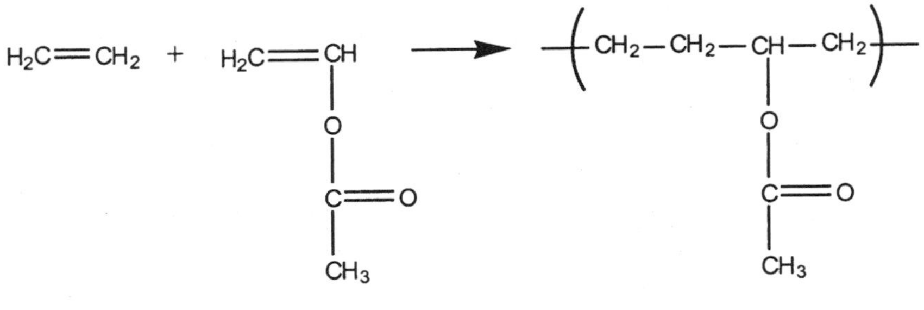

Figure 1.35 Polymerization of EVA.

Vinyl acetate is a low-cost co-monomer that is nontoxic, which allows for this copolymer to be used in many food packaging applications. These films are soft and tacky and therefore appropriate for cling-wrap applications (they are more thermally stable than the PVDC films often used as cling wrap) as well as interlayers in coextruded and laminated films.

EVA copolymers with approximately 11 mole percent vinyl acetate are widely used in the hot-melt coatings and adhesives arena, where the additional intermolecular bonding promoted by the polarity of the vinyl acetate ether and carbonyl linkages enhances melt strength while still enabling low melt processing temperatures. At 15 mole percent vinyl acetate, a copolymer with very similar mechanical properties to plasticized PVC is formed. There are many advantages to an inherently flexible polymer for which there is no risk of plasticizer migration, and PVC alternatives is the area of largest growth opportunity. These copolymers have higher moduli than standard elastomers and are preferable in that they are more easily processed without concern for the need to vulcanize.

1.5.15.6 Ethylene-vinyl alcohol (EVOH) copolymers. Poly(vinyl alcohol) is prepared through alcoholysis of poly(vinyl acetate). PVOH is an atactic polymer, but, since the crystal lattice structure is not disrupted by hydroxyl groups, the presence of residual acetate groups greatly diminishes the crystal formation and the degree of hydrogen bonding. Polymers that are highly hydrolyzed (have low residual acetate content) have a high tendency to crystallize and to undergo hydrogen bonding. As the degree of hydrolysis increases, the molecules will very readily crystallize, and hydrogen bonds will keep them associated if they are not fully dispersed prior to dissolution. At degrees of hydrolysis above 98 percent, manufacturers recommend a minimum temperature of 96°C to ensure that the highest-molecular-weight components have enough thermal energy to go into solution. Polymers with low degrees of residual acetate have high humidity resistance.

1.5.16 Modified Polyethylenes

The properties of PE can be tailored to meet the needs of a particular application by a variety of different methods. Chemical modification, copolymerization, and compounding can all dramatically alter specific properties. The homopolymer itself has a range of properties depending upon the molecular weight, the number and length of side branches, the degree of crystallinity, and the presence of additives such as fillers or reinforcing agents. Further modification is possible by chemical substitution of hydrogen atoms; this occurs preferentially at the tertiary carbons of a branching point and primarily involves chlorination, sulfonation, phosphorylination, and intermediate combinations.

1.5.16.1 Chlorinated polyethylene (CPE). The first patent on the chlorination of PE was awarded to ICI in 1938.[245] CPE is polymerized by substituting select hydrogen atoms on the backbone of either HDPE or LDPE with chlorine. Chlorination can occur in the gaseous phase, in solution, or as an emulsion. In the solution phase, chlorination is random, while the emulsion process can result in uneven chlorination due to the crystalline regions. The chlorination process generally occurs by a free-radical mechanism, shown in Fig. 1.36, where the chlorine free radical is catalyzed by ultraviolet light or initiators.

Interestingly, the properties of CPE can be adjusted to almost any intermediary position between PE and PVC by varying the properties of the parent PE and the degree and tacticity of chlorine substitution. Since the introduction of chlorine reduces the regularity of the PE, crystallinity is disrupted and, at up to a 20 percent chlorine level, the modified material is rubbery (if the chlorine was randomly substituted). When the level of chlorine

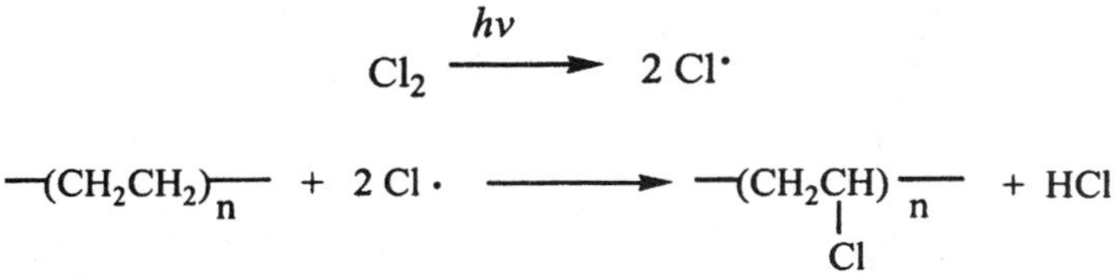

Figure 1.36 Chlorination process of CPE.

reaches 45 percent (approaching PVC), the material is stiff at room temperature. Typically, HDPE is chlorinated to a chlorine content of 23 to 48 percent.[246] Once the chlorine substitution reaches 50 percent, the polymer is identical to PVC, although the polymerization route differs. The largest use of CPE is as a blending agent with PVC to promote flexibility and thermal stability for increased ease of processing. Blending CPE with PVC essentially plasticizes the PVC without adding double-bond unsaturation prevalent with rubber-modified PVCs and results in a more UV-stable, weather-resistant polymer. While rigid PVC is too brittle to be machined, the addition of as little as three to six parts per hundred CPE in PVC allows extruded profiles such as sheets, films, and tubes to be sawed, bored, and nailed.[247] Higher CPE content blends result in improved impact strength of PVC and are made into flexible films that do not have plasticizer migration problems. These films find applications in roofing, water and sewage-treatment pond covers, and sealing films in building construction.

CPE is used in highly filled applications, often using $CaCO_3$ as the filler, and finds use as a homopolymer in industrial sheeting, wire and cable insulations, and solution applications. When PE is reacted with chlorine in the presence of sulfur dioxide, a chlorosulfonyl substitution takes place, yielding an elastomer.

1.5.16.2 Chlorosulfonated polyethylenes (CSPE). Chlorosulfonation introduces the polar, cross-linkable SO_2 group onto the polymer chain, with the unavoidable introduction of chlorine atoms as well. The most common method involves exposing LDPE, which has been solubilized in a chlorinated hydrocarbon, to SO_2 and Cl in the presence of UV or high-energy radiation.[248] Both linear and branched PEs are used, and CSPEs contain 29 to 43 percent chlorine and 1 to 1.5 percent sulfur.[249] As in the case of CPEs, the introduction of Cl and SO_2 functionalities reduces the regularity of the PE structure, hence reducing the degree of crystallinity, and the resultant polymer is more elastomeric than the unmodified homopolymer. CSPE is manufactured by DuPont under the trade name Hypalon and is used in protective coating applications such as the lining for chemical processing equipment; as the liners and covers for waste-containment ponds; as cable jacketing and wire insulation, spark plug boots, power steering pressure hoses; and in the manufacture of elastomers.

1.5.16.3 Phosphorylated polyethylenes. Phosphorylated PEs have higher ozone and heat resistance than ethylene propylene copolymers due to the fire retardant nature provided by phosphor.[250]

1.5.16.4 Ionomers. Acrylic acid can be copolymerized with polyethylene to form an ethylene acrylic acid copolymer (EAA) through addition or chain growth polymerization. It is structurally similar to ethylene vinyl acetate but with acid groups off the backbone.

The concentration of acrylic acid groups is generally in the range of 3 to 20 percent.[251] The acid groups are then reacted with a metal containing base, such as sodium methoxide or magnesium acetate, to form the metal salt as depicted in Fig. 1.37.[252] The ionic groups can associate with each other, forming a cross-link between chains. The resulting materials are called *ionomers* in reference to the ionic bonds formed between chains. They were originally developed by DuPont under the trade name of Surlyn.

The association of the ionic groups forms a thermally reversible cross-link that can be broken when exposed to heat and shear. This allows ionomers to be processed on conventional thermoplastic processing equipment while still maintaining some of the behavior of a thermoset at room temperature.[253] The association of ionic groups is generally believed to take two forms: multiplets and clusters.[254] Multiplets are considered to be a small number of ionic groups dispersed in the matrix, while clusters are phase-separated regions containing many ion pairs and also hydrocarbon backbone.

A wide range of properties can be obtained by varying the ethylene/methacrylic acid ratios, molecular weight, and the amount and type of metal cation used. Most commercial grades use either zinc or sodium for the cation. Materials using sodium as the cation generally have better optical properties and oil resistance, while those using zinc usually have better adhesive properties, lower water absorption, and better impact strength.[255]

The presence of the co-monomer breaks up the crystallinity of the polyethylene, so ionomer films have lower crystallinity and better clarity as compared with polyethylene.[256] Ionomers are known for their toughness and abrasion resistance, and the polar nature of the polymer improves both its bondability and paintability. Ionomers have good low-temperature flexibility and resistance to oils and organic solvents. Ionomers show a yield point with considerable cold drawing. In contrast to PE, the stress increases with strain during cold drawing, giving a very high energy to break.[257]

Ionomers can be processed by most conventional extrusion and molding techniques using conditions similar to those used for other olefin polymers. For injection molding, the melt temperatures are in the range 210 to 260°C.[258] The melts are highly elastic, due to the presence of the metal ions. Increasing the temperatures rapidly decreases the melt viscosity, with the sodium- and zinc-based ionomers showing similar rheological behavior. Typical commercial ionomers have melt index values between 0.5 and 15.[259] Both unmodified and glass-filled grades are available.

Ionomers are used in applications such as golf ball covers and bowling pin coatings, where their good abrasion resistance is important.[260] The puncture resistance of films allows these materials to be widely used in packaging applications. One of the early applications was the packaging of fish hooks.[261] They are often used in composite products as an outer heat-seal layer. Their ability to bond to aluminum foil is also utilized in packaging applications.[262] Ionomers also find application in footwear for shoe heels.[263]

1.5.17 Polyimide (PI)

Thermoplastic polyimides are linear polymers noted for their high-temperature properties. Polyimides are prepared by condensation polymerization of pyromellitic anhydrides and

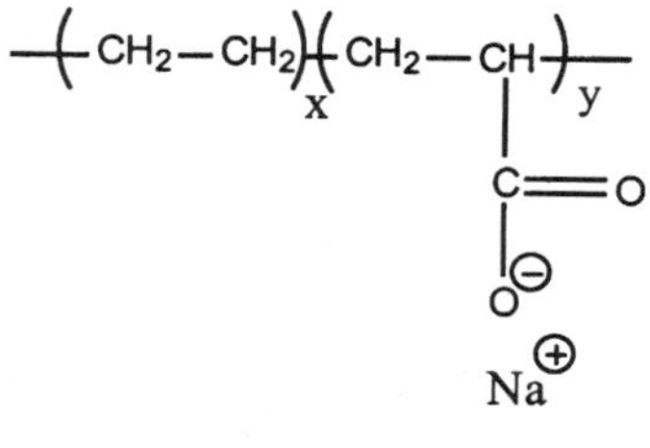

Figure 1.37 Structure of an ionomer.

primary diamines. A polyimide contains the structure –CO-NR-CO as a part of a ring structure along the backbone. The presence of ring structures along the backbone as depicted in Fig. 1.38 gives the polymer good high-temperature properties.[264] Polyimides are used in high-performance applications as replacements for metal and glass. The use of aromatic diamines gives the polymer exceptional thermal stability. An example of this is the use of di-(4-amino-phenyl) ether, which is used in the manufacture of Kapton (DuPont).

Although called *thermoplastics,* some polyimides must be processed in precursor form, because they will degrade before their softening point.[265] Fully imidized injection molding grades are available along with powder forms for compression molding and cold forming. However, injection molding of polyimides requires experience on the part of the molder.[266] Polyimides are also available as films and preformed stock shapes. The polymer may also be used as a soluble prepolymer, where heat and pressure are used to convert the polymer into the final, fully imidized form. Films can be formed by casting soluble polymers or precursors. It is generally difficult to form good films by melt extrusion. Laminates of polyimides can also be formed by impregnating fibers such as glass or graphite.

Polyimides have excellent physical properties and are used in applications where parts are exposed to harsh environments. They have outstanding high-temperature properties, and their oxidative stability allows them to withstand continuous service in air at temperatures of 260°C.[267] Polyimides will burn, but they have self-extinguishing properties.[268] They are resistant to weak acids and organic solvents but are attacked by bases. The polymer also has good electrical properties and resistance to ionizing radiation.[269] A disadvantage of polyimides is their hydrolysis resistance. Exposure to water or steam above 100°C may cause parts to crack.[270]

The first application of polyimides was for wire enamel.[271] Applications for polyimides include bearings for appliances and aircraft, seals, and gaskets. Film versions are used in flexible wiring and electric motor insulation. Printed circuit boards are also fabricated with polyimides.[272]

1.5.18 Polyketones

The family of aromatic polyether ketones includes structures that vary in the location and number of ketonic and ether linkages on their repeat unit and therefore include polyether ketone (PEK), polyether ether ketone (PEEK), polyether ether ketone ketone (PEEKK), as well as other combinations. Their structures are as shown in Fig. 1.39. All have very high

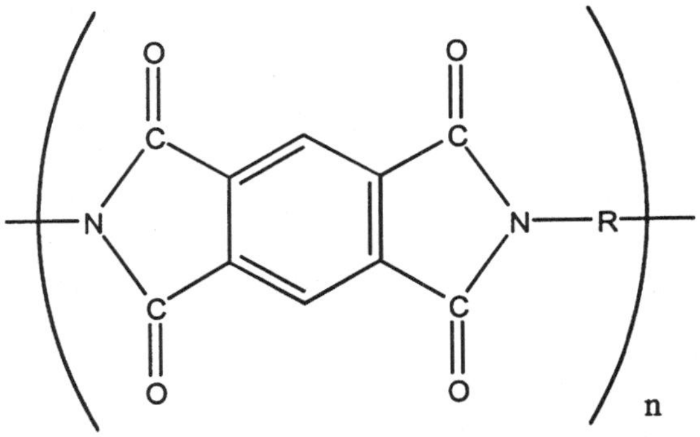

Figure 1.38 Structure of a polyimide.

PEK

PEEK

PEEKK

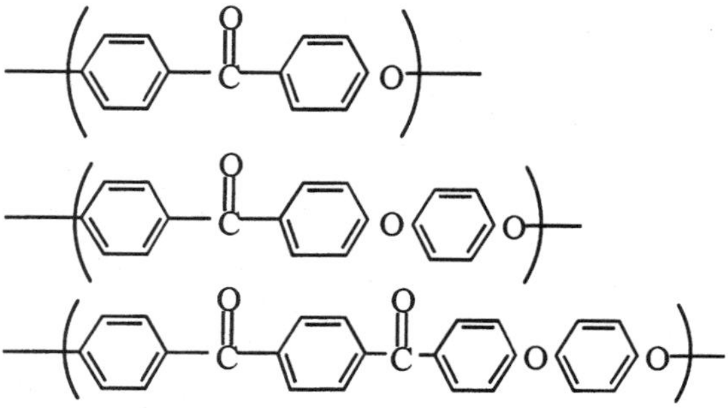

Figure 1.39 Structures of PEK, PEEK, and PEEKK.

thermal properties due to the aromaticity of their backbones and are readily processed via injection molding and extrusion, although their melt temperatures are very high—370°C for unfilled PEEK and 390°C for filled PEEK and both unfilled and filled PEK. Mold temperatures as high as 165°C are also used.[273] Their toughness (surprisingly high for such high-heat-resistant materials), high dynamic cycles and fatigue resistance capabilities, low moisture absorption, and good hydrolytic stability lend these materials to applications such as parts found in nuclear plants, oil wells, high-pressure steam valves, chemical plants, and airplane and automobile engines.

One of the two ether linkages in PEEK is not present in PEK, and the ensuing loss of some molecular flexibility results in PEK having an even higher T_m and heat distortion temperature than PEEK. A relatively higher ketonic concentration in the repeat unit results in high ultimate tensile properties as well. A comparison of different aromatic polyether ketones is given in Table 1.11.[274,275] As these properties are from different sources, strict comparison between the data is not advisable due to the likelihood that differing testing techniques were employed.

Glass and carbon fiber reinforcements are the most important fillers for all of the PEK family. While elastic extensibility is sacrificed, the additional heat resistance and moduli improvements allow glass- or carbon-fiber formulations entry into many applications.

PEK is polymerized either through self-condensation of structure (a) shown in Fig. 1.40, or via the reaction of intermediates shown in (b) below. Since these polymers can crystallize and tend therefore to precipitate from the reactant mixture, they must be reacted in high-boiling solvents close to the 320°C melt temperature.[276]

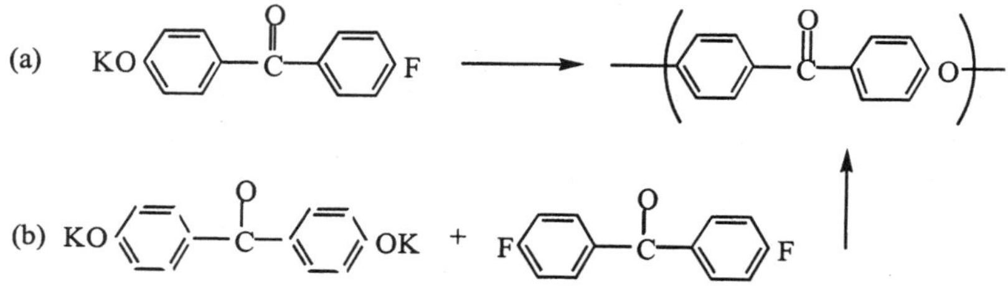

Figure 1.40 Routes for PEK synthesis.

TABLE 1.11 Comparison of Selected PEK, PEEK, and PEEKK Properties

	PEK unfilled	30% glass-filled PEK	PEEK unfilled	30% glass-filled PEEK	PEEKK unfilled	30% glass-filled PEEKK
T_m, °C	323–381	329–381	334	334	365	–
Tensile modulus, MPa	3,585–4,000	9,722–12,090	–	8,620–11,030	4000	13,500
Ultimate elongation, %	50	2.2–3.4	30–150	2–3	–	–
Ultimate tensile strength, MPa	103	–	91	–	86	168
Specific gravity	1.3	1.47–1.53	1.30–1.32	1.49–1.54	1.3	1.55
Heat deflection temperature, °C, 264 lb/in^2	162–170	326–350	160	288–315	160	>320

1.5.19 Poly(methylmethacrylate) (PMMA)

Poly(methylmethacrylate) is a transparent thermoplastic material of moderate mechanical strength and outstanding outdoor weather resistance. It is available as sheet, tubes, and rods that can be machined, bonded, and formed into a variety of different parts, and in bead form, which can be conventionally processed via extrusion or injection molding. The sheet form material is polymerized *in situ* by casting a monomer that has been partly prepolymerized by removing any inhibitor, heating, and adding an agent to initiate the free radical polymerization. This agent is typically a peroxide. This mixture of polymer and monomer is then poured into the sheet mold, and the plates are brought together and reinforced to prevent bowing to ensure that the final product will be of uniform thickness and flatness. This bulk polymerization process generates such high-molecular-weight material that the sheet or rod will decompose prior to melting. As such, this technique is not suitable for producing injection-molding-grade resin, but it does aid in producing material that has a large rubbery plateau and has high enough elevated temperature strength to allow for band sawing, drilling, and other common machinery practices as long as the localized heating does not reach the polymer's decomposition temperature.

Suspension polymerization provides a final polymer with low enough molecular weight to allow for typical melt processing. In this process, methyl methacrylate monomer is suspended in water, to which the peroxide is added along with emulsifying/suspension agents, protective colloids, lubricants, and chain transfer agents to aid in molecular weight control. The resultant bead can then be dried and is ready for injection molding or it can be further compounded with any desired colorants, plasticizers, and rubber modifier, as required.[277] Number-average molecular weights from the suspension process are approximately 60,000 g/mol, while the bulk polymerization process can result in number average molecular weights of approximately 1 million g/mol.[278]

Typically applications for PMMA optimize use of its clarity, with up to 92 percent light transmission, depending on the thickness of the sample. Again, because it has such strong weathering behavior, it is well suited for applications such as automobile tail light housings, lenses, aircraft cockpits, helicopter canopies, dentures, steering wheel bosses, and windshields. Cast PMMA is used extensively as bathtub materials, in showers, and in whirlpools.[279]

Since the homopolymer is fairly brittle, PMMA can be toughened via copolymerization with another monomer (such as polybutadiene) or blended with an elastomer in the same way that high-impact polystyrene is used to enable better stress distribution via the elastomeric domain.

1.5.20 Polymethylpentene (PMP)

Polymethylpentene was introduced in the mid 1960s by ICI and is now marketed under the same trade name, TPX, by Mitsui Petrochemical Industries. The most significant commercial polymerization method involves the dimerization of propylene, as shown in Fig. 1.41.

As a polyolefin, this material offers chemical resistance to mineral acids, alkaline solutions, alcohols, and boiling water. It is not resistant to ketones or aromatic and chlorinated hydrocarbons. Like polyethylene and polypropylene, it is susceptible to environmental

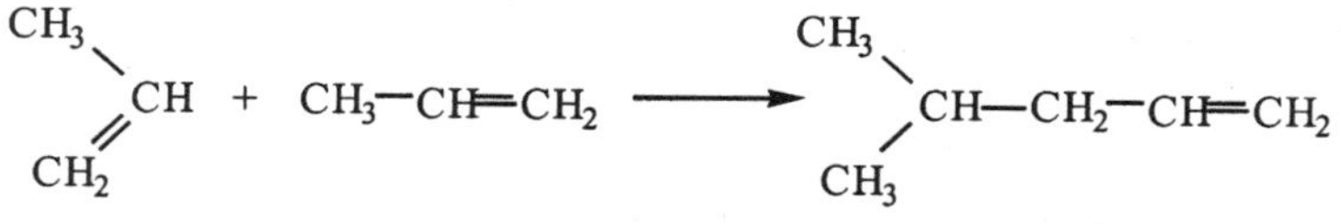

Figure 1.41 Polymerization route for polymethylpentene.

stress cracking[280] and requires formulation with antioxidants. Its use is primarily in injection molding and thermoforming applications, where the additional cost incurred as compared with other polyolefins is justified by its high melt point (245°C), transparency, low density, and good dielectric properties. The high degree of transparency of polymethylpentene is attributed both to the similarities of the refractive indices of the amorphous and crystalline regions and to the large coil size of the polymer due to the bulky branched four-carbon side chain. The free volume regions are large enough to allow light of visible region wavelengths to pass unimpeded. This degree of free volume is also responsible for the 0.83 g/cm^3 low density. As typically cooled, the polymer achieves about 40 percent crystallinity but, with annealing, it can reach 65 percent crystallinity.[281] The structure of the polymer repeat unit is shown in Fig. 1.42.

Voids are frequently formed at the crystalline/amorphous region interfaces during injection molding, rendering an often undesirable lack of transparency. To counter this, polymethylpentene is often copolymerized with hex-1-ene, oct-1-ene, dec-1-ene, and octadec-1-ene, which reduces the voids and concomitantly reduces the melting point and degree of crystallinity.[282] Typical products made from polymethylpentene include transparent pipes and other chemical plant applications, sterilizable medical equipment, light fittings, and transparent housings.

1.5.21 Polyphenylene Oxide (PPO)

The term *polyphenylene oxide (PPO)* is a misnomer for a polymer that is more accurately named poly-(2,6-dimethyl-*p*-phenylene ether). In Europe, it is more commonly known as a polymer covered by the more generic term *polyphenylene ether (PPE)*. This engineering polymer has high temperature properties due to the large degree of aromaticity on the backbone, with dimethyl-substituted benzene rings joined by an ether linkage, as shown in Fig. 1.43.

The stiffness of this repeat unit results in a heat-resistant polymer with a T_g of 208°C and a T_m of 257°C. The fact that these two thermal transitions occur within such a short temperature span of each other means that PPO does not have time to crystallize while it cools before reaching a glassy state and as such is typically amorphous after processing.[283] Commercially available as PPO from General Electric, the polymer is sold in mo-

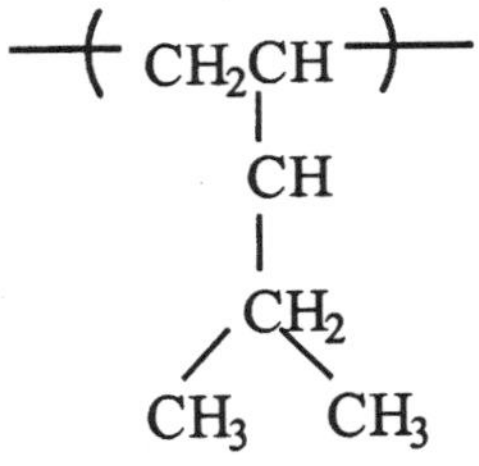

Figure 1.42 Repeat structure of polymethylpentene.

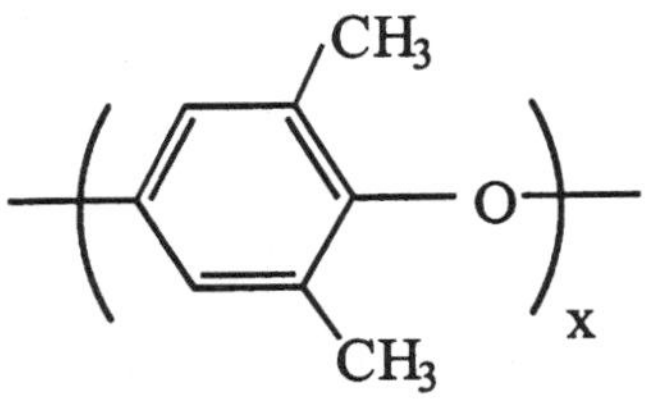

Figure 1.43 Repeat structure of PPO.

lecular weight ranges of 25,000 to 60,000 g/mol.[284] Properties that distinguish PPO from other engineering polymers are its high degree of hydrolytic and dimensional stabilities, which enable it to be molded with precision; however, high processing temperatures are required. It finds application as television tuner strips, microwave insulation components, and transformer housings, which take advantage of its strong dielectric properties over wide temperature ranges. It is also used in applications that benefit from its hydrolytic stability including pumps, water meters, sprinkler systems, and hot water tanks.[285] Its greater use is limited by the often-prohibitive cost, and General Electric responded by commercializing a PPO/PS blend marketed under the trade name Noryl. GE sells many grades of Noryl, based on different blend ratios and specialty formulations. The styrenic nature of PPO leads one to surmise very close compatibility (similar solubility parameters) with PS, although strict thermodynamic compatibility is questioned due to the presence of two distinct T_g peaks when measured by mechanical rather than calorimetric means.[286] The blends present the same high degree of dimensional stability, low water absorption, excellent resistance to hydrolysis, and good dielectric properties offered by PS, yet with the elevated heat distort temperatures that result from PPO's contribution. These polymers are more cost-competitive than PPO and are used in moldings for dishwashers, washing machines, hair dryers, cameras, instrument housings, and television accessories.[287]

1.5.22 Polyphenylene Sulphide (PPS)

The structure of PPS, shown in Fig. 1.44, clearly indicates high temperature, high strength, and high chemical resistance due to the presence of the aromatic benzene ring on the backbone linked with the electronegative sulfur atom. In fact, the melt point of PPS is 288°C, and the tensile strength is 70 MPa at room temperature. The brittleness of PPS, due to the highly crystalline nature of the polymer, is often overcome by compounding with glass fiber reinforcements. Typical properties of PPS and a commercially available 40 percent glass-filled polymer blend are shown in Table 1.12.[288] The mechanical properties of PPS are similar to those of other engineering thermoplastics such as polycarbonate and polysulfones, except that, as mentioned, the PPS suffers from the brittleness arising from its crystallinity but does however offer improved resistance to environmental stress cracking.[289]

PPS is of most significant commercial interest as a thermoplastic, although it can be cross-linked into a thermoset system. Its strong inherent flame retardance puts this polymer in a fairly select class of polymers, including polyethersulfones, liquid crystal polyesters, polyketones, and polyetherimides.[290] As such, PPS finds application in electrical components, printed circuits, and contact and connector encapsulation. Other uses take advantage of the low mold shrinkage values and strong mechanical properties even at elevated temperatures. These include pump housings, impellers, bushings, and ball valves.[291]

1.5.23 Polyphthalamide (PPA)

Polyphthalamides were originally developed for use as fibers and later found application in other areas. They are semi-aromatic polyamides based on the polymerization of tereph-

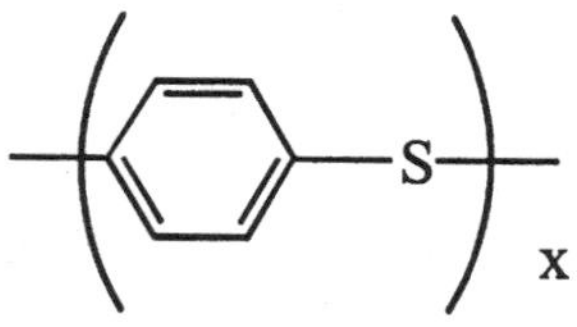

Figure 1.44 Repeat structure of polyphenylene sulphide.

TABLE 1.12 Selected Properties of PPS and GF PPS

Property, units	PPS	40 percent glass-filled PPS
T_g, °C	85	–
Heat distortion temperature, Method A, °C	135	265
Tensile strength		
21°C MPa	64–77	150
204°C, MPa	33	33
Elongation at break, %	3	2
Flexural modulus, MPa	3900	10,500
Limiting oxygen index, %	44	47

thalic acid or isophthalic acid and an amine.[292] Both amorphous and crystalline grades are available. Polyphthalamides are polar materials with a melting point near 310°C and a glass transition temperature of 127°C.[293] The material has good strength and stiffness along with good chemical resistance. Polyphthalamides can be attacked by strong acids or oxidizing agents and are soluble in cresol and phenol.[294] Polyphthalamides are stronger, less moisture sensitive, and possess better thermal properties as compared with the aliphatic polyamides such as nylon 6/6. However, polyphthalamide is less ductile than nylon 6/6, although impact grades are available.[295] Polyphthalamides will absorb moisture, decreasing the glass transition temperature and causing dimensional changes. The material can be reinforced with glass and has extremely good high-temperature performance. Reinforced grades of polyphthalamides are able to withstand continuous use at 180°C.[296]

The crystalline grades are generally used in injection molding, while the amorphous grades are often used as barrier materials.[297] The recommended mold temperatures are 135 to 165°C, with recommended melt temperatures of 320 to 340°C.[298] The material should have a moisture content of 0.15 percent or less for processing.[299] Because mold temperature is important to surface finish, higher mold temperatures may be required for some applications.

Both crystalline and amorphous grades are available under the trade name Amodel (Amoco); amorphous grades are available under the names Zytel (Dupont) and Trogamid (Dynamit Nobel). Crystalline grades are available under the trade name Arlen (Mitsui).[300]

Polyphthalamides are used in automotive applications where their chemical resistance and temperature stability are important.[301] Examples include sensor housings, fuel line components, headlamp reflectors, electrical components, and structural components. Electrical components attached by infrared and vapor phase soldering are applications utilizing PPA's high-temperature stability. Switching devices, connectors, and motor brackets are often made from PPA. Mineral-filled grades are used in applications that require plating, such as decorative hardware and plumbing. Impact modified grades of unreinforced PPA are used in sporting goods, oil field parts, and military applications.

1.5.24 Polypropylene (PP)

Polypropylene is a versatile polymer used in applications from films to fibers with a worldwide demand of over 21 million pounds.[302] It is similar to polyethylene in structure, except for the substitution of one hydrogen group with a methyl group on every other car-

bon. On the surface, this change would appear trivial, but this one replacement now changes the symmetry of the polymer chain. This allows for the preparation of different stereoisomers, namely, syndiotactic, isotactic, and atactic chains. These configurations are shown previously in Fig. 1.6.

Polypropylene (PP) is synthesized by the polymerization of propylene, a monomer derived from petroleum products through the reaction shown in Fig. 1.45. It was not until Ziegler-Natta catalysts became available that polypropylene could be polymerized into a commercially viable product. These catalysts allowed the control of stereochemistry during polymerization to form polypropylene in the isotactic and syndiotactic forms, both capable of crystallizing into a more rigid, useful polymeric material.[303] The first commercial method for the production of polypropylene was a suspension process. Current methods of production include a gas phase process and a liquid slurry process.[304] New grades of polypropylene are now being polymerized using metallocene catalysts.[305] The range of molecular weights for PP is M_n = 38,000 to 60,000 and M_w = 220,000 to 700,000. The molecular weight distribution (M_n/M_w) can range from 2 to about 11.[306]

Different behavior can be found for each of the three stereoisomers. Isotactic and syndiotactic polypropylene can pack into a regular crystalline array, giving a polymer with more rigidity. Both materials are crystalline; however, syndiotactic polypropylene has a lower T_m than the isotactic polymer.[307] The isotactic polymer is the most commercially used form with a melting point of 165°C. Atactic polypropylene has a very small amount of crystallinity (5 to 10 percent), because its irregular structure prevents crystallization; thus, it behaves as a soft flexible material.[308] It is used in applications such as sealing strips, paper laminating, and adhesives.

Unlike polyethylene, which crystallizes in the planar zigzag form, isotactic polypropylene crystallizes in a helical form because of the presence of the methyl groups on the chain.[309] Commercial polymers are about 90 to 95 percent isotactic. The amount of isotacticity present in the chain will influence the properties. As the amount of isotactic material (often quantified by an isotactic index) increases, the amount of crystallinity will also increase, resulting in increased modulus, softening point, and hardness.

Although, in many respects, polypropylene is similar to polyethylene, both being saturated hydrocarbon polymers, they differ in some significant properties. Isotactic polypropylene is harder and has a higher softening point than polyethylene, so it is used where higher-stiffness materials are required. Polypropylene is less resistant to degradation (particularly high-temperature oxidation) than polyethylene, but it has better environmental stress cracking resistance.[310] The decreased degradation resistance of PP is due to the presence of a tertiary carbon in PP, allowing for easier hydrogen abstraction compared with PE.[311] As a result, antioxidants are added to polypropylene to improve the oxidation resistance. The degradation mechanisms of the two polymers are also different. PE cross-links on oxidation, while PP undergoes chain scission. This is also true of the polymers when exposed to high-energy radiation, a method commonly used to cross-link PE.

Polypropylene is one of the lightest plastics, with a density of 0.905.[312] The nonpolar nature of the polymer gives PP low water absorption. Polypropylene has good chemical resistance, but liquids such as chlorinated solvents, gasoline, and xylene can affect the material. Polypropylene has a low dielectric constant and is a good insulator. Difficulty in

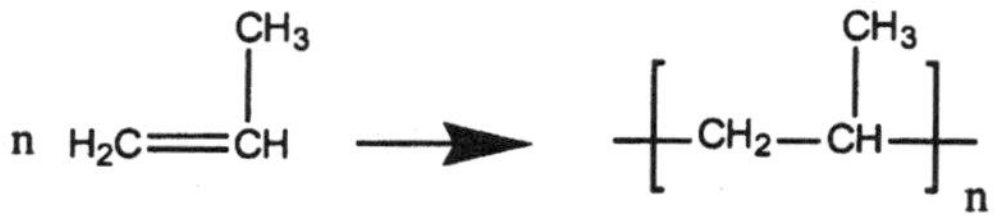

Figure 1.45 The reaction to prepare polypropylene.

bonding to polypropylene can be overcome by the use of surface treatments to improve the adhesion characteristics.

With the exception of UHMWPE, polypropylene has a higher T_g and melting point than polyethylene. Service temperature is increased, but PP needs to be processed at higher temperatures. Because of the higher softening, PP can withstand boiling water and can be used in applications requiring steam sterilization.[313] Polypropylene is also more resistant to cracking in bending than PE and is preferred in applications that require tolerance to bending. This includes applications such as ropes, tapes, carpet fibers, and parts requiring a living hinge. Living hinges are integral parts of a molded piece that are thinner and allow for bending.[314] One weakness of polypropylene is its low-temperature brittleness behavior, with the polymer becoming brittle near 0°C.[315] This can be improved through copolymerization with other polymers such as ethylene.

Comparing the processing behavior of PP with PE, it is found that polypropylene is more non-Newtonian than PE and that the specific heat of PP is lower than polyethylene.[316] The melt viscosity of PE is less temperature sensitive than PP.[317] Mold shrinkage is generally less than for PE but is dependent on the actual processing conditions.

Unlike many other polymers, an increase in molecular weight of polypropylene does not always translate into improved properties. The melt viscosity and impact strength will increase with molecular weight, but often with a decrease in hardness and softening point. A decrease in the ability of the polymer to crystallize as molecular weight increases is often offered as an explanation for this behavior.[318]

The molecular weight distribution (MWD) has important implications for processing. A PP grade with a broad MWD is more shear sensitive than a grade with a narrow MWD. Broad MWD materials will generally process better in injection molding applications. In contrast, a narrow MWD may be preferred for fiber formation.[319] Various grades of polypropylene are available tailored to particular application. These grades can be classified by flow rate, which depends on both average molecular weight and MWD. Lower flow rate materials are used in extrusion applications. In injection molding applications, low flow rate materials are used for thick parts and high-flow-rate materials are used for thin-wall molding.

Polypropylene can be processed by methods similar to those used for PE. The melt temperatures are generally in the range of 210 to 250°C.[320] Heating times should be minimized to reduce the possibility of oxidation. Blow molding of PP requires the use of higher melt temperatures and shear, but these conditions tend to accelerate the degradation of PP. Because of this, blow molding of PP is more difficult than for PE. The screw metering zone should not be too shallow to avoid excessive shear. For a 60-mm screw, the flight depths are typically about 2.25 mm, and 3.0 mm for a 90-mm screw.[321]

In film applications, film clarity requires careful control of the crystallization process to ensure that small crystallites are formed. This is accomplished in blown film by extruding downward into two converging boards. In the Shell TQ process, the boards are covered with a film of flowing, cooling water. Oriented films of PP are manufactured by passing the PP film into a heated area and stretching the film both transversely and longitudinally. To reduce shrinkage, the film may be annealed at 100°C while under tension.[322] Highly oriented films may show low transverse strength and a tendency to fibrillate. Other manufacturing methods for polypropylene include extruded sheet for thermoforming applications and extruded profiles.

If higher stiffness is required, short glass reinforcement can be added. The use of a coupling agent can dramatically improve the properties of glass-filled PP.[323] Other fillers for polypropylene include calcium carbonate and talc, which can also improve the stiffness of PP.

Other additives, such as pigments, antioxidants, and nucleating agents, can be blended into polypropylene to give the desired properties. Carbon black is often added to polypro-

pylene to impart UV resistance in outdoor applications. Antiblocking and slip agents may be added for film applications to decrease friction and prevent sticking. In packaging applications, antistatic agents may be incorporated.

The addition of rubber to polypropylene can lead to improvements in impact resistance. One of the most commonly added elastomers is ethylene-propylene rubber. The elastomer is blended with polypropylene, forming a separate elastomer phase. Rubber can be added in excess of 50 percent to give elastomeric compositions. Compounds with less than 50 percent added rubber are of considerable interest as modified thermoplastics. Impact grades of PP can be formed into films with good puncture resistance.

Copolymers of polypropylene with other monomers are also available, the most common monomer being ethylene. Copolymers usually contain between 1 and 7 weight percent of ethylene randomly placed in the polypropylene backbone. This disrupts the ability of the polymer chain to crystallize, giving more flexible products. This improves the impact resistance of the polymer, decreases the melting point, and increases flexibility. The degree of flexibility increases with ethylene content, eventually turning the polymer into an elastomer (ethylene propylene rubber). The copolymers also exhibit increased clarity and are used in blow molding, injection molding, and extrusion.

Polypropylene has many applications. Injection molding applications cover a broad range from automotive uses such as dome lights, kick panels, and car battery cases to luggage and washing machine parts. Filled PP can be used in automotive applications such as mounts and engine covers. Elastomer modified PP is used in the automotive area for bumpers, fascia panels, and radiator grills. Ski boots are another application for these materials.[324] Structural foams, prepared with glass-filled PP, are used in the outer tank of washing machines. New grades of high-flow PPs are allowing manufacturers to mold high-performance housewares.[325]

Polypropylene films are used in a variety of packaging applications. Both oriented and non-oriented films are used. Film tapes are used for carpet backing and sacks. Foamed sheet is used in a variety of applications including thermoformed packaging. Fibers are another important application for polypropylene, particularly in carpeting, because of its low cost and wear resistance. Fibers prepared from polypropylene are used in both woven and nonwoven fabrics.

1.5.25 Polyurethane (PUR)

Polyurethanes are very versatile polymers. They are used as flexible and rigid foams, elastomers, and coatings. Polyurethanes are available as both thermosets and thermoplastics, in addition, their hardnesses span the range from rigid material to elastomer. Thermoplastic polyurethanes will be the focus of this section. The term polyurethane is used to cover materials formed from the reaction of isocyanates and polyols.[326] The general reaction for a polyurethane produced through the reaction of a diisocyanate with a diol is shown in Fig. 1.46.

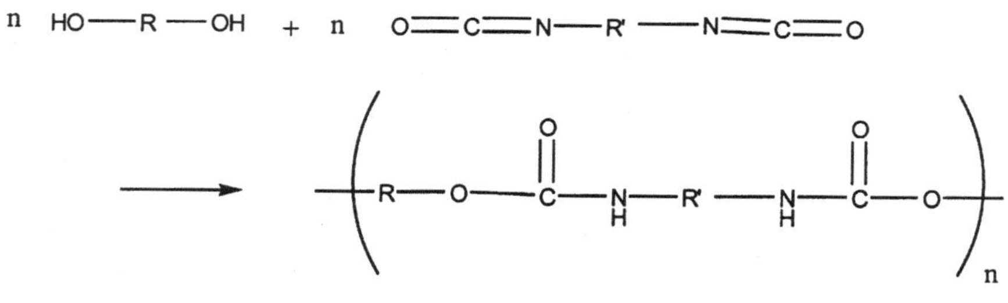

Figure 1.46 Polyurethane reaction.

[—A—B—A—B—A—B—]$_n$

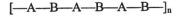

Figure 1.47 Block structure of polyurethanes.

Polyurethanes are phase separated block copolymers as depicted in Fig. 1.47, where the A and B portions represent different polymer segments. One segment, called the *hard segment,* is rigid, while the other, the *soft segment,* is elastomeric. In polyurethanes, the soft segment is prepared from an elastomeric long chain polyol, generally a polyester or polyether, but other rubbery polymers end-capped with a hydroxyl group could be used. The hard segment is composed of the diisocyanate and a short chain diol called a *chain extender.* The hard segments have high interchain attraction due to hydrogen bonding between the urethane groups. In addition, they may be capable of crystallizing.[327] The soft elastomeric segments are held together by the hard phases, which are rigid at room temperature and act as physical cross-links. The hard segments hold the material together at room temperature, but at processing temperatures the hard segments can flow and be processed.

The properties of polyurethanes can be varied by changing the type or amount of the three basic building blocks of the polyurethane: diisocyanate, short chain diol, or long chain diol. Given the same starting materials, the polymer can be varied simply by changing the ratio of the hard and soft segments. This allows the manufacturer a great deal of flexibility in compound development for specific applications. The materials are typically manufactured by reacting a linear polyol with an excess of diisocyanate. The polyol is end-capped with isocyanate groups. The end-capped polyol and free isocyanate are then reacted with a chain extender, usually a short chain diol to form the polyurethane.[328]

There are a variety of starting materials available for use in the preparation of polyurethanes, some of which are listed below.

- Diisocyanates
  - 4,4´-diphenylmethane diisocyanate (MDI)
  - Hexamethylene diisocyanate (HDI)
  - Hydrogenated 4,4´-diphenylmethane diisocyanate (HMDI)
- Chain extenders
  - 1,4 butanediol
  - Ethylene glycol
  - 1,6 hexanediol
- Polyols
  - Polyesters
  - Polyethers

Polyurethanes are generally classified by the type of polyol used—for example, polyester polyurethane or polyether polyurethane. The type of polyol can affect certain properties. For example, polyether polyurethanes are more resistant to hydrolysis than polyester-based urethanes, while the polyester polyurethanes have better fuel and oil resistance.[329] Low-temperature flexibility can be controlled by proper selection of the long chain polyol. Polyether polyurethanes generally have lower glass transition temperatures than polyester polyurethanes. The heat resistance of the polyurethane is governed by the hard segments. Polyurethanes are noted for their abrasion resistance, toughness, low-temperature impact strength, cut resistance, weather resistance, and fungus resistance.[330] Specialty polyurethanes include glass-reinforced products, fire-retardant grades, and UV-stabilized grades.

Polyurethanes find application in many areas. They can be used as impact modifiers for other plastics. Other applications include rollers or wheels, exterior body parts, drive belts,

and hydraulic seals.[331] Polyurethanes can be used in film applications such as textile laminates for clothing and protective coatings for hospital beds. They are also used in tubing and hose, in both unreinforced and reinforced forms, because of their low-temperature properties and toughness. Their abrasion resistance allows them to be used in applications such as athletic shoe soles and ski boots. Polyurethanes are also used as coatings for wire and cable.[332]

Polyurethanes can be processed by a variety of methods, including extrusion, blow molding, and injection molding. They tend to pick up moisture and must be thoroughly dried prior to use. The processing conditions vary with the type of polyurethane; higher hardness grades usually require higher processing temperatures. Polyurethanes tend to exhibit shear sensitivity at lower melt temperatures. Post-mold heating in an oven, shortly after processing, can often improve the properties of the finished product. A cure cycle of 16 to 24 hr at 100°C is typical.[333]

1.5.26 Styrenics

The styrene family is well suited for applications where rigid, dimensionally stable molded parts are required. PS is a transparent, brittle, high-modulus material with a multitude of applications, primarily in packaging, disposable cups, and medical ware. When the mechanical properties of the PS homopolymer are modified to produce a tougher, more ductile blend, as in the case of rubber-modified high-impact grades of PS (HIPS), a far wider range of applications becomes available. HIPS is preferred for durable molded items including radio, television, and stereo cabinets as well as compact disc jewel cases. Copolymerization is also used to produce engineering-grade plastics of higher performance as well as higher price, with acrylonitrile-butadiene-styrene (ABS) and styrene-acrylonitrile (SAN) plastics being of greatest industrial importance.

1.5.26.1 Acrylonitrile butadiene styrene (ABS) terpolymer. As with any copolymers, there is tremendous flexibility in tailoring the properties of ABS by varying the ratios of the three monomers, acrylonitrile, butadiene, and styrene. The acrylonitrile component contributes heat resistance, strength, and chemical resistance. The elastomeric contribution of butadiene imparts higher impact strength, toughness, low-temperature property retention, and flexibility, while the styrene contributes rigidity, glossy finish, and ease of processability. As such, worldwide usage of ABS is surpassed only by that of the "big four" commodity thermoplastics (polyethylene, polypropylene, polystyrene, and polyvinyl chloride). Primary drawbacks to ABS include opacity, poor weather resistance, and poor flame resistance. Flame retardance can be improved by the addition of fire-retardant additives or by blending ABS with PVC, with some reduction in ease of processability.[334] As it is widely used as equipment housings (such as telephones, televisions, and computers), these disadvantages are tolerated. Figure 1.48 shows the repeat structure of ABS.

Most common methods of manufacturing ABS include graft polymerization of styrene and acrylonitrile onto a polybutadiene latex, blending with a styrene-acrylonitrile latex, and then coagulating and drying the resultant blend. Alternatively, the graft polymer of styrene, acrylonitrile, and polybutadiene can be manufactured separately from the styrene acrylonitrile latex, and the two grafts blended and granulated after drying.[335]

Its ease of processing by a variety of common methods (including injection molding, extrusion, thermoforming, compression molding, and blow molding), combined with a good economic value for the mechanical properties achieved, results in widespread use of ABS. It is commonly found in under-hood automotive applications, refrigerator linings, radios, computer housings, telephones, business machine housings, and television housings.

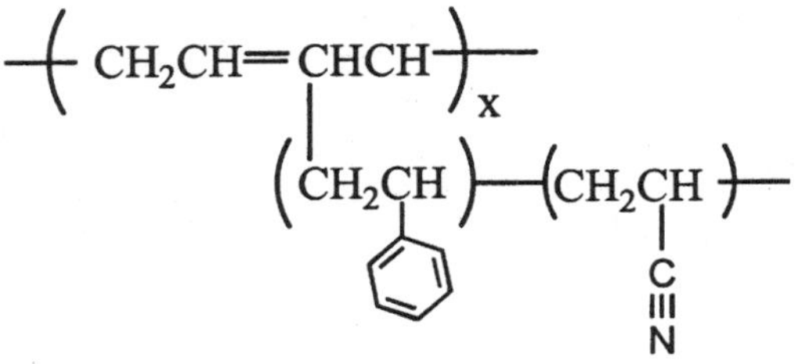

Figure 1.48 Repeat structure of ABS.

1.5.26.2 Acrylonitrile-chlorinated polyethylene-styrene (ACS) terpolymer. While ABS itself can be readily tailored by modifying the ratios of the three monomers and by modifying the lengths of each grafted segment, several companies are pursuing the addition of a fourth monomer, such as alpha-methylstyrene for enhanced heat resistance, and methylmethacrylate to produce a transparent ABS. One such modification involves using chlorinated polyethylene in place of the butadiene segments. This terpolymer, ACS, has very similar properties to the engineering terpolymer ABS, but the addition of chlorinated polyethylene imparts improved flame retardance, weatherability, and resistance to electrostatic deposition of dust, without the addition of antistatic agents. The addition of the chlorinated olefin requires more care when injection molding to ensure that the chlorine does not dehydrohalogenate. Mold temperatures are recommended to be kept at between 190 and 210°C and not to exceed 220°C, and, as with other chlorinated polymers such as polyvinyl chloride, that residence times be kept relatively short in the molding machine.[336]

Applications for ACS include housings and parts for office machines such as desktop calculators, copying machines, and electronic cash registers, and as housings for television sets and videocassette recorders.[337]

1.5.26.3 Acrylic styrene acrylonitrile (ASA) terpolymer. Like ACS, ASA is a specialty product with similar mechanical properties to ABS, but it offers improved outdoor weathering properties. This is due to the grafting of an acrylic ester elastomer onto the styrene-acrylonitrile backbone. Sunlight usually combines with atmospheric oxygen to result in embrittlement and yellowing of thermoplastics, and this process takes a much longer time in the case of ASA; therefore, ASA finds applications in gutters, drain pipe fittings, signs, mailboxes, shutters, window trims, and outdoor furniture.[338]

1.5.26.4 General-purpose polystyrene (PS). PS is one of the four plastics whose combined usage accounts for 75 percent of the worldwide usage of plastics.[339] These four commodity thermoplastics are PE, PP, PVC, and PS. Although it can be polymerized via free-radical, anionic, cationic, and Ziegler mechanisms, commercially available PS is produced via free-radical addition polymerization. PS's popularity is due to its transparency, low density, relatively high modulus, excellent electrical properties, low cost, and ease of processing. The steric hindrance caused by the presence of the bulky benzene side groups results in brittle mechanical properties, with ultimate elongations only around 2 to 3 percent, depending upon molecular weight and additive levels. Most commercially available

PS grades are atactic and, in combination with the large benzene groups, result in an amorphous polymer. The amorphous morphology provides not only transparency, but the lack of crystalline regions also means that there is no clearly defined temperature at which the plastic melts. PS is a glassy solid until its T_g of ~100°C is reached, whereupon further heating softens the plastic gradually from a glass to a liquid. Advantage is taken of this gradual transition by molders who can eject parts that have cooled to beneath the relatively high Vicat temperature. Also, the lack of a heat of crystallization means that high heating and cooling rates can be achieved. These reduce cycle time and also promote an economical process. Lastly, upon cooling, PS does not crystallize the way PE and PP do. This gives PS low shrinkage values (0.004 to 0.005 mm/mm) and high dimensional stability during molding and forming operations.

Commercial PS is segmented into easy-flow, medium-flow, and high-heat-resistance grades. Comparison of these three grades is made in Table 1.13. The easy-flow grades are the lowest in molecular weight, to which 3 to 4 percent mineral oil has been added. The mineral oil reduces melt viscosity, which is well suited for increased injection speeds while molding inexpensive thin-walled parts such as disposable dinnerware, toys, and packaging. The reduction in processing time comes at the cost of a reduced softening temperature and a more brittle polymer. The medium-flow grades are slightly higher in molecular weight and contain only 1 to 2 percent mineral oil. Applications include injection-molded tumblers, medical ware, toys, injection-blow-molded bottles, and extruded food packaging. The high-heat-resistance plastics are the highest in molecular weight and have the least level of additives such as extrusion aids. These products are used in sheet extrusion and thermoforming and extruded film applications for oriented food packaging.[340]

TABLE 1.13 Properties of Commercial Grades of General-Purpose PS[456]

Property	Easy-flow PS	Medium-flow PS	High-heat-resistance PS
M_w	218,000	225,000	300,000
M_n	74,000	92,000	130,000
Melt flow index, g/10 min	16	7.5	1.6
Vicat softening temperature, °C	88	102	108
Tensile modulus, MPa	3,100	2,450	3,340
Ultimate tensile strength, MPa	1.6	2.0	2.4

1.5.26.5 Styrene-acrylonitrile copolymers (SANs). Styrene-acrylonitrile polymers are copolymers prepared from styrene and acrylonitrile monomers. The polymerization can be done under emulsion, bulk, or suspension conditions.[341] The polymers generally contain between 20 and 30 percent acrylonitrile.[342] The acrylonitrile content of the polymer influences the final properties with tensile strength, elongation, and heat distortion temperature, increasing as the amount of acrylonitrile in the copolymer increases.

SAN copolymers are linear, amorphous materials with improved heat resistance over pure polystyrene.[343] The polymer is transparent but may have a yellow color as the acrylonitrile content increases. The addition of a polar monomer, acrylonitrile, to the backbone gives these polymers better resistance to oils, greases, and hydrocarbons as compared with

polystyrene.[344] Glass-reinforced grades of SAN are available for applications requiring higher modulus combined with lower mold shrinkage and lower coefficient of thermal expansion.[345]

As the polymer is polar, it should be dried before processing. It can be processed by injection molding into a variety of parts. SAN can also be processed by blow molding, extrusion, casting, and thermoforming.[346]

SAN competes with polystyrene, cellulose acetate, and polymethyl methacrylate. Applications for SAN include injection molded parts for medical devices, PVC tubing connectors, dishwasher-safe products, and refrigerator shelving.[347] Other applications include packaging for the pharmaceutical and cosmetics markets, automotive equipment, and industrial uses.

1.5.26.6 Olefin-modified SAN. SAN can be modified with olefins, resulting in a polymer that can be extruded and injection molded. The polymer has good weatherability and is often used as a capstock to provide weatherability to less expensive parts such as swimming pools, spas, and boats.[348]

1.5.26.7 Styrene-butadiene copolymers. Styrene-butadiene polymers are block copolymers prepared from styrene and butadiene monomers. The polymerization is performed using sequential anionic polymerization.[349] The copolymers are better known as *thermoplastic elastomers,* but copolymers with high styrene contents can be treated as thermoplastics. The polymers can be prepared as either a star block form or as a linear, multiblock polymer. The butadiene exists as a separate dispersed phase in a continuous matrix of polystyrene.[350] The size of the butadiene phase is controlled to be less than the wavelength of light, resulting in clear materials. The resulting amorphous polymer is tough, with good flex life and low mold shrinkage. The copolymer can be ultrasonically welded, solvent welded, or vibration welded. The copolymers are available in injection-molding grades and thermoforming grades. The injection-molding grades generally contain a higher styrene content in the block copolymer. Thermoforming grades are usually mixed with pure polystyrene.

Styrene-butadiene copolymers can be processed by injection molding, extrusion, thermoforming, and blow molding. The polymer does not need to be dried prior to use.[351]

Styrene-butadiene copolymers are used in toys, housewares, and medical applications.[352] Thermoformed products include disposable food packaging such as cups, bowls, "clam shells," deli containers, and lids. Blister packs and other display packaging also use styrene-butadiene copolymers. Other packaging applications include shrink wrap and vegetable wrap.[353]

1.5.27 Sulfone-Based Resins

Sulfone resins refers to polymers containing SO_2 groups along the backbone as depicted in Fig. 1.49. The R groups are generally aromatic. The polymers are usually yellowish, trans-

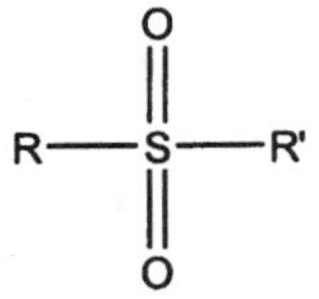

Figure 1.49 General structure of a polysulfone.

parent, amorphous materials and are known for their high stiffness, strength, and thermal stability.[354] The polymers have low creep over a large temperature range. Sulfones can compete against some thermoset materials in performance, while their ability to be injection molded offers an advantage.

The first commercial polysulfone was Udel (Union Carbide, now Amoco), followed by Astrel 360 (Minnesota Mining and Manufacturing), which is termed a *polyarylsulfone,* and, finally, Victrex (ICI), a polyethersulfone.[355] Current manufacturers also include Amoco, Carborundum, and BASF, among others. The different polysulfones vary by the spacing between the aromatic groups, which in turn affects their T_g values and their heat distortion temperatures. Commercial polysulfones are linear with high T_g values in the range of 180 to 250°C, allowing for continuous use from 150 to 200°C.[356] As a result, the processing temperatures of polysulfones are above 300°C.[357] Although the polymer is polar, it still has good electrical insulating properties. Polysulfones are resistant to high thermal and ionizing radiation. They are also resistant to most aqueous acids and alkalis but may be attacked by concentrated sulfuric acid. The polymers have good hydrolytic stability and can withstand hot water and steam.[358] Polysulfones are tough materials, but they do exhibit notch sensitivity. The presence of the aromatic rings causes the polymer chain to be rigid. Polysulfones generally do not require the addition of flame retardants and usually emit low smoke.

The properties of the main polysulfones are generally similar, although polyethersulfones have better creep resistance at high temperatures and higher heat distortion temperature, but more water absorption and higher density than the Udel-type materials.[359] Glass-fiber-filled grades of polysulfone are available as are blends of polysulfone with ABS.

Polysulfones may absorb water, leading to potential processing problems such as streaks or bubbling.[360] The processing temperatures are quite high, and the melt is very viscous. Polysulfones show little change in melt viscosity with shear. Injection molding melt temperatures are in the range of 335 to 400°C, and mold temperatures are in the range of 100 to 160°C. The high viscosity necessitates the use of large cross-sectional runners and gates. Purging should be done periodically, as a layer of black, degraded polymer may build up on the cylinder wall, yielding parts with black marks. Residual stresses may be reduced by higher mold temperatures or by annealing. Extrusion and blow molding grades of polysulfones are higher molecular weight with blow molding melt temperatures in the range of 300 to 360°C and mold temperatures between 70 and 95°C.

The good heat resistance and electrical properties of polysulfones allows them to be used in applications such as circuit boards and TV components.[361] Chemical and heat resistance are important properties for automotive applications. Hair dryer components can also be made from polysulfones. Polysulfones find application in ignition components and structural foams.[362] Another important market for polysulfones is microwave cookware.[363]

1.5.27.1 Polyaryl sulfone (PAS). This polymer differs from the other polysulfones in the lack of any aliphatic groups in the chain. The lack of aliphatic groups gives this polymer excellent oxidative stability, as the aliphatic groups are more susceptible to oxidative degradation.[364] Polyaryl sulfones are stiff, strong, and tough polymers with very good chemical resistance. Most fuels, lubricants, cleaning agents, and hydraulic fluids will not affect the polymer.[365] However, methylene chloride, dimethyl acetamide, and dimethyl formamide will dissolve the polymer.[366] The glass transition temperature of these polymers is about 210°C, with a heat deflection temperature of 205°C at 1.82 MPa.[367] PAS also has good hydrolytic stability. Polyarylsulfone is available in filled and reinforced grades as well as both opaque and transparent versions.[368] This polymer finds application in electrical applications for motor parts, connectors, and lamp housings.[369]

The polymer can be injection molded, provided the cylinder and nozzle are capable of reaching 425°C.[370] It may also be extruded. The polymer should be dried prior to processing. Injection molding barrel temperatures should be 270 to 360°C at the rear, 295 to 390°C in the middle, and 300 to 395°C at the front.[371]

1.5.27.2 Polyether sulfone (PES). Polyether sulfone is a transparent polymer with high temperature resistance and self-extinguishing properties.[372] It gives off little smoke when burned. Polyether sulfone has the basic structure as shown in Fig. 1.50.

Polyether sulfone has a T_g near 225°C and is dimensionally stable over a wide range of temperatures.[373] It can withstand long-term use up to 200°C and can carry loads for long times up to 180°C.[374] Glass-fiber-reinforced grades are available for increased properties. It is resistant to most chemicals with the exception of polar, aromatic hydrocarbons.[375]

Polyether sulfone can be processed by injection molding, extrusion, blow molding, or thermoforming.[376] It exhibits low mold shrinkage. For injection molding, barrel temperatures of 340 to 380°C, with melt temperatures of 360°C, are recommended.[377] Mold temperatures should be in the range or 140 to 180°C. For thin-walled molding, higher temperatures may be required. Unfilled PES can be extruded into sheets, rods, films, and profiles.

PES finds application in aircraft interior parts due to its low smoke emission.[378] Electrical applications include switches, integrated circuit carriers, and battery parts.[379] The high-temperature oil and gas resistance allows polyether sulfone to be used in the automotive markets for water pumps, fuse housings, and car heater fans. The ability of PES to endure repeated sterilization allows PES to be used in a variety of medical applications, such as parts for centrifuges and root canal drills. Other applications include membranes for kidney dialysis, chemical separation, and desalination. Consumer uses include cooking equipment and lighting fittings. PES can also be vacuum metallized for a high-gloss mirror finish.

1.5.27.3 Polysulfone (PSU). Polysulfone is a transparent thermoplastic prepared from bisphenol A and 4,4′-dichlorodiphenylsulfone.[380] The structure is shown below in Fig. 1.51. It is self-extinguishing and has a high heat-distortion temperature. The polymer has a glass transition temperature of 185°C.[381] Polysulfones have impact resistance and ductility below 0°C. Polysulfone also has good electrical properties. The electrical and mechanical properties are maintained to temperatures near 175°C. Polysulfone shows good chemical resistance to alkali, salt, and acid solutions.[382] It has resistance to oils, detergents, and alcohols, but polar organic solvents and chlorinated aliphatic solvents may attack the polymer. Glass and mineral filled grades are available.[383]

Properties such as physical aging and solvent crazing can be improved by annealing the parts.[384] This also reduces molded-in stresses. Molded-in stresses can also be reduced by

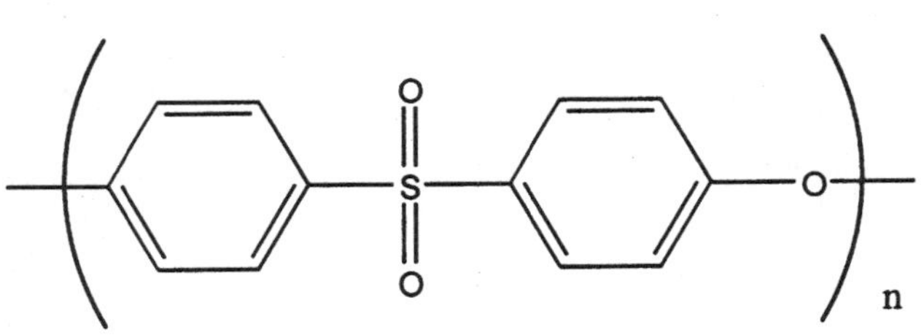

Figure 1.50 Structure of polyether sulfone.

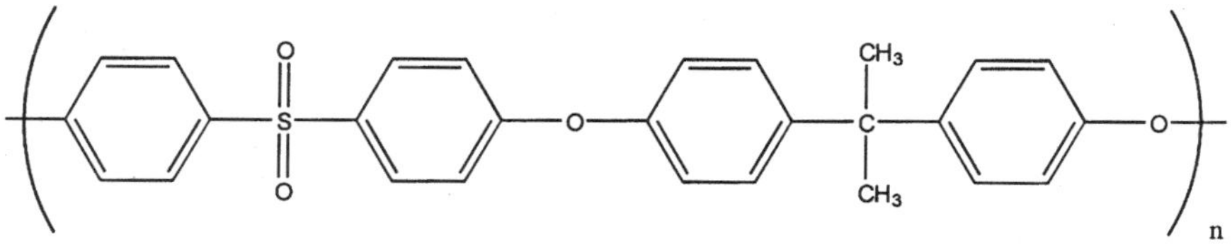

Figure 1.51 Structure of polysulfone.

using hot molds during injection molding. As mentioned above, runners and gates should be as large as possible due to the high melt viscosity. The polymer should hit a wall or pin shortly after entering the cavity of the mold, as polysulfone has a tendency toward jetting. For thin-walled or long parts, multiple gates are recommended.

For injection molding, barrel temperatures should be in the range of 310 to 400°C, with mold temperatures of 100 to 170°C.[385] In blow molding, the screw type should have a low compression ratio, 2.0/1 to 2.5/1. Higher compression ratios will generate excessive frictional heat. Mold temperatures of 70 to 95°C with blow air pressures of 0.3 to 0.5 MPa are generally used. Polysulfone can be extruded into films, pipe, or wire coatings. Extrusion melt temperatures should be from 315 to 375°C. High-compression-ratio screws should not be used for extrusion. Polysulfone shows high melt strength, allowing for good draw down and the manufacture of thin films. Sheets of polysulfone can be thermoformed, with surface temperatures of 230 to 260°C recommended. Sheets may be bonded by heat sealing, adhesive bonding, solvent fusion, or ultrasonic welding.

Polysulfone is used in applications requiring good high-temperature resistance such as coffee carafes, piping, sterilizing equipment, and microwave oven cookware.[386] The good hydrolytic stability of polysulfone is important in these applications. Polysulfone is also used in electrical applications for connectors, switches, and circuit boards and in reverse osmosis applications as a membrane support.[387]

1.5.28 Vinyl-Based Resins

1.5.28.1 Polyvinyl chloride (PVC). Polyvinyl chloride polymers (PVCs), generally referred to as *vinyl resins,* are prepared by the polymerization of vinyl chloride in a free radical addition polymerization reaction. Vinyl chloride monomer is prepared by reacting ethylene with chlorine to form 1,2-dichloroethane.[388] The 1,2 dichloroethane is then cracked to give vinyl chloride. The polymerization reaction is depicted in Fig. 1.52.

The polymer can be made by suspension, emulsion, solution, or bulk polymerization methods. Most of the PVC used in calendering, extrusion, and molding is prepared by suspension polymerization. Emulsion polymerized vinyl resins are used in plastisols and organosols.[389] Only a small amount of commercial PVC is prepared by solution polymerization. The microstructure of PVC is mostly atactic, but a sufficient quantity of syndiotactic portions of the chain allow for a low fraction of crystallinity (about 5 percent). The polymers are essentially linear, but a low number of short chain branches may exist.[390] The monomers are predominantly arranged head to tail along the backbone of the

$$n\,CH_2{=}CHCl \rightarrow -(CH_2{-}CHCl)_n-$$

Figure 1.52 Synthesis of polyvinyl chloride.

chain. Due to the presence of the chlorine group, PVC polymers are more polar than polyethylene. The molecular weights of commercial polymers are M_w = 100,000 to 200,000; M_n= 45,000 to 64,000.[391] M_w/M_n = 2 for these polymers.

The polymeric PVC is insoluble in the monomer; therefore, bulk polymerization of PVC is a heterogeneous process.[392] Suspension PVC is synthesized by suspension polymerization. These are suspended droplets approximately 10 to 100 nm in diameter of vinyl chloride monomer in water. Suspension polymerizations allow control of particle size, shape, and size distribution by varying the dispersing agents and stirring rate. Emulsion polymerization results in much smaller particle sizes than suspension polymerized PVC, but soaps used in the emulsion polymerization process can affect the electrical and optical properties.

The glass transition temperature of PVC varies with the polymerization method but falls within the range of 60 to 80°C.[393] PVC is a self-extinguishing polymer and therefore has application in the field of wire and cable. PVC's good flame resistance results from removal of HCl from the chain, releasing HCl gas.[394] Air is restricted from reaching the flame, because HCl gas is more dense than air. Because PVC is thermally sensitive, the thermal history of the polymer must be carefully controlled to avoid decomposition. At temperatures above 70°C, degradation of PVC by loss of HCl can occur, resulting in the generation of unsaturation in the backbone of the chain. This is indicated by a change in the color of the polymer. As degradation proceeds, the polymer changes color from yellow to brown to black, visually indicating that degradation has occurred. The loss of HCl accelerates the further degradation and is called *autocatalytic decomposition*. The degradation can be significant at processing temperatures if the material has not been heat stabilized, so thermal stabilizers are often added at additional cost to PVC to reduce this tendency. UV stabilizers are also added to protect the material from ultraviolet light, which may also cause the loss of HCl.

There are two basic forms of PVC: rigid and plasticized. Rigid PVC, as its name suggests, is an unmodified polymer and exhibits high rigidity.[395] Unmodified PVC is stronger and stiffer than PE and PP. Plasticized PVC is modified by the addition of a low-molecular-weight species (plasticizer) to flexibilize the polymer.[396] Plasticized PVC can be formulated to give products with rubbery behavior.

PVC is often compounded with additives to improve the properties. A wide variety of applications for PVC exist, because one can tailor the properties by proper selection of additives. As mentioned above, one of the principal additives are stabilizers. Lead compounds are often added for this purpose, reacting with the HCl released during degradation.[397] Among the lead compounds commonly used are basic lead carbonate or white lead and tribasic lead sulfate. Other stabilizers include metal stearates, ricinoleates, palmitates, and octoates. Of particular importance are the cadmium-barium systems with synergistic behavior. Organo-tin compounds are also used as stabilizers to give clear compounds. In addition to stabilizers, other additives, such as fillers, lubricants, pigments, and plasticizers, are used. Fillers are often added to reduce cost and include talc, calcium carbonate, and clay.[398] These fillers may also impart additional stiffness to the compound.

The addition of plasticizers lowers the T_g of rigid PVC, making it more flexible. A wide range of products can be manufactured by using different amounts of plasticizer. As the plasticizer content increases, there is usually an increase in toughness and a decrease in the modulus and tensile strength.[399] Many different compounds can be used to plasticize PVC, but the solvent must be miscible with the polymer. A compatible plasticizer is considered a nonvolatile solvent for the polymer. The absorption of solvent may occur automatically at room temperature or may require the addition of slight heat and mixing. PVC plasticizers are divided into three groups depending on their compatibility with the polymer: primary plasticizers, secondary plasticizers, and extenders. Primary plasticizers are compatible (have similar solubility parameters) with the polymer and should not exude. If

the plasticizer and polymer have differences in their solubility parameters, they tend to be incompatible or have limited compatibility and are called *secondary plasticizers.* Secondary plasticizers are added along with the primary plasticizer to meet a secondary performance requirement (cost, low-temperature properties, permanence). The plasticizer can still be used in mixtures with a primary plasticizer, provided the mixture has a solubility parameter within the desired range. Extenders are used to lower the cost and are generally not compatible when used alone. Common plasticizers for PVC include dioctyl phthalate, di-iso-octyl phthalate, dibutyl phthalate among others.[400]

The plasticizer is normally added to the PVC before processing. Since the plasticizers are considered to be solvents for PVC, they will normally be absorbed the polymer with only a slight rise in temperature.[401] This reduces the time the PVC is exposed to high temperatures and potential degradation. In addition, the plasticizer reduces the T_g and T_m, thereby lowering the processing temperatures and thermal exposure. Plasticized PVC can be processed by methods such as extrusion and calendering into a variety of products.

Rigid PVC can be processed using most conventional processing equipment. Because HCl can be given off in small amounts during processing, corrosion of metal parts is a concern. Metal molds, tooling, and screws should be inspected regularly. Corrosion resistant metals and coatings are available but add to the cost of manufacturing.

Rigid PVC products include house siding, extruded pipe, and thermoformed and injection molded parts. Rigid PVC is calendered into credit cards. Plasticized PVC is used in applications such as flexible tubing, floor mats, garden hose, shrink wrap, and bottles.

PVC joints can be solvent welded rather than heated to fuse the two part together. This can be an advantage when heating the part is not feasible.

1.5.28.2 Chlorinated PVC. Post chlorination of PVC was practiced during World War II.[402] Chlorinated PVC (CPVC) can be prepared by passing chlorine through a solution of PVC. The chlorine adds to the carbon that does not already have a chlorine atom present. Commercial materials have chlorine contents around 66 to 67 percent. The materials have a higher softening point and higher viscosity than PVC. They are known for good chemical resistance. Compared with PVC, chlorinated PVC has higher modulus and tensile strength. Compounding processes are similar to those for PVC but are more difficult.

Chlorinated PVC can be extruded, calendered, or injection molded.[403] The extrusion screw should be chrome plated or stainless steel. Dies should be streamlined. Injection molds should be chrome or nickel plated or stainless steel. CPVC is used for water distribution piping, industrial chemical liquid piping, outdoor skylight frames, automotive interior parts, and a variety of other applications.

1.5.28.3 Copolymers. Vinyl chloride can be copolymerized with vinyl acetate, giving a polymer with a lower softening point and better stability than pure PVC.[404] The compositions can vary from 5 to 40 percent vinyl acetate content. This material has application in areas where PVC is too rigid and the use of plasticized PVC is unacceptable. Flooring is one application for these copolymers. Copolymers with about 10 percent vinylidene chloride and copolymers with 10 to 20 percent diethyl fumarate or diethyl maleate are also available.

1.5.28.4 Dispersion PVC. If sufficient quantity of solvent is added to PVC, it can become suspended in the solvent, giving a fluid that can be used in coating applications.[405]

This form of PVC is called a *plastisol* or *organisol*. PVC in the fluid form can be processed by methods such as spread coating, rotational casting, dipping, and spraying. The parts are then dried with heat to remove any solvent and fuse the polymer. Parts such as handles for tools and vinyl gloves are produced by this method.

The plastisol or organisols are prepared from PVC produced through emulsion polymerization.[406] The latex is then spray dried to form particles from 0.1 to 1 micron. These particles are then mixed with plasticizers to make plastisols or with plasticizers and other volatile organic liquids to make organisols. Less plasticizer is required with the organisols, so harder coatings can be produced. The polymer particles are not dissolved in the liquid but remain dispersed until the material is heated and fused. Other additives such as stabilizers and fillers may be compounded into the dispersion.

As plasticizer is added, the mixture goes through different stages as the voids between the polymer particles are filled.[407] Once all the voids between particles have been filled, the material is considered to be a paste. In these materials, the size of the particle is an important variable. If the particles are too large, they may settle out, so small particles are preferred. Very small particles have the disadvantage that the particles will absorb the plasticizer with time, giving a continuous increase in viscosity of the mixture. Paste polymers have particle sizes in the range of 0.2 to 1.5 μm. Particle size distribution will also affect the paste. It is usually better to have a wide particle size distribution so that particles can pack efficiently. This reduces the void space that must be filled by the plasticizer, and any additional plasticizer will act as lubricant. For a fixed particle/plasticizer ratio, a wide distribution will generally have lower viscosity than for a constant particle size. In some cases, very large particles are added to the paste, as they will take up volume, again reducing the amount of plasticizer required. These particles are made by suspension polymerization. With the mixture of particle sizes, these larger particles will not settle out as they would if used alone. Plastisols and organisols require the addition of heat to fuse. Temperatures in the range of 300 to 410°F are used to form the polymer.

1.5.28.5 Polyvinylidene chloride (PVDC).

Polyvinylidene chloride (PVDC) is similar to PVC except that two chlorine atoms are present on one of the carbon groups.[408] Like PVC, PVDC is also polymerized by addition polymerization methods. Both emulsion and suspension polymerization methods are used. The reaction is shown below in Fig. 1.53. The emulsion polymers are either used directly as a latex or dried for use in coatings or melt processing.

This material has excellent barrier properties and is frequently used in food packaging applications. Films made from PVDC have good cling properties, which is an advantage for food wraps. Commercial polymers are all copolymers of vinylidene chloride with vinyl chloride, acrylates, or nitriles. Copolymerization of vinylidene chloride with other monomers reduces the melting point to allow easier processing. Corrosion-resistant materials should be considered for use when processing PVDC.

1.5.29 Additives

There is a broad range of additives for thermoplastics. Some of the more important additives include plasticizers, lubricants, anti-aging additives, colorants, flame retardants,

$$n\ CH_2{=}CCl_2 \rightarrow (\text{-}CH_2\text{-}CCl_2\text{-})_n$$

Figure 1.53 Preparation of vinylidene chloride polymers.

blowing agents, cross-linking agents, and UV protectants. Fillers are also considered additives but are covered separately above.

Plasticizers are considered nonvolatile solvents.[409] They act to soften a material by separating the polymer chains, allowing them to be more flexible. As a result, the plasticized polymer is softer, with greater extensibility. Plasticizers reduce the melt viscosity and glass transition temperature of the polymer. For the plasticizer to be a "solvent" for the polymer, it is necessary for the solubility parameter of the plasticizer to be similar to the polymer. As a result, the plasticizer must be selected carefully so it is compatible with the polymer. One of the primary applications of plasticizers is for the modification of PVC. In this case, the plasticizers are divided into three classes: primary and secondary plasticizers and extenders.[410] Primary plasticizers are compatible, can be used alone, and will not exude from the polymer. They should have a solubility parameter similar to the polymer. Secondary plasticizers have limited compatibility and are generally used with a primary plasticizer. Extenders have limited compatibility and will exude from the polymer if used alone. They are usually used along with the primary plasticizer. Plasticizers are usually in the form of high-viscosity liquids. The plasticizer should be capable of withstanding the high processing temperatures without degradation and discoloration that would adversely affect the end product. The plasticizer should be capable of withstanding any environmental conditions that the final product will see. This might include UV exposure, fungal attack, or water. In addition, it is important that the plasticizer show low volatility and migration so that the properties of the plasticized polymer will remain relatively stable over time. There is a wide range of plasticizer types. Some typical classes include phthalic esters, phosphoric esters, fatty acid esters, fatty acid esters, polyesters, hydrocarbons, aromatic oils, and alcohols.

Lubricants are added to thermoplastics to aid in processing. High-molecular-weight thermoplastics have high viscosity. The addition of lubricants acts to reduce the melt viscosity to minimize machine wear and energy consumption.[411] Lubricants may also be added to prevent friction between molded products. Examples of these types of lubricants include graphite and molybdenum disulphide.[412] Lubricants that function by exuding from the polymer to the interface between the polymer and machine surface are termed *external lubricants.* Their presence at the interface between the polymer and metal walls acts to ease the processing. They have low compatibility with the polymer and may contain polar groups so that they have an attraction to metal. Lubricants must be selected based on the thermoplastic used. Lubricants may cause problems with clarity, ability to heat seal, and printing on the material. Examples of these lubricants include stearic acid or other carboxylic acids, paraffin oils, and certain alcohols and ketones for PVC. Low-molecular-weight materials that do not affect the solid properties but act to enhance flow in the melt state are termed *internal lubricants*. Internal lubricants for PVC include amine waxes, montan wax ester derivatives, and long chain esters. Polymeric flow promoters are also examples of internal lubricants. They have solubility parameters similar to the thermoplastic but lower viscosity at processing temperatures. They have little effect on the mechanical properties of the solid polymer. An example is the use of ethylene-vinyl acetate copolymers with PVC.

Anti-aging additives are incorporated to improve the resistance of the formulation. Examples of aging include attack by oxygen, ozone, dehydrochlorination, and UV degradation. Aging often results in changes in the structure of the polymer chain such as cross-linking, chain scission, addition of polar groups, or the addition of groups that cause discoloration. Additives are used to help prevent these reactions. Antioxidants are added to the polymer to stop the free radical reactions that occur during oxidation. Antioxidants include compound such as phenols and amines. Phenols are often used because they have less of a tendency to stain.[413] Peroxide decomposers are also added to improve the aging properties of thermoplastics. These include mercaptans, sulfonic acids, and zinc dialkylth-

iophosphate. The presence of metal ions can act to increase the oxidation rate, even in the presence of antioxidants. Metal deactivators are often added to prevent this from taking place. Chelating agents are added to complex with the metal ion.

The absorption of ultraviolet light by a polymer may lead to the production of free radicals. These radicals react with oxygen, resulting in what is termed *photodegradation*. This leads to the production of chemical groups that tend to absorb ultraviolet light, increasing the amount photodegradation. To reduce this effect, UV stabilizers are added. One way to accomplish UV stabilization is by the addition of UV absorbers such as benzophenones, salicylates, and carbon black.[414] They act to dissipate the energy in a harmless fashion. Quenching agents react with the activated polymer molecule. Nickel chelates and hindered amines can be used as quenching agents. Peroxide decomposers may be used to aid in UV stability.

In certain applications, flame resistance can be important. In this case, flame retarders may be added.[415] They act by one of four possible mechanisms. They may

1. Act to chemically interfere with the propagation of flame
2. React or decompose to absorb heat
3. Form a fire-resistant coating on the polymer
4. Produce gases that reduce the supply of air

Phosphates are an important class of flame retarders. Tritolyl phosphate and trixylyl phosphate are often used in PVC. Halogenated compounds such as chlorinated paraffins may also be used. Antimony oxide is often used in conjunction to obtain better results. Other flame retarders include titanium dioxide, zinc oxide, zinc borate, and red phosphorus. As with other additives, the proper selection of a flame retarder will depend on the particular thermoplastic.

Colorants are added to produce color in the polymeric part. They are separated into pigments and dyes. Pigments are insoluble in the polymer, while dyes are soluble in the polymer. The particular color desired and the type of polymer will affect the selection of the colorants.

Blowing agents are added to the polymer to produce a foam or cellular structure.[416] They may be chemical blowing agents that decompose at certain temperatures and release a gas, or they may be low-boiling liquids that become volatile at the processing temperatures. Gases may be introduced into the polymer under pressure and expand when the polymer is depressurized. Mechanical whipping and the incorporation of hollow glass spheres can also be used to produce cellular materials.

Peroxides are often added to produce cross-linking in a system. Peroxides can be selected to decompose at a particular temperature for the application. Peroxides can be used to cross-link saturated polymers.

1.5.30 Polymer Blends

There is considerable interest in polymer blends. This is driven by consideration of the difficulty in developing new polymeric materials from monomers. In many cases, it can be more cost-effective to tailor the properties of a material through the blending of existing materials. One of the most basic questions in blends is whether the two polymers are miscible or exist as a single phase. In many cases, the polymers will exist as two separate phases. In this case, the morphology of the phases is of great importance. In the case of a miscible single-phase blend, there is a single T_g, which is dependent on the composition of the blend.[417] Where two phases exist, the blend will exhibit two separate T_g values—one for each of the phases present. In the case where the polymers can crystallize, the crystal-

line portions will exhibit a melting point (T_m), even in the case where the two polymers are a miscible blend.

Although miscible blends of polymers exist, most blends of high-molecular-weight polymers exist as two-phase materials. Control of the morphology of these two-phase systems is critical to achieve the desired properties. A variety of morphologies exist, such as dispersed spheres of one polymer in another, lamellar structures, and co-continuous phases. As a result, the properties depend in a complex manner on the types of polymers in the blend, the morphology of the blend, and the effects of processing, which may orient the phases by shear.

Miscible blends of commercial importance include PPO-PS, PVC-nitrile rubber, and PBT-PET. Miscible blends show a single T_g that is dependent on the ratios of the two components in the blend and their respective T_g values. In immiscible blends, the major component has a large effect on the final properties of the blend. Immiscible blends include toughened polymers in which an elastomer is added, existing as a second phase. The addition of the elastomer phase dramatically improves the toughness of the resulting blend as a result of the crazing and shear yielding caused by the rubber phase. Examples of toughed polymers include high-impact polystyrene (HIPS), modified polypropylene, ABS, PVC, nylon, and others. In addition to toughened polymers, a variety of other two phase blends are commercially available. Examples include PC-PBT, PVC-ABS, PC-PE, PP-EPDM, and PC-ABS.

1.6 Processing of Thermoplastics

Processing involves the conversion of the solid polymer into a desirable size and shape. There are a number of methods to shape the polymer, including injection molding, extrusion, thermoforming, blow molding, and rotational molding. The plastic material is heated to the appropriate temperature for it to flow, and the material is shaped and then cooled to preserve the desired shape.

1.6.1 Extrusion

Extruders are used to continuously produce rods or sheets; this is the method used to produce PVC pipes, profiled PVC window moldings, and sheets of all sorts of plastics such as PC (Lexan®) and PMMA (Plexiglass®) for applications including storm doors. Extruders can continuously produce shapes of any cross section and are thus used to manufacture garden hoses, gutters, floor tiles, sealing strips for car windows and doors, and to coat wire insulation, among other applications. This method was adapted from metallurgists who use a similar form of extrusion to process molten aluminum, and it was first adapted in 1845 by Bewley and Brooman to extrude rubber around cable as a coating.[418] As a most basic description, extruders are a machine with a drive system (motor, gearbox, and thrust bearing), a plasticating unit (with at least one Archimedes-type screw, a barrel, and a temperature control system) and a control cabinet with control devices and the power supply.[419] Figure 1.54 shows a single-screw extruder.[420]

Plastic pellets are fed into the cavity between the screw and extruder barrel and, as the screw rotates, the pellets are dragged forward, compressed, heated by conduction through the barrel walls and through frictional heating, melted, and forced under pressure through a die that forms the molten plastic. Figure 1.55[421] shows a single-screw extruder with three zones along the screw: a feed or compaction zone, a transition zone, and a metering zone. This is typical of single-screw design, where the first zone is used to compact the pellets together and to begin transporting them along the screw channel, the transition zone starts to compress and melt the pellets, and the metering zone homogenizes the melt and brings it to its final temperature. The screw channel depth is constant in both the feed

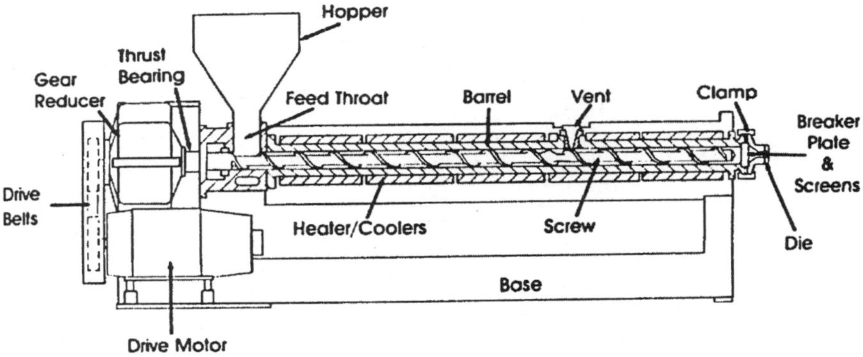

Figure 1.54 Single-screw extruder.[420]

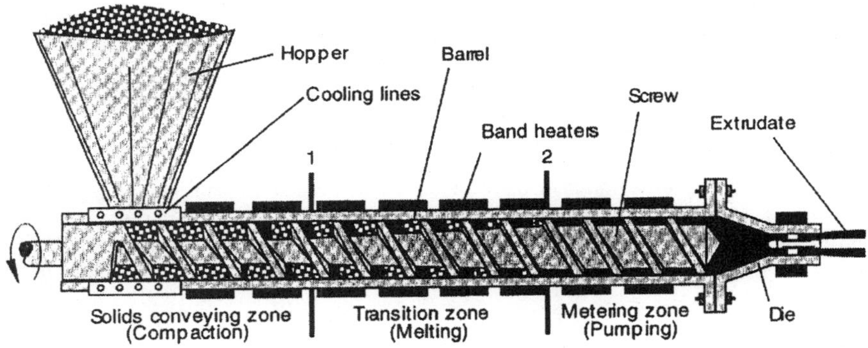

Figure 1.55 The zones of an extruder screw.[421]

and metering zones (being deeper in the feed zone) and varies in the transition zone to begin generating pressure and to force the pellets to begin to melt. The length of each zone in a screw's design varies according to the type of plastic being processed. For polymers that melt gradually, such as LDPE, the overall length of the screw is roughly divided into three even zones, while plastics such as nylon, which have a sharp melting point, may have a screw designed such that the transition zone only consists of one turn of the screw flight. Polymers, such as PVC, that are prone to thermal degradation and that melt very gradually may be processed with a screw whose entire length is composed of a compression zone, sometimes with the addition of a metering zone. Figure 1.56 shows some common screw design variations.[422] There are many other configurations designed to improve the physical mixing of the polymer melt as well as homogenization of the temperature and pressure throughout the melt. These include the addition of mixing pins on the barrel of the screw, ring barriers, and modified designs that involve very large screw diameters so as to force molten polymer through a small clearance between the mixing head and the inside of the barrel wall. Venting may also be required along the length of the barrel to enable entrapped moisture, air, or other volatiles to escape from the melt.

Typical extruder dimensions include a screw length/diameter ratio of 20 or less for melt extruders, with standard diameters varying from 20 mm (0.75 in) to 600 mm (24 in). They

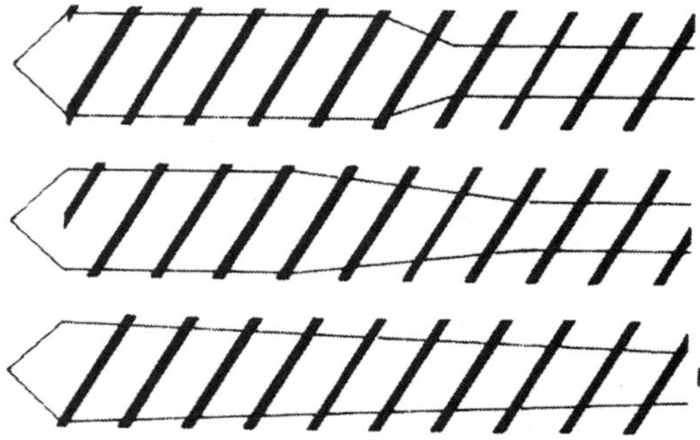

Figure 1.56 Screw design variations.[422]

typically operate at 1 to 2 r/s (60 to 120 r/min) for large extruders and 1 to 5 r/s (60 to 300 r/min) for small extruders.[423] Output varies as a function of processing parameters (particularly screw speed and pressure), the thermal and mechanical properties of the polymer, and the design and geometry of the screw. A 600-mm diameter single-screw extruder is capable of delivering 29 metric tons of product an hour, whereas the smallest 20-mm diameter single-screw extruders have a throughput capacity of 5 kg/h.[424] Operating pressures as high as 69 MPa (10,000 lb/in^2) are common.

Intermeshing twin screw extruders also exist; the screws are either co-rotating or counter-rotating. Co-rotating twin screw extruders are typically used in applications where mixing and compounding need to be accomplished in addition to the molding of the plastic melt. They are highly capable of dispersing small agglomerates such as carbon black and can be used, for example, to blend the components of duct tape adhesive as well as coat the finished adhesive onto the tape backing. Counter-rotating screws impart very little shear to the polymer and are thus used to extrude shear sensitive materials such as rigid PVC into pipe profiles, particularly from a powder feedstock.[425]

Die design is also critical to avoid "dead spots" where the polymer melt can become stagnant and risk thermal degradation. It is also important that the polymer molecules be allowed to return to an equilibrium position to the greatest extent possible to minimize the orientation as a result of flow. Laminar flow is desired, and finite element analysis is used to design dies that enable laminar flow to the greatest extent. Dies known as *coat hanger* dies are used for producing sheets and films. These dies are designed to promote uniform cross-die melt flow.

Modifications to the extrusion process include coextrusion, where multiple single-screw extruders can individually feed melt into a sheet die to form a multilayer sheet or film. This is a common technique for producing multilayer packaging films, where each layer provides a particular feature. For example, garbage bags are often multilaminate constructions, as are packaging films where a PVDC layer may be incorporated for moisture or oxygen barrier properties, and HDPE may be used as a less expensive, relatively strong, outer layer. EVA is a common "bonding layer" between different plastic layers. As many as eight or more extruders may be used to form highly specialized multilayer films.

Common defects encountered with extrusion include effects associated with the viscoelastic nature of plastic melts. As the melt is extruded from the die for example, it may exhibit shark-skinning, melt fracture, and die swell. Diagrams of these defects are shown in Fig. 1.57.[426] Shark-skinning and melt fracture occur when the stresses being applied to the plastic melt exceed its tensile strength. Die swell occurs due to the elastic component of the polymer melt's response to stress and is the result of the elastic rebound of the polymer as it leaves the constraints of the die channel prior to cooling.

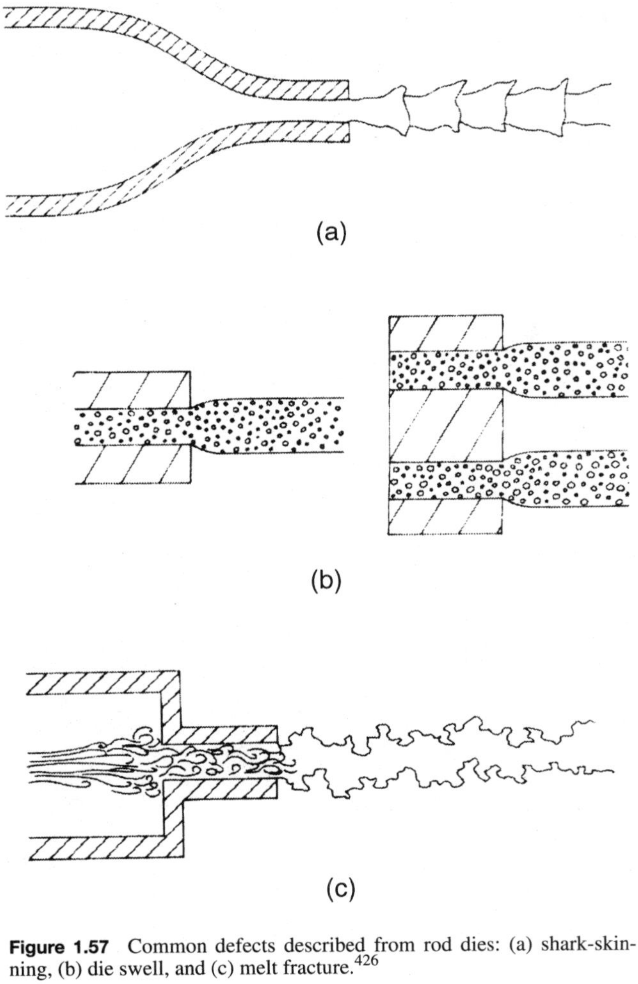

Figure 1.57 Common defects described from rod dies: (a) shark-skinning, (b) die swell, and (c) melt fracture.[426]

An extrusion process known as *blown film* involves passing the plastic melt upward through an annular die. Hot air is blown into the annulus through a hole in the die mandrel to increase the diameter of the "tube" of film being formed, while pullers are used to stretch the film further upward. This biaxial orientation, thinning of the tube of film through the internal pressurization of the bubble, combined with the thinning of the film as it is stretched upward, results in a strong, biaxially oriented film. Stretching continues to the freezing line, at which point the film has cooled off to such an extent as to provide a high enough modulus to resist further deformation. Crystallization also enables the orientation to be maintained. Figure 1.58[427] shows a diagram of the blown film process, where a pair of collapsing rolls are used to flatten the bubble and allow the film to then be wound into a master roll for later converting processes such as slitting. The blown film process can be used to blow multilaminate films when several extruders are used to feed polymer melt into the annular die.

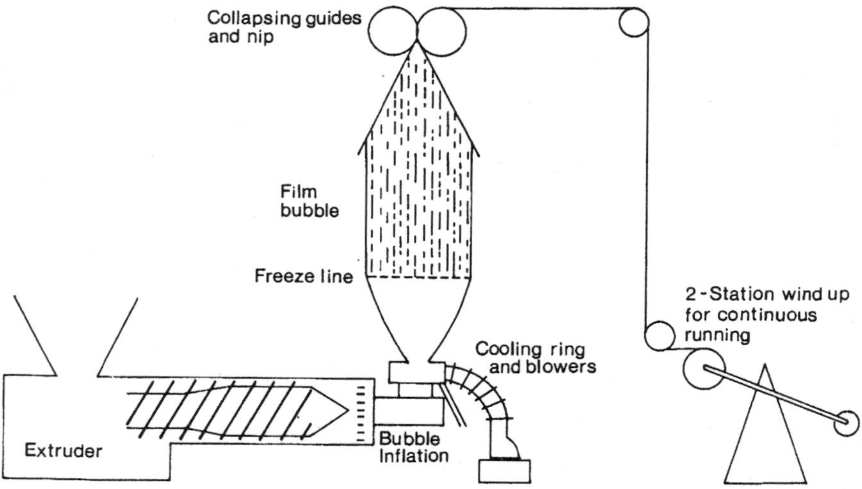

Figure 1.58 Blown film process.[427]

1.6.2 Injection Molding

Injection molding is a widely used process to produce parts with variable dimensions. An injection molding machine consists of the following four components:[428]

- Injection unit
- Control system
- Drive system
- Clamping unit

The purpose of the injection unit is to heat and melt the polymer, inject the melt into the cavity, and apply pressure during the cooling phase. The most common type of injection molding machine is the reciprocating screw. In this type of machine, the screw rotates to plasticize the polymer, moving backward to deposit a volume of polymer melt ahead of the screw (shot). Once the correct shot size has been built, the screw then moves forward to inject the melt into the mold. Injection molding is a discontinuous process, and the clamping unit allows for the mold to open and close for part removal and to provide pressure as the cavity is filled. This is depicted in Fig. 1.59.

The purpose of an injection mold is to give the shape of the part (cavity), distribute the polymer melt to the cavities through a runner system, cool the part, and eject the part. During the injection molding cycle, the polymer flows from the nozzle on the injection unit through the sprue, then to the runners, which distribute the melt to each of the cavities. The entrance to the cavity is called the *gate* and is usually small so that the runner system can be easily removed from the part. A typical feed system for injection molding is shown in Fig. 1.60. Figure 1.61 depicts a number of gate configurations. Molding conditions for a wide variety of thermoplastics are given in Table 1.14.

The molding process itself can have a large influence on the final properties of the part. The polymer chains undergo orientation in the flow direction during the mold-filling phase of the injection cycle as shown in Fig. 1.62. The amount of orientation in the final part depends on how much orientation was induced during filling minus the amount that was re-

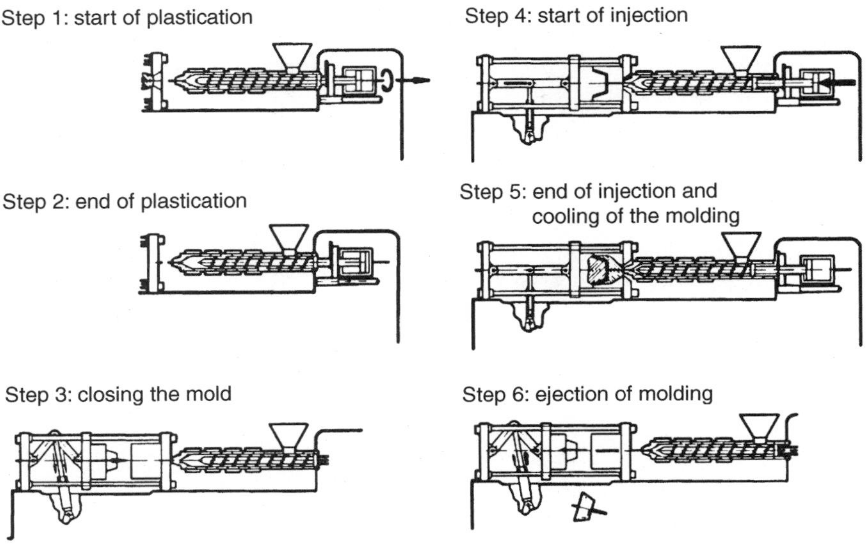

Figure 1.59 Injection molding.[461]

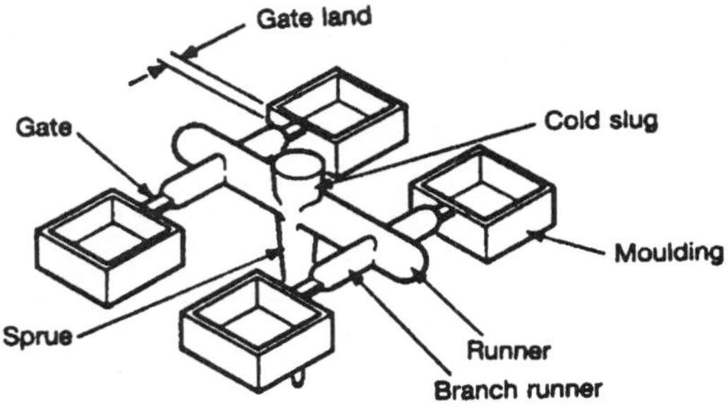

Figure 1.60 Injection molding feed system.[462]

moved through molecular relaxation.[429] This is particularly true of the surface of the part, where hot material reaches the cool walls of the mold with rapid solidification, coupled with the highest shear and induced orientation at the mold surface. Orientation can result in parts with anisotropic properties and should be accounted for during part design. Mechanical properties are thus typically higher in the direction of orientation.[430]

For semicrystalline polymers, the injection molding process parameters can have a large impact on the degree of crystallinity. As the cooling rate increases, the degree of crystallinity will decrease.[431] Cooling rate effects can cause a gradient of crystallinity across the thickness of a part, where interior portions of the part may have higher crystallinity due to their slower cooling rates compared to the surface. The crystalline morphology will also be influenced by the cooling behavior. Slower cooling rates result in larger spherulites, while more rapid cooling rates result in a larger number of smaller spherulites.[432]

A number of specialized injection molding processes also exist and are outlined below.

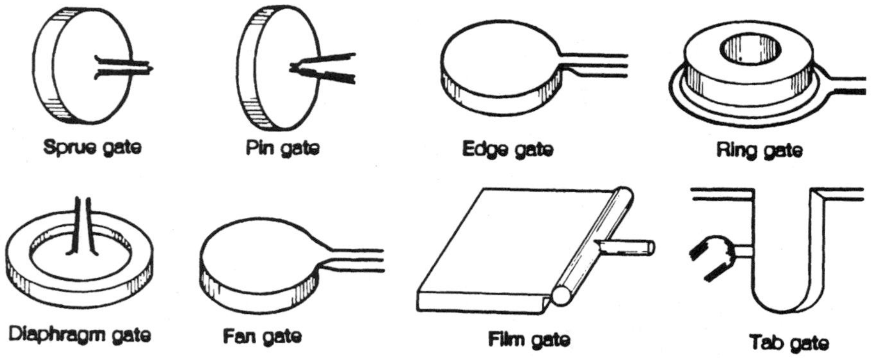

Figure 1.61 Injection molding gate types.[463]

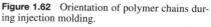

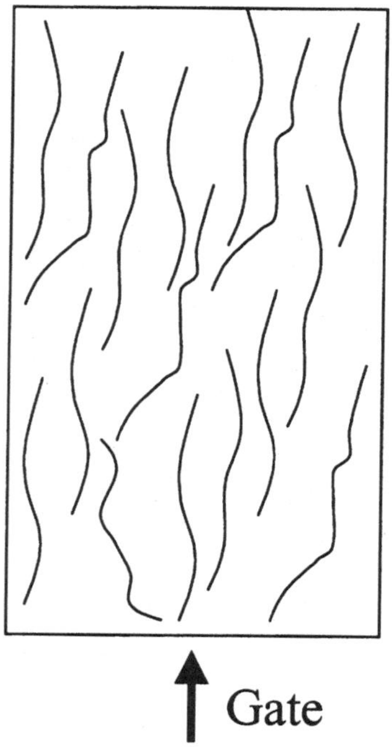

Figure 1.62 Orientation of polymer chains during injection molding.

1.6.2.1 Injection-compression molding. Injection-compression molding refers to the process whereby the cavity is not completely filled during injection of the resin.[428] In this process, the resin is injected while the mold is slightly open. The two halves of the mold then close, distributing the resin and filling the cavity. This process is useful for products that require high surface replication, such as compact discs or optical parts. Thin-walled parts can also be molded by this process, as the pressure losses are reduced, and there is less risk of premature resin solidification. Figure 1.63 illustrates this process.

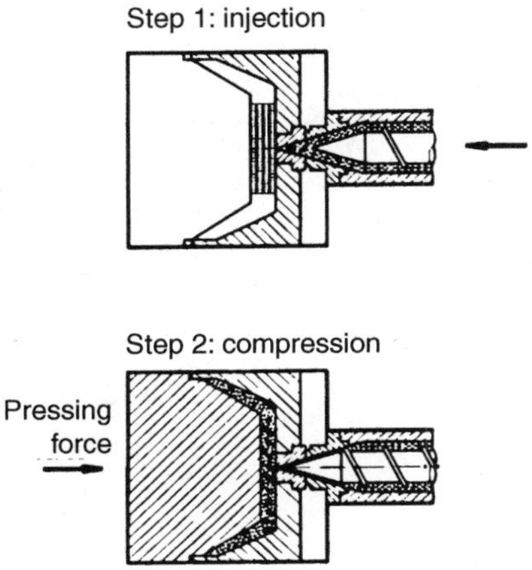

Figure 1.63 Injection compression process.[464]

1.6.2.2 Lost-core process. Products that are hollow or contain complex undercuts can be fabricated using the lost core process as illustrated in Fig. 1.64. Core materials are typically low-melting alloys (around 150°C) that are removed by heating the part. Before each molding cycle, a core is inserted into the mold, and the part is injection molded. The core is ejected with the part and then melted, resulting in the finished product. It is important that the core material melt at temperatures low enough that the plastic material is not affected by the heating cycle. Air manifolds for automotive and pump parts are often fabricated using this method.

1.6.2.3 Gas-assisted injection molding. In gas-assisted injection molding, the mold is partially filled with polymer, followed by a gas, which presses the polymer out to the surface of the mold, resulting in a hollow part. This process can be used for producing lighter weight parts, often with reduced cycle times as a result of less material to cool.

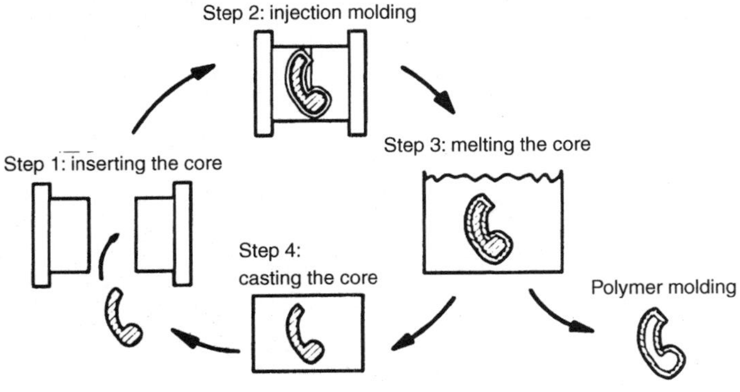

Figure 1.64 Lost-core injection molding process.[465]

TABLE 1.14 Injection Molding Guidelines for Unfilled Materials[457]

Material	Melt temp. range, °C	Mold temp. range, °C	Drying temp., °C	Drying times, hr
ASA	250–265	40–80	80–85	2–4
ABS	220–260	60–90	80–85	2–4
BDS	190–230	10–60	60	1
CA/CAB/CAP	160–230	40–80	55–85	3–4
FEP	300–380	200–240	150	2–4
HIPS/TPS	200–270	10–80	65–70	3–4
PA6	230–280	60–90	80–105	12–16
PA66	260–290	20–100	85–105	5–12
PA11/PA12	240–300	30–100	85	3– 5
PBT	220–260	20–110	120–150	2.5–5.5
PC	280–320	80–120	120	2– 4
PEBA	185–220	20–40	70–80	2–6
PEEL	195–255	10–70	90–120	10
PEEK	360–380	160–170	150	3
PE-HD	205–280	10–60	65	3
PE-LD	180–280	10–60	65	3
PE-LLD	160–280	10–60	65	3
PES	350–380	140–160	135–150	3–4
PET/PETP	365–295	120–140	135–165	2–4
PMMA	210–270	60–90	75	2–4
POM-H	190–215	40–120	110	2–3
POM-CO	175–220	40–120	110	2–3
PPO-M	260–300	60–110	100	2
PPS	300–360	135–160	150	3–6
PP	220–275	30–80	80	2–3
GPPS	200–250	10–80	70	2–3
PSU	350–380	100–150	135–150	3–4
PVDF	180–300	30–120	80	2–4
SAN	200–270	40–80	70–75	3–4
TPU/PUR	180–230	15–70	80	3
UPVC	185–205	30–60	65	2–3
PPVC	175–200	30–50	65	2
EVA	140–225	15–40	50–60	8

Thick-walled parts can be produced with fewer surface imperfections, such as sink marks, but equipment costs will be higher. Figure 1.65 shows the gas-assist injection molding method.

1.6.2.4 Coinjection molding. Coinjection molding refers to a process whereby two materials are injected into the same cavity.[428] The first material is injected into the cavity and then followed by the second material as depicted in Fig. 1.66. In this process, the first material goes to the outside of the mold and forms the skin, and the second material forms

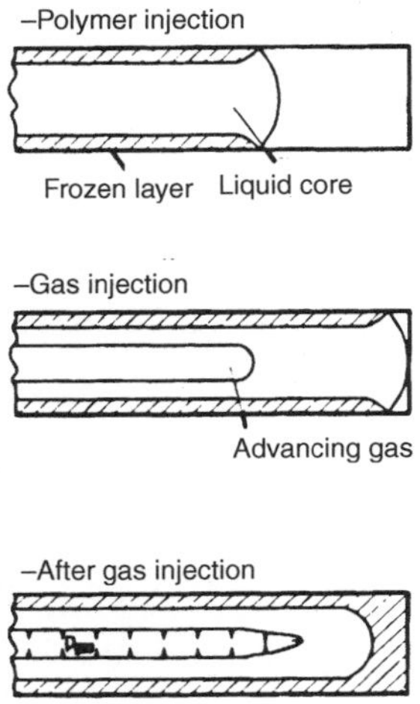

Figure 1.65 Gas-assist injection molding.[466]

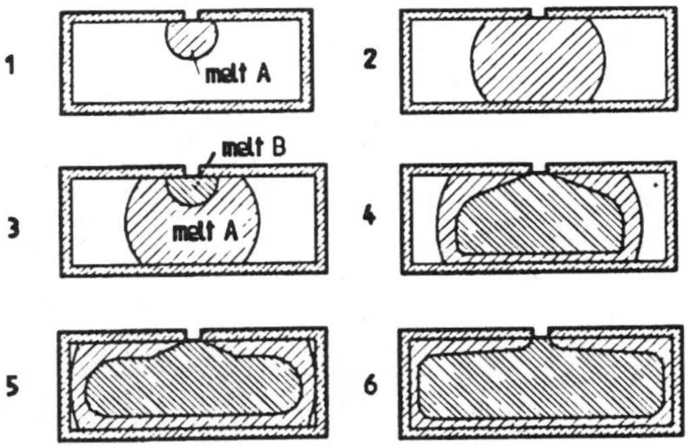

Figure 1.66 Coinjection molding.[467]

the interior of the part. This is often referred to as *sandwich molding*. Materials may be injected either sequentially or simultaneously. Applications include the use of an expensive outer layer and a cheaper core material or with fiber reinforced materials, where a skin material is used for improved surface quality.

1.6.2.5 Two-shot injection molding. Two-shot or overmolding refers to a process wherein either different colors or different materials are molded into one part. In this process, the first material or color is injected, then the mold is rotated, and the second shot is injected as depicted in Fig. 1.67. An alternative method is to use a retractable core.[468] In this case, the first material is injected and cooled to solidify, and then the core is retracted to allow injection of the second material as shown in Fig. 1.68. Bonding is accomplished through either strictly mechanical means or by adhesion between the two components through diffusion of the chains. This can result in parts with two combined materials without the need for an additional adhesive bonding step.[433] In the case in which direct adhesion of the materials is desired, proper selection of compatible materials is required. Table

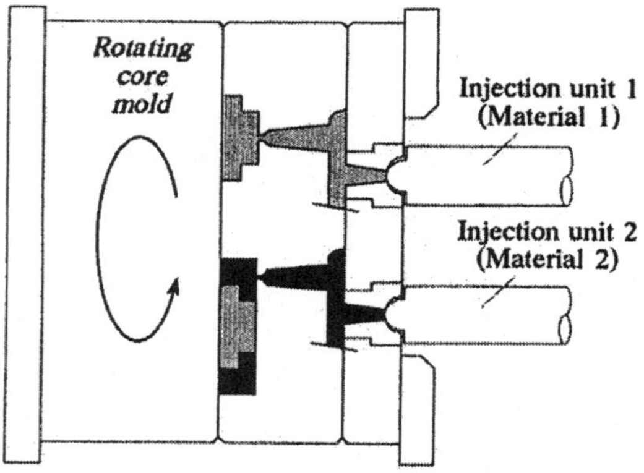

Figure 1.67 Two-shot molding with rotating mold.[468]

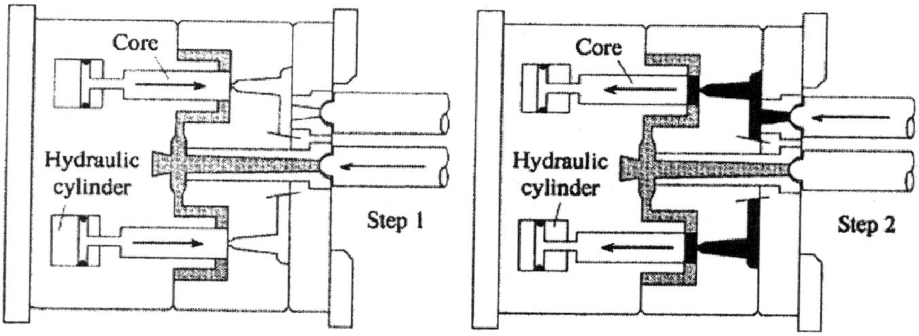

Figure 1.68 Two-shot injection molding with retractable core.[468]

1.15 shows the bonding strength for a number of thermoplastic combinations for use in multicomponent injection molding.

1.6.3 Thermoforming

Thermoforming is the heating of a thermoplastic sheet until it is soft and stretchable, and then forcing the hot sheet against the contours of a mold using mechanical force (plug assist), vacuum, pressure, or a combination of all three. After cooling, the plastic sheet retains the mold's shape and detail.[434] Thermoforming is still a rapidly growing processing method because of the range of products that can be formed and the relatively low cost of required tooling and equipment.[435] Thermoformed products include dinnerware, cups, automotive parts, egg cartons, and blister packaging.[436]

There are a wide variety of processes for thermoforming. One-step processes include[437]

- Drape forming
- Vacuum forming
- Pressure forming

Table 1.15 Bonding for Thermoplastic Combinations in Multicomponent Injection Molding[458]

Material	ABS	ASA	CA	EVA	PA6	PA6,6	PC	HDPE	LDPE	PMMA	POM	PP	PPO mod.	PS–GP	PS–HI	PBTP	TPU	PVC (soft)	SAN	TPR	PETP	PVAC	PSU	PC–PBTP	PC–ABS
ABS	+	+	+	+	+	+	+	–	–	+	–	–	–	N	N	+	+	+	+	N	+		+	+	+
ASA	+	+	+	+	+	+	+	–	–	+	–	–	–	N	–	+	+	+	+	N	+		+	+	+
CA	+	+	+	N				–	–		–	–	–	–	–	+	+	+	+	–					
EVA	+	+	N	+				+	+			+		+	+			–	+						
PA 6	+	+			+	+	+	N	N		–	N	–	–	–	+	+		+	–	–			+	+
PA 6,6	+	+			+	+	N	N	N		–	–	–	–	–	+	+		+	–	+			+	+
PC	+	+			+	N	+	–	–		–	–	–	–	–	+	+		+	–	–		+	+	+
HDPE	–	–	–	+	N	N	–	+	+	N	N	–	–	–	–	–	–	–	–	N	–		–	–	–
LDPE	–	–	–	+	N	N	–	+	+	N	N	+	–	N	–	–	–	–	–	N	–		–	–	–
PMMA	+	+						N	N	+		N	–	–	–	–		+	+		–			+	+
POM	–	–	–		–	–	–	N	N		+	–	–	–	–	–			–		–			–	–
PP	–	–	–	+	–	–	–	–	+	N	–	+	–	–	–	–	–	–	–	+			–	–	–
PPO mod.	–	–	–		–	–	–	–	–	–	–	–	+	+	+	–	–	–	N	+	–		–	–	–
PSGP	N	N	–	+	–	–	–	–	N	–	–	–	+	+	+	–	–	–	–	–	–		–	–	–
PS–HI	N	–	–	+	–	–	–	–	–	–	–	–	+	+	+	–	–	–	–	N	–		–	–	–
PBTP	+	+	+		+	+	+	–	–	–	–	–	–	–	–	+	+	+	+	–	+		+	+	+
TPU	+	+	+		+	+	+	–	–			–	–	–	–	+	+	+	+	–	+			+	+
PVC (soft)	+	+	+	–				–	–	+		–	–	–	–	+	+	+	+					+	+
SAN	+	+	+	+	+	+	+	–	–	+		–	N	–	–	+		+	+					+	+
TPR	N	N	–		–	–	–	N	N			+	+	–	N					+	–		–	–	–
PETP	+	+					+	–	–	–	–		–	–	–	+				–	+		+	+	+
PVAC																						+			
PSU	+	+					+	–	–			–	–	–	–	+				–	+		+	+	+
PC–PBTP	+	+			+	+	+	–	–	+	–	–	–	–	–	+	+	+	+	–	+		+	+	+
PC–ABS	+	+			+	+	+	–	–	+	–	–	–	–	–	+	+	+	+	–	+		+	+	+

+ = good bonding, – = poor bonding, N = no bonding, blank = not evaluated.

- Free blowing
- Matched die molding

Drape forming, as shown in Fig. 1.69, involves the either lowering the heated sheet onto a male mold or raising the mold into the sheet. Usually, either vacuum or pressure is used to force the sheet against the mold. In vacuum forming (Fig. 1.70), the sheet is clamped to the edges of a female mold, then vacuum is applied to force the sheet against the mold. Pressure forming is similar to vacuum forming, except that air pressure is used to form the part (Fig. 1.71). In free blowing, the heated sheet is stretched by air pressure into shape, and the height of the bubble is controlled using air pressure. As the sheet expands outward, it cools into a free-form shape as shown in Fig. 1.72. This method was originally developed for aircraft gun enclosures. Matched die molding (Fig. 1.73) uses

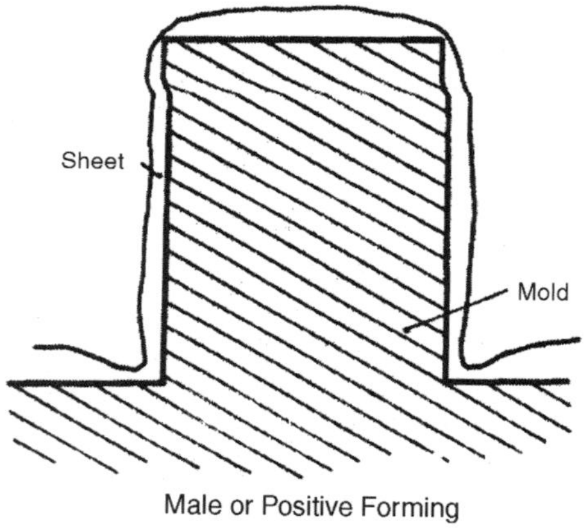

Figure 1.69 Drape forming process.[469]

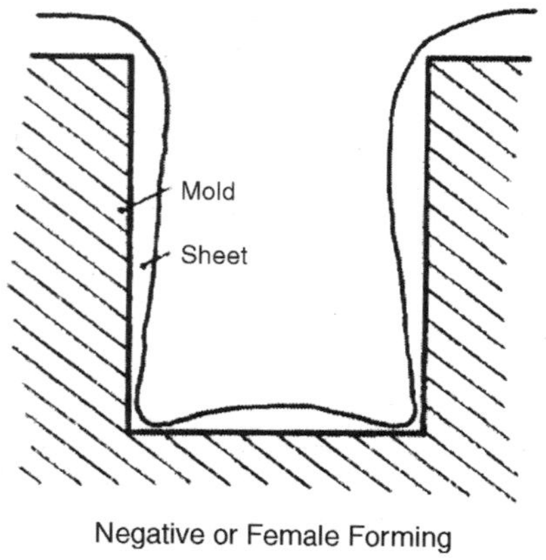

Figure 1.70 Vacuum forming process.[469]

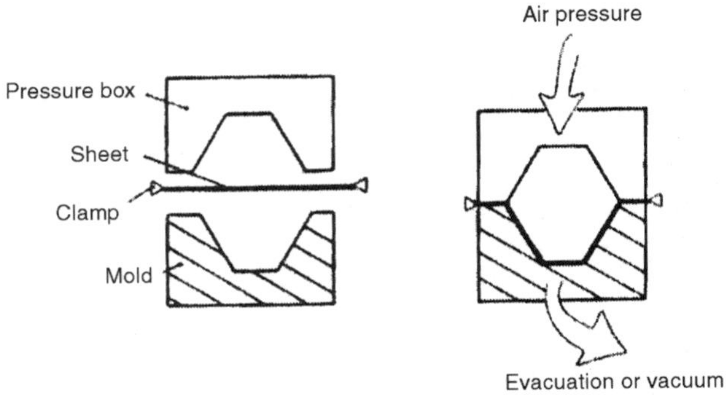

Figure 1.71 Pressure forming.[470]

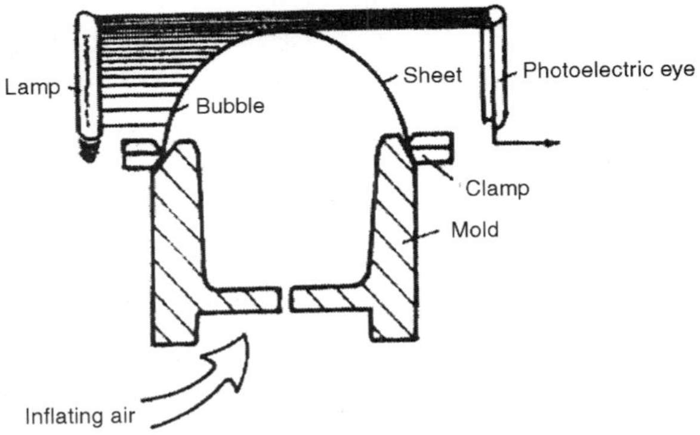

Figure 1.72 Free-blowing process.[470]

two mold halves to form the heated sheet. This method is often used to form relatively stiff sheets.

Multi-step forming is used in applications for thicker sheets or complex geometries with deep draw. In this type of thermoforming, the first step involves prestretching the sheet by techniques such as billowing or plug assist. After prestretching, the sheet is pressed against the mold. Multi-step forming includes[438]

- Billow drape forming
- Billow vacuum forming
- Vacuum snap-back forming
- Plug assist vacuum forming
- Plug assist pressure forming
- Plug assist drape forming

Billow drape forming consists of a male mold pressed into a sheet prestretched by the billowing process (Fig. 1.74). A similar process is billow vacuum forming, wherein a female mold is used (Fig. 1.75). In vacuum snap-back forming, vacuum is used to prestretch the sheet, then a male mold is pressed into the sheet, and, finally, pressure is used to force

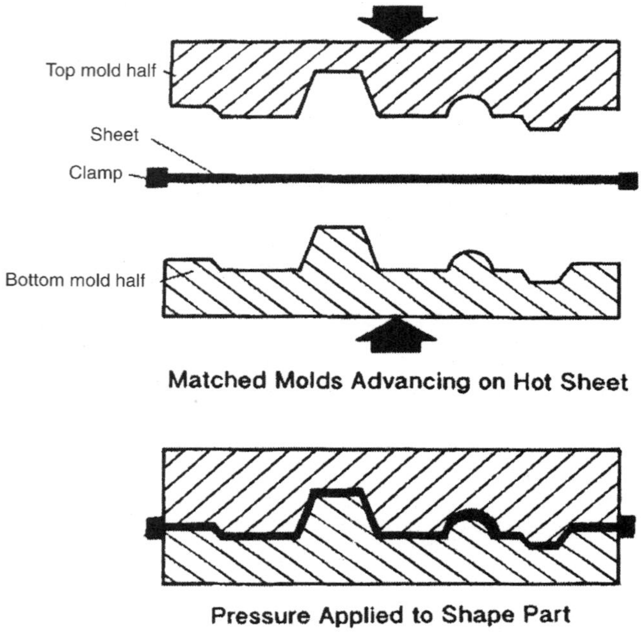

Figure 1.73 Matched die thermoforming.[471]

the sheet against the mold as seen in Fig. 1.76. In plug assist, a plug of material is used to prestretch the sheet. Either vacuum or pressure is then used to force the sheet against the walls of the mold as shown in Figs. 1.77 and 1.78. Plug assist drape forming is used to force a sheet into undercuts or corners (Fig. 1.79). The advantage of prestretching the sheet is more uniform wall thickness.

Materials suitable for thermoforming must be compliant enough to allow for forming against the mold yet not produce excessive flow or sag while being heated.[439] Amorphous materials generally exhibit a wider process window than semicrystalline materials. Processing temperatures are typically 30 to 60°C above T_g for amorphous materials, and usually just above T_m in the case of semicrystalline polymers.[440] Amorphous materials that are thermoformed include PS, ABS, PVC, PMMA, PETP, and PC. Semicrystalline materials that can be successfully thermoformed include PE and nucleated PETP. Nylons typically do not have sufficient melt strength to be thermoformed. Table 1.16 shows processing temperatures for thermoforming a number of thermoplastics.

1.6.4 Blow Molding

Blow molding is a technique for forming nearly hollow articles and is very commonly practiced in the formation of PET soft-drink bottles. It is also used to make air ducts, surfboards, suitcase halves, and automobile gasoline tanks.[441] Blow molding involves taking a parison (a tubular profile) and expanding it against the walls of a mold by inserting pressurized air into it. The mold is machined to have the negative contour of the final desired finished part. The mold, typically a mold split into two halves, then opens after the part has cooled to the extent that the dimensions are stable, and the bottle is ejected. Molds are

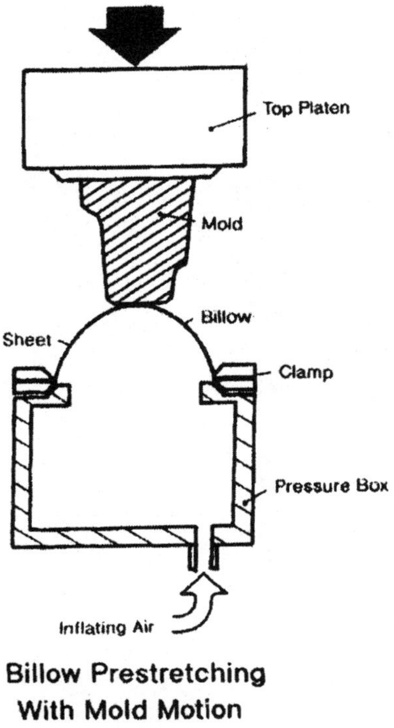

Figure 1.74 Billow drape forming.[472]

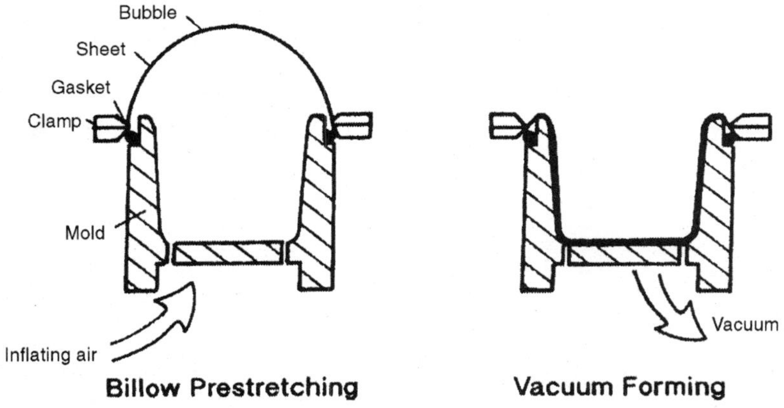

Figure 1.75 Billow vacuum process.[473]

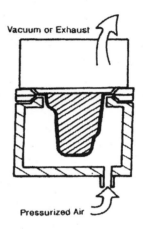

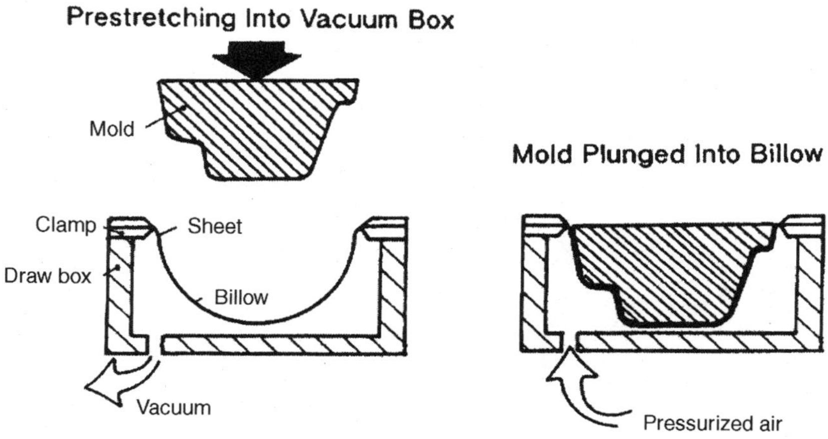

Figure 1.76 Vacuum snap-back process.[473]

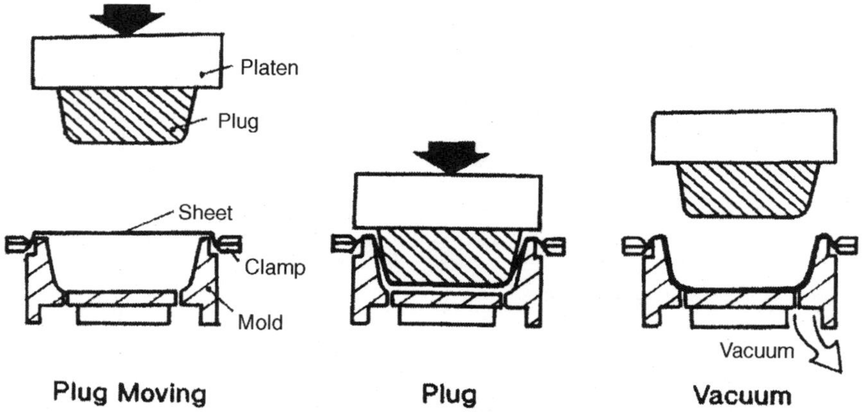

Figure 1.77 Plug assist vacuum forming.[474]

commonly made out of aluminum, as molding pressures are relatively low, and aluminum has high thermal conductivity to promote rapid cooling of the part. The parison can either be made continuously with an extruder or it can be injection molded; the method of parison production governs whether the process is called *extrusion blow molding* or *injection blow molding*. Figure 1.80 shows both the extrusion and injection blow molding processes.[442] Extrusion blow molding is often done with a rotary table so that the parison is extruded into a two-plate open mold; the mold closes as the table rotates another mold under the extruder's die. The closing of the mold cuts off the parison and leaves the characteristic weld-line on the bottom of many bottles as evidence of the pinch-off. Air is then blown into the parison to expand it to fit the mold configuration, and the part is then cooled and ejected before the position rotates back under the die to begin the process again. The

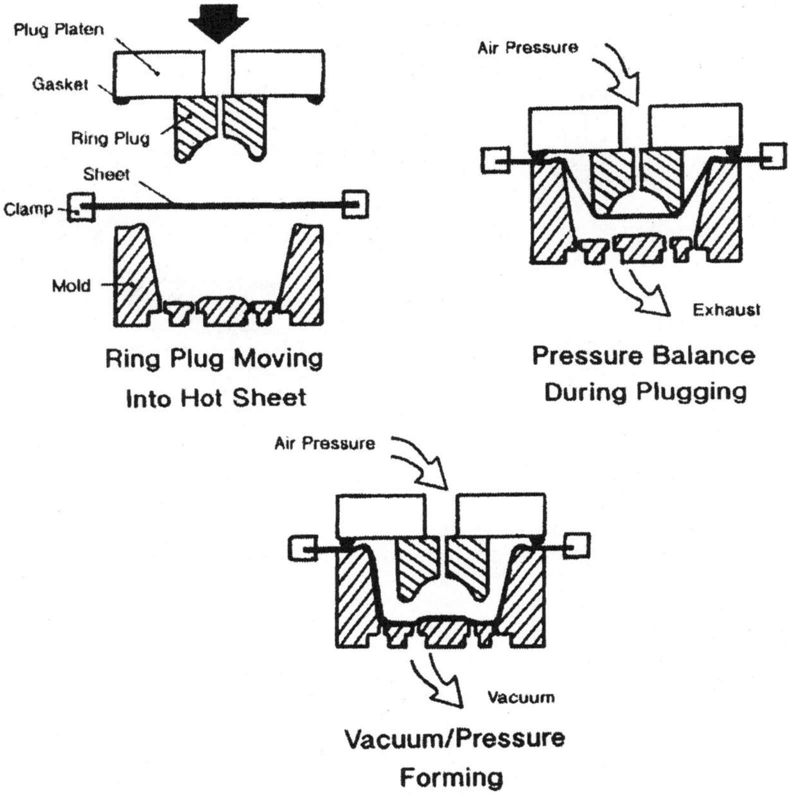

Figure 1.78 Plug assist pressure forming.[475]

blowing operation imparts radial and longitudinal orientation to the plastic melt, strengthening it through biaxial orientation. A container featuring this biaxial orientation is more optically clear, has increased mechanical properties, and has reduced permeability, which is important in maintaining carbonation in soft drinks.

Injection blow molding has very similar treatment of the parison, but the parison itself is injection molded rather than extruded continuously. There is evidence of the gate on the bottom of the bottles rather than having a weld line where the parison was cut off. The parison can be blown directly after molding while it is still hot, or it can be stored and reheated for the secondary blowing operation. An advantage of injection blow molding is that the parison can be molded to have finished threads. Cooling time is the largest part of this cycle and is the rate-limiting step. HDPE, LDPE, PP, PVC, and PET are commonly used in blow molding operations.

1.6.5 Rotational Molding

Rotational molding, also known as *rotomolding* or *centrifugal casting,* involves filling a mold cavity, generally with powder, and rotating the entire heated mold along two axes to uniformly distribute the plastic along the mold walls. This method is commonly used for

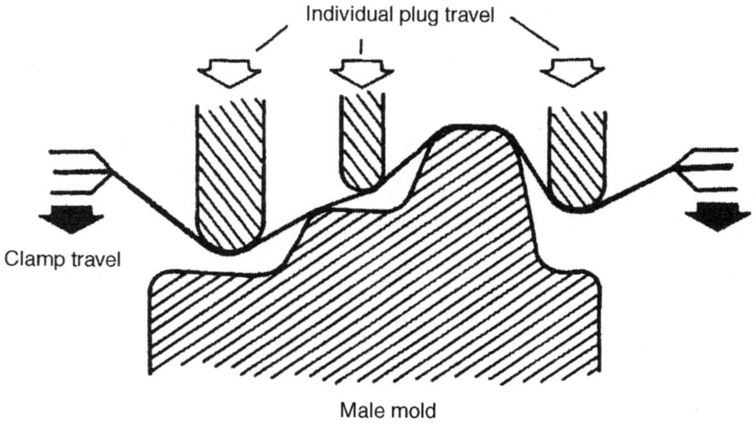

Figure 1.79 Plug assist drape forming.[475]

making hollow parts, like blow molding, but is used either when the parts are very large (as in the case of kayaks, outdoor portable toilets, phone booths, and large chemical storage drums) or when the part requires very low residual stresses. Also, rotomolding is well suited, as compared with blow molding, if the desired part design is complex or requires uniform wall thicknesses. Part walls produced by this method are very uniform as long as neither of the rotational axes corresponds to the centroid of the part design. The rotomolding operation imparts no shear stresses to the plastic, and the resultant molded article is therefore less prone to stress cracking, environmental attack, or premature failures along stress lines. Molded parts also are free of seams. Figure 1.81 shows a diagram of a typical rotational molding process.[443]

This is a relatively low-cost method, as molds are inexpensive, and energy costs are low, thus making it suitable for short-run products. The drawback is that the required heating and cooling times are long, and therefore the cycle time is correspondingly long. High melt flow index PEs are often used in this process.

1.6.6 Foaming

The act of foaming a plastic material results in products with a wide range of densities. These materials are often termed *cellular plastics*. Cellular plastics can exist in two basic structures: closed-cell or open-cell. Closed-cell materials have individual voids or cells that are completely enclosed by plastics, and gas transport takes place by diffusion through the cell walls. In contrast, open-cell foams have cells that are interconnected, and fluids may pass easily between the cells. The two structures may exist together in a material so that it may be a combination of open and closed cells.

Blowing agents are used to produce foams, and they can be classified as either physical or chemical. Physical blowing agents include

- Incorporation of glass or resin beads (syntactic foams)
- Inclusion of an inert gas, such as nitrogen or carbon dioxide, into the polymer at high pressure, which expands when the pressure is reduced
- Addition of low-boiling liquids, which volatilize on heating, forming gas bubbles when pressure is released

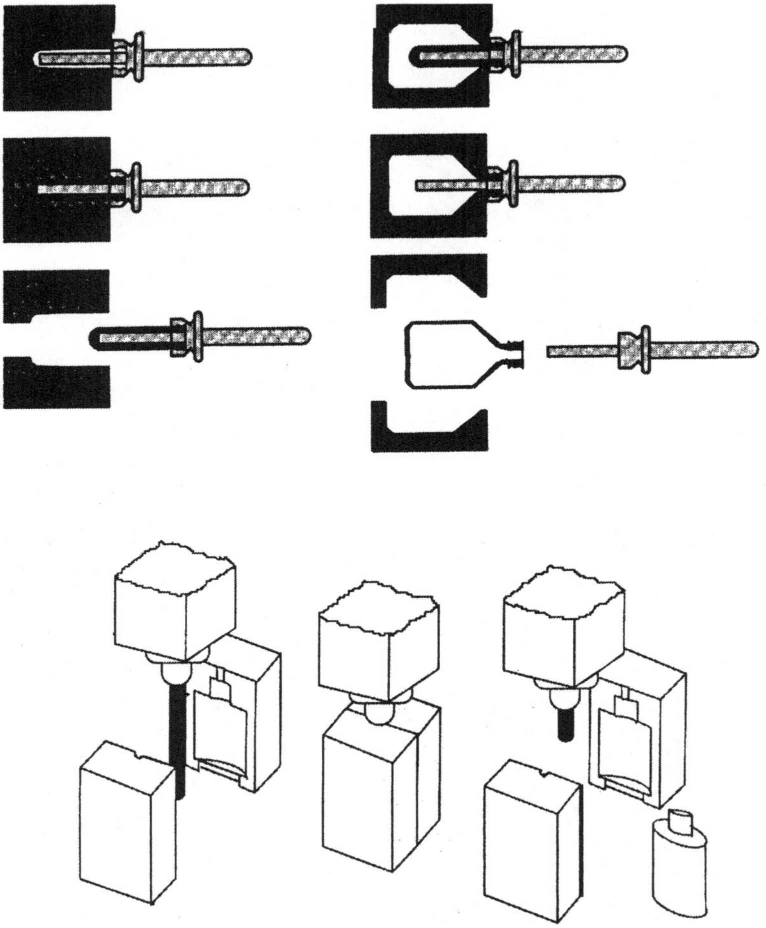

Figure 1.80 Extrusion and injection blow molding processes.[442]

Chemical blowing agents include

- Addition of compounds that decompose over a suitable temperature range with the evolution of gas
- Chemical reaction between components

The major types of chemical blowing agents include the azo compounds, hydrazine derivatives, semicarbazides, tetrazoles, and benzoxazines.[444] Table 1.17 shows some of the common blowing agents, their decomposition temperature, and primary uses.

A wide range of thermoplastics can be converted into foams. Some of the most common materials include polyurethanes, polystyrene, and polyethylene. Polyurethanes are a popular and versatile material for the production of foams and may be foamed by either physical or chemical methods. In the physical reaction, an inert low-boiling chemical is

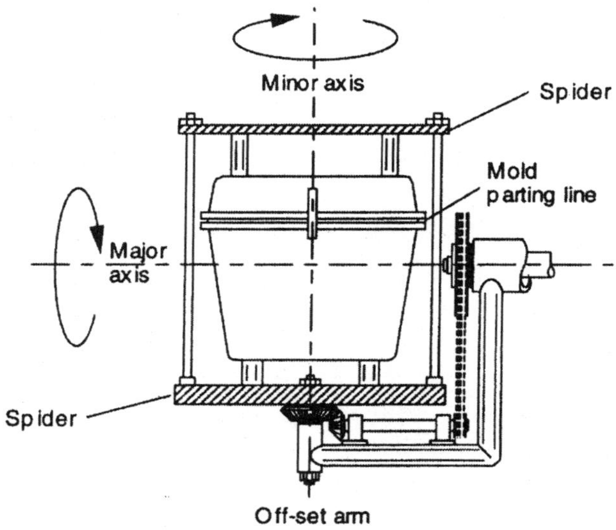

Figure 1.81 The rotational molding process.[443]

added to the mixture, which volatilizes as a result of the heat produced from the exothermic chemical reaction to produce the polyurethane (reaction of isocyanate and diol). Chemical foaming can be done through the reaction of the isocyanate groups with water to produce carbamic acid, which decomposes to an amine and carbon dioxide gas.[445]

Rigid polyurethane foams can be formed by pour, spray, and froth.[446] Liquid polyurethane is poured into a cavity and allowed to expand in the pour process. In the spray method, heated two-component spray guns are used to apply the foam. This method is suitable for application in the field. The froth technique is similar to the pour technique, except that the polyurethane is partially expanded before molding. A two-step expansion is used for this method using a low-boiling agent for preparation of the froth and a second higher-boiling agent for expansion once the mold is filled.

Polyurethane foams can also be produced by reaction injection molding or RIM.[447] This process combines low-molecular-weight isocyanate and polyol, which are accurately metered into the mixing chamber and then injected into the mold. The resulting structure consists of a solid skin and a foamed core.

Polystyrene foams are typically considered either as extruded or expanded bead.[448] Extruded polystyrene foam is produced by extrusion of polystyrene containing a blowing agent and allowing the material to expand into a closed cell foam. This product is used extensively as thermal insulation. Molded expanded polystyrene is produced by exposing polystyrene beads containing a blowing agent to heat.[449] If the shape is to be used as loose-fill packaging, then no further processing steps are needed. If a part is to be made, the beads are then fused in a heated mold to shape the part. Bead polystyrene foam is used in thermal insulation applications, flotation devices, and insulated hot and cold drink cups.

Polyethylene foams are produced using chemical blowing agents and are typically closed cell foams.[450] Cellular polyethylene offers advantages over solid polyethylene in terms of reduced weight and lower dielectric constant. As a result, these materials find application in electrical insulation markets. Polyethylene foams are also used in cushioning applications to protect products during shipping and handling.

TABLE 1.16 Thermoforming Process Temperatures for Selected Materials[459]

Material	Mold and set temperature, °C	Lower processing limit, °C	Normal forming temperature, °C	Upper temperature limit, °C
ABS	85	127	149	182
Acetate	71	127	149	182
Acrylic	85	149	177	193
Butyrate	79	127	146	182
Polycarbonate	138	168	191	204
Polyester (PETG)	77	121	149	166
Polyethersulfone	204	274	316	371
Polyethersulfone-glass filled	210	279	343	382
HDPE	82	127	146	182
PP	88	129	154–166	166
PP-glass filled	91	129	204+	232
Polysulfone	163	190	246	302
Polystyrene	85	127	149	182
FEP	149	232	288	327
PVC–rigid	66	104	138–141	154

References

1. F. W. Billmeyer, *Textbook of Polymer Science,* 2/e, John Wiley & Sons, Inc., New York, 1971.
2. A.W. Birley, B. Haworth, and J. Batchelor, *Physics of Plastics,* Carl Hanser Verlag, Munich, 1992.
3. M.L. Williams, R.F. Landel, and J.D. Ferry, *J. Am. Chem. Soc.,* 77, 3701(1955).
4. P.C. Powell, *Engineering with Polymers,* Chapman and Hall, London, 1983.
5. A.W. Birley, B. Haworth, and J. Batchelor, *Physics of Plastics,* Carl Hanser Verlag, Munich, 1992, pp. 283–284.
6. L.E. Nielsen and R.F. Landel, *Mechanical Properties of Polymers and Composites,* Marcel Dekker, New York, 1994, pp. 342–352.
7. A.W. Bosshard and H.P. Schlumpf, Fillers and Reinforcements, in *Plastics Additives,* 2/e, R. Gachter and H. Muller, Eds., Hanser Publishers, New York, 1987, p. 397.
8. Brydson, J.A., *Plastics Materials,* 6/e, Butterworth-Heinemann, Oxford, 1995, p. 122.

TABLE 1.17 Common Chemical Blowing Agents[460]

Blowing agent	Decomposition temp., °C	Gas yield, ml/g	Polymer applications
Azodicarbonamide	205–215	220	PVC, PE, PP, PS, ABS, PA
Modified azodicarbonamide	155–220	150–220	PVC, PE, PP, EVA, PS, ABS
4,4´–oxybis (benzenesulfohydrazide)	150–160	125	PE, PVC, EVA
Diphenylsulfone–3,3´–disulfo-hydrazide	155	110	PVC, PE, EVA
Trihydrazinotriazine	275	225	ABS, PE PP, PA
p-toliuylenesulfonyl semicarbazide	228–235	140	ABS, PE, PP, PA, PS
5-phenyltetrazole	240–250	190	ABS, PPE, PC, PA, PBT, LCP
Isatoic anhydride	210–225	115	PS, ABS, PA, PPE, PBT, PC

9. A.W. Bosshard and H.P. Schlumpf, Fillers and Reinforcements, in *Plastics Additives* 2/e, R. Gachter and H. Muller, Eds., Hanser Publishers, New York, 1987, p. 407.
10. A.W. Bosshard and H.P. Schlumpf, Fillers and Reinforcements, in *Plastics Additives* 2/e, R. Gachter and H. Muller, Eds., Hanser Publishers, New York, 1987, p. 420.
11. Carraher, C.E., *Polymer Chemistry, An Introduction,* 4/e, Marcel Dekker, Inc., New York, 1996, p. 238.
12. Brydson, J.A., *Plastics Materials,* 6/e, Butterworth-Heinemann, Oxford, 1995, p. 516.
13. Kroschwitz, J.I., *Concise Encyclopedia of Polymer Science and Engineering,* John Wiley and Sons, New York, 1990, p. 4.
14. Brydson, J.A., *Plastics Materials,* 6/e, Butterworth-Heinemann, Oxford, 1995, p. 517.
15. Brydson, J.A., *Plastics Materials,* 6/e, Butterworth-Heinemann, Oxford, 1995, p. 518.
16. Billmeyer, F.W., Jr., *Textbook of Polymer Science,* 2/e, John Wiley & Sons, Inc., New York, 1962, p. 439.
17. Brydson, J.A., *Plastics Materials,* 6/e, Butterworth-Heinemann, Oxford, 1995, p. 519.
18. Berins, M.L., *Plastics Engineering Handbook of the Society of the Plastics Industry,* 5/e, Chapman and Hall, New York, 1991, p. 61.
19. Brydson, J.A., *Plastics Materials,* 6/e, Butterworth-Heinemann, Oxford, 1995, p. 521.
20. Brydson, J.A., *Plastics Materials,* 6/e, Butterworth-Heinemann, Oxford, 1995, p. 523.
21. Brydson, J.A., *Plastics Materials,* 6/e, Butterworth-Heinemann, Oxford, 1995, p. 524.
22. Berins, M.L., *Plastics Engineering Handbook of the Society of the Plastics Industry,* 5/e, Chapman and Hall, New York, 1991, p. 62.
23. Strong, A.B., *Plastics: Materials and Processing,* Prentice-Hall, NJ, 1996, p. 193.
24. Brydson, J.A., *Plastics Materials,* 6/e, Butterworth-Heinemann, Oxford, 1995, p. 525.

25. *Modern Plastics,* January 1998, p. 76.
26. Brydson, J.A., *Plastics Materials,* 6/e, Butterworth-Heinemann, Oxford, 1995, p. 527.
27. Carraher, C.E., *Polymer Chemistry, An Introduction,* 4/e, Marcel Dekker, Inc., New York, 1996, p. 524.
28. Carraher, C.E., *Polymer Chemistry, An Introduction,* 4/e, Marcel Dekker, Inc., New York, 1996, p. 524.
29. McCarthy, S.P. Biodegradable Polymers for Packaging, in *Biotechnological Polymers*, C.G. Gebelein, Ed.
30. Carraher, C.E., *Polymer Chemistry, An Introduction,* 4/e, Marcel Dekker, Inc., New York, 1996, p. 525.
31. Brydson, J.A., *Plastics Materials,* 6/e, Butterworth-Heinemann, Oxford, 1995, p. 858.
32. Brydson, J.A., *Plastics Materials,* 6/e, Butterworth-Heinemann, Oxford, 1995, p. 858.
33. Brydson, J.A., *Plastics Materials,* 6/e, Butterworth-Heinemann, Oxford, 1995, p. 859.
34. McCarthy, S.P. Biodegradable Polymers for Packaging, in *Biotechnological Polymers*, C.G. Gebelein, Ed.
35. McCarthy, S.P. Biodegradable Polymers for Packaging, in *Biotechnological Polymers*, C.G. Gebelein, Ed.
36. Brydson, J.A., *Plastics Materials,* 6/e, Butterworth-Heinemann, Oxford, 1995, p. 608.
37. Byrom, D., Miscellaneous biomaterials, in *Biomaterials,* D. Byrom, Ed., Stockton Press, New York, 1991, p. 341.
38. Byrom, D., Miscellaneous biomaterials, in *Biomaterials*, D. Byrom, Ed., Stockton Press, New York, 1991, p. 341.
39. Byrom, D., Miscellaneous biomaterials, in *Biomaterials,* D. Byrom, Ed., Stockton Press, New York, 1991, p. 343.
40. Brydson, J.A., *Plastics Materials,* 6/e, Butterworth-Heinemann, Oxford, 1995, p. 859.
41. Brydson, J.A., *Plastics Materials,* 6/e, Butterworth-Heinemann, Oxford, 1995, p. 718.
42. McCarthy, S.P. Biodegradable Polymers for Packaging, in *Biotechnological Polymers*, C.G. Gebelein, Ed.
43. Brydson, J.A., *Plastics Materials,* 6/e, Butterworth-Heinemann, Oxford, 1995, p. 860.
44. Byrom, D., Miscellaneous biomaterials, in *Biomaterials,* D. Byrom, Ed., Stockton Press, New York, 1991, p. 338.
45. Brydson, J.A., *Plastics Materials,* 6/e, Butterworth-Heinemann, Oxford, 1995, p. 860.
46. Brydson, J.A., *Plastics Materials,* 6/e, Butterworth-Heinemann, Oxford, 1995, p. 862.
47. McCarthy, S.P. Biodegradable Polymers for Packaging, in *Biotechnological Polymers*, C.G. Gebelein, Ed.
48. Byrom, D., Miscellaneous biomaterials, in *Biomaterials,* D. Byrom, Ed., Stockton Press, New York, 1991, p. 351.
49. Byrom, D., Miscellaneous biomaterials, in *Biomaterials,* D. Byrom, Ed., Stockton Press, New York, 1991, p. 353.
50. *Encyclopedia of Polymer Science and Engineering,* 2/e, vol. 3, Mark, Bilkales, Overberger, Menges, Kroschwitz, Eds., Wiley Interscience, 1986, p. 60.

51. *Encyclopedia of Polymer Science and Engineering,* 2/e, vol. 3, Mark, Bilkales, Overberger, Menges, Kroschwitz, Eds., Wiley Interscience, 1986, p. 68
52. *Encyclopedia of Polymer Science and Engineering,* 2/e, vol. 3, Mark, Bilkales, Overberger, Menges, Kroschwitz, Eds., Wiley Interscience, 1986, p. 92.
53. *Encyclopedia of Polymer Science and Engineering,* 2/e, vol. 3,Mark, Bilkales, Overberger, Menges, Kroschwitz, Eds., Wiley Interscience, 1986, p. 182.
54. *Encyclopedia of Polymer Science and Engineering,* 2/e, vol. 3, Mark, Bilkales, Overberger, Menges, Kroschwitz, Eds., Wiley Interscience, 1986, p. 182.
55. *Plastics Materials*, 5/e, J.A. Brydson, Butterworths, 1989, p. 583.
56. *Plastics Materials*, 5/e, J.A. Brydson, Butterworths, 1989, p. 187.
57. Williams, R.W., Cellulosics, in *Modern Plastics Encyclopedia Handbook,* McGraw-Hill, Inc., 1994, p. 8.
58. Brydson, J.A., *Plastics Materials,* 6/e, Butterworth-Heinemann, Oxford, 1995, p. 349.
59. Brydson, J.A., *Plastics Materials,* 6/e, Butterworth-Heinemann, Oxford, 1995, p. 349.
60. Berins, M.L., *Plastics Engineering Handbook of the Society of the Plastics Industry,* 5/e, Chapman and Hall, New York, 1991, p. 62.
61. Billmeyer, F.W., Jr., *Textbook of Polymer Science,* 2/e, John Wiley & Sons, Inc., New York, 1962, p. 423.
62. Berins, M.L., *Plastics Engineering Handbook of the Society of the Plastics Industry,* 5/e, Chapman and Hall, New York, 1991, p. 63.
63. Carraher, C.E., *Polymer Chemistry, An Introduction,* 4/e, Marcel Dekker, Inc., New York, 1996, p. 319.
64. Brydson, J.A., *Plastics Materials,* 6/e, Butterworth-Heinemann, Oxford, 1995, p. 359.
65. Billmeyer, F.W., Jr., *Textbook of Polymer Science,* 2/e, John Wiley & Sons, Inc., New York, 1962, p. 426.
66. Brydson, J.A., *Plastics Materials,* 6/e, Butterworth-Heinemann, Oxford, 1995, p. 359.
67. Brydson, J.A., *Plastics Materials,* 6/e, Butterworth-Heinemann, Oxford, 1995, p. 359.
68. Berins, M.L., *Plastics Engineering Handbook of the Society of the Plastics Industry,* 5/e, Chapman and Hall, New York, 1991, p. 63.
69. Berins, M.L., *Plastics Engineering Handbook of the Society of the Plastics Industry,* 5/e, Chapman and Hall, New York, 1991, p. 63.
70. Berins, M.L., *Plastics Engineering Handbook of the Society of the Plastics Industry,* 5/e, Chapman and Hall, New York, 1991, p. 63.
71. Brydson, J.A., *Plastics Materials,* 6/e, Butterworth-Heinemann, Oxford, 1995, p. 360.
72. Billmeyer, F.W., Jr., *Textbook of Polymer Science,* 2/e, John Wiley & Sons, Inc., New York, 1962, p. 427.
73. Berins, M.L., *Plastics Engineering Handbook of the Society of the Plastics Industry,* 5/e, Chapman and Hall, New York, 1991, p. 62.
74. Billmeyer, F.W., Jr., *Textbook of Polymer Science,* 2/e, John Wiley & Sons, Inc., New York, 1962, p. 428.
75. Brydson, J.A., *Plastics Materials,* 6/e, Butterworth-Heinemann, Oxford, 1995, p. 361.
76. Berins, M.L., *Plastics Engineering Handbook of the Society of the Plastics Industry,* 5/e, Chapman and Hall, New York, 1991, p. 62.
77. Billmeyer, F.W., Jr., *Textbook of Polymer Science,* 2/e, John Wiley & Sons, Inc., New York, 1962, p. 423.

78. Brydson, J.A., *Plastics Materials,* 6/e, Butterworth-Heinemann, Oxford, 1995, p. 352.
79. Billmeyer, F.W., Jr., *Textbook of Polymer Science,* 2/e, John Wiley & Sons, Inc., New York, 1962, p. 424.
80. Brydson, J.A., *Plastics Materials,* 6/e, Butterworth-Heinemann, Oxford, 1995, p. 351.
81. Billmeyer, F.W., Jr., *Textbook of Polymer Science,* 2/e, John Wiley & Sons, Inc., New York, 1962, p. 425.
82. Brydson, J.A., *Plastics Materials,* 6/e, Butterworth-Heinemann, Oxford, 1995, p. 355.
83. Brydson, J.A., *Plastics Materials,* 6/e, Butterworth-Heinemann, Oxford, 1995, p. 353.
84. Berins, M.L., *Plastics Engineering Handbook of the Society of the Plastics Industry,* 5/e, Chapman and Hall, New York, 1991, p. 62.
85. Brydson, J.A., *Plastics Materials,* 6/e, Butterworth-Heinemann, Oxford, 1995, p. 356.
86. Brydson, J.A., *Plastics Materials,* 6/e, Butterworth-Heinemann, Oxford, 1995, p. 357.
87. Brydson, J.A., *Plastics Materials,* 6/e, Butterworth-Heinemann, Oxford, 1995, p. 357.
88. Brydson, J.A., *Plastics Materials,* 6/e, Butterworth-Heinemann, Oxford, 1995, p. 357.
89. Billmeyer, F.W., Jr., *Textbook of Polymer Science,* 2/e, John Wiley & Sons, Inc., New York, 1962, p. 426.
90. Billmeyer, F.W., Jr., *Textbook of Polymer Science,* 2/e, John Wiley & Sons, Inc., New York, 1962, p. 428.
91. Brydson, J.A., *Plastics Materials,* 6/e, Butterworth-Heinemann, Oxford, 1995, p. 362.
92. Carraher, C.E., *Polymer Chemistry, An Introduction,* 4/e, Marcel Dekker, Inc., New York, 1996, p. 319.
93. Brydson, J.A., *Plastics Materials,* 6/e, Butterworth-Heinemann, Oxford, 1995, p. 363.
94. Berins, M.L., *Plastics Engineering Handbook of the Society of the Plastics Industry,* 5/e, Chapman and Hall, New York, 1991, p. 63.
95. Berins, M.L., *Plastics Engineering Handbook of the Society of the Plastics Industry,* 5/e, Chapman and Hall, New York, 1991, p. 63.
96. Brydson, J.A., *Plastics Materials,* 6/e, Butterworth-Heinemann, Oxford, 1995, p. 362.
97. Billmeyer, F.W., Jr., *Textbook of Polymer Science,* 2/e, John Wiley & Sons, Inc., New York, 1962, p. 434.
98. Modern Plastics, January 1998, p. 76.
99. Berins, M.L., *Plastics Engineering Handbook of the Society of the Plastics Industry,* 5/e, Chapman and Hall, New York, 1991, p. 64.
100. Brydson, J.A., *Plastics Materials,* 6/e, Butterworth-Heinemann, Oxford, 1995, p. 462.
101. Berins, M.L., *Plastics Engineering Handbook of the Society of the Plastics Industry,* 5/e, Chapman and Hall, New York, 1991, p. 64.
102. Billmeyer, F.W., Jr., *Textbook of Polymer Science,* 2/e, John Wiley & Sons, Inc., New York, 1962, p. 433.
103. Deanin, R.D., *Polymer Structure, Properties and Applications,* Cahners Publishing Company, Inc., York, PA, 1972 p. 455.

104. Brydson, J.A., *Plastics Materials,* 6/e, Butterworth-Heinemann, Oxford, 1995, p. 470.
105. Strong, A.B., *Plastics: Materials and Processing,* Prentice-Hall, NJ, 1996, p. 190.
106. Berins, M.L., *Plastics Engineering Handbook of the Society of the Plastics Industry,* 5/e, Chapman and Hall, New York, 1991, p. 64.
107. Brydson, J.A., *Plastics Materials,* 6/e, Butterworth-Heinemann, Oxford, 1995, p. 477.
108. Strong, A.B., *Plastics: Materials and Processing,* Prentice-Hall, NJ, 1996, p. 191.
109. Brydson, J.A., *Plastics Materials,* 6/e, Butterworth-Heinemann, Oxford, 1995, p. 471.
110. Carraher, C.E., *Polymer Chemistry, An Introduction,* 4/e, Marcel Dekker, Inc., New York, 1996, p. 233.
111. Brydson, J.A., *Plastics Materials,* 6/e, Butterworth-Heinemann, Oxford, 1995, p. 472.
112. Brydson, J.A., *Plastics Materials,* 6/e, Butterworth-Heinemann, Oxford, 1995, p. 472.
113. Galanty, P.G. and Bujtas, G.A., Nylon, in *Modern Plastics Encyclopedia Handbook,* McGraw-Hill, Inc., 1994, p. 12.
114. Brydson, J.A., *Plastics Materials,* 6/e, Butterworth-Heinemann, Oxford, 1995, p. 473.
115. Brydson, J.A., *Plastics Materials,* 6/e, Butterworth-Heinemann, Oxford, 1995, p. 472.
116. Brydson, J.A., *Plastics Materials,* 6/e, Butterworth-Heinemann, Oxford, 1995, p. 473.
117. Strong, A.B., *Plastics: Materials and Processing,* Prentice-Hall, NJ, 1996, p. 190.
118. Brydson, J.A., *Plastics Materials,* 6/e, Butterworth-Heinemann, Oxford, 1995, p. 484.
119. Brydson, J.A., *Plastics Materials,* 6/e, Butterworth-Heinemann, Oxford, 1995, p. 484.
120. Berins, M.L., *Plastics Engineering Handbook of the Society of the Plastics Industry,* 5/e, Chapman and Hall, New York, 1991, p. 64.
121. Brydson, J.A., *Plastics Materials,* 6/e, Butterworth-Heinemann, Oxford, 1995, p. 461.
122. Billmeyer, F.W., Jr., *Textbook of Polymer Science,* 2/e, John Wiley & Sons, Inc., New York, 1962, p. 435.
123. Billmeyer, F.W., Jr., *Textbook of Polymer Science,* 2/e, John Wiley & Sons, Inc., New York, 1962, p. 436.
124. Billmeyer, F.W., Jr., *Textbook of Polymer Science,* 2/e, John Wiley & Sons, Inc., New York, 1962, p. 437.
125. Brydson, J.A., *Plastics Materials,* 6/e, Butterworth-Heinemann, Oxford, 1995, p. 486.
126. Brydson, J.A., *Plastics Materials,* 6/e, Butterworth-Heinemann, Oxford, 1995, p. 480.
127. Strong, A.B., *Plastics: Materials and Processing,* Prentice-Hall, NJ, 1996, p. 191.
128. Brydson, J.A., *Plastics Materials,* 6/e, Butterworth-Heinemann, Oxford, 1995, p. 492.
129. Brydson, J.A., *Plastics Materials,* 6/e, Butterworth-Heinemann, Oxford, 1995, p. 492.
130. Brydson, J.A., *Plastics Materials,* 6/e, Butterworth-Heinemann, Oxford, 1995, p. 494–495.
131. Brydson, J.A., *Plastics Materials,* 6/e, Butterworth-Heinemann, Oxford, 1995, p. 493.

132. Brydson, J.A., *Plastics Materials,* 6/e, Butterworth-Heinemann, Oxford, 1995, p. 496.
133. Brydson, J.A., *Plastics Materials,* 6/e, Butterworth-Heinemann, Oxford, 1995, p. 497.
134. Strong, A.B., *Plastics: Materials and Processing,* Prentice-Hall, NJ, 1996, p. 192.
135. Brydson, J.A., *Plastics Materials,* 6/e, Butterworth-Heinemann, Oxford, 1995, p. 400.
136. Billmeyer, F.W., Jr., *Textbook of Polymer Science,* 2/e, John Wiley & Sons, Inc., New York, 1962, p. 414.
137. Kroschwitz, J.I., *Concise Encyclopedia of Polymer Science and Engineering,* John Wiley and Sons, New York, 1990, p. 28.
138. Kroschwitz, J.I., *Concise Encyclopedia of Polymer Science and Engineering,* John Wiley and Sons, New York, 1990, p. 29.
139. Brydson, J.A., *Plastics Materials,* 6/e, Butterworth-Heinemann, Oxford, 1995, p. 401.
140. Brydson, J.A., *Plastics Materials,* 6/e, Butterworth-Heinemann, Oxford, 1995, p. 402.
141. Kroschwitz, J.I., *Concise Encyclopedia of Polymer Science and Engineering,* John Wiley and Sons, New York, 1990, p. 29.
142. Billmeyer, F.W., Jr., *Textbook of Polymer Science,* 2/e, John Wiley & Sons, Inc., New York, 1962, p. 413.
143. Kroschwitz, J.I., *Concise Encyclopedia of Polymer Science and Engineering,* John Wiley and Sons, New York, 1990, p. 23.
144. Brydson, J.A., *Plastics Materials,* 6/e, Butterworth-Heinemann, Oxford, 1995, p. 507.
145. Berins, M.L., *Plastics Engineering Handbook of the Society of the Plastics Industry,* 5/e, Chapman and Hall, New York, 1991, p. 65.
146. Carraher, C.E., *Polymer Chemistry, An Introduction,* 4/e, Marcel Dekker, Inc., New York, 1996, p. 533.
147. Johnson, S.H., Polyamide-imide, in *Modern Plastics Encyclopedia Handbook,* McGraw-Hill, Inc., 1994, p. 14.
148. Johnson, S.H., Polyamide-imide, in *Modern Plastics Encyclopedia Handbook,* McGraw-Hill, Inc., 1994, p. 14.
149. Berins, M.L., *Plastics Engineering Handbook of the Society of the Plastics Industry,* 5/e, Chapman and Hall, New York, 1991, p. 65.
150. Brydson, J.A., *Plastics Materials,* 6/e, Butterworth-Heinemann, Oxford, 1995, p. 507.
151. Berins, M.L., *Plastics Engineering Handbook of the Society of the Plastics Industry,* 5/e, Chapman and Hall, New York, 1991, p. 65.
152. Brydson, J.A., *Plastics Materials,* 6/e, Butterworth-Heinemann, Oxford, 1995, p. 507.
153. Brydson, J.A., *Plastics Materials,* 6/e, Butterworth-Heinemann, Oxford, 1995, p. 507
154. Brydson, J.A., *Plastics Materials,* 6/e, Butterworth-Heinemann, Oxford, 1995, p. 708.
155. Berins, M.L., *Plastics Engineering Handbook of the Society of the Plastics Industry,* 5/e, Chapman and Hall, New York, 1991, p. 66.
156. Dunkle, S.R. and Dean, B.D., Polyarylate, in *Modern Plastics Encyclopedia Handbook,* McGraw-Hill, Inc., 1994, p. 15.
157. Berins, M.L., *Plastics Engineering Handbook of the Society of the Plastics Industry,* 5/e, Chapman and Hall, New York, 1991, p. 66.

158. Dunkle, S.R. and Dean, B.D., Polyarylate, in *Modern Plastics Encyclopedia Handbook,* McGraw-Hill, Inc., 1994, p. 16.
159. DiSano, L., Polybenzimidazole, in *Modern Plastics Encyclopedia Handbook,* McGraw-Hill, Inc., 1994, p. 16.
160. Carraher, C.E., *Polymer Chemistry, An Introduction,* 4/e, Marcel Dekker, Inc., New York, 1996, p. 236.
161. Kroschwitz, J.I., *Concise Encyclopedia of Polymer Science and Engineering,* John Wiley and Sons, New York, 1990, p. 772.
162. DiSano, L., Polybenzimidazole, in *Modern Plastics Encyclopedia Handbook,* McGraw-Hill, Inc., 1994, p. 16.
163. DiSano, L., Polybenzimidazole, in *Modern Plastics Encyclopedia Handbook,* McGraw-Hill, Inc., 1994, p. 16.
164. Kroschwitz, J.I., *Concise Encyclopedia of Polymer Science and Engineering,* John Wiley and Sons, New York, 1990, p. 773.
165. DiSano, L., Polybenzimidazole, in *Modern Plastics Encyclopedia Handbook,* McGraw-Hill, Inc., 1994, p. 17.
166. Brydson, J.A., *Plastics Materials,* 6/e, Butterworth-Heinemann, Oxford, 1995, p. 259.
167. Kroschwitz, J.I., *Concise Encyclopedia of Polymer Science and Engineering,* John Wiley and Sons, New York, 1990, p. 100.
168. Berins, M.L., *Plastics Engineering Handbook of the Society of the Plastics Industry,* 5/e, Chapman and Hall, New York, 1991, p. 55.
169. Brydson, J.A., *Plastics Materials,* 6/e, Butterworth-Heinemann, Oxford, 1995, p. 259.
170. Kroschwitz, J.I., *Concise Encyclopedia of Polymer Science and Engineering,* John Wiley and Sons, New York, 1990, p. 100.
171. Brydson, J.A., *Plastics Materials,* 6/e, Butterworth-Heinemann, Oxford, 1995, p. 259.
172. Berins, M.L., *Plastics Engineering Handbook of the Society of the Plastics Industry,* 5/e, Chapman and Hall, New York, 1991, p. 55.
173. Brydson, J.A., *Plastics Materials,* 6/e, Butterworth-Heinemann, Oxford, 1995, p. 260.
174. Berins, M.L., *Plastics Engineering Handbook of the Society of the Plastics Industry,* 5/e, Chapman and Hall, New York, 1991, p. 55.
175. Kroschwitz, J.I., *Concise Encyclopedia of Polymer Science and Engineering,* John Wiley and Sons, New York, 1990, p. 101.
176. Modern Plastics, January 1999, p. 64 vol. 79, no. 1.
177. *Plastics Materials*, 5/e, J.A. Brydson, Butterworths, 1989, p. 525.
178. *Plastics for Engineers, Materials, Properties, Applications*, by H. Domininghaus, Hanser Publishers, New York, 1988, p. 423.
179. *Plastics for Engineers, Materials, Properties, Applications*, by H. Domininghaus, Hanser Publishers, New York, 1988, p. 424.
180. *Plastics for Engineers, Materials, Properties, Applications*, by H. Domininghaus, Hanser Publishers, New York, 1988, p. 426.
181. Brydson, J.A., *Plastics Materials,* 6/e, Butterworth-Heinemann, Oxford, 1995, p. 711.
182. Berins, M.L., *Plastics Engineering Handbook of the Society of the Plastics Industry, Inc.*, 5/e, Chapman and Hall, New York, 1991, p. 67.
183. Brydson, J.A., *Plastics Materials,* 6/e, Butterworth-Heinemann, Oxford, 1995, p. 707.
184. *Concise Polymer Handbook,* p. 477.

185. Brydson, J.A., *Plastics Materials,* 6/e, Butterworth-Heinemann, Oxford, 1995, p. 712.
186. Brydson, J.A., *Plastics Materials,* 6/e, Butterworth-Heinemann, Oxford, 1995, p. 713.
187. Brydson, J.A., *Plastics Materials,* 6/e, Butterworth-Heinemann, Oxford, 1995, p. 714.
188. Brydson, J.A., *Plastics Materials,* 6/e, Butterworth-Heinemann, Oxford, 1995, p. 712.
189. McChesney, C.E., in *Engineering Plastics*, vol. 2, Engineering Materials Handbook, ASM International, Metals Park, OH, 1988, p. 181.
190. Brydson, J.A., *Plastics Materials,* 6/e, Butterworth-Heinemann, Oxford, 1995, p. 712.
191. McChesney, C.E., in *Engineering Plastics*, vol. 2, Engineering Materials Handbook, ASM International, Metals Park, OH, 1988, p. 181.
192. Brydson, J.A., *Plastics Materials,* 6/e, Butterworth-Heinemann, Oxford, 1995, p. 713.
193. *Modern Plastics Encyclopedia Handbook,* McGraw-Hill, Inc., 1994. p. 20.
194. McChesney, C.E, in *Engineering Plastics*, vol. 2, *Engineering Materials Handbook,* ASM International, Metals Park, OH, 1988, p. 181.
195. Berins, M.L., *Plastics Engineering Handbook of the Society of the Plastics Industry, Inc.*, 5/e, Chapman and Hall, New York, 1991, p. 67.
196. *Modern Plastics,* January 1999, vol. 76, no. 1, McGraw-Hill, New York, p. 65.
197. *Plastics Materials*, 5/e, J.A. Brydson, Butterworths, Boston, 1989, p. 681.
198. *Modern Plastics,* January 1999, vol. 76, no. 1, McGraw-Hill Publication, New York, p. 75.
199. *Plastics Materials*, 5/e, J.A. Brydson, Butterworths, Boston, 1989, p. 677.
200. *Modern Plastics Encyclopedia Handbook,* McGraw-Hill, Inc., 1994. p. 23.
201. *Modern Plastics Encyclopedia Handbook,* McGraw-Hill, Inc., 1994. p. 23.
202. *Modern Plastics,* January 1999, vol. 76, no. 1, McGraw-Hill, New York, pp. 74, 75.
203. *Principles of Polymerization*, 2/e, G. Odian, John Wiley & Sons, Inc., New York, 1981, p. 103.
204. *Plastics Materials*, 5/e, J.A. Brydson, Butterworths, Boston, 1989, p. 675.
205. *Encyclopedia of Polymer Science & Engineering,* Vol. 12, John Wiley & Sons, New York, 1985, p. 223.
206. *Principles of Polymerization*, 2/e, G. Odian, John Wiley & Sons, Inc., New York, 1981, p. 103.
207. *Encyclopedia of Polymer Science & Engineering,* Vol. 12, John Wiley & Sons, New York, 1985, p. 223.
208. *Principles of Polymerization*, 2/e, G. Odian, John Wiley & Sons, Inc., New York, 1981, p. 105.
209. *Encyclopedia of Polymer Science & Engineering,* Vol. 12, John Wiley & Sons, New York, 1985, p. 223.
210. *Encyclopedia of Polymer Science & Engineering,* Vol. 12, John Wiley & Sons, New York, 1985, p. 222.
211. *Encyclopedia of Polymer Science & Engineering,* Vol. 12, John Wiley & Sons, New York, 1985, p. 223.
212. *Encyclopedia of Polymer Science & Engineering,* Vol. 12, John Wiley & Sons, New York, 1985, p. 195.
213. *Encyclopedia of Polymer Science & Engineering,* Vol. 12, John Wiley & Sons, New York, 1985, p. 228.
214. *Encyclopedia of Polymer Science & Engineering,* Vol. 12, John Wiley & Sons, New York, 1985, p. 194.

215. *Encyclopedia of Polymer Science & Engineering,* Vol. 12, John Wiley & Sons, New York, 1985, pp. 195, 204–209.
216. *Encyclopedia of Polymer Science & Engineering,* Vol. 12, John Wiley & Sons, New York, 1985, p. 197.
217. *Encyclopedia of Polymer Science & Engineering,* Vol. 12, John Wiley & Sons, New York, 1985, p. 213.
218. Berins, M.L., *Plastics Engineering Handbook of the Society of the Plastics Industry,* 5/e, Chapman and Hall, New York, 1991, p. 67.
219. Kroschwitz, J.I., *Concise Encyclopedia of Polymer Science and Engineering,* John Wiley and Sons, New York, 1990, p. 327.
220. Brydson, J.A., *Plastics Materials,* 6/e, Butterworth-Heinemann, Oxford, 1995, p. 508.
221. Brydson, J.A., *Plastics Materials,* 6/e, Butterworth-Heinemann, Oxford, 1995, p. 508.
222. Berins, M.L., *Plastics Engineering Handbook of the Society of the Plastics Industry,* 5/e, Chapman and Hall, New York, 1991, p. 68.
223. Berins, M.L., *Plastics Engineering Handbook of the Society of the Plastics Industry,* 5/e, Chapman and Hall, New York, 1991, p. 68.
224. Brydson, J.A., *Plastics Materials,* 6/e, Butterworth-Heinemann, Oxford, 1995, p. 508.
225. Berins, M.L., *Plastics Engineering Handbook of the Society of the Plastics Industry,* 5/e, Chapman and Hall, New York, 1991, p. 68.
226. *Plastics for Engineers, Materials, Properties, Applications*, by H. Domininghaus, Hanser Publishers, New York, 1988, p. 24.
227. *Modern Plastics Encyclopedia,* 1998, vol. 74, no. 13, McGraw-Hill Incorporated, p. B-4.
228. *Plastics Materials*, 5/e, J.A. Brydson, Butterworths, 1989, p. 217.
229. *Plastics for Engineers, Materials, Properties, Applications*, by H. Domininghaus, Hanser Publishers, New York, 1988, p. 55.
230. *Encyclopedia of Polymer Science and Engineering,* 2/e, vol. 6, edited by: Mark, Bilkales, Overberger, Menges, Kroschwitz, Wiley Interscience, 1986, p. 383.
231. *McGraw-Hill Encyclopedia of Science & Technology,* 5/e, vol. 10, 19826, p. 647.
232. *Encyclopedia of Polymer Science and Engineering,* 2/e, vol. 6, edited by: Mark, Bilkales, Overberger, Menges, Kroschwitz, Wiley Interscience, 1986, p. 385.
233. *Modern Plastics Encyclopedia,* 1998, p. A-15.
234. *Encyclopedia of Polymer Science and Engineering,* 2/e, vol. 6, Mark, Bilkales, Overberger, Menges, Kroschwitz, Eds., Wiley Interscience, 1986, p. 486.
235. *Encyclopedia of Polymer Science and Engineering,* 2/e, vol. 6, Mark, Bilkales, Overberger, Menges, Kroschwitz, Eds., Wiley Interscience, 1986, p. 493.
236. *Plastics Materials*, 5/e, J.A. Brydson, Butterworths, 1989, p. 262.
237. *Encyclopedia of Polymer Science and Engineering,* 2/e, vol. 6, Mark, Bilkales, Overberger, Menges, Kroschwitz, Eds., Wiley Interscience, 1986, p. 422.
238. Kung, D.M., Ethylene-ethyl acrylate, in *Modern Plastics Encyclopedia Handbook,* McGraw-Hill, Inc., 1994, p. 38.
239. Kung, D.M., Ethylene-ethyl acrylate, in *Modern Plastics Encyclopedia Handbook,* McGraw-Hill, Inc., 1994, p. 38.
240. *Plastics Materials*, 5/e, J.A. Brydson, Butterworths 1989, p. 262.
241. Kung, D.M., Ethylene-ethyl acrylate, in *Modern Plastics Encyclopedia Handbook,* McGraw-Hill, Inc., 1994, p. 38.
242. Baker, G., Ethylene-methyl acrylate, in *Modern Plastics Encyclopedia Handbook,* McGraw-Hill, Inc., 1994, p. 38.

243. *Encyclopedia of Polymer Science and Engineering,* 2/e, vol. 6, Mark, Bilkales, Overberger, Menges, Kroschwitz, Eds., Wiley Interscience, 1986, p. 422.
244. *Plastic materials*, J.A. Brydson, 5/e, Butterworths, London, 1989, p. 261.
245. *Plastic materials*, J.A. Brydson, 5/e, Butterworths, London, 1989, p. 229.
246. *Concise Encyclopedia of Polymer Science and Engineering,* Jacqueline Kroschwitz, Ex. Ed., Wiley-Interscience Publication, New York, 1990, p. 357.
247. *Plastics for Engineers, Materials, Properties, Applications*, by H. Domininghaus, Hanser Publishers, New York, 1988, p. 65.
248. *Plastics for Engineers, Materials, Properties, Applications*, by H. Domininghaus, Hanser Publishers, New York, 1988, p. 67.
249. *Plastic materials*, J.A. Brydson, 5/e, Butterworths, London, 1989, p. 284.
250. *Plastics for Engineers, Materials, Properties, Applications*, by H. Domininghaus, Hanser Publishers, New York, 1988, p. 68.
251. Strong, A.B., *Plastics: Materials and Processing*, Prentice-Hall, NJ, 1996, p. 165.
252. Brydson, J.A., *Plastics Materials,* 6/e, Butterworth-Heinemann, Oxford, 1995, p. 268.
253. Brydson, J.A., *Plastics Materials,* 6/e, Butterworth-Heinemann, Oxford, 1995, p. 268.
254. MacKnight, W.J. and Lundberg, R.D., in *Thermoplastic Elastomers* 2/e, Holden, G., Legge, N.R., Quirk, R.P., and Schroeder, H.E., Eds., Hanser Publishers, New York, 1996, p. 279.
255. Kroschwitz, J.I., *Concise Encyclopedia of Polymer Science and Engineering,* John Wiley and Sons, New York, 1990, p. 126.
256. Strong, A.B., *Plastics: Materials and Processing*, Prentice-Hall, NJ, 1996, p. 165.
257. Rees, R. W., Ionomers, in *Engineering Plastics*, vol. 2, *Engineering Materials Handbook,* ASM International, Metals Park, OH, 1988, p. 120.
258. Rees, R. W., Ionomers, in *Engineering Plastics*, vol. 2, *Engineering Materials Handbook,* ASM International, Metals Park, OH, 1988, p. 122.
259. Rees, R. W., Ionomers, in *Engineering Plastics*, vol. 2, *Engineering Materials Handbook,* ASM International, Metals Park, OH, 1988, p. 123.
260. Strong, A.B., *Plastics: Materials and Processing*, Prentice-Hall, NJ, 1996, p. 165.
261. Rees, R.W., in *Thermoplastic Elastomers* 2/e, Holden, G., Legge, N.R., Quirk, R.P., and Schroeder, H.E., Eds., Hanser Publishers, New York, 1996, p. 263.
262. Brydson, J.A., *Plastics Materials,* 6/e, Butterworth-Heinemann, Oxford, 1995, p. 268.
263. Brydson, J.A., *Plastics Materials,* 6/e, Butterworth-Heinemann, Oxford, 1995, p. 269.
264. Kroschwitz, J.I., *Concise Encyclopedia of Polymer Science and Engineering,* John Wiley and Sons, New York, 1990, p. 827.
265. Berins, M.L., *Plastics Engineering Handbook of the Society of the Plastics Industry,* 5/e, Chapman and Hall, New York, 1991, p. 69.
266. Albermarle, Polyimide, Thermoplastic, in *Modern Plastics Encyclopedia Handbook,* McGraw-Hill, Inc., 1994, p. 43.
267. Berins, M.L., *Plastics Engineering Handbook of the Society of the Plastics Industry,* 5/e, Chapman and Hall, New York, 1991, p. 69.
268. Kroschwitz, J.I., *Concise Encyclopedia of Polymer Science and Engineering,* John Wiley and Sons, New York, 1990, p. 827.
269. Berins, M.L., *Plastics Engineering Handbook of the Society of the Plastics Industry,* 5/e, Chapman and Hall, New York, 1991, p. 69.
270. Brydson, J.A., *Plastics Materials,* 6/e, Butterworth-Heinemann, Oxford, 1995, p. 504.

271. Brydson, J.A., *Plastics Materials,* 6/e, Butterworth-Heinemann, Oxford, 1995, p. 501.
272. Berins, M.L., *Plastics Engineering Handbook of the Society of the Plastics Industry,* 5/e, Chapman and Hall, New York, 1991, p. 69.
273. Brydson, J.A., *Plastics Materials,* 5/e, Butterworth-Heinemann, Oxford, 1989, p. 565.
274. *Modern Plastics Encyclopedia 1998,* mid-November 1997 issue, vol. 74, no. 13, McGraw-Hill Companies, pp. B-162, B-163.
275. Brydson, J.A., *Plastics Materials,* 6/e, Butterworth-Heinemann, Oxford, 1995, p. 586.
276. Brydson, J.A., *Plastics Materials,* 6/e, Butterworth-Heinemann, Oxford, 1995, p. 564.
277. Brydson, J.A., *Plastics Materials,* 6/e, Butterworth-Heinemann, Oxford, 1995, p. 389.
278. Brydson, J.A., *Plastics Materials,* 6/e, Butterworth-Heinemann, Oxford, 1995, p. 391.
279. *Plastics for Engineers, Materials, Properties, Applications*, by H. Domininghaus, Hanser Publishers, New York, 1988, p. 280.
280. *Plastics for Engineers, Materials, Properties, Applications*, H. Domininghaus, Hanser Publishers, New York 1988, p. 122.
281. Brydson, J.A., *Plastics Materials*, 6/e, Butterworth-Heinemann, Oxford, 1995, p. 261.
282. Brydson, J.A., *Plastics Materials*, 6/e, Butterworth-Heinemann, Oxford, 1995, p. 263.
283. Brydson, J.A., *Plastics Materials*, 6/e, Butterworth-Heinemann, Oxford, 1995, p. 567.
284. Brydson, J.A., *Plastics Materials*, 6/e, Butterworth-Heinemann, Oxford, 1995, p. 568.
285. Brydson, J.A., *Plastics Materials*, 6/e, Butterworth-Heinemann, Oxford, 1995, p. 570.
286. Brydson, J.A., *Plastics Materials*, 6/e, Butterworth-Heinemann, Oxford, 1995, p. 570.
287. *Plastics for Engineers, Materials, Properties, Applications*, H. Domininghaus, Hanser Publishers, New York 1988, p. 490.
288. Brydson, J.A., *Plastics Materials*, 6/e, Butterworth-Heinemann, Oxford, 1995, p. 575.
289. Brydson, J.A., *Plastics Materials*, 6/e, Butterworth-Heinemann, Oxford, 1995, p. 576.
290. Brydson, J.A., *Plastics Materials*, 6/e, Butterworth-Heinemann, Oxford, 1995, p. 575.
291. *Plastics for Engineers, Materials, Properties, Applications*, H. Domininghaus, Hanser Publishers, New York 1988, p. 529.
292. Harris, J.H. and Reksc, J.A., Polyphthalamide, in *Modern Plastics Encyclopedia Handbook,* McGraw-Hill, Inc., 1994, p. 47.
293. Brydson, J.A., *Plastics Materials,* 6/e, Butterworth-Heinemann, Oxford, 1995, p. 499.
294. Harris, J.H. and Reksc, J.A., Polyphthalamide, in *Modern Plastics Encyclopedia Handbook,* McGraw-Hill, Inc., 1994, p. 47.
295. Harris, J.H. and Reksc, J.A., Polyphthalamide, in *Modern Plastics Encyclopedia Handbook,* McGraw-Hill, Inc., 1994, p. 47.
296. Brydson, J.A., *Plastics Materials,* 6/e, Butterworth-Heinemann, Oxford, 1995, p. 499.

297. Harris, J.H. and Reksc, J.A., Polyphthalamide, in *Modern Plastics Encyclopedia Handbook,* McGraw-Hill, Inc., 1994, p. 47.
298. Brydson, J.A., *Plastics Materials,* 6/e, Butterworth-Heinemann, Oxford, 1995, p. 499.
299. Harris, J.H. and Reksc, J.A., Polyphthalamide, in *Modern Plastics Encyclopedia Handbook,* McGraw-Hill, Inc., 1994, p. 47.
300. Harris, J.H. and Reksc, J.A., Polyphthalamide, in *Modern Plastics Encyclopedia Handbook,* McGraw-Hill, Inc., 1994, p. 48.
301. Harris, J.H. and Reksc, J.A., Polyphthalamide, in *Modern Plastics Encyclopedia Handbook,* McGraw-Hill, Inc., 1994, p. 48.
302. Modern Plastics, Jan. 1998 p. 58.
303. Brydson, J.A., *Plastics Materials,* 6/e, Butterworth-Heinemann, Oxford, 1995, p. 244.
304. Cradic, G.W., PP Homopolymer, in *Modern Plastics Encyclopedia Handbook,* McGraw-Hill, Inc., 1994, p. 49.
305. Colvin, R., Modern Plastics, May 1997, p. 62.
306. Brydson, J.A., *Plastics Materials,* 6/e, Butterworth-Heinemann, Oxford, 1995, p. 245.
307. Odian, G., *Principles of Polymerization,* 2/e, John Wiley & Sons, Inc., New York, 1981, p. 581.
308. Brydson, J.A., *Plastics Materials,* 6/e, Butterworth-Heinemann, Oxford, 1995, p. 258.
309. Brydson, J.A., *Plastics Materials,* 6/e, Butterworth-Heinemann, Oxford, 1995, p. 244.
310. Strong, A.B., *Plastics: Materials and Processing,* Prentice-Hall, NJ, 1996, p. 168.
311. Billmeyer, F.W., Jr., *Textbook of Polymer Science,* 2/e, John Wiley & Sons, Inc., New York, 1962, p. 388.
312. Billmeyer, F.W., Jr., *Textbook of Polymer Science,* 2/e, John Wiley & Sons, Inc., New York, 1962, p. 387.
313. Brydson, J.A., *Plastics Materials,* 6/e, Butterworth-Heinemann, Oxford, 1995, p. 256.
314. Strong, A.B., *Plastics: Materials and Processing,* Prentice-Hall, NJ, 1996, p. 169.
315. Brydson, J.A., *Plastics Materials,* 6/e, Butterworth-Heinemann, Oxford, 1995, p. 245.
316. Brydson, J.A., *Plastics Materials,* 6/e, Butterworth-Heinemann, Oxford, 1995, p. 246.
317. Brydson, J.A., *Plastics Materials,* 6/e, Butterworth-Heinemann, Oxford, 1995, p. 248.
318. Brydson, J.A., *Plastics Materials,* 6/e, Butterworth-Heinemann, Oxford, 1995, p. 245.
319. Cradic, G.W., PP Homopolymer, in *Modern Plastics Encyclopedia Handbook,* McGraw-Hill, Inc., 1994, p. 49.
320. Brydson, J.A., *Plastics Materials,* 6/e, Butterworth-Heinemann, Oxford, 1995, p. 253.
321. Brydson, J.A., *Plastics Materials,* 6/e, Butterworth-Heinemann, Oxford, 1995, p. 254.
322. Brydson, J.A., *Plastics Materials,* 6/e, Butterworth-Heinemann, Oxford, 1995, p. 255.
323. Brydson, J.A., *Plastics Materials,* 6/e, Butterworth-Heinemann, Oxford, 1995, p. 251.
324. Brydson, J.A., *Plastics Materials,* 6/e, Butterworth-Heinemann, Oxford, 1995, p. 257.

325. Leaversuch, R.D., *Modern Plastics,* Dec. 1996, p. 52
326. Brydson, J.A., *Plastics Materials,* 6/e, Butterworth-Heinemann, Oxford, 1995, p. 756.
327. Brydson, J.A., *Plastics Materials,* 6/e, Butterworth-Heinemann, Oxford, 1995, p. 767.
328. Brydson, J.A., *Plastics Materials,* 6/e, Butterworth-Heinemann, Oxford, 1995, p. 768.
329. Sardanopoli, A.A., Thermoplastic Polyurethanes, in *Engineering Plastics*, vol. 2, *Engineering Materials Handbook,* ASM International, Metals Park, OH, 1988, p. 203.
330. Sardanopoli, A.A., Thermoplastic Polyurethanes, in *Engineering Plastics*, vol. 2, *Engineering Materials Handbook,* ASM International, Metals Park, OH, 1988, p. 206.
331. Sardanopoli, A.A., Thermoplastic Polyurethanes, in *Engineering Plastics*, vol. 2, *Engineering Materials Handbook,* ASM International, Metals Park, OH, 1988, p. 205.
332. Sardanopoli, A.A., Thermoplastic Polyurethanes, in *Engineering Plastics*, vol. 2, *Engineering Materials Handbook,* ASM International, Metals Park, OH, 1988, p. 205.
333. Sardanopoli, A.A., Thermoplastic Polyurethanes, in *Engineering Plastics*, vol. 2, *Engineering Materials Handbook,* ASM International, Metals Park, OH, 1988, p. 207.
334. Brydson, J.A., *Plastics Materials*, 6/e, Butterworth-Heinemann, Oxford, 1995, p. 427.
335. *Plastics for Engineers, Materials, Properties, Applications*, H. Domininghaus, Hanser Publishers, New York 1988, p. 226.
336. Akane, J., ACS, in *Modern Plastics Encyclopedia Handbook,* McGraw-Hill, Inc., 1994, p. 54.
337. Akane, J., ACS, in *Modern Plastics Encyclopedia Handbook,* McGraw-Hill, Inc., 1994, p. 54.
338. Ostrowski, S., Acrylic-styrene-acrylonitrile, in *Modern Plastics Encyclopedia Handbook,* McGraw-Hill, Inc., 1994, p. 54.
339. *Principles of Polymer Engineering,* 2/e, McCrum, Buckley and Bucknall, Oxford Science Publications, p. 372.
340. *Encyclopedia of Polymer Science and Engineering,* 2/e, vol. 16, Mark, Bilkales, Overberger, Menges, Kroschwitz, Eds., Wiley Interscience, 1986, p. 65.
341. Kroschwitz, J.I., *Concise Encyclopedia of Polymer Science and Engineering,* John Wiley and Sons, New York, 1990, p. 30.
342. Brydson, J.A., *Plastics Materials,* 6/e, Butterworth-Heinemann, Oxford, 1995, p. 426.
343. Berins, M.L., *Plastics Engineering Handbook of the Society of the Plastics Industry,* 5/e, Chapman and Hall, New York, 1991, p. 57.
344. Brydson, J.A., *Plastics Materials,* 6/e, Butterworth-Heinemann, Oxford, 1995, p. 426.
345. Brydson, J.A., *Plastics Materials,* 6/e, Butterworth-Heinemann, Oxford, 1995, p. 426.
346. Kroschwitz, J.I., *Concise Encyclopedia of Polymer Science and Engineering,* John Wiley and Sons, New York, 1990, p. 30.
347. Berins, M.L., *Plastics Engineering Handbook of the Society of the Plastics Industry,* 5/e, Chapman and Hall, New York, 1991, p. 57.
348. Berins, M.L., *Plastics Engineering Handbook of the Society of the Plastics Industry,* 5/e, Chapman and Hall, New York, 1991, p. 57.
349. Brydson, J.A., *Plastics Materials,* 6/e, Butterworth-Heinemann, Oxford, 1995, p. 435.
350. Salay, J.E. and Dougherty, D.J., Styrene-butadiene copolymers, in *Modern Plastics Encyclopedia Handbook,* McGraw-Hill, Inc., 1994, p. 60.

351. Salay, J.E. and Dougherty, D.J., Styrene-butadiene copolymers, in *Modern Plastics Encyclopedia Handbook,* McGraw-Hill, Inc., 1994, p. 60.
352. Brydson, J.A., *Plastics Materials,* 6/e, Butterworth-Heinemann, Oxford, 1995, p. 435.
353. Salay, J.E. and Dougherty, D.J., Styrene-butadiene copolymers, in *Modern Plastics Encyclopedia Handbook,* McGraw-Hill, Inc., 1994, p. 60.
354. Strong, A.B., *Plastics: Materials and Processing,* Prentice-Hall, NJ, 1996, p. 205.
355. Brydson, J.A., *Plastics Materials,* 6/e, Butterworth-Heinemann, Oxford, 1995, p. 577.
356. Kroschwitz, J.I., *Concise Encyclopedia of Polymer Science and Engineering,* John Wiley and Sons, New York, 1990, p. 886.
357. Brydson, J.A., *Plastics Materials,* 6/e, Butterworth-Heinemann, Oxford, 1995, p. 580.
358. Kroschwitz, J.I., *Concise Encyclopedia of Polymer Science and Engineering,* John Wiley and Sons, New York, 1990, p. 886.
359. Brydson, J.A., *Plastics Materials,* 6/e, Butterworth-Heinemann, Oxford, 1995, p. 582.
360. Brydson, J.A., *Plastics Materials,* 6/e, Butterworth-Heinemann, Oxford, 1995, p. 582.
361. Brydson, J.A., *Plastics Materials,* 6/e, Butterworth-Heinemann, Oxford, 1995, p. 583.
362. Carraher, C.E., *Polymer Chemistry, An Introduction,* 4/e, Marcel Dekker, Inc., New York, 1996, p. 240.
363. Kroschwitz, J.I., *Concise Encyclopedia of Polymer Science and Engineering,* John Wiley and Sons, New York, 1990, p. 888.
364. Berins, M.L., *Plastics Engineering Handbook of the Society of the Plastics Industry,* 5/e, Chapman and Hall, New York, 1991, p. 71.
365. Berins, M.L., *Plastics Engineering Handbook of the Society of the Plastics Industry,* 5/e, Chapman and Hall, New York, 1991, p. 71.
366. Sauers, M.E., Polyaryl Sulfones, in *Engineering Plastics, vol. 2, Engineered Materials Handbook,* ASM International, Metals Park, OH, 1988, p. 146.
367. Sauers, M.E., Polyaryl Sulfones, in *Engineering Plastics, vol. 2, Engineered Materials Handbook,* ASM International, Metals Park, OH, 1988, p. 145.
368. Berins, M.L., *Plastics Engineering Handbook of the Society of the Plastics Industry,* 5/e, Chapman and Hall, New York, 1991, p. 72.
369. Berins, M.L., *Plastics Engineering Handbook of the Society of the Plastics Industry,* 5/e, Chapman and Hall, New York, 1991, p. 72.
370. Berins, M.L., *Plastics Engineering Handbook of the Society of the Plastics Industry,* 5/e, Chapman and Hall, New York, 1991, p. 71.
371. Sauers, M.E., Polyaryl Sulfones, in *Engineering Plastics,* vol. 2, *Engineered Materials Handbook,* ASM International, Metals Park, OH, 1988, p. 146.
372. Berins, M.L., *Plastics Engineering Handbook of the Society of the Plastics Industry,* 5/e, Chapman and Hall, New York, 1991, p. 72.
373. Watterson, E.C., Polyether Sulfones, in *Engineering Plastics,* vol. 2, *Engineered Materials Handbook,* ASM International, Metals Park, OH, 1988, p. 161.
374. Watterson, E.C., Polyether Sulfones, in *Engineering Plastics,* vol. 2, *Engineered Materials Handbook,* ASM International, Metals Park, OH, 1988, p. 160.
375. Berins, M.L., *Plastics Engineering Handbook of the Society of the Plastics Industry,* 5/e, Chapman and Hall, New York, 1991, p. 72.
376. Berins, M.L., *Plastics Engineering Handbook of the Society of the Plastics Industry,* 5/e, Chapman and Hall, New York, 1991, p. 72.

377. Watterson, E.C., Polyether Sulfones, in *Engineering Plastics, vol. 2, Engineered Materials Handbook,* ASM International, Metals Park, OH, 1988, p. 161.
378. Berins, M.L., *Plastics Engineering Handbook of the Society of the Plastics Industry,* 5/e, Chapman and Hall, New York, 1991, p. 72.
379. Watterson, E.C., Polyether Sulfones, in *Engineering Plastics,* vol. 2, *Engineered Materials Handbook,* ASM International, Metals Park, OH, 1988, p. 159.
380. Dunkle, S.R., Polysulfones, in *Engineering Plastics,* vol. 2, *Engineered Materials Handbook,* ASM International, Metals Park, OH, 1988, p. 200.
381. Dunkle, S.R., Polysulfones, in *Engineering Plastics,* vol. 2, *Engineered Materials Handbook,* ASM International, Metals Park, OH, 1988, p. 200.
382. Berins, M.L., *Plastics Engineering Handbook of the Society of the Plastics Industry,* 5/e, Chapman and Hall, New York, 1991, p. 71.
383. Berins, M.L., *Plastics Engineering Handbook of the Society of the Plastics Industry,* 5/e, Chapman and Hall, New York, 1991, p. 71.
384. Dunkle, S.R., Polysulfones, in *Engineering Plastics,* vol. 2, *Engineered Materials Handbook,* ASM International, Metals Park, OH, 1988, p. 200.
385. Dunkle, S.R., Polysulfones, in *Engineering Plastics,* vol. 2, *Engineered Materials Handbook,* ASM International, Metals Park, OH, 1988, p. 201.
386. Berins, M.L., *Plastics Engineering Handbook of the Society of the Plastics Industry,* 5/e, Chapman and Hall, New York, 1991, p. 71.
387. Dunkle, S.R., Polysulfones, in *Engineering Plastics,* vol. 2, *Engineered Materials Handbook,* ASM International, Metals Park, OH, 1988, p. 200.
388. Brydson, J.A., *Plastics Materials,* 6/e, Butterworth-Heinemann, Oxford, 1995, p. 301.
389. Billmeyer, F.W., Jr., *Textbook of Polymer Science,* 2/e, John Wiley & Sons, Inc., New York, 1962, p. 420.
390. Brydson, J.A., *Plastics Materials,* 6/e, Butterworth-Heinemann, Oxford, 1995, p. 304.
391. Brydson, J.A., *Plastics Materials,* 6/e, Butterworth-Heinemann, Oxford, 1995, p. 307.
392. Brydson, J.A., *Plastics Materials,* 6/e, Butterworth-Heinemann, Oxford, 1995, p. 302–304.
393. Strong, A.B., *Plastics: Materials and Processing,* Prentice-Hall, NJ, 1996, p. 171.
394. Strong, A.B., *Plastics: Materials and Processing,* Prentice-Hall, NJ, 1996, p. 170.
395. Billmeyer, F.W., Jr., *Textbook of Polymer Science,* 2/e, John Wiley & Sons, Inc., New York, 1962, p. 420.
396. Strong, A.B., *Plastics: Materials and Processing,* Prentice-Hall, NJ, 1996, p. 172.
397. Brydson, J.A., *Plastics Materials,* 6/e, Butterworth-Heinemann, Oxford, 1995, p. 314–316.
398. Strong, A.B., *Plastics: Materials and Processing,* Prentice-Hall, NJ, 1996, p. 171.
399. Strong, A.B., *Plastics: Materials and Processing,* Prentice-Hall, NJ, 1996, p. 172.
400. Brydson, J.A., *Plastics Materials,* 6/e, Butterworth-Heinemann, Oxford, 1995, p. 317–319.
401. Strong, A.B., *Plastics: Materials and Processing,* Prentice-Hall, NJ, 1996, p. 173.
402. Brydson, J.A., *Plastics Materials,* 6/e, Butterworth-Heinemann, Oxford, 1995, p. 346.
403. Martello, G.A., Chlorinated PVC, in *Modern Plastics Encyclopedia Handbook,* McGraw-Hill, Inc., 1994, p. 71.
404. Brydson, J.A., *Plastics Materials,* 6/e, Butterworth-Heinemann, Oxford, 1995, p. 341.
405. Strong, A.B., *Plastics: Materials and Processing,* Prentice-Hall, NJ, 1996, p. 173.

406. Hurter, D., Dispersion PVC, in *Modern Plastics Encyclopedia Handbook,* McGraw-Hill, Inc., 1994, p. 72.
407. Brydson, J.A., *Plastics Materials,* 6/e, Butterworth-Heinemann, Oxford, 1995, p. 309.
408. Brydson, J.A., *Plastics Materials,* 6/e, Butterworth-Heinemann, Oxford, 1995, p. 450.
409. Brydson, J.A., *Plastics Materials,* 6/e, Butterworth-Heinemann, Oxford, 1995, p. 127.
410. I.W. Sommer, Plasticizers, in *Plastics Additives* 2/e, R. Gachter and H. Muller, Eds., Hanser Publishers, New York, 1987, p. 253–255.
411. W. Brotz, Lubricants and Related Auxiliaries for Thermoplastic Materials, in *Plastics Additives* 2/e, R. Gachter and H. Muller, Eds., Hanser Publishers, New York, 1987, p. 297.
412. Brydson, J.A., *Plastics Materials,* 6/e, Butterworth-Heinemann, Oxford, 1995, p. 129.
413. Brydson, J.A., *Plastics Materials,* 6/e, Butterworth-Heinemann, Oxford, 1995, p. 136.
414. Brydson, J.A., *Plastics Materials,* 6/e, Butterworth-Heinemann, Oxford, 1995, p. 130–141.
415. Brydson, J.A., *Plastics Materials,* 6/e, Butterworth-Heinemann, Oxford, 1995, p. 141–145.
416. Brydson, J.A., *Plastics Materials,* 6/e, Butterworth-Heinemann, Oxford, 1995, p. 146–149.
417. Kroschwitz, J.I., *Concise Encyclopedia of Polymer Science and Engineering,* John Wiley and Sons, New York, 1990, p. 830–835.
418. Osswald, T.A., *Polymer Processing Fundamentals*, Hanser/Gardner Publications, New York, 1998, p. 67.
419. Michaeli, W., *Plastics Processing, An Introduction*, Hanser/Gardner Publications, New York,1992, p. 86.
420. Berins, M.L., *Plastics Engineering Handbook of the Society of the Plastics Industry,* 5/e, Chapman and Hall, New York,1991, p. 81.
421. Osswald, T.A., *Polymer Processing Fundamentals*, Hanser/Gardner Publications, New York,1998, p. 69.
422. Morton-Jones, D.H., *Polymer Processing*, Chapman and Hall, New York,1989, p. 77.
423. Osswald, T.A., *Polymer Processing Fundamentals*, Hanser/Gardner Publications, New York,1998, p. 70.
424. *Encyclopedia of Polymer Science and Engineering,* 2/e, vol. 6, Mark, Bilkales, Overberger, Menges, Kroschwitz, Eds., Wiley Interscience, 1986, p. 571.
425. Berins, M.L., *Plastics Engineering Handbook of the Society of the Plastics Industry,* 5/e, Chapman and Hall, New York,1991, p. 92.
426. Morton-Jones, D.H., *Polymer Processing*, Chapman and Hall, New York, 1989, pp. 107,110, and 111.
427. Morton-Jones, D.H., *Polymer Processing*, Chapman and Hall, New York, 1989, p. 118.
428. G. Pötsch and W. Michaeli, *Injection Molding,* Hanser Publishers, Munich, Germany, 1995.
429. R.A. Malloy, *Plastic Part Design for Injection Molding*, Carl Hanser Verlag, Munich, 1994, p. 20.
430. G. Pötsch and W. Michaeli, *Injection Molding,* Hanser Publishers, Munich, Germany, 1995, p. 115.
431. G. Pötsch and W. Michaeli, *Injection Molding,* Hanser Publishers, Munich, Germany, 1995, p. 133.

432. G. Pötsch and W. Michaeli, *Injection Molding,* Hanser Publishers, Munich, Germany, 1995, p. 135.
433. R.A. Malloy, *Plastic Part Design for Injection Molding*, Carl Hanser Verlag, Munich, 1994, p. 120.
434. A.B. Strong., *Plastics: Materials and Processing,* Prentice-Hall, NJ, 1996.
435. J.L. Throne, *Technology of Thermoforming,* Carl Hanser Verlag, Munich, 1996.
436. M.L. Berins, *Plastics Engineering Handbook of the Society of the Plastics Industry,* 5/e, Chapman and Hall, New York, 1991, p. 383.
437. J.L. Throne, *Technology of Thermoforming,* Carl Hanser Verlag, Munich, 1996, pp. 17–19.
438. J.L. Throne, *Technology of Thermoforming,* Carl Hanser Verlag, Munich, 1996, pp. 19–22.
439. A.W. Birley, B. Haworth, and J. Batchelor, *Physics of Plastics,* Carl Hanser Verlag, Munich, 1992, p. 229.
440. A.W. Birley, B. Haworth, and J. Batchelor, *Physics of Plastics,* Carl Hanser Verlag, Munich, 1992, p. 230.
441. Michaeli, W., *Plastics Processing, An Introduction*, Hanser/Gardner Publications, New York,1992, p. 102.
442. Osswald, T.A., *Polymer Processing Fundamentals*, Hanser/Gardner Publications, New York,1998, pp. 149, 151.
443. Osswald, T.A., *Polymer Processing Fundamentals*, Hanser/Gardner Publications, New York,1998, p. 176.
444. H. Hurnik, Chemical Blowing Agents, in *Plastics Additives,* 4/e, R. Gächter and H. Müller, Eds., Carl Hanser Verlag, Munich, 1993.
445. Berins, M.L., *Plastics Engineering Handbook of the Society of the Plastics Industry,* 5/e, Chapman and Hall, New York, 1991, p. 553.
446. Berins, M.L., *Plastics Engineering Handbook of the Society of the Plastics Industry,* 5/e, Chapman and Hall, New York, 1991, p. 555.
447. Berins, M.L., *Plastics Engineering Handbook of the Society of the Plastics Industry,* 5/e, Chapman and Hall, New York, 1991, p. 559.
448. M.L. Berins, *Plastics Engineering Handbook of the Society of the Plastics Industry,* 5/e, Chapman and Hall, New York, 1991, p. 593.
449. M.L. Berins, *Plastics Engineering Handbook of the Society of the Plastics Industry,* 5/e, Chapman and Hall, New York, 1991, pp. 593–599.
450. M.L. Berins, *Plastics Engineering Handbook of the Society of the Plastics Industry,* 5/e, Chapman and Hall, New York, 1991, pp. 600–605.
451. L.H. Van Vlack, *Elements of Materials Science and Engineering*, 3/e, Addison-Wesley, Reading, MA, 1975.
452. R.R. Maccani, Characteristics Crucial to the Application of Engineering Plastics, in *Engineering Plastics*, vol. 2, *Engineering Materials Handbook,* ASM International, Metals Park, OH, 1988, p. 69.
453. Berins, M.L., *Plastics Engineering Handbook of the Society of the Plastics Industry,* 5/e, Chapman and Hall, New York, 1991, p. 48–49.
454. H.P. Schlumpf, Fillers and Reinforcements, in *Plastics Additives,* 4/e, R. Gächter and H. Müller, Eds., Carl Hanser Verlag, Munich, 1993.
455. *Encyclopedia of Polymer Science and Engineering,* 2/e, vol. 6, Mark, Bilkales, Overberger, Menges, Kroschwitz, Eds., Wiley Interscience, 1986, p. 433.
456. *Encyclopedia of Polymer Science and Engineering,* 2/e, vol. 16, Mark, Bilkales, Overberger, Menges, Kroschwitz, Eds., Wiley Interscience, 1986, p. 65.
457. T. Whelan and J. Goff, *The Dynisco Injection Molders Handbook,* 1/e, Dynisco, copyright T. Whelan and J. Goff, 1991.

458. http://www.mgstech.com/multishot_molding/materials/battenfeld_plastic_bonding_chart.gif.
459. Berins, M.L., *Plastics Engineering Handbook of the Society of the Plastics Industry,* 5/e, Chapman and Hall, New York, 1991, p. 405.
460. H. Hurnik, Chemical Blowing Agents, in *Plastics Additives,* 4/e, R. Gächter and H. Müller, Eds., Carl Hanser Verlag, Munich, 1993.
461. G. Pötsch and W. Michaeli, *Injection Molding,* Hanser Publishers, Munich, Germany, 1995, p. 2.
462. N.G. McCrum, C.P. Buckley, and C.B. Bucknall, *Principles of Polymer Engineering,* 2/e, Oxford University Press, New York, 1997, p. 334.
463. N.G. McCrum, C.P. Buckley, and C.B. Bucknall, *Principles of Polymer Engineering,* 2/e, Oxford University Press, New York, 1997, p. 338.
464. G. Pötsch and W. Michaeli, *Injection Molding,* Hanser Publishers, Munich, Germany, 1995, p. 172.
465. G. Pötsch and W. Michaeli, *Injection Molding,* Hanser Publishers, Munich, Germany, 1995, p. 173.
466. G. Pötsch and W. Michaeli, *Injection Molding,* Hanser Publishers, Munich, Germany, 1995, p. 178.
467. G. Pötsch and W. Michaeli, *Injection Molding,* Hanser Publishers, Munich, Germany, 1995, p. 177.
468. R.A. Malloy, *Plastic Part Design for Injection Molding*, Carl Hanser Verlag, Munich, 1994, p. 396.
469. J.L. Throne, *Technology of Thermoforming,* Carl Hanser Verlag, Munich, 1996, p. 17.
470. J.L. Throne, *Technology of Thermoforming,* Carl Hanser Verlag, Munich, 1996, p. 18.
471. J.L. Throne, *Technology of Thermoforming,* Carl Hanser Verlag, Munich, 1996, p. 19.
472. J.L. Throne, *Technology of Thermoforming,* Carl Hanser Verlag, Munich, 1996, p. 20.
473. J.L. Throne, *Technology of Thermoforming,* Carl Hanser Verlag, Munich, 1996, p. 21.
474. J.L. Throne, *Technology of Thermoforming,* Carl Hanser Verlag, Munich, 1996, p. 22.
475. J.L. Throne, *Technology of Thermoforming,* Carl Hanser Verlag, Munich, 1996, p. 23.

Chapter

2

Thermosets, Reinforced Plastics, and Composites

Ralph E. Wright

R. E. Wright Associates
Yarmouth, Maine

2.1 Resins

One definition of *resin* is "any class of solid, semi-solid, or liquid organic material, generally the product of natural or synthetic origin with a high molecular weight and with no melting point." The 10 basic thermosetting resins all possess a commonality in that they will, upon exposure to elevated temperature from ambient to upward of 450°F, undergo an irreversible chemical reaction often referred to as *polymerization* or cure. Each family member has its own set of individual chemical characteristics based on its molecular makeup and its ability to either homopolymerize, copolymerize, or both.

This transformation process represents the line of demarcation separating the thermosets from the thermoplastic polymers. Crystalline thermoplastic polymers are capable of a degree of crystalline cross-linking, but there is little, if any, of the chemical cross-linking that occurs during the thermosetting reaction. The important beneficial factor here lies in the inherent enhancement of thermoset resins in their physical, electrical, thermal, and chemical properties due to that chemical cross-linking polymerization reaction which, in turn, also contributes to their ability to maintain and retain these enhanced properties when exposed to severe environmental conditions.

2.2 Thermosetting Resin Family

2.2.1 Allyls: Diallyl ortho phthalate (DAP) and Diallyl iso phthalate (DAIP)

2.2.1.1 Chemical characteristics.[1] The most broadly used allyl resins are prepared from the prepolymers of either DAP or DAIP, which have been condensed from dibasic acids. The diallyl phthalate monomer is an ester produced by the esterification process in-

volving a reaction between a dibasic acid (phthalic anhydride) and an alcohol (allyl alcohol), which yields the DAP ortho monomer, as shown in Fig. 2.1. Similar reactions with dibasic acids will yield the DAIP (iso) prepolymer. Both prepolymers are white, free-flowing powders and are relatively stable whether catalyzed or not, with the DAP being more stable and showing negligible change after storage of several years in temperatures up to 90°C.

The monomer is capable of cross-linking and will polymerize in the presence of certain peroxide catalysts such as

- Dicumyl peroxide (DICUP)
- t-Butyl perbenzoate (TBP)
- t-Butylperoxyisopropyl carbonate (TBIC)

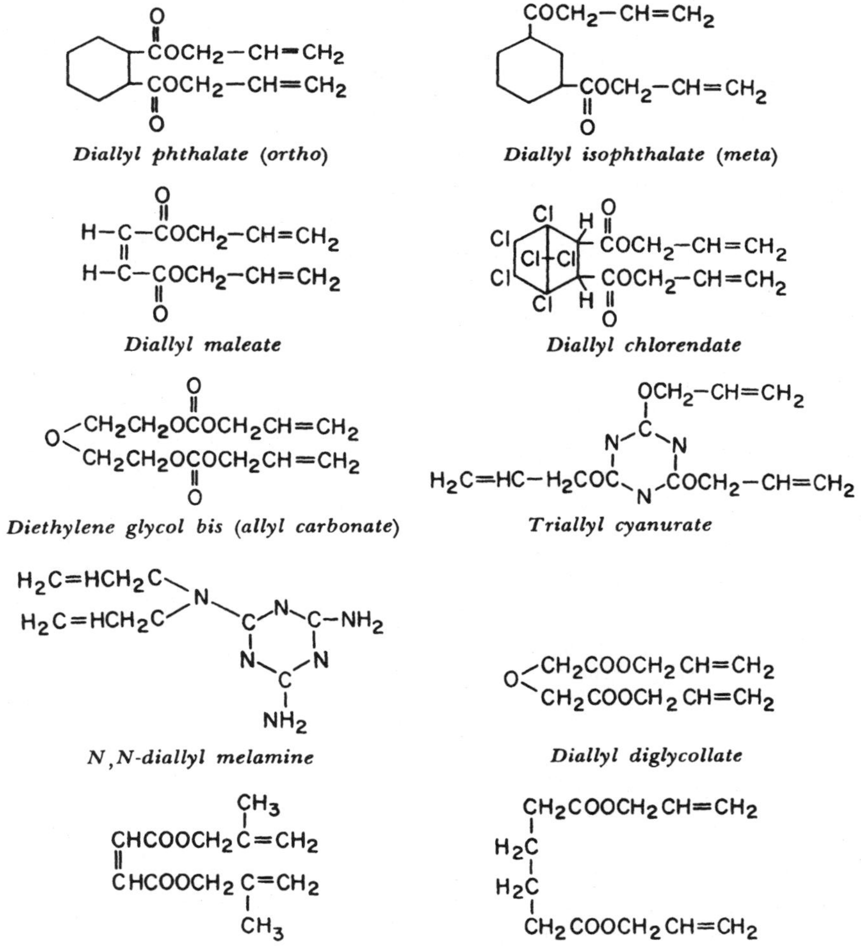

Figure 2.1 Structural formula for allylic resins.

2.2.2 Aminos: Urea and Melamine

2.2.2.1 Chemical characteristics.[2] Both resins will react with formaldehyde to initiate and form monomeric addition products. Six molecules of formaldehyde added to one single molecule of melamine will form hexamethylol melamine, whereas a single molecule of urea will combine with two molecules of formaldehyde to form dimethylolurea. If carried on further, these condensation reactions produce an infusible polymer network. The urea/formaldehyde and melamine/formaldehyde reactions are illustrated in Fig. 2.2.

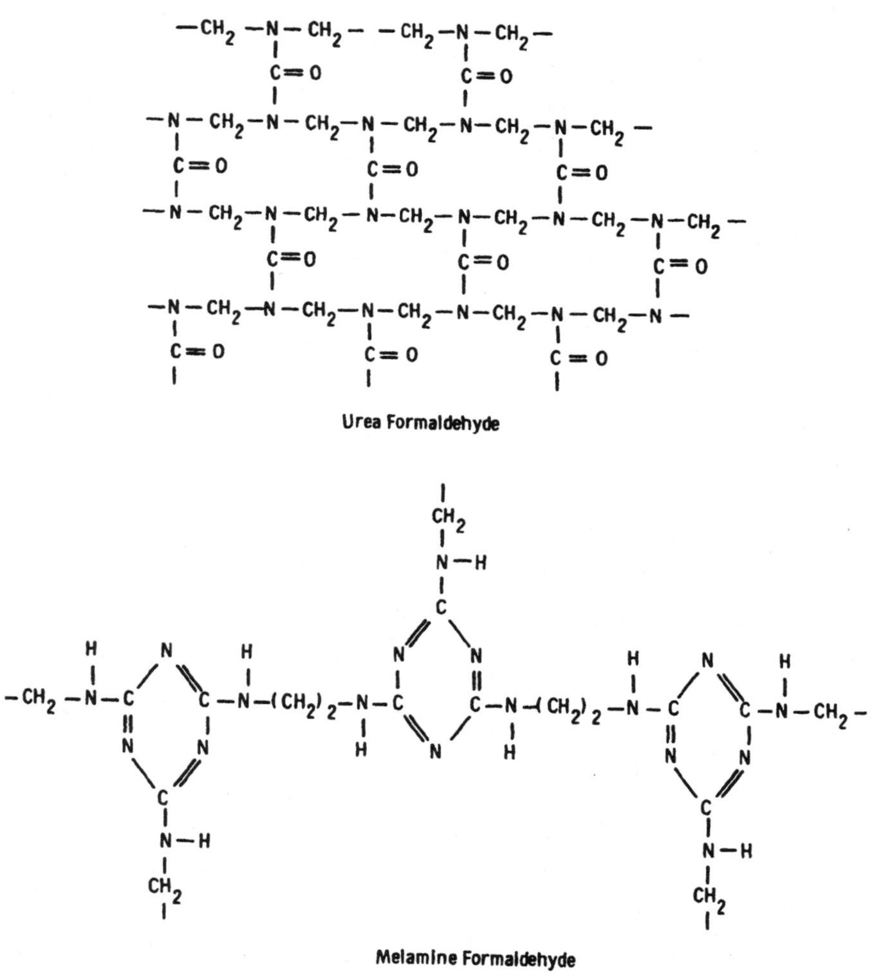

Figure 2.2 Structural formula for amino resins. *(Source: Charles A. Harper,* Handbook of Plastics, Elastomers, and Composites, *3d ed., McGraw-Hill, New York, 1996, p. 1.29)*

2.2.3 Bismaleimides (BMIs)

2.2.3.1 Chemical characteristics.[3] The bismaleimides (BMIs) are generally prepared by the condensation reaction of a diamine with maleic anhydride. A typical BMI based on methylene dianiline (MDA) is illustrated in Fig. 2.3.

2.2.4 Epoxies

2.2.4.1 Chemical characteristics.[4] *Epoxy resins* are a group of cross-linking polymers and are sometimes known as the *oxirane group,* which is reactive toward a broad range of curing agents. The curing reactions convert the low-molecular-weight resins into three-dimensional thermoset structures exhibiting valuable properties. The standard epoxy resins used in molding compounds meeting Mil-M-24325 (ships) are based on bisphenol A and epichiorohydrin as raw materials with anhydride catalysts, as illustrated in Fig. 2.4. The low-pressure encapsulation compounds consist of an epoxy-novolac resin system using amine catalysts.

2.2.5 Phenolics: Resoles and Novolacs

2.2.5.1 Chemical characteristics.[5] Phenol and formaldehyde, when reacted together, will produce condensation products when there are free positions on the benzene ring—*ortho* and *para* to the hydroxyl group. Formaldehyde is by far the most reactive and is used almost exclusively in commercial applications. The product is greatly dependent upon the type of catalyst and the mole ratio of the reactants. Although there are four major reactions in the phenolic resin chemistry, the resole (single stage) and the novolac (two stage) are the two primarily used in the manufacture of phenolic molding compounds.

Novolacs (two-stage). In the presence of acid catalysts, and with a mole ratio of formaldehyde to phenol of less than 1, the methylol derivatives condense with phenol to form, first, dihydroxydiphenyl methane, as shown in Fig. 2.5. On further condensation and with methyl bridge formation, fusible and soluble linear low polymers called *novolacs* are formed with the structure where the *ortho* and *para* links occur at random, as shown in Fig. 2.6.

The novolac (two-stage) resins are made with an acid catalyst, and only part of the necessary formaldehyde is added to the reaction kettle, producing a mole ratio of about 0.81. The rest is added later: a hexamethylenetetramine (hexa), which decomposes in the final

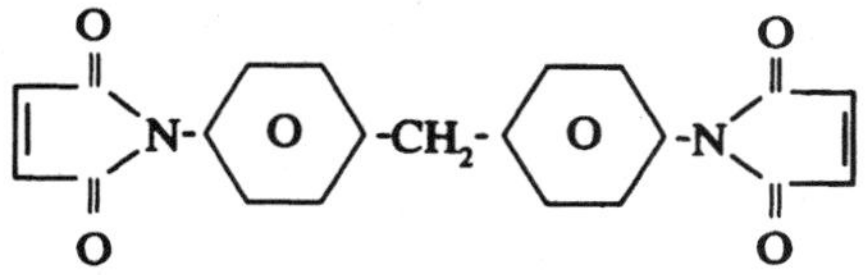

Figure 2.3 Structural formulas for bismaleimide resins. *(From* Plastics Handbook, *Modern Plastics Magazine, McGraw-Hill, New York, 1994, p. 76)*

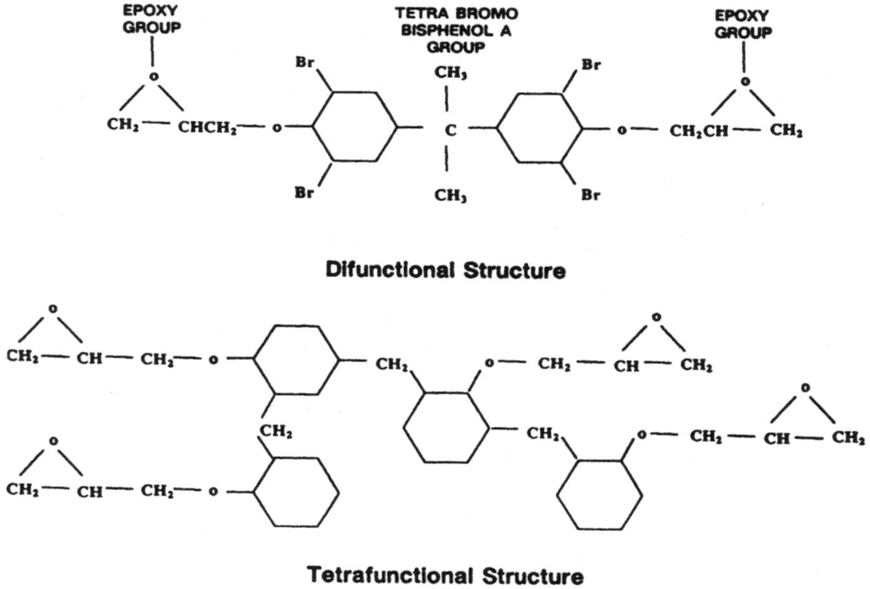

Figure 2.4 Structural formulas for epoxy resins.

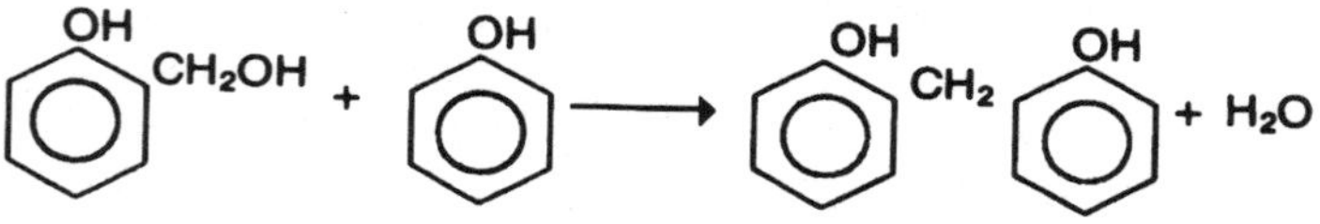

Figure 2.5 Structural formulas for single-stage phenolic resins.

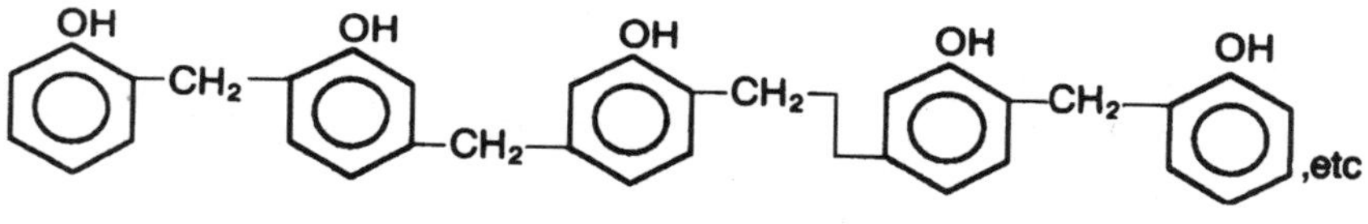

Figure 2.6 Structural formulas for single-stage novolac phenolic resins.

curing step, with heat and moisture present, to yield formaldehyde and ammonia, which act as the catalyst for curing.

Resole (single-stage). In the presence of alkaline catalysts and with more formaldehyde, the methylol phenols can condense either through methylene linkages or through ether linkages. In the latter case, subsequent loss of formaldehyde may occur with methylene bridge formation. Products of this type, soluble and fusible but containing alcohol groups, are called *resoles* and are shown in Fig. 2.7. If the reactions leading to their forma-

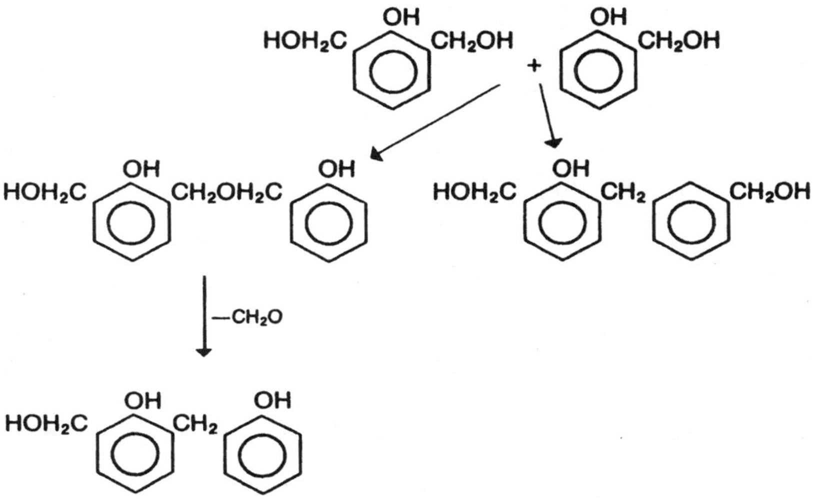

Figure 2.7 Structural formula for two-stage resole phenolic resin.

tion are carried further, large numbers of phenol nuclei can condense to give a network formation.

In the production of a single-stage phenolic resin, all the necessary reactants for the final polymer (phenol, formaldehyde, and catalyst) are charged into the resin kettle and reacted together. The ratio of formaldehyde to phenol is about 1.25:1, and an alkaline catalyst is used. Resole resin-based molding compounds will not outgas ammonia during the molding process as do the novolac compounds, with the result that the resole compounds are used for applications where their lack of outgassing is most beneficial and, in same cases, essential. The chief drawback lies in their sensitivity to temperatures in excess of 45°F (7.2°C). This makes it imperative to produce these compounds in air-conditioned rooms and to store the finished goods in similar conditions in sealed containers.

The novolac (two-stage) resin based compounds are the most widely used despite the ammonia outgassing, which can be controlled through proper mold venting. They have an outstanding property profile, can be readily molded in all thermosetting molding techniques, and are among the least costly compounds available. Their chief drawback has been their limited color range, but that has been improved with the advent of phenolic-melamine alloying.

2.2.6 Polyesters (Thermosetting)

2.2.6.1 Chemical characteristics.[6] Thermosetting polyesters can be produced from phthalic or maleic anhydrides and polyfunctional alcohols with catalyzation achieved by the use of free radical-producing peroxides. The chemical linkages shown in Fig. 2.8 will form a rigid, cross-linked, thermosetting molecular structure. Thermosetting polyesters are derived from the condensation effects of combining an unsaturated dibasic acid (maleic anhydride with a glycol (propylene glycol, ethylene glycol, and dipropylene glycol). When utilized as a molding compound, the unsaturated polyester resin is dis-

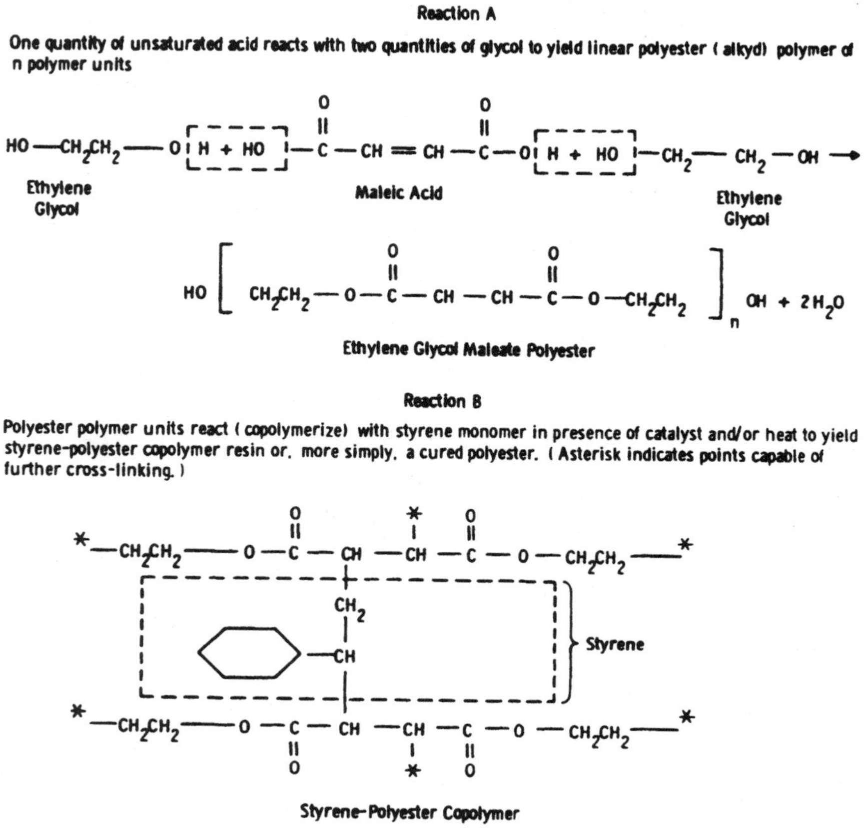

Reaction A
One quantity of unsaturated acid reacts with two quantities of glycol to yield linear polyester (alkyd) polymer of n polymer units
Ethylene Glycol
Maleic Acid
Ethylene Glycol
Ethylene Glycol Maleate Polyester
Reaction B
Polyester polymer units react (copolymerize) with styrene monomer in presence of catalyst and/or heat to yield styrene-polyester copolymer resin or, more simply, a cured polyester. (Asterisk indicates points capable of further cross-linking.)
Styrene
Styrene-Polyester Copolymer

Figure 2.8 Simplified diagrams showing how cross-linking reactions produce polyester resin (styrene-polyester copolymer) from basic chemicals. *(Source: Charles A. Harper,* Handbook of Plastics, Elastomers, and Composites, *3d ed., McGraw-Hill, New York, 1996, p. 1.3)*

solved in a cross-linking monomer (styrene) with the addition of an inhibitor (hydroquinone) to prevent the cross-linking until the compound is ready for use in the molding process. The compounds are further enhanced with additives, such as chiorendic anhydride for flame retardance, isophthalic acid for chemical resistance, and neopentyl glycol for weathering resistance.

The compounder uses the free radical addition to polymerize the resin. The catalyst (organic peroxide) becomes the source for the free radicals, and with elevated temperature the heat decomposes the peroxide, producing free radicals. Peroxyesters and benzoyl peroxide are the organic peroxides primarily used at elevated temperatures.

2.2.7 Polyimides (Thermosetting)

2.2.7.1 Chemical characteristics.[7] *Polyimides* are heterocyclic polymers with a noncarbon atom of nitrogen in one of the rings in the molecular chains, as shown in Fig. 2.9.

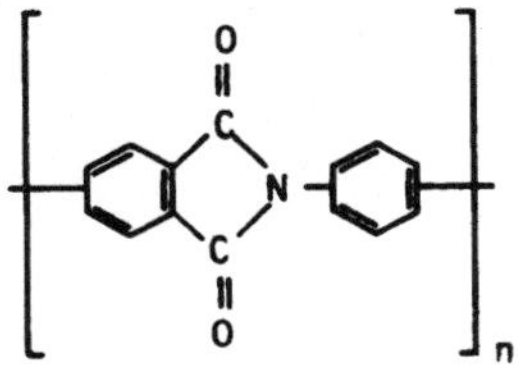

Figure 2.9 Structural formula for polyimide resin. *(Source: Charles A. Harper,* Handbook of Plastics, Elastomers, and Composites, *3d ed., McGraw-Hill, New York, 1996, p. 1.63)*

The fused rings provide chain stiffness essential to high-temperature strength retention. The low concentration of hydrogen provides oxidation resistance by preventing thermal degradative fracture of the chain.

Poly(amide-imide) resins contain aromatic rings and the characteristic nitrogen linkages, as shown in Fig. 2.10. There are two basic types of polyimides:

1. *Condensation resin* is based upon a reaction of an aromatic diamine with an aromatic dianhydride, producing a fusible polyamic acid intermediate which is converted by heat to an insoluble and infusible polyimide, with water being given off during the cure.
2. *Addition resins* are based on short, preimidized polymer chain segments similar to those comprising condensation polyimides. The prepolymer chains, which have unsaturated aliphatic end groups, are capped by termini which polymerize thermally without the loss of volatiles.

While condensation polyimides are available as both thermosetting and thermoplastic, the addition polyimides are available only as thermosets.

2.2.8 Polyurethanes

2.2.8.1 Chemical characteristics.[8] The true foundation of the polyurethane industry is the *isocyanate,* an organic functional group that is capable of an enormously diverse

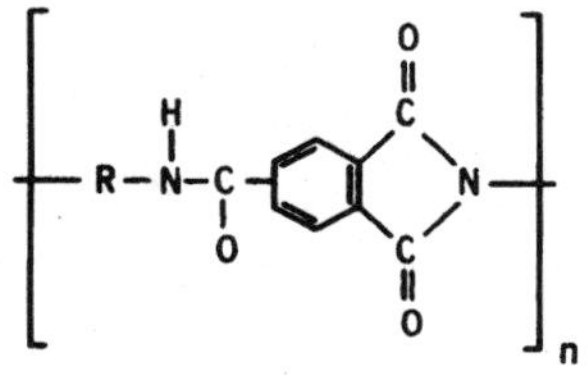

Figure 2.10 Structural formulas for poly(amide-imide) resin. *(Source: Charles A. Harper,* Handbook of Plastics, Elastomers, and Composites, *3d ed., McGraw-Hill, New York, 1996, p. 1.64)*

range of chemical reactions. The reactions of isocyanates fall into two broad categories, with the most important category being the reactions involving *active hydrogens*. This category of reactions requires at least one co-reagent containing one or more hydrogens that are potentially exchangeable under conditions of reaction. The familiar reaction of isocyanates with polyols, to form polycarbamates, is of the "active hydrogen" type. The active groups are, in this case, the hydroxyl groups on the polyol. *Nonactive hydrogen reactions* constitute the second broad category, including cycloaddition reactions and linear polymerizations (which may or may not involve co-reagents).

The diversity of isocyanate chemistry, combined with the availability of selective catalysts, has made it possible to "select" reactions that "fit" desired modes and rates of processing. Table 2.1 lists some of the known polymer-forming reactions of the isocyanates.

2.2.9 Silicones

2.2.9.1 Chemical characteristics.[9] The silicone family consists of three forms, as shown in Fig. 2.11 and described below:

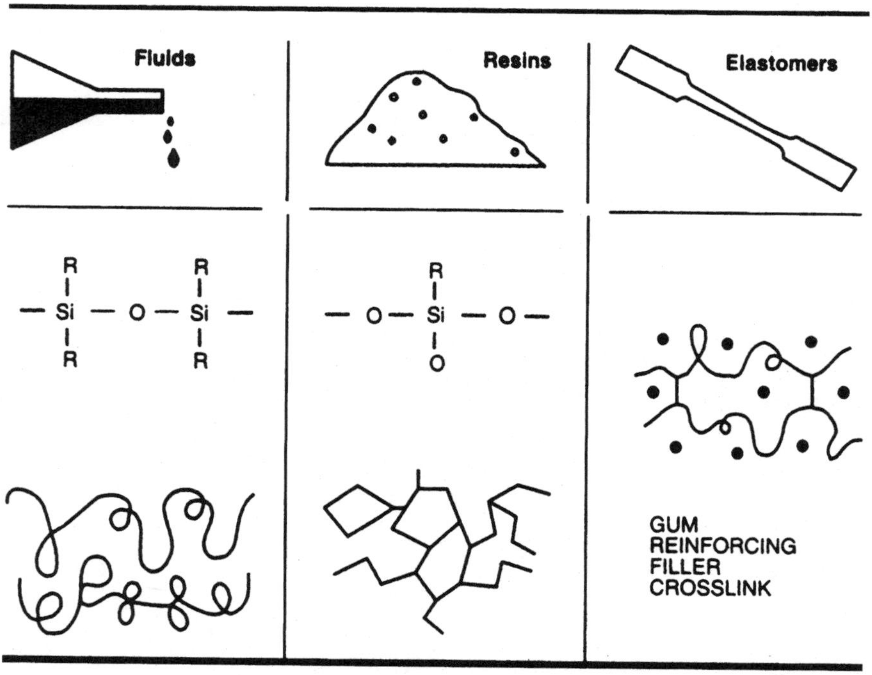

Figure 2.11 Fluids, resins, and elastomers. Silicone fluids are linear chains where molecular weight determines viscosity; silicone resins are branched polymers and thus glass-like solids; and silicone elastomers are composed of long, linear polysiloxane chains reinforced with an inorganic filler and cross-linked. *(From* Plastics Handbook, *Modern Plastics Magazine, McGraw-Hill, New York, 1994, p. 86)*

TABLE 2.1 Examples of Polymer-Forming Reactions of Isocyanates

Co-reactant	Product(s)	Catalysts	Category
Alcohols	Carbamates	3° amines Tin soaps Alkalai soaps	Active-H
Carbamates	Allophanates	3° amines Tin soaps Alkalai soaps	Active-H
Primary amines	Ureas	Carboxylic acids	Active-H
Secondary amines	Ureas	Carboxylic acids	Active-H
Ureas	Biurates	Carboxylic acids	Active-H
Imines	Ureas Amides Triazines	Carboxylic acids	Either category
Enamines	Amides Heterocycles	Carboxylic acids	Either category
Carboxylic acids	Amides	Phospholene oxides	Active-H
Amides	Acyl ureas	Acids/bases	Active-H
H_2O	Ureas CO_2	3° amines Alkalai soaps	Active-H
Anhydrides (cyclic)	Imides	Phospholene oxides	Nonactive hydrogen
Ketones	CO_2		
	Imines	Alkalai soaps	Either category
	Heterocycles CO_2	Alkalai soaps	Either category
Aldehydes	Imines	Alkalai soaps	Either category
	Heterocycles CO_2	Alkalai soaps	Either category
Active methylene (active methine) compounds	Amides Heterocycles	Bases	Either category
Isocyanates	Carbodiimides	Phosphorous-oxygen compounds	Nonactive Nonactive Nonactive
	Dimers	Pyridines	Nonactive
	Trimers	Alkalai soaps	Nonactive
	Polymers	Strong bases	Nonactive
Carbodiimides	Uretonimines	—	Nonactive
Epoxides	Oxazolidones	Organoantimony iodides	Either category
2 or B hydroxy acids (esters)	Heterocycles ROH or H_2O	Acids/bases	Active-H
2 or B amino acids (esters)	Heterocycles ROH or H_2O	Acids/bases	Active-H
Cyanohydrins	Heterocycles	Acids/bases	Active-H
2-Cyano amines	Heterocycles	Acids/bases	Active-H

TABLE 2.1 Examples of Polymer-Forming Reactions of Isocyanates (*Continued*)

Co-reactant	Product(s)	Catalysts	Category
2-Amino esters	Heterocycles ROH	Acids/bases	Active-H
Orthoformates	Heterocycles ROH	Acids/bases	Active-H
Oxazolines; imidazolines	Various heterocycles	Acids/bases	Either category
Cyclic carbonates	Heterocycles CO2	Bases	Either category
Acetylenes	Various heterocycles	Various	Either category
Pyrroles	Amides	Acids/bases	Active-H
Carbamic acid	Ureas	—	Active-H
Amine salts	CO_2		

Fluids. The fluids are linear chains of dimethyl siloxane whose molecular weights determine their viscosity. They are supplied as both neat fluids and as water emulsions, and as nonreactive and reactive fluids.

Resins. The *resins* are highly branched polymers that cure to solids. They resemble glass but are somewhat softer and usually soluble in solvent until cured. Their degree of hardness when cured depends on the extent of cross-linking.

Elastomers. The *elastomers* are prepared from linear silicone oils or gums and reinforced with a filler and then vulcanized (cured or cross-linked). The base resin for silicone molding compounds are the result of reacting silicone monomer with methylchloride to produce methyl chlorosilane:

$$\text{Si} + \text{RCl (catalyst/heat)} \rightarrow \text{RSiCl}$$

2.3 Resin Characteristics

1. *A stage.* Resin is still soluble in certain liquids and still fusible.
2. *B stage.* An intermediate stage in the reaction process when the material softens with heat and swells in contact with certain liquids but does not entirely fuse or dissolve.
3. *C stage.* Final stage of the reaction when the resin becomes completely and solidly cured and is relatively insoluble and infusible.

2.4 Resin Forms

1. *Liquid.* Naked for coatings.
2. *Liquid.* Catalyzed for castings, foundry resins, and encapsulation.
3. *Liquid.* In molding compounds.
4. *Solid.* Flakes, granules, powders in molding compounds, and/or foundry resins.

2.5 Liquid Resin Processes[10]

2.5.1 Potting Process

The processing of liquid potting compounds commences with dispersing curing agents and other ingredients with a simple propeller mixer, producing potting compounds capable of a variety of properties achieved by changes in formulations to gain stiffness, color, strength, and electrical properties meeting specific needs and applications. The liquid potting compound is poured into a receptacle or jacket where the compound cures in place, either with added heat or at ambient temperature. The receptacle or jacket remains as the permanent outer skin of the final product. Proper compounding can produce articles with negligible shrinkage and optical clarity along with a minimum of internal stress. Part and mold design, coupled with the proper placement of the reinforcement, are calculated with sophisticated computer software.

2.5.2 Potting Resin Selection

Tables 2.2 through 2.5 provide helpful information for guidance in the selection process for potting resins.

2.5.3 Casting Process

The process for casting fluid monomeric resin compounds involves pouring the premixed compound into a stationary mold (metal or glass) and then allowing the compound to cure at ambient or elevated temperatures. The mold may contain objects that will have been prepositioned in the mold and they become embedded by the resin as it cures. This technique can be done with minimal equipment and delicate inserts can he embedded, or they can he introduced into the compound prior to the curing stage. The chief disadvantages are that high-viscosity resins are difficult to handle and the occurrence of voids or bubbles can present problems.

2.5.3.1 Cast epoxies. *Cast epoxy resins* have proven to be very popular in a wide variety of applications because of their versatility, excellent adhesion, low cure shrinkage, good electrical properties, compatibility with many other materials, resistance to weathering and chemicals, dependability, and ability to cure under adverse conditions. Some of their application fields are adhesives, coatings, castings, pottings, building construction, chemical resistant equipment, and marine applications. The most widely used epoxy resins in the casting field are the epi-bis and cycloaliphatic epoxies. Table 2.6 lists the properties of typical cured epi-bis resins with a variety of curing agents, and Table 2.7 provides information on the properties of blends of cycloaliphatic epoxy resins.

Novolac epoxy resins, phenolic or cresol novolacs, are reacted with epichiorohydrin to produce these novolac epoxy resins which cure more rapidly than the epi-bis epoxies and have higher exotherms. These cured novolacs have higher heat-deflection temperatures than the epi-bis resins as shown in Table 2.8. The novolacs also have excellent resistance to solvents and chemicals when compared with that of an epi-bis resin as seen in Table 2.9.

2.5.3.2 Cast polyesters. General purpose polyester, when blended with a monomer such as polystyrene and then cured, will produce rigid, rapidly curing transparent castings exhibiting the properties shown in Table 2.10. Other monomers, in conjunction with polystyrene, such as alpha methyl styrene, methyl methacrylate, vinyl toluene, diallyl phtha-

TABLE 2.2 Comparison of Properties of Liquid Resins

Material	Cure shrinkage	Adhesion	Thermal shock	Electrical properties	Mechanical properties	Handling properties	Cost
Epoxy:							
Room temperature cure	Low	Good	Fair*	Fair	Fair	Good	Moderate
High temperature	Low	Good	Fair*	Good	Good	Fair to good	Moderate
Flexible	Low	Excellent	Good	Fair	Fair	Good	Moderate
Polyesters:							
Rigid	High	Fair	Fair*	Fair to good	Good	Fair	Low to moderate
Flexible	Moderate	Fair	Fair	Fair	Fair	Fair	Low to moderate
Silicones:							
Rubbers	Very low	Poor	Excellent	Good	Poor	Good	High
Rigid	High	Poor	Poor	Excellent	Poor	Fair	High
Polyurethanes:							
Solid	Low	Excellent	Good	Fair	Good	Fair	Moderate
Foams	Variable	Good	Good	Fair	Good for density	Fair	Moderate
Butyl LM	Low	Good	Excellent	Good	Poor	Fair	Low to moderate
Butadienes	Moderate	Fair	Poor*	Excellent	Fair	Fair	Moderate

*Depends on filler.
SOURCE: From Ref. 10, p. 4.3.

TABLE 2.3 Properties of Various Liquid Resins

Property	ASTM	Rigids (silica-filled)			Flexibilized epoxies			Liquid elastomers				Rigid polyurethane
		Styrene polyester	Epoxy	Silicone	Epoxy-polysulfide (50–50)	Epoxy-polyamide (50–50)	Epoxy-polyurethane (50–50, diamine cure)	Poly-sulfide	Silicone	Poly-urethane-(diamine cure)	Plas-tisol	Prepolymer
Electrical Properties												
Dielectric strength, V/mil*	D 149	425	425	350	350	430	640	340	400	350	200	
Dielectric constant:												
60 Hz	D 150	3.7	3.8	3.7	5.6	3.2		7.2	3	8.2	5.6	1.05
10^3 Hz	D 150	3.7	3.6	3.6	5.4	3.2	4.9	7.2	3	7.3	4.9	1.06
10^6 Hz	D 150	3.6	3.4	3.6	4.8	3.1		7.2	3		3.6	1.04
Dissipation factor:												
60 Hz	D 150	0.01	0.02	0.008	0.02	0.01		0.01	0.005	0.08	0.12	0.004
10^3 Hz	D 150	0.02	0.02	0.004	0.02	0.01	0.04	0.01	0.004	0.09	0.1	0.003
10^6 Hz	D 150	0.02	0.03	0.01	0.06	0.02		0.02	0.003		0.12	0.003
Surface resistivity, Ω/square:												
Dry	D 257	10^{15}	10^{15}	10^{15}	10^{12}	10^{14}		10^{11}	10^{13}		10^{10}	$>10^{12}$
After 96 h at 95°F, 90% RH	D 257	10^{11}	10^{13}	10^{15}	10^{10}			10^{10}	10^{12}		10^{10}	$>10^{12}$
Volume resistivity (dry), Ω-cm	D 257	$>10^{15}$	$>10^{15}$	10^{15}	10^{12}	10^{14}	10^{14}	10^{11}	10^{14}	10^{12}	10^{10}	$>10^{14}$
Mechanical Properties												
Tensile strength, lb/in²	D 638, D 412	10,000	9,000	4,000	1,800	4,600	6,000	800	275	4,000	2,400	110
Elongation, %	D 412				30		10	450	200	450	300	
Compression strength, lb/in²	D 695, D 575	25,000	16,000	13,000		7,000						85
Flexural modulus of rupture, lb/in²	D 790	10,000	12,000	8,000		8,300						
Izod impact strength (notched), ft-lb	D 256	0.3	0.4	0.3								
Shore hardness	D 676, D 1484				40D		80D	45A	35A	90A	80A	
Penetration at 77°F, mils	D 5											
Shrinkage on cooling or curing, % by vol		6	3.5	8	3	3	6	1	2	4.5		

		Rigids (silica-filled)			Flexibilized epoxies			Liquid elastomers				Rigid polyurethane
Property	ASTM	Styrene polyester	Epoxy	Silicone	Epoxy-polysulfide (50–50)	Epoxy-polyamide (50–50)	Epoxy-polyurethane (50–50, diamine cure)	Poly-sulfide	Silicone	Poly-urethane-(diamine cure)	Plas-tisol	Prepolymer
Physical Properties												
Specific gravity	D 71, D 792	1.6	1.6	1.8	1.2	1.0	1.2	1.2	1.1	1.1	1.2	0.1†
Minimum cold flow, °F												
Softening or drip point, °F	D 36											
Heat-distortion temp, °F	D 648	230	165	300			100					
Maximum continuous service temp, °F		250	250	480	225	175	250	200	350	210	150	165
Coefficient of thermal expansion per $°F \times 10^{-6}$	D 696	26	22	44	44	44	110		128	110		19
Thermal conductivity, $Btu/(ft^2)(h)(°F/ft)$		0.193	0.29	0.23			0.77		0.13		0.09	0.02
Cost, $/lb		0.25	0.50	2.75	0.90	0.80	1.10	0.85	4.25	1.30	0.30	1.75

*Short-time test on $^1/_8$-in specimen.
†6 lb/ft^3.
SOURCE: C. V. Lundberg, "A Guide to Potting and Encapsulation Materials," *Material Engineering*, May 1960.

TABLE 2.4 Characteristics Influencing Choice of Resins in Electrical Applications

Resin	Cure and handling characteristics	Final part properties
Epoxies	Low shrinkage, compatible with a wide variety of modifiers, very long storage stability, moderate viscosity, cure under adverse conditions	Excellent adhesion, high strength, available clear, resistant to solvents and strong bases, sacrifice of properties for high flexibility
Polyesters	Moderate to high shrinkage, cure cycle variable over wide range, very low viscosity possible, limited compatibility, low cost, long pot life, easily modified, limited shelf life, strong odor with styrene	Fair adhesion, good electrical properties, water-white, range of flexibilities
Polyurethanes	Free isocyanate is toxic, must be kept water-free, low cure shrinkage, solid curing agent for best properties	Wide range of hardness, excellent wear, tear, and chemical resistance, fair electricals, excellent adhesion, reverts in humidity
Silicones (flexible)	Some are badly cure-inhibited, some have uncertain cure times, low cure shrinkage, room temperature cure, long shelf life, adjustable cure times, expensive	Properties constant with temperature, excellent electrical properties, available from soft gels to strong elastomers, good release properties
Silicones (rigid)	High cure shrinkage, expensive	Brittle, high-temperature stability, electrical properties excellent, low tensile and impact strength
Polybutadienes	High viscosity, reacted with isocyanates, epoxies, or vinyl monomers, moderate cure shrinkage	Excellent electrical properties and low water absorption, lower strength than other materials
Polysulfides	Disagreeable odor, no cure exotherm, high viscosity	Good flexibility and adhesion, excellent resistance to solvents and oxidation, poor physical properties
Depolymerized rubber	High viscosity, low cure shrinkage, low cost, variable cure times and temperature, one-part material available	Low strength, flexible, low vapor transmission, good electrical properties
Allylic resins	High viscosity for low-cure-shrinkage materials, high cost	Excellent electrical properties, resistant to water and chemicals

SOURCE: From Ref. 10, p. 4.6.

TABLE 2.5 Characteristics of Liquid Resins for Nonelectrical Applications

Resin	Handling characteristics	Final part properties	Typical applications
Cast acrylics	Low to high viscosity, long cure times, bubbles a problem, special equipment necessary for large parts	Optical clarity, excellent weathering, resistance to chemicals and solvents	In glazing, furniture, embedments, impregnation
Cast nylon	Complex casting procedure, very large parts possible	Strong, abrasion-resistant, wear-resistant, resistance to chemicals and solvents, good lubricity	Gears, bushings, wear plates, stock shapes, bearings
Phenolics	Acid catalyst used, water given off in cure	High density unfilled, brittle, high temperature, brilliance	Billiard balls, beads
Vinyl plastisol	Inexpensive, range of viscosities, fast set, cure in place, one-part system, good shelf life	Same as molded vinyls, flame-resistant, range of hardness	Sealing gaskets, hollow toys, foamed carpet backing
Epoxies	Cures under adverse conditions, over a wide temperature range, compatible with many modifiers, higher filler loadings	High strength, good wear, chemical and abrasion resistance, excellent adhesion	Tooling, fixtures, road and bridge repairs, chemical-resistant coating, laminates, and adhesives
Polyesters	Inexpensive, low viscosity, good pot life, fast cures, high exotherm, high cure shrinkage, some cure inhibition possible, wets fibers easily	Moderate strength, range of flexibilities, water-white available, good chemical resistance, easily made fire retardant	Art objects, laminates for boats, chemical piping, tanks, aircraft, and building panels

TABLE 2.5 Characteristics of Liquid Resins for Nonelectrical Applications (continued)

Polyurethanes	Free isocyanate is toxic, must be kept water-free, low cure shrinkage, cast hot	Wide range of hardnesses, excellent wear, tear and, chemical resistance, very strong	Press pads, truck wheels, impellers, shoe heels and soles
Flexible soft silicone	High-strength materials, easily inhibited, low cure shrinkage, adjustable cure time, no exotherm	Good release properties, flexible and useful over wide temperature range, resistant to many chemicals, good tear resistance	Casting molds for plastics and metals, high-temperature seals
Allylic resins	Long pot life, low vapor pressure monomers, high cure temperature, low viscosity for monomers	Excellent clarity, abrasion resistant,- color stability, resistant to solvents and acids	Safety lenses, face shields, casting impregnation, as monomer in polyester
Depolymerized rubber	High viscosity, low cost, adjustable cure time	Low strength, low vapor transmission, resistant to reversion in high humidity	Roofing coating, sealant, in reservoir liners
Polysulfide	High viscosity, characteristic odor, low cure exotherm	Good flexibility and adhesion, excellent resistance to solvents and oxidation, poor physical properties	Sealants, leather impregnation
Epoxy vinyl esters	Low viscosity, high cure shrinkage, wets fibers easily, fast cures	Excellent corrosion resistance, high impact resistance, excellent electrical insulation properties	Absorption towers, process vessels, storage tanks, piping, hood scrubbers, ducts and exhaust stacks
Cyanate esters	Low viscosity at room temperature or heated, rapid fiber wetting, high temperature cure	High operating temperature, good water resistance, excellent adhesive properties and low dielectric constant	Structural fiber-reinforced products, high temperature film adhesive, pultrusion and filament winding

SOURCE: From Ref. 10, p. 4.7.

TABLE 2.6 Properties of Epi-Bis Resin with Various Hardeners

Hardener	phr*	Tensile strength, lb/in² at 25°C	Tensile modulus, lb/in² × 10^{-6}	Tensile elongation, %	Dielectric constant			Dielectric strength S/T, V/mil	Dissipation factor, 60 Hz at				Volume resistivity, MΩ-cm	Arc resistance, s (ASTM D 495)
					25°C	60 Hz at 50°C	100°C		25°C	50°C	100°C	150°C		
Aliphatic amines:														
Diethyl amino propylamine	7	8,500												58
Aminoethyl piperazine	20	10,000		5–6										62
Aromatic amines:														
Mixture of aromatic amines	23	12,600		6–5										80
p,p'-Methylene dianiline	27	9,500	0.4	4–5	4.4		4.6	420	0.007		0.002		$>1 \times 10^{10}$	83
m-Phenylene diamine	14.5	13,000	0.5	4	4.4		4.7	483	0.007		0.003		$>1 \times 10^{10}$	78
Diaminodiphenyl sulfone	30	7,000	0.47		4.22		4.71	410	0.004		0.068		$>2 \times 10^{8}$	65
Anhydrides:														
Methyl tetrahydrophthalic anhydride	84	12,200	0.44	6–4	3.0	3–0	3.0	377	0.006	0.005	0.005	0.09	$>2 \times 10^{9}$	110
Phthalic anhydride	78	8,500	0.5		3.5		3.8	390	0.004		0.006		$>5 \times 10^{9}$	95
"Nadic" methyl anhydride	80	11,400	0.40	5–6	3.3	3–4	3.4	443	0.003	0.002	0.004		$>4.7 \times 10^{10}$	110
BF_3MEA	3	5,900	0.4	1–8	3.6		4.1	480	0.004		0.049		13×10^{9}	120

*Parts hardener per hundred parts resin.

SOURCE: "Bahelite Liquid Epoxy Resins and Hardeners," Technical Bulletin, Union Carbide Corporation.

TABLE 2.7 Cast-Resin Data on Blends of Cycloaliphatic Epoxy Resins

Resin ERL-4221,* parts	100	75	50	25	0
Resin ERR-4090,* parts	0	25	50	75	100
Hardener, hexa-hydrophthalic (HHPA), phr†	100	83	65	50	34
Catalyst (BDMA), phr†	1	1	1	1	1
Cure, h/°C	2/120	2/120	2/120	2/120	2/120
Postcure, h/°C	4/160	4/160	4/160	4/160	4/160
Pot life, h/°C	>8/25	>8/25	>8/25	>8/25	>8/25
HDT (ASTM D 648), °C	190	155	100	30	−25
Flexural strength (D 790), lb/in^2	14,000	17,000	13,500	5,000	Too soft
Compressive strength (D 695) lb/in^2	20,000	23,000	20,900	Too soft	Too soft
Compressive yield (D 695), lb/in^2	18,800	17,000	12,300	Too soft	Too soft
Tensile strength (D 638), lb/in^2	8,000–10,000	10,500	8,000	4,000	500
Tensile elongation (D 638), %	2	6	27	70	115
Dielectric constant (D 150), 60 Hz:					
25°C	2.8	2.7	2.9	3.7	5.6
50°C	3.0	3.1	3.4	4.5	6.0
100°C	2.7	2.8	3.2	4.9	Too high
150°C	2.4	2.6	3.3	4.6	Too high
Dissipation factor:					
25°C	0.008	0.009	0.010	0.020	0.090
100°C	0.007	0.008	0.030	0.30	Too high
150°C	0.003	0.010	0.080	0.80	
Volume resistivity (D 257), Ω-cm	1×10^{13}	1×10^{12}	1×10^{11}	1×10^{8}	1×10^{6}
Arc resistance (D 495), s	>150‡	>150	>150	>150	>150

*Union Carbide Corp.
†Parts per 100 resin.
‡Systems started to burn at 120 s. All tests stopped at 150 s.
SOURCE: "Bakelite Cycloaliphatic Epoxides," Technical Bulletin, Union Carbide Corporation.

late, triallyl cyanurate, divinyl benzene, and chlorostyrene, can be blended to achieve specific property enhancements. The reactivity of the polyester used, as well as the configuration of the product, affect the choice of systems.

Flexible polyester resins are available that are tougher and slower curing and produce lower exotherms and less cure shrinkage. They absorb more water and are more easily scratched but show more abrasion resistance than the rigid type. Their property profile is shown in Table 2.11. The two types, rigid and flexible, can be blended to produce intermediate properties, as shown in Table 2.12.

2.5.3.3 Cast polyurethanes. *Polyurethanes* are reaction products of an isocyanate, a polyol, and a curing agent. Because of the hazards involved in handling free isocyanate, prepolymers of the isocyanate and the polyol are generally used.

TABLE 2.8 Heat-Deflection Temperatures* of Blends of Novolac Epoxy and Epi-Bis Resins

Hardener	D. E. N. 438†	75/25	50/50	25/75	D. E. R. 332‡
TETA	§	§	133	126	127
MPDA	202	192	180		165
MDA	205	193	190	186	168
5% BF_3MEA	235			204	160
HET	225	213	205	203	196

*Heat-distortion temperature, °C (stoichiometric amount of curing agent—except BF_3 MEA—cured 15 h at 180°C).
†Novolac resin, Dow Chemical Co.
‡Epi-bis resin, Dow Chemical Co.
§The mixture reacts too quickly to permit proper mixing by hand.
SOURCE: "Dow Epoxy Novolac Resins," Technical Bulletin, Dow Chemical Company.

TABLE 2.9 Comparison of Chemical Resistance for Cured Epoxy Novolac and Epi-Bis Resin

	Weight gain,* %	
Chemical	D.E.N 438†	D.E.R. 331‡
Acetone	1.9	12.4
Ethyl alcohol	1.0	1.5
Ethylene dichloride	2.6	6.5
Distilled water	1.6	1.5
Glacial acetic acid	0.3	1.0
30% sulfuric acid	1.9	2.1
3% sulfuric acid	1.6	1.2
10% sodium hydroxide	1.4	1.2
1% sodium hydroxide	1.6	1.3
10% ammonium hydroxide	1.1	1.3

[1]One-year immersion at 25°C; cured with methylene dianiline; gelled 16 h at 25°C, postcured $4^{1}/_{2}$ h at 166°C; D.E.N. 438 cured additional $3^{1}/_{2}$ h at 204°C.
*Sample size $^{1}/_{2}$ by $^{1}/_{2}$ by 1 in.
†Novolac epoxy resin, Dow Chemical Co.
‡Epi-bis resin, Dow Chemical Co.
SOURCE: "Dow Epoxy Novolac Resins," Technical Bulletin, Dow Chemical Company.

The choice of curing agent influences the curing characteristics and final properties. Diamines are the best general-purpose curing agent, as shown by Table 2.13. The highest physical properties are produced using MOCA 4,4-methyl-bis (2chloroaniline). The other major class of curing agents, the polyols, are more convenient to use, but the final products have lower physical properties. By providing good abrasion resistance and a low coefficient of friction, polyurethanes find application in roller coatings and press pads as well as gaskets, casting molds, timing belts, wear strips, liners, and heels and soles.

2.5.3.4 Cast phenolics. *Phenolic casting resins* are available as syrupy liquids produced in huge kettles by the condensation of formaldehyde and phenol at high temperature in the presence of a catalyst and the removal of excess moisture by vacuum distillation. These resins, when blended with a chemically active hardener, can be cast and cured solid

TABLE 2.10 Properties of Typical Polyester Producing Rigid Castings

Product specifications at 25°C	
Flash point, Seta closed cup, °F	89
Shelf life, minimum, months	3
Specific gravity	1.10–1.20
Weight per gallon, lb	9.15–10.0
% styrene monomer	31–35
Viscosity, Brookfield model LVF, #3 spindle at 60 r/min, cP	650–850
Gel time	
150–190°F, min	4–7
190°F to peak exotherm, min	1–3
Peak exotherm, °F	385–425
Color	Amber clear
Typical physical properties (clear casting)	
Barcol hardness (ASTM D 2583)	47
Heat-deflection temperature, °C (°F) (ASTM D 648)	87 (189)
Tensile strength, lb/in^2 (ASTM D 638)	8000
Tensile modulus, 10^5 lb/in^2 (ASTM D 638)	5.12
Flexural strength, lb/in^2 (ASTM D 790)	13,500
Flexural modulus, 10^5 lb/in^2 (ASTM D 790)	6.0
Compressive strength (ASTM D 695)	22,000
Tensile elongation, % at break (ASTM D 638)	1.5
Dielectric constant (ASTM D 150)	
At 60 Hz	2.97
At 1 MHz	2.87
Power factor	
At 1 kHz	0.005
At 1 MHz	0.017
Loss tangent at 1 MHz	0.017

SOURCE: From Ref. 10, p. 4.24.

TABLE 2.11 Properties of Typical Polyester Producing Flexible Material

Product specifications at 25°C	
Flash point, Seta closed cup, °F	89
Shelf life, minimum, months	3
Specific gravity	1.13–1.25
Weight per gallon, lb	9.4–10.4
% styrene monomer	18–22
Viscosity, Brookfield model LVF, #3 spindle at 60 r/min, cP	1100–1400
Gel time	
SPI 150–190 °F, min	6–8
190°F to peak exotherm, min	4.5–6.0
Peak temperature, °C (°F)	102–116 (215–240)
Color	Amber clear
Typical physical properties (clear casting)	
Tensile strength, lb/in^2 (ASTM D 638)	50
Tensile elongation, % (ASTM D 638)	10
Flexural strength, lb/in^2 (ASTM D 790)	Yields
Hardness, Shore D (ASTM D 2240)	15

SOURCE: From Ref. 10, p. 4.25.

TABLE 2.12 Properties of Blend of Rigid and Flexible Polyesters

Flexible polyester*	30%	20%	10%	5%	—
Rigid polyester†	70%	80%	90%	95%	100%
Tensile strength, lb/in^2 (ASTM D 638)	5200	8100	7300	6800	6500
Tensile elongation, % (ASTM D 638)	10.0	4.8	1.7	1.3	1.0
Flexural strength, lb/in^2 (ASTM D 790)	8100	13,200	15,600	14,600	13,500
Flexural modulus, 10^5 lb/in^2 (ASTM D 790)	2.40	3.75	5.60	5.80	6.00
Barcol hardness (ASTM D 2583)	0–5	20–25	30–35	35–40	40–45
Heat-deflection temperature, °C (°F) (ASTM D 648)	51 (124)	57.5 (136)	63 (145)	85 (185)	88 (190)

*Polylite 31-820, Reichhold Chemicals, Inc.
†Polylite 31-000, Reichhold Chemicals, Inc.
SOURCE: From Ref. 10, p. 4.25.

TABLE 2.13 Properties of Polyurethane Resins Cured with Diamines and Polyols

Property	Diamine cures	Polyol cures
Resilience	Medium	Medium to high
Reactivity	Medium to high	Low to medium
Modulus	Medium to high	Low
Tensile strength	Medium to high	Low to medium
Ultimate elongation	Medium to high	High
Tear strength	High	Low to medium
Abrasion resistance:		
Sliding	High to very high	Low to medium
Impact	Medium to high	High to very high
Compression set	Medium	Low
Hardness	High	Low to medium

SOURCE: "Adiprene," Technical Bulletin, Elastomers Chemicals Dept., du Pont de Nemours & Company.

in molds constructed from various materials and of a variety of mold designs. They will exhibit a broad-based property profile as described in Table 2.14. The mold designs available are draw molds, split molds, cored molds, flexible molds, and plaster molds.

2.5.3.5 Cast allylics.[11]

The *allylic ester resins* possess excellent clarity, hardness, and color stability and thus can be cast to form optical parts. These castings can be either homo or co-polymers. The free radical addition polymerization of the allylic ester presents some casting difficulties such as exotherm control, monomer shrinkage during curing and the interaction between the exotherm, the free radical source, and the environmental heat required to decompose the peroxide and initiate the reaction.

A simple casting formulation is as follows:

Prepolymer:	60 parts/wt
Monomer:	40 parts/wt

TABLE 2.14 Properties of Cured Cast-Phenolic Resins

Property	Test method	Test values
Inherent properties:		
Specific gravity	D 792	1.30–1.32
Specific volume, in^3/lb	D 792	21.3–20.9
Refractive index, n_D	D 542	1.58–1.66
Cured properties:		
Tensile strength, lb/in^2	D 638–D 651	6,000–9,000
Elongation, %	D 638	1.5–2.0
Modulus of elasticity in tension, 10^5 lb/in^2	D 638	4–5
Compressive strength, lb/in^2	D 695	12,000–15,000
Flexural strength, lb/in^2	D 790	11,000–17,000
Impact strength, ft-lb/in notch ($^1/_2 \times {}^1/_2$-in notched bar, Izod test)	D 256	0.25–0.40
Hardness, Rockwell	D 785	M93–M120
Thermal conductivity, 10^4 cal/(s)(cm^2)(°C)(cm)	C 177	3–5
Specific heat, cal/(°C)(g)		0.3–0.4
Thermal expansion, 10^{-5} per °C	D 696	6–8
Resistance to heat (continuous), °F		160
Heat-distortion temperature, °F	D 648	165–175
Volume resistivity (50% RH and 23°C), Ω-cm	D 257	$10^{12}-10^{13}$
Dielectric strength ($^1/_8$-in thickness), V/mil:		
Short time	D 149	350–400
Step-by-step	D 149	250–300
Dielectric constant:		
60 Hz	D 150	5.6–7.5
10^5 Hz	D 150	5.5–6.0
10^6 Hz	D 150	4.0–5.5
Dissipation (power) factor:		
60 Hz	D 150	0.10–0.15
10^3 Hz	D 150	0.01–0.05
10^6 Hz	D 150	0.04–0.05
Arc resistance, s	D 495	200–250
Water absorption (24 h, $^1/_8$-in thickness), %	D 570	0.3–0.4
Burning rate	D 635	Very low
Effect of sunlight		Colors may fade
Effect of weak acids	D 543	None to slight, depending on acid
Effect of strong acids	D 543	Decomposed by oxidizing acids; reducing and organic acids, none to slight effect
Effect of weak alkalies	D 543	Slight to marked, depending on alkalinity
Effect of strong alkalies	D 543	Decomposes
Effect of organic solvents	D 543	Attacked by some
Machining qualities		Excellent
Clarity		Transparent, translucent, opaque

SOURCE: "Plastics Property Chart," *Modern Plastics,* MPE Supplement, 1970–1971.

Tert-butyl perbenzoate:	2 parts/wt
Tert-butylcatechol:	0.1 parts/wt

The function of the catechol is to retard the polymerization and the exothermic heat over a longer time period, allowing the heat to dissipate and minimize cracking. Monomer catalyzed with benzoyl peroxide or tert-butyl perbenzoate may be stored at room temperature from 2 weeks to 1 year, but, at 120°F (49°C), the catalyzed resin will gel in a few hours.

Molds for casting allylics may be ground and polished metal or glass, with glass being the preferred type, since it is scratch resistant and able to take a high polish.

Cast allylics are noted for their hardness, heat resistance, electrical properties, and chemical resistance as shown in Tables 2.15 and 2.16. They do lack strength, so their usage is confined to optical parts and some small electrical insulators.

Allylic monomers are sometimes used with alkyds to produce polyesters, with the orthophthalate resin being the most widely used because of its lower cost and very low water vapor pressures. Alkyd-diallylphthalate copolymers have significantly lower exotherm than an alkyd-styrene copolymer. The electrical properties of allylic resins are excellent, and the variations of dissipation factor, dielectric constant, and dielectric strength with temperature and frequency are given in Figs. 2.12 and 2.13. The surface and volume resistivities remain high after prolonged exposure to high humidity. Resistance to solvents and acids is excellent, along with good resistance to alkalies. Tables 2.17 and 2.18 compare the chemical resistance of some plastics with the chemical resistance of the DAIP formulations.

2.6 Laminates[12]

2.6.1 Laminates

Laminates can be defined as combinations of liquid thermosetting resins with reinforcing materials that are bonded together by the application of heat and pressure, forming an infusible matrix. Plywood is a good example of a thermosetting laminate with the phenolic resin serving as the binder to bond the layers of wood sheets together when compressed with heat in a molding press.

2.6.2 Resins

The resin systems primarily used for laminates are bismaleimides, epoxies, melamines, polyesters, polyimides, silicones, and phenolics and cyanate esters. These resins are described in Sec. 2.7.8, and the reinforcements are described in Sec. 2.6.3.

2.6.3 Reinforcements

Glass fibers are the most commonly used laminate reinforcement and are available in some six formulations, with the E-glass providing excellent moisture resistance, which, in turn, results in superior electrical properties along with other valuable properties. Properties of reinforced plastics using a variety of reinforcing fibers are shown in Table 2.19. Compositions of the major types of glass fibers are shown in Table 2.20. The glass fibers are available in filaments, chopped strands, mats, and fabrics in a wide variety of diameters, as described in Table 2.21.

Glass fabrics: Many different fabrics are made for reinforced plastics, with E-glass being the most common. Filament laminates using glass types of D, G, H, and K are also

TABLE 2.15 Properties of Allyl Ester Monomers

Property	Diallyl phthalate	Diallyl iso-phthalate	Diallyl maleate	Diallyl chloren-date	Diallyl adipate	Diallyl diglycollate	Triallyl cyanurate	Diethylene glycol bis-(allyl carbonate)
Density, at 20°C	1.12	1.12	1.076	1.47	1.025	1.113	1.113	1.143
Molecular wt	246.35	246.35	196	462.76	226.14	214.11	249.26	274.3
Boiling pt, °C, 4 mm Hg	160	181	111		137	135	162	160
Freezing pt, °C	−70	−3	−47	29.5	−33		27	−4
Flash pt, °C	166	340	122	210	150	146	>80	177
Viscosity at 20°C, cP	12	16.9	4.5	4.0	4.12	7.80	12	9
Vapor pressure at 20°C, mm Hg	27							
Surface tension at 20°C, dynes/cm	34.4	35.4	33		32.10	34.37		35
Solubility in gasoline, %	24	100	23		100	4.8	80	
Thermal expansion, in/(in)(°C)	0.00076							

SOURCE: From Ref. 11.

TABLE 2.16 Properties of Two Cured Allyl Esters

Property by ASTM procedures	Diallyl phthalate	Diallyl isophthalate
Dielectric constant:		
25°C and 60 Hz	3.6	3.5
25°C and 10^6 Hz	3.4	3.2
Dissipation factor:		
25°C and 60 Hz	0.010	0.008
25°C and 10^6 Hz	0.011	0.009
Volume resistivity, Ω-cm at 25°C	1.8×10^{16}	3.9×10^{17}
Volume resistivity, Ω-cm at 25°C (wet)*	1.0×10^{14}	
Surface resistivity, Ω at 25°C	9.7×10^{15}	8.4×10^{12}
Surface resistivity, Ω at 25°C (wet)*	4.0×10^{13}	
Dielectric strength, V/mil at 25°C†	450	422
Arc resistance, s	118	123–128
Moisture absorption, %, 24 h at 25°C	0.09	0.1
Tensile strength, lb/in^2	3,000–4,000	4,000–4,500
Specific gravity	1.270	1.264
Heat-distortion temperature, °C at 264 lb/in^2	155 (310°F)	238‡ (460°F)
Heat-distortion temperature, °C at 546 lb/in^2	125 (257°F)	184–211 (364–412°F)
Chemical resistance, % gain in wt		
After 1 month immersion at 25°C in:		
Water	0.9	0.8
Acetone	1.3	−0.03
1% NaOH	0.7	0.7
10% NaOH	0.5	0.6
3% H_2SO_4	0.8	0.7
30% H_2SO_4	0.4	0.4

*Tested in humidity chamber after 30 days at 70°C (158°F) and 100% relative humidity.

†Step by step.

‡No deflection.

SOURCE: Directory/Encyclopedia Issue, *Insulation/Circuits,* June/July 1972.

common with the filaments combined into strands, and the strands plied into yarns. These yarns can be woven into fabrics on looms.

2.6.4 Processes

The liquid resins are poured onto the reinforcement material and the combined resin-reinforcement sheet is then placed in a "horizontal treater," as shown in Fig. 2.12, where the resin is dissolved in a solvent to achieve optimum wetting or saturation of the resin into the reinforcement. The sheets come out of the treater and are sheared to size and stacked

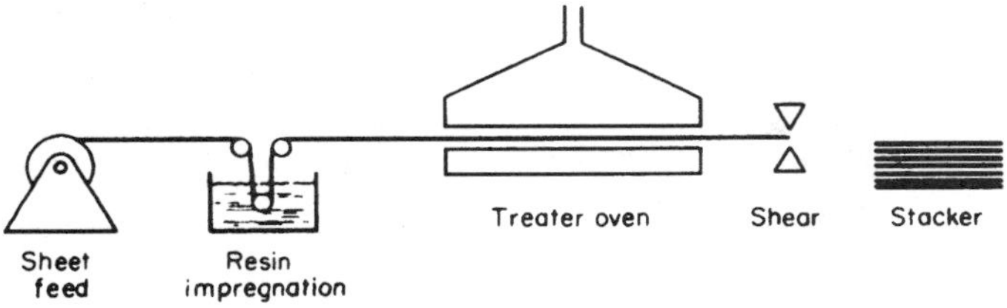

(a)

(b)

Figure 2.12 (a) Horizontal treater and (b) decorative laminate treater. *(Source: Charles A. Harper,* Handbook of Plastics, Elastomers, and Composites, *3d ed., McGraw-Hill, New York, 1996, p. 2.2)*

and stored in temperature- and humidity-controlled rooms. The preimpregnated, stacked sheets are further stacked into packs with each pack containing 10 laminates. The stacked units are then placed in between two polished steel plates, and the press is closed. This process involves molding temperatures ranging from 250 to 400°F (122 to 204°C), with molding pressures in the 200- to 3000-lb/in^2 range.

The press may have 24 platens, making a press load of some 240 laminates each cycle. After molding, the laminates are trimmed to final size and sometimes postcured by heating them in ovens.

2.6.5 Properties

2.6.5.1 Physical and mechanical. The use of reinforcements in combination with thermosetting resins will produce laminates exhibiting higher tensile, compressive, and

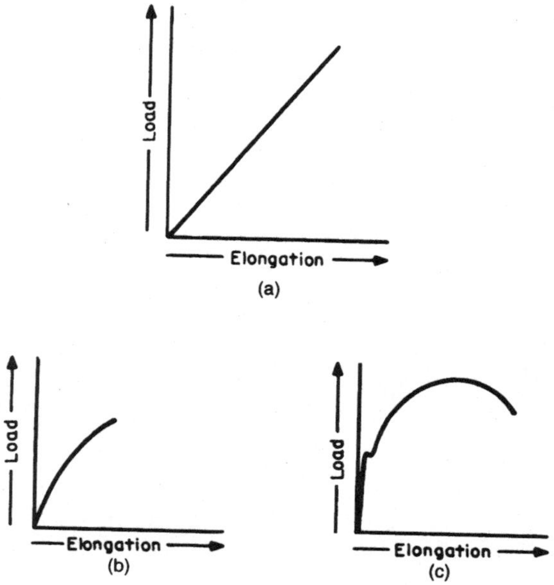

Figure 2.13 Stress-strain curves for various materials. (a) Reinforced plastics, (b) wood and most metals, and (c) steel. *(Source: Charles A. Harper,* Handbook of Plastics, Elastomers, and Composites, *3d ed., McGraw-Hill, New York, 1996, p. 2.3)*

flexural strengths due to the polymerization of the resin with its property enhancement contribution to the matrix. The impact strength sometimes improves by a factor of 10, and the laminates have no clear yield point, as shown in Fig. 2.13 (stress and strain).

Reinforced thermosetting laminates are anisotropic, with their properties differing depending upon the direction of measurement. The properties of a fabric laminate are controlled by the weave of the fabric and the number and density of the threads in the warp and woof directions, with these values differing from the values in the z of thickness direction. Both thermal expansion and conductivity properties are also anisotropic, as displayed in Table 2.22.

2.6.5.2 Electrical. The dielectric strength of laminates will decrease with increasing thickness and is highly dependent upon the direction of the electric field stress. This property will show higher when tested across the sample's thickness, whereas end-to-end testing will show lower values. Laminates with higher resin content will show better electrical properties but poorer physical properties than laminates with lower resin content.

2.6.5.3 Dimensional stability. Laminates using thermosetting resins are superior in stability, thermal resistance, and electrical properties, with dimensional stability the most important due to stresses incurred within the laminate during the molding and curing pro-

TABLE 2.17 Chemical Resistance of Some Plastics (Change in wt on exposure, % after 24 hr)

	Melamine cellulose-filled	Poly-styrene high impact	Styrene polyester unfilled medium unsaturation	Epoxy resin unfilled	Diallyl ortho-phthalate
Acetone at 25°C	−0.75	*	+2.5	*	+0.05
Benzene at 25°C	+0.05	*	−0.02	+0.15	+0.01
Carbon tetrachloride at 25°C	−0.11	*	−0.01	−0.09	+0.01
95% Ethanol at 25°C	+0.013	+0.32	+0.22	−0.03	+0.06
Heptane at 25°C	+0.48	+70.0*(5)	−0.02	+0.21	+0.03
n-Butyl acetate at 70°C	−3.3	*	+2.82	+10.9	−3.3
1% Sodium hydroxide at 70°C	+3.10	+1.00	+1.49	−2.67	+1.07
20% Sodium hydroxide at 70°C	+2.10	+0.16	+0.89	−2.90	+0.43
1% Sulfuric acid at 70°C	+3.50	+1.00	+1.63	+0.10	+0.99
30% Sulfuric acid at 70°C	+4.6	+0.17	+1.12	+0.05	+0.61
1% Nitric acid at 70°C	+3.40	+1.0	+1.85	+0.66	+1.10
10% Nitric acid at 70°C	+8.52	+0.98	+1.73	−1.9	+1.0
3% Citric acid at 70°C	+3.1	+0.9	+1.53	+0.49	+1.1
Chromic acid at 70°C	+2.8	+1.1	+1.52	+0.65	+1.1
Mazola oil at 70°C	−1.89	+0.91	−0.25	−0.08	+0.05

*Disintegrated.
SOURCE: From Ref. 11.

TABLE 2.18 Chemical Resistance of Cast Diallyl Metaphthalate

Material	Time at 25°C	% wt Gain
Water	24 h	0.2
	7 days	0.4
	1 month	0.8
	Reconditioned*	0.6
Acetone	24 h	−0.03
	7 days	−0.03
	1 month	−0.03
	Reconditioned*	−0.07
1% NaOH	24 h	0.0
	7 days	0.3
	1 month	0.7
	Reconditioned*	0.6
10% NaOH	24 h	0.1
	7 days	0.3
	1 month	0.6
	Reconditioned*	0.5
3% H_2SO_4	24 h	0.1
	7 days	0.3
	1 month	0.7
	Reconditioned*	0.5
30% H_2SO_4	24 h	0.1
	7 days	0.2
	1 month	0.4
	Reconditioned*	0.4

*Followed after the immersion period by 2 h drying at 70°C.
SOURCE: From Ref. 11.

cess. Laminate thickness is limited by the thermal conductivity of the polymerizing resin and the thermal conductivity of the cured laminate. The laminate derives its heat from the press platens to begin the curing action, and the resin, as it polymerizes, will produce substantial reactive heat, making it difficult to produce very thick laminates. The normal thickness range for laminates will range from 0.002 to about 2.000 in with special grades produced at 4 to 10 in.

2.7 Molding Compounds

2.7.1 Resin Systems

Thermosetting resin systems are the backbone of a large, versatile, and important family of molding compounds. This family provides the industrial, military, and commercial markets with plastic molding materials exhibiting exceptional electrical, mechanical, thermal, and chemical properties. These important property values enable the product designer, manufacturing engineer, and research and design engineer to select from a wide choice of products, and enable them to choose the most suitable molding compound to meet their specific needs and requirements. These versatile materials are covered by military, industrial, and commercial specifications that are designed to ensure the quality of the molded articles utilizing thermosetting molding compounds.

TABLE 2.19 Properties of Reinforced Plastic and of Reinforcing Fiber

	Laminate properties									
Reinforcing fiber	Mechanical strength	Electrical properties	Impact resistance	Chemical resistance	Machining and punching	Heat resistance	Moisture resistance	Abrasion resistance	Low cost	Stiffness
Glass strands	X		X	X		X	X		X	X
Glass fabric	X	X	X	X		X	X			X
Glass mat			X	X		X	X		X	X
Asbestos		X	X			X				
Paper		X			X				X	
Cotton/linen	X	X	X		X				X	
Nylon		X	X	X				X		
Short inorganic fibers	X		X						X	
Organic fibers	X	X	X	X	X			X		
Ribbons		X					X			
Metals	X		X				X			X
Polyethylene	X		X		X					X
Aramid	X	X	X			X		X		X
Boron	X		X			X				X
Carbon/graphite	X		X			X				X
Ceramic	X		X						X	

SOURCE: From Ref. 12, p. 2.5.

TABLE 2.20 Composition of Glass Fibers

Ingredient	Glass type*					
	A	C	D	E	R	S and S2
SiO_2	72	65	74	52–56	60	65
CaO	10	14	0.5	16–25	9	—
Al_2O_3	0.6	4	0.3	12–16	25	25
MgO	2.5	3	—	0–5	6	10
B_2O_3	—	6	22	5–10	—	—
TiO_2	—	—	—	0–1.5	—	—
Na_2O	14.2	8	1.0	0–2	—	—
K_2O	—	—	1.5	—	—	—
Fe_2O_3	—	0.2	—	0–0.8	—	—
So_3	0.7	0.1	—	—	—	—
F_2	—	—	—	0–1.0	—	—

*Values represent weight percent of total.
SOURCE: From Ref. 12, p. 2.6.

The following sections will take each of the family members, in alphabetical order, and outline their history; specifications; and reinforcements, fillers, and data sheet value applications.

2.7.2 Allyls: Diallyl-ortho-phthalate (DAP) and Diallyl-iso-phthalate (DAIP)

2.7.2.1 History.[13] The most sophisticated work on saturated polyesters is usually traced to W. H. Carothers, who, from 1928 to 1935 with DuPont, studied polyhydroxy condensates of carbolic acids. Unable to achieve suitable heat and chemical resistance from these esters, he turned to polyamide-carboxylic acid reaction products, which later became *nylon*. Carothers' concepts on saturated polyesters yielded several other polyester products such as Terylene, which became patented and introduced in the United States as *dacron* and *mylar*.

In 1937, Carlton Ellis found that unsaturated polyesters, which are condensation products of unsaturated dicarboxylic acids and dihydroxy alcohols, would freely co-polymerize with monomers that contained double bond unsaturation, yielding rigid thermosetting resins. The allylic resins became commercial resins suitable for compounding into a broad range of products with exceptional electrical, mechanical, thermal, and chemical properties.

2.7.2.2 Physical forms. Molding compounds using either the DAP or the DAIP resin systems are available in free-flowing granular form and also in high-bulk factor flake form. The resins are in a white powder form, which makes it possible to provide a broad opaque color range. Both compounds in their granular form are readily molded, preformed, or preplasticated automatically, whereas the high-bulk-factor compounds generally require auxiliary equipment for such operations.

2.7.2.3 Reinforcements. Compounds of either DAP or DAIP resin systems utilize a variety of reinforcement materials ranging from the granular compounds with mineral,

TABLE 2.21 Glass Fiber Reinforcement Forms

Form of fiberglass reinforcement	Definition and description	Range of grades available	General types of sizing applied	General usage in RP/C (and secondary uses if any)
Twisted yarns	Single-end fiberglass strands twisted on standard textile tube-drive machinery.	B to K fiber, S or Z twist, 0.25–10.0 twists/in, many fiber and yardage variations.	Starch.	Into single and plied yarns for weaving; many other industrial uses (and decorative uses).
Plied yarns	Twisted yarns plied with reverse twist on standard textile ply frames.	B to K fiber, up to $^4/_{18}$ ply—many fiber and yardage variations.	Starch.	Weaving industrial fabrics and tapes in many different cloth styles; also heavy cordage.
Fabrics	Yarns woven into a multiplicity of cloth styles with various thicknesses and strength orientations.	D to K fibers, 2.5–40 oz/yd^2 in weight.	Starch size removed and compatible finish; applied after weaving	Wet lay-up for open molding, prepreg, high pressure lamination, and also some press molding.
Chopped strands	Filament bundles (strands) bonded by sizings, subsequently cured and cut or chopped into short lengths. Also the reverse, i.e., chopped and cured.	G to M fiber, $^1/_8$- to $^1/_2$-in or longer lengths, various yardages.	LSB and RTP.	Compounding, compression, transfer and injection press molding.
Roving	Gathered bundle of one or more continuous strands wound in parallel and in an untwisted manner into a cylindrical package.	G or T fiber used. Roving yields 1800 to 28 yd/lb and packages 15- to 450-lb weight (size of up to 24 × 24 in).	HSB, LSB, and RTP.	Used in all phases of RP/C. Some (HSB) are used in continuous form, e.g., filament winding, and others (LSB) are chopped, as in sheet molding compound for compression molding.
Woven roving	Coarse fabric, bidirectional reinforcement, mostly plain weave, but some twill. Uni- and multidirectional nonwoven rovings are also produced.	K to T fiber, fabric weights 10–48 oz/yd^2.	HSB.	Mostly wet lay-up, but some press molding.

Chopped strand mats	Strands from forming packages chopped and collected in a random pattern with additional binder applied and cured; some "needled" mat produced with no extra binder required.	G to K fiber, weights 0.75–6.0 oz/ft^2.	HSB and LSB.	Both wet lay-up and press molding.
Mats, continuous strand (swirl)	Strands converted directly into mat form without cutting with additional binder applied and cured, or needled.	Nominally M to R fiber diameter, weights 0.75–4.5 oz/ft^2.	LSB.	Compression and press molding, resin transfer molding, and pultrusion.
Mat-woven roving combinations	Chopped strand mat and woven roving combined into a drapable reinforcement by addition of binder or by stitching.	30–62 oz/yd^2.	HSB.	Wet lay-up to save time in handling.
Three- dimensional reinforcements	Woven, knitted, stitched, or braided strands or yarns in bulky, continuous shapes.	—	HSB.	Molding, pultrusion.
Milled fibers	Fibers reduced by mechanical attrition to short lengths in powder or nodule form.	Screened from $^1/_{32}$–$^1/_4$ in. Actual lengths range 0.001–$^1/_4$ in. Several grades.	None, HSB, and RTP.	Casting, potting, injection molding, reinforced reaction injection molding (RRIM).
Related forms: Glass beads	Small solid or hollow spheres of glass.	Range 1–53 μm in diameter, bulk densities; hollow = 0.15–0.38 g/cm^3, solid = 1.55 g/cm^3.	Usually treated with cross-linking additives.	Used as filler, flow aid, or weight reduction medium in casting, lamination, and press molding.
Glass flake	Thin glass platelets of controlled thickness and size.	0.0001-in and up thick.	None, or treated with doupling agent.	Used as a barrier in, or to enhance abrasion resistance of, linear resins; coatings used for corrosion-resistance applications. Also used in RRIM for increased dimensional stability.

SOURCE: C. V. Lundberg, "A Guide to Potting and Encapsulation Materials," *Material Engineering*, May 1960.

TABLE 2.22 Thermal Coefficient of Expansion of Epoxy Glass Fabric Laminate and of Kevlar Fiber

Material	Direction	Coefficient of thermal expansion, 10^{-6}/°C
Epoxy	*x*	16.0
	y	12.7
	z	200
Kevlar fiber	*x*	−2
	z	60

SOURCE: From Ref. 12, p. 2.4.

glass, and synthetic fibers to the high-strength, high-bulk factor types employing long fibers of cotton flock, glass, acrylic, and polyethylene terephthalate.

2.7.2.4 Specifications. DAP and DAIP molding compounds meet the requirements of several commercial and military specifications as well as those of certain industrial and electronic firms who will often use the Mil spec in conjunction with their own special requirements.

Military

Mil-M-14 DAP (ortho) compounds:

MDG: Mineral filler, general purpose

SDG: Short glass filler, general purpose

SDG-F: Short glass filler, flame resistant

SDI-5: Orlon filler, impact value of 0.5 ft-lb

SDI-:30: Dacron filler, impact value of 3.0 ft-lb

GDI-30: Long glass filler, impact value of 3.0 ft-lb

GDI-30F: Long glass filler, impact value of 3.0 ft-lb, flame resistant

GDI-300: Long glass filler, impact value of 30 ft-lb

GDI-300F: Long glass filler, impact value of 30 ft-lb, flame resistant

DAIP (iso) compounds. Mil-M-14 H lists designations for the (iso) compounds and refers to them as heat-resistant compounds:

MIG: Mineral filler, heat resistant

MIG-F: Mineral filler, heat and flame resistant

SIG: Short glass filler, heat resistant

SIG-F: Short glass filler, heat and flame resistant

GII-30: Long glass filler, heat resistant, impact value of 3.0 ft-lb

GII-30-F: Long glass filler, heat and flame resistant, impact value of 30 ft-lb

Note: All of these DAIP compounds will perform without substantial property loss during exposure to elevated temperatures of 490°F (200°C).

ASTM. Under D-1636-75 A, both DAP and DAIP are listed by type, class and grade as shown below:

Type	Class	Grade
I	A Ortho (DAP)	1. Long glass
	B Ortho FR (DAP)	2. Medium glass
	C Iso (DAIP)	3. Short glass
II	A Ortho (DAP)	1. Mineral
	B Ortho FR (DAP)	2. Mineral and organic fiber
	C Iso (DAIP)	1. Mineral
		2. Mineral and organic fiber
III	A Ortho (DAP)	1. Acrylic fiber
		2. Polyester fiber, long
		3. Polyester fiber, milled

Underwriters Laboratory (UL). UL rates DAP and DAIP thermally at 266°F (130°C), with a flame resistance rating of 94 VO in a 1/16-in section. Neither the DAP nor DAIP compounds have been submitted to UL's long-term thermal testing. However, MIL-M-14 H does recognize their excellent long-term heat resistance and ability to retain their initial property profile under adverse thermal conditions over lengthy time spans.

Data sheet values. Table 2.23 includes a typical property value sheet for a wide range of allyl molding compounds, and Table 2.24 summarizes the allyl molding compound properties.

Applications. The DAP and DAIP molding compounds are used primarily in applications requiring superior electrical and electronic insulating properties with high reliability under the most severe environmental conditions. A few such applications are military and commercial connectors, potentiometer housings, insulators, switches, circuit boards and breakers, x-ray tube holders, and TV components.

Molded articles of either the DAP or the DAIP compounds have a nil lifetime shrinkage after their cooling-off shrinkage. This feature has been one of the main reasons for the use of DAP and DAIP compounds in the military connector field. Parts molded today will fit parts molded years ago.

Suppliers

Rogers Corporation

Cosmic Plastics, Inc.

(See App. C for supplier addresses.)

2.7.3 Aminos (Urea Melamine)

2.7.3.1 History[14]

Urea. Urea formaldehyde resins came into being through work done by Fritz Pollack and Kurt Hitter along with Carlton Ellis in the 1920s. This was followed up by the introduction of a white urea compound by Dr. A. M. Howald, named *Plaskon,* which was sold by Toledo Synthetic Products Company, which, in turn, became a part of Allied Chemical Company.

TABLE 2.23 Allyl Molding Compound Data Sheet Values

		Grade and Mil-M-14 type						
Property*	ASTM test method	73-70-70 C SDG-F	73-70-70 R SDG-F	FS-10 VO SDG-F	52-20-30 GDI-30	FS-4 GDI-30	52-40-40 GDI-30F	FS-80 GDI-30F
Reinforcement	—	Short glass	Short glass	Short glass	Long glass	Long glass	Long glass	Long glass
Resin-isomer (DAP)	—	Ortho	Ortho	Iso	Ortho	Iso	Ortho	Iso
Form	—	Granular	Granular	Granular	Flake	Flake	Flake	Flake
Bulk factor	—	2.4	2.3	2.3	6	6	6	6
Specific gravity, g/cm^3	D-792A	1.91	1.85	1.91	1.72	1.64	1.79	1.74
Shrinkage (comp), in/in	D-955	0.002–0.004	0.002–0.004	0.002–0.004	0.0025	0.0025	0.0025	0.0025
Izod impact, ft/lb/in	D-256A	0.6	0.6	0.6	4.0	3.6	3.9	3.4
Flexural strength, lb/in^2	D-790	13,000	14,000	13,000	15,000	15,000	15,000	15,000
Flexural modulus, lb/in^2 × 10^6	D-790	2.0	1.9	2.0	1.3	1.4	1.3	1.4
Tensile strength, lb/in^2	D-638	7,000	6,500	8,000	—	—	—	—
Compressive strength, lb/in^2	D-695	—	—	—	—	—	—	—
Water absorption, %, 24 h at 100°C and 48 h at 50°C in H_2O	D-570	0.25	0.30	0.35	0.35	0.35	0.32	0.35
CTE, in/in °C × 10^{-5}	D-696	—	—	—	—	—	—	—
Deflection temperature, °F	D-648A	500	450	500	500	500	500	500
Flammability (ign./burn), s	Fed. 2023	90A 90	90A 90	90A 90	—	—	90A 90	90A 90
UL flammability rating								
$^1/_8$ in	UL 94	94V-O	94V-O	94V-O	—	—	—	—
$^1/_{16}$ in	UL 94	94V-O	94V-O	94V-O	—	—	—	—
Comparative tracking index, s	—	600+	600+	600+	600+	600+	600+	600+
Dielectric strength 60 Hz ST/SS wet, V/mil	D-149	350	350	350	350	350	350	340
Dielectric constant, 1 kHz/1 MHz wet	D-150	4.4	4.2	4.6	4.2	4.0	4.3	4.1
Dissipation factor, 1 kHz/1 MHz wet	D-150	0.014	0.015	0.016	0.016	0.014	0.016	0.015
Arc resistance, s	D-495	130	125	175	130	138	133	145

TABLE 2.23 Allyl Molding Compound Data Sheet Values *(Continued)*

Property*	ASTM test method	Grade and Mil-M-14 type: RX 2-520 SDG	RX 1-520 RX 1310 SDG	RX 3-2-520F SDG-F	RX 3-1-525 RX 1366FR SDG-F	RX-1-530 GDI-30	RX-2-530 GDI-30	RX 3-1-530 GDI-30F	RX-2-530 GDI-30F
Reinforcement	—	Short glass	Short glass	Short glass	Short glass	Long glass	Long glass	Long glass	Long glass
Resin-isomer (DAP)	—	Ortho	Ortho	Iso	Ortho	Ortho	Iso	Ortho	—
Form	—	Granular	Granular	Granular	Granular	Flake	Flake	Flake	Flake
Bulk factor	—	2.5	2.4	2.4	2.3	3.5	3.5	3.5	—
Specific gravity, g/cm^3	D-792A	1.75	1.83	1.90	1.87	1.74	1.73	1.76	1.00
Shrinkage (comp), in/in	D-955	0.001–0.003	0.001–0.003	0.001–0.003	0.001–0.003	0.001–0.004	0.001–0.004	0.001–0.004	0.000–0.000
Izod impact, ft/lb/in	D-256A	0.80	0.70	0.80	0.80	3.0–7.0	3.0–7.0	3.0–7.0	3.0–0.0
Flexural strength, lb/in^2	D-790	22,000	22,000	23,000	23,000	23,000	23,000	23,000	23,000
Flexural modulus, lb/in^2 × 10^6	D-790	2.6	2.6	2.6	2.6	1.1	2.0	2.0	0.0
Tensile strength, lb/in^2	D-638	12,000	11,000	12,000	12,000	10,000	10,000	10,000	10,000
Compressive strength, lb/in^2	D-695	25,000	25,000	25,000	25,000	27,000	27,000	27,000	27,000
Water absorption, %, 24 h at 100°C and 48 h at 50°C in H_2O	D-570	0.25	0.25	0.25	0.25	0.35	0.35	0.35	0.00
CTE, in in °C × 10^{-5}	D-696	1.7	1.5	1.2	1.5	1.6	1.5	2.2	0.0
Deflection temperature, °F	D-648A	525	400	525	400	500	550	500	550
Flammability (ign./burn), s	Fed. 2023	—	—	1120 40	110 40	—	—	110 40	110 00
UL flammability rating									
$^1/_8$ in	UL 94	94 HB	—	94 VO	94 VO	—	—	94 VO	
$^1/_{16}$ in	UL 94	94 HB	—	94 VO	94 VO	—	—	94 VO	
Comparative tracking index, s	—	—	—	600+	600+	—	—	600+	
Dielectric strength 60 Hz ST/SS wet, V/mil	D-149	450/400	450/400	450/400	450/400	400/375	400/375	400/375	400/375
Dielectric constant, 1 kHz/1 MHz wet	D-150	3.9/3.7	4.0/3.5	4.1/3.9	4.2/3.5	0.0/0.0	3.8/3.6	4.1/3.5	4.1/0.0
Dissipation factor, 1 kHz/1 MHz wet	D-150	0.011–0.015	0.009–0.016	0.010–0.013	0.010–0.016	0.011–0.019	0.010–0.019	0.010–0.018	0.010–0.019
Arc resistance, s	D-495	150	130	150	130	135	135	130	135

*Certification to Mil-M-14 requires batch testing. All testing in accordance with Mil-M-14. Values are typical and not statistical minimums.
SOURCE: FROM REF. 11.

TABLE 2.24 Summary of Properties of Allyls

Summary of properties of allyls
Physical:
Excellent long-term dimensional stability
Virtually no postmold shrinkage
Chemically inert
Mechanical:
Excellent flexural strength
High impact resistance
Highly resistant to sudden, extreme jolts and severe stresses
Electrical:
Retains high insulation resistance at elevated temperatures
Retains performance characteristics at high ambient humidity
Thermal:
Successful performance in vapor-phase soldering environments—419° and 487°F (215° and 253°C)
FS-10VO:200°C UL temperature index
Isophthalate materials are recommended for very high temperature applications
Chemical:
Highly resistant to solvents, acids, alkalies, fuels, hydraulic fluids, plating chemicals, and sterilizing solutions

SOURCE: From Ref. 11.

Melamines. Melamine was isolated in 1834, and it wasn't until 1933 that Palmer Griffith produced dicyanamide and found that it contained melamine. The addition of formaldehyde produced a resin that could be compounded into a desirable molding compound. This new compound had a number of desirable qualities superior to phenolics and ureas of that time. The colorability and surface hardness led to its use in molded dinnerware along with some very important military and electrical applications.

2.7.3.2 Physical forms. The urea molding compounds are available as free-flowing granular products that are readily preformable and can be preheated and preplasticated prior to molding. They mold very easy in all thermosetting molding methods. The melamine compounds are available as both free-flowing granular products and as high-strength, high-bulk factor materials. The bulky materials require auxiliary equipment for preforming or preplasticating.

2.7.3.3 Reinforcements. The reinforcements are available as purified cellulose fibers, minerals, chopped cotton flock, wood flour, and glass fibers (short and long).

2.7.3.4 Specifications

Military (Mil-M-14)

Urea compounds are not listed.

Melamine

CMG: Cellulose filler, general purpose

CMI-5: Cellulose filler, impact value of 0.5 ft-lb

CMI- 10: Cellulose filler, impact value of 1.0 ft-lb

MMD: Mineral filler, general purpose

MMI-5: Glass filler, impact value of 0.5 ft-lb

MMI-30: Glass filler, impact value of 3.0 ft-lb

ASTM

Urea D-705

Melamine D-704

UL

	Thermal index	Flame resistance
Urea	221°F (105°C) maximum	74 VO in 1/16-in section
Melamine	220 to 292°F (130 to 150°C)	74 VO in 1/16-in section

2.7.3.5 Data sheet values. Table 2.25 provides a list of property values for a range of urea compounds, and Table 2.26 shows a similar list for melamine compounds.

2.7.3.6 Applications

Urea. Molded urea articles will have a high-gloss finish that is scratch resistant and used in the following applications: closures, control housings, wiring devices, control buttons, electric shaver housings, and knobs.

Melamines. Melamine molded products are used in commercial, industrial, and military applications, taking advantage of the variety of reinforcements available, scratch resistance, and very wide color range. Typical military uses are for connector bodies and

TABLE 2.25 Amino Molding Compound Data Sheet Values—Urea

	Filler				
	α Cellulose	Mineral	Chopped cotton fabric	Cellulose	Glass
Specific gravity	1.49	1.78	1.5	1.5	1.97
Flex strength, lb/in^2	10,000	7,600	12,000	7,000	10,500
Impact strength, ft-lb	0.25	0.3	0.6	0.25	3.0–4.0
Water absorption, % gain (24 h at 25°C)	0.4	0.15	0.5	0.65	0.2
Dielectric strength, V/mil					
Short time	300	350	250	350	170
Step × step	250	250	100	200	170
Arc resistance, s	120	120	125	70	180

SOURCE: From Ref. 11.

TABLE 2.26 Amino Molding Compound Data Sheet Values—Melamine

	Filler	
	α Cellulose	Cellulose
Specific gravity	1.49	1.49
Flex strength, lb/in^2	8000	7500
Impact strength, ft-lb	0.2–0.3	0.2–0.3
Water absorption, % gain (24 h at 25°C)	0.6	0.6
Dielectric strength, V/mil		
Short time	350	350
Step × step	275	275
Arc resistance, s	120	120

SOURCE: From Ref. 17.

circuit breaker housings. Typical commercial uses are for dinnerware, shavers, knobs, buttons, ashtrays, and connector bodies.

Note: Combinations of phenolic and melamine resins in molding compounds have produced materials with excellent color stability, ease of moldability, and good heat resistance. Their usage is in appliance components such as pot or pan handles.

2.7.3.7 Suppliers

Amino compounds

American Cyanamide Company, Polymer Products Division

ICI/Fiberite

Plenco Engineering Company

(See App. C for supplier addresses.)

2.7.4 Epoxies

2.7.4.1 History. The Shell Chemical Corporation introduced the *epoxy resin systems* into the United States in 1941, and their good property profile has been utilized in a wide range of applications. The molding compounds are available in extreme soft flows and long gelation times, which make them very adaptable for encapsulation molding techniques in the encapsulation of electronic components such as integrated circuits, resistors, diode capacitors, relays, and bobbins.

The compounds also found a market in the commercial and military industrial areas for connector bodies, potting shells, printed circuit boards, coils, and bobbins with the higher-strength and higher filler and pressure compounds that compete with the phenolics and DAPs.

2.7.4.2 Physical forms. The molding compounds are available in a free-flowing granular form suitable for automatic preforming or preplasticating and are readily mold-

able in all thermosetting molding techniques. The higher-impact, bulkier compounds mold readily but require special auxiliary equipment for either preforming or preplasticating.

2.7.4.3 Reinforcements. The standard compounds (Mil-M-24325) are available with mineral fillers and also with short and long glass fibers. The low-pressure encapsulation compounds have fused or crystalline silica reinforcements and are in a free-flowing granulate state.

2.7.4.4 Specifications

Military (Mil-M-24325)

MEC: Mineral filler, encapsulation grade, low pressure

MEE: Mineral filler, electrical grade, low pressure

MEG: Mineral filler. general purpose

MEH: Mineral filler, heat resistant

GEl: Glass filler, impact value of 0.5 ft-lb

GEI-20: Glass filler, impact value of 2.0 ft-lb

GEl- 100: Glass filler, impact value of 10.0 ft-lb

ASTM. These specifications are listed under D-3013-77 and D-1763. UL. These specifications have a thermal index rating of 266°F (130°C) and a flammability rating of 94 VO.

2.7.4.5 Data sheet values. Table 2.27 contains a list of property values for a wide range of epoxy molding compounds—both for the low-pressure encapsulation types and the high-pressure types.

2.7.4.6 Applications. Due to retention of their excellent electrical, mechanical, and chemical properties at elevated temperature and very high moisture resistance, the high-pressure Mil spec molding compounds have found their market niche in high-performance military and commercial applications such as connectors, potting shells, relays, printed circuit boards, switches, coils, and bobbins.

The low-pressure encapsulation compounds, which exhibit the same high-performance characteristics as the high-pressure compounds, have become the primary insulating medium for the encapsulation of components such as integrated circuits, resistors, coils, diodes, capacitors, relays, and bobbins.

2.7.4.7 Suppliers

Cytec Fiberite

Dexter Electronics, Materials Division

Morton Chemical

(See App. C for supplier addresses.)

TABLE 2.27 Epoxy Molding Compound Data Sheet Values

Property*	ASTM test method	Class: Hardware grade 1904B	1906	2004B	1907	1908 1908B	1914
Reinforcement		Mineral glass	Mineral glass	Mineral glass	Short-glass	Short-glass	Short-glass
Bulk factor	—	2.4	2.3	2.3	2.2	2.3	2.1
Form	—	Granular	Granular	Granular	Granular	Granular	Granular
Specific gravity, g/cm^3	D-792	2.05	1.90	1.95	1.95	1.85	1.94
Mold shrinkage, in/in	D-955	0.001–0.003	0.001–0.003	0.002–0.004	0.002–0.004	0.002–0.004	0.001–0.003
Impact strength, ft/lb/in	D-256	0.61	0.40	0.60	0.58	0.55	0.61
Flexural strength, RT, lb/in^2	D-790	17,000	15,000	16,000	15,000	17,000	17,000
Flexural modulus, $lb/in^2 \times 10^6$	D-790	1.9	2.1	1.8	1.9	1.9	2.2
Tensile strength, lb/in^2	D-651	10,000	10,000	10,000	10,000	9,000	10,500
Compressive strength, lb/in^2	D-695	32,000	32,000	32,000	32,000	33,000	31,000
Water absorption, %, 48 h at 122°F (50°C)	D-570	0.2	0.2	0.2	0.3	0.3	0.3
Barcol hardness	D-2583	72	70	72	73	72	70
Deflection temperature, °F	D-648	500	340	500	450	500	500
Coefficient of linear thermal expansion, in/in/°C $\times 10^{-6}$ (−30°C±30°C) (−30°C−+30°C)	D-696	28	37	33	47	37	41
Coefficient of thermal conductivity, $cal/s/cm^3$/°C/cm$\times 10^{-4}$	—	17.5	24	14.5	14	17.5	14

TABLE 2.27 Epoxy Molding Compound Data Sheet Values *(Continued)*

Property*	ASTM test method	Class: Electrical encapulation grade 1960B	1961B	2060B	2061B
Reinforcement		Mineral glass	Mineral glass	Mineral glass	Mineral glass
Bulk factor	—	2.5	2.3	2.5	2.3
Form	—	Granular	Granular	Granular	Granular
Specific gravity, g/cm^3	D-792	1.75	1.90	1.80	1.95
Mold shrinkage, in/in	D-955	0.004–0.006	0.002–0.001	0.004–0.006	0.002–0.001
Impact strength, ft/lb/in	D-256	0.56	0.63	0.60	0.61
Flexural strength, RT, lb/in^2	D-790	15,000	14,000	15,000	14,000
Flexural modulus, lb/in^2 × 10^6	D-790	1.7	2.0	1.7	2.0
Tensile strength, lb/in^2	D-651	9,000	10,500	9,000	11,000
Compressive strength, lb/in^2	D-695	30,000	31,500	30,500	31,000
Water absorption, %, 48 h at 122°F (50°C)	D-570	0.3	0.2	0.3	0.3
Barcol hardness	D-2583	65	65	65	70
Deflection temperature, °F	D-648	350	425	375	450
Coefficient of linear thermal expansion, in/in/°C × 10^{-6} (−30°C±30°C) (−30°C−+30°C)	D-696	38	38	35	42
Coefficient of thermal conductivity, cal/s/cm^3/°C/cm × 10^{-4}	—	14	16	17	13
Dielectric constant (1 MHz/24 h at 23°C H_2O)	D-150	4.2	4.5	4.0	4.2
Dissipation factor (1 MHz/24 h at 23°C H_2O)	D-150	0.02	0.02	0.02	0.02
Dielectric strength (S/S) (VPM/48 h at 50°C H_2O)	D-149	325	325	320	320
Arc resistance, s	D-495	185	186	170	170
Insulation resistance (MΩ/30 days at 150°F, 95% RH)	D-257	10	10	10	10
Oxygen index	D-2863	50	3208	—	45.0
UL flammability rating					
$^1/_4$ in	UL 94	—	—	—	—
$^1/_{16}$ in	UL 94	—	—	—	94 VO
Mil-M-14, type	—	—	—	—	

*Certification to Mil-M-14 requires batch testing. All testing in accordance with Mil-M-14. Values are typical and not statistical minimums.

SOURCE: From Ref. 17.

2.7.5 Phenolics

2.7.5.1 History. *Phenolic resins* came into being when Dr. Leo Baekeland, in the early 1900s, discovered that a successful reaction between phenol and formaldehyde in a heated pressure kettle produced an amber-colored liquid thermosetting resin. This resin became the foundation of the entire thermosetting molding compound industry, and an entire family of thermosetting resins and compounds was developed over the next several decades. Phenolic molding compounds became the primary insulating material for a wide and diversified range of applications for industrial, commercial, and military applications.

2.7.5.2 Physical forms. The novolac-based molding compounds are available as free-flowing granular powders or pellets in their general-purpose grades and in a variety of large pellets and flakes in the high-strength bulky grades. The general purpose grades are easily preformed and preplasticated while the bulky products often require auxiliary equipment for such operations.

The resole-based compounds are generally only available as granular powder or small pellets. Both resin system compounds are easily molded in all thermosetting molding procedures.

2.7.5.3 Reinforcements. Both the resole and novolac compounds use a broad array of reinforcements to meet the demands of the marketplace: wood flour, cotton flock, minerals, chopped fabric, Teflon, glass fibers (long and short), nylon, rubber, and kevlar. Asbestos, which had been a widely used filler in many thermosetting compounds, has been replaced over the past 20 years with glass fiber-reinforced phenolic compounds in many applications.

2.7.5.4 Specifications

Military (Mil-M-14)

CFG: Cellulose filler, general purpose

CFI-5: Cellulose filler, impact value of 0.5 ft-lb

CFI-10: Cellulose filler, impact value of 1.0 ft-lb

CFI-20: Cellulose filler, impact value of 2.0 ft-lb

CFI-30: Cellulose filler, impact value of 3.0 ft-lb

CFI-40: Cellulose filler, impact value of 4.0 ft-lb

MFE: Mineral filler, best electrical grade

MFH: Mineral filler, heat resistant

GPG: Glass filler, general purpose

GPI-10: Glass filler, impact value of 1.0 ft-lb

GPI-20: Glass filler, impact value of 2.0 ft-lb

GPI-30: Glass filler, impact value of 3.0 ft-lb

GPI-40: Glass filler, impact value of 4.0 ft-lb

ASTM. The ASTM specifications can be found in D-700-65 and D-4617.

UL. The UL specification has a thermal index rating of 155°C (378°F) for most compounds, but the glass-reinforced grades and some phenolic alloys carry a 185°C (417°F) rating. The flame retardant rating is at 94 VO in a 1/16-in section.

2.7.5.5 Data sheet values. Table 2.28 lists the property values for a broad range of phenolic molding compounds, including both the resole (single-stage and the novolac (two-stage) compounds.

2.7.5.6 Applications. The property profile, which includes a broad range of reinforcements, provides design engineers with great flexibility in their compound selection process. Phenolic molding compounds have found worldwide acceptance in such diverse market areas as

- Automotive (as seen in Figs. 2.14*a, b,* and *c*)
- General transportation
- Electronics
- Aeronautics
- Aerospace
- Electrical
- Appliances
- Business equipment

2.7.5.7 Suppliers

Cytec Fiberite
Occidental Chemical
Plastics Engineering Company
Plaslok Corporation
Rogers Corporation
Resinoid Engineering Corporation
Valentine Sugars Inc., Valite Division

(See App. C for supplier addresses.)

2.7.6 Thermoset Polyesters

2.7.6.1 History. Polyester-based thermosetting molding compounds have been an important component of the thermosetting molding industry for many years, but the past decade has seen a marked increase in their use in many market areas. This increase has come about because of their low cost, wide range of colors, high strength/weight ratio, and, more importantly, because of the introduction of molding equipment capable of injection

TABLE 2.28 Phenolic Molding Compound Data Sheet Values

Grade	Special self-lube RX 342	RX 525	RX 448	RX 431	RX 475	RX 466	RX 468
Reinforcement type	Cellulose	Cellulose	Cellulose	Cellulose	Cellulose	Asbestos	Asbestos
Phenolic resin type	Two-step	Two-step	Two-step	Two-step	Two-step	Two-step	Two-step
Form	Nodular	Nodular	Nodular	Nodular	Nodular	Nodular	Nodular
Bulk factor	2.8	2.7	2.6	3.5	3.3	2.1	2.2
Specific gravity, g/cm^3	1.44	1.39	1.45	1.41	1.40	1.69	1.72
Shrinkage, in/in	0.006	0.006	0.006	0.005	0.005	0.002	0.002
Izod impact, ft-lb/in notch	0.60	0.50	0.70	1.2	1.8	0.70	0.75
Flexural strength, lb/in^2	10,000	10,000	11,000	10,000	10,500	11,000	10,500
Flexural modulus, lb/in^2	1.3×10^6	1.3×10^6	1.3×10^6	1.3×10^6	1.3×10^6	1.8×10^6	1.8×10^6
Tensile strength, lb/in^2	5,500	6,500	6,000	5,700	6,000	6,500	6,000
Compressive strength, lb/in^2	24,000	32,000	28,000	26,000	27,000	24,000	26,000
Water absorption, %	0.4	0.4	0.3	1.0	1.0	0.5	0.35
Deflection temperature @ 264 lb/in^2	350°F	325°F	325°F	335°F	330°F	525°F	500+°F
Continuous use temperature	300°F	300°F	300°F	300°F	310°F	—	—
UL flammability rating @ $^1/_8$ in	—	94 VO	94 HB	94 HB	—	94 VO	94 VO
Dielectric strength 60Hz ST/SS, V/mil	—	325/250	270/175	250/190	250/190	150/100	150/100
Arc resistance, s	—	90	130	125	75	165	175

TABLE 2.28 Phenolic Molding Compound Data Sheet Values *(Continued)*

Grade	RX 655	RX 630	RX 611	RX 660	RX 862	RX 865	RX 867	Ammonia Free RX 640
Reinforcement type	Glass	Glass	Glass	Glass	Glass	Glass	Glass	Glass
Phenolic resin type	Two-step	Two-step	Two-step	Two-step	Two-step	Two-step	Two-step	One-step
Form	Granular	Granular	Granular	Granular	Coarse granular	Coarse granular	Nodular	Granular
Bulk factor	2.1	2.7	2.3	2.2	2.5	2.4	2.7	2.2
Specific gravity, g/cm^3	2.08	1.75	1.75	1.79	1.88	1.88	1.78	1.73
Shrinkage, in/in	0.0015	0.001	0.0015	0.0015	0.0015	0.001	0.0015	0.0015
Izod impact, ft-lb/in notch	0.45	1.20	0.90	0.75	0.90	1.3	0.70	0.90
Flexural strength, lb/in^2	12,000	23,000	18,000	18,000	12,000	15,000	11,500	16,000
Flexural modulus, lb/in^2	2.7×10^6	2.2×10^6	2.0×10^6	2.0×10^6	2.3×10^6	2.5×10^6	2.0×10^6	1.8×10^6
Tensile strength, lb/in^2	6,400	12,000	10,000	8,500	6,500	7,500	6,000	9,000
Compressive strength, lb/in^2	30,000	40,000+	40,000	40,000	28,000	33,000	28,000	35,000
Water absorption, %	0.04	0.07	0.07	0.08	0.07	0.05	0.17	0.15
Deflection temperature @ 264 lb/in^2	550+°F	450°F	440°F	410°F	500°F	550°F	500°F	500°F
Continuous use temperature	385°F	390°F	380°F	375°F	380°F	390°F	350°F	380°F
UL flammability rating @ $^1/_8$ in	—	94 VO	94 VO	94 VO	94 VO	94 VO	94 VO	—
Dielectric strength 60Hz ST/SS, V/mil	—	500/440	450/425	450/425	300/250	300/250	300/230	475/425
Arc resistance, s	—	180	181	180	183	185	184	180

SOURCE: From Ref. 19.

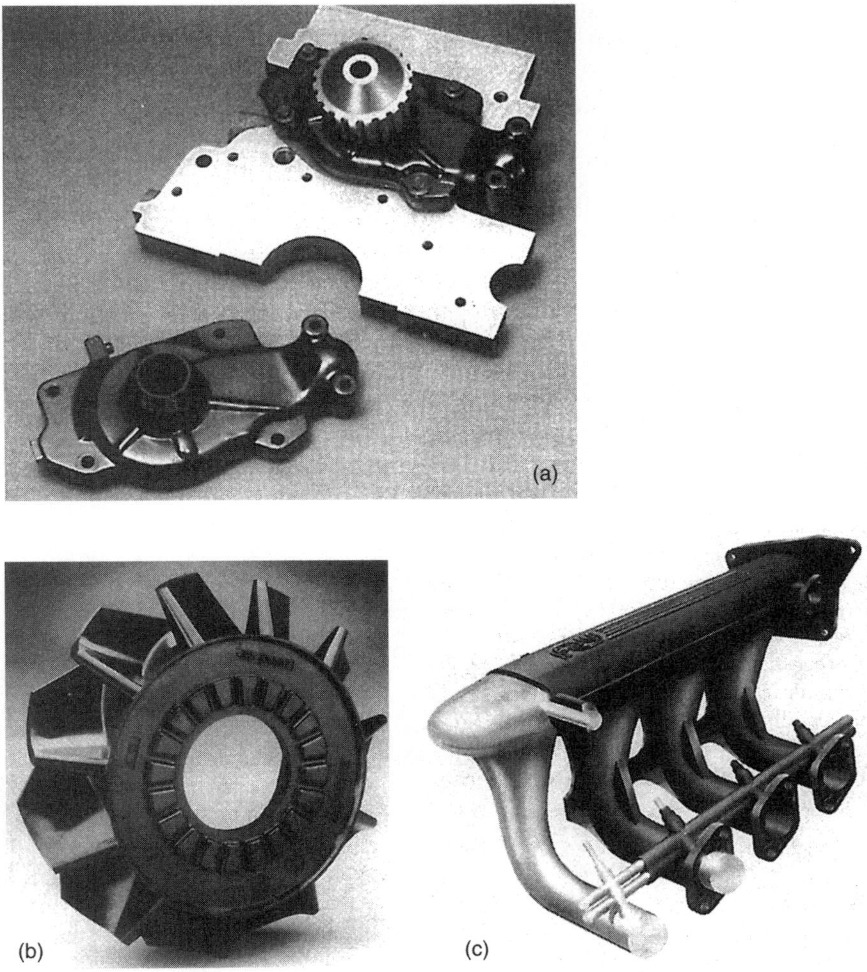

Figure 2.14 (a) Water pump housing, (b) transmission stator and reactor, and (c) manifold housing.

molding these bulky, dough-like compounds by using stuffing mechanisms that augment the passage of the compound from the hopper into the barrel for delivery into the mold. Lower molding costs and reduced finishing costs are among the benefits derived with the use of these versatile molding compounds with the injection molding process.

2.7.6.2 Physical forms. Thermoset polyester molding compounds are available in several physical forms: free-flowing granules, pelletized (PMG), putty or rope-type extrudates, sheet molding compound (SMC), high-bulk molding compound (BMC), and thick molding compound (TMC) forms. The molding compounds, regardless of reinforcement type, are all readily moldable in all thermosetting processes. When compression or trans-

fer molded, the preforming and preplasticating operation will necessitate the use of auxiliary equipment, especially the bulky, dough-like material.

2.7.6.3 Reinforcements. The types of reinforcements available are minerals, long and short glass fibers, and organic fibers.

2.7.6.4 Specifications

Military (MIl-M-14)

MAG: Mineral filler, general purpose

MAI-30: Glass filler, impact value of 3.0 ft-lb

MAT 60: Glass filler, impact value of 6.0 ft-lb

MAT 30: Glass filler, impact value of 3.0 ft-lb

ASTM. The specifications are covered under D-1201-80 and D-1201-62 9 (reapproved 1975).

UL. The thermoset polyester compounds generally carry a 365°F (180 °C) thermal index rating, with a flammability rating of 94 VO in 1/16-in sections.

2.7.6.5 Data sheet values. Table 2.29 includes a list of the property values for a range of thermoset polyester compounds.

2.7.6.6 Applications. The applications are automotive components, circuit breaker housings, brush holders, commercial connectors, battery racks, business machine housings, marine structures, and household articles.

2.7.6.7 Suppliers

American Cyanamid Company BMC Inc.

(See App. *C* for supplier addresses.)

2.7.7 Silicones

2.7.7.1 History. Silicone fluids, resins, and elastomers have been in use for over 50 years, originating with the discovery by E. Rochow (GE Company) in 1940 of what was designated the "direst process," in which elemental silicon was obtained by the reduction of silicon dioxide in an electric furnace. The resultant silicon was then pulverized and reacted with gaseous methyl chloride in the presence of a copper catalyst.

2.7.7.2 Physical forms. The molding compound will consist of 20 to 25% resin (phenyl and methyl siloxanes), 75% filler (glass fiber and fused silica mix), a lead-based

TABLE 2.29 Thermoset Polyester Molding Compound Data Sheet Values

Sheet molding compounds (typical ASTM data)	ASTM test method	UL recognized electrical		
		4000-( )-( )	42000-( )-( )	47000-( )-( )
Water absorption, %	D-570	0.2	0.18	0.15
Specific gravity	—	1.85	1.75	1.85
UL 94 flame classification—UL File E-27875	—	94 VO	94 VO	94 VO, 945-V
Tensile strength, lb/in^2	D-638	14,000	10,000	15,000
Flexural strength, lb/in^2	D-790	30,000	25,000	30,000
Compressive strength, flatwise, lb/in^2	D-695	30,000	28,000	30,000
Impact strength, izod, edgewise, ft/lb/in notch	D-256	13.5	18.5	20.0
Arc resistance, s	D-495	180	180	180
Track resistance, min	D-2303	450	>600	500
Dielectric strength, perpendicular, short time, V/mil in oil	D-149	420	425	420
Dielectric constant at 60 Hz	D-150	4.7	4.4	4.4
Dissipation factor at 60 Hz	D-150	0.96	0.96	0.95
Glass content, % for data shown	—	22	22	33
Range of glass content available, %	—	15–35	15–35	33
Mold shrinkage, in/in	—	0.002	0.001	0.002

Sheet molding compounds (typical ASTM data)	ASTM test method	UL recognized electrical		
		13600-( )-( )	14100-( )-( )	14100-( )-( )
Water absorption, %	D-570	0.35	0.1	0.10
Specific gravity	—	1.94	1.89	1.9
UL 94 flame classification—UL File E-27875	—	94 VO	94 VO	94 VO, 945-V
Tensile strength, lb/in^2	D-638	6,000	6,000	7,000
Flexural strength, lb/in^2	D-790	15,000	12,500	15,000
Compressive strength, flatwise, lb/in^2	D-695	15,000	13,000	20,000
Impact strength, izod, edgewise, ft/lb/in notch	D-256	3.5	3.5	5.0
Arc resistance, s	D-495	190	192	180
Track resistance, min	D-2303	>600	>600	600
Dielectric strength, perpendicular, short time, V/mil in oil	D-149	375	400	400
Dielectric constant at 60 Hz	D-150	5.8	5.0	5.0
Dissipation factor at 60 Hz	D-150	295	3.25	3.0
Glass content, % for data shown	—	15	15	20
Range of glass content available, %	—	5–35	5–35	20
Mold shrinkage, in/in	—	0.002	0.0001	0.002

SOURCE: From Ref. 10.

catalyst pigment, and lubricants. The compounds are free-flowing granular in form and are available in opaque colors (mostly red). They are readily moldable in compression, transfer, and injection molding processes.

2.7.7.3 Reinforcements.

The reinforcements available are quartz, "E"-type glass fibers, and fused silica.

2.7.7.4 Specifications

Military (MIl-M-14)

MSG: Mineral filler, heat resistant

MSI-30: Glass filler, impact value of 3.0 ft-lb, heat resistant

UL. The UL ratings are (1) thermal index rating of 464°F (240°C) and(2) flammability rating of 94 VO in 1/16-in sections.

2.7.7.5 Data sheet values.

Table 2.30 provides a property value list for the silicone family of molding compounds.

2.7.7.6 Applications.

The silicones are nonconductors of either heat or electricity; have good resistance to oxidation, ozone, and ultra-violet radiation (weatherability); and are generally inert. They have a constant property profile of tensile, modulus, and viscosity values over a broad temperature range 60 to 390°F (13 to 166°C). They also have a low glass transition temperature (T_g) of –185°F. Encapsulation of semiconductor devices such as microcircuits, capacitors, and resistors, electrical connectors seals, gaskets, O-rings, and terminal and plug covers all take advantage of these excellent properties.

2.7.7.7 Suppliers

Dow Corning Corporation

Cytec Fiberite

General Electric Company Silicone Products Division

(See App. C for supplier addresses.)

2.7.8 Composites

2.7.8.1 History.[15]

The introduction of fiberglass-reinforced structural applications in 1949 brought a new plastics application field which began with the consumption of 10 lb and burgeoned into the annual usage of over 1 to 2 billion lb over the next several decades. This usage has been, and still is, taking place in application areas that take advantage of the extraordinarily low-weight/high-strength ratio inherent in these composite materials.

TABLE 2.30 Silicone Molding Compound Data Sheet Values

Properties[a]	Dow Corning silicone molding compound 306	307	308	480	1-5021	Test method CTM[b]	ASTM
Physical							
Specific gravity	1.88	1.85	1.86	1.92	1.92	0540	D-792
Flexural strength (lb/in^2 × 10^3)	8.8±0.7	7.5±0.8	9.2±1.4	9.1±1.1	8.9±1.1	0491A	D-790
Flexural modulus (lb/in^2 × 10^6)	1.2	0.7	1.2	1.5	1.5	0491A	D-790
Compressive strength, lb/in^2	14,000	10,000	14,000	18,000	18,000	0533	D-695
Tensile strength, lb/in^2	5,000	4,500	5,200	5,000	5,000	—	D-638
Izod impact strength, ft/lb/in notch[c]	0.34	0.34	0.26	0.24	0.24	0498	D-256
Water absorption, %	0.1	0.1	0.1	0.1	0.1	0248	D-570
Flammability, UL 94, 0.0625 in[d]	V-0	V-0	V-0	V-0	V-0	—	—
Thermal conductivity, cal/s-°C-cm × 10^{-4}	15	14	16	18	18	Colora	—
Thermal expansion,[e]							
Measured parallel to flow							
50–150°C, in/in°C × 10^{-6}	22	22	24	25	25 0563	—	
150–250°C, in/in°C × 10^{-6}	27	29	28	26	28 0563	—	
Measured perpendicular to flow							
50–150°C, in/in°C × 10^{-6}	40	43	38	34	35 0563	—	
150–250°C, in/in°C × 10^{-6}	46	52	39	35	38 0563	—	
Stability to 70°C/93% RH after 10,000 h weight change, %	+1.4	+1.5	—	—	—	—	—
Dimension change, %	+0.14	+0.16	—	—	—	—	—
Electrical							
Arc resistance, s	290	270	280	300	—	0171	D-495
Dielectric strength, in oil, V/mil[f]	320	310	290	300	370	0114	D-149
Volume resistivity							
Condition A[g] Ω-cm × 10^{15}	2.8	2.7	2.8	2.8	2.6	0249	D-257
Condition D[h] Ω-cm × 10^{15}	2.7	2.7	2.8	2.8	2.5	0249	D-257
Dielectric constant at 10^6 Hz							
Condition A[g]	3.55	3.61	3.68	3.60	3.49	0543	D-150
Condition D[h]	3.60	3.64	3.72	3.65	3.53	0543	D-150
Dissipation factor at 10^6 Hz							
Condition A[g]	0.0016	0.0018	0.0018	0.0014	0.0012	0543	D-150
Condition D[h]	0.0019	0.0027	0.0019	0.0019	0.0015	0543	D-150

[a]All values shown are determined periodically and must be considered typical. Should a property be judged to be pertinent in a given application, verification must be made by the user prior to use.

[b]Dow Corning Test Method. Similar to ASTM method shown. Copies available upon request.

[c]Not corrected for energy to toss.

[d]UL yellow card listed.

[e]Test specimens 1.5 in × 0.25 in × 0.10 in. Transfer molded bars postcured 2 h at 200°C.

[f]Tested under 1/4-in electrodes, 500-V/s rise, 1/8-in-thick specimen.

[g]Condition A = as received.

[h]Condition D = after 24-h immersion in distilled water at 23°C (75°F).

SOURCE: From Ref. 10.

A *thermosetting matrix* is defined as a composite matrix capable of curing at some temperature from ambient to several hundred degrees of elevated temperature and cannot be reshaped by subsequent reheating. In general, thermosetting polymers contain two or more ingredients—a resinous matrix with a curing agent that causes the matrix to polymerize (cure) at room temperature, or a resinous matrix and curing agent that, when subjected to elevated temperatures, will commence to polymerize and cure.

2.7.8.2 Resins (matrices). The available resins are polyester and vinyl esters, polyureas, epoxy, bismaleimides, polyimides, cyanate ester, and phenyl triazine.

Polyester and vinyl esters. Polyester matrices have had the longest period of use, with wide application in many large structural applications (see Table 2.31). They will

TABLE 2.31 Neat Resin Casting Properties of Polyester-Related Matrices

Material	Barcol hardness	Tensile strength		Tensile modulus		Elongation, %	Flexural strength		Flexural modulus		Compressive strength		Heat deflection temperature	
		MPa	kips/in^2	10^{-2} Pa	10^{-5} kips/in^2		MPa	kips/in^2	10^{-2} Pa	10^{-5} kips/in^2	MPa	kips/in^2	°C	°F
Orthophthalic	—	55	8	34.5	5.0	2.1	80	12	34.5	5.0	—	—	80	175
Isophthalic	40	75	11	33.8	4.9	3.3	130	19	35.9	5.2	120	17	90	195
BPA fumarate	34	40	6	28.3	4.1	1.4	110	16	33.8	4.9	100	15	130	265
Chlorendic	40	20	3	33.8	4.9	—	120	17	39.3	5.7	100	15	140	285
Vinylester	35	80	12	35.9	5.2	4.0	140	20	37.2	5.4	—	—	100	212

SOURCE: Charles D. Dudgeon in *Engineering Materials Handbook*, vol. 1, Theodore Reinhart, Tech. Chairman, ASM International, 1987, p. 91.

cure at room temperature with a catalyst (peroxide), which produces an exothermic reaction. The resultant polymer is nonpolar and very water resistant, making it an excellent choice in the marine construction field. The isopolyester resins, regarded as the most water-resistant polymers in the polymer group, have been chosen as the prime matrix materials for use on a fleet of U.S. Navy mine hunters.

Epoxy. The most widely used matrices for advanced composites are the epoxy resins, even though they are more costly and do not have the high-temperature capability of the bismaleimides or polyimide. The advantages listed in Table 2.32 show why they are widely used.

Bismaleimides (BMIs). The *bismaleimide resins* have found their niche in high-temperature aircraft design applications where temperature requirements are in the 177°C (350°F) range. BMI is the primary product and is based on the reaction product from methylene dianiline (MDA) and maleic anhydride: bis (4 maleimidophynyl) methane (MDA BMI). Variations of this polymer with compounded additives to improve impregnation are now on the market and can be used to impregnate suitable reinforcements to result in high-temperature mechanical properties (Table 2.33).

Polyimides. *Polyimides* are the highest-temperature polymers in the general advanced composite, with a long-term upper temperature limit of 232 to 316°C (450 to 600°F). Table 2.34 is a list of commercial polyimides being used in structural composites.

Polyureas. Polyureas involve the combination of novel MDI polymers and either amine or imino-functional polyether polyols. The resin systems can be reinforced with milled glass fibers, flaked glass, Wollastanite, or treated mica, depending on the compound requirements as too processability or final product.

TABLE 2.32 Epoxy Resin Selection Factors

Advantages
- Adhesion to fibers and to resin
- No by-products formed during cure
- Low shrinkage during cure
- Solvent and chemical resistance
- High or low strength and flexibility
- Resistance to creep and fatigue
- Good electrical properties
- Solid or liquid resins in uncured state
- Wide range of curative options

Disadvantages
- Resins and curatives somewhat toxic in uncured form
- Absorb moisture
  Heat distortion point lowered by moisture absorption
  Change in dimensions and physical properties due to moisture absorption
- Limited to about 200°C upper temperature use (dry)
- Difficult to combine toughness and high temperature resistance
- High thermal coefficient of expansion
- High degree of smoke liberation in a fire
- May be sensitive to ultraviolet light degradation
- Slow curing

SOURCE: From Ref. 15, p. 3.15.

TABLE 2.33 Approximation of Mechanical Properties of BMI Composites

Property	Unreinforced homopolymer	Glass-reinforced homopolymer	Carbon-reinforced homopolymer	Carbon-reinforced homopolymer
Reinforcement, vol %	0	60	60	70
Service temperature, °C (°F)	260 (500)	177–232 (350–450)	177–232 (350–450)	149–204 (300–400)
Flexural strength, MPA (kips/in^2)				
At room temperature	210 (30)	480 (70)	2000 (290)	725 (105)
At 230°C (450°F)	105 (15)	290 (42)	1340 (194)	—
Flexural modulus, GPa (10^6 lb/in^2)				
At room temperature	4.8 (0.7)	17.2 (2.5)	126 (18.4)	71 (10.3)
At 230°C (450°F)	3.4 (0.5)	15.1 (2.2)	57 (8.2)	—
Interlaminar shear strength, MPa (kips/in^2)				
At room temperature	—	—	117 (17)	—
At 230°C (450°F)	—	—	59 (8.6)	—
Tensile strength, MPa (kips/in^2)				
At room temperature	97 (14)	—	1725 (250)	570 (83)
At 230°C (450°F)	76 (11)	—	—	—
Tensile modulus, GPa (10^6 lb/in^2)				
At room temperature	4.1 (0.6)	—	148 (21.5)	15.9 (2.3)
At 230°C (450°F)	2.8 (0.4)	—	—	—

SOURCE: James A. Harvey in *Engineering Materials Handbook,* vol. 1, Theodore Reinhart, Tech. Chairman, ASM International, 1987, p. 256

TABLE 2.34 Commercial Polyimides Used for Structural Composites

	Upper temperature capability	
	°C	°F
Condensation		
Monsanto Skybond 700, 703	316	600
DuPont NR-150B2 (Avimid N)	316	600
LARC TPI	300	572
Avimid K-III	225	432
Ultem	200	400
Addition		
PMR-15 (Reverse Diels-Adler nadic end-capped)	316	600
LARC 160 (Reverse Diels-Adler nadic end-capped)	316	600
Thermid 600 (Acetylene end-capped)	288	550
BMIs (Bismaleimides, maleimide end-capped)	232	450

SOURCE: D. A. Scola in *Engineering Materials Handbook,* J. N. Epel et al. (eds.), vol. 2, ASM International, 1988, p. 241.

Cyanate ester and phenolic triazine (PT). The cyanate ester resins have shown superior dielectric properties and much lower moisture absorption than any other structural resin for composites. The physical properties of cyanate ester resins are compared to those of a representative BMI resin in Table 2.33. The PT resins also possess superior elevated-temperature properties, along with excellent properties at cryogenic temperatures. They are available in several viscosities, ranging from a viscous liquid to powder, which facilitates their use in applications that use liquid resins such as filament winding and transfer molding.

2.7.8.3 Reinforcements

Fiberglass. Fiberglass possesses high tensile strength and strain to failure, but the real benefits of its use relate to its heat and fire resistance, chemical resistance, moisture resistance, and very good thermal and electrical values. Some important properties of glass fibers are shown in Table 2.35.

Graphite. Graphite fibers have the widest variety of strength and moduli and also the greatest number of suppliers. These fibers start out as organic fiber, rayon, polyacrylonitrile, or pitch called the *precursor.* The precursor is stretched, oxidized, carbonized, and graphitized. The relative amount of exposure to temperatures from 2500 to 3000°C will then determine the graphitization level of the fiber. A higher degree of graphitization will usually result in a stiffer (higher modulus) fiber with greater electrical and thermal conductivities. Some important properties of carbon and graphite fibers are shown in Table 2.36.

Aramid. The organic fiber kevlar 49, an aramid, essentially revolutionized pressure vessel technology because of its great tensile strength and consistency coupled with low density, resulting in much more weight-effective designs for rocket motors.

Boron. Boron fibers, the first fibers to be used in production aircraft, are produced as individual monofilaments upon a tungsten or carbon substrate by pyrolytic reduction of boron trichloride (BCL) in a sealed glass chamber. Some important properties of boron fibers are shown in Table 2.37.

2.7.9 Molding Compound Production[16]

2.7.9.1 Introduction. The selection of production equipment and processes for thermosetting molding compounds commences with the compound designer's formulation, which designates the type and quantity of the various ingredients that make up the compound. These molding compounds are a physical mixture of resin, reinforcement or filler, catalyst, lubricant, and color. The resin, by far, is the *key* component in any thermosetting molding compound, since it is the only component that actually goes through the chemical reaction known as *polymerization* or *cure* during the molding process. Also, because of this curing quality, the resin, production process, and equipment are governed by the need to understand this chemical reaction with its effects on the production process and/or equipment. Also, the resin is the primary flow promoter and chief provider of the desired electrical insulating properties of the final molded product.

The next most important component is the reinforcement or filler, because the type and quantity of either will determine the manufacturing process and equipment. This will be seen in the following sections, which describe the various processes and their equipment.

TABLE 2.35 Glass Fibers in Order of Ascending Modulus Normalized to 100% Fiber Volume (Vendor Data)

Type	Nominal tensile modulus, GPa ($kips/in^2 \times 10^6$) strand	Nominal tensile strength, MPa ($kips/in^2 \times 10^6$) strand	Ultimate strain, %	Fiber density, kg/m^3 (lb/in^3)	Typical suppliers
E	72.5 (10.5)	3447 (500)	4.8	2600 (0.093)	Pittsburgh Plate Glass, Manville Co., Owens Corning Fiberglass
R	86.2 (12.5)	4400 (638)	5.1	2530 (0.089)	Vetrotex St. Gobain, Certainteed
Te	84.3 (12.2)	4660 (675)	5.5	2530 (0.089)	Nittobo
S-2, S	88 (12.6)	4600 (665)	5.2	2490 (0.090)	Owens Corning

SOURCE: S. T. Peters, W. D. Humphrey, and R. F. Foral, *Filament Winding: Composite Structure Fabrications*, SAMPE Publishers, Covina, Calif., 1991.

TABLE 2.36 Carbon and Graphite Fibers in Order of Ascending Modulus Normalized to 100 Percent Fiber Volume (Vendor Data)

Class of fiber	Nominal tensile modulus GPa ($lb/in^2 \times 10^6$) strand	Nominal tensile strength MPa ($lb/in^2 \times 10^3$) strand	Ultimate strain (%)	Fiber density kg/m^3 (lb/in^3)	Suppliers/ typical products
High tensile strength	227 (33)	3996 (580)	1.60	1750 (0.063)	Amoco, T-300; Hercules, AS-4
High strain	234 (34)	4100 (594)	1.95	1790 (0.064)	Courtaulds Grafil, 33-600
Intermediate modulus	275 (40)	5133 (745)	1.75	1740 (0.062)	Hercules, IM-6; Amoco, T-40; Courtaulds Grafil, 42-500
Very high strength	289 (42)	7027 (1020)	1.82	1820 (0.066)	Toray, T-1000
High modulus	358	2482 (360)	0.70	1810 (0.065)	Amoco, T-50; Celanese, G-50
High modulus (pitch)	379 (55)	2068 (300)	0.50	2000 (0.072)	Amoco, P-55
Ultrahigh modulus (pan)	517 (75)	1816 (270)	0.36	1960 (0.070)	Celanese, GY-70
Ultrahigh modulus (pitch)	517 (75)	2068 (300)	0.40	2000 (0.072)	Amoco, P-75
Extremely high modulus (pitch)	689 (100)	2240 (325)	0.31	2150 (0.077)	Amoco, P-100

SOURCE: S. T. Peters, W. D. Humphrey, and R. F. Foral, *Filament Winding: Composite Structure Fabrications,* SAMPE Publishers, Covina, Calif., 1991.

TABLE 2.37 Boron and Ceramic Fibers, Normalized to 100 Percent Fiber Volume (Vendor Data)

Type	Nominal tensile modulus, GPa ($lb/in^2 \times 10^6$)	Nominal tensile strength, MPa ($lb/in^2 \times 10^3$)	Fiber density kg/m^3 (lb/in^3)	Suppliers
Boron	400 (58)	3520 (510)	2.55–3.30 (0.093)	Textron, Huber, Nippon, Tokai
Silicon carbide	425 (62)	611	3.56 (0.125)	Textron, Dow Corning, Nippon Carbon
Silicon nitride	300	2500	2.5	Tonen
Silica	66		1.80–2.50	Enka, Huber
Alumina	345 (50)	1380 (200)	3.71 (0.134)	DuPont, FP
Alumina boria silica	(27)	300	2.71 (0.098)	3MS, Nexel 312

The compounds utilizing "fillers" as opposed to "reinforcements" can be processed with either the "dry" or the "wet" (solvent) process, since the compound formulations include free-flowing granular fillers that are not as susceptible to degradation when exposed to the hot roll mill phase of the operation. The catalyst in each compound serves as the *reaction controller,* with the type and quantity of catalyst acting to either accelerate or inhibit the curing rate in both the production and the molding phases.

Lubricants, which provide a measure of flow promotion, mold release, and barrel life during molding, are generally internally supplied but are occasionally provided as an external addition. All thermosetting molding compound colors are opaque, with the pigments or dyes heat stable within the molding process temperature range of 200 to 400°. Coloring does have a large effect on the manufacturing process when the product line includes a wide variety of colors such as are common to the DAP, melamine, urea, and thermoset polyester compounds. Choice of production equipment has to be designed to meet the need for quick and easy color changes.

2.7.9.2 Production processes

Dry process (batch and blend). The *dry* or *nonsolvent process,* illustrated in Fig. 2.15, employs low-strength, low-cost, free-flowing granular fillers and involves the use of ribbon or conical mixers to homogenize the dry ingredients prior to feeding the mix onto the heated roll mills where the mix is compounded (worked) for a specific time and temperature. Once the mix or hatch has been worked to the proper consistency and temperature, it is then fed onto a three-roll calendering mill where it is shaped to a specific width and thickness to allow the sheet to pass into a grinder, and then onto screens to obtain the desired granulation and for dust removal. The thickness and temperature of the calendered sheet is controlled for ease of granulation. If the sheet is too warm, it will not cut cleanly; if the sheet is too cool, it will be too fragile to produce clean, even size particles.

An individual batch is generally 200 lb, which eventually is blended with other hatches into 2500- to 5000-lb blends that are ready for shipment to the customer. The batch and

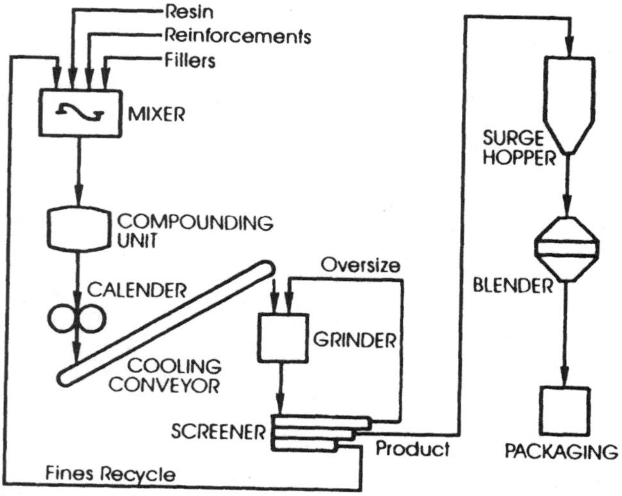

Figure 2.15 Batch and blend dry production method.

blend process is employed for short runs, especially where the production schedules call for a variety of colors. The equipment must he such that it allows for relatively quick color changeovers.

Wet process (pelletized). The wet process, shown in Fig. 2.16, when low-strength, low-cost, free-flowing granular fillers are used, can produce free-flowing pelletized material, with water as the solvent and the ingredients thoroughly mixed in a kneader and then auger fed to a heated extruder. The extruder screw densifies the wet mass, forcing it out of the extruder head, which contains many small through orifices that determine the diameter of the pellets, and the fly-cutters working across the face of the extruder head determine the pellet length. The extruded pellets then require a drying operation prior to final blending for shipment.

Wet process (high strength). As can he seen in Fig. 2.17, this process involves the use of mixers, mobile carts, air-drying rooms, prebreakers, hammer mills, lenders, and extruders. The basic purpose of the entire process is to provide minimum reinforcement degradation so as to maintain sufficient fiber integrity to meet the various mechanical strength requirements which are generally set by military or commercial specifications.

The process begins with the mixing of all the ingredients, except the reinforcement, into a suitable solvent. Once the mix has been properly dispersed, the reinforcement is added, keeping the mixing time to a minimum to preserve the fiber integrity. Solvent recovery is possible during this phase of the operation as well as later on in the drying phase. The wet mass is removed from the mixer and spread onto wire trays capable of holding about 25 lb. These trays are loaded into a mobile cart and placed in a drying room, generally overnight. After drying, the now-hardened slabs of compound are fed into a "prebreaker," which tears the slab into pieces that are then sent to a hammer mill for particle size reduction. The eventual particle size is determined by the size of the openings at the bottom of the mill.

There is no need for blending, since the ultimate flow properties of the compound are governed by the resin, reinforcement, and catalyst mix. There has been little, if any, temperature imposed upon the materials that might affect the flow properties.

High volume (general purpose). The method for producing large volume runs of granular compounds, generally of a single color, flow, and granulation, shown in Fig. 2.18*a*, involves the use of extruders or kneaders into which the compound mix (wet or dry) is fed for "working" or homogenizing prior to being extruded out the exit end of the unit.

The open buss kneader (Fig. 2.18*b*) works with an external screw bearing in the production of pastel colors in the urea and melamine compounds. When processing epoxy, polyesters (with or without glass fibers), and phenolics, the external bearing is not required. Both the open buss kneader and the Werner Pfliederer compounding unit will process the compounds and also control the granulation and flow properties.

Sheet process (SMC). Almost regardless of the specific resin used in the SMC process, the compound manufacturing technique is the same as that shown in Fig. 2.19, with the doughy material and its reinforcement being covered on both lower and upper surfaces with a thin film of polyethylene. The finished product is then conveyed onto a rotating mandrel and wound up until it reaches a preset weight, and then it is cut off. The sheets ready for use or for shipment generally weigh about 50 lb, are 4 ft wide, and are approximately 0.075 to 0.250 in thick. Formulations generally consist of an unsaturated polyester resin (20 to 30 percent), chopped glass rovings (40 to 50 percent), fine particle size calcium carbonate, filler, catalyst, pigment, and modifiers. The resin system can be epoxy, polyester, or vinyl ester to meet the need of the marketplace.

Sheet process (TMC). The production of TMC, shown in Fig. 2.20, differs from that of SMC in that the glass fibers are wetted between the impregnating rollers before being

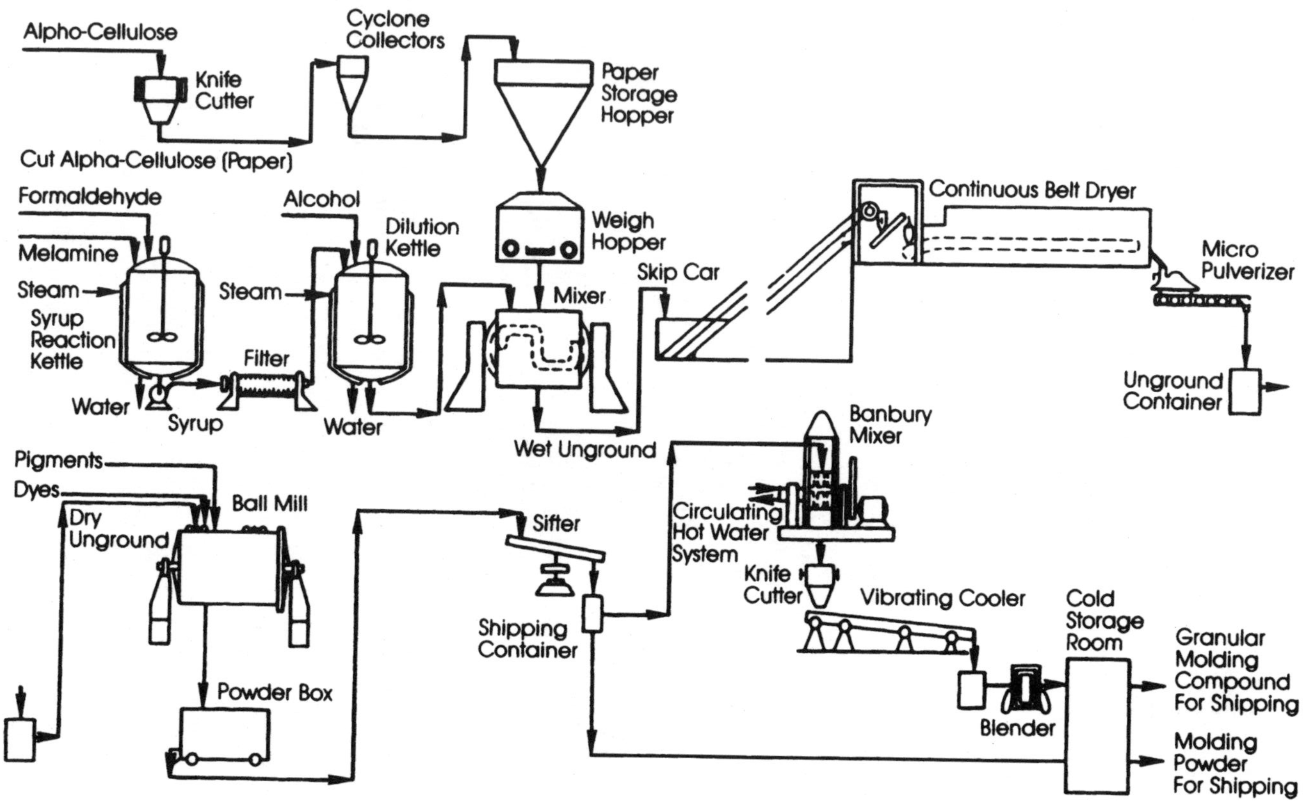

Figure 2.16 Continuous wet high-production method.

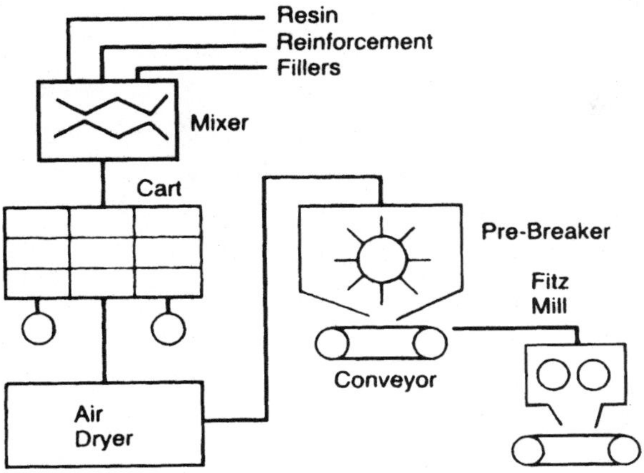

Figure 2.17 Batch and blend wet high-strength method.

deposited onto the moving film. TMC sheets can be produced in thicknesses of up to 2 in with glass lengths of 1 in/min at loading levels of 20 to 30 percent. These sheets are generally compression molded using matched metal-hardened steel molds. Packaging and shipment of the unmolded product is similar to that of the SMC products.

High strength (BMC). Bulky high-strength compounds are produced with the batch method during which the resin, lubricants, catalyst, and chopped glass fibers (1/8 to 1/2 in long) are all compounded in relatively low-intensity mixers. The mixing procedure is carefully monitored to achieve the highest possible mechanical properties with the least amount of fiber degradation. The finished product is shipped in bulk form using vapor barrier cartons with the compound in a sealed polyethylene bag. These compounds are also available in a "rope" form in any length or diameter specified by the customer. The shape is attained by feeding the doughy material through an extruder, with the extruder nozzle providing the desired shape and length.

Quality assurance. Thermosetting compound production uses comprehensive quality assurance programs to ensure that customers receive products that meet their specific specifications, regardless of the process or equipment employed. Formulation and processing specifications require that the incoming raw materials meet specific quality standards and that the manufacturing processes be carefully monitored to make certain that the final compound meets a designated property profile.

The formulation requirements are based on meeting certain standards relating to the following property standards:

- Flow properties
- Electrical (insulation)
- Mechanical (strength)
- Chemical resistance
- Weather resistance
- Thermal resistance
- Flame resistance

(a)

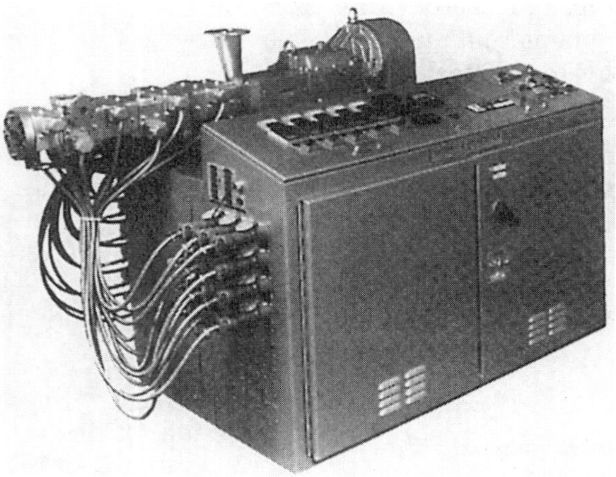

(b)

Figure 2.18 (a) Continuous dry high-volume method and (b) open bus kneader.

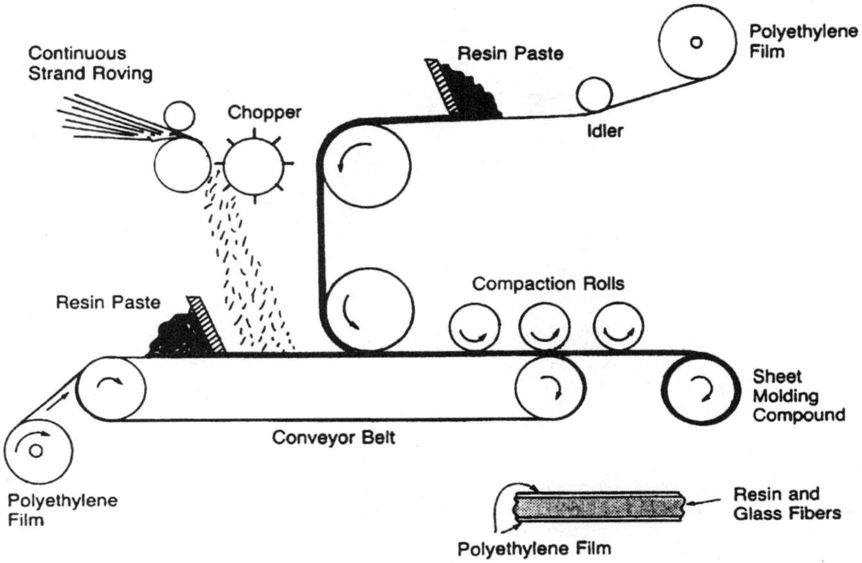

Figure 2.19 Sheet molding compound (SMC) process.

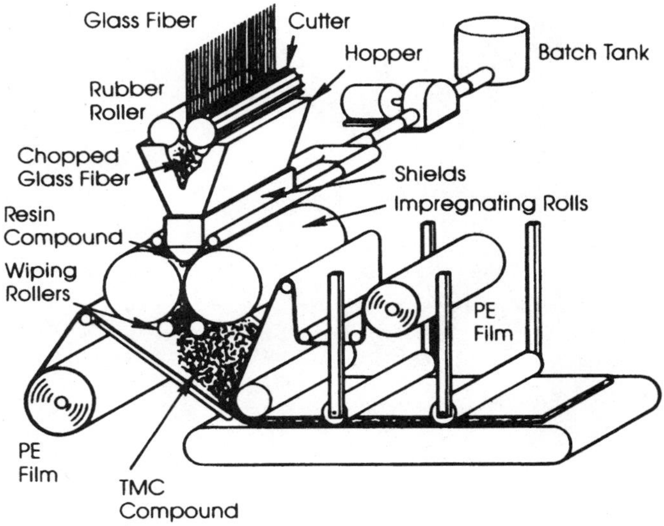

Figure 2.20 Thick molding compound (TMC) process.

- Surface finish
- Color

These specifications are established by either the military or commercial users or by ASTM and UL standards. The compound designer's *primary* task is one of selecting the individual ingredients that, within a cost range, will not only create the appropriate compound but will also function properly in the manufacturing process. Process procedures are governed by a set of standards that spell out check points along the line to aid in the process control. The formulation will furnish the necessary information relating to the acquisition of raw materials along with the basic standards by which each ingredient is accepted for incorporation into the compound mixture.

The specific ingredients are resin reinforcement, pigments or dyes, lubricants, and solvents. Each of these ingredients has specific characteristics that are measurable and controllable, and the compound producer can, and often will, supply the molder with test data on each production blend the customer has received. The special characteristics of each of these ingredients are as follows:

Resin. Viscosity, gel time, cure rate, and solubility

Reinforcements and fillers. Aspect ratio, moisture content, fiber size and length, purity, and color

Pigments and dyes. Solubility, coatability, and thermal stability

Catalysts. Solubility, reaction temperature, and purity

Lubricants. Solubility, melting point, and purity

Solvents. Purity and toxicity

Compounds manufactured to meet Mil specifications will be subjected to a certification process involving the documentation of actual property values derived from the testing of the compound both during, and upon completion of the manufacturing procedure. The values called for are as follows:

- Specific gravity
- Shrinkage
- Arc resistance
- Dielectric constant
- Dielectric strength
- Impact strength
- Flexural strength
- Volume and surface resistivity

Certification will cover a blend of 2500 lb or more, with the actual values recorded and furnished to the customer on request. The alternate certification process involves the compounder furnishing a letter of compliance that confirms that the actual blend involved does meet all the requirements of the specification.

2.7.9.3 Testing equipment and procedures. Most compounders employ a variety of testing tools to control and monitor their manufacturing processes, as well as their R&D programs, and for troubleshooting problems encountered in the field. A list of such equipment follows.

Unitron metallograph. This is a sophisticated metal detection device used to check production as well as troubleshoot areas.

Scanning electron microscope (SEM). An SEM is employed to examine surfaces from low magnification up to a 100,000× enlargement. It is an excellent research and problem-solving tool.

Dynamic mechanical spectrophotometer (DMS). The DMS measures the viscoelastic properties of polymers, thus determining a compound's viscosity and elastic modulus, following the change of these properties over time and changes in temperature.

Infrared spectrophotometer. The *infrared spectrophotometer* can provide information concerning the composition of a compound such as degradation, replacement for fillers and reinforcing agents, and it can be used to evaluate the purity of resins, fillers, catalysts, reinforcements, solvents, and lubricants.

Capillary rheometer. The Monsanto *capillary rheometer* measures the viscosity properties of polymers and provides a direct measure of viscosity and the change in viscosity with time and flow rate at plastication temperatures. The capillary orifice simulates the gate and runner system of actual molding conditions, thus providing valuable flow information for molding compounds.

Brabender plasticorder. The plasticorder is a small mixer capable of measuring the viscosity and the gel time of thermosetting molding compounds with results that can be correlated to the performance of a compound during molding conditions.

High-pressure liquid chromatograph. This device is available in two forms:

1. Gel *permeation.* Measures the molecular weight distribution and average molecular weight of the molecules in a sample of a compound.
2. *Liquid.* Detects and measures the amount of chemical constituents present in a compound.

Thermal analysis (TA). DuPont's TA equipment is available in four modules that provide information regarding the effect of temperature on a compound's physical properties.

1. *Differential scanning calorimeter (DSC).* This measures heat uptake or heat release of a compound as the temperature is raised and also the heat effects associated with material transitions such as melting.
2. *Thermal analyzer (TMA).* A thermal analyzer measures the variation in the length of a sample as temperature is increased. It is good for comparing this property with a sample of another compound. TMA also measures thermal transition points by predicting the point and rate at which a compound will melt as well as determining the temperature at which blistering will occur if a molded part has not been properly postbaked.
3. *Dynamic mechanical analyzer (DMA).* Measures a compound's modulus (stiffness) as its temperature is raised. This instrument has provided interesting insights into the properties of phenolics as well as those of DAPs, thermoset polyesters, silicones, and epoxies, by indicating the ability of thermosets to retain their modulus at elevated temperatures.
4. *Thermogravimetric analyzer (TGA).* TGA measures weight changes in a sample as the temperature is varied, providing a useful means to determine degradative processes and heat resistance in polymeric compounds.

Particle size analyzer (PSA). A particle size analyzer is an accurate and automatic development tool that allows for a very rapid measurement of particle size distribution in powder or slurry compounds.

Humidity chamber. A humidity chamber is employed to measure the effects of temperature and humidity cycles on molded parts.

Instron testers. *Instron testers* will measure flexural, tensile, and compressive strength as well as stress-strain curves at ambient temperatures, and, when fitted with an environmental chamber, the flexural and tensile tests can run at elevated temperatures.

2.7.9.4 Rheology (flow testing). Easily the foremost characteristic of a thermosetting molding compound is its ability to flow under pressure within the confines of the heated mold. This property value is of utmost importance in the eyes of the molder and will vary according to the molding method, mold design, molding equipment, and certainly the configuration of the molded part.

Since the molding compound is subjected to elevated temperature and pressure in the molding process, its ability to flow is greatly affected by the chemical reaction taking place as a result of these conditions. As a general rule, the speed of this chemical reaction will double with every 10° increase in temperature. In every thermosetting molding cycle, regardless of the type of compound, mold, or molding press used, the molding compound will go through the typical thermosetting reaction curve shown in Fig. 2.21. At the top left side (A), the compound is at room temperature and 0 pressure. With pressure applied and the compound increasing in temperature, its viscosity decreases as shown on the slope at B. This decrease in viscosity continues along the B slope as the compound temperature increases until the compound reaches its peak of flow at C, just prior to a rapid acceleration of the reaction as the curve turns upward to D, thus completing the cure.

Every individual molding compound possesses its own flow characteristics, which are affected by the resin, catalyst, and reinforcement ratio as well as the type and content of all the ingredients that make up the complete formulation. The *desired* rheology or flow property requirements of a specific compound will be determined by

Molding method. Compression, transfer, injection, mold designs, part configuration, number of cavities and their location in the mold, and size and location of the runner and gate system.

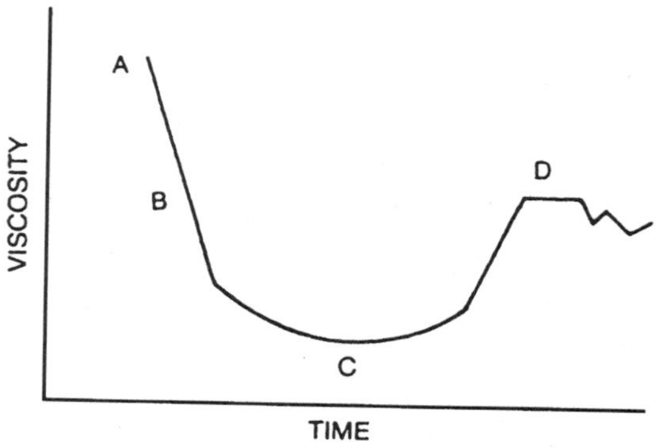

Figure 2.21 Typical thermosetting curve.

Flow specifications. Generally identified as stiff, medium, or soft or by a designated cure rate or flow time. All thermosetting molding compounds possess flow characteristics that are both measurable and controllable, with the important characteristics being the rate of curing, speed of the flow, distance of the flow, and finally the amount of compound used during the flow time.

2.7.9.5 Flow testing procedures. The five most widely used flow testing procedures have one main purpose—to provide specific and detailed information based on the compound's intended use. Compounds that are designated for use in compression molds will have decidedly different flow requirements than if the intended use is in either a transfer or an injection molding process.

In all of the following flow tests, with the exception of the Brabender, three elements are always kept constant during the testing procedure:

- *Amount of compound.* Charge weight
- *Mold temperature.* Usually 300°F
- *Molding pressure.* Usually 1000 lb/in^2

Cup closing (test and mold) (Fig. 2.22). With the mold set at 300°F (148°C) and a molding pressure of 1000 $lb/in,^2$ a room-temperature charge of compound is placed in the lower half of the mold, and the mold is closed. The time required for the mold to completely close is recorded in seconds. The longer the time, the stiffer the compound; the shorter the time, the shorter the flow. Generally speaking, stiff flow compounds will be 15 s or more, whereas the medium flow compounds will be in the 8- to 14-s range, and the soft flow compounds closing in less than 8 s.

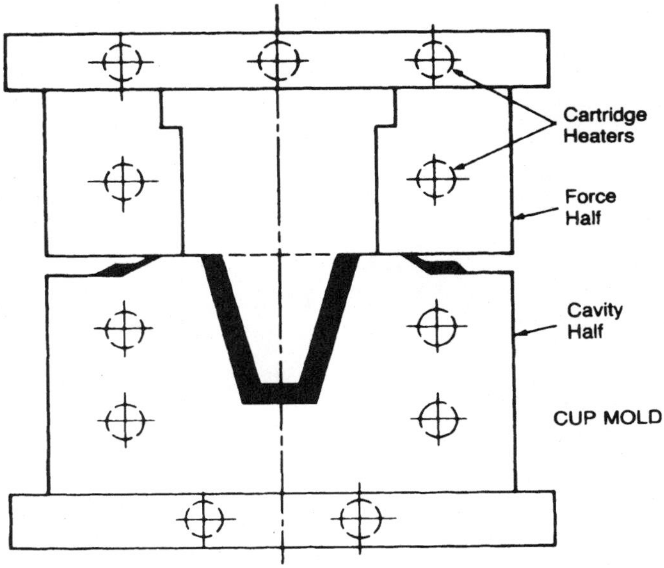

Figure 2.22 Cup closing test and mold.

Disk flow I (test and mold) (Fig. 2.23). With the mold in the open position, a measured amount of room-temperature compound is placed on the lower half of the mold, and the mold is closed and then reopened as soon as the compound is cured. The molded disk is then measured for diameter and thickness. The thinner the disk, the softer the flow; the thicker the disk, the stiffer the flow.

Disk flow II (test and mold) (Fig. 2.24). With the diameter of the disk being used as the gauge, the molded disk is placed on a target of concentric circles numbered 1 to 5. A disk matching the #1 circle on the target will he designated as 1S flow, whereas a disk matching the #5 circle will carry a 5S How. The higher the number, the softer the flow.

Orifice flow I (test and mold) (Fig. 2.25). This flow test involves the use of a mold with a lower plate containing a cavity into which a measured quantity of room-temperature compound is placed. The upper plate has a plunger with two small orifices cut into the outer circumference, as shown in Fig. 2.25. The test generally uses a charge of 12 to 15 g and a mold temperature of 300°F (148°C). The molding pressure employed can be 600, 900, 1800, or 2700 lb/in^2, depending on the molding process to be used. With the charged mold on, the heated platens of the press close, and the compound is forced out of the two orifices. The mold is kept closed until the compound has stopped flowing and is cured. On completion of the molding cycle, the cured compound remaining in the mold is extracted and weighed to determine the percentage of the flow. For example, if 30 percent of the

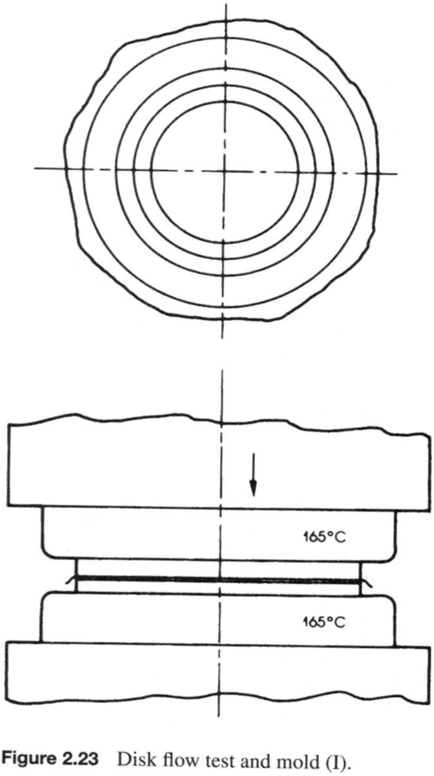

Figure 2.23 Disk flow test and mold (I).

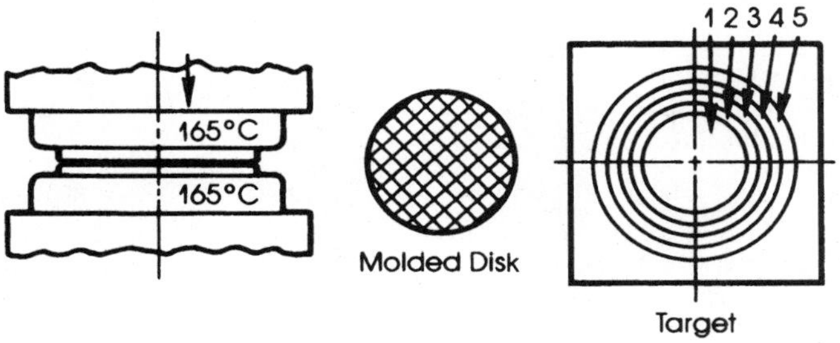

Figure 2.24 Disk flow test and mold (II).

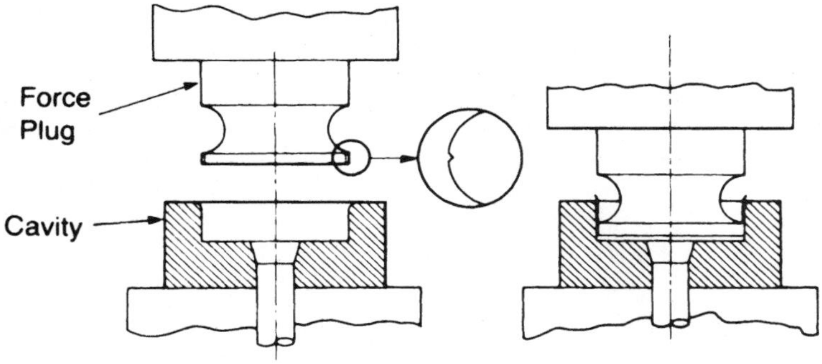

Figure 2.25 Orifice flow test and mold (I).

original shot weight is left in the mold, and the molding pressure was 900 lb/in^2, the compound would be designated as 30 percent @ 900 lb/in^2. The use of varying pressures reflects the need to consider the flow types normally used for either compression, transfer, or injection molding.

Orifice flow II (test and mold) (Fig. 2.26). As shown, the mold has a pot in the lower half and a plunger affixed to the top half. The pot block is designed with a sprue hole in the bottom of the cavity, which feeds the molding compound into a runner that comes out one side of the mold. The pot is charged with 90 g of compound, and the mold is closed under 1000 lb/in^2 with a mold temperature of 300°F (148°C). The compound exiting from the runner can he weighed once the flow has ceased. If the time of flow from the start has been timed from start to finish, the rate of flow can he calculated as x g/s. This is a very useful tool for use in transfer or injection molding of thermosets.

If desired, the extrudate from the mold can be cut off every 20 s, and each segment can then be weighed. The result will provide a time-weight ratio that depicts the decrease in viscosity and increase in weight as the compound reacts to the increased mold temperature.

Spiral flow (test and mold). There are two types of spiral flow molds—one for the very soft flow encapsulation compound generally associated with the encapsulation grades of the epoxy family of compounds, and a spiral flow mold, which is used when testing the high-pressure phenolic, DAP, melamine, urea, epoxy, and thermoset polyester compounds.

Figure 2.27 illustrates the *Emmi mold,* which is used for testing the very soft flow encapsulation compounds. It contains a runner system cut in an Archimedes spiral starting

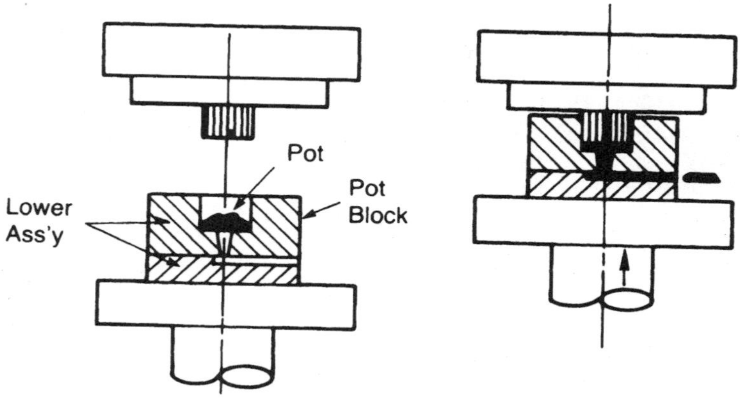

Figure 2.26 Orifice flow test and mold (II).

from the center of the mold. The runner configuration is 1 /8-in half-round, 100 in long. The mold is heated to 300°F and run @ 300 lb/in^2. The runner is marked in linear inches, and the test results are recorded as a flow of x in.

Figure 2.28 depicts the *Mesa flow mold,* which has a different runner configuration. The runner is cut 0.250 in wide by 0.033 in deep and is 50 linear inches long. The molding cycle is run with the molding pressure set at 1000 lb/in^2, and the mold temperature is 300°F.

The flow measurement results from either the Emmi or Mesa molds can be enhanced by using a timer in conjunction with the transfer stroke during the molding cycle, thus gathering additional data relating to the quantity of the compound used as well as the distance of the compound flow.

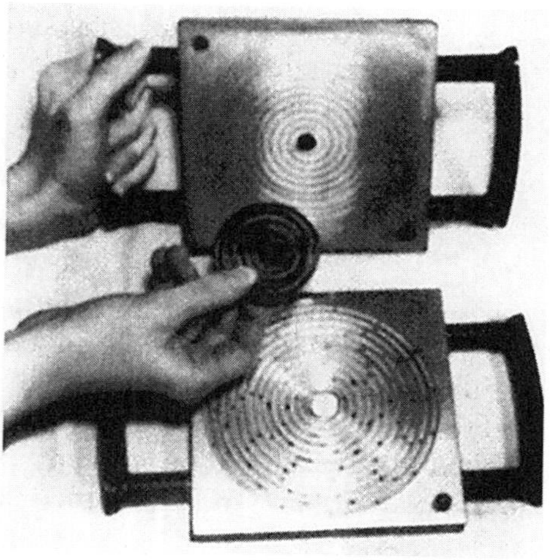

Figure 2.27 Spiral flow test and mold (Emmi).

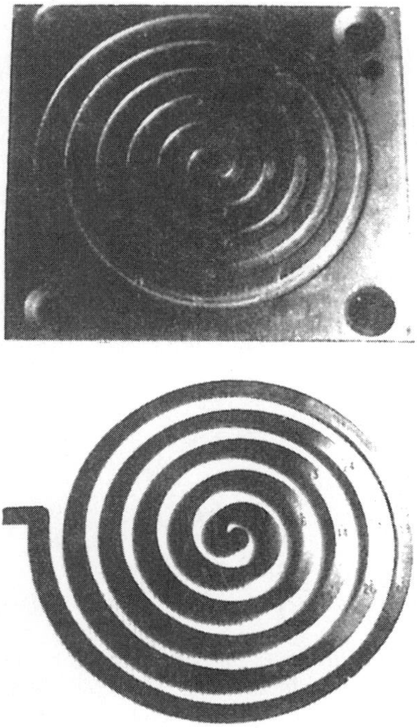

Figure 2.28 Spiral flow test and mold (Mesa).

Brabender Plasticorder (Figs. 2.29 and 2.30). The introduction of the extruder screw as an integral component of the so-called *closed mold method (injection)* brought the need for a more sophisticated means for measuring the rheology of thermosetting compounds when exposed to the different molding conditions encountered with this molding technique. There is a need to know the duration of a compound's flow life when exposed to both the initial barrel temperatures as the compound is prepared for its movement into the mold, and when the compound enters the much hotter mold. Generally, the compound temperature within the barrel or "reservoir" will not exceed 250°F and, once it moves into the mold, it will be met with temperatures in the 325 to 400°F range. The compound will also begin to gather more heat as it flows at great speed through the runner system and then through a small gate area into the mold cavities. This "shear" action will raise the compound temperature considerably with the effect of greatly increasing the cure rate and diminishing the How life of the compound. The Brabender Plasticorder* as shown in Fig. 2.29 will provide meaningful data for a specific compound's flow life or duration when exposed to the thermal conditions previously described (Fig. 2.30). The instrument supplier, Brabender Instruments, Inc., should be contacted for complete technical details.

2.7.9.6 Cure characteristics. The degree of cure accomplished during the molding cycle is the criterion for determining the ultimate mechanical, electrical, chemical, and

* Tradename of Brabender Instruments, Inc., P.O. Box 2127, South Hackensack, NJ, 07606.

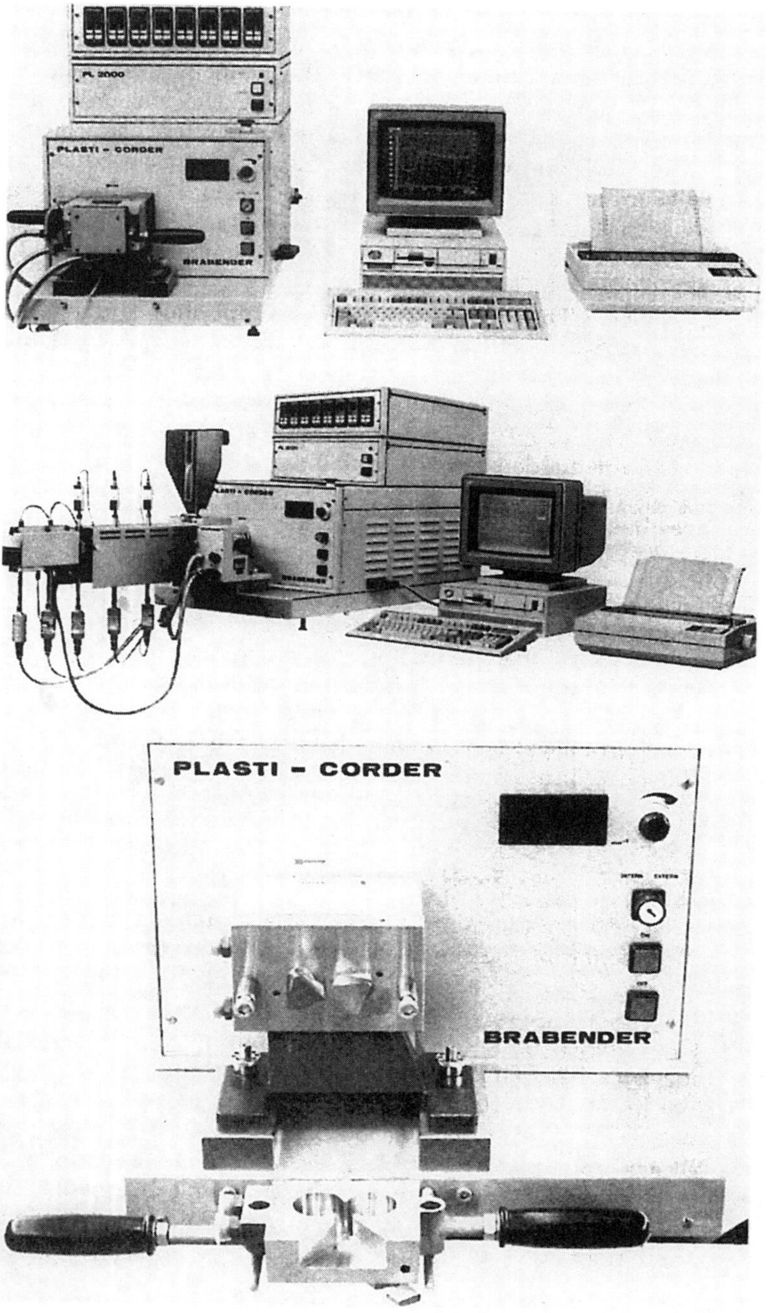

Figure 2.29 Brabender Plasticorder.

BRABENDER

Data-Processing Plasti-Corder PL2000 and Mixer Measuring Head
Flow-Curing Behav. of Crosslink. Polymers acc. to DIN suggestion

Test-Conditions

Order	: BRABENDER	Mixer-Temp.	:	140 °C
Operator	: EICKMEIER	Speed	:	30 1/min
Check-Date	: 16. MAY '88	Meas. Range	:	50 Nm
PL-Type	: 2000-3	Zero-Suppr.	:	0 %
Mixer-Type	: MB 30	Damping	:	3
Load. Chute	: MANUAL + 5 KG	Test-Time	:	5.0 min
Sample	: THERMOSETTING	Sample-Weight	:	24.00 g
Additive	:	Codenumber	:	1
		Start-Temp.	:	137 °C

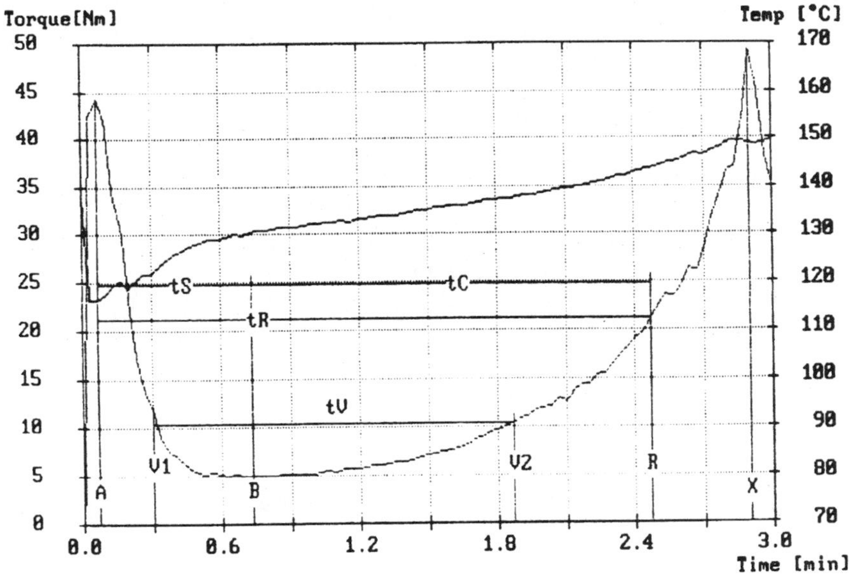

Value		at Time	Torque [Nm]	Stocktemp.[°C]
Loading Peak	A	00:00:04	44.3	117
Minimum	B	00:00:44	5.0	130
Maximum	X	00:02:54	49.3	149
Start Delay Time	V1	00:00:18	11.7	122
Stop Delay Time	V2	00:01:52	10.4	138
Stop Reaction	R	00:02:28	21.2	144

Integration / Energy

- Load.Peak to Minimum	A - B :	W1 =	1.6 [kNm]
- Minimum to Reaction	B - R :	W2 =	3.1 [kNm]
- Delay Time	tV :	W3 =	1.9 [kNm]
- Load.Peak to Reaction	A - R :	W4 =	4.7 [kNm]
- Specific Energy (W4/Sample-Weight)	:	W5 =	0.2 [kNm/g]

Results

- Delay Time	f(B+ 5.0 Nm)	V1- V2:	tV =	00:01:34
- Melt Time		A - B :	tS =	00:00:40
- Curing Time		B - R :	tC =	00:01:44
- Reaction Time	f(B+ 15.0 Nm)	A - R :	tR =	00:02:24

Figure 2.30 Brabender Plasticorder data output.

thermal properties of the molded article. The individual compound, depending on its resin component, will exhibit its very best property profile when the molded article has achieved maximum molded density and has been fully cured. The degree of cure can be ascertained by one or more of the specific testing procedures based on the resin used as described in the following subsections.

Allyls (DAP and DAIP). The reflux apparatus described in Fig. 2.31 is the generally accepted cure testing procedure for allyls. Sections of a molded specimen are refluxed in boiling chloroform for 3 hr, after which the specimen is visually examined for evidence of a serious problem such as cracking, swelling, fissuring, or actual crumbling of the surface. Surface hardness can be checked with a hardness tester, but this should not be done until

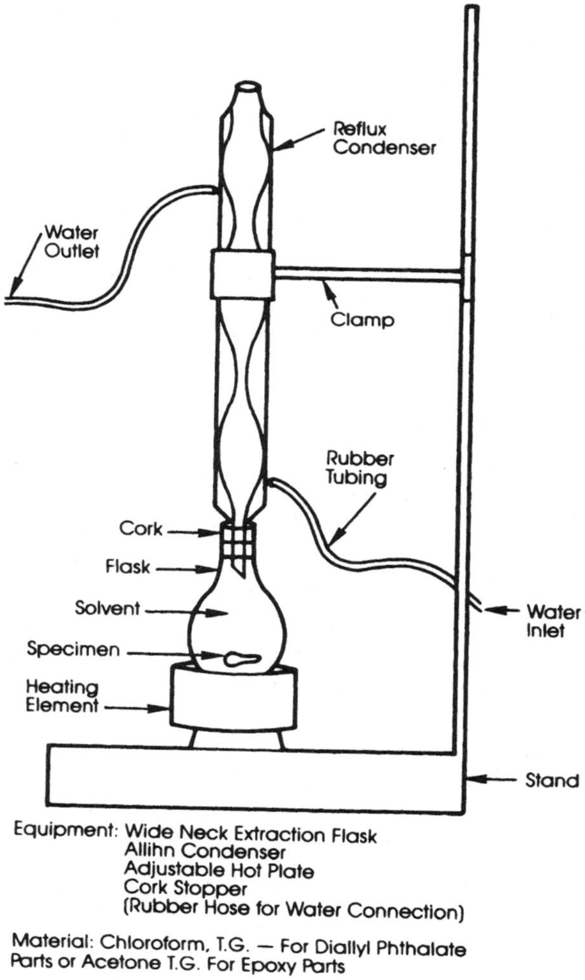

Figure 2.31 Cure test reflux apparatus.

some 12 to 14 hr have elapsed after the removal of the specimen from the boiling chloroform.

Phenolics and epoxies. ASTM D-494 describes the acetone extraction test that is commonly used for these two resin systems.

Thermoset polyesters. A Barcol hardness test, employed before and after boiling a section of the molded specimen in water for 1 hr, will detect any sign of undercure. Note that none of these chemical tests will accurately determine the actual effects of the degree of undercure on the ultimate integrity of a molded article. Often, what might be considered "undercured" by one of these tests will be blister free, rigid enough to be ejected from the mold, warpage free, and dimensionally correct while exhibiting electrical, mechanical, chemical, and thermal properties that are highly acceptable for commercial applications.

The integrity of a molded article may be further examined if the end-use application requires it, through more sophisticated thermoanalytical equipment such as

- Differential scanning calorimetry) DSC)
- Thermo-mechanical analysis (TMA)
- Thermo-gravimetric analysis (TGA)

These techniques involve the analysis of the changes in the chemical and physical properties of a molded specimen as a function of temperature. They are not generally used for production or quality control but are quite useful in resolving problems that could be the result of insufficient cure and that are not readily detected by the more common visual or chemical tests. There is a "rule-of-thumb" test procedure that is a fairly reliable cure test, and it is based upon a visual inspection of the molded article for obvious defects such as

- Lack of rigidity when ejected from the mold
- Cracks, surface porosity, and/or blisters and swelling
- Lack of flatness and/or signs of warpage that affect the ultimate dimensional stability

2.7.9.7 Postbaking. Thermosetting molded articles are frequently postcured by afterbaking at recommended times and temperatures for the purpose of enhancing mechanical and thermal properties, particularly when the end-use application requires optimal performance. The usual end results of postbaking are

- Improved creep resistance
- Reduction of molded-in volatiles and/or stresses
- Improved dimensional stability

In a recent study on postcuring, the following observations were made:

1. The starting temperature should be well below the glass transition *(T_g)* of the compound involved. A temperature below the actual molding temperature will suffice.
2. Parts of uneven cross-sectional thickness will exhibit uneven shrinkage.
3. Postbaking should be done with a multistage temperature cycle as shown here.
4. The reinforcement system of the compound will, to a large extent, distort the temperature and time cycles. Parts molded with organic reinforcements must be postcured at lower temperatures than those molded with glass and mineral reinforcement systems.

Postcure cycles1

Phenolics

One-step resole *compounds:* At 1/8 in or less thick—2 hr @ 280°F (138°C), 4 hr @ 330°F (166°C), and 4 hr @ 375°F (191°C)

Two-step novolac compounds: At 1 in or less thick—2 hr @ 300°F (149°C), 4 hr @ 350°F (177°C, and 4 hr @ 375°F (191°C)

For parts exceeding the 1/8-in thickness, it is recommended that the time be doubled for each 1/16 in of added thickness for best results. Longer times are more effective than increased temperature.

Allyls (DAP and DAIP). The recommendations are 8 hr at a range starting at 275°F (135°C) and raising the temperature 20°F/hr up to 415°F (213°C).

Epoxies. The recommendations are 8 hr at a range starting at 275°F (135°C) and raising the temperature 20°F/hr up to 415°F (213°C).

Thermoset polyesters. The recommendations are 8 hr at 250°F (121°C) and raising the temperature 20°F up to a maximum of 350°F (177°C).

Note that much of the information on postbaking procedures have come from a study conducted by Bruce Fitts and David Daniels of the Rogers Corporation.[18]

References

1. R. E. Wright, Chap. 1 in *Molded Thermosets,* Hanser Verlag, Munich, 1991.
2. B. J. Shupp, *Plastics Handbook, Modern Plastics Magazine,* McGraw-Hill, New York, 1994.
3. M. A. Chaudhari, *Plastics Handbook, Modern Plastics Magazine,* McGraw-Hill, New York, 1994.
4. J. Gannon, *Plastics Handbook, Modern Plastics Magazine,* McGraw-Hill, New York, 1994.
5. W. Ayles, *Plastics Handbook, Modern Plastics Magazine,* McGraw-Hill, New York, 1994.
6. W. McNeil, *Plastics Handbook, Modern Plastics Magazine,* McGraw-Hill, New York 1994.
7. R. Ray Patrylak, *Plastics Handbook, Modern Plastics Magazine,* McGraw-Hill, New York, 1994.
8. H. R. Gillis, *Plastics Handbook, Modern Plastics Magazine,* McGraw-Hill, New York, 1994.
9. R. Bruce Frye, *Plastics Handbook, Modern Plastics Magazine,* McGraw-Hill, New York, 1994.
10. Charles A Harper, Chap. 4 in *Handbook of Plastics, Elastomers, and Composites,* McGraw-Hill, New York, 1996.
11. Harry Raech, Jr., Chap. 1 in *Allylic Resins and Monomers,* Reinhold Publishing, New York, 1965.
12. Charles A. Harper, Chap. 2 in *Handbook of Plastics, Elastomers, and Composites,* McGraw Hill, New York, 1996.
13. Harry Raech, Jr., Chap. 1 in *Allylic Resins and Monomers,* Reinhold Publishing, New York, 1965.
14. Harry Dubois, *Plastics History USA,* Cahners, Boston, 1972.
15. Charles A. Harper, Chap. 3 in *Handbook of Plastics, Elastomers, and Composites,* McGraw-Hill, New York, 1996.

16. R. E. Wright, Chap. 1 in *Molded Thermosets,* Hanser Verlag, Munich, 1991.
17. Charles A. Harper, Chap. 1 in *Handbook of Plastics, Elastomers, and Composites,* McGraw-Hill, New York 1996.
18. Bruce Fitts and David Daniels, Applications Report, Rogers Corporation, Rogers, Conn., 1991.
19. Rogers Corporation, *Molding Compound Product Bulletin,* Rogers, Conn.

Chapter

3

Elastomeric Materials and Processes*

James M. Margolis

Montreal, Province of Quebec, Canada

3.1 Introduction

Another important group of polymers is the group that is elastic or rubberlike, known as *elastomers*. This chapter discusses this group of materials, including TPEs, MPRs, TPVs, synthetic rubbers, and natural rubber.

3.2 Thermoplastic Elastomers

Worldwide consumption of thermoplastic elastomer (TPE) for the year 2000 is estimated to be about 2.5 billion pounds, primarily due to new polymer and processing technologies, with an annual average growth rate of about 6 percent between 1996 and 2000. About 40 percent of this total is consumed in North America.[4]

TPE grades are often characterized by their hardness, resistance to abrasion, cutting, scratching, local strain, and wear. A conventional measure of hardness is Shore A and Shore D, shown in Fig. 3.1. Shore A is a softer, and Shore D is a harder TPE, with ranges from as soft as Shore A 28 to as hard as Shore D 82. Durometer hardness (ASTM D 2240) is an industry standard test method for rubbery materials, covering two types of durometers, A and D. The *durometer* is the hardness measuring apparatus, and the term *durometer hardness* is often used with Shore hardness values. There are other hardness test methods such as Rockwell hardness for plastics and electrical insulating materials (ASTM D785 and ISO 2039), and Barcol hardness (ASTM D2583) for rigid plastics. While hardness is

---

* The chapter author, editors, publisher, and companies referred to are not responsible for the use or accuracy of information in this chapter, such as property data, processing parameters, and applications.

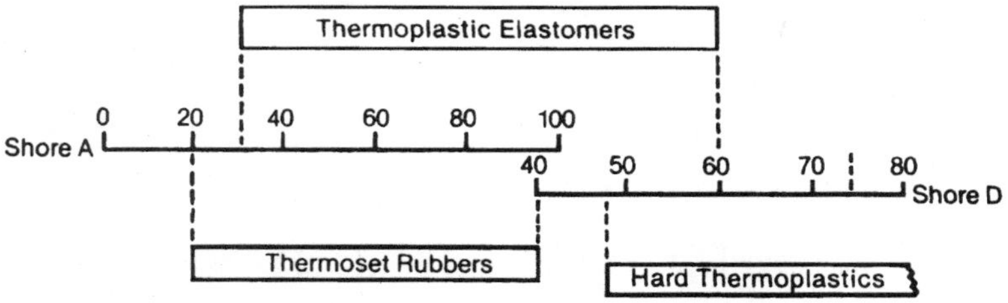

Figure 3.1 TPEs bridge the hardness ranges of rubbers and plastics. *(Source: Ref. 10, p. 5.2)*

often a quantifying distinction between grades, it does not indicate comparisons between physical/mechanical, chemical, and electrical properties.

Drying times depend on moisture absorption of a given resin. TPE producers suggest typical drying times and processing parameters. Actual processing temperature and pressure settings are determined by resin melt temperatures and rheological properties, mold cavity design, and equipment design such as screw configuration.

Performance property tables provided by suppliers usually refer to compounded grades containing property enhancers (additives) such as stabilizers, modifiers, and flame retardants. Sometimes, the suppliers' property tables refer to a polymer rather than a formulated compound.

3.2.1 Styrenics

Styrene block copolymers are the most widely used TPEs, accounting for close to 45 percent of total TPE consumption worldwide at the close of the twentieth century.[1] They are characterized by their molecular architecture, which has a "hard" thermoplastic segment (block) and a "soft" elastomeric segment (block) (see Fig. 3.2). Styrenic TPEs are usually styrene butadiene styrene (SBS), styrene ethylene/butylene styrene (SEBS), and styrene isoprene styrene (SIS). Styrenic TPEs usually have about 30 to 40 percent (wt) bound styrene; certain grades have a higher bound styrene content. The polystyrene endblocks create a network of reversible physical cross links that allow thermoplasticity for melt processing or solvation. With cooling or solvent evaporation, the polystyrene domains reform and harden, and the rubber network is fixed in position.[2]

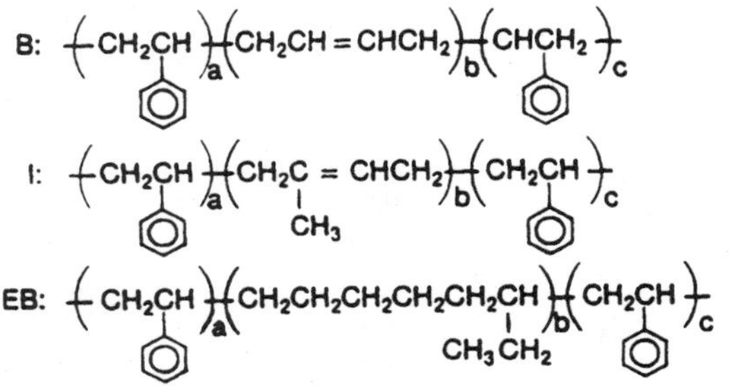

Figure 3.2 Structures of three common styrenic block copolymer TPEs: *a* and *c* = 50 to 80; *b* = 20 to 100. *(Source: Ref. 10, p. 5.12)*

Principal styrenic TPE markets are: molded shoe soles and other footwear; extruded film/sheet and wire/cable covering; pressure-sensitive adhesives (PSA) and hot-melt adhesives; and viscosity index (VI) improver additives in lube oils, resin modifiers, and asphalt modifiers. They are also popular as grips (bike handles), kitchen utensils, clear medical products, and personal care products.[1,4] Adhesives and sealants are the largest single market.[1] Styrenic TPEs are useful in adhesive compositions in web coatings.[1]

Styrenic block copolymer (SBC) thermoplastic elastomers were produced by Shell Chemical (KRATON®) and are available from Firestone Synthetic Rubber and Latex, Division of Bridgestone/Firestone (Stereon®*), Dexco Polymers (Vector®†), and EniChem Elastomers (Europrene®‡). SBC properties and processes are described for these four SBCs.

KRATON§ TPEs are usually SBS, SEBS, and SIS, as are SEP (styrene ethylene/propylene) and SEB (styrene ethylene/butylene).[2] The polymers can be precisely controlled during polymerization to meet property requirements for a given application.[2]

Two KRATON types are chemically distinguished: KRATON G and (raton D, as described below. A third type, KRATON Liquid®, poly(ethylene/butylene), is described thereafter. KRATON G and D have different performance and processing properties. KRATON G polymers have saturated midblocks with better resistance to oxygen, ozone and ultraviolet (UV) radiation, and higher service temperatures, depending on load, up to 350°F (177°C) for certain grades.[2] They can be steam sterilized for reusable hospital products. KRATON D polymers have unsaturated midblocks with service temperatures up to 150°F (66°C).[2] SBC upper service temperature limits depend on the type and weight percent (wt%) thermoplastic and type and wt% elastomer, and the addition of heat stabilizers. A number of KRATON G polymers are linear SEBS, while several KRATON D polymers are linear SIS.[2] KRATON G polymer compounds' melt process is similar to polypropylene; KRATON D polymer compounds' process is comparable to polystyrene (PS).[2]

Styrenic TPEs have strength properties equal to those of vulcanized rubber, but they do not require vulcanization.[2] Properties are determined by polymer type and formulation. There is wide latitude in compounding to meet a wide variety of application properties.[2] According to application-driven formulations, KRATONs are compounded with a hardness range from Shore A 28 to 95 (Shore A 95 is approximately equal to Shore D 40), sp gr from 0.90 to 1.18, tensile strengths from 150 to 5000 lb/in^2 (1.03 to 34.4 MPa), and flexibility down to –112°F (80°C) (see Table 3.1).[2]
KRATONs are resistant to acids, alkalis, and water, but long soaking in hydrocarbon solvents and oils deteriorates the polymers.[2]

Automotive applications range from window seals and gasketing to enhanced noise/vibration attenuation.[1] The polymers are candidates for automotive seating, interior padded trim and insulation, hospital padding, and topper pads.[1] SEBS is extruded/blown into 1-mil films for disposable gloves for surgical/hospital/dental, food/pharmaceutical, and household markets.[1]

KRATONs are used in PSAs, hot-melt adhesives, sealants, solution-applied coatings, flexible oil gels, modifiers in asphalt, thermoplastics, and thermosetting resins.[2] When KRATONs are used as an impact modifier in nylon 66, notched Izod impact strength can be increased from 0.8 ft-lb/in for unmodified nylon 66 to 19 ft-lb/in. Flexural modulus may

---

* Stereon is a registered trademark of Firestone Synthetic Rubber and Latex Company, Division of Bridgestone/Firestone.

† Vector is a registered trademark of Dexco, A Dow/Exxon Partnership.

‡ Europrene is a registered trademark of EniChem Elastomers.

§ KRATON styrenics were developed by Shell Chemical. In March 2001, the company formally announced finalization of sale to Ripplewood Holdings. For information, contact Kate Hill, Shell International Media Relations, telephone 44(0)2079342914.

TABLE 3.1 Typical Properties of a KRATON D and KRATON G Polymer for Use as Formulation Ingredients and as Additive (U.S. FDA Compliance)

Property [74°F (23°C)]	Kraton D D1101 (linear SBS)	Kraton G G1650 (linear SEBS)
Specific gravity, g/cm^3	0.94	0.91
Hardness, Shore A	71	75
Tensile strength, lb/in^2 (MPa)	4600 (32)	5000 (34)
300% modulus, lb/in2 (MPa)	400 (2.7)	800 (5.5)
Elongation, %	880	550
Set @ break, %	10	—
Melt flow index, g/10 min	<1	—
Brookfield viscosity, Hz @ 77°F (25°C), toluene solution	4000	8000
Styrene/rubber ratio	31/69	29/71

decrease from 44,000 lb/in^2 (302 MPa) for unmodified nylon 66 to about 27,000 lb/in^2 (186 MPa) for impact-modified nylon 66.

SBCs are injection molded, extruded, blow molded, and compression molded.[2]

KRATON Liquid polymers are polymeric diols with an aliphatic, primary OH^- group on each terminal end of the poly(ethylene/butylene) elastomer. They are used in formulations for adhesives, sealants, coatings, inks, foams, fibers, surfactants, and polymer modifiers.[13]

Two large markets for Firestone's styrenic block copolymer SBS Stereon TPEs are (1) impact modifiers (enhancers) for flame-retardant polystyrene and polyolefin resins and (2) PSA and hot-melt adhesives. Moldable SBS block copolymers possess high clarity and gloss, have good flex cycle stability for "living hinge" applications, and come in FDA-compliant grades for food containers and medical/hospital products.[1] Typical mechanical properties are 4600-lb/in^2 (31.7-MPa) tensile strength, 6000-lb/in^2 (41.4-MPa) flexural strength, and 200,000-lb/in^2 (1.4-GPa) flexural modulus.[1]

Stereon stereospecific butadiene styrene block copolymer is used as an impact modifier in PS, high-impact polystyrene (HIPS), polyolefin sheet and films, such as blown film grade linear low-density polyethylene (LLDPE), to achieve downgauging and improve tear resistance and heat sealing.[1] Blown LLDPE film modified with 7.5 percent stereospecific styrene block copolymers has a Dart impact strength of 185°F per 50 g, compared with 135°F per 50 g for unmodified LLDPE film. These copolymers also improve environmental stress crack resistance (ESCR) (especially to fats and oils for meat/poultry packaging trays), increase melt flow rates, increase gloss, and meet U.S. FDA 21 CFR 177.1640 (PS and rubber modified PS) with at least 60 percent PS for food contact packaging.[1] When used with thermoformable foam PS, flexibility is improved without sacrificing stiffness, allowing deeper draws.[1] The stereospecific butadiene block copolymer TPEs are easily dispersed and improve blendability of primary polymer with scrap for recycling.

Vector SBS, SIS, and SB styrenic block copolymers are produced as diblock-free and diblock copolymers.[29] The company's process to make linear SBCs yields virtually no diblock residuals. Residual styrene butadiene and styrene isoprene require endblocks at both ends of the polymer to have a load-bearing segment in the elastomeric network.[29] However, diblocks are blended into the copolymer for certain applications.[29] Vector SBCs are injection molded, extruded, and formulated into pressure-sensitive adhesives for tapes and labels, hot-melt product-assembly adhesives, construction adhesives, mastics, sealants, and asphalt modifiers.[29] The asphalts are used to make membranes for single-ply roofing and waterproofing systems, binders for pavement construction and repair, and

sealants for joints and cracks.[29] Vector SBCs are used as property enhancers (additives) to improve the toughness and impact strength at ambient and low temperatures of engineering thermoplastics, olefinic and styrenic thermoplastics, and thermosetting resins.[29] The copolymers meet applicable U.S. FDA food additive 21 CFR 177.1810 regulations and United States Pharmacopoeia (USP) (Class VI medical devices) standards for health-care applications.[29]

The company's patented hydrogenation techniques are developed to improve SBC heat resistance as well as ultraviolet resistance.[29]

EniChem Europrene SOL T products are styrene butadiene and styrene isoprene linear and radial block copolymers.[1] They are solution-polymerized using anionic type catalysts.[33] The molecules have polystyrene endblocks with central elastomeric polydiene (butadiene or isoprene) blocks.[33] The copolymers are *(S-B)nX* type where S = polystyrene, B = polybutadiene and polyisoprene, and X = a coupling agent. Both configurations have polystyrene (PS) endblocks, with bound styrene content ranging from 25 to 70 percent (wt).[1] Polystyrene contributes styrene hardness, tensile strength, and modulus; polybutadiene and polyisoprene contribute high resilience and flexibility, even at low temperatures.[1] Higher molecular weight (MW) contributes a little to mechanical properties but decreases melt flow characteristics and processibility.

The polystyrene and polydiene blocks are mutually insoluble, and this shows with two T_g peaks on a cartesian graph with tan δ (y axis) versus temperature (x axis): one T_g for the polydiene phase and a second T_g, for the polystyrene phase. A synthetic rubber, such as SBR, shows one T_g.[33] The two phases of a styrenic TPE are chemically bound, forming a network with the PS domains dispersed in the polydiene phase. This structure accounts for mechanical/elastic properties and thermoplastic processing properties.[33] At temperatures up to about 167°F (80°C), which is below PS T_g of 203 to 212°F (95 to 100°C), the PS phase is rigid.[33] Consequently, the PS domains behave as cross-linking sites in the polydiene phase, similar to sulfur links in vulcanized rubber.[33] The rigid PS phase also acts as a reinforcement, as noted here.[33] Crystal PS, HIPS, poly-alpha-methylstyrene, ethylene vinyl acetate (EVA) copolymers, low-density polyethylene (LDPE), and high-density polyethylene (HDPE) can be used as organic reinforcements. $CaCO_3$, clay, silica, and silicates act as inorganic fillers, with little reinforcement, and they can adversely affect melt flow if used in excessive amounts.[33]

The type of PS, as well as its percent content, affect properties. Crystal PS, which is the most commonly used, and HIPS increase hardness, stiffness, and tear resistance without reducing melt rheology.[1] High-styrene copolymers, especially Europrene SOL S types produced by solution polymerization, significantly improve tensile strength, hardness, and plasticity, and they enhance adhesive properties.[33] High styrene content does not decrease the translucency of the compounds.[33] Poly-alpha-methylstyrene provides higher hardness and modulus, but abrasion resistance decreases.[33] EVA improves resistance to weather, ozone, aging, and solvents, retaining melt rheology and finished product elasticity. The highest Shore hardness is 90 A, the highest melt flow is 16 g/10 min, and specific gravity is 0.92–0.96.[1]

Europrene compounds can be extended with plasticizers that are basically a paraflinic oil containing specified amounts of naphthenic and aromatic fractions.[33] Europrenes are produced in both oil-extended and dry forms.[1] Oils were specially developed for optimum mechanical, aging, processing, and color properties.[33] Increasing oil content significantly increases melt flow properties, but it reduces mechanical properties. Oil extenders must be incompatible with PS so as to avoid PS swelling, which would decrease mechanical properties even more.[33]

The elastomers are compounded with antioxidants to prevent thermal and photo-oxidation, which can be initiated through the unsaturated zones in the copolymers.[33] Oxidation can take place during melt processing and during the life of the fabricated product.[33] Phe-

nolic, or phosphitic antioxidants, and dilauryldithiopropionate as a stabilizer during melt processing, are recommended.[33] Conventional UV stabilizers are used such as benzophenone and benzotriazine.[33] Depending on the application, the elastomer is compounded with flow enhancers such as low MW polyethylene (PE), microcrystalline waxes or zinc stearate, pigments, and blowing agents.[33]

Europrene compounds, especially oil-extended grades, are used in shoe soles and other footwear.[1] Principal applications are impact modifiers in PS, HDPE, LDPE, polypropylene (PP), other thermoplastic resins and asphalt; extruded hose, tubing, O-rings, gaskets, mats, swimming equipment (eye masks, snorkels, fins, "rubberized" suits) and rafts; and pressure-sensitive adhesives (PSA) and hot melts.[1] SIS types are used in PSA and hot melts; SBS types are used in footwear.[1]

The copolymer is supplied in crumb form, and mixing is done by conventional industry practices, with an internal mixer or low-speed room temperature premixing and compounding with either a single- or twin-screw extruder.[33] Low-speed premixing/extrusion compounding is the process of choice.

Europrenes have thermoplastic polymer melt processing properties and characteristics of TPEs. At melt processing temperatures, they behave as thermoplastics, and below the PS T_g of 203 to 212°F (95 to 100°C) the copolymers act as cross-linked elastomers, as noted earlier. Injection-molding barrel temperature settings are from 284 to 374°F (140 to 190°C). Extrusion temperature at the head of the extruder is maintained between 212 and 356°F (100 and 180°C).

3.2.2 Olefenics and TPO Elastomers

Thermoplastic polyolefin (TPO) elastomers are typically composed of ethylene propylene rubber (EPR) or ethylene propylene diene "M" (EPDM) as the elastomeric segment and polypropylene thermoplastic segment.[18] LDPE, HDPE, and LLDPE; copolymers ethylene vinyl acetate (EVA), ethylene ethylacrylate (EEA), ethylene, methyl-acrylate (EMA); and polybutene-1 can be used in TPOs.[18] Hydrogenation of polyisoprene can yield ethylene propylene copolymers, and hydrogenation of 1,4- and 1,2-stereoisomers of S-B-S yields ethylene butylene copolymers.[1]

TPO elastomers are the second most widely used TPEs on a tonnage basis, accounting for about 25 percent of total world consumption at the close of the twentieth century (according to what TPOs are included as thermoplastic elastomers).

EPR and polypropylene can be polymerized in a single reactor or in two reactors. With two reactors, one polymerizes propylene monomer to polypropylene, and the second copolymerizes polypropylene with ethylene propylene rubber (EPR) or EPDM. Reactor grades are (co)polymerized in a single reactor. Compounding can be done in the single reactor.

Montell's in-reactor Catalloy®* ("catalytic alloy") polymerization process alloys propylene with comonomers, such as EPR and EPDM, yielding very soft, very hard, and rigid plastics, impact grades, or elastomeric TPOs, depending on the EPR or EPDM percent content. The term *olefinic* for thermoplastic olefinic elastomers is arguable because of the generic definition of olefinic. TPVs are composed of a continuous thermoplastic polypropylene phase and a discontinuous vulcanized rubber phase, usually EPDM, EPR, nitrile rubber, or butyl rubber.

Montell describes TPOs as flexible plastics, stating "TPOs are not TPEs."[32a] The company's Catalloy catalytic polymerization is a cost-effective process, used with propylene monomers that are alloyed with comonomers, including the same comonomers with dif-

---

* Catalloy is a registered trademark of Montell North America Inc., wholly owned by the Royal Dutch/Shell Group.

ferent molecular architecture. Catalloy technology uses multiple gas-phase reactors that allow the separate polymerization of a variety of monomer streams.[32] Alloyed or blended polymers are produced directly from a series of reactors that can be operated independently from each other to a degree.[32]

Typical applications are flexible products such as boots, bellows, drive belts, conveyor belts, diaphragms, and keypads; connectors, gaskets, grommets, lip seals, O-rings, and plugs; bumper components, bushings, dunnage, motor mounts, sound deadening; and casters, handle grips, rollers, and step pads.[9]

Insite®* technology is used to produce Affinity®† polyolefin plastomers (POPs), which contain up to 20 wt% octene co-monomer.[14] Dow Chemical's 8-carbon octene polyethylene technology produces the company's ULDPE Attane®‡ ethylene-octene-1 copolymer for cast and blown films. Alternative copolymers are 6-carbon hexene and 4-carbon butene, for heat-sealing packaging films. Octene copolymer POP has lower heat-sealing temperatures for high-speed form-fill-seal lines and high hot-tack strength over a wide temperature range.[14] Other benefits cited by Dow Chemical are toughness, clarity, and low taste/odor transmission.[14]

Insite technology is used for homogeneous single-site catalysts that produce virtually identical molecular structure, such as branching, co-monomer distribution, and narrow molecular weight distribution (MWD).[14] Solution polymerization yields Affinity polymers with uniform, consistent structures, resulting in controllable, predictable performance properties.[14]

Improved performance properties are obtained without diminishing processibility, because the Insite process adds long-chain branching onto a linear short-chain, branched polymer.[14] The addition of long-chain branches improves melt strength and flow.[14] Long-chain branching results in polyolefin plastomers processing at least as smooth as LLDPE and ultra-low-density polyethylene (ULDPE) film extrusion.[14] Polymer design contributes to extrusion advantages such as enhanced shear flow, drawdown, and thermoformability.

For extrusion temperature and machine design, the melt temperature is 450 to 550°F (232 to 288°C), the feed zone temperature setting is 300 to 325°F (149 to 163°C), 24/1 to 32/1 *L/D;* for the sizing gear box, use 5 lb/hr/hp (1.38 kg/hr/kW) to estimate power required to extrude POP at a given rate; for single-flight screws, line draw over the length of the line, 10 to 15 ft/min (3 to 4.5 m/min) maximum. Processing conditions and equipment design vary according to the resin selection and finished product. For example, a melt temperature of 450 to 550°F (232 to 288°C) applies to cast, nip-roll fabrication using an ethylene alpha-olefin POP[14a]), while 350 to 450°F (177 to 232°C) is recommended for extrusion/blown film for packaging, using an ethylene aipha-olefin POP.[14b]

POP applications are sealants for multilayer bags and pouches to package cake mixes, coffee, processed meats and cheese, and liquids; overwraps; shrink films; skin packaging; heavy-duty bags and sacks; and molded storage containers and lids.[14]

Engage®§ polyolefin elastomers (POEs), ethylene octene copolymer elastomers, produced by DuPont Dow Elastomers, use Insite catalytic technology.[5] Table 3.2 shows their low density and wide range of physical/mechanical properties (using ASTM test methods).[5]

The copolymer retains toughness and flexibility down to 40°F (–40°C).[5] When cross-linked with peroxide or silane, or by radiation, heat resistance and thermal aging increase to >302°F (>150°C).[5] Cross-linked copolymer is extruded into covering for low- and me-

---

* Insite is a registered trademark of DuPont Dow Elastomers LLC.

† Affinity is a registered trademark of The Dow Chemical Company.

‡ Attane is a registered trademark of The Dow Chemical Company.

§ Engage is a registered trademark of DuPont Dow Elastomers LLC.

TABLE 3.2 Typical Property Profile for Engage Polyolefin Elastomers (Unfilled Polymer, Room Temperature Except where Indicated)

Property	Values
Specific gravity, g/cm^3	0.857–0.91
Flexural modulus, lb/in^2 (MPa), 2% secant	435–27,55 (3–190)
100% modulus, lb/in^2(MPa)	145–>725 (1–>5)
Elongation, %	700+
Hardness, Shore A	50–95
Haze, %, 0.070 in (1.8 mm) injection-molded plaque	<10–20
Low-temperature brittleness, °F (°C)	<–104 (<–76)
Melt flow index, g/10 min	0.5–30
Melting point, °F (°C)	91–225 (33–107)

dium-voltage cables. POE elastomers have a saturated chain, providing inherent UV stability.[5] Ethylene octene copolymers are used as impact modifiers, for example, in polypropylene. Typical products are foams and cushioning components, sandal and slipper bottoms, sockliners and midsoles, swim fins, and winter and work boots; TPO bumpers, interior trim and rub strips, automotive interior air ducts, mats and liners, extruded hose and tube, interior trim, NVH applications, primary covering for wire and cable voltage insulation (low and medium voltage), appliance wire, semiconductive shields, nonhalogen flame-retardant and low smoke emission jackets, and bedding compounds.[5]

Union Carbide elastomeric polyolefin flexomers combine flexibility, toughness, and weatherability, with properties midrange between polyethylene and EPR.

3.3 Polyurethane Thermoplastic Elastomers (TPUs)

TPUs are the third most widely used TPEs, accounting for about 15 percent of TPE consumption worldwide.

Linear polyurethane thermoplastic elastomers can be produced by reacting a diisocyanate [methane diisocyanate (MDI) or toluene diisocyanate (TDI)] with long-chain diols such as liquid polyester or polyether polyols and adding a chain extender such as 1,4-butanediol.[17,18c] The diisocyanate and chain extender form the hard segment, and the long-chain diol forms the soft segment.[18c] For sulfur curing, unsaturation is introduced, usually with an allyl ether group.[17] Peroxide curing agents can be used for cross linking.

The two principal types of TPUs are polyether and polyester. Polyethers have good low-temperature properties and resistance to fungi; polyesters have good resistance to fuel, oil, and hydrocarbon solvents.

BASF Elastollan®* TPU elastomer property profiles show typical properties of polyurethane thermoplastic elastomers (see Table 3.3).

Shore hardness can be as soft as 70 A and as hard as 74 D, depending on the hard/soft segment ratio. Specific gravity, modulus, compressive stress, load-bearing strength, and tear strength are also hard/soft ratio dependent.[18c] TPU thermoplastic elastomers are tough, tear resistant, abrasion resistant, and exhibit low-temperature properties.[4]

* Elastopllan is a registered trademark of BASF Corporation.

TABLE 3.3 Mechanical Property Profile of Elastollan

Property	Value
Specific gravity, g/cm^3	1.11–1.21
Hardness, Shore	70 A–74 D
Tensile strength, lb/in^2 (MPa)	4600–5800 (31.7–40.0)
Tensile strength, lb/in^2 (MPa):	
100% elongation	770–1450 (5.3–10.0)
300% elongation	1300–1750 (9.0–12.0)
Elongation @ break, %	550–700
Tensile set @ break, %	45–50
Tear strength, Die C pli	515–770
Abrasion-resistance, mg loss (Tabor)	25

Dow Plastics Pellethane®* TPU elastomers are based on both polyester and polyether soft segments.[3]

Five series indicate typical applications:

1. Polyester polycaprolactones for injection-molded automotive panels, painted (without primer) with urethane and acrylic enamels, or water-based elastomeric coating
2. Polyester polycaprolactones for seals, gaskets, and belting
3. Polyester polyadipates extruded into film, sheet, and tubing
4. Polytetramethylene glycol ethers with excellent dielectrics for extruded wire and cable covering, and also for films, tubing, belting, and caster wheels
5. Polytetramethylene glycol ethers for healthcare applications[3]

Polyether-polyester hybrid specialty compounds are the softest nonplasticized TPU (Shore hardness 70 A), and they are used as impact modifiers.

Polycaprolactones possess good low-temperature impact strength for paintable body panels, good fuel and oil resistance, and hydrolytic stability for seals, gaskets, and belting.[3] Polycaprolactones have fast crystallization rates and high crystallinity, and they are generally easily processed into complex parts.

Polyester polyadipates have improved oil and chemical resistance but slightly lower hydrolytic stability than polycaprolactones, which are used for seals, gaskets, and beltings.

Polytetramethylene glycol ether resins for wire/cable covering have excellent resistance to hydrolysis and microorganisms, compared with polyester polyurethanes. Healthcare grade polyether TPUs are resistant to fungi, have low levels of extractable ingredients, offer excellent hydrolysis resistance, and can be sterilized for reuse by gamma irradiation, ethylene oxide, and dry heat—but not with pressurized steam (autoclave).[3] Polyether TPUs are an option for sneakers and athletic footwear components such as outer soles.

Bayer Bayflex®† elastomeric polyurethane reaction injection molding (RIM) is a two-component diphenylmethane diisocyanate- (MDI)-based liquid system produced in unreinforced, glass-reinforced, and mineral-/microsphere-reinforced grades.[15] They possess a wide stiffness range, relatively high impact strength, and quality molded product surface, and they can be in-mold coated. Room temperature properties are as follows:

---

* Pellethane is a registered trademark of The Dow Chemical Company.

† Bayflex is a registered trademark of Bayer Corporation.

- Specific gravity, 0.95 to 1.18
- Ultimate tensile strength, 2300 to 4200 lb/in^2 (16 to 29 MPa)
- Flexural modulus, 5000 to 210,000 lb/in^2 (34 to 1443 MPa)
- Tear strength, Die C, 230 to 700 lb/in (40 to 123 kN/m)[15]

Related Bayer U.S. patents are TPU-Urea Elastomers, U.S. Patent 5,739,250 assigned to Bayer AG, April 14, 1998; and RIM Elastomers Based on Prepolymers of Cyoloaliphatic Diisocyanates, U.S. Patent 5,738,253 assigned to Bayer Corp., April 14, 1998.

Representative applications are tractor body panels and doors, automotive fascia, body panels, window encapsulation, heavy-duty truck bumpers, and recreation vehicle (RV) panels.

Bayer's Texin®* polyester and polyether TPU and TPU/polycarbonate (PC) elastomers were pioneer TPEs in the early development of passenger car fronts and rear bumpers. PC imparts Izod impact strength toughness required for automotive exterior body panels. Extrusion applications include film/sheet, hose, tubing, profiles, and wire/cable covering. Hardness ranges from Shore A 70 to Shore D 75. Texin can be painted without a primer.

Morton International's Morthane®† TPU elastomers are classified into four groups: polyesters, polyethers, polycaprolactones, and polyblends. Polyester polyurethanes have good tear and abrasion resistance, toughness, and low-temperature flexibility, and Shore hardness ranges from 75 A to 65 D. They are extruded into clear film and tubing and fuel line hose. Certain grades are blended with acrylonitrile buta-diene styrene (ABS), styrene acrylonitrile (SAN), nylon, PC, polyvinyl chloride (PVC), and other thermoplastic resins. Polyethers possess hydrolytic stability, resilience, toughness, good low-temperature flexibility, easy processibility and fast cycles. They also have tensile strength up to 7500 lb/in^2 (52 MPa) and melt flow ranges from about 5 to 60 g/10 min. Hardnesses are in the Shore A range up to 90. Certain grades can be used in medical applications. Aliphatic polyester and polyether grades provide UV resistance for pipes, tubing, films, and liners. They can be formulated for high clarity.

The polyblends are polyester TPUs blended with ABS, SAN, PC, nylon, PVC, and other thermoplastics for injection molding and extrusion. A 10 to 20 percent loading into PVC compositions can increase mechanical properties 30 to 40 percent.[1] Although the elastomeric TPUs are inherently flexible, plasticizers may be recommended, for example, in films.

TPU elastomers are processed on rubber equipment, injection molded, extruded, compression molded, transfer molded, and calendered. To be fabricated into products, such as athletic shoe outer soles, the elastomer and ingredients are mixed in conventional rubber equipment (two-roll mills, internal mixers) and compounded.[17] Subsequently, the compound is processed; for example, injection molded.[17]

Typical melt processing practices are described with Pellethane TPU (see Table 3.4). The moisture content is brought to <0.02 percent before molding or extruding.[3] Desiccant, dehumidifying hopper dryers that can produce a –40°F (40°C) dew point at the air inlet are suggested. A dew point of –20°F (–29°C) or lower is suggested for TPU elastomers.[3] The suggested air-inlet temperature range is 180 to 230°F (82 to 110°C)[3]: 180 to 200°F (82 to 93°C) for the softer Shore A elastomers, and 210 to 230°F (99 to 110°C) for harder Shore D elastomers.

Drying time to achieve a given moisture content for resin used directly from sealed bags is shown in Fig. 3.3: about 4 hr to achieve <0.02 percent moisture content @ 210°F (99°C)

---

* Texin is a registered trademark of Bayer Corporation.

† Morthane is a registered trademark of Morton International Inc.

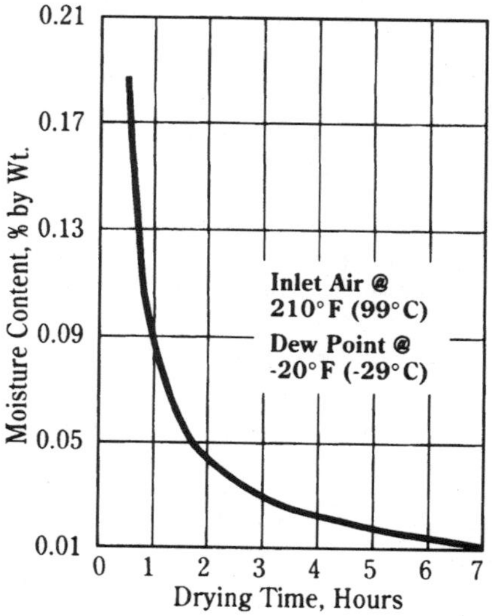

Figure 3.3 Typical drying curve for Pellethane elastomers. *(From Ref. 3, p. 6)*

TABLE 3.4 Typical Injection-Molding Settings* for Pellethane TPU

Temperature, °F (°C)	Shore A 80	Shore D 55
Melt temperature max	415 (213)	435 (224)
Cylinder zone		
Rear (feed)	350–370 (177–188)	360–380 (183–193)
Middle (transition)	360–380 (183–193)	370–390 (188–198)
Front (metering)	370–390 (188–199)	390–410 (204–210)
Nozzle	390–410 (199–210)	400–410 (204–210)
Mold temperature	80–140	(27–60)

*Typical temperature and pressure settings are based on Ref. 3. Settings are based on studies using a reciprocating screw, general-purpose crew, clamp capacity of 175 tons, and rated shot capacity of 10oz (280 g). Molded specimen thicknesses ranged from 0.065 to 0.125 in (1.7 to 3.2 mm).

air-inlet temperature and –20°F (–29°C) dew point.[3] When TPU elastomers are exposed to air just prior to processing, the pellets are maintained at 150 to 200°F (65 to 93°C), a warmer temperature than the ambient air.[3] A polymer temperature that is warmer than the ambient air reduces the ambient moisture absorption.[3]

Melt temperature is determined by T_m of resin and processing and equipment specifications, including machine capacity, rated shot size, screw configuration (*L/D*, flight number, and design), part design, mold design (gate type and runner geometry), and cycle time.[3] Shear energy created by the reciprocating screw contributes heat to the melt, causing the actual melt temperature to be 10 to 20°F (6 to 10°C) higher than the barrel temperature settings.[3] Temperature settings should take shear energy into account. To ensure maximum product quality, the processor should discuss processing parameters, specific

machinery, equipment, and tool and product data with the resin supplier. For example, if it is suspected that an improper screw design will be used, melt temperature gradient may be reversed. Instead of increasing temperature from rear to front, it may be reduced from rear to front.[3]

A higher mold temperature favors a uniform melt cooling rate, minimizing residual stresses, and improves the surface finish, mold release, and product quality. The mold cooling rate affects finished product quality. Polyether type TPU can set up better and release better.

High pressures and temperatures fill a high surface-to-volume ratio mold cavity more easily, but TPU melts can flash fairly easily at high pressures (Table 3.5). Pressure can be

TABLE 3.5 Typical Temperature Pressure Settings* for Pellethane TPU

	Pressure, lb/in^{2} † (MPa)
Injection pressure	
First stage	8000–15,000 (55.0–103)
Second stage	5000–10,000 (34.5–69.60)
Back pressure	0–100 (0–0.69)
Cushion, in/mm	0.25 (6.4)
Screw speed, r/min	50–75
Cycle time, s (injection, relatively slow to avoid flash, etc.)	3–10

*Typical temperature and pressure settings are based on Ref. 3. Settings are based on studies using a reciprocating screw, general-purpose screw, clamp capcity of 175 tons, and rated shot capacity of 10 oz (280 g). Molded speciment thickness ranged from 0.065 to 0.125 in (1.7– to 3.2–mm).

†U.S. units refer to line pressures; metric units are based on the pressure on the (average) cross-sectional area of the screw.

carefully controlled to achieve a quality product by using higher pressure during quick-fill, followed by lower pressure.[3] The initial higher pressure may reduce mold shrinkage by compressing the elastomeric TPU.[3]

The back pressure ranges from 0 to 100 lb/in^2 (0 to 0.69 MPa). TPU elastomers usually require very little or no back pressure.[3] When additives are introduced by the processor prior to molding, back pressure will enhance mixing, and when the plastication rate of the machine is insufficient for shot size or cycle time, a back pressure up to 200 lb/in^2 (1.4 MPa) can be used.[3]

Product quality is not as sensitive to screw speed as it is to process temperatures and pressures. The rotating speed of the screw, along with flight design, affects mixing (when additives have been introduced) and shear energy. Higher speeds generate more shear energy (heat). A speed above 90 r/min can generate excessive shear energy, creating voids and bubbles in the melt, which remain in the molded part.[3]

Cycle times are related to TPU hardness, part design, temperatures, and wall thickness. Higher temperature melt and a hot mold require longer cycles, when the cooling gradient is not too steep. The cycle time for thin-wall parts, <0.125 in (<3.2 mm), is typically about 20 s.[3] The wall thickness for most parts is less than 0.125 in (3.2 mm), and a wall thickness as small as 0.062 in (1.6 mm) is not uncommon. When the wall thickness is 0.250 in (6.4 mm), the cycle time can increase to about 90 s.[3]

Mold shrinkage is related to TPU hardness and wall thickness, part and mold designs, and processing parameters (temperatures and pressures). For a wall thickness of 0.062 in (1.6 mm) for durometer hardness Shore A 70, the mold shrinkage is 0.35 percent. Using the same wall thickness for durometer hardness Shore A 90, the mold shrinkage is 0.83 percent.[3]

Purging when advisable is accomplished with conventional purging materials, polyethylene, or polystyrene. Good machine maintenance includes removing and cleaning the screw and barrel mechanically with a salt bath or with a high-temperature fluidized sand bath.[3]

Reciprocating screw injection machines are usually used to injection mold TPU, and these are the preferred machines, but ram types can be successfully used. Ram machines are slightly oversized to avoid (1) incomplete melting and (2) steep temperature gradients during resin melting and freezing. Oversizing applies especially to TPU durometers harder than Shore D 55.[3]

Molded and extruded TPU have a wide range of applications, including:

- Automotive: body panels (tractors) and RVs, doors, bumpers (heavy-duty trucks), fascia, and window encapsulations
- Belting
- Caster wheels
- Covering for wire and cable
- Film/sheet
- Footwear and outer soles
- Seals and gaskets
- Tubing

3.3.1 Copolyesters

Thermoplastic copolyester elastomers are segmented block copolymers with a polyester hard crystalline segment and a flexible soft amorphous segment with a very low T_g.[35] Typically, the hard segments are composed of short-chain ester blocks such as tetramethylene terephthalate, and the soft segments are composed of aliphatic polyether or aliphatic polyester glycols, their derivatives, or polyetherester glycols. The copolymers are also called *thermoplastic etheresterelastomers (TEEEs)*.[35] The terms *COPE* and *TEEE* are used interchangeably (see Fig. 3.4).

TEEEs are typically produced by condensation polymerization of an aromatic dicarboxylic acid or ester with a low MW aliphatic diol and a polyakylene ether glycol.[35] Reaction of the first two components leads to the hard segment, and the soft segment is the product of the diacid or diester with a long-chain glycol.[35] This can be described as a melt transesterification of an aromatic dicarboxylic acid, or preferably its dimethyl ester, with a low MW poly(alklylene glycol ether) plus a short-chain diol.[35]

An example is melt phase polycondensation of a mixture of dimethyl terephthaate (DMT) + poly(tetramethylene oxide) glycol + an excess of tetramethylene glycol. A wide range of properties can be built into the TEEE by using different mixtures of isomeric phthalate esters, different polymeric glycols, and varying MW and MWD.[36] Antioxidants,

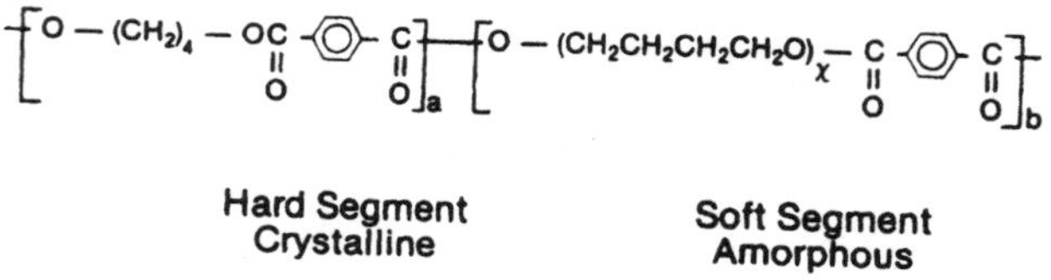

Figure 3.4 Structure of a commercial COPE TPE: a = 16 to 40, x = 10 to 50, and b = 16 to 40. *(Source: Ref. 10, p. 5.14)*

such as hindered phenols or secondary aromatic amines, are added during polymerization, and the process is carried out under nitrogen, because the polyethers are subject to oxidative and thermal degradation.[35]

Hytrel®* TEEElastomer block copolymers' property profile is given in Table 3.6.

TABLE 3.6 Typical Hytrel Property Profile[1]—ASTM Test Methods

Property	Value
Specific gravity, g/cm^3	1.01–1.43
Tensile strength @ break, lb/in^2 (MPa)	1,400–7,000 (10–48)
Tensile elongation @ break, %	200–700
Hardness Shore D	30–82
Flexural modulus, lb/in^2 (MPa)	
–40°F (–40°C)	9,000–440,000 (62–3,030)
73°F (22.8°C)	4,700–175,000 (32–1,203)
212°F (100°C)	1,010–37,000 (7.0–255)
Izod impact strength, ft-lb/in (J/m), notched	
–40°F (–40°C)	No break–0.4 (No break–20)
–73°F (22.8°C)	No break–0.8 (No break–40)
Tabor abrasion, mg/1000 rev	
CS–17 wheel	0–85
H–18 wheel	20–310
Tear resistance, lb/in, initial Die C	210–1,440
Vicat softening temperature, °F (°C)	169–414 (7601–212)
Melt point, °F (°C)	302–433 (150–223)

The mechanical properties are between rigid thermoplastics and thermosetting hard rubber.[35] Mechanical properties and processing parameters for Hytrel, and for a number of other materials in this chapter, can be found on the producers' Internet home pages.

Copolymer properties are largely determined by the soft/hard segment ratio; as with any commercial resin, properties are determined with compound formulations.

TEEEs combine flexural fatigue strength, low-temperature flexibility, good apparent modulus (creep resistance), DTUL and heat resistance, resistance to hydrolysis, and good chemical resistance to nonpolar solvents at elevated temperatures. A tensile stress/percent elongation curve reveals an initial narrow linear region.[19] COPEs are attacked by polar solvents at elevated temperatures. The copolymers can be completely soluble in metacresol, which can be used for dilute solution polymer analysis.[19]

TEEEs are processed by conventional thermoplastic melt-processing methods, injection molding, and extrusion, requiring no vulcanization.[35] They have sharp melting transitions and rapid crystallization (except for softer grades with higher amount of amorphous segment), and apparently melt viscosity decreases slightly with shear rate (at low shear rates).[35] The melt behaves like a Newtonian fluid.[35] In a true Newtonian fluid, the coefficient of viscosity is independent of the rate of deformation. In a non-Newtonian fluid, the apparent viscosity is dependent on shear rate and temperature.

TPE melts are typically highly non-Newtonian fluids, and their apparent viscosity is a function of shear rate.[10] TPE's apparent viscosity is much less sensitive to temperature

* Hytrel is a registered trademark of DuPont for its brand of thermoplastic polyester elastomer.

than it is to shear rate.[10] The apparent viscosity of TPEs as a function of apparent shear rate and as a function of temperature are shown in Figs. 3.5 and 3.6.

TEEEs can be processed successfully by low-shear methods such as laminating, rotational molding, and casting.[35] Standard TEEElastomers are usually modified with viscosity enhancers for improved melt viscosity for blow molding.[35]

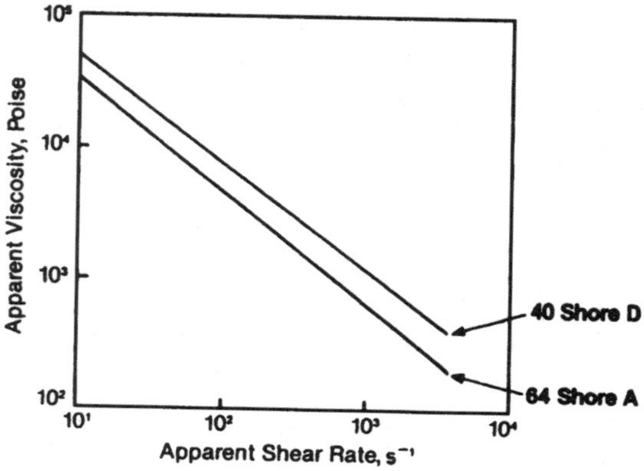

Figure 3.5 Viscosity as a function of shear rate for hard and soft TPEs. *(Source: Ref. 10, p. 5.31)*

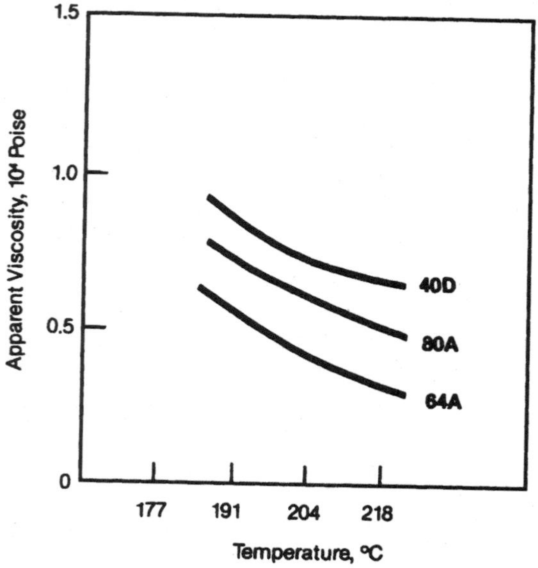

Figure 3.6 TPE (with different hardnesses) viscosity as a function of temperature. *(Source: Ref. 10, p. 5.31)*

Riteflex®* copolyester elastomers have high fatigue resistance, chemical resistance, good low-temperature [–40°F (–40°C)] impact strength, and service temperatures up to 250°F (121°C). Riteflex grades are classified according to hardness and thermal stability. The typical hardness range is Shore D 35 to 77. They are injection molded, extruded, and blow molded. The copolyester can be used as a modifier in other polymer formulations. Applications for Riteflex copolyester and other compounds that use it as a modifier include bellows, hydraulic tubing, seals, wire coating, and jacketing; molded air dams, automotive exterior panel components (fender extensions, spoilers), fascia and fascia coverings, radiator panels; extruded hose, belting, and cable covering; and spark plug and ignition boots.

Arnitel®† TPEs are based on polyether ester or polyester ester, including specialty compounds as well as standard grades.[34] Specialty grades are classified as (1) flame-retardant UL 94 V/0 @ 0.031 in (0.79 mm), (2) high modulus glass-reinforced, (3) internally lubricated with polytetrafluoro ethylene (PTFE) or silicone for improved wear resistance, and (4) conductive, compounded with carbon black, carbon fibers, nickel-coated fibers, stainless-steel fibers, for ESD applications.

Standard grades have a hardness range of about Shore D 38 to 74 for injection molding, extrusion, and powder rotational molding.[1] Arnitels have high impact strength, even at subzero temperatures, near-constant stiffness over a wide temperature range, and good abrasion.[34] They have excellent chemical resistance to mineral acids, organic solvents, oils, and hydraulic fluids.[34] They can be compounded with property enhancers (additives) for resistance to oxygen, light, and hydrolysis.[34] Glass fiber-reinforced grades, like other thermoplastic composites, have improved DTUL, modulus, and coefficient of linear thermal expansion (CLTE).[34]

Typical products are automotive exterior trim, fascia components, spoilers, window track tapes, boots, bellows, underhood wire covering, connectors, hose, and belts; appliance seals, power tool components, ski boots, and camping equipment.[1]

Like other thermoplastics, processing temperatures and pressures and machinery/tool designs are adjusted to the compound and application.

The following conditions apply to Arnitel COPE compounds, for optimum product quality: melt temperature range, 428 to 500°F (220 to 260°C); cylinder (barrel) temperature setting range, 392 to 482°F (200 to 250°C); mold temperature range for thin-wall products, 122°F (50°C) and for thick-wall products, 68°F (20°C).

Injection pressure is a function of flow length, wall thickness, and melt rheology, and it is calculated to achieve uniform mold filling. The Arnitel injection pressure range is <5000 to >20,000 lb/in^2 (<34 to >137 MPa). Thermoplastic elastomers may not require back pressure, and when back pressure is applied, it is much lower than for thermoplastics that are not elastomeric. Back pressure for Arnitel is about 44 to 87 lb/in^2 (0.3 to 0.6 MPa). Back pressure is used to ensure a homogeneous melt with no bubbles.

The screw configuration is as follows: thread depth ratio, approximately 1:2, and *L/D* ratio, 17/1 to 23/1 (standard three-zone screws: feed, transition or middle, and metering or feed zones).[34] Screws are equipped with a nonreturn valve to prevent backflow.[34] Decompression-controlled injection-molding machines have an open nozzle.[34] A short nozzle with a wide bore (3-mm minimum) is recommended to minimize pressure loss and heat due to friction.[34] Residence time should be as short as possible, and this is accomplished with barrel temperatures at the lower limits of recommended settings.[34]

Tool design generally follows conventional requirements for gates and runners. DSM recommends trapezoidal gates or, for wall thickness more than 3 to 5 mm, full sprue

---

* Riteflex is a registered trademark of Ticona.

† Arnitel is a registered trademark of DSM.

gates.[34] Vents approximately 1.5 × 0.02 mm are located in the mold at the end of the flow patterns, either in the mold faces or through existing channels around the ejector pins and cores.[34] Ejector pins and plates for thermoplastic elastomers must take into account the molded product's flexibility. Knock-out pins/plates for flexible products should have a large enough face to distribute evenly the minimum possible load. Prior to ejection, the part is cooled, carefully following the resin supplier's recommendation. The cooling system configuration in the mold base, and the cooling rate, are critical to optimum cycle time and product quality. The product is cooled as fast as possible without causing warpage. Cycle times vary from about 6 s for a wall thickness of 0.8 to 1.5 mm to 40 s for a wall thickness of about 5 to 6 mm. Drying temperatures and times range from 3 to 10 hr at 194 to 248°F (90 to 120°C).

In general, COPEs can require drying for 4 hr @ 225°F (107°C) in a dehumidifying oven to bring the pellet moisture content to 0.02 percent max.[1] The melt processing range is typically about 428 to 448°F (220 to 231°C); however, melt processing temperatures can be as high as 450 to 500°F (232 to 260°C). A typical injection-molding grade has a T_m of 385°F (196°C).[1] The mold temperature is usually between 75 and 125°F (24 and 52°C).

Injection-molding screws have a gradual transition (center) zone to avoid excess shearing of the melt and high metering (front) zone flight depths [0.10 to 0.12 in (2.5 to 3.0 mm)], a compression ratio of 3.0:1 to 3.5:1, and an *L/D* of 18/1 min (24/1 for extrusion).[18a] Barrier screws can provide more efficient melting and uniform melt temperatures for molding very large parts and for high-speed extrusions.[18a] When Hytrel is injection molded, molding pressures range from 6000 to 14,000 lb/in^2 (41.2 to 96.2 MPa). When pressures are too high, over-packing and sticking to the mold cavity wall can occur.[18a] Certain mold designs are recommended: large knock-out pins and stripper plates, and generous draft angles for parts with cores.[18a]

3.4 Polyamides

Polyamide TPEs are usually either polyester-amides, polyetherester-amide block copolymers, or polyether block amides (PEBA) (see Fig. 3.7). PEBA block copolymer molecular architecture is similar to typical block copolymers.[10] The polyamide is the hard (thermoplastic) segment, whereas the polyester, polyetherester, and polyether segments are the soft (elastomeric) segment.[10]

Polyamide TPEs can be produced by reacting a polyamide with a polyol such as polyoxyethylene glycol or polyoxypropylene glycol, a polyesterification reaction.[1] Relatively

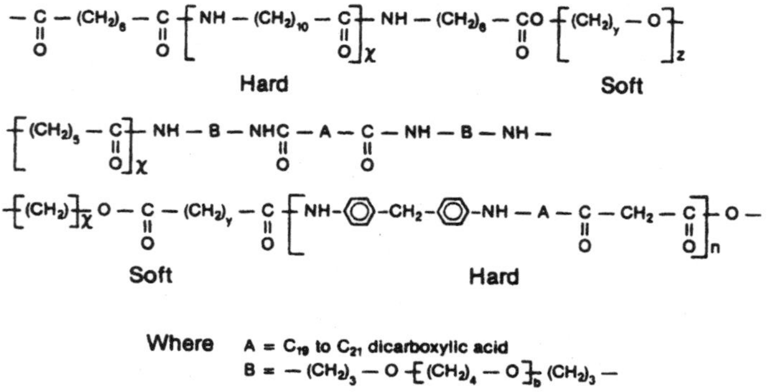

Figure 3.7 Structure of three PEBA TPEs. *(Source: Ref. 10, p. 5.17)*

high aromaticity is achieved by esterification of a glycol to form an acid-terminated soft segment, which is reacted with a diisocyanate to produce a polyesteramide. The polyamide segment is formed by adding diacid and diisocyanate.[1] The chain extender can be a dicarboxylic acid.[1] Polyamide TPEs can be composed of lauryl lactam and ethylene-propylene rubber (EPR).

Polyamide thermoplastic elastomers are characterized by their high service temperature under load, good heat aging, and solvent resistance.[1] They retain serviceable properties >120 hr @ 302°F (150°C) without adding heat stabilizers.[1] Addition of a heat stabilizer increases service temperature. Polyesteramides retain tensile strength, elongation, and modulus to 347°F (175°C).[1] Oxidative instability of the ether linkage develops at 347°F (175°C). The advantages of polyether block amide copolymers are their elastic memory, which allows repeated strain (deformation) without significant loss of properties, lower hysteresis, good cold-weather properties, hydrocarbon solvent resistance, UV stabilization without discoloration, and lot-to-lot consistency.[1]

The copolymers are used for waterproof/breathable outerwear; air-conditioning hose; underhood wire covering; automotive bellows; flexible keypads; decorative watch faces; rotationally molded basketballs, soccer balls, and volleyballs; and athletic footwear soles.[1] They are insert-molded over metal cores for nonslip handle covers (for video cameras) and coinjected with polycarbonate core for radio/TV control knobs.[1]

Pebax®* polyether block amide copolymers consist of regular linear chains of rigid polyamide blocks and flexible polyether blocks. They are injection molded, extruded, blow molded, thermoformed, and rotational molded.

The property profile is as follows: specific gravity about 1.0; Shore hardness range about 73 A to 72 D; water absorption, 1.2 percent; flexural modulus range, 2600 to 69,000 lb/in^2 (18.0 to 474 MPa); high torsional modulus from –40° to 0°C; Izod impact strength (notched), no break from –40to 68°F (–40 to 20°C); abrasion resistance; long wear life; elastic memory, allowing repeated strain under severe conditions without permanent deformation; lower hysteresis values than many thermoplastics and thermosets with equivalent hardness; flexibility temperature range, –40to 178°F (–40 to 81°C), and flexibility temperature range is achieved without plasticizer (it is accomplished by engineering the polymer configuration); lower temperature increase with dynamic applications; chemical resistance similar to polyurethane (PUR); good adhesion to metals; small variation in electrical properties over service temperature range and frequency (Hz) range; printability and colorability; tactile properties, such as good "hand," feel; and nonallergenic.[1]

The T_m for polyetheresteramides is about 248 to 401°F (120 to 205°C) and about 464°F (240°C) for aromatic polyesteramides.[18b]

Typical Pebax applications are one-piece, thin-wall soft keyboard pads; rotationally molded, high-resiliency, elastic memory soccer balls, basketballs, and volleyballs; flexible, tough mouthpieces for respiratory devices, scuba equipment, frames for goggles, and ski and swimming breakers; and decorative watch faces. Pebax offers good nonslip adhesion to metal and can be used for coverings over metal housings for hand-held devices such as remote controls, electric shavers, camera handle covers; coinjected over polycarbonate for control knobs; and employed as films for waterproof, breathable outerwear.[1]

Polyamide/ethylene-propylene, with higher crystallinity than other elastomeric polyamides, has improved fatigue resistance and improved oil and weather resistance.[1] T_m and service temperature usually increase with higher polyamide crystallinity.[1]

Polyamide/acrylate graft copolymers have a Shore D hardness range from 50 to 65, and continuous service temperature range from –40 to 329°F (–40 to 165°C). The markets are

* Pebax is a registered trademark of Elf Atochem.

underhood hose and tubing, seals and gaskets, and connectors and optic fiber sheathing, snap-fit fasteners.[1] Nylon 12/nitrile rubber blends were commercialized by Denki Kagaku Kogyo as part of the company's overall nitrile blend development.[1]

3.5 Melt Processable Rubber (MPR)

MPRs are amorphous polymers, with no sharp melt point,[1] which can be processed in both resin melt and rubber processing machines, injection molded, extruded, blow molded, calendered, and compression molded.[*1] Flow properties are more similar to rubber than to thermoplastics.[1] The polymer does not melt by externally applied heat alone but becomes a high-viscosity, intractable semifluid. It must be subjected to shear to achieve flowable melt viscosities, and shear force applied by the plasticating screw is necessary. Without applied shear, melt viscosity and melt strength increase too rapidly in the mold. Even with shear and a hot mold, as soon as the mold is filled and the plasticating screw stops or retracts, melt viscosity and melt strength increase rapidly.

Melt rheology is illustrated with Alcryn®.[†] The combination of applied heat and shear-generated heat brings the melt to 320 to 330°F (160 to 166°C). The melt temperature should not be higher than 360°F (182°C). New grades have been introduced with improved melt processing.

Proponents of MPR view its rheology as a processing cost benefit by allowing faster demolding and lower processing temperature settings, significantly reducing cycle time.[1] High melt strength can minimize or virtually eliminate distortion and sticking, and cleanup is easier.[1] MPR is usually composed of halogenated (chlorinated) polyolefins, with reactive intermediate-stage ethylene interpolymers that promote H^+ bonding.

Alcryn is an example of single-phase MPR with overall midrange performance properties, supplementing the higher-price COPE thermoplastic elastomers. Polymers in single-phase blends are miscible, but polymers in multiple-phase blends are immiscible, requiring a compatibilizer for blending. Alcryns are partially cross-linked halogenated polyolefin MPR blends.[1] The specific gravity ranges from 1.08 to 1.35.[1] MPRs are compounded with various property enhancers (additives), especially stabilizers, plasticizers, and flame retardants.[1]

The applications are automotive window seals and fuel filler gaskets, industrial door and window seals and weatherstripping, wire/cable covering, and hand-held power tool housing/handles. Nonslip soft-touch hand-held tool handles provide weather and chemical resistance and vibration absorption.[16] Translucent grade is extruded into films for face masks and tube/hosing and injection-molded into flexible keypads for computers and telephones.[1] Certain grades are paintable without a primer. Typical durometer hardnesses are Shore A 60, 76, and 80.

The halogen content of MPRs requires corrosion-resistant equipment and tool cavity steels along with adequate venting. Viscosity and melt strength buildup are taken into account with product design, equipment, and tooling design: wall thickness gradients and radii, screw configuration (flights, *L/D,* length), gate type and size, and runner dimensions.[1] The processing temperature and pressure setting are calculated according to rheology.[1]

To convert solid pellet feed into uniform melt, moderate screws with some shallow flights are recommended. Melt flow is kept uniform in the mold with small gates (which maximize shear), large vents, and large sprues for smooth mold filling.[1] Runners should be balanced and radiused for smooth, uniform melt flow.[1] Recommendations, such as bal-

---

* MPR is a trademark of Advanced Polymer Alloys, Division of Ferro Corporation.

† Alcryn is a registered trademark of Advanced Polymer Alloys Division of Ferro Corporation.

anced, radiused runners, are conventional practice for any mold design, but they are more critical for certain melts such as MPRs. Molds have large knock-out pins or plates to facilitate stripping the rubbery parts during demolding. Molds may be chilled to 75°F (24°C). Mold temperatures depend on grades and applications; hot molds are used for smooth surfaces and to minimize orientation.[1]

Similar objectives of the injection-molding process apply to extrusion and blow molding, namely, creating and maintaining uniform, homogeneous, and properly fluxed melt. Shallow-screw flights increase shear and mixing. Screws that are 4.5 in (11.4 cm) in diameter with *L/D* 20/1 to 30/1 are recommended for extrusion. Longer barrels and screws produce more uniform melt flux, but *L/D* ratios can be as low as 15/1. The temperature gradient is reversed. Instead of the temperature setting being increased from the rear (feed) zone to the front (metering) zone, a higher temperature is set in the rear zone, and a lower temperature is set at the front zone and at the adapter (head).[1] Extruder dies are tapered, with short land lengths, and die dimensions are close to the finished part dimension.[1] Alcryns have low to minimum die swell.

The polymer's melt rheology is an advantage in blow molding during parison formation, because the parison is not under shear, and it begins to solidify at about 330°F (166°C). High melt viscosity allows blow ratios up to 3:1 and significantly reduces demolding time.

MPRs are thermoformed and calendered with similar considerations described for molding and extrusion. Film and sheet can be calendered with thicknesses from 0.005 to 0.035 in (0.13 to 0.89 mm).

3.6 Thermoplastic Vulcanizate (TPV)

TPVs are composed of a vulcanized rubber component, such as EPDM, nitrile rubber, and butyl rubber, in a thermoplastic olefinic matrix. TPVs have a continuous thermoplastic phase and a discontinuous vulcanized rubber phase. TPVs are dynamically vulcanized during a melt-mixing process in which vulcanization of the rubber polymer takes place under conditions of high temperatures and high shear. Static vulcanization of thermoset rubber involves heating a compounded rubber stock under zero shear (no mixing), with subsequent cross-linking of the polymer chains.

Advanced Elastomer Systems' Santoprene®* thermoplastic vulcanizate is composed of polypropylene and finely dispersed, highly vulcanized EPDM rubber. Geolast®† TPV is composed of polypropylene and nitrile rubber, and the company's Trefsin®‡ is a dynamically vulcanized composition of polypropylene plus butyl rubber.

EPDM particle size is a significant parameter for Santoprene's mechanical properties, with smaller particles providing higher strength and elongation.[1] Higher cross-link density increases tensile strength and reduces tension set (plastic deformation under tension).[1] Santoprene grades can be characterized by EPDM particle size and cross-link density.[1]

These copolymers are rated as midrange with overall performance generally between the Tower cost styrenics and the higher-cost TPUs and copolyesters.[1] The properties of Santoprene, according to its developer (Monsanto), are generally equivalent to the properties of general purpose EPDM, and oil resistance is comparable to that of neoprene.[1] Geolast has higher fuel/oil resistance and better hot oil aging than Santoprene (see Tables 3.7, 3.8, and 3.9).

* Santoprene is a registered trademark of Advanced Elastomer Systems LP.

† Geolast is a registered trademark of Advanced Elastomer Inc. Systems LP.

‡ Trefsin is a registered trademark of Advanced Elastomer Systems LP.

TABLE 3.7 Santoprene Mechanical Property Profile—ASTM Test Methods—Durometer Hardness Range, Shore 55A to 50 D

Property	Shore hardness		
	55A	80A	50D
Specific gravity, g/cm^3	0.97	0.97	0.94
Tensile strength, lb/in^2 (MPa)	640 (4.4)	600 (11)	4000 (27.5)
Ultimate elongation, %	330	450	600
Compression set, %, 168 hr	23	29	41
Tension set, %	6	20	61
Tear strength, pli			
77°F (25°C)	108 (42)	194 (90)	594 (312)
212°F (100°C)	42 (5.6)	75 (24)	364 (184)
Flex fatigue megacycles to failure	>3.4	—	—
Brittle point, °F (°C)	<–76 (<–60)	–81 (–63)	–29 (–34)

TABLE 3.8 Santoprene Mechanical Property Profile—Hot Oil Aging*/Hot Air Aging—Durometer Hardness Range Shore 55 A to 55 D *(from Ref. 1)*

Property	Shore hardness		
	55A	80A	50D
Tensile strength, ultimate			
lb/in^2 (MPa)	470 (3.2)	980 (6.8)	2620 (18.10)
Percent retention	77	73	70
Ultimate elongation, %	320	270	450
Percent retention	101	54	69
100% modulus, lb/in^2 (MPa)	250 (1.7)	610 (4.2)	1500 (10.3)
Percent retention	87	84	91

*Hot oil aging (IRM 903), 70 hr @ 257°F (125°C).

TABLE 3.9 Santoprene Mechanical Property Profile—Hot Oil Aging/Hot Air Aging*—Durometer Hardness Range Shore 55 A to 55 D *(from Ref. 1)*

Property	Shore hardness		
	55A	80A	50D
Tensile strength, ultimate			
lb/in^2 (MPa)	680 (4.7)	1530 (10.6)	3800 (26.2)
Percent retention	104	109	97
Ultimate elongation, %	370	400	560
Percent retention	101	93	90
100% modulus, lb/in^2 (MPa)	277 (1.9)	710 (4.9)	1830 (12.6)
Percent retention	105	111	117

*Hot air aging, 168 hr @ 257°F (125°C).

Tensile stress-strain curves for Santoprene at several temperatures for Shore 55 A and 50 D hardnesses are shown in Fig.3.8.[8]

Generally, tensile stress decreases with temperature increase, while elongation at break increases with temperature. Tensile stress at a given strain increases with hardness from the softer Shore A grades to the harder Shore D grades. For a given hardness, the tensile stress-strain curve becomes progressively more rubber-like with increasing temperature. For a given temperature, the curve is progressively more rubberlike with decreasing hardness. Figure 3.9 shows dynamic mechanical properties for Shore 55 A and 50 D hardness grades over a wide range of temperatures.[8]

TPVs composed of polypropylene and EPDM have a service temperature range from –75 to 275°F (–60 to 135°C) for more than 30 days and 302°F (150°C) for short times (up to 1 week). Reference 8 reports further properties, including tensile and compression set, fatigue resistance, and resilience and tear strength. Polypropylene/nitrile rubber high/low service temperature limits are 257°F (125°C)/–40°F (–40°C).

Santoprene automotive applications include air ducts, body seals, boots (covers), bumper components, cable/wire covering, weatherstripping, underhood and other automotive hose/tubing, and gaskets. Appliance uses include diaphragms, handles, motor mounts, vibration dampers, seals, gaskets, wheels, and rollers. Santoprene rubber is used in building/construction for expansion joints, sewer pipe seals, valves for irrigation, weatherstripping, and welding line connectors. Prominent electrical uses are in cable jackets, motor shaft mounts, switch boots, and terminal plugs. Business machines, power tools, and plumbing/hardware provide TPVs with numerous applications. In healthcare applications, it is used in disposable bed covers, drainage bags, pharmaceutical packaging, wound dressings (U.S. Pharmacopoeia Class VI rating for biocompatibility). Special-purpose Santoprene grades meet flame retardance, outdoor weathering, and heat aging requirements.

Santoprene applications of note are a nylon-bondable grade for the General Motors GMT 800 truck air-induction system; driveshaft boot in Ford-F Series trucks, giving easier assembly, lighter weight, and higher temperature resistance than the material it replaced; and Santoprene cover and intermediate layers of tubing assembly for hydraulic oil hose. Nylon-bondable Santoprene TPV is coextruded with an impact modified (or pure) nylon 6 inner layer.

Polypropylene/EPDM TPVs are hygroscopic, requiring drying at least 3 hr at 160°F (71°C) and avoiding exposure to humidity.[1] They are not susceptible to hydrolysis.[1] Moisture in the resin can create voids, disturbing processing and finished product performance properties. Moisture precautions are similar to those for polyethylene or polypropylene.[1]

Typical of melts with a relatively low melt flow index (0.5 to 30 g/10 min for Santoprene), gates should be small, and runners and sprues should be short; long plasticating screws are used with an *L/D* ratio typically 24/1 or higher.[1] The high viscosity at low shear rates (see Fig. 3.10) provides good melt integrity and retention of design dimensions during cooling.[1]

Similar injection-molding equipment design considerations apply to extrusion equipment such as long plasticating screws with 24/1 or higher *L/D* ratios and approximately 3:1 compression ratios.[1]

Equipment/tool design, construction, and processing of TPVs differ from that of other thermoplastics. EPDM/polypropylene is thermally stable up to 500°F (260°C), and it should not be processed above this temperature.[1] It has a flash ignition temperature above 650°F (343°C).

TPV's high shear sensitivity allows easy mold removal; thus, sprays and dry powder mold release agents are not recommended.

Geolast TPVs are composed of polypropylene and nitrile rubber. Table 3.10 profiles the mechanical properties for these TPVs with Shore hardness range of 70 A to 45 D.

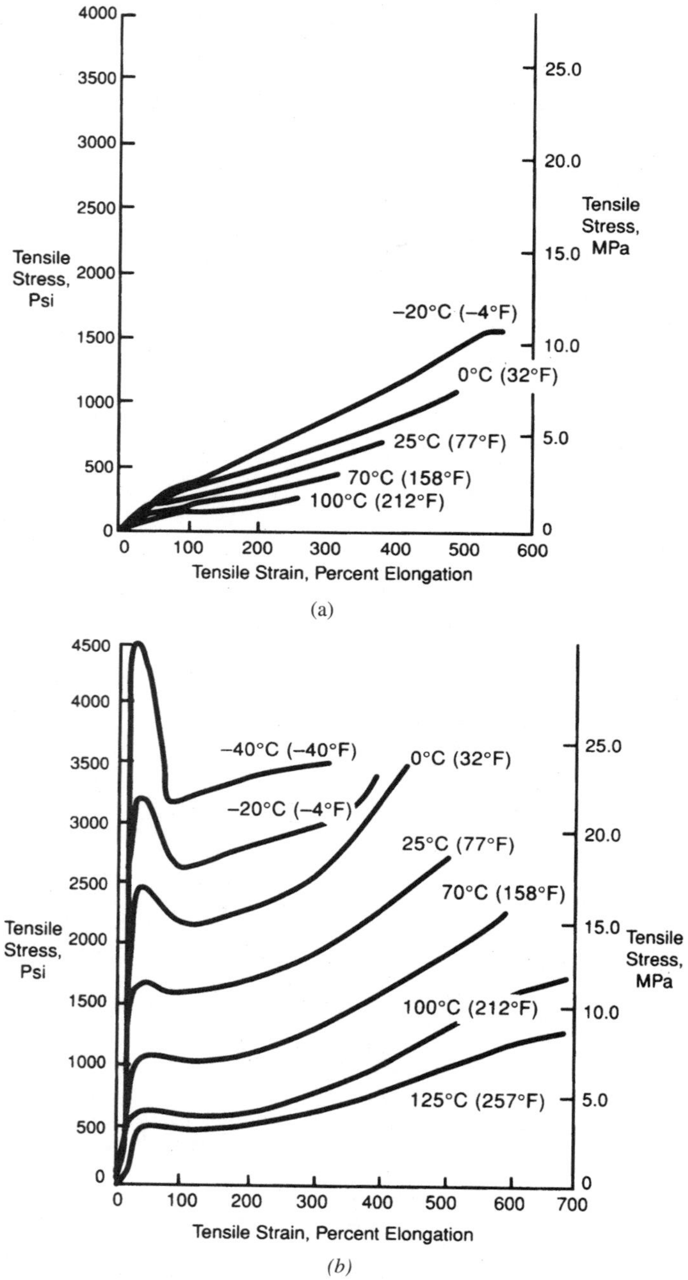

Figure 3.8 Tensile stress-strain curves for Santoprene at several temperatures for different hardness grades. *(a)* 55 Shore A grades (ASTM D 412), *(b)* 50 Shore D grades (ASTM D 412). *(Source: Ref. 8, pp. 3–4)*

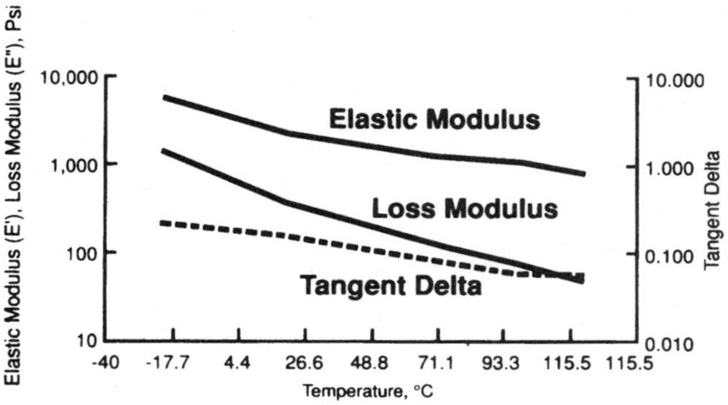

(a)

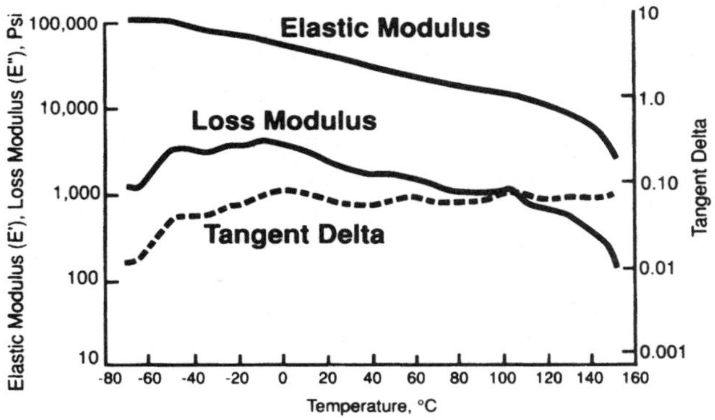

(b)

Figure 3.9 Dynamic mechanical properties for different hardness grades over a range of temperatures. *(a)* 55 Shore A grades, *(b)* 50 Shore D grades. *(Source: Ref. 8, pp. 12–13)*

Geolast (polypropylene plus nitrile rubber) has a higher resistance than Santoprene (polypropylene plus EPDM) to oils (such as IRM 903) and fuels, plus good hot-oil/hot-air aging.[1] Geolast applications include molded fuel filler gasket (Cadillac Seville), carburetor components, hydraulic lines, and engine parts such as mounts and tank liners.

Three property distinctions among Trefsin grades are (1) heat aging; (2) high energy attenuation for vibration damping applications such as automotive mounts, energy absorbing fascia and bumper parts, and sound deadening; and (3) moisture and 0_2 barrier. Other applications are soft bellows; basketballs, soccer balls, and footballs; calendered textile coatings; and packaging seals. Since Trefsin is hygroscopic, it requires drying before processing. Melt has low viscosity at high shear rates, providing fast mold filling. High viscosity at low shear during cooling provides a short cooling time. Overall, cycle times are reduced.

Advanced Elastomer Systems L.P. (AES) is the beneficiary of Monsanto Polymers' TPE technology and business, which included Monsanto's earlier acquisition of BP Performance Polymers' partially vulcanized EPDM/polypropylene (TPR), and Bayer's partially vulcanized EPDM/polyolefin TPEs in Europe.

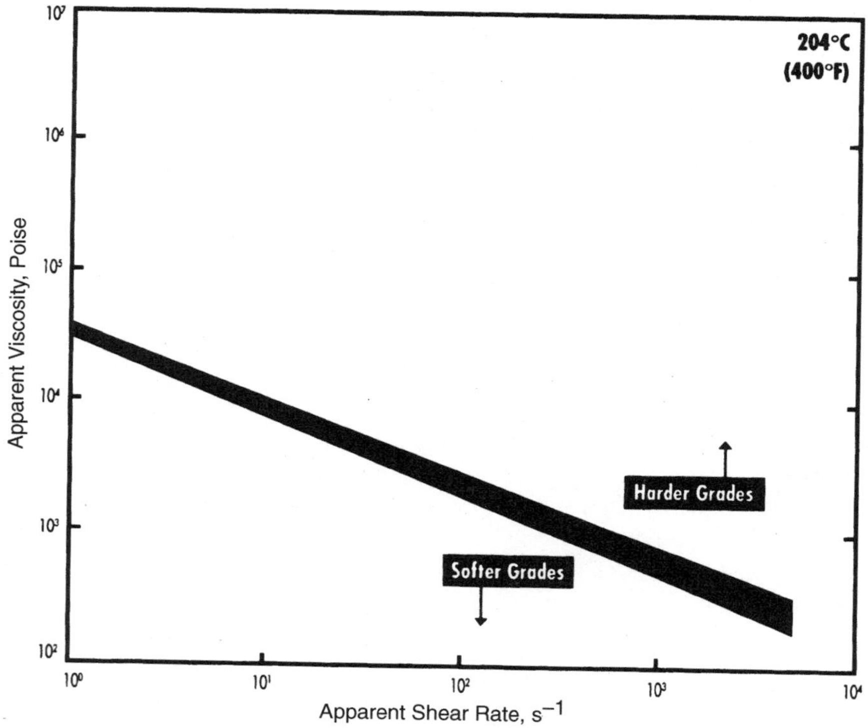

Figure 3.10 Apparent viscosity vs. apparent shear rate @ 400°F (04°C). *(Source: Ref. 7, pp. 36)*

TABLE 3.10 Geolast Mechanical Property Profile—ASTM Test Methods—Room Temperature—Durometer Hardness Range, Shore 70 A to 45 D

	Shore hardness		
Property	70 A	87 A	45 D
Specific gravity, g/cm^3	1.00	0.98	0.97
Tensile strength, lb/in^2 (MPa)	900 (6.2)	1750 (12)	2150 (15)
Ultimate elongation, %	265	380	350
Compression set @ 22 hr, %			
212°F (100°C)	28	39	52
257°F (125°C)	37	48	78
Tension set (%)	10	24	40
Tear strength, pli			
73°F (23°C)	175 (79)	350 (177)	440 (227)
212°/F (100°C)	52 (11)	150 (66)	220 (104)
Brittle point, °F (°C)	–40 (–40)	–33 (–28)	–31 (–36)

3.7 Synthetic Rubbers (SRs)

A second major group of elastomers is that group known as *synthetic rubbers.* Elastomers in this group, discussed in detail in this section, are

- Acrylonitrile butadiene copolymers (NBR)
- Butadiene rubber (BR)
- Butyl rubber (IIR)
- Chlorosulfonated polyethylene (CSM)
- Epichiorohydrin (ECH, ECO)
- Ethylene propylene diene monomer (EPDM)
- Ethylene propylene monomer (EPM)
- Fluoroelastomers (FKM)
- Polyacrylate (ACM)
- Polybutadiene (PB)
- Polychloroprene (CR)
- Polyisoprene (IR)
- Polysulfide rubber (PSR)
- Silicone rubber (SiR)
- Styrene butadiene rubber (SBR)

Worldwide consumption of synthetic rubber can be expected to be about 11 million metric tons in 2000 and about 12 million metric tons in 2003, based on earlier reporting (1999) by the International Institute of Synthetic Rubber Producers.[26] About 24 percent is consumed in North America.[1] Estimates depend on which synthetic rubbers are included and reporting sources from world regions.

New synthetic rubber polymerization technologies replacing older plants and increasing world consumption are two reasons new production facilities are being built around the world. Goodyear Tire & Rubber's 110,000-metric tons/y butadiene-based solution polymers went onstream in 2000 in Beaumont Texas.[25] Goodyear's 18,200-metric tons/y polyisoprene unit went onstream in 1999 in Beaumont.[25] Sumitomo Sumika AL built a 15,000-metric tons/y SBR plant in Chiba, Japan, adding to the company's 40,000-metric tons/y SBR capacity at Ehime.[25] Haldia Petrochemical Ltd. of India is constructing a 50,000-metric tons/y SBR unit and a 50,000-metric tons/y PB unit using BASF technology.[25]

Bayer Corporation added a 75,000-metric tons/y SBR and PB capacity at Orange, Texas, in 1999, converting a lithium PB unit to produce solution SBR and neodymium PB.[25] Bayer AG increased SBR and PB capacity from 85,000 to 120,000 metric tons/y at Port Jerome, France, in 1999.[25] Bayer AG will complete a worldwide butadiene rubber capacity increase from 345,000 metric tons/y in 1998 to more than 600,000 metric tons/y by 2001.[25] Bayer AG increased EPDM capacity at Orange, Texas, using slurry polymerization and at Marl, Germany, using solvent polymerization in 1999.[25] Bayer Inc. added 20,000–metric tons/y butyl rubber capacity at Sarnia, Ontario, to the company's 70,000-metric tons/y butyl rubber capacity and 50,000-metric tons/y halo-butyl capacity at Sarnia. Bayer's 90,000-metric tons/y halo- or regular butyl capacity was scheduled to be restarted in 2000.[25]

Mitsui Chemicals goes onstream with a 40,000-metric tons/y metallocene EPDM in Singapore in 2001.[25] The joint venture Nitrilo SA between Uniroyal Chemical subsidiary

of Crompton and Girsa subsidiary of Desc SA (Mexico) went onstream with a 28,000-metric tons/y NBR at Altamira, Mexico, in 1999.[25] Uniroyal NBR technology and Girsa process technology were joined.[25] Chevron Chemical went onstream in 1999 with a 60,000-metric tons/y capacity polyisobutylene (PIB) at Belle Chase, Louisiana, licensing technology from BASF.[25] BASF is adding a 20,000–metric tons/y medium MW PIB at its Lufwigshafen complex, which will double the unit's capacity to 40,000 metric tons/y. This addition will be completed in 2001. BASF has 70,000–metric tons/y low MW PIB capacity. The company is using its own selective polymerization technology, which allows MW to be controlled.[25]

BST Elastomers, a joint venture of Bangkok Synthetics, Japan Synthetic Rubber (JSR), Nippon Zeon, Mitsui, and Itochu, went onstream in 1998–1999 with 40,000–metric tons/y PB capacity and 60,000–metric tons/y SBR at Map Ta Phut, Rayong, Thailand.[25] Nippon Zeon completed adding 25,000–metric tons/y SBR capacity to its existing 30,000-metric tons/y at Yamaguchi, Japan, in 1999, and the company licensed its solution polymerization technology to Buna Sow Leuna Olefinverbund (BSLO).[25] BSLO will start 60,000-metric tons/y SBR capacity at Schkopau, Germany, in 2000.[25]

Sinopec, China's state-owned petrochemical and polymer company, is increasing synthetic rubber capacity across the board, including butyls, SBRs, nitrile, and chloroprene. Sinopec is starting polyisoprene and EPR production, although the company did not produce polyisoprene or EPR prior to 1999.[25] Total synthetic rubber capacity will be 1.15 million metric tons/y by 2000. China's synthetic rubber consumption is forecast by the company to be almost 7 million metric tons/y in 2000.[25] Dow chemical purchased Shell's E-SBR and BR. Dow is the fastest growing SR producer with the broadest portfolio of SBR and BR.

Synthetic rubber is milled and cured prior to processing such as injection molding. Processing machinery is designed specifically for synthetic rubber.

Engel (Guelph, Ontario) ELAST®* technology includes injection-molding machines designed specifically for molding cross-linked rubbers.[31] Typical process temperature settings, depending on the polymer and finished product, are 380 to 425°F (193 to 218°C). Pressures, which also depend on polymer and product, are typically 20,000 to 30,000 lb/in^2 (137 to 206 MPa). A vertical machine's typical clamping force is 100 to 600 U.S. tons, while a horizontal machine's typical clamping force is 60 to 400 U.S. tons. They have short flow paths, allowing injection of rubber very close to the cross-linking temperature.[31] The screw *L/D* can be as small as 10/1.31 ELAST technology includes tiebarless machines for small and medium capacities and proprietary state-of-the-art computer controls.[31]

3.7.1 Acrylonitrile Butadiene Copolymers

Nitrile butadiene rubbers (NBRs) are poly(acrylonitrile-co-1,3-butadiene) copolymers of butadiene and acrylonitrile.[23] Resistance to swelling caused by oils, greases, solvents, and fuels is related to percent bound acrylonitrile (ACN) content, which usually ranges from 20 to 46 percent.[6] Higher ACN provides higher resistance to swelling but diminishes compression set and low temperature flexibility.[6] ACN properties are related to percent acrylo and percent nitrile content. Nitrile increases compression set, flex properties, and processing properties.[1] The rubber has good barrier properties due to the polar nitrile groups.[23] Continuous-use temperature for vulcanized NBR is up to 248°F (120°C) in air and up to 302°F (150°C) immersed in oil.[23]

NBR curing, compounding, and processing are similar to those for other synthetic rubbers.[6]

---

* ELAST is a registered trademark of Engel Canada.

Fine-powder NBR grades are ingredients in PVC/nitrile TPEs and in other polar thermoplastics to improve melt processibility; reduce plasticizer blooming (migration of plasticizer to the surface of a finished product); and improve oil resistance, compression set, flex properties, feel, and finish of the plastic product.[1] Chemigum®* fine powder is blended with PVC/ABS and other polar thermoplastics.[1] The powders are typically less than 1 mm in diameter (0.5 nominal diameter particle size), containing 9 percent partitioning agent. Partitioning agents may be SiO_2, $CaCO_3$, or PVC. Their structures may be linear, linear/cross linked, and branched/cross linked (see Table 3.11).

TABLE 3.11 Typical Chemigum/PVC Formulations and Properties for General-Purpose and Oil/Fuel-Resistant Hose *(from Ref. 1)*

	General-purpose	Oil/fuel resistant
Ingredient, parts by weight		
PVC	100	100
Chemigum	25	100
DOP plasticizer	78	40
$CaCO_3$	—	20
Epoxidized soya oil	3	5
Stabilizer, lubricant	2	3
Properties		
Specific gravity, g/cm^3	1.17	1.20
Tensile strength, lb/in^2 (MPa)	2566 (17.6)	1929 (13.3)
Elongation @ break, %	390	340
Hardness Shore A	57	73

Nitrile rubber applications are belting, sheeting, cable jacketing, hose for fuel lines and air conditioners, sponge, gaskets, arctic/aviation O-rings and seals, precision dynamic abrasion seals, and shoe soles. Nitrile rubbers are coextruded as the inner tube with chlorinated polyethylene outer tube for automotive applications.[10d] Nitrile provides resistance to hydrocarbon fluids, and chlorinated polyethylene provides ozone resistance.[10d] Other automotive applications are engine gaskets, fluid- and vapor-resistant tubing, fuel filler neck inner hose, fuel system vent inner hose, oil, and grease seals. Nitrile powder grades are used in window seals, appliance gasketing, footwear, cable covering, hose, friction material composites such as brake linings, and food contact applications.[1,6]

Blends based on nitrile rubbers are used in underground wire/cable covering; automotive weatherstripping, spoiler extensions, foam-integral skin core-cover armrests, and window frames; footwear; and flexible, lay-flat, reinforced, rigid, and spiral hose for oils, water, food, and compressed air.

3.7.2 Butadiene Rubber (BR) and Polybutadiene (PB)

Budene®† solution polybutadiene (solution polymerized) is cis-1,4-poly(butadiene) produced with stereospecific catalysts which yield a controlled MWD, which is essentially a

* Chemigum is a registered trademark of Goodyear Tire and Rubber Company.
† Budene is a registered trademark of Goodyear Tire and Rubber Company.

linear polymer.[6] Butadiene rubber, polybutadiene, is solution-polymerized to stereospecific polymer configurations[10a] by the additional polymerization of butadiene monomer. The following cis- and trans-1,4-polybutadiene isomers can be produced: cis-1,4-polybutadiene with good dynamic properties, low hysteresis, and good abrasion resistance; trans isomers are tougher, harder, and show more thermoplasticity.[10,23] Grades are oil and non-oil extended and vary according to their cis-content.[6]

The applications are primarily tire tread and carcass stock, conveyor belt coverings, V-belts, hose covers, tubing, golf balls, shoe soles and heels, sponges, and mechanical goods.[6] They are blended with SBR for tire treads to improve abrasion and wear resistance.[10a] The tread is the part of the tire that contacts the road, requiring low rolling resistance, abrasion and wear resistance, and good traction and durability.[22]

Replacement passenger car shipments in the United States are expected to increase from 185.5 million units in 1998 to 199 million units in 2004, according to the Rubber Manufacturers Association. Synthetic rubber choices for tires and tire treads are related to tire design. Composition and design for passenger cars, sport utility vehicles (SUV), pickup trucks, tractor trailers, and snow tires are continually under development, as illustrated by the following sampling of U.S. Patents.[16] Tire Having Silica Reinforced Rubber Tread Containing Carbon Fibers to Goodyear Tire & Rubber, U.S. Patent 5,718,781, February 17, 1998; Silica Reinforced Rubber Composition to Goodyear Tire & Rubber, U.S. Patent 5,719,208, February 17, 1998; and Silica Reinforced Rubber Composition and Tire With Tread to Goodyear Tire & Rubber, U.S. Patent 5,719,207, February 17, 1998.

Other patents include "Ternary Blend of Polyisoprene, Epoxidized Natural Rubber and Chlorosulfonated Polyethylene" to Goodyear Tire & Rubber, U.S. Patent 5,736,593, April 7, 1998, and "Truck Tire With Cap/Base Construction Tread" to Goodyear Tire & Rubber, U.S. Patent 5,718,782, February 17, 1998; "Tread Of Heavy Duty Pneumatic Radial Tire" to Bridgestone Corporation, U.S. Patent 5,720,831, February 24, 1998; "Pneumatic Tire With Asymmetric Tread Profile" to Dunlop Tire, U.S. Patent 5,735,979, April 7, 1998; and "Tire Having Specified Crown Reinforcement" to Michelin, U.S. Patent 5,738,740, April 14, 1998. Silica improves a passenger car's tread rolling resistance and traction when used with carbon black.[28] High dispersible silica (HDS) in high vinyl solution polymerized SER compounds show improved processing and passenger car tread abrasion resistance.[28] Precipitated silica with carbon black has been used in truck tire tread compounds which are commonly made with natural rubber (NR).[28]

Modeling is the method of choice for analyzing passenger car cord-reinforced rubber composite behavior. Large-scale three-dimensional finite element analysis (FEA) improves understanding of tire performance, including tire and tread behavior when "the rubber meets the road."

BR is extruded and calendered. Processing properties and performance properties are related to polymer configuration: cis- or trans- stereoisomerism, MW and MWD, degree of crystallization (DC), degree of branching, and Mooney viscosity.[23] Broad MWD and branched BR tend to mill and process more easily than narrow MWD and more linear polymer.[23] Lower Mooney viscosity enhances processing.[23] BR is blended with other synthetic rubbers such as SBR to combine BR properties with millability and extrudability.

3.7.3 Butyl Rubber

Butyl rubber (IIR) is an isobutylene-based rubber, which includes copolymers of isobutylene and isoprene, halogenated butyl rubbers, and isobutylene/p-methylstyrene/bromo-p-methylstyrene terpolymers.[22] IIR can be slurry polymerized from isobutylene copolymerized with small amounts of isoprene in methyl chloride diluent at –130 to –148°F (–90 to –100°C). Halogenated butyl is produced by dissolving butyl rubber in a hydrocarbon solvent and introducing elemental halogen in gas or liquid state.[23] Cross-linked terpolymers are formed with isobutylene + isoprene + divinylbenzene.

Most butyl rubber is used in the tire industry. Isobutylene-based rubbers are used in underhood hose for the polymer's low permeability and temperature resistance, and high damping, resilient butyl rubbers are used for NVH (noise, vibration, harshness) applications such as automotive mounts for engine and vehicle/road NVH attenuation.[22]

Butyl rubber is ideal for automotive body mounts that connect the chassis to the body, damping road vibration.[10d] Road vibration generates low vibration frequencies. Butyl rubber can absorb and dissipate large amounts of energy due to its high mechanical hysteresis over a useful temperature range.[10d]

Low-MW "liquid" butyls are used for sealants, caulking compounds, potting compounds, and coatings.[23] Depolymerized virgin butyl rubber is high viscosity and is used for reservoir liners, roofing coatings, and aquarium sealants.[10b] It has property values similar to conventional butyl rubber: extremely low VTR (vapor transmission rate); resistance to degradation in hot, humid environments; excellent electrical properties; and resistance to chemicals, oxidation, and soil bacteria.[10b] To make high-viscosity depolymerized butyl rubber pourable, solvents or oil is added.[10b]

Chlorobutyl provides flex resistance in the blend chlorobutyl rubber/EPDM rubber/NR for white sidewall tires and white sidewall coverstrips.[22] An important application of chlorobutyl rubber in automotive hose is extruded air conditioning hose to provide barrier properties to reduce moisture gain and minimize refrigerant loss.[22] The polymer is used in compounds for fuel line and brake line hoses.[22] Brominated isobutylene-p-methylstyrene (BIMS) was shown to have better aging properties than halobutyl rubber for underhood hose and comparable aging properties to peroxide-cured EPDM, depending on compound formulations.[22] Bromobutyls demonstrate good resistance to brake fluids for hydraulic brake lines and to methanol and methanol/gasoline blends.[22]

3.7.4 Chlorosulfonated Polyethylene (CSM)

Chlorosulfonated polyethylene is a saturated chlorohydrocarbon rubber produced from Cl_2, SO_2, and a number of polyethylenes, and contains about 20 to 40 percent chlorine and 1 to 2 percent sulfur as sulfonyl chloride.[23] Sulfonyl chloride groups are the curing or cross-linking sites.[23] CSM properties are largely based on initial polyethylene (PE) and percent chlorine. A free-radical-based PE with 28 percent chlorine and 1.24 percent S has a dynamic shear modulus range from 1000 to 300,000 lb/in^2 (7 MPa to 2.1 GPa).[23] Stiffness differs for free-radical-based PE and linear PE, with chlorine content: at about 30 percent, Cl_2 free-radical-based PE stiffness decreases to minimum value, and at about 35 percent, Cl_2 content linear PE stiffness decreases to minimum value.[23] When the Cl_2 content is increased more than 30 and 35 percent, respectively, the stiffness (modulus) increases.[23]

Hypalon®* CSMs are specified by their Cl_2, S contents, and Mooney viscosity.[23] CSM has an excellent combination of heat and oil resistance and oxygen and ozone resistance. CSM, like other polymers, is compounded to meet specific application requirements. Hypalon is used for underhood wiring and fuel hose resistance.

3.7.5 Epichlorohydrin (ECH, ECO)

ECH and ECO polyethers are homo- and copolymers, respectively: chloromethyloxirane homopolymer and chloromethyloxirane copolymer with oxirane.[23] Chloromethyl side chains provide sites for cross-linking (curing and vulcanizing). These chlorohydrins are

* Hypalon is a registered trademark of DuPont Dow Elastomers LLC.

chemically 1-chloro-2,3-epoxypropane. They have excellent resistance to swelling when exposed to oils and fuels; good resistance to acids, alkalis, water, and ozone; and good aging properties.[10a] Aging can be ascribed to environments such as weathering (UV radiation, oxygen, ozone, heat, and stress).[10a] High chlorine content provides inherent flame retardance,[10a] and, like other halogenated polymers, flame-retardant enhancers (additives) may be added to increase UL 94 flammability rating.

ECH and ECO can be blended with other polymers to increase high-and low-temperature properties and oil resistance.[23] Modified polyethers have potential use for new, improved synthetic rubbers. ECH and ECO derivatives, formed by nucleophilic substitution on the chloromethyl side chains, may provide better processing.

3.7.6 Ethylene Propylene Copolymer (EPDM)

EPM [poly(ethylene-co-propylene)] and EPDM [poly(ethylene-copropylene-co-5-ethylidene-2-norbornene)][23] can be metallocene catalyst polymerized. Metallocene catalyst technologies include (1) Insite, a constrained geometry group of catalysts used to produce Affinity polyolefin plastomers (POP), Elite®* PE, Nordel®† EPDM, and Engage polyolefin elastomers (POP) and (2) Exxpol®‡ ionic metallocene catalyst compositions used to produce Exact§ plastomer octene copolymers.[24] Insite technology produces EPDM-based Nordel IP with property consistency and predictability[16] (see Sec. 3.10.2).

Mitsui Chemical reportedly has developed "FI" catalyst technology, called a *phenoxy-cyimine complex,* with 10 times the ethylene polymerization activity of metallocene catalysts, according to *Japan Chemical Weekly* (summer, 1999).[25]

EPM and EPDM can be produced by solution polymerization, while suspension and slurry polymerization are viable options. EPDM can be gas-phase 1,-4 hexadiene polymerized using Ziegler-Natta catalysts. Union Carbide produces ethylene propylene rubber (EPR) using modified Unipol® low-pressure gas-phase technology.

The letter "M" designates that the ethylene propylene has a saturated polymer chain of the polymethylene type, according to the ASTM.[12] EPM (copolymer of ethylene and propylene) rubber and EPDM (terpolymer of ethylene, propylene, and a nonconjugated diene) with residual side chain unsaturation, are subclassified under the ASTM "M" designation.[12]

The diene ethylidiene norbornene in Vistalon®** EPDM allows sulfur vulcanization (see Table 3.12).[12] 1,4-Hexadiene and dicyclopentadiene (DCPD) are also used as curing agents.[18] The completely saturated polymer "backbone" precludes the need for antioxidants that can bleed to the surface (bloom) of the finished product and cause staining.[12] Saturation provides inherent ozone and weather resistance, good thermal properties, and a low compression set.[12] Saturation also allows a relatively high-volume addition of low-cost fillers and oils in compounds while retaining a high level of mechanical properties.[12] The ethylene/propylene monomer ratio also affects the properties.

EPM and EPDM compounds, in general, have excellent chemical resistance to water, ozone, radiation, weather, brake fluid (nonpetroleum based), and glycol.[12]

EPM is preferred for dynamic applications, because its age resistance retains initial product design over time and environmental exposure.[12] EPDM is preferred for its high resilience.[12] EPM is resistant to acids, bases (alkalis), and hot detergent solution. EPM and EPDM are resistant to salt solutions, oxygenated solvents, and synthetic hydraulic fluids.[12] Properties are determined by the composition of the base compound. A typical for-

* Elite is a registered trademark of Dow Chemical Company.

† Nordel is a registered trademark of DuPont Dow Elastomers LLC.

‡ Exxpol is a registered trademark of Exxon Mobil Corporation.

§ Exact is a registered trademark of Exxon Mobil Corporation.

** Vistalon is a registered trademark of Exxon Chemical Company, Division of Exxon Corporation.

TABLE 3.12 Typical Properties of EPM/EPDM Compounds[12] Based on Vistalon

Property	Value
Hardness Shore A	35–90
Tensile strength, lb/in^2 (MPa)	580–3200 (4–22)
Compression set (%), 70 hr @ 302°F (150°C)	15–35
Elongation, %	150–180
Tear strength, lb/in (kN/m)	86–286 (15–50)
Continuous service temperature, °F (°C)	302 (150) max
Intermediate service temperature, °F (°C)	347 (175) max
Resilience (Yerzley), %	75
Loss tangent (15 Hz), % (dynamic)	0.14
Elastic spring rate, lb/in (kN/m) (15 Hz)	3143 (550)
Dielectric constant, %	2.8
Dielectric strength, kV/mm	26
Power factor, %	0.25
Volume resistivity, Ω–cm	1×10^{16}

mulation includes Vistalon EP(D)M, carbon black, process oil, zinc oxide, stearic acid, and sulfur. [12]

EPDM formulations are increasingly popular for medium-voltage, up to 221°F (105°C) continuous-use temperature wire and cable covering.[20] Thinner walls, yet lower (power) losses and better production rates, are sought by cable manufacturers.[20] Low-MW (Mooney viscosity, ML), high-ethylene-content copolymers and terpolymers are used In medium-voltage cable formulations.[20] With ethylidiene norbornene or hexadiene, EPDMs are good vulcanizates, providing improved wet electrical properties.[20] When the diene vinyl norbornene is incorporated on the EPDM backbone by a gel-free process, a significantly improved EPDM terpolymer is obtained for wire/cable applications.[20] Other applications are automotive body seals, mounts, weatherstripping, roofing, hose, tubing, ducts, and tires. Molded EPDM rubber is used for bumpers and fillers to dampen vibrations around the vehicle, such as deck-lid over-slam bumpers, for its ozone and heat resistance.[10d] EPDM can be bonded to steels, aluminum, and brass, with modified poly(acrylic acid) and polyvinylamine water-soluble coupling agents.[27]

EPDM is a favorable selection for passenger car washer-fluid tubes and automotive body seals, and it is used for automotive vacuum tubing.[10d] EPDM has good water-alcohol resistance for delivering fluid from the reservoir to the spray nozzle and good oxygen and UV resistance.[10d] Random polymerization yields a liquid with a viscosity of 100,000 centipoise (cP) @ 203°F (95°C), room temperature-cured with para-quinone dioxime systems or two-component peroxide systems, or cured at an elevated temperature with sulfur.[10b] They can be used as automotive and construction sealants, waterproof roofing membranes, and for encapsulating electrical components.[10b]

3.7.7 Fluoroelastomers (FKM)

Fluoroelastomers can be polymerized with copolymers and terpolymers of tetrafluoroethylene, hexafluoroethylene, and vinylidene fluoride. The fluorine content largely determines chemical resistance and T_g, which increases with increasing fluorine content. Low-temperature flexibility decreases with increasing fluorine content.[1] The fluorine content is typically 57 wt%.[11]

TFE/propylene copolymers can be represented by Aflas®* TFE, produced by Asahi Glass. They are copolymers of tetrafluoroethylene (TFE) + propylene, and terpolymers of TFE + propylene + vinylidene fluoride. Fluoroelastomer dipolymer and terpolymer gums are amine- or bisphenol-cured and peroxide-cured for covulcanizable blends with other peroxide curable elastomers. They can contain cure accelerators for faster cures, and they are divided into three categories: (1) gums with incorporated cures, (2) gums without incorporated cures, and (3) specialty master batches used with other fluoroelastomers.

Aflas products are marketed in five categories according to their MW and viscosities.[11] The five categories possess similar thermal, chemical, and electrical resistance properties but different mechanical properties.[11] The lowest viscosity is used for chemical process industry tank and valve linings, gaskets for heat exchangers and pipe/flanges, flue duct expansion joints, flexible and spool joints, and viscosity improver additives in other Aflas grades.[11]

The second-lowest viscosity grade is high-speed extruded into wire/cable coverings, sheet, and calendered stock.[11] Wire and cable covering are a principal application, especially in Japan. The third grade, general-purpose, is molded, extruded, and calendered into pipe connector gaskets, seals, and diaphragms in pumps and valves.[11] The fourth grade, with higher MW, is compression molded into O-rings and other seal applications.

The fifth grade, with the highest MW, is compression molded into oil field applications requiring resistance to high-pressure gas blistering.[11] It is used for down-hole packers and seals in oil exploration and production. Oilfield equipment seals are exposed to short-term temperatures from 302 to 482°F (150 to 250°C) and pressures above 10,000 lb/in^2 (68.7 MPa) in the presence of aggressive hydrocarbons H_2S, CH_4, $C0_2$, amine-containing corrosion inhibitors, and steam and water.[11]

Synthetic rubbers, EPM/EPDM, nitrile, polychloroprene (neoprene), epichlorohydrin, and polyacrylate have good oil resistance, heat stability, and chemical resistance. Fluoropolymers are used in oil and gas wells 20,000 ft (6096 m) deep. These depths can have pressures of 20,000 lb/in^2 (137.5 MPa), which cause "extrusion" failures of down-hole seals by forcing the rubber part out of its retaining gland. TFE/propylene jackets protect down-hole assemblies, which consist of stainless steel tubes that deliver corrosion-resistant fluid into the well.

Aircraft jet engine O-rings require fluoropolymer grades for engine cover gaskets that are resistant to jet fuel, turbine lube oils, and hydraulic fluids.

Dyneon®† BREs (base-resistant elastomers) are used in applications exposed to automotive fluids such as ATF, gear lubricants, engine oils shaft seals, O-rings, and gaskets.

DuPont Dow Elastomers fuel-resistant Viton®‡ fluoroelastomers are an important source for the applications described previously. The company's Kalrez®§ perfluoroelastomers with reduced contamination are widely used with semiconductors and other contamination-sensitive applications. Contamination caused by high alcohol content in gasoline can cause fuel pump malfunction. The choice of polymer can determine whether an engine functions properly.

The three principal Viton categories are (1) Viton A dipolymers composed of vinylidene fluoride (VF_2) and hexafluoropropylene (HFP) to produce a polymer with 66 percent (wt%) fluorine content, (2) Viton B terpolymers of VF_2 + HFP + tetrafluoroethylene (TFE) to produce a polymer with 68 percent fluorine, and (3) Viton F terpolymers composed of VF_2 + HFP + TFE to produce a polymer with 70 percent fluorine.[36] The three categories

---

* Aflas is a registered trademark of Asahi Glass Company.

† Dyneon and Dyneon BRE are registered trademarks of Dyneon LLC.

‡ Viton is a registered trademark of DuPont Dow Elastomers LLC.

§ Kalrez is a registered trademark of DuPont Dow Elastomers LLC.

are based on their resistance to fluids and chemicals.[36] Fluid resistance generally increases but low-temperature flexibility decreases with higher fluorine content.[36] Specialty Viton grades are made with additional or different principal monomers in order to achieve specialty performance properties.[36] An example of a specialty property is low-temperature flexibility.

Compounding further yields properties to meet a given application.[36] Curing systems are an important variable affecting properties.

DuPont Dow Elastomers developed curing systems during the 1990s, and the company should be consulted for the appropriate system for a given Viton grade.

FKMs are coextruded with lower-cost (co)polymers such as ethylene acrylic copolymer.[10d] They can be modified by blending and vulcanizing with other synthetic rubbers such as silicones, EPR and EPDM, epichlorohydrin, and nitriles. Fluoroelastomers are blended with modified NBR to obtain an intermediate performance/cost balance. These blends are useful for underhood applications in environments outside the engine temperature zone such as timing chain tensioner seals.

Fluoroelastomers are blended with fluorosilicones and other high-temperature polymers to meet engine compartment environments and cost/performance balance. Fine-particle silica increases hardness, red iron oxide improves heat resistance, and zinc oxide improves thermal conductivity. Hardness ranges from about Shore 35 A to 70 A. Fluorosilicones are resistant to nonpolar and nominally polar solvents, diesel and jet fuel, and gasoline, but not to solvents such as ketones and esters.

Typical applications are exhaust gas recirculating and seals for engine valve stems and cylinders, crankshaft, speedometers, and O-rings for fuel injector systems.

FKMs are compounded in either water-cooled internal mixers or two-roll mills. A two-pass mixing is recommended for internally mixed compounds with the peroxide curing agent added in the second pass.[11] Compounds press-cured 10 min @ 350°F (177°C) can be formulated to possess more than 2100-lb/in^2 (14.4-MPa) tensile strength, 380 percent elongation, 525 percent @ 100 percent modulus, and higher values when postcured 16 hr @ 392°F (200°C).[11] Processing temperatures are >392°F (200°C).[30]

3.7.8 Polyacrylate Acrylic Rubber (ACM)

Acrylic rubber can be emulsion- and suspension-polymerized from acrylic esters such as ethyl, butyl, and/or methoxyethyl acetate to produce polymers of ethyl acetate and copolymers of ethyl, butyl, and methoxyl acetate. Polyacrylate rubber, such as Acron®* from Cancarb Ltd., Alberta, Canada, possesses heat resistance and oil resistance between nitrile and silicone rubbers[23] Acrylic rubbers retain properties in the presence of hot oils and other automotive fluids, and they resist softening or cracking when exposed to air up to 392°F (200°C). The copolymers retain flexibility down to –40°F (–40°C). Automotive seals and gaskets constitute a major market.[23] These properties and inherent ozone resistance are largely due to the polymer's saturated "backbone" (see Table 3.13). Polyacrylates are vulcanized with sulfur or metal carboxylate, with a reactive chlorine-containing monomer to create a cross-linking site.[23]

Copolymers of ethylene and methyl acrylate, and ethylene acrylics, have a fully saturated "backbone," providing heat-aging resistance and inherent ozone resistance.[23] They are compounded in a Banbury mixer and fabricated by injection molding, compression molding, resin transfer molding, extrusion, and calendering.

---

* Acron is a registered trademark of Cancarb Ltd.

TABLE 3.13 Property Profile of Polyacrylic Rubbers[23]

Property at room temperature*	Value
Tensile strength, lb/in^2 (MPa)	2212 (15.2)
100% modulus	1500 (10.3)
Compression set, % [70 hr @ 302°F (150°C)]	28
Hardness Shore A	80

*Unless indicated otherwise.

3.7.9 Polychloroprene (Neoprene) (CR)

Polychloroprene is produced by free-radical emulsion polymerization of primarily trans-2-chloro-2-butenylene moieties.[23] Chloroprene rubber possesses moderate oil resistance, very good weather and oil resistance, and good resistance to oxidative chemicals.[10a] Performance properties depend on compound formulation, with the polymer providing fundamental properties. This is typical of any polymer and its compounds. Chloride imparts inherent self-extinguishing flame retardance.

Crystallization contributes to high tensile strength, elongation, and wear resistance in its pure gum state before CR is extended or hardened.[10a]

3.7.10 Polyisoprene (IR)

Polymerization of isoprene can yield high-purity cis-1,4-polyisoprene and trans-1,4-polyisoprene. Isoprene is 2-methyl-1,3-butadiene, 2-methyldivinyl, or 2-methylerythrene.[23] Isoprene is polymerized by 1,4 or vinyl addition, the former producing cis-1,4 or trans-1,4 isomer.[23]

Synthetic polyisoprene, isoprene rubber (IR), was introduced in the 1950s as odorless rubber with virtually the same properties as natural rubber. Isoprene rubber product and processing properties are better than natural rubber in a number of characteristics. MW and MWD can be controlled for consistent performance and processing properties.

Polyisoprene rubber products are illustrated by Natsyn®,* which is used to make tires and tire tread (cis isomer). Tires are the major cispolyisoprene product. Trans-polyisoprene can be used to make golf ball covers, hot-melt adhesives, and automotive and industrial products.

Depolymerized polyisoprene liquid is used as a reactive plasticizer for adhesive tapes, hot melts, brake linings, grinding wheels, and wire and cable sealants.[10b]

3.7.11 Polysulfide Rubber (PSR)

PSR is highly resistant to hydrocarbon solvents, aliphatic fluids, and aliphatic-aromatic blends.[10a] It is also resistant to conventional alcohols, ketones, and esters used in coatings and inks and to certain chlorinated solvents.[10d] With these attributes, PSR is extruded into hose to carry solvents and printing rolls, and, due to its good weather resistance, it is useful in exterior caulking compounds.[10a] Its limitations, compared with nitrile, are relatively poor tensile strength, rebound, abrasion resistance, high creep under strain, and odor.[10a]

---

* Natsyn is registered trademark of Goodyear Tire and Rubber Company.

Liquid PSR is oxidized to rubbers with service temperatures from –67 to 302°F (–55 to 150°C), excellent resistance to most solvents, and good resistance to water and ozone.[10b] It has very low selective permeability rates to a number of highly volatile solvents and gases and odors. Compounds formulated with liquid PSR can be used as a flexibilizer in epoxy resins, and epoxy-terminated polysulfides have better underwater lap shear strength than toughened epoxies.[10b] Other PSR applications are aircraft fuel tank sealant, seals for flexible electrical connections, printing rollers, protective coatings on metals, binders in gaskets, caulking compound ingredient, adhesives, and to provide water and solvent resistance to leather.

3.7.12 Silicone Rubber (SiR)

Silicone rubber polymers have the more stable Si atom compared with carbon. Silicone's property signature is its combined (1) high-temperature resistance [>500°F (260°C)], (2) good flexibility at <–100°F (–73°C), (3) good electrical properties, (4) good compression set, and (5) tear resistance and stability over a wide temperature range.[10a] When exposed to decomposition level temperature, the polymer forms SiO_2, which can continue to serve as an electrical insulator.[10a] Silicone rubber is used for high-purity coatings for semiconductor junctions, high-temperature wire, and cable coverings.[1]

RTV (room temperature vulcanizing) silicones cure in about 24 hr.[10b] They can be graded according to their room-temperature viscosities, which range from as low as 1500 cP (general-purpose soft) up to 700,000 cP (high-temperature paste). Most, however, are between 12,000 and 40,000 cP.[10b] RTV silicone has a low modulus over a wide temperature range from –85 to 392°F (–65 to 200°C), making them suitable for encapsulating electrical components during thermal cycling and shock.[10b] Low modulus minimizes stress on the encapsulated electrical components.[10b]

High-consistency rubber (HCR) from Dow Corning is injection-molded into high-voltage insulators, surge arrestors, weather sheds, and railway insulators. The key properties are wet electrical performance and high tracking resistance.

Liquid silicone rubbers (LSRs) are two-part grades that can be coinjection-molded with thermoplastics to make door locks and flaps for vents.[10b] LSRs can be biocompatible and have low compression set, low durometer hardness, and excellent adhesion.[10b] One-part silicones are cured by ambient moisture. They are used for adhesives and sealants with plastic, metal, glass, ceramic, and silicone rubber substrates.[10b] A solventless, clear silicone/PC has been developed that requires no mixing and can be applied without a primer.[1]

3.7.13 Styrene Butadiene Rubber (SBR)

SBR is emulsion- and solution-polymerized from styrene and butadiene, plus small volumes of emulsifiers, catalysts and initiators, endcapping agents, and other chemicals. It can be sulfur-cured SBR types are illustrated with Plioflex®* emulsion SBR (emulsion polymerized) and Solflex®† solution SBR.[6] Emulsion SBR is produced by hot polymerization for adhesives and by cold polymerization for tires and other molded automotive and industrial products.[6] Solution SBR is used for tires.

SBR is a low-cost rubber with slightly better heat aging and wear resistance than NR for tires.[10a] SBR grades are largely established by the bound styrene/butadiene ratio, polymerization conditions such as reaction temperature, and auxiliary chemicals added during polymerization.

---

* Plioflex is a registered trademark of Goodyear Tire and Rubber Company.

† Solflex is a registered trademark of Goodyear Tire and Rubber Company.

SBR/PVC blends that employ nitrile rubber (NBR) as a compatibilizer show improved mechanical properties at lower cost than NBI/PVC.[21] This was the conclusion of studies using a divinylbenzene cross-linked, hot-polymerized emulsion polymer with 30 percent bound styrene and a cold-polymerized emulsion polymer with 23 percent bound styrene; PVC with inherent viscosity from 0.86 to 1.4; NBR with Mooney viscosity from 30 to 86; acrylonitrile content of 23.5, 32.6, and 39.7 percent; and ZnO, stabilizers, sulfur, and accelerators.[21]

3.8 Natural Rubber (NR)

Natural rubber, the original elastomer, still plays an important role among elastomers. Worldwide consumption of NR in 2000 is expected to be about 7 million metric tons/y, based on earlier reporting by the International Rubber Study Group. Chemically, natural rubber is cis1,4-polyisoprene and occurs in Hevea rubber trees. NR tapped from other rubber trees (gutta-percha and balata) is the trans isomer of polyisoprene.[23] NR's principal uses are automotive tires, tire tread, and mechanical goods. Automotive applications are always compounded with carbon black to impart UV resistance and to increase mechanical properties.[10d] Latex concentrate is used for dipped goods, adhesives, and latex thread.[23] Latex concentrate is produced by centrifuge-concentrating field latex tapped from rubber trees. The dry rubber content is subsequently increased from 30 to 40 to 60 percent minimum.[23]

Vulcanization is the most important NR chemical reaction.[23] Most applications require cross-linking via vulcanization to increase resiliency and strength. Exceptions are crepe rubber shoe soles and rubber cements.[23] There are a number of methods for sulfur vulcanization, with certain methods producing polysulfidic cross-linking and other methods producing more monosulfidic cross-links.[10d]

NR is imported from areas such as Southeast Asia to the world's most industrial regions, North America, Europe, and Japan, since it is not indigenous to these regions. The huge rubber trees require about 80 to 100 in/y (200 to 250 cm/y) rainfall, and they flourish at an altitude of about 1000 ft (300 m).[23] As long as NH is needed for tires, industrial regions will be import dependent.

NR has good resilience; high tensile strength; low compression set; resistance to wear and tear, cut-through and cold flow; and good electrical properties.[10a] Resilience is the principal property advantage compared with synthetic rubbers.[10a] For this reason, NH is usually used for engine mounts, because NR isolates vibrations caused when an engine is running. NH is an effective decoupler, isolating vibrations such as engine vibration from being transmitted to another location such as the passenger compartment.[10d] With decoupling, vibration is returned to its source instead of being transmitted through the rubber.[10d] Polychloroprene is used for higher under-hood temperatures above NR service limits; butyl rubber is used for body mounts and for road vibration frequencies, which occur less frequently than engine vibrations or have low energy; EPDM is often used for molded rubber bumpers and fillers throughout the vehicle, such as deck-lid over-slam bumpers.

Degree of crystallinity (DC) can affect NH properties, and milling reduces MW. MW is reduced by mastication, typically with a Banbury mill, adding a peptizing agent during milling to further reduce MW, which improves NR solubility after milling.[23] NR latex grades are provided to customers in low (0.20 wt%) and high (0.75 wt%), with ammonia added as a preservative.[23] Low NH_4 has reduced odor and eliminates the need for deammoniation.[23]

Properties of polymers are improved by compounding with enhancing agents (additives), and NR is not an exception. Compounding NR with property enhancers improves resistance to UV oxygen, and ozone, but formulated TPEs and synthetic rubbers overall

have better resistance than compounded NR to UV, oxygen, and ozone.[10a] NR does not have satisfactory resistance to fuels, vegetable, and animal oils, while TPEs and synthetic rubbers can possess good resistance to them.[10a] NR has good resistance to acids and alkalis.[10a] It is soluble in aliphatic, aromatic, and chlorinated solvents, but it does not dissolve easily because of its high MW. Synthetic rubbers have better aging properties; they harden over time, while NR softens over time (see Table 3.14).[10a]

TABLE 3.14 Typical Thermal and Electrical Property Profile of NR[23]

Property	Value
Specific gravity	
@ 32°F (0°C)	0.950
@ 68°F (20°C)	0.934
T_g, °F (°C)	–98 (–72)
Specific heat	0.502
Heat of combustion, cal/g (J/g)	10,547 (44,129)
Thermal conductivity, (BTU-in) (hr-ft^2-°F)	0.90
W/(m · K)	0.13
Coefficient of cubical expansion, in^3/°C	0.00062
Dielectric strength, V/mm	3.937
Dielectric constant	2.37
Power factor @ 1000 cycles	0.15–0.20
Volume resistivity, Ω · cm	1015
Cohesive energy density, cal/cm^3 (J/cm^3)	64 (266.5)
Refractive index	
68°F (20°C) RSS*	1.5192
68°F (20°C) pale crepe	1.5218

*RSS = ribbed smoked sheet.

There are several visually graded latex NRs, including ribbed smoked sheets (RSS) and crepes such as white and pale, thin and thick brown latex, etc.[23] Two types of raw NR are field latex and raw coagulum, and these two types comprise all NR ("downstream") grades.[23]

Depolymerized NR is used as a base for asphalt modifiers, potting compound, and cold-molding compounds for arts and crafts.[10b]

3.9 Conclusion

Producers can engineer polymers and copolymers, and compounders can formulate recipes for a range of products that challenges the designers' imaginations. Computer variable-controlled machinery, tools, and dies can meet the designers' demands. Processing elastomeric materials is not as established as the more traditional thermoplastic and thermosetting polymers. Melt rheology, more than just viscosity, is the central differentiating characteristic for processing elastomeric materials. Processing temperature and pressure settings are not fixed ranges; they are dynamic, changing values from the hopper to the demolded product. Operators and management of future elastomeric materials processing plants will be educated to the finesse of melt processing these materials. Elastomeric materials industries, welcome to the twenty-first century.

References

1. James M. Margolis, Elastomeric Polymers 2000 to 2010: Properties, Processes and Products Report, 2000.
2. *KRATON Polymers and Compounds, Typical Properties Guide,* Shell Chemical Company, Houston, Texas, 1997.
3. *Products, Properties and Processing for PELLETHANE Thermoplastic Polyurethane Elastomers,* Dow Plastics, The Dow Chemical Company Midland, Michigan, ca. 1997.
4. *Modern Plastics Encyclopedia* '99, McGraw-Hill, New York, 1999, pp. B-51, B-52.
5. Engage, A Product of DuPont Dow Elastomers, Wilmington, Delaware, December 1998.
6. *Product Guide,* Goodyear Chemical, Goodyear Tire & Rubber Company, Akron, Ohio, October 1996.
7. *Injection Molding Guide for Thermoplastic Rubber–Processing, Mold Design, Equipment,* Advanced Elastomer Systems LP, Akron, Ohio, 1997.
8. *Santoprene Rubber Physical Properties Guide,* Advanced Elastomer Systems LP, Akron, Ohio, ca. 1998.
9. Hifax MXL 55A01 (1998), FXL 75A01 (1997) and MXL 42D01 Developmental Data Sheets secured during product development and subject to change before final commercialization. Montell Polyolefins Montell North America Inc., Wilmington, Delaware.
10. Charles B. Rader, Thermoplastic Elastomers, in *Handbook of Plastics, Elastomers, and Composites,* 3d ed., Charles A. Harper, ed., McGraw-Hill, New York, 1996.
11. Joseph F. Meier, Fundamentals of Plastics and Elastomers, in *Handbook of Plastics, Elastomers, and Composites,* 3d ed., Charles A. Harper, ed., McGraw-Hill, New York, 1996.
12. Leonard S. Buchoff, Liquid and Low-Pressure Resin Systems, in *Handbook of Plastics, Elastomers, and Composites,* 3d ed., Charles A. Harper, ed., McGraw-Hill, New York, 1996.
13. Edward M. Petrie. Joining of Plastics, Elastomers, and Composites, in *Handbook* of *Plastics, Elastomers, and Composites,* 3d ed., Charles A. Harper, ed., McGraw-Hill, New York, 1996.
14. Ronald Toth, Elastomers and Engineering Thermoplastics for Automotive Applications, in *Handbook of Plastics, Elastomers, and Composites,* 3d ed., Charles Harper, ed., McGraw-Hill, New York, 1996.
15. Aflas TFE Elastomers Technical Information and Performance Profile Data Sheets, Dyneon LLC, A 3M-Hoechst Enterprise, Oakdale, Minnesota, 1997.
16. Vistalon User's Guide, Properties of Ethylene-Propylene Rubber, Exxon Chemical Company, Houston, Texas, Division of Exxon Corporation, ca. 1996.
17. KRATON Liquid L-2203 Polymer, Shell Chemical Company, Houston, Texas, 1997.
18. Affinity Polyolefin Plastomers, Dow Plastics, The Dow Chemical Company, Midland, Michigan, 1997.
19. Affinity HF-1030 Data Sheet, Dow Plastics, The Dow Chemical Company Midland, Michigan, 1997.
20. Affinity PF 1140 Data Sheet, Dow Plastics, The Dow Chemical Company Midland, Michigan, 1997.
21. Bayer Engineering Polymers Properties Guide, Thermoplastics and Polyurethanes, Bayer Corporation, Pittsburgh, Pennsylvania, 1998.
22. *Rubber World Magazine,* monthly, 1999.
23. Jim Ahnemiller, PU Rubber Outsoles for Athletic Footwear, *Rubber World,* December 1998.

24. Charles D. Shedd, Thermoplastic Polyolefin Elastomers, in *Handbook of Thermoplastic Elastomers,* 2d ed., Benjamin M. Walker and Charles P. Rader, eds., Van Nostrand Reinhold, New York, 1988.
25. Thomas W. Sheridan, Copolyester Thermoplastic Elastomers, *Handbook* of *Thermoplastic Elastomers,* 2d ed., Benjamin M. Walker and Charles P. Rader, eds., Van Nostrand Reinhold, New York, 1988.
26. William J. Farrisey Polyamide Thermoplastic Elastomers, in *Handbook of Thermoplastic Elastomers,* 2d ed., Benjamin M. Walker and Charles P. Rader, eds., Van Nostrand Reinhold, New York, 1988.
27. Eric C. Ma, Thermoplastic Polyurethane Elastomers, in *Handbook of Thermoplastic Elastomers,* 2d ed., Benjamin M. Walker and Charles P. Rader, eds., Van Nostrand Reinhold, New York, 1988.
28. N. R. Legge, G. Holden, and H. E. Schroeder, eds., *Thermoplastic Elastomers, A Comprehensive Review,* Hanser Publishers, Munich, Germany, 1987.
29. P S. Ravisbanker, Advanced EPDM for W & C Applications, *Rubber World,* December 1998.
30. Junling Zbao, G. N. Chebremeskel, and J. Peasley SBR/PVC Blends With NBR As Compatibilizer, *Rubber World,* December 1998.
31. John E. Rogers and Walter H. Waddell, A Review of Isobutylene-Based Elastomers Used in Automotive Applications, *Rubber World,* February 1999.
32. *Kirk-Othmer Concise Encyclopedia of Chemical Technology,* John Wiley & Sons, New York, 1999.
33. *PetroChemical News (PCN),* weekly, William F. Bland Company Chapel Hill, North Carolina, September 14,1998.
34. *PeroChemical News (PCN),* weekly, William F. Bland Company, Chapel Hill, North Carolina, 1998 and 1999.
35. *PetroChemical News (PCN),* weekly William F. Bland Company, Chapel Hill, North Carolina, February 22, 1999.
36. C. P. J. van der Aar, et al., Adhesion of EPDMs and Fluorocarbons to Metals by Using Water-Soluble Polymers, *Rubber World,* November 1998.
37. Larry R. Evans and William C. Fultz, Tread Compounds with Highly Dispersible Silica, *Rubber World,* December 1998.
38. Vector Styrene Block Copolymers, Dexco Polymers, A Dow/Exxon Partnership, Houston, Texas, 1997.
39. *Fluoroelastomers Product Information Manual* (1997), *Product Comparison Guide* (1999), Dyneon LLC, A 3M-Hoechst Enterprise, Oakdale, Minnesota, 1997.
40. Engel data sheets and brochures, Guelph, Ontario, 1998.
41. Catalloy Process Resins, Montell Polyolefins, Wilmington, Delaware.
42. Catalloy Process Resins, Montell Polyolefins, Wilmington, Delaware, p.7.
43. EniChem Europrene SOL T Thermoplastic Rubber, styrene butadiene types, styrene isoprene types, EniChem Elastomers Americas Inc., Technical Assistance Laboratory, Baytown, Texas.
44. Arnitel Guidelines for the Injection Molding of Thermoplastic Elastomer TPE-E, DSM Engineering Plastics, Evansville, Ind., ca, 1998.
45. Correspondence from DuPont Engineering Polymers, July 1999.
46. Correspondence from DuPont Dow Elastomers, Wilmington, Delaware, August 1999.

Chapter

4

Composite Materials and Processes

S. T. Peters

Process Research
Mountain View, California

4.1 Introduction

There are two general types of composites, distinguished by the type of materials that are used in construction and by the general market in which they can be found. The more prevalent composites, such as used in printed circuit boards, shower enclosures, and pleasure boats, are generally reinforced with fiberglass fabric, use a type of polyester resin as the matrix, and can be referred to as *commodity* composites. Large overlaps exist for the two types; for instance, there is a significant weight percent of fiberglass-reinforced plastic in most commercial airliners, and carbon/graphite or aramid have been used in reinforcing laminated truss beams for home building. Modern structural composites, frequently referred to as *advanced composites,* can be distinguished from *commodity* composites because of their frequent use of more exotic or expensive matrix materials and higher-priced reinforcements such as carbon/graphite, and they can be found in more structurally demanding locations that have a greater need for weight savings. They are a blend of two or more components. One component is made up of stiff, long fibers, and the other, for polymeric composites, is a resinous binder or *matrix* that holds the fibers in place. The fibers are strong and stiff relative to the matrix and are generally orthotropic (having different properties in two different directions). These properties are most evident when the components are shown in a breakdown view as in Fig. 4.1.

The fiber for advanced structural composites is long, with length-to-diameter ratios >100. Predominately, for advanced structural composites, the fiber has been continuous, but there is an increased awareness that discontinuous fibers allow potentially huge savings in manufacturing costs, so there are now efforts to incorporate them in areas previously reserved for continuous fiber. The fiber's strength and stiffness are much greater—many times more than the matrix material. For instance, the tensile strength quoted for 3502 resin is 4.8 ksi, and the longitudinal elastic modulus (Young's modulus) is 526 ksi. The "B" basis unidirectional tensile strength of the AS-4/3502 lamina is 205 ksi,

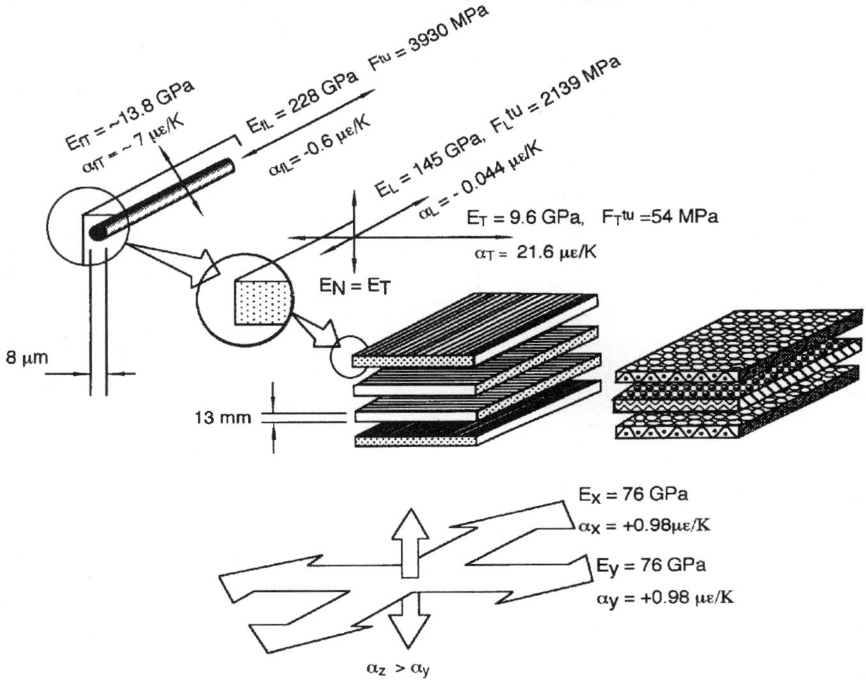

Figure 4.1 The anatomy of a composite laminate.

and the fiber (tow) tensile strength and modulus are 580 ksi and 36 msi, respectively. Thus, for this combination of "typical" advanced composite materials, the fiber is >100 times the strength of the resin and >50 times the modulus.[1–3] (These data will change over time due to manufacturing improvements, and more detailed data, including design allowables, will be shown later in this chapter.) When the fiber and matrix are joined to form a composite, they both retain their individual identities, and both directly influence the composite's final properties. The resulting composite is composed of layers (laminae) of the fibers and matrix stacked to achieve the desired properties in one or more directions.

Designers of aircraft structures have been quick to realize that the high strength-to-weight or modulus-to-weight ratios of composites could result in lighter structural components with lower operating costs and better maintenance histories. The first high-performance aircraft to use advanced composites (boron/epoxy horizontal stabilizers, 1500 built) in a production contract was the F-14A. The use of composites has continued and has resulted in a preponderance of the structure being fabricated from composite materials in one aircraft, the Bell-Boeing V-22, of which approximately 60% of the weight of the craft is composite (Fig. 4.2). Some of the advantages in the use of composites are shown in Table 4.1. These advantages translate not only into aircraft production but also into everyday activities. For example, a carbon/graphite-shafted golf club produces longer drives, because more of the mass is concentrated at the club head. Tennis players also experience less fatigue and pain, because a carbon/graphite composite tennis racquet is lighter and has some inherent vibration damping. Generally, the advantages shown in Table 4.1 can be realized for most fiber/composite combinations, and the disadvantages are more obvious

with some. These advantages have now resulted in composite applications far outside the aircraft industry, with many more reasons for use as shown in Table 4.2. Proper design and material selection can circumvent many of the disadvantages.

4.2 Material Systems

An advanced or commodity composite laminate can be tailored so that the directional dependence of strength and stiffness matches that of the loading environment. To do that, layers of unidirectional material called lamina, or woven fabric with fibers predominately in the expected loading directions, are oriented to satisfy the strength or stiffness requirements. These laminae and fabrics contain both fibers and a matrix. Because of the use of directional laminae, the tensile, flexural, and torsional shear properties of a structure can be disassociated from each other to some extent, and a golf shaft, for example, can be changed in torsional stiffness without changing the flexural or tensile stiffness. This allows for almost infinite variations in the shafts to accommodate individual needs. It also allows for altering the stiffness of a forward-swept aircraft wing to respond to the incoming loads (Fig. 4.3). This is not an option with isotropic metal.

4.2.1 Fibers

Fibers can be of the same material within a lamina or several fibers mixed (hybrid). The common commercially available fiber classes are as follows:

Figure 4.2 The Bell-Boeing V-22. *(Courtesy of the Boeing Company)*

Table 4.1 Reasons for Using Composites

Reason for use	Material selected	Application/driver
Lighter, stiffer, stronger	Boron, all carbon/graphites, some aramid	Military aircraft, better performance; commercial aircraft, operating costs
Controlled or zero thermal expansion	Very high modulus carbon/graphite	Spacecraft with high positional accuracy requirements for optical sensors
Environmental resistance	Fiberglass, vinyl esters, Bisphenol-A Fumarates, Chlorendic Resins	Tanks and piping, corrosion resistance to industrial chemicals, crude oil, gasoline at elevated temperatures
Lower inertia, faster startups, less deflection	High-strength carbon/graphite, epoxy	Industrial rolls, for paper, films
Light weight, damage tolerance	High-strength carbon/graphite, fiberglass, (hybrids), epoxy	CNG tanks for "green" cars, trucks and busses to reduce environmental pollution
More reproducible complex surfaces	High-strength or high-modulus carbon graphite/epoxy	High-speed aircraft; metal skins cannot be formed accurately
Less pain and fatigue	Carbon/graphite/epoxy	Tennis, squash and racquetball racquets; metallic racquets no longer available
Reduces logging in "old growth" forests	Aramid, carbon/graphite	Laminated "new" growth wooden support beams with high-modulus fibers incorporated
Reduces need for intermediate support and resists constant 100% humidity atmosphere	High-strength carbon/graphite-epoxy	Cooling tower driveshafts
Tailorability of bending and twisting response	Carbon/graphite-epoxy	Golf club shafts, fishing rods
Transparency to radiation	Carbon/graphite-epoxy	X-ray tables
Crashworthiness	Carbon/graphite-epoxy	Racing cars
Higher natural frequency, lighter	Carbon/graphite-epoxy	Automotive and industrial driveshafts
Water resistance	Fiberglass (woven fabric), polyester, or isopolyester	Commercial boats
Ease of field application	Carbon/graphite, fiberglass- epoxy, tape, and fabric	Freeway support structure repair after earthquake

Figure 4.3 NASA-Grumman forward-swept wing, X-29 aircraft. *(Courtesy of Dryden Flight Research Center)*

Table 4.2 Advantages and Disadvantages of Advanced Composites

Advantages	Disadvantages
Weight reduction	Cost of raw materials and fabrication
High strength- or stiffness-to-weight ratio)	Transverse properties may be weak
Tailorable properties: can tailor strength or stiffness to be in the load direction	Matrix weakness, low toughness
Redundant load paths (fiber to fiber)	Matrix subject to environmental degradation
Longer life (no corrosion)	Difficult to attach
Lower manufacturing costs because of lower part count	Analysis for physical properties and mechanical properties difficult, analysis for damping efficiency has not reached a consensus
Inherent damping	Nondestructive testing tedious
Increased (or decreased) thermal or electrical conductivity	Acceptable methods for evaluation of residual properties have not reached a consensus
Better fatigue life	

- Carbon/graphite
- Fiberglass
- Organic
  - Aramid
  - Polyethylene
  - PBO

- Boron
- Silicon carbide
- Silicon nitride, silica, alumina, alumina silica

4.2.1.1 Fiberglass. The most widely used fiber for commodity composites, and the fiber that has had the longest period for development, is fiberglass, which has been marketed in several grades in the United States for more than 40 years. During that time, many different glasses were developed, including, leaded glass, and beryllium high-modulus glass, all of which are no longer produced or are in limited supply. The various types of glass that continue to be useful for composite structures are shown in Table 4.3.[4]

Table 4.3 Glass Fibers*

Type	Nominal tensile modulus, GPa ($lb/in^2 \times 10^6$)	Nominal tensile strength, MPa ($lb/in^2 \times 10^3$)	Ultimate strain,%	Fiber density, Mg/m^3 (lb/in^3)	Suppliers
E	72.5 (10.5)	3447 (500)	4.8	2600 (0.093)	PPG Manville Co. Owens Corning Fiberglass
R	85.2 (12.5)	2068 (300)	5.1	2491 (0.089)	Vetrotex Certainteed
Te	84.3 (12.2)	4660 (675)	5.5	2491 (0.089)	Nittobo
S-2	86.9 (12.6)	4585 (665)	5.4	2550 (0.092)	Owens Corning
Zentron high silica	94 (13.5)	3970 (575)		2460 (0.089)	Owens Corning

*In order of ascending modulus normalized to 100% fiber volume (vendor data) (see Ref. 4, p. 2–3).

The table shows the common description for the glass and the nominal tensile strength and tensile modulus of strands and composite. The composite data are average values generated from a series of tests by the manufacturer of the fiber, and they reflect the "ideal" or maximum strength and stiffness of a single fiber or tow or thin unidirectional laminated composite. A tensile test by the user will generally not reflect these values for strength. The maximum number of fibers per strand is usually important information for applications (e.g., filament winding or pultrusion) that use dry fibers mixed with the matrix resin at the point of fiber laydown, since it influences handling ease and per ply thickness. The fiber density is included so that the rule of mixtures equations involving fiber volume and resin volume can be used to evaluate void volume and theoretical mechanical properties. Compressive strength of glass-reinforced composites is relatively high and has led to their selection for use in underwater deep diving applications. The electrical properties of glass-reinforced composites have allowed their use as radomes and printed circuit boards, and in many other areas that require high dielectric strength.

Fiberglass is a product of silica sand, limestone, boric acid, and other ingredients that are dry-mixed, melted (at approximately 1260°C), and then drawn into fibers. Fiberglass,

for structural use, is marketed in the form of fiber strand, unidirectional fabric, woven and knitted fabrics, and as preform shapes. It is by far the most widely used fiber, primarily because of its low cost, but its mechanical properties and specific mechanical properties (i.e., its strength or modulus to weight ratio) are not comparable with other structural fibers. Fiber properties for any of the materials, by themselves, are of little use to most designers, given that they prefer to work with laminate or lamina data. However, with the equations shown later, the designer can use these fiber values to compute the preliminary laminate properties, without the benefit of a computer program or extensive laminate data, which may not be available for newer fibers.

Glass fibers, like all other continuous composite reinforcements, are coated with a thin coating, somewhat like a paint and called a *finish* or *sizing*, that forms a bond between the fiber and the matrix material, improves the handling characteristics, and protects the fiber in the composite from some environmental effects. Each fiber type has a series of unique finishes that are formulated specifically for that type of fiber. Finishes are of little interest to the composite manufacturer, but they are somewhat critical to some processes that use dry strands or tow. For instance, if the fiber finish on the strand or tow for wet filament winding is changed, the spread of the fiber will be changed. If there are no other changes in the process, the resulting width of the applied band will be wider or narrower, and there will be overlaps or gaps in the laminae. Usually, the operator of the machine will adjust to remove laps and gaps (they are usually covered in a specification). Thus, the final laminate thickness will be less than specified and could result in a problem such as premature burst of a pressure vessel.

4.2.1.2 Carbon/graphite. Carbon/graphite fibers have demonstrated the widest variety of strengths and moduli and have the greatest number of suppliers (Tables 4.4a and 4.4b).[4] The carbon/graphite fiber manufacturers, as a result of the turmoil in the composite

Table 4.4a Carbon/Graphite Fibers (Pan) (from Ref. 4, p. 2-7)

Class of fiber	Nominal tensile modulus, GPa ($lb/in^2 \times 10^6$)	Nominal tensile strength, MPa ($lb/in^2 \times 10^3$)	Ultimate strain,%	Fiber density, Mg/m^3 (lb/in^3)	Suppliers/typical products
High tensile strength	227 (33)	3996 (580)	1.60	1.750 (0.063)	Amoco T-300 Hexcel AS-4
High strain	234 (34) 248 (36)	4500 (650) 4550 (660)	1.9 1.7	1.800 (0.064) 1.77	Mitsubishi Grafil 34-700 Amoco T-650-35
Intermediate modulus	275 (40) 290	5133 (745) 5650	1.75 1.8	1740 (0.062) 1.81	Hexcel IM-6 Amoco T-40
Very high strength	289 (42.7)	6370 (924)	2.1	1800 (0.06)	Toray T-1000G
High modulus	390 (57) 436	2900 (420) 4.210	0.70 1.0	1810 (0.065) 1.84	Amoco T-50 Toray M-46J
Very high modulus	540	3920	0.7	1.750	Toray M55-J

Table 4.4b Carbon/Graphite Fibers (Pitch) (from Ref. 4, p. 2-7)

Class of fiber	Nominal tensile modulus, GPa ($lb/in^2 \times 10^6$)	Nominal tensile strength, MPa ($lb/in^2 \times 10^3$)	Ultimate strain,%	Fiber density, Mg/m^3 (lb/in^3)	Thermal conductivity,*,† K W/m-K (Btu/h/°F	CTE,† ppm/K (10^{-6}/°F)	Suppliers/typical products
High modulus (55–65 msi)	379 (55)	2068 (300)	0.50	2000 (0.072)	120	−1.3	Amoco P-55
							Nippon Granoc XN-40
	441		0.76	2080	n/a‡	n/a	Mitsubishi-Kasei Dialead K-1334U
Very high modulus (75–85 msi)	517 (75)	2068 (300)	0.40	2000 (0.072)	185	−1.4	Mitsubishi K-9354U Amoco P-75
Ultra high modulus (90–100 msi)	689 (100)	2240 (325)	0.31	2150 (0.077)	520	−1.6	Granoc XN-70 Mitsubishi K1374U Amoco P-100
Extreme high modulus (>100 msi)	895–999 (130–145)	2411–3789 (350–550)	n/a	2150–2250 (0.077–0.081)	1000	−1.6	Amoco Thornel K-1100X
	780	3530	0.5	n/a	n/a		Granoc XN 80
	826	2239	0.3	2180	640	−1.45	Amoco P-120K
	900 (130)	3800 (550)	0.42	2200	620	−1.2	Mitsubishi Dialead K13C2U

*K and CTE for copper are 400 and 17, respectively.
†Unidirectional composite property, not fiber.
‡Not available.

industry induced by the severe decline in the U.S. defense demand for advanced composites, have gone through drastic price declines, inducing the departure of several manufacturers during the past 10 years. The fibers begin as an organic fiber, such as rayon, polyacrylonitrile (PAN), or pitch (a derivative of crude oil) that is called the *precursor*. The precursor is then stretched, oxidized, carbonized, and graphitized. There are many ways to produce these fibers,[5] but the relative amount of exposure at temperatures from 2500–3000°C results in greater or less graphitization of the fiber. The degree of graphitization in most high-strength or intermediate fibers is low (less than 10%), so the appellation *graphite* fiber is a misnomer and has been promulgated as a marketing tool for sporting goods. Higher degrees of graphitization usually result in a stiffer fiber (higher modulus) with greater electrical and thermal conductivities. Pitch fibers above 689 GPa (100 msi) tensile modulus have thermal conductivity greater than copper and have been used in spacecraft for thermal control as well as for structural applications. Figures 4.4 and 4.5[5,6] show the effect of processing temperature on tensile strength, tensile modulus, and thermal and electrical conductivity.

4.2.1.3 Aramid fibers. Table 4.5[4] shows the properties of several organic reinforcing fibers. Aramid fibers were introduced by DuPont in 1972. Kevlar 49, one aramid, essentially revolutionized pressure vessel technology because of its great tensile strength and consistency coupled with low density, resulting in much more weight-effective designs for rocket motors. The specific tensile strength of Kevlar was, at its introduction, the highest of any fiber. Carbon/graphite fibers, because of advances in processing, have the highest values now. (Specific strength and modulus based on fiber values are simply tensile strength or tensile modulus/density and are a good measure of structural efficiency of a fiber for airborne applications.) The values for tensile modulii may be near to those developed in a composite structure, but the tensile strength values for fibers may be quite different because of factors such as translation efficiency, possibility of flaws, processing damage, or incorrect fiber orientations. Aramid composites are still widely used for pressure vessels but have been largely supplanted by the very high-strength carbon/graphite fibers. Aramids have outstanding toughness and creep resistance, and their failure mode in

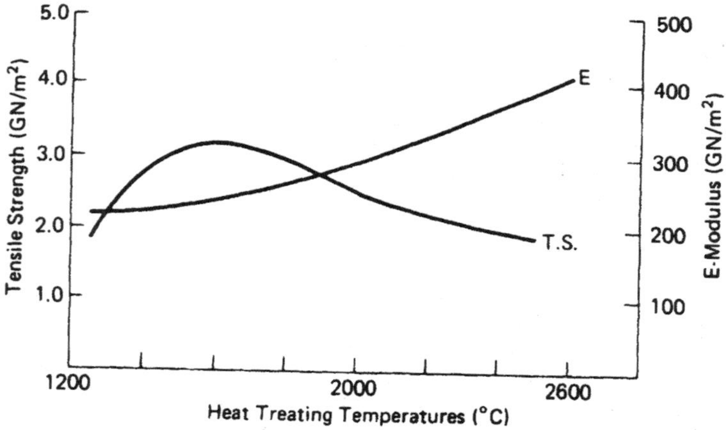

Figure 4.4 Graphitization temperature vs. modulus and tensile strength for carbon graphite fibers.[6]

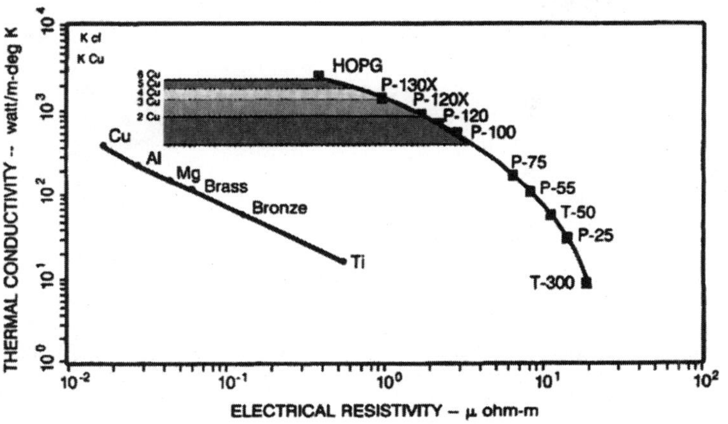

Figure 4.5 Electrical and thermal conductivity of carbon fibers and metals.[7]

Table 4.5 Organic Fibers (from Ref. 4, p. 2-5)

Type	Nominal tensile modulus, GPa ($lb/in^2 \times 10^6$)	Nominal tensile strength, MPa ($lb/in^2 \times 10^3$)	Ultimate strain,%	Fiber density, Mg/m^3 (lb/in^3)	Suppliers/ products
Aramid (medium modulus)	62 (9.0)	3617 (525)	4.0	1440 (0.052)	DuPont Kevlar 29
	80 (11.6)	3150 (457)	3.3	1440	Enka Twaron
	70 (10.1)	3000 (440)	4.4	1390	Teijin Technora
Oriented polyethylene	117 (17)	2585 (375)	3.5	968 (0.035)	Allied Fibers Spectra 900
Aramid (intermediate modulus)	121 (18) 121(18)	3792 (550) 3150 (457)	2.9 2.0	1440 (0.052) 1450	DuPont Kevlar 49 Enka Twaron HM
Oriented polyethylene	172 (25)	3274 (471)	2.7	968 (0.035)	Allied Fibers Spectra 1000
Aramid (high modulus)	186(27)	3445 (500)	1.8	1440 (0.052)	DuPont Kevlar 149
PBO	180 (26)	5800 (840)	2.5	1540 (0.056)	Toyobo Zylon AS*

*High-modulus grade also available.

compression, shear, or flexure is not in a brittle manner and requires a relatively great deal of work. Aramid composites have relatively poor shear and compression properties; careful design is required for their use in structural applications that involve bending.

4.2.1.4 PBO fibers. A new fiber, poly(p-phenylene-2-6-benzobisoxazole) (PBO) was introduced in the 1990s. It was developed by Dow Chemical Co. under U.S. Air Force funding and was tested in pressure vessels by Lincoln Composites. It has shown great promise due to its high strength (5.5 GPa,798 ksi) and low density(1560 kg/m^3, 0.056 lb/in^3). Pressure vessels (Fig. 4.6) fabricated with the fiber and a proprietary Brunswick Composites resin system, LRF-0092, demonstrated a performance factor 30% better than test vessels fabricated with the highest-performing carbon/graphite fibers.[7] There are two grades of the fiber, with tensile strengths and moduli almost double those of p-aramid fibers, along with attractive strain to failure and low moisture regain, making them competitive with carbon high-strength fibers for pressure vessels. The performance factor of the fiber, using strength and density values quoted by the vendor, shows that the PBO fiber is 3% less efficient than T-1000 carbon/graphite fiber. Pressure vessel testing may result in a higher efficiency factor because of the higher strain-to-failure and other considerations. Development ended in the U.S., and now the fiber is manufactured in Japan and marketed by Toyobo under the trade name *Zylon*.

4.2.1.5 Polyethylene fibers. The polyethylene fibers have the same shear and compression property drawbacks as the aramids, but they also suffer from a low melting temperature that limits their use to composites that cure or operate below 149°C (300°F) and a susceptibility to degradation by ultraviolet light exposure. Both the aramids and the polyethylene fibers have wide use in personal protective armor, and the polyethylene fibers have found wide use as ropes and lines for boating and sailing due to their high strength and low density. They float on water and have a pleasant feel or *hand* as a rope or line. In spite of the drawbacks, both of these fibers are enjoying strong worldwide growth.

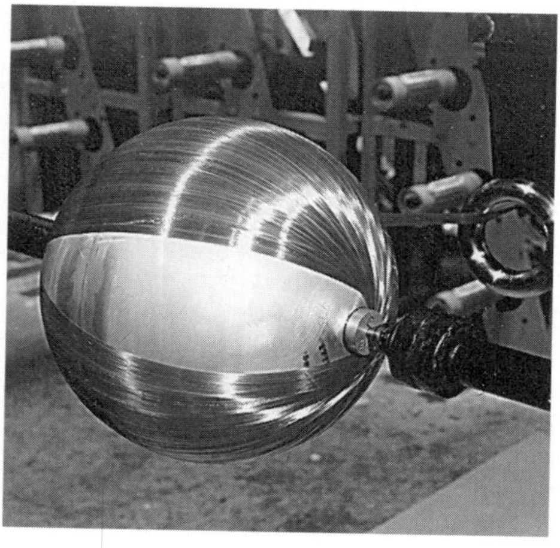

Figure 4.6 Winding PBO spherical test vessel. *(Courtesy of Lincoln Composites)*

4.2.1.6 Boron fibers. Boron fibers, the first fibers to be used on production aircraft (rudders for USAF F-14A fighter, and horizontal stabilizers for the F-111 in approximately 1964–1970), are produced as individual monofilaments on a tungsten or carbon substrate by pyrolylic reduction of boron trichloride (BCl_3) in a sealed glass chamber. (Fig. 4.7). Because the fiber is made as a single filament rather than as a group or tow, the manufacturing process is slower, and the prices are, and will continue to be, higher than for most carbon/graphite fibers. The relatively large-cross-section fiber is used today primarily in polymeric composites that undergo significant compressive stresses (combat aircraft control surfaces) or in composites that are processed at temperatures that would attack carbon/graphite fibers (i.e., metal matrix composites). The carbon/graphite core is protected by the unreactive boron (Table 4.6).[4]

4.2.1.7 Ceramic fibers. The other fibers shown in Table 4.6[4] have varying uses, and several are still in development. Silicon carbide continuous fiber is produced in a chemical vapor deposition (CVD) process similar to that for boron, and it has many mechanical properties identical to those of boron. The other fibers show promise in metal matrix composites, as high-temperature polymeric ablative reinforcements, in ceramic-ceramic composites, and in microwave transparent structures (radomes or microwave printed wiring boards).

4.2.2 Matrix Materials

If parallel and continuous fibers are combined with a suitable matrix and cured properly, unidirectional composite properties such as those shown in Table 4.7 are the result. The functions for and requirements of the matrix are to:

- Help to distribute or transfer loads
- Protect the filaments, both in the structure and before and during structure fabrication
- Control the electrical and chemical properties of the composite
- Carry interlaminar shear

The requirements of and for the matrix, which will vary somewhat with the purpose of the structure, are as follows. It must achieve the following:

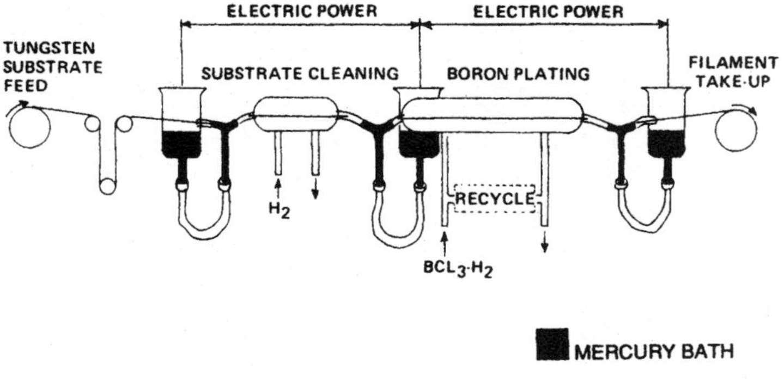

Figure 4.7 Production of boron fiber. *(From Ref. 8)*

Table 4.6 Boron and Ceramic Fibers (from Ref. 4, p. 2-4)

Class of fiber	Nominal tensile modulus, GPa (lb/in^2 × 10^6)	Nominal tensile strength, MPa (lb/in^2 × 10^3)	Ultimate strain,%	Fiber density, Mg/m^3 (lb/in^3)	Thermal conductivity,* K W/m-K (Btu/h/°F	CTE, ppm/K (10^{-6}/°F)	Suppliers/typical products
Alumina	206 (30)	1760 (255)	n/a	3200	1.32	n/a	Sumitomo Altex
	150 (22)	1700 (250)	1.2	2700	0.06	3	3M Nextel 312
SiC	167 (24.3)	2962 (430)	1.4–1.5	2300–2400	n/a	3.1	UBE Tyranno
	186 (27)	2962 (430)	1.6	2360	n/a	n/a	Nippon (DC) HVR Nicolon
SiO_2	69 (10)	3600 (530)	n/a	2200	n/a	n/a	J.P. Stevens Astroquartz II
	72 (10)	3600 (530)	n/a	2200	n/a	n/a	Quartz Products Quartzel

*Unidirectional composite property, not. fiber.

Table 4.7 Properties of Typical Unidirectional Graphite/Epoxy Composites (Fiber Volume Fraction, V_f = 0.60) (from Ref. 10)

	High strength	High modulus
Elastic constants, GPa ($lb/in^2 \times 10^6$)		
Longitudinal modulus, E_L	145 (21)	220 (32)
Transverse modulus, E_T	10 (1.5)	6.9 (1.0)
Shear modulus, G_{LT}	4.8 (0.7)	4.8 (0.7)
Poisson's ratio (dimensionless) υ_{LT}	0.25	0.25
Strength properties, MPa (10^3 lb/in^2)		
Longitudinal tension, $F^{tu}{}_L$	1240 (180)	760 (110)
Transverse tension, $F^{tu}{}_T$	41 (6)	28 (4)
Longitudinal compression, $F^{cu}{}_L$	1240 (180)	690 (100)
Transverse compression, $F^{cu}{}_T$	170 (25)	170 (25)
In-plane shear, $F^{su}{}_{LT}$	80 (12)	70 (10)
Interlaminar shear, F^{Lsu}	90 (13)	70 (10)
Ultimate strains, %		
Longitudinal tension, $\in^{tu}{}_L$	0.9	0.3
Transverse tension, $\in^{tu}{}_T$	0.4	0.4
Longitudinal compression, $\in^{cu}{}_L$	0.9	0.3
Transverse compression, $\in^{cu}{}_T$	1.6	2.8
In-plane shear	2.0	—
Physical properties		
Specific gravity	1.6	1.7
Density (lb/in^3)	0.056	0.058
Longitudinal CTE, 10^{-6} in/in/°F (10^{-6} m/m/°C)	–0.2	–0.3
Transverse CTE, 10^{-6} m/m/°C (10^{-6} in/in/°F)	32 (18)	32 (18)

- Minimize moisture absorption
- Have low shrinkage
- Wet and bond to fiber
- Have a low coefficient of thermal expansion
- Flow to penetrate the fiber bundles completely and eliminate voids during the compacting/curing process
- Have reasonable strength, modulus, and elongation (elongation should be greater than fiber)

- Be elastic to transfer load to fibers
- Have strength at elevated temperature (depending on application)
- Have low-temperature capability (depending on application)
- Have excellent chemical resistance (depending on application)
- Be easily processable into the final composite shape
- Have dimensional stability (maintain its shape)

There are two alternates in matrix selection, thermoplastic and thermoset, and there are many matrix choices available within the two main divisions. The basic difference between the two is that thermoplastic materials can be repeatedly softened by heat, and thermosetting resins cannot be changed after the chemical reaction to cause their cure has been completed. The two alternatives differ profoundly in terms of manufacture, processing, physical and mechanical properties of the final product, and the environmental resistance of the resultant composite.

4.2.2.1 Thermoplastic matrices. Several thermoplastic matrices were developed to increase hot-wet use temperature and the fracture toughness of aerospace, continuous-fiber composites. There are also many thermoplastic matrices, such as polyethylene, ABS, and nylon, that are common to the *commodity* plastics arena. Although continuous-fiber, high-performance "aerospace" thermoplastic composites are still not in general usage, their properties are well documented because of sponsorship of development programs by the U.S. Air Force. Table 4.8 shows the relative advantages and disadvantages of both thermoplastics and thermoset matrices. Thermoplastic matrix choices range from nylon and polypropylene in the commodity arena to those matrices selected for extreme resistance to high temperature and aggressive solvents encountered in the commercial aircraft daily environment, such as the polyether-ether-ketone (PEEK) resins. There is a decided difference in the costs of the commodity resins and the resins that would be used for aerospace use—in a similar order as the differences in fiber prices, for instance, (~U.S.$1.00/lb for polypropylene to >U.S.$100.00/lb for PEEK). Some manufacturers have elected to propose the use of a commodity approach to manufacturing aerospace structures such as small aircraft with polypropylene/glass.[8] The aerospace, high-performance thermoplastic composites have a relatively high potential advantage, because their large-scale use is still in the future. Some special considerations must be made for thermoplastics, as follows:

- Because high temperatures (up to 300°C) are required for processing the higher-performance matrices, special autoclaves, processes, ovens, and bagging materials may be needed.
- The fiber finishes used for thermosetting resins may not be compatible with thermoplastic matrices, requiring alternative treatment.
- Thermoplastic composites can have greater or much less solvent resistance than a thermoset material. If the stressed matrix of the composite is not resistant to the solvent, the attack and destruction of the composite may be nearly instantaneous. (This is due to *stress corrosion cracking,* a common concern for commodity thermoplastics. Thermoplastic liquid detergent bottle materials must undergo rigorous testing to verify their resistance to stress cracking with the contained material, and the addition of fibers into the matrix aggravates the propensity to crack).

4.2.2.2 Thermoset matrices. Thermoset matrices do not necessarily have the same stress corrosion problems but have a completely different and just as extensive set of envi-

Table 4.8 Composite Matrix Trade-Offs

Property	Thermoset	Thermoplastic	Notes
Resin cost	Low to medium-high, based on resin requirements	Low to high. Premium thermoplastic prepregs are more than thermoset prepregs	Will decrease for thermoplastics as volume increases
Formulation	Complex	Simple	
Melt viscosity	Very low	High	High melt viscosity interferes with fiber impregnation
Fiber impregnation	Easy	Difficult	
Prepreg tack/drape	Good	None	Simplified by co-mingled fibers
Prepreg stability	Poor	Good	
Composite voids	Good (low)	Good to excellent	
Processing cycles	Long	Short to long (long processing degrades polymer)	
Fabrication costs	High for aerospace, low for pipes and tanks with glass fibers	Low (potentially); some shapes still cannot be processed economically	
Composite mechanical properties	Fair to good	Good	
Interlaminar fracture toughness	Low	High	
Resistance to fluids/ solvents	Good	Poor to excellent; choose matrix well	Thermoplastics stress craze
Damage tolerance	Poor to excellent	Fair to good	
Resistance to creep	Good	Not known	
Data base	Very large	Small	
Crystallinity problems	None	Possible	Crystallinity affects solvent resistance
Other		Thermoplastics can be reformed to make an interference joint	

ronmental and physical-mechanical concerns. To provide solutions for these potential problems, a great number of matrices have been under development for over 50 years.

The common thermoset matrices for composites include the following:

- Polyester and vinylesters
- Epoxy
- Bismaleimide
- Polyimide
- Cyanate ester and phenolic triazine

Each of the resin systems has some drawbacks that must be accounted for in design and manufacturing plans. Polyester matrices have been in use for the longest period, and they are used in the widest variety and greatest number of structures. These structures have included storage tanks with fiberglass and many types of watercraft, ranging from small fishing or speed boats to large minesweepers. The usable polymers can contain up to 50% by weight of unsaturated monomers and solvents such as styrene. These can cause a significant shrinkage on matrix cure. Polyesters cure via a catalyst (usually a peroxide), which results in an exothermic reaction. This reaction can be initiated at room temperature. Because of the large shrinkage with the polyester-type matrices, they are generally not used with the high-modulus fibers.

The most widely used matrices for advanced composites have been the epoxy resins. These resins cost more than polyesters and do not have the high-temperature capability of the bisimalimides or polyimides; but, because of the advantages shown in Table 4.9, they are widely used.

Table 4.9 Epoxy Resin Selection Factors

Advantages	Disadvantages
Adhesion to fibers and resin	Resins and curatives somewhat toxic in uncured form
No by-products formed during cure	Moisture absorption:
Low shrinkage during cure	Heat distortion point lowered by moisture absorption
Solvent and chemical resistance	Change in dimensions and physical properties due to moisture absorption
High or low strength and flexibility	Limited to about 200°C upper temperature use (dry)
Resistance to creep and fatigue	Difficult to combine toughness and high temperature resistance
Good electrical properties	High thermal coefficient of expansion
Solid or liquid resins in uncured state	High degree of smoke liberation in a fire
Wide range of curative options	May be sensitive to UV light degradation
	Slow curing

There are two resin systems in common use for higher temperatures, bismaleimides and polyimides. New designs for aircraft demand a 177°C (350°F) operating temperature that is not met by the other common structural resin systems. The primary bismaleimide (BMI) in use is based on the reaction product from methylene dianiline (MDA) and maleic anhydride: bis (4 maleimidophenyl) methane (MDA BMI).

Two newer resin systems have been developed and have found applications in widely diverse areas. The cyanate ester resins, marketed by Ciba-Geigy, have shown superior dielectric properties and much lower moisture absorption than any other structural resin for composites. The dielectric properties have enabled their use as adhesives in multilayer mi-

crowave printed circuit boards and the low moisture absorbance have caused them to be the resin of universal choice for structurally stable spacecraft components.

The PT resins also have superior elevated temperature properties, along with excellent properties at cryogenic temperatures. Their resistance to proton radiation under cryogenic conditions was a prime cause for their choice for use in the superconducting supercollider, subsequently canceled by the U.S. Congress. They are still available from the Lonza Company.

Polyimides are the highest-temperature polymer in general advanced composite use, with a long-term upper temperature limit of 232°C (450°F) or 316°C (600°F). Two general types are *condensation* polyimides, which release water during the curing reaction, and *addition* type polyimides, with somewhat easier process requirements.

Several problems consistently arise with thermoset matrices and prepregs that do not apply to thermoplastic composite starting materials. Because of the problems shown below, if raw material and processing costs were comparable for the two matrices, the choice would probably always be thermoplastic composites, without regard to the other advantages resulting in the composite. These problems lead to a great increase in quality control efforts that may result in the bulk of final composite structure costs. They are as follows:

Problems Associated with Thermoset Matrices

1. Frequent variations from batch to batch
   - Effects of small amounts of impurities
   - Effects of small changes in chemistry
   - Change in matrix component vendor or manufacturing location
2. Void generation, caused by
   - Premature gelation
   - Premature pressure application
   - Effects on interlaminar shear and flexural modulus because of water absorption
3. Change in processing characteristics
   - Absorbed water in prepreg
   - Length of time under refrigeration
   - Length of time out before cure
   - Loss of solvent in wet systems

Some other resins that are in general commercial and aerospace use are not treated here, because they are not in wide use with the modern fibers.

The following general notes are more or less applicable to all thermoset matrices:

- The higher the service temperature limitation the less strain to failure.
- The greater the service temperature, the more difficult the processing that may be due to:
  1. Volatiles in matrix
  2. Higher melt viscosity
  3. Longer heating curing cycles
- The greater the service temperature or the greater the curing temperature, the greater the chance for development of color in the matrix.

- Higher service temperatures and higher curing temperatures may sometimes result in better flame resistance (although this is not evident for epoxies with curing temperatures between 250°F and 350°F).

4.2.3 Fiber Matrix Systems

The end-user sees a composite structure. Someone else, probably a prepregger, combined the fiber and the resin system, and someone else caused the cure and compaction to result in a laminated structure. A schematic of the steps is shown in Fig. 4.8. In many cases, the end-user of the structure has fabricated the composite from prepreg. The three types of continuous fibers, roving or tow, tape, and woven fabric available as prepregs give the end user many options in terms of design and manufacture of a composite structure. Although the use of dry fibers and impregnation at the work (i.e., filament winding pultrusion or hand layup) is very advantageous in terms of raw material costs, there are many advantages to the use of prepregs, as shown in Table 4.10, particularly for the manufacture of modern composites. In general, fabricators skilled in manufacturing from prepreg will not care to use wet processes.

Table 4.10 Advantages of Prepregs over Wet Impregnation

Prepregs reduce the handling damage to dry fibers.
They improve laminate properties via better dispersion of short fibers.
Prepregs allow the use of hard-to-mix or proprietary resin systems.
They allow more consistency, because there is a chance for inspection before use.
Heat curing provides more time for the proper laydown of fibers and for the resin to move and degas before cure.
Increasing curing pressure reduces voids and improves fiber wetting.
Most prepregs have been optimized as individual systems to improve processing.

The prepreg process for thermoset matrices is accomplished by feeding the fiber continuous tape, woven fabric, strands, or roving through a resin-rich solvent solution and then removing the solvent by hot tower drying. The excess resin is removed via a doctor blade or metering rolls, and then the product is staged to the cold-stable prepreg form (B stage). The newer technique, the hot-melt procedure for prepregs, has substantially replaced the solvent method because of environmental concerns and the need to exert better control over the amount of resin on the fiber. A film of resin that has been cast hot onto release paper is fed, along with the reinforcement, through a series of heaters and rollers to force the resin into the reinforcement. Two layers of resin are commonly used so that a resin film is on both sides of the reinforcement; one of the release papers is removed, and the prepreg is then trimmed, rolled, and frozen. The two types of prepregging techniques, solvent and film are shown in Figs. 4.9 and 4.10.[9]

4.2.4 Unidirectional Ply Properties

The manufacturer of the prepreg reports an areal weight for the prepreg and a resin percentage, by weight. Since fiber volume is used to relate the properties of the manufactured composites, the following equations can be used to convert between weight fraction and fiber volume.

$$W_f = \frac{w_f}{w_c} = \frac{\rho_f V_f}{\rho_c V_f} = \frac{\rho_f}{\rho_c} V_f \qquad (4.1)$$

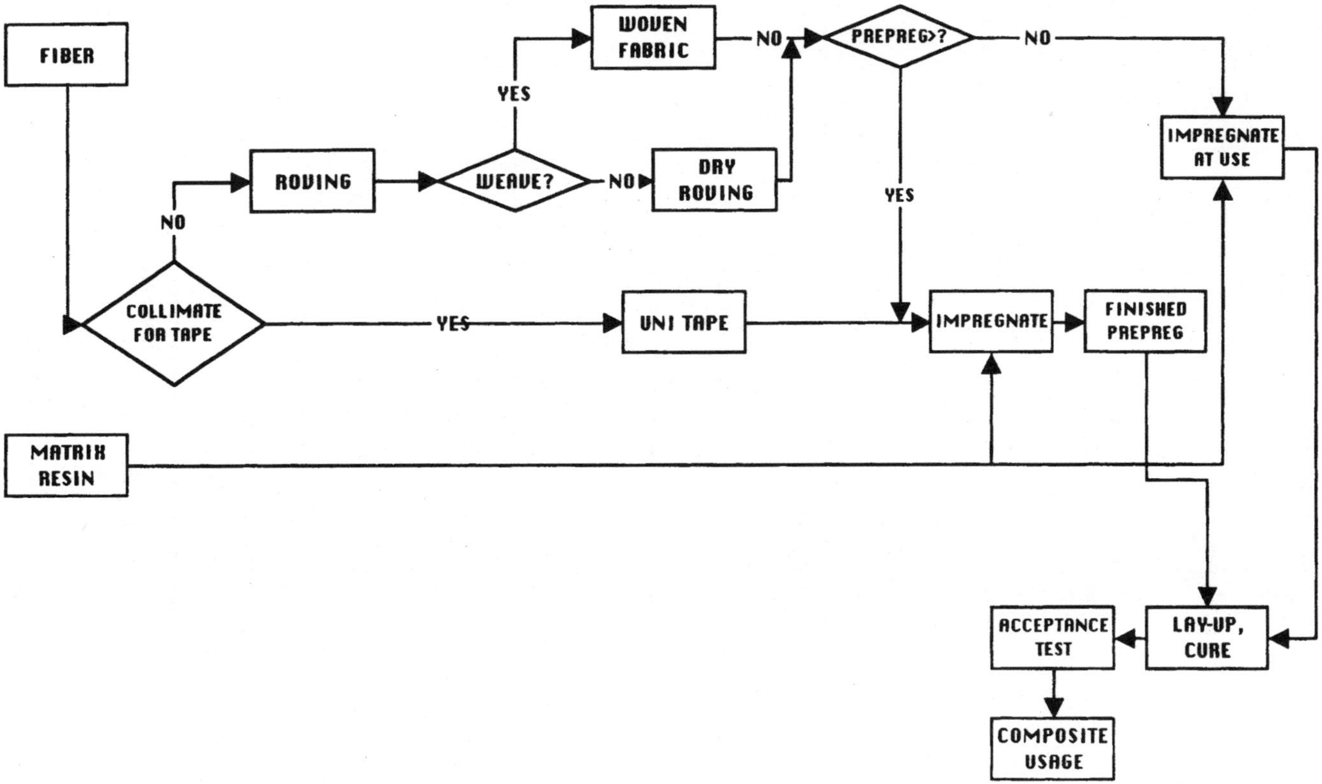

Figure 4.8 The manufacturing steps in composite structure fabrication.

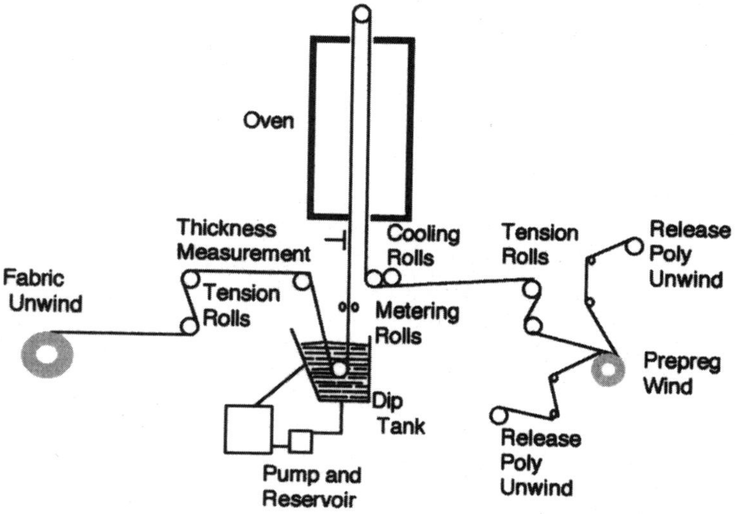

Figure 4.9 Schematic of the typical solution prepregging process.

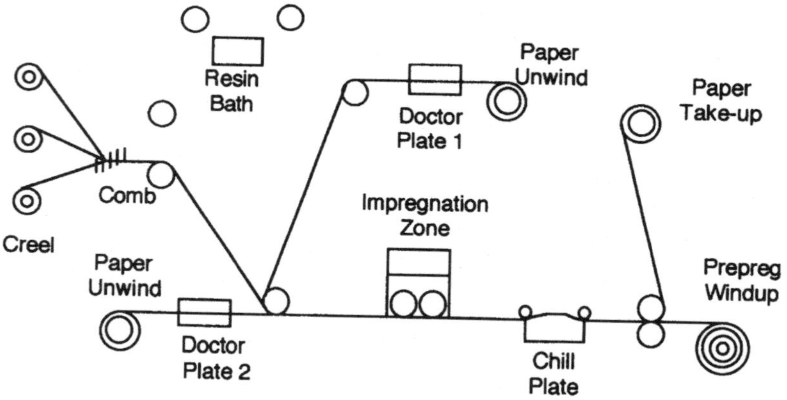

Figure 4.10 Schematic of the typical film prepregging process.

$$V_f = \frac{\rho_c}{\rho_f} W_f = 1 - V_m \tag{4.2}$$

where W_f = weight fraction of fiber
w_f = weight of fiber
w_c = weight of composite
ρ_f = density of fiber

ρ_c = density of composite
V_f = volume fraction of fiber
V_m =volume fraction of matrix
ρ_m = density of matrix

A percentage fiber that is easily achievable and repeatable in a composite and convenient for reporting mechanical and physical properties for several fibers is 60%. The properties of unidirectional fiber laminates are shown in Tables 4.7, 4.11, 4.12, and 4.13.[10]

Table 4.11 Properties of Typical Unidirectional Glass/Epoxy Composites (Fiber Volume Fraction, V_f = 0.60); Elastic Constants, Strengths, Strains, and Physical Properties (from Ref. 10)

	E-glass	S-glass
Elastic constants, GPa (lb/in² × 10⁶)		
Longitudinal modulus, E_L	45 (6.5)	55 (8.0)
Transverse modulus, E_T	12 (1.8)	16 (2.3)
Shear modulus, G_{LT}	5.5 (0.8)	7.6 (1.1)
Poisson's ratio (dimensionless) υ_{LT}	0.19	0.28
Strength properties, MPa (10³ lb/in²)		
Longitudinal tension, $F^{tu}{}_L$	1020 (150)	1620 (230)
Transverse tension, $F^{tu}{}_T$	40 (7)	40 (7)
Longitudinal compression, $F^{cu}{}_L$	620 (90)	690 (100)
Transverse compression, $F^{cu}{}_T$	140 (20)	140 (20)
In-plane shear, $F^{su}{}_{LT}$	60 (9)	60 (9)
Interlaminar shear, F^{Lsu}	60 (9)	80 (12)
Ultimate strains, %		
Longitudinal tension, $\in^{tu}{}_L$	2.30.9	2.9
Transverse tension, $\in^{tu}{}_T$	0.4	0.3
Longitudinal compression, $\in^{cu}{}_L$	1.4	1.3
Transverse compression, $\in^{cu}{}_T$	1.1	1.9
In-plane shear	—	3.2
Physical properties		
Specific gravity	2.1	2.0
Density (lb/in³)	0.075	0.72
Longitudinal CTE, 10^{-6} in/in/°F (10^{-6} m/m/°C)	3.7 (6.6)	3.5 (6.3)
Transverse CTE, 10^{-6} m/m/°C (10^{-6} in/in/°F)	30 (17)	32 (18)

Table 4.12 Properties of Unidirectional Aramid/Epoxy Composites (Fiber Volume Fraction, $V_f = 0.60$) (from Ref. 10)

	Kevlar 49
Elastic constants, GPa (lb/in^2 $\times$ 10^6)	
Longitudinal modulus, E_L	76 (11)
Transverse modulus, E_T	5.5 (0.8)
Shear modulus, G_{LT}	2.1 (0.3)
Poisson's ratio (dimensionless) υ_{LT}	0.34
Strength properties, MPa (10^3 lb/in^2)	
Longitudinal tension, $F^{tu}{}_L$	1380 (200)
Transverse tension, $F^{tu}{}_T$	30 (4.3)
Longitudinal compression, $F^{cu}{}_L$	280 (40)
Transverse compression, $F^{cu}{}_T$	140 (20)
In-plane shear, $F^{su}{}_{LT}$	60 (9)
Interlaminar shear, F^{Lsu}	60 (9)
Ultimate strains, %	
Longitudinal tension, $\in^{tu}{}_L$	1.8
Transverse tension, $\in^{tu}{}_T$	0.5
Longitudinal compression, $\in^{cu}{}_L$	2.0
Transverse compression, $\in^{cu}{}_T$	2.5
In-plane shear	—
Physical properties	
Specific gravity	1.4
Density (lb/in^3)	0.050
Longitudinal CTE, 10^{-6} in/in/°F (10^{-6} m/m/°C)	–4 (–2.2)
Transverse CTE, 10^{-6} m/m/°C (10^{-6} in/in/°F)	70 (40)

These values are for individual lamina or for a unidirectional composite, and they represent the theoretical maximum (for that fiber volume) for longitudinal in-plane properties. Transverse, shear, and compression properties will show maximums at different fiber volumes and for different fibers, depending on how the matrix and fiber interact. These properties are not reflected in strand data. These values may also be used to calculate the properties of a laminate that has fibers oriented in several directions. Using the techniques shown in Sec. 4.5.1, the methods of description for ply orientation must be introduced.

Table 4.13 Properties of Typical Unidirectional Boron/Epoxy Composites (from Ref. 10)

	Boron
Elastic constants, GPa (lb/in^2 × 10^6)	
Longitudinal modulus, E_L	207 (30)
Transverse modulus, E_T	19 (2.7)
Shear modulus, G_{LT}	4.8 (0.7)
Poisson's ratio (dimensionless) υ_{LT}	0.21
Strength properties, MPa (10^3 lb/in^2)	
Longitudinal tension, $F^{tu}{}_L$	1320 (192)
Transverse tension, $F^{tu}{}_T$	72 (10.4)
Longitudinal compression, $F^{cu}{}_L$	2430 (350)
Transverse compression, $F^{cu}{}_T$	276 (40)
In-plane shear, $F^{su}{}_{LT}$	105 (15)
Interlaminar shear, F^{Lsu}	90 (13)
Ultimate strains, %	
Longitudinal tension, $\in^{tu}{}_L$	0.6
Transverse tension, $\in^{tu}{}_T$	0.4
Longitudinal compression, $\in^{cu}{}_L$	—
Transverse compression, $\in^{cu}{}_T$	—
In-plane shear	—
Physical properties	
Specific gravity	2.0
Density (lb/in^3)	0.072
Longitudinal CTE, 10^{-6} in/in/°F (10^{-6} m/m/°C)	4.1 (2.3)
Transverse CTE, 10^{-6} m/m/°C (10^{-6} in/in/°F)	19 (11)

4.3 Ply Orientations, Symmetry, and Balance

4.3.1 Ply Orientations

One of the advantages of using a modern composite is the potential to orient the fibers to respond to the load requirements. This means that the composite designer must show the material, the fiber orientations in each ply, and how the plies are arranged (ply stackup). A

shorthand "code" (Fig. 4.11*b*) for ply fiber orientations has been adapted for use in layouts and studies.

Each ply (lamina) is shown by a number representing the direction of the fibers in degrees, with respect to a reference (x) axis. 0° fibers of both tape and fabric are normally aligned with the largest axial load (axis) (Fig. 4.11*a*).

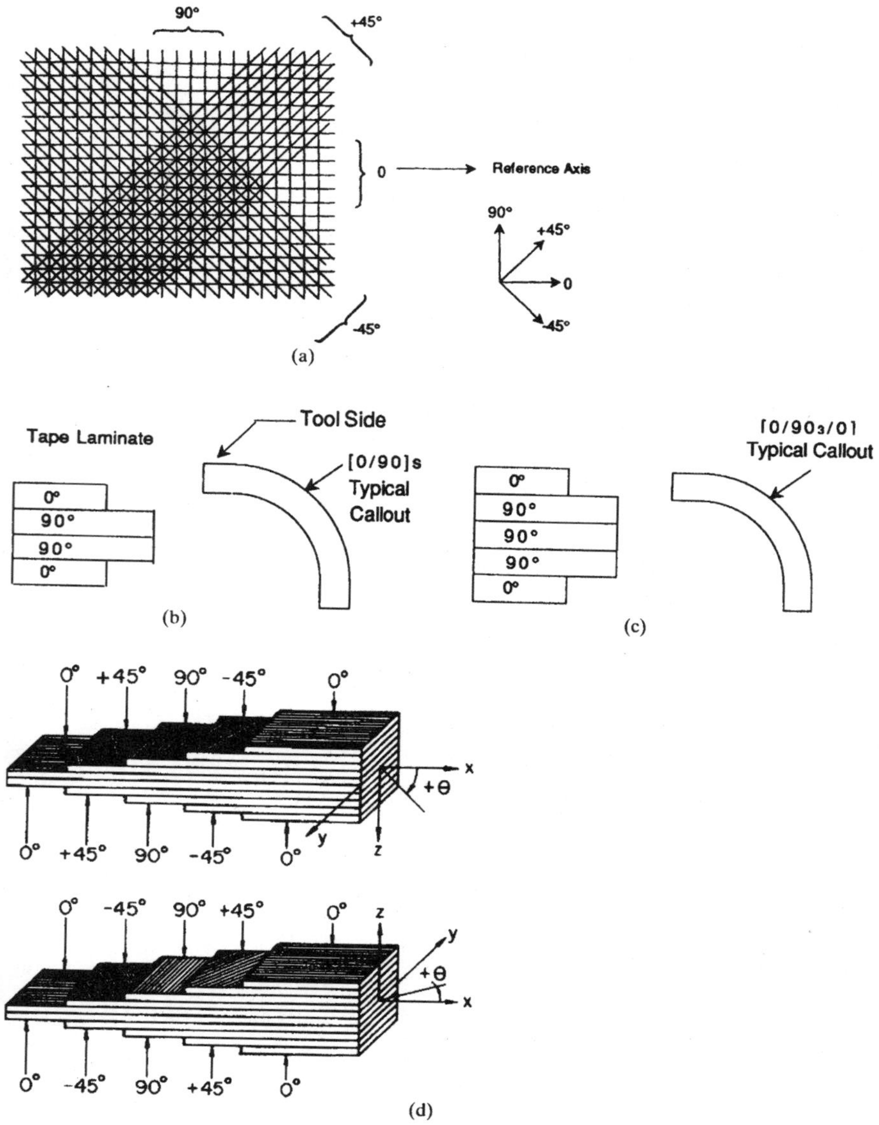

Figure 4.11 Ply orientations, symmetry, and balance.

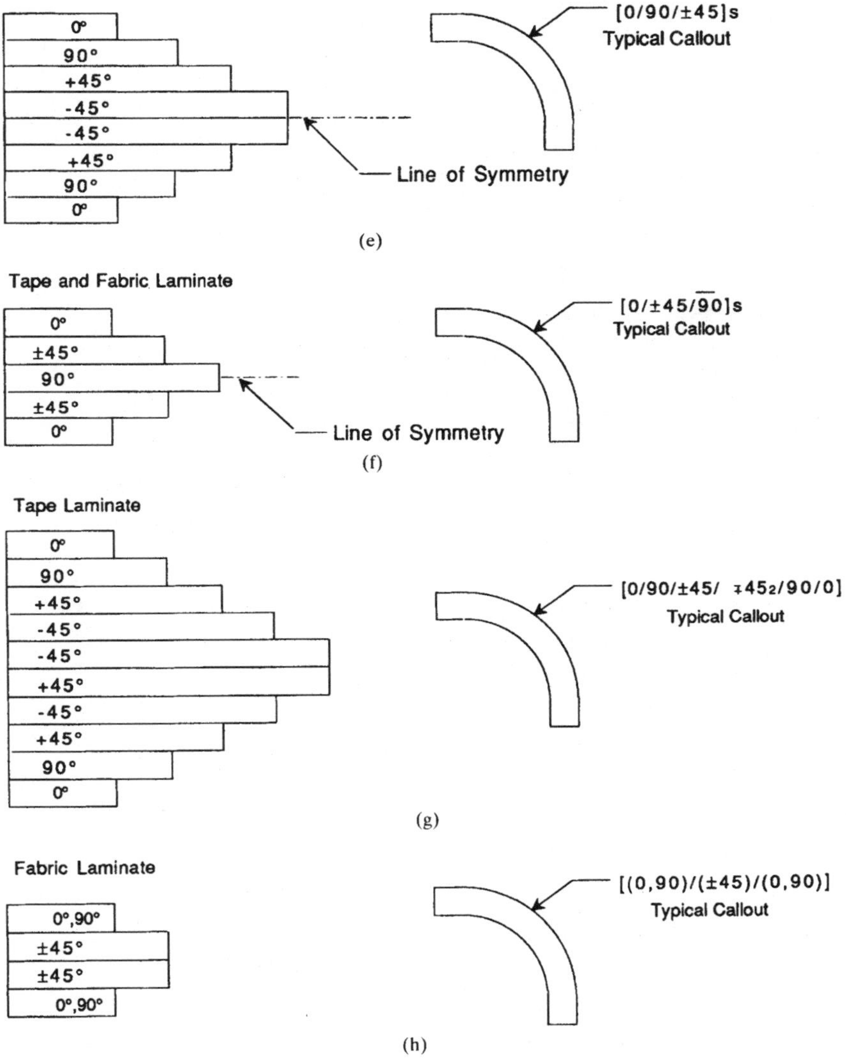

Figure 4.11 Ply orientations, symmetry, and balance *(continued).*

Individual adjacent plies are separated by a slash in the code if their angles are different (Fig. 4.11*b*).

The plies are listed in sequence, from one laminate face to the other, starting with the ply first on the tool and indicated by the code arrow with brackets indicating the beginning and end of the code. Adjacent plies of the same angle of orientation are shown by a numerical subscript (Fig. 4.11*c*).

When tape plies are oriented at angles equal in magnitude but opposite in sign, (+) and (–) are used. Each (+) or (–) sign represents one ply. A numerical subscript is used only

when there are repeating angles of the same sign. Positive and negative angles should be consistent with the coordinate system chosen. An orientation shown as positive in one right-handed coordinate system may be negative in another. If the Y and Z axis directions are reversed, the ±45 plies are reversed (Fig. 4.11*d*).

Symmetric laminates with an even number of plies are listed in sequence, starting at one face and stopping at the midpoint. A subscript "S" following the bracket indicates only one half of the code is shown (Fig. 4.11*e*).

Symmetric laminates with an odd number of plies are coded as a symmetric laminate except that the center ply, listed last, is overlined to indicate that half of it lies on either side of the plane of symmetry (Fig. 4.11*f–h*).

4.3.2 Symmetry

The geometric midplane is the reference surface for determining if a laminate is symmetrical. In general, to reduce out-of-plane strains, coupled bending and stretching of the laminate, and complexity of analysis, symmetric laminates should be used. However, some composite structures (e.g., filament wound pressure vessels) are geometrically symmetric, so symmetry through a single laminate wall is not necessary if it constrains manufacture. To construct a midplane symmetric laminate, for each layer above the midplane there must exist an identical layer (same thickness, material properties, and angular orientation) below the midplane (Fig. 4.11*e*).

4.3.3 Balance

All laminates should be balanced to achieve in-plane orthotropic behavior. To achieve balance, for every layer centered at some positive angle $+\theta$, there must exist an identical layer oriented at $-\theta$ with the same thickness and material properties. If the laminate contains only 0° and/or 90° layers, it satisfies the requirements for balance. Laminates may be midplane symmetric but not balanced, and vice versa. Figure 4.11*e* is symmetric and balanced, whereas Fig. 4.11*g* is balanced but unsymmetric.

4.4 Quasi-isotropic Laminate

The goal of composite design is to achieve the lightest, most efficient structure by aligning most of the fibers in the direction of the load. Many times, there is a need, however, to produce a composite that has some isotropic properties, similar to metal, because of multiple or undefined load paths or for a more conservative design. A *quasi-isotropic* laminate layup accomplishes this for the x and y planes only; the z, or through-the-laminate thickness plane, is quite different and lower. Most laminates produced for aircraft applications have been, with few exceptions, quasi-isotropic. One exception was the X-29 (Fig. 4.3). As designers become more confident and have access to a greater database with fiber-based structures, more applications will evolve. For a quasi-isotropic (QI) laminate, the following are requirements:

- It must have three layers or more.
- Individual layers must have identical stiffness matrices and thicknesses
- The layers must be oriented at equal angles. For example, if the total number of layers is *n*, the angle between two adjacent layers should be 360°/*n*. If a laminate is constructed from identical sets of three or more layers each, the condition on orientation must be satisfied by the layers in each set, for example: $[0°/\pm60°]_S$ or $[0°/\pm45°/90]_S$ [Ref. 11, p. 199].

Table 4.14[12] shows mechanical values for several composite laminates with the high-strength fiber of Table 4.4 and a typical resin system. The first and second entries are for simple 0/90 laminates and show the effect of changing the position of the plies. The effect of increasing the number of 0° plies is shown next, and the final two laminates demonstrate the effect of ±45° plies on mechanical properties, particularly the shear modulus. The last entry is a quasi-isotropic laminate. These laminates are then compared to a typical aluminum alloy. This effectively shows that there is a strength and modulus penalty that goes with the conservatism of the use a QI laminate.

Table 4.14 High-Strength Carbon Graphite Laminate Properties

Laminate	Longitudinal modulus, E_{11}, GPa	Bending modulus, E_B, GPa	Shear modulus, G_{xy}, GPa
$[0/90_2/0]$	76.5	126.8	5.24
$[90/0_2/90]$	76.5	26.3	5.24
$[0_2/90_2/0]$	98.5	137.8	5.24
$[0_2/\pm45_2/0]$	81.5	127.5	21.0
$[0/\pm45/90]_s$	55.0	89.6	21.0
Aluminum	41.34	41.34	27.56

When employing the data extracted from tables, there are some cautions that should be observed by the reader. The values seen in many tables of data may not always be consistent for the same materials or the same group of materials from several sources for the following reasons:

1. Manufacturers have been refining their production processes so that newer fibers may have greater strength or stiffness. These new data may not be reflected in the compiled data.
2. The manufacturer may not be able to change the value quoted for the fiber because of government or commercial restrictions imposed by the specification process of his customers.
3. Many different high-strength fibers are commercially available. Each manufacturer has optimized its process to maximize the mechanical properties, and each of the processes may by different from that of the competitor, so all vendor values in a generic class may differ widely.
4. Most tables of values are presented as "typical values." Those values and the values that are part of the menu of many computer analysis programs should be used with care. Each user must find the most appropriate set of values for design, develop useful design allowables, and apply appropriate "knock down" factors, based on the operating environments expected in service.

4.5 Analysis

4.5.1 Micromechanical Analysis

A number of methods are in common use for the analysis of composite laminates. The use of micromechanics, i.e., the application of the properties of the constituents to arrive at the properties of the composite ply, can be used to achieve the following:

1. Arrive at "back of the envelope" values to determine if a composite is feasible
2. Arrive at values for insertion into computer programs for laminate analysis or finite element analysis
3. Check on the results of computer analysis

The rule of mixtures holds for composites. The micromechanics formula to arrive at the Young's modulus for a given composite is

$$E_c = V_f E_f + V_m E_m$$

and

$$V_f + V_m = 1$$

$$E_c = V_f E_f + E_m(1 - V_f) \tag{4.3}$$

where E_c = composite or ply Young's modulus in tension for fibers oriented in direction of applied load
V = volume fraction of fiber (f) or matrix (m)
E = Young's modulus of fiber (f) or matrix (m)

But, since the fiber has much higher Young's modulus than the matrix, (Table 4.7 vs. the value for the 3502 matrix shown in Fig. 4.1), the second part of the equation can be ignored.

$$E_f \gg E_m$$

$$E_c = E_f V_f \tag{4.4}$$

This is the basic rule of mixture and represents the highest Young's modulus composite, where all fibers are aligned in the direction of load. The minimum Young's modulus for a reasonable design (other than a preponderance of fibers being orientated transverse to the load direction) is the quasi-isotropic composite and can be approximated by

$$E_c \cong \frac{3}{8} E_f V_f \tag{4.5}$$

Note: the quasi-isotropic modulus, E, of a composite laminate is

$$\frac{3}{8}E_{11} + \frac{5}{8}E_{22} \text{ (see Ref. 13)}$$

where E_{11} is the modulus of the lamina in the fiber direction and E_{22} is the transverse modulus of the lamina. The transverse modulus for polymeric-based composites is a small fraction of the longitudinal modulus (see E_T in Table 4.7) and can be ignored for preliminary estimates, resulting in a slightly lower-than-theoretical value for E_c for a quasi-isotropic laminate. This approximate value for quasi-isotropic modulus represents the lower bound of composite modulus. It is useful for comparisons of composite properties to those of metals and to establish if a composite is appropriate for a particular application.

The following formulas also can be used to obtain important data for unidirectional composites:

Density, $$\rho_c = V_f\rho_f + V_m\rho_m \tag{4.6}$$

Poisson's ratio, $$\nu_{12} = \nu_f V_f + \nu_m V_m \tag{4.7}$$

Transverse Young's modulus, $$E_2 = \frac{E_{2m}(1 + \xi\eta_2 V_f)}{1 - \eta_2 V_f} \tag{4.8}$$

and values for η_2 and ξ can be seen in Ref. 14 and Ref. 11, pp. 76–78. The matrix is isotropic.

4.5.2 Carpet Plots

The analysis of a multilayered composite, if attempted by hand calculations, is not trivial. Fortunately, there are a significant number of computer programs to perform the matrix multiplications and the transformations.[14–16] However, the use of carpet plots is still in practice in U.S. industry, and these plots are useful for preliminary analysis. The carpet plot shows graphically the range of properties available with a specific laminate configuration. For example, if the design options include $[\pm 0/90]_S$ laminates, a separate carpet plot for each value of θ would show properties attainable by varying percentage of $\pm\theta$ plies versus 90° plies. A sequence of these charts would display attainable properties over a range of θ values. The computer programs described above can be programmed to produce such charts for arbitrary laminates.

Figure 4.12 shows a sample carpet plot[17] of extensional modulus of elasticity E_x for Kevlar 49/epoxy with $[0/\pm 45/90]_s$ construction. As expected, the chart shows $E_x = 76$ GPa (11×10^6 lb/in^2) with all 0° plies, and $E_x = 5.5$ GPa (0.8×10^6 lb/in^2) with all 90s. With all 45s, an axial modulus is only slightly higher, 8 GPa (1.1×10^6 lb/in^2), than the all 90s value predicted for this material. A quasi-isotropic laminate (Sec. 4.5.2) with 25% 0s, 50% ±45s, and 25% 90s, produces an intermediate value of $E_x = 29$ GPa (4.2×10^6 lb/in^2).

4.6 Composite Failure and Design Allowables

4.6.1 Failure[18–20]

Composite failure modes are different from those of isotropic materials such as metals. Because of the fibers, they do not tend to fail in only one area, they do not have the strain-bearing capacity of most metals, and they are prone to premature failure if stressed in a direction that was not anticipated in the design. Useful structures nearly always have been constructed from ductile materials such as steel or aluminum, with fairly well defined strengths. This allows designers to accurately comprehend and specify safety factors that provide some assurance that the structures will not fail in service.

It has became necessary, in the practical design of structures for demanding environments, to use brittle materials such as glass and ceramics to take advantage of special properties such as high-temperature strength. When brittle materials are employed in practical structures, the designer still has the need to ensure that the structure will not fail prematurely.

The data that provide the background for the design confidence can be obtained from various sources. They can be derived from previous designs that have proven reliable and resulted in data being published in a reference work such as Mil-Handbook-5 for Aerospace Metals[21] or industry journals. Or the data can be obtained through testing conducted

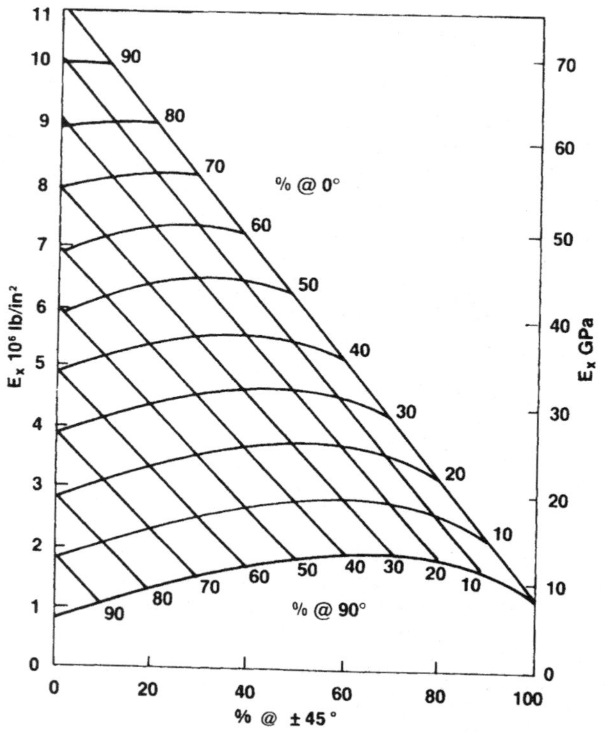

Figure 4.12 Predicted axial modulii for [0/±45/90] Kevlar® epoxy laminates.

by the designer's own organization. Typically, on the basis of laboratory experiments on a statistically determined number of small specimens tested in simple tension or bending, the probability of failure can be calculated for structural members of other sizes and shapes, often under completely different loading conditions. The tool for accomplishing this is *statistical fracture theory.*

To predict strength of the ply with the laminate, it is usually assumed that knowledge of failure of a ply by itself under simple tension, compression, or shear will allow prediction of failure of that ply under combined loading in the laminate.

The matrix plays a special role in the failure of the composite. The matrix is extremely weak compared to the fibers (particularly if they are the *advanced composite* fibers) and cannot carry primary loads, but it efficiently allows the transfer of the loads in the composite. This is demonstrated by the experimental observation that the strength of matrix-impregnated fiber bundles can be on the order of a factor of 2 higher than the measured tensile strength of dry fiber bundles without matrix impregnation. The key to this apparently contradictory evidence lies in a synergistic effect between fiber and matrix. The first and primary design rule for composites of this type is that the fibers must be oriented to carry the primary loads. A comparison of the tensile strengths illustrates this point. High-strength carbon fibers have tensile strengths that approach 1×10^6 lb/in^2 (6900 MPa), while the tensile strength of typical polymer matrices may be on the order of 3×10^4 lb/in^2 (200 MPa) or less. Clearly, the tensile strength of the matrix is insignificant in comparison.

A number of investigators have provided an explanation for the above observation. It can be explained by noting that the strength of individual brittle fibers varies widely because of a statistical distribution of flaws. The fibers can be considered to be brittle and sensitive to surface imperfections randomly distributed over the fiber length. The strength of individual fibers varies widely and will decrease with increasing length. These are characteristics that are typical of brittle materials failing at random defects, and they are changed dramatically through the addition of the matrix. The matrix acts to almost double the apparent strength of a fiber bundle, and it significantly reduces the variability.

In a dry fiber bundle, when a fiber breaks, it loses all of its load-carrying ability over its entire length, and the load is shifted to the remaining fibers. When enough of the weaker fibers fail, the strength of the remaining fibers is exceeded, and the bundle fails. In matrix-impregnated fiber bundles, the matrix acts to bridge around individual fiber breaks so that adjacent fibers quickly pick up the load. Thus, the adjacent fibers have to carry an increased load over only a small axial distance. Statistical distribution of fiber defects makes it unlikely that each fiber would be weakest at the same axial location, so failure will occur at a higher load value after enough fibers have failed in adjacent locations. Because of the small diameter of individual fibers (5.7 mm for some typical carbon fibers), there are many millions of fibers in a structure. This makes statistical effects important.

4.6.2 Failure Theories

For over three decades, there has been a continuous effort to develop a more universal failure criterion for unidirectional fiber composites and their laminates. A recent FAA publication lists 21 of these theories.[22] The simplest choices for failure criteria are maximum stress or maximum strain. With the maximum stress theory, the ply stresses, in-plane tensile, out-of-plane tensile, and shear are calculated for each individual ply using lamination theory and compared with the allowables. When one of these stresses equals the allowable stress, the ply is considered to have failed. Other theories use more complicated (e.g., quadratic) parameters, which allow for interaction of these stresses in the failure process.

Although long-fiber composites typically fail at low tensile strains, they are generally not considered to be brittle, i.e., in the realm of glass or ceramics. The fibers do have strain to failure, and the failures can be predicted. A bundle of fibers bound together by a matrix does not usually fail when the first fiber ruptures. Instead, the final failure is preceded by a period of progressive damage.

The basic assumption of statistical fracture theory is that the reason for the variations in strength of nominally identical specimens is their varying content of randomly distributed (and generally invisible) flaws. The strength of a specimen thus becomes the strength of its weakest flaw, just as the strength of a chain is that of its weakest link.

Since it is not possible to obtain strengths in all possible lamina orientations or for all combinations of lamina, a means must be established by which these characteristics can be determined from basic layer data. Theories of failure are hypotheses concerning the limit of load-carrying ability under different load combinations. Using expressions derived from these theories, it is possible to construct failure envelopes or, if in three dimensions, failure surfaces that represent the limit of usefulness of the material as a load-bearing component, i.e., if a given loading condition is within the envelope, the material will not fail. The suitability of any proposed criterion is determined by a number of factors, the most important of which has to do with the nature of the failure mode. As a result, it is important that proposed failure criteria be accompanied by a definition of material behavior.

4.6.3 Design Allowables

The design of composites involves knowledge of a significantly greater number of material properties than those needed for conventional isotropic metals. As mentioned previ-

ously, these data are not always conveniently available from a single source of data such as a handbook. The data at the maximum, for the design of aerospace structures, takes the form of those shown in Table 4.15.[23]

Data for design use requires statistical significance with a known confidence level. The old MIL-Hdbk-17B[24] provides a guide concerning the number and type of tests sufficient to establish statistically based material properties along with some limited data that is now somewhat out of date. The (new) Composites Materials Handbook, Mil-17,[25] in preparation, has somewhat enhanced statistical treatment approaches. Three classes of allowables, pertinent to current usage of composites for many applications, are:

- *A-Basis Allowable.* The value above which 99% of the population of values is expected to fall, with a confidence of 95%.
- *B-Basis Allowable.* The value above which 90% of the population of values is expected to fall, with a confidence of 95%.
- *S-Basis Allowable.* The value that is usually the specified minimum value of the appropriate government specification.

For most flightworthy composites, material properties are usually required to be either A-Basis or B-Basis allowables.

The effort is still in progress to provide a new family of design allowables including the most advanced fiber composites and the background guidance for their use. This is a reprise of the Mil-Hdbk-17 effort initiated in 1972, which had property values primarily for fiberglass fibers. The new Mil-Hdbk-17 committee has published the first interim report on the effort. The original Mil-Hdbk-17 treated unidirectional and woven fabric laminates, and two typical entries are shown in Tables 4.16 and 4.17. The first table shows the properties of 3M XP 2515 fiberglass epoxy with 100% unidirectional fibers. Although the resin portion of this laminate is no longer produced, and the data has not been upgraded, this is one of the few areas in which one can see statistically significant data for unidirectional fiberglass that is of a quality suitable for inclusion in computer analysis programs for laminate analysis. The second table is for materials that still exist in the marketplace and was continued in the 1999 edition of the handbook. This table shows the data for a woven fabric with a heat-curing epoxy resin. The 7781 fabric is a reasonably balanced fabric with good drape qualities in wide use. Tables 4.18 and 4.19, presented for carbon/graphite/epoxy materials, are extracted from the new MIL-Hdbk. They show the "B" basis allowables for strength and modulus of a unidirectional and a woven fabric laminate for three temperatures of interest. If the values of the fiber and the resin as shown by the vendor are used for contrast, a full picture of the laminate materials emerges and further simplifies analysis of laminates made with these materials or similar materials for extrapolation

4.7 Composite Fabrication Techniques

4.7.1 Choosing the Manufacturing Method

There is a history of choosing the composite manufacturing technique for the wrong reasons. Sometimes the choice is good regardless of the method, but often the end product or the schedule suffers and, in turn, the customer is unhappy. The rationales for choice have historically been as outlined below.

Design needs. This is the best reason for choosing a manufacturing method. The key to attaining a good composite design with a manufacturing process that can operate with

Table 4.15 Lamina Properties and Equations Used To Calculate Material Properties[23]

Lamina material properties	Definition	Equation used to calculate material property
Elastic		
E_1	Elastic modulus in the fiber direction	Property based on test data
E_2	Elastic modulus transverse to the fiber direction	Property based on test data
E_3	Elastic modulus through-the-thickness	Transverse isotropy: $E_3 = E_2$
G_{12}	Shear modulus in the 1-2 plane	Property based on test data
G_{23}	Shear modulus in the 2-3 plane	$G_{23} = \frac{E_3}{2(1+\nu_{23})}$
G_{13}	Shear modulus in the 1-3 plane	Transverse isotropy: $G_{13} = G_{12}$
ν_{12}	Poisson's ratio in 1-3 plane	Property based on test data
ν_{23}	Poisson's ratio in 2-3 plane	$\nu_{23} = \nu_f V_f + \nu_m(1 - V_f)\left[\frac{1 + \nu_m - \nu_{12}\frac{E_m}{E_1}}{1 - \nu_m^2 + \nu_m \nu_{12}\frac{E_m}{E_1}}\right]$
ν_{13}	Poisson's ratio in 1-3 plane	Transverse isotropy: $\nu_{13} = \nu_{12}$
Strength		
σ_1	Tensile strength in the fiber direction	Property based on test data
$-\sigma_1$	Compressive strength in the fiber direction	Property based on test data
σ_2	Tensile strength transverse to the fiber	Property based on test data
$-\sigma_2$	Compressive strength transverse to the fiber	Property based on test data
σ_3	Tensile strength through-the-thickness	$\sigma_3 = \sigma_2$
$-\sigma_3$	Compressive strength through-the-thickness	$-\sigma_3 = -\sigma_2$
τ_{12}	Shear strength in 1-2 plane (in-plane)	Property based on test data
τ_{13}	Shear strength in the 1-3 plane (interlaminar)	Property based on test data
τ_{23}	Shear strength in the 2-3 plane (interlaminar)	$\tau_{23} = \tau_{13}$

TABLE 4.16 Unidirectional Fiberglass Laminate Data

Fabrication		Layup: unidirectional		Vacuum		Pressure: 50 psi		Bleedout: vertical		Cure: 2 hr/300°F 4 hr & 350°F		Postcure:		Plies: 6	
Physical properties		Weight percent resin: 21.4				Avg. specific gravity: 1.96		Avg. percent voids: Figure 4.32.5				Avg. thickness: 0.045 inch			
Test methods		Tension: see text		Compression: see text		Shear: rail		Flexure: ASTM – D790		Bearing: ASTM – D953		Interlaminar shear: short beam			
Temperature		−65°F				75°F				160°F				300°F	
Condition		Dry		Wet		Dry		Wet		Dry		Wet		Dry	
		Avg	SD	Avg	SD	Avg	SD	Avg	SD	Avg	SD	Avg	SD	Avg	SD
Tension															
ultimate stress, ksi	0°	288.6	15.5	304.3	11.9	276.0	11.2	274.0	8.2	247.8	7.2	231.2	16.6	248.2	11.3
	90%	12.5	1.4	11.6	1.4	9.0	1.6	9.2	0.8	7.0	1.55	5.2	1.1	2.2	0.3
ultimate stress,%	0°	3.85	0.29	4.38	0.18	3.57	0.22	3.83	0.12	3.82	0.31	3.27	0.36	3.13	0.26
	90°	0.49	0.07	0.46	0.05	0.35	0.07	0.37	0.04	0.41	0.10	0.29	0.12	0.61	0.13
proportional limit, ksi	0°	247.6	16.9	237.0	19.8	201.8	15.8	213.4	16.0	195.2	11.4	167.9	18.8	187.2	22.6
	90°	10.6	0.8	10.6	1.2	5.7	0.70	6.2	0.36	4.7	0.73	3.7	0.46	0.8	0.1
initial modulus, 10^6 psi	0°	7.64	0.29	7.64	0.21	7.99	0.28	7.69	0.26	7.62	0.18	7.60	0.18	8.30	0.44
	90°	2.59	0.16	2.57	0.09	2.69	0.14	2.65	0.13	1.91	0.10	2.03	0.28	0.69	0.15
secondary modulus, 10^6 psi	0°														
	90°														
Compression															
ultimate stress, ksi	0°	115.3	5.8	119.9	10.8	100.0	7.7	92.4	10.0	76.0	5.5	70.0	4.1	48.6	4.1
	90°	41.4	2.2	37.5	2.8	29.3	0.6	28.0	0.7	22.0	0.8	19.8	0.7	8.4	0.8
ultimate stress,%	0°	1.66	0.22	1.89	0.21	1.36	0.06	1.41	0.11	1.06	0.09	1.08	0.07	0.76	0.06
	90°	2.36	0.27	2.03	0.29	1.95	0.20	1.94	0.19	1.99	0.32	2.04	0.34	4.08	0.33
proportional limit, ksi	0°	75.6	5.6	87.0	10.2	84.2	5.5	66.3	6.1	68.5	4.2	54.2	3.4	31.7	5.0
	90°	14.8	1.6	14.8	1.5	10.4	0.8	11.1	1.1	8.4	0.8	8.0	0.4	2.8	0.3
initial modulus, 10^6 psi	0°	7.73	0.20	7.07	0.24	7.52	0.18	6.78	0.22	7.24	0.30	6.68	0.26	7.50	0.47
	90°	2.60	0.10	2.53	0.09	2.69	0.09	2.49	0.07	2.21	0.11	2.08	0.08	0.57	0.16
secondary modulus, 10^6 psi	0°														
	90°														

TABLE 4.16 Unidirectional Fiberglass Laminate Data *(Continued)*

Fabrication	Layup: unidirectional	Vacuum	Pressure: 50 psi	Bleedout: vertical	Cure: 2 hr/300°F 4 hr & 350°F	Postcure:	Plies: 6	
Physical properties	Weight percent resin: 21.4		Avg. specific gravity: 1.96	Avg. percent voids: Figure 4.32.5		Avg. thickness: 0.045 inch		
Test methods	Tension: see text	Compression: see text	Shear: rail	Flexure: ASTM – D790	Bearing: ASTM – D953	Interlaminar shear: short beam		

Temperature		−65°F				75°F				160°F				300°F	
Condition		Dry		Wet		Dry		Wet		Dry		Wet		Dry	
		Avg	SD	Avg	SD	Avg	SD	Avg	SD	Avg	SD	Avg	SD	Avg	SD
Shear ultimate stress, ksi	0–90°					9.9				8.2					
	90–0°					13.4				12.0					

		−65°F dry			75°F dry			160°F dry		
		Avg	Max	Min	Avg	Max	Min	Avg	Max	Min
Flexure										
ultimate stress, ksi	0°	262.2	271.5	251.9	219.6	224.2	216.1	193.2	196.7	198.2
proportional limit, ksi	0°	177.6	181.3	173.8	181.8	187.9	170.6	165.0	173.8	154.7
initial modulus, 106 psi	0°	7.00	7.11	6.85	6.78	6.82	6.75	6.95	7.18	6.80
Bearing										
ultimate stress, ksi	0°	77.3	80.9	74.8	68.8	71.4	65.3	53.6	54.6	52.1
stress at 4% elong. ksi	0°	20.5	21.5	19.1	17.7	19.7	17.2	17.0	18.5	15.8
Interlaminar shear ultimate stress, ksi	0°	14.6	15.2	13.7	11.9	12.6	11.0	14.2	14.5	13.9

TABLE 4.17 Woven Fiberglass (Style 7781) Laminate Data

	Summary of Mechanical Properties of U.S. Polymeric E-720E/7781 (ECDE–1/0–550) Fiberglass Epoxy													
Fabrication	Layup: parallel	Vacuum: none	Pressure: 55–65 psi	Bleedout: edge & vertical	Cure: 2 hr/350°F		Postcure: 4 hr/400°F		Postcure: 4 hr/400°F		Plies: 8			
Physical properties	Weight percent resin: 34.9		Avg. specific gravity: 1.78		Avg. specific gravity: 1.78		Avg. percent voids: 2.0			Avg. thickness: 0.082 inches				
Test methods	Tension: ASTM D 638 TYPE–1		Compression: MIL–HDBK–17		Shear: Rail		Flexure: ASTM D 790		Bearing: ASTM D 953		Interlaminar shear: short beam			
Temperature	–65°F				75°F				160°F				400°F	
Condition	Dry		Wet		Dry		Wet		Dry		Wet		Dry	
	Avg	SD	Avg	SD	Avg	SD	Avg	SD	Avg	SD	Avg	SD	Avg	SD
Tension														
ultimate stress, ksi 0°	69.2	1.6	69.1	1.7	60.4	1.7	55.7	1.5	52.5	1.0	42.9	0.8	44.8	2.0
90°	56.0	2.0	56.5	2.0	49.0	1.8	45.9	1.4	42.3	1.2	36.9	1.1	34.9	1.6
ultimate strain,% 0°	2.93	0.08	2.70	0.11	2.43	0.14	2.12	0.08	2.05	0.08	1.61	0.06	1.80	0.20
90°	2.92	0.22	2.54	0.19	2.33	0.09	2.04	0.09	1.98	0.08	1.70	0.13	1.72	0.22
proportional limit, ksi 0°														
90°														
initial modulus, 10^6 psi 0°	3.30		3.38		3.12		3.12		2.95		2.76		2.60	
90°	2.90		3.02		2.82		2.78		2.50		2.65		2.30	
secondary modulus, 10^6 psi 0°	2.30		2.85		2.45		2.50		2.46		2.37			
90°	1.90		1.74		2.05		2.19		2.01		1.97			
Compression														
ultimate stress, ksi 0°	77.1	4.0	75.0	3.7	64.8	2.9	57.3	3.8	54.0	1.4	46.2	1.4	23.8	2.2
90°	57.2	2.7	53.9	2.7	50.2	2.9	45.2	2.4	40.8	2.9	36.2	3.1	14.7	1.6
ultimate strain,% 0°	2.48	0.16	2.44	0.15	2.14	0.11	1.99	0.09	1.86	0.08	1.62	0.06	1.12	0.22
90°	1.93	0.16	1.81	0.19	1.70	0.14	1.58	0.14	1.46	0.17	1.37	0.15	0.91	0.08
proportional limit, ksi 0°														
90°														
initial modulus, 10^6 ksi 0°	3.50		3.45		3.25		3.10		3.15		3.03		2.45	
90°	3.20		3.26		3.21		3.03		2.99		2.85		1.85	

TABLE 4.17 Woven Fiberglass (Style 7781) Laminate Data *(Continued)*

	Summary of Mechanical Properties of U.S. Polymeric E-720E/7781 (ECDE–1/0–550) Fiberglass Epoxy													
Fabrication	Layup: parallel	Vacuum: none	Pressure: 55–65 psi	Bleedout: edge & vertical	Cure: 2 hr/350°F		Postcure: 4 hr/400°F		Postcure: 4 hr/400°F		Plies: 8			
Physical properties	Weight percent resin: 34.9		Avg. specific gravity: 1.78		Avg. specific gravity: 1.78		Avg. percent voids: 2.0				Avg. thickness: 0.082 inches			
Test methods	Tension: ASTM D 638 TYPE–1		Compression: MIL–HDBK–17		Shear: Rail		Flexure: ASTM D 790		Bearing: ASTM D 953		Interlaminar shear: short beam			
Temperature	–65°F				75°F				160°F				400°F	
Condition	Dry		Wet		Dry		Wet		Dry		Wet		Dry	
	Avg	SD	Avg	SD	Avg	SD	Avg	SD	Avg	SD	Avg	SD	Avg	SD
Shear ultimate stress, ksi 0–90°	17.5				14.3	0.6			11.2					
Flexure ultimate stress, ksi0	115.6	119.4	111.5	91.7	93.4	90.3	69.4	71.1	67.2					
proportional limit, ksi 0°	88.1	100.7	77.5	32.5	36.2	30.8	56.2	62.8	49.4					
initial modulus, 10^6 ksi 0°	2.87	2.91	2.74	3.21	3.36	3.03	2.81	2.87	2.76					
	–65°F			75°F			160° dry							
	Avg	Max	Min	Avg	Max	Min	Avg	Max	Min					
Bearing ultimate stress, ksi 0°	74.1	78.4	70.7	60.8	64.4	58.2	50.0	53.0	47.9					
stress at 4% elong., ksi 0°	32.1	34.8	29.1	23.9	34.2	20.1	18.1	21.5	15.9					
Interlaminar shear ultimate stress, ksi 0°	7.09	7.36	6.80	5.90	6.07	5.72	6.05	6.16	5.91					

TABLE 4.18 Unidirectional Carbon/Graphite Laminate Data (AS4/3502)

Material:	AS4 12k/3502 unidirectional tape
Resin content:	30–33 wt%
Fiber volume:	59–61%
Ply thickness:	0.0049–0.0071 in
Comp. density:	1.56–1.59 g/cm^3
Void content:	0.0–1.0%
Test method:	ASTM D 3039-76
Modulus calculation:	Linear portion of curve
Normalized by:	Ply thickness to 0.0055 in. (58%)

Table 4.2.8(a)
C/Ep 147–UT
AS4/3502
Tension, 1-axis
$[0]_s$
75/A, –65/A, 180/W
Fully approved

Temperature, °F		75		–65		180	
Moisture content,%		ambient		ambient		1.1–1.3	
Equilibrium at T, RH						(1)	
Source code		49		49		49	
		Normalized	Measured	Normalized	Measured	Normalized	Measured
F_1^{tu} ksi	Mean	258	253	231	227	261	255
	Minimum	191	196	162	151	140	135
	Maximum	317	302	285	280	317	315
	C.V.,%	9.83	9.13	13.4	13.6	14.8	15.2
	B-value	204		173		200	
	Distribution	Weibull		Weibull		Weibull	
	C_1	269		244		276	
	C_2	11.2		8.82		9.39	
	No. specimens	36		38		40	
	No. batches	5		5		5	
	Approval class	Fully approved		Fully approved		Fully approved	
E_1^t msi	Mean	19.3		19.2		19.7	
	Minimum	15.6		16.8		15.1	
	Maximum	21.0		23.2		23.3	
	C.V.,%	5.74		6.31		6.87	
	No. specimens	35		38		40	
	No. batches	5		5		5	
	Approval class	Fully approved		Fully approved		Fully approved	

TABLE 4.19 Woven Fabric Carbon/Graphite Laminate Data (AS4/3502)

Material:	AS4 6k/3502 5-harness satin weave
Resin content:	36–37 wt%
Fiber volume:	56–57%
Ply thickness:	0.0146–0.0157 in
Comp. density:	1.55–1.56 g/cm^3
Void content:	0.0–1.00.2
Test method:	BMS 8-168D
Modulus calculation:	Linear portion of curve
Normalized by:	Ply thickness to 0.0145 in. (57%)

Table 4.2.16(a)
C/Ep 365–5HS
AS4/3502
Tension, 1-axis
$[/90/0/90/90/0]_f$
75/A, –65/A, 180/W
Fully approved

Temperature, °F		75		–65		180	
Moisture content,%		ambient		ambient		1.1–1.3	
Equilibrium at T, RH						(1)	
Source code		49		49		49	
		Normalized	Measured	Normalized	Measured	Normalized	Measured
F_1^{tu} ksi	Mean	114	109	105	101	117	111
	Minimum	126	91.9	87.9	84.0	102	96.1
	Maximum	6.87	121	116	112	128	123
	C.V.,%		7.01	5.33	5.53	5.29	5.36
	B-value	91.9		95.0		102	
	Distribution	ANOVA		Normal		ANOVA	
	C_1	8.15		104.9		6.31	
	C_2	2.70		5.59		2.33	
	No. specimens	30		30		30	
	No. batches	5		5		5	
	Approval class	Fully approved		Fully approved		Fully approved	
E_1^t msi	Mean	9.61		9.67		10.5	
	Minimum	9.29		9.09		9.74	
	Maximum	10.4		10.1		10.9	
	C.V.,%	3.08		2.35		2.75	
	No. specimens	30		30		30	
	No. batches	5		5		5	
	Approval class	Fully approved		Fully approved		Fully approved	

minimum dysfunction is the choice of the method based on the design of the composite component. Thus, the manufacturing process must be kept in mind during the component design phase and must also be a consideration in laminate design.

Part configuration. This must have a great influence on manufacturing technique. In no other manufacturing endeavor does the finished configuration of the structure play such an important role. Some component configurations, such as pressure vessels, drive the process decision to only one method (e.g., filament winding), while others, such as driveshafts, have more processing options (e.g., filament winding, pultrusion, roll wrapping, or RTM).

Prior experience. Experience provides a good guide to manufacturing method selection, but it is not the best. We tend to stay in the same track because of inertia; however, it imperative that each new design be thoroughly analyzed to select the optimum and not the wrong manufacturing method.

Facilities available or underutilized. This is considered the least effective reason for manufacturing choice, but it regrettably seems to be the most prevalent. The history of the composite manufacturing business has been that an independent composite manufacturer would never be expected to reject a contract because of the lack of the optimum facilities or experienced personnel. The independent contractor would find a way to build it or would, in turn, subcontract. The same comment can be extended to most larger companies, such as the large aerospace manufacturers with adequate, but not infinite, facilities in house. Usually, the capital cost and overhead rate of a particular machine will be a strong consideration in the choosing process; manufacturers do not want to see valuable machinery idle.

This approach to composite manufacturing was demonstrated in the last decade when a commercial aircraft supplier selected two companies to fabricate composite fuselage prototypes with different approaches. One manufacturer chose a filament-wound isogrid, and the other chose a filament-wound honeycomb-stiffened structure with carbon/graphite-epoxy skins. Both of these approaches were viable and very cost effective, and the stiffened honeycomb structure approach was chosen for advanced development. Seven fuselages were then built, and one was flown on a subscale test aircraft. However, the autoclave that had been a long-lead acquisition for the aircraft manufacturer, and that was the reason for considering the subcontracts, became available earlier than anticipated. The subcontracts were cancelled, and the manufacturing was taken in house. Subsequently, a few years ago, the manufacturer acquired a fiber placement machine. It is difficult to say whether the fuselage by either method is lighter, stiffer, or stronger, but we know that the filament-wound structure did fly, and we also know that the fiber-placed structure has to cost considerably more than the filament-wound structure, because there is up to a 10:1 acquisition cost differential for the fiber placement machine vs. a suitable filament winding machine. The cost of most fiber placement machines is in the realm of \$3 million to \$5 million, and the cost of most CNC filament winding machines with 5 to 7 axes of motion is \$300 thousand to \$700 thousand. The machine, in this case, used to wind the fuselages was a two-axis winder that had been modified with a tracer-lathe-type mechanism to add a third axis of movement. The cost of the two-axis filament winding machine is unknown but would be expected to be less than \$100 thousand. We also know that the differential in cost of the materials was greater than 3:1. The fuselage was wet wound; the fiber placed fuselage uses prepreg tow or slit prepreg tape.

This is the reason why all participants in the design or acquisition process for composites should have some familiarity with the composite manufacturing processes. All participants in the design process, at a minimum, should include the program manager, the materials engineer, the manufacturing engineer, and the purchasing agent. In smaller com-

panies, many of these tasks are combined. This knowledge can be used for the following objectives:

- For help in vendor selection
- To have a knowledge of the limitations of the different manufacturing processes
- To influence the design so as to simplify manufacture
- To exert the proper control over vendors

The goals any composite manufacturing process are to

- Achieve a consistent product by controlling:
  - composite thickness
  - fiber volume
  - fiber directions
- Minimize voids
- Reduce internal residual stresses
- Process in the least costly manner

The procedures to reach these goals involve iterative processes to select the three key components:

- Composite material and its configuration
- Tooling
- Process

Composite material selection. Composite material selection will be dictated by the requirements of the structure and may be constrained by the manufacturing process. It involves both the selection of the material and the form of material. The forms of material can be described as dry fiber (with wet resin), prepreg, and fiber preform (prepreg or dry). Often, the form-of-material choice is also constrained by the manufacturing process selected; i.e., the wet process hand-layup technique is constrained to use woven fabric instead of unidirectional tape, and filament winding uses tow or strand rather than fabric (fabric can be used for local reinforcement). Because these choices for any significant structure are not trivial, this is another reason for concurrent engineering. The reinforcement configurations and their relation to the several manufacturing processes are shown in Table 4.20.

4.7.2 Tooling

Once material selection has been completed, the first step leading to the acceptable composite structure is the selection of tooling, which is intimately tied to process and material. For all curing techniques, the tool must be:

- Strong and stiff enough to resist the pressure exerted during cure
- Dimensionally stable through repeated heating and cooling cycles
- Light enough to respond reasonably quickly to the changes in cure cycle temperature and to be moved in the shop (thermal conductivity may also be a concern)
- Leakproof so that the vacuum and pressure cycles are consistent

Table 4.21 shows some thermal, physical, and mechanical properties for different tool materials.

Table 4.20 Common Reinforcement Configuration for Manufacturing Process

Reinforcement configuration	Prepreg tape	Prepreg or (dry) tow	Prepreg or (dry) woven or nonwoven fabric	Other woven preforms, chopped fibers
Hand layup	X		X, (X)	X
Automatic tape laydown	X			
Filament winding		X, (X)	X, (X)	X, (X)
Resin transfer molding		(X)	(X)	X
Pultrusion		(X)		
Fiber placement		X		

Table 4.21 Typical Tooling Materials for Advanced Composites

Material	Density, kg/m^3	Young's modulus in tension E, GPa	Temp. limitation, °C	Coefficient of thermal expansion, m/m°C × 10^{-6}
Silicone rubber	1605		260	81–360
Aluminum alloy	2714	68.9	204	22.5
Steel	7833		Note*	12.5
Electroless nickel	–	–	Note*	12.6
Cast iron	7474	165	Note*	10.8
Fiberglass	1950	20	177†	11.7–13.1
Carbon fiber epoxy (T-300)	1577	66	177†	2.8–3.6
Cast ceramic	3266	10	Note*	0.81
Monolithic graphite	1522	13	Note*	2.7–3
Low-expansion nickel alloys	8137	144	Note*	1.4–1.7
Carbon fiber epoxy pitch 55	1720	227	177†	<1

*Above composite
†Limited by resin matrix

The tool face is commonly the surface imparted to the outer surface of the composite and must be smooth, particularly for aerodynamic surfaces. The other surface frequently may be of lower finish quality and is imparted by the disposable or reusable vacuum bag. This surface can be improved by the use of a supplemental metal tool known as a *caul plate*. (Press curing, resin transfer molding, injection molding, and pultrusion require a fully closed or two-sided mold.) Figure 4.13 shows the basic components of the tooling for vacuum bag or autoclave processed components. Table 4.22 shows the function of each

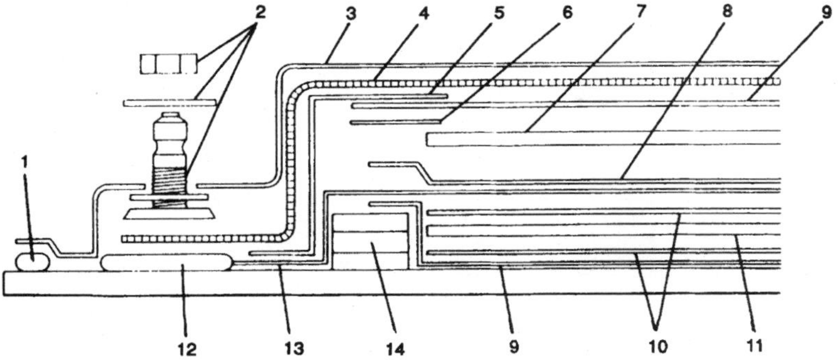

Figure 4.13 Typical vacuum bag layup components.[25]

part of the system. Tooling options have been augmented by the introduction of elastomeric tooling wherein the thermal expansion of an elastomer provides some or all of the pressure curing cure, or a rubber blanket is used as a reusable vacuum bag. The volumetric expansion of an elastomer can be used to fill a cavity between the uncured composite and an outer mold. The use of elastomeric tooling can provide the means for fabricating complex box-like structures such as integrally stiffened skins with a co-cured substructure in a single curing operation.[26]

Tooling and the configuration of the reinforcement have a great influence on the curing process selected, and vice-versa. The choice between unidirectional tape and woven fabric frequently has been made on the basis of the greater strength and modulus attainable with the tape, particularly in applications which compression strength is important. There are other factors that should be included in the trade, as shown on Table 4.23.

4.7.3 Layup Technique

4.7.3.1 Automation considerations. The layup of a composite, the actual placement of plies in their expected final position in the laminate, cannot be separated from the total manufacturing process in some procedures, e.g., filament winding or pultrusion. In other techniques, e.g., automatic tape laydown or RTM, the layup process is a separate *batch* process and is completely separate from the compaction and cure phase. All composite manufacturing processes can be automated to some extent, and the amount of automation depends on how amenable the optimum manufacturing technique is to total or partial automation, the capital costs of the automation machinery, eventual total number of piece parts to be manufactured, the expected cost of each, the time frame available, and a host of other factors.

Layup techniques, along with composite cure control, have received the greatest attention in terms of cost control. In efforts to reduce labor costs of composite fabrication, to which layup has traditionally been cast as the largest contributor, mechanically assisted, controlled tape laying and automated integrated manufacturing systems have been developed. Table 4.24 shows some of the considerations for choosing a layup technique. In addition to any cost savings by the use of automated layup technique for long production runs, there are two key quality assurance factors that favor the use of them. These are the greatly reduced chance that release paper or film could be retained, which would destroy shear and compressive strength if undetected, and the reduced probability of the addition

Table 4.22 Functions of Vacuum Bag Components

Component of process	Functions
1. Bag sealant*	Temporarily bonds vacuum bag to tool
2. Vacuum fitting and hardware	Exhausts air, provides convenient connection to vacuum pump
3. Bagging film	Encloses part, allows for vacuum and pressure
4. Open-weave breather mat	Allows air or vacuum transfer to all of part
5. Polyester tape (wide)	Holds other components of bag in place
6. Polyester tape (narrow)	Holds components in place
7. Caul sheet	Imparts desired contour and surface finish to composite
8. Perforated release film	Allows flow of resin or air without adhesion
9. Nonperforated release film	Prevents adhesion of laminate resin to tool surface
10. Peel ply	Imparts a bondable surface to cured laminate
11. Laminate	
12. 1581-style glass breather manifold	Allows transfer of air or vacuum
13. 1581-style glass bleeder ply	Soaks up excess resin
14. Stacked silicone rubber edge dam	Forces excess resin to flow vertically, increasing fluid pressure

*Numbers refer to Fig. 4.13.

or loss of an angle ply that would cause warping due to the laminate's lack of symmetry and balance. New ATL machines also have a laser ply mapping accessory that can verify the position of the surface ply.

The U.S. Air Force (USAF) was quick to realize that the composite materials that added so much performance were also very costly. The USAF supported, and still supports, many development programs to identify the excessive cost contributors and to find techniques to reduce them. A recent presentation showed the present accretion of the cost elements. These costs, referred to as the *composite cost chain,* have been fairly consistent over the past 30 years for aircraft and aerospace applications and are broken down as follows:[28]

Material or operation	Est. cost, $U.S.
Polyacrylonitrile (PAN) fiber precursor	$2/lb
Conversion PAN to carbon fiber	$18/lb
Woven into fabric	$22/lb
Preimpregnation	$200/lb
Assembly	$300/lb

Table 4.23 Fabric vs. Tape Reinforcement

Tape advantages	*Tape disadvantages*
Best modulus and strength efficiency	Poor drape on complex shapes
High fiber volume achievable	Cured composite more difficult to machine
Low scrap rate	Lower impact resistance
No discontinuities	Multiple plies required for balance and symmetry
Automated layup possible	Higher costs for hand layup
Available in thin plies	
Lowest cost prepreg form	
Less tendency to trap volatiles	
Fabric advantages	*Fabric disadvantages*
Better drape for complex shapes	Fiber discontinuities (splices)
Single ply is balanced and may be essentially symmetric	Less strength and modulus efficient
Can be laid up without resin	Lower fiber volume than tape
Plys stay in line better during cure	Greater scrap rates
Cured parts easier to machine	Warp and fill properties differ
Better impact resistance	Fabric distortion can cause part warpage
Many forms available	

As the largest user and the principal supporter of programs to quantify and lower composite costs, the USAF has tried to lower costs through automation and a number of other ways. One of these other ways was to lower the costs of the raw materials. They attempted to do this by ensuring that there were always multiple suppliers for carbon fibers. However, the "buy America" programs, the loss of U.S. suppliers because of the turndown in aerospace demand, and the extensive quality assurance and specification requirements have ensured that the fibers supplied to Air Force programs have always been at the high end of the fiber cost spectrum. This was in spite of the fact that some commercial fibers are sharply decreasing in cost. (It is expected that there will be a high-strength, high-tow-count carbon fiber on the market at $5.00/lb within three years.) A more successful approach that the Air Force has been taking to reduce these costs has been to stress the use of automation in the composite manufacturing process, and to include the life cycle costs in the ultimate cost considerations. The Air Force, along with NASA, has funded many programs to quantify and/or lower the costs of composite structures. These programs usually had a common goal of reducing composite acquisition costs by 25% minimum. They have all had some success, but the acquisition costs for high-performance military aircraft remain high and close to the estimate above.

These automation efforts have traditionally been directed toward the prepreg tape-laydown—autoclave cure processes that were necessary for complex aircraft shapes. This effort has resulted in the emergence of fiber placement systems available from two vendors (Cincinnati Machines and Ingersoll); contour tape layers from Cincinnati Machines; and flat tape layers from Cincinnati Machines, Ingersoll, and Brisard Machine (Fig. 4.14).[27] To adequately take advantage of these expensive machines (flat tape layers have been estimated to cost U.S.$2–4 million, and contour tape layers U.S.$3–5 million),[28] a series of support machines are needed. These can include cutting machines for prepreg and core materials; robots and conveyor material handling systems and automated methods of edge trimming and drilling; such as water jet or abrasive jet; and better autoclave controls. Also, in conjunction with some of the support machines mentioned above, there are automated methods that can assist hand layup, the principal method being laser projection. Finally, there are faster techniques that can come into play when the starting material forms are not

predisposed. Thus, if the form of fiber to be considered can be expanded to discontinuous fibers, other automation techniques become apparent, as discussed later and as shown in Table 4.25. Obviously, these same machines can be used in support of the other process techniques that have inherent automation but are not widely used in the aircraft industry, such as filament winding and pultrusion.

The goals in the curing-compacting process are the same for all techniques, namely,

- Composite with no or minimal voids and porosity
  - This is to ensure the preservation of matrix-dominated properties and matrix-influenced properties such as compression.
- Full cure of the composite
  - This results in the most environmentally stable composite as well as maximizing matrix-dominated properties.
- Reproducible, consistent fiber volumes
  - This assures consistent composite mechanical response along with constant or anticipated thickness of the composite.

When these goals are not met, or are only partially met, the result is production slowdowns, excessive rework and repair, and extensive nondestructive testing, in addition to

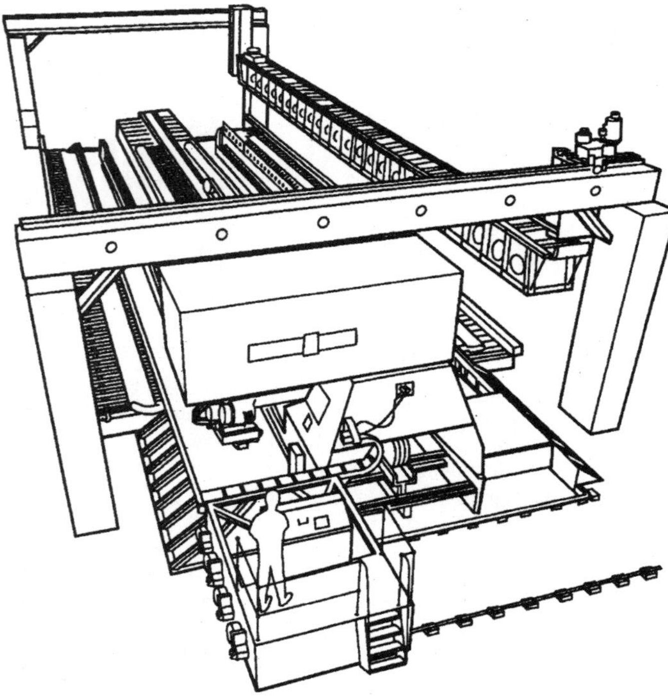

Figure 4.14 Typical contour tape layout machine layout. *(From Sarl, B., Moore, B., and Riedell, J.,* Proc. 40th International SAMPE Symposium, *May 1995, p. 384)*

Table 4.24 Considerations in Composite Layup Technique

	Manual	Flat tape	Contoured tape
Orientation accuracy	Least accurate	Automatic Dependent on tape	Somewhat dependent on tape accuracy and computer program
Ply count	Dependent on operator, count mylars, and separator films	Dependent on operator	Program records
Release film retention	Up to operator	Automatic	Automatic removal
Labor costs	High	86% improvement	Slight additional improvement
Machine costs	N/A	Some costs	Approximately $2–3 million or greater
Production rate	Low (1.5 lb/h)	10 lb/h	Up to 30 lb/h quoted
Machine uptime	N/A	Not a consideration	Complex program and machine make this a consideration
Varying tape widths	Not a concern	Easily changed	Difficulty in changing
Tape lengths	Longer tapes more difficult	Longer is more economical	Longer tape is more economical
Cutting waste	Scrap on cutting	Less scrap	Least scrap due to back-and-forth laydown
Compaction pressure	No pressure	Fewer voids because of compaction at laydown	Fewer voids because of compaction at laydown
Programming	N/A	N/A	Necessary

the production of expensive scrap composite. Obviously, there have been many studies to determine the critical parameters in the process and for the autoclave curing of thin, composite aircraft components. There are now recommendations dealing with prepreg properties such as tack and drape, layup environment (moisture), laminate thickness, ply orientation and drop-offs, and bleeder amount and placement.[29]

The cure/consolidation techniques are somewhat different for thermosets and thermoplastics. The process with thermosets is generally to squeeze the excess resin out of the composite. An exception is resin transfer molding (RTM) processes, where the procedure is to infiltrate the proper resin content into the dry fiber preform. Thermoplastics are completely "cured" in the prepreg, and the process to form them to the final shape involves melting or softening and forming to the final desired shape. Because of the much higher melt viscosity, thermoplastic matrices must have close to the final composite resin content within the layup, with very little resin bleed. There are a number of unique routes to the formation of a thermoplastic composite; these are shown in Table 4.26.[30] All curing techniques use heat and pressure to cause the matrix to flow and wet out all the fibers before the matrix solidifies or cools.

Table 4.25 Types of Automation Based on Starting Fiber Form

Reinforcement	Fiber type	Proximate use	Technique for use	Application (potential)
Continuous	Glass	Preforms	Matched metal dies, VARTM, SCRIMP™	Commercial, aerospace, automotive
		Direct	Prepregs, wet layup (fabrics), pultrusion, filament winding	Aerospace, commercial, (tanks and pipes), automotive
	High-performance fiber	Preforms	SCRIMP™, autoclave	Aerospace (Navy ship structure), commercial
		Direct	Prepregs, filament winding, pultrusion	Aerospace, commercial, sporting goods
Discontinuous	Glass	Preforms	P-4 process	Automotive, commercial
		Direct	Glass fabrics, dry and prepreg sprayup	Commercial, automotive
	High-performance fiber	Preforms	P-4 process	Aerospace
		Direct	Metal matrix composites, thermoplastics	Aerospace, commercial, (automotive)

Generally, for thermosets, the percent matrix weight is higher before cure initiation; the matrix flows out of the laminate and takes the excess resin with the potential voids. The matrix exhibits a low-viscosity phase during the cure of a thermoset and then advances in viscosity rapidly after a gel period at an intermediate temperature (Fig. 4.15). Because of the many changes going on in the matrix, determining the actual point at which cure occurs is very difficult, and cure techniques (by almost all methods) have evolved into a stepped cure with gradual application of heat to avoid formation of voids. An arbitrary 1% void limit has been adopted for most autoclaved composites; filament-wound and pultruded composites may have higher void volumes, which may be acceptable in several product applications (i.e., 3–4% is acceptable in many filament-wound pressure vessel applications) depending on the situation.

4.7.3.2 Curing. Each technique for compaction and cure is unique in terms of types of structures for which it can be used, and also in terms of the forms of starting materials. For instance, a unidirectional prepreg is optimum for automated laydown methods with autoclave cure but, because of potential fiber washout by low viscosity resin movement, which can move fibers, and for better handling, a prepreg fabric may be more appropriate for a press-cured laminate. The advantages and disadvantages of fabric vs. unidirectional tape fiber forms for composite manufacture were shown earlier in Table 4.23.

An autoclave is essentially a closed, pressurized oven; the most common epoxy laminates are cured at an upper temperature of 177°C (350°F) and 6 MPa (100 psig). Figure 4.16 shows a typical cure schedule. Autoclaves are still the primary tool in advanced composite curing and have been built up to 16m (55 ft) long at 6.1 m (20 ft) diameter. Since

Table 4.26 Manufacturing Routes for Thermoplastic Composites Based on Processing Methods

Manufacturing route	Outline of fabrication and processing methods
	Open-mold processes
1. Autoclave	Unidirectional or woven fibers pre-impregnated by the resin (prepreg) are used. Other forms of prepreg have reinforcing fibers in combination with the resin as fibers or as powder. The prepreg layers are stacked on the mold surface and covered with a flexible bag. Consolidation is obtained by external pressure applied in an autoclave at elevated temperature.
2. Filament winding	Prepreg tape or tape with the resin as fibers or powder are wound onto a mandrel at pre-determined angles. Heat and pressure are applied to the tape in order to continuously weld it onto the underlying material.
3. Folding	Preconsolidated sheets are heated. Simple fixtures are then used to shape the sheets into the desired geometry.
	Closed-mold processes
4. Injection molding (short fibers, 0.1–10 mm)	A mixture of molten thermoplastic and short fibers is injected into a colder metal mold at very high pressure. The component is allowed to solidify and is automatically ejected.
5. Compression molding (short fibers, 5–50 mm)	Semi-finished sheets of glass mat thermoplastics are heated and placed in the lower part of the mold in a fast press. The press is quickly closed and pressure is applied so that the material can flow to fill the mold. Technology is also available where the hot molding compound reaches the mold from an extruder.
6. Compression molding (continuous fibers)	The same principle as for short fiber materials. Continuous fibers require special clamping fixtures for the sheets and can primarily be used for simple geometries.
7. Diaphragm forming	A stack of prepreg is placed in between two diaphragms (superplastic aluminium or polymer film). The diaphragms are fixed whereas the prepreg can move freely. The material is slowly deformed by external pressure and the mold.
8. Pultrusion	Prepreg tape or tape with the resin as fibers or powder is pulled through a heated die to form beams or similar continuous structures with constant cross-section geometry. The material is allowed to cool and solidify.
9. Resin injection	Dry reinforcing fibers are placed in the mold. Monomers and/or low-molecular-weight polymer with low viscosity are injected, the reinforcement is impregnated. Polymerization to a high molecular weight thermoplastic occurs by mixing of reactive components and/or thermal activation.

SOURCE: Bergland, Lars, in S.T. Peters, ed., *Handbook of Composites,* 2nd ed., p. 116, Chapman & Hall, London, 1998.

autoclaves are so expensive to build and operate, other methods of curing and compacting composites have been promoted for large aircraft components.

Each resin fiber combination has one or more optimum cure techniques, depending on the proposed environment. The prepreg manufacturer supplies a time-temperature cycle that may have been used to manufacture the test specimens employed in generating preliminary mechanical and physical properties. Frequently, it has been necessary to modify the preliminary cure cycle because of part configuration (thickness) or because of production economics.[31] There are several developments that help to inject science into the otherwise hit-or-miss, time-consuming procedure of cure cycle optimization. They are

- Development of dielectric sensors and signal processing to gain a knowledge of the resin condition (viscosity) at any point during the cure process
- Use of thermocouple data, transducer outputs, and dielectric information to interactively control the curing process within an autoclave

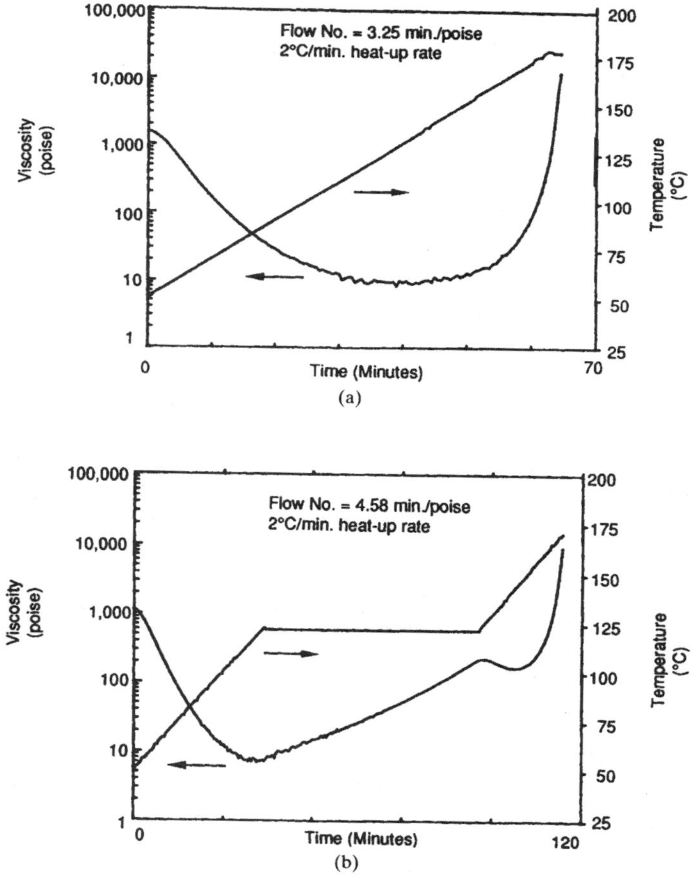

Figure 4.15 Matrix changes during cure. *(From Ref. 27)*

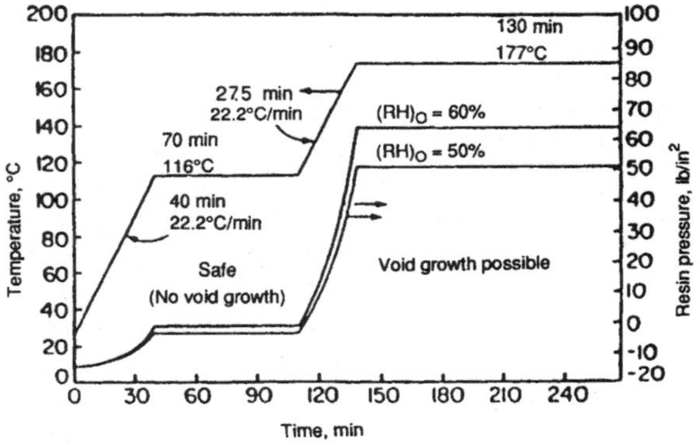

Figure 4.16 Time-temperature cure cycle for autoclave cure to prevent moisture-induced void growth.[28]

- Development of computer programs supported by test results to predict the physical changes that will occur during cure
- Analytical techniques, such as rheometric dynamic scanning and differential scanning calorimetry, that give information on changes in viscosity and the amount of heat absorbed or liberated during cure

4.7.4 Resin Transfer Molding

Previous discussions have centered on moving resin out of the laminate to reduce voids. Resin transfer involves the placement of dry fiber reinforcement into a closed mold and then injecting a catalyzed resin into the mold to encapsulate the reinforcement and form a composite. The impetus for the use of this process comes from the large cost reductions that can be realized in raw materials and layup. The process can utilize low injection pressures, i.e., 55 MPa (80 psig); therefore, the tooling can be lower-cost plastic rather than metal. The process is most appropriately used for non-aerospace composites but has been extended to many advanced applications. RTM manufacturing considerations are shown in Table 4.27. The advantages of RTM include the possibility of producing very large (Fig. 4.17) and complex shapes efficiently and inexpensively, and reducing production times with the ability to include inserts in the composite. Effective for large structures such as boats, the SCRIMP™ process uses a vacuum bag on one side of the laminate instead of the two plates of a typical mold. Advantages quoted for this technique are one sided mold, better control of and higher fiber volume, and lower porosity in the composite. It has been used for structures that need higher quality than can be obtained by other RTM processes. Table 4.28 shows the range of applications for the RTM technique. RTM is also a way of preparing a composite structure from a knitted preform. Knitting and braiding and sewn tridimensionally reinforced preforms offer complex shapes that are not attainable by other techniques. The techniques can possibly lower costs due to reduction of labor. The product may also gain increased impact resistance due to the multiple, interlocked directions of fiber.

4.7.5 Fiber Placement

Fiber placement was invented by Hercules Aerospace Co. Now, machines are marketed by Cincinnati Machines and Ingersoll in the U.S.A. The process is a cross between filament

Figure 4.17 RTM (SCRIMP™) process for injecting a boat hull. *(Copyright Billy Black, courtesy of Seemann Composites)*

Table 4.27 RTM Manufacturing Considerations

Materials	Tooling	Technology
Fiber type	Mold material	Resin viscosity
Preform complexity	Mold surface finish	Flow modeling
Preform cost	Resin pump type	Gating and vent design
Inserts	Integral or oven heating	Composite strength, stiffness, fiber volume, transverse properties
	Clamping method	Vacuum assist or not
	Tool durability	

winding and automatic tape laydown, retaining many of the advantages of both processes. Fiber placement, the natural outgrowth of adding multiple axes of control to filament winding machines, results in control of the fiber laydown so that nonaxisymmetric surfaces can be wound. This involves the addition of a modified tape laydown head to the filament winding machine and other sub-machine additions that include in-process compaction, individual tow cut/start capabilities, a resin tack control system, differential tow payout, low tension on fiber, and enhanced off-line programming. Several manufacturers now use slit prepreg tape rather than tow. The machines are capable of winding the shapes shown in Fig. 4.18 and can change fiber paths such as shown on Fig. 4.19. Table 4.29[32] shows the advantages quoted for the technique. The present disadvantages are the cost of the machines (very high when compared to some filament winding machines), the dependence on computers and electronics rather than mechanical means of directing fiber laydown, and the cost and complexity of mandrels.

Table 4.28 RTM Composite End Uses

Composite use	Part
Industrial	Solar collectors Electrostatic precipitator plates Fan blades Business machine cabinetry Water tanks
Recreational	Canoe paddles Large yachts Television antennas Snowmobile bodies
Construction	Seating Baths and showers Roofing
Aerospace	Airplane wing ribs Cockpit hatch covers Speed brakes Escape doors
Automotive	Crash members Leaf springs Car bodies Bus shelters

4.7.6 Filament Winding

Filament winding is a process by which continuous reinforcements in the form of rovings or tows (gathered strands of fiber) are wound over a rotating mandrel. The mandrel can be cylindrical, round, or any other shape as long as it does not have re-entrant curvature. Special machines (Fig. 4.20) traversing a wind eye at speeds synchronized with the mandrel rotation, control winding angle of the reinforcement and the fiber lay-down rate. The reinforcement may be wrapped in adjacent bands or in repeating bands that are stepped the width of the band and that eventually cover the mandrel surface. Local reinforcement can be added to the structure using circumferential windings or local helical bands, or by the use of woven or unidirectional cloth. The wrap angle can be varied from low-angle helical to high-angle circumferential or *hoop,* which allows winding from about 4–90° relative to the mandrel axis for older mechanical machines; newer machines can *place* fiber at 0°.

There are advantages and disadvantages to filament winding as compared to other methods. The most obvious advantages, summarized in Table 4.30, are cost savings (both capital and recurring labor) and the ability to build a structure that is larger than autoclave capacity. The disadvantages of filament winding, in most cases, can be worked around by innovative engineering and manufacture. Fabricators of large rocket motors have used plaster mandrels that can be stripped, reduced in size, and passed out through the relatively small port. Reverse curvature can be formed into a positive curvature by the addition of oriented fibers or mats or, if the curvature is necessary to the design, such as on an airfoil, it can be accomplished by removing the uncured structure from the mandrel and using alternate means of compaction to form the composite. This is the fabrication of a continuous multisided preform.

Figure 4.18 Versatility of shapes fabricated by fiber placement. *(Courtesy of Hercules Aerospace Co., now Alliant Techsystems)*

Newer filament winding machines have the capacity to change the wind angle at any point over the part surface. This gives the option of actually winding the fiber into a reverse curvature by selecting the wind angle that will follow a hyperboliodal path into a smooth recess, without bridging or slipping. This technique has been used to wind in an *in-situ* metallic end ring for a composite-to-metal joint, eliminating the need for further bolting or pinning and providing a measure of fail-safe operation The fiber path can be altered by pins or sawtooth to avoid slipping or bridging. Mandrels are less expensive than two-sided molds, and their reusability contributes to the cost-effectiveness of filament winding. The expansion of the mandrel during the cure cycle provides the compaction necessary to result in a fully compacted, dense laminate, without external compacting measures such as autoclave. Filament wound laminates, without special curing conditions, can have void contents on the order of 3 to 7%. Because there is no external mold, there is a poor external or bumpy surface that can be improved by mild compaction as exerted by shrink tape. The poor external surface can be smoothed somewhat by proper selection of resin and fiber, use of surfacing mat or filled smoothing compounds at some weight penalty, or by compaction and cure in a female die mold using vacuum bag or autoclave pressure. Figure 4.21 shows the Beech Starship carbon/graphite epoxy fuselage that was filament wound then expanded into tooling during cure to form a smooth outer skin.

Thermoset resins generally have been used as the binders for the reinforcements. These resins can be applied to the dry roving at the time of winding (wet winding) or applied previously and gelled to a "B" stage as prepreg. The fiber can be impregnated and rerolled without B staging and used promptly or refrigerated. Prepreg and wet rerolled materials are useful because of the opportunity to perform quality control checks early.

The cure of the filament wound composite is generally conducted at elevated temperatures without the addition of any process for composite compaction. The filament winding

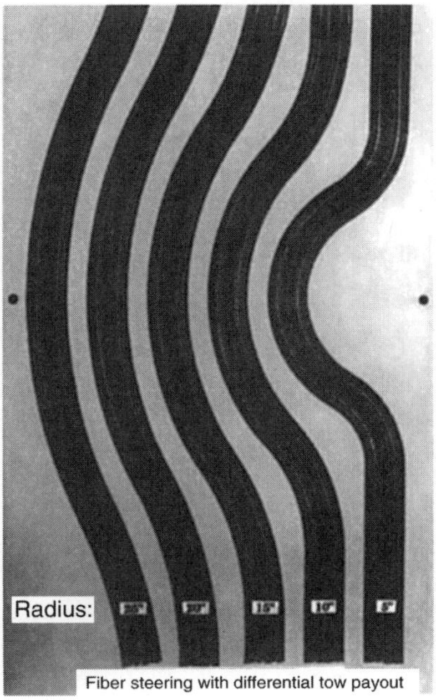

Figure 4.19 Versatility of shapes fabricated by fiber placement. *(Courtesy of Hercules Aerospace Co., now Alliant Techsystems)*

Table 4.29 Fiber Placement Processing Advantages (adapted from Ref. 32)

Flexibility	Compaction	Material usage
Full range of fiber orientations Non-geodesic path generation Constant ply thickness over complex shapes Localized reinforcements Continuous fibers over three-dimensional shapes (without joints) Large structures	Continuous in-process debulking Complex surface fabrications: ▪ Concave/convex ▪ Nonaxisymmetric and axisymmetric shapes	Wide use of advanced thermoset material systems Near prepreg tape equivalent

process, like pultrusion, can employ wet resin systems to result in potentially lower cost composite structures.

Handling guidelines that are unique to wet filament winding for a wet resin system are:

- Viscosity should be 2 PaS or lower.
- Pot life should be as long as possible (preferably >6 hr).
- Toxicity should be low.

The other approach is to use a wet-rerolled system, essentially a wet resin that is applied to the fiber beforehand and then kept in a freezer until use.

Figure 4.20 Filament winding. *(Courtesy of Plastrex)*

Fibers are used in tow (carbon/graphite) or (roving) glass and Kevlar. The two terms define a gathered parallel bunch of fibers with essentially no twist. The fibers that are made in individual processes (e.g., boron) have not been used extensively in filament winding applications due to their stiffness.

4.7.7 Pultrusion Braiding and Weaving

4.7.7.1 Pultrusion. Pultrusion is an automated process for the manufacture of constant-volume/shape profiles from composite materials. The composite reinforcement is continuously pulled through a heated die and shaped (Fig. 4.22) and cured simultaneously. If the cross-sectional shape is conducive to the process, it is the fastest and most economical method of composite production. Straight and cured configurations can be fabricated with square, round, hat-shaped, angled *I*, or *T*-shaped cross sections from vinylester, polyester, or epoxy matrices with E and S-glass, Kevlar and carbon/graphite reinforcements. Some of the available cross-sectional shapes and limitations are shown in Table 4.31.[33] The curing is effected by combinations of dielectric preheating and microwave or induction (with conductive reinforcements like carbon graphite) while the shape traverses the die. The resin systems, predominantly polyester, can be wet or prepreg, but the cure rates will be much more rapid than for processes that use thermal conduction for heat transfer into the laminate. The process lends itself to long component lengths with the need for reinforcement in the 0° direction. Many uses have been found for the process; pultrusion is the primary fabrication technique for reinforced plastic booms, ladder components, light poles, boundary stakes for snow roads, and conduits. The primary reasons for the use of the technique are design considerations driven by commercial uses, such as cost, weight, electrical properties, and environmental resistance, but the process has also been used to produce some components from advanced composite materials. An application that uses the cost-effective technique for high-volume production is the pultruded car-

Table 4.30 Comparison of Filament Winding with Other Fiber Deposition, Compacting, and Curing Processes

Advantages
Filament winding is highly repetitive and accurate in fiber placement (from part to part and from layer to layer).
It can use continuous fibers over the whole component area (without joints); can orient fibers easily in load direction. This simplifies the fabrication of aircraft fuselages and reduces the joints.
It avoids capital expense and the recurring expense for inert gas of autoclave.
Large and thick-walled structures can be built—larger than any autoclave.
Mandrel costs can be lower than other tooling costs. There is only one tool, the male mandrel, which sets the inside diameter and the inner surface finish. The outer surface is uncontrolled and may be rough.
The cost is lower for large numbers of components, since there is less labor than in many other processes.
Material costs are relatively low, since fiber and resin can be used in their lowest-cost form rather than as prepreg, and no preforming is necessary. (Preforming is necessary for RTM and may be a significant recurring expense.)
Disadvantages
The shape of the component must permit mandrel removal. Long, tubular shapes will generally have a taper. Different mandrel materials, because of differing thermal expansion and differing laminate layup percentages of hoops versus helical plies, will demonstrate varying amounts of difficulty in removal of the part from the mandrel.
One generally cannot wind reverse curvature. To wind a reverse curvature, wind the exact shape on a positive dummy mandrel insert and then remove the insert and place the fiber.
One cannot change fiber path easily (in one lamina). It can be done via the use of pins or slip of the tow. Fiber placement is the only fabrication method capable of "steering" the fiber.
The process requires a mandrel, which sometimes can be complex or expensive. Usually, the mandrel is less costly than the dies or molds for forming methods other than pultrusion or RTM.
Generally poor external surfaces are produced, which may hamper aerodynamics. A better outside surface can be obtained by ▪ Use of outer clamshell molds ▪ External hoop plies or thinner tows on last ply ▪ Shrink tape or porous TFE-glass tape overwrap

bon/graphite-reinforced drive shaft that replaces two metal shafts, universal joints, and hangers and results in a quieter product because of inherent composite damping (Fig. 4.23). The process has some limitations:

- The part must be of constant cross section over its length; it cannot be tapered.
- Transverse strength will be somewhat lower than for other manufacturing methods.
- Curved shapes require special machines.
- Thick sections are difficult because of exotherm with rapidly curing resins.
- Cure shrinkage may cause dimensional and mechanical problems.
- There generally will be a cut edge, and this must be cut carefully to reduce delamination and then sealed to prevent moisture or other attack to fiber or interface.
- Joints are more difficult because of fiber orientation and lack of section changes that would allow geometric locking.

Figure 4.21 Filament wound aircraft fuselage with smooth skin. *(Courtesy of Fibertec Div. of Alcoa)*

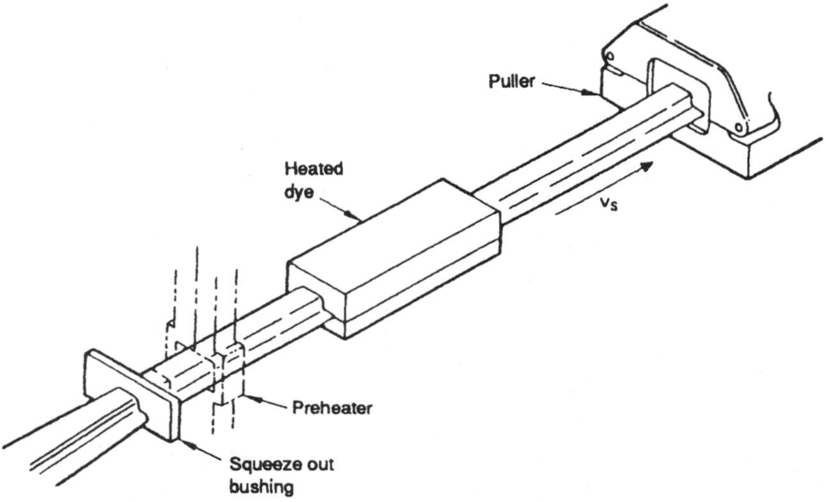

Figure 4.22 Schematic of pultrusion elements. *(From Ref. 34)*

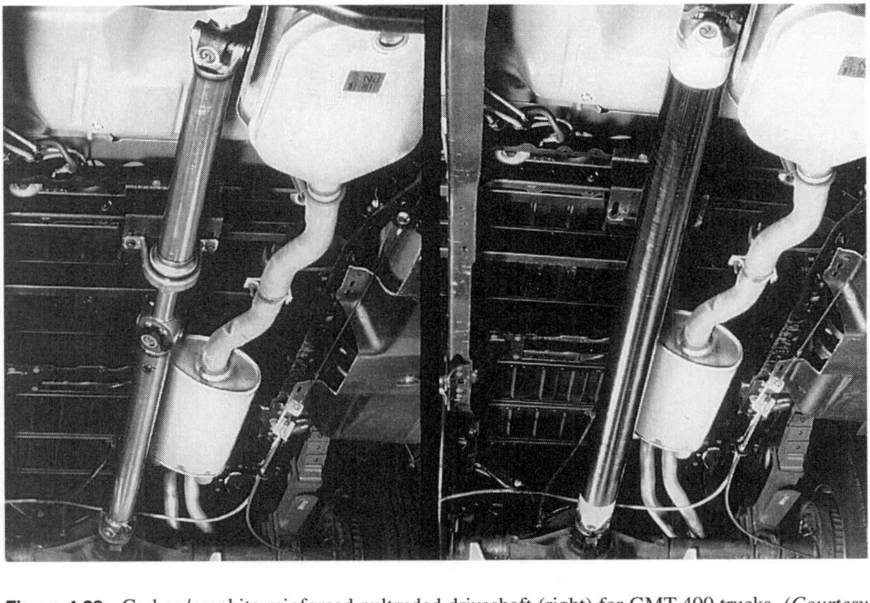

Figure 4.23 Carbon/graphite reinforced pultruded driveshaft (right) for GMT-400 trucks. *(Courtesy of Spicer Div. of Dana Corp.)*

TABLE 4.31 Pultrusion Design Guidelines

Minimum inside radius		Roving, 0.79 mm mat, 1.6 mm	Corrugated sections		Yes, longitudinal
Molded-in holes	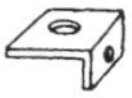	No	Metal inserts		No
Trimmed in mold		Yes	Bosses		No
Core pull and slides	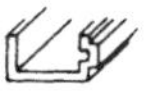	Yes	Ribs		Yes, longitudinal
Undercuts		Yes	Molded-in labels		Yes, but not recessed
Minimum recommended draft		No limitation	Raised numbers		No
Minimum practical thickness	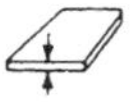	Roving, 1.0 mm Mat, 1.5 mm	Finished surfaces (reproduces mold surface)	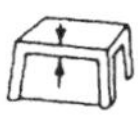	Two
Maximum practical thickness	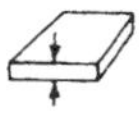	Roving, 75 mm Mat, 25 mm	Hollow sections	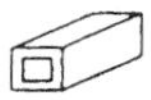	Yes, longitudinal
Normal thickness variation		See ASTM D 3917[22]	Wire inserts	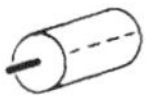	Yes, longitudinal
Maximum thickness buildup	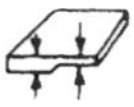	As required	Embossed surface		No

SOURCE: From J. D. Martin and J. E. Sumerak in Reinhart,[33] p. 542.

Pultrusion has also been combined with filament winding to achieve high transverse properties with advanced composite starting materials as shown in Fig. 4.24.

4.7.7.2 Braiding, weaving, and other preform techniques. Braiding, weaving, knitting, and stitching (with low- or high-modulus fibers) represent methods of forming a shape, generally referred to as a *preform* (a complete fiber layup representing the total laminate thickness with a small amount of resin or transverse fiber to hold it in place before resin infusion). The shape may be the final product or some intermediate form such as a woven fabric. New techniques allow prepregs to be used, and the introduction of three-dimensional braids has extended braiding to airborne structural components that meet high fracture toughness requirements with high damage tolerance. The braiding process is continuous and is amenable to round or rectangular shapes or smooth curved surfaces, and it can transition easily from one shape to another. Resin systems are generally epoxies or polyester, and the fiber options are similar to those for filament winding; the single stiff fibers such as boron and ceramics cannot endure the tight bend radii.

Figure 4.24 Filament-wound/protruded graphite epoxy bridge beam. *(Courtesy of Sci. Div. of Harsco)*

In the process, the mandrel is fed through the center of the machine at a controlled rate, and fibers from a large number of rolls are deposited at multiple angles, usually 45° and 90° (Fig. 4.25). 0° plies can be laid from tape or placed by a triaxial braider.

Braiding, in some applications, has turned out to be almost half the cost of filament winding because of labor savings in assembly and simplification of design.[34] The fiber volume of a braided composite will generally be lower than for other methods.

The other fabric preforming techniques are weaving, knitting, and the nonstructural stitching of unidirectional tapes. The weaving and knitting are compared to braiding in Table 4.32.[34] Stitching simply uses a nonstructural thread such as nylon or Dacron or a stiff fiber such as Kevlar to hold dry tapes or fibers at selected fiber angles. Preforming in this manner results in a higher-cost raw material but saves labor costs. The stitched preform has known, stable fiber orientations similar to woven fabric, without the crossovers.

4.7.7.3 Short fiber composites. Short fiber composites have only recently arrived in the advanced composite field. There is an increased interest in short fiber composites

Figure 4.25 Continuous braiding machine. *(Courtesy of B.C.M.E. Ltd.)*

Table 4.32 Comparison of Fabric Formation Techniques (from Ref. 34)

	Braiding	Weaving	Knitting
Basic direction of yarn introduction	One (machine direction)	Two (0°/90°) (warp and fill)	One (0° or 90°) (warp or fill)
Basic formation technique	Intertwining (position displacement)	Interlacing (by selective insertion of 90° yarns into 0° yarn system)	Interlooping (by drawing loops of yarns over previous loops)

because of the advantages in lay-down speed inherent in SCRIMP™ and other RTM resin infusion techniques. The P-4 process *(programmable powdered preforming process)* has great promise for automotive uses with the use of a chopper gun to lay down the fiber preform of a thermoplastic composite. Short fiber composites have been dismissed for advanced composites because of the weight, stiffness, and strength penalties inherent with the use of short fibers.

The speed of fabricating a composite short fiber preform is substantially increased when the P-4 technology is applied. This technology was jointly developed by Owens Corning (Battice, Belgium) and Applicator System AB (Mölnlycke, Sweden) and is now under further development at the National Composites Center (Kettering, OH). The process uses chopped fiberglass with a thermoplastic matrix for applications such as a structural automotive components at high production rates of up to 50,000 units per year for a pickup truck box. This high production rate has been the driving force in the process development. A parallel interest is for the quick and lower-cost manufacture of structural advanced composites with discontinuous carbon/graphite fibers. Besides the control of physical and mechanical properties exerted by fiber strength, stiffness, direction, and position in the laminate, a new consideration has to be added for advanced structural applications: fiber length. There are equations for predicting the properties the result from varying the fiber lengths,[34] and studies have determined some relationships empirically. Figure 4.26[35] shows the relation between fiber length and ultimate tensile strength for a discontinuous, unidirectional, glass fiber composite. Note that the composite has a low fiber volume. Another penalty is evident when the fibers are not aligned. Thus, one of the main development tasks for aerospace applications is the need to align the fibers.

4.7.8 Machining Techniques

The machining of composite materials also poses some challenges. The unique machining requirements for composites starts with the need for harder rather than high-speed drills. PCD-coated (polycrystalline diamond), carbide, or some other hard coating is necessary because of tool wear and the high temperatures generated by dull drills in the relatively low thermal conductivity of composite materials. Of practical interest, it is recommended that high-speed steel not be used. Drilling polymeric composites can result in delamination from too high or low feed-to-speed ratios (Figs. 4.27[36] and 4.28[37]), from heat buildup (composites generally have lower thermal conductivities than metallic materials), and because of inadequate backup at the exit point. The hole quality can also suffer, resulting in fiber pullout and fuzzing, matrix cratering, thermal alterations, and delaminations. The recommendations are even more relevant in the case of aramid composites. These composites require a different drill technique, such as the *spade* drill, also hard coated, because the fiber is so hard to shear—it tends to recede back into the composite matrix. In general, aramid composites can be expected to generate copious quantities of fluff under most machining inputs—this is natural and in most cases does not represent any significant

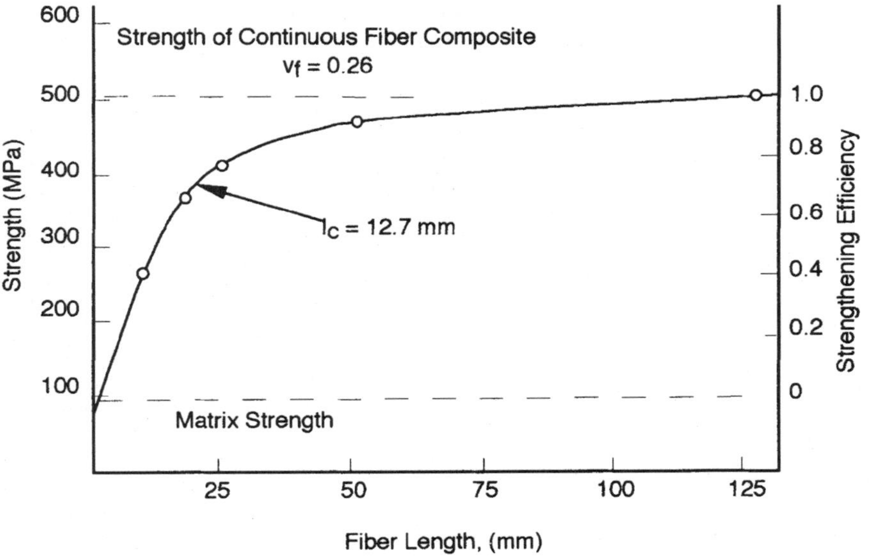

Figure 4.26 Relationship between short fiber length and composite strength. *(After Hancock, P., and Cuthbertson, R.C.,* J. Mat. Sci., *5, 76–768, 1970)*

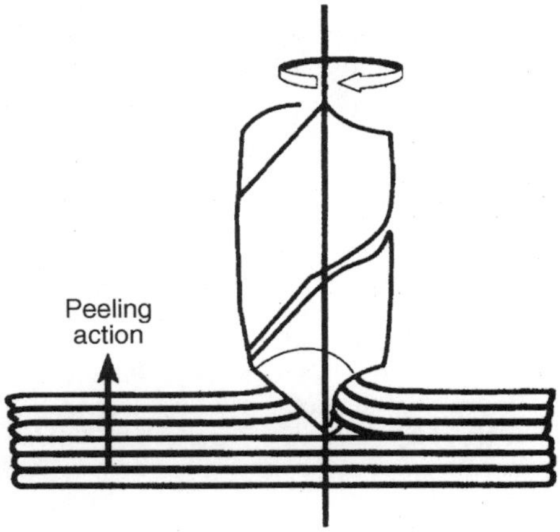

Figure 4.27 Typical composite drilling problem: peelup delamination at entrance. *(Kohkonen, K.E., and Potdar, N., in S.T. Peters, ed.,* Handbook of Composites, *2nd ed., Chapman & Hall, p. 598, London, 1998)*

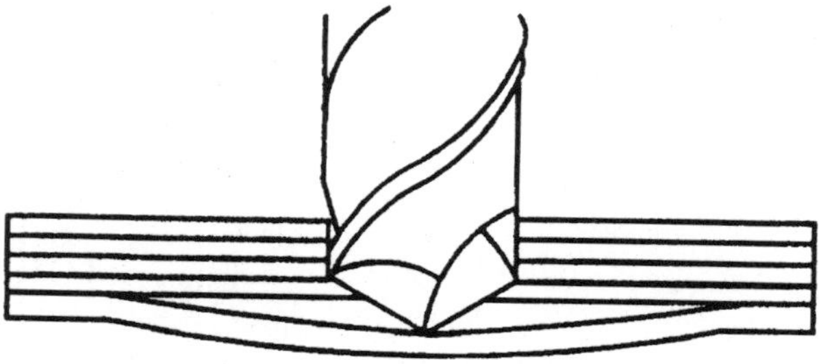

Figure 4.28 Typical composite drilling problem: pushout delamination at exit. *(Abrate, S., in P.K. Mallick, ed.,* Composites Engineering Handbook, *Marcel Dekker, New York, 1997, p. 783)*

degradation of the remaining composite. A summary of some relevant parameters for drilling of several composite materials with different drilling conditions is shown in Table 4.33.[37]

Table 4.33 Typical Machining Parameters for Drilling Polymeric Composite Materials[37]

Workpiece material	Tool material	Hole dia., mm	Material thickness, mm	Cutting speed, m/min	Feed rate, mm/rev
Unidirectional graphite-epoxy	Carbide PCD	4.85–7.92 4.85–7.92	0–12.7 0–12.7	42.7 61.0	0.0254–0.0508 0.0508–0.0889
Multidirectional graphite-epoxy	PCD Carbide	4.85–7.924 85–7.92	0–12.7 0–12.7	61.0 68.6	0.0254–0.0508 0.0508–0.0889
Glass-epoxy	HSS	3	10	33.0	0.05
Boron-epoxy	PCD	6.35 6.35	2.0 25.4	91–182 91–182	25.4*
Boron-epoxy	PCD	6.35	10.4	79	41.91*
Kevlar-epoxy	Carbide	5.6	—	158	0.05

*Units = mm/min

The precautions described for drilling merit repeating for machining with the additional proviso that, when machining unidirectional composites, the operator must attend to the fiber direction to avoid excessive fuzzing and potential splitting of the laminate. Table 4.34[37] shows some machining parameters for two typical composites. Grinding is generally recommended as an approach to avoid many of the foregoing problems. Grinding does generate fine airborne dust that poses an environmental hazard and should be controlled or contained.

There are many other approaches to composite machining that involve some significant capital outlays and should be justified by a reasonably long production run. These include water jet or abrasive jet, and laser cutting.

Table 4.34 Typical Machining Parameters for Sawing Polymeric Composites[37]

Material	Thickness, mm	Cutting speed, m/s	Feed rate, mm/min	Type of saw
Graphite-epoxy	4	2–12	50.8	Circular (HSS)
	0–25.4	15.24	50.8	Band saw (PCD)
Boron-epoxy	2.0	30.48	254	Circular* (PCD)
	25.4	15.24	50.8	Band saw* (PCD)

*With coolant.

4.8 Analysis

4.8.1 Overview of Mechanics of Composite Materials

The 1, 2, and 3 axes in Fig. 4.29 are special and are called the *ply axes,* or *material axes.* The 1 axis is in the direction of the fibers and is called the *longitudinal axis* or the *fiber axis.* The longitudinal axis typically has the highest stiffness and strength of any direction. Any direction perpendicular to the fibers (in the 2, 3 plane) is called a *transverse direction.* Sometimes, to simplify analysis and test requirements, ply properties are assumed to be the same in any transverse direction. This is the transverse isotropy assumption; it is approximately satisfied for most unidirectional composite plies.

These properties are typically modified by transformation relative to the laminate axis, where these may not be the same as the ply axes. These calculations involve matrix manipulations. For the simplest ply, stresses along the ply axes, the strains produced by the stresses are

$$\begin{bmatrix} \varepsilon_1 \\ \varepsilon_2 \\ \gamma_{12} \end{bmatrix} = \begin{bmatrix} \dfrac{1}{E} & -\dfrac{\upsilon_{21}}{E_2} & 0 \\ -\dfrac{\gamma_{12}}{E_1} & \dfrac{1}{E_2} & 0 \\ 0 & 0 & \dfrac{1}{G_{12}} \end{bmatrix} \begin{bmatrix} \sigma_1 \\ \sigma_2 \\ \tau_{12} \end{bmatrix} \tag{4.9}$$

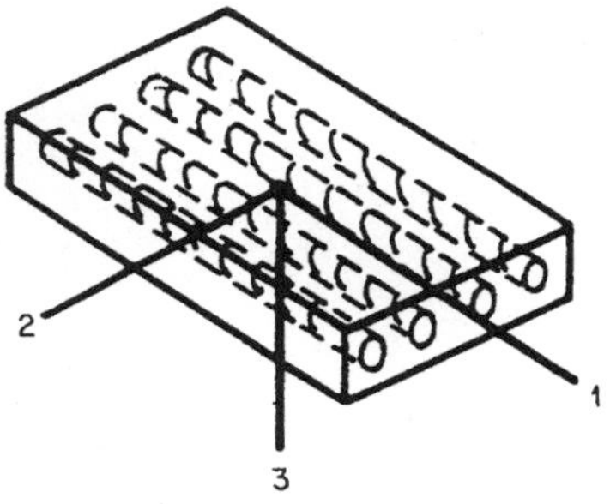

Figure 4.29 The unidirectional ply.

where ε = strain
υ = Poisson's ratio
E = Young's modulus
G = shear modulus
σ = tensile stress
τ = shear stress
γ = shear strain

In a multidirectional laminate, there can be as many as 21 stiffness constants. Strength predictions are equally complicated (a) because of directional differences, i.e., compression is not always equal to tension, and (b) because of the complexity of the several failure theories. As the complexity of the matrix calculations increase, it becomes evident that errorless mathematical manipulations are impossible without the aid of computers. Fortunately, there are several software packages that can accomplish these manipulations. The output of the computer, however, is only as good as the input information. Users must view the data in handbooks and computer material property files as "preliminary" and should verify the necessary constants and failure properties of the composite materials and processes by subscale and full-scale tests. Since there are no standard, accepted "design allowables" for composite materials, most organizations that extensively use the materials have had to develop their own data to supply "A" or "B" basis allowables. Because of the possible numbers of permutations of resins, fibers, and curing techniques, it will be some time before standardized strength and modulii values can be published as is done now for most metallic structural materials.

4.9 Design of Composite Structures

4.9.1 Composite Laminate Design

The design process for composites involves both laminate design and component design and must also include considerations of manufacturing process and eventual environmental exposure. The following are some steps that can simplify the process.

1. Take advantage of the orthotropic nature of the fiber composite ply.
   - To carry in-plane tensile or compressive loads, align the fibers in the directions of these loads.
   - For in-plane shear loads, align most fibers at ±45° to these shear loads.
   - For combined normal and shear in-plane loading, provide multiple or intermediate ply angles for a combined load capability.
2. Intersperse the ply orientations.
   - If a design requires a laminate with 16 plies at ±45°, 16 plies at 0°, and 16 plies at 90°, use the interspersed design $[90_2/\pm45_2/0_2]_{4S}$ rather than $[90_8/\pm45_8/0_8]_S$. Concentrating plies at nearly the same angle (0° and 90° in the above example) provides the opportunity for large matrix cracks to form. These produce lower laminate allowables, probably because large cracks are more injurious to the fibers and more readily form delaminations than the finer cracks occurring in interspersed laminates.
   - If a design requires all 0° plies, some 90° plies (and perhaps some off-angle plies) should be interspersed in the laminate to provide some biaxial strength and stability and to accommodate unplanned loads. This improves handling characteristics and also serves to prevent large matrix cracks from forming.

- Locally reinforce with fabric or mat in areas of concentrated loading. (This technique is used to locally reinforce pressure vessel domes.)
- Use fabric, particularly fiberglass or Kevlar, as a surface ply to restrict surface (handling) damage.
- Ensure that the laminate has sufficient fiber orientations to avoid dependence on the matrix for stability. A minimum coverage of 6 to 10% of total thickness in 0, ±45, 90 directions is recommended.

3. Select the layup to avoid a mismatch of laminate properties with those of the adjoining structures, or provide a shear/separator ply.

- Poisson's ratio—if the transverse strain of a laminate greatly differs from that of adjoining structure, large interlaminar stresses are produced under load. Large Poisson's ratios can be achieved with some carbon/graphite-reinforced laminates. This effect is graphically portrayed in Fig. 4.30. The solution to the problem is to analyze the effects of induced loads on the metal/composite couple and to consider using 90° plies to reduce the Poisson's ratio.
- Coefficient of thermal expansion—temperature change can produce large interlaminar stresses if the coefficient of thermal expansion of the laminate differs greatly from that of adjoining structure. If the adjacent structure is metallic, try to adjust the laminate CTE to more closely approximate the CTE of the metal. If 90° plies are not effective, hybridizing may solve the problem. Fiberglass woven fabric has been successfully used to reduce the CTE of a carbon/graphite-epoxy joint to titanium. If the

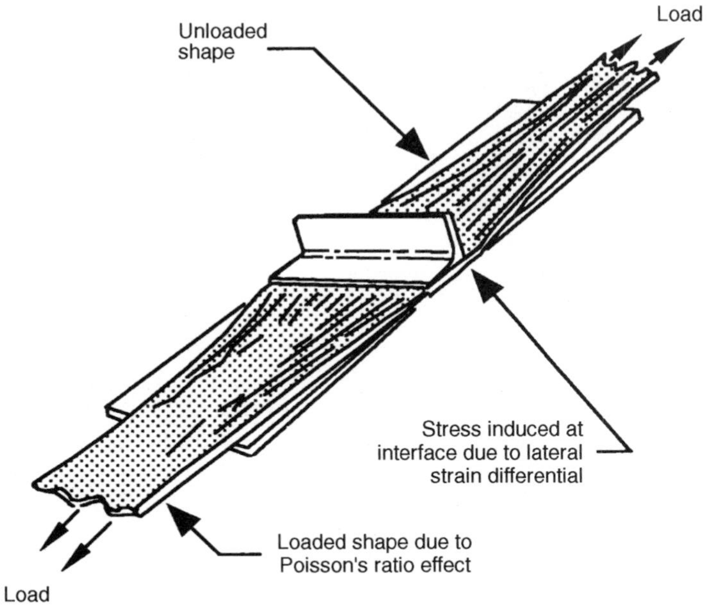

Figure 4.30 Poisson's ratio problem.

composite and the metal's CTE cannot be reasonably matched, allow for the differences in the joint. Note that the composite will generally have the lower CTE.

- Composite curing stresses, called *residual stresses* are significant and, in the case of high-modulus composites, can cause part failure on cool-down from cure. These stresses will also cause Poisson's ratio strains.
- The ply layer adjacent to most bonded joints should not be perpendicular to the direction of loading. Reduce composite section thickness in the joint area, soften the composite by adding fiberglass or angle plies, and select the highest-strain-capability adhesive.
- Hygrothermal stresses, induced in the laminate when it absorbs moisture, should also be considered. These are generally factored in by most computer analysis programs.

4. Use multiple ply angles.

- Typical composite laminates are constructed from multiple unidirectional or fabric layers that are positioned at angular orientations in a specified stacking sequence. From many choices, experience suggests a rather narrow range of practical construction from which the final laminate configuration is usually selected. The multiple layers are usually oriented in at least two different angles, and possibly three or four ($\pm\theta$, $0°/\pm\theta$, or $0°/\pm\theta/90°$ cover most applications, with θ between 30 and 60°). Unidirectional laminates are rarely used except when the basic composite material is only mildly orthotropic (e.g., certain metal matrix applications) or when the load path is absolutely known or carefully oriented parallel to the reinforcement (e.g., stiffener caps or cryogenic vessel support straps).
- An important observation concerning practical laminate construction is the concept of *midplane symmetry* with respect to stacking sequence, i.e., uncoupled response, and *balanced construction* with regard to the in-plane angular orientations, i.e., orthotropic behavior as opposed to anisotropic. If a laminate does not exhibit midplane symmetry, the stretching and bending behavior can be highly coupled. The severity of this coupling is inversely proportional to the number of layers within the laminate. The fewer the layers, the worse the problem. This heterogeneous nature should be avoided. In general, seek a symmetric laminate to satisfy the design. When symmetric and balanced laminates are used, bending, twisting, and warping effects are reduced.
- When unsymmetrical laminates are used, such as for the faces of a sandwich structure or the sides of a cylinder, the entire laminate is generally made symmetric because of geometry. If an unsymmetric laminate must be used alone, it is very prudent practice to avoid those of less than eight layers. Problems inherent with unsymmetric laminates may be more severe with higher-modulus fibers

5. Use midplane symmetry.

- The usual reference surface for determining if a laminate is symmetrical is the geometric midplane. A midplane symmetric laminate is highly desirable and to be preferred. Extremely few cases exist where an unsymmetrical stacking sequence should even be considered. To construct a midplane symmetric laminate, for each layer above the midplane there must exist an identical layer the same distance below the midplane, with a repetition of thickness, material properties, and angular orientation. Thus, the lamination stacking sequence will possess a mirror image about the geo-

metric midsurface. As Figs. 4.11*e* and 4.11*f* show, both even- and odd-number-ply laminates can satisfy these criteria. Note that the use of an identical woven fabric at identical distances from the midplane may not confer symmetry. One of the layers will have to "flipped," because the fabric may not identical on both surfaces.

6. Observe the effects of stacking sequence.

– Once the orientations of a laminate are specified, the stacking sequence will control the flexural rigidities of a laminated plate. Thus, the stability, vibration, and static bending behavior are controlled by the relative dispersion and thickness distribution of these oriented layers.

– Interlaminar stresses (normal and shear) are influenced by the laminate stacking sequence. For example, a $(\pm45°_2/0°_2)_S$ laminate generates compressive normal stresses at the laminate boundaries that are 10 times as high as the compressive stresses produced by a $(45°/0°/\pm45°/0°/-45°)_S$ laminate, when each is loaded in compression. Tensile free-edge stresses would result if these laminates were subjected to axial tensile loads. However, if the 0° layers were stacked on the outside, tensile (delamination) stresses would be induced at the edges as a result of compressive loads. The interlaminar normal stresses can be minimized, particularly for fatigue applications, by optimizing the stacking sequence in reference to the direction of load. The interlaminar stress problem is normally referred to as the free-edge effect. Such stresses are usually dominant at the free-edges of laminates (Fig. 4.31). Their magnitude can be significant, especially for fatigue conditions such as thermal cycling. The major free-edge-effect width is over a very narrow region (approximately equal to the laminate thickness). Furthermore, the magnitudes of such stresses are proportional to the thicknesses of nondispersed oriented layers. Hence, avoid stacks of layers at the same orientation; alternate or disperse the layers.

– If plies must be terminated (in the laminate), provide for equal steps of 0.10 or greater and cover the step area with an additional ply (preferably fabric) to avoid in-

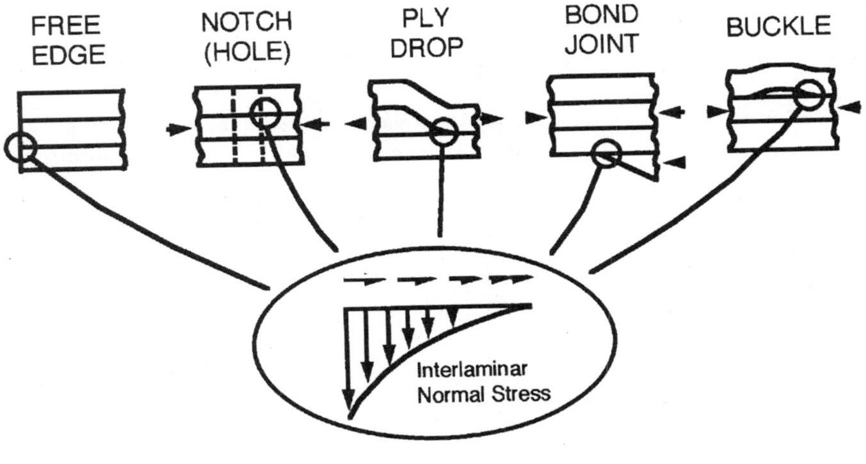

Figure 4.31 Sources of out-of-plane loads.

terlaminar shear failure, to provide some load redistribution, and to avoid ply edge peeling.

- Adjacent layers, if possible, should be positioned with a maximum orientation differential of 60°. Potential macrocracking of layers, from induced thermal stress during cool-down from curing, is sufficient reason to avoid this problem. As the adjacent-orientation differential increases, so do thermal stresses (e.g., ±45° or 0°/90° laminates are representative of the worst-case situations).
- Macrocracking of layers within a laminate is an irreversible process. Although it may not affect static strengths, it can lead to reduced fatigue life. Generally, the frequency of macrocracking increases with thermal cycling and can quickly lead to reduced laminate stiffnesses and changes in Poisson's ratio and coefficients of thermal expansion. In addition, layer cracking may interact with free-edge stresses that can trigger delamination. Fabric layers (as opposed to unidirectional) are very resistant to cracking.

4.9.2 Component Design Recommendations

These recommendations relate to material and processes interactions with environmental concerns and cost-effective fabrication.

- Consider electrode potential; corrosive galvanic cells can be produced whenever two materials of different electrode potential are in electrical contact in the presence of an electrolyte. This is an important consideration for carbon/graphite/epoxy laminates, which are electrically conductive. Many metals are anodic with respect to carbon/graphite/epoxy and are corroded when they are part of a galvanic cell with carbon/graphite/epoxy. An electrolyte, a fluid capable of transferring electrons, is very difficult to avoid in practice. In designs with carbon/graphite/epoxy laminates, observe the following:
  1. Avoid magnesium and magnesium alloys because of their high potential relative to carbon/graphite/epoxy. Aluminum has been used in conjunction with carbon/graphite/epoxy composites with a designed separation to ensure that there is no conductive path from the composite to the aluminum. Titanium has been generally recommended as a replacement for aluminum alloys in areas where galvanic corrosion is expected.
  2. Carefully investigate the potential of other metals and provide special inter-surface protection, such as a fiberglass ply or other nonconductor, as required.
  3. Consider the use of a nonconductive structural fiber composite such as aramid or glass rather than carbon/graphite/epoxy in areas where the presence of electrolytes is expected.

4.9.3 Component Fabrication Recommendations

- Consider fabrication requirements during preliminary design.
- Avoid selecting a design that might be technically elegant but could not be built consistently or reliably. The design of a composite structure requires early concurrent engineering to ensure that the composite can be manufactured.
- Review the design carefully to avoid out-of-plane loads that can cause high levels of interlaminar shear. These, in turn, can cause delaminations.
- Select the simplest design to manufacture.

- If there is a hard choice between ease of analysis and ease of manufacture, choose the latter.

  Design for a composite structure rather than using composite as a substitute for a metal.

  The designer/composites technologist should perform preliminary analysis.

  Many tools are available for preliminary composite sizing and layout, including excellent computer programs for use on microcomputers. Use these computer programs and closed form analyses early, and defer finite element analysis.

- Reduce the part count.

  Consolidate as many substructures as possible, co-cure adhesive bonded assemblies, and build in stiffeners. This is the primary method of reducing costs. (Co-curing is not universally accepted in the industry due to the inability to perform an inspection on the co-cured joint and the possible cost ramifications if the joint is found to be defective.)

- Use the greatest tow size and thickest tape or fiber as practical.

  Larger tow sizes will generally be much less expensive on a weight basis and will tend to decrease fabrication time, but they may have a negative effect on physical properties and surface finish.

- Use the most appropriate fabrication method for the part.

  Note the constraints associated with each technique, e.g., filament winding results in non-smooth outer surface; pultrusion does not normally result in a quasi-isotropic laminate. For large structures, the relative cost of the filament winding is less than one-fourth that of hand layup and less than one-half that of the best tape laying machine.[38]

- Use the least expensive form of composite raw materials.

  Use a wet resin and dry fiber, if appropriate to the process. This method reduces the cost of filament winding. In Ref. 38, the cost for the large MX launch canister wet filament wound by Hercules in the 1990–95 time period was $45.00/lb—less than the cost of just the prepreg tape material for many composite parts.

- Consider the use of fabric, if appropriate to the structure.

  Although raw material costs are higher (some estimates are 18 to 20%), and some physical properties are reduced, woven fabric shortens layup times, avoids tape spread-out problems at abrupt contour changes, remains in place during cure (no distortion or fiber washout), and is frequently easier to drill or machine.

- Use composite and metals together to obtain a synergistic part.

  Avoid making the component of either metal or composite only when the judicious use of both will make a better, cheaper part.

- Make in-situ joints if possible.

  A metal-to-composite in-situ joint, made concurrently with the composite structure, saves costs in terms of machining, surface preparation quality assurance, and adhesive application. Generally, the technique will require the use of a film adhesive. This is similar to co-curing, but one side of the joint can be metal or composite, and the joint will generally have a mechanical locking feature.

- Select verification methods carefully.

  For coupon testing, avoid extensive tests for matrix-dominated properties when loads are almost totally reacted by fiber, and vice-versa. Make witness panels as a part of the component if possible. Coupon testing does not remove the need for full-scale component testing, and vice versa.

- Water absorption will change the mechanical and physical properties of a composite laminate. Matrix-dominated shear, compression and bending modulii, and strength may be reduced, and coefficient of expansion may be increased. Account for these changes in design.

- Carbon/graphite composites have poor impact resistance. If impact is a consideration, add glass plies to the exterior surfaces.

4.10 Damage Tolerance

One of the primary hurdles for the widespread use of composites, in addition to cost, has been concern about the damage tolerance. Composite materials show no yield behavior, defects are hard to find, and their effects are not always predictable.

Almost all applications for advanced composites, except one-time-use structures like rocket motors, involve repeated cyclic application of stresses. Defects can be introduced by the manufacturing processes or by unplanned occurrences such as impacts. The effects of these defects must be evaluated, because their growth during cyclic stresses could cause delaminations or other failures of the structure. Thus, the structure must be damage tolerant to ensure that it will not fail catastrophically during its operating life. Three suggestions for safety of advanced composite structures are:

1. Undetectable defects or damage should not affect the life of the structure.
2. Detectable damage should be sustainable by the composite structure for a period of time until it is found.
3. Damage during a mission or flight should be sustainable to as great an extent as possible to complete the mission.

The operational stresses in the composite structure can be kept low by increasing wall thicknesses, but that approach would negate the obvious weight advantage of composites and would also make the product more expensive. Also, some stresses, such as interlaminar or transverse, will not be reduced but may be actually increased by additional plies. Increased attention to processing controls can reduce or eliminate many defects (e.g., voids, delaminations, and ply buckles) but not all (e.g., ply drops and free edges). Costs escalate when non-standard controls are exercised in production. The advantage of low-maintenance costs of composites over product lifetime is enhanced because of good fracture-toughness properties of the composite.

4.10.1 Reducing Damage Concerns

Probably the easiest way to reduce concerns over damage to composites is to impose a ply of an impact absorbing material on the exposed surfaces. Thus, carbon/graphite laminates have received a measure of protection from impacts and other environmental insults by the addition of a fiberglass or aramid ply on the exterior surfaces. This can also have the effect of increasing the laminate allowables by reducing the *knockdown* values for many environmental insults. Another advantage of a thin fiberglass external ply is the telltale stress whitening that happens with external impact when the ply is added to the exterior a carbon/graphite composite.[39] Other ways are as follows:[40]

- Use higher strain fibers and tougher matrices.

  This has been the approach used by the U.S. Air Force to increase the fracture toughness and impact resistance of carbon/graphite laminates for new high-performance aircraft. The bulk of effort has centered on higher-strain carbon/graphite fibers and advanced thermoplastic matrices with their higher levels of toughness.

- Modify the stacking sequence.

  Modify the stacking sequence of the laminate plies to reduce interlaminar stresses or

damage openings. One of the layup arrangements studied has been $(30/-30_2/30/90)_S$. This layup places the 90° ply at the midplane of the laminate, which may be a better position than others in terms of longitudinal tensile strength and modulus. However, it must be considered in light of the following potential problems:

Ply thickness. Since the failure may originate more predominately at a 90° ply, reduce the thickness of the 90° plies and do not stack them together. Possibly move them away from the midplane of the laminate.

Stitching. The benefit of through-the-laminate-thickness stitching has already been demonstrated to lock the laminate plies together and to reduce the propensity for transverse tensile failure

Braiding. The elimination of a weak ply interface can be accomplished by braiding. It also has some drawbacks, namely a complicated manufacturing process and a loss of the ability to tailor other properties of the laminate.

Edge cap reinforcement. A layer of aramid or fiberglass can act to reduce interlaminar normal stresses and increase both static and fatigue strengths at a cut edge of the laminate. Capping will also increase bending rigidity, manufacturing, and NDT costs.

Notched edges. Some investigators have found that narrow notches along the edges of a thin laminate seem to disrupt the edge load path. It can result in stress concentrations, also.

Critical ply termination. When a laminate with a 90° ply at mid-plane is subjected to tension, peak interlaminar stresses are generated at the interface between the ply or within it. This is because the Poisson's ratio is so different from the surrounding plies. One solution has been to terminate the 90° ply away from the free edge.

Discrete critical ply. If there is a material mismatch at the midplane, such as a 90° ply, it can cause delamination, but if the discrete ply is cut, the delamination will be contained

Hybridization. Replace the 90° ply with a softer material, such as fiberglass. Do not use an aramid, due to the poorer shear strength.

Adhesive layer interleave. The adhesive layer softens the center of the laminate and also increases the impact resistance but, in turn, it decreases the bending stiffness dramatically.

The fracture toughness of the laminate material can be measured by a number of methods, some of which are shown in Fig. 4.32. The results will vary with different layup orientations. Other methods may be similar to the test specimens shown in ASTM-E-399 and may involve impact. Presently, the most widely accepted technique for evaluating the fracture toughness of a laminate or structure is a compression test after impact. Compression is a valid indicator; without matrix support, the fibers will buckle, while the fibers do not need as much matrix support for tensile loading.

Fracture toughness depends on matrix and fiber properties, fiber orientations, and stacking sequence. The emphasis on thermoplastic composites by the U.S. Air Force has been directed toward increasing the fracture toughness of composites.

4.11 Composite Repairs

There will be damage to composite structures, particularly in aircraft, that may not be visible but are of concern. The damage to a composite structure may be more than that incurred by a comparable metal structure under an identical impact, due to the lack of strain

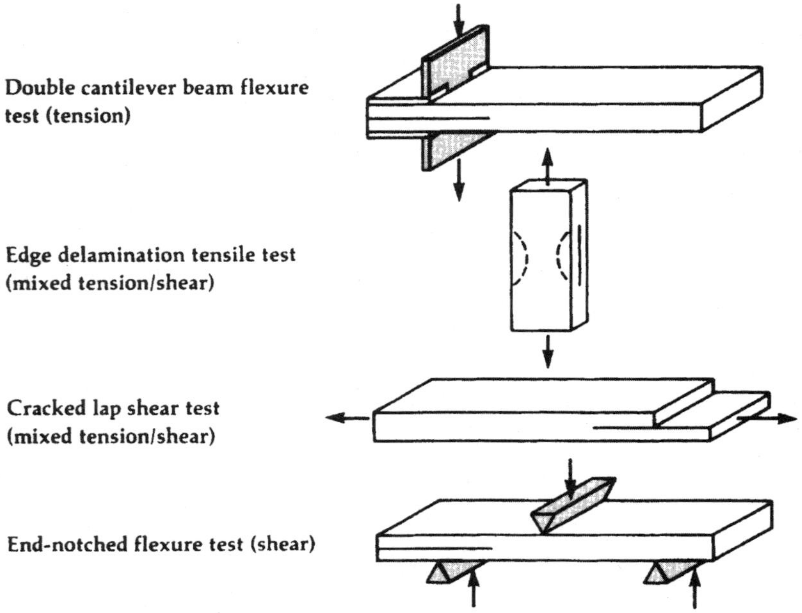

Figure 4.32 Typical methods of measuring fracture toughness of composites. *(From Ref. 12)*

capability in the composite. The metal may yield to show the impact site, but the composite may delaminate within and not reveal the damaged area. In-plane tensile strength may not always be compromised by impact damage to the matrix, but damaged matrix cannot stabilize fibers under compressive bending or shear stresses, and compression is usually the critical loading mode in aircraft structures. A modest impact to a composite can result in severe undetected internal damage in the form of delamination.

The objectives of most repair/maintenance programs are as follows:

1. Investigate and quantify the extent of damage.
2. Determine the comprehensiveness of the repair effort and where the repairs will be performed.
   - Repair to full properties (with all flaws detected and corrected), or
   - Repair to acceptable properties that will allow the structure to be usable to some high, predetermined percentage of full-scale operation, or
   - Repair to some emergency level that will allow use at a low percentage of operational effectiveness (i.e., fly the damaged aircraft to depot maintenance base).
3. Select the repair configuration.
4. Define the materials and processes to be used, i.e., adhesives, cure cycle limitations.

The repair options after detection and quantification are shown schematically in Fig. 4.33.[41] Most repairs to aircraft composites involve damaged and possibly wet honeycomb

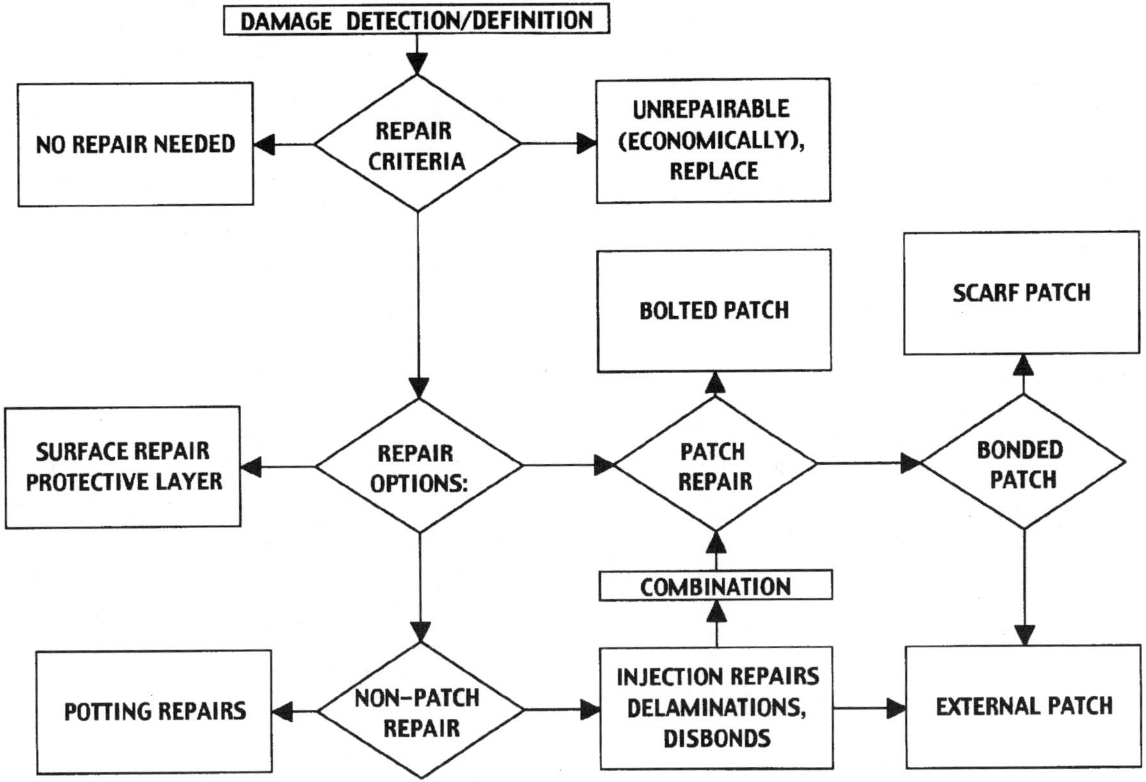

Figure 4.33 Flow diagram for repair options. *(From Baker, A., in Mallik, P.K., ed.,* Composites Engineering Handbook, *Marcel Dekker, New York, 1997, p. 747)*

in addition to the composite. A nomograph for determining acceptable conditions for these repairs has been prepared (Fig. 4.34).[42]

The first priority is preparing the composite for repair and may involve drilling holes for a bolt-on patch or reducing the moisture to a level that will allow heating of the composite and the adhesive for cure.

Repair can involve the simple injection of a wet resin through drilled holes into the delaminations then adding blind fasteners to defer the onset of local buckling, or the use of wet or dry prepregs in conjunction with adhesives to form a plug or flush on bonded-on patch. A summary of the available techniques for repair of solid and honeycomb laminates is shown in Table 4.35.[43] Several of the most important points for consideration, summarized in Ref. 43, are:

1. Match the modulus of the original material and the fiber direction as closely as possible.
2. Repair techniques are sufficiently complex that a high degree of technician skill and training are necessary.
3. For aircraft, nonmetallic repairs must incorporate lightning strike protection.

4.12 Adhesive Bonding and Mechanical Fastening

There are overwhelming reasons to adhesive bond composites to themselves and to metals, and conversely there are substantial reasons for using mechanical methods such as bolts or rivets (Table 4.36).

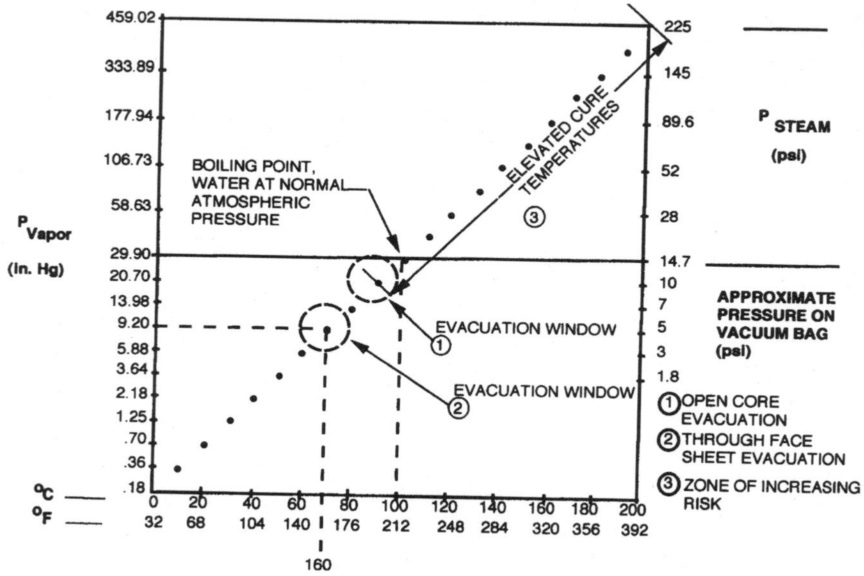

Figure 4.34 Pressure and evacuation guidelines for honeycomb core repair.

Table 4.35 Summary of Available Composite Repair Techniques[42]

Method	Advantages	Disadvantages	Ease of repair	Structural integrity
Bolted patch	No surface treatment; no refrigeration, heating blankets, or vacuum bags required	Bolt holes weaken structure; bolts can pull out	Fast	Low
Bonded patch	Flat or curved surface; field repair	Not suitable for high temperatures or critical parts	Fast, but depends on cure cycle of adhesive	Low to medium
Flush aerodynamic	Restores full design strength; high-temp capability	Time consuming; usually limited to depot; requires refrigeration	Time consuming	High
Resin injection	Quick; may be combined with an external patch	May cause plies to separate further	Fast	Low
Honeycomb, fill in with body filler	Fast; restores aerodynamic shape	Limited to minimal damage	Fast	Low
Honeycomb, remove damage, replace with synthetic foam	Restores aerodynamic shape and full compressive strength	Some loss of impact strength, gain in weight	Relatively quick	High
Honeycomb, remove damage, replace with another piece of honeycomb	Restores full strength with nominal weight gain	Time consuming; requires spare honeycomb	More difficult	High

Table 4.36 Reasons For and Against Adhesive Bonding

For	Against
Higher strength-to-weight ratio	Sometimes difficult surface preparation techniques cannot be verified as 100% effective
Lower manufacturing cost	Formulation may change
Better distribution of stresses	May require heat and pressure
Electrical isolation of components	Must track shelf life and out-time
Minimal strength reduction of composite	Adhesives change values with temperature
Reduced maintenance costs	May be attacked by solvents or cleaners
Reduced corrosion of metal adherend (no drilled holes)	Common attitude: "I won't ride in a glued-together airplane"
Better sonic fatigue resistance	

The studies on adhesives, surface preparation, test specimen preparation, and design of bonded joints reported for the PABST Program[44] gave much more credibility to the concept of a bonded aircraft and provided reliable methods of transferring loads between composites and metals or other composites.

A number of decisions must be made before bonding to a composite is attempted. These are as follows:

1. Surface preparation technique
   - *Composite.* Peel ply or manual abrasion or, if possible, co-cure.
   - *Metal:*
     Aluminum—Phosphoric anodize is accepted, preferred airframe method.
     Titanium—Several types of etches are available.
     Steel—Time between surface preparation and bonding is concern.
2. Type of adhesive
   - *Film adhesive.* Reproducible chemistry; early, frequent quality assurance (because resin is premixed by the manufacturer); requires special storage, and it generally is more difficult to accommodate varying bond line thicknesses.
   - *Paste adhesive.* Longer shelf life, no or less changes in storage, accommodates varying bond line thicknesses, but mixing errors are possible, and quality assurance of the adhesive (mix ratio) is generally performed after the structure is bonded.
3. Cure temperature of adhesive
   The maximum cure temperature of adhesives should be below the composite cure temperature unless they are co-cured. Cure temperature or upper use temperature of the adhesive may dictate the maximum environmental exposure temperature of the component. There are several cure temperature ranges:
   - Room temperature to 225°F. Generally for paste adhesives for noncritical structures.
   - 225°F to 285°F. For nonaircraft critical structures. Cannot be used for aircraft generally, because moisture absorption may lower HDT (heat deflection temperature) below environmental operating temperature.
   - 350°F and above for aircraft structural bonding. Higher cure temperature may mean more strain discontinuity between adhesive and both mating surfaces (residual stresses in bond).
4. Joint design
   The primary desired method of load transfer through an adhesive bonded joint is by shear. This means that the design must avoid peel, cleavage, and normal tensile stresses. One practical application of a method avoiding other than shear stresses is the use of a rivet bonded construction (Fig. 4.35). The direction of the fibers in the outer ply of the composite (against the adhesive) should not be 90° to the expected load path.

4.12.1 Design Guidelines for a Bolted or Bonded Joint

Observe the unique design requirements of a bolted or bonded joint from a composite to a metal or other composite: composites are different from metals. Metals can be relied on to reduce high stress concentrations because of the metal's plasticity. In a brittle material (composite), bolt tolerances and hole drilling are very important. Inadequate attention can result in a few bolts being forced to carry the load. Brittle failure at a low strain can happen at the bolts before the load can be redistributed. Some general design rules for a bolted joint are as follows:

- Design the joint to be critical in bearing with a non-catastrophic failure (Fig. 4.36).[45]
- Fasteners must have sufficient diameter and strength. Locally reinforce as required. Maintain fastener-to-edge distance (e/d) and spacing distance (s/d) as follows: e/d = 2.5 to 3 and s/d = 6 vs. 1.7 to 2 and 4, respectively, for aluminum alloy. Figure 4.37 shows these physical dimension suggestions.

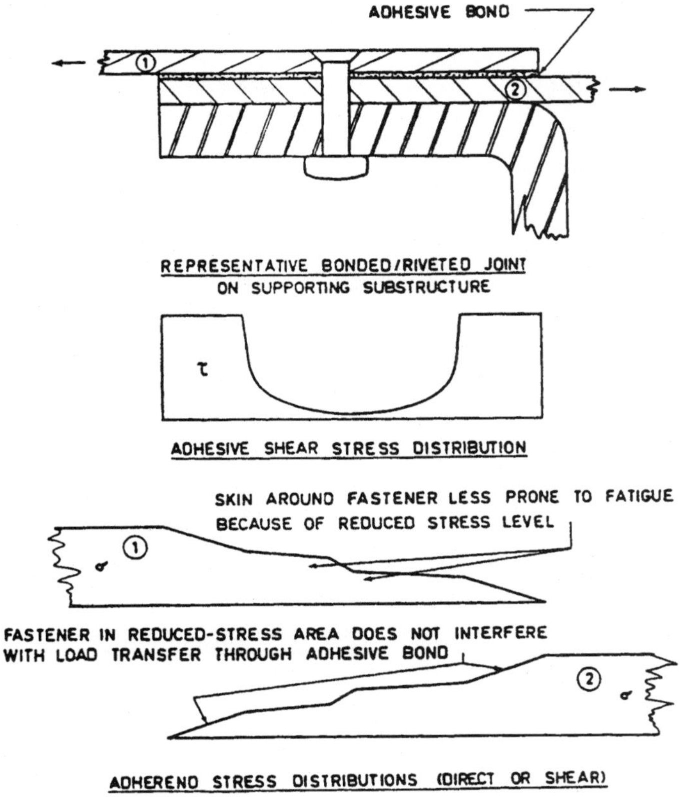

Figure 4.35 Load transfer in rivet-bonded construction. *(From Ref. 45)*

- Keep nearest row of pins at least three fastener diameters from the free edge. Note that multiple rows of holes do not share the load equally. This suggests that two rows of holes would be more efficient than, and preferred to, multiple rows. Keep bearing and tensile values well within the elastic range of expected laminate properties.
- Eliminate, as much as possible, the joint eccentricity. Do not introduce bending into the joint. Design for tight pin-to-hole tolerance. Ensure a smooth fastener shank loading against composite. Fastener-to-composite thickness ratio should be greater than ~0.70.
- Increase radial clamping for higher bearing allowables. Finally, and this could be the most important consideration: All of the above recommendations are predicated on the use of a quasi-isotropic laminate. Any other laminate type must be structurally evaluated, and additional laminate must be added to the joint area if necessary.

Rules for a bonded joint are as follows:

- Do not allow tensile loads into the joint, only shear loads. For a tensile load, use a bolted joint.

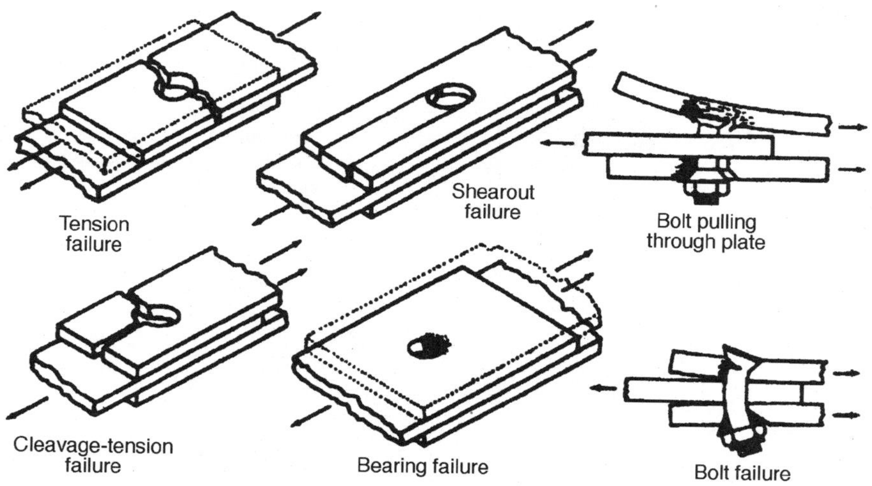

Figure 4.36 Failure modes for bolted joints. *(From Nelson, W.D., Bunin, B.L., and Hart-Smith, L.J., in* Proc. 4th Conf. Fibrous Composites in Structural Design, *Army Materials and Mechanics Research Center, Manuscript Report AMMRC MS 83-2, 1983, pp. U-2 through II-38)*

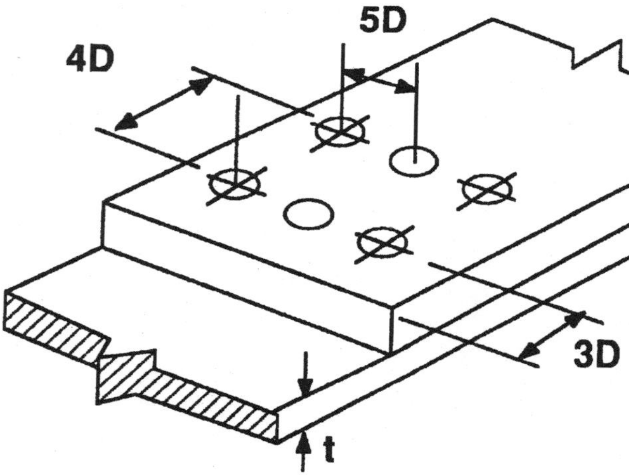

Figure 4.37 Recommended fastener spacings and edge margins for bolted joints.

- Short bonded joints are more efficient. The peak stress occurs at the ends of the joint, and the allowable stress decreases with length.
- Avoid eccentricity in the joint, because it produces peel stresses. This means the thickness of the adhesive should not be excessive. Most manufacturers recommend joint thickness on the order of 0.005–0.010 in. Besides changing of the stress type, thick ad-

hesives generally are weaker. Thick laminates are more affected than thin. If a thick laminate has to be bonded, chamfer. There are concerns for both bolting and bonding; they are summarized in Table 4.37.[46] Further details are provided in Ref. 44.

Table 4.37 Comparison of Bolting and Bonding (from Ref. 46)

Advantages	Disadvantages
Bonded joints	
Small stress concentration in adherends	Limits to thickness that can be joined with simple joint configuration
Stiff connection	Inspection difficult other than for gross flaws
Excellent fatigue properties	Prone to environmental degradation
No fretting problems	Sensitive to peel and through-the-thickness stresses
Sealed against corrosion	Residual stress problems when joining to metals
Smooth surface contour	Cannot be disassembled
Relatively lightweight	May require costly tooling and facilities
Damage tolerant	Requires high degree of quality control
	May be environmental concerns
Bolted joints	
Positive connection, low initial risk	Considerable stress concentration
Can be disassembled	Prone to fatigue cracking in metallic component
No thickness limitations	Hole formation can damage composite
Simple joint configuration	In composites, relatively poor bearing properties
Simple manufacturing process	Prone to fretting in metal
Simple inspection procedure	Prone to corrosion in metal
Not environmentally sensitive	May require extensive shimming with thick composites
Provides through-the-thickness reinforcement, not sensitive to peel stresses	
No major residual stress problem	

4.13 Environmental Effects

The primary environmental concern for all polymeric matrix composites is the effect of moisture intrusion both into the raw materials and into the finished components. Moisture in the raw materials causes voids and delaminations of the composite, and moisture in the finished composite structure can result in the effects shown in Table 4.38. Moisture enters the composite through cracks or voids in the matrix, and diffuses through the resin or through the fiber-to-matrix interface (if open to atmosphere). Except for fiberglass and Kevlar, the advanced composite fibers do not transport water, and none of the fibers except for fiberglass are substantially affected by water intrusion. Figure 4.38 shows typical degradation techniques for fiberglass and Kevlar.[47] Most of the effects of moisture are reversible, and the affected property will be restored when the composite is dried. It may take a long time to dry, since the moisture must diffuse out.

Many users of composites require a systematic evaluation of the reaction of each composite (fiber and resin) to a broad list of environments. Many of these environments are listed in test documents such as MIL-STD-810. The evaluation can result in the approval of a composite structure, because it is the same or very close to one already tested (similarity), because it is theoretically unresponsive (analysis), or because the actual structure or a

Table 4.38 Possible Effects of Absorbed Moisture on Polymeric Composites

- Plasticization of epoxy matrix
  - Reduction of glass transition temperature reduction of usable range
- Change in dimensions due to matrix swelling
- Enhanced creep and stress relaxation
  - Increased ductility
- Change in coefficient of expansion
- Reduction in ultimate strength and stiffness properties in matrix-dominated properties
  - Transverse tension
  - In-plane shear
  - Interlaminar shear
  - Longitudinal compression
  - Fatigue properties (change may be beneficial)
- Fiber-dominated properties are generally not affected
- Change in microwave transmissibility properties

series of coupons has been tested by the composite manufacturer (test). It is in the best interest of both the fabricator and the user whenever the resolution of the composites' response to an environment can be obtained at minimum cost, i.e., verification by similarity. Some of these environments, the predicted effects, and references are shown in Table 4.39.[47]

4.14 Composite Testing

To ensure consistent, reproducible components, three levels of testing are employed: incoming materials testing, in-process testing, and control and final structure verification.

4.14.1 Incoming Materials Testing

Incoming materials testing seeks to verify the conformance of the raw materials to specifications and to ensure processibility. The levels of knowledge of composite raw materials do not approach those for metals, which can be bought to several consensus specifications and will appear to be generally identical, even if purchased from many manufacturers. Although there are fewer suppliers for composite raw materials, the numbers of permutations of resins, fibers, and manufacturers prevents the kind of standardization necessary to be able to buy composite raw materials as if they were alloys.

The fabricators of composites will rely on specifications for control of fiber, resin and/or the prepreg as shown in Table 4.40. Many prepreg resin and fiber vendors will certify only to their own specifications, which may differ from those shown; users should consult the vendors to determine what certification limits exist before committing to specification control.

The purpose of incoming testing is to achieve a consistently reliable product that can be verified to meet the user's requirements and processing techniques; thus, the testing should reflect, if possible, the individuality of the processing method.

As part of raw materials verification, composite design effort, and final product verification, mechanical testing of composite test specimens will be performed. The testing of composite materials offers unique challenges because of the special characteristics of composites. Factors not considered important in metals testing are very important in testing composites. For example, composites are anisotropic, with properties that depend on the direction in which they are measured. Strain and load rate must be carefully monitored, specimen conditioning (drying, storage, etc.) can substantially affect results, and

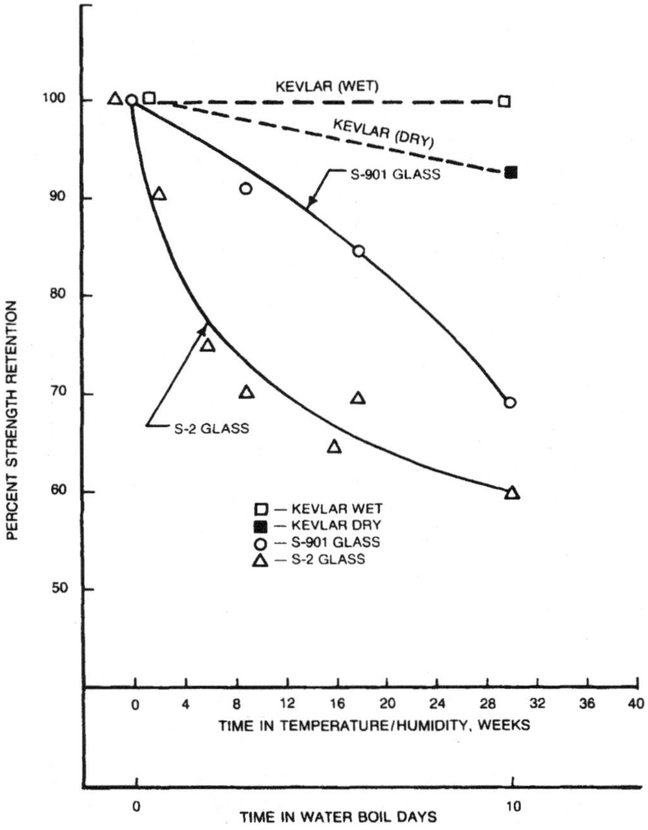

Figure 4.38 Composite response to humidity. *(From Ref. 46)*

even humidity conditions at the time of specimen fabrication or test can significantly affect the material. Fiber content and void content, which can also vary with manufacturing conditions, have important effects on material properties. All personnel involved in the generation and use of test results, from fabrication of the material to final data interpretation, should be familiar with these factors and their influence on test results. References 47 (pp. 7-1 through 7-31) and 48 provide reference monographs on composites testing; Refs. 49 and 50 provide excellent descriptions of the factors that influence test results.

4.14.2 In-Process Testing

In-process testing seeks to ensure the repeatability of the process so as to reach a consistent fiber volume, resin content and fiber angle, and correct layer placement. Since the processing from start of the cure is a continuous part of the batch processing, there can be no test sample testing. The control is essentially interrogation of the curing process, modification of the process to accommodate the batch to-batch inconsistencies of the input materials, and correlation of the effects of changes in input materials and processes to arrive at

Table 4.39 Environmental Effects on Composite Structures (adapted from Ref. 47, pp. 9-2 through 9-44)

Environment	Effects (matrix of fiber)	Comments
Moisture related		
Rain	Matrix, softens and swells	Rain may cause water intrusion into composite and joints
Humidity	Matrix, softens and swells	Effect can be aggravated by high, >1%, void content.
Salt fog	Matrix, softens and swells	Corrosion of metal attached to graphite epoxy composite can be increased.
Deep submergence	Matrix	Interface may be affected, low void content is requirement for compressive inputs.
Rain and sand erosion	Both	Protect surfaces with paint or elastomer, composite response is significantly different from metals.
Galvanic corrosion	Neither	Graphite composites can cause corrosion of metals; composite is unaffected.
Radiation		
Solar radiation (Earth)	Matrix	Unprotected aramid or polyethylene is affected.
Ultraviolet	Matrix	Unprotected aramid or polyethylene is affected.
Nonionizing space	Both	Composites can be destroyed, most metals are unaffected.
Solar (LEO)	Both	Composites can be destroyed, most metals are unaffected.
Temperature		
High	Matrix	Fibers (except polyethylene) are generally more resistant than matrix.
Low	Matrix	
Extreme (ablation)	Both	
Thermal cycling (shock)	Matrix	
Miscellaneous environments		
Solvents, fuels	Matrix	Determine for each matrix, stressed and unstressed; effects may be aggravated by voids.
Vacuum	Matrix	
Fungus	Matrix	Each material must be evaluated. May affect organic fibers.
Fatigue	Both	Effects may be diminished by voids.

Table 4.40 Test Methods for Composite Raw Materials

Component	Test	Test method	Reference
Resin	Moisture content	Karl Fisher Titration*	ARP† 1610
	Comparison identity	Chromatography (several types)*	ARP 1610
	Chlorine (epoxies)	Hydrolyzable chlorine*	ARP1610
	Comparison	Color*	ASTM D-1544
	Comparison	Density	ASTM D-1475
	Comparison	Refractometric	User defined
	Comparison	Infrared identity*	ARP 1610
	Comparison	Viscosity	ASTM D-445, 1545
	Reactivity	Weight per epoxide	ASTM D-1652
Fiber	Strand mechanicals	Strand tensile strength and modulus	AMS 3892
	Yield	Weight/length	AMS 3892
	Sizing	Extraction	ARP 1610, AMS 3901, AMS 3892
	Fuzz		User defined
	Ribbonization		User defined
	Packaging		AMS 3901, AMS 3894
	Density		ASTM D-3800
Prepreg	Fiber mechanicals	Tow test or unidirectional tensile	ASTM D-2343, D-3379, D-4018, D-3039
	Resin content	Extraction	ASTM C-613, D-1652, D-3529
	Resin identity	Infrared identity	ARP 1610
	Curative identity,%	Extraction, infrared	ARP 1610
	Nonvolatiles		ASTM D-3530
	Resin flow		ASTM D-3531
	Gel time		ASTM D-3532
	Tack		AMS 3894
	Visual imperfection	Optical	AMS 3894
	Areal weight	Weight	ASTM D-3529, D-3539

*Also applicable to prepreg.
†ARP = Aerospace Recommended Practice, Society of Automotice Engineers (SAE), 400 Commercial Dr., Warrendale, PA 15096. AMS = Aerospace Materials Specification, SAE. ASTM = Americal Society for Tesging and Materials, 1916 Race St., Philadelphia, PA 19193.

limits on the process variables and inputs. There have been frequent problems of resin exotherm and cure temperature overshoot with large or thick structures that could ruin a component. New autoclaves have been provided with interactive controls and "smart" operating software to accommodate changes in cure dynamics. An interactive "smart" curing machine can respond to the curing temperature overshoot automatically and rapidly. Table 4.41[48] shows techniques for in-process monitoring.

Table 4.41 In-Process Monitoring of Advanced Composites (see Ref. 48 for further details)

Parameter controlled (measured)	Desired location control*	Method	Notes
Temperature	In laminate, top bottom center on tool, air temp.	Thermocouple	Includes rate of heating and cooling†
Vacuum	Composite surface	Gage (transducer)	Note†
Pressure	Composite surface in composite vessel	Gage (transducer) Springback go/no-go gage	Note†
Cure rate	In composite	Dielectric analysis Optical (fluorescence)‡ FTIR spectra‡	Note†
Viscosity	In composite	Dielectric analysis Ultrasonic‡ Acoustic Waveguides‡	Note†

*Frequency and number of monitored factors will decrease as product confidence is gained.
†Information is part of adaptive control system.
‡Not in general use yet.

Since composite structures can be large and complex to cost-effectively replace metal assemblies, the labor input value of these structures can be considerable. The benefit of testing the incoming materials and of verifying the processing is obvious. Ideally, with feed-back control and complete control over the incoming materials, there would be reduced need for final product inspection.

4.14.3 Final Product Verification

Rarely does composite testing include actual mechanical verification of the final product. The exceptions are filament-wound pressure vessels and some small composite structures, such as golf shafts, that are tested in bending. Many pressure vessels experience some deterioration due to the proof pressure test, which causes transverse stress in the hoop plies that ends up fracturing the matrix in the outer plies and results in a *first-ply* failure. This is one reason why another failure criterion is invoked for pressure vessels, namely the *last-ply* failure criterion. However, the failure in the outermost plies reduces the composite's resistance to water intrusion, and consequently the product may have to be recoated with a moisture-resistant paint to satisfy some environmental fears.

Nearly all final composite structure verification includes mechanical tests of tag end or coprocessed coupons and nondestructive tests on the structural laminate. Mechanical test-

ing is also used to verify the materials and processes prior to committing to the layup and cure of a larger composite structure. Table 4.42, adapted from Ref. 50, shows the key data and test methods that are needed to support design or production of a composite structure.

It can be seen that there are several tests in current use to determine some properties. Many of the tests are still evolving and will eventually become standardized through ASTM or other consensus organizations. Also, this list does not exhaust the available test methods, and nonstandard tests abound for composites fabricated by alternative techniques (e.g., filament winding, Ref. 47).

Table 4.42 Advanced Composite Test Methods in Current Use (adapted from Ref. 50)

Test	Test method(s)
Tensile strength and modulus axial and transverse	ASTM D-3039
Compression	ASTM D-695 (modified for high-modulus composites); Celanese,* ASTM D-3410; IITRI, ASTM D-3410; sandwich beam, ASTM D3410
Shear	Iosipescu, ASTM D-3518 Rail shear, ASTM D-4255 Short beam, ASTM D-2344
Flexure	ASTM D-790
Fracture toughness	ASTM E-399 NASA 1092* End notched flexure mode II Cracked lap shear (mixed modes I and II) Edge delamination NASA 1092
Impact	Instrumented drop-weight impact Tensile impact Compression after impact (Boeing BSS-7260)
Fatigue	ASTM D-3479, ASTM D-671
Coefficient of thermal expansion	User defined
Coefficient of moisture expansion	User defined
Single fiber tension	ASTM D-3379
Single fiber tensile creep	Note†
Open-hole tension, compression	Boeing BMS 8-276
Thermal conductivity	User defined
Bolt bearing	ASTM D-953

*Refers to the original design agency or the agency that prepared the reference specification or to the commercial company whose internal specification was used throughout the industry.
†No general consensus specification exists.

Nondestructive test techniques are preferred to verify the final structure, because coprocessed or tag end test specimens suffer from edge effects and may not mirror the fiber angles, resin, or fiber content. The component may also have discrepancies such as fiber kinking and washout that may not be reflected in the test specimens. Confidence is gained in a structure because of careful control of incoming materials, controlled processing, adherence of tag end or co-processed test specimens to required test values, and the meeting of NDI standards. The simplest NDI techniques, which involved coin tapping and visual observance, have evolved into those shown in Table 4.43 (adapted from Ref. 51). They can be used individually or, in some cases, concurrently.

Table 4.43 NDE Test Methods for Composites (adapted from Ref. 51)

	Principal characteristic detected	Advantages	Limitations
Radiography	Differential absorption of penetrating radiation	Film provides record of inspection, extensive data base	Expensive, depth of defect not indicated, rad. safety
Computer tomography	Conventional x-ray technology with computer digital processing	Pinpoint defect location; image display is computer controlled	Very expensive, thin wall structure might give problems
Ultrasonics	Changes in acoustic impedance caused by defects	Can penetrate thick materials, can be automated	Water immersion or couplant needed
Acoustic emission	Defects in part stressed generate stress waves	Remote and continuous surveillance	Requires application of stress for defect detection
Acoustic ultrasonics	Uses pulsed ultrasound stress wave stimulation	Portable, quantitative, automated, graphic imaging	Surface, contact surface geometry critical
Thermography	Mapping of temperature distribution over the test area	Rapid, remote measurement, need not contact part, quantitative	Poor resolution for thick specimens
Optical holography	3-D imaging of a diffusely reflecting object	No special surface preparation or coating required	Vibration-free environment required, heavy base needed

4.15 Safety Issues with Composite Materials

Composite materials, particularly in the raw material form, present toxic hazards to users and should be handled carefully. Essentially, there is a need for management inputs and controls on the following aspects of composite fabrication:

- Material handling
- Training
- Gaining awareness of hazards and proper use of toxic materials
- Isolation of some operations
- Use of personal protective equipment (if necessary)
- Personal hygiene
- Significance of warnings and labels

- Housekeeping
- Dispensing and storage
- Emergency instructions

The typical materials that are encountered are shown on Table 4.44. This table is not exhaustive, and the user should consult the Material Safety Data Sheets (MSDS) and more in-depth references (e.g., Ref. 52). The liquid resins and catalysts are the highest noted hazards, because there are so many avenues of attack to our bodies, and because there are many different effects. Once the composite structure is fabricated, the fibers and resins are rendered essentially innocuous, and machining, drilling, etc. pose minimal hazards. Machining dusts should be contained and properly disposed of to reduce the nuisance.

Table 4.44 Commonly Encountered Hazards with Advanced Composite Materials

Hazard	Form	Reported hazards
Epoxy resins	Liquid or as prepreg	Dermatitis, some may be potential skin carcinogens
Epoxy curing agents		
Aromatic amines	Liquids, but primarily in prepreg	Liver, kidney damage, jaundice, some are carcinogens, anemia
Aliphatic amines	Liquids	Severe bases and irritants and visual disturbances
Polyaminoamides	Liquids	Somewhat less irritating, may cause sensitization
Amide		Slight irritant
Anhydride	Liquid, solid, prepreg	Severe eye irritants, strong skin irritants
Phenolic and amino resins	Usually as dry prepreg	Free Phenol, also released on cure. Formaldehyde is strong eye, skin and respiratory irritant
Bismaleimide resins	Liquid or paste, generally as prepreg	Skin irritant and sensitization, not fully characterized yet
Thermoplastic resins	Solids or prepreg	Exercise care with molten materials & provide good ventilation. Consult MSDS
Reinforcing materials		
Graphite fibers	Airborne dusts	Skin irritation, contain dusts, prevent exposure and protect skin
Aramid fibers		Contain airborne dusts
Fiberglass		Mechanical irritation to skin, eyes, nose, and throat

References

1. *Mil-Hdbk-17/2D.*
2. Hercules Data Sheet, H050-377/GF.
3. *Carbon and High Performance Fibres*, 6th ed., Chapman and Hall, London, 1995.
4. Peters, S.T., Humphrey, W.D. and Foral, R.F., *Filament Winding, Composite Structure Fabrication,* 2nd ed., SAMPE Publishers, Covina, CA, 1999.
5. Delmonte, John, *Technology of Carbon and Graphite Fiber Composites*, Van Nostrand Reinhold, New York, 1981, p. 59.
6. Bacon, R., Towne, M., Amoco Performance Products, personal communication.
7. Humphrey, W. Donald and Vedula, Murali, *SAMPE Proceedings,* Vol. 40, May 1995, pp. 14785–7.
8. Goldsworthy, W.B. and Hiel, C., *Proc. 44th SAMPE Symposium,* May 1999, pp. 931–942.
9. Mayorga, G.D., in Lee, S.M., ed. *International Encyclopedia of Composites,* Vol. 4, VCH Publishers, New York, NY.
10. Foral, Ralph F., and Peters, Stanley T., Composite Structures and Technology Seminar Notes, 1989.
11. Agarwal, Bhagwan D., and Broutman, Lawrence J., *Analysis and Performance of Fiber Composites*, 2nd ed., Wiley Interscience, New York, 1990, p. 199.
12. Peters, S.T., in S.T. Peters, ed. *Handbook of Composites*, 2nd ed., p. 16, Chapman & Hall, London, 1998.
13. Tsai, Stephen W. and Pagano, Nicholas J., in *Composite Materials Workshop*, S. W. Tsai, J.C. Halpin and Nicholas J. Pagano, eds., Technomic Publishing Co., Lancaster, PA, 1978 p. 249.
14. Tsai, Steven W., *Theory of Composites Design*, 1st ed., Think Composites, Dayton, Ohio, 1992.
15. Brown, Richard T., in *Engineered Materials Handbook,* Vol. 1, Composites, Theodore Reinhart, Tech. Chairman, ASM International, 1987, p. 268–274.
16. *Genlam*, General Purpose Laminate Program, Think Composites, P.O. Box 581, Dayton, Ohio 45419.
17. DuPont, *Data Manual for Kevlar 49 Aramid*, May 1986, p. 95.
18. G. Eckold, *Design and Manufacture of Composite Structures*, pp. 76–96, McGraw-Hill, New York, NY, 1994.
19. S. Swanson, in P.K. Mallick, ed. *Composites Engineering Handbook*, Marcel Dekker, pp. 1183 et seq, New York, NY, 1997.
20. S.B. Batdorf, in S.M. Lee, ed., *International Encyclopedia of Composites*, VCH Publishers, New York, NY, 1989.
21. *Military Handbook 5 for Aerospace Metals.*
22. Sun, C.T, Quinn, B.J., Tao, J., and Oplinger, D.W., *Comparative Evaluation of Failure Analysis Methods for Composite Laminates,* DOT/FAA/AR-95/109, May 1996.
23. Kirchner–Lapp in *Handbook of Composites*, 2nd ed., S. T. Peters, ed., Chapman and Hall, London, 1998, p. 760.
24. Military Handbook 17 B, *Plastics for Aerospace Vehicles Part 1, Notice-1,* 1973.
25. *The Composites Materials Handbook-Mil 17*, Technomic Publishing Co. Inc., Lancaster, PA, 1999.
26. Foston, Marvin and Adams, R.C., in *Engineered Materials Handbook,* Vol. 1, Composites, Theodore Reinhart, Tech. Chairman, ASM International, 1987, pp. 591–601.
27. B. Sarh, B. Moore, and J. Riedell, 40th International SAMPE Symposium, May 8–11, 1995, p. 381.
28. Habermeier, J., Verbal Presentation at SAMPE 44th ISSS&E, May 1999.
29. Campbell, Flake C., et al, *J. Adv. Materials*, July 1995, pp.18–33.

30. Bergland, Lars in S.T. Peters, ed. *Handbook of Composites*, 2nd ed., p. 116, Chapman & Hall, London, 1998.
31. Loos, Alfred C. and Springer, George, S.J., *Comp. Mater.*, March 17, 1983, pp. 135–169.
32. Enders, Mark L. and Hopkins, Paul C., *Proc. SAMPE International Symposium and Exhibition,* Vol. 36, April 1981, pp. 778–790.
33. Martin, Jeffrey D. and Sumerak, Joseph E., Composite Structures and Technology Seminar Notes, 1989, p. 542.
34. Ko, Frank K., in *Engineered Materials Handbook,* Vol. 1, Composites, Theodore Reinhart, Tech. Chairman, ASM International, 1987, p. 519.
35. Hancock, P. and Cuthbertson, R.C., *J. Mat Sci.,* 5, 762–768, 1970.
36. Kohkonen, K.E. and Potdar, N., in S.T. Peters, ed. *Handbook of Composites*, 2nd ed., Chapman and Hall, p. 598, London, 1998.
37. Abrate, S., in P.K. Mallick, ed., *Composites Engineering Handbook*, Marcel Dekker, New York, NY, 1997, p. 783.
38. Freeman, W.T. and Stein, B. A., *Aerospace America*, Oct. 1985, pp. 44–49.
39. Heil, C., Dittman, D., and Ishai, O., *Composites* (24) no. 5, 1993, pp. 447–450.
40. Chan, W.S., in P.K. Mallick, ed., *Composites Engineering Handbook*, Marcel Dekker, New York, NY, 1997, pp. 357–364.
41. Baker, A., P.K. Mallick, ed., *Composites Engineering Handbook,* Marcel Dekker, p. 747, New York, NY 1997.
42. Seidl, A.L., in *Handbook of Composites*, 2nd ed., S. T. Peters, ed., Chapman and Hall, London, 1999, p. 864.
43. Seidl, A.L., Repair of Composite Structures on Commercial Aircraft, 15th Annual Advanced Composites Workshop, Northern California Chapter of SAMPE, 27 Jan. 1989.
44. Potter, D.L., *Primary Adhesively Bonded Structure Technology (PABST) Design Handbook for Adhesive Bonding*, Douglas Aircraft Co., McDonnell Douglas Corporation, Long Beach, CA, Jan. 1979.
45. Nelson, W.D., Bunin, B.L., and Hart-Smith, L.J., Critical Joints in Large Composite Aircraft Structure, in *Proc. 6th Conf. Fibrous Composites in Structural Design,* Army Materials and Mechanics Research Center Manuscript Report AMMRC MS 83-2 (1983), pp. U-2 through II-38.
46. Baker, A., P.K. Mallick, ed., *Composites Engineering Handbook,* Marcel Dekker, p. 674, New York, NY, 1997.
47. Peters, S.T., Humphrey, W. D., and Foral, R., *Filament Winding, Composite Structure Fabrication,* 2nd ed., SAMPE Publishers, 1999, Covina, CA, pp. 9–13.
48. Kranbuehl, David E., in *International Encyclopedia of Composites*, Vol. 1, pp. 531–543, Stuart M. Lee, ed., VCH Publishers, New York.
49. Whitney, J.M., Daniel, I.M., and Pipes, R.B., *Experimental Mechanics of Fiber Reinforced Composite Materials*, SESA Monograph No. 4, The Society for Experimental Stress Analysis, Brookfield Center, Connnecticut, 1982.
50. Carlsson, L.A. and Pipes, R.B., *Experimental Characterization of Advanced Composite Materials*, Prentice-Hall, Englewood Cliffs, N.J., 1987.
51. Munjal, A., *SAMPE Quarterly*, Jan. 1986.
52. Safe Handling of Advanced Composite Materials Components, Health Association, Arlington, VA, April 1989.

Chapter

5

Metal Matrix Composites, Ceramic Matrix Composites, Carbon Matrix Composites, and Thermally Conductive Polymer Matrix Composites

Carl Zweben

Devon, Pennsylvania

5.1 Introduction

Chapter 4 discusses polymer matrix composites (PMCs) used in structural applications. This chapter covers PMCs used in thermal management and electronic packaging. It also provides an overview of metal matrix composites (MMCs), ceramic matrix composites (CMCs), and carbon matrix composites (CAMCs).

The development of composite materials and the related design and manufacturing technologies is one of the most important advances in the history of materials. Composites are multifunctional materials having unprecedented mechanical and physical properties that can be tailored to meet the requirements of a particular application. Some composites also exhibit great resistance to high-temperature corrosion, oxidation, and wear. These unique characteristics provide the engineer with design opportunities not possible with conventional monolithic (unreinforced) materials. Composites technology also makes possible the use of an entire class of solid materials, ceramics, in applications for which monolithic versions are unsuited because of their great strength scatter and poor resistance to mechanical and thermal shock. Furthermore, many manufacturing processes for composites are well adapted to the fabrication of large, complex structures. This allows consolidation of parts, which can reduce manufacturing costs.

In recent years, carbon fibers with thermal conductivities much greater than that of copper have been developed. These reinforcements are being used in polymer, metal, and carbon matrices to create composites with high thermal conductivities that are being used in applications for which thermal management is important. Discontinuous versions of these fibers are also being incorporated in thermoplastic injection molding compounds, improving their thermal conductivity by as much as two orders of magnitude or more. This greatly expands the range of products for which injection molded polymers can be used.

Composites are important materials that are now used widely, not only in the aerospace industry, but also in a large and increasing number of commercial applications, including

- Internal combustion engines
- Machine components
- Thermal control and electronic packaging
- Automobile, train, and aircraft structures
- Mechanical components, such as brakes, drive shafts, and flywheels
- Tanks and pressure vessels
- Dimensionally stable components
- Process industries equipment requiring resistance to high-temperature corrosion, oxidation, and wear
- Offshore and onshore oil exploration and production
- Marine structures
- Sports and leisure equipment
- Biomedical devices
- Civil engineering structures

The resulting increases in production volumes have helped to reduce material prices, increasing their attractiveness in cost-sensitive applications.

It should be noted that biological structural materials occurring in nature are typically composites. Common examples are wood, bamboo, bone, teeth, and shell. Furthermore, use of artificial composite materials is not new. Bricks made from straw-reinforced mud were employed in biblical times. This material also has been widely used in the American Southwest for centuries, where it is known as *adobe*. In current terminology, it would be called an *organic fiber-reinforced ceramic matrix composite.*

To put things in perspective, it is important to consider that modern composites technology is only several decades old. This is an extremely short period of time compared with other materials, such as metals, which go back millennia. In the future, improved and new materials and processes can be expected. It is also likely that new concepts will emerge, such as greater functionality, including integration of electronics, sensors, and actuators.

There is no universally accepted definition of a composite material. A good description of a composite is a material consisting of two or more distinct materials bonded together.[1] This differentiates composites from materials such as alloys.

Solid materials can be divided into four categories (polymers, metals, ceramics, and carbon). We consider carbon as a separate class because of its unique characteristics. We find both reinforcements and matrix materials in all four categories. This results in the potential for a limitless number of new material systems having unique properties that cannot be obtained with any single monolithic material. Table 5.1 shows the types of material combinations that are now in use.

Composites are usually classified by the type of material used for the matrix. The four primary categories of composites are polymer matrix composites (PMCs), metal matrix composites (MMCs), ceramic matrix composites (CMCs), and carbon matrix composites (CAMCs). The last category, CAMCs, includes carbon/carbon composites (CCCs), which consist of carbon matrices reinforced with carbon fibers. For decades, CCCs were the only significant type of CAMC. However, there are now other types of composites utilizing a carbon matrix. Notable among these is silicon carbide fiber-reinforced carbon, which is being used in military aircraft gas turbine engine components.

The characteristics of the four classes of matrix materials used in composites differ radically. Table 5.2 presents properties of selected matrix materials from the four classes.

TABLE 5.1 Types of Composite Materials

	Reinforcement			
Matrix	Polymer	Metal	Ceramic	Carbon
Polymer	✔	✔	✔	✔
Metal	✔	✔	✔	✔
Ceramic	✔	✔	✔	✔
Carbon		✔	✔	✔

TABLE 5.2 Properties of Selected Matrix Materials

Material	Class	Density g/cm^3 (Pci)	Modulus GPa (Msi)	Tensile strength MPa (Ksi)	Tensile failure strain (%)	Thermal conductivity W/mK (Btu/hr·ft·°F)	Coefficient of thermal expansion ppm/K (ppm/°F)
Epoxy	Polymer	1.8 (0.065)	3.5 (0.5)	70 (10)	3	0.1 (0.06)	60 (33)
Aluminum (6061)	Metal	2.7 (0.098)	69 (10)	300 (43)	10	180 (104)	23 (13)
Titanium (6Al-4V)	Metal	4.4 (0.16)	105 (15.2)	1100 (160)	10	16 (9.5)	9.5 (5.3)
Silicon carbide	Ceramic	2.9 (0.106)	520 (75)	–	<0.1	81 (47)	4.9 (2.7)
Alumina	Ceramic	3.9 (0.141)	380 (55)	–	<0.1	20 (12)	6.7 (3.7)
Glass (borosilicate)	Ceramic	2.2 (0.079)	63 (9)	–	<0.1	2 (1)	5 (3)
Amorphous carbon	Carbon	1.8 (0.065)	20 (3)	–	<0.1	5–90 (3–50)	2 (1)

Note that the densities, moduli, strengths, and failure strains differ greatly. These and other differences result in composite materials that have very dissimilar characteristics.

Composites are now important commercial and aerospace materials.[2] At this time, PMCs are the most widely used composites. MMCs are employed in a significant and increasing number of commercial and aerospace applications, such as automobile engines, electronic packaging, cutting tools, circuit breaker contact pads, high-speed and precision machinery, and aircraft structures. CCCs are used in high-temperature, lightly loaded applications, such as aircraft brakes, rocket nozzles, glass processing equipment, and heat treatment furnace support fixtures and insulation. Although CMCs are not as widely used at this time, there are notable applications that are indicative of their great potential.

The main types of reinforcements used in composite materials include aligned continuous fibers, discontinuous fibers, whiskers (elongated single crystals), particles, and numer-

ous forms of fibrous architectures produced by textile technology, such as fabrics and braids.[3–5]

Increasingly, designers are using hybrid composites that combine different types of reinforcements and reinforcement forms to achieve greater efficiency and reduce cost.[6,7] For example, fabrics and unidirectional tapes are often used together in structural components. In addition, carbon fibers are combined with glass or aramid to improve impact resistance. Laminates combining composites and metals, such as "Glare," which consists of layers of aluminum and glass fiber-reinforced epoxy, are being used in aircraft structures to improve fatigue resistance.

Composites are strongly heterogeneous materials. That is, the properties of a composite vary considerably from point to point in the material, depending on the material phase in which the point is located. Monolithic ceramics, metallic alloys, and intermetallic compounds are usually considered to be homogeneous materials, as a first approximation.

Many artificial composites, especially those reinforced with fibers, are anisotropic, which means their properties vary with direction (the properties of isotropic materials are the same in every direction). This is a characteristic they share with a widely used natural fibrous composite, wood. As for wood, when structures made from artificial fibrous composites are required to carry load in more than one direction, they are typically used in laminated form. It is worth noting that the strength properties of some metals also vary with direction. This is typically related to the manufacturing process, such as rolling.

With the exception of MMCs, composites do not display plastic behavior as monolithic metals do, which makes composites more sensitive to stress concentrations. However, the absence of plastic deformation does not mean that composites should be considered brittle materials like monolithic ceramics. The heterogeneous nature of composites results in complex failure mechanisms that impart toughness. Fiber-reinforced materials have been found to produce durable, reliable structural components in countless applications.[2] For example, PMCs have been used in production boats, electrical equipment, and solid rocket motors since the 1950s, and in aircraft since the early 1970s. The technology has progressed to the point where the entire empennage (tail section) of the Boeing 777 is made of carbon/epoxy.

5.2 Comparative Properties of Composite Materials

There are a large and increasing number of materials that fall into each of the four types of composites, so generalization is difficult. However, as a class of materials, composites tend to have the following characteristics:

- Tailorable mechanical and physical properties
- High strength
- High modulus
- Low density
- Excellent resistance to fatigue, creep, creep rupture, corrosion, and wear

Composites are available with tailorable thermal and electrical conductivities that range from very low to very high. Composites are available with tailorable coefficients of thermal expansion (CTEs) ranging from –2 to + 60 ppm/K (1 to 30 ppm/°F).

As for monolithic materials, each of the four classes of composites has its own particular attributes. For example, CMCs tend to have particularly good resistance to corrosion, oxidation, and wear, along with high-temperature capability.

There are many types of fibrous and particulate reinforcements used in composite materials, including carbon, aramid, glasses, oxides, boron, and so on.[8–18] Carbon, glass, and aramid fibers are probably the most important at this time.

There are dozens of different types of commercial carbon fibers. The stiffest versions have moduli of 965 GPa (140 Msi). Strengths top out at 7 GPa (1,000,000 lb/in^2). Carbon fibers are made from several types of precursor materials, polyacrylonitrile (PAN), petroleum pitch, coal tar pitch, and rayon. Except for a few applications initially developed many years ago, rayon-based carbon fibers are no longer of great importance. Characteristics of the two types of pitch-based fibers tend to be similar but very different from those made from PAN. The key types of carbon fibers are standard modulus (SM) PAN, intermediate modulus (IM) PAN, ultrahigh modulus (UHM) PAN, and ultrahigh modulus (UHM) pitch. The strongest UHS carbon fibers are forms of intermediate modulus (IM) fibers. Carbon fiber cost varies greatly. The least expensive industrial versions are now available for about USD 10/kg (USD 5/lb).

The outstanding mechanical properties of composite materials has been a key reason for their extensive use in structures. However, composites also have important physical properties, especially low, tailorable coefficient of thermal expansion (CTE) and high thermal conductivity, which are key reasons for their selection in an increasing number of applications. Key examples are electronic packaging and thermal management.[19–21]

Many composites, such as PMCs reinforced with carbon and aramid fibers, and silicon carbide particle-reinforced aluminum, have low CTEs, which are advantageous in applications requiring dimensional stability. Examples include spacecraft structures, instrument structures, and optical benches.[22] By appropriate selection of reinforcements and matrix materials, it is possible to produce composites with near-zero CTEs.

Coefficient of thermal expansion tailorability provides a way to minimize thermal stresses and distortions that often arise when dissimilar materials are joined. For example, the CTE of silicon carbide particle-reinforced aluminum depends on particle content. By varying the amount of reinforcement, it is possible to match the CTEs of a variety of key engineering materials, such as steel, titanium, and alumina (aluminum oxide).[23]

The ability to tailor CTE is important in many applications. For example, titanium fittings are often used with carbon/epoxy (C/Ep) structures instead of aluminum, because the latter has a much larger CTE that can cause high thermal stresses under thermal cycling. Another application for which CTE is important is electronic packaging, because thermal stresses can cause failure of ceramic substrates, semiconductors, and solder joints.

A unique and increasingly important property of some composites is exceptionally high thermal conductivity. This is leading to increasing use of composites in applications for which heat dissipation is a key design consideration. In addition, the low densities of composites make them particularly advantageous in thermal control applications for which weight is important. An important recent breakthrough is the development of injection molded PMCs with thermal conductivities as high as 100 W/m·K (58 Btu/hr·ft·°F). This is discussed in Sec. 5.5.

There are a large and increasing number of thermally conductive PMCs, MMCs, and CAMCs. One of the most important types of reinforcements for these materials is pitch fibers.[11] PAN-based fibers have relatively low thermal conductivities.[10] However, pitch-based fibers with thermal conductivities more than twice that of copper are commercially available. These ultrahigh thermal conductivity (UHK) reinforcements also have very high stiffnesses and low densities. Fibers made by chemical vapor deposition (CVD), also called *vapor-grown fibers,* have reported thermal conductivities as high as 2000 W/m·K (1160 Btu/hr·ft·°F), about five times that of copper.[24] Fibers made from another form of carbon, diamond, also have the potential for thermal conductivities in this range. PMCs and CCCs reinforced with UHK carbon fibers are being used in a wide range of applica-

tions, including spacecraft radiators, battery sleeves, electronic packaging, and motor enclosures. Applications for specific materials are discussed later in Secs. 5.5 through 5.9.

5.3 Overview of Mechanical and Physical Properties

Initially, the excellent mechanical properties of composites were the main reason for their use.[25] However, there are an increasing number of applications for which the unique and tailorable physical properties of composites are key considerations. For example, the extremely high thermal conductivity and tailorable CTE of some composite material systems are leading to their increasing use in electronic packaging. Similarly, the extremely high stiffness, near-zero CTE, and low density of carbon fiber-reinforced polymers have made these composites the materials of choice in a variety of applications, including spacecraft structures, antennas, and optomechanical system components such as telescope metering structures and optical benches.

Composites are complex, heterogeneous, and often anisotropic material systems. Their properties are affected by many variables, including *in situ* constituent properties; reinforcement form, volume fraction, and geometry; properties of the interphase, the region where the reinforcement and matrix are joined (also called the *interface*); and void content. The process by which the composite is made affects many of these variables.[26] Composites containing the same matrix material and reinforcements, when combined by different processes, may have very different properties.

Several other important things must be kept in mind when considering composite properties. For one, most composites are proprietary material systems made by proprietary processes. There are few industry or government specifications for composites as there are for many structural metals. However, this is also the case for many monolithic ceramics and polymers, which are widely used engineering materials. Despite their inherently proprietary nature, there are some widely used composite materials made by a number of manufacturers that have similar properties. Notable examples are standard modulus (SM) and intermediate modulus (IM) carbon fiber-reinforced epoxy.

Another critical issue is that properties are sensitive to the test methods by which they are measured, and there are many different test methods used throughout the industry.[27,28] Furthermore, test results are very sensitive to the skill of the technician performing them. Because of these factors, it is very common to find significant differences in reported properties of what is nominally the same composite material.

There is often a great deal of confusion among those unfamiliar with composites about the effect of reinforcement form. The properties of composites are very sensitive to reinforcement form, volume fraction, and internal reinforcement geometry.

It is important to keep in mind that one of the key problems with using discontinuous fiber reinforcement is that it is often difficult to control fiber orientation. For example, material flow can significantly align fibers in some regions. This affects all mechanical and physical properties, including modulus, strength, CTE, thermal conductivity, etc. For example, if there is significant flow in a region, strength properties perpendicular to the fiber direction in this area may be low. This has been a frequent source of component failures.

Traditional fabric reinforcements have fibers oriented at 0 and 90°. For the sake of completeness, we note that triaxial fabrics, which have fibers at 0°, +60°, and –60°, are now commercially available. Composites using a single layer of this type of reinforcement are approximately quasi-isotropic, which means that they have the same in-plane elastic (but not strength) properties in every direction. Their thermal conductivity and CTE are also approximately isotropic in the plane of the fabric.

5.4 Manufacturing Considerations

Composites also offer a number of significant manufacturing advantages over monolithic metals and ceramics. For example, fiber-reinforced polymers and ceramics can be fabricated in large, complex shapes that would be difficult or impossible to make with other materials.[26,29–31] The ability to fabricate complex shapes allows consolidation of parts, which reduces machining and assembly costs. Some processes allow fabrication of parts in their final shape (net shape) or close to their final shape (near-net shape), which also produces manufacturing cost savings. The relative ease with which smooth shapes can be made is a significant factor in the use of composites in boats, aircraft, and other applications for which aerodynamic considerations are important. Manufacturing considerations for each of the four classes of composites are discussed in the following sections, along with their properties.

5.5 Polymer Matrix Composites

Chapter 4 presents a thorough discussion of polymer matrix composites focused primarily on structural applications. This chapter emphasizes physical properties, but mechanical properties are presented for completeness.

The high thermal conductivities of some PMCs has led to their increasing use in applications like spacecraft structures and electronic packaging components, e.g., printed circuit board heat sinks, heat spreaders, and heat sinks used to cool microprocessors. The addition of thermally conductive carbon fibers and ceramic particles to thermoplastics has opened the door to use of injection molded parts for which plastics previously could not be used because of their low thermal conductivities. We consider some examples in this section.

There are important issues that must be discussed before presenting composite properties. The traditional structural materials are primarily metal alloys for most of which there are industry and government standards. The situation is very different for composites. Most reinforcements and matrices are proprietary materials for which there are no standards. In addition, many processes are proprietary. This is similar to the current situation for most polymers and ceramics. The matter is further complicated by the fact that there are many test methods in use to measure mechanical and physical properties.[27,28] As a result, there are often conflicting material property data in the usual sources used by engineers, published papers, and manufacturers' literature. The data presented in this chapter represent a carefully evaluated distillation of information from many sources. However, in view of the uncertainties discussed, the properties presented in this chapter should be considered approximate values.

Polymers are relatively weak, low-stiffness materials with low thermal conductivities and high coefficients of thermal expansion. To obtain materials with mechanical properties that are acceptable for structural applications, it is necessary to reinforce them with continuous or discontinuous fibers. The addition of ceramic or metallic particles to polymers results in materials that have increased modulus. As a rule, strength typically does not increase significantly and may actually decrease. However, there are many particle-reinforced polymers used in electronic packaging, primarily because of their physical properties. For these applications, ceramic particles such as alumina, aluminum nitride, boron nitride, and even diamond are added to obtain an electrically insulating material with higher thermal conductivity and lower CTE than the monolithic base polymer. Metallic particles such as silver and aluminum are added to create materials that are both electrically and thermally conductive. These materials have replaced lead-based solders in some applications. There are also magnetic composites made by incorporating ferrous or permanent magnet particles in various polymers. A common example is magnetic tape used to record audio and video.

Polymer matrices generally are relatively weak, low-stiffness, viscoelastic materials. They also have very low thermal and electrical conductivities. The strength and stiffness of PMCs come primarily from the fiber phase.

In a vacuum, resins outgas water and organic and inorganic chemicals, which can condense on surfaces with which they come in contact. This can be a problem in optical systems and electronic packaging. Outgassing can result in corrosion and affect surface properties critical for thermal control, such as absorptivity and emissivity. Outgassing can be controlled by resin selection and baking out the component, followed by storage in a dry environment.

For a wide range of applications, composites reinforced with continuous fibers are the most efficient structural materials at low to moderate temperatures. Consequently, we focus on them. Table 5.3 presents room-temperature mechanical properties of unidirectional polymer matrix composites reinforced with key fibers: E-glass, aramid, boron, standard modulus (SM) PAN (polyacrylonitrile) carbon, intermediate modulus (IM) PAN carbon, ultrahigh modulus (UHM) PAN carbon, ultrahigh modulus (UHM) pitch carbon, and ultrahigh thermal conductivity (UHK) pitch carbon. The fiber volume fraction is 60 percent, a typical value.

The properties presented in Table 5.3 are representative of what can be obtained at room temperature with a well made PMC employing an epoxy matrix. Epoxies are widely used, provide good mechanical properties, and can be considered a reference matrix material. Properties of composites using other resins may differ from these. This has to be examined on a case-by-case basis.

The properties of PMCs, especially strengths, depend strongly on temperature. The temperature dependence of polymer properties differs considerably. This is also true for different epoxy formulations, which have different cure and glass transition temperatures.

The properties shown in Table 5.3 are axial, transverse and shear moduli, Poisson's ratio, tensile and compressive strengths in the axial and transverse directions, and in-plane shear strength. The Poisson's ratio presented is called the *major Poisson's ratio*. It is defined as the ratio of the magnitude of transverse strain divided by the magnitude of axial strain when the composite is loaded in the axial direction. Note that transverse moduli and strengths are much lower than corresponding axial values.

Carbon fibers display nonlinear stress-strain behavior. Their moduli increase under increasing tensile stress and decrease under increasing compressive stress. This makes the method of calculating modulus critical. Various tangent and secant definitions are used throughout the industry, contributing to the confusion in reported properties. For example, on one program, it was found that the fiber supplier, prepreg supplier, and end user were all using different definitions of modulus, resulting in significantly different values.

The moduli presented in Table 5.3 are based on tangents to the stress-strain curves at the origin. Using this definition, tensile and compressive moduli are usually very similar. However, this is not the case for moduli computed using various secants. These typically produce compression moduli that are significantly lower than tensile moduli, because the stress-strain curves are nonlinear.

As a result of the low transverse strengths of unidirectional laminates, they are rarely used in structural applications. The design engineer selects laminates with layers in several directions to meet requirements for strength, stiffness, buckling, etc. There are an infinite number of laminate geometries that can be selected. For comparative purposes, it is useful to consider quasi-isotropic laminates, which have the same elastic properties in all directions in the plane.

Laminates have quasi-isotropic elastic properties when they have the same percentage of layers every $180/n$ degrees, where $n \geq 3$. The most common quasi-isotropic laminates have layers that repeat every 60°, 45°, or 30°. We note, however, that strength properties in the plane are not isotropic for these laminates, although they tend to become more uniform

TABLE 5.3 Representative Mechanical Properties of Selected Unidirectional Polymer Matrix Composites, Nominal Fiber Volume Fraction = 60 Percent

Fiber	Axial modulus GPa (Msi)	Transverse modulus GPa (Msi)	In-plane shear modulus GPa (Msi)	Poisson's ratio	Axial tensile strength MPa (Msi)	Transverse tensile strength MPa (Msi)	Axial compressive strength MPa (Msi)	Transverse compressive strength MPa (Msi)	In-plane shear strength MPa (Msi)
E-glass	45 (6.5)	12 (1.8)	5.5 (0.8)	0.28	1020 (150)	40 (7)	620 (90)	140 (20)	70 (10)
Aramid	76 (11)	5.5 (0.8)	2.1 (0.3)	0.34	1240 (180)	30 (4.3)	280 (40)	140 (20)	60 (9)
Boron	210 (30)	19 (2.7)	4.8 (0.7)	0.25	1240 (180)	70 (10)	3310 (480)	280 (40)	90 (13)
Standard modulus carbon (PAN)	145 (21)	10 (1.5)	4.1 (0.6)	0.25	1520 (220)	41 (6)	1380 (200)	170 (25)	80 (12)
Ultrahigh modulus carbon (PAN)	310 (45)	9 (1.3)	4.1 (0.6)	0.20	1380 (200)	41 (6)	760 (110)	170 (25)	80 (12)
Ultrahigh modulus carbon (pitch)	480 (70)	9 (1.3)	4.1 (0.6)	0.25	900 (130)	20 (3)	280 (40)	100 (15)	41 (6)

as the angle of repetition becomes smaller. Laminates have quasi-isotropic CTE and coefficient of expansion when they have the same percentage of layers in every $180/m$ degrees, where $m \geq 2$. For example, laminates with equal numbers of layers at 0° and 90° have quasi-isotropic thermal properties.

Table 5.4 presents the mechanical properties of quasi-isotropic laminates having equal numbers of layers at 0°, +45°, -45°, and 90°. The elastic moduli of all quasi-isotropic laminates are the same for a given material. Note that the moduli and strengths are much lower than the axial properties of unidirectional laminates made of the same material. In many applications, laminate geometry is such that the maximum axial modulus and tensile and compressive strengths fall somewhere between axial unidirectional and quasi-isotropic values.

Table 5.5 presents physical properties of selected unidirectional composite materials having a typical fiber volume fraction of 60 percent. The densities of all of the materials are considerably lower than that of aluminum, and some are lower than that of magnesium. This reflects the low densities of both fibers and matrix materials. The low densities of most polymers give PMCs a significant advantage over most MMCs and CMCs, all other things being equal.

As Table 5.5 shows, all of the composites have relatively low axial CTEs. This results from the combination of low fiber axial CTE, high fiber stiffness, and low matrix stiffness. The CTE of most polymers is very high. Note that the axial CTEs of PMCs reinforced with aramid fibers and some carbon fibers are negative. This means that, contrary to the general behavior of most monolithic materials, they contract when heated. The transverse CTEs of the composites are all positive, and their magnitudes are much larger than the magnitudes of the corresponding axial CTEs. This results from the high CTE of the matrix and a Poisson effect caused by constraint of the matrix in the axial direction and lack of constraint in the transverse direction. The transverse CTE of aramid composites is particularly high, in part because the fibers have a relatively high positive radial CTE.

The axial thermal conductivities of composites reinforced with glass, aramid, boron, and a number of the carbon fibers are relatively low. In fact, E-glass and aramid PMCs are often used as thermal insulators. As Table 5.5 shows, most PMCs have low thermal conductivities in the transverse direction, as a result of the low thermal conductivities of the matrix and the fibers in the radial direction. Through-thickness conductivities of laminates tend to be similar to the transverse thermal conductivities of unidirectional composites.

Table 5.6 shows the in-plane thermal conductivities and CTEs of quasi-isotropic laminates made from the same materials as in Table 5.5. Here again, a fiber volume fraction of 60 percent is assumed.

Note that the CTEs of the quasi-isotropic composites are higher than the axial values of corresponding unidirectional composites. However, the CTEs of quasi-isotropic composites reinforced with aramid and carbon fibers are still very small. By appropriate selection of fiber, matrix, and fiber volume fraction, it is possible to obtain quasi-isotropic materials with CTEs very close to zero. The through-thickness CTEs of these laminates are typically positive and relatively large. However, this is not a significant issue for most applications. One exception is optical mirrors, for which through-thickness CTE can be an important issue.

The in-plane thermal conductivity of quasi-isotropic laminates reinforced with UHM pitch carbon fibers is similar to that of aluminum alloys, while UHK pitch carbon fibers provide laminates with a conductivity over 50 percent higher. Both materials have densities about 35 percent lower than aluminum.

As mentioned earlier, through-thickness thermal conductivities of laminates tend to be similar to the transverse thermal conductivities of unidirectional composites, which are relatively low. If laminate thickness is small, this may not be a severe limitation. However, low through-thickness thermal conductivity can be a significant issue for thick laminates and for very high thermal loads. This issue needs to be addressed on a case-by-case basis.

TABLE 5.4 Mechanical Properties of Selected Quasi-isotropic Polymer Matrix Composites, Fiber Volume Fraction = 60 Percent

Fiber	Axial modulus GPa (Msi)	Transverse modulus GPa (Msi)	In-plane modulus GPa (Msi)	Poisson's ratio	Axial tensile strength MPa (Ksi)	Transverse tensile strength MPa (Ksi)	Axial compressive strength MPa (Ksi)	Transverse compressive strength MPa (Ksi)	In-plane Shear Strength MPa (Ksi)
E-glass	23 (3.4)	23 (3.4)	9.0 (1.3)	0.28	550 (80)	550 (80)	330 (48)	330 (48)	250 (37)
Aramid	29 (4.2)	29 (4.2)	11 (1.6)	0.32	460 (67)	460 (67)	190 (28)	190 (28)	65 (9.4)
Boron	80 (11.6)	80 (11.6)	30 (4.3)	0.33	480 (69)	480 (69)	1100 (160)	1100 (160)	360 (52)
SM carbon (PAN)	54 (7.8)	54 (7.8)	21 (3.0)	0.31	580 (84)	580 (84)	580 (84)	580 (84)	410 (59)
IM carbon (PAN)	63 (9.1)	63 (9.1)	21 (3.0)	0.31	1350 (200)	1350 (200)	580 (84)	580 (84)	410 (59)
UHM carbon (PAN)	110 (16)	110 (16)	41 (6.0)	0.32	490 (71)	490 (71)	270 (39)	270 (39)	205 (30)
UHM carbon (pitch)	165 (24)	165 (24)	63 (9.2)	0.32	310 (45)	310 (45)	96 (14)	96 (14)	73 (11)
UHK carbon (pitch)	165 (24)	165 (24)	63 (9.2)	0.32	310 (45)	310 (45)	96 (14)	96 (14)	73 (11)

TABLE 5.5 Physical Properties of Selected Unidirectional Polymer Matrix Composites, Fiber Volume Fraction = 60 Percent

Fiber	Density g/cm^3 (lb/in^3)	Axial CTE $10^{-6}/K$ ($10^{-6}/°F$)	Transverse CTE $10^{-6}/K$ ($10^{-6}/°F$)	Axial thermal conductivity W/m·K (Btu/hr-ft-°F)	Transverse thermal conductivity W/m·K (Btu/hr-ft-°F)
E-glass	2.1 (0.075)	6.3 (3.5)	22 (12)	1.2 (0.7)	0.6 (0.3)
Aramid	1.38 (0.050)	–4.0 (–2.2)	58 (32)	1.7 (1.0)	0.1 (0.08)
Boron	2.0 (0.073)	4.5 (2.5)	23 (13)	2.2 (1.3)	0.7 (0.4)
SM carbon (PAN)	1.58 (0.057)	0.9 (0.5)	27 (15)	5 (3)	0.5 (0.3)
IM carbon (PAN)	1.61 (0.058)	0.5 (0.3)	27 (15)	10 (6)	0.5 (0.3)
UHM carbon (PAN)	1.66 (0.060)	–0.9 (–0.5)	40 (22)	45 (26)	0.5 (0.3)
UHM carbon (pitch)	1.80 (0.065)	–1.1 (–0.6)	27 (15)	380 (220)	10 (6)
UHK carbon (pitch)	1.80 (0.065)	–1.1 (–0.6)	27 (15)	660 (380)	10 (6)

TABLE 5.6 Physical Properties of Selected Quasi-isotropic Polymer Matrix Composites, Fiber Volume Fraction = 60 Percent

Fiber	Density g/cm^3 (lb/in^3)	Axial CTE $10^{-6}/K$ ($10^{-6}/°F$)	Transverse CTE $10^{-6}/K$ ($10^{-6}/°F$)	Axial thermal conductivity W/m-K (Btu/hr-ft-°F)	Transverse thermal conductivity W/m-K (Btu/hr-ft-°F)
E-glass	2.1 (0.075)	10 (5.6)	10 (5.6)	0.9 (0.5)	0.9 (0.5)
Aramid	1.38 (0.050)	1.4 (0.8)	1.4 (0.8)	0.9 (0.5)	0.9 (0.5)
Boron	2.0 (0.073)	6.5 (3.6)	6.5 (3.6)	1.4 (0.8)	1.4 (0.8)
SM carbon (PAN)	1.58(0.057)	3.1 (1.7)	3.1 (1.7)	2.8 (1.6)	2.8 (1.6)
IM carbon (PAN)	1.61 (0.058)	2.3 (1.3)	2.3 (1.3)	6 (3)	6 (3)
UHM carbon (PAN)	1.66 (0.060)	0.4 (0.2)	0.4 (0.2)	23 (13)	23 (13)
UHM carbon (pitch)	1.80 (0.065)	–0.4 (–0.2)	–0.4 (–0.2)	195 (113)	195 (113)
UHK carbon (pitch)	1.80 (0.065)	–0.4 (–0.2)	–0.4 (–0.2)	335 (195)	335 (195)

A significant recent advance in PMC technology is the development of injection-moldable materials with much higher thermal conductivities than those available in the past. Unreinforced polymers have thermal conductivities in the range of 0.2 W/m·K (0.1 Btu/hr·ft·°F). A number of commercially available PMCs consisting of thermoplastic matrices reinforced with discontinuous carbon fibers have reported thermal conductivities ranging from 2 W/m·K (1.2 Btu/hr·ft·°F) to as high as 100 W/m·K (58 Btu/hr·ft·°F).[33] Matrices include PPS, nylon 6, polycarbonate and liquid crystal polymers. These materials are electrically conductive. Electrically insulating PMCs reinforced with thermally conductive discontinuous ceramic reinforcements have reported thermal conductivities of up to 15 W/m·K (8.8 Btu/hr·ft·°F).

Thermally conductive injection molding compounds are now being used in a significant and increasing number of applications for which unreinforced polymers and traditional injection molding compounds, because of their low thermal conductivities, are not acceptable. As discussed earlier, the mechanical and physical properties of fiber-reinforced injection molded PMCs are affected by fiber orientation induced by material flow. In addition to the ability to dissipate heat, another advantage of these PMCs is that their use results in lower temperatures and thermal gradients, which tends to reduce distortion. On the negative side, as reinforcement volume fraction increases, fracture toughness decreases. Nevertheless, these materials present the design engineer with a greater range of options for injection molded parts than in the past.

Figure 5.1 shows a stepper motor having an injection molded carbon fiber-reinforced PPS combination enclosure-heat sink.[34] In this application, the PMC replaces die-cast aluminum. Although the composite thermal conductivity of 20 W/m·K (12 Btu/hr·ft·°F), is lower than that of die cast aluminum, it meets the required maximum operating temperature of 60°C, which is acceptable for this application. The reason is that the thermal design is convection limited. The lower composite thermal conductivity only increases operating temperature by 2 to 5°C. Injection molded thermally conductive thermoplastic PMCs are also being used in electronic packaging. Figure 5.2 shows a microprocessor heat spreader.

5.6 Metal Matrix Composites

MMCs consist of metals reinforced with a variety of ceramic and carbon fibers, whiskers, and particles.[35] There are wide ranges of materials that fall in this category. An important example is a material consisting of tungsten carbide particles embedded in a cobalt matrix, which is used extensively in cutting tools and dies. This composite, often referred to as a

Figure 5.1 Thermally conductive carbon fiber-reinforced PPS injection molded stepper motor enclosure—heat sink. *(Courtesy of Cool Polymers, Inc.)*

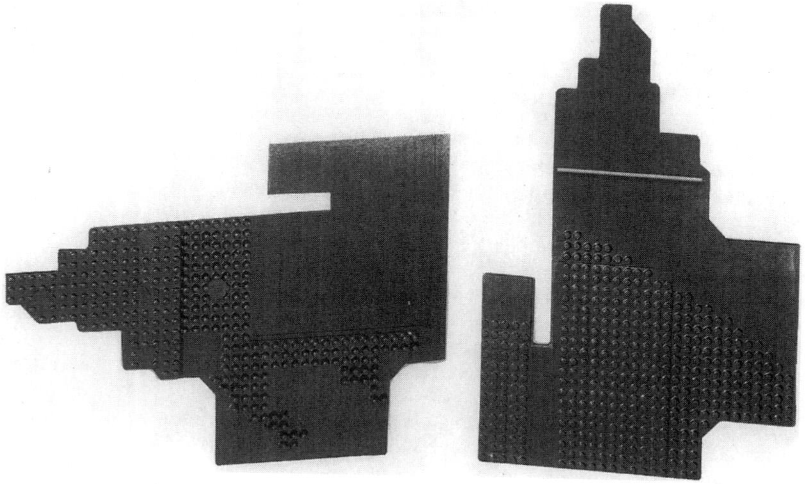

Figure 5.2 Injection-molded carbon fiber-reinforced thermoplastic microprocessor heat spreader. *(Courtesy of Cool Polymers, Inc.)*

cermet, cemented carbide, or simply (but incorrectly) *tungsten carbide,* has much better fracture toughness than monolithic tungsten carbide, which is a brittle ceramic material. Another interesting MMC, tungsten carbide particle-reinforced silver, is a key circuit breaker contact pad material. Here, the composite provides good electrical conductivity and much greater hardness and wear resistance than monolithic silver, which is too soft to be used in this application. Ferrous alloys reinforced with titanium carbide particles have been used for many years in numerous aerospace and commercial production applications, including dies, engine valves, and aircraft fuel pumps. Compared to the monolithic base metals, they offer better wear resistance, higher stiffness, and lower density.

Another notable commercial MMC application is the Honda Prelude engine block, which has cylinder walls reinforced with a combination of aluminum oxide (alumina) and carbon fibers, enabling elimination of cast iron cylinder liners.[36] MMCs are also being used in high-speed electronics manufacturing equipment and in photolithography tables and other equipment used for production of microprocessor chips. Other applications include aircraft structures, aircraft engine fan exit guide vanes, and automobile and train brake rotors.[37]

One of the most important uses for MMCs is in electronic packaging and thermal management.[24,36,38] For example, silicon carbide particle-reinforced aluminum, often called *Al/SiC* in the electronics industry, is being used in high-volume production parts such as microprocessor lids and power modules for hybrid electric vehicles such as the Toyota Prius. Other MMCs used in packaging are carbon fiber reinforced-aluminum and copper, beryllium oxide (beryllia) particle-reinforced beryllium and silicon/aluminum. Here, the advantages are high stiffness, high thermal conductivity, and low density and CTE. Two traditional packaging materials, copper/tungsten and copper/molybdenum, can also be considered MMCs. The CTEs of these composites can be tailored by varying the ratio of the two constituent metals. A major drawback is that both have high densities.

Monolithic metallic alloys are among the most widely used structural materials. By reinforcing them with continuous fibers, discontinuous fibers, whiskers, and particles, new materials are created with enhanced or modified properties, such as higher strength and

stiffness, better wear resistance, lower CTE, etc. In some cases, the improvements are dramatic.

The greatest increases in strength and modulus are achieved with continuous fibers, at least in the direction parallel to the fibers, called the *axial* or *longitudinal* direction. As for PMCs, transverse properties are dominated by the properties of the matrix and interface. The latter is more properly referred to as the *interphase* region. However, because the matrices are in themselves structural materials, transverse strength properties are frequently great enough to permit use of unidirectional MMCs in some structural applications, which is usually not possible for PMCs. One example is the boron fiber-reinforced aluminum struts used on the Space Shuttle Orbiter.[39]

One of the major advantages of MMCs reinforced with continuous fibers over PMCs is that many, if not most, unidirectional MMCs have much greater transverse strengths that allow them to be used in a unidirectional configuration. In general, the axial moduli and strengths of the composites are much greater than those of the monolithic base metals used for the matrices. However, MMC transverse strengths are often somewhat lower than those of the parent matrix materials.

The key particle-reinforced MMCs, include titanium carbide-reinforced steel, aluminum reinforced with silicon carbide and with alumina particles, titanium carbide particle-reinforced titanium, and titanium boride-reinforced titanium.

Aluminum reinforced with silicon carbide particles, the focus of this section, is arguably the most important of the newer types of MMCs. The low cost of the aluminum matrix and silicon carbide particle makes these composites of particular interest. There are wide ranges of materials falling in this category. They are made by a variety of processes, which are discussed later in this section. Properties depend on the type of particle, particle volume fraction, matrix alloy, and the process used to make them. Table 5.7 presents representative composite properties for three particle volume fractions, 25, 55, and 70 percent. In general, as particle volume fraction increases, modulus and yield strength increase, and fracture toughness, tensile ultimate strain, and CTE decrease. Particle reinforcement also improves elevated temperature strength properties and fatigue resistance. The ability to tailor CTE by varying particle volume fraction is a key attribute of these materials.

There are a variety of processes to make silicon carbide particle-reinforced aluminum, including powder metallurgy, stir casting, and pressure and pressureless infiltration. The last two, as well as remelt casting, can make net shape or near-net shape parts.

5.7 Carbon Matrix Composites

Carbon matrix composites (CAMCs) consist of a carbon matrix reinforced with any combination of fibers, whiskers, or particles.[19–21] For many years, the only significant CAMCs were carbon/carbon composites (CCCs), in which the reinforcements are discontinuous or continuous carbon fibers. In the last few years, a new proprietary carbon matrix material system was developed that has a silicon carbide fiber reinforcement. This material is now being used for engine flaps on a military aircraft engine. One of the key reported advantages of this new material is that it has a higher CTE than CCCs, reducing the tendency of protective ceramic coatings to crack. The focus of this section is on CCCs.

CCCs are used in a variety of applications, including electronic packaging, spacecraft radiator panels, rocket nozzles, reentry vehicle nose tips, the Space Shuttle Orbiter leading edges and nose cap, aircraft brakes, heat treating furnaces, and glass-making equipment.

As for PMCs, there are many different CCC materials having widely different mechanical and physical properties. The primary advantages of CCCs are

- High strength compared with competing materials at very high temperatures
- High stiffness

TABLE 5.7 Properties of Silicon Carbide Particle-Reinforced Aluminum

Property	Aluminum 6061-T6	Titanium 6Al-4V	Steel 4340	Composite particle volume fraction (%)		
				25	55	70
Modulus, GPa (Msi)	69 (10)	113 (6.5)	200 (29)	114 (17)	186 (27)	265 (38)
Tensile yield strength, MPa (Ksi)	275 (40)	1000 (145)	1480 (215)	400 (58)	495 (72)	–
Tensile ultimate strength, MPa (Ksi)	310 (45)	1100 (160)	1790 (260)	485 (70)	530 (77)	225 (33)
Elongation (%)	15	5	10	3.8	0.6	0.1
Specific modulus, GPa	5	26	26	40	63	88
CTE, 10^{-6}/K (10^{-6}/°F)	23 (13)	9.5 (5.3)	12 (6.6)	16.4 (9.1)	10.4 (5.8)	6.2 (3.4)
Density, g/cm^3 (lb/in^3)	2.77 (0.10)	4.43 (0.16)	7.76 (0.28)	2.88 (0.104)	2.96 (0.107)	3.00 (0.108)

- Ablation resistance
- High thermal conductivity (for some systems)
- Low CTEs
- Low density
- Absence of outgassing

In addition, CCCs are less brittle than monolithic carbon.

The primary disadvantages are

- Susceptibility to oxidation at temperatures above about 370 to 500°C (700 to 930°F)
- Low interlaminar (through-thickness) tensile and shear strengths for materials with 2D reinforcement
- Microcracking at low stresses in some directions for 3D composites
- High cost of many systems

The variables affecting properties include type of fiber, reinforcement form, geometry, and volume fraction and matrix characteristics. Because of the low interlaminar strength properties of CCCs, many applications, particularly those with thick walls, often use three-dimensional reinforcement.

As mentioned earlier, one of the most significant limitations of CCCs is oxidation. Addition of oxidation inhibitors to the matrix and protective coatings raises the threshold substantially. In inert atmospheres, CCCs retain their properties to temperatures as high as 2400°C (4300°F).

Carbon matrices are typically weak, brittle, low-stiffness materials. As a result, transverse and through-thickness elastic moduli and strength properties of unidirectional CCCs are low. Because of this, three-dimensional reinforcement forms are used in some applications.

As for all composites, properties of CCCs depend on those of the reinforcement, matrix, fiber-matrix interphase, and the process by which they are made. Table 5.8 presents mechanical and physical properties of CCCs reinforced with unbalanced fabrics having a warp-to-fill ratio of 4:1. The reinforcements are two types of carbon fibers that have very high thermal conductivities, P120, and K1100. For comparison, copper has a thermal conductivity of about 400 W/m·K (230 Btu/hr·ft·°F). The combination of high thermal conductivity and low density makes CCCs attractive candidates for thermal management and electronic packaging. In addition, CCCs have very low CTEs, leading to their use as thermal doublers with carbon fiber-reinforced PMC structures. The unique combination of properties possessed by CCCs, combined with a lack of outgassing, also makes them attractive for optical subsystems. There are a variety of fibrous carbonaceous materials at various stages of development that have even higher thermal conductivities than the CCCs reported here. For example, highly oriented pyrolytic graphite, also called *thermal pyrolytic graphite,* has a reported in-plane thermal conductivity of as high as 1700 W/m·K (980 Btu/hr·ft·°F).

There are two basic types of processes used to make CAMCs.[19–21] The first is chemical vapor infiltration (CVI). CVI is a process in which gaseous chemicals are reacted or decomposed, depositing a solid material on a fibrous preform. In the case of CAMCs, hydrocarbon gases like methane and propane are broken down, and the material deposited is the carbon matrix. The second class of processes involves infiltration of a preform with polymers or pitches, which are then converted to carbon by pyrolysis (heating in an inert atmosphere).[19–21] After pyrolysis, the composite is heated to high temperatures to graphitize the matrix. To minimize porosity, the process is repeated until a satisfactory density is achieved. This is called *densification.* Common matrix precursors are phenolic and furan resins, and pitches derived from coal tar and petroleum.

TABLE 5.8 Properties of Carbon/Carbon Composites Reinforced with Unbalanced Fabrics Having a Warp-to-Fill Ratio of 4:1 (*Source:* BF Goodrich)

Fiber	Density g/cm^3 (lb/in^3)	Warp modulus GPa (Msi)	Warp tensile strength MPa (Ksi)	Through-thickness tensile strength MPa (Ksi)	Warp compression strength MPa (Ksi)	Interlaminar shear strength MPa (Ksi)	Warp thermal conductivity W/m.K (Btu/hr·ft·°F)	Through-thickness thermal conductivity W/m·K (Btu/hr·ft·°F)
P-120	1.6–1.8 (0.058–0.065)	290–390 (42–57)	255–475 (37–69)	2.3–4.7 (0.33–0.68)	–	9.0–26 (1.3–3.8)	349–412 (202–238)	24–47 (14–27)
K-1100	1.8 (0.065)	303–344 (44–50)	338–420 (49–61)	2.8–3.9 (0.40–0.57)	124–145 (18–21)	9.6–16 (1.4–2.3)	400–483 (231–279)	56–72 (32–42)

5.8 Ceramic Matrix Composites

Ceramic matrix composites (CMCs) can be thought of as an improved form of carbon matrix composite in which the carbon matrix is replaced with ceramics that are stronger and much more resistant to oxidation.[40–42] CMCs employ a variety of reinforcements including continuous fibers, discontinuous fibers, whiskers, and particles. Continuous fibers provide the best properties. There are many different types of CMCs, and they are at various stages of development.

The key advantage of ceramic matrix composites is that, when properly designed and manufactured, they have many of the advantages of monolithic ceramics, such as much lower density than high-temperature metals, but are much more durable. That is, CMCs have higher effective fracture toughnesses, so they are less susceptible to failure when subjected to mechanical and thermal shock. As a consequence, it is possible to consider CMCs for applications where they are subjected to moderate tensile loads. However, CMCs are the most complex of all types of composites, and CMC technology is less developed than that of PMCs, MMCs, and CAMCs.

CMCs are being used in a number of commercial production applications. One of the most successful is silicon carbide whisker-reinforced alumina cutting tool inserts, which have greater fracture toughness and are therefore more durable than monolithic ceramics. Another application is silicon carbide whisker-reinforced aluminum nitride crucibles, which are used for casting molten aluminum. In this application, the key advantage of the CMC over monolithic ceramics is thermal shock resistance. Silicon carbide particle-reinforced alumina is being used in slurry pumps because of the good durability and wear resistance of this material. In this application, the process makes it possible to fabricate reliable, complex parts that would be hard to make out of monolithic ceramics. Other high-temperature CMC applications include coal-fired power plant candle filters used for particulate removal, natural gas burner elements, and U-tubes. In addition, there is a wide variety of candidate applications, including stationary gas turbine combustor liners and shrouds, abradable rim seals, reverberatory screens, particle separators, tube shields, recuperators, turbine tip shoes, pipe hangers, heat treating furnace fans, hot gas filters, and natural gas burner elements.

Aerospace applications of ceramic matrix composites to date have been limited. Perhaps the most significant are the aircraft engine flaps used on a French fighter. There are two types. Both use silicon carbide matrices. One is reinforced with carbon fibers, and the other with a multifilament silicon carbide fiber. Another application is a missile diverter thruster made of carbon fiber-reinforced silicon carbide. Again, the process used to make this part is CVI. The Space Shuttle Orbiter thermal protection system (TPS) makes extensive use of tiles composed of a three-dimensional network of discontinuous oxide fibers with silicate surface layers. While there is no continuous matrix for most of the tile, the surface region is a form of CMC. In a sense, this can be considered to be a type of functionally graded material.

As a class of materials, monolithic ceramics are characterized by high stiffness and hardness; resistance to wear, corrosion, and oxidation; and high-temperature operational capability. However, they also have serious deficiencies that have severely limited their use in applications that are subjected to significant tensile stresses. The fundamental problem is that ceramics have very low fracture toughnesses, which makes them very sensitive to the presence of small flaws. This results in great strength scatter and poor resistance to thermal and mechanical shock. Civil engineers recognized this deficiency long ago, and, in construction, ceramic materials like stone and concrete are rarely used to carry tensile loads. In concrete, this function has been relegated to reinforcing bars made of steel or, more recently, PMCs. An important exception has been in lightly loaded structures where dispersed reinforcing fibers of asbestos, steel, glass, and carbon allow modest tensile stresses to be supported.

The addition of continuous fibers to a ceramic matrix can significantly change the failure mode. Monolithic ceramics have linear stress-strain curves and fail catastrophically at low strain levels. However, CMCs display nonlinear stress-strain behavior with much more area under the curve, indicating that more energy is absorbed during failure and that the material has a less catastrophic failure mode.

Reinforcements that have been used for CMCs include continuous fibers, discontinuous fibers, whiskers, and particles. Key continuous fibers used in CMCs include carbon, silicon carbide-based, alumina-based, alumina-boria-silica, quartz, and alkali-resistant glass. Steel wires are also used. Discontinuous CMC fibers are primarily silica based. Silicon carbide is the key whisker reinforcement. Particulate reinforcements include silicon carbide, zirconium carbide, hafnium carbide, hafnium diboride, and zirconium diboride.

A large number of ceramics have been considered for matrix materials, including alumina, glass, glass-ceramic, mullite (aluminum silicate), cordierite (magnesium aluminosilicate), Yttrium alumina garnet (YAG), barium aluminosilicate (BAS), barium magnesium aluminosilicate (BMAS), calcium aluminosilicate (CAS), barium and strontium aluminosilicate (BSAS, or celsian), "Blackglas" (silicon oxycarbide or Si-O-C), silicon nitride, silicon carbide, silicon nitride-bonded silicon carbide, silicon carbide and silicon, hafnium carbide, tantalum carbide, zirconium carbide, hafnium diboride, zirconium diboride, and molybdenum disilicide.

The most mature CMCs consist of silicon carbide matrices reinforced with silicon carbide-based fibers (SiC/SiC) and silicon carbide reinforced with carbon fibers (C/SiC).

Table 5.9 presents room-temperature properties of one of these materials, enhanced silicon carbide fiber-reinforced silicon carbide, which has a SiC matrix containing proprietary additives that improve oxidation resistance. The composite is reinforced with a plain weave fabric woven from CG "Nicalon" silicon carbide fibers.

TABLE 5.9 Properties of Enhanced Silicon Carbide Fiber-Reinforced Silicon Carbide Ceramic Matrix Composites (*Source:* Allied Signal)

Property	Units	
Density	g/cc (Pci)	2.30 (0.083)
Initial axial tensile modulus	GPa (Msi)	140 (20)
In-plane shear modulus	GPa (Msi)	70 (10)
Through-thickness shear modulus	GPa (Msi)	28 (4)
Axial tensile strength	MPa (Ksi)	225 (33)
Axial compressive strength	MPa (Ksi)	500 (73)
In-plane shear strength	MPa (Ksi)	180 (26)
Interlaminar shear strength	MPa (Ksi)	30 (4.3)
In-plane CTE	PPM/°C (ppm/°F)	5.0 (2.8)
In-plane thermal conductivity	W/m·K (BTU/hr·ft·°F)	17 (10)

As a consequence of the low fiber modulus and matrix porosity, the composite modulus is actually lower than that of good quality monolithic SiC. The key benefit of the reinforcement is that it improves fracture toughness, preventing catastrophic failure by crack propagation, which is a limitation of monolithic ceramics. The in-plane strength properties

are in the range of those of aluminum alloys. However, the through-thickness strengths are relatively low, as they are for PMCs and CAMCs. To put things in perspective, the densities of C/SiC and SiC/SiC are low, compared with those of high-temperature metals, so that these materials are still of considerable interest where characteristics such as resistance to mechanical and thermal shock, corrosion resistance, and weight reduction are important.

As for other classes of composite materials, there are many processes that can be used to make CMCs.[42] Key considerations in process selection are porosity and reactions among reinforcements, reinforcement coatings, and matrices. The most important processes for making CMCs at this time are chemical vapor infiltration, melt infiltration, preceramic polymer infiltration and pyrolysis (PIP), slurry infiltration, sol-gel, hot pressing, and hot isostatic pressing. In addition, there are a number of reaction-based processes, which include reaction bonding and direct metal oxidation ("Dimox").

5.9 Acknowledgements

The author is grateful for helpful discussions with Dr. James Miller of Cool Polymers, Inc. Portions of this article were taken from Ref. 25 and appear courtesy of the publisher.

References

1. Introduction, *Concise Encyclopedia of Composite Materials*, Revised Edition, A. Kelly, Ed., Pergamon Press, Oxford, 1994.
2. *Comprehensive Composite Materials,* Vol. 6: Design and Applications, M. G. Bader, Keith K. Kedward, and Yoshihiro Sawada, Vol. Eds., Anthony Kelly and Carl Zweben, Editors-in-Chief, Pergamon Press, Elsevier Science Ltd., Oxford, 2000.
3. F. K. Ko, Advanced Textile Structural Composites, *Advanced Topics in Materials Science and Engineering,* J. I. Moran-Lopez and J. M. Sanchez, Eds., Plenum Press, New York, 1993.
4. W. S. Smith and Carl Zweben, Properties of Constituent Materials, Section 2, *Engineered Materials Handbook*, Vol. 1, Composites, ASM International, Materials Park, Ohio, 1987, pp. 43–104.
5. *Comprehensive Composite Materials*, Vol. 1: Fiber Reinforcements and General Theory of Composites, Tsu-Wei Chou, Vol. Ed., Anthony Kelly and Carl Zweben, Editors-in-Chief, Pergamon Press, Elsevier Science Ltd., Oxford, 2000.
6. J.C. Norman and C. Zweben, Kevlar® 49/Thornel® 300 Hybrid Fabric Composites for Aerospace Applications, *SAMPE Quarterly,* Vol. 7, No. 4, July 1976, pp. 1–10.
7. A. Afaghi-Khatibi, L. Ye and Y.-W. Mai, Hybrids and Sandwiches, *Comprehensive Composite Materials*, Vol. 2: Polymer Matrix Composites, Ramesh Talreja, Vol. Ed., Anthony Kelly and Carl Zweben, Editors-in-Chief, Pergamon Press, Elsevier Science Ltd., Oxford, 2000.
8. J. J. Pigliacampi, Inorganic Fibers, *Engineered Materials Handbook*, Vol. 1, Composites, ASM International, Materials Park, Ohio, 1987, pp. 43–104.
9. D. D. Johnson and H. G. Sowman, Ceramic Fibers, *Engineered Materials Handbook*, Vol. 1, Composites, ASM International, Materials Park, Ohio, 1987, pp. 60–65.
10. A. Shindo, Polyacrylonitrile (PAN)-based Carbon Fibers, *Comprehensive Composite Materials*, Vol. 1: Fiber Reinforcements and General Theory of Composites, Tsu-Wei Chou, Vol. Ed., Anthony Kelly and Carl Zweben, Editors-in-Chief, Pergamon Press, Elsevier Science Ltd., Oxford, 2000.
11. R. J. Diefendorf, Pitch Precursor Carbon Fibers, *Comprehensive Composite Materials*, Vol. 1: Fiber Reinforcements and General Theory of Composites, Tsu-Wei Chou, Vol.

Ed., Anthony Kelly and Carl Zweben, Editors-in-Chief, Pergamon Press, Elsevier Science Ltd., Oxford, 2000.

12. F. E. Wawner, Boron and Silicon Carbide Fibers (CVD), *Comprehensive Composite Materials*, Vol. 1: Fiber Reinforcements and General Theory of Composites, Tsu-Wei Chou, Vol. Ed., Anthony Kelly and Carl Zweben, Editors-in-Chief, Pergamon Press, Elsevier Science Ltd., Oxford, 2000.
13. H. Ichikawa and T. Ishikawa, Silicon Carbide Fibers (Organometallic Pyrolysis), *Comprehensive Composite Materials*, Vol. 1: Fiber Reinforcements and General Theory of Composites, Tsu-Wei Chou, Vol. Ed., Anthony Kelly and Carl Zweben, Editors-in-Chief, Pergamon Press, Elsevier Science Ltd., Oxford, 2000.
14. M. H. Berger and A. R. Bunsell, Oxide Fibers, *Comprehensive Composite Materials*, Vol. 1: Fiber Reinforcements and General Theory of Composites, Tsu-Wei Chou, Vol. Ed., Anthony Kelly and Carl Zweben, Editors-in-Chief, Pergamon Press, Elsevier Science Ltd., Oxford, 2000.
15. A. Parvizi-Majidi, Whiskers and Particulates, *Comprehensive Composite Materials*, Vol. 1: Fiber Reinforcements and General Theory of Composites, Tsu-Wei Chou, Vol. Ed., Anthony Kelly and Carl Zweben, Editors-in-Chief, Pergamon Press, Elsevier Science Ltd., Oxford, 2000.
16. H. H. Yang, Aramid Fibers, *Comprehensive Composite Materials*, Vol. 1: Fiber Reinforcements and General Theory of Composites, Tsu-Wei Chou, Vol. Ed., Anthony Kelly and Carl Zweben, Editors-in-Chief, Pergamon Press, Elsevier Science Ltd., Oxford, 2000.
17. D. W. Dwight, Glass Fiber Reinforcements, *Comprehensive Composite Materials*, Vol. 1: Fiber Reinforcements and General Theory of Composites, Tsu-Wei Chou, Vol. Ed., Anthony Kelly and Carl Zweben, Editors-in-Chief, Pergamon Press, Elsevier Science Ltd., Oxford, 2000.
18. T. Peijs, M. J. N. Jacobs and P. J. Lemstra, High Performance Polyethylene Fibers, *Comprehensive Composite Materials*, Vol. 1: Fiber Reinforcements and General Theory of Composites, Tsu-Wei Chou, Vol. Ed., Anthony Kelly and Carl Zweben, Editors-in-Chief, Pergamon Press, Elsevier Science Ltd., Oxford, 2000.
19. G. Savage, *Carbon-Carbon Composites*, Chapman & Hall, London, 1993.
20. L. E. McAllister, Carbon-Carbon Composites, *Concise Encyclopedia of Composite Materials*, Revised Edition, A. Kelly, ed., Pergamon Press, Oxford, 1994.
21. J. Jortner, Carbon/Carbon Composites, *Comprehensive Composite Materials,* Vol. 6: Design and Applications, M. G. Bader, Keith K. Kedward, and Yoshihiro Sawada, Vol. Eds., Anthony Kelly and Carl Zweben, Editors-in-Chief, Pergamon Press, Elsevier Science Ltd., Oxford, 2000.
22. C. Zweben, Advanced Composites in Spacecraft and Launch Vehicles, *Launchspace,* June/July 1998.
23. K. Schmidt and C. Zweben, Mechanical and Thermal Properties of Silicon Carbide Particle-Reinforced Aluminum, *Thermal and Mechanical Behavior of Metal Matrix and Ceramic Matrix Composites*, ASTM STP 1080, L. M. Kennedy, H.H. Moeller and W. S. Johnson, Eds., American Society for Testing and Materials, Philadelphia, 1989.
24. D. D. L. Chung and C. Zweben, Composites for Electronic Packaging and Thermal Management, *Comprehensive Composite Materials,* Vol. 6: Design and Applications, M. G. Bader, Keith K. Kedward, and Yoshihiro Sawada, Vol. Eds., Anthony Kelly and Carl Zweben, Editors-in-Chief, Pergamon Press, Elsevier Science Ltd., Oxford, 2000.
25. C. Zweben, Composite Materials And Mechanical Design, *Mechanical Engineers' Handbook,* Second Edition, Myer Kutz, Ed, John Wiley & Sons, Inc., New York, 1998.
26. J.-A. E. Månson, Manufacturing Process—An Overview, *Comprehensive Composite Materials*, Vol. 2: Polymer Matrix Composites, Ramesh Talreja, Vol. Ed., Anthony

Kelly and Carl Zweben, Editors-in-Chief, Pergamon Press, Elsevier Science Ltd., Oxford, 2000.

27. D. F. Adams, Test Methods for Mechanical Properties, *Comprehensive Composite Materials*, Vol. 5: Test Methods, Nondestructive Evaluation, and Smart Materials, Leif Carlsson, Robert L. Crane and K. Uchino, Vol. Eds., Anthony Kelly and Carl Zweben, Editors-in-Chief, Pergamon Press, Elsevier Science Ltd., Oxford, 2000.
28. M. J. Parker, Test Methods for Physical Properties, Vol. 5: Test Methods, Nondestructive Evaluation, and Smart Materials, Leif Carlsson, Robert L. Crane and K. Uchino, Vol. Eds., Anthony Kelly and Carl Zweben, Editors-in-Chief, Pergamon Press, Elsevier Science Ltd., Oxford, 2000.
29. F. Langlais, Chemical Vapor Infiltration Processing of Ceramic Matrix Composites, *Comprehensive Composite Materials*, Vol. 4: Carbon/Carbon, Cement and Ceramic Matrix Composites, Richard Warren, Vol. Ed., Anthony Kelly and Carl Zweben, Editors-in-Chief, Pergamon Press, Elsevier Science Ltd., Oxford, 2000.
30. F. F. Lange, F. W. Zok and C. G. Levi, Processing and Properties of Porous Ceramic Matrix Composites, *Comprehensive Composite Materials*, Vol. 4: Carbon/Carbon, Cement and Ceramic Matrix Composites, Richard Warren, Vol. Ed., Anthony Kelly and Carl Zweben, Editors-in-Chief, Pergamon Press, Elsevier Science Ltd., Oxford, 2000.
31. A. R. Bhatti and P. Ferris, Preparation of Fiber Reinforced Dense Ceramic Matrix Composites, *Comprehensive Composite Materials*, Vol. 4: Carbon/Carbon, Cement and Ceramic Matrix Composites, Richard Warren, Vol. Ed., Anthony Kelly and Carl Zweben, Editors-in-Chief, Pergamon Press, Elsevier Science Ltd., Oxford, 2000.
32. R. Talreja and J.-A. E. Månson, Introduction, *Comprehensive Composite Materials*, Vol. 2: Polymer Matrix Composites, Ramesh Talreja, Vol. Ed., Anthony Kelly and Carl Zweben, Editors-in-Chief, Pergamon Press, Elsevier Science Ltd., Oxford, 2000.
33. Cool Polymers product data sheets, www.coolpolymers.com.
34. J. Ogando, Thermally conductive plastics beat the heat, *Design News,* September 17, 2001.
35. T. W. Clyne, An Introductory Overview of MMC Systems, Types, and Applications, *Comprehensive Composite Materials*, Vol. 3: Metal Matrix Composites, T. W. Clyne, Vol. Ed., Anthony Kelly and Carl Zweben, Editors-in-Chief, Pergamon Press, Elsevier Science Ltd., Oxford, 2000.
36. Hayashi, H. Ushio and M Ebisawa, *The Properties of Hybrid Fiber Reinforced Metal and Its Application for Engine Block,* SAE Technical Paper No. 890557, 1989.
37. Warren Hunt, Metal Matrix Composites, *Comprehensive Composite Materials,* Vol. 6: Design and Applications, M. G. Bader, Keith K. Kedward, and Yoshihiro Sawada, Vol. Eds., Anthony Kelly and Carl Zweben, Editors-in-Chief, Pergamon Press, Elsevier Science Ltd., Oxford, 2000.
38. C. Zweben, Thermal Management and Electronic Packaging Applications, *ASM Handbook,* Vol. 21, *Composites*, ASM International, Materials Park, Ohio, 2002.
39. C. Zweben, Metal Matrix Composites: Aerospace Applications, *Encyclopedia of Advanced Materials*, M.C. Flemings, et al., Eds., Pergamon Press, Oxford 1994.
40. R. Warren, Ed., *Ceramic-Matrix Composites*, Chapman and Hall, New York, 1992.
41. *Advanced Materials by Design*, OTA-E-351 U.S. Congress Office of Technology Assessment, U.S. Government Printing Office, Washington, DC, June 1988).
42. C. Zweben, *Material Selection and Manufacturing For Spacecraft And Launch Vehicles, A State-of-the-Art Report,* Advanced Materials and Processes Technology Information Analysis Center, Rome, NY, in press.

Chapter

6

Plastics in Coatings and Finishes

Carl P. Izzo

Industrial Paint Consultant
Export, Pennsylvania

6.1 Introduction

Coatings and finishes are composed of film-forming resins, pigments, solvents, and additives.[1] Although their existence predates plastics by several millennia, it is important to note that the resins used in modern coatings and finishes are the same as those used in plastics. For example, looking back to the 1930s, linseed-oil-impregnated webs called, *oil cloth,* were used in home decorating. They were also used to make floor coverings. At the same time, Linseed oil was the main resin used in industrial and trade sales paints. Today, vinyl resins have replaced linseed oil in decorative cloth, floor coverings, and many paints.

After coatings and finishes are applied, they form films and cure. Curing mechanisms can be as simple as solvent evaporation or as complicated as free-radical polymerization. Basically, coatings and finishes can be classified as baking or air drying, which usually means room-temperature curing. The curing method and times are important in coating selection, because they must be considered for optimizing equipment and production schedule choices.

Coatings and finishes today are engineering materials. They evolved from cave dwellers decorating their walls with earth pigments ground in egg whites into factory workers protecting products with E-coat primers and urethane acrylic enamels. Unlike Noah (who received divine instructions to paint his ark with pitch), the Egyptians (who mixed amber with oils to make decorative varnishes), or the Romans (who mixed white lead with amber and with pitch), scientists and engineers using plastic resins today must design coatings and finishes for specific applications. Significant changes have been made in the film-forming resinous portions of the coatings and finishes.[2] Because of concern for environmental regulations, safety in the workplace, performance requirements, and costs, the design of these coatings and finishes is being optimized.

The first 50 years of the twentieth century were the decades of discovery. Significant changes were made in the resins, which are the liquid portions of the coatings and finishes composed of binder and thinner.[2] The binder is also called *vehicle*. Since the 1900s and the introduction of phenolic synthetic resin vehicles, coatings and finishes have been designed

to increase production and meet performance requirements at lower costs. These developments were highlighted by the introduction of nitrocellulose lacquers for the automotive and furniture industries; followed by the alkyds, epoxies, vinyls, polyesters, acrylics, and a host of other resins; and finally the polyurethanes.

In the 1950s, the decade of expansion, manufacturers built new plants to supply the coatings and finishes demand for industrial and consumer products. Coatings and finishes were applied at low solids using inefficient, conventional air-atomized spray guns. The atmosphere was polluted with volatile organic compounds (VOCs), but no one cared as long as finished products were shipped out of the factories. Coatings and finishes suppliers were fine tuning formulations to provide faster curing and improved performance properties. In 1956, powder coatings were invented. By 1959, there were several commercial conveyorized lines applying powder coatings by the fluidized bed process.

In the 1960s, the decade of technology, just as coatings and finishes were becoming highly developed, another variable—environmental impact—was added to the equation. Someone finally noticed that the solvents, which were used in coatings and finishes for viscosity and flow control and which and evaporated during application and cure, were emitted into the atmosphere. Los Angeles County officials, who found that VOC emissions were a major source of air pollution, enacted Rule 66 to control the emission of solvents that cause photochemical smog. To comply with Rule 66, the paint industry reformulated its coatings and finishes using exempt solvents, which presumably did not produce smog. California's Rule 66 was followed by other local air-quality standards and finally by the establishment of the U.S. Environmental Protection Agency (EPA), whose charter, under the law, was to improve air quality by reducing solvent emissions. During this decade, there were three notable developments that would eventually reduce VOC emissions in coating operations: (1) electrocoating, (2) electrostatically sprayed powder coatings, and (3) radiation curable coatings.

In the 1970s, the decade of energy conservation, a crisis resulted in terms of shortages and price increases for solvents and coating and finish materials. Also affected was the distribution of natural gas, the primary fuel for curing ovens, which caused shortages and price increases. In response to those pressures, the coatings and finishes industry developed low-temperature curing coatings and finishes in an effort to reduce energy consumption. Of greater importance to suppliers and end users of coatings and finishes was the establishment of the Clean Air Act of 1970. An important development during this period was radiation curable coatings, mostly clears, for flat line applications using electron beam (EB) and ultraviolet (UV) radiation sources.

In the 1980s, the decade of restriction, the energy crisis was ending, and the more restrictive air-quality standards were beginning. However, energy costs remained high. The importance of transfer efficiency (the percentage of an applied coating that actually coats the product) was recognized by industry and the EPA. This led to the development and use of coatings application equipment and coating methods that have higher transfer efficiencies. The benefits of using higher transfer efficiency coating methods are threefold: lower coatings material usage, lower solvent emissions, and lower costs. The automotive industry switched almost exclusively to electrocoating for the application of primers. Powder coating material and equipment suppliers worked feverishly to solve the problems associated with automotive topcoats. Other powder coatings applications saw rapid growth. Radiation-curable pigmented coatings and finishes for three-dimensional products were developed.

In the 1990s, the decade of compliance, resin and coatings suppliers developed compliance coatings: electrocoating, high solids, powder, radiation curable, and waterborne. Equipment suppliers developed devices to apply and cure these new coatings. These developments were in response to the 1990 amendments to the Clean Air Act of 1970. The amendments established a national permit program that made the law more enforceable, ensuring better compliance and calling for nationwide regulation of VOC emissions from

all organic finishing operations. The amendments also established Control Technique Guidelines to allow state and local governments to develop Attainment Rules. Electrocoating was the process of choice for priming many industrial and consumer products. Powder coatings were used in a host of applications where durability was essential. UV-curable coatings and finishes were applied to three-dimensional objects. Equipment suppliers developed more efficient application equipment.

In the 2000s, the beginning of the green millennium, coating and finish equipment suppliers' investments in research and development will pay dividends. Improvements in coatings materials and application equipment have enabled end users to comply with the air-quality regulations. Primers are applied by electrocoating. One-coat finishes are replacing two coats in many cases. High-solids and waterborne liquid coatings and finishes are replacing conventional solvent-thinned ones. Powder coatings use has increased dramatically. Radiation-cured coatings and finishes are finding more applications. The use of coating and finish materials and solvent use, as well as application costs, are being reduced. Air-quality standards are being met. Suppliers of coatings and finishes and related equipment, as well as end users, are recognizing the cost savings bonus associated with attainment. Compliance coatings applied by more efficient painting methods will reduce coating and finish materials and solvent use, effecting cost savings.

Since coatings and finishes today are considered to be engineering materials, their performance characteristics must not only match service requirements, they must also meet governmental regulations and production cost considerations. In the past, the selection of a coating depended mainly on the service requirements and application method. Now, more than ever before, worker safety, environmental impact, and economics must be considered. For this reason, compliance coatings—electrocoating, high-solids, powders, radiation-cured, and waterborne coatings—are the most sensible choices.

Coatings and finishes are applied to most industrial products by spraying. Figure 6.1 shows a typical industrial spray booth. In 1890, Joseph Binks invented the cold-water

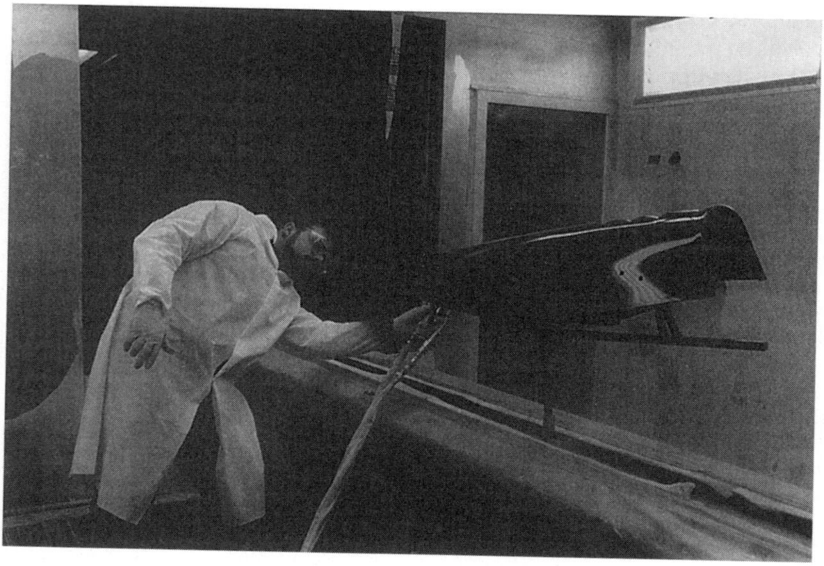

Figure 6.1 A typical industrial spray booth used for applying industrial coatings. *(Courtesy of George Koch Sons, LLC)*

paint-spraying machine, the first airless sprayer, which was used to apply whitewash to barns and other building interiors. In 1924, Thomas DeVilbiss used a modified medical atomizer, the first air-atomizing sprayer, to apply a nitrocellulose lacquer on the Oakland automobile. Since then, these tools have remained virtually unchanged, and, until the enactment of the air quality standards, they were used to apply coatings and finishes at 25 to 50 percent volume solids at transfer efficiencies of 30 to 50 percent. Using this equipment, the remainder of the nonvolatile material, the overspray, coated the floor and walls of spray booths and became hazardous or nonhazardous waste, while the solvents, the VOCs, evaporated from the coating during application and cure to become air pollutants. Today, finishes are applied by highly transfer-efficient application equipment. The choice of application equipment must be optimized.

Even the best coatings and finishes will not perform their function if they are not applied on properly prepared substrates. For this reason, surfaces must first be cleaned to remove oily soils, corrosion products, and particulates, and then pretreated before applying any coatings and finishes.

The purposes of this chapter are threefold.

1. To aid the designer in optimizing the selection of coating and finishing materials and application equipment
2. To acquaint the reader with surface preparation, coating materials, application equipment, and curing methods
3. To stress the importance of environmental compliance in coating operations

6.2 Environment and Safety

In the past, changes in coating materials and coating application lines were discussed only when lower prices, novel products, new coating lines, or new plants were considered. Today, with rising material costs, rising energy costs, and more restrictive governmental regulations, they are the subject of frequent discussions. During these discussions, both in-house and with suppliers, choices of coating materials and processing equipment are optimized. Coating material and solvent costs, which are tied to the price of crude oil, have risen since the 1970s, as has the cost of natural gas, which is the most frequently used fuel for coating and finish bake ovens. The EPA has imposed restrictive air-quality standards. The Occupational Safety and Health Act (OSHA) and the Toxic Substances Control Act (TSCA) regulate the environment in the workplace and limit workers' contact with hazardous materials. These factors have increased finishing costs and the awareness of product finishers. To meet the challenge, they must investigate and use alternative coating materials and processes for compliance and cost effectiveness.

Initial attempts to control air pollution in the late 1940s resulted in smoke-control laws to reduce airborne particulates. The increased use of the automobile and industrial expansion during that period caused a condition called *photochemical smog* (smog created by the reaction of chemicals exposed to sunlight in the atmosphere) in major cities throughout the United States. Los Angeles County officials recognized that automobile exhaust and VOC emissions were major sources of smog, and they enacted an air pollution regulation called Rule 66. Rule 66 forbade the use of specific solvents that produced photochemical smog and published a list of exempt solvents for use in coatings. Further study by the EPA has shown that, if given enough time, even the Rule 66-exempt solvents will produce photochemical smog in the atmosphere.

The Clean Air Act of 1970 and its 1990 amendments, formulated by the EPA, established national air-quality standards that regulate the amount of solvents emitted. The EPA divided the 50 states into 250 air-quality regions, each of which is responsible for the im-

plementation of the national air quality standards. It is important to recognize that many of the local standards are more stringent than the national ones. For this reason, specific coatings and finishes that comply with the air-quality standards of one district may not comply with another's. Waterborne, high-solids, powder, electrophoretic, and radiation-cured coatings and finishes will comply. The use of precoated metal can eliminate all the compliance problems.

Not only because the EPA mandates the reduction of VOC emissions, but also because of economic advantages, spray painting, which is the most widely used application method, must be done more efficiently. The increased efficiency will reduce the amount of expensive coatings and finishes materials and solvents used, thereby reducing production costs.

6.3 Surface Preparation

The most important step in any coating operation is surface preparation, which includes cleaning and pretreatment. For coatings and finishes to adhere, surfaces must be free from oily soils, corrosion products, and loose particulates. New wood surfaces are often coated without cleaning. Old wood and coated wood must be cleaned to remove oily soils and loose, flaky coatings. Plastics are cleaned by using solvents and chemicals to remove mold release. Metals are cleaned by media blasting, sanding, brushing, and by solvents or aqueous chemicals. The choice of a cleaning method depends on the substrate type and the size and shape of the object.

After cleaning, pretreatments are applied to enhance coating adhesion and, in the case of metals, corrosion resistance. Some wood surfaces require no pretreatment, while others require priming of knots and filling of nail holes. Cementitious and masonry substrates are pretreated, using acids, to remove loosely adhering contaminants and to passivate the surfaces. Metals, still the most common industrial substrates, are generally pretreated using phosphates, chromates, and oxides to passivate their surfaces and provide corrosion resistance. Plastics, second only to steel, are gaining rapidly in use as industrial substrates. Some are paintable after cleaning to remove mold release and other contaminants, while others require priming, physical treatments, or chemical etching to ensure coating adhesion. Since most of the industrial substrates coated are metals and plastics, their cleaning and pretreatment are described in the next sections. Because of their complexities, detailed descriptions of cleaning and pretreatment processes are beyond the scope of this chapter. Enough detail will be given to allow the reader to make a choice. As with the choice of a cleaning method, the choice of a pretreatment method depends on the composition, size, and shape of the product.

6.3.1 Metal Surface Cleaning

Oily soils must be removed before any other surface preparation is attempted. Otherwise, these soils may be spread over the surface. These soils can also contaminate abrasive cleaning media and tools. Oily soils can be removed faster using liquid cleaners that impinge on the surface or in agitated immersion baths. It is often necessary to heat liquid cleaners to facilitate soil removal.

6.3.1.1 Abrasive cleaning. After removal of the oily soils, surfaces are abrasive-cleaned to remove rust and corrosion by media blasting, hand or power sanding, and hand or power brushing. Media blasting consists of propelling materials, such as sand, metallic shot, nut shells, plastic pellets, and dry ice crystals, by gases under pressure, so that they

impinge on the surfaces to be cleaned. High-pressure water-jet cleaning is similar to media blasting.

6.3.1.2 Alkaline cleaning. To remove oily soils, aqueous solutions of alkaline phosphates, borates, and hydroxides are applied to metals by immersion or spray. After cleaning, the surfaces are rinsed with clear water to remove the alkali. These materials are not effective for removing rust and corrosion.

6.3.1.3 Detergent cleaning. Aqueous solutions of detergents are used to remove oily soils in much the same way as alkaline cleaners. Then they are rinsed with cold water to flush away the soils.

6.3.1.4 Emulsion cleaning. Heavy oily soils and greases are removed by aqueous emulsions of organic solvents such as mineral spirits and kerosene. After the emulsified solvent has dissolved the oily soils, they are flushed away using a hot-water rinse. Any remaining oily residue must be removed using clean solvent, alkaline, or detergent cleaners.

6.3.1.5 Solvent cleaning. Immersion, hand wiping, and spraying using organic solvents are effective methods for removing oily soils. Since these soils will contaminate solvents and wipers, it is important to change them frequently. Otherwise, oily residues will remain on substrates. Safe handling practices must be followed because of the hazardous nature of most organic solvents.

6.3.1.6 Manual spray cleaning. For large products, detergent and alkaline cleaners applied using steam cleaners are a well known degreasing method. In addition to oily soils, the impingement of the steam and the action of the chemicals will dissolve and flush away heavy greases and waxes. Hot-water spray cleaning using chemicals is nearly as effective as steam cleaning.

6.3.1.7 Vapor degreasing. Vapor degreasing has been a very popular cleaning method for removing oily soils. Boiling solvent condenses on the cool surface of the product and flushes away oily soils but does not remove particulates. Since this process uses chlorinated solvents, which are under regulatory scrutiny by government agencies, its popularity is declining. However, closed-loop systems are still available.

6.3.2 Metal Surface Pretreatment

Cleaning metals will remove oily soils but will generally not remove rust and corrosion from substrates to be coated. Abrasive cleaning will remove corrosion products, and for this reason it is also considered a pretreatment, because the impingement of blasting media and the action of abrasive pads and brushes roughen the substrate and therefore enhance adhesion. The other pretreatments use aqueous chemical solutions, which are applied by immersion or spray techniques. Pretreatments for metallic substrates used on industrial products are discussed in this section. Because they provide corrosion protection to fer-

rous and nonferrous metals, chromates are used in pretreatment stages and as conversion coatings. They are being replaced by nonchromate chemicals.

6.3.2.1 Aluminum. Aluminum is cleaned by solvents and chemical solutions to remove oily soils and corrosion products. Cleaned aluminum is pretreated using chromate conversion coating and anodizing. Phosphoric acid-activated vinyl wash primers, which are also considered pretreatments, must be applied directly to metal and not over other pretreatments.

6.3.2.2 Copper. Copper is cleaned by solvents and chemicals and then abraded to remove corrosion. Bright dipping in acids will also remove corrosion. Cleaned surfaces are often pretreated using chromates and vinyl wash primers.

6.3.2.3 Galvanized steel. Galvanized steel must be cleaned to remove the oil or wax that is applied at the mill to prevent white corrosion. After cleaning, the surfaces are pretreated using chromates and phosphates. Vinyl wash primer pretreatments can also be applied on galvanized steel surfaces having no other pretreatments.

6.3.2.4 Steel. Steel surfaces are cleaned to remove oily soils and, if necessary, pickled in acid to remove rust. Clean steel is generally pretreated with phosphates to provide corrosion resistance. Other pretreatments for steel are chromates and wash primers.

6.3.2.5 Stainless steel. Owing to its corrosion resistance, stainless steel is usually not coated. Otherwise the substrate must be cleaned to remove oily soils and then abraded to roughen the surface. Wash primers will enhance adhesion.

6.3.2.6 Titanium. Cleaned titanium is pretreated like stainless steel.

6.3.2.7 Zinc and cadmium. Zinc and cadmium substrates are pretreated like galvanized steel.

6.3.3 Plastic Surface Cleaning

6.3.3.1 Alkaline cleaning. Aqueous solutions of alkaline phosphates, borates, and hydroxides are applied to plastics by immersion or spray to remove oily soils and mold release agents. After cleaning, the surfaces are rinsed with clear water to remove the alkali.

6.3.3.2 Detergent cleaning. Aqueous solutions of detergents are used to remove oily soils and mold release agents in much the same way as with alkaline cleaners. Then they are rinsed with cold water to flush away the soils.

6.3.3.3 Emulsion cleaning. Heavy, oily soils, greases, and mold release agents are removed by aqueous emulsions of organic solvents such as mineral spirits and kerosene. After the emulsified solvent has dissolved the oily soils, they are flushed away using a hot-water rinse. The remaining oily residue must be removed using clean solvent, alkaline, or detergent cleaners.

6.3.3.4 Solvent cleaning. Immersion, hand wiping, and spraying, using organic solvents, are effective methods for removing oily soils and mold release agents. Since these soils will contaminate solvents and wipers, it is important to change them frequently. Otherwise, oily residues will remain on substrates. Compatibility of cleaning solvents with the plastic substrates is extremely important. Solvents that affect plastics are shown in Table 6.1. Suppliers of mold release agents are the best source for information on solvents which will remove their materials. Safe handling practices must be followed because of the hazardous nature of most organic solvents.

TABLE 6.1 Solvents that Affect Plastics

Resin	Heat-distortion point, °F	Solvents that affect surface
Acetal	338	None
Methyl methacrylate	160–105	Ketones, esters, aromatics
Modified acrylic	170–190	Ketones, esters, aromatics
Cellulose acetate	110–209	Ketones, some esters
Cellulose propionate	110–250	Ketones, esters, aromatics, alcohols
Cellulose acetate butyrate	115–227	Alcohols, ketones, esters, aromatics
Nylon	260–360	None
Polyethylene:		
High density	140–180	
Medium density	120–150	None
Low density	105–121	
Polypropylene	210–230	None
Polycarbonate	210–290	Ketones, esters, aromatics
Polystyrene (GP high-heat)	150–195	Some aliphatics, ketones, esters, aromatics
Polystyrene (impact, heat-resistant)	148–200	Ketones, esters, aromatics, some aliphatics
Acrylonitrile butadiene styrene (ABS)	165–225	Ketones, esters, aromatics, alcohol

6.3.3.5 Manual spray cleaning. Detergent and alkaline cleaners applied using steam and hot-water spray cleaners are a well known degreasing method. It can also be used for removing mold release agents. The impingement of the steam and hot water and the action of the chemicals will dissolve and flush away the contaminants. Manual spray cleaning is used for large products.

6.3.4 Plastic Surface Pretreatment

Cleaning will remove oily soils and mold release agents, but additional pretreatment may be needed on certain plastic surfaces to ensure adhesion. Many of the plastic substrates are chemically inert and will not accept coatings and finishes because of their poor wettability.

Depending on their chemical composition, they will require mechanical, chemical, and physical pretreatment or priming to enhance coating adhesion. Since mechanical pretreatment consists of abrasion, its effect on the substrate must be considered. Chemical pretreatments involve corrosive materials which etch the substrates and can be hazardous. Therefore, handling and disposal must be considered. Physical pretreatments consist of plasma, corona discharge, and flame impingement. Process control must be considered.

6.3.4.1 Abrasive cleaning. After removal of the oily soils, surfaces are abrasive pretreated to roughen the substrate by media blasting, hand or power sanding, and hand or power brushing. Media blasting consists of propelling materials, such as sand, metallic shot, nut shells, plastic pellets, and dry ice crystals, by gases under pressure, so that they impinge on the surfaces to be pretreated.

6.3.4.2 Chemical etching. Chemical pretreatments use solutions of corrosive chemicals, which are applied by immersion or spray techniques, to etch the substrate

6.3.4.3 Corona discharge. During corona discharge pretreatment the plastic is bombarded by gasses directed toward its surface

6.3.4.4 Flame treating. During the flame pretreatment, an open flame impinging on the surface of the plastic product causes alterations in the surface chemistry.

6.3.4.5 Plasma pretreatment. Low-pressure plasma pretreatment is conducted in a chamber while atmospheric plasma pretreatment is done in the open. In both cases, ablation alters the surface chemistry and causes changes in surface roughness.

6.3.4.6 Laser pretreatment. Laser pretreatment ablates the plastic substrate causing increased surface roughness and changes in the surface chemistry.

6.3.5 Priming

Priming involves the application of a coating on the surface of the plastic product to promote adhesion or to prevent attack by the solvents in a subsequent protective or decorative coating. In some cases, priming can be done after cleaning. In others, it must be done after pretreatment.

6.4 Coatings and Finishes Selection

To aid in their selection, coatings and finishes will be classified by use in finish systems, physical state, and resin type. Coatings and finishes are also classified by their use as electrical insulation. It is not the intention of this chapter to instruct the reader in the chemistry of organic coating but rather to aid in selection of coatings and finishes for specific applications. Therefore, the coating resin's raw material feed stock and polymerization reac-

tions will not be discussed. On the other hand, generic resin types, curing, physical states, and application methods are discussed.

6.4.1 Selection by Finish Systems

Finish systems can be one-coat or multicoat that use primers, intermediate coats, and topcoats. Primers provide adhesion, corrosion protection, passivation, and solvent resistance to substrates. Topcoats provide weather, chemical, and physical resistance, and generally determine the performance characteristics of finish systems. Performance properties for coatings formulated with the most commonly used resins are shown in Table 6.2.

TABLE 6.2 Performance Properties* of Common Coating Resins[3]

Resin type	Humidity resistance	Corrosion resistance	Exterior durability	Chemical resistance	Mar resistance
Acrylic	E	E	E	G	E
Alkyd	F	F	P	G	G
Epoxy	E	E	G	E	E
Polyester	E	G	G	G	G
Polyurethane	E	G	E	G	E
Vinyl	E	G	G	G	G

*Note: E = excellent, G = good, F = fair, P = poor.

In coatings and finishes selection, intended service conditions must be considered. To illustrate this point, consider the differences between service conditions for toy boats and for battleships. Table 6.3. shows the use of industrial finish systems in various service conditions.

TABLE 6.3 Typical Industrial Finish Systems[3]

Service conditions	Primer	One–coat enamel	Intermediate coat	Topcoat
Interior				
Light duty		X		
Heavy duty	X			
Exterior				X
Light duty		X		
Heavy duty	X			X
Extreme duty	X		X	X

6.4.2 Selection by Physical State

A resin's physical state can help determine the application equipment required. Solid materials can be applied by powder coating methods. Table 6.4 lists plastics applied as powder coatings. Liquids can be applied by most of the other methods, which are discussed later. Many of the coating resins exist in several physical states. Table 6.5 lists the physical states of common coating resins.

TABLE 6.4 Plastics Used in Powder Coatings

Resin	Fluidizing conditions: Preheat temperature, °F	Fluidizing conditions: Cure or fusion Temperature, °F	Fluidizing conditions: Cure or fusion Time, min	Fluidized bed powder: Maximum operating temperature, °F	Fluidized bed powder: Adhesion	Fluidized bed powder: Weather resistance
Epoxy	250–450	250–450	1–60	200–400	Excellent	Good
Vinyl	450–550	400–600	1–3	225	Poor	Good
Cellulose acetate butyrate	550–600	400–550	1–3	225	Poor	Good
Nylon	550–800	650–700	1	300	Poor	Fair
Polyethylene	500–600	400–600	1–5	225	Fair	Good
Polypropylene	500–700	400–600	1–3	260	Poor	Good
Penton	500–650	450–600	1–10	350	Poor	Good
Teflon	800–1000	800–900	1–3	500	Poor	Good

TABLE 6.5 Physical States of Common Coating Resins[3]

Resin	Conventional solvent	Waterborne	High-solids	Powder coating	100% solids liquid	Two-component liquid
Acrylic	x	x	x	x		
Alkyd	x	x	x			
Epoxy	x	x	x	x	x	x
Polyester	x		x	x	x	x
Polyurethane	x	x	x	x	x	x
Vinyl	x	x	x	x	x	

6.4.3 Selection by Resin Type

Since resin type determines the performance properties of a coating, it is used most often. Table 6.6 shows the physical, environmental, and film-forming characteristics of coatings and finishes by polymer (resin) type. It is important to realize that, in selecting coatings, tables of performance properties of generic resins must be used only as guides, because coatings and finishes of one generic type, such as acrylic, epoxy, or polyurethane, are often modified using one or more of the other generic types. Notable examples are acrylic alkyds, acrylic urethanes, acrylic melamines, epoxy esters, epoxy polyamides, silicone alkyds, silicone epoxies, silicone polyesters, vinyl acrylics, and vinyl alkyds. While predicting specific coating performance properties of unmodified resins is simple, predicting the properties of modified resins is difficult if not impossible. Parameters causing these difficulties are resin modification percentages and modifying methods such as simple blending or copolymerization. The performance of a 30 percent copolymerized silicone alkyd is not necessarily the same as one that was modified by blending. These modifications can change the performance properties subtly or dramatically.[3]

There are nearly 1500 coating manufacturers in the United States, each having various formulations that could number in the hundreds. Further complicating the coatings and finishes selection difficulty is the well known practice of a few manufacturers who add small amounts of a more expensive, better performing resin to a less expensive, poorer performing resin and call the product by the name of the former. An unsuspecting person,

TABLE 6.6 Properties of Coatings by Polymer Type

Coating type	Electrical properties: Volume resistivity, Ω-cm (ASTM D 257)	Dielectric strength, V/mil	Dielectric constant	Dissipation factor	Physical characteristics: Maximum continuous service temperature, °F	Adhesion to metals	Flexibility	Approximate Sward hardness (higher number is harder)	Abrasion resistance
Acrylic	10^{14}–10^{15}	450–550	2.7–3.5	0.02–0.06	180	Good	Good	12–24	Fair
Alkyd	10^{14}	300–350	4.5–5.0	0.003–0.06	200 250 TS	Excellent	Fair to good Low temperature—poor	3–13 (air dry) 10–24 (bake)	Fair
Cellulosic (nitrate butyrate)		250–400	3.2–6.2		180	Good	Good Low temperature—poor	10–15	
Chlorinated polyether (Penton*)	10^{15}	400	3.0	0.01	250	Excellent	Good		
Epoxy-amine cure	10^{14} at 30°C 10^{10} at 105°C	400–550	3.5–5.0	0.02–0.03 at 30°C	350	Excellent	Fair to good Low temperature—poor	26–36	Good to excellent
Epoxy-anhydride, dicy		650–730	3.4–3.8	0.01–0.03	400	Excellent	Good to excellent Low temperature—poor	20	Good to excellent
Epoxy-polyamide	10^{14} at 30°C 10^{10} at 105°C	400–500	2.5–3.0	0.008–0.02	350	Excellent	Good to excellent Low temperature—poor	20	Fair to good
Epoxy-phenolic	10^{12}–10^{13}	300–450			400	Excellent	Good Low temperature—fair		Good to excellent
Fluorocarbon TFE	10^{18}	430	2.0–2.1	0.0002	500	Can be excellent; primers required	Excellent		
FEP	10^{18}	480	2.1	0.0003–0.0007	400				
CTFE	10^{18}	500–600	2.3–2.8	0.003–0.004	400	Can be excellent; primers required	Excellent		
Parylene (polyxylylenes)	10^{16}–10^{17}	700	2.6–3.1	0.0002–0.02	240°F (air) 510°F (inert atmosphere)	Good	Good		
Phenolics	10^{9}–10^{12}	100–300	4–8	0.005–0.5	350	Excellent	Poor to good Low temperature—poor	30–38	Fair

Phenolic-oil varnish					250	Excellent	Good Low temperature—fair		Poor to fair
Phenoxy	10^{13}–10^{14}	500	3.7–4.0	0.001	180	Excellent	Excellent		
Polyamide (nylon)	10^{13}–10^{15}	400–500	2.8–3.6	0.01–0.1	225–250	Excellent			
Polyester	10^{12}–10^{14}	500	3.3–8.1	0.008–0.04	200	Good on rough surfaces; poor to polished metals	Fair to excellent	25–30	Good
Chlorosulfonated (polyethylene Hypalon)†			6–10	0.03–0.07	250	Good	Elastomeric	Less than 10	
Polyimide	10^{16}–10^{18}	400 3000 (10 mil)	3.4–3.8	0.003	500	Good	Fair to excellent		
Polystyrene	10^{10}–10^{19}	500–700	2.4–2.6	0.0001–0.0005	140–180		Poor to fair		Good
Polyurethane	10^{12}–10^{13}	450–500 3800 (1 mil)	6.8 (1 kHz) 4.4 (1 MHz)	0.02–0.08	250	Often poor to metals (excellent to most nonmetals)	Good to excellent Low temperature—poor	10–17 (castor oil) 50–60 (polyester)	
Silicone	10^{14}–10^{16}	550	3.0–4.2	0.001–0.008	500	Varies, but usually needs primer for good adhesion	Excellent Low temperature—excellent	12–16	Fair to excellent
Vinyl chloride (poly-)	10^{11}–10^{15}	300–800	3–9	0.04–0.14	150	Excellent, if so formulated	Excellent Low temperature—fair to good	5–10	
Vinyl chloride (plastisol, organisol)	10^{9}–10^{16}	400	2.3–9	0.10–0.15	150	Requires adhesive primer	Excellent Low temperature—fair to good	3–6	
Vinyl fluoride	10^{13}–10^{14}	260 1200 (8 mil)	6.4–8.4	0.05–0.15	300	Excellent, if fused on surface	Excellent Low temperature—excellent		
Vinyl formal (Formvar‡)	10^{13}–10^{15}	850–1000	3.7	0.007–0.2	200	Excellent			

*Trademark of Hercules Powder Co., Inc., Wilmington, Del.
†Trademark of E. I. du Pont de Nemours & Co., Wilmington, Del.
‡Trademark of Monsanto Co., St. Louis, Mo.

TABLE 6.6 Properties of Coatings by Polymer Type *(Continued)*

	Resistance to environmental effects						Film formation			
Coating type	Chemical and solvent resistance	Moisture and humidity resistance	Weatherability	Resistance to micro-organisms	Flammability	Repairability	Method of cure	Cure schedule	Application method	Typical uses
Acrylic		Good	Excellent resistance to UV and weather	Good	Medium	Remove with solvent	Solvent evaporation	Air dry or low-temperature bake	Spray, brush, dip	Coatings for circuit boards. Quick dry protection for markings and color coding.
Alkyd	Solvents—poor Alkalies—poor Dilute acids—poor to fair	Poor	Good to excellent	Poor	Medium	Poor	Oxidation or heat	Air dry or baking types	Most common methods	Painting of metal parts and hardware.
Cellulosic (nitrate butyrate)	Solvents—good Alkalies—good Acids—good	Fair		Poor to good	High	Remove with solvents	Solvent evaporation	Air dry or low-temperature bake	Spray, dip	Lacquers for decoration and protection. Hot-metal coatings.
Chlorinate polyether (Penton*)		Good			Low		Powder or dispersion fuses	High-temperature fusion	Spray, dip, fluid bed	Chemically resistant coatings.
Epoxy-amine cure	Solvents—good to excellent Alkalies—good Dilute acids—fair	Good	Pigmented—fair; clear—poor (chalks)	Good	Medium	No	Cured by catalyst reaction	Air dry to medium bake	Spray, dip, fluid bed	Coatings for circuit boards. Corrosion-protective coatings for metals.
Epoxy-anhydride, Dicy	Solvents—good Alkalies— Dilute acids—	Good		Good	Medium	No	Cured by chemical reaction	High bakes 300 to 400°F	Spray, dip, fluid bed, impregnate	High-bake, high-temperature-resistant dielectric and corrosion coatings.
Epoxy-polyamide	Solvents—fair Alkalies—good Dilute acids—poor	Good		Good	Medium	No	Cured by coreactant	Air dry or medium bake	Spray, dip	Coatings for circuit boards. Filleting coating.
Epoxy-phenolic	Solvents—excellent Alkalies—fair Dilute acids—good	Excellent	Pigmented—fair; clear—poor	Good	Medium	No	Cured by coreactant	High bakes 300 to 400°F	Spray, dip	High-bake solvent and chemical resistant coating.
Fluorocarbon TFE	Solvents—excellent Alkalies—good Dilute acids—excellent	Excellent		Good	None	No	Fusion from water or solvent dispersion	Approximately 750°F	Spray, dip	High-temperature resistant insulation for wiring.
FEP CTFE		Excellent		Good	None	No	Fusion from water or solvent dispersion	500–600°F	Spray, dip	High-temperature-resistant insulation. Extrudable.
Parylene (polyxylylenes)		Excellent			None		Vapor phase deposition and polymerization requiring special license from Union Carbide.			Very thin, pinhole-free coatings, possible semiconductable coating.
Phenolics	Solvents—good to excellent Alkalies—poor Dilute acids—good to excellent	Excellent	Fair	Poor to good	Medium	No	Cured by heat	Bake 350–500°F	Spray, dip	High-bake chemical and solvent-resistant coatings.

Phenolic-oil varnish	Solvents—poor Alkalies—poor Dilute acids—good to excellent	Good	Good	Poor, unless toxic—additive	Medium	Poor	Oxidation or heat		Spray, brush, dip-impregnate	Impregnation of electronic modules, quick protective coating.
Phenoxy		Good		Good			Cured by heat			Chemical resistant coating.
Polyamide (nylon)		Fair				Fairly solderable				Wire coating.
Polyester	Solvents—poor Alkalies—poor to fair Dilute acids—good	Fair	Very good	Good	Medium	Poor	Cured by heat or catalyst	Air dry or bake 100–250°F	Spray, brush, dip	
Chlorosulfonated polyethylene (Hypalon†)		Good		Good	Low		Solvent evaporation	Air dry or low-temperature bake	Spray, brush	Moisture and fungus proofing of materials.
Polyimide	Solvents—excellent Alkalies—poor to fair Dilute acids—good	Good	Good	Good	Low	Poor	Cured by heat	High bake	Dip, impregnate, wire coater	Very high temperature resistant with insulation.
Polystyrene		Good		Good	High	Dissolve with solvents	Solvent evaporation	Air dry or low bake	Spray, dip	Coil coating, low dielectric constant, low loss in radar uses.
Polyurethane	Solvents—good Dilute alkalies—fair Dilute acids—good	Good		Poor to good	Medium	Excellent; melts, solder-through properties	Coreactant or moisture cure	Air dry to medium bake	Spray, brush, dip	Conformal coating of circuitry, solderable wire insulation.
Silicone	Solvents—poor Alkalies—good (dilute) poor (concentrated) Dilute acids—good	Excellent	Excellent	Good	Very low (except in O_2 atmosphere)	Fair to excellent. Cut and peel	Cured by heat or catalyst	Air dry (RTV) to high bakes	Spray, brush, dip	Heat-resistant coating for electronic circuitry. Good moisture resistance.
Vinyl chloride (poly-)	Solvents—alcohol, good Alkalies—good	Good	Pigmented—fair to good Clear—poor	Poor to good (depends on plasticizer)	Very low	Dissolve with solvents	Solvent evaporation	Air dry or elevated temperature for speed	Spray, dip, roller coat	Wire insulation. Metal protection (especially magnesium, aluminum).
Vinyl chloride (plastisol, organisol)		Good		Poor to good (depends on plasticizer)	Low	Poor	Fusion of liquid to gel	Bake 250–350°F	Spray, dip, reverse roll	Soft-to-hard thick coatings, electroplating racks, equipment.
Vinyl fluoride		Good	Excellent	Good	Very low	Poor	Fusion from solvent dispersion	Bake 400–500°F	Spray, roller coat	Coatings for circuitry. Long-life exterior finish.
Vinyl formal (Form-var‡)		Good		Good	Medium	Poor	Cured by heat	Bake 300–500°F	Roller coat, wire coater	Wire insulation (thin coatings) coil impregnation.

*Trademark of Hercules Powder Co., Inc., Wilmington, Del.
†Trademark of E. I. du Pont de Nemours & Co., Wilmington, Del.
‡Trademark of Monsanto Co., St. Louis, Mo.
SOURCE: This table has been reprinted from *Machine Design*, May 25, 1967. Copyright, 1967, by The Penton Publishing Company, Cleveland, Ohio.

whose choice of such a coating is based on properties of the generic resin, can be greatly disappointed. Instead, selections must be made on the basis of performance data for specific coatings or finish systems. Performance data are generated by the paint and product manufacturing industries when conducting standard paint evaluation tests. Test methods for coating material evaluation are listed in Table 6.7.

6.4.4 Selection by Electrical Properties

Electrical properties of organic coatings and finishes vary by resin (also referred to as *polymer*) type. When selecting insulating varnishes, insulating enamels, and magnet wire enamels, the electrical properties and physical properties determine the choice.

Table 6.8 shows electric strengths, Table 6.9 shows volume resistivities, Table 6.10 shows dielectric constants, and Table 6.11 shows dissipation factors for coatings using most of the available resins. Magnet wire insulation is an important use for organic coatings. National Electrical Manufacturer's Association (NEMA) standards and manufacturers' trade names for various wire enamels are shown in Table 6.12. This information can be used to guide the selection of coatings. However, it is important to remember the aforementioned warnings about blends of various resins and the effects on performance properties.

6.5 Coating and Finishing Materials

Since it is the resin in the coating's vehicle that determines its performance properties, coatings and finishes can be classified by their resin types. The most widely used resins for manufacturing modern coatings and finishes are acrylics, alkyds, epoxies, polyesters, polyurethanes, and vinyls.[3] In the following section, the resins used in coatings and finishes are described.

6.5.1 Common Coating Resins

6.5.1.1 Acrylics. Acrylics are noted for color and gloss retention in outdoor exposure. Acrylics are supplied as solvent-containing, high-solids, waterborne, and powder coatings. They are formulated as lacquers, enamels, and emulsions. Lacquers and baking enamels are used as automotive and appliance finishes. Both these industries use acrylics as topcoats for multicoat finish systems. Thermosetting acrylics have replaced alkyds in applications requiring greater mar resistance such as appliance finishes. Acrylic lacquers are brittle and therefore have poor impact resistance, but their outstanding weather resistance allowed them to replace nitrocellulose lacquers in automotive finishes for many years. Acrylic and modified acrylic emulsions have been used as architectural coatings and finishes and also on industrial products. These medium-priced resins can be formulated to have excellent hardness, adhesion, abrasion, chemical, and mar resistance. When acrylic resins are used to modify other resins, their properties are often imparted to the resultant resin system.

Uses. Acrylics, both lacquers and enamels, were the topcoats of choice for the automotive industry from the early 1960s to the mid 1980s. Thermosetting acrylics are still used by the major appliance industry. Acrylics are used in electrodeposition and have largely replaced alkyds. The chemistry of acrylic-based resins allows them to be used in radiation curing applications alone or as monomeric modifiers for other resins. Acrylic-modified polyurethane coatings and finishes have excellent exterior durability.

TABLE 6.7 Specific Test Methods for Coatings*

Test	ASTM	Federal Std. 141a, method	MIL-Std.-202, method	Federal Std. 406, method	Others
Abrasion	D968	6191 (Falling sand) 6192 (Taber)		1091	Fed. Std. 601, 14111
Adhesion	D2197	6301.1 (Tape test, wet) 6302.1 (Microknife) 6303.1 (Scratch adhesion) 6304.1 (Knife test)		1111	Fed. Std. 601, 8031
Arc resistance	D495		303	4011	
Dielectric constant	D150		301	4021	
Dielectric strength (breakdown voltage)	D149 D115			4031	Fed. Std. 101, 303 Fed. Std. 601, 13311
Dissipation factor	D150			4021	
Drying time	D1640 D115	4061.1			
Electrical insulation resistance	D229 D257		302	4041	MIL-W-81044, 4.7.5.2
Exposure (interior)	D1014	6160 (On metals) 6161.1 (Outdoor rack)			
Flash point	D56, D92 D1310 (Tag open cup)	4291 (Tag Closed Cup) 4294 (Cleveland Open Cup)			Fed. Std. 810, 509
Flexibility	D1737 D522	6221 (Mandrel) 6222 (Conical Mandrel)		1031	Fed. Std. 601, 11041
Fungus resistance	D1924				MIL-E-5272, 4.8 MIL-STD-810, 508.1 MIL-T-5422, 4.8
Hardness	D1474	6211 (Print Hardness) 6212 (Indentation)			
Heat resistance	D115 D1932	6051			
Humidity	D2247	6071 (100% RH) 6201 (Continuous Condensation)	103 106A		MIL-E-5272, proc. 1 Fed. Std. 810, 507

TABLE 6.7 Specific Test Methods for Coatings* *(Continued)*

Test	ASTM	Federal Std. 141a, method	MIL-Std.-202, method	Federal Std. 406, method	Others
Impact resistance	D2794	6226 (G.E. Impact)	1074		
Moisture-vapor permeability	E 96 D1653	6171	7032		
Nonvolatile content		4044			
Salt spray (fog)	B117	6061	101C	6071	MIL-STD-810, 509.1 MIL-E-5272, 4.6 Fed. Std. 151, 811.1 Fed. Std. 810, 509
Temperature-altitude					MIL-E-5272, 4.14 MIL-T-5422, 4.1 MIL-STD-810, 504.1
Thermal conductivity	D1674 (Cenco Fitch) C177 (Guarded hot plate)				MIL-I-16923, 4.6.9
Thermal shock			107		MIL-E-5272, 4.3 MIL-STD-810,503.1
Thickness (dry film)	D1005 D1186	6181 (Magnetic Gage) 6183 (Mechanical Gage)		2111, 2121, 2131, 2141, 2151	Fed. Std. 151, 520, 521.1
Viscosity	D1545 D562 D1200 D88	4271 (Gardner Tubes) 4281 (Krebs-Stormer) 4282 (Ford Cup) 4285 (Saybolt) 4287 (Brookfield)			
Weathering (accelerated)	D822	6151 (Open Arc) 6152 (Enclosed Arc)		6024	

*Note: A more complete compilation of test methods is found in J.J. Licari, *Plastic Coatings for Electronics,* McGraw-Hill, New York, 1970. The major collection of complete test methods for coatings is *Physical and Chemical Examination of Paints, Varnishes, Lacquers, and Colors,* by Gardner and Sward, Gardner Laboratory, Bethesda, MD. This has gone through many editions.

TABLE 6.8 Electric Strengths of Coatings

Material	Dielectric strength, V/mil	Comments*	Source of information
Polymer coatings:			
Acrylics	450–550	Short-time method	a
	350–400	Step-by-step method	a
	400–530		b
	1700–2500	2-mil-thick samples	Columbia Technical Corporation, Humiseal Coatings
Alkyds	300–350		b
Chlorinated polyether	400	Short-time method	a
Chlorosulfonated polyethylene	500	Short-time method	a
Diallyl phthalate	275–450		h
	450	Step-by-step method	a
Diallyl isophthalate	422	Step-by-step method	a
Depolymerized rubber (DPR)	360–380		H. V. Hardman Company, DPR Subsidiary
Epoxy	650–730	Cured with anhydride–castor oil adduct	Autonetics, Division of North American Rockwell
Epoxy	1300	10-mil-thick dip coating	
Epoxies, modified	1200–2000	2-mil-thick sample	Columbia Technical Corporation, Humiseal Coatings
Neoprene	150–600	Short-time method	a
Phenolic	300–450		b
Polyamide	780	106 mil thick sample	
Polyamide-imide	2700		
Polyesters	250–400	Short-time method	a
	170	Step-by-step method	a
Polyethylene	480		b
	300	60-mil-thick sample	c
	500	Short-time method	a
Polyimide	3000	Pyre-ML, 10 mils thick	d
	4500–5000	Pyre-ML (RC-675)	e
	560	Short-time method, 80 mils thick	
Polypropylene	750–800	Short-time method	a
Polystyrene	500–700	Short-time method	a
	400–600	Step-by-step method	a
	450	60-mil-thick sample	c
Polysulfide	250–600	Short-time method	a
Polyurethane (single component)	3800	1-mil-thick sample	f
Polyurethane (two components)/castor oil cured	530–1010		g
Polyurethane (two components, 100% solids)	275	125-mil-thick sample	Products Research & Chemical Corporation (PR-1538)
	750	25-mil thick sample	
Polyurethane (single component)	2500	2-mil-thick sample	Columbia Technical Corporation, Humiseal 1A27
Polyvinyl butyral	400		
Polyvinyl chloride	300–1000	Short-time method	
	275–900	Step-by-step method	b
Polyvinyl formal	860–1000		
Polyvinylidene fluoride	260	Short-time, 500-V/s, $^1/_8$-in sample	h
	1280	Short-time, 500-V/s, 8-mil sample	h
	950	Step by step (1-kV steps)	h
Polyxylylenes:			
Parylene N	6000	Step by step	Union Carbide Corporation
	6500	Short time	Union Carbide Corporation
Parylene C	3700	Short time	Union Carbide Corporation
	1200	Step by step	Union Carbide Corporation
Parylene D	5500	Short time	Union Carbide Corporation
	4500	Step by step	Union Carbide Corporation
Silicone	500	Sylgard 182	Dow Corning Corporation
Silicone	550–650	RTV types	General Electric & Stauffer Chemical Company bulletins
Silicone	800	Flexible dielectric gel	Dow Corning Corporation
Silicone	1500	2-mil-thick sample	Columbia Technical Corporation, Humiseal 1H34
TFE fluorocarbons	400	60-mil-thick sample	c
	480	Short-time method	a
	430	Step by step	a
Teflon TFE dispersion coating	3000–4500	1- to 4-mil-thick sample	E. I. du Pont de Nemours & Company
Teflon FEP dispersion coating	4000	1.5-mil-thick sample	E. I. du Pont de Nemours & Company

TABLE 6.8 Electric Strengths of Coatings *(Continued)*

Material	Dielectric strength, V/mil	Comments*	Source of information
Other materials used in electronic assemblies:			
Alumina ceramics	200–300		[b]
Boron nitride	900–1400		
Electrical ceramics	55–300		[b]
Forsterite	250		[b]
Glass, borosilicate	4500	40-mil sample	[c]
Steatite	145–280		[b]

*All samples are standard 125 mils thick unless otherwise specified.
[a] *Insulation*, Directory Encyclopedia Issue, no. 7, June–July 1968.
[b] *Material Engineering*, Materials Selector Issue, vol. 66, no. 5, Chapman-Reinhold Publication, mid-October 1967–1968.
[c] W. H. Kohl, *Handbook of Materials and Techniques for Vacuum Devices*, p. 586, Reinhold Publishing Corporation, New York, 1967, p. 586.
[d] J. R. Learn and M. P. Seegers, "Teflon-Pyre-M. L. Wire Insulation System," *13th Symposium on Technical Progress in Commun. Wire and Cables*, Atlantic City, NJ, December 2–4, 1964.
[e] J. T. Milek: Polyimide Plastics: A State of the Art Report, *Hughes Aircraft Report*, S-8, October, 1965.
[f] *Hughson Chemical Co.* Bulletin 7030A.
[g] *Spencer-Kellogg* (Division of Textron, Inc.) Bulletin TS-6593.
[h] *Pennsalt Chemicals Corp. Prod. Sheet* KI-66a, Kynar Vinylidene Fluoride Resin, 1967.

6.5.1.2 Alkyds.

Alkyd resin-based coatings and finishes were introduced in the 1930s as replacements for nitrocellulose lacquers and oleoresinous coatings. They offer the advantage of good durability at relatively low cost. These low- to medium-priced coatings and finishes are still used for finishing a wide variety of products, either alone or modified with oils or other resins. The degree and type of modification determine their performance properties. They were used extensively by the automotive and appliance industries through the 1960s. Although alkyds are used in outdoor applications, they are not as durable in long-term exposure, and their color and gloss retention is inferior to that of acrylics.

Uses. Once the mainstay of organic coatings and finishes, alkyds are still used for finishing metal and wood products. Their durability in interior exposures is generally good, but their exterior durability is only fair. Alkyd resins are used in fillers, sealers, and caulks for wood finishing because of their formulating flexibility. Alkyds have also been used in electrodeposition as replacements for the oleoresinous vehicles. They are still used for finishing by the machine tool and other industries. Alkyds have also been widely used in architectural and trade sales coatings. Alkyd-modified acrylic latex paints are excellent architectural finishes.

6.5.1.3 Epoxies.

Epoxy resins can be formulated with a wide range of properties. These medium- to high-priced resins are noted for their adhesion, make excellent primers, and are used widely in the appliance and automotive industries. Their heat resistance permits them to be used for electrical insulation. When epoxy topcoats are used outdoors, they tend to chalk and discolor because of inherently poor ultraviolet light resistance. Other resins modified with epoxies are used for outdoor exposure as topcoats, and properties of many other resins can be improved by their addition. Two-component epoxy coatings and finishes are used in environments with extreme corrosion and chemical conditions. Flexibility in formulating two-component epoxy resin-based coatings and finishes results in a wide range of physical properties.

Uses. Owing to their excellent adhesion, they are used extensively as primers for most coatings and finishes over most substrates. Epoxy coatings and finishes provide excellent chemical and corrosion resistance. They are used as electrical insulating coatings and finishes because of their high electric strength at elevated temperatures. Some of the

TABLE 6.9 Volume Resistivities of Coatings

Material	Volume resistivity at 25°C, Ω-cm	Source of information
Acrylics	10^{14}–10^{15}	[a]
	$>10^{14}$	[b]
	7.6×10^{14}–1.0×10^{15}	Columbia Technical Corporation, Humiseal
Alkyds	10^{14}	[a]
Chlorinated polyether	10^{15}	[b]
Chlorosulfonated polyethylene	10^{14}	[b]
Depolymerized rubber	1.3×10^{13}	H. V. Hardman, DPR Subsidiary
Diallyl phthalate	10^{8}–2.5×10^{10}	[a]
Epoxy (cured with DETA)	2×10^{16}	[c,d]
Epoxy polyamide	1.1–1.5×10^{14}	
Phenolics	6×10^{12}–10^{13}	[a]
Polyamides	10^{13}	
Polyamide-imide	7.7×10^{16}	[e]
Polyethylene	$>10^{16}$	
Polyimide	10^{16}–10^{18}	[f]
Polypropylene	10^{10}–$>10^{16}$	[a]
Polystyrene	$>10^{16}$	[b]
Polysulfide	2.4×10^{11}	[g]
Polyurethane (single component)	5.5×10^{12}	[h]
Polyurethane (single component)	2.0×10^{12}	[h]
Polyurethane (single component)	4×10^{13}	Columbia Technical Corporation
Polyurethane (two components)	1×10^{13}	Products Research & Chemical
	5×10^{9}(300°F)	Corporation (PR-1538)
Polyvinyl chloride	10^{11}–10^{15}	[b]
Polyvinylidene chloride	10^{14}–10^{16}	[a]
Polyvinylidene fluoride	2×10^{14}	[h]
Polyxylylenes (parylenes)	10^{16}–10^{17}	Union Carbide Corporation
Silicone (RTV)	6×10^{14}–3×10^{15}	Stauffer Chemical Company, Si-O-Flex SS 831, 832, & 833
Silicone, flexible dielectric gel	1×10^{15}	Dow Corning Corporation
Silicone, flexible, clear	2×10^{15}	Dow Corning Corporation
Silicone	3.3×10^{14}	Columbia Technical Corporation Humiseal 1H34
Teflon TFE	$>10^{18}$	[b]
Teflon FEP	$>2 \times 10^{18}$	[b]

[a] *Materials Engineering,* Materials Selector Issue, vol. 66, no. 5, Chapman-Reinhold Publication, mid-October 1967–1968.
[b] *Insulation,* Directory Encyclopedia Issue, no. 7, June–July 1968.
[c] H. Lee and K. Neville, *Epoxy Resins,* McGraw-Hill, New York, 1966.
[d] Tucker, Cooperman, and Franklin, Dielectric Properties of Casting Resins, *Electronics Equipment,* July 1956.
[e] J. H. Freeman, "A New Concept in Flat Cable Systems," *5th Annual Symposium on Advanced Technology for Aircraft Electrical Systems,* Washington, D.C., October 1964.
[f] J. T. Milek, "Polyimide Plastics: A State of the Art Report," *Hughes Aircraft Rep.* S-8, October 1965.
[g] L. Hockenberger, *Chem.-Ing. Tech.,* vol. 36, 1964.
[h] *Hughson Chemical Company Technical Bulletin* 7030A; *Pennsalt Chemicals Corporation Product Sheet* KI-66a, Kynar Vinylidene Fluoride Resin, 1967.

TABLE 6.10 Dielectric Constants of Coatings

Coating	60–100 Hz	10^6 Hz	>10^6 Hz	Reference source
Acrylic		2.7–3.2		*a*
Alkyd			3.8 (10^{10} Hz)	*b*
Asphalt and tars			3.5 (10^{10} Hz)	*b*
Cellulose acetate butyrate		3.2–6.2		*a*
Cellulose nitrate		6.4		*a*
Chlorinated polyether	3.1	2.92		*a*
Chlorosulfonated polyethylene (Hypalon)	6.19 7–10 (10^3 Hz)	~5		E. I. du Pont de Nemours & Company
Depolymerized rubber (DPR)	4.1–4.2	3.9–4.0		H. V. Hardman, DPR Subsidiary
Diallyl isophthalate	3.5	3.2	3 (10^8 Hz)	*a,c*
Diallyl phthalate	3–3.6	3.3–4.5		*a,c*
Epoxy-anhydride—castor oil adduct	3.4	3.1	2.9 (10^7 Hz)	Autonetics, Division of North American Rockwell
Epoxy (one component)	3.8	3.7		Conap Inc.
Epoxy (two components)	3.7			Conap Inc.
Epoxy cured with methyl nadic anhydride (100:84 pbw)	3.31			*d*
Epoxy cured with dodecenylsuccinic anhydride (100:132 pbw)	2.82			*d*
Epoxy cured with DETA	4.1	4.2	4.1	*e*
Epoxy cured with *m*-phenylenediamine	4.6	3.8	3.25 (10^{10} Hz)	*e*
Epoxy dip coating (two components)	3.3	3.1		Conap Inc.
Epoxy (one component)	3.8	3.5		Conap Inc.
Epoxy-polyamide (40% Versamid* 125, 60% epoxy)	3.37	3.08		*e*
Epoxy-polyamide (50% Versamid 125, 50% epoxy)	3.20	3.01		*e*
Fluorocarbon (TFE, Teflon)	2.0–2.08	2.0–2.08		E. I. du Pont de Nemours & Company
Phenolic		4–11		*a*
Phenolic	5–6.5	4.5–5.0		*c*
Polyamide	2.8–3.9	2.7–2.96		
Polyamide-imide	3.09	3.07		
Polyesters	3.3–8.1	3.2–5.9		*c*
Polyethylene		2.3		*a*
Polyethylenes	2.3	2.3		*c*
Polyimide-Pyre-M.L.† enamel	3.8	3.8		*f*
Polyimide-Du Pont RK-692 varnish	3.8			*g*
Polyimide-Du Pont RC-B-24951	3.0(10^3 Hz)			*g*
Polyimide-Du Pont RC-5060	2.8(10^3 Hz)			*g*
Polypropylene		2.1		*a*

TABLE 6.10 Dielectric Constants of Coatings *(Continued)*

Coating	60–100 Hz	10^6 Hz	>10^6 Hz	Reference source
Polypropylene	2.22–2.28	2.22–2.28		[c]
Polystyrene	2.45–2.65	2.4–2.65	2.5(10^{10} Hz)	[b]
Polysulfides	6.9			[h]
Polyurethane (one component)	4.10	3.8		Conap Inc.
Polyurethane (two components—castor oil cured)	6.8(10^3 Hz)	2.98–3.28		[i]
Polyurethane		4.4		Products Research & (two components) Chemical Corporation (PR-1538)
Polyvinyl butyral	3.6	3.33		[a]
Polyvinyl chloride	3.3–6.7	2.3–3.5		[a]
Polyvinyl chloride—vinyl acetate copolymer	3–10			[a]
Polyvinyl formal	3.7	3.0		[a]
Polyvinylidene chloride		3–5		[a]
Polyvinylidene fluoride	8.1	6.6		[i]
Polyvinylidene fluoride	8.4	6.43	2.98(10^9 Hz)	[k]
p-Polyxylylene:				
Parylene N	2.65	2.65		Union Carbide Corporation
Parylene C	3.10	2.90		Union Carbide Corporation
Parylene D	2.84	2.80		Union Carbide Corporation
Shellac (natural, dewaxed)	3.6	3.3	2.75(10^9 Hz)	[l]
Silicone (RTV types)	3.3–4.2	3.1–4.0		General Electric and Stauffer Chemical Companies
Silicone (Sylgard‡ type)	2.88	2.88		Dow Corning Corporation
Silicone, flexible	3.0			Dow Corning dielectric gel Corporation
FEP dispersion coating	2.1(10^3 Hz)			E. I. du Pont de Nemours & Company
TFE dispersion coating	2.0–2.2(10^3 Hz)			E. I. du Pont de Nemours & Company
Wax (paraffinic)	2.25	2.25	2.22(10^{10} Hz)	[l]

*Trademark of General Mills, Inc., Kankakee, Ill.
†Trademark of E. I. du Pont de Nemours & Company, Wilmington, Del.
‡Trademark of Dow Corning Corporation, Midland, Mich.
[a]*Materials Engineering*, Materials Selector Issue, vol. 66, no. 5, Chapman-Reinhold Publication, mid-October 1967–1968.
[b]M. C. Volk, J. W. Lefforge, and R. Stetson, *Electrical Encapsulation*, Reinhold Publishing Corporation, New York, 1962.
[c]*Insulation*, Directory Encyclopedia Issue, no. 7, June–July 1968.
[d]C. F. Coombs, ed., *Printed Circuits Handbook*, McGraw-Hill, New York, 1967.
[e]H. Lee and K. Neville, *Handbook of Epoxy Resins*, McGraw-Hill, New York, 1967.
[f]J. R. Learn and M. P. Seeger, Teflon-Pyre-M.L. Wire Insulation System, *13th Symposium of Technical Progress in Communications Wire and Cable*, Atlantic City, NJ, Dec. 2–4, 1964.
[g]*Du Pont Bulletin* H65-4, Experimental Polyimide Insulating Varnishes, RC-B-24951 and RC-5060, January 1965.
[h]L. Hockenberger, *Chem.-Ing. Tech.* vol. 36, 1964.
[i]*Spencer-Kellogg* (division of Textron, Inc.) Bull. TS-6593.
[j]W. S. Barnhart, R. A. Ferren, and H. Iserson, 17th ANTEC of SPE, January 1961.
[k]*Pennsalt Chemicals Corporation Product Sheet* KI-66a, Kynar Vinylidene Fluoride Resin, 1967.
[l]A. R. Von Hippel, ed., *Dielectric Materials and Applications*, Technology Press of MIT and John Wiley & Sons, Inc., New York, 1961.

TABLE 6.11 Dissipation Factors of Coatings

Coating	60–100 Hz	10^6 Hz	$>10^6$ Hz	Reference source
Acrylics	0.04–0.06	0.02–0.03		*a*
Alkyds	0.003–0.06			*a*
Chlorinated polyether	0.01	0.01		*a*
Chlorosulfonated polyethylene	0.03	0.07(10^3 Hz)		*b*
Depolymerized rubber (DPR)	0.007–0.013	0.0073–0.016		H. V. Hardman, DPR Subsidiary
Diallyl phthalate	0.010	0.011	0.011	*b*
Diallyl isophthalate	0.008	0.009	0.014(10^{10} Hz)	*b*
Epoxy dip coating (two components)	0.027	0.018		Conap Inc.
Epoxy (one component)	0.011	0.004		Conap Inc.
Epoxy (one component)	0.008	0.006		Conap Inc.
Epoxy polyamide (40% versamid 125, 60% epoxy)	0.0085	0.0213		
Epoxy polyamide (50% Versamid 115, 50% epoxy)	0.009	0.0170		
Epoxy cured with anhydride–castor oil adduct	0.0084	0.0165	0.0240	Autonetics, North American Rockwell
Phenolics	0.005–0.5	0.022		*a*
Polyamide	0.015	0.022–0.097		
Polyesters	0.008–0.041			*a*
Polyethylene (linear)	0.00015	0.00015	0.0004(10^{10} Hz)	*d*
Polymethyl methacrylate	0.06	0.02	0.009(10^{10} Hz)	*d*
Polystyrene	0.0001–0.0005	0.0001–0.0004		
Polyurethane (two component, castor oil cure)		0.016–0.036		
Polyurethane (one component)	0.038–0.039	0.068–0.074		Conap Inc.
Polyurethane (one component)	0.02			Conap Inc.
Polyvinyl butyral	0.007	0.0065		
Polyvinyl chloride	0.08–0.15	0.04–0.14		*a*
Polyvinyl chloride, plasticized	0.10	0.15	0.01(10^{10} Hz)	*d*
Polyvinyl chloride-vinyl acetate copolymer	0.6–0.10			

Polyvinyl formal	0.007	0.02		
Polyvinylidene fluoride	0.049	0.17		[f]
	0.049	0.159	0.110	[g]
Polyxylylenes:				
Parylene N	0.0002	0.0006		Union Carbide Corporation
Parylene C	0.02	0.0128		Union Carbide Corporation
Parylene D	0.004	0.0020		Union Carbide Corporation
Silicone (Sylgard 182)	0.001	0.001		Dow Corning Corporation
Silicone, flexible dielectric gel	0.0005			Dow Corning Corporation
Silicone, flexible, clear	0.001			Dow Corning Corporation
Silicone (RTV types)	0.011–0.02	0.003–0.006		General Electric
Teflon FEP dispersion coating	0.0002–0.0007			E. I. du Pont de Nemours & Company
Teflon FEP	<0.0003	<0.0003		[a]
Teflon TFE	<0.0003			[a]
Teflon TFE	0.00012	0.00005		Union Carbide Corporation
Other materials:				
Alumina (99.5%)		0.0001		[h]
Beryllia (99.5%)		0.0003		[h]
Glass silica	0.0006	0.0001	0.00017(10^{10} Hz)	[d]
Glass, borosilicate		0.013–0.016		Corning Glass Works
Glass, 96% silica		0.0015–0.0019		Corning Glass Works

[a] *Machine Design*, Plastics Reference Issue, vol. 38, no. 14, Penton Publishing Co., 1966.
[b] *Insulation*, Directory, Encyclopedia Issue, no. 7, June–July 1968.
[c] H. Lee and K. Neville, *Handbook of Epoxy Resins*, McGraw-Hill, New York, 1967.
[d] K. Mathes, Electrical Insulation Conference, 1967.
[e] *Spencer-Kellogg* (Division of Textron, Inc.) *Technical Bulletin* TS-6593.
[f] W. S. Barnhart, R. A. Ferren, and H. Iserson, 17th ANTEC of SPE, January 1961.
[g] *Pennsalt Chemicals Corporation Product Sheet* KI-66a, Kynar Vinylidene Fluoride Resin, 1967.
[h] *Machine Design*, Design Guide, September 28, 1967.

TABLE 6.12 NEMA Standards and Manufacturers' Trade Names for Magnet Wire Insulation

Manufacturer	Plain enamel	Polyvinyl formal	Polyvinyl formal modified	Polyvinyl formal with nylon overcoat	Polyvinyl formal with butyral overcoat	Polyamide	Acrylic	Epoxy
Thermal class	105°C	105°C	105°C	105°C	105°C	105°C	105°C	130°C
NEMA Standard*	MW 1	MW 15	MW 27	MW 17	MW 19	MW 6	MW 4	MW 9
Anaconda Wire & Cable Company	Plain enamel	Formvar	Hermetic Formvar	Nyform	Cement-coated Formvar	....	....	Epoxy, epoxy-cement-coated
Asco Wire & Cable Company	Enamel	Formvar	....	Nyform	Formbond	Nylon	Acrylic	Epoxy
Belden Manufacturing Company	Beld enamel	Formvar	....	Nyclad	....	....	....	Epoxy
Bridgeport Insulated Wire Company	....	Formvar	....	....	Quickbond	Quick-Sol	....	....
Chicago Magnet Wire Corporation	Plain enamel	Formvar	....	Nyform	Bondable Formvar	Nylon	Acrylic	Epoxy
Essex Wire Corporation	Plain enamel	Formvar	Formetex	Nyform	Bondex	....	Ensolex/ ESX	Epoxy
General Cable Corporation	Plain enamel	Formvar	Formetic	Formlon	Formese	....	Solder-able acrylic	Epoxy
General Electric Corporation	....	Formex	....	....	....	Nylon	....	....
Haveg-Super Temp Division	....	....	....	....	....	....	....	....
Hitemp Wires Company Division Simplex Wire & Cable Company	....	....	....	....	....	....	....	....
Hudson Wire Company	Plain enamel	Formvar	....	Nyform	Formvar AVC	Ezsol	....	....
New Haven Wire & Cable, Inc.	Plain enamel	....	....	....	....	....	....	....
Phelps Dodge Magnet Wire Corporation	Enamel	Formvar	Hermeteze	Nyform	Bondeze	....	....	....
Rea Magnet Wire Company, Inc.	Plain enamel	Formvar	Hermetic Formvar special	Nyform	Koilset	Nylon	....	Epoxy
Viking Wire Company, Inc.	Enamel	Formvar	....	Nyform	F-Bondall	Nylon	....	....

*National Electrical Manufacturers Association.
SOURCE: Courtesy of Rea Magnet Wire Company, Inc.

original work with powder coating was done using epoxy resins, and they are still applied using this method. Many of the primers used for coil coating are epoxy resin-based.

6.5.1.4 Polyesters. Polyesters are used alone or modified with other resins to formulate coatings and finishes ranging from clear furniture finishes (replacing lacquers) to industrial finishes (replacing alkyds). These moderately priced finishes permit the same formulating flexibility as alkyds but are tougher and more weather resistant. There are basi-

Teflon	Poly-urethane	Polyure-thane with friction surface	Polyure-thane with nylon overcoat	Polyure-thane with butyral overcoat	Polyure-thane with nylon and butyral overcoat	Polyester	Polyester with overcoat	Poly-imide	Polyester polyimide	Ceramic, ceramic-Teflon, ceramic-silicon
200°C MW 10	105°C MW 2	105°C	130°C MW 28	105°C MW 3 (PROP)	130°C MW 29 (PROP)	155°C MW 5	155°C MW 5	220°C MW 16	180°C	180°C+ MW 7
....	Analac	....	Nylac	Cement-coated analac	Cement-coated nylac	....	Anatherm D Anatherm 200	AL 220 ML	Anatherm N	
....	Poly	....	Nypol	Asco bond-P	Asco bond	Ascotherm	Isotherm 200	ML	Anamid M (amide-imide),	
....	Beldure	....	Beldsol	....	....	Isonel	Polyther-malese	ML	Ascomid	
....	Polyure-thane	Uniwind	Poly-nylon	Polybond	....	Isonel 200				
....	Soderbrite	....	Nysod	Bondable polyurethene	....	Polyester 155				
....	Soderex	....	Soderon	Soder-bond	Soder-bond N	Thermalex F	Polyther-malex/ PTX 200	Allex		
....	Enamel "G"	....	Genlon	....	....	Gentherm	Polyther-maleze 200			
....	....	....	....	....	....	Alkanex				
Teflon	....	....	....	....	....	Isonel	....			
Temprite										
....	Hudsol	Gripon	Nypoly	Hudsol AVC	Nypoly AVC	Isonel 200	Isonel 200-A	ML	Isomid	
....	Impsol	....	Impsolon	....	....	....	Imp-200	....	....	
....	Sodereze	Gripeze	Nyleze	....	S-Y Bondeze	Polyther-maleze 200 II	ML			
....	Solvar	....	Nylon solvar	Solvar Koilset	....	Isonel 200	Polyther-maleze 200	Pyre ML	Isomid	Ceroc
....	Polyure-thane	....	Polynylon	P-Bondall	....	Isonel 200	Iso-poly	ML	Isomid Isomid-P	

cally two types of polyesters: two-component and single-package. Two-component polyesters are cured using peroxides, which initiate free-radical polymerization, while single-package polyesters, sometimes called *oil-free alkyds,* are self-curing, usually at elevated temperatures. It is important to realize that, in both cases, the resin formulator can adjust properties to meet most exposure conditions. Polyesters are also applied as powder coatings.

Uses. Two-component polyesters are well known as gel coats for glass-reinforced plastic bathtubs, lavatories, boats, and automobiles. Figure 6.2 shows tub and shower units

Figure 6.2 Polyester gel coats are used to give a decorative and protective surface to tub shower units that are made out of glass fiber-reinforced plastics. *(Courtesy of Owens-Corning Fiberglas Corporation)*

using a polyester gel coat. High-quality one-package polyester finishes are used on furniture, appliances, automobiles, magnet wire, and industrial products. Polyester powder coatings are used as high-quality finishes in indoor and outdoor applications for anything from tables to trucks. They are also used as coil coatings.

6.5.1.5 Polyurethanes. Polyurethane resin-based coatings and finishes are extremely versatile. They are higher in price than alkyds but lower than epoxies. Polyurethane resins are available as oil-modified, moisture-curing, blocked, two-component, and lacquers. Table 6.13. is a selection guide for polyurethane coatings. Two-component polyurethanes can be formulated in a wide range of hardnesses. They can be abrasion resistant, flexible, resilient, tough, chemical resistant, and weather resistant. Abrasion resistance of organic coatings is shown in Table 6.14. Polyurethanes can be combined with other resins to reinforce or adopt their properties. Urethane-modified acrylics have excellent outdoor weathering properties. They can also be applied as air-drying, forced-dried, and baking liquid finishes as well as powder coatings.

Uses. Polyurethanes have become very important finishes in the transportation industry, which includes aircraft, automobiles, railroads, trucks, and ships. Owing to their chemical resistance and ease of decontamination from chemical, biological, and radiological warfare agents, they are widely used for painting military land vehicles, ships, and aircraft. They are used on automobiles as coatings and finishes for plastic parts and as clear topcoats in the basecoat-clearcoat finish systems. Low-temperature baking polyurethanes are used as mar-resistant finishes for products that must be packaged while still warm.

TABLE 6.13 Guide to Selecting Polyurethane Coatings

	One-component				
Property	Urethane Oil	Moisture	Blocked	Two-component	Lacquer
Abrasion resistance	Fair–good	Excellent	Good–excellent	Excellent	Fair
Hardness	Medium	Medium–hard	Medium–hard	Soft–very hard	Soft–medium
Flexibility	Fair–good	Good–excellent	Good	Good–excellent	Excellent
Impact resistance	Good	Excellent	Good–excellent	Excellent	Excellent
Solvent resistance	Fair	Poor–fair	Good	Excellent	Poor
Chemical resistance	Fair	Fair	Good	Excellent	Fair–good
Corrosion resistance	Fair	Fair	Good	Excellent	Good–Excellent
Adhesion	Good	Fair–good	Fair	Excellent	Fair–good
Toughness	Good	Excellent	Good	Excellent	Excellent
Elongation	Poor	Poor	Poor	Excellent	Excellent
Tensile	Fair	Good	Fair–good	Good–excellent	Excellent
Weatherability:					
Aliphatic	Good	Poor–fair		Good–excellent	Good
Conventional	Poor–fair	Poor–fair	Poor–fair	Poor–fair	Poor
Pigmented gloss	High	High	High	High	Medium
Cure rate	Slow	Slow	Fast	Fast	None 150–225°F
Cure temperature	Room temperature	Room temperature	300–390°F	212°F	225°F
Work life	Infinite	1 y	6 months	1–24 hr	Infinite

TABLE 6.14 Abrasion Resistance of Coatings

Coating	Taber ware index, mg/1000 rev.
Polyurethane type 1	55–67
Polyurethane type 2 (clear)	8–24
Polyurethane type 2 (pigmented)	31–35
Polyurethane type 5	60
Urethane oil varnish	155
Alkyd	147
Vinyl	85–106
Epoxy-amine-cured varnish	38
Epoxy-polyamide enamel	95
Epoxy-ester enamel	196
Epoxy-polyamide coating (1:1)	50
Phenolic spar varnish	172
Clear nitrocellulose lacquer	96
Chlorinated rubber	200–220
Silicone, white enamel	113
Catalyzed epoxy, air-cured (PT-401)	208
Catalyzed epoxy, Teflon-filled (PT-401)	122
Catalyzed epoxy, bake-Teflon-filled (PT-201)	136
Parylene N	9.7
Parylene C	44
Parylene D	305
Polyamide	290–310
Polyethylene	360
Alkyd TT-E-508 enamel (cured for 45 min at 250°F)	51
Alkyd TTLE-508 (cured for 24 hr at room temperature)	70

Polyurethanes are used in an increasing number of applications. They are also used in radiation curable coatings.

6.5.1.6 Polyvinyl chloride. Polyvinyl chloride (PVC) coatings, commonly called *vinyls,* are noted for their toughness, chemical resistance, and durability. They are available as solutions, dispersions, and lattices. Properties of vinyl coatings are listed in Table 6.15. They are applied as lacquers, plastisols, organisols, and lattices. PVC coating powders have essentially the same properties as liquids. PVC organisol, plastisol, and powder coatings have limited adhesion and require primers.

Uses. Vinyls have been used in various applications, including beverage and other can linings, automobile interiors, and office machine exteriors. They are also used as thick film liquids and as powder coatings for electrical insulation. Owing to their excellent chemical resistance, they are used as tank linings and as rack coatings in electroplating shops. Typical applications for vinyl coatings are shown in Fig. 6.3. Vinyl-modified acrylic

TABLE 6.15 Properties of Vinyl Coatings

		Mechanical properties[a]					Color and gloss[a]			Weathering[a] properties	
Coating type	Outstanding characteristics	Hardness	Abrasion resistance	Adhesion	Flexibility	Toughness	Film color	Color retention	Gloss	Weather resistance	Gloss retention
Solution[b]	Excellent color, flexibility, chemical resistance; tasteless, odorless	F	E	F to G	E	E	E[f]	E	G	E[f]	E
Plastisol[c]	Toughness; resilience; abrasion resistance; can be applied without solvents	F	E	E to cloth[e]	E	E	E[f]	E	F	E[f]	F to G
Organosol[d]	High solids content; excellent color, flexibility; tasteless, odorless	F	E	E to cloth[e]	E	E	E[f]	E	P to G	E[f]	G

[a]E = excellent; G = good; F = fair; P = poor.
[b]Vinyl chloride acetate copolymers; resins vary widely in compatibility with other materials.
[c]Vinyl chloride acetate copolymer and vinyl chloride resins.
[d]Vinyl chloride acetate copolymers; require grinding for good dispersions.
[e]Requires primer for use on metal.
[f]Pigmented.

Figure 6.3 Vinyl plastisols and organisols are used extensively for dip coating of wire products. The coatings can be varied from very hard to very soft. *(Courtesy of M & T Chemicals)*

latex trade sale paints are used as trim enamels for exterior applications and as semigloss wall enamels for interior applications.

6.5.2 Other Coating Resins

In addition to the aforementioned materials, there are a number of other important resins used in formulating coatings. These materials, used alone or as modifiers for other resins, provide coating vehicles with diverse properties.

6.5.2.1 Aminos. Resins of this type, such as urea formaldehyde and melamine, are used in modifying other resins to increase their durability. Notable among these modified resins are the super alkyds used in automotive and appliance finishes.

Uses. Melamine and urea formaldehyde resins are used as modifiers for alkyds and other resins to increase hardness and accelerate cure.

6.5.2.2 Cellulosics. Nitrocellulose lacquers are the most important of the cellulosics. They were introduced in the 1920s and used as fast-drying finishes for a number of manufactured products. Applied at low solids using expensive solvents, they will not meet air-quality standards. By modifying nitrocellulose with other resins such as alkyds and ureas, the VOC content can be lowered, and performance properties can be increased. Other important cellulosic resins are cellulose acetate butyrate and ethyl cellulose.

Uses. Although no longer used extensively by the automotive industry, nitrocellulose lacquers are still used by the furniture industry because of their fast-drying and hand-rubbing properties. Cellulose acetate butyrate has been used for coating metal in numerous applications. In 1959, one of the first conveyorized powder coating lines in the United States coated distribution transformer lids and hand-hole covers with a cellulose acetate butyrate powder coating.

6.5.2.3 Chlorinated rubber. Chlorinated rubber coatings and finishes are used as swimming pool paints and traffic paints.

6.5.2.4 Fluorocarbons. These high-priced coatings and finishes require high processing temperatures and therefore are limited in their usage. They are noted for their lubricity or nonstick properties due to low coefficients of friction, and also for weatherability. Table 6.16. gives the coefficients of friction of typical coatings.

TABLE 6.16 Coefficients of Friction of Typical Coatings

Coating	Coefficient of friction, μ	Information source
Polyvinyl chloride	0.4–0.5	a
Polystyrene	0.4–0.5	a
Polymethyl methacrylate	0.4–0.5	a
Nylon	0.3	a
Polyethylene	0.6–0.8	a
Polytetrafluoroethylene (Teflon)	0.05–0.1	a
Catalyzed epoxy air-dry coating with Teflon filler	0.15	b
Parylene N	0.25	c
Parylene C	0.29	c
Parylene D	0.31–0.33	c
Polyimide (Pyre–ML)	0.17	d
Graphite	0.18	d
Graphite–molybdenum sulfide:		
Dry-film lubricant	0.02–0.06	e
Steel on steel	0.45–0.60	e
Brass on steel	0.44	e
Babbitt on mild steel	0.33	e
Glass on glass	0.4	e
Steel on steel with SAE no. 20 oil	0.044	e
Polymethyl methacrylate to self	0.8 (static)	e
Polymethyl methacrylate to steel	0.4–0.5 (static)	e

[a] R.P. Bowder, *Endeavor,* Vol. 16, No. 61, 1957, p. 5.
[b] Product Techniques Incorporated, Bulletin on PT–401 TE, October 17, 1961.
[c] Union Carbide data.
[d] *DuPont Technical Bulletin 19,* Pyre–ML Wire Enamel, August 1967.
[e] Electrofilm, Inc. data.

Uses. Fluorocarbons are used as chemical-resistant coatings and finishes for processing equipment. They are also used as nonstick coatings and finishes for cookware, friction-reducing coatings and finishes for tools, and as dry lubricated surfaces in many other consumer and industrial products, as shown in Fig. 6.4. Table 6.17 compares the properties of four fluorocarbons.

6.5.2.5 Oleoresinous. Oleoresinous coatings, based on drying oils such as soybean and linseed, are slow curing. For many years prior to the introduction of synthetic resins,

TABLE 6.17 Properties of Four Fluorocarbons

Property	Polyvinyl fluoride (PVF) $(CH_2–CHF)_n$	Polyvinyl-idene fluoride $(CH_2–CF_2)_n$	Polytrifluorochloroethylene (PTFC1) $(CC1F–CF_2)_n$	Polytetrafluoroethylene (PTFE) $(CF_2–CF_2)_n$
Physical properties:				
Density	1.4	1.76	2.104	2.17–2.21
Fusing temperature, °F	300	460	500	750
Maximum continuous service and temperature, °F	225	300	400	550
Coefficient of friction	0.16	0.16	0.15	0.1
Flammability	Burns	Nonflammable	Nonflammable	Nonflammable
Mechanical properties:				
Tensile strength, lb/in^2	7000	7000	5000	2500–3500
Elongation, %	115–250	300	250	200–400
Izod impact, ft-lb/in		3.8	5	3
Durometer hardness		80	74–78	50–65
Yield strength at 77°F, lb/in^2	6000	5500	4500	1300
Heat-distortion temperature at 66 lb/in^2 °F	NA	300	265	250
Coefficient of linear expansion	2.8×10^5	8.5×10^5	15×10^5	8×10^5
Modulus (tension) $\times 10^5$ lb/in^2	2.5–3.7	1.2	1.9	0.6
Electrical properties:				
Dielectric strength, V/mil		260 (0.125)	500 (0.063)	600 (0.060)
Short time, V/mil (in)	3400 (0.002)			
Dielectric constant, 10^3 Hz	8.5	7.72	2.6	2.1
Arc resistance (77°F) ASTM D 495	NA	60	300	300
Volume resistivity Ω-cm at 50% RH, 77°F	10^{12}	10^{14}	10^{16}	10^{18}
Dissipation factor, 100 Hz	1.6	0.05	0.022	0.0003

Figure 6.4 Nonstick feature of fluorocarbon finishes makes them useful for products such as saws, fan and blower blades, door-lock parts, sliding- and folding-door hardware, skis, and snow shovels. *(Courtesy of E. I. DuPont de Nemours & Company)*

they were used as the vehicles in most coatings. They still find application alone or as modifiers to other resins.

Uses. Oleoresinous vehicles are used in low-cost primers and enamels for structural, marine, architectural, and, to a limited extent, industrial product finishing.

6.5.2.6 Phenolics. Introduced in the early 1900s, phenolics were the first commercial synthetic resins. They are available as 100 percent phenolic baking resins, oil-modified, and phenolic dispersions. Phenolic resins, used as modifiers, will improve the heat and chemical resistance of other resins. Baked phenolic resin-based coatings and finishes are well known for their corrosion, chemical, moisture, and heat resistance.

Uses. Phenolic coatings and finishes are used on heavy-duty air-handling equipment, on chemical equipment, and as insulating varnishes. Phenolic resins are also used as binders for electrical and decorative laminated plastics.

6.5.2.7 Polyamides. One of the more notable polyamide resins is nylon, which is tough, wear resistant, and has a relatively low coefficient of friction. It can be applied as a powder coating by fluidized bed, electrostatic spray, or flame spray. Table 6.18 compares the properties of three types of nylon polymers used in coatings. Nylon coatings and finishes generally require a primer. Polyamide resins are also used as curing agents for two-component epoxy resin coatings. Film properties can be varied widely by polyamide selection.

Uses. Applied as a powder coating, nylon provides a high degree of toughness and mechanical durability to office furniture. Other polyamide resins are used as curing agents in two-component epoxy resin-based primers and topcoats, adhesives, and sealants.

TABLE 6.18 Properties of Nylon Coatings

	Nylon 11	Nylon 6/6	Nylon 6
Elongation (73°F), %	120	90	50–200
Tensile strength (73°F), lb/in^2	8,500	10,500	10,500
Modulus of elasticity (73°F), lb/in^2	178,000	400,000	350,000
Rockwell hardness	R 100.5	R 118	R 112–118
Specific gravity	1.04	1.14	1.14
Moisture absorption, ASTM D 570	0.4	1.5	1.6–2.3
Thermal conductivity, Btu/(ft^2) (h × °F/in)	1.5	1.7	1.2–1.3
Dielectric strength (short time), V/mil	430	385	440
Dielectric constant (10 Hz)	3.5	4	4.8
Effect of:			
Weak acids	None	None	None
Strong acids	Attack	Attack	Attack
Strong alkalies	None	None	None
Alcohols	None	None	None
Esters	None	None	None
Hydrocarbons	None	None	None

6.5.2.8 Polyolefins. These coatings, which can be applied by flame spraying, hot melt, or powder coating methods, have limited usage.

Uses. Polyethylene is used for impregnating or coating packaging materials such as paper and aluminum foil. Certain polyethylene-coated composite packaging materials are virtually moisture proof. Table 6.19 compares the moisture vapor transmission rates of various coatings and films. Polyethylene powder coatings are used on chemical processing and food-handling equipment.

6.5.2.9 Polyimides. Polyimide coatings and finishes have excellent long-term thermal stability, wear, mar and moisture resistance, and electrical properties. They are high in price.

Uses. Polyimide coatings and finishes are used in electrical applications as insulating varnishes and magnet wire enamels in high-temperature, high-reliability applications. They are also used as alternatives to fluorocarbon coatings and finishes on cookware, as shown in Fig. 6.5.

6.5.2.10 Silicones. Silicone resins are high in price and are used alone or as modifiers to upgrade other resins. They are noted for their high-temperature resistance, moisture resistance, and weatherability. They can be hard or elastomeric, baking, or room temperature curing.

Uses. Silicones are used in high-temperature coatings and finishes for exhaust stacks, ovens, and space heaters. Figure 6.6 shows silicone coatings and finishes on fireplace equipment. They are also used as conformal coatings for printed wiring boards, moisture repellents for masonry, weather-resistant finishes for outdoors, and thermal control coatings for space vehicles. The thermal conductivities of coatings are listed in Table 6.20.

TABLE 6.19 Moisture-Vapor Transmission Rates per 24-hr Period of Coatings and Films in g/(mil) (in²)

Coating or Film	MVTR	Information Source
Epoxy-anhydride	2.38	Autonetics data (25°C)
Epoxy-aromatic amine	1.79	Autonetics data (25°C)
Neoprene	15.5	Baer (39°C)
Polyurethane (Magna X–500)	2.4	Autonetics data (25°C)
Polyurethane (isocyanate-polyester)	8.72	Autonetics data (25°C)
Olefane,* polypropylene	0.70	Avisum data
Cellophane (type PVD uncoated film)	134	DuPont
Cellulose acetate (film)	219	DuPont
Polycarbonate	10	FMC data
Mylar†	1.9	Baer (39°C)
	1.8	DuPont data
Polystyrene	8.6	Baer (39°C)
	9.6	Dow data
Polyethylene film	0.97	Dow data (1-mil film)
Saran resin (F120)	0.097–0.45	Baer (39°C)
Polyvinylidene chloride	0.15	Baer (2-mil sample, 40°C)
Polytetrafluoroethylene (PTFE)	0.32	Baer (2-mil sample 40°C)
PTFE, dispersion cast	0.2	DuPont data
Fluorinated ethylene propylene (FEP)	0.46	Baer (40°C)
Polyvinyl fluoride	2.97	Baer (40°C)
Teslar	2.7	DuPont data
Parylene N	14	Union Carbide data (2-mil sample)
Parylene C	1	Union Carbide data (2-mil sample)
Silicone (RTV 521)	120.78	Autonetics data
Methyl phenyl silicone	38.31	Autonetics data
Polyurethane (ABO130–002)	4.33	Autonetics data
Phenoxy	3.5	Lee, Stoffey, and Neville
Alkyd-silicone (DC–1377)	6.47	Autonetics data
Alkyd-silicone (DC–1400)	4.45	Autonetics data
Alkyd-silicone	6.16–7.9	Autonetics data
Polyvinyl fluoride (PT–207)	0.7	Product Techniques Incorporated

*Trademark of Avisun Corporation, Philadelphia, PA
†Trademark of E.I. DuPont de Nemours & Company, Wilmington, DE

6.6 Application Methods

The selection of an application method is as important as the selection of the coating itself. Basically, the application methods for industrial liquid coatings and finishes are dipping, flow coating, and spraying, although some are applied by brushing, rolling, printing, and silk screening. The application methods for powder coatings and finishes are fluidized beds, electrostatic fluidized beds, and electrostatic spray outfits. In these times of environmental awareness, regulation, and compliance, it is mandatory that coatings and finishes be applied in the most efficient manner.[3] Not only will this help meet the air-quality standards, it will also reduce material costs. The advantages and disadvantages of various coating application methods are given in Table 6.21.

Figure 6.5 Polyimide coating is used as a protective finish on the inside of aluminum, stainless steel, and other cookware. *(Courtesy of Mirro Aluminum Company)*

Figure 6.6 Silicone coatings are used as heat-stable finishes for severe high-temperature applications such as fireplace equipment, exhaust stacks, thermal control coatings for spacecraft, and wall and space heaters. *(Courtesy of Copper Development Assn.)*

TABLE 6.20 Thermal Conductivities of Coatings

Material	k value,* cal/(s)(cm^2)(°C/cm) × 10^4	Source of information
Unfilled plastics:		
Acrylic	4–5	a
Alkyd	8.3	a
Depolymerized rubber	3.2	H. V. Hardman, DPR Subsidiary
Epoxy	3–6	b
Epoxy (electrostatic spray coating)	6.6	Hysol Corporation, DK-4
Epoxy (electrostatic spray coating)	2.9	Minnesota Mining & Manufacturing, No. 5133
Epoxy (Epon† 828, 71.4% DEA, 10.7%)	5.2	
Epoxy (cured with diethylenetriamine)	4.8	c
Fluorocarbon (Teflon TFE)	7.0	DuPont
Fluorocarbon (Teflon FEP)	5.8	DuPont
Nylon	10	d
Polyester	4–5	a
Polyethylenes	8	a
Polyimide (Pyre-ML enamel)	3.5	e
Polyimide (Pyre-ML varnish)	7.2	f
Polystyrene	1.73–2.76	g
Polystyrene	2.5–3.3	a
Polyurethane	4–5	n
Polyvinyl chloride	3–4	a
Polyvinyl formal	3.7	a
Polyvinylidene chloride	2.0	a
Polyvinylidene fluoride	3.6	h
Polyxylylene (Parylene N)	3	Union Carbide
Silicones (RTV types)	5–7.5	Dow Corning Corporation
Silicones (Sylgard types)	3.5–7.5	Dow Corning Corporation
Silicones (Sylgard varnishes and coatings)	3.5–3.6	Dow Corning Corporation
Silicone (gel coating)	3.7	Dow Corning Corporation
Silicone (gel coating)	7 (150°C)	Dow Corning Corporation
Filled plastics:		
Epon 828/diethylenetriamine = A	4	b
A + 50% silica	10	b
A + 50% alumina	11	b
A + 50% beryllium oxide	12.5	b
A + 70% silica	12	b
A + 70% alumina	13	b
A + 70% beryllium oxide	17.8	b
Epoxy, flexibilized = B	5.4	i
B + 66% by weight tabular alumina	18.0	i
B + 64% by volume tabular alumina	50.0	i
Epoxy, filled	20.2	Emerson & Cuming, 2651 ft
Epoxy (highly filled)	15–20	Wakefield Engineering Company
Polyurethane (highly filled)	8–11	International Electronic Research Company

TABLE 6.20 Thermal Conductivities of Coatings *(Continued)*

Material	k value,* cal/(s)(cm^2) (°C/cm) × 10^4	Source of information
Other materials used in electronic assemblies:		
Alumina ceramic	256–442 (20–212°F)	*a*
Aluminum	2767–5575	*a*
Aluminum oxide (alumina), 96%	840	*i*
Beryllium oxide, 99%	5500	*i*
Copper	8095–9334	*a*
Glass (Borosill, 7052)	28	*k*
Glass (pot-soda-lead, 0120)	18	*k*
Glass (silica, 99.8% SiO_2)	40	*l*
Gold	7104 (20–212°F)	*a*
Kovar	395	*m*
Mica	8.3–16.5	*a*
Nichrome‡	325	*m*
Silica	40	*k*
Silicon nitride	359	*m*
Silver	9995 (20–212°F)	*a*
Zircon	120–149	*a*

*All values are at room temperature unless otherwise specified.

†Trademark of Shell Chemical Company, New York, N.Y.

‡Trademark of Driver-Harris Company, Harrison, N.J.

[a]*Materials Engineering,* Materials Selector Issue, vol. 66, no. 5, Chapman-Reinhold Publication, mid-October 1967.

[b]D. C. Wolf, *Proceedings, National Electronics and Packaging Symposium,* New York, June 1964.

[c]H. Lee and K. Neville, *Handbook of Epoxy Resins,* McGraw-Hill, New York, 1966.

[d]R. Davis, *Reinforced Plastics,* October 1962.

[e]*DuPont Technical Bulletin* 19, Pyre-ML Wire Enamel, August 1967.

[f]*DuPont Technical Bulletin* 1, Pyre-ML Varnish RK-692, April 1966.

[g]W. C. Teach and G. C. Kiessling, *Polystyrene,* Reinhold Publishing Corporation, New York, 1960.

[h]W. S. Barnhart, R. A. Ferren, and H. Iserson, 17th ANTEC of SPE, January 1961.

[i]A. J. Gershman and J. R. Andreotti, *Insulation,* September 1967.

[j]*American Lava Corporation* Chart 651.

[k]E. B. Shand, *Glass Engineering Handbook,* McGraw-Hill, 1958.

[l]W. D. Kingery, "Oxides for High Temperature Applications," *Proceedings, International Symposium,* Asilomar, Calif., October 1959, McGraw-Hill, New York, 1960.

[m]W. H. Kohl, *Handbook of Materials and Techniques for Vacuum Devices,* Reinhold Publishing Company, New York, 1967.

[n]*Modern Plastics Encyclopedia,* McGraw-Hill, New York, 1968.

TABLE 6.21 Application Methods for Coatings

	Advantages	Limitations	Typical applications
Spray	Fast, adaptable to varied shapes and sizes. Equipment cost is low.	Difficult to completely coat complex parts and to obtain uniform thickness and reproducible coverage.	Motor frames and housings, electronic enclosures, circuit boards, electronic modules, automobiles, and appliances
Dip	Provides thorough coverage, even on complex parts such as tubes and high-density electronic modules.	Viscosity and pot life of dip must be monitored. Speed of withdrawal must be regulated for consistent coating thickness	Small- and- medium sized parts, castings, moisture and fungus proofing of modules, temporary protection of finished machined parts.
Brush	Brushing action provides good "wetting" of surface, resulting in good adhesion. Cost of equipment is lowest.	Poor thickness control; not for precise applications. High labor cost.	Coating of individual components, spot repairs, or maintenance.
Roller	High-speed continuous process; provides excellent control on thickness.	Large runs of flat sheets or coil stock required to justify equipment cost and setup time. Equipment cost is high.	Metal decorating of sheet to be used for aluminum siding, gutter and downspouts, and to fabricate cans, boxes.
Impregnation	Results in complete coverage of intricate and closely spaced parts. Seals fine leaks or pores.	Requires vacuum or pressure cycling or both. Special equipment usually required.	Coils, transformers, field and armature windings, metal castings, and sealing of porous structures.
Fluidized bed	Thick coatings can be applied in one dip. Uniform coating thickness on exposed surfaces. Dry materials are used, saving cost of solvents.	Requires preheating of part to above fusion temperature of coating. This temperature may be too high for some parts.	Motor stators; heavy-duty electrical insulation on castings, metal substrates for circuit boards, heat sinks.
Screen-on	Deposits coating in selected areas through a mask. Provides good pattern deposition and controlled thickness.	Requires flat or smoothly curved surface. Preparation of screens is time-consuming.	Circuit boards, artwork, labels, masking against etching solution, spot insulation between circuitry layers or under heat sinks or components.
Electrocoating	Provides good control of thickness and uniformity. Parts wet from cleaning need not be dried before coating.	Limited number of coating types can be used; compounds must be specially formulated ionic polymers. Often porous, sometimes nonadherent.	Primers for automobile frames and bodies, complex casting such as open work, motor end bells.
Vacuum deposition	Ultrathin, pinhole-free films possible. Selective deposition can be made through masks.	Thermal instability of most plastics; decomposition occurs on products. Vacuum control needed.	Experimental at present. Potential use is in microelectronics, capacitor dielectrics.
Electrostatic spray	High efficient coverage and use of paint on complex parts. Successfully automated.	High equipment cost. Requires specially formulated coatings.	Heat dissipators, electronic enclosures, open-work grills, complex parts, autos, and appliances.

Liquid spray coating equipment can be classified by its atomizing method: air, hydraulic, or centrifugal. These can be subclassified into air atomizing, airless, airless electrostatic, air-assisted airless electrostatic, rotating electrostatic disks and bells, and high-volume, low-pressure types. While liquid dip coating equipment is usually simple, electrocoating equipment is fairly complex, using electrophoresis as the driving force. Other liquid coating methods include flow coating, which can be manual or automated, roller coating, curtain coating, and centrifugal coating. Equipment for applying powder coatings is not as diversified as for liquid coatings. It can only be classified as fluidized bed, electrostatic fluidized bed, and electrostatic spray.

It is important to note that environmental and worker safety regulations can be met, hazardous and nonhazardous wastes can be reduced, and money can be saved by using compliance coatings (those that meet the VOC emission standards) in equipment having the highest transfer efficiency (the percentage of the coating used which actually coats the product and is not otherwise wasted). The theoretical transfer efficiencies (TEs) of coating application equipment are indicated in the text and in Table 6.22, where they are listed in descending order.[4] The aforementioned transfer efficiencies are meant to be used only as guidelines. Actual transfer efficiencies are dependant on a number of factors that are unique to each coating application line.

TABLE 6.22 Theoretical Transfer Efficiencies (TEs) of Coating Application Methods

Coating method	TE, %
Autodeposition	95–100
Centrifugal coating	95–100
Curtain coating	95–100
Electrocoating	95–100
Fluidized-bed powder	95–100
Electrostatic fluidized-bed powder	95–100
Electrostatic-spray powder	95–100
Flow coating	95–100
Roller coating	95–100
Dip coating	95–100
Rotating electrostatic disks and bells	80–90
Airless electrostatic spray	70–80
Air-assisted airless electrostatic spray	70–80
Air electrostatic spray	60–70
Airless spray	50–60
High-volume, low-pressure spray	40–60
Air-assisted airless spray	40–60
Multicomponent spray	30–70
Air-atomized spray	30–40

In the selection of a coating method and equipment, the product's size, configuration, intended market, and appearance must be considered. To aid in the selection of the most efficient application method, each will be discussed in greater detail.

6.6.1 Dip Coating

Dip coating (95–100 percent TE) is a simple coating method in which products are dipped in a tank of coating material, withdrawn, and allowed to drain in the solvent-rich area

above the coating's surface and then allowed to dry. The film thickness is controlled by viscosity, flow, percent solids by volume, and rate of withdrawal. This simple process can also be automated with the addition of a drain-off area, which allows excess coating material to flow back to the dip tank.

Dip coating is a simple, quick method that does not require sophisticated equipment. The disadvantages of dip coating are film thickness differential from top to bottom, resulting in the so-called *wedge effect;* fatty edges on lower parts of products; and runs and sags. Although this method coats all surface areas, solvent reflux can cause low film build. Light products can float off the hanger and hooks and fall into the dip tank. Solvent-containing coatings and finishes in dip tanks and drain tunnels must be protected by fire extinguishers and safety dump tanks. The fire hazard can be eliminated by using waterborne coatings.

6.6.2 Electrocoating

Electrocoating (95–100 percent TE) is a sophisticated dipping method commercialized in the 1960s to solve severe corrosion problems in the automotive industry. In principle, it is similar to electroplating, except that organic coatings, rather than metals, are deposited on products from an electrolytic bath. Electrocoating can be either anodic (deposition of coatings on the anode from an alkaline bath) or cathodic (deposition of coatings on the cathode from an acidic bath). The bath is aqueous and contains very little VOCs. The phenomenon called *throwing power* causes inaccessible areas to be coated with uniform film thicknesses. Electrocoating has gained a significant share of the primer and one-coat enamel market.

Advantages of the electrocoating method include environmental acceptability owing to decreased solvent emissions and increased corrosion protection to inaccessible areas. It is less labor intensive than other methods, and it produces uniform film thickness from top to bottom and inside and outside on products with a complex shape. Disadvantages are high capital equipment costs, high material costs, and more thorough pretreatment. Higher operator skills are required.

6.6.3 Spray Coating

Spray coating (30–90 percent TE), which was introduced to the automobile industry in the 1920s, revolutionized industrial painting. The results of this development were increased production and improved appearance. Electrostatics, which were added in the 1940s, improved transfer efficiency and reduced material consumption. Eight types of spray-painting equipment are discussed in this section. The transfer efficiencies listed are theoretical. The actual transfer efficiency depends on many variables, including the size and configuration of the product and the airflow in the spray booth.

6.6.3.1 Rotating electrostatic disks and bell spray coating. Rotating electrostatic spray coaters (80–90 percent TE) rely on centrifugal force to atomize droplets of liquid as they leave the highly machined, knife-edged rim of an electrically charged rotating applicator. The new higher-rotational-speed applicators will atomize high-viscosity, high-solids coatings and finishes (65 percent volume solids and higher). Disk-shaped applicators are almost always used in the automatic mode, with vertical reciprocators, inside a loop in the conveyor line. Bell-shaped applicators are used in automated systems in the same configurations as spray guns and can also be used manually.

An advantage of rotating disk and bell spray coating is its ability to atomize high-viscosity coating materials. A disadvantage is maintenance of the equipment.

6.6.3.2 High-volume, low-pressure spray coating. High-volume, low-pressure (HVLP) spray coaters (40–60 percent TE) are a development of the early 1960s that has been upgraded. Turbines rather than pumps are now used to supply high volumes of low-pressure heated air to the spray guns. Newer versions use ordinary compressed air. The air is heated to reduce the tendency to condense atmospheric moisture and to stabilize solvent evaporation. Low atomizing pressure results in lower droplet velocity, reduced bounce-back, and reduced overspray.

The main advantage of HVLP spray coating is the reduction of overspray and bounce-back and the elimination of the vapor cloud usually associated with spray painting. A disadvantage is the poor appearance of the cured film

6.6.3.3 Airless electrostatic spray coating. The airless electrostatic spray coating method (70– 80 percent TE) uses airless spray guns with the addition of a dc power source that electrostatically charges the coating droplets. Its advantage over airless spray is the increase in transfer efficiency due to the electrostatic attraction of charged droplets to the product.

6.6.3.4 Air-assisted airless electrostatic spray coating. The air-assisted airless electrostatic spray coating method (70–80 percent TE) is a hybrid of technologies. The addition of atomizing air to the airless spray gun allows the use of high-viscosity, high-solids coatings. Although the theoretical transfer efficiency is in a high range, it is lower than that of airless electrostatic spray coating because of the higher droplet velocity.

The advantage of using the air-assisted airless electrostatic spray method is its ability to handle high-viscosity materials. An additional advantage is better spray pattern control.

6.6.3.5 Air electrostatic spray coating. The air electrostatic spray coating method (60–70 percent TE) uses conventional equipment with the addition of electrostatic charging capability. The atomizing air permits the use of most high-solids coatings.

Air electrostatic spray equipment has the advantage of being able to handle high-solids materials. This is overshadowed by the fact that it has the lowest transfer efficiency of the electrostatic spray coating methods.

6.6.3.6 Airless spray coating. When it was introduced, airless spray coating (50–60 percent TE) was an important paint-saving development. The coating material is forced by hydraulic pressure through a small orifice in the spray gun nozzle. As the liquid leaves the orifice, it expands and atomizes. The droplets have low velocities because they are not propelled by air pressure as in conventional spray guns. To reduce the coating's viscosity without adding solvents, in-line heaters were added.

Advantages of airless spray coating are: less solvent is used, less overspray, less bounce back, and compensation for seasonal ambient air temperature and humidity changes. A disadvantage is its slower coating rate.

6.6.3.7 Multicomponent spray coating. Multicomponent spray coating equipment (30–70 percent TE) is used for applying fast-curing coating system components simultaneously. Since they can be either hydraulic or air-atomizing, their transfer

efficiencies vary from low to medium. They have two or more sets of supply and metering pumps to transport components to a common spray head.

Their main advantage, the ability to apply fast-curing multicomponent coatings, can be overshadowed by disadvantages in equipment cleanup, maintenance, and low transfer efficiency.

6.6.3.8 Air-atomized spray coating. Air-atomized spray coating equipment (30–40 percent TE) has been used to apply coatings and finishes to products since the 1920s. A stream of compressed air mixes with a stream of liquid coating material, causing it to atomize or break up into small droplets. The liquid and air streams are adjustable, as is the spray pattern, to meet the finishing requirements of most products. This equipment is still being used.

The advantage of the air-atomized spray gun is that a skilled operator can adjust fluid flow, air pressure, and the coating's viscosity to apply a high-quality finish on most products. The disadvantages are its low transfer efficiency and ability to spray only low-viscosity coatings, which emit great quantities of VOCs to the atmosphere.

6.6.4 Powder Coating

Powder coating (95–100 percent TE), developed in the 1950s, is a method for applying finely divided, dry, solid resinous coatings and finishes by dipping products in a fluidized bed or by spraying them electrostatically. The fluidized bed is essentially a modified dip tank. During the electrostatic spraying method, charged powder particles are applied, and they adhere to grounded parts until fused and cured. In all cases, the powder coating must be heated to its melt temperature, where a phase change occurs, causing it to adhere to the product and fuse to form a continuous coating film. Elaborate reclaiming systems to collect and reuse oversprayed material in electrostatic spray powder systems boost transfer efficiency. Since the enactment of air-quality standards, this method has grown markedly.

6.6.4.1 Fluidized bed powder coating. Fluidized bed powder coating (95–100 percent TE) is simply a dipping process that uses dry, finely divided plastic materials. A fluidized bed is a tank with a porous bottom plate, as illustrated in Fig. 6.7. The plenum below the porous plate supplies low-pressure air uniformly across the plate. The rising air surrounds and suspends the finely divided plastic powder particles, so the powder-air mixture resembles a boiling liquid. Products that are preheated above the melt temperature of the coating material are dipped in the fluidized bed, where the powder melts and fuses into a continuous coating. Thermosetting powders often require additional heat to cure the film on the product. The high transfer efficiency results from little dragout and no dripping. This method is used to apply heavy coats, 3 to 10 mil, in one dip, uniformly, to complex-shaped products. The film thickness is dependent on the powder chemistry, preheat temperature, and dwell time. It is possible to build a film thickness of 100 mils using higher preheat temperatures and multiple dips. An illustration of film build is presented in Fig. 6.8.

Advantages of fluidized bed powder coating are uniform and reproducible film thicknesses on all complex-shaped product surfaces. Another advantage is a heavy coating in one dip. A disadvantage of this method is the 3-mil minimum thickness required to form a continuous film.

6.6.4.2 Electrostatic fluidized bed powder coating. An electrostatic fluidized bed (95 to 100 percent TE) is essentially a fluidized bed with a high-voltage dc grid in-

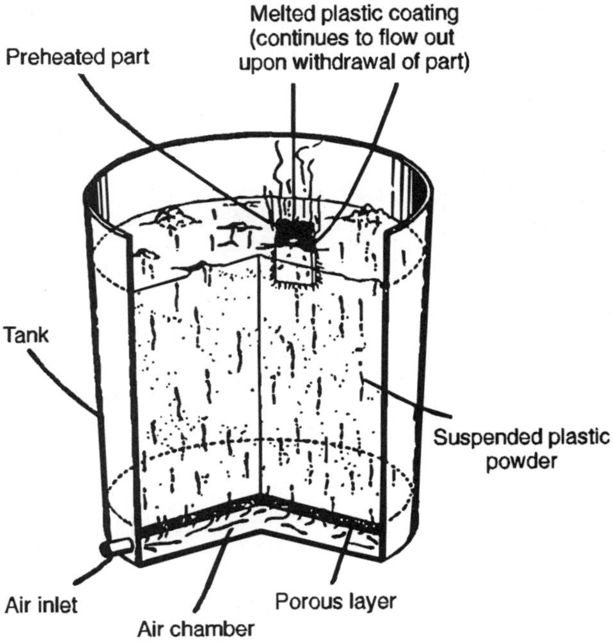

Figure 6.7 Illustration of fluidized-bed process principle.

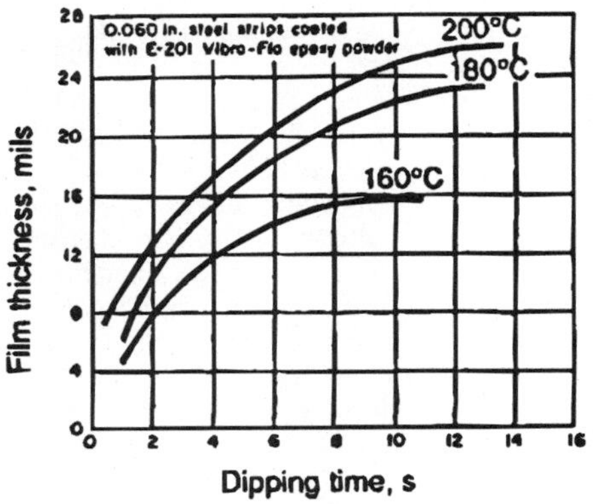

Figure 6.8 Effect of preheat temperature and dipping time on film build in coating a steel bar with epoxy resin.

stalled above the porous plate to charge the finely divided particles. Once charged, the particles are repelled by the grid and repel each other, forming a cloud of powder above the grid. These electrostatically charged particles are attracted to and coat products that are at ground potential. Film thicknesses of 1.5 to 5 mil are possible on cold parts, and of 20 to 25 mil are possible on heated parts.

The advantage of the electrostatic fluidized bed is that small products, such as electrical components, can be coated uniformly and quickly. The disadvantages are that the product size is limited, and inside corners have low film thicknesses owing to the well known Faraday cage effect.

6.6.4.3 Electrostatic spray powder coating. Electrostatic spray powder coating (95–100 percent TE) is a method for applying finely divided, electrostatically charged plastic particles to products that are at ground potential. A powder-air mixture from a small fluidized bed in the powder reservoir is supplied by a hose to the spray gun, which has a charged electrode in the nozzle fed by a high-voltage dc power pack. In some cases, the powder is electrostatically charged by friction. The spray guns can be manual or automatic, fixed or reciprocating, and mounted on one or both sides of a conveyorized spray booth. Electrostatic spray powder coating operations use collectors to reclaim overspray. Film thicknesses of 1.5 to 5 mil can be applied on cold products. If the products are heated slightly, 20- to 25-mil-thick coatings can be applied on these products. As with other coating methods, electrostatic spray powder coating has limitations. Despite these limitations, powder coatings are replacing liquid coatings and finishes in a growing number of cases. A variation of the electrostatic spray powder coater is the electrostatic disk.

The advantage of this method is that coatings, using many of the resin types, can be applied in low (1.5- to 3-mil) film thicknesses with no VOC emission at extremely high transfer efficiency. Disadvantages include the difficulty in obtaining less than a 1-mil-thick continuous coating and, owing to the complex powder reclaiming systems, color changes are more difficult than with liquid spray systems.

6.6.5 Other Application Methods

6.6.5.1 Autodeposition coating. Autodeposition (95–100 percent TE) is a dipping method where coatings are applied on the product from an aqueous solution. Unlike electrocoating, there is no electric current applied. Instead, the driving force is chemical, because the coating reacts with the metallic substrate.

The advantages of autophoretic coating are no VOC emissions, no metal pretreatment other than cleaning, and uniform coating thickness. This technique requires 30 percent less floor space than electrocoating, and capital equipment costs are 25 percent lower than for electrocoating. Disadvantages of autophoretic coatings are that they are available only in dark colors, and corrosion resistance is lower than for electrocoated products.

6.6.5.2 Centrifugal coating. A centrifugal coater (95–100 percent TE) is a self-contained unit. It consists of an inner basket, a dip coating tank, and exterior housing. Products are placed in the inner basket, which is dipped into the coating tank. The basket is withdrawn and spun at a speed high enough to remove the excess coating material by centrifugal force. This causes the coating to be flung onto the inside of the exterior housing, from which it drains back into the dip coating tank.

The advantage of centrifugal coating is that large numbers of small parts can be coated at the same time. The disadvantage is that the appearance of the finish is a problem, because the parts touch each other.

6.6.5.3 Flow coating. In a flow coater (95–100 percent TE), the coating material is pumped through hoses and nozzles onto the surfaces of the product, from which the excess drains into a reservoir to be recycled. Flow coaters can be either automatic or manual. Film thickness is controlled by the viscosity and solvent balance of the coating material. A continuous coater is an advanced flow coater using airless spray nozzles mounted on a rotating arm in an enclosure.

Advantages of flow coating are high transfer efficiency and low volume of paint in the system. Products will not float off hangers, and extremely large products can be painted. As with dip coating, the disadvantages of flow coating are coating thickness control and solvent refluxing.

6.6.5.4 Curtain coating. Curtain coating (95–100 percent TE), which is similar to flow coating, is used to coat flat products on conveyorized lines. The coating falls from a slotted pipe or flows over a weir in a steady stream or curtain while the product is conveyed through it. Excess material is collected and recycled through the system. Film thickness is controlled by coating composition, flow rates, and line speed.

The advantage of curtain coating is uniform coating thickness on flat products with high transfer efficiency. The disadvantage is the inability to uniformly coat three-dimensional objects.

6.6.5.5 Roller coating. Roller coating (95–100 percent TE), which is used mainly by the coil coating industry for prefinishing metal coils that will later be formed into products, has seen steady growth. It is also used for finishing flat sheets of material. There are two types of roller coaters, direct and reverse, depending on the direction of the applicator roller relative to the direction of the substrate movement. Roller coating can apply multiple coats to the front and back of coil stock with great uniformity.

The advantages of roller coating are consistent film thickness and elimination of painting operations at a fabricating plant. The disadvantages are limited metal thickness, limited bend radius, and corrosion of unpainted cut edges.

6.7 Curing

No dissertation on organic coatings and finishes is complete without mentioning film formation and cure. It is not the intent of this chapter to fully discuss the mechanisms, which are more important to researchers and formulators than to designers and end users, but rather to show that differences exist and to aid the reader in making selections. Most of the organic coating resins are liquid, which cure or dry to form solid films. They are classified as *thermoplastic* or *thermosetting*. Thermoplastic resins dry by solvent evaporation and will soften when heated and harden when cooled. Thermosetting resins will not soften when heated after they are cured. Another classification of coatings and finishes is by their various film-forming mechanisms, such as solvent evaporation, coalescing, phase change, and conversion. Coatings and finishes are also classified as room-temperature curing, sometimes called air drying; or heat curing, generally called baking, or force drying,

which uses elevated temperatures to accelerate air drying. Thermoplastic and thermosetting coatings and finishes can be both air drying and baking.

6.7.1 Air Drying

Air drying coatings and finishes will form films and cure at room or ambient temperatures (20°C) by the mechanisms described in this section.

6.7.1.1 Solvent evaporation. Thermoplastic coating resins that form films by solvent evaporation are shellac and lacquers such as nitrocellulose, acrylic, styrene-butadiene, and cellulose acetate butyrate.

6.7.1.2 Conversion. In these coatings, films are formed as solvents evaporate, and they cure by oxidation, catalyzation, or cross-linking. Thermosetting coatings and finishes cross-link to form films at room temperature by oxidation or catalyzation. Oxidative curing of drying oils and oil-modified resins can be accelerated by using catalysts. Monomeric materials can form films and cure by cross-linking with polymers in the presence of catalysts, as in the case of styrene monomers and polyester resins. Epoxy resins will cross-link with polyamide resins to form films and cure. In the moisture curing polyurethane resin coating systems, airborne moisture starts a reaction in the vehicle, resulting in film formation and cure.

6.7.1.3 Coalescing. Emulsion or latex coatings, such as styrene-butadiene, acrylic ester, and vinyl acetate acrylic, form films by coalescing and dry by solvent evaporation.

6.7.2 Baking

Baking coatings and finishes will form films at room temperature, but require elevated temperatures (150 to 200°C) to cure. Most coatings and finishes are baked in gas-fired ovens, although oil-fired ovens are also used. Steam-heated and electric ovens are used on a limited basis. Both electric and gas-fired infrared elements are used as heat sources in paint bake ovens.

6.7.2.1 Conversion. The cure of many oxidative thermosetting coatings and finishes is accelerated by heating. In other resins systems, such as thermosetting acrylics and alkyd melamines, the reactions do not occur below temperature thresholds of 135°C or higher. Baking coatings and finishes (those that require heat to cure) are generally tougher than air-drying coatings. In some cases, the cured films are so hard and brittle that they must be modified with other resins.

6.7.2.2 Phase change. Thermoplastic coatings and finishes that form films by phase changes, generally from solid to liquid then back to solid, are polyolefins, waxes, and polyamides. Plastisols and organisols undergo phase changes from liquid to solid during film formation. Fluidized bed powder coatings, both thermoplastic and thermosetting, also undergo phase changes from solid to liquid to solid during film formation and cure.

6.7.3 Radiation Curing

Films are formed and cured by bombardment with ultraviolet and electron beam radiation with little increase in surface temperature. Infrared radiation, on the other hand, increases the surface temperature of films and is therefore a baking process. The most notable radiation curing is UV curing. This process requires the use of specially formulated coatings. They incorporate photoinitiators and photosensitizers that respond to specific wavelengths of the spectrum to cause a conversion reaction. Curing is practically instantaneous with little or no surface heating. It is, therefore, useful in coating plastic and other temperature-sensitive substrates. Since the coatings and finishes are 100 percent solids, there are no VOCs. UV curable powder coatings are also commercially available. Although most UV coating are clears (unpigmented), paints can also be cured. Figure 6.9 shows radiation curing equipment.

6.7.4 Force Drying

In many cases, the cure rate of thermoplastic and thermosetting coatings and finishes can be accelerated by exposure to elevated temperatures that are below those considered to be baking temperatures.

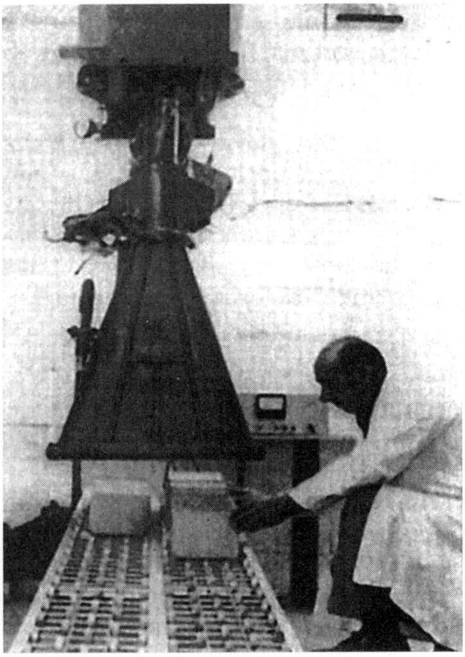

Figure 6.9 Radiation curing is fast, allowing production-line speeds of 2000 ft/min. This technique takes place at room temperature, and heat-sensitive wooden and plastic products and electronic components and assemblies can be given the equivalent of a baked finish at speeds never before possible. *(Courtesy of Radiation Dynamics)*

6.7.5 Vapor Curing

Vapor curing is essentially a catalyzation or cross-linking conversion method for two-component coatings. The product is coated with one component of the coating in a conventional manner. It is then placed in an enclosure filled with the other component, the curing agent, in vapor form. It is in this enclosure that the reaction occurs.

6.7.6 Reflowing

Although not actually a curing process, certain thermoplastic films will soften and flow to become smooth and glossy at elevated temperatures. This technique is used on acrylic lacquers by the automotive industry to eliminate buffing.

6.8 Summary

The purpose of this chapter is to aid the reader in selecting surface preparation methods, coating materials, and application methods. It also acquaints readers with curing methods and helps them to comply with environmental regulations.

Coating and finish selection is not easy, owing to the use and formulating versatility of modern plastic resins. This versatility also contributes to one of their faults, which is the possible decline in one performance property when another is enhanced. Because of this, the choice of a coating and finish must be based on specific performance properties and not on generalizations. This choice is further complicated by the need to comply with governmental regulations. Obviously, the choice of compliance coatings, electrocoating, waterborne, high solids, powder, radiation curable, and vapor cure is well advised.

To apply coatings and finishes in the most effective manner, the product's size, shape, ultimate appearance, and end use must be addressed. This chapter emphasizes the importance of transfer efficiency in choosing a coating method.

To meet these requirements, product designers, coaters, and finishers have all the tools at their disposal. They can choose coatings and finishes that apply easily, coat uniformly, cure rapidly and efficiently, and comply with governmental regulations at lower costs. By applying compliance coatings using methods that have high transfer efficiencies, they will not only comply with air- and water-quality standards, they will also provide a safe workplace and decrease the generation of hazardous wastes.[4] Reducing hazardous wastes, using less material, and emitting fewer VOCs can also effect significant cost savings.

References

1. C. P. Izzo, Today's Paint Finishes: Better than Ever, *Products Finishing Magazine,* vol. 51, no. 1, October 1986, 48–52.
2. *Coatings Encyclopedic Dictionary,* Federation of Societies for Coatings Technology, 1995.
3. C. Izzo, How Are Coatings Applied, in *Products Finishing Directory and Technology Guide,* Gardner Publications, Cincinnati, OH, 2000.
4. C. Izzo, Overview of Industrial Coating Materials, in *Products Finishing Directory and Technology Guide,* Gardner Publications, Cincinnati, OH, 2000.

Chapter

7

Plastics and Elastomers in Adhesives

Edward M. Petrie

ABB Electric Systems Technology Institute
Raleigh, North Carolina

7.1 Introduction to Adhesives

Adhesives were first used many thousands of years ago, and most were derived from naturally occurring vegetable, animal, or mineral substances. Synthetic polymeric adhesives displaced many of these early products due to stronger adhesion and greater resistance to operating environments. These modern plastic and elastomer based adhesives are the principal subject of this chapter.

An adhesive is a substance capable of holding substrates (adherends) together by surface attachment. A material merely conforming to this definition does not necessarily ensure success in an assembly process. For an adhesive to be useful, it must not only hold materials together but also withstand operating loads and last the life of the product.

The successful application of an adhesive depends on many factors. Anyone using an adhesive faces a complex task of selecting the proper adhesive and the correct processing conditions that allow a bond to form. One must also determine the substrate-surface treatment which will permit an acceptable degree of permanence and bond strength. The adhesive joint must be correctly designed to avoid stresses within the joint that could cause premature failure. Also, the physical and chemical stability of the bond must be forecast with relation to its service environment. This chapter is intended to guide the adhesives user through these numerous considerations.

7.1.1 Advantages and Disadvantages of Adhesive Bonding

Adhesive bonding presents several distinct advantages over conventional mechanical methods of fastening. There are also some disadvantages that may make adhesive bonding impractical. These are summarized in Table 7.1.

TABLE 7.1 Advantages and Disadvantages of Adhesive Bonding

Advantages	Disadvantages
1. It provides large stress-bearing area.	1. Surfaces must be carefully cleaned.
2. It provides excellent fatigue strength.	2. Long cure times may be needed.
3. It damps vibration and absorbs shock.	3. There is a limitation on upper continuous operating temperature (generally 350°F).
4. It minimizes or prevents galvanic corrosion between dissimilar metals.	4. Heat and pressure may be required.
5. It joins all shapes and thicknesses.	5. Jigs and fixtures may be needed.
6. It provides smooth contours.	6. Rigid process control is usually necessary.
7. It seals joints.	7. Inspection of the finished joint is difficult.
8. It joins any combination of similar or dissimilar materials.	8. Useful life depends on environment.
9. Often, it is less expensive and faster than mechanical fastening.	9. Environmental, health, and safety considerations are necessary.
10. Heat, if required, is too low to affect metal parts.	10. Special raining is sometimes required.
11. It provides an attractive strength-to-weight ratio.	

The design engineer must consider and weigh these factors before deciding on a method of fastening. However, in many applications, adhesive bonding is the only practical method for assembly. In the aircraft industry, for example, adhesives make the use of thin metal and honeycomb structures, which is feasible because stresses are transmitted more effectively by adhesives than by rivets or welds. Plastics and elastomers can also be more reliably joined with adhesives than by other methods.

7.1.1.1 Mechanical advantages. The most common methods of structural fastening are shown in Fig. 7.1. Because of the uniformity of an adhesive bond, certain mechanical advantages can be provided as shown. The stress-distribution characteristics and inherent toughness of polymeric adhesives provide bonds with superior fatigue resistance, as shown in Fig. 7.2. Generally, in well designed joints, the adherends will fail in fatigue before the adhesive.

7.1.1.2 Design advantages. Adhesives offer certain design advantages that are often valuable.

- Unlike rivets or bolts, adhesives produce smooth contours that are aerodynamically and cosmetically beneficial.

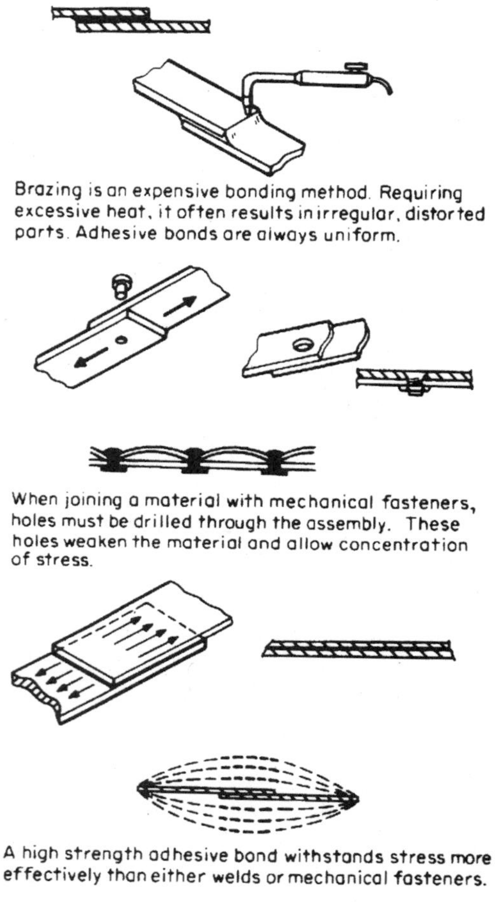
Brazing is an expensive bonding method. Requiring excessive heat, it often results in irregular, distorted parts. Adhesive bonds are always uniform.
When joining a material with mechanical fasteners, holes must be drilled through the assembly. These holes weaken the material and allow concentration of stress.
A high strength adhesive bond withstands stress more effectively than either welds or mechanical fasteners.

Figure 7.1 Common methods of structural fastening.[1]

- Adhesives also offer a better strength/weight ratio than mechanical fasteners.
- Adhesives can join any combination of solid materials, regardless of shape or thickness. Materials such as plastics, elastomers, ceramics, and wood can be joined more economically and efficiently by adhesive bonding than by other methods.
- Adhesive bonding is frequently faster and less expensive than conventional fastening methods. As the size of the area to be joined increases, the time and labor saved by using adhesives instead of mechanical fasteners become progressively greater, because the entire joint area can be assembled in one operation.

7.1.1.3 Other advantages. Adhesives can be made to function as electrical and thermal insulators. The degree of insulation can be varied with different adhesive formulations and fillers. Adhesives can even be made electrically and thermally conductive with silver

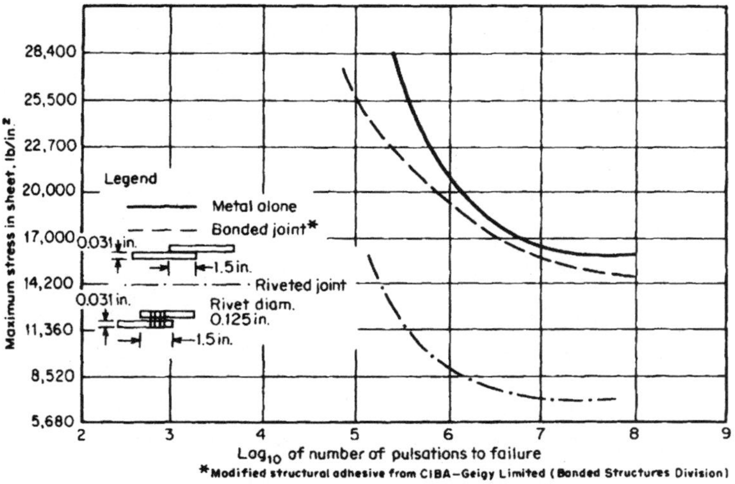

Figure 7.2 Fatigue strengths of aluminum alloy specimens under pulsating tensile load.[2]

and boron nitride fillers, respectively. Adhesives can also perform sealing functions, offering a barrier to the passage of fluids and gases. Adhesives may also act as vibration dampers to reduce the noise and oscillation encountered in assemblies.

Frequently, adhesives may be called upon to do multiple functions. In addition to being a mechanical fastener, an adhesive may also be used as a sealant, vibration damper, insulator, and gap filler in the same application.

7.1.1.4 Mechanical limitations. The most serious limitation to the use of modern polymeric adhesives is their time-dependent strength in degrading service environments such as moisture, high temperatures, or chemicals.

There are polymeric adhesives that perform well at temperatures between –60 and 350°F. But only a few adhesives can withstand operating temperatures outside that range. Adhesives can also be degraded by chemical environments and outdoor weathering. The rate of strength degradation may be accelerated by continuous stress or elevated temperatures.

7.1.1.5 Design limitations. The adhesive joint must be carefully designed for optimal performance. Design factors must include the type of stress, environmental influences, and production methods that will be used. The strength of the adhesive joint is dependent on the type and direction of stress. Generally, adhesives perform better when stressed in shear or tension than when exposed to cleavage or peel forces.

Since nearly every adhesive application is somewhat unique, the adhesive manufacturers often do not have data concerning the aging characteristics of their adhesives in specific environments. Thus, before any adhesive is incorporated into production, a thorough evaluation should be made in a simulated operating environment. Time must also be allowed to train personnel in what can be a rather complex and critical manufacturing process.

7.1.1.6 Production limitations. All adhesives require clean surfaces to attain optimal results. Depending on the type and condition of the substrate and the bond strength desired, surface preparations ranging from a simple solvent wipe to chemical etching are necessary.

If the adhesive has multiple components, the parts must be carefully weighed and mixed. The setting operation often requires heat and pressure. Lengthy set time could make assembly jigs and fixtures necessary.

Adhesives may be composed of materials that present personnel hazards, including flammability and dermatitis, in which case necessary precautions must be considered. Finally, the inspection of finished joints for quality control is very difficult. This necessitates strict control over the entire bonding process to ensure uniform bond quality.

Although the material cost is relatively low, some adhesive systems may require metering, mixing, and dispensing equipment as well as curing fixtures, ovens, and presses. Capital equipment investment must be included in any economic evaluation. The following items contribute to a "hidden cost" of using adhesives, and they also could lead to serious production difficulties:

1. The storage life (shelf life) of the adhesive may be unrealistically short; some adhesives require refrigerated storage.
2. The adhesive may begin to solidify or gel too early in the bonding process.
3. Waste, safety, and environmental concerns can be essential cost factors.
4. Cleanup is a cost factor, especially where misapplied adhesive may ruin the appearance of a product.
5. Once bonded, samples cannot easily be disassembled; if misalignment occurs and the adhesive cures, usually the part must be scrapped.

7.1.2 Theories of Adhesion

Various theories attempt to describe the phenomena of adhesion. No single theory explains adhesion in a general way. However, knowledge of adhesion theories can assist in understanding the basic requirements for a good bond.

7.1.2.1 Mechanical theory. The surface of a solid material is never truly smooth but consists of a maze of microscopic peaks and valleys. According to the mechanical theory of adhesion, the adhesive must penetrate the cavities on the surface and displace the trapped air at the interface.

Such mechanical anchoring appears to be a prime factor in bonding many porous substrates. Adhesives also frequently bond better to abraded surfaces than to natural surfaces. This beneficial effect may be due to

1. Mechanical interlocking
2. Formation of a clean surface
3. Formation of a more reactive surface
4. Formation of a larger surface area

7.1.2.2 Adsorption theory. The adsorption theory states that adhesion results from molecular contact between two materials and the surface forces that develop. The process

of establishing intimate contact between an adhesive and the adherend is known as *wetting*. Figure 7.3 illustrates good and poor wetting of a liquid spreading over a surface.

For an adhesive to wet a solid surface, the adhesive should have a lower surface tension than the solid's critical surface tension. Tables 7.2 and 7.3 list surface tensions of common adherends and liquids.

TABLE 7.2 Critical Surface Tension of Common Plastics and Metals

Materials	Critical surface tension, dyn/cm
Acetal	47
Acrylonitrile-butadiene-styrene	35
Cellulose	45
Epoxy	47
Fluoroethylene propylene	16
Polyamide	46
Polycarbonate	46
Polyethylene	31
Polyethylene terephthalate	43
Polyimide	40
Polymethylmethacrylate	39
Polyphenylene sulfide	38
Polystyrene	33
Polysulfone	41
Polytetrafluoroethylene	18
Polyvinyl chloride	39
Silicone	24
Aluminum	≈500
Copper	≈1000

TABLE 7.3 Surface Tension of Common Adhesives and Liquids

Material	Surface tension, dyn/cm
Epoxy resin	47
Fluorinated epoxy resin*	33
Glycerol	63
Petroleum lubricating oil	29
Silicone oils	21
Water	73

*Experimental resin, developed to wet low-energy surfaces. (*Note:* low surface tension relative to most plastics.)

Most organic adhesives easily wet metallic solids. But many solid organic substrates have surface tensions less than those of common adhesives. From Tables 7.2 and 7.3, it is apparent that epoxy adhesives will wet clean aluminum or copper surface. However, epoxy resin will not wet a substrate having a critical surface tension significantly less than 47 dynes/cm. Epoxies will not, for example, wet either a metal surface contaminated with silicone oil or a clean polyethylene substrate.

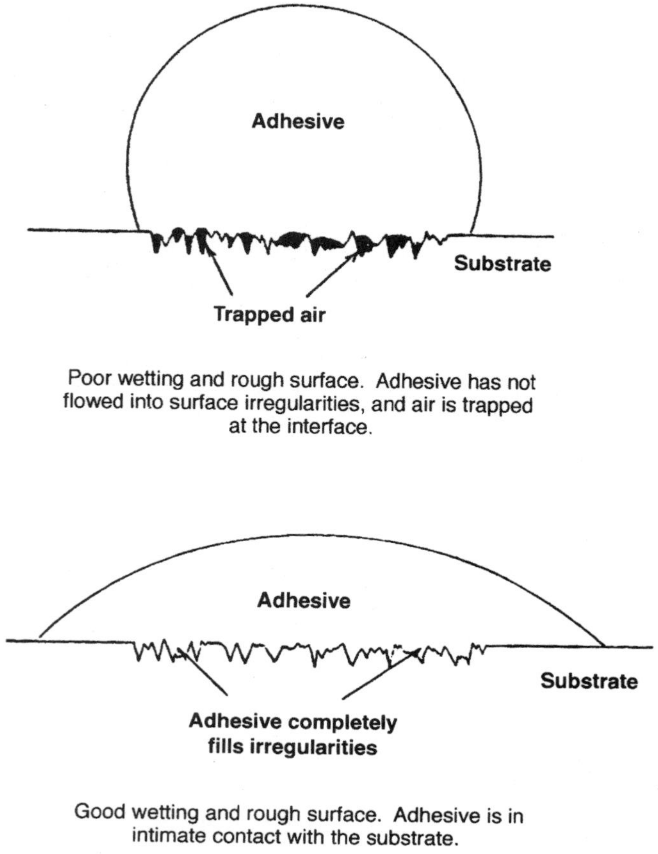
Adhesive
Substrate
Trapped air
Poor wetting and rough surface. Adhesive has not flowed into surface irregularities, and air is trapped at the interface.
Adhesive
Substrate
Adhesive completely fills irregularities
Good wetting and rough surface. Adhesive is in intimate contact with the substrate.

Figure 7.3 Illustration of good and poor wetting by adhesive spreading over a surface.

After intimate contact is achieved between adhesive and adherend through wetting, it is believed that adhesion results primarily through forces of molecular attraction. Four general types of chemical bonds are recognized: electrostatic, covalent, and metallic, which are referred to as primary bonds, and van der Walls forces, which are referred to as secondary bonds. The adhesion between adhesive and adherend is thought to be primarily due to van der Walls forces of attraction.

7.1.2.3 Electrostatic and diffusion theories. The electrostatic theory states that electrostatic forces in the form of an electrical double layer are formed at the adhesive–adherend interface. These forces account for resistance to separation. The theory gathers support from the fact that electrical discharges have been noticed when an adhesive is peeled from a substrate.

The fundamental concept of the diffusion theory is that adhesion arises through the inter-diffusion of molecules in the adhesive and adherend. The diffusion theory is primarily

applicable when both the adhesive and adherend are polymeric, having long-chain molecules capable of movement. Bonds formed by solvent or heat welding thermoplastics result from the diffusion of molecules.

7.1.2.4 Weak-boundary-layer theory. According to the weak-boundary-layer theory, when bond failure seems to be at the interface, a cohesive break of a weak boundary layer usually is the real event.[3] Weak boundary layers can originate from the adhesive, the adherend, the environment, or a combination of any of the three.

Weak boundary layers can occur on the adhesive or adherend if an impurity concentrates near the bonding surface and forms a weak attachment to the substrate. When bond failure occurs, it is the weak boundary layer that fails, although failure seems to occur at the adhesive–adherend interface.

Two examples of a weak boundary layer effect are polyethylene and metal oxides. Conventional grades of polyethylene have weak, low-molecular-weight constituents evenly distributed throughout the polymer. These weak elements are present at the interface and contribute to low failing stress when polyethylene is used as an adhesive or adherend. Certain metal oxides are weakly attached to their base metals.

Failure of adhesive joints made with these adherends will occur cohesively within the weak oxide. Weak boundary layers can be removed or strengthened by various surface treatments.

Weak boundary layers formed from the shop environment are very common. When the adhesive does not wet the substrate as shown in Fig. 7.3, a weak boundary layer of air is trapped at the interface, causing lowered joint strength. Moisture from the air may also form a weak boundary layer on hydrophilic adherends.

7.1.3 Requirements of a Good Bond

The basic requirements for a good adhesive bond are cleanliness, wetting, solidification, and proper selection of adhesive and joint design.

7.1.3.1 Cleanliness. To achieve an effective adhesive bond, one must start with a clean surface. Foreign materials such as dirt, oil, moisture, and weak oxide layers must be removed from the substrate surface, or the adhesive will bond to these weak boundary layers rather than the actual substrate. There are various surface preparations that remove or strengthen the weak boundary layer. These treatments generally involve physical or chemical processes or a combination of both. Surface-preparation methods for specific substrates will be discussed in a later section.

7.1.3.2 Wetting. While it is in the liquid state, the adhesives must "wet" the substrate. Examples of good and poor wetting have been explained. The result of good wetting is greater contact area between adherend and adhesive over which the forces of adhesion may act.

7.1.3.3 Solidification. The liquid adhesive, once applied, must be capable of conversion into a solid. The process of solidifying can be completed in different ways (e.g., chemical reaction by any combination of heat, pressure, and curing agents; cooling from a molten liquid to a solid; and drying due to solvent evaporation). The method by which solidification occurs depends on the adhesive.

7.1.3.4 Adhesive choice. Factors most likely to influence adhesive selection are listed in Table 7.4. With regard to the controlling factors involved, the many adhesives available can usually be narrowed to a few candidates that are most likely to be successful. The general areas of concern to the design engineer when selecting adhesives should be the material to be bonded, service requirements, production requirements, and overall cost.

TABLE 7.4 Factors Influencing Adhesive Selection

Stress	Tension Shear Impact Peel Cleavage Fatigue
Chemical factors	External (service-related) Internal (effect of adherend on adhesives)
Exposure	Weathering Light Oxidation Moisture Salt spray
Temperature	High Low Cycling
Biological factors	Bacteria or mold Rodents or vermin
Working properties	Application Bonding time and temperature range Tackiness Curing rate Storage stability Coverage

7.1.3.5 Joint design. The adhesive joint should be designed to optimize the forces of adhesion. Such design considerations will be discussed in the next section. Although adequate adhesive-bonded assemblies have been made from joints designed for mechanical fastening, maximum benefit can be obtained only in assemblies specifically designed for adhesive bonding.

7.1.4 Mechanism of Bond Degradation

Adhesive joints may fail adhesively or cohesively. *Adhesive* failure is interfacial bond failure between the adhesive and adherend. *Cohesive* failure occurs when the adhesive fractures allowing a layer of adhesive to remain on both substrates. When the adherend fails before the adhesive, it is known as a cohesive failure of the adherend. The various modes of possible bond failures are shown in Fig. 7.4.

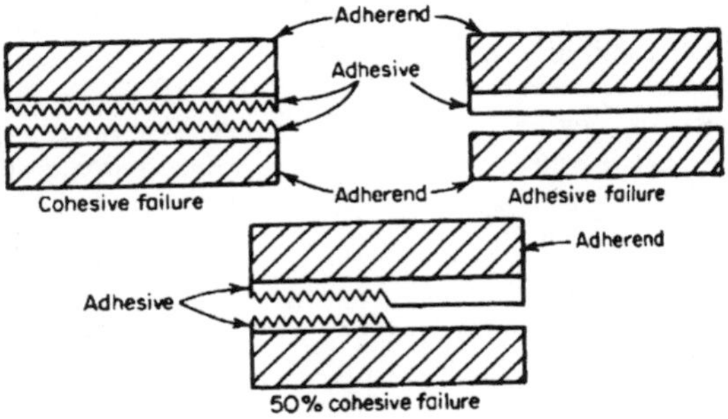

Figure 7.4 Cohesive and adhesive bond failures.

Cohesive failure within the adhesive or one of the adherends is the ideal type of failure, because the maximum strength of the materials in the joint has been reached. However, failure mode should not be used as a criterion for a useful joint. Some adhesive-adherend combinations may fail in adhesion but provide sufficient strength margin to be practical. An analysis of failure mode can be an extremely useful tool to determine if a failure is due to a weak boundary layer or improper surface preparation.

The exact cause of adhesive failure is very hard to determine, because so many factors in adhesive bonding are interrelated. However, there are certain common factors at work when an adhesive bond is made that contribute to the weakening of all bonds. The influences of these factors are qualitatively summarized in Fig. 7.5.

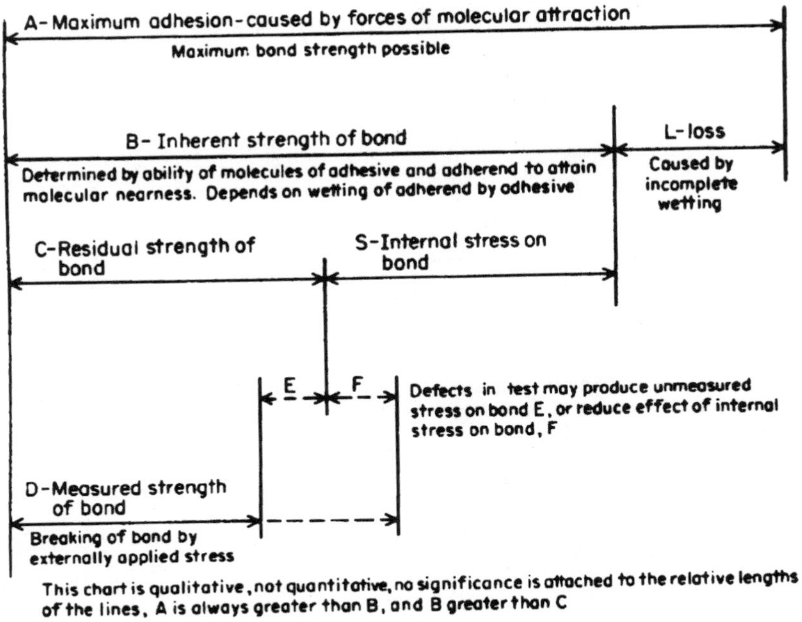

Figure 7.5 Relation between factors involved in adhesion.[4]

If the adhesive does not wet the surface of the substrate, the joint will be inferior. It is also important to allow the adhesive enough time to wet the substrate effectively before gelation occurs.

Internal stresses occur in the adhesive joint because of a natural tendency of the adhesive to shrink during solidification and because of differences in physical properties between adhesive and substrate. The coefficient of thermal expansion of adhesive and adherend should be as close as possible to limit stresses that may develop during thermal cycling or after cooling from an elevated-temperature cure. Polymeric adhesives generally have a thermal-expansion coefficient an order of magnitude greater than that of metals. Adhesives can be formulated with various fillers to modify their thermal-expansion characteristics and limit internal stresses. A relatively elastic adhesive capable of accommodating internal stress may also be useful when thermal-expansion differences are of concern.

Once an adhesive bond is placed in service, other forces are at work weakening the bond. The type of stress, its orientation to the joint, and the rate in which the stress is applied are important. Sustained loads can cause premature failure in service, even though similar unloaded joints may exhibit adequate strength when tested after aging. Most adhesives have poor strength when stresses are acting to peel or cleave the adhesive from the substrate. Many adhesives are sensitive to the rate in which the joint is stressed. Rigid, brittle adhesives sometimes have excellent tensile or shear strength but stand up very poorly under an impact test.

Operating environments are capable of degrading an adhesive joint in various ways. The adhesive may have to withstand temperature variation, weathering, oxidation, moisture, and other exposure conditions. If more than one of these factors are present in the operating environment, their synergistic effect could cause a rapid decline in adhesive strength.

7.1.5 Adhesive Classification

Adhesives may be classified by many methods. The most common methods are by function, chemical composition, mode of application and setting, and end use.

The functional classification defines adhesives as being structural or nonstructural. Structural adhesives are materials of high strength and permanence. Their primary function is to hold structures together and be capable of resisting high loads. Nonstructural adhesives are not required to support substantial loads. They merely hold lightweight materials in place or provide a seal without having a high degree of strength.

The chemical composition classification broadly describes adhesives as thermosetting, thermoplastic, elastomeric, or combinations of these. There are then many chemical types within each classification. They are described in Table 7.5.

Adhesives are often classified by their mode of application. Depending on viscosity, adhesives are sprayable, brushable, or trowelable. Heavily bodied adhesive pastes and mastics are considered extrudable; they are applied by syringe, caulking gun, or pneumatic pumping equipment.

Another distinction between adhesives is the manner in which they flow or solidify. Some adhesives solidify simply by losing solvent while others harden as a result of heat activation or chemical reaction. Pressure-sensitive adhesives flow under pressure and are stable when the pressure is removed.

Adhesives may also be classified according to their end use. Thus, metal adhesives, wood adhesives, and vinyl adhesives refer to the substrates they bond; and acid-resistant adhesives, heat-resistant adhesives, and weatherable adhesives indicate the environments for which each is suited.

TABLE 7.5 Adhesives Classified by Chemical Composition (from Ref. 5)

Classification	Thermoplastic	Thermosetting	Elastomeric	Alloys
Types within group	Cellulose acetate, cellulose acetate butyrate, cellulose nitrate, polyvinyl acetate, vinyl vinylidene, polyvinyl acetals, polyvinyl alcohol, polyamide, acrylic, phenoxy	Cyanoacrylate, polyester, urea formaldehyde, melamine formaldehyde, resorcinol and phenol-resorcinol formaldehyde, epoxy, polyimide, polybenzimidazole, acrylic, acrylate acid diester	National rubber, reclaimed rubber, butyl, polyisobutylene, nitrile, styrene-butadiene, polyurethane, polysulfide, silicone, neoprene	Epoxy-phenolic, epoxy-polysulfide, epoxy-nylon, nitrile-phenolic, neoprene-phenolic, vinyl-phenolic
Most used form	Liquid, some dry film	Liquid, but all forms common	Liquid, some film	Liquid, paste, film
Common further classifications	By vehicle (most are solvent dispersions or water emulsions)	By cure requirements (heat and/or pressure most common but some are catalyst types)	By cure requirements (all are common); also by vehicle (most are solvent dispersions or water emulsions)	By cure requirements (usually heat and pressure except some epoxy types); by vehicle (most are solvent dispersions or 100% solids); and by type of adherends or end-service conditions
Bond characteristics	Good to 150–200°F; poor creep strength; fair peel strength	Good to 200–500°F; good creep strength; fair peel strength	Good to 150–400°F; never melt completely; low strength; high flexibility	Balanced combination of properties of other chemical groups depending on formulation; generally higher strength over wider temp range
Major type of use	Unstressed joints; designs with caps, overlaps, stiffeners	Stressed joints at slightly elevated temp	Unstressed joints on light-weight materials; joints in flexure	Where highest and strictest end-service conditions must be met; sometimes regardless of cost, as military uses
Materials most commonly bonded	Formulation range covers all materials, but emphasis on nonmetallics–esp. wood, leather, cork, paper, etc.	For structural uses of most materials	Few used "straight" for rubber, fabric, foil, paper, leather, plastics films; also as tapes. Most modified with synthetic resins	Metals, ceramics, glass, thermosetting plastics; nature of adherends often not as vital as design or end-service conditions (i.e., high strength, temp)

7.1.6 Adhesive Bonding Process

A typical flow chart for the adhesive bonding process is shown in Fig. 7.6. The elements of the bonding process are as important as the adhesive itself for a successful end product.

Many of the adhesive problems that develop are not due to a poor choice of material or joint design but are directly related to faulty production techniques. The adhesive user must obtain the proper processing instructions from the manufacturer and follow them closely and consistently to ensure acceptable results. Adhesive production involves four basic steps.

1. Design of joints and selection of adhesive
2. Preparation of adherends
3. Applying and curing the adhesive
4. Inspection of bonded parts

The remainder of this chapter will discuss these steps in detail.

7.2 Design and Testing of Adhesive Joints

7.2.1 Types of Stress

Four basic types of loading stress are common to adhesive joints: tensile, shear, cleavage, and peel. Any combination of these stresses, illustrated in Fig. 7.7, may be encountered in an adhesive application.

Tensile stress develops when forces acting perpendicular to the plane of the joint are distributed uniformly over the entire bonded area. Adhesive joints show good resistance to tensile loading because all the adhesive contributes to the strength of the joint. In practical applications, however, loads are rarely axial, and unwanted cleavage or peel stresses tend to develop.

Shear stress results when forces acting in the plane of the adhesive try to separate the adherends. Joints that depend on the adhesive's shear strength are relatively easy to design and offer favorable properties. Adhesive joints are strong when stressed in shear because all the bonded area contributes to the strength.

Cleavage and peel stresses are undesirable. Cleavage occurs when forces at one end of a rigid bonded assembly act to split the adherends apart. Peel stress is similar to cleavage but

Figure 7.6 Basic steps in bonding process.

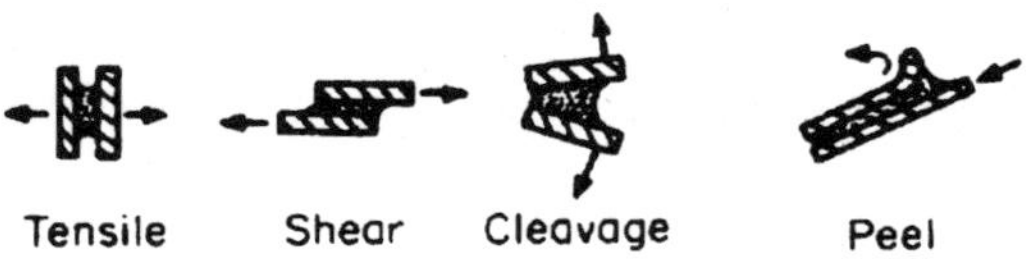

Figure 7.7 Four basic types of adhesive stress.

applies to a joint where one or both of the adherends are flexible. Joints loaded in peel or cleavage offer much lower strength then joints loaded in shear, because the stress is concentrated on only a very small area of the total bond. The remainder of the bonded area makes no contribution to the strength of the joint. Peel and cleavage stresses should be avoided where possible.

7.2.2 Joint Efficiency

To avoid concentration of stress, the joint designer should take into consideration the following rules:

1. Keep the stress on the bond line to a minimum.
2. Design the joint so that the operating loads will stress the adhesive in shear.
3. Peel and cleavage stresses should be minimized.
4. Distribute the stress as uniformly as possible over the entire bonded area.
5. Adhesive strength is directly proportional to bond width. Increasing width will always increase bond strength; increasing the depth does not always increase strength.
6. Generally, rigid adhesives are better in shear, and flexible adhesives are better in peel.

Brittle adhesives are particularly weak in peel, because the stress is localized at only a very thin line at the edge of the bond, as shown in Fig. 7.8. Tough, flexible adhesives distribute the peeling stress over a wider bond area and show greater resistance to peel.

For a given adhesive and adherend, the strength of a joint stressed in shear depends primarily on the width and depth of the overlap and the thickness of the adherend. Adhesive shear strength is directly proportional to the width of the joint. Strength can sometimes be increased by increasing the overlap depth, but the relationship is not linear. Since the ends of the bonded joint carry a higher proportion of the load than the interior area, the most efficient way of increasing joint strength is by increasing the width of the bonded area.

In a shear joint made from thin, relatively flexible adherends, there is a tendency for the bonded area to distort because of eccentricity of the applied load. This distortion, illustrated in Fig. 7.9, causes cleavage stress on the ends of the joint, and the joint strength may be considerably impaired. Thicker adherends are more rigid, and the distortion is not as

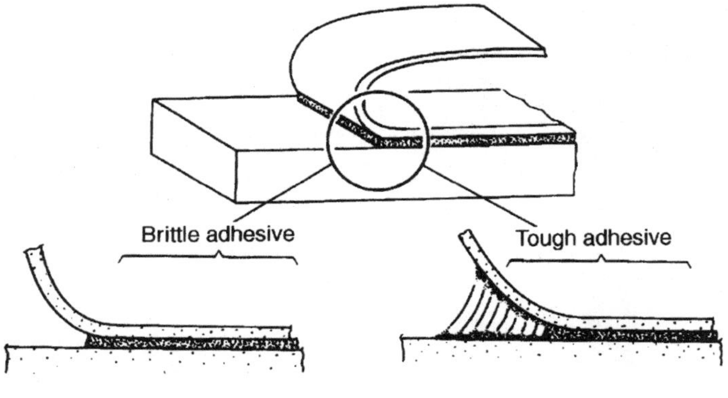

Figure 7.8 Tough, flexible adhesives distribute peel stress over a larger area.[6]

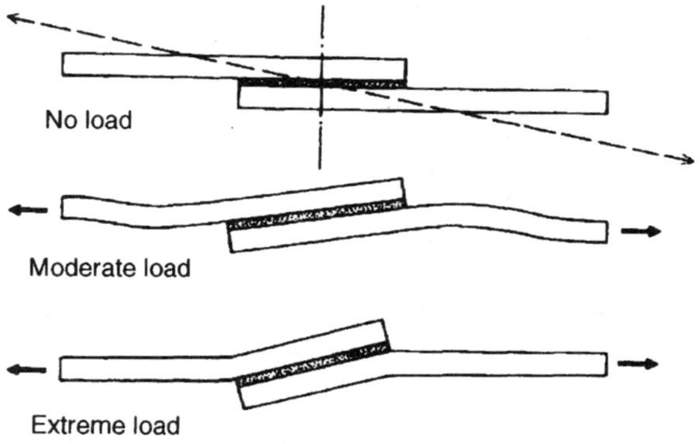

Figure 7.9 Distortion caused by loading can introduce cleavage stresses and must be considered in joint design.[5]

much a problem as with thin-gage adherends. Figure 7.10 shows the general interrelationship between failure load, depth of overlap, and adherend thickness for a specific metallic adhesive joint.

Since the stress distribution across the bonded area is not uniform and depends on joint geometry, the failure load of one specimen cannot be used to predict the failure load of another specimen with different joint geometry. The results of a particular shear test pertain only to joints that are exact duplicates. To characterize overlap joints more closely, the ratio of overlap length to adherend thickness *l/t* can be plotted against shear strength. A set of *l/t* curves for aluminum bonded with a nitrile-rubber adhesive is shown in Fig. 7.11.

The strength of an adhesive joint also depends on the thickness of the adhesive. Thin adhesive films offer the highest strength, provided that the bonded area does not have

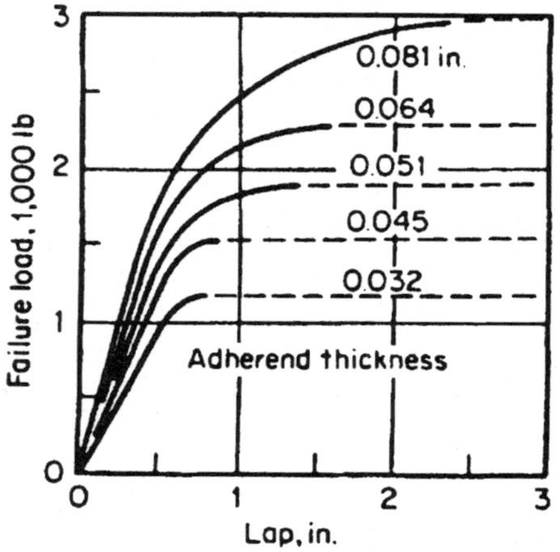

Figure 7.10 Interrelation of failure loads, depth of lap, and adherend thickness for lap joints with a specific adhesive and adherend.[7]

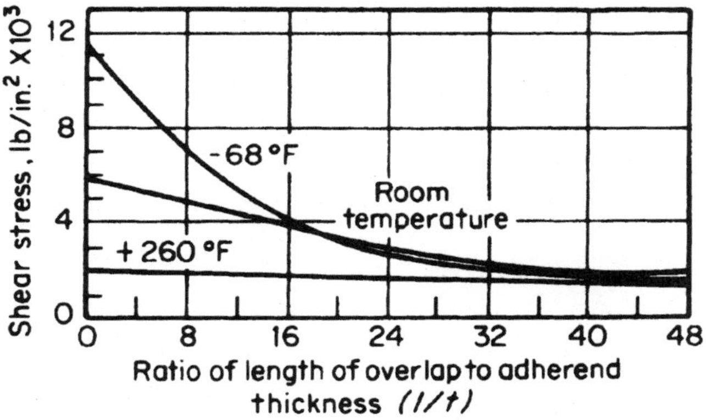

Figure 7.11 *l/t* curves at three test temperatures for aluminum joints bonded with nitrile-rubber adhesive.[8]

"starved" areas where all the adhesive has been forced out. Excessively heavy adhesive-film thicknesses cause greater internal stresses during cure and concentration of stress under load at the ends of a joint. Optimal adhesive thickness for maximum shear strength is generally considered to be between 2 and 10 mils. Strength does not vary significantly with bond-line thickness in this range.

7.2.3 Joint Design

A favorable stress can be applied by using proper joint design. However, joint designs may be impractical, expensive to make, or hard to align. The design engineer will often have to weigh these factors against optimum joint performance.

7.2.3.1 Flat adherends. The simplest joint to make is the plain butt joint. However, butt joints cannot withstand bending forces, because the adhesive would experience cleavage stress. The butt joint can be improved by redesigning in a number of ways, as shown in Fig. 7.12.

Lap joints are commonly used because they are simple to make, are applicable to thin adherends, and stress the adhesive in its strongest direction. Tensile loading of a lap joint causes the adhesive to be stressed in shear. However, the simple lap joint is offset, and the shear forces are not in line, as was illustrated in Fig. 7.9. Modifications of lap-joint design include

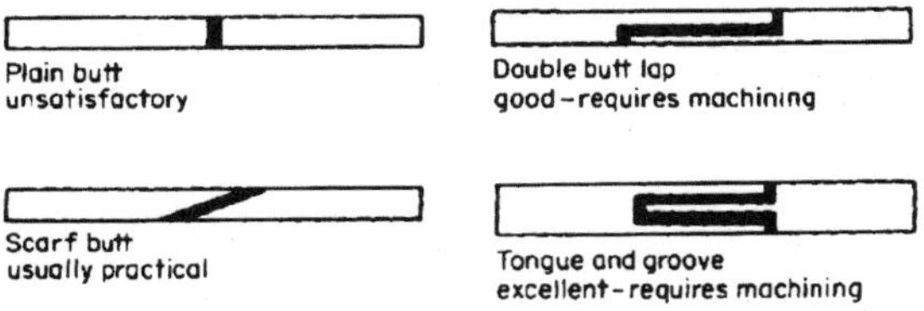

Figure 7.12 Butt connections.

1. Redesigning the joint to bring the load on the adherends in line
2. Making the adherends more rigid (thicker) near the bond area (see Fig. 7.10)
3. Making the edges of the bonded area more flexible for better conformance, thus minimizing peel

Modifications of lap joints are shown in Fig. 7.13.

Strap joints keep the operating loads aligned and are generally used where overlap joints are impractical because of adherend thickness. Strap-joint designs are shown in Fig. 7.14. Like the lap joint, the single strap is subjected to cleavage stress under bending forces.

When thin members are bonded to thicker sheets, operating loads generally tend to peel the thin member from its base, as shown in Fig. 7.15*a*. The subsequent illustrations show what can be done to decrease peeling tendencies in simple joints.

7.2.3.2 Cylindrical adherends. Several recommended designs for rod and tube joints are illustrated in Fig. 7.16. These designs should be used instead of the simpler butt joint. Their resistance to bending forces and subsequent cleavage is much better, and the bonded area is larger. Unfortunately, most of these joint designs require a machining operation.

7.2.3.3 Angle and corner joints. A butt joint is the simplest method of bonding two surfaces that meet at an angle. Although the butt joint has good resistance to pure tension

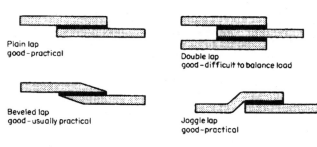

Figure 7.13 Lap connections.

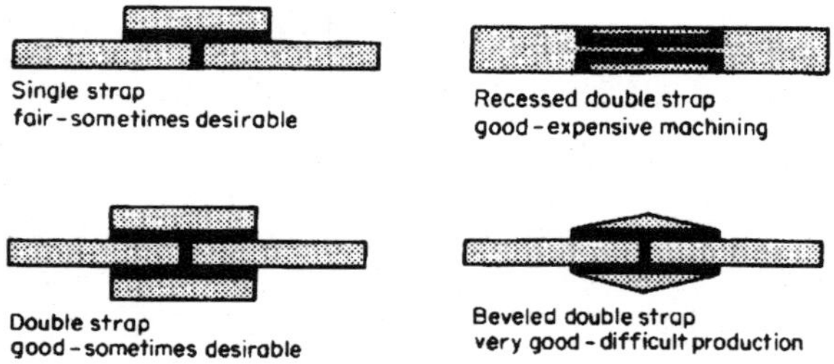

Figure 7.14 Strap connections.

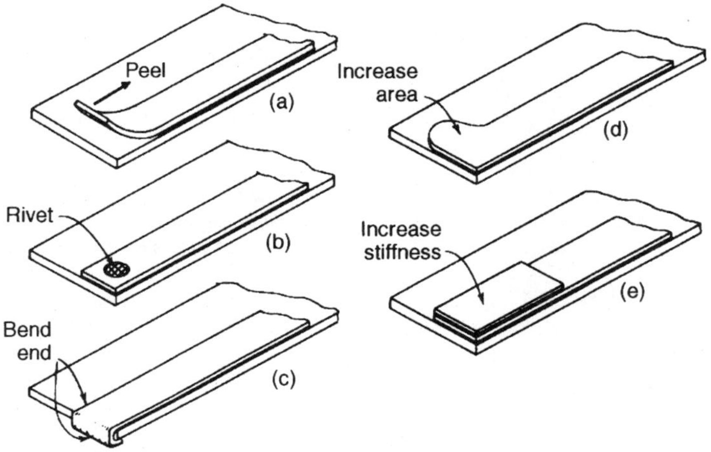

Figure 7.15 Minimizing peel in adhesive joints.[9]

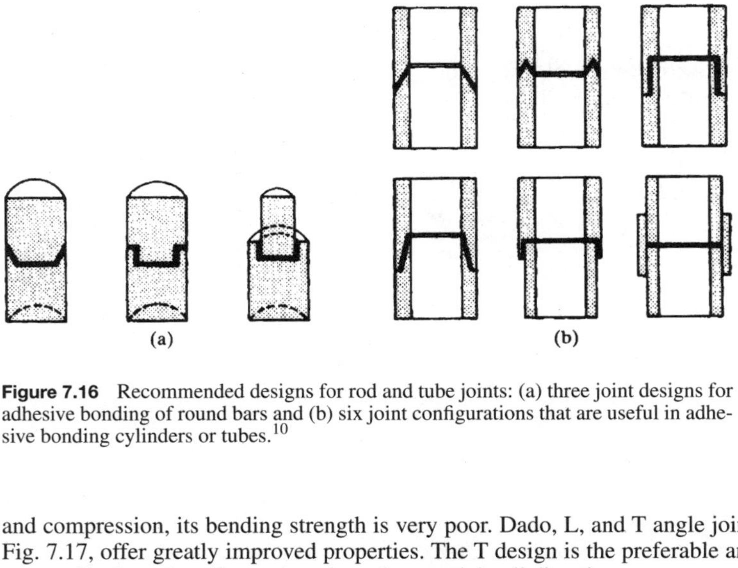

Figure 7.16 Recommended designs for rod and tube joints: (a) three joint designs for adhesive bonding of round bars and (b) six joint configurations that are useful in adhesive bonding cylinders or tubes.[10]

and compression, its bending strength is very poor. Dado, L, and T angle joints, shown in Fig. 7.17, offer greatly improved properties. The T design is the preferable angle joint because of its large bonding area and good strength in all directions.

Corner joints made of relatively flexible adherends such as sheet metal should be designed with reinforcements for support. Various corner-joint designs are shown in Fig. 7.18.

7.2.3.4 Plastic and elastomeric joints. The design of joints for plastic and elastomeric substrates follows the same practice as for metal. However, certain characteristics of these materials require special consideration.

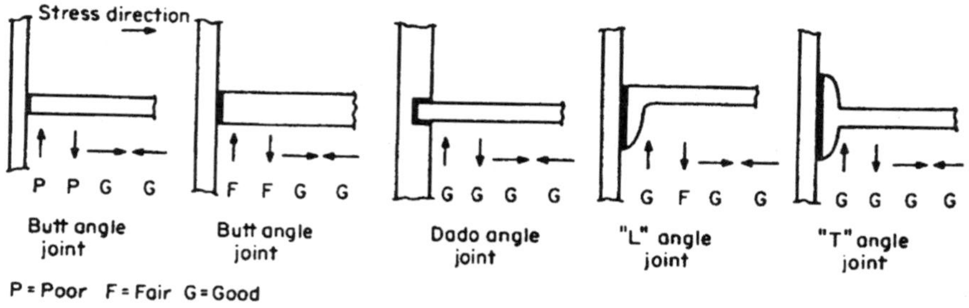

Figure 7.17 Types of angle joints and methods of reducing cleavage.[9]

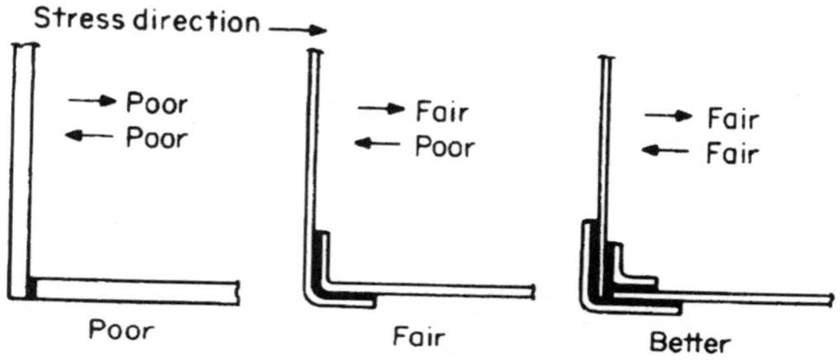

Figure 7.18 Reinforcement of bonded corners.[9]

Flexible plastics and elastomers. Thin or flexible polymeric substrates may be joined using a simple or modified lap joint. The double strap joint is best, but it is also the most time consuming to fabricate. The strap material should be made out of the same material as the parts to be joined or at least have approximately equivalent strength, flexibility, and thickness. The adhesive should have the same degree of flexibility as the adherends. If the sections to be bonded are relatively thick, a scarf joint is acceptable. The length of the scarf should be at least four times the thickness; sometimes, larger scarfs may be needed.

When bonding elastomers, forces on the substrate during setting of the adhesive should be carefully controlled, since excess pressure will cause residual stresses at the bond interface.

As with all joint designs, polymeric joints should be designed to avoid peel stress. Figure 7.19 illustrates methods of bonding flexible substrates so that the adhesive will be stressed in its strongest direction.

Rigid plastics. Rigid unreinforced plastics can be bonded using the joint design principles for metals. However, reinforced plastics are often anisotropic materials. This means that their strength properties are directional. Joints made from anisotropic substrates should be designed to stress both the adhesive and adherend in the direction of greatest strength. Laminates, for example, should be stressed parallel to the laminations. Stresses normal to the laminate may cause the substrate to delaminate. Single and joggle lap joints are more likely to cause delamination than scarf or beveled lap joints. The strap-joint variations are useful when bending loads are expected.

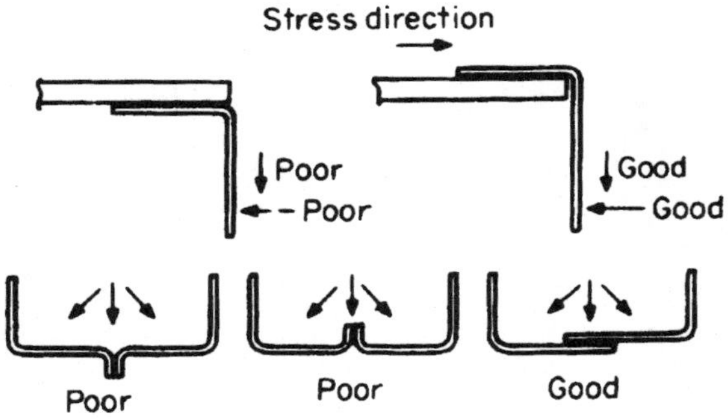

Figure 7.19 Methods of joining flexible plastic or rubber.[9]

7.2.4 Test Methods

7.2.4.1 Standard ASTM test methods. A number of standard tests for adhesive bonds have been specified by the American Society for Testing and Materials (ASTM). Selected ASTM standards are presented in Table 7.6. The properties usually reported by adhesive suppliers are ASTM tensile-shear and peel strength.

7.2.4.2 Lap-shear tests. The lap-shear or tensile-shear test measures the strength of the adhesive in shear. It is the most common adhesive test because the specimens are inexpensive, easy to fabricate, and simple to test. This method is described in ASTM D 1002, and the standard test specimen is shown in Fig. 7.20*a*. The specimen is loaded in tension, causing the adhesive to be stressed in shear until failure occurs. Since the test calls for a sample population of five, specimens can be made and cut from large test panels, illustrated in Fig. 7.20*b*.

7.2.4.3 Tensile tests. The tensile strength of an adhesive joint is seldom reported in the adhesive suppliers' literature, because pure tensile stress is not often encountered in actual production. Tensile test specimens also require considerable machining to ensure parallel surfaces.

ASTM tension tests are described in D 897 and D 2095 and employ bar- or rod-shaped butt joints. The maximum load at which failure occurs is recorded in pounds per square inch of bonded area. Test environment, joint geometry, and type of failure should also be recorded.

A simple cross-lap specimen to determine tensile strength is described in ASTM D 1344. This specimen has the advantage of being easy to make, but grip alignment and adherend deflection during loading can cause irreproducibility. A sample population of at least ten is recommended for this test method.

7.2.4.4 Peel test. Because adhesives are notoriously weak in peel, tests to measure peel resistance are very important. Peel tests involve stripping away a flexible adherend

TABLE 7.6 ASTM Adhesive Standards

Aging	Resistance of Adhesives to Cyclic Aging Conditions, Test for (D1183) Bonding Permanency of Water- or Solvent-Soluble Liquid Adhesives for Labeling Glass Bottles, Test for (D 1581) Bonding Permanency of Water- or Solvent-Soluble Liquid Adhesives for Automatic Machine Sealing Top Flaps of Fiber Specimens, Test for (D1713) Permanence of Adhesive-Bonded Joints in Plywood under Mold Conditions, Test for (D 1877) Accelerated Aging of Adhesive Joints by the Oxygen-Pressure Method, Practice for (D 3632)
Amylaceous matter	Amylaceous Matter in Adhesives, Test for (D 1488)
Biodeterioration	Susceptibility of Dry Adhesive Film to Attack by Roaches, Test for (D 1382) Susceptibility of Dry Adhesive Film to Attack by Laboratory Rats, Test for (D 1383) Permanence of Adhesive-Bonded Joints in Plywood under Mold Conditions, Test for (D 1877) Effect of Bacterial Contamination of Adhesive Preparations and Adhesive Films, Test for (D 4299) Effect of Mold Contamination on Permanence of Adhesive Preparation and Adhesive Films, Test for (D 4300)
Blocking point	Blocking Point of Potentially Adhesive Layers, Test for (D1146)
Bonding permanency	(See Aging.)
Chemical reagents	Resistance of Adhesive Bonds to Chemical Reagents, Test for (D 896)
Cleavage	Cleavage Strength of Metal-to-Metal Adhesive Bonds, Test for (D 1062)
Cleavage/peel strength	Strength Properties of Adhesives in Cleavage Peel by Tension Loading (Engineering Plastics-to-Engineering Plastics), Test for (D 3807)
Corrosivity	Determining Corrosivity if Adhesive Materials, Practice for (D 3310)
Creep	Conducting Creep Tests of Metal-to-Metal Adhesives, Practice for (D 1780) Creep Properties of Adhesives in Shear by Compression Loading (Metal-to-Metal), Test for (D 2293) Creep Properties of Adhesives in Shear by Tension Loading, Test for (D 2294)
Cryogenic temperatures	Strength Properties of Adhesives in Shear by Tension Loading in the Temperature Range from –267.8 to –55°C (–450 to –67°F), Test for (D 2557)
Density	Density of Adhesives in Fluid Form, Test for (D 1875)
Durability (including weathering)	Effect of Moisture and Temperature on Adhesive Bonds, Test for (D 1151) Atmospheric Exposure of Adhesive-Bonded Joints and Structures, Practice for (D 1828) Determining Durability of Adhesive Joints Stressed in Peel, Practice for (D 2918) Determining Durability of Adhesive Joints Stressed in Shear by Tension Loading, Practice for (D 2919) (See also Wedge Test)

TABLE 7.6 ASTM Adhesive Standards *(Continued)*

Electrical properties	Adhesives Relative to Their Use as Electrical Insulation, Testing (D 1304)
Electrolytic corrosion	Determining Electrolytic Corrosion of Copper by Adhesives, Practice for (D 3482)
Fatigue	Fatigue Properties of Adhesives in Shear by Tension Loading (Metal/ Metal), Test for (D 3166)
Filler content	Filler content of Phenol, Resorcinol, and Melamine Adhesives, Test for (D 3166)
Flexibility	(See Flexural strength)
Flexural strength	Flexural Strength of Adhesive Bonded Laminated Assemblies, Test for (D 1184) Flexibility Determination of Hot Melt Adhesives by Mandrel Bend Test Methods, Practice for (D 3111)
Flow properties	Flow Properties of Adhesives, Test for (D 2183)
Fracture strength in cleavage	Fracture Strength in Cleavage of Adhesives in Bonded Joints, Practice for (D 3433)
Gap-filling adhesive bonds	Strength of Gap Filling Adhesive Bonds in Shear by Compression Loading, Practice for (D 3931)
High-temperature effects	Strength Properties of Adhesives in Shear by Tension Loading at Elevated Temperatures (Metal-to-Metal), Test for (D 2295)
Hydrogen-ion concentration	Hydrogen Ion Concentration, est for (D 1583)
Impact strength	Impact Strength of Adhesive Bonds, Test for (D 950)
Light exposure	(See Radiation Exposure)
Low and cryogenic temperatures	Strength Properties of Adhesives in Shear by Tension Loading in the Temperature Range from –267.8 to –55°C (–450 to –67°F), Test for (D 2557)
Nonvolatile content	Nonvolatile Content of Aqueous Adhesives, Test for (D 1489) Nonvolatile Content of Urea-Formaldehyde Resin Solutions, Test for (D 1490) Nonvolatile Content of Phenol, Resorcinol, and Melamine Adhesives, Test for (D 1582)
Odor	Determination of the Odor of Adhesives, Test for (D 4339)
Peel strength (stripping strength)	Peel or Stripping Strength of Adhesive Bonds, Test for (D 903) Climbing Drum Peel Test for Adhesives, Method for (D 1781) Peel Resistance of Adhesives (T-Peel Test), Test for (D 1876) Evaluating Peel Strength of Shoe Sole Attaching Adhesives, Test for (D 2558) Determining Durability of Adhesive Joints Stressed in Peel, Practice for (D 2918) Floating Roller Peel Resistance, Test for (D 3167)

TABLE 7.6 ASTM Adhesive Standards *(Continued)*

Penetration	Penetration of Adhesives, Test for (D 1916)
pH	(See Hydrogen-Ion Concentration)
Radiation exposure (including light)	Exposure of Adhesive Specimens to Artificial (Carbon-Arc Type) and Natural Light, Practice for (D 904) Exposure of Adhesive Specimens to High-Energy Radiation, Practice for (D 1879)
Rubber cement tests	Rubber Cements, Testing of (D 816)
Salt spray (fog) testing	Salt Spray (Fog) Testing, Method of (B 117) Modified Salt Spray (Fog) Testing, Practice for (G 85)
Shear Strength (Tensile Shear Strength)	Shear Strength and Shear Modulus of Structural Adhesives, Test for (E 229) Strength Properties of Adhesive Bonds in Shear by Compression Loading, Test for (D 905) Strength Properties of Adhesives in Plywood Type Construction in Shear by Tension Loading, Test for (D 906) Strength Properties of Adhesives in Shear by Tension Loading (Metal-to-Metal), Test for (D 1002) Determining Strength Development of Adhesive Bonds, Practice for (D 1144) Strength Properties of Metal-to-Metal Adhesives by Compression Loading (Disk Shear), Test for (D 2181) Strength Properties of Adhesives in Shear by Tension Loading at Elevated Temperatures (Metal-to-Metal), Test for (D 2295) Strength Properties of Adhesives in Two-Ply Wood construction in Shear by Tension Loading, Test for (D 2339) Strength Properties of Adhesives in Shear by Tension Loading in the Temperature Range from –267.8 to –55°C (–450 to –67°F), Test for (D 2557) Determining Durability of Adhesive Joints Stressed in Shear by Tension Loading, Practice for (D 2919) Determining the Strength of Adhesively Bonded Rigid Plastic Lap-Shear Joints in Shear by Tension Loading, Practice for (D 3163) Determining the Strength of Adhesively Bonded Plastic Lap-Shear Sandwich Joints in Shear by Tension Loading, Practice for (D 3164) Strength Properties of Adhesives in Shear by Tension Loading of Laminated Assemblies, Test for (D 3165) Fatigue Properties of Adhesives in Shear by Tension Loading (Metal/Metal), Test for (D 3166) Strength Properties of Double Lap Shear Adhesive Joints by Tension Loading, Test for (3528) Strength of Gap-Filling Adhesive Bonds in Shear by Compression Loading, Practice for (D 3931) Measuring Strength and Shear Modulus of Nonrigid Adhesives by the Thick Adherend Tensile Lap Specimen, Practice for (D 3983) Measuring Shear Properties of Structural Adhesives by the Modified-Rail Test, Practice for (D 4027)
Specimen Preparation	Preparation of Bar and Rod Specimens for Adhesion Tests, Practice for (D 2094)
Spot-Adhesion Test	Qualitative Determination of Adhesion of Adhesives to Substrates by Spot Adhesion Test Method, Practice for (D 3808)

TABLE 7.6 ASTM Adhesive Standards *(Continued)*

Spread (Coverage)	Applied Weight per Unit Area of Dried Adhesive Solids, Test for (D 898) Applied Weight per Unit Area of Liquid Adhesive, Test for (D 899)
Storage Life	Storage Life of Adhesives by Consistency and Bond Strength, Test for (D 1337)
Strength Development	Determining Strength Development of Adhesive Bonds, Practice for (D 1144)
Stress-Cracking Resistance	Evaluating the Stress Cracking of Plastics by Adhesives Using the Bent Beam Method, Practice for (D 3929)
Stripping Strength	(See Peel Strength)
Surface Preparation	Preparation of Surfaces of Plastics Prior to Adhesive Bonding, Practice for (D 2093) Preparation of Metal Surfaces for Adhesive Bonding, Practice for (D 2651) Analysis of Sulfochromate Etch Solution Used in Surface Preparation of Aluminum, Methods of (D 2674) Preparation of Aluminum Surfaces for Structural Adhesive Bonding (Phosphoric Acid Anodizing), Practice for (D 3933)
Tack	Pressure Sensitive Tack of Adhesives Using an Inverted Probe Machine, Test for (D 2979) Tack of Pressure-Sensitive Adhesives by Rolling Ball, Test for (D 3121)
Tensile Strength	Tensile Properties of Adhesive Bonds, Test for (D 897) Determining Strength Development of Adhesive Bonds, Practice for (D 1144) Cross-Lap Specimens for Tensile Properties of Adhesives, Testing of (D 1344) Tensile Strength of Adhesives by Means of Bar and Rod Specimens, Method for (D 2095)
Torque Strength	Determining the Torque Strength of Ultraviolet (UV) Light-Cured Glass/Metal Adhesive Joints, Practice for (D 3658)
Viscosity	Viscosity of Adhesives, Test for (D 1084) Apparent Viscosity of Adhesives Having Shear-Rate-Dependent Flow Properties, Test for (D 2556) Viscosity of Hot Melt Adhesives and Coating Materials, Test for (D 3236)
Volume Resistivity	Volume Resistivity of Conductive Adhesives, Test for (D 2739)
Water Absorptiveness (of Paper Labels)	Water Absorptiveness of Paper Labels, Test for (D 1584)
Weathering	(See Durability)
Wedge Test	Adhesive Bonded Surface Durability of Aluminum (Wedge Test) (D 3762)
Working Life	Working Life of Liquid or Paste Adhesive by Consistency and Bond Strength, Test for (D 1338)

*The latest revisions of ASTM standards can be obtained from the American Society for Testing and Materials, 100 Barr Harbor Dr. West, Conshohocken, PA 19482.

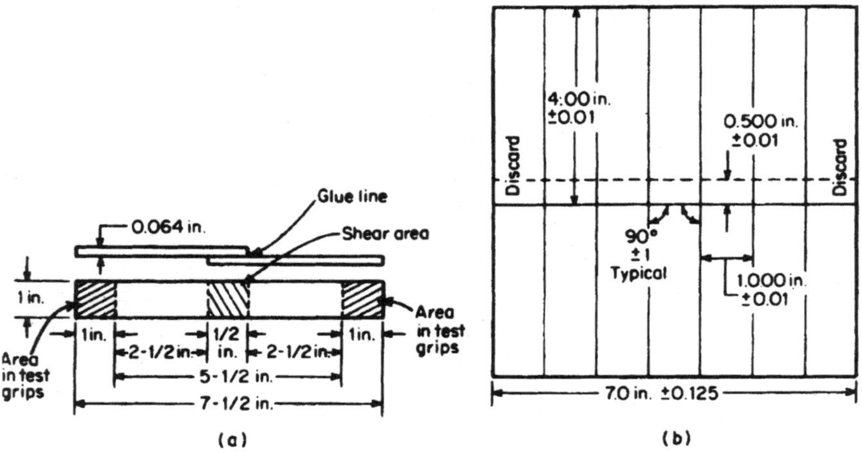

Figure 7.20 Standard lap-shear test specimen design: (a) form and dimensions of lap-shear test specimen and (b) standard test panel of five lap-shear specimens. *(From ASTM D 1002)*

from another adherend that may be flexible or rigid. The specimen is usually peeled at an angle of 90 or 180°.

The most common types of peel test are the T-peel, Bell, and climbing-drum methods. Representative test specimens are shown in Fig. 7.21. The values resulting from each test method can be substantially different; hence, it is important to specify the test method employed.

Peel values are recorded in pounds per inch of width of the bonded specimen. They tend to fluctuate more than other adhesive test results because of the extremely small area at which the stress is localized during loading.

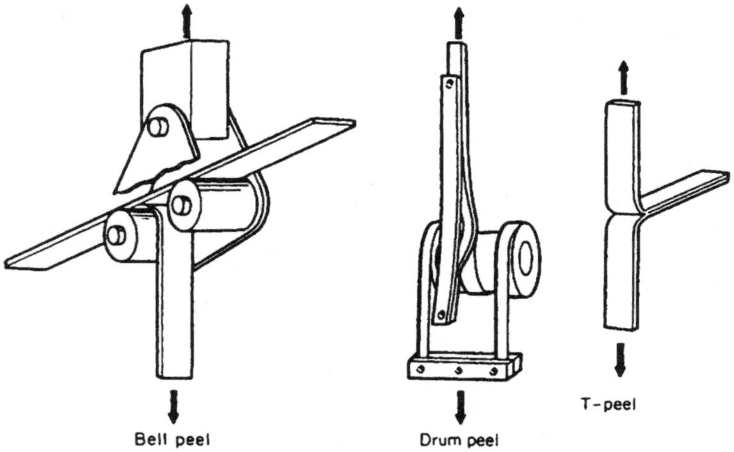

Figure 7.21 Common types of adhesive peel tests.[11]

The T-peel test is described in ASTM D 1876 and is the most common of all peel tests. The T-peel specimen is shown in Fig. 7.22. Generally, this test method is used when both adherends are flexible.

A 90° peel test, such as the Bell peel (ASTM D 3167), is used when one adherend is flexible and the other is rigid. The flexible member is peeled at a constant 90° angle through a spool arrangement. Thus, the values obtained are generally more reproducible.

The climbing-drum peel specimen is described in ASTM D 1781. This test method is intended for determining peel strength of thin metal facings on honeycomb cores, although it can be generally used for joints where at least one member is flexible.

A variation of the T-peel test is a 180° stripping test illustrated in Fig. 7.23 and described in ASTM D 903. It is commonly used when one adherend is flexible enough to permit a 180° turn near the point of loading. This test offers more reproducible results than the T-peel test, because the angle of peel is maintained constant.

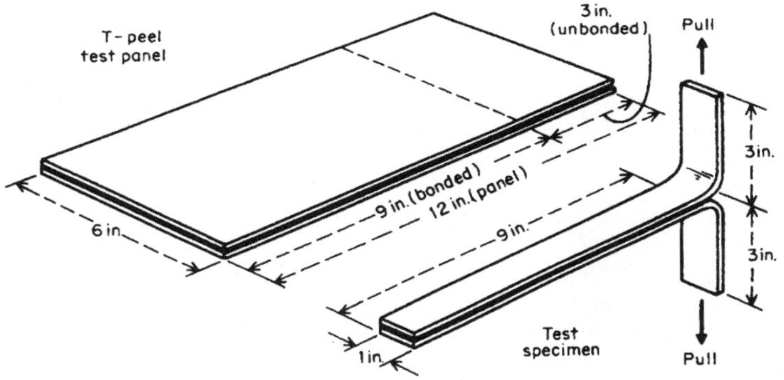

Figure 7.22 Test panel and specimen for T peel. *(From ASTM D 1876)*

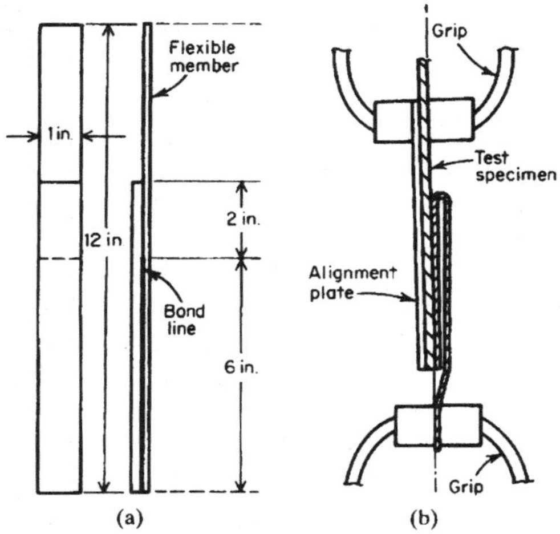

Figure 7.23 180° peel test specimens: (a) specimen design and (b) specimen under test. *(From ASTM D 903)*

7.2.4.5 Cleavage test. Cleavage tests are conducted by prying apart one end of a rigid bonded joint and measuring the load necessary to cause rupture. The test method is described in ASTM D 1062. A standard test specimen is illustrated in Fig. 7.24. Cleavage values are reported in pounds per inch of adhesive width. Because cleavage test specimens involve considerable machining, peel tests are usually preferred.

7.2.4.6 Fatigue test. Fatigue testing places a given load repeatedly on a bonded joint. Standard lap-shear specimens are tested on a fatiguing machine capable of inducing cyclic loading (usually in tension) on the joint. The fatigue strength of an adhesive is reported as the number of cycles of a known load necessary to cause failure.

Fatigue strength is dependent on adhesive, curing conditions, joint geometry, mode of stressing, magnitude of stress, and duration and frequency of load cycling.

7.2.4.7 Impact test. The resistance of an adhesive to impact can be determined by ASTM D 950. The specimen is mounted in a grip shown in Fig. 7.25 and placed in a standard impact machine. One adherend is struck with a pendulum hammer traveling at 11 ft/s, and the energy of impact is reported in pounds per square inch of bonded area.

7.2.4.8 Creep test. The dimensional change occurring in a stressed adhesive over a long time period is called *creep*. Creep data are seldom reported in the adhesive suppliers'

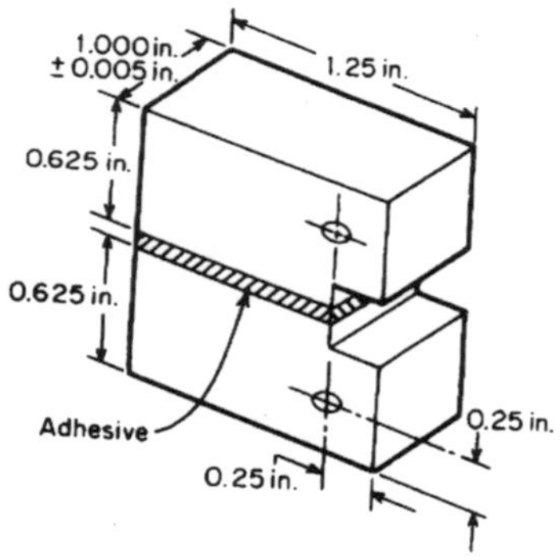

Figure 7.24 Cleavage test specimen. *(From ASTM D 1062)*

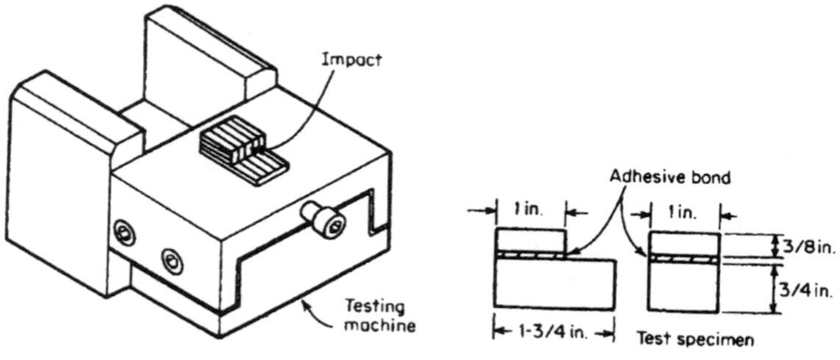

Figure 7.25 Impact test specimen and holding fixture. *(From ASTM D 950)*

literature, because the tests are time-consuming and expensive. This is very unfortunate, since sustained loading is a common occurrence in adhesive applications. All adhesives tend to creep, some much more than others. With weak adhesives, creep may be so extensive that bond failure occurs prematurely. Certain adhesives have also been found to degrade more rapidly when aged in a stressed rather than an unstressed condition.

Creep tests are made by loading a specimen with a predetermined stress and measuring the total deformation as a function of time or measuring the time necessary for complete failure of the specimen. Depending on the adhesive, loads, and testing conditions, the time required for a measurable deformation may be extremely long. ASTM D 2294 defines a test for creep properties of adhesives utilizing a spring-loaded apparatus to maintain constant stress.

7.2.4.9 Environmental tests. Strength values determined by short-term tests do not give an adequate indication of an adhesive's permanence during continuous environmental exposure. Laboratory-controlled aging tests seldom last longer than a few thousand hours. To predict the permanence of an adhesive over a 20-year product life requires accelerated test procedures and extrapolation of data. Such extrapolations are extremely risky, because the causes of adhesive-bond deterioration are many and not well understood.

7.2.4.10 Federal test specifications. A variety of federal and military specifications describing adhesives and test methods have been prepared. Selected government specifications are described in Table 7.7.

TABLE 7.7 Government Adhesive Specifications

	Military Specifications
MIL-A-928	Adhesive; metal to work, structural
MIL-A-1154	Adhesive; bonding, vulcanized synthetic rubber to steel
MIL-C-1219	Cement; iron and steel
MIL-C-3316	Adhesive; fire resistant, thermal insulation
MIL-C-4003	Cement; general purpose, synthetic base
MIL-A-5092	Adhesive; rubber (synthetic and reclaimed rubber base)
MIL-A-5534	Adhesive; high-temperature setting resin (phenol, melamine, and resorcinol base)
MIL-C-5339	Cement; natural rubber
MIL-A-5540	Adhesive; polychloroprene
MIL-A-8576	Adhesive; acrylic monomer base, for acrylic plastic
MIL-A-8623	Adhesive; epoxy resin, metal-to-metal structural bonding
MIL-A-9117	Adhesive; sealing, for aromatic fuel cells and general repair
MIL-C-10523	Cement, gasket, for automobile applications
MIL-S-11030	Sealing compound; noncuring polysulfide base
MIL-S-11031	Sealing compound, adhesive; curing, polysulfide base
MIL-A-11238	Adhesive; cellulose nitrate
MIL-C-12850	Cement, rubber

TABLE 7.7 Government Adhesive Specifications *(Continued)*

	Military Specifications
MIL-A-13554	Adhesive for cellulose nitrate film on metals
MIL-C-13792	Cement, vinyl acetate base solvent type
MIL-A-13883	Adhesive, synthetic rubber (hot or cold bonding)
MIL-A-14042	Adhesive, epoxy
MIL-C-14064	Cement; grinding disk
MIL-P-14536	Polyisobutylene binder
MIL-I-15126	Insulation tape, electrical, pressure-sensitive adhesive and pressure-sensitive thermosetting adhesive
MIL-C-l8726	Cement, vinyl alcohol-acetate
MIL-A-22010	Adhesive, solvent type, polyvinyl chloride
MIL-A-22397	Adhesive, phenol and resorcinol resin base
MIL-A-22434	Adhesive, polyester, thixotropic
MIL-C-22608	Compound insulating, high temperature
MIL-A-22895	Adhesive, metal identification plate
MIL-C-23092	Cement, natural rubber
MIL-A-25055	Adhesive; acrylic monomer base, for acrylic plastics
MIL-A-25457	Adhesive, air-drying silicone rubber
MIL-A-25463	Adhesive, metallic structural honeycomb construction
MIL-A-46050	Adhesive, special; rapid room-temperature curing, solventless
MIL-A-46051	Adhesive, room-temperature and intermediate-temperature setting resin (phenol, resorcinol, and melamine base)
MIL-A-52194	Adhesive, epoxy (for bonding glass reinforced polyester)
MIL-A-9067C	Adhesive bonding, process and inspection requirements for
MIL-C-7438	Core material, aluminum, for sandwich construction
MIL-C-8073	Core material, plastic honeycomb, laminated glass fabric base, for aircraft structural applications
MIL-C-21275	Core material, metallic, heat-resisting, for structural sandwich construction
MIL-H-9884	Honeycomb material, cushioning, paper
MIL-S-9041A	Sandwich construction, plastic resin, glass fabric base, laminated facings for aircraft structural applications
	Federal Specifications
MMM-A-l8l	Adhesive, room-temperature and intermediate-temperature setting resin (phenol, resorcinol, and melamine Base)
MMM-A-00185	Adhesive, rubber
MMM-A-00187	Adhesive, synthetic, epoxy resin base, paste form, general purpose
MMM-A-132	Adhesives, Heat-resistant, airframe structural, metal-to-metal
Federal Test Method l75	Adhesives; methods of testing
MIL-STD-401	Sandwich construction and core materials; general test methods

7.3 Surface Preparation

7.3.1 Importance of Surface Preparation

Surface preparation of adherends prior to bonding is one of the most important factors in the adhesive-bonding process. Initial bond strength and joint permanence are greatly dependent on the quality of surface that is in contact with the adhesive. Prebond treatments are intended to remove weak boundary layers and provide easily wettable surfaces. As a general rule, all adherends must be treated in some manner prior to bonding.

Surface preparations can range from simple solvent wiping to a combination of mechanical abrading, chemical cleaning, and acid etching. In many low- to medium-strength applications extensive surface preparation may be unnecessary. But, where maximum bond strength, permanence, and reliability are required, carefully controlled surface-treating processes are necessary. The following factors should be considered in the selection of a surface preparation process:

1. The ultimate initial bond strength required
2. The degree of permanence necessary and the service environment
3. The degree and type of contamination on the adherend
4. The type of adherend and adhesive

Table 7.8 shows the effect of various metallic-surface preparations on adhesive-joint strength.

Surface preparations enhance the quality of a bonded joint by performing one or more of the following functions: (1) remove contaminants, (2) control adsorbed water, (3) control oxide formation, (4) poison surface atoms that catalyze adhesive breakdown, (5) protect the adhesive from the adherend and vice versa, (6) match the adherend crystal structure to the adhesive molecular structure, and (7) control surface roughness. Thus, surface preparations can affect the permanence of the joint as well as its initial strength.

Plastic and elastomeric adherends are even more dependent than metals on surface preparation. Many of these surfaces are contaminated with mold-release agents or processing additives. Such contaminants must be removed before bonding. Because of their low surface energy, polytetrafluoroethylene, polyethylene, and certain other polymeric materials are completely unsuitable for adhesive bonding in their natural state. The surfaces of these materials must be chemically or physically altered prior to bonding to improve wetting.

7.3.2 General Surface Preparation Methods

Listed here are several methods of preparing both metal and polymer substrates for adhesive bonding. The chosen method will ultimately be the process that yields the necessary strength and permanence with the least cost.

7.3.2.1 Solvent wiping. Where loosely held dirt, grease, and oil are the only contaminants, simple solvent wiping will provide surfaces for weak- to medium-strength bonds. Solvent wiping is widely used, but it is the least effective substrate treatment. Volatile solvents such as acetone and trichloroethylene are acceptable. Trichloroethylene is often favored because of its nonflammability. A clean cloth should be saturated with the solvent and wiped across the area to be bonded until no signs of residue are evident on the cloth or

TABLE 7.8 Effect of Substrate Pretreatment on Strength of Adhesive-Bonded Joints

Adherend	Treatment	Adhesive	Shear strength, lb/in^2	Ref.
Aluminum	As received	Epoxy	444	12
	Vapor degreased		837	
	Grit blast		1,751	
	Acid etch		2,756	
Aluminum	As received	Vinyl-phenolic	2,442	13
	Degreased		2,741	
	Acid etch		5,173	
Stainless steel	As received	Vinyl-phenolic	5,215	13
	Vapor degreased		6,306	
	Acid etch		7,056	
Cold-rolled steel	As received	Epoxy	2,900	14
	Vapor degreased		2,910	
	Grit blast		4,260	
	Acid etch		4,470	
Copper	Vapor degreased	Epoxy	1,790	15
	Acid etch		2,330	
Titanium	As received	Vinyl-phenolic	1,356	13
	Degreased		3,180	
	Acid etch		6,743	
Titanium	Acid etch	Epoxy	3,183	16
	Liquid pickle		3,317	
	Liquid hone		3,900	
	Hydrofluorosilicic acid etch		4,005	

substrate. Special precautions are necessary to prevent the solvent from becoming contaminated. For example, the wiping cloth should never touch the solvent container, and new wiping cloths must be used often. After cleaning, the parts should be air dried in a clean, dry environment before being bonded.

7.3.2.2 Vapor degreasing. Vapor degreasing is a reproducible form of solvent cleaning that is attractive when many parts must be prepared. It consists of suspending the adherends in a container of solvent vapor such as trichloroethylene or perchloroethylene. When the hot vapors come into contact with the relatively cool substrate, solvent condensation occurs on the surface of the part, which dissolves the organic contaminants. Vapor degreasing is preferred to solvent wiping, because the surfaces are continuously being washed in distilled, uncontaminated solvent. The vapor degreaser must be kept clean, and a fresh supply of solvent must be used when the contaminants in the solvent trough lower the boiling point significantly.

Modern vapor degreasing equipment is available with ultrasonic transducers built into the solvent rinse tank. The parts are initially cleaned by vapor and then subjected to ultrasonic scrubbing. The cleaning solutions and processing parameters must be optimized by test.

7.3.2.3 Abrasive cleaning. Mechanical methods for surface preparation include sandblasting, wire brushing, and abrasion with sandpaper, emery cloth, or metal wool. These methods are most effective for removing heavy, loose particles such as dirt, scale, tarnish, and oxide layers.

The parts should always be degreased prior to abrasive treatment to prevent contaminants from being rubbed into the surface. Solid particles left on the surfaces after abrading can be removed by blasts of clean, oil free, dry air or solvent wiping.

Each metal reacts favorably with a specific range of abrasive sizes. It has been found that joint strength generally increases with the degree of surface roughness provided that the adhesive wets the adherend (Table 7.9). However, excessively rough surfaces increase the probability tat voids will be left at the interface, causing stress risers detrimental to the joint.

TABLE 7.9 Effect of Surface Roughness in Butt Tensile Strength of DER 332-Versamid 140 (60/40)* Epoxy Joints (from Ref. 17)

Adherend		Adherend surface†	Butt tensile strength, lb/in^2
6061	Al	Polished	4,720 ± 1,000
6061	Al	0.005-in grooves	6,420 ± 500
6061	Al	0.005-in grooves, sandblasted	7,020 ± 1,120
6061	Al	Sandblasted (40–50 grit)	7,920 ± 530
6061	Al	Sandblasted (10–20 grit)	7,680 ± 360
304	SS	Polished	4,030 ± 840
304	SS	0.010-in grooves	5,110 ± 1,020
304	SS	0.010-in grooves, sandblasted	5,510 ± 770
304	SS	Sandblasted (40–50 grit)	7,750 ± 840
304	SS	Sandblasted (10–20 grit)	9,120 ± 470

*74°C/16 hr cure.
†Adherend surfaces were chromate etched.

7.3.2.4 Chemical cleaning. Strong detergent solutions are used to emulsify surface contaminants on both metallic and nonmetallic substrates. These solutions are usually heated. Parts for cleaning are generally immersed in a well agitated solution maintained at 150 to 210°F for approximately 10 min. The surfaces are then rinsed immediately with deionized water and dried. Chemical cleaning is often used in combination with other surface treatments. Chemical cleaning by itself will not remove heavy or strongly attached contaminants such as rust or scale.

Alkaline detergents recommended for prebond cleaning are combinations of alkaline salts such as sodium metasilicate and tetrasodium pyrophosphate with surfactants included. Many commercial detergents are available.

7.3.2.5 Other cleaning methods. Vapor-honing and ultrasonic cleaning are efficient treating methods for small, delicate parts. In cases where the substrate is so delicate that usual abrasive treatments may be too rough, contaminants can be removed by vapor honing. This method is similar to grit blasting except that very fine abrasive particles are suspended in a high-velocity water or steam spray. Thorough rinsing after vapor honing is usually not required.

Ultrasonic cleaning employs a bath of cleaning liquid or solvent that is ultrasonically activated by a high-frequency transducer. The part to be cleaned is immersed in the liquid, which carries the sonic waves to the surface of the part. High-frequency vibrations then dislodge the contaminants. Commercial ultrasonic cleaning units are available from a number of manufacturers.

7.3.2.6 Alteration of surfaces. Certain treatments of substrate surfaces change the physical and chemical properties of the surface to produce greater wettability and/or a stronger surface. Specific processes are required for each substrate material. The part or area to be bonded is usually immersed in an active solution for a matter of minutes. The parts are then immediately rinsed with deionized water and dried.

Chemical solutions must be changed regularly to prevent contamination and ensure repeatable concentration. Tank temperature and agitation must also be controlled. Personnel need to be trained in the safe handling and use of chemical solutions and must wear the proper clothing.

Some paste-type acid etching products are available that simultaneously clean and chemically treat surfaces. They react at room temperature and need only be applied to the specific area to be bonded. However, these paste etchants generally require much longer treatment time than hot acid-bath processes.

Treatment of certain polymeric surfaces with excited inert gases greatly improves the bond strength of adhesive joints prepared from these materials. With this technique, called *plasma treatment,* a low-pressure inert gas is activated by an electrode-less radio-frequency discharge or microwave excitation to produce metastable species which react with the polymeric surface. The type of plasma gas can be selected to initiate a wide assortment of chemical reactions. In the case of polyethylene, plasma treatment produces a strong, wettable, cross-linked skin. Commercial instruments are available that can treat polymeric materials in this manner. Table 7.10 presents bond strength of various plastic joints pretreated with activated gas and bonded with an epoxy adhesive.

7.3.2.7 Combined methods. More than one cleaning method is usually required for optimal adhesive properties. A three-step process that is recommended for most substrates consists of (1) degreasing, (2) mechanical abrasion, and (3) chemical treatment. Table 7.11 shows the effect of various combinations of aluminum-surface preparations on lap-shear strength.

7.3.3 Surface Treatment of Polymers

Treating of plastic surfaces usually consists of one or a combination of the following processes: solvent cleaning, abrasive treatment and etching, flame, hot air, electric discharge, or plasma treatments. The purpose of the treatment is to either remove or strengthen the weak boundary layer or to increase the critical surface tension.

Abrasive treatments consist of scouring, machining, hand sanding, and dry and wet abrasive blasting. The choice is generally determined by available production facilities and

TABLE 7.10 Typical Adhesive-Strength Improvement with Plasma Treatment; Aluminum-Plastic Shear Specimen Bonded with Epon 828–Versamide 140 (70/30) Epoxy Adhesive (from Ref. 18)

	Strength of bond, lb/in^2	
Material	Control	After plasma treatment
High-density polyethylene	315 + 38	>3,125 + 68
Low-density polyethylene	372 + 52	>1,466 ± 106
Nylon 6	846 ± 166	>3,956 ± 195
Polystyrene	566 ± 17	>4,015 ± 85
Mylar* A	530 ± 51	1,660 ± 40
Mylar D	618 ± 25	1,185 ± 83
Polyvinylfluoride (Tedlar†)	278 ± 2	>1280 ± 73
Lexan*	410 ± 10	928 ± 66
Polypropylene	370 ±	3,080 ± 180
Teflon* TFE	75	750

*Trademark of E.I. du Pont de Nemours & Company, Inc.
†Trademark of General Electric Company.

cost. Laminates can be prepared by either abrasion or the tear-ply technique. In the tear-ply design, the laminate is manufactured so that one ply of heavy fabric, such as Dacron or equivalent, is attached at the bonding surface. Just prior to bonding, the tear-ply is stripped away, and a fresh, clean, bondable surface is exposed.

Chemical etching treatments vary with the type of plastic surface. The plastic resin supplier is the best guide to the appropriate etching chemicals and process. Etching processes can involve the use of corrosive and hazardous materials. The most common processes are sulfuric-dichromate etch (polyethylene and polypropylene) and sodium-naphthalene etch (fluorocarbons).

Flame, hot air, electrical discharge, and plasma treatments physically and chemically change the nature of polymeric surfaces. The plasma treating process has been found to be very successful on most hard-to-bond plastic surfaces. Table 7.10 shows that plasma treatment results in improved plastic joint strength with common epoxy adhesives. Plasma treatment requires high vacuum and special processing equipment.

Better adhesion can be obtained if parts are formed, treated, and coated with an adhesive in a continuous operation. The sooner an article can be bonded after surface treatment, the better will be the adhesion. After the part is treated, handling and exposure to shop environments should be kept to a minimum. Some surface treatments, such as plasma, have a long effective shelf life. However, some treating processes, such as electrical discharge and flame treating, will become less effective with a longer time between surface preparation and bonding.

TABLE 7.11 Surface Preparation of Aluminum Substrates vs. Lap–Shear Strength (from Ref. 12)

Group treatment	X, lb/in^2	s, lb/in^2	Cu, %
1. Vapor degrease, grit blast 90–mesh grit, alkaline clean, $Na_2Cr_2O_7$-H_2SO_4, distilled water	3091	103	3.5
2. Vapor degrease, grit blast 90-mesh grit, alkaline clean, $Na_2Cr_2O_7$-H_2SO_4, tap water	2929	215	7.3
3. Vapor degrease, alkaline clean, $Na_2Cr_2O_7$-H_2SO_4, distilled water	2800	307	10.96
4. Vapor degrease, alkaline clean, $Na_2Cr_2O_7$-H_2SO_4, tap water	2826	115	4.1
5. Vapor degrease, alkaline clean, chromic-H_2SO_4, deionized water	2874	163	5.6
6. Vapor degrease, Na_2Cr_2O7-H2SO4, tap water	1756	363	1.3
7. Unsealed anodized	1935	209	10.8
8. Vapor degrease, grit blast 90-mesh grit	1751	138	7.9
9. Vapor degrease, wet and dry sand, 100 + 240 mesh grit, N_2 blown	1758	160	9.1
10. Vapor degrease, wet and dry sand, wipe off with sandpaper	1726	60	3.4
11. Solvent wipe, wet and dry sand, wipe off with sandpaper (done rapidly)	1540	68	4
12. Solvent wipe, sand (not wet and dry), 120 grit	1329	135	1.0
13. Solvent wipe, wet and dry sand, 240 grit only	1345	205	15.2
14. Vapor degrease, aluminum wood	1478		
15. Vapor degrease, 15% NaOH	1671		
16. Vapor degrease	837	72	8.5
17. Solvent wipe (benzene)	353		
18. As received	444	232	52.2

X = average value. s = standard deviation. C_v = coefficient of variation.
Resin employed is EA 934 Hysol Division, Dexter Corp.; cured 16 hr at 75°F plus 1 hr at 180°F.

7.3.4 Evaluation of Surface Preparation before and after Bonding

On many nonporous surfaces, a useful and quick method for testing the effectiveness of the surface preparation is the "water-break test." If distilled water beads up when sprayed on the surface and does not wet the substrate, the surface-preparation steps should be repeated. If the water wets the surface in a uniform film, an effective surface operation has been achieved.

The objective of surface treating is to obtain a joint where the weakest link is the adhesive layer and not the interface. Thus, destructively tested joints should be examined for mode of failure. If failure is cohesive (within the adhesive layer or adherend), the surface treatment is the optimum for that particular combination of adherend, adhesive, and testing condition. However, it must be realized that specimens could exhibit cohesive failure initially and interfacial failure after aging. Both adhesive and surface preparations need to be tested with respect to the intended service environment.

7.3.5 Substrate Equilibrium

After the surface-preparation process has been completed, the substrates may have to be stored before bonding. During storage, the surface could regain its oxide or weak boundary layer and become exposed to contaminating environments. Typical storage life for various metals subjected to different treatments is shown in Table 7.12. Because of the relatively short storage life of many treated materials, bonding should be conducted as soon after the surface preparation as possible.

TABLE 7.12 Maximum Allowable Time between Surface Preparation and Bonding or Priming (from Ref. 19)

Metal	Surface	Time
Aluminum	Wet-abrasive-blasted	72 hr
Aluminum	Sulfuric-chromic acid etched	6 days
Aluminum	Anodized	30 days
Stainless steel	Sulfuric acid etched	30 days
Steel	Sandblasted	4 hr
Brass	Wet-abrasive-blasted	8 hr

If prolonged storage is necessary, a compatible primer may be used to coat the treated substrates after surface preparation. The primer will protect the surface during storage and interact with the adhesive during bonding. Many primer systems are sold with adhesives for this purpose.

7.3.6 Primers and Adhesion Promoters

Primers are applied and cured onto the adherend prior to adhesive bonding. They serve four primary functions singly or in combination.

1. Protection of surfaces after treatment (primers can be used to extend the time between preparing the adherend surface and bonding)
2. Developing tack for holding or positioning parts to be bonded
3. Inhibiting corrosion during service
4. Serving as an intermediate layer to enhance the physical properties of the joint and improve bond strength

Some primers have been found to provide corrosion resistance for the joint during service. The primer protects the adhesive–adherend interface and lengthens the service life of the bonded joint. Representative data are shown in Fig. 7.26.

Primers may also be used to modify the characteristics of the joint. For example, elastomeric primers are used with rigid adhesives to provide greater peel or cleavage resistance.

Primers can also chemically react with the adhesive and adherend to provide greater joint strengths. This type of primer is referred to as an *adhesion promoter*. The use of reactive silane to improve the adhesion of resin to glass fibers in polymeric laminates is well known in the plastic industry.

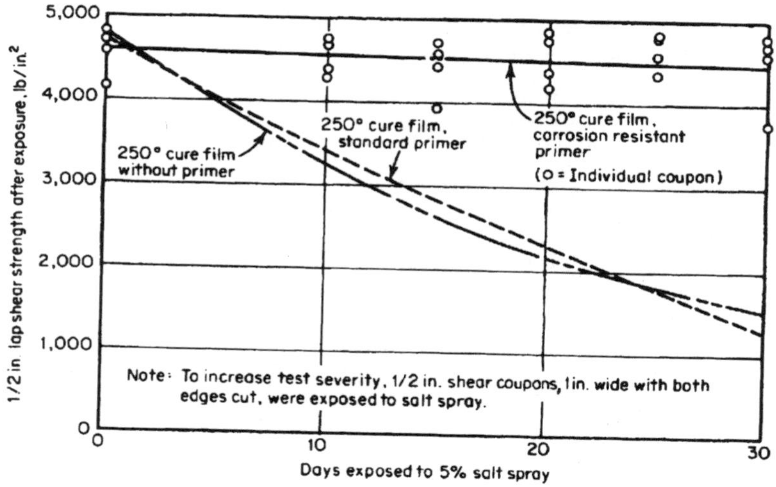

Figure 7.26 Effect of primer on lap-shear strength of aluminum joints exposed to 5 percent salt spray.[20]

7.3.7 Specific Surface Treatments and Characteristics

7.3.7.1 Metallic adherends. Table 7.13 lists common surface-treating procedures for metallic adherends. The general methods previously described are all applicable to me-

TABLE 7.13 Surface Preparation for Metals

Adherend	Degreasing solvent	Method of treatment	Remarks
Aluminum and aluminum alloys	Trichloroethylene	1. Sandblast or 100-grit emery cloth followed by solvent degreasing	Medium- to high-strength bonds, suitable for noncritical applications
		2. Immerse for 10 min at 77 ± 6°C in a commercial alkaline cleaner or *Parts by wt.* Sodium metasilicate.. 30.0 Sodium hydroxide.... 1.5 Sodium pyrophosphate 1.5 Nacconol NR (Allied Chemical Co.)...... 0.5 Water (distilled)...... 128.0 Wash in water below 65°C and etch for 10 min at 68±3°C in *Parts by wt.* Sodium dichromate... 1 Sulfuric acid (96%, sp. gr. 1.84)......... 10 Water (distilled)...... 30 Rinse in distilled water after washing in tap water, and dry in air	Optimum bond strength, specified in ASTM D 2651 and MIL-A-9067. Solvent degrease may replace alkaline cleaning

TABLE 7.13 Surface Preparation for Metals *(Continued)*

Adherend	Degreasing solvent	Method of treatment	Remarks
		3. Vapor degrease or solvent wipe. Immerse for 5 min at 71–82°C in Sulfuric acid (96%, sp. gr. 1.84) .... 1 gal Chromic acid .... 45 oz Distilled water .... 9 gal Rinse in distilled water and dry in air	Alternative to sulfuric-dichromate etch
		4. Degrease with 50/50 solution of methyl ethyl ketone and chlorothene. Abrade lightly with mildly abrasive cleaner. Rinse in deionized water; wipe; or air dry. Etch 20 min at RT in *Parts by wt.* Sodium dichromate ... 2 Sulfuric acid (96%, sp. gr. 1.84) .... 7 Rinse thoroughly in deionized water; dry at 70°C for 30 min	Room-temp etch
		5. Form a paste using sulfuric acid–dichromate solution and finely divided silica or fuller's earth. Apply; do not permit paste to dry. Time depends on degree of contamination (usually greater than 10 min at RT). Wash very thoroughly with deionized water, and air-dry	Paste form of acid etch useful when part cannot be immersed. ASTM D 2651
Beryllium	Trichloro-ethylene	1. Scrub with a nonchlorinated cleaner. Rinse with deionized water. Force dry at 250–300°F. Check for water break and repeat procedure if necessary	
		2. Immerse for 5–10 min at 20°C in *Parts by wt.* Sodium hydroxide ..... 20–30 Water (distilled) ...... 170–180 Rinse in distilled water after washing in tap water, and oven-dry for 10 min at 121–177°C	
		3. Same procedure as 2, except immerse for 3–4 min at 75–80°C	
Brass and bronze (see also Copper and copper alloys)	Trichloro-ethylene	1. Etch for 5 min at 20°C in *Parts by wt.* Zinc oxide .... 20 Sulfuric acid (96%, sp. gr. 1.84) 460	Temperatures must not exceed 65°C when washing and drying

TABLE 7.13 Surface Preparation for Metals *(Continued)*

Adherend	Degreasing solvent	Method of treatment	Remarks
		Nitric acid (69%, sp. gr. 1.41) .... 360 Rinse in water, below 65°C, and re-etch in the acid solution for 5 min at 49°C. Rinse in distilled water after washing, and dry in air	
Cadmium	Trichloroethylene	1. Abrasion. Grit or vapor blast, or 100-grit emery cloth, followed by solvent degreasing	Suitable for general-purpose bonding
		2. Electroplate with nickel or silver	For maximum bond strength
Chromium	Trichloroethylene	1. Abrasion. Grit or vapor blast, or 100-grit emery cloth, followed by solvent degreasing	Suitable for general-purpose bonding
		2. Etch for 1–5 min at 90–95°C in *Parts by wt.* Hydrochloric acid (37%) .... 17 Water .... 20 Rinse in distilled water after cold/hot-water washing, and dry in hot air	For maximum bond strength
Copper and copper alloys	Trichloroethylene	1. Abrasion. Sanding, wire brushing, or 100-grit emery cloth, followed by vapor or solvent degreasing	Suitable for general-purpose bonding. Use 320-grit emery cloth for foil
		2. Etch for 10 min at 66°C in *Parts by wt.* Ferric sulfate .... 1.0 Sulfuric acid (96%) .... 0.75 Water .... 8.0 Wash in water at 20°C, and etch in cold solution of *Parts by wt.* Sodium dichromate .... 5 Sulfuric acid (96%) .... 10 Water .... 85 Etch until a bright clean surface has been obtained. Rinse in water, dip in ammonium hydroxide (sp. gr. 0.88), and wash in tap water. Rinse in distilled water, and dry in warm air	For maximum bond strength. Suitable for brass and bronze. ASTM D 2651
		3. Etch for 1–2 min at 20°C in *Parts by wt.* Ferric chloride (42% w/w solution) .... 0.75 Nitric acid (sp. gr. 1.41) .... 1.5 Water .... 10.0 Rinse in distilled water after cold-water wash and dry in air stream at 20°C	Room-temp etch. ASTM D 2651

TABLE 7.13 Surface Preparation for Metals *(Continued)*

Adherend	Degreasing solvent	Method of treatment	Remarks
		4. Etch for 30 s at 20°C in *Parts by wt.* Ammonium persulfate ... 1 Water ... 4 Rinse in distilled water after cold-water wash, and dry in air stream at 20°C	Alternative etching solution to above where fast processing is required
		5. Solvent degrease. Immerse 30 s at 20°C in *Parts by vol.* Nitric acid (69%) ... 10 Deionized water ... 90 Rinse in running water and transfer immediately to next solution; immerse for 1–2 min at 98°C in Ebonol C (Ethone, Inc., New Haven, Conn.) 24 oz and equivalent water to make 1 gal Rinse in deionized water, and air-dry	For copper alloys containing over 95% copper. Stable surface for hot bonding. ASTM D 2651
Germanium	Trichloro-ethylene	Abrasion. Grit or vapor blast followed by solvent degreasing	
Gold	Trichloro-ethylene	Solvent or vapor degrease after light abrasion with a fine emery cloth	
Iron		See Steel (mild)	
Lead and solders	Trichloro-ethylene	Abrasion. Grit or vapor blast, or 100-grit emery cloth, followed by solvent degreasing	
Magnesium and magnesium alloys	Trichloro-ethylene	1. Abrasion with 100-grit emery cloth followed by solvent degreasing	Apply the adhesive immediately after abrasion
		2. Vapor degrease. Immerse for 10 min at 60–70°C in *Parts by wt.* Deionized water ... 95 Sodium metasilicate ... 2.5 Trisodium pyrophosphate ... 1.1 Sodium hydroxide ... 1.1 Nacconal NR (Allied Chemical Co.) ... 0.3 Rinse in water and dry below 60°C	Medium to high bond strength. ASTM D 2651
		3. Vapor degrease. Immerse for 10 min at 71–88°C *Parts by wt.* Water ... 4 Chromic acid ... 1 Rinse in water, and dry below 60°C	High bond strength. ASTM D 2651

TABLE 7.13 Surface Preparation for Metals *(Continued)*

Adherend	Degreasing solvent	Method of treatment	Remarks
		4. Vapor degrease. Immerse for 5–10 min at 63–80°C in *Parts by wt.* Water............... 12 Sodium hydroxide.... 1 Rinse in water. Immerse for 5–15 min at RT in *Parts by wt.* Water............... 123 Chromic acid........ 24 Calcium nitrate...... 1.8 Rinse in water, and dry below 60°C	Alternative to procedure 2
		5. Light anodic treatment and various corrosion-preventive treatments have been developed by magnesium producers (Dow 17 and Dow 7, Dow Chemical Co.)	Dow 17 preferred under extreme environmental conditions
Nickel	Trichloroethylene	1. Abrasion with 100-grit emery cloth followed by solvent degreasing	For general-purpose bonding
		2. Etch for 5 s at 20°C in nitric acid (69%, sp.gr.1.41). Wash in cold and hot water followed by a distilled-water rinse, and air-dry at 40°C	For general-purpose bonding
Platinum	Trichloroethylene	Solvent or vapor degrease after light abrasion with a 320-grit emery cloth	
Silver	Trichloroethylene	Abrasion with 320-grit emery cloth followed by solvent degreasing	
Steel (stainless)	Trichloroethylene	1. Abrasion with 100-grit emery cloth, grit or vapor blast, followed by solvent degreasing	
		2. Solvent degrease and abrade with grit paper. Degrease again. Immerse for 10 min at 71–82°C in *Parts by wt.* Sodium metasilicate.. 3 Tetrasodium pyrophosphate.......... 1.5 Sodium hydroxide.... 1.5 Nacconol NR (Allied Chemical Co.)...... 0.5 Distilled water....... 138.0 Rinse in deionized water; dry in air, at 93°C. Immerse for 10 min at 85–91°C in *Parts by wt.* Oxalic acid.......... 1 Sulfuric acid (sp. gr. 1.84)............... 1 Distilled water....... 8 Rinse in deionized water; dry at 93°C for 10–15 min	Heat-resistant bond. Alkaline clean alone sufficient for general bonding. Commercial alkaline cleaners (Prebond 700, American Cyanamid) available

TABLE 7.13 **Surface Preparation for Metals *(Continued)***

Adherend	Degreasing solvent	Method of treatment	Remarks
		3. Etch for 15 min at 63°C in *Parts by wt.* Sodium dichromate (saturated solution) .. 0.30 Sulfuric acid .... 10.0 Remove carbon residue with nylon brush while rinsing. Rinse in distilled water and dry in warm air, at 93°C	ASTM D 2651
		4. Etch for 2 min at 93°C in *Parts by wt.* Hydrochloric acid (37%) .... 20 Orthophosphoric acid (85%) .... 3 Hydrofluoric acid (35%) .... 1 Rinse in warm water with a final rinse in distilled water. Dry in air below 93°C	For maximum resistance to heat and environment. ASTM D 2651
		5. Vapor degrease for 10 min and pickle for 10 min at 20°C in *Parts by vol.* Nitric acid (69%, sp. gr. 1.41) .... 10 Hydrofluoric acid (48%) .... 2 Water .... 88 Dry in air under 70°C	Room-temperature etch. Treatment may be followed by passivation for 20 min in 5–10% w/v chromic acid (CrO_3) solution
Steel (mild), iron, and ferrous metals other than stainless	Trichloroethylene	1. Abrasion. Grit or vapor blast followed by solvent degreasing with water-free solvents	Xylene or toluene is preferred to acetone and ketone, which may be moist enough to cause rusting
		2. Etch for 5–10 min at 20°C in *Parts by wt.* Hydrochloric acid (37%) .... 1 Water .... 1 Rinse in distilled water after cold-water wash, and dry in warm air for 10 min at 93°C	Bonding should follow immediately after etching treatment since ferrous metals are prone to rusting. Abrasion is more suitable for procedure where bonding is delayed
		3. Etch for 10 min at 60°C in *Parts by wt.* Orthophosphoric acid (85%) .... 1 Ethyl alcohol (denatured) .... 2 Brush off carbon residue with nylon brush while washing in running water. Rinse with distilled water, and heat for 1 h at 120°C	For maximum strength
Tin	Trichloroethylene	Solvent or vapor degrease after light abrasion with a fine emery cloth (320-grit)	

TABLE 7.13 Surface Preparation for Metals *(Continued)*

Adherend	Degreasing solvent	Method of treatment	Remarks
Titanium and titanium alloys	Trichloroethylene	1. Abrasion. Grit or vapor blast, or 100-grit emery cloth, followed by solvent degrease; or scour with a nonchlorinated cleaner, rinse, and dry	For general-purpose bonding
		2. Etch for 5–10 min at 20°C in *Parts by wt.* Sodium fluoride...... 2 Chromium trioxide... 1 Sulfuric acid (96%, sp. gr. 1.84)......... 10 Water............... 50 Rinse in water and distilled water. Dry in air at 93°C	
		3. Etch for 2 min at 20°C in *Parts by vol.* Hydrofluoric acid (60%).............. 63 Hydrochloric acid (37%).............. 841 Orthosphosphoric acid (85%).............. 89 Rinse in water and distilled water. Dry in air at 93°C	Suitable for alloys to be bonded with polybenzimidazole adhesives. Bond within 10 min of treatment. ASTM D 2651
		4. Etch for 10–15 min at 38–52°C in *Parts by vol.* Nitric acid (69%).... 6 Hydrofluoric acid (60%).............. 1 Water............... 20 Rinse with water and distilled water. Dry in oven at 71–82°C for 15 min	Alternative etch for alloys to be bonded with polyimide adhesives is nitric: hydrofluoric: water in ratio 5:1:27 by wt. Etch 30 s at 20°C
		5. Commercial etching liquids and pastes (Plasa-Jell 107C available from Semco, South Hoover, Los Angeles, Calif. 18881)	
Tungsten and tungsten alloys	Trichloroethylene	Etch for 1–5 min at 20°C in *Parts by wt.* Nitric acid (69%, sp. gr. 1.41)............ 6 Hydrofluoric acid (60%).............. 1 Sulfuric acid (96%, sp. gr. 1.84)......... 10 Water............... 3 Add a few drops of hydrogen peroxide (20%). Rinse with water and distilled water. Dry in air at 71–82°C for 15 min	
Uranium	Trichloroethylene	Abrasion of the metal in a pool of liquid adhesive (epoxy resins have been used)	Prevents oxidation of metal

TABLE 7.13 Surface Preparation for Metals *(Continued)*

Adherend	Degreasing solvent	Method of treatment	Remarks
Zinc and zinc alloys	Trichloroethylene	1. Abrasion. Grit or vapor blast, or 100-grit emery cloth, followed by solvent degreasing	For general-purpose bonding
		2. Etch for 2–4 min at 20°C in *Parts by vol.* Hydrochloric acid (37%) .... 10–20 Water .... 90–80 Rinse with warm water and distilled water. Dry in air at 66–71°C for 30 min	Glacial acetic acid is an alternative to hydrochloric acid
		3. Etch for 3–6 min at 38°C in *Parts by wt.* Sulfuric acid (96%, sp. gr. 1.84) .... 2 Sodium dichromate (crystalline) .... 1 Water 8	Suitable for freshly galvanized metal

Source: Based on Refs. 11, 21–24, 26

tallic surfaces, but the processes listed in Table 7.13 have been specifically found to provide reproducible structural bonds and fit easily into the bonding operation. The metals most commonly used in bonded structures and their respective surface treatments are described more fully in the following sections.

Aluminum and aluminum alloys. The effects of various aluminum surface treatments have been studied extensively. The most widely used process for high-strength, environment-resistant adhesive joints is the sodium dichromate-sulfuric acid etch, developed by Forest Product Laboratories and known as the FPL etch process. Abrasion or solvent degreasing treatments result in lower bond strengths, but these simpler processes are more easily placed into production. Table 7.14 qualitatively lists the bond strengths that can be realized with various aluminum treatments.

Copper and copper alloys. Surface preparation of copper alloys is necessary to remove weak oxide layers attached to the copper surface. This oxide layer is especially troublesome, because it forms very rapidly. Copper specimens must be bonded or primed as quickly as possible after surface preparation. Copper also has a tendency to form brittle surface compounds when used with certain adhesives that are corrosive to copper.

One of the better surface treatments for copper, utilizing a commercial product named Ebonol C (Enthane, Inc. New Haven, CT), does not remove the oxide layer but creates a deeper and stronger oxide formation. This process, called *black oxide,* is commonly used when bonding requires elevated temperatures; for example, laminating copper foil. Chromate conversion coatings are also used for high strength copper joints.

Magnesium and magnesium alloys. Magnesium is one of the lightest metals. The surface is very sensitive to corrosion, and chemical products are often formed at the adhesive–metal interface during bonding. Preferred surface preparations for magnesium develop a strong surface coating to prevent corrosion. Proprietary methods of producing such coatings have been developed by magnesium producers.

Steel and stainless steel. Steels are generally easy to bond provided that all rust, scale, and organic contaminants are removed. This may be accomplished easily by a combination of mechanical abrasion and solvent cleaning. Table 7.15 shows the effect of vari-

TABLE 7.14 Surface Treatment for Adhesive Bonding Aluminum (from Ref. 25)

Surface treatment	Type of bond
Solvent wipe (MEK, MIBK, trichloroethylene)	Low to medium strength
Abrasion of surface, plus solvent wipe (sandblasting, coarse sandpaper, etc.)	Medium to high strength
Hot-vapor degrease (trichloroethylene)	Medium strength
Abrasion of surface, plus vapor degrease	Medium to high strength
Alodine treatment	Low strength
Anodize	Medium strength
Caustic etch*	High strength
Chromic acid etch (sodium dichromate-surface acid)†	

*A good caustic etch is Oakite 164 (Oakite Products, Inc., 19 Rector Street, New York).
†Recommended pretreatment for aluminum to achieve maximum bond strength and weatherability:
1. Degrease in hot trichloroethylene vapor (160°F).
2. Dip in the following chromic acid solutionfor 10 min at 160°F:
   Sodium dischromate (Na_2Cr_2)H · $2H_2O$ 1 part/wt
   Cone, sulfuric acid (sp. gr. 1.86) 10 parts/wt
   Distilled water 30 parts/wt
3. Rinse thoroughly in cold, running, distilled or deionized water.
4. Air dry for 30 min, followed by 10 min at 150°F.

TABLE 7.15 Effect of Pretreatment on the Shear Strength of Steel Joints Bonded with a Polyvinyl Formal Phenolic Adhesive (from Ref. 26)

	Martensitic steel		Austenitic steel		Mild steel	
Pretreatment	*M*	*C*	*M*	*C*	*M*	*C*
Grit blast + vapor degreasing	5120	13.1	4100	4.7	4360	7.8
Vapor blast + vapor degreasing	6150	5.6	4940	7.1	4800	5.9
The following treatments were preceded by vapor degreasing:						
Cleaning in metasilicate solution	4360	5.7	3550	7.8	4540	6.1
Acid-dichromate etch	5780	5.8	2150	22.5	4070	4.0
Vapor blast + acid dichromate etch	6180	4.1				
Hydrochloric acid etch + phosphoric acid etch	3700	17.9	950	20.2	3090	20.7
Nitric/hydrofluoric acid etch	6570	7.5	3210	15.2	4050	8.4

M = mean failing load, lb/in^2. C = coefficient of variation, %.

ous surface treatments on the tensile shear strength of steel joints bonded with a vinyl-phenolic adhesive.

Prepared steel surfaces are easily oxidized. Once processed, hey should be kept free of moisture and primed or bonded within 4 hr. Stainless surfaces are not as sensitive to oxidation as carbon steels, and a slightly longer time between surface preparation and bonding is acceptable.

Titanium alloys. Because of the usual use of titanium at high temperatures, most surface preparations are directed at improving the thermal resistance of titanium joints. Like magnesium, titanium can also react with the adhesive during cure and create a weak boundary layer.

7.3.7.2 Plastic adherends. Many plastics and plastic composites can be treated by simple mechanical abrasion or alkaline cleaning to remove surface contaminants. In some cases, it is necessary that the polymeric surface be physically or chemically modified to achieve acceptable bonding. This applies particularly to crystalline thermoplastics such as the polyolefins, linear polyesters, and fluorocarbons. Methods used to improve the bonding characteristics of these surfaces include

1. Oxidation via chemical treatment or flame treatment
2. Electrical discharge to leave a more reactive surface
3. Ionized inert gas, which strengthens the surface by cross-linking and leaves it more reactive
4. Metal-ion treatment

Table 7.16 lists common recommended surface treatments for plastic adherends. These treatments are necessary when plastics are to be joined with adhesives. Solvent and heat welding are other methods of fastening plastics that do not require chemical alteration of the surface. Welding procedures were discussed in the previous chapter.

As with metallic substrates, the effects of plastic surface treatments decrease with time. It is necessary to prime or bond soon after the surfaces are treated. Listed below are some common plastic materials that require special physical or chemical treatments to achieve adequate surfaces for adhesive bonding.

Fluorocarbons. Fluorocarbons such as polytetrafluoroethylene (TFE), polyfluoroethylene propylene (FEP), polychlorotrifluoroethylene (CFE), and polymonochlorotrifluoroethylene (Kel-F) are notoriously difficult to bond because of their low surface tension. However, epoxy and polyurethane adhesives offer moderate strength if the fluorocarbon is treated prior to bonding.

The fluorocarbon surface may be made more "wettable" by exposing it for a brief moment to a hot flame to oxidize the surface. The most satisfactory surface treatment is achieved by immersing the plastic in a bath consisting of sodium-naphthalene dispersion in tetrahydrofuran. This process is believed to remove fluorine atoms, leaving a carbonized surface which can be wet easily. Fluorocarbon films pretreated for adhesive bonding are available from most suppliers. A formulation and description of the sodium-naphthalene process may be found in Table 7.16. Commercial chemical products for etching fluorocarbons are also listed.

Polyethylene terephthalate (Mylar®). A medium-strength bond can be obtained with polyethylene terephthalate plastics and films by abrasion and solvent cleaning. However, a stronger bond can be achieved by immersing the surface in a warm solution of sodium hydroxide or in an alkaline cleaning solution for 2 to 10 min.

TABLE 7.16 Surface Preparations for Plastics

Adherend	Degreasing solvent	Method of treatment	Remarks
Acetal (copolymer)	Acetone	1. Abrasion. Grit or vapor blast, or medium-grit emery cloth followed by solvent degreasing	For general-purpose bonding
		2. Etch in the following acid solution: Parts by wt.: Potassium dichromate 75; Distilled water 120; Concentrated sulfuric acid (96%, sp. gr. 1.84) 1,500; for 10 s at 25°C. Rinse in distilled water, and dry in air at RT	For maximum bond strength. ASTM D 2093
Acetal (homopolymer)	Acetone	1. Abrasion. Sand with 280A-grit emery cloth followed by solvent degreasing	For general-purpose bonding
		2. "Satinizing" technique. Immerse the part in Parts by wt.: Perchloroethylene 96.85; 1,4-Dioxane 3.00; *p*-Toluenesulfonic acid 0.05; Cab-o-Sil (Cabot Corp.) 0.10; for 5–30 s at 80–120°C. Transfer the part immediately to an oven at 120°C for 1 min. Wash in hot water. Dry in air at 120°C	For maximum bond strength. Recommended by du Pont
Acrylonitrile butadiene styrene	Acetone	1. Abrasion. Grit or vapor blast, or 220-grit emery cloth, followed by solvent degreasing	
		2. Etch in chromic acid solution for 20 min at 60°C	Recipe 2 for methyl pentane
Cellulosics: Cellulose, cellulose acetate, cellulose acetate butyrate, cellulose nitrate, cellulose propionate, ethyl cellulose	Methanol, isopropanol	1. Abrasion. Grit or vapor blast, or 220-grit emery cloth, followed by solvent degreasing	For general bonding purposes
		2. After procedure 1, dry the plastic at 100°C for 1 h, and apply adhesive before the plastic cools to room temperature	
Diallyl phthalate, diallyl isophthalate	Acetone, methyl ethyl ketone	Abrasion. Grit or vapor blast, or 100-grit emery cloth, followed by solvent degreasing	Steel wool may be used for abrasion

TABLE 7.16 Surface Preparations for Plastics *(Continued)*

Adherend	Degreasing solvent	Method of treatment	Remarks
Epoxy resins	Acetone, methyl ethyl ketone	Abrasion. Grit or vapor blast, or 100-grit emery cloth, followed by solvent degreasing	Sand or steel shot are suitable abrasives
Ethylene vinyl acetate	Methanol	Prime with epoxy adhesive and fuse into the surface by heating for 30 min at 100°C	
Furane	Acetone, methyl ethyl ketone	Abrasion. Grit or vapor blast, or 100-grit emery cloth, followed by solvent degreasing	
Ionomer	Acetone, methyl ethyl ketone	Abrasion. Grit or vapor blast, or 100-grit emery cloth, followed by solvent degreasing	Alumina (180-grit) is a suitable abrasive
Melamine resins	Acetone, methyl ethyl ketone	Abrasion. Grit or vapor blast, or 100-grit emery cloth, followed by solvent degreasing	
Methyl pentene	Acetone	1. Abrasion. Grit or vapor blast, or 100-grit emery cloth, followed by solvent degreasing	For general-purpose bonding
		2. Immerse for 1 h at 60°C in Parts by wt. Sulfuric acid (96%, sp. gr. 1.84) .... 26 Potassium chromate .. 3 Water .... 11 Rinse in water and distilled water. Dry in warm air	
		3. Immerse for 5–10 min at 90°C in potassium permanganate (saturated solution), acidified with sulfuric acid (96%, sp. gr. 1.84). Rinse in water and distilled water. Dry in warm air	
		4. Prime surface with lacquer *based on urea-formaldehyde* resin diluted with carbon tetrachloride	Coatings (dried) offer excellent *bonding surfaces without further* pretreatment
Phenolic resins, phenolic melamine resins	Acetone, methyl ethyl ketone, detergent	1. Abrasion. Grit or vapor blast, or abrade with 100-grit emery cloth, followed by solvent degreasing 2. Removal of surface layer of one ply of fabric previously placed on surface before curing. Expose fresh bonding surface by tearing off the ply prior to bonding	Steel wool may be used for abrasion. Sand or steel shot are suitable abrasives. Glass-fabric decorative laminates may be degreased with detergent solution

TABLE 7.16 Surface Preparations for Plastics *(Continued)*

Adherend	Degreasing solvent	Method of treatment	Remarks
Polyamide (nylon)	Acetone, methyl ethyl ketone, detergent	1. Abrasion. Grit or vapor blast, or abrade with 100-grit emery cloth, followed by solvent degreasing	Sand or steel shot are suitable abrasives
		2. Prime with a spreading dough based on the type of rubber to be bonded in admixture with isocyanate	Suitable for bonding polyamide textiles to natural and synthetic rubbers
		3. Prime with resorcinol-formaldehyde adhesive	Good adhesion to primer coat with epoxy adhesives in metal-plastic joints
Polycarbonate, allyl diglycol carbonate	Methanol, isopropanol, detergent	Abrasion. Grit or vapor blast, or 100-grit emery cloth, followed by solvent degreasing	Sand or steel shot are suitable abrasives
Fluorocarbons: Polychlorotrifluoroethylene, polytetrafluoroethylene, polyvinyl fluoride, polymonochlorotrifluoroethylene	Trichloroethylene	1. Wipe with solvent and treat with the following for 15 min at RT: Naphthalene (128 g) dissolved in tetrahydrofuran (1 liter) to which is added sodium (23 g) during a stirring period of 2 h. Rinse in deionized water, and dry in warm air	Sodium-treated surfaces must not be abraded before use. Hazardous etching solutions requiring skillful handling. Proprietary etching solutions are commercially available (see 2). PTFE available in etched tape. ASTM D 2093
		2. Wipe with solvent and treat as recommended in one of the following commercial etchants: Bond aid.....W. S. Shamban and Co. 11617 W. Jefferson Blvd. Culver City, Calif. Fluorobond...Joclin Mfg. Co. 15 Lufbery Ave. Wallingford, Conn. Fluoroetch....Action Associates 1180 Raymond Blvd. Newark, N.J. Tetraetch....W. L. Gore Associates 487 Paper Mill Rd. Newark, Del.	
		3. Prime with epoxy adhesive, and fuse into the surface by heating for 10 min at 370°C followed by 5 min at 400°C	
		4. Expose to one of the following gases activated by corona discharge: Air (dry) for 5 min; Air (wet) for 5 min; Nitrous oxide for 10 min; Nitrogen for 5 min	Bond within 15 min of pretreatment
		5. Expose to electric discharge from a tesla coil (50,000 V ac) for 4 min	Bond within 15 min of pretreatment

TABLE 7.16 Surface Preparations for Plastics *(Continued)*

Adherend	Degreasing solvent	Method of treatment	Remarks
Polyesters, polyethylene terephthalate (Mylar)	Detergent, acetone, methyl ethyl ketone	1. Abrasion. Grit or vapor blast, or 100-grit emery cloth, followed by solvent degreasing	For general-purpose bonding
		2. Immerse for 10 min at 70–95°C in *Parts by wt.* Sodium hydroxide.... 2 Water............... 8 Rinse in hot water and dry in hot air	For maximum bond strength. Suitable for linear polyester films (Mylar)
Chlorinated polyether	Acetone, methyl ethyl ketone	Etch for 5–10 min at 66–71°C in *Parts by wt.* Sodium dichromate...... 5 Water.................. 8 Sulfuric acid (96%, sp. gr. 1.84)............... 100 Rinse in water and distilled water. Dry in air	Suitable for film materials such as Penton. ASTM D 2093
Polyethylene, polyethylene (chlorinated), polyethylene terephthalate (see polyesters), polypropylene, polyformaldehyde	Acetone, methyl ethyl ketone	1. Solvent degreasing	Low-bond-strength applications
		2. Expose surface to gas-burner flame (or oxyacetylene oxidizing flame) until the substrate is glossy	
		3. Etch in the following: *Parts by wt.* Sodium dichromate... 5 Water............... 8 Sulfuric acid (96%, sp. gr. 1.84)......... 100 Polyethylene and polypropylene 60 min at 25°C or 1 min at 71°C Polyformaldehyde......10 s at 25°C	For maximum bond strength. ASTM D 2093
		4. Expose to following gases activated by corona discharge: Air (dry)............For 15 min Air (wet)............For 5 min Nitrous Oxide.......For 10 min Nitrogen............For 15 min	Bond within 15 min of pretreatment. Suitable for polyolefins.
		5. Expose to electric discharge from a tesla coil (50,000 V ac) for 1 min	Bond within 15 min of pretreatment. Suitable for polyolefins.
Polymethyl methacrylate, methacrylate butadiene styrene	Acetone, methyl ethyl ketone, detergent, methanol, trichloroethylene, isopropanol	Abrasion. Grit or vapor blast, or 100-grit emery cloth, followed by solvent degreasing	For maximum strength relieve stresses by heating plastic for 5 h at 100°C

TABLE 7.16 Surface Preparations for P astics *(Continued)*

Adherend	Degreasing solvent	Method of treatment	Remarks
Poly-phenylene	Trichloro-ethylene	Abrasion. Grit or vapor blast, or 100-grit emery cloth, followed by solvent degreasing	
Poly-phenylene oxide	Methanol	Solvent degrease	Plastic is soluble in xylene and may be primed with adhesive in xylene solvent
Polysty-rene	Methanol, isopro-panol, deter-gent	Abrasion, Grit or vapor blast, or 100-grit emery cloth, followed by solvent degreasing	Suitable for rigid plastic
Poly-sulfone	Methanol	Vapor degrease	
Polyure-thane	Acetone, methyl ethyl ketone	Abrade with 100-grit emery cloth and solvent degrease	
Polyvinyl chloride, polyvinyl-idene chloride polyvinyl fluoride	Trichloro-ethylene, methyl ethyl ketone	1. Abrasion. Grit or vapor blast, or 100-grit emery cloth followed by solvent degreasing 2. Solvent wipe with ketone	Suitable for rigid plastic. For maximum strength, prime with nitrile-phenolic adhesive Suitable for plasticized material
Styrene acrylo-nitrile	Trichloro-ethylene	Solvent degrease	
Urea for-malde-hyde	Acetone, methyl ethyl ketone	Abrasion. Grit or vapor blast, or 100-grit emery cloth, followed by solvent degreasing	

Source: **Based on Refs. 11, 21–24, 26.**

Polyolefins, polyformaldehyde, polyether. These materials can be effectively bonded only if the surface is first located. Polyethylene and polypropylene can be prepared for bonding by holding the flame of an oxyacetylene torch over the plastic until it becomes glossy, or else by heating the surface momentarily with a blast of hot air. It is important not to overheat the plastic, thereby causing deformation. The treated plastic must be bonded as quickly as possible after surface preparation.

Polyolefins, such as polyethylene, polypropylene, and polymethylpentene, as well as polyformaldehyde and polyether, may be more effectively treated with a sodium dichromate- sulfuric acid solution. This treatment oxidizes the surface, allowing better wetting. Activated gas plasma treatment, described in the general section on surface treatments is also an effective treatment for these plastics. Table 7.17 shows the tensile-shear strength of bonded polyethylene pretreated by these various methods.

7.3.7.3 Elastomeric adherends. Vulcanized-rubber joints are often contaminated with mold release and plasticizers or extenders that can migrate to the surface. As shown in Table 7.18, solvent washing and abrading are common treatments for most elastomers,

TABLE 7.17 Effects of Surface Treatments on Bonding to Polyethylene with Various Types of Adhesives (from Ref. 27)

Specimen No.	Control	Flame treated	Sanded	Acid treated, oven-dried at 90°C	Acid treated, oven-dried at 71°C	Acid treated, wiped, air-dried at 22°C	Acid treated, acetone-dried	Plasma treatment: Helium (30 s)	Plasma treatment: Helium (30 min)	Plasma treatment: Oxygen (30 s)	Plasma treatment: Oxygen (30 min)
Epoxy											
1	40	464*	186	454	428*	500*	516*	468*	...	423*	490*
2	48	480*	166	480*	440*	524*	490*	470*	...	463*	424*
3	24	452*	182	472*	532*	500*	502*	470*	...	484*	439*
4	58	486*	220	462*	460*	524*	500*	450*	...	495*	424*
5	58	520*	216	506*	424*	448*	476*				
Avg lb/in.²	46	480	195	475	457	499	497	464	...	466	445
Polyester											
1	74	502*	214	290	300	294	452*	196	284	264	480*
2	102	472*	146	290	288	416*	462*	230	396*	246	514*
3	70	430*	170	230	322	412	392	214	380	320	300
4	70	364*	188	268	318	426*	464*	160	400*	240	370
5	108	400*	178	308	256	236	200	148	372	244	484*
Avg lb/in.²	85	434	175	277	297	357	394	190	346	263	430
Nitrile-Rubber-Phenolic											
1	42	196	54	102	100	124	106	...	166	...	210
2	38	120	54	100	92	128	136	...	110	...	170
3	46	88	52	88	64	120	102	...	276	...	220
4	46	120	64	96	158	88	54	...	112	...	110
5	48	166	56	96	124	88	164	...	224	...	170
Avg lb/in.²	44	138	56	96	108	110	112	...	178	...	176

* Adherend failed rather than bond. All values in this table are based upon lap-shear strength calculated as lb/in.².

TABLE 7.18 Surface Preparation for Elastomers

Adherend	Degreasing solvent	Method of treatment	Remarks
Natural rubber	Methanol, isopropanol	1. Abrasion followed by brushing. Grit or vapor blast, or 280-grit emery cloth, followed by solvent wipe	For general-purpose bonding
		2. Treatment surface for 2–10 min with sulfuric acid (sp. gr. 1.84) at RT. Rinse thoroughly with cold water/hot water. Dry after rinsing in distilled water. (Residual acid may be neutralized by soaking for 10 min in 10% ammonium hydroxide after hot-water washing)	Adequate pretreatment is indicated by the appearance of hairline surface cracks on flexing the rubber. Suitable for many synthetic rubbers when given 10–15 min etch at room temperature. Unsuitable for use on butyl, polysulfide, silicone, chlorinated polyethylene, and polyurethane rubbers
		3. Treat surface for 2–10 min with paste made from sulfuric acid and barium sulfate. Apply paste with stainless-steel spatula, and follow procedure 2, above	
		4. Treat surface for 2–10 min in *Parts by vol.*: Sodium hypochlorite .... 6; Hydrochloric acid (37%) .... 1; Water .... 200. Rinse with cold water and dry	Suitable for those rubbers amenable to treatments 2 and 3
Butadiene styrene	Toluene	1. Abrasion followed by brushing. Grit or vapor blast, or 280-grit emery cloth, followed by solvent wipe 2. Prime with butadiene styrene adhesive in an aliphatic solvent. 3. Etch surface for 1–5 min at RT, following method 2 for natural rubber	Excess toluene results in swollen rubber. A 20-min drying time will restore the part to its original dimensions
Butadiene nitrile	Methanol	1. Abrasion followed brushing. Grit or vapor blast, or 280-grit emery cloth, followed by solvent wipe 2. Etch surface for 10–45 s at RT, following method 2 for natural rubber	
Butyl	Toluene	1. Solvent wipe	For general-purpose bonding
		2. Prime with butyl-rubber adhesive in an aliphatic solvent	For maximum strength
Chlorosulfonated polyethylene	Acetone or methyl ethyl ketone	Abrasion followed by brushing. Grit or vapor blast, or 280-grit emery cloth, followed by solvent wipe	General-purpose bonding

TABLE 7.18 Surface Preparation for Elastomers *(Continued)*

Adherend	Degreasing solvent	Method of treatment	Remarks
Ethylene propylene	Acetone or methyl ethyl ketone	Abrasion followed by brushing. Grit or vapor blast, or 280-grit emery cloth, followed by solvent wipe	General-purpose bonding
Fluorosilicone	Methanol	Application of fluorosilicone primer (A 4040) to metal where intention is to bond unvulcanized rubber	Primer available from Dow Corning
Polyacrylic	Methanol	Abrasion followed by brushing. Grit or vapor blast, or 100-grit emery cloth followed by solvent wipe	General-purpose bonding
Polybutadiene	Methanol	Solvent wipe	General-purpose bonding
Poly-chloroprene	Toluene, methanol, isopropanol	1. Abrasion followed by brushing. Grit or vapor blast, or 100-grit emery cloth, followed by solvent wipe 2. Etch surface for 5–30 min at RT, following method 2 for natural rubber	Adhesion improved by abrasion with 280-grit emery cloth followed by acetone wipe
Polysulfide	Methanol	Immerse overnight in strong chlorine water, wash and dry	
Polyurethane	Methanol	1. Abrasion followed by brushing. Grit or vapor blast, or 280-grit emery cloth followed by solvent wipe	
		2. Incorporation of a chlorosilane into the adhesive-elastomer system. 1% w/w is usually sufficient	Chlorosilane is available commercially. Addition to adhesive eliminates need for priming and improves adhesion to glass, metals. Silane may be used as a surface primer
Silicone	Acetone or methanol	1. Application of primer, Chemlok 607, in solvent (dries 10–15 min) 2. Expose to oxygen gas activated by corona discharge for 10 min	Primer available from Hughson Chemical Company

Source: Based on Refs. 11, 21–24, and 26

but chemical treatment may be required for maximum properties. Synthetic and natural rubbers may require "cyclizing" with concentrated sulfuric acid until hairline fractures are evident on the surface.

7.3.7.4 Other adherends. Table 7.19 provides surface treatments for a variety of materials not covered in the preceding tables. Bonding to painted or plated parts requires special consideration. The resulting adhesive bond is only as strong as the adhesion of the paint or plating to the base material.

7.4 Types of Adhesives

7.4.1 Adhesive Composition

Modern-day adhesives are often fairly complex formulations of components that perform specialty functions. The adhesive base or binder is the primary component of an adhesive. The binder is generally the resinous component from which the name of the adhesive is derived. For example, an epoxy adhesive may have many components, but the primary material is epoxy resin.

A hardener is a substance added to an adhesive formulation to initiate the curing reaction and take part in it. Two-component adhesive systems have one component, which is the base, and a second component, which is the hardener. Upon mixing, a chemical reaction ensues that causes the adhesive to solidify. A catalyst is sometimes incorporated into an adhesive formulation to speed the reaction between base and hardener.

Solvents are sometimes needed to lower viscosity or to disperse the adhesive to a spreadable consistency. Often, a mixture of solvents is required to achieve the desired properties.

A reactive ingredient added to an adhesive to reduce the concentration of binder is called a *diluent*. Diluents are principally used to lower viscosity and modify processing conditions of some adhesives. Diluents react with the binder during cure, become part of the product, and do not evaporate as does a solvent.

Fillers are generally inorganic particulates added to the adhesive to improve working properties, strength, permanence, or other qualities. Fillers are also used to reduce material cost. By selective use of fillers, the properties of an adhesive can be changed tremendously. Thermal expansion, electrical and thermal conduction, shrinkage, viscosity, and thermal resistance are only a few properties that can be modified by use of selective fillers.

A carrier or reinforcement is usually a thin fabric used to support a semi-cured (B-staged) adhesive to provide a product that can be used as a tape or film. The carrier can also serve as a spacer between the adherends and reinforcement for the adhesive.

Adhesives can be broadly classified as being thermoplastic, thermosetting, elastomeric, or alloy blend. These four adhesive classifications can be further subdivided by specific chemical composition as described in Tables 7.20 through 7.23. The types of resins that go into the thermosetting and alloy adhesive classes are noted for high strength, creep resistance, and resistance to environments such as heat, moisture, solvents, and oils. Their physical properties are well suited for structural-adhesive applications.

Elastomeric and thermoplastic adhesive classes are not used in applications requiring continuous load because of their tendency to creep under stress. They are also degraded by many common service environments. These adhesives find greatest use in low-strength applications such as pressure-sensitive tape, sealants, and hot-melt products.

7.4.2 Structural Adhesives

7.4.2.1 Epoxy. Epoxy adhesives offer a high degree of adhesion to all substrates except some untreated plastics and elastomers. Cured epoxies have thermosetting molecular structures. They exhibit excellent tensile-shear strength but poor peel strength unless modified with a more resilient polymer. Epoxy adhesives offer excellent resistance to oil, moisture, and many solvents. Low cure shrinkage and high resistance to creep under prolonged stress are characteristics of epoxy resins.

Epoxy adhesives are commercially available as liquids, pastes, and semi-cured (B-staged) film and solids. Epoxy adhesives are generally supplied as a 100 percent solids

(non-solvent) formulation, but some sprayable epoxy adhesives are available in solvent systems. Epoxy resins have no evolution of volatiles during cure and are useful in gap-filling applications.

Depending on the type of curing agent, epoxy adhesives can cure at room or elevated temperatures. Higher strengths and better heat resistance are usually obtained with the heat-curing types. Room-temperature-curing epoxies can harden in as little as 1 min at room temperature, but most systems require from 18 to 72 hr. The curing time is greatly temperature-dependent, as shown in Fig. 7.27.

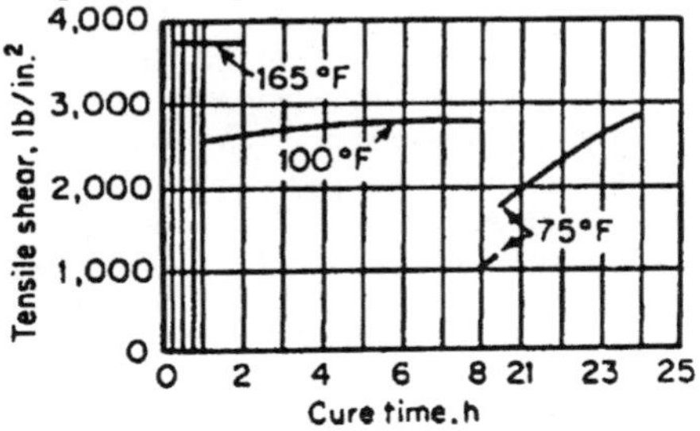

Figure 7.27 Characteristics of particular epoxy adhesive under different curing time and temperature relationships.[28]

Epoxy resins are the most versatile of structural adhesives, because they can be cured and co-reacted with many different resins to provide widely varying properties. Table 7.24 describes the influence of curing agents on the bond strength of epoxy to various adherends. The type of epoxy resin used in most adhesives is derived from the reaction of bisphenol A and epichlorohydrin. This resin can be cured with amines or polyamides for room-temperature-setting systems; anhydrides for elevated-temperature cure; or latent curing agents such as boron trifluoride complexes for use in one-component, heat-curing adhesives. Polyamide curing agents are used in most "general-purpose" epoxy adhesives. They provide a room-temperature cure and bond well to many substrates including plastics, glass, and elastomers. The polyamide-cured epoxy also offers a relatively flexible adhesive with fair peel and thermal-cycling properties.

7.4.2.2 Epoxy alloy or hybrids. A variety of polymers can be blended and co-reacted with epoxy resins to provide certain desired properties. The most common of these are phenolic, nylon, and polysulfide resins.

Epoxy-phenolic. Adhesives based on epoxy-phenolic blends are good for continuous high-temperature service in the 350°F range or intermittent service as high as 500°F. They retain their properties over a very wide temperature range, as shown in Fig. 7.28. Shear strengths of up to 3,000 lb/in^2 at room temperature and 1,000 to 2,000 lb/in^2 at 500°F are available. Resistance to oil, solvents, and moisture is very good. Because of their rigid nature, epoxy-phenolic adhesives have low peel strength and limited thermal-shock resistance.

These adhesives are available as pastes, solvent solutions, and B-staged film supported on glass fabric. Cure generally requires 350°F for 1 hr under moderate pressure. Epoxy-phenolic adhesives were developed primarily for bonding metal joints in high-temperature applications.

Epoxy-nylon. Epoxy-nylon adhesives offer both excellent shear and peel strength. They maintain their physical properties at cryogenic temperatures but are limited to a maximum service temperature of 180°F.

Epoxy-nylon adhesives are available as unsupported B-staged film or in solvent solutions. A moderate pressure of 25 lb/in^2 and temperature of 350°F are generally required for 1 hr to cure the adhesive. Because of their excellent filleting properties and high peel strength, epoxy-nylon adhesives are used to bond aluminum skins to honeycomb core in aircraft structures.

Epoxy-polysulfide. Polysulfide resins combine with epoxy resins to provide adhesives with excellent flexibility and chemical resistance. These adhesives bond well to many different substrates. Shear strength and elevated temperature properties are poor, but resistance to peel forces and low temperatures is very good. The epoxy-polysulfide alloy is supplied as a two-part, flowable paste that cures to a rubbery solid at room temperature. A common application for epoxy-polysulfide adhesives is as a sealant.

7.4.2.3 Resorcinol and phenol resorcinol. Resorcinol adhesives are primarily used for bonding wood, plastic skins to wood cores, and primed metal to wood. Adhesive bonds as strong as wood itself are obtainable. Resorcinol adhesives are resistant to boiling water, oil, many solvents, and mold growth. Their service temperature ranges from –300 to 350°F. Because of high cost, resorcinol resins are often modified by the addition of phenolic resins to form phenol resorcinol.

Resorcinol and phenol resorcinol adhesives are available in liquid form and are mixed with a powder hardener before application. These adhesives are cold setting, but they can also be pressed at elevated temperatures for faster production.

7.4.2.4 Melamine formaldehyde and urea formaldehydes. Melamine formaldehyde resins are colorless adhesives for wood. Because of high cost, they are sometimes blended with urea formaldehyde. Melamine formaldehyde is usually supplied in powder form and reconstituted with water; a hardener is added at the time of use. Temperatures of about 200°F are necessary for cure. Adhesive strength is greater than the strength of the wood substrate.

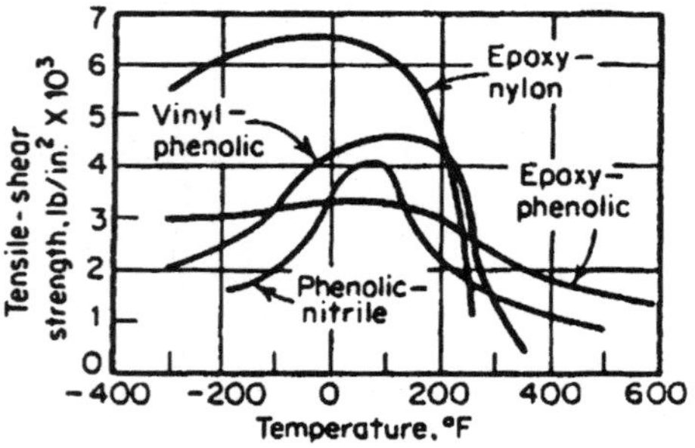

Figure 7.28 Effect of temperature on tensile-shear strength of adhesive alloys (substrate material is aluminum).[28]

TABLE 7.20 Thermosetting Adhesives

Adhesive	Description	Curing method	Special characteristics	Usual adherends	Price range
Cyanoacrylate	One-part liquid	Rapidly at RT in absence of air	Fast setting; good bond strength; low viscosity; high cost; poor heat and shock resistance; will not bond to acidic surfaces	Metals, plastics, glass	Very high
Polyester	Two-part liquid or paste	RT or higher	Resistant to chemicals, moisture, heat, weathering; good electrical properties; wide range of strengths; some resins do not fully cure in presence of air; isocyanate-cured system bonds well to many plastic films	Metals, foils, plastics, plastic laminates, glass	Low–med
Urea formaldehyde	Usually supplied as two-part resin and hardening agent; extenders and fillers used	Under pressure	Not as durable as others but suitable for fair range of service conditions; generally low cost and ease of application and cure; pot life limited to 1 to 24 hr	Plywood	Low
Melamine formaldehyde	Powder to be mixed with hardening agent	Heat and pressure	Equivalent in durability and water resistance (including boiling water) to phenolics and resorcinols; often combined with ureas to lower cost; higher service temp than ureas	Plywood, other wood products	Med
Resorcinol and phenol-resorcinol formaldehyde	Usually alcohol-water solutions to which formaldehyde must be added	RT or higher with moderate pressure	Suitable for exterior use; unaffected by boiling water, mold, fungus, grease, oil, most solvents; bond strength equals or betters strength of wood; do not bond directly to metal	Wood, plastics, paper, fiberboard, plywood	Med

Epoxy	Two-part liquid or paste; one-part liquid, paste or solid; solutions	RT or higher	Most versatile adhesive available; excellent tensile-shear strength; poor peel strength; excellent resistance to moisture and solvents; low cure shrinkage; variety of curing agents/hardeners results in many variations	Metals, plastic, glass, rubber, wood, ceramics	Med
Polyimide	Supported film, solvent solution	High temp	Excellent thermal and oxidation resistance; suitable for continuous use at 550°F and short-term use to 900·F; expensive	Metals, metal foil, honeycomb core	Very high
Polybenzimidazole	Supported film	Long, high-temp cure	Good strength at high temperatures; suitable for continuous use at 450°F and short-term use at 1000°F; volatiles released during cure; deteriorates at high temperatures on exposure to air; expensive	Metals, metal foil, honeycomb core	Very high
Acrylic	Two-part liquid or paste	RT	Excellent bond to many plastics, good weather resistance, fast cure, catalyst can be used as a substrate primer; poor peel and impact strength	Metals, many plastics, wood	Very high
Acrylate acid diester	One-part liquid or paste	RT or higher in absence of air	Chemically blocked, anaerobic type; excellent wetting ability; useful temperature range of –65 to 300°F; withstands rapid thermal cycling; high-tensile-strength grade requires cure at 250°F, cures in minutes at 280°F	Metals, plastics, glass, wood	Very high

TABLE 7.21 Thermoplastic Adhesives

Adhesive	Description	Curing method	Special characteristics	Usual adherends	Price range
Cellulose acetate, cellulose acetate butyrate	Solvent solutions	Solvent evaporation	Water-clear, more heat resistant but less water resistant than cellulose nitrate; cellulose acetate butyrate has better heat and water resistance than cellulose acetate and is compatible with a wider range of plasticizers	Plastics, leather, paper, wood, glass, fabrics	Low
Cellulose nitrate	Solvent solutions	Evaporation of solvent	Tough, develops strength rapidly, water resistant; bonds to many surfaces; discolors in sunlight; dried adhesive is flammable	Glass, metal, cloth, plastics	Low
Polyvinyl acetate	Solvent solutions and water emulsions, plasticized or unplasticized, often containing fillers and pigments; also dried film that is light stable, water-white, transparent	On evaporation of solvent or water; film by heat and pressure	Bond strength of several thousand $lb/in.^2$ but not under continuous loading; the most versatile in terms of formulations and uses; tasteless, odorless; good resistance to oil, grease, acid; fair water resistance	Emulsions particularly useful with porous materials like wood and paper; solutions used with plastic films, mica, glass, metal, ceramics	Low
Vinyl vinylidene	Solutions in solvents like methyl ethyl ketone	Evaporation of solvent	Tough, strong, transparent, and colorless; resistant to hydrocarbon solvents, greases, oils	Particularly useful with textiles; also porous materials, plastics	Med

Polyvinyl acetals	Solvent solutions, film, and solids	Evaporation of solvent; film and solid by heat and pressure	Flexible bond; modified with phenolics for structural use; good resistance to chemicals and oils; includes polyvinyl formal and polyvinyl butyral types	Metals, mica, glass, rubber, wood, paper	Med
Polyvinyl alcohol	Water solutions, often extended with starch or clay	Evaporation of water	Odorless, tasteless, and fungus-resistant (if desired); excellent resistance to grease and oils; water soluble	Porous materials such as fiberboard, paper, cloth	Low
Polyamide	Solid hot-melt, film, solvent solutions	Heat and pressure	Good film flexibility; resistant to oil and water; used for heat-sealing compounds	Metals, paper, plastic films	Med
Acrylic	Solvent solutions, emulsions, and mixtures requiring added catalysts	Evaporation of solvent; RT or elevated temp (two-part)	Good low-temperature bonds; poor heat resistance; excellent resistance to ultraviolet; clear; colorless	Glass, metals, paper, textiles, metallic foils, plastics	Med
Phenoxy	Solvent solutions, film, solid hot-melt	Heat and pressure	Retains high strength from 40 to 180°F; resists creep up to 180°F; suitable for structural use	Metals, wood, paper, plastic film	Med

TABLE 7.22 Elastomeric Adhesives

Adhesive	Description	Curing method	Special characteristics	Usual adherends	Price range
Natural rubber	Solvent solutions, latexes, and vulcanizing type	Solvent evaporation, vulcanizing type by heat or RT (two-part)	Excellent tack, good strength; shear strength 30–180 lb/in.^{2}; peel strength 0.56 lb/in. width; surface can be tack-free to touch and yet bond to similarly coated surface	Natural rubber, masonite, wood, felt, fabric, paper, metal	Med
Reclaimed rubber	Solvent solutions, some water dispersions; most are black, some gray and red	Evaporation of solvent	Low cost, widely used; peel strength higher than natural rubber; failure occurs under relatively low constant loads	Rubber, sponge rubber, fabric, leather, wood, metal, painted metal, building materials	Low
Butyl	Solvent system, latex	Solvent evaporation, chemical cross-linking with curing agents and heat	Low permeability to gases, good resistance to water and chemicals, poor resistance to oils, low strength	Rubber, metals	Med
Polyisobutylene	Solvent solution	Evaporation of solvent	Sticky, low-strength bonds; copolymers can be cured to improve adhesion, environmental resistance, and elasticity; good aging; poor thermal resistance; attacked by solvents	Plastic film, rubber, metal foil, paper	Low
Nitrile	Latexes and solvent solutions compounded with resins, metallic oxides, fillers, etc.	Evaporation of solvent and/or heat and pressure	Most versatile rubber adhesive; superior resistance to oil and hydrocarbon solvents; inferior in tack range, but most dry tack-free, an advantage in precoated assemblies; shear strength of 150–2000 lb/in.^{2}, higher than neoprene, if cured	Rubber (particularly nitrile), metal, vinyl plastics	Med

Styrene butadiene	Solvent solutions and latexes; because tack is low, rubber is compounded with tackifiers and plasticizing oils	Evaporation of solvent	Usually better aging properties than natural or reclaimed; low dead load strength; bond strength similar to reclaimed; useful temp range from –40 to 160°F	Fabrics, foils, plastic film laminates, rubber and sponge rubber, wood	Low
Polyurethane	Two-part liquid or paste	RT or higher	Excellent tensile-shear strength from –400 to 200°F; poor resistance to moisture before and after cure; good adhesion to plastics	Plastics, metals, rubber	Med
Polysulfide	Two-part liquid or paste	RT or higher	Resistant to wide range of solvents, oils, and greases; good gas impermeability; resistant to weather, sunlight, ozone; retains flexibility over wide temperature range; not suitable for permanent load-bearing applications	Metals, wood, plastics	High
Silicone	Solvent solution; heat or RT curing and pressure sensitive; and RT vulcanizing pastes	Solvent evaporation, RT or elevated temp	Of primary interest is pressure-sensitive type used for tape; high strengths for other forms are reported from –100 to 500°F; limited service to 700°F; excellent dielectric properties	Metals; glass; paper; plastics and rubber, including silicone and butyl rubber and fluorocarbons	High–very high
Neoprene	Latexes and solvent solutions, often compounded with resins, metallic oxides, fillers, etc.	Evaporation of solvent	Superior to other rubber adhesives in most respects—quickness; strength; max temp (to 200°F, sometimes 350°F); aging; resistant to light, weathering, mild acids, oils	Metals, leather, fabric, plastics, rubber (particularly neoprene), wood, building materials	Med

TABLE 7.23 Alloy Adhesives

Adhesive	Description	Curing method	Special characteristics	Usual adherends	Price range
Epoxy-phenolic	Two-part paste, supported film	Heat and pressure	Good properties at moderate cures; volatiles released during cure; retains 50% of bond strength at 500°F; limited shelf life; low peel strength and shock resistance	Metals, honeycomb core, plastic laminates, ceramics	Med
Epoxy-polysulfide	Two-part liquid or paste	RT or higher	Useful temperature range –70 to 200°F, greater resistance to impact, higher elongation, and less brittleness than epoxies	Metals, plastic, wood, concrete	Med
Epoxy-nylon	Solvent solutions, supported and unsupported film	Heat and pressure	Excellent tensile-shear strength at cryogenic temperature; useful temperature range –423 to 180°F; limited shelf life	Metals, honeycomb core, plastics	Med
Nitrile-phenolic	Solvent solutions, unsupported and supported film	Heat and pressure	Excellent shear strength; good peel strength; superior to vinyl and neoprene—phenolics; good adhesion	Metals, plastics, glass, rubber	Med
Neoprene-phenolic	Solvent solutions, unsupported and supported film	Heat and pressure	Good bonds to a variety of substrates; useful temp range –70 to 200°F, excellent fatigue and impact strength	Metals, glass, plastics	Med
Vinyl-phenolic	Solvent solutions and emulsions, tape, liquid, and coreacting powder	Heat and pressure	Good shear and peel strength; good heat resistance; good resistance to weathering, humidity, oil, water, and solvents; vinyl formal and vinyl butyral forms available, vinyl formal–phenolic is strongest	Metals, paper, honeycomb core	Low–med

TABLE 7.19 Surface Preparations for Materials Other than Metals, Plastics, and Elastomers

Adherend	Degreasing solvent	Method of treatment	Remarks
Asbestos (rigid)	Acetone	1. Abrasion. Abrade with 100-grit emery cloth, remove dust, and solvent degrease 2. Prime with diluted adhesive or low-viscosity rosin ester	Allow the board to stand for sufficient time to allow solvent to evaporate off
Brick and fired non-glazed building materials	Methyethyl ketone	Abrade surface with a wire brush; remove all dust and contaminants	
Carbon graphite	Acetone	Abrasion. Abrade with 220-grit emery cloth and solvent degrease after dust removal	For general-purpose bonding
Glass and quartz (nonoptical)	Acetone, detergent	1. Abrasion. Grit blast with carborundum and water slurry, and solvent degrease. Dry for 30 min at 100°C. Apply the adhesive before the glass cools to RT	For general-purpose bonding. Drying process improves bond strength
		2. Immerse for 10–15 min at 20°C in *Parts by wt.* Sodium dichromate......... 7 Water.............................7 Sulfuric acid (96%, sp. gr. 1.84)...............400 Rinse in water and distilled water. Dry thoroughly	For maximum strength
Glass (optical)	Acetone, detergent	Clean in an ultrasonically agitated detergent bath. Rinse; dry below 38°C	
Ceramics and porcelain	Acetone	1. Abrasion. Grit blast with carborundum and water slurry, and solvent degrease	Suitable for unglazed ceramics such as alumina, silica
		2. Solvent degrease or wash in warm aqueous detergent. Rinse and dry	For glazed ceramics such as porcelain
		3. Immerse for 15 min at 20°C in *Parts by wt.* Sodium dichromate........... 7 Water.................................7 Sulfuric acid (95% sp.gr. 1.84).................. 400 Rinse in water and distilled water. Oven-dry at 66°C	For maximum strength bonding of small ceramic (glazed) artefacts

Table 7.24 The Influence of Epoxy Curing Agent on the Bond Strength Obtained with Various Base Materials (from Ref. 29)

			Tensile—shear strength, lb/in²						
Curing agent	Amount*	Cure cycle, hr at °F	Polyester glass-mat laminate	Polyester glass-cloth laminate	Cold-rolled steel	Aluminum	Brass	Copper
Triethylamine	6	24 at 75, 4 at 150	1,850	2,100	2,456	1,810	1,765	655
Trimethylamine	6	24 at 75, 4 at 150	1,054	1,453	1,385	1,543	1,524	1,745
Triethylenetetramine	12	24 at 75, 4 at 150	1,150	1,632	1,423	1,675	1,625	1,325
Pyrrolidine	5	24 at 75, 4 at 150	1,250	1,694	1,295	1,733	1,632	1,420
Polyamid amine equivalent 210–230	35–65	24 at 75, 4 at 150	1,200	1,450	2,340	3,120	2,005	1,876
Metaphenylenidiamine	12.5	4 at 350	780	640	2,150	2,258	2,150	1,650
Diethylenetriamine	11	24 at 75, 4 at 150	1,010	1,126	1,350	1,420	1,135	1,236
Boron trifluoride monoethylamine	3	3 at 375	—	—	1,732	1,876	1,525	1,635
Dicyandiamide		4 at 350	530	432	2,680	2,785	2,635	2,550
Methy nadic anhydride	85	6 at 350	600	756	2,280	2,165	1,955	1,835

*Per 100 parts by weight of resin.
Epoxy resin used was derived from bisphenol A and epichlorohydrin and had an epoxide equivalent of 180 to 195; the adhesives contained no filler.

TABLE 7.19 Surface Preparations for Materials Other than Metals, Plastics, and Elastomers *(Continued)*

Adherend	Degreasing solvent	Method of treatment	Remarks
Concrete, granite, stone	Perchloro-ethylene, detergent	1. Abrasion. Abrade with a wire brush, degrease with detergent, and rinse with hot water before drying	For general-purpose bonding
		2. Etch with 15%hydrochloric acid until effervescence ceases. Wash with water until surface is litmus-neutral. *Rinse with 1% ammonia and water. Dry thoroughly before bonding	Applied by still-bristle brush. Acid should be prepared in a polyethylene pail 10–12% hydrochloric or sulfuric acids are alternative etchants. 10%w/w sodium bicarbonate may be used instead of ammonia for acid neutralization
Wood, plywood		Abrasion. Dry wood is smoothed with a suitable emery paper. Sand plywood along the direction of the grain	For general-purpose bonding
Painted surface	Detergent	1. clean with detergent solution, abrade with a medium emery cloth, final wash with detergent	Bond generally as strong as the paint
		2. Remove paint by solvent or abrasion, and pretreat exposed base	For maximum adhesion

Source: Based on Refs. 22–26

Urea formaldehyde adhesives are not as strong or as moisture resistant as resorcinol adhesives. However, they are inexpensive, and both hot- and cold-setting types are available. Maximum service temperature of urea adhesive is approximately 140°F. Cold-water resistance is good, but boiling-water resistance may be improved by the addition of melamine formaldehyde or phenol resorcinol resins. Urea-based adhesives are used in plywood manufacture.

7.4.2.5 Modified phenolics. Phenolic or phenol formaldehyde is also used as an adhesive for bonding wood. However, because of its brittle nature, this resin is unsuitable alone for more extensive adhesive applications. By modifying phenolic resin with various synthetic rubbers and thermoplastic materials, flexibility is greatly improved. The modified adhesive is well suited for structural bonding of many materials.

Nitrile-phenolic. Certain blends of phenolic resins with nitrile rubber produce adhesives useful to 300°F. On metals, nitrile phenolics offer shear strength in excess of 4,000 lb/in^2 and excellent peel properties. Good bond strengths can also be achieved on rubber, plastics, and glass. These adhesives have high impact strength and resistance to creep and fatigue. Their resistance to solvent, oil, and water is also good.

Nitrile-phenolic adhesives are available as solvent solutions and as supported and unsupported film. They require heat curing at 300 to 500°F under pressure of up to 200 lb/in^2. The nitrile-phenolic systems with the highest curing temperature have the greatest resistance to elevated temperatures during service. Because of good peel strength and elevated-temperature properties, nitrile-phenolic adhesives are commonly used for bonding linings to automobile brake shoes.

Vinyl-phenolic. Vinyl-phenolic adhesives are based on a combination of phenolic resin with polyvinyl formal or polyvinyl butyral resins. They have excellent shear and peel strength. Room-temperature shear strength as high as 5,000 lb/in^2 is available. Maximum operating temperature, however, is only 200°F, because the thermoplastic constituent softens at elevated temperatures. Chemical resistance and impact strength are excellent.

Vinyl-phenolic adhesives are supplied in solvent solutions and as supported and unsupported film. The adhesive cures rapidly at elevated temperatures under pressure. They are used to bond metals, elastomers, and plastics to themselves or each other. A major application of vinyl-phenolic adhesive is the bonding of copper sheet to plastic laminate in printed circuit board manufacture.

Neoprene-phenolic. Neoprene-phenolic alloys are used to bond a variety of substrates. Normal service temperature is –70 to 200°F. Because of high resistance to creep and most service environments, neoprene-phenolic joints can withstand prolonged stress. Fatigue and impact strengths are also excellent. Shear strength, however, is lower than that of other modified phenolic adhesives.

Temperatures over 300°F and pressure greater than 50 lb/in^2 are needed for cure. Neoprene-phenolic adhesives are available as solvent solutions and film. During cure, these adhesives are quite sensitive to atmospheric moisture, surface contamination, and other processing variables.

7.4.2.6 Polyaromatics. Polyimide and polybenzimidazole resins belong to the aromatic heterocycle polymer family, which is noted for its outstanding thermal resistance. These two highly cross-linked adhesives are the most thermally stable systems commercially available. The polybenzimidazole (PBI) adhesive has shear strength on steel of 3,000 lb/in^2 at room temperature and 2,500 lb/in^2 at 700°F. The polyimide adhesive offers a shear strength of approximately 3,000 lb/in^2 at room temperature, but it does not have the excellent strength at 700 to 1,000°F that is characteristic of PBI. Polyimide adhesives offer better elevated-temperature aging properties than PBI. The maximum continuous operating temperature for a polyimide adhesive is 600 to 650°F, whereas PBI adhesives oxidize rapidly at temperatures over 500°F.

Both adhesives are available as supported film, and polyimide resins are also available in solvent solution. During cure, temperatures of 550 to 650°F and high pressure are required. Volatiles are released during cure, which contribute to a porous, brittle bond line with relatively low peel strength.

Polyester. Polyesters are a large class of synthetic resins having widely varying properties. They may be divided into two distinct groups: saturated and unsaturated.

Unsaturated polyesters are fast-curing, two-part systems that harden by the addition of catalysts, usually peroxides. Styrene monomer is generally used as a reactive diluent for polyester resins. Cure can occur at room or elevated temperature depending on the type of catalyst. Accelerators such as cobalt naphthalene are sometimes incorporated into the resin to speed cure. Unsaturated polyester adhesives exhibit greater shrinkage during cure and poorer chemical resistance than epoxy adhesives. Certain types of polyesters are inhibited from curing by the presence of air, but they cure fully when enclosed between two

substrates. Depending on the type of resin, polyester adhesives can be quite flexible or very rigid. Uses include patching kits for the repair of automobile bodies and concrete flooring. Polyester adhesives also have strong bond strength to glass-reinforced polyester laminates.

Saturated polyester resins exhibit high peel strength and are used to laminate plastic films such as polyethylene terephthalate (Mylar). They also offer excellent clarity and color stability. These polyester types, in both solution and solid form, can be chemically cross-linked with curing agents such as the isocyanates for improved thermal and chemical stability.

Polyurethane. Polyurethane-based adhesives form tough bonds with high peel strength. Generally supplied as a two-part liquid, polyurethane adhesives can be cured at room or elevated temperatures. They have exceptionally high strength at cryogenic temperatures, but only a few formulations offer operating temperatures greater than 250°F. Like epoxies, urethane adhesives can be applied by a variety of methods and form strong bonds to most surfaces. Some polyurethane adhesives degrade substantially when exposed to high-humidity environments.

Polyurethane adhesives bond well to many substrates, including hard-to-bond plastics. Since they are very flexible, polyurethane adhesives are often used to bond films, foils, and elastomers. Moisture curing one-part urethanes are also available. These adhesives utilize the humidity in the air to activate their curing mechanism.

Anerobic adhesives. Acrylate acid diester and cyanoacrylate resins are called "anaerobic" adhesives because they cure when air is excluded from the resin. Anaerobic resins are noted for being simple to use, one-part adhesives, having fast cure at room temperature and high cost. However, the cost is moderate when considering a bonded-area basis because only a small volume of adhesive is required. Most anaerobic adhesives do not cure when gaps between adherend surfaces are greater than 10 mils, although some monomers have been developed to provide for thicker bond-lines.

The acrylate acid diester adhesives are available in various viscosities. They cure in minutes at room temperature when a special primer is used or in 3 to 10 min at 250°F without the primer. Without the primer, the adhesive requires 3 to 4 hr at room temperature to cure.

The cyanoacrylate adhesives are more rigid and less resistant to moisture than acrylate acid diester adhesives. They are available only as low-viscosity liquids that cure in seconds at room temperature without the need of a primer. The cyanoacrylate adhesives bond well to a variety of substrates, as shown in Table 7.25, but have relatively poor thermal resistance. Modifications of the original cyanoacrylate resins have been introduced to provide faster cures, higher strengths with some plastics, and greater thermal resistance.

Thermosetting Acrylics. Thermosetting acrylic adhesives are newly developed two-part systems which provide high shear strength to many metals and plastics, as shown in Table 7.26. These acrylics retain their strength to 200°F. They are relatively rigid adhesives with poor peel strength. These adhesives are particularly noted for their weather and moisture resistance as well as fast cure at room temperature.

One manufacturer has developed an acrylic adhesive system in which the hardener is applied to the substrate as a primer solution. The substrate can then be dried and stored for up to six months. When the parts are to be bonded, only the acrylic resin need be applied between the already primed substrates. Cure can occur in minutes at room temperature, depending on the type of acrylic resin used. Thus, this system offers the user a fast-reacting, one-part adhesive (with primer) with long shelf life.

7.4.3 Nonstructural Adhesives

TABLE 7.25 Performance of Cyanoacrylate Adhesives on Various Substrates (from Ref. 30)

Substrate	Age of bond	Shear strength, lb/in^2, of adhesive bonds
Steel–steel	10 min 48 hr	1920 3300
Aluminum–aluminum	10 min 48 hr	1480 2270
Butyl rubber–butyl rubber	10 min	150*
SBR rubber–SBR rubber	10 min	130
Neoprene rubber–neoprene rubber	10 min	100*
SBR rubber–phenolic	10 min	110*
Phenolic–phenolic	10 min 48 hr	930 940*
Phenolic–aluminum	10 min 48 hr	650 920*
Aluminum–nylon	10 min 48 hr	500 950
Nylon–nylon	10 min 48 hr	330 600
Neoprene rubber–polyester glass	10 min	110*
Polyester glass–polyester glass	10 min	680
Acrylic–acrylic	10 min 48 hr	810* 790*
ABS–ABS	10 min 48 hr	640* 710*
Polystyrene–polystyrene	10 min	330*
Polycarbonate–polycarbonate	10 min 48 hr	790 850*

*Substrate failure

Nonstructural adhesives are characterized by low shear strength (usually less than 1000 lb/in^2) and poor creep resistance at slightly elevated temperatures. The most common nonstructural adhesives are based on elastomers and thermoplastics. Although these systems have low strength, they usually are easy to use and fast setting. Most nonstructural adhesives are used in assembly-line fastening operations or as sealants and pressure-sensitive tapes.

7.4.3.1 Elastomer-based adhesives. Natural- or synthetic-rubber-based adhesives usually have excellent peel strength but low shear strength. Their resiliency provides good fatigue and impact properties. Temperature resistance is generally limited to 150 to 200°F, and creep under load occurs at room temperature.

The basic types of rubber-based adhesives used for nonstructural applications are shown in Table 7.27. These systems are generally supplied as solvent solutions, latex ce-

TABLE 7.26 Tensile-Shear Strength of Various Substrates Bonded with Thermosetting Acrylic Adhesives (from Ref. 31)

	Average lap shear, lb/in^2, at 77°F		
Substrate*	Adhesive A	Adhesive B	Adhesive C
Alclad aluminum, etched	4430	4235	5420
Bare aluminum, etched	4305	3985	5015
Bare aluminum, blasted	3375	3695	4375
Brass, blasted	4015	3150	4075
302 stainless steel, blasted	4645	4700	5170
302 stainless steel, etched	2840	4275	2650
Cold-rolled steel, blasted	2050	3385	2135
Copper, blasted	2915	2740	3255
Polyvinyl chloride, solvent wiped	1375†	1250†	1250†
Polymethyl methacrylate, solvent wiped	1550†	1160†	865†
Polycarbonate, solvent wiped	2570†	960	2570†
ABS, solvent wiped	1610†	1635	1280†
Alclad aluminum–PVC	1180†		
Plywood, 5/8-in exterior glued (lb/in)	802†	978†	
AFG-01 gap fill (1/16-in) (lb/in)		1083†	

*Metals solvent cleaned and degreased before etching or blasting.
†Substrate failure.

ments, and pressure-sensitive tapes. The first two forms require driving the solvent or water vehicle from the adhesive before bonding. This is accomplished by either simple ambient air evaporation or forced heating. Some of the stronger and more environmental-resistant rubber-based adhesives require an elevated-temperature cure. Generally, only slight pressure is required to achieve a substantial bond.

Pressure-sensitive adhesives are permanently tacky and flow under pressure to provide intimate contact with the adherend surface. Pressure-sensitive tapes are made by placing these adhesives on a backing material such as rubber, vinyl, canvas, or cotton cloth. After pressure is applied, the adhesive tightly grips the part being mounted as well as the surface to which it is affixed. The ease of application and the many different properties that can be obtained from elastomeric adhesives account for their wide use.

In addition to pressure-sensitive adhesives, elastomers go into mastic compounds that find wide use in the construction industry. Neoprene and reclaimed-rubber mastics are used to bond gypsum board and plywood flooring to wood-framing members. Often, the

TABLE 7.27 Properties of Elastomeric Adhesives Used in Nonstructural Applications (from Ref. 29)

Adhesive	Application	Advantages	Limitations
Reclaimed rubber	Bonding paper, rubber, plastic and ceramic tile, plastic films, fibrous sound insulation and weatherstripping; also used for the adhesive on surgical and electrical tape	Low cost, applied very easily with roller coating, spraying, dipping, or brushing, gains strength very rapidly after joining, excellent moisture and water resistance	Becomes quite brittle with age, poor resistance to organic solvents
Natural rubber	Same as reclaimed rubber; also used for bonding leather and rubber sides to shoes	Excellent resilience, moisture and water resistance	Becomes quite brittle with age; poor resistance to organic solvents; does not bond well to metals
Neoprene rubber	Bonding weather stripping and fibrous soundproofing materials to metal; used extensively in industry; bonding synthetic fibers, i.e., Dacron	Good strength to 150°F, fair resistance to creep	Poor storage life, high cost; small amounts of hydrochloric acid evolved during aging that may cause corrosion inclosed systems; poor resistance to sunlight
Nitrile rubber	Bonding plastic films to metals, and fibrous materials such as wood and fabrics to aluminum, brass, and steel; also, bonding nylon to nylon and other materials	Most stable synthetic-rubber adhesive, excellent oil resistance, easily modified by addition of thermosetting resins	Does not bond well to natural rubber or butyl rubber
Polyioso-butylene	Bonding rubber to itself and plastic materials; also, bonding polyethylene terephthalate film to itself, aluminum foil and other plastic films	Good aging characteristics	Attacked by hydrocarbons; poor thermal resistance
Butyl	Bonding rubber to itself and metals; forms good bonding with most plastic films such as polyethylene terephthalate and polyvinylidene chloride	Excellent aging characteristics; chemically cross-linked materials have good thermal properties	Metals should be treated with an appropriate primer before bonding; attacked by hydrocarbons

adhesive bond is much stronger than the substrate. These mastic systems cure by evaporation of solvent through the porous substrates.

Silicone. Silicone pressure-sensitive adhesives have low shear strength but excellent peel strength and heat resistance. Silicone adhesives can be supplied as solvent solutions

for pressure-sensitive application. The adhesive reaches maximum physical properties after being cured at elevated temperature with an organic peroxide catalyst. A lesser degree of adhesion can also be developed at room temperature. Silicone adhesives retain their qualities over a wide temperature range and after extended exposure to elevated temperature. Table 7.28 shows typical adhesive-strength properties of a silicone pressure-sensitive tape prepared with aluminum-foil backing.

TABLE 7.28 Effect of Temperature and Aging on Silicone Pressure-Sensitive Tape (Aluminum Foil Backing) (from Ref. 32)

Testing temperature		Adhesive strength, oz/in	
°C	°F	Uncatalyzed adhesive	Catalyzed adhesive (1% benzoyl peroxide)
–70	–94	>100*	>100*
–20	–4	>100*	>100*
100	212	60	48
150	302	60	45
200	392	Cohesive failure	40
250	482	Cohesive failure	35
Heat-aging cycle prior to testing (tested at 25°C, 77°F):			
No aging		60	52
1 hr at 150°C (302°F)		55	50
24 hr at 150°C (302°F)		60	50
7 days at 150°C (302°F)		70	50
1 hr at 250°C (482°F)		60	50
24 hr at 250°C (482°F)		65	65
7 days at 250°C (482°F)		Cohesive failure	65

*Maximum limit of testing equipment.

Room-temperature-vulcanizing (RTV) silicone-rubber adhesives and sealants form flexible bonds with high peel strength to many substrates. These resins are one-component pastes that cure by reacting with moisture in the air. Because of this unique curing mechanism, nonporous substrates should not overlap by more than 1 in.

RTV silicone materials cure at room temperature in about 24 hr. Fully cured adhesives can be used for extended periods up to 450°F and for shorter periods up to 500°F. Figure 7.29 illustrates the peel strength of an RTV adhesive on aluminum as a function of heat aging. With most RTV silicone formulations, acetic acid is released during cure. Consequently, corrosion of metals such as copper and brass in the bonding area may be a

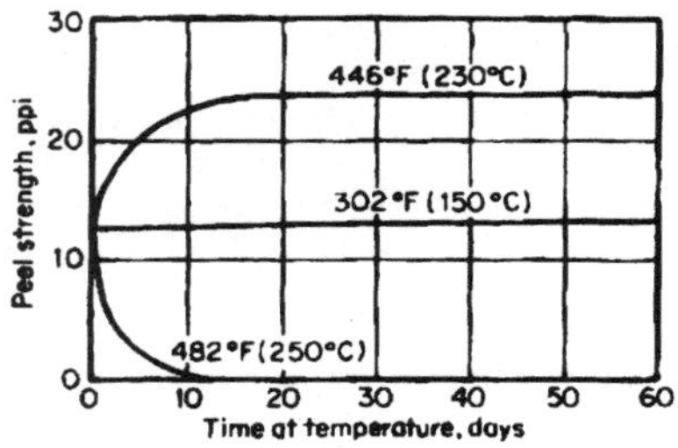

Figure 7.29 Peel strength of RTV silicone rubber bonded to aluminum as a function of heat aging.[33]

problem. However, special formulations are available that liberate methanol instead of acetic acid during cure. Silicone rubber bonds to clean metal, glass, wood, silicone resin, vulcanized silicone rubber, ceramic, and many plastic surfaces.

7.4.3.2 Thermoplastic adhesives. Table 7.21 describes the most common types of thermoplastic adhesives. These adhesives are useful in the –20 to 150°F temperature range. Their physical properties vary with chemical type. Some resins like the polyamides are quite tacky and flexible, while others are very rigid.

Thermoplastic adhesives are generally available as solvent solutions, water-based emulsions, and hot melts. The first two systems are useful in bonding porous materials such as wood, plastic foam, and paper. Water-based systems are especially useful for bonding foams that could be affected adversely by solvents. When hardened, thermoplastic adhesives are very nonresistant to the solvent in which they are originally supplied.

Hot-melt systems are usually flexible and tough. They are used extensively for sealing applications involving paper, plastic films, and metal foil. Table 7.29 offers a general comparison of hot-melt adhesives. Hot melts can be supplied as (1) tapes or ribbons, (2) films, (3) granules, (4) pellets, (5) blocks, or (6) cards, which are melted and pressed between the substrate. The rate at which the adhesive cools and sets is dependent on the type of substrate and whether it is preheated. Table 7.30 lists the advantages and disadvantages associated with the use of water-based, solvent-based, and hot-melt thermoplastic adhesives.

7.5 Selecting an Adhesive

7.5.1 Factors Affecting Selection

There is no general-purpose adhesive. The best adhesive for a particular application will depend on the materials to be bonded, the service and assembly requirements, and the economics of the bonding operation. By using these factors as criteria for selection, the many commercially available adhesives can be narrowed down to a few possible candidates. All the desired properties are often not available in a single adhesive system. In these cases, a compromise adhesive can usually be chosen by deciding which properties are of major and minor importance.

7.5.1.1 Adhesive substrates. The materials to be bonded are a prime factor in determining which adhesive to use. Some adherends, such as stainless steel or wood, can be successfully bonded with a great many adhesive types; other adherends, such as nylon, can be bonded with only a few. Typical adhesive-adherend combinations are listed in Table 7.31. A number in a given column indicates the particular adhesive that will bond to a

TABLE 7.29 General Comparison of Common Hot-Melt Adhesives (from Ref. 34)

Property	Ethylene vinyl acetate	Ethylene ethyl acetate	Ethylene acrylic acid	Ionomer	Phenoxy	Poly-amide	Polyester	Poly-ethylene	Polyvinyl acetate	Polyvinyl butyral
Softening point, °C	40	60	70	75	100	100	.....	.....	65–195	
Melting point, °C	95	90 (80–100)	.....	.....	.....	.....	267	137	.....	130
Crystallinity	L	L	M	L	L	L	H	H or L	L	L
Melt index	6	3 (2–20)	3 (0.5–400)	2	2.5	2	5	5 (0.5–20)		
Tensile strength, lb/in.²	2,750	2,000	3,300	4,000	9,500	2,000	4,500	2,000	5,000	6,500
% elongation	800	700	600	450	75	300	500	150	10	100
Lap shear on metal, lb/in.²	.....	.....	1,700	.....	3,500	1,050				
Film peel, lb/in.	1	4.5	12	12	.....	.....	5	2		
Cost	M	M	M	M	H	M	H	L	M–L	M
Usage	H	M	L	L	L	H	H	L	M	H

H = high. M = medium. L = low.

TABLE 7.30 Advantages and Disadvantages of Thermoplastic Adhesive Forms (from Ref. 35)

Form	Advantages	Disadvantages
Water base	Lower cost Nonflammable Nontoxic solvent Wide range of solids content Wide range of viscosity High concentration of high-molecular-weight material can be used Penetration and wetting can be varied	Poorer water resistance Subject to freezing Shrinks fabrics Wrinkles or curls paper Can be contaminated by some metals used for storage and application Corrosive to some metals Slow drying Poorer electrical properties
Solvent base	Water-resistant Wide range of drying rates and open times Develop high early bond strength Easily wets some difficult surfaces	Fire and explosive hazard Health hazard Special explosion proof and ventilating equipment needed
Hot melt	Lower package and shipping cost per pound of solid material Will not freeze Drying and equipment for drying are unnecessary Impervious surfaces easily bonded Fast bond-strength development Good storage stability Provides continuous water-vapor impermeable and water-resistant adhesive film	Special application equipment needed Limited strength because of viscosity and temperature limitations Degrades on continuous heating Poorer coating weight control Preheating of adherends may be necessary

cross-referenced substrate. If two different materials are to be bonded, the recommended adhesives in Table 7.31 are those showing identical numbers under both substrates.

This information is intended only as a guideline to show common adhesives that have been used successfully in various applications. The adhesive selections are listed without regard to strength or service requirements. Lack of a suggested adhesive for a particular substrate does not necessarily mean that a poor bond will result, only that information is not commonly available concerning that particular combination.

7.5.1.2 Service requirements. Adhesives must also be selected with regard to the type of stress and environmental conditions to which they will be exposed. These factors will further limit the number of candidate adhesives to be tested. Information on the environmental resistance of various adhesive classifications will appear in the next section.

The chosen adhesive should have strength great enough to resist the maximum stress during any time in service with reasonable safety factor. Overspecifying could result in certain adhesives being overlooked that can do the job at lower cost and with less demanding curing conditions.

7.5.1.3 Production requirements. Adhesives require time, pressure, and heat, singly or in combination, to harden or set. Production requirements are often a severe restrict-

Table 7.31 Selecting Adhesives for Use with Various Substrates (based on Reference 23)

Adhesives (number key given in Table 7.32)

Adherend																																											
Metals, alloys																																											
Aluminum	1	2	3		5				9	10	11			14	15						21		23	24	25		27	28	29	30	31	32	33	34	35	36	37	38	39	40	41	42	43
Brass	1				5				9		11			14	15						21		23	24						30		32			35	36	37	38	39	40	41	42	43
Bronze	1				5				9		11										21		23	24						30		32			35	36							
Cadmium									9	10																																	
Chromium	1	2	3						9	10	11										21			24					29			32											
Copper	1	2			5				9	10	11			14	15						21		23	24						30		32		34	35	36			39	40	41	42	43
Germanium									9		11																																
Gold		2			5					10	11																		29	30							37						
Lead		2	3						9		11																																
Magnesium	1				5				9	10	11								19		21		23	24			27	28	29	30	31	32	33	34	35	36	37		39	40	41	42	43
Nickel					5				9	10	11																		29							36							
Platinum									9		11												23															38					
Silver					5				9	10	11																									36							
Steel (mild), iron	1	2	3		5				9	10	11	12	13	14	15	16	17	18	19		21		23	24	25		27	28	29	30	31	32	33	34	35	36	37		39	40	41		43
Steel (stainless)	1	2	3		5				9	10	11	12	13	14	15	16	17	18	19		21		23	24	25		27	28	29	30	31	32	33	34	35	36	37		39	40	41		43
Tin									9	10	11																		29	30	31	32	33										
Titanium	1	2			5				9	10	11	12	13										23						29	30	31	32	33	34	35	36	37		39	40	41	42	43
Tungsten									9	10																				30													
Zinc	1	2	3		5					10	11								19				23	24			27	28	29	30	31	32	33	34	35	36				40			
Uranium		2									11																																
Plastics																																											
Acetals	1	2									11																			30						36				40			
Acrylonitrile butadiene styrene	1	2	3						9		11			14	15									24			27		29	30		32				36				40			
Allyl diglycol carbonate																								24																			
Cellulose																16		18																									
Cellulose acetate	1	2														16	17	18	19				23			26			29	30						36				40			
Cellulose acetate-butyrate	1	2														16	17	18	19							26			29	30						36				40			
Cellulose nitrate																		18					23																				
Cellulose propionate	1	2		4					9							16	17	18	19				23			26			29	30						36				40			
Diallyl isophthalate				4					9																						31					36				40		42	
Diallyl phthalate				4					9																						31					36				40		42	

Table 7.31 Selecting Adhesives for Use with Various Substrates *(Continued)* (based on Reference 23)

Adherend	Adhesives (number key given in Table 7.32)
Epoxy resins	2 5 9 11 27 28 29 30 32 35 36 37 40
Ethyl cellulose	18
Methacrylate butadiene styrene	31
Melamine formaldehyde	4 7 9 11 19 36 40 43
Methyl pentene	9 30
Phenol formaldehyde	4 7 9 11 19 36 40 43
Phenolic–melamine	2 5 7 9 10 11 14 23 27 28 29 30 31 32 35 36 37 39 41 43
Phenoxy	11
Polyamide (nylon)	1 2 7 8 11 15 23 30 32 40
Polycarbonate	1 2 3 9 11 14 15 32
Polychlorotrifluoroethylene	9 10 11 29 34 35 40
Polyester (fiber composite)	2 3 9 11 14 24 27 28 29 30 32 36 40
Polyether (chlorinated)	30 36
Polyethylene	9 11 23 30 32 36 40
Polyethylene (film)	2 9 11 23 29 30 32 35 36 40
Polyethylene (chlorinated)	2 5 30 35 36
Polyethylene terephthalate (Mylar)	2 5 10 19 23 29 30 35
Polyformaldehyde	2 9 11 19 24 30 40
Polyimide	9 10 11 12 30 34 35 36 37 39
Polymethyl metacrylate	1 5 11 14 15 19 24 30 34 35 36 40
Polyphenylene	11
Polyphenylene oxide	2 3 9 11 14 29 30 32 36 40
Polypropylene	11 30
Polypropylene (film)	2 11 23 30 35
Polystyrene	1 3 14 15 23 24
Polystyrene (film)	1 15 19
Polysulfone	14 30
Polytetrafluoroethylene	9 10 11 29 34 35 40
Polyurethane	11 30 32 36
Polyvinyl chloride	1 2 9 11 14 15 19 24 29 30 32 36 40
Polyvinyl chloride (film)	2 19 24 30 32 35 40

Table 7.31 Selecting Adhesives for Use with Various Substrates *(Continued)* (based on Reference 23)

Adherend	Adhesives (number key given in Table 7.32)
Polyvinyl fluoride	2 3 11 30 32 33 40
Polyvinylidene chloride	2 23 30 32
Silicone	2 34 35
Styrene acrylonitrile	11 14 30
Urea formaldehyde	4
Foams	
Epoxy	9 11
Latex	1 2 9 11 26 27 30 32
Phenol formaldehyde	2 5 9 11 19 27 28 32 36 41 43
Polyethylene	11 29 30
Polyethylene–cellulose acetate	9
Polyphenylene oxide	2 9 11 14 21 28 30
Polystyrene	2 9 11 14 19 30 31 36
Polyurethane	2 5 9 11 15 24 28 30 32 35
Polyvinylchloride	2 3 11 19 30 32 33 36
Silicone	34
Urea formaldehyde	4 7
Rubbers	
Butyl	1 15 28 32 35
Butadiene nitrile	1 2 7 9 11 15 30 32 35
Butadiene styrene	1 2 9 10 15 30 32 35 40 43
Chlorosulfonated polyethylene	30 31 32 35 40 43
Ethylene propylene	11
Fluorocarbon	1 9 10 11 15 34 35
Fluorosilicone	34
Polyacrylic	24
Polybutadiene	2 11 31
Polychloroprene (neoprene)	1 2 8 10 11 15 28 30 31 32 33 35 36 40 41 43
Polyisoprene (natural)	1 2 7 9 10 11 15 26 27 30 31 32 34 35 40 43
Polysulfide	33
Polyurethane	1 2 11 15 28 29 30 32 35 36

Table 7.31 Selecting Adhesives for Use with Various Substrates *(Continued)* (based on Reference 23)

Adherend	Adhesives (number key given in Table 7.32)
Silicone	34 35
Wood, allied materials	
Cork	2 3 4 5 6 7 8 9 11 16 17 18 19 21 23 26 27 28 30 31 32 33 34 36 40 41
Hardboard, chipboard	4 7 31
Wood	2 3 4 5 6 7 8 9 11 14 15 16 17 18 19 21 23 25 26 27 28 29 30 31 32 33 34 36 40
Wood (laminates)	2 3 4 5 6 7 8 9 11 16 17 18 19 21 25 26 28 29 30 31 32 33 34 36 40
Fiber products	
Cardboard	7 8 9 11 16 17 18 19 21 22 23 24 26 27 28 29 30 31 32 33 36 37
Cotton	2 3 4 5 7 9 11 16 17 18 19 23 24 26 27 30 31 32 33 34 35 36 40 41
Felt	3 4 5 7 9 11 16 17 18 19 23 24 26 27 30 31 32 33 34 35 36 40 41
Jute	2 3 4 5 7 9 11 16 17 19 23 24 26 27 30 31 32 33 34 36 40 41
Leather	2 3 5 9 11 16 17 18 19 21 23 24 26 27 28 29 30 31 32 33 34 36 40 41
Paper (bookbinding)	16 17 18 19 21 22 23 26 27 28 29 31 32 33 36
Paper (labels)	16 17 18 19 22 23 24 26 27 29 30 31 32 33 34 36
Paper (packaging)	2 16 17 18 19 21 22 23 24 26 27 28 29 30 31 32 33 36
Rayon	2 3 7 9 11 16 17 18 19 22 24 26 27 28 29 30 31 32 33 34 35 36 40
Silk	2 3 4 5 7 9 11 16 17 18 19 23 24 26 27 28 30 31 32 33 34 35 36 40 41 43
Wool	2 3 4 5 7 9 11 16 17 18 19 23 24 26 27 28 30 31 32 33 34 35 36 40 41
Inorganic materials	
Asbestos	2 3 5 7 8 9 10 11 14 16 18 19 21 27 28 29 30 31 32 33 34 35 37 40 43
Carbon	9 11
Carborundum	9 11
Ceramics (porcelain, vitreous)	1 2 3 4 7 8 9 10 11 14 15 16 17 18 19 21 27 28 29 30 31 32 33 34 35 36 37
Concrete, stone, granite	2 3 4 7 8 11 14 35 40 41
Ferrite	9 11
Glass	1 2 3 5 9 10 11 14 15 16 17 18 21 28 30 31 32 34 35 37 40
Magnesium fluoride	11
Mica	35 40 43
Quartz	9 11 14
Sodium chloride	11
Tungsten carbide	9 11

TABLE 7.32 Selecting Adhesives with Respect to Form and Processing Factors

	Common forms available					Method of cure				Processing conditions			
Adhesive type	Solid	Film	Paste	Liquid	Solvent sol, emulsion	Solvent release	Fusion on heating	Pressure-sensitive	Chemical reaction	Room-temperature	Elevated-temperature	Bonding pressure required	Bonding pressure not required
Thermosetting adhesives													
1. Cyanoacrylate	–	–	–	X	–	–	–	–	X	X	–	X	–
2. Polyester + isocyanate	–	X	–	–	X	X	–	–	X	X	X	X	–
3. Polyester + monomer	–	–	X	X	–	–	–	–	X	X	X	–	X
4. Urea formaldehyde	X	–	–	X	–	–	–	–	X	X	X	X	–
5. Melamine formaldehyde	X	–	–	–	–	–	–	–	X	–	X	X	–
6. Urea-melamine formaldehyde	X	–	–	X–	–	–	–	–	X	X	X	X	–
7. Resorcinol formaldehyde	–	–	–	X	–	–	–	–	X	X	X	X	–
8. Phenol-resorcinol formaldehyde	–	–	–	X	–	–	–	–	X	X	X	X	–
9. Epoxy (+ polyamine)	–	–	X	X	–	–	–	–	X	X	X	–	X
10. Epoxy (+ polyanhydride)	X	X	X	X	–	–	–	–	X	–	X	–	X
11. Epoxy (+ polyamide)	–	–	X	X	–	–	–	–	X	X	X	–	X
12. Polyimide	–	X	–	–	X	–	–	–	X	–	X	X	–
13. Polybenzimidazole	–	X	–	–	–	–	–	–	X	–	X	X	–
14. Acrylic	–	–	X	X	–	–	–	–	X	X	–	–	X
15. Acrylate acid diester	–	–	X	X	–	–	–	–	X	X	X	X	–
Thermoplastic adhesives													–
16. Cellulose acetate	–	–	–	–	X	X	–	–	–	X	–	–	X
17. Cellulose acetate butyrate	–	–	–	–	X	X	–	–	–	X	–	–	X
18. Cellulose nitrate	–	–	–	–	X	X	–	–	–	X	–	–	X
19. Polyvinyl acetate	–	X	–	–	X	X	X*	–	–	X	X*	X*	X
20. Vinyl vinylidene	–	–	–	–	X	X	–	–	–	X	–	–	X
21. Polyvinyl acetal	X	X	–	–	X	X	X*	–	–	X	X*	X*	X
22. Polyvinyl alcohol	–	–	–	–	X	X	–	–	–	X	–	–	X

TABLE 7.32 Selecting Adhesives with Respect to Form and Processing Factors

	Common forms available					Method of cure				Processing conditions			
Adhesive type	Solid	Film	Paste	Liquid	Solvent sol, emulsion	Solvent release	Fusion on heating	Pressure-sensitive	Chemical reaction	Room-temperature	Elevated-temperature	Bonding pressure required	Bonding pressure not required
23. Polyamide	X	X	–	–	–	–	X	–	–	–	X	X	–
24. Acrylic	–	–	–	–	X	X	–	–	X	X	X	X	–
25. Phenoxy	X	X	–	–	–	–	X	–	–	–	X	X	–
Elastomer adhesives													–
26. Natural rubber	–	–	–	–	X	X	–	X	X	X	X	X	–
27. Reclaimed rubber	–	–	–	–	X	X	–	X	–	X	–	X	–
28. Butyl	–	–	–	–	X	X	–	X	X	X	X	X	–
29. Polyisobutylene	–	–	–	–	X	X	–	X	–	X	–	X	–
30. Nitrile	–	–	–	–	X	X	–	X	–	X	–	X	–
31. Styrene butadiene	X	–	–	–	X	X	X*	X	–	X	X*	X	–
32. Polyurethane	–	–	X	X	X	X	–	–	X	X	–	–	X
33. Polysulfide	–	–	X	X	–	–	–	–	X	X	–	–	X
34. Silicone (RTV)	–	–	X	–	–	–	–	–	X	X	–	–	X
35. Silicone resin	–	–	–	–	X	–	–	X	X	–	X	X	–
36. Neoprene	–	–	–	–	X	X	–	X	–	X	–	X	–
Alloy adhesives													–
37. Epoxy-phenolic	–	X	X	–	–	–	–	–	X	–	X	X	–
38. Epoxy-polysulfide	–	X	X	–	–	–	–	–	X	X	–	–	X
39. Epoxy-nylon	–	X	–	–	X	–	–	–	X	–	X	X	–
40. Phenolic-nitrile	–	X	–	–	X	–	–	–	X	–	X	X	–
41. Phenolic-neoprene	–	X	–	–	–	–	–	–	X	–	X	X	–
42. Phenolic-polyvinyl butyral	–	X	–	–	–	–	–	–	X	–	X	X	–
43. Phenolic-polyvinyl formal	–	X	–	–	–	–	–	–	X	–	X	X	–

*Heat and pressure required for hot–melt types.

ing factor in the selection of an adhesive. Typical factors involved in assembly are the equipment available, allowable cure time, pressure required, necessary bonding temperature, degree of substrate preparation required, and physical form of the adhesive. Table 7.32 lists available forms and processing requirements for various types of adhesives.

7.5.1.4 Cost requirements. Economic analysis of the bonding operation should consider not only raw-materials cost but also the processing equipment necessary, time and labor required, and cost incurred by wasted adhesive and rejected parts. Final cost per bonded part is a realistic criterion for selection of an adhesive.

7.5.2 Adhesives for Metal

The chemical types of structural adhesives for metal bonding were described in the preceding section. Since organic adhesives readily wet most metallic surfaces, the adhesive selection does not depend as much on the type of metal substrate as on other bonding requirements.

Selecting a specific adhesive from a table of general properties is difficult, because formulations within one class of adhesive may vary widely in physical properties. General physical data for structural metal adhesives are presented in Table 7.33. This table may prove useful in making preliminary selections or eliminating obviously unsuitable adhesives. Nonstructural adhesives for metals include elastomeric and thermoplastic resins. These are generally used as pressure-sensitive or hot-melt adhesives. They are noted for fast production, low cost, and low to medium strength. Typical adhesives for nonstructural bonding applications were previously described. Most pressure-sensitive and hot-melt cements can be used on any clean metal surface and on many plastics and elastomers.

7.5.3 Adhesives for Plastics

The physical and chemical properties of both the solidified adhesive and the plastic substrate affect the quality of the bonded joint. Major elements of concern are the thermal expansion coefficient and glass transition temperature of the substrate relative to the adhesive. Special consideration is also required of polymeric surfaces that can change during normal aging or exposure to operating environments.

Significant differences in thermal expansion coefficient between substrates and the adhesive can cause serious stress at the plastic joint's interface. These stresses are compounded by thermal cycling and low-temperature service requirements. Selection of a resilient adhesive or adjustments in the adhesive's thermal expansion coefficient via fillers or additives can reduce such stress.

Structural adhesives must have a glass transition temperature higher than the operating temperature to avoid a cohesively weak bond and possible creep problems. Modern engineering plastics, such as polyimide or polyphenylene sulfide, have very high glass transition temperatures. Most common adhesives have a relatively low glass transition temperature so that the weakest thermal link in the joint may often be the adhesive.

Use of an adhesive too far below the glass transition temperature could result in low peel or cleavage strength. Brittleness of the adhesive at very low temperatures could also manifest itself in poor impact strength.

Plastic substrates could be chemically active, even when isolated from the operating environment. Many polymeric surfaces slowly undergo chemical and physical change. The plastic surface, at the time of bonding, may be well suited to the adhesive process. However, after aging, undesirable surface conditions may present themselves at the interface,

Table 7.33 Properties of Structural Adhesives Used to Bond Metals

Adhesive	Service temp. °F		Shear strength, lb/in.^{2}	Peel strength	Impact strength	Creep resistance	Solvent resistance	Moisture resistance	Type of bond
	Max	Min							
Epoxy-amine	150	–50	3000–5000	Poor	Poor	Good	Good	Good	Rigid
Epoxy-polyamide	150	–60	2000–4000	Medium	Good	Good	Good	Medium	Tough and moderately flexible
Epoxy-anhydride	300	–60	3000–5000	Poor	Medium	Good	Good	Good	Rigid
Epoxy-phenolic	350	–423	3200	Poor	Poor	Good	Good	Good	Rigid
Epoxy-nylon	180	–423	6500	Very good	Good	Medium	Good	Poor	Tough
Epoxy-polysulfide	150	–100	3000	Good	Medium	Medium	Good	Good	Flexible
Nitrile-phenolic	300	–100	3000	Good	Good	Good	Good	Good	Tough and moderately flexible
Vinyl-phenolic	225	–60	2000–5000	Very Good	Good	Medium	Medium	Good	Tough and moderately flexible
Neoprene-phenolic	200	–70	3000	Good	Good	Good	Good	Good	Tough and moderately flexible
Polyimide	600	–423	3000	Poor	Poor	Good	Good	Medium	Rigid
Polybenzimidazole	500	–423	2000–3000	Poor	Poor	Good	Good	Good	Rigid
Polyurethane	150	–423	5000	Good	Good	Good	Medium	Poor	Flexible
Acrylate acid diester	200	–60	2000–4000	Poor	Medium	Good	Poor	Poor	Rigid
Cyanoacrylate	150	–60	2000	Poor	Poor	Good	Poor	Poor	Rigid
Phenoxy	180	–70	2500	Medium	Good	Good	Poor	Good	Tough and moderately flexible
Thermosetting acrylic	250	–60	3000–4000	Poor	Poor	Good	God	Good	Rigid

displace the adhesive, and result in bond failure. These weak boundary layers may come from the environment or from within the plastic substrate itself.

Moisture, solvent, plasticizers, and various gases and ions can compete with the cured adhesive for bonding sites. The process where a weak boundary layer preferentially displaces the adhesive at the interface is called *desorption*. Moisture is the most common desorbing substance, being present both in the environment and within many polymeric substrates.

Solutions to the desorption problem consist of eliminating the source of the weak boundary layer or selecting an adhesive that is compatible with the desorbing material. Excessive moisture can be eliminated from a plastic part by post curing or drying the part before bonding. Additives that can migrate to the surface can possibly be eliminated by reformulating the plastic resin. Also, certain adhesives are more compatible with oils and plasticizers than others. For example, the migration of plasticizer from flexible polyvinyl chloride can be counteracted by using a nitrile-based adhesive. Nitrile adhesive resins are capable of absorbing the plasticizer without degrading.

7.5.3.1 Thermoplastics. Many thermoplastics can be joined by solvent or heat welding as well as with adhesives. These alternative joining processes are discussed in detail in another chapter. The plastic manufacturer is generally the leading source of information on the proper methods of joining a particular plastic.

7.5.3.2 Thermosetting plastics. Thermosetting plastics cannot be heat or solvent welded. They are easily bonded with many adhesives, some of which have been listed in Table 7.31. Abrasion is generally recommended as a surface treatment.

7.5.3.3 Reinforced plastics. Adhesives that give satisfactory results on the resin matrix alone may also be used to bond reinforced plastics. Surface preparation of reinforced thermosetting plastics consists of abrasion and solvent cleaning. A degree of abrasion is desired so that the reinforcing material is exposed to the adhesive.

Reinforced thermoplastic parts are generally abraded and cleaned prior to adhesive bonding. However, special surface treatment such as used on the thermoplastic resin matrix may be necessary for optimal strength. Care must be taken so that the treatment chemicals do not wick into the substrate and cause degradation. Certain reinforced thermoplastics may also be solvent cemented or heat welded. However, the percentage of filler in the substrate must be limited, or the bond will be starved of resin.

7.5.3.4 Plastic foams. Some solvent cements and solvent-containing pressure-sensitive adhesives will collapse thermoplastic foams. Water-based adhesives, based on styrene butadiene rubber (SBR) or polyvinyl acetate, and 100 percent solid adhesives are often used. Butyl, nitrile, and polyurethane adhesives are often used for flexible polyurethane foam. Epoxy adhesives offer excellent properties on rigid polyurethane foam.

7.5.4 Adhesives for Elastomers

7.5.4.1 Vulcanized elastomers. Bonding of vulcanized elastomers to themselves and other materials is generally accomplished by using a pressure-sensitive adhesive derived from an elastomer similar to the one being bonded. Flexible thermosetting adhesives such as epoxy-polyamide or polyurethane also offer excellent bond strength to most elastomers. Surface treatment consists of washing with a solvent, abrading, or acid cyclizing as described in Table 7.18.

Elastomers vary greatly in formulation from one manufacturer to another. Fillers, plasticizers, antioxidants, etc., may affect the adhesive bond. Adhesives should be thoroughly tested on a specific elastomer and then re-evaluated if the elastomer manufacturer or formulation is changed.

7.5.4.2 Unvulcanized elastomers. Unvulcanized elastomers may be bonded to metals and other rigid adherends by priming the adherend with a suitable air- or heat-drying adhesive before the elastomer is molded against the adherend. The most common elastomers to be bonded in this way include nitrile, neoprene, urethane, natural rubber, SBR, and butyl rubber. Less common unvulcanized elastomers such as the silicones, fluorocarbons, chlorosulfonated polyethylene, and polyacrylate are more difficult to bond.

However, recently developed adhesive primers improve the bond of these elastomers to metal. Surface treatment of the adherend before priming should be according to good standards.

7.5.5 Adhesives for Wood

Resorcinol-formaldehyde resins are cold-setting adhesives for wood structures. Urea-formaldehyde adhesives, commonly modified with melamine formaldehyde, are used in the production of plywood and in wood veneering for interior applications. Phenol-formaldehyde and resorcinol-formaldehyde adhesive systems have the best heat and weather resistance.

Polyvinyl acetates are quick-drying, water-based adhesives commonly used for the assembly of furniture. This adhesive produces bonds stronger than the wood itself, but it is not resistant to moisture or high temperature. Table 7.34 describes common adhesives used for bonding wood.

7.5.6 Adhesives for Glass

Glass adhesives are generally transparent, heat-setting resins that are water-resistant to meet the requirements of outdoor applications. Adhesives generally used to bond glass, and their physical characteristics, are presented in Table 7.35.

7.6 Effect of the Environment

For an adhesive bond to be useful, it not only must withstand the mechanical forces acting on it; it must also resist the service environment. Adhesive strength is influenced by many common environments, including temperature, moisture, chemical fluids, and outdoor weathering. Table 7.36 summarizes the relative resistance of various adhesive types to common environments.

7.6.1 High Temperature

TABLE 7.34 Properties of Common Wood Adhesives (from Ref. 36)

Resin type used	Resin solids in glue mix, %	Principal use	Method of application	Principal property	Principal limitation
Urea formaldehyde	23–30	Wood-to-wood interior	Spreader rolls	Bleed-through-free; good adhesion	Poor durability
Phenol formaldehyde	23–27	Plywood exterior	Spreader rolls	Durability	Comparatively long cure times
Melamine formaldehyde	68–72	Wood-to-wood, splicing, patching, scarfing	Sprayed, combed	Adhesion, color, durability	Relative cost; poor washability; needs heat to cure
Melamine urea 1/1	55–60	End and edge gluing exterior	Applicator	Colorless, durability and speed	Cost
Resorcinol formaldehyde	50–56	Exterior wood-to-wood (laminating)	Spreader rolls	Cold sets durability	Cost, odor
Phenol-resorcinol 10/90	50–56	Wood-to-wood exterior (laminating)	Spreader rolls	Warm-set durability	Cost, odor
Polyvinyl acetate emulsion	45–55	Wood-to-wood interior	Brushed, sprayed, spreader rolls	Handy	Lack of H_2O and heat resistance

All polymeric materials are degraded to some extent by exposure to elevated temperatures. Not only are physical properties lowered at high temperatures, they also degrade due to thermal aging. Newly developed polymeric adhesives have been found to withstand 500 to 600°F continuously. To use these materials, the designer must pay a premium in adhesive cost and also be capable of providing long, high-temperature cures.

For an adhesive to withstand elevated-temperature exposure, it must have a high melting or softening point and resistance to oxidation. Materials with a low melting point, such as many of the thermoplastic adhesives, may prove excellent adhesives at room temperature. However, once the service temperature approaches the glass transition temperature of these adhesives, plastic flow results in deformation of the bond and degradation in cohesive strength. Thermosetting materials, exhibiting no melting point, consist of highly cross-linked networks of macromolecules. Many of these materials are suitable for high-temperature applications. When considering thermoset adhesives, the critical factor is the rate of strength reduction due to thermal oxidation and pyrolysis.

Thermal oxidation initiates progressive chain scission of molecules resulting in losses of weight, strength, elongation, and toughness within the adhesive. Figure 7.30 illustrates the effect of oxidation by comparing adhesive joints aged in both high-temperature air and inert-gas environments. The rate of strength degradation in air depends on the temperature, the adhesive, the rate of airflow, and even the type of adherend. Certain metal–adhesive interfaces are capable of accelerating the rate of oxidation. For example, many structural adhesives exhibit better thermal stability when bonded to aluminum than when bonded to stainless steel or titanium (Fig. 7.30).

High-temperature adhesives are usually characterized by a rigid polymeric structure, high softening temperature, and stable chemical groups. The same factors also make these

TABLE 7.35 Commercial Adhesives Most Desirable for Glass (from Ref. 37)

		Bond characteristics		
Trade name	Chemical type	Strength, lb/in^2	Type of failure	Weathering quality
Butacite, Butvar	Polyvinyl butyral	2000–4000	Adhesive	Fair
Bostik 7026, FM–45, FM–46	Phenolic butyral	2000–5500	Glass	Excellent
EC826, EC776			Adhesion and glass	
N–199, Scotchweld	Phenolic nitrile	1000–1200		Excellent
Pliobond M–20, EC847	Vinyl nitrile	1200–3000	Adhesion and glass	Fair to good
EC711, EC882				
EC870	Neoprene	800–1200	Adhesion and cohesive	Fair
EC801, EC612	Polysulfide	200–400	Cohesive	Excellent
EC526, R660T, EC669	Rubber base	200–800	Adhesive	Fair to poor
Siliastic	Silicone	200–300	Cohesive	Excellent
Res–N–Glue, du Pont 5459	Cellulose vinyl	1000–1200	Adhesive	Fair
Vinylite AYAF, 28–18	Vinyl acetate	1500–2000	Adhesive	Poor
Araldite, Epon L–1372, ERL–2774, R–313, C–14, SH–1, J–1152	Epoxy	600–2000	Adhesive	Fair to good

adhesives very difficult to process. Only epoxy-phenolic-, polyimide-, and polybenzimidazole-based adhesives can withstand long-term service temperatures greater than 350°F.

7.6.1.1 Epoxy. Epoxy adhesives are generally limited to continuous applications below 300°F. Figure 7.31 illustrates the aging characteristics of a typical epoxy adhesive at elevated temperatures. Certain epoxy adhesives are able to withstand short terms at 500°F and long-term service at 300 to 350°F. These systems were formulated especially for thermal environments by incorporation of stable epoxy co-reactants, high-temperature curing agents, and antioxidants into the adhesive.

One successful epoxy co-reactant system is an epoxy-phenolic alloy. The excellent thermal stability of the phenolic resins is coupled with the adhesion properties of epoxies to provide an adhesive capable of 700°F short-term operation and continuous use at 350°F. The heat-resistance and thermal-aging properties of an epoxy-phenolic adhesive are compared with those of other high-temperature adhesives in Fig. 7.32.

Table 7.36 Relative Resistance of Synthetic Adhesives to Common Service Environment (from Ref. 38)

Adhesive type	Shear	Peel	Heat	Cold	Water	Hot water	Acid	Alkali	Oil, grease	Fuels	Alcohols	Ketones	Esters	Aromatics	Chlorinated solvents
Thermosetting adhesives															
1. Cyanoacrylate	2	6	5	–	6	6	6	6	3	3	5	5	5	4	4
2. Polyester + isocyanate	2	2	3	2	1	3	3	2	2	2	3	2	2	6	2
3. Polyester + monomer	2	6	5	3	3	6	3	6	2	2	2	6	6	6	6
4. Urea formaldehyde	2	6	3	3	2	6	2	2	2	2	2	2	2	2	2
5. Melamine formaldehyde	2	6	2	2	2	5	2	2	2	2	2	2	2	2	2
6. Urea-melamine formaldehyde	2	6	2	2	2	2	1	1	2	2	2	2	2	2	2
7. Resorcinol formaldehyde	2	6	2	2	2	2	2	2	2	2	2	2	2	2	2
8. Phenol-resorcinol formaldehyde	2	6	2	2	2	2	2	2	2	2	2	2	2	2	2
9. Epoxy (+ polyamine)	2	5	3	5	2	2	2	2	2	3	1	6	6	1	
10. Epoxy (+ polyanhydride)	2	5	1	4	3	3	2	2	–	2	2	6	6	2	
11. Epoxy (+ polyamide)	2	2	6	2	2	6	3	6	2	2	1	6	6	3	
12. Polyimide	2	4	1	1	2	4	2	2	2	2	2	2	2	2	2
13. Polybenzimidazole	2	4	1	1	2	4	2	2	2	2	2	2	2	2	2
14. Acrylic	2	6	5	3	1	3	2	2	2	2	2	2	2	2	2
15. Acrylate acid diester	2	5	3	3	4	4	6	6	3	3	5	5	5	4	4
Thermoplastic adhesives															
16. Cellulose acetate	2	6	2	3	1	6	1	2	–	2	4	6	6	6	6
17. Cellulose acetate butyrate	2	3	3	3	2	–	3	2	–	–	6	6	6	6	6
18. Cellulose nitrate	2	6	3	3	3	3	3	6	2	2	6	6	6	6	6
19. Polyvinyl acetate	2	6	6	–	3	6	3	3	2	2	6	6	6	6	6
20. Vinyl vinylidene	2	3	3	3	3	3	–	–	2	2	2	2	2	–	–
21. Polyvinyl acetal	2	6	5	2	2	–	6	3	2	2	3	3	6	3	2
22. Polyvinyl alcohol	–	2	3	–	6	6	5	5	2	1	3	1	1	1	1

Table 7.36 Relative Resistance of Synthetic Adhesives to Common Service Environment *(Continued)* (from Ref. 38)

Adhesive type	Shear	Peel	Heat	Cold	Water	Hot water	Acid	Alkali	Oil, grease	Fuels	Alcohols	Ketones	Esters	Aromatics	Chlorinated solvents
23. Polyamide	2	3	5	–	5	6	6	2	2	2	6	2	2	2	6
24. Acrylic	2	2	4	3	3	3	–	–	2	–	–	4	4	–	4
25. Phenoxy	2	3	4	3	3	4	3	2	3	5	5	–	–	6	–
Elastomer adhesives															
26. Natural rubber	2	3	3	–	3	–	3	3	6	6	2	4	4	6	6
27. Reclaimed rubber	2	3	3	–	2	–	3	3	6	6	2	4	4	6	6
28. Butyl	3	6	6	3	2	6	1	2	6	6	2	2	2	6	6
29. Polyisobutylene	6	6	6	3	2	6	2	2	6	6	2	2	2	6	6
30. Nitrile	2	3	3	3	2	5	5	6	2	2	3	6	6	3	6
31. Styrene butadiene	3	6	3	3	1	–	3	2	–	5	2	6	6	6	6
32. Polyurethane	2	3	3	2	2	3	3	3	2	2	2	5	5	–	5
33. Polysulfide	3	2	6	2	1	6	2	2	2	2	2	6	6	2	6
34. Silicone (RTV)	3	5	1	1	2	2	3	3	2	3	3	3	3	3	3
35. Silicone resin	2	2	1	2	2	2	–	2	2	2	2	4	4	3	6
36. Neoprene	2	3	3	3	2	–	2	2	2	2	3	6	6	6	6
Alloy adhesives															
37. Epoxy-phenolic	1	6	1	3	2	2	2	2	3	3	2	6	6	2	
38. Epoxy-polysulfide	2	2	6	2	1	6	2	2	2	2	2	6	6	2	6
39. Epoxy-nylon	1	1	6	2	2	6	–	–	–	2	3	6	6	6	6
40. Phenolic-nitrile	2	2	2	3	2	2	2	2	2	2	2	6	6	6	6
41. Phenolic-neoprene	2	3	3	2	2	–	3	2	2	2	3	6	6	6	6
42. Phenolic-polyvinyl butyral	2	3	3	3	2	3	4	2	2	2	4	6	6	6	6
43. Phenolic-polyvinyl formal	2	3	6	6	2	6	6	4	2	2	2	4	6	6	6

Key: 1. Excellent; 2. Good; 3. Fair; 4. Poor; 5. Very poor; 6. Extremely poor

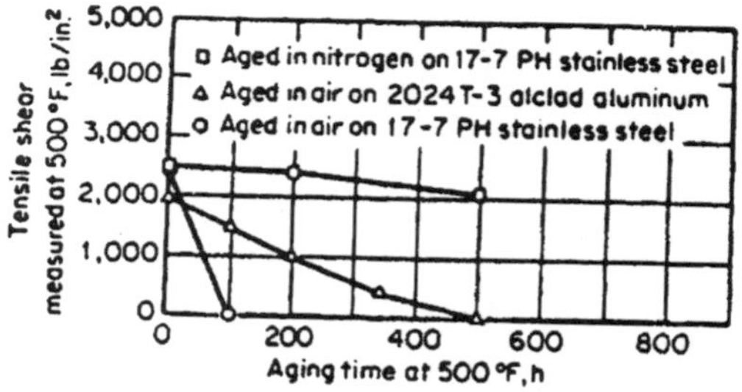

Figure 7.30 The effect of 500°F aging in air and nitrogen on an epoxy-phenolic adhesive (HT-424).[39]

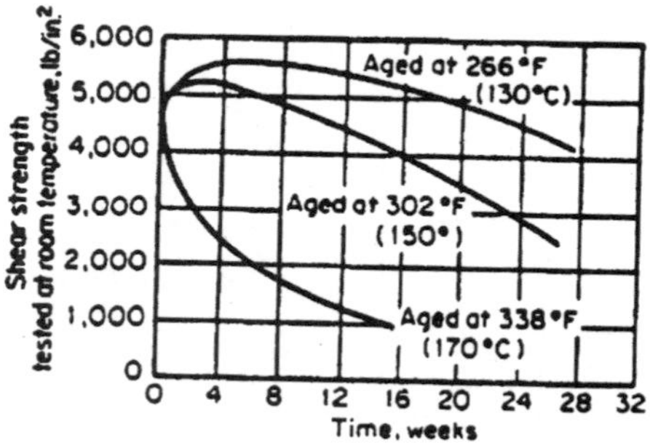

Figure 7.31 Effect of temperature aging on typical epoxy adhesive in air. Strength measured at room temperature.[40]

Anhydride curing agents give unmodified epoxy adhesives greater thermal stability than most other epoxy curing agents. Phthalic anhydride, pyromellitic dianhydride, and chlorendic anhydride allow greater cross-linking and result in short-term heat resistance to 450°F. Long-term thermal endurance, however, is limited to 300°F. Typical epoxy formulations cured with pyromellitic dianhydride offer 1,200 to 2,600 lb/in^2 shear strength at 300°F and 1,000 lb/in^2 at 450°F.

7.6.1.2 Modified phenolics. Of the common modified phenolic adhesives, the nitrile-phenolic blend has the best resistance to shear at elevated temperatures. Nitrile phenolic adhesives have high shear strength up to 250 to 350°F, and the strength retention on aging at these temperatures is very good. The nitrile phenolic adhesives are also extremely tough and provide high peel strength.

7.6.1.3 Silicone. Silicone adhesives have very good thermal stability but low strength. Their chief application is in nonstructural applications such as high-temperature pressure-sensitive tape.

Attempts have been made to incorporate silicones with other resins such as epoxies and phenolics, but long cure times and low strength have limited their use.

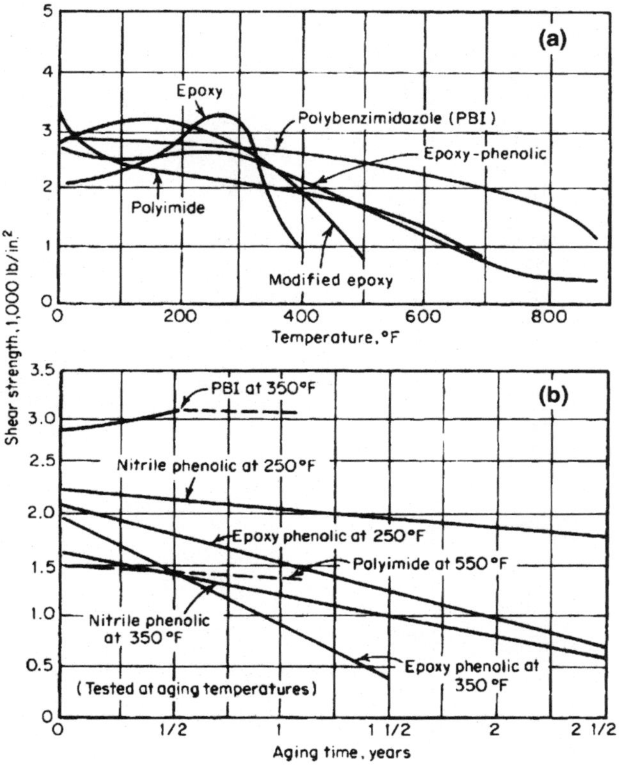

Figure 7.32 Comparison of (a) heat resistance and (b) thermal aging of high-temperature structural adhesives.[41]

7.6.1.4 Polyaromatics. The most common polyaromatic resins, polyimide and polybenzimidazole, offer greater thermal resistance then any other commercially available adhesive. The rigidity of their molecular chains decreases the possibility of chain scission caused by high temperatures. The aromaticity of these structures provides high bond-dissociation energy and acts as an "energy sink" to the thermal environment.

Polyimide. The strength retention of polyimide adhesives for short exposures to 1000°F is slightly better than that of an epoxy-phenolic alloy. However, the thermal endurance of polyimides at temperatures greater than 500°F is unmatched by other commercially available adhesives.

Polyimide adhesives are usually supplied as a glass-fabric-reinforced film having a limited shelf life. A cure of 90 min at 500 to 600°F and 15 to 200 lb/in^2 pressure is usually necessary for optimal properties. High-boiling volatiles can be released during cure, which causes a somewhat porous adhesive layer. Because of the inherent rigidity of this material, peel strength is low.

Polybenzimidazole. As illustrated in Fig. 7.32, polybenzimidazole (PBI) adhesives offer the best short-term performance at elevated temperatures. However, PBI resins oxidize rapidly and are not recommended for continuous use at temperatures over 450°F.

PBI adhesives require a cure at 600°F. Release of volatiles during cure contributes to a porous adhesive bond. Supplied as a very stiff, glass-fabric-reinforced film, this adhesive is expensive, and applications are limited by a long, high-temperature curing cycle.

7.6.2 Low Temperature

The factors that determine the strength of an adhesive at very low temperatures are (1) the difference in coefficient of thermal expansion between adhesive and adherend, (2) the elastic modulus, and (3) the thermal conductivity of the adhesive. The difference in thermal expansion is very important, especially since the elastic modulus of the adhesive generally decreases with falling temperature. It is necessary that the adhesive retain some resiliency if the thermal expansion coefficients of adhesive and adherend cannot be closely matched. The adhesive's coefficient of thermal conductivity is important in minimizing transient stresses during cooling. This is why thinner bonds have better cryogenic properties than thicker ones.

Low-temperature properties of common structural adhesives used for cryogenic applications are illustrated in Fig. 7.33.

Epoxy-polyamide adhesives can be made serviceable at very low temperatures by the addition of appropriate fillers to control thermal expansion. But the epoxy-based systems are not as attractive as some others because of brittleness and corresponding low peel and impact strength at cryogenic temperatures.

Epoxy-phenolic adhesives are exceptional in that they have good adhesive properties at both elevated and low temperatures. Vinyl-phenolic adhesives maintain fair shear and peel strength at –423°F, but strength decreases with decreasing temperature. Nitrile-phenolic adhesives do not have high strength at low service temperatures, because of rigidity.

Polyurethane and epoxy-nylon systems offer outstanding cryogenic properties. Polyurethane adhesives are easily processable and bond well to many substrates. Peel strength ranges from 22 lb/in at 75° to 26 lb/in at –423°F, and the increase in shear strength at

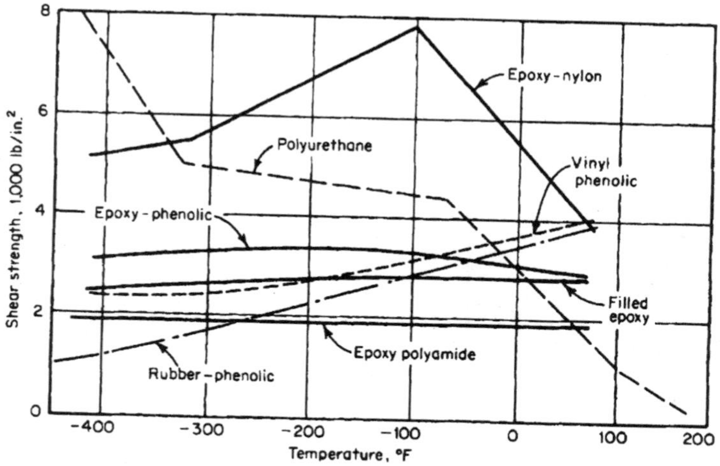

Figure 7.33 Properties of cryogenic structural adhesive systems.[41]

–423°F is even more dramatic. Epoxy-nylon adhesives also retain flexibility and yield 5,000 lb/in^2 shear strength in the cryogenic-temperature range.

Heat-resistant polyaromatic adhesives also have shown promising low-temperature properties. The shear strength of a polybenzimidazole adhesive on stainless-steel substrates is 5,690 lb/in^2 at a test temperature of –423°F, and polyimide adhesives have exhibited shear strength of 4,100 lb/in^2 at –320°F. These unique properties show the applicability of polyaromatic adhesives on structures seeing both very high and low temperatures.

7.6.3 Humidity and Water Immersion

Moisture can affect adhesive strength in two significant ways. Some polymeric materials, notably ester-based polyurethanes, will *revert*—i.e., lose hardness, strength, and (in the worst cases) turn fluid during exposure to warm, humid air. Water can also permeate the adhesive and preferentially displace the adhesive at the bond interface. This later mechanism is the most common cause of adhesive-strength reduction in moist environments.

The rate of reversion or hydrolytic instability depends on the chemical structure of the base adhesive, the type and amount of catalyst used, and the flexibility of the adhesive. Certain chemical linkages, such as ester, urethane, amide, and urea, can be hydrolyzed. The rate of attack is fastest for ester-based linkages. Ester linkages are present in certain types of polyurethanes and anhydride-cured epoxies. Generally, amine-cured epoxies offer better hydrolytic stability than anhydride-cured types. Figure 7.34 illustrates the hydrolytic stability of various polymeric materials determined by a hardness measurement. Reversion is usually much faster in flexible materials because water permeates more easily.

Structural adhesives not susceptible to the reversion phenomenon are also likely to lose adhesive strength when exposed to moisture. The degradation curves shown in Fig. 7.35 are typical for an adhesive exposed to moist, high-temperature environments. The mode of failure in the initial stages of aging is usually truly cohesive. After aging, the failure becomes one of adhesion. It is expected that water vapor permeates the adhesive through its exposed edges and concentrates in weak boundary layers at the interface. This effect is greatly dependent on the type of adhesive and substrate.

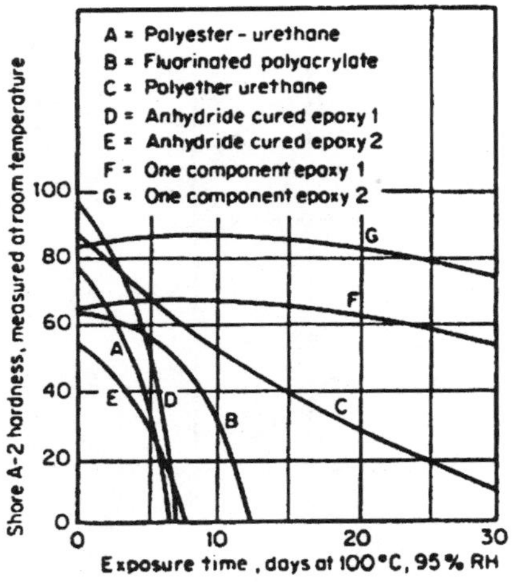

Figure 7.34 Hydrolytic stability of potting compounds. Materials showing rapid hardness loss will soften similarly after two to four years at ambient temperatures in a high-humidity, tropical climate.[42]

Stress accelerates the effect of environments on the adhesive joint. Little data are available on this phenomenon because of the time and expense associated with stress-aging tests. However, it is known that moisture, as an environmental burden, markedly decreases the ability of an adhesive to bear prolonged stress. Figure 7.36 illustrates the effect of stress aging on specimens exposed to relative humidity cycling from 90 to 100 percent and simultaneous temperature cycling from 80 to 120°F. The loss of load-bearing ability of a certain flexibilized epoxy adhesive (Fig. 7.36) is exceptional. The stress on this particular adhesive had to be reduced to 13 percent of its original strength for the joint to last a little more than 44 days in the test environment.

7.6.4 Outdoor Weathering

The most detrimental factors influencing adhesives aged outdoors are heat and humidity. Thermal cycling, ultraviolet radiation, and cold are relatively minor factors. The reasons why warm, moist climates degrade adhesive joints were presented in the previous section.

When exposed to weather, structural adhesives rapidly lose strength during the first six months to one year. After two to three years, the rate of decline usually levels off, depending on the climate zone, adherend, adhesive, and stress level. Figure 7.37 shows the weathering characteristics of unstressed epoxy adhesives to the Richmond, VA, climate.

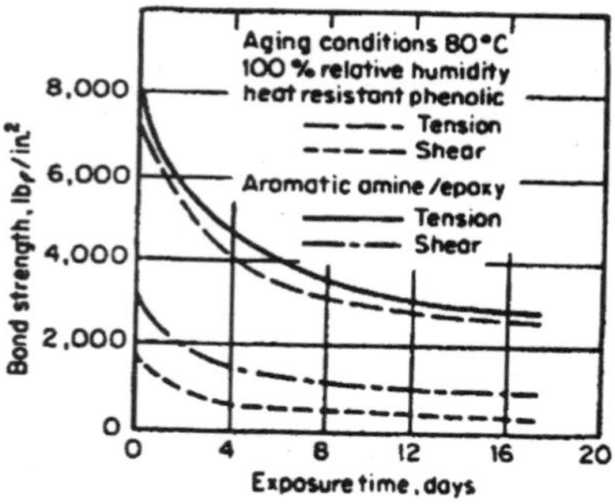

Figure 7.35 Effect of humidity on adhesion of two structural adhesives to stainless steel.[43]

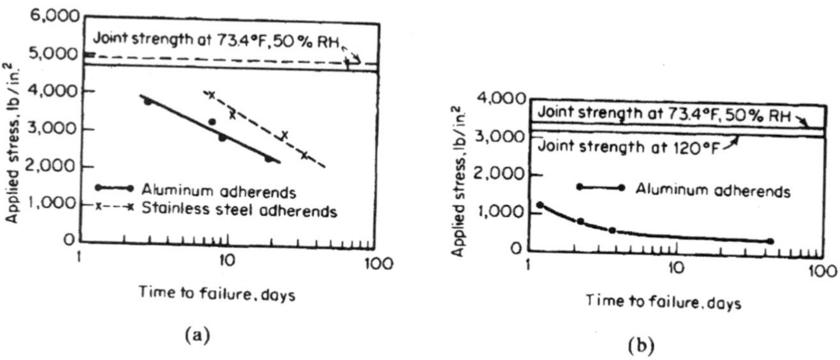

Figure 7.36 Time to failure vs. stress for two adhesives in a warm, high-humidity environment: (a) adhesive = one-part, heat-curing, modified epoxy; (b) adhesive = flexibilized, amine-cured epoxy.[44]

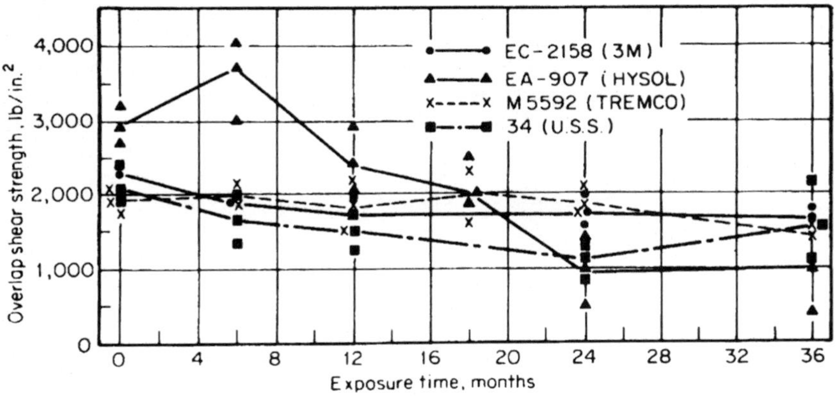

Figure 7.37 Effect of outdoor weathering on typical aluminum joints made with four different two-part epoxies cured at room temperature.[25]

The following generalizations are of importance in designing an adhesive joint for outdoor service:

1. The most severe locations are those with high humidity and warm temperatures.
2. Stressed panels deteriorate more rapidly than unstressed panels.
3. Stainless-steel panels are more resistant than aluminum panels because of corrosion.
4. Heat-cured adhesive systems are generally more resistant than room-temperature-cured systems.
5. With the better adhesives, unstressed bonds are relatively resistant to severe outdoor weathering, although all joints will eventually exhibit some strength loss.

MIL-STD-304 is a commonly used accelerated-exposure technique to determine the effect of weathering and high humidity on adhesive specimens. Adhesive comparisons can be made with this type of test. In this procedure, bonded panels are exposed to alternating cold (–65°F), dry heat (160°F), and heat and humidity (160°F, 95 percent RH) for 30 days.

The effect of MIL-STD-304 conditioning on the joint strength of common structural adhesives is presented in Table 7.37.

TABLE 7.37 Effect of MIL–STD–304 Aging on Bonded Aluminum Joints (from Ref. 45)

Adhesive	Shear, lb/in², 73°F		Shear, lb/in², 160°F	
	Control	Aged	Control	Aged
Room-temperature cured:				
Epoxy-polyamide	1800	2100	2700	1800
Epoxy-polysulfide	1900	1640	1700	6070
Epoxy-aromatic amine	2000	Failed	720	Failed
Epoxy-nylon	2600	1730	220	80
Resorcinol epoxy-polyamide	3500	3120‘	3300	2720
Epoxy-anhydride	3000	920	3300	1330
Polyurethane	2600	1970	1600	1560
Cured 45 min at 330°F, epoxy-phenolic	2900	2350	2900	2190
Cured 1 hr at 350°F:				
Modified epoxy	4900	3400	4100	3200
Nylon-phenolic	4600	3900	3070	2900

7.6.5 Chemicals and Solvents

Most organic adhesives tend to be susceptible to chemicals and solvents, especially at elevated temperatures. Standard test fluids and immersion conditions are used by adhesive suppliers and are defined in MMM-A-132. Unfortunately exposure tests lasting less than 30 days are not applicable to many requirements. Practically all adhesives are resistant to these fluids over short time periods and at room temperatures. Some epoxy adhesives even show an increase in strength during aging in fuel or oil. This effect is possibly due to a post-curing or plasticizing of the epoxy by oil.

Epoxy adhesives are generally more resistant to a wide variety of liquid environments than other structural adhesives. However, the resistance to a specific environment is greatly dependent on the type of epoxy curing agent used. Aromatic amine, such as metaphenylene diamine, cured systems are frequently preferred for long-term chemical resistance.

There is no "best adhesive" for universal chemical environments. As an example, maximum resistance to bases almost axiomatically means poor resistance to acids. It is relatively easy to find an adhesive that is resistant to one particular chemical environment. It becomes more difficult to find an adhesive that will not degrade in two widely differing chemical environments. Generally, adhesives that are most resistant to high temperatures have the best resistance to chemicals and solvents.

The temperature of the immersion medium is a significant factor in the aging properties of the adhesive. As the temperature increases, more fluid is generally absorbed by the adhesive, and the degradation rate increases.

From the rather limited information reported in the literature, it may be summarized that

1. Chemical-resistance tests are not uniform in concentrations, temperature, time, properties measured.

2. Generally, chlorinated solvents and ketones are severe environments.
3. High-boiling solvents, such as dimethylformamide and dimethyl sulfoxide, are severe environments.
4. Acetic acid is a severe environment.
5. Amine curing agents for epoxies are poor in oxidizing acids.
6. Anhydride curing agents are poor in caustics.

7.6.6 Vacuum

The ability of an adhesive to withstand long periods of exposure to a vacuum is of primary importance for certain applications. Loss of low-molecular-weight constituents such as plasticizers or diluents could result in hardening and porosity of cured adhesives or sealants.

Since most structural adhesives are relatively high-molecular- weight polymers, exposure to pressures as low as 10^{-9} torr is not harmful. However, high temperatures, radiation, or other degrading environments may cause the formation of low-molecular-weight fragments that tend to bleed out of the adhesive in a vacuum.

Epoxy and polyurethane adhesives are not appreciably affected by 10^{-9} torr for seven days at room temperature. However, polyurethane adhesives exhibit significant outgassing when aged under 10^{-9} torr at 225°F.

7.6.7 Radiation

High-energy particulate and electromagnetic radiation, including neutron, electron, and gamma radiation, have similar effects on organic adhesives. Radiation causes molecular-chain scission of the adhesive, which results in weakening and embrittlement of the bond. This degradation is worsened when the adhesive is simultaneously exposed to elevated temperatures and radiation.

Figure 7.38 illustrates the effect of radiation dosage on the tensile-shear strength of structural adhesives. Generally, heat-resistant adhesives have been found to resist radiation better than less thermally stable systems. Fibrous reinforcement, fillers, curing agents, and reactive diluents affect the radiation resistance of adhesive systems. In epoxy-based adhesives, aromatic curing agents offer greater radiation resistance than aliphatic-type curing agents.

7.7 Processing and Quality Control of Adhesive Joints

Processing and quality control are usually the final considerations in the design of an adhesive-bonding system. These decisions are very important, however, because they alone may (1) restrict the degrees of freedom in designing the end product, (2) determine the types and number of adhesives that can be considered, (3) affect the quality and reproducibility of the joint, and (4) affect the total assembly cost.

7.7.1 Measuring and Mixing

When a multiple-part adhesive is used, the concentration ratios have a significant effect on the quality of the joint. Strength differences caused by varying curing-agent concentration are most noticeable when the joints are tested at elevated temperatures or after exposure to

water or solvents. Exact proportions of resin and hardener must be weighed out on an accurate balance or in a measuring container for best adhesive quality and reproducibility.

The weighed-out components must be mixed thoroughly. Mixing should be continued until no color streaks or density stratifications are noticeable. Caution should be taken to prevent air from being mixed into the adhesive through overagitation. This can cause foaming of the adhesive during heat cure, resulting in porous bonds. If air does become mixed into the adhesive, vacuum degassing may be necessary before application.

Only enough adhesive should be mixed that can be used before the adhesive begins to cure. Working life of an adhesive is defined as the period of time during which an adhesive remains suitable for use after mixing with catalyst. Working life is decreased as the ambient temperature increases and as the batch size becomes larger. One-part, and some heat-curing, two-part, adhesives have very long working lives at room temperature, and application and assembly speed or batch size are not critical.

For a large-scale bonding operation, hand mixing is costly, messy, and slow, and repeatability is entirely dependent on the operator. Equipment is available that can meter, mix, and dispense multicomponent adhesives on a continuous or shot basis.

7.7.2 Application of Adhesives

The selection of an application method depends primarily on the form of the adhesive: liquid, paste, powder, or film. Table 7.38 describes the advantages and limitations realized in using each of the four basic forms. Other factors influencing the application method are the size and shape of parts to be bonded, the total area where the adhesive is to be applied, and production volume and rate.

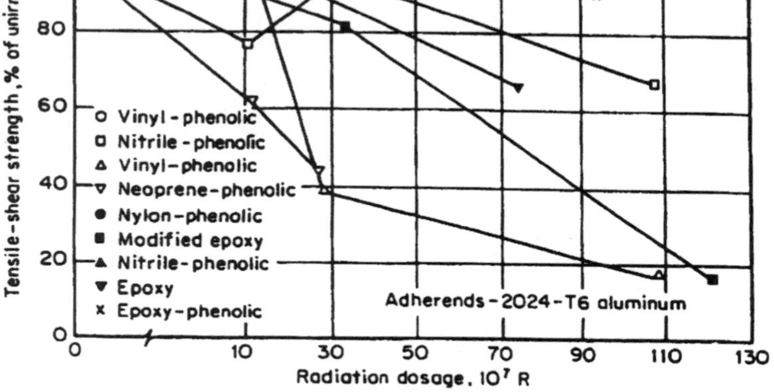

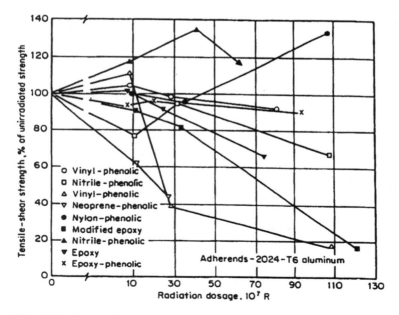

Figure 7.38 Percent change of initial tensile-shear strength caused by nuclear radiation dosage.[46]

Table 7.38 Characteristics of Various Adhesive-Application Methods (from Ref. 47)

Application method	Viscosity	Operator skill	Production rate	Equipment cost	Coating uniformity	Material loss
Liquid						
Manual, brush or roller	Low to medium	Little	Low	Low	Poor	Low
Roll coating, reverse, gravure	Low	Moderate	High	High	Good	Low
Spray, manual, automatic, airless, or external mix	Low to high	Moderate to high	Moderate to high	Moderate to high	Good	Low to high
Curtain coating	Low	Moderate	High	High	Good to excellent	Low
Bulk						
Paste and mastic	High	Little	Low to moderate	Low	Fair	Low to high
Powder						
Dry or liquid primed	—	Moderate	Low	High	Poor to fair	Low
Dry Film	—	Moderate to high	Low to high	Low to high	Excellent	Lowest

7.7.2.1 Liquids. Liquids, the most common form of adhesive, can be applied by a variety of methods. Brushes, simple rollers, and glue guns are manual methods that provide simplicity, low cost, and versatility. Spray, dipping, and mechanical roll coaters are generally used on large production runs. Mechanical roller methods are commonly used to apply a uniform layer of adhesive to a continuous roll or coil. Such automated systems are used with adhesives that have a long working life and low viscosity. Spray methods can be used on both small and large production runs. The spray adhesive is generally in solvent solution, and sizable amounts of adhesive may be lost from overspray. Two-component adhesives are usually mixed prior to placing in the spray-gun reservoir. Application systems are available, however, that meter and mix the adhesive in the spray-gun barrel. This is ideal for fast-reacting systems.

7.7.2.2 Pastes. Bulk adhesives such as pastes and mastics are the simplest and most reproducible adhesives to apply. These systems can be troweled or extruded through a caulking gun. Little operator skill is required. Since the thixotropic nature of the paste prevents it from flowing excessively, application is usually clean, and little waste is generated.

7.7.2.3 Powders. Powder adhesives can be applied in three ways.

1. They may be sifted onto a preheated substrate. The powder that falls onto the substrate melts and adheres. The assembly is then mated and cured according to recommended processes.
2. A preheated substrate could also be dipped into a fluidized bed of the powder and then extracted with an attached coating of adhesive. This method helps to ensure an even distribution of powder.
3. The powder can be melted into a paste or liquid and applied by conventional means.

Powder adhesives are generally one-part, epoxy-based systems that require heat and pressure to cure. They do not require metering and mixing but often must be refrigerated for extended shelf life.

7.7.2.4 Films. Dry adhesive films have the following advantages:

1. High repeatability—no mixing or metering, constant thickness.
2. Easy to handle—low equipment cost, relatively hazard-free, clean operating.
3. Very little waste—preforms can be cut to size.
4. Excellent physical properties—wide variety of adhesive types available.

Films are limited to flat surfaces or simple contours. Application requires a relatively high degree of care to ensure nonwrinkling and removal of separator sheets. Characteristics of available film adhesives vary widely, depending on the type of adhesive used.

Film adhesives are made as both unsupported and supported types. The carrier for supported films is generally fibrous fabric or mat. Film adhesives are supplied in heat-activated, pressure-sensitive, or solvent-activated forms. Solvent-activated adhesives are made tacky and pressure-sensitive by wiping with solvent. They are not as strong as the other types but are well suited for contoured, curved, or irregularly shaped parts. Manual sol-

vent-reactivation methods should be closely monitored so that excessive solvent is not used. Solvent-activated films include neoprene, nitrile, and butyral phenolics. Decorative trim and nameplates are usually fastened onto a product with solvent-activated adhesives.

7.7.3 Bonding Equipment

After the adhesive is applied, the assembly must be mated as quickly as possible to prevent contamination of the adhesive surface. The substrates are held together under pressure and heated, if necessary, until cure is achieved. The equipment required to perform these functions must provide adequate heat and pressure, maintain the prescribed pressure during the entire cure cycle, and distribute pressure uniformly over the bond area. Of course, many adhesives require only simple contact pressure at room temperature, and extensive bonding equipment is not necessary.

7.7.3.1 Pressure equipment. Pressure devices should be designed to maintain constant pressure on the bond during the entire cure cycle. They must compensate for thickness reduction from adhesive flow or thermal expansion of assembly parts. Thus, screw-actuated devices like C-clamps and bolted fixtures are not acceptable when a constant pressure is important. Spring pressure can often be used to supplement clamps and compensate for thickness variations. Dead-weight loading may be applied in many instances; however, this method is sometimes impractical, especially when heat cure is necessary.

Pneumatic and hydraulic presses are excellent tools for applying constant pressure. Steam or electrically heated platen presses with hydraulic rams are often used for adhesive bonding. Some units have multiple platens, thereby permitting the bonding of several assemblies at one time.

Large bonded areas such as on aircraft parts are usually cured in an autoclave. The parts are mated first and covered with a rubber blanket to provide uniform pressure distribution. The assembly is then placed in an autoclave, which can be pressurized and heated. This method requires heavy capital-equipment investment.

Vacuum-bagging techniques can be an inexpensive method of applying pressure to large parts. A film or plastic bag is used to enclose the assembly, and the edges of the film are sealed airtight. A vacuum is drawn on the bag, enabling atmospheric pressure to force the adherends together. Vacuum bags are especially effective on large areas, because size is not limited by equipment.

7.7.3.2 Heating equipment. Many structural adhesives require heat as well as pressure. Most often, the strongest bonds are achieved by an elevated-temperature cure. With many adhesives, trade-offs between cure times and temperature are permissible. But, generally, the manufacturer will recommend a certain curing schedule for optimum properties.

If, for example, a cure of 60 min at 300°F is recommended, this does not mean that the assembly should be placed in a 300°F oven for 60 min. It is the bond line that should be at 300°F for 60 min. Total oven time would be 60 min plus whatever time is required to bring the assembly up to 300°F. Large parts act as a heat sink and may require substantial time for an adhesive in the bond line to reach the necessary temperature. Bond-line temperatures are best measured by thermocouples placed very close to the adhesive. In some cases, it may be desirable to place the thermocouple in the adhesive joint for the first few assemblies being cured.

Oven heating is the most common source of heat for bonded parts, even though it involves long curing cycles because of the heat-sink action of large assemblies. Ovens may

be heated with gas, oil, electricity, or infrared units. Good air circulation within the oven is mandatory to prevent nonuniform heating.

Heated-platen presses are good for bonding flat or moderately contoured panels when faster cure cycles are desired. Platens are heated with steam, hot oil, or electricity and are easily adapted with cooling-water connections to further speed the bonding cycle.

7.7.3.3 Adhesive-thickness control. It is highly desirable to have a uniformly thin (2- to 10-mil) adhesive bond line. Starved adhesive joints, however, will yield exceptionally poor properties. Three basic methods are used to control adhesive thickness. The first method is to use mechanical shims or stops which can be removed after the curing operation. Sometimes it is possible to design stops into the joint.

The second method is to employ a film adhesive that becomes highly viscous during the cure cycle preventing excessive adhesive flow-out. With supported films, the adhesive carrier itself can act as the *shims*. Generally, the cured bond-line thickness will be determined by the original thickness of the adhesive film. The third method of controlling adhesive thickness is to use trial and error to determine the correct pressure-adhesive viscosity factors that will yield the desired bond thickness.

7.7.4 Quality Control

A flow chart of a quality-control system for a major aircraft company is illustrated in Fig. 7.39. This system is designed to ensure reproducible bonds and, if a substandard bond is detected, to make suitable corrections. Quality control should cover all phases of the bonding cycle from inspection of incoming material to the inspection of the completed assembly. In fact, good quality control will start even before receipt of materials.

7.7.4.1 Prebonding conditions. The human element enters the adhesive-bonding process more than in other fabrication techniques. An extremely high percentage of defects can be traced to poor workmanship. This generally prevails in the surface-preparation steps but may also arise in any of the other bonding steps. This problem can be largely overcome by proper motivation and education. All employees, from design engineer to laborer to quality-control inspector, should be somewhat familiar with adhesive-bonding technology and be aware of the circumstances that can lead to poor joints.

The plant's bonding area should be as clean as possible prior to receipt of materials. The basic approach to keeping the assembly area clean is to segregate it from the other manufacturing operations either in a corner of the plant or in isolated rooms. The air should be dry and filtered to prevent moisture or other contaminants from gathering at a possible interface. The cleaning and bonding operations should be separated from each other. If mold release is used to prevent adhesive flash from sticking to bonding equipment, it is advisable that great care be taken to ensure that the release does not contaminate the adherends. Spray mold releases, especially silicone release agents, have a tendency to migrate to undesirable areas.

7.7.4.2 Quality control of adhesive and surface treatment. Acceptance tests on adhesives should be directed toward assurance that incoming materials are identical from lot to lot. The tests should be those that can quickly and accurately detect deficiencies in the adhesive's physical or chemical properties. ASTM lists various test methods that are commonly used for adhesive acceptance. Actual test specimens should also be

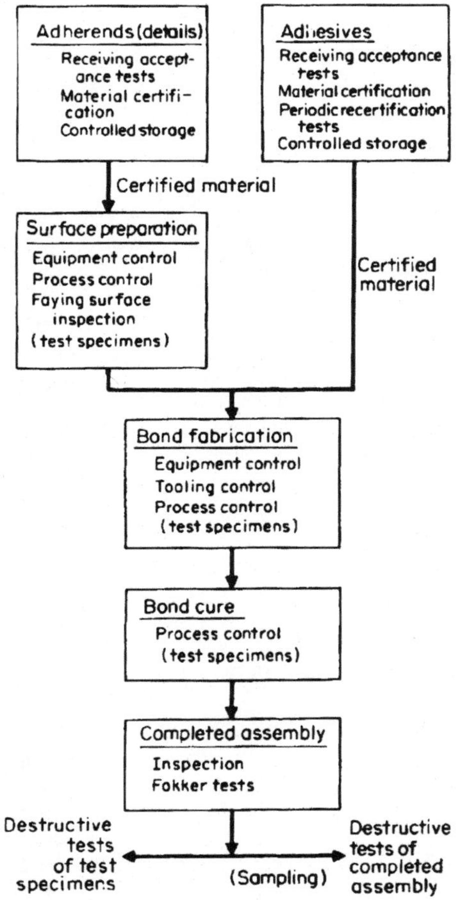

Figure 7.39 Flowchart of a quality-control system for adhesive bonding.[48]

made to verify strength of the adhesive. These specimens should be stressed in directions that are representative of the forces that the bond will see in service, i.e., shear, peel, tension, or cleavage. If possible, the specimens should be prepared and cured in the same manner as actual production assemblies. If time permits, specimens should also be tested in simulated service environments, e.g., high temperature or humidity.

Surface preparations must be carefully controlled for reliable production of adhesive-bonded parts. If a chemical surface treatment is required, the process must be monitored for proper sequence, bath temperature, solution concentration, and contaminants. If sand or grit blasting is employed, the abrasive must be changed regularly. An adequate supply of clean wiping cloths for solvent cleaning is also mandatory. Checks should be made to determine if cloths or solvent containers have become contaminated.

The specific surface preparation can be checked for effectiveness by the water-break free test. After the final treating step, the substrate surface is checked for a continuous film of water that should form when deionized water drops are placed on the surface.

After the adequacy of the surface treatment has been determined, precautions must be taken to ensure that the substrates are kept clean and dry until bonding. The adhesive or primer should be applied to the treated surface as quickly as possible.

7.7.4.3 Quality control of the bonding process. The adhesive metering and mixing operation should be monitored by periodically sampling the mixed adhesive and testing it for adhesive properties. A visual inspection can also be made for air entrapment and degree of mixing. The quality-control engineer should be sure that the oldest adhesive is used first and that the specified shelf life has not been exceeded.

During the actual assembly operation, the cleanliness of the shop and tools should be verified. The shop atmosphere should be controlled as closely as possible. Temperature in the range of 65 to 90°F and relative humidity from 20 to 65 percent is best for almost all bonding operations.

The amount of the applied adhesive and the final bond-line thickness must also be monitored, because they can have a significant effect on joint strength. Curing conditions should be monitored for heat-up rate, maximum and minimum temperature during cure, time at the required temperature, and cool-down rate.

7.7.4.4 Bond inspection. After the adhesive is cured, the joint can be inspected to detect gross flaws or defects. This inspection procedure can be either destructive or nondestructive in nature. Destructive testing generally involves placing samples in simulated or accelerated service and determining if they have similar properties to a specimen that is known to have a good bond and adequate service performance. The causes and remedies for faults revealed by such mechanical tests are described in Table 7.39. Nondestructive testing (NDT) is far more economical, and every assembly can be tested if desired. Great amounts of energy are now being devoted to improve NDT techniques.

7.7.4.5 Nondestructive testing procedures

Visual inspection. A trained eye can detect a surprising number of faulty joints by close inspection of the adhesive around the bonded area. Table 7.40 lists the characteristics of faulty joints that can be detected visually. The most difficult defect to be found by any way are those related to improper curing and surface treatment. Therefore, great care and control must be given to surface-preparation procedures and shop cleanliness.

Sonic Inspection. Sonic and ultrasonic methods are, at present, the most popular NDT techniques for use on adhesive joints. Simple tapping of a bonded joint with a coin or light hammer can indicate an unbonded area. Sharp, clear tones indicate that adhesive is present and adhering to the substrate in some degree; dull, hollow tones indicate a void or unattached area. Ultrasonic testing basically measures the response of the bonded joint to loading by low-power ultrasonic energy.

Other NDT methods. Radiography (x-ray) inspection can be used to detect voids or discontinuities in the adhesive bond. This method is more expensive and requires more skilled experience than ultrasonic methods.

Thermal-transmission methods are relatively new techniques for adhesive inspection. Liquid crystals applied to the joint can make voids visible if the substrate is heated. This test is simple and inexpensive, although materials with poor heat-transfer properties are

TABLE 7.39 Faults Revealed by Mechanical Tests

Fault	Cause	Remedy
Thick, uneven glue line	Clamping pressure too low	Increase pressure. Check that clamps are seating properly
	No follow-up pressure	Modify clamps or check for freedom of moving parts
	Curing temperature too low	Use higher curing temperature. Check that temperature is above the minimum specified throughout the curing cycle
	Adhesive exceeded its shelf life, resulting in increased viscosity	Use fresh adhesive
Adhesive residue has spongy appearance or contains bubbles	Excess air stirred into adhesive	Vacuum-degas adhesive before application
	Solvents not completely dried out before bonding	Increase drying time or temperature. Make sure drying area is properly ventilated
	Adhesive material contains volatile constituent	Seek advice from manufacturers
	A low-boiling constituent boiled away	Curing temperature is too high
Voids in bond (i.e., areas that are not bonded), clean bare metal exposed, adhesive failure at interface	Joint surfaces not properly treated	Check treating procedure; use clean solvents and wiping rags. Wiping rags must not be made from synthetic fiber. Make sure cleaned parts are not touched before bonding. Cover stored parts to prevent dust from settling on them
	Resin may be contaminated	Replace resin. Check solids content. Clean resin tank
	Uneven clamping pressure	Check clamps for distortion
	Substrates distorted	Check for distortion; correct or discard distorted components. If distorted components must be used, try adhesive with better gap-filling ability
Adhesive can be softened by heating or wiping with solvent	Adhesive not properly cured	Use higher curing temperature or extend curing time. Temperature and time must be above the minimum specified throughout the curing cycle. Check mixing ratios and thoroughness of mixing. Large parts act as a heat sink, necessitating larger cure times

difficult to test, and the joint must be accessible from both sides. An infrared inspection technique has also been developed for detection of internal voids and nonbonds. This technique is somewhat expensive, but it can accurately determine the size and depth of the flaw.

The science of holography has also been used for NDT of adhesive bonds. Holography is a method of producing photographic images of flaws and voids using coherent light

TABLE 7.40 Visual Inspection for Faulty Bonds

Fault	Cause	Remedy
No appearance of adhesive around edges of joint or adhesive bond line too thick	Clamping pressure too low	Increase pressure. Check that clamps are seating properly
	Starved joint	Apply more adhesive
	Curing temperature too low	Use higher curing temperature. Check that temperature is above the minimum specified
Adhesive bond line too thin	Clamping pressure too high	Lessen pressure
	Curing temperature too high	Use lower curing temperature
	Starved joint	Apply more adhesive
Adhesive flash breaks easily away from substrate	Improper surface treatment	Check treating procedure; use clean solvents and wiping rags. Make sure cleaned parts are not touched before bonding
Adhesive flash is excessively porous	Excess air stirred into adhesive	Vacuum-degas adhesive before application
	Solvent not completely dried out before bonding	Increase drying time or temperature
	Adhesive material contains volatile constituent	Seek advice from manufacturers
Adhesive flash can be softened by heating or wiping with solvent	Adhesive not properly cured	Use higher curing temperature or extend curing time. Temperature and time must be above minimum specified. Check mixing

such as that produced by a laser. The major advantage of holography is that it photographs successive "slices" through the scene volume. A true three-dimensional image of a defect or void can then be reconstructed.

7.7.5 Environmental and Safety Concerns

Four primary safety factors must be considered in all adhesive bonding operations: toxicity, flammability, hazardous incompatibility, and equipment.

All adhesives, solvents, chemical treatments, etc. must be handled in a manner that prevents toxic exposure to the work force. Methods and facilities must be provided to ensure that the maximum acceptable concentrations of hazardous materials are never exceeded. These values are prominently displayed on the material's Material Safety Data Sheet (MSDA), which must be maintained and available for the workforce.

Where flammable solvents and adhesives are used, they must be stored, handled, and used in a manner that prevents any possibility of ignition. Proper safety containers, storage areas, and well ventilated workplaces are required.

Certain adhesive materials are hazardous when mixed together. Epoxy and polyester catalysts, especially, must be well understood prior to departing from the manufacturers' recommended procedure for mixing. Certain unstabilized solvents, such as trichloroethylene and perchloroethylene, are subject to chemical reaction on contact with oxygen or moisture. Only stabilized grades of solvents should be used.

Certain adhesive systems, such as heat-curing epoxy and room-temperature-curing polyesters, can develop very large exothermic reactions on mixing. The temperature generated during this exotherm is dependent on the mass of the material being mixed. Exotherm temperatures can get so high that the adhesive will catch fire and burn. Adhesive products should always be applied in thin bond lines to minimize the exotherm until the chemistry of the product is well understood.

Safe equipment and proper operation are, of course, crucial to a workplace. Sufficient training and safety precautions must be installed in the factory before the bonding process is established.

References

1. Structural Adhesives, Minnesota Mining and Manufacturing Co., Adhesives Coatings and Sealess Division, Technical Bulletin.
2. Powis, C. N., Some Applications of Structural Adhesives, in D. J. Alner, Ed., *Aspects of Adhesion,* vol. 4, University of London Press, Ltd., London 1968.
3. Bikerman, J. J., Causes of Poor Adhesion, *Ind. Eng. Chem.*, September 1967.
4. Reinhart, F. W., Survey of Adhesion and Types of Bonds Involved, in J. E. Rutzler and R. L. Savage, Eds., *Adhesion and Adhesives Fundamentals and Practices,* Society of Chemical Industry, London, 1954.
5. Merriam, J. C., Adhesive Bonding, *Mater. Des. Eng.*, September 1959.
6. Rider, D. K., Which Adhesives for Bonded Metal Assembly, *Prod. Eng.,* May 25, 1964.
7. Perry, H. A., Room Temperature Setting Adhesives for Metals and Plastics, in J. E. Rutzler and R. L. Savage, Eds., *Adhesion and Adhesives Fundamentals and Practices,* Society of Chemical Industry, London, 1954.
8. Lunsford, L. R., Design of Bonded Joints, in M. J. Bodnar, Ed., *Symposium on Adhesives for Structural Applications,* Interscience, New York, 1962.
9. Koehn, G. W., Design Manual on Adhesives, *Machine Design,* April 1954.
10. *Adhesive Bonding Alcoa Aluminum,* Aluminum Co. of America, 1967.
11. DeLollis, N. J., *Adhesives for Metals Theory and Technology,* Industrial Press, New York, 1970.
12. Chessin, N., and V. Curran, Preparation of Aluminum Surface for Bonding, in M. J. Bodnar, Ed., *Structural Adhesive Bonding,* Interscience, New York, 1966.
13. Muchnick, S. N., Adhesive Bonding of Metals, *Mech. Eng.*, January 1956.
14. Vazirani, H. N., Surface Preparation of Steel and Its Alloys for Adhesive Bonding and Organic Coatings, *J. Adhesion*, July 1969.
15. Vazirani, H. N., Surface Preparation of Copper and Its Alloys for Adhesive Bonding and Organic Coatings, *J. Adhesion,* July 1969.
16. Walter, R. E., D. L. Voss, and M. S. Hochberg, Structural Bonding of Titanium for Advanced Aircraft, *Nat. SAMPE Tech. Conf. Proc.,* vol 2, Aerospace Adhesives and Elastomers, 1970.
17. Jennings, C. W., Surface Roughness and Bond Strength of Adhesives, *J. Adhesion,* May 1972.
18. Bersin, R.L., How to Obtain Strong Adhesive Bonds via Plasma Treatment, *Adhesive Age*, March 1972.
19. Rogers, N. L., Surface Preparation of Metals, in M. J. Bodnar, Ed., *Structural Adhesive Bonding,* Interscience, New York, 1966.

20. Krieger, R. B., Advances in Corrosion Resistance of Bonded Structures, *Nat. SAMPE Tech. Conf. Proc.,* vol. 2, Aerospace Adhesives and Elastomers, 1970.
21. Cagle, C. V., *Adhesive Bonding Techniques and Applications,* McGraw-Hill, New York, 1968.
22. Landrock, A. H., *Adhesives Technology Handbook,* Noyes Publications, New Jersey, 1985.
23. Guttmann, W. H., *Concise Guide to Structural Adhesives,*" Reinhold, New York, 1961.
24. *Preparing the Surface for Adhesive Bonding,* Hysol Div., Dexter Corp., Bull. G1-600.
25. *Adhesive Bonding Aluminum,* Reynolds Metal Co., 1966.
26. Schields, J., *Adhesives Handbook,* CRC Press, 1970.
27. Devine, A. T., and M. J. Bodnar, Effects of Various Surface Treatments on Adhesive Bonding of Polyethylene, *Adhesives Age,* May 1969.
28. Austin, J. E., and L. C. Jackson, Management: Teach Your Engineers to Design Better With Adhesives, *SAE J.,* October 1961.
29. Burgman, H. A., Selecting Structural Adhesive Materials, *Electrotechnol.,* June 1965.
30. *Three New Cyanoacrylate Adhesives,* Eastman Chemical Co., Leaflet R-206A.
31. TAME, A New Concept in Structural Adhesives, *B. F. Goodrich General Products Co. Bull.* GPC-72-AD-3.
32. Information about Pressure Sensitive Adhesives, *Dow Corning, Bull.* 02-032.
33. Information about Silastic RTV Silicone Rubber, *Dow Corning, Bull.* 61-015a.
34. Bruno, E. J., Ed., *Adhesives in Modern Manufacturing,* p. 29, Society of Manufacturing Engineers, 1970.
35. Lichman, J., Water-based and Solvent-based Adhesives, in I. Skeist, Ed., *Handbook of Adhesives,* Reinhold, New York, 1977.
36. Hemming, C. B., Wood Gluing, in I. Skiest, Ed., *Handbook of Adhesives,* 1st ed., Reinhold, New York, 1962.
37. Moser, F., Bonding Glass, in I. Skiest, Ed., *Handbook of Adhesives,* 1st ed., Reinhold, New York, 1962.
38. Weggemans, D. M., Adhesive Charts, in *Adhesion and Adhesives,* vol. 2, Elsevier, Amsterdam, 1967.
39. Krieger, R. B. and R. E. Politi, High Temperature Structural Adhesives, in D. J. Alner, Ed., *Aspects of Adhesion,* vol. 3, University of London Press, London, 1967.
40. Burgman, H. A., The Trend in Structural Adhesives, *Electrotechnol.,* November 21, 1963.
41. Kausen, R. C., Adhesives for High and Low Temperature, *Mater. Eng.,* August-September 1964.
42. Bolger, J. C., New One-Part Epoxies are Flexible and Reversion Resistant, *Insulation,* October 1969.
43. Falconer, D. J., et. al., The Effect of High Humidity Environments on the Strength of Adhesive Joints, *Chem. Ind.,* July 4, 1964.
44. Sharpe, L. H., Aspects of the Permanence of Adhesive Joints, in M. J. Bodnar, Ed., *Structural Adhesive Bonding,* Interscience, New York, 1966.
45. Tanner, W. C., Adhesives and Adhesion in Structural Bonding for Military Material, in M. J. Bodnar, Ed. *Structural Adhesive Bonding,* Interscience, New York, 1966.
46. Arlook, R. S., and D. G. Harvey, *Effects of Nuclear Radiation on Structural Adhesive Bonds,* Wright Air Development Center, Report WADC-TR-46-467.
47. Carroll, K. W., How to Apply Adhesive, *Prod. Eng.,* November 22, 1965.
48. Smith, D. F., and C. V. Cagle, A Quality Control System for Adhesive Bonding Utilizing Ultrasonic Testing, in M. J. Bodnar, Ed., *Structural Adhesive Bonding,* Interscience, New York, 1966.

Chapter

8

Plastics Joining

Edward M. Petrie

Materials and Process Center
ABB Transmission Technology Institute
Raleigh, North Carolina

8.1 Introduction

To fabricate large plastic assemblies, the most cost-effective method often involves producing smaller subsections and joining them together. In such cases, the fabricator has a variety of joining options to consider, including

- Adhesive bonding
- Thermal welding
- Solvent cementing
- Mechanical fastening

The strength of the plastic assembly relies heavily on the characteristics of the joint. In most applications, the joint must be nearly as strong as the substrate.throughout the expected life of the product. The method chosen to join plastics should be carefully evaluated. In addition to strength and permanence, consideration must be given to tooling cost, labor and energy cost, production time, appearance of the final part, and disassembly requirements. Some plastic materials will be better suited for certain joining processes than others due to their physical and chemical characteristics.

A consideration in plastic assembly that is usually not dominant when joining other substrates is the time it takes to complete the joining operation. Plastic products generally require very fast, high-volume assembly processes in industries such as consumer products, automotive, and packaging. Speed, simplicity, and reliability are key concerns in most of these high-volume assembly processes.

Speed and simplicity are usually considered to be of greater value than reliability or durability when bonding commodity plastic substrates. Because of the nature of the polymeric substrate and the type of applications for which such materials are best suited, exceedingly high strength and durability in hash environments are not generally necessary.

These properties will gladly be sacrificed for faster, lower-cost production methods. Often, there is not enough time for critical surface preparation or nondestructive testing of every part.

In certain industries, such as the automotive industry, plastic materials are often chosen because of their fast joining ability. Thus, thermoplastics are often preferred over thermosets, because they can be joined via thermal welding processes in a few seconds.

Although adhesive bonding often proves to be an effective method for bonding plastics, there are various other ways of joining plastics to themselves or to other materials. Thermal welding, solvent cementing, and mechanical fastening are usually faster than adhesive bonding and as a result are often preferred in high-volume assembly operations. These non-adhesive methods of joining will be the subject of this chapter. Adhesive bonding methods for plastics as well as other substrates had be considered in the previous chapter.

The sections that follow will describe the various processes that are can be used for joining plastic materials. Information will be provided on how to choose the most appropriate process for a specific substrate and application. The plastic materials that are best suited for each process will be identified. Important process parameters and test results will also be reviewed. Recommendations regarding joining processes for specific types of plastic materials (e.g., polyethylene, glass reinforced epoxy, polysulfone) will also be given.

8.2 General Types of Plastic Materials

There are many types of plastic materials with a range of properties that depend on the base polymer and the additives used. Plastics are used routinely in many commodity items such as packaging, pipe, clothing, appliances, and electronics. They are also increasingly being used in structural and engineering applications such as aerospace, building, and automotive.

All plastics can be classified into two categories: thermoplastics and thermosets. Thermoplastics are not cross-linked, and the polymeric molecules making up the thermoplastic can easily slip by one another. This slip or flow can be caused by thermal energy, by solvents or other chemicals, and by the application of continuous stress. Thermoplastics can be repeatedly softened by heating and hardened on cooling. Hence, they can be welded by the application of heat. Thermoplastics can also be dissolved in solvents, so it is also possible to join thermoplastic parts by solvent cementing. Typical thermoplastics are polyethylene, polyvinyl chloride, polystyrene, polypropylene, nylon, and acrylic.

Engineering thermoplastics are a class of thermoplastic materials that have exceptionally high temperature and chemical resistance. They have properties very similar to those of thermoset plastics and metals. As a result, they are not as easy to weld thermally or to solvent cement as are the more conventional thermoplastics. Typical engineering thermoplastics are polysulfone, acetal, amide-imide, and thermoplastic polyimide.

Thermoset plastics are cross-linked by chemical reaction so that their molecules cannot slip by one another. They are rigid when cool and cannot be softened by the action of heat. If excessive heat is applied, thermoset plastics will degrade. Consequently, they are not weldable. Because of their chemical resistance, they cannot be solvent cemented. Thermoset plastics are usually joined by either adhesive bonding or mechanical fastening. Typical thermosetting plastics are epoxy, urethane, phenolic, and melamine formaldehyde.

There are but a few typical basic polymer resin suppliers. However, there are many companies that formulate filled plastic systems from these basic resins. These smaller companies are quite often the ultimate suppliers to the manufacturer of the end-product. Both the basic resin manufacturer and the formulator have considerable influence on the joining characteristics of the final material. Because of this, they should be considered to be the primary source of information regarding joining processes and expected end results.

Plastics offer several important advantages compared with traditional materials, and their joining process should not detract from these benefits. The greatest advantage is low processing cost. Low weight is also a major advantage. Relative densities of most unfilled plastics materials range from 0.9–1.4, compared with 2.7 for aluminum and 7.8 for steel. Relative densities for highly filled plastic compounds can rise to 3.5.

Other advantages of plastics are their low frictional resistance, good corrosion resistance, insulation properties, and ease at which they can be fabricated into various shapes. The chemical resistance of plastics is specific to the type of plastic. Some plastics have chemical resistance comparable to that of metals; others are attacked by acids, and still others are attacked by solvents and oils. Certain plastics are also attacked by moisture, especially at moderately elevated temperatures. The nature of this attack is generally first a swelling of the substrate and then, finally, dissolution. Some plastics are transparent, and others are translucent. These plastics can pass UV radiation, so the adhesive or joint could be affected by UV. In certain cases, the plastic itself can be degraded by UV. Resistance to UV degradation can be improved by the addition of UV absorbers such as carbon black. All plastics can all be colored by the addition of pigments. Addition of pigments generally eliminates the need for painting.

The most serious disadvantage of plastics are that, compared with metals, they have low stiffness and strength and are not suitable for use at high temperatures. Some plastics are unusable at temperatures above 50°C. A very few are capable of resisting short-term exposure to temperatures up to 400°C. Surface hardness of plastics is generally low. This can lead to indentation and compression under local stress. The elastic modulus of plastic materials is relatively high compared with metals. They can be elongated and compressed with applied stress. The thermal expansion coefficient of plastics is generally an order of magnitude greater than metals. This results in internal joint stress when joined to a substrate having a much lower coefficient of expansion.

8.3 Types of Plastic Joining Processes

The joining of plastics with adhesives is generally made difficult because of the low surface energy, poor wettability, and presence of weak boundary layers associated with these substrates. Adhesive bonding is a relatively slow process that could be a significant drawback in many industries that produce high-volume plastic assemblies. However, with plastic substrates, the designer has a greater choice of joining techniques than with many other substrate materials. Thermoset plastics must usually be bonded with adhesives or mechanically fastened, but many thermoplastics can be joined by solvent or heat welding as well as by adhesives or mechanical fasteners.

In the thermal or solvent welding processes, the plastic resin that makes up the substrate itself acts as the adhesive. These processes require that the surface region of the substrate be made fluid so that it can wet the mating substrate. If the mating substrate is also a polymer, both substrate surfaces can be made fluid so that the resin can molecularly diffuse into the opposite interfaces. This fluid interface region is usually achieved by thermally heating the surface areas of one or both substrates or by dissolving the surfaces in an appropriate solvent. Once the substrate surface is in a fluid condition, they are then brought together and held in place with moderate pressure. At this point, the molecules of substrate A and substrate B will diffuse into one another and form a very tight bond. The fluid polymer mix then returns back to the solid state, usually by the dissipation of solvent or by cooling from the molten condition.

Thermal or solvent welding can be effectively used when at least one substrate is polymeric. In the case of thermal welding, the molten polymer surface wets the nonpolymeric substrate and acts as a hot melt type of adhesive. Internal stresses that occur on cooling the interface from the molten condition are the greatest detriment to this method of bonding.

Solvent welding can also be used only if the nonpolymeric substrate is porous. If it is not porous, the solvent may become entrapped at the bond line and cause very weak joints.

Table 8.1 indicates the most common joining methods that are suitable for various plastic materials. Descriptions of these joining techniques are summarized both in Table 8.2 and in the section to follow. In general, non-adhesive joining methods for plastics can be classified as follows:

1. Welding by direct heating (heated tool, hot gas, resistance wire, infrared, laser, extrusion)
2. Induced heating (induction, electrofusion, dielectric)
3. Frictional (ultrasonic, vibration, spin)
4. Solvent welding
5. Mechanical fastening

Equipment costs for each method vary considerably, as do the amount of labor involved and the speed of the operation. Most techniques have limitations regarding the design of the joint and the types of plastic materials that can be joined.

8.4 Direct Heat Welding

Welding by the direct application of heat provides an advantageous method of joining many thermoplastic materials that do not degrade rapidly at their melt temperatures. The principal methods of direct heat welding are

1. Heated tool
2. Hot gas
3. Resistance wire
4. Laser
5. Infrared

These methods are generally capable of joining thermoplastics to themselves and other thermoplastics, and in certain cases they may also be used to weld thermoplastics to nonplastic substrates.

8.4.1 Heated-Tool Welding

Fusion or heated-tool welding is an excellent method of joining many thermoplastics. In this method, the surfaces to be fused are heated by holding them against a hot surface. When the plastic becomes molten and a flash about half the thickness of the substrate is visible, the parts are removed from the hot surface. They are then immediately joined under slight pressure (5 to 15 lb/in^2) and allowed to cool and harden. The molten polymer acts as a hot-melt adhesive, providing a bond between the substrates.

This method is often used in high-volume operations where adhesive bonding time is objectionably long. It is also often used to join low-surface-energy materials such as polypropylene where the cost and complexity required for substrate treatment and adhesive bonding cannot be tolerated. Surface treatment, other than simple cleaning, is usually not required for thermal welding. Heated-tool welding is a simple, economical technique in which high-strength joints can be achieved with large and small parts. Hermetic seals

TABLE 8.1 Assembly Methods for Plastics (from Ref. 1)

Plastics	Common assembly methods							
	Adhesives	Dielectric welding	Induction bonding	Mechanical fastening	Solvent welding	Spin welding	Thermal welding	Ultrasonic welding
Thermoplastics								
ABS	X	—	X	X	X	X	X	X
Acetals	X	—	X	X	X	X	X	X
Acrylics	X	—	X	X	X	X	—	X
Cellulosics	X	—	—	—	X	X	—	—
Chlorinated polyether	X	—	—	X	—	—	—	—
Ethylene copolymers	—	X	—	—	—	—	X	—
Fluoroplastics	X	—	—	—	—	—	—	—
Ionomer	—	X	—	—	—	—	X	—
Methylpentene	—	—	—	—	—	—	—	X
Nylons	X	—	X	X	X	—	—	X
Phenylene oxide–based materials	X	—	—	X	X	X	X	X
Polyesters	X	—	—	X	X	X	—	X
Polyamide–imide	X	—	—	X	—	—	—	—
Polyaryl ether	X	—	—	X	—	—	—	X
Polyaryl sulfone	X	—	—	X	—	—	—	X
Polybutylene	—	—	—	—	—	—	—	X
Polycarbonate	X	—	X	X	X	X	X	X
Polycarbonate/ABS	X	—	—	—	X	X	X	X
Polyethylene	X	X	X	X	—	X	X	X
Polyimide	X	—	—	—	—	—	—	—
Polyphenylene sulfide	X	—	—	—	—	—	—	—
Polypropylenes	X	X	X	X	—	X	X	X
Polystyrene	X	—	X	X	X	X	X	X

TABLE 8.1 Assembly Methods for Plastics (from Ref. 1) (continued)

	Common assembly methods							
Plastics	Adhesives	Dielectric welding	Induction bonding	Mechanical fastening	Solvent welding	Spin welding	Thermal welding	Ultrasonic welding
Polysulfone	X	—	—	—	X	—	—	X
Polypropylene copolymers	X	X	X	X	—	X	X	—
PVC/acrylic alloy	X	—	X	X	—	—	—	X
PVC/ABS alloys	X	X	—	—	—	—	X	—
Styrene acrylonitrile	X	—	X	X	X	X	X	X
Vinyl	X	X	X	X	X	—	X	—
Thermosets								
Alkyd	X	—	—	—	—	—	—	—
Allyl diglycol carbonate	X	—	—	—	—	—	—	—
Diallyl phthalate	X	—	—	X	X	—	—	—
Epoxies	X	—	—	X	X	—	—	—
Melamines	X	—	—	—	—	—	—	—
Phenolics	X	—	—	X	X	—	—	—
Polybutadienes	X	—	—	—	—	—	—	—
Polyesters	X	—	—	X	X	—	—	—
Silicones	X	—	—	X	X	—	—	—
Ureas	X	—	—	—	—	—	—	—
Urethanes	X	—	—	X	X	—	—	—

can also be achieved. Heated-tool welding does not introduce foreign materials into the part and, as a result, plastic parts are more easily recycled.

Success in heated-tool welding depends primarily on having the proper temperature at the heating surface and on the timing of the various steps in the process. These periods include time for application of heat, time between removal of heat and joining of parts, and time the parts are under pressure. The tool should be hot enough to produce sufficient flow for fusion within 10 s. The parts are generally pressed against the heated tool with a certain degree of pressure. However, to avoid strain, the pressure on the parts should be released for a period of time before they are removed from contact with the heated tool.

TABLE 8.2 Bonding or Joining Plastics; What Techniques are Available and What Do They have to Offer? (from Ref. 2)

Technique	Description	Advantages	Limitations	Processing Considerations
Solvent cementing and dopes	Solvent softens the surface of an amorphous thermoplastic; mating takes place when the solvent has completely evaporated. Bodied cement with small percentage of parent material can give more workable cement, fill in voids in bond area. Cannot be used for polyolefins and acetal homopolymers.	Strength, up to 100% of parent materials, easily and economically obtained with minimal equipment requirements.	Long evaporation times required; solvent may be hazardous; may cause crazing in some resins.	Equipment ranges from hypodermic needle or just a wiping media to tanks for dip and soak. Clamping devices are necessary, and air dryer is usually required. Solvent-recovery apparatus may be necessary or required. Processing speeds are relatively slow because of drying times. Equipment costs are low to medium.
		Thermal Bonding		
Ultrasonics	High-frequency sound vibrations transmitted by a metal horn generate friction at the bond area of a thermoplastic part, melting plastics just enough to permit a bond. Materials most readily weldable are acetal, ABS, acrylic, nylon, PC, polyimide, PS, SAN, phenoxy.	Strong bonds for most thermoplastics; fast, often less than 1 s. Strong bonds obtainable in most thermal techniques if complete fusion is obtained.	Size and shape limited. Limited applications to PVSs, polyolefins.	Converter to change 20 kHz electrical into 20 kHz mechanical energy is required along with stand and horn to transmit energy to part. Rotary tables and high-speed feeder can be incorporated.
Hot-plate and hot-tool welding	Mating surfaces are heated against a hot surface, allowed to soften sufficiently to produce a good bond, then clamped together while bond sets. Applicable to rigid thermoplastics.	Can be very fast, e.g., 4–10 s in some cases; strong.	Stresses may occur in bond area.	Use simple soldering guns and hot irons, relatively simple hot plates attached to heating elements up to semiautomatic hot-plate equipment. Clamps needed in all cases.

TABLE 8.2 Bonding or Joining Plastics; What Techniques are Available and What Do They have to Offer? (from Ref. 2) (continued)

Technique	Description	Advantages	Limitations	Processing Considerations
Hot-gas welding	Welding rod of the same material being joined (largest application is vinyl) is softened by hot air or nitrogen as it is fed through a gun that is softening part surface simultaneously. Rod fills in joint area and cools to effect a bond.	Strong bonds, especially for large structural shapes.	Relatively slow; not an "appearance" weld.	Requires a hand gun, special welding tips, an air source and welding rod. Regular hand-gun speeds run 6 in/min; high-speed hand-held tool boosts this to 48–60 in/min.
Spin welding	Parts to be bonded are spun at high speed, developing friction at the bond area; when spinning stops, parts cool in fixture under pressure to set bond. Applicable to most rigid thermoplastics.	Very fast (as low as 1–2 s); strong bonds.	Bond area must be circular.	Basic apparatus is a spinning device, but sophisticated feeding and handling devices are generally incorporated to take advantage of high-speed operation.
Dielectries	High-frequency voltage applied to film or sheet causes material to melt at bonding surfaces. Material cools rapidly to effect a bond. Most widely used with vinyls.	Fast seal with minimum heat applied.	Only for film and sheet.	Requires rf generator, dies, and press. Operation can range from hand-fed to semiautomatic with speeds depending on thickness and type of product being handled. 3–35 kW units are most common.
Induction	A metal insert or screen is placed between the parts to be welded, and energized with an electromagnetic field. As the insert heats up, the parts around it melt, and when cooled form a bond. For most thermoplastics.	Provides rapid heating of solid sections to reduce change of degradation.	Since metal is embedded in plastic, stress may because at bond.	High-frequency generator, heating coil, and inserts (generally 0.02–0.04 in thick). Hooked up to automated devices, speeds are high. Work coils, water cooling for electronics, automatic timers, multiple-positions stations may also be required.

TABLE 8.2 Bonding or Joining Plastics; What Techniques are Available and What Do They have to Offer? (from Ref. 2) (continued)

Technique	Description	Advantages	Limitations	Processing Considerations
		Adhesives*		
Liquids solvent, water base, anaerobic	Solvent-and-water-based liquid adhesives, available in a wide number of bases—e.g., polyester, vinyl—in one-or two-part form fill bonding needs ranging from high-speed lamination to one-of-a-kind joining of dissimilar plastics parts. Solvents provide more bite but cost much more than similar base water-type adhesive. Anaerobics are a group of adhesives that cure in the absence of air.	Easy to apply; adhesives available to fit most applications.	Shelf and pot life often limited. Solvents may cause pollution problems; water-base not as strong; anaerobics toxic.	Application techniques range from simply brushing on to spraying and roller coating lamination for very high production. Adhesive application techniques, often similar to decorating equipment, from hundreds to thousands of dollars with sophisticated laminating equipment costing in the tens of thousands of dollars. Anaerobics are generally applied a drop at a time from a special bottle or dispenser.
Pastes, mastics	Highly viscous single- or two-component materials that cure to a very hard or flexible joint depending on adhesive type.	Does not run when applied.	Shelf and pot life are often limited.	Often applied via a trowel, knife, or gun-type dispenser; one-component systems can be applied directly from a tube. Various types of roller coaters are also used. Metering-type dispensing equipment in the $2,500 range has been used to some extent.
Hot melts	100% solids adhesives that become flowable when heat is applied. Often used to bond continuous flat surfaces.	Fast application; clean operation.	Virtually no structural hot melts for plastics.	Hot melts are applied at high speeds via heating the adhesive, then extruding (actually squirting) it onto a substrate, roller coating, using a special dispenser or roll to apply dots or simply dipping.

TABLE 8.2 Bonding or Joining Plastics; What Techniques are Available and What Do They have to Offer? (from Ref. 2) (continued)

Technique	Description	Advantages	Limitations	Processing Considerations
Film	Available in several forms including hotmelts, these are sheets of solid adhesive. Mostly used to bond film or sheet to substrate.	Clean, efficient.	High cost.	Film adhesive is reactivated by a heat source; production costs are in the medium to high range, depending on heat source used.
Pressure-sensitive	Tacky adhesives used in a variety of commercial applications (e.g., cellophane too). Often used with polyolefins.	Flexible.	Bonds not very strong.	Generally applied by spray with bonding affected by light pressure.
Mechanical bonding				
Mechanical fasteners (staples, screws, molded-in inserts, snap fits and variety of proprietary fasteners)	Typical mechanical fasteners are listed at left. Devices are made of metal or plastic. Type selected will depend on how strong the end product must be, appearance factors. Often used to join dissimilar plastics or plastics to nonplastics.	Adaptable to many materials; low to medium costs; can be used for parts that must be disassembled.	Some have limited pull-out strength; molded-in inserts may result in stresses.	Nails and staples are applied by simply hammering or stapling. Other fasteners may be inserted by drill press, ultrasonics, air or electric gun, hand tool. Special molding (i.e., molded-in-hole) may be required.

*Typical adhesives in each class are: Liquids: 1. Solvent—polyester, vinyl, phenolics acrylics, rubbers, epoxies, polyamide; 2. Water—acrylics, rubber-casein; 3. Anaerobics—cyanoacrylate; mastics—rubbers, epoxies; hot melts—polyamides, PE, PS PVA; film—epoxies, polyamide, phenolics; pressure, sensitive—rubbers.

While some rules of thumb can apply, the final process settings for temperature, duration of heating and cooling times, and pressures will depend on the polymer. Adjustments will be required until the desired bond quality is achieved. The thickness of the molten layer is an important determinant of weld strength. Dimensions are usually controlled through the incorporation of displacement stops at both the heating and mating steps in the process. If welds are wider than 1/4 in, the heated parts should be glided across each other during the mating step to prevent air entrapment in the joint.

Heated-tool welding is suitable for almost any thermoplastic, but it is most often used for softer, semicrystalline thermoplastics. Common plastic substrates that are suitable for heated-tool welding include polyethylene, polypropylene, polystyrene, ABS, PVC, and acetals. It is usually not suitable for nylon or other materials that have long molecular chains. Dissimilar yet chemically compatible materials that have different melting temperatures can be welded in heated-tool welding by using two heated platens, each heated to the melting temperature of the part to be welded.

Heated-tool welding can be accomplished with either no surface treatment or very minor surface preparation (degreasing and removal of mold release), depending on the strength and reliability dictated by the application. Generally, surface degreasing to remove mold release or other organic contaminants is the only prebond treatment necessary. Mechanical roughening or chemical treatment of the surface provides no advantage, since the surface will be melted, and a new surface will be formed. Plastic parts that have significant degree of internal moisture may have to be dried before heated-tool welding, or else the moisture will tend to escape the molten surface in the form of vapor bubbles.

Electric strip heaters, soldering irons, hot plates, and resistance blades are common methods of providing heat locally. Usually, the heating platen is coated with a fluorocarbon such as PTFE for non-stick characteristics. A simple hot plate has been used extensively with many plastics. Figure 8.1 illustrates an arrangement for direct heat welding consisting of heated platens and fixturing. The parts are held on the hot plate until sufficient fusible material has been developed. Table 8.3 lists typical hot-plate temperatures for a variety of plastics. A similar technique involves butting flat plastic sheets on a flat table against a heated blade that runs the length of the sheet. Once the plastic begins to soften, the blade is raised, and the sheets are pressed together and fused.

TABLE 8.3 Hot Plate Temperatures to Weld Plastics (from Ref. 3)

Plastic	Temperature, °F
ABS	450
Acetal	500
Phenoxy	550
Polyethylene LD	360
HD	390
Polycarbonate	650
PPO	650
Noryl*	525
Polypropylene	400
Polystyrene	420
SAN	450
Nylon 6, 6	475
PVC	450

*Trademark of General Electric Co.

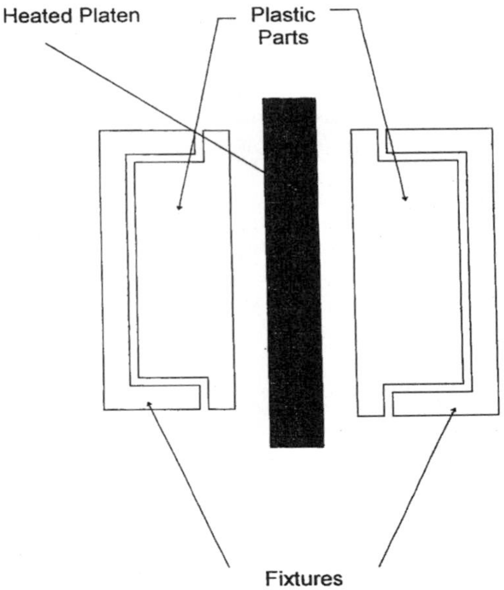

Figure 8.1 In direct hot-plate welding, two fixtures press components into a hot moving platen, causing the plastic to melt at the interface.

The direct heat welding operation can be completely manual, as in the case of producing a few prototypes, or it can be semiautomatic or fully automatic for fast, high-volume production. For automated assembly, rotary machines are often used where there is an independent station for each of the following processes:

1. Clamping into fixtures
2. Heating
3. Joining and cooling
4. Unloading

Heated wheels or continuously moving heated bands are common tools used to bond thin plastic sheet and film. This is commonly used for sealing purposes such as packaging of food. Care must be taken, especially with thin film, not to apply excessive pressure or heat. This could result in melting through the plastic. Table 8.4 provides heat-sealing temperature ranges for common plastic films.

Heated-tool welding is commonly used in medium- to high-volume industries that can make use of the simplicity and speed of these joining processes. Industries that commonly use this fastening method include appliance and automotive. Welding times range from 10–20 s for small parts, to up to 30 min for larger parts such as heavy-duty pipe. Typical cycle times are less than 60 s. Although heated-tool welding is faster than adhesive bonding, it is not as fast as other welding methods such as ultrasonic and induction welding.

The heated tool process is extremely useful for pipe and duct work, rods and bars, and for continuous seals in films. However, irregular surfaces are difficult to heat unless complicated tools are provided. Special tooling configuration can be used for bonding any

TABLE 8.4 Heat Sealing Temperatures for Plastic Films (from Ref. 4)

Film	Temperature, °F
Coated cellophane	200–350
Cellulose acetate	400–500
Coated polyester	490
Poly (chlorotrifluoroethylene)	415–450
Polyethylene	250–375
Polystyrene (oriented)	220–300
Poly (vinyl alcohol)	300–400
Poly (vinyl chloride) and copolymers (nonrigid)	200–400
Poly (vinyl chloride) and copolymers (rigid)	260–400
Poly (vinyl chloride)–nitrile rubber blend	220–350
Poly (vinylidene chloride)	285
Rubber hydrochloride	225–350
Fluorinatedthylene–propylene copolymer	600–750

structural profile to a flat surface. In certain applications, direct heating can also be used to shape the joint. With pipe, for example, a technique called *groove welding* is generally employed. Groove welding involves two heating elements. One element melts a groove in one substrate that is the exact shape of the mating part, and the other element heats the edge of the mating substrate. The heated part is quickly placed into the heated groove and allowed to cool.

8.4.2 Hot-Gas Welding

An electrically heated welding gun can be used to bond many thermoplastic materials. An electrical heating element in the welding gun is capable of heating either compressed air or an inert gas to 425–700°F and forcing the heated gas onto the substrate surface. The pieces to be joined are beveled and positioned with a small gap between them. A welding rod made of the same plastic that is being bonded is laid in the joint with a steady pressure. The heat from the gun is directed to the tip of the rod, where it fills the gap, as shown in Fig. 8.2. Several passes may be necessary with the rod to fill the pocket. Thin sheets that are to be butt welded together, as in the case of tank linings, use a flat strip instead of a rod. The strip is laid over the joint and is welded in place in a single pass. Usually, the parts to be joined are held by fixtures so that they do not move during welding or while the weld is cooling. Alternatively, the parts can be first tacked together using a tool similar to a soldering iron.

Hot-gas welding is usually a manual operation in which the quality of the joint corresponds to the skill and experience of the operator. However, automatic welding machines are available and are used for overlap welding of seams or membranes. In either case, bond strength approximately 85 percent of the strength of the bulk material can be achieved. Hot-gas welding is a relatively fast operation. It can be used to weld a 1-in wide tank seam at rates up to 60 in/min. It can also be used to do temporary tack work and to repair faulty or damaged joints that are made by gas welding or other joining processes.

Hot-gas welding can be used to join most thermoplastics, including polypropylene, polyethylene, acrylonitrile butadiene styrene, polyvinyl chloride, thermoplastic polyurethane, high density polyethylene, polyamide, polycarbonate, and polymethylmethacrylate. For polyolefins and other plastics that are easily oxidized, the heated gas must be inert (e.g., nitrogen or argon), since hot air will oxidize the surface of the plastic.

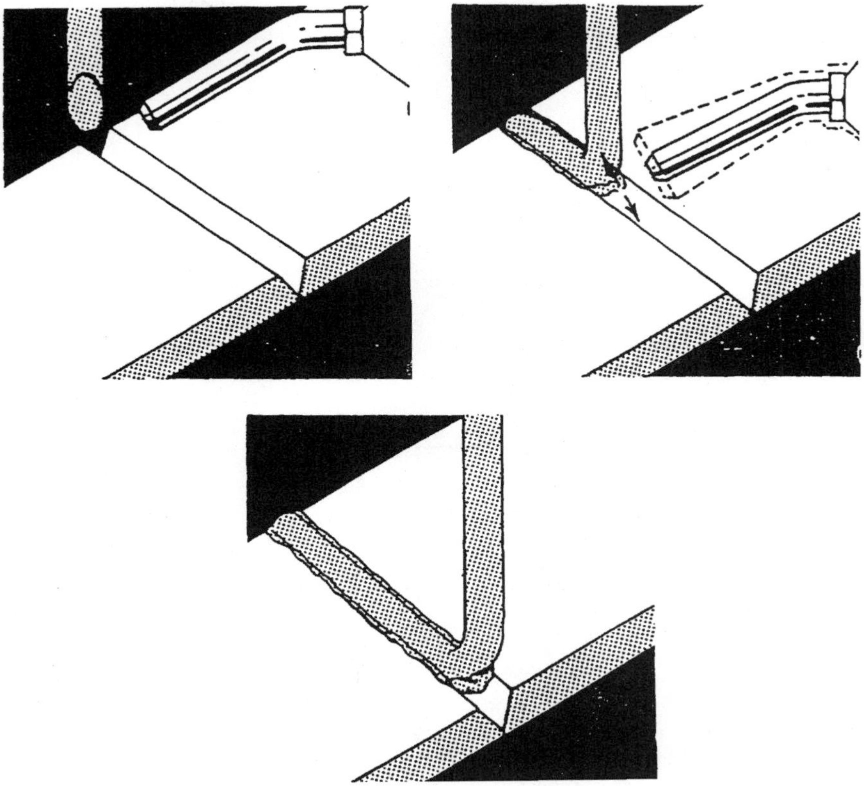

Thermoplastic Welding Chart						
	PVC	H.D. polyethylene	Polypropylene	Penton	ABS	Plexiglas®
Welding temperature	525	550	575	600	500	575
Forming temperature	300	300	350	350	300	300
Welding gas	Air	WPN*	WPN	Air	WPN	Air

*WPN = water-pumped nitrogen

Figure 8.2 Hot-gas welding apparatus, method of application, and thermoplastic welding parameters.[5]

Process parameters that are responsible for the strength of a hot-gas weld include the type of plastic being welded, the temperature and type of gas, the pressure on the rod during welding, the preparation of the material before welding, and the skill of the welder. After welding, the joint should not be stressed for several hours. This is particularly true for polyolefins, nylons, and polyformaldehyde. Hot-gas welding is not recommended for

filled materials or substrates that are less than 1/16 in thick. Conventional hot-gas welding joint designs are shown in Fig. 8.3.

Ideally, the welding rod should have a triangular cross section to match the bevel in the joint. A joint can be filled in one pass using triangular rod, saving time and material. Plastic welding rods of various types and cross sections are commercially available. However, it is also possible to cut welding rod from the sheet of plastic that is being joined. Although this may require multiple passes for filling, and the chance of air pockets is greater, the welding rod is very low in cost, and the user is guaranteed material compatibility between the rod and the plastic being joined.

Hot-gas welding can be used in a wide variety of welding, sealing, and repair applications. Applications are usually large structural assemblies. Hot-gas welding is used very often in industrial applications such as chemical storage tank repair, pipe fittings, etc. It is an ideal system for a small fabricator or anyone looking for an inexpensive welding system. Welders are available for several hundred dollars. The weld may not be as cosmetically attractive as other joining methods, but fast processing and tensile strengths of 85 percent of the parent material can be obtained easily.

Another form of hot-gas welding is extrusion welding. In this process, an extruder is used instead of a hot-gas gun. The molten welding material is expelled continuously from the extruder and fills a groove in the preheated weld area. A welding shoe follows the application of the hot extrudate and actually molds the seam in place. The main advantage with extrusion welding is the pressure that can be applied to the joint. This adds to the quality and consistency of the joint.

8.4.3 Resistance Wire Welding

The resistance wire welding method of joining employs an electrical resistance heating element laid between mating substrates to generate the needed heat of fusion. Once the element is heated, the surrounding plastic melts and flows together. Heating elements can be anything that conducts current and can be heated through Joule heating. This includes nichrome wire, carbon fiber, woven graphite fabric, and stainless steel foil. Figure 8.4 shows an example of such a joint where a nichrome wire is used as the heating element. After the bond has been made, the resistance element that is exterior to the joint is cut off. Implant materials should be compatible with the intended application, since they will remain in the bond line for the life of the product.

Like hot-plate welding, resistance welding has three steps: heating, pressing, and maintaining contact pressure as the joint gels and cures. The entire cycle takes 30 s to several minutes. Resistance welders can be automated or manually operated. Processing parameters include power (voltage and current), weld pressure, peak temperature, dwell time at temperature, and cooling time.

With resistance wire welding, surface preparation steps are necessary only when one of the substrates cannot be melted (e.g., thermosets and metals). Standard adhesive joining surface preparation processes such as those suggested in the next chapter can be used with these substrates.

The resistance heating process can be performed at either constant power or at constant temperature. When using constant power, a particular voltage and current is applied and held for a specified period of time. The actual temperatures are not controlled and are difficult to predict. In constant-temperature resistance wire welding, temperature sensors monitor the temperature of the weld and automatically adjust the current and voltage to maintain a predefined temperature. Accurate control of heating and cooling rates is important when welding some plastics such as semicrystalline thermoplastics or when welding substrates having significantly different melt temperatures or thermal expansion coefficients. This heating and cooling control can be used to minimize internal stresses in the joint due to thermal effects.

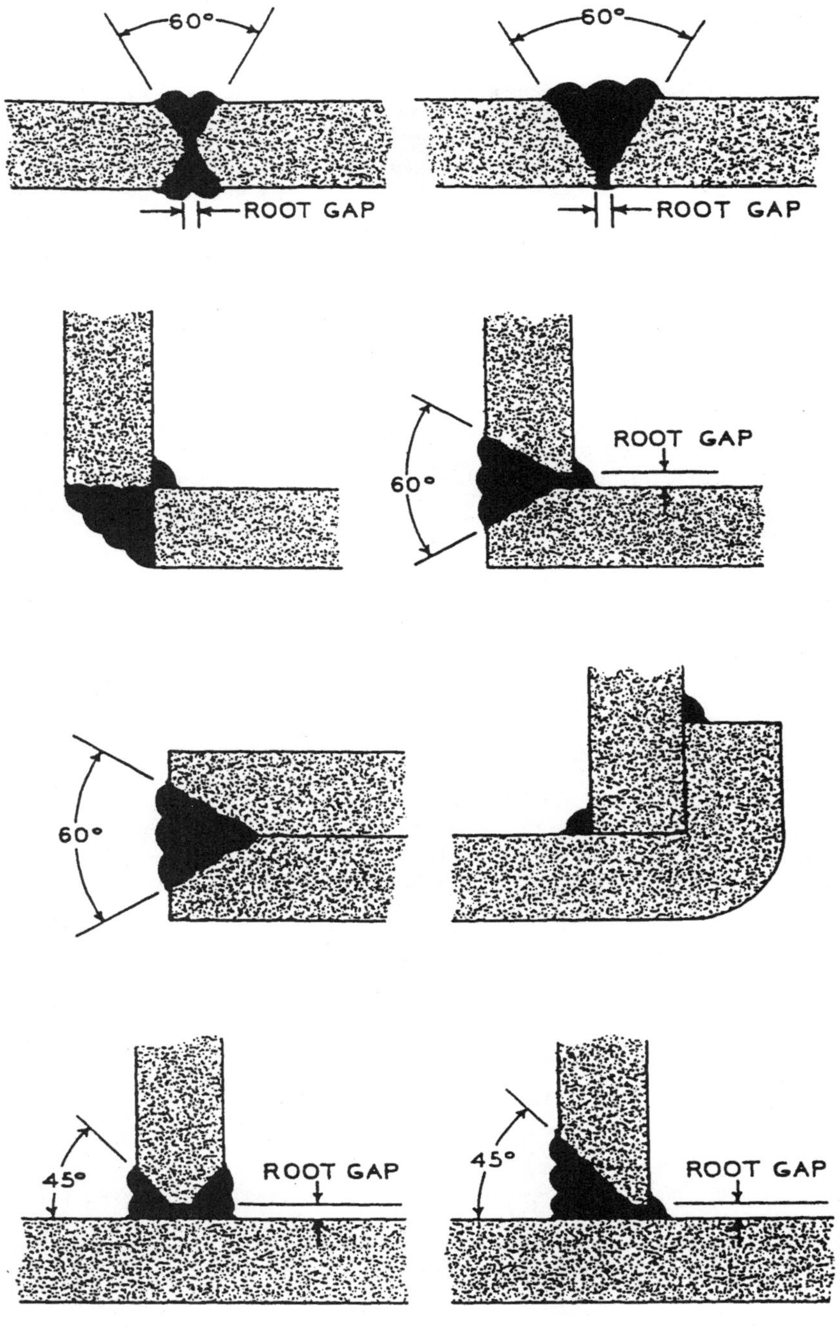

Figure 8.3 Conventional hot-gas welding joint designs.[5]

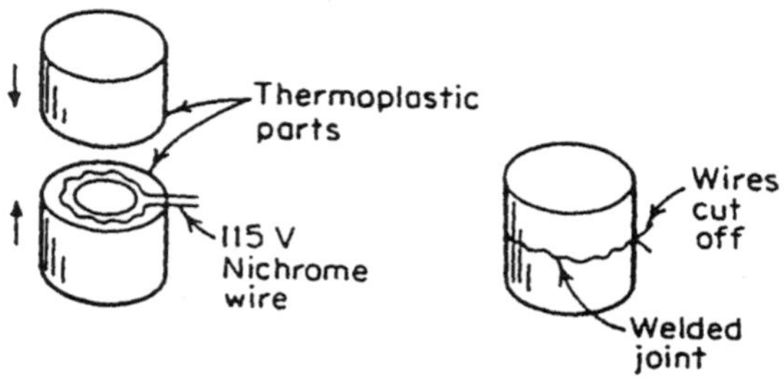

Figure 8.4 Resistance wire welding of thermoplastic joints.[6]

Resistance wire welding can be used to weld dissimilar materials, including thermoplastics, thermoplastic composites, thermosets, and metal, in many combinations. When the substrate is not the source of the adhesive melt, such as when bonding two aluminum strips together, then a thermoplastic film with an embedded heating element can be used as the adhesive. Large parts can require considerable power. Resistance welding has been applied to complex joints in automotive applications (including vehicle bumpers and panels), joints in plastic pipe, and medical devices. Resistance wire welding is not restricted to flat surfaces. If access to the heating element is possible, repair of badly bonded joints is possible, and joints can be disassembled in a reverse process to which they were made. A similar type of process can be used to cure thermosetting adhesives when the heat generated by the resistance wire is used to advance the cure.

8.4.4 Laser Welding

Laser welding of plastic parts has been available for the last 30 years. However, only recently have the technology and cost allowed these joining techniques to be considered broadly.[7] Laser welders produce small beams of photons and electrons, respectively. The beams are focused onto the workpiece. Power density varies from a few to several thousand W/mm^2, but low-power lasers (less than 50 W/mm^2) are generally used for plastic parts.

Laser welding is a high-speed, noncontact process for welding thermoplastics. It is expected to find applications in the packaging and medical products industries.[8] Thermal radiation absorbed by the work piece forms the weld. Sold state Nd:YAG and CO_2 lasers are most commonly used for welding. Laser radiation, in the normal mode of operation, is so intense and focused that it very quickly degrades thermoplastics. However, lasers have been used to butt weld polyethylene by pressing the unwelded parts together and tracking a defocused laser beam along the joint area. High-speed laser welding of polyethylene films has been demonstrated at weld speeds of 164 ft/min using carbon dioxide and Nd:YAG lasers. Weld strengths are very near the strength of the parent substrate.

Processing parameters that have been studied in laser welding are the power level of the laser, shielding gas flow rate, offset of the laser beam from a focal point on the top surface of the weld interface, travel speed of the beam along the interface, and welding pressure.[9] Butt joint designs can be laser welded; lap joints can be welded by directing the beam at the edges of the joint.

Lasers have been used primarily for welding polyethylene and polypropylene. Usually, laser welding is applied only to films or thin-walled components. The least powerful beams, around 50 W, with the widest weld spots are used for fear of degrading the polymer substrate. The primary goal in laser welding is to reach a melt temperature where the parts can be joined quickly before the plastic degrades. To avoid material degradation, ac-

curate temperature measurement of the weld surfaces and temperature control by varying laser strength are essential.

Lasers have been primarily used for joining delicate components that cannot stand the pressure of heated tool or other thermal welding methods. Applications exist in the medical, automotive, and chemical industries. Perhaps the greatest opportunity for this process will be for the high-speed joining of films.

Laser welding has also been used for filament winding of fiber reinforced composite materials using a thermoplastic prepreg. A defocused laser beam is directed on the area where the prepreg meets the winding as it is being built up. With suitable control over the winding speed, applied pressure, and the temperature of the laser, excellent reinforced structures of relatively complex shape can be achieved.

Laser welding requires a high investment in equipment and creates the need for a ventilation system to remove hazardous gaseous and particulate materials resulting from the vaporization of polymer degradation products. Of course, suitable precautions must also be taken to protect the eyesight of anyone in the vicinity of a laser welding operation.

8.4.5 Infrared Welding

Infrared radiation is a noncontact alternative for hot-plate welding. Infrared is particularly promising for higher-melting polymers, since the parts do not contact and stick to the heat source. Infrared radiation can penetrate into a polymer and create a melt zone quickly. By contrast, hot-plate welding involves heating the polymer surface and relying on conduction to create the required melt zone.

Infrared welding is at least 30 percent faster than heated-tool welding. High reproducibility and bond quality can be obtained. Infrared welding can be easily automated, and it can be used for continuous joining. Often heated-tool welding equipment can be modified to accept infrared heating elements.

Infrared radiation can be supplied by high-intensity quartz heat lamps. The lamps are removed after melting the polymer, and the parts are forced together as with hot-plate welding. The depth of the melt zone depends on many factors, including minor changes in polymer formulation. For example, colorants and pigments will change a polymer's absorption properties and will affect the quality of the infrared welding process. Generally, the darker the polymer, the less infrared energy is transferred down through a melt zone, and the more likely is surface degradation to occur through overheating.

8.5 Indirect Heating Methods

Many plastic parts may be joined by indirect heating. With these methods, the materials are heated by external energy sources. The heat is induced within the polymer or at the interface. The most popular indirect heating methods are

- Induction welding
- Dielectric welding

For induction welding, the energy source is an electromagnetic field; for dielectric welding, the energy source is an electric field of high frequency.

Indirect heat joining is possible for almost all thermoplastics; however, it is most often used with the newer engineering thermoplastics. The engineering thermoplastics generally have greater heat and chemical resistance than the more conventional plastics. In many applications, engineering plastics are reinforced to improve structural characteristics. They are generally stronger than other plastics and have excellent strength-to-weight ratios.

However, many of the engineering plastics are not well suited to joining by direct heat, because of the high melt temperatures. Indirect heating methods and frictional heating methods must be used to obtain fast, high-quality bonds with these useful plastic materials.

8.5.1 Induction Welding

The electromagnetic induction field can be used to heat a metal grid or insert placed between mating thermoplastic substrates. Radio-frequency energy from the electromagnetic field induces eddy currents in the conductive material, and the material's resistance to these currents produces heat. When the joint is positioned between induction coils, the hot insert causes the plastic to melt and fuse together. Slight pressure is maintained as the induction field is turned off and the joint hardens. The main advantage is that heating occurs only where the electromagnetic insert is applied. The bulk substrate remains at room temperature, avoiding degradation and distortion.

Induction welding is very much like resistance wire welding. An implant is heated to melt the surrounding polymer. Rather than heating the implant restively, in induction welding, the implant is heated with an electromagnetic field. More popular forms of induction welding have been developed that use a bonding agent consisting of a thermoplastic resin filled with metal particles. This bonding agent melts in the induction field and forms the adhesive joint. The advantage of this method is that stresses caused by large metal inserts are avoided.

The bonding agent should be similar to the substrates. When joining polyethylene, for example, the bonding agent may be a polyethylene resin containing 0.5 to 0.6 percent by volume magnetic iron oxide powder. Electromagnetic adhesives can be made from iron-oxide-filled thermoplastics. These adhesives can be shaped into gaskets or film that will melt in the induction field. The step-by-step EMABOND thermoplastic assembly system is illustrated in Fig. 8.5.

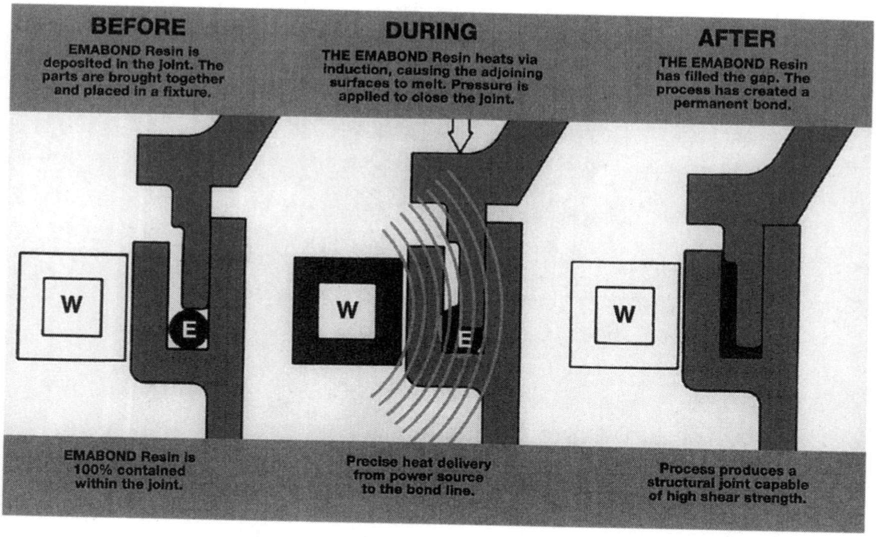

Figure 8.5 Schematic of EMABOND® thermoplastic assembly system.[10] *(Courtesy of Ashland Specialty Chemical Company. EMABOND is a registered trademark of Ashland Inc.)*

The electromagnetic welding process comprises four basic components.

1. An induction generator that converts the 60-Hz electrical supply to 3- to 40-MHz output frequency and output power from 1 to 5 kW
2. An induction heating coil consisting of water-cooled copper tubing, usually formed into hairpin-shaped loops
3. Fixturing used to hold parts in place
4. A bonding material, in the form of molded or extruded preforms, which becomes an integral part of the welded product

Induction heating coils should be placed as close as possible to the joint. For complex designs, coils can be contoured to the joint. Electromagnetic welding systems can be designed for semiautomatic or completely automatic operation. With automated equipment, a sealing rate of up to 150 parts/min can be achieved. Equipment costs are generally in the range of 10,000 to hundreds of thousands of dollars, depending on the degree of automation required.

The bonding agent is usually produced for a particular application to ensure compatibility with the materials being joined. However, induction welding equipment suppliers also offer proprietary compounds for joining dissimilar materials. The bonding agent is often shaped into a profile to match the joint design (i.e., gaskets, rings, ribbon). The fillers used in the bonding agents are micron-size ferromagnetic powders. They can be metallic, such as iron or stainless steel, or a ceramic ferrite material.

Quick bonding rates are generally obtainable, because heating occurs only at the interface. Heat does not have to flow from an outside source or through the substrate material to the point of need. Polyethylene joints can be made in as little as 3 s with electromagnetic welding. Depending on the weld area, most plastics can be joined by electromagnetic welding in 3- to 12-s cycle times.

Plastics that are readily bonded with induction methods include all grades of ABS, nylon, polyester, polyethylene, polypropylene, and polystyrene, as well as those materials often considered more difficult to bond such as acetals, modified polyphenylene oxide, and polycarbonate. Reinforced thermoplastics with filler levels up to 65 percent have been joined successfully.[10] Many combinations of dissimilar materials can be bonded with induction welding processes. Table 8.5 shows compatible plastic combinations for electromagnetic adhesives. Thermoset and other nonmetallic substrates can also be electromagnetically bonded. In these applications the bonding agent acts as a hot-melt adhesive.

Advantages of induction welding include the following:

- Heat damage, distortion and, over-softening of the parts are reduced.
- Squeeze-out of fused material form the bond line is limited.
- Hermetic seals are possible.
- Control is easily maintained by adjusting the output of the power supply.
- No pretreatment of the substrates is required.
- Bonding agents have unlimited storage life.

The ability to produce hermetic seals is cited as one of the prime advantages in certain applications, such as in medical equipment. Welds can also be disassembled by placing the bonded article in an electromagnetic field and remelting the joint. There are few limitations on part size or geometry. The only requirement is that the induction coils can be designed to apply a uniform field. The primary disadvantages of electromagnetic bonding

TABLE 8.5 Compatible Plastic Combinations* for Thermoplastics Bonded by the EMABOND Electromagnetic Bonding Process[12]

	ABS	Acetals	Acrylic	Cellulosics	Ionomer (Surlyn)	Nylon 6.6, 11, 12	Polybutylene	Polycarbonate	Polyethylene	Polyphenylene Oxide (Noryl)	Polypropylene	Polystyrene	Polysulfone	Polyvinyl Chloride	Polyurethane	SAN	Thermoplastic Polyester	Thermoplastic elastomers: Copolyester	Styrene bl. copolymer	Olefin type
ABS	●		●					●								●				
Acetals		●																		
Acrylic	●		●					●				●				●				
Cellulosics				●																
Ionomer (Surlyn)					●															
Nylon 6.6, 11, 12						●														
Polybutylene							●													
Polycarbonate	●		●					●				●	●			●				
Polyethylene									●										●	
Polyphenylene oxide (Noryl)										●		●								
Polypropylene											●									●
Polystyrene			●					●		●		●				●				
Polysulfone								●					●							
Polyvinyl chloride														●						
Polyurethane															●					
SAN	●		●					●				●				●				
Thermoplastic polyester																	●			
Thermoplastic elastomers																				
Copolyester																		●		●
Styrene bl. copolymer									●										●	
Olefin type											●							●		●

*● = compatible

are that the metal inserts remain in the finished product, and they represent an added cost. The cost of induction welding equipment is high. The weld is generally not as strong as those obtained by other welding methods.

Induction welding is frequently used for high-speed bonding of many plastic parts. Production cycles are generally faster than with other bonding methods. It is especially useful on plastics that have a high melt temperature, such as the modern engineering plastics. Thus, induction welding is used in many under-the-hood automotive applications. It is also frequently used for welding large or irregularly shaped parts.

Electromagnetic induction methods have also been used to quickly cure thermosetting adhesives such as epoxies. Metal particle fillers or wire or mesh inserts are used to provide the heat source. These systems generally have to be formulated so that they cure with a low internal exotherm.Otherwise, the joint will overheat, and the adhesive will thermally degrade.

8.5.2 Dielectric Heating

Dielectric sealing can be used on most thermoplastics except those that are relatively transparent to high-frequency electric fields. This method is used mostly to seal vinyl

sheeting such as automobile upholstery, swimming pool liners, and rainwear. An alternating electric field is imposed on the joint, which causes rapid reorientation of polar molecules. As a result, heat is generated within the polymer by molecular friction. The heat causes the polymer to melt, and pressure is applied to the joint. The field is then removed, and the joint is held until the weld cools. The main difficulty in using dielectric heating as a bonding method is in directing the heat to the interface. Generally, heating occurs in the entire volume of the polymer that is exposed to the electric field.

Variables in the bonding operation are the frequency generated, dielectric loss of the plastic, the power applied, pressure, and time. The materials most suitable for dielectric welding are those that have strong dipoles. These can often be identified by their high electrical dissipation factors. Materials most commonly welded by this process include polyvinyl chloride, polyurethane, polyamide, and thermoplastic polyester. Since the field intensity decreases with distance from the source, this process is normally used with thin polymer films.

Dielectric heating can also be used to generate the heat necessary for curing polar, thermosetting adhesives, and it can be used to quickly evaporate water from a water-based adhesive formulation. Dielectric-processing water-based adhesives are commonly used in the furniture industry for very fast drying of wood joints in furniture. Common white glues, such as polyvinyl acetate emulsions, can be dried in seconds using dielectric heating processes.

There are basically two forms of dielectric welding: radio frequency welding and microwave welding. Radio frequency welding uses high frequencies (13–100 MHz) to generate heat in polar materials, resulting in melting and weld formation after cooling. The electrodes are usually designed into the platens of a press. Microwave welding uses high-frequency (2–20 GHz) electromagnetic radiation to heat a susceptor material located at the joint interface. The generated heat melts thermoplastic materials at the joint interface, producing a weld upon cooling. Heat generation occurs in microwave welding through absorption of electrical energy similar to radio frequency welding.

Polyaniline doped with an aqueous acid is used as a susceptor in microwave welding. This introduces polar groups and a degree of conductivity into the molecular structure. It is these polar groups that preferentially generate heat when exposed to microwave energy. These doped materials are used to produce gaskets that can be used as an adhesive in dielectric welding.

Dielectric welding is commonly used for sealing thin films such as polyvinyl chloride for lawn waste bags, inflatable articles, liners, and clothing. It is used to produce high-volume stationery items such as loose-leaf notebooks and checkbook covers. Because of the cost of the equipment and the nature of the process, industries of major importance for dielectric welding are the commodity industries.

8.6 Friction Welding

In friction welding, the joint interface alone is heated via mechanical friction caused by one substrate surface contacting and sliding over another substrate surface. The frictional heat generated is sufficient to create a melt zone at the interface. Once a melt zone is created the relative movement is stopped, and the parts are held together under slight pressure until the melt cools and sets. Common friction welding processes include

- Spin welding
- Ultrasonic welding
- Vibration welding

8.6.1 Spin Welding

Spin welding uses frictional forces to provide the heat of fusion at the interface. One substrate is rotated very rapidly while in touch with the other substrate, which is fixed in a stationary position. The surfaces melt by frictional heating without heating or otherwise damaging the areas outside of the joint. Sufficient pressure is applied during the process to force out a slight amount of resinous flash along with excess air bubbles. Once the rotation is stopped, position and pressure are maintained until the weld sets. The rotation speed and pressure are dependent on the type of thermoplastic being joined.

Spin welding is an old and uncomplicated technique. The equipment required can be as simple as lathes or modified drills. Spin welding has a lower capital cost than other welding methods. The base equipment required is comparatively inexpensive; however, auxiliary equipment, such as fixtures, part feeders, and unloaders, can drive up the cost of the system. Depending on the geometry and size of the part, the fixture that attaches the part to the rotating motor may be complex. A production rate of 300 parts/min is possible on simple circular joints with an automated system containing multiple heads.

The main advantages of spin welding are its simplicity, high weld quality, and the wide range of possible materials that can be joined with this method. Spin welding is capable of very high throughput. Heavy welds are possible with spin welding. Actual welding times for most parts are only several seconds. A strong hermetic seal can be obtained that is frequently stronger than the material substrate itself. No foreign materials are introduced into the weld, and no environmental considerations are involved. The main disadvantage of this process is that spin welding is used primarily on parts where at least one substrate is circular.

When considering a part as a candidate for spin welding, there are three items that must be considered.

1. The type of material and the temperature at which it starts to become tacky
2. The diameter of the parts
3. How much flash will develop and what to do with the flash

The parts that are to be welded must be structurally stiff enough to resist the pressure required. Joint areas must be circular, and a shallow matching groove is desirable to index the two parts and provide a uniform bearing surface. In addition, the tongue-and-groove type joint is useful in hiding the flash that is generated during the welding process. However, a flash "trap" will usually lower the ultimate bond strength. It is generally more desirable either to remove the flash or to design the part so that the flash accumulates on the inside of the joint and is hidden from view. Figure 8.6 shows conventional joint designs used in spin welding.

Since the heating generated at the interface depends on the relative surface velocity, the outside edges of circular components will see higher temperatures by virtue of their greater diameter and surface velocity. This will cause a thermal differential that could result in internal stress in the joint. To alleviate this affect, joints with hollow section and thin walls are preferred.

The larger the part, the larger the motor required to spin the part, as more torque is required to spin the part and obtain sufficient friction. Parts with diameters of 1–5 in have been spin welded using motors from 1/4 to 3 hp.[14] The weld can be controlled by the rotational speed of the motor and somewhat by the pressure on the piece being joined, the timing of the pressure during spin and during joining, and the cooling time and pressure. In commercial rotation welding machines, speeds can range from 200–14,000 r/min. Welding times range from tenths of a second to 20 s, and cool-down times are in the range of 0.5 s. A typical complete process time is two seconds.[13] Axial pressure on the part ranges

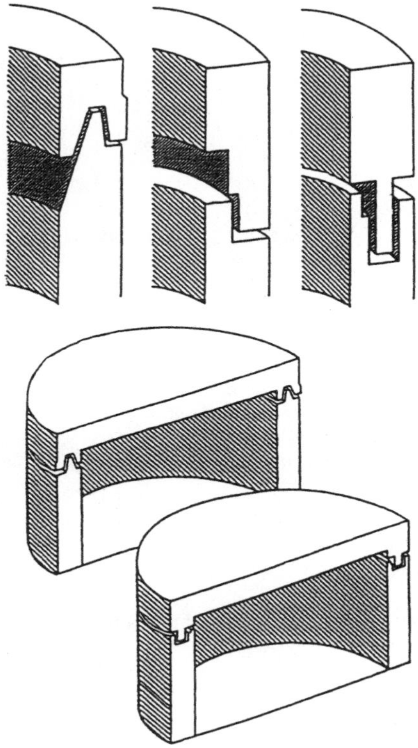

Figure 8.6 Common joints used in the spin welding process.[13]

from 150–1000 lb/in^2. A prototype appraisal is usually completed to determine the optimal parameters of the process for a particular material and joint design.

Table 8.6 shows the temperature at which tackiness starts for most thermoplastics that can be spin welded. This data is useful for all forms of thermal welding—not only spin welding. The tackiness temperature can be used as a guide to determine r/min or s·ft/m (surface feet per minute) required at the part. Rotational speeds from 200–14,000 r/min are often used. An unfilled 1-in diameter polyethylene part may be spun at 1000 r/min (260 s·ft/m) to reach the tackiness temperature of 280°F. As the amount of inert filler increases in the part, the speed needs to be increased. Effects of increasing rotations per minute are similar to those of increased pressure.

Typical applications include small parts such as fuel filters, check valves, aerosol cylinders, tubes, and containers. Spin welding is also a popular method of joining large-volume products such as packaging and toys. Spin welding can also be used for attaching studs to plastic parts.

8.6.2 Ultrasonic Welding

Ultrasonic welding is also a frictional process that can be used on many thermoplastic parts. Frictional heat in this form of welding is generated by high-frequency vibration. The basic parts of a standard ultrasonic welding device are shown in Fig. 8.7. During ultrasonic welding, a high-frequency electrodynamic field is generated that resonates a metal

TABLE 8.6 Tack Temperatures of Common Thermoplastics (from Ref. 14)

Plastic	Tackiness temperature, °F
Ethylene, vinyl, acetate	150
PVC	170
Polystyrene, high-impact	180
ABS, high-impact	200
Acetal	240
Polyurethane, thermoplastic	245
SAN, CAB	250
Polypropylene, Noryl®	260
Cellulose acetate	270
Polycarbonate	275
Polyethylene	280
Acetal	290
Acrylic	320
Polysulfone	325
PET	350
PES	430
Fluorocarbon, melt-processable	630

horn that is in contact with one substrate. The horn vibrates the substrate with sufficient speed, relative to a fixed substrate, that significant heat is generated at the interface. With pressure and subsequent cooling, a strong bond can be obtained. The stages of the ultrasonic welding process are shown in Fig. 8.8.

The frequency generally used in ultrasonic assembly is 20 kHz, because the vibration amplitude and power necessary to melt thermoplastics are easy to achieve. However, this power can produce a great deal of mechanical vibration, which is difficult to control, and tooling becomes large. Higher frequencies (40 kHz) that produce less vibration are possible and are generally used for welding the engineering thermoplastics and reinforced polymers. Higher frequencies also more appropriate for smaller parts and for parts where lower material degradation is required.

Ultrasonic welding is clean, fast (20–30 parts per minute), and usually results in a joint that is as strong as the parent material. The method can provide hermetically sealed components if the entire joint can be welded at one time. Large parts generally are too massive to be joined with one continuous bond, so spot welding is necessary. It is difficult to obtain completely sealed joint with spot welding. Materials handling equipment can be easily interfaced with the ultrasonic system to further improve rapid assembly.

Rigid plastics with a modulus of elasticity are best. Rigid plastics readily transmit the ultrasonic energy, whereas softer plastics tend to dampen the energy before it reaches the critical joint area. Excellent results generally are obtainable with polystyrene, SAN, ABS, polycarbonate, and acrylic plastics. PVC and the cellulosics tend to attenuate energy and

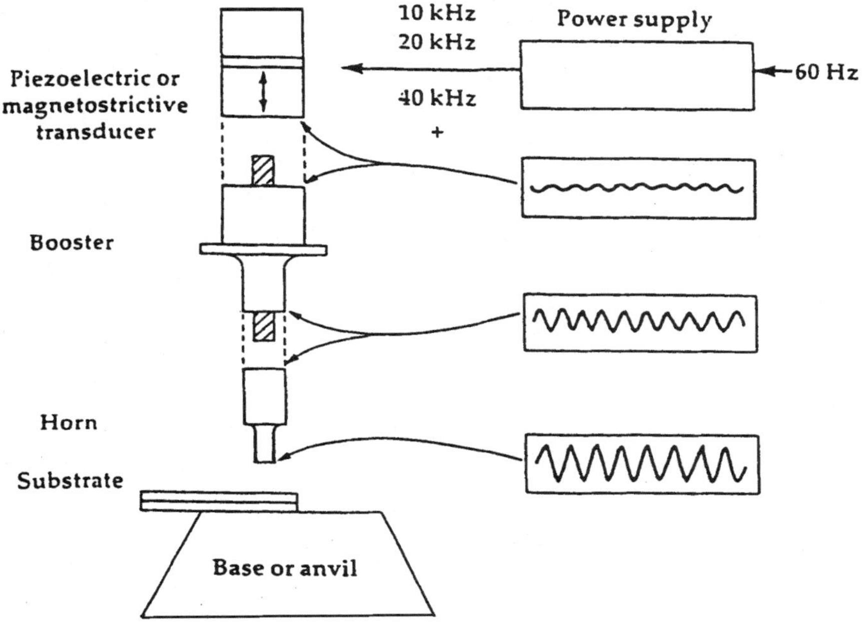

Figure 8.7 Equipment used in a standard ultrasonic welding process.[13]

deform or degrade at their surfaces. Figure 8.9 shows an index for the ultrasonic weldability of conventional thermoplastics. Dissimilar plastics may be joined if they have similar melt temperatures and are chemical compatible. The plastic compatibility chart for ultrasonic welding is shown in Table 8.7. Materials such a polycarbonate and nylon must be dried before welding, otherwise their high level of internal moisture will cause foaming and interfere with the joint.

Common ultrasonic welding joint designs are shown in Fig. 8.10. The most common design is a butt joint that uses an "energy director." This design is appropriate for most amorphous plastic materials. The wedge design concentrates the vibrational energy at the tip of the energy director. A uniform melt then develops where the volume of material formed by the energy director becomes the material that is consumed in the joint. Without the energy director, a butt joint would produce voids along the interface, resulting in stress and a low strength joint. Shear and scarf joints are employed for crystalline polymeric materials. They are usually formed by designing an interference fit.

Ultrasonic welding can also be used to stake plastics to other substrates and for inserting metal parts. It can also be used for spot welding two plastic components. Figure 8.11 illustrates ultrasonic insertion, swaging, stacking, and spot welding operations. In ultrasonic spot welding, the horn tip passes through the top sheet to be welded. The molten plastic forms a neat raised ring on the surface that is shaped by the horn tip. Energy is also released at the interface of the two sheets producing frictional heat. As the tip penetrates the bottom substrate, displaced molten plastic flows between the sheets into the preheated area and forms a permanent bond.

Ultrasonic heating is also applicable to hot-melt and thermosetting adhesives.[20] In these cases, the frictional energy is generated by the substrate contacting an adhesive film be-

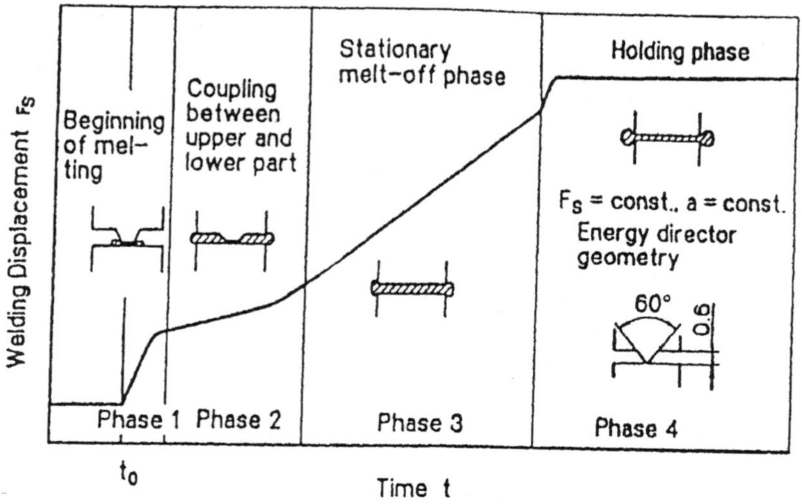

Figure 8.8 Stages in the ultrasonic welding process. In Phase 1, the horn is placed in contact with the part, pressure is applied, and vibratory motion is started. Heat generation due to friction melts the energy director, and it flows into the joint interface. The weld displacement begins to increase as the distance between the parts decreases. In Phase 2, the melting rate increases, resulting in increased weld displacement, and the part surfaces meet. Steady-state melting occurs in Phase 3, as a constant melt layer thickness is maintained in the weld. In Phase 4, the holding phase, vibrations cease. Maximum displacement is reached, and intermolecular diffusion occurs as the weld cools and solidifies.[16]

tween the two substrates. The frictional energy generated is sufficient either to melt the hot-melt adhesive or to cure the thermosetting adhesive.

8.6.3 Vibration Welding

Vibration welding is similar to ultrasonic welding in that it uses the heat generated at the surface of two parts rubbing together. This frictional heading produces melting in the interfacial area of the joint. Vibration welding is different from ultrasonic welding, however, in that it uses lower frequencies of vibration—120–240 Hz rather than 20–40 kHz as used for ultrasonic welding. With lower frequencies, much larger parts can be bonded because of less reliance on the power supply. Figure 8.12 shows the joining and sealing of a two-part plastic tank design of different sizes using vibration welding.

There are two types of vibration welding: linear and axial. Linear vibration welding is most commonly used. Friction is generated by a linear, back-and-forth motion. Axial or orbital vibration welding allows irregularly shaped plastic parts to be vibration welded. In axial welding, one component is clamped to a stationary structure, and the other component is vibrated using orbital motion.

Vibration welding fills a gap in the spectrum of thermoplastic welding in that it is suitable for large, irregularly shaped parts. Vibration welding has been used successfully on large thermoplastic parts such as canisters, pipe sections, and other parts that are too large to be excited with an ultrasonic generator and ultrasonically welded. Vibration welding is also capable of producing strong, pressure-tight joints at rapid rates. The major advantage is its application to large parts and to non-circular joints, provided that a small relative motion between the parts in the welding plane is possible.

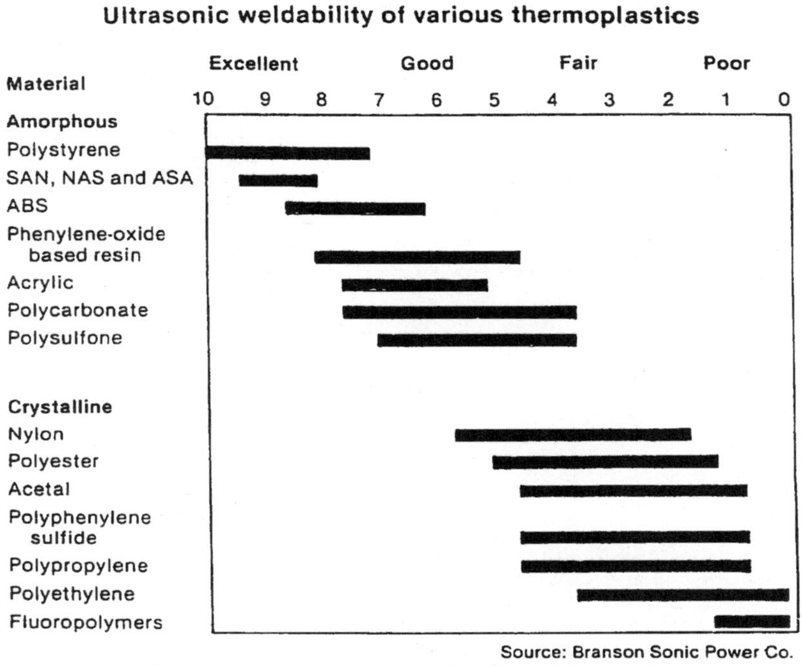

Figure 8.9 Ultrasonic weldability index for common thermoplastics.[17]

Usually, the same manufacturers of ultrasonic welding equipment will also provide vibration welding equipment. Vibration welding equipment can be either electrically driven (variable frequency) or hydraulically driven (constant frequency). Capital cost is generally higher than with ultrasonic welding.

Process parameters to control in vibration welding are the amplitude and frequency of motion, weld pressure, and weld time. Most industrial vibration welding machines operated at frequencies of 120–240 Hz. The amplitude of vibration is usually less than 0.2 in. Lower weld amplitudes are used with higher frequencies. Lower amplitudes are necessary when welding parts into recessed cavities. Lower amplitudes (0.020 in) are used for high-temperature thermoplastics. Joint pressure is held in the rage of 200–250 lb/in^2, although much higher pressures are required at times. High mechanical strength can usually be obtained at shorter weld times by decreasing the pressure during the welding cycle. Vibration welding equipment has been designed to vary the pressure during the welding cycle to improve weld quality and decrease cycle times. This also allows more of the melted polymer to remain in the bond area, producing a wider weld zone.

Vibration welding times depend on the melt temperature of the resin and range from 1–10 s with solidification times of less than 1 s. Total cycle times typically range from 6–15 s. This is slightly longer than typical spin welding and ultrasonic welding cycles, but much shorter than hot-plate welding and solvent cementing.

A number of factors must be considered when vibration welding larger parts. Clearances must be maintained between the parts to allow for movement between the halves. The fixture must support the entire joint area, and the parts must not flex during welding. Vibration welding is applicable to a variety of thermoplastic parts with planar or slightly

TABLE 8.7 Compatibility* of Plastics for Ultrasonic Welding (from Ref. 15)

	ABS	ABS/polycarbonate alloy	ABS, PVC alloy (Cycoloy 800)	Acetal	Acrylics	Acrylic multipolymer	Acrylic/PVC alloy (Kydex)	ASA	Butyrates	Cellulosics	Modified phenylene oxide	Nylon	Polycarbonate	Polyethylene	Polyimide	Polypropylene	Polystyrene	Polysulfone	PPO	PVC	SAN-NAS
ABS	▨	▨	■	□	▨	■	■	■	■	□	□	□	□	□	□	□	■	□	□	□	■
ABS/polycarbonate allow	▨	▨	■	□	■	■	■	■	□	□	□	□	▨	□	□	□	□	□	□	□	■
ABS, PVC alloy (Cycovin)	■	■	▨	□	■	■	■	■	□	□	□	□	□	□	□	□	□	□	□	■	□
Acetal	□	□	□	▨	□	□	□	□	□	□	□	□	□	□	□	□	□	□	□	□	□
Acrylics	▨	■	■	□	▨	■	■	■	□	□	□	□	■	□	□	□	□	□	□	□	■
Acrylic multipolymer	■	■	■	□	■	▨	■	■	□	□	□	□	□	□	□	□	■	□	□	□	■
Acrylic/PVC alloy (Kydex)	■	■	■	□	■	■	▨	■	□	□	□	□	□	□	□	□	□	□	□	■	□
ASA	■	■	■	□	■	■	■	▨	□	□	□	□	□	□	□	□	■	□	□	□	■
Butyrates	■	□	□	□	□	□	□	□	▨	□	□	□	□	□	□	□	□	□	□	□	□
Cellulosics	□	□	□	□	□	□	□	□	□	▨	□	□	□	□	□	□	□	□	□	□	□
Modified phenylene oxide	□	□	□	□	□	□	□	□	□	□	▨	□	□	□	□	□	▨	□	▨	□	■
Nylon	□	□	□	□	□	□	□	□	□	□	□	▨	□	□	□	□	□	□	□	□	□
Polycarbonate	□	▨	□	□	■	□	□	□	□	□	□	□	▨	□	□	□	□	□	□	□	□
Polyethylene	□	□	□	□	□	□	□	□	□	□	□	□	□	▨	□	□	□	□	□	□	□
Polyimide	□	□	□	□	□	□	□	□	□	□	□	□	□	□	▨	□	□	□	□	□	□
Polypropylene	□	□	□	□	□	□	□	□	□	□	□	□	□	□	□	▨	□	□	□	□	□
Polystyrene	■	□	□	□	□	■	□	■	□	□	▨	□	□	□	□	□	▨	□	□	□	■
Polysulfone	□	□	□	□	□	□	□	□	□	□	□	□	□	□	□	□	□	▨	□	□	□
PPO	□	□	□	□	□	□	□	□	□	□	▨	□	□	□	□	□	□	□	▨	□	□
PVC	□	□	■	□	□	□	■	□	□	□	□	□	□	□	□	□	□	□	□	▨	□
SAN-NAS	■	■	□	□	■	■	□	■	□	□	■	□	□	□	□	□	■	□	□	□	▨

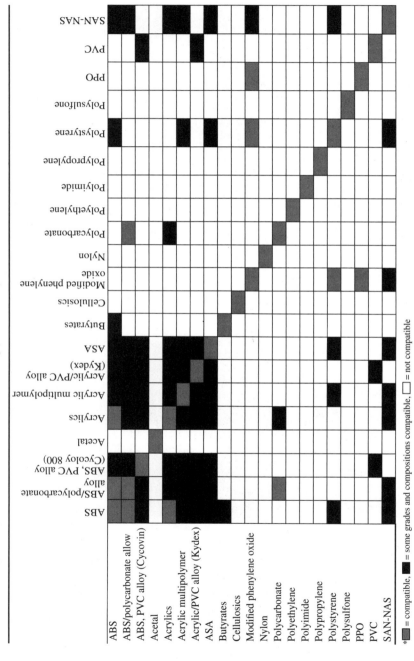

*▨ = compatible, ■ = some grades and compositions compatible, □ = not compatible

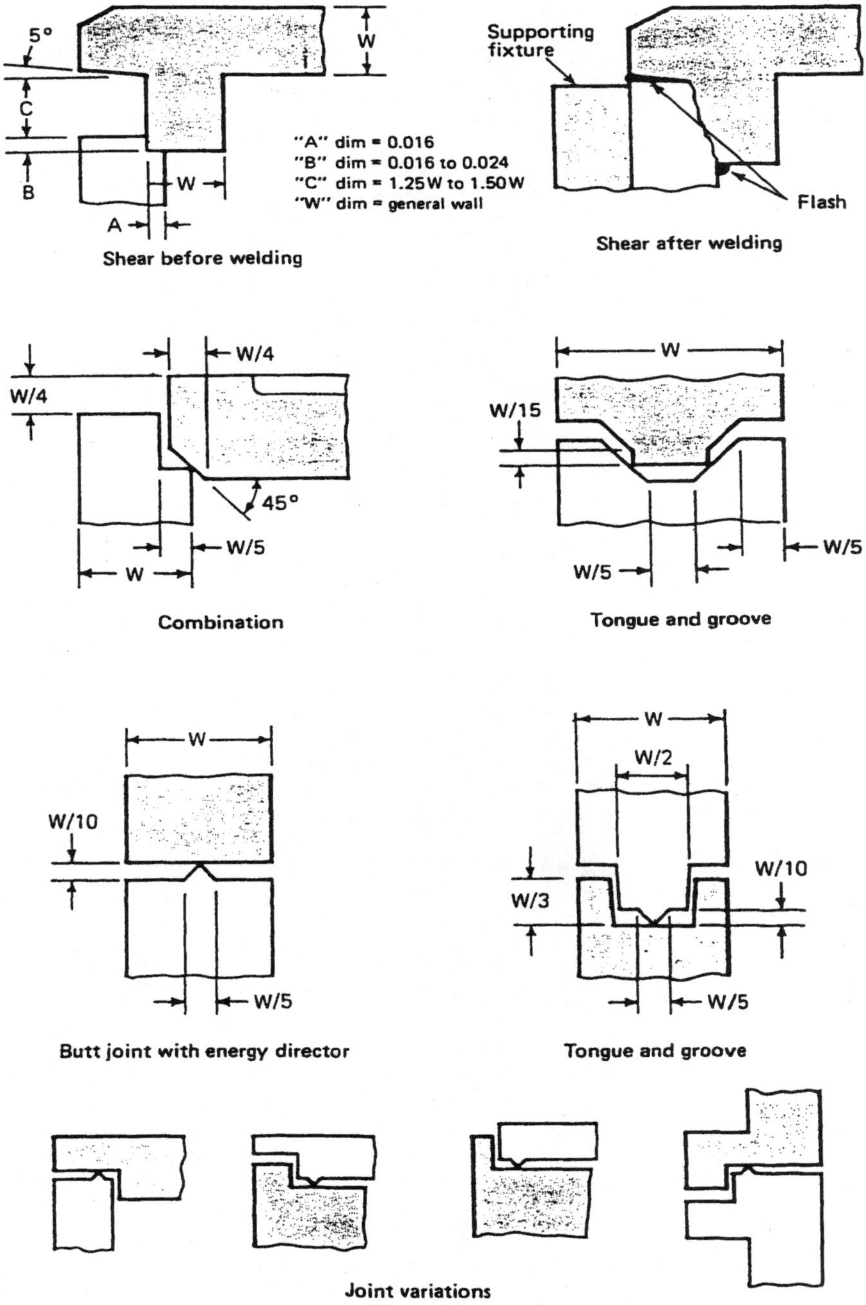

Figure 8.10 Ultrasonic welding joints for amorphous and crystalline polymeric materials.[18]

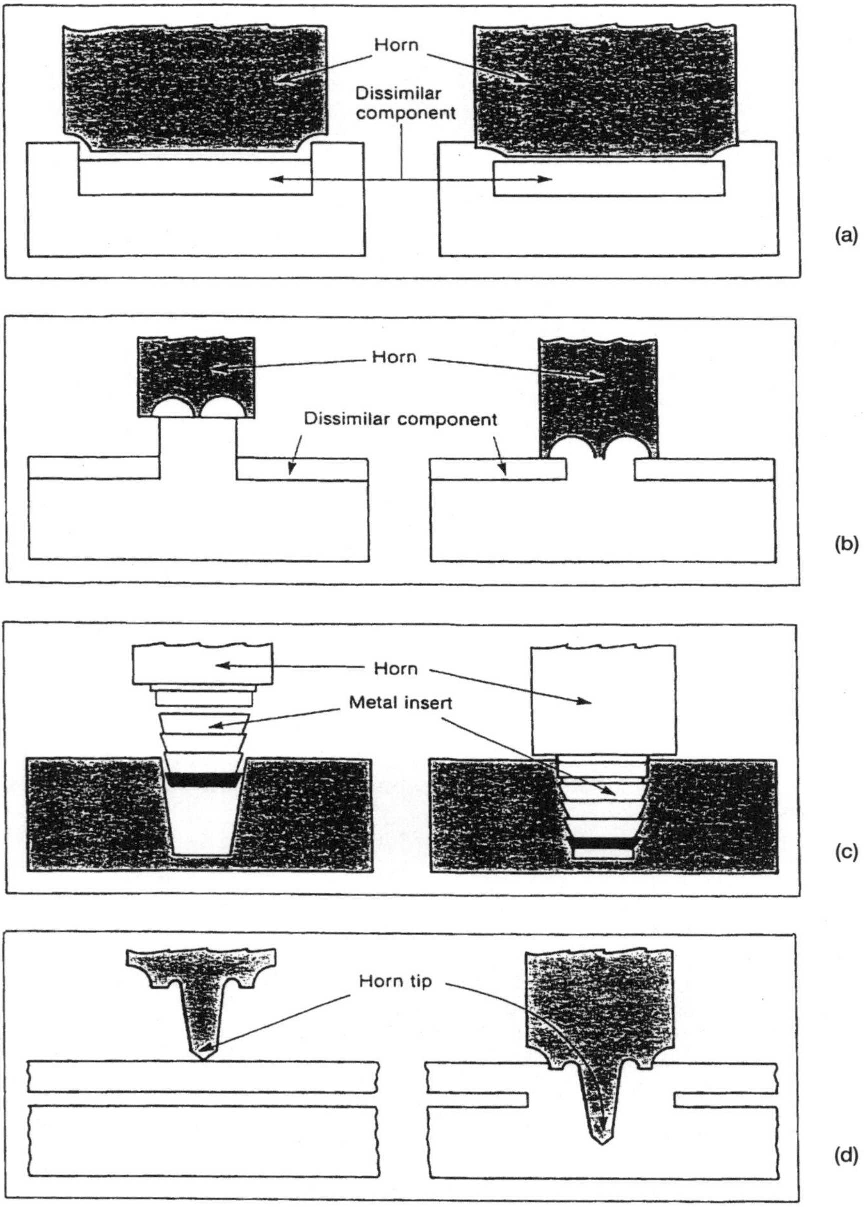

Figure 8.11 Ultrasonic joining operating. (a) Swaging: the plastic ridge is melted and reshaped (left) by ultrasonic vibrations to lock another part into place. (b) Staking: ultrasonic vibrations melt and reform a plastic stud (left) to lock a dissimilar component into place (right). (c) Insertion: a metal insert (left) is embedded in a preformed hole in a plastic part by ultrasonic vibration (right). (d) Spot welding: two plastic components (left) are joined at localized points (right).[19]

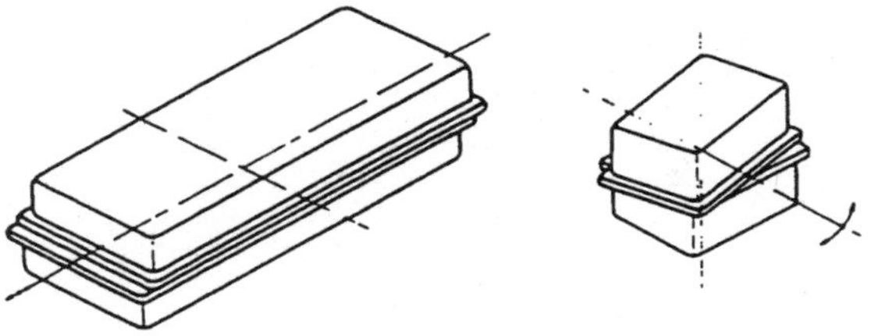

Linear vibration (left) is employed where the length to width ratio precludes the use of axial welding (right) where the axial shift is still within the width of the welded edge.

Figure 8.12 Linear and axial vibration welding of a two-part container.[21]

curved surfaces. The basic joint is a butt joint, but, unless parts have thick walls, a heavy flange is generally required to provide rigidity and an adequate welding surface. Typical joint designs for vibration welds are shown in Fig. 8.13.

Vibration welding is ideally suited to parts injection molded or extruded in engineering thermoplastics as well as acetal, nylon, polyethylene, ionomer, and acrylic resins. Almost any thermoplastic can be vibration welded. Unlike other welding methods, vibration welding is applicable to crystalline or amorphous or filled, reinforced, or pigmented materials. Vibration welding also can be utilized with fluoropolymers and polyester elastomers, none of which can be joined by ultrasonic welding. By optimizing welding parameters and glass fiber loadings, nylon 6 and nylon 66 butt joints can be produced having up to 17 percent higher strength than the base resin.[23] Any pair of dissimilar materials that can be ultrasonically joined can also be vibration welded.

Vibration welding techniques have found several applications in the automobile industry, including emission control canisters, fuel pumps and tanks, head and tail light assemblies, heater valves, air intake filters, water pump housings, and bumper assemblies. They have also been used for joining pressure vessels and for batteries, motor housings, and butane gas lighter tanks.

8.7 Solvent Cementing

Solvent cementing is the simplest and most economical method of joining noncrystalline thermoplastics. In solvent cementing, the application of the solvent softens and dissolves the substrate surfaces being bonded. The solvent diffuses into the surface, allowing increased freedom of movement of the polymer chains. As the parts are then brought together under pressure, the solvent-softened plastic flows. Van der Walls attractive forces are formed between molecules from each part, and polymer chains from each part intermingle and diffuse into one another. The parts then are held in place until the solvent evaporates from the joint area.

Solvent-cemented joints of like materials are less sensitive to thermal cycling than joints bonded with adhesives, because there is no stress at the interface due to differences in thermal expansion between the adhesive and the substrate. When two dissimilar plastics are to be joined, adhesive bonding is generally desirable because of solvent and polymer compatibility problems. Solvent-cemented joints are as resistant to degrading environments as the parent plastic. Bond strength greater than 85 percent of the parent plastic generally can be obtained. Solvents provide high strength bonds quickly due to rapid evaporation rates.

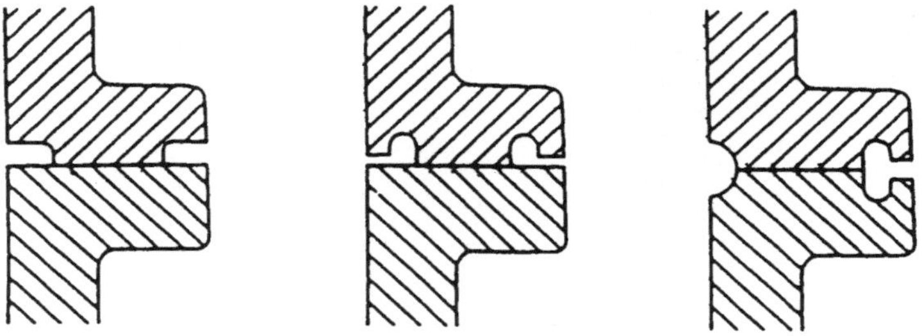

Unless the parts have thick walls, it is necessary to form a flange for the butt weld. In practice, melt traps are usually formed so that the molten resin does not create a flashing.

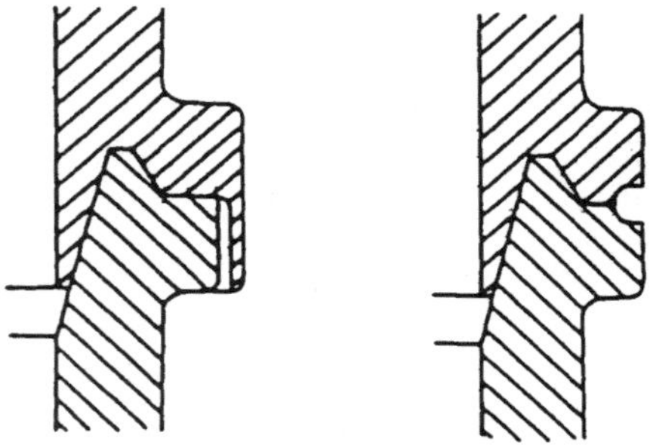

Circular parts can be joined more effectively with tongue-in-groove joints, which retain most of the melt and create a joint of high integrity.

Figure 8.13 Typical vibration welding joint designs.[21]

Solvent bonding is suitable for all amorphous plastics. It is used primarily on ABS, acrylics, cellulosics, polycarbonates, polystyrene, polyphenylene oxide, and vinyls. Solvent welding is not suitable for crystalline thermoplastics. It is not effective on polyolefins, fluorocarbons, or other solvent-resistant polymers. Solvent welding is moderately effective on nylon and acetal polymers. Solvent welding cannot be used to bond thermosets. It can be used to bond soluble plastics to unlike porous surfaces, including wood and paper, through impregnation and encapsulation of the fibrous surface.

The major disadvantage of solvent cementing is the possibility of stress cracking in certain plastic substrates. Stress cracking or *crazing* is the formation of microcracks on the surface of a plastic part that has residual internal stresses due to its molding process. The contact with a solvent will cause the stresses to release uncontrollably, resulting in stress cracking of the part. When this is a problem, annealing of the plastic part at a temperature slightly below its glass transition temperature will usually relieve the internal stresses and reduce the stress cracking probability. Annealing time must be sufficiently long to allow the entire part to come up to the annealing temperature. Another disadvantage of solvent welding is that many solvents are flammable and/or toxic and must be handled accordingly. Proper ventilation must be provided when bonding large areas or with high-volume production.

Solvent cements should be chosen with approximately the same solubility parameter as the plastic to be bonded. Table 8.8 lists typical solvents used to bond major plastics. Solvents used for bonding can be a single pure solvent, a combination of solvents, or a solvent(s) mixed with resin. It is common to use a mixture of a fast-drying solvent with a less volatile solvent to prevent crazing. The solvent cement can be bodied up to 25 percent by weight with the parent plastic to increase viscosity. These bodied solvent cements can fill gaps and provide less shrinkage and internal stress than if only pure solvent is used.

The parts to be bonded should be unstressed and annealed if necessary. For solvent bonding, surfaces should be clean and should fit together uniformly throughout the joint. Close-fitting edges are necessary for good bonding. The solvent cement is generally applied to the substrate with a syringe or brush. In some cases, the surface may be immersed in the solvent. However, solvent application generally must be carefully controlled, since a small difference in the amount of solvent applied to a substrate greatly affects joint strength. After the area to be bonded softens, the parts are mated and held under light pressure until dry. Pressure should be low and uniform so that the joint will not be stressed. After the joint hardens, the pressure is released, and an elevated-temperature cure may by necessary, depending on the plastic and desired joint strength. Exact processing parameters for solvent welding are usually determined by trial and error. They will be dependent on the exact polymer, ambient conditions, and type of solvent used.

The bonded part should not be packaged or stressed until the solvent has adequate time to escape from the joint. Complete evaporation of solvent may not occur for hours or days. Some solvent joined parts may have to be "cured" at elevated temperatures to encourage the release of solvent prior to packaging.

8.8 Methods of Mechanical Joining

There are instances when adhesive bonding, thermal welding, and solvent cementing are not practical joining methods for plastic assembly. This usually occurs because the optimal joint design is not possible, the cost and complexity is too great, or the skill and resources are not present to attempt these forms of fastening. Another common reason for forgoing bonding or welding is when repeated disassembly of the product is required. Fortunately, when these situations occur, the designer can still turn to mechanical fastening as a possible solution.

There are basically two methods of mechanical assembly for plastic parts. The first uses fasteners, such as screws or bolts, and the second uses interference fit such as press fit or snap fit and is generally used in thermoplastic applications. This latter method of fastening is also called design for assembly or self-fastening. If possible, the designer should try to design the entire product as a one-part molding or with the capability of being press-fit or snap-fit together, because it will eliminate the need for a secondary assembly operation. However, mechanical limitations often will make it necessary to join one part to another using a fastening device. Fortunately, there are a number of mechanical fasteners designed for metals that are also generally suitable with plastics, and there are many other fasteners specifically designed for plastics. Typical of these are thread-forming screws, rivets, threaded inserts, and spring clips.

As in adhesive bonding or welding, special considerations must be given to mechanical fastening because of the nature of the plastic material. Care must be taken to avoid overstressing the parts. Mechanical creep can result in loss of preload in poorly designed systems. Reliable mechanically fastened plastic joints require the following:

- A firm strong connection
- Materials that are stable in the environment

TABLE 8.8 Typical Solvents for Solvent Cementing of Plastics (from Ref. 23)

Solvents for bonding plastics	Acetic Acid (Glacial)	Acetone	Acetone: ethyl acetate: R (40:40:20)	Acetone: ethyl lactate (90:10)	Acetone: ethyl lactate (80:20)	Acetone: methyl acetate (70:30)	Butyl acetate: acetone: methyl acetate (50:30:20)	Butyl acetate: methyl methacrylate (40:60)	Chloroform	Ethyl acetate	Ethyl acetate: ethyl alcohol (80:20)	Ethyl dichloride	Ethyl dichloride: methylene chloride (50:50)	Methyl acetate	Methylene chloride	Methylene chloride: methyl methacrylate (60:40)	Methylene chloride: methyl methacrylate (50:50)	Methylene chloride: trichloro-ethylene (85:15)	Methyl ethyl ketone	Methyl isobutyl ketone	Methyl methacrylate	Tetrachloroethylene	Tetrachloroethane	Tetrahydrofuran: cyclohexanone (80:20)	Toluene	Toluene: ethyl alcohol (90:10)	Toluene: Methyl Ethyl Ketone (50:50)	Trichloroethane	Trichloroethylene	Xylene	Xylene: Methyl Isobutyl Ketone (25:75)
ABS																			■	■							■				■
Cellulose acetate		■		■	■	■	■			■				■																	
Cellulose acetate butrate		■	■	■	■	■	■			■				■																	
Cellulose propionate						■																									
Ethyl cellulose											■															■		■			
Nylon	■																														
Polycarbonate																															
Polymethyl methacrylate												■	■			■						■									
Polyphanylene oxide									■			■			■			■											■		■
Polystyrene										■					■				■		■				■				■		
Polysulfone															■																
Polyvinyl chloride																			■	■				■						■	
SAN								■		■									■												
Styrene butadiene		■																	■	■											

- Stable geometry
- Appropriate stresses in the parts including a correct clamping force

In addition to joint strength, mechanically fastened joints should prevent slip, separation, vibration, misalignment, and wear of parts. Well designed joints provide the above without being excessively large or heavy, or burdening assemblers with bulky tools. Designing plastic parts for mechanical fastening will depend primarily on the particular plastic being joined and the functional requirements of the application.

8.8.1 Mechanical Fasteners

A large variety of mechanical fasteners can be used for joining plastic parts to themselves and to other materials. These include

- Machine screws and bolts
- Self-threading screws
- Rivets
- Spring fasteners and clips

In general, when repeated disassembly of the product is anticipated, mechanical fasteners are used. Metal fasteners of high strength can overstress plastic parts, so torque-controlled tightening or special design provisions are required. Where torque cannot be controlled, various types of washers can be used to spread the compression force over larger areas.

8.8.1.1 Machine screws, bolts, etc. Parts molded of thermoplastic resin are sometimes assembled with machine screws or with bolts, nuts, and washers, especially if it is a very strong plastic. Machine screws are generally used with threaded inserts, nuts, and clips. They rarely are used in pretapped holes. Figure 8.14 shows correct and incorrect methods of mechanical fastening of plastic parts using this hardware.

Inserts into the plastic part can be used effectively to provide the female part of the fastener. Inserts that are used for plastic assembly consist of molded-in inserts and post-molded inserts.

Molded-in inserts represent inserts that are placed in the mold before the plastic resin is injected. The resin provides is then shaped to the part geometry and locks the insert into its body. Molded-in inserts provide very high-strength assemblies and relatively low unit cost. However, molded-in inserts could increase part cycle time while the inserts are manually placed in the mold. When the application involves infrequent disassembly, molded-in threads can be used successfully. Coarse threads can also be molded into most materials. Threads of 32 or finer pitch should be avoided, along with tapered threads, because of excessive stress on the part. If the mating connector is metal, overtorque will result in part failure.

Post-molded inserts come in four types: press-in, expansion, self-tapping, and thread forming, and inserts that are installed by some method of heating (e.g., ultrasonic). Metal inserts are available in a wide range of shapes and sizes for permanent installation. Inserts are typically installed in molded bosses, designed with holes to suit the insert to be used. Some inserts are pressed into place, and others are installed by methods designed to limit the stress and increase strength. Generally, the outside of the insert is provided with projections of various configurations that penetrate the plastic and prevent movement under normal forces exerted during assembly.

Whatever mechanical fastener is used, particular attention should be paid to the head of the fastener. Conical heads, called *flat heads,* produce undesirable tensile stresses and

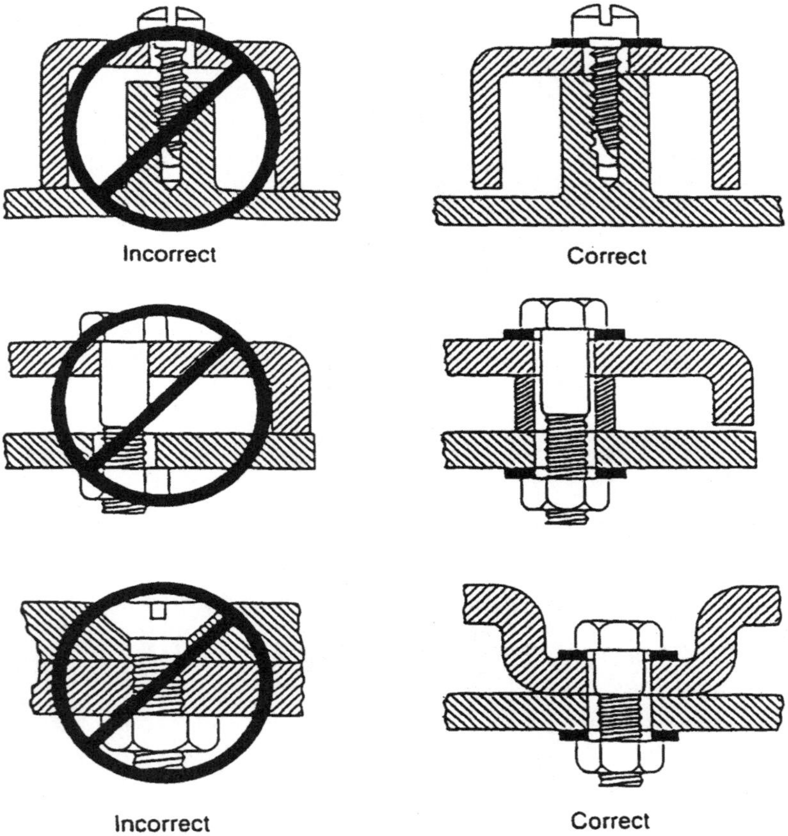

Figure 8.14 Mechanical fastening with (a) self-tapping screws and (b) bolts, nuts, and washers.[24]

should not be used. Bolt or screw heads with a flat underside [such as pan heads, round heads, and so forth (Fig. 8.15)] are preferred, because the stress produced is more compressive. Flat washers are also suggested and should be used under both the nut and the fastener head. Sufficient diametrical clearance for the body of the fastener should always be provided in the parts to be joined. This clearance can nominally be 0.25 mm (0.010 in).

8.8.1.2 Self-threading screws. Self-threading screws can be either thread cutting or thread forming. To select the correct screw, the designer must know which plastic will be used and its modulus of elasticity. The advantage of using these types of screws are as follows:

- They are generally off-the-shelf items.
- They are low in cost.

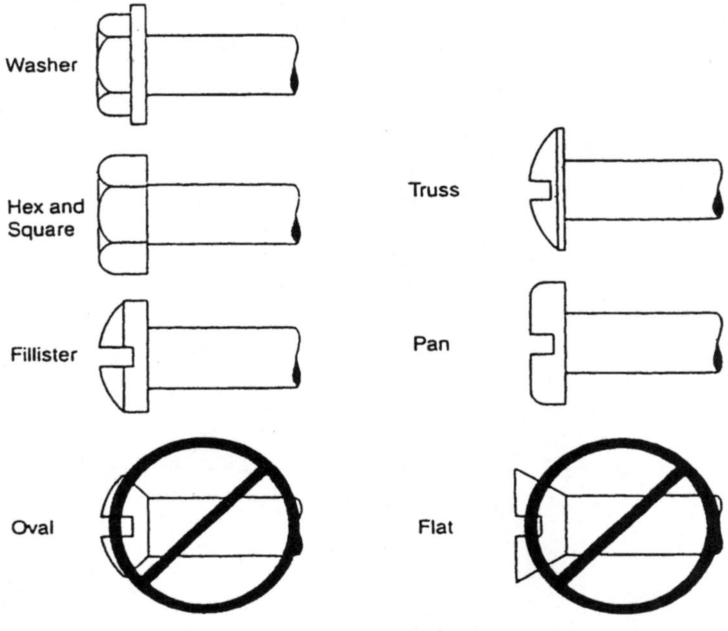

Figure 8.15 Common head systems of screws and bolts. Flat underside of head is preferred.[24]

- They allow high production rates.
- They require minimal tooling investment.

The principal disadvantage of these screws is limited reuse; after repeated disassembly and assembly, these screws will cut or form new threads in the hole, eventually destroying the integrity of the assembly.

Thread-forming screws are used in the softer, more ductile plastics with moduli below 1380 MPa (200,000 lb/in^2). There are a number of fasteners especially designed for use with plastics (Fig. 8.16). Thread-forming screws displace plastic material during the threading operation. This type of screw induces high stress levels in the part and is not recommended for parts made of weak resins.

Assembly strengths using thread-forming screws can be increased by reducing hole diameter in the more ductile plastics, by increasing screw thread engagement, or by going to a larger-diameter screw when space permits. The most common problem encountered with these types of screws is *boss cracking*. This can be minimized or eliminated by increasing the size of the boss, increasing the diameter of the hole, decreasing the size of the screw, changing the thread configuration of the screw, or changing the part to a more ductile plastic.

Thread-cutting screws are used in harder, less-ductile plastics. Thread-cutting screws remove material as they are installed, thereby avoiding high stress. However, these screws should not be installed and removed repeatedly.

8.8.1.3 Rivets. Rivets provide permanent assembly at very low cost. Clamp load must be limited to low levels to prevent distortion of the part. To distribute the load, rivets with

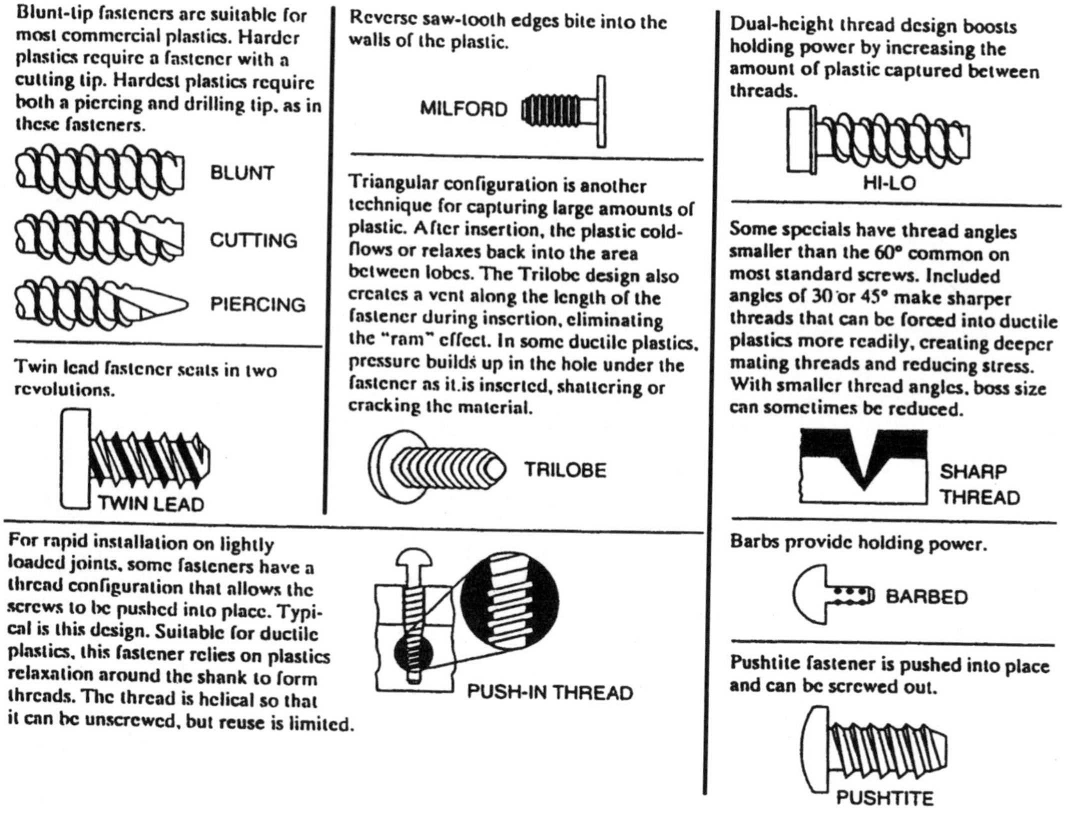
Blunt-tip fasteners are suitable for most commercial plastics. Harder plastics require a fastener with a cutting tip. Hardest plastics require both a piercing and drilling tip, as in these fasteners.
BLUNT
CUTTING
PIERCING
Twin lead fastener seats in two revolutions.
TWIN LEAD
Reverse saw-tooth edges bite into the walls of the plastic.
MILFORD
Triangular configuration is another technique for capturing large amounts of plastic. After insertion, the plastic cold-flows or relaxes back into the area between lobes. The Trilobe design also creates a vent along the length of the fastener during insertion, eliminating the "ram" effect. In some ductile plastics, pressure builds up in the hole under the fastener as it is inserted, shattering or cracking the material.
TRILOBE
For rapid installation on lightly loaded joints, some fasteners have a thread configuration that allows the screws to be pushed into place. Typical is this design. Suitable for ductile plastics, this fastener relies on plastics relaxation around the shank to form threads. The thread is helical so that it can be unscrewed, but reuse is limited.
PUSH-IN THREAD
Dual-height thread design boosts holding power by increasing the amount of plastic captured between threads.
HI-LO
Some specials have thread angles smaller than the 60° common on most standard screws. Included angles of 30 or 45° make sharper threads that can be forced into ductile plastics more readily, creating deeper mating threads and reducing stress. With smaller thread angles, boss size can sometimes be reduced.
SHARP THREAD
Barbs provide holding power.
BARBED
Pushtite fastener is pushed into place and can be screwed out.
PUSHTITE

Figure 8.16 Thread-forming fasteners for plastics.[25]

large heads should be used with washers under the flared end of the rivet. The heads should be three times the shank diameter.

Riveted composite joints should be designed to avoid loading the rivet in tension. Generally, a hole 1/64 in larger than the rivet shank is satisfactory for composite joints. A number of patented rivet designs are commercially available for joining aircraft and aerospace structural composites.

8.8.1.4 Spring steel fasteners. Push-on spring steel fasteners (Fig. 8.17) can be used for holding light loads. Spring steel fasteners are simply pushed on over a molded stud. The stud should have a minimum 0.38 mm (0.015 in) radius at its base. Too large a radius could create a thick section, resulting in sinks or voids in the plastic molding.

8.8.2 Design for Self-Assembly

It is often possible and desirable to incorporate fastening mechanisms in the design of the molded part itself. The two most common methods of doing this are by interference fit (including press-fit or shrink- fit) and by snap-fit. Whether these methods can be used will depend heavily on the nature of the plastic material and the freedom one has in part design.

8.8.2.1 Press fit. In press or interference fits, a shaft of one material is joined with the hub of another material by a dimensional interference between the shaft's outside diameter

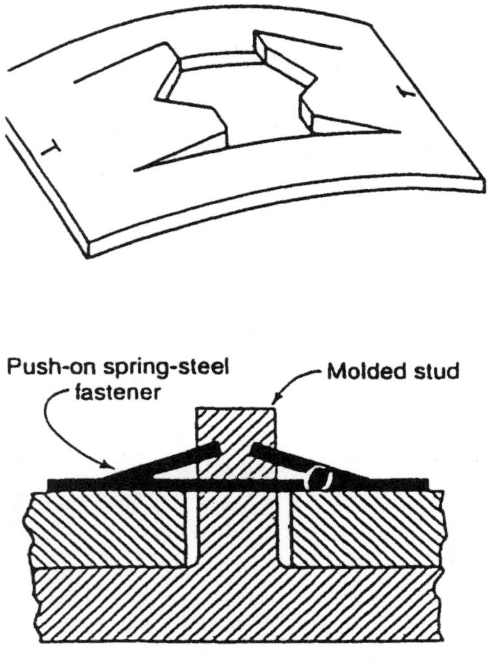

Figure 8.17 Push-on spring steel fasteners.[24]

and the hub's inside diameter. This simple, fast assembly method provides joints with high strength and low cost. Press fitting is applicable to parts that must be joined to themselves or to other plastic and nonplastic parts. The advisability of its use will depend on the relative properties of the two materials being assembled. When two different materials are being assembled, the harder material should be forced into the softer. For example, a metal shaft can be press-fitted into plastic hubs. Press-fit joints can be made by simple application of force or by heating or cooling one part relative to the other.

Press-fitting produces very high stresses in the plastic parts. With brittle plastics, such as thermosets, press-fit assembly may cause the plastic to crack if conditions are not carefully controlled.

Where press-fits are used, the designer generally seeks the maximum pullout force using the greatest allowable interference between parts that is consistent with the strength of the plastic. Figure 8.18 provides general equations for interference fits (when the hub and shaft are made of the same materials and for when they are a metal shaft and a plastic hub). Safety factors of 1.5–2.0 are used in most applications.

For a press-fit joint, the effect of thermal cycling, stress relaxation, and environmental conditioning must be carefully evaluated. Testing of the factory-assembled parts under expected temperature cycles, or under any condition that can cause changes to the dimensions or modulus of the parts, is obviously indicated. Differences in coefficient of thermal expansion can result in reduced interference due either to one material shrinking or expanding away from the other, or it can cause thermal stresses as the temperature changes. Since plastic materials will creep or stress-relieve under continued loading, loosening of the press fit, at least to some extent, can be expected during service. To counteract this, the designer can knurl or groove the parts. The plastic will then tend to flow into the grooves and retain the holding power of the joint.

8.8.2.2 Snap fit. In all types of snap-fit joints, a protruding part of one component, such as a hook, stud, or bead, is briefly deflected during the joining operation, and it is made to catch in a depression (undercut) in the mating component. This method of assembly is uniquely suited to thermoplastic materials due to their flexibility, high elongation, and ability to be molded into complex shapes. However, snap-fit joints cannot carry a load in excess of the force necessary to make or break the snap fit. Snap-fit assemblies are usually employed to attach lids or covers that are meant to be disassembled or that will be lightly loaded. The design should be such that, after assembly, the joint will return to a stress-free condition.

The two most common types of snap fits are those with flexible cantilevered lugs (Fig. 8.19) and those with a full cylindrical undercut and mating lip (Fig. 8.20). Cylindrical snap fits are generally stronger but require deformation for removal from the mold. Materials with good recovery characteristics are required.

To obtain satisfactory results, the undercut design must fulfill certain requirements.

- The wall thickness should be kept uniform.
- The snap fit must be placed in an area where the undercut section can expand freely.
- The ideal geometric shape is circular.
- Ejection of an undercut core from the mold is assisted by the fact that the resin is still at relatively high temperatures.
- Weld lines should be avoided in the area of the undercut.

In the cantilevered snap-fit design, the retaining force is essentially a function of the bending stiffness of the resin. Cantilevered lugs should be designed in a way as not to ex-

General equation for interference

$$I = \frac{S_d D_s}{W}\left[\frac{W\mu_h}{E_h} + \frac{1-\mu_s}{E_s}\right]$$

in which

$$W = \frac{1+\left(\frac{D_s}{D_h}\right)^2}{1-\left(\frac{D_s}{D_h}\right)^2}$$

I = Diametral interference, mm (in.)

S_d = Design stress limit or yield strength of the polymer, generally in the hub, MPa (psi) (A typical design limit for an interference fit with thermoplastics is 0.5% strain at 73°C.)

D_h = Outside diameter of hub, mm (in.)

D_s = Diameter of shaft, mm (in.)

E_h = Modulus of elasticity of hub, MPa (psi)

E_s = Elasticity of shaft, MPa (psi)

μ_h = Poisson's ratio of hub material

μ_s = Poisson's ratio of shaft material

W = Geometric factor

If the shaft and hub are of the same material, $E_h = E_s$ and $\mu_h = \mu_s$. The above equation simplifies to:

Shaft and hub of same material

$$I = \frac{S_d D_s}{W} \times \frac{W+1}{E_h}$$

If the shaft is a high modulus metal or other material, with $E_s > 34.4 \times 10^3$ MPa, the last term in the general interference equation is negligible, and the equation simplifies to:

Metal shaft, plastic hub

$$I = \frac{S_d D_s}{W} \times \frac{W+\mu_h}{E_h}$$

Figure 8.18 General calculation of interference fit between a shaft and a hub.[26]

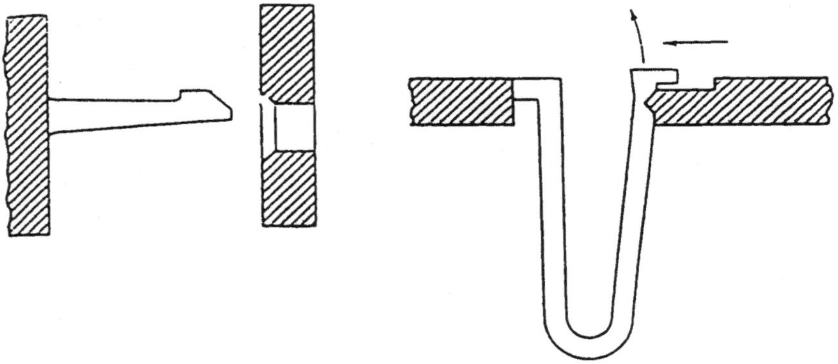

Figure 8.19 Snap-fitting cantilevered arms.[24]

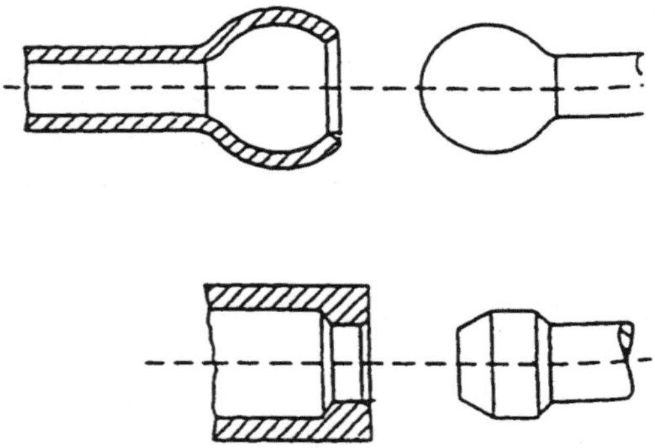

Figure 8.20 Undercuts for snap joints.[24]

ceed allowable stresses during assembly. Cantilevered snap fits should be dimensioned to develop constant stress distribution over their length. This can be achieved by providing a slightly tapered section or by adding a rib. Special care must be taken to avoid sharp corners and other possible stress concentrations. Cantilever design equations have been recently developed to allow for both the part and the snap fit to flex.[28] Many more designs and configurations can be used with snap-fit configuration than only cantilever or snap-fit joints. The individual plastic resin suppliers are suggested for design rules and guidance on specific applications.

8.9 Recommended Assembly Processes for Common Plastics

When decisions are to be made relative to assembly methods (mechanical fastening, adhesive bonding, thermal welding, or solvent cementing), special considerations must be taken because of the nature of the substrate and possible interactions with the adhesive or the environment. The following sections identify some of these considerations and offer an

assembly guide to the various methods of assemblies that have been found appropriate for specific plastics.

8.9.1 Acetal Homopolymer and Acetal Copolymer

Parts made of acetal homopolymer and copolymer are generally strong and tough, with a surface finish that is the mirror image of the mold surface. Acetal parts are generally ready for end-use or further assembling with little or no postmold finishing.

Press fitting has been found to provide joints of high strength at minimum cost. Acetal copolymer can be used to provide snap-fit parts. Use of self-tapping screws may provide substantial cost savings by simplifying machined parts and reducing assembly costs.

Epoxies, isocyanate cured polyester, and cyanoacrylates are used to bond acetal copolymer. Generally, the surface is treated with a sulfuric-chromic acid treatment. Epoxies have shown 150–500 lb/in^2 shear strength on sanded surfaces and 500–1000 lb/in^2 on chemically treated surfaces. Plasma treatment has also shown to be effective on acetal substrates. Acetal homopolymer surfaces should be chemically treated prior to bonding. This is accomplished with a sulfuric-chromic acid treatment followed by a solvent wipe. Epoxies, nitrile, and nitrile-phenolics can be used as adhesives.

Thermal welding and solvent cementing are commonly used for bonding this material to itself. Heated-tool welding produces exceptionally strong joints with acetal homopolymers and copolymers. With the homopolymer, a temperature of the heated surface near 550°F and a contact time of 2–10 s are recommended. The copolymer can be hot-plate welded from 430–560°F. Annealing acetal copolymer joints is claimed to strengthen them further. Annealing can be done by immersing the part in 350°F oil. Acetal resin can be bonded by hot-wire welding. Pressure on the joint, duration of the current, and wire type and size must be varied to achieve optimal results. Shear strength on the order of 150–300 lb/in or more have been obtained with both varieties, depending on the wire size, energizing times (wire temperature), and clamping force.

Hot-gas welding is used effectively on heavy acetal sections. Joints with 50 percent of the tensile strength of the acetal resin have been obtained. Conditions of joint design and rod placement are similar to those presented for ABS. A nitrogen blanket is suggested to avoid oxidation. The outlet temperature of the welding gun should be approximately 630°F for the homopolymer and 560°F for the copolymer. For maximum joint strength, both the welding rod and parts to be welded must be heated so that all surfaces are melted.

Acetal components can easily be joined by spin welding, which is a fast and generally economical method to obtain joints of good strength. Spin-welded acetal joints can have straight 90° mating surfaces, or surfaces can be angles, molded in a V-shape, or flanged.

Although not a common practice, acetal copolymer can be solvent welded at room temperature with full strength hexafluoroacetone sesquihydrate (Allied Chemical Corp., Morristown, N.J.). The cement has been found to be an effective bonding agent for adhering to itself, nylon, or ABS. Bond strengths in shear are greater than 850 lb/in^2 using "as molded" surfaces. Hexafluoroacetone sesquihydrate is a severe eye and skin irritant. Specific handling recommendations and information on toxicity should be requested from Allied Chemical Corp. Because of its high solvent resistance, acetal homopolymer cannot be solvent cemented.

8.9.2 Acrylonitrile-Butadiene-Styrene (ABS)

ABS parts can be designed for snap-fit assembly using a general guideline of 5 percent allowable strain during the interference phase of the assembly. Thread-cutting screws are frequently recommended for non-foamed ABS, and thread-forming screws for foamed grades. Depending on the application, the use of bosses and boss caps may be advantageous.

The best adhesives for ABS are epoxies, urethanes, thermosetting acrylics, nitrile-phenolics, and cyanoacrylates. These adhesives have shown joint strength greater than that of the ABS substrates being bonded. ABS substrates do not require special surface treatments other than simple cleaning and removal of possible contaminants.

ABS can also be bonded to itself and to certain other thermoplastics by either solvent cementing or any of the heat welding methods. For bonding acrylonitrile butadiene styrene (ABS) to itself, it is recommended that the hot-plate temperatures be between 430–550°F. Lower temperatures will result in sticking of the materials to the heated platens, while temperatures above 550°F will increase the possibility of thermal degradation of the surface. In joining ABS, the surfaces should be in contact with the heated tool until they are molten, then brought carefully and quickly together and held with minimum pressure. If too much pressure is applied, the molten material will be forced from the weld and result in poor appearance and reduced weld strength. Normally, if a weld flash greater than 1/8 in occurs, too much joining pressure has been used.

Hot-gas welding has been used to join ABS thermoplastic with much success. Joints with over 50 percent the strength of the parent material have been obtained. The ABS welding rod should be held approximately at a 90° angle to the base material; the gun should be held at a 45° angle with the nozzle 1/4–1/2 in from the rod. ABS parts to be hot-gas welded should be bonded at 60° angles. The welding gun, capable of heating the gas to 500 to 600°F, must be moved continuously in a fanning motion to heat both the welding rod and bed. Slight pressure must be maintained on the rod to ensure good adhesion.

Spin welded ABS joints can have straight 90° mating surfaces, or the surface can be angled, molded in a V-shape, or flanged. The most important factor in the quality of the weld is the joint design. The area of the spinning part should be as large as possible, but the difference in linear velocity between the maximum and minimum radii should be as small as feasible.

One of the fastest methods of bonding ABS and acetal thermoplastics is induction welding. This process usually takes 3–10 s but can be done in as little as 1 s. During welding, a constant pressure of at least 100 lb/in^2 should be applied on the joint to minimize the development of bubbles; this pressure should be maintained until the joint has sufficiently cooled. When used, metal inserts should be 0.02–0.04 in thick. Joints should be designed to enclose completely the metal insert. Inserts made of carbon steel require less power for heating, although other types of metal can be used. The insert should be located as close as possible to the electromagnetic generator coil and centered within the coil to assure uniform heating.

The solvents recommended for ABS are methyl ethyl ketone, methyl isobutyl ketone, tetrahydrofuran, and methylene chloride. The solvent used should be quick drying to prevent moisture absorption yet slow enough to allow assembly of the parts. The recommended cure time is 12 to 24 hr at room temperature. The time can be reduced by curing at 130–150°F. A cement can be made by dissolving ABS resin in solvent up to 25 percent solids. This type of cement is very effective in joining parts that have irregular surfaces or areas that are not readily accessible. Because of the rapid softening actions of the solvent, the pressure and amount of solvent applied should be minimal.

8.9.3 Cellulosics (Cellulose Acetate, Cellulose Acetate Butyrate, Cellulose Nitrate, Ethyl Cellulose, Etc.)

Cellulosic materials can be mechanically fastened by a number of methods. However, their rigidity and propensity to have internal molding stresses must be carefully considered.

Adhesives commonly used are epoxies, urethanes, isocyanate cured polyesters, nitrile-phenolic, and cyanoacrylate. Only cleaning is required prior to applying the adhesive. A

recommended surface cleaner is isopropyl alcohol. Cellulosic plastics may contain plasticizers. The extent of plasticizer migration and the compatibility with the adhesive must be evaluated. Cellulosics can be stress cracked by uncured cyanoacrylate and acrylic adhesives. Any excess adhesive should be removed from the surface immediately.

Cellulosic materials can also be solvent cemented. Where stress crazing is a problem, adhesives are a preferred method of assembly.

8.9.4 Fluorocarbons (PTFE, CTFE, FEP, Etc.)

Because of the lower ductility of the fluorocarbon materials, snap-fit and press-fit joints are seldom used. Rivets or studs can be used in forming permanent mechanical joints. These can be provided with thermal techniques on the melt-processable grades. Self-tapping screws and threaded inserts are used for many mechanical joining operations. In bolted connections, some stress relaxation may occur the first day after installation. In such cases, mechanical fasteners should be tightened; thereafter, stress relaxation is negligible.

The combination of properties that makes fluorocarbons highly desirable engineering plastics also makes them nearly impossible to heat or solvent weld and very difficult to bond with adhesives without proper surface treatment. The most common surface preparation for fluorocarbons is a sodium naphthalene etch, which is believed to remove fluorine atoms from the surface to provide better wetting properties. A formulation and description of the sodium naphthalene process can be found in another chapter. Commercial chemical products for etching fluorocarbons are also listed.

Another process for treating fluorocarbons, as well as some other hard-to-bond plastics (notably polyolefins), is plasma treating. Plasma surface treatment has been shown to increase the tensile shear strength of Teflon® bonded with epoxy adhesive from 50 to 1000 lb/in^2. The major disadvantage of plasma treating is that it is a batch process, which involves large capital equipment expense, and part size is often limited because of available plasma treating vessel volume. Epoxies and polyurethanes are commonly used for bonding treated fluorocarbon surfaces.

Melt-processable fluorocarbon parts have been successfully heat welded, and certain grades have been spin welded and hermetically sealed with induction heating. However, because of the extremely high temperatures involved and the resulting weak bonds, these processes are seldom used for structural applications.

Fluorocarbon parts cannot be solvent welded because of their great resistance to all solvents.

8.9.5 Polyamide (Nylon)

Because of their toughness, abrasion resistance, and generally good chemical resistance parts made from polyamide or resin (or nylon) are generally more difficult to finish and assemble than other plastic parts. However, nylons are used virtually in every industry and market. The number of chemical types and formulations of nylon available also provide difficulty in selecting fabrication and finishing processes.

Nylon parts can be mechanically fastened by most of the methods described in this chapter. Mechanical fastening is usually the preferred method of assembly, because adhesives bonding and welding often show variable results mainly due to the high internal moisture levels in nylon. Nylon parts can contain a high percentage of absorbed water. This water can create a weak boundary layer under certain conditions. Generally, parts are dried to less than 0.5 percent water before bonding.

Some epoxy, resorcinol formaldehyde, phenol resorcinol, and rubber based adhesives have been found to produce satisfactory joints between nylon and metal, wood, glass, and

leather. The adhesive tensile shear strength is about 250 to 1000 lb/in^2. Adhesive bonding is usually considered inferior to heat welding or solvent cementing. However, priming of nylon adherends with compositions based on resorcinol formaldehyde, isocyanate modified rubber, and cationic surfactant have been reported to provide improved joint strength. Elastomeric (nitrile, urethane), hot-melt (polyamide, polyester), and reactive (epoxy, urethane, acrylic, and cyanoacrylate) adhesives have been used for bonding nylon.

Induction welding has also been used for nylon and polycarbonate parts. Because of the variety of formulation available and their direct effect on heat welding parameters, the reader is referred to the resin manufacturer for starting parameter for use in these welding methods. Both nylon and polycarbonate resins should be predried before induction welding.

Recommended solvent systems for bonding nylon to nylon are aqueous phenol, solutions of resorcinol in alcohol, and solutions of calcium chloride in alcohol. These solvents are sometimes bodied by adding nylon resin.

8.9.6 Polycarbonate

Polycarbonate parts lend themselves to all mechanical assembly methods. Polycarbonate parts can be easily joined by solvents or thermal welding methods; they can also be joined by adhesives. However, polycarbonate is soluble in selected chlorinated hydrocarbons. It also exhibits crazing in acetone and is attacked by bases.

When adhesives are used, epoxies, urethanes, and cyanoacrylates are chosen. Adhesive bond strengths with polycarbonate are generally 1000 to 2000 lb/in^2. Cyanoacrylates, however, are claimed to provide over 3000 lb/in^2 when bonding polycarbonate to itself. No special surface preparation is required of polycarbonate other than sanding and cleaning. Polycarbonates can stress crack in the presence of certain solvents. When cementing polycarbonate parts to metal parts, a room-temperature curing adhesive is suggested to avoid stress in the interface caused by differences in thermal expansion.

Polycarbonate film is effectively heat sealed in the packaging industry. The sealing temperature is approximately 425°F. For maximum strength, the film should be dried at 250°F to remove moisture before bonding. The drying time varies with the thickness of the film or sheet. A period of approximately 20 min is suggested for a 20-mil thick film and 6 hr for a 1/4-in thick sheet. Predried films and sheets should be sealed within two hours after drying. Hot-plate welding of thick sheets of polycarbonate is accomplished at about 650°F. The faces of the substrates should be butted against the heating element for 2–5 s or until molten. The surfaces are then immediately pressed together and held for several seconds to make the weld. Excessive pressure can cause localized strain and reduce the strength of the bond. Pressure during cooling should not be greater than 100 lb/in^2.

Polycarbonate parts having a thickness of at least 40 mils can be successfully hot gas welded. Bond strengths in excess of 70 percent of the parent resin have been achieved. Equipment should be used capable of providing gas temperature of 600–1200°F. As prescribed for the heated tool process, it is important to adequately predry (250°F) both the polycarbonate parts and welding rods. The bonding process should occur within minutes of removing the parts from the predrying oven.

For spin welding, tip speeds of 30–50 ft/min create the most favorable conditions to get polycarbonate resin surfaces to their sealing temperature of 435°F. Contact times as short as 1/2 s are sufficient for small parts. Pressures of 300–400 lb/in^2 are generally adequate. For the best bonds, parts should be heat treated for stress relief at 250°F for several hours after welding. However, this stress relief step is often unnecessary and may lead to degraded impact properties of the parent plastics.

Methylene chloride is a very fast solvent cement for polycarbonate. This solvent is recommended only for temperature climate zones and on small areas. A mixture of 60 percent

methylene chloride and 40 percent ethylene chloride is slower drying and the most common solvent cement used. Ethylene chloride is recommended in very hot climate. These solvents can be bodied with 1–5 percent polycarbonate resin where gap filling properties are important. A pressure of 200 lb/in^2 is recommended.

8.9.7 Polyethylene, Polypropylene, and Polymethyl Pentene

Because of their ductility, polyolefin parts must be carefully assembled using mechanical fasteners. These assembly methods are normally used on the materials having higher modulus such as high molecular weights of polyethylene and polypropylene.

Epoxy and nitrile-phenolic adhesives have been used to bond these plastics after surface preparation. The surface can be etched with a sodium sulfuric-dichromate acid solution at elevated temperature. Flame treatment and corona discharge have also been used. However, plasma treatment has proven to be the optimal surface process for these materials. Shear strengths in excess of 3000 lb/in^2 have been reported on polyethylene treated for 10 min in an oxygen plasma and bonded with an epoxy adhesive. Polyolefin materials can also be thermally welded, but they cannot be solvent cemented.

Polyolefins can be thermally welded by almost any technique. However, they cannot be solvent welded because of their resistance to most solvents.

8.9.8 Polyethylene Terephthalate and Polybutylene Terephthalate

These materials can be joined by mechanical self fastening methods or by mechanical fasteners. Polyethylene terephthalate (PET) and polybutylene terephthalate (PBT) parts are generally joined by adhesives. Surface treatments recommended specifically for PBT include abrasion and solvent cleaning with toluene. Gas plasma surface treatments and chemical etch have been used where maximum strength is necessary. Solvent cleaning of PET surfaces is recommended. The linear film of polyethylene terephthalate (Mylar®) surface can be pretreated by alkaline etching or plasma for maximum adhesion, but often a special treatment is unnecessary. Commonly used adhesives for both PBT and PET substrates are isocyanate cured polyesters, epoxies, and urethanes. Polyethylene terephthalate cannot be solvent cemented or heat welded.

Ultrasonic welding is the most common thermal assembly process used with polybutylene terephthalate parts. However, heated-tool welding and other welding methods have proven satisfactory joints when bonding PET and PBT to itself and to dissimilar materials.

Solvent cementing is generally not used to assemble PET or PBT parts because of their solvent resistance.

8.9.9 Polyetherimide (PEI), Polyamide-imide, Polyetheretherketone (PEEK), Polyaryl Sulfone, and Polyethersulfone (PES)

These high-temperature thermoplastic materials are generally joined mechanically or with adhesives. The high modulus, low creep strength, and superior fatigue resistance make these materials ideal for snap-fit joints.

They are easily bonded with epoxy or urethane adhesives; however, the temperature resistance of the adhesives do not match the temperature resistance of the plastic part. No special surface treatment is required other than abrasion and solvent cleaning. Polyetherimide (ULTEM®), polyamide-imide (TORLON®), and polyethersulfone can be solvent cemented, and ultrasonic welding is possible.

These plastics can also be welded using vibration and ultrasonic thermal processes. Solvent welding is also possible with selected solvents and processing conditions.

8.9.10 Polyimide

Polyimide parts can be joined with mechanical fasteners. Self-tapping screws must be strong enough to withstand distortion when they are inserted into the polyimide resin that is very hard.

Polyimide parts can be bonded with epoxy adhesives. Only abrasion and solvent cleaning is necessary to treat the substrate prior to bonding. The plastic part will usually have higher thermal rating than the adhesive. Thermosetting polyimides cannot be heat welded or solvent cemented.

8.9.11 Polymethylmethacrylate (Acrylic)

Acrylics are commonly solvent cemented or heat welded. Because acrylics are a noncrystalline material, they can be welded with greater ease than semicrystalline parts. Ultrasonic welding is the most popular process for welding acrylic parts. However, because they are relatively brittle materials, mechanical fastening processes must be carefully chosen.

Epoxies, urethanes, cyanoacrylates, and thermosetting acrylics will result in bond strengths greater than the strength of the acrylic part. The surface needs only to be clean of contamination. Molded parts may stress crack when in contact with an adhesive containing solvent or monomer. If this is a problem, an anneal (slightly below the heat distortion temperature) is recommended prior to bonding.

8.9.12 Polyphenylene Oxide (PPO)

Polystyrene modified polyphenylene oxide can be joined with almost all techniques described in this chapter. Snap-fit and press-fit assemblies can be easily made with this material. Maximum strain limit of 8 percent is commonly used in the flexing member of PPO parts. Metal screw and bolts are commonly used to assemble PPO parts or for attaching various components.

Epoxy, polyester, polyurethane, and thermosetting acrylic have been used to bond modified PPO to itself and other materials. Bond strengths are approximately 600–1500 lb/in^2 on sanded surfaces and 1000–2200 lb/in^2 on chromic acid-etched surfaces.

Polystyrene modified polyphenylene oxide (PPO) or Noryl® can be hot-plate welded at 500–550°F and 20–30 s contact time. Unmodified PPO can be welded at hot-plate temperatures of 650°F. Excellent spin welded bonds are possible with modified polyphenylene oxide (PPO), because the low thermal conductivity of the resin prevents head dissipation from the bonding surfaces. Typical spin welding condones are rotational speed of 40–50 ft/min and a pressure of 300–400 lb/in^2. Spin time should be sufficient to ensure molten surfaces.

Polyphenylene oxide joints must mate almost perfectly; otherwise, solvent welding provides a weak bond. Very little solvent cement is needed. Best results are obtained by applying the solvent cement to only one substrate. Optimum holding time has been found to be 4 min at approximately 400 lb/in^2. A mixture of 95 percent chloroform and 5 percent carbon tetrachloride is the best solvent system for general-purpose bonding, but very good ventilation is necessary. Ethylene dichloride offers a slower rate of evaporation for large structures or hot climates.

8.9.13 Polyphenylene Sulfide (PPS)

Being a semicrystalline thermoplastic, PPS is not ideally suited to ultrasonic welding. Because of its excellent solvent resistance, PPS cannot be solvent cemented. PPS assemblies can be made by a variety of mechanical fastening methods as well as by adhesive bonding.

Adhesives recommended for polyphenylene sulfide include epoxies and urethanes. Joint strengths in excess of 1000 lb/in^2 have been reported for abraded and solvent-cleaned surfaces. Somewhat better adhesion has been reported for machined surfaces over as-molded surfaces. The high heat and chemical resistance of polyphenylene sulfide plastics make them inappropriate for solvent cementing or heat welding.

Polyimide and polyphenylene sulfide (PPS) resins present a problem in that their high temperature resistance generally requires that the adhesive have similar thermal properties. Thus, high-temperature epoxies adhesive are most often used with polyimide and PPS parts. Joint strength is superior (greater than 1000 lb/in^2), but thermal resistance is not better than the best epoxy systems (300–400°F continuous).

8.9.14 Polystyrene

Polystyrene parts are conventionally solvent cemented or heat welded. However, urethanes, epoxies, unsaturated polyesters, and cyanoacrylates will provide good adhesion to abraded and solvent-cleaned surfaces. Hot-melt adhesives are used in the furniture industry. Polystyrene foams will collapse when in contact with certain solvents. For polystyrene foams, a 100 percent solids adhesive or a water-based contact adhesive is recommended.

Polystyrene can be joined by either thermal or solvent welding techniques. Preference is generally given to ultrasonic methods because of its speed and simplicity. However, heated-tool welding and spin welding are also commonly used.

8.9.15 Polysulfone

Polysulfone parts can be joined with all the processes described in this chapter. Because of their inherent dimensional stability and creep resistance, polysulfone parts can be press fitted with ease. Generally, the amount of interference will be less than that required for other thermoplastics. Self-tapping screws and threaded inserts have also been used.

Urethane and epoxy adhesives are recommended for bonding polysulfone substrates. No special surface treatment is necessary. Polysulfones can also be easily joined by solvent cementing or thermal welding methods.

Direct thermal welding of polysulfone requires a heated tool capable of attaining 700°F. Contact time should be approximately 10 s, and then the parts must be joined immediately. Polysulfone parts should be dried 3 to 6 hr at 250°F before attempting to heat seal. Polysulfone can also be joined to metal, since polysulfone resins have good adhesive characteristics. Bonding to aluminum requires 700°F. With cold-rolled steel, the surface of the metal first must be primed with 5 to 10 percent solution of polysulfone and baked for 10 min at 500°F. The primed piece then can be heat welded to the polysulfone part at 500–600°F.

A special tool has been developed for hot-gas welding of polysulfone. The welding process is similar to standard hot-gas welding methods but requires greater elevated temperature control. At the welding temperature, great care must be taken to avoid excessive application of heat, which will result in degradation of the polysulfone resin.

For polysulfone, a 5 percent solution of polysulfone resin in methylene chloride is recommended as a solvent cement. A minimal amount of cement should be used. The assembled pieces should be held for 5 min under 500 lb/in^2. The strength of the joint will improve over a period of several weeks as the residual solvent evaporates.

8.9.16 Polyvinyl Chloride (PVC)

Rigid polyvinyl chloride can be easily bonded with epoxies, urethanes, cyanoacrylates, and thermosetting acrylics. Flexible polyvinyl chloride parts present a problem because of

plasticizer migration over time. Nitrile adhesives are recommended for bonding flexible polyvinyl chloride because of compatibility with the plasticizers used. Adhesives that are found to be compatible with one particular polyvinyl chloride plasticizer may not work with another formulation. Solvent cementing and thermal welding methods are also commonly used to bond both rigid and flexible polyvinyl chloride parts.

8.9.17 Thermoplastic Polyesters

These materials may be bonded with epoxy, thermosetting acrylic, urethane, and nitrile-phenolic adhesives. Special surface treatment is not necessary for adequate bonds. However, plasma treatment has been reported to provide enhanced adhesion. Solvent cementing and certain thermal welding methods can also be used with thermoplastic polyester.

Thermoplastic polyester resin can be solvent cemented using hexafluoroisopropanol or hexafluoroacetone sesquihydrate. The solvent should be applied to both surfaces and the parts assembled as quickly as possible. Moderate pressure should be applied as soon as the parts are assembled. Pressure should be maintained for at least 1 to 2 min; maximum bond strength will not develop until at least 18 hr at room temperature. Bond strengths of thermoplastic polyester bonded to itself will be in the 800–1500 lb/in^2 range.

8.9.18 Thermosetting Plastics (Epoxies; Diallyl Phthalate; Polyesters; Melamine, Phenol, and Urea Formaldehyde; Polyurethanes; Etc.)

Thermosetting plastics are joined either mechanically or by adhesives. Their thermosetting nature prohibits the use of solvent or thermal welding processes; however, they are easily bonded with many adhesives.

Abrasion and solvent cleaning are generally recommended as the surface treatment. Surface preparation is generally necessary to remove contaminant, mold release, or gloss from the part surface. Simple solvent washing and abrasion is a satisfactory surface treatment for bonds approaching the strength of the parent plastic. An adhesive should be selected that has a similar coefficient of expansion and modulus as the part being bonded. Rigid parts are best bonded with rigid adhesives based on epoxy formulations. More flexible parts should be bonded with adhesives that are flexible in nature after curing. Epoxies, thermosetting acrylics, and urethanes are the best adhesives for the purpose.

8.10 More Information on Joining Plastics

Additional details on joining plastics by adhesive bonding, direct heat welding, indirect heat welding, frictional welding, solvent cementing, or mechanical fastening can be found in numerous places. The best source of information is often the plastic resin manufacturers themselves. They often freely offer related recipes and processes, because it is in their interest that their materials successfully find implementation in joined components.

Another source of information is the equipment manufacturers. The manufacturers of induction bonding, ultrasonic bonding, spin welding, vibration welding, and such equipment will often provide guidance on the correct parameter to be used for specific materials and joint designs. Many of these equipment suppliers will have customer service laboratories where prototype parts can be tried and guidance provided regarding optimal processing parameters.

Of course, adhesive suppliers and mechanical fastener suppliers can provide detailed information on their products and guidance as to which substrate is most appropriate. They can generally provide complete processes and specifications relative to the assembly operation. They usually also have moderate amounts of test data to provide an indication of strength and durability.

Finally, a very useful source of information is the technical literature, conference publications, books, and handbooks relative to the subject of joining plastics. The following works are especially recommended for anyone requiring detailed information in this area:

- *Handbook of Plastics Joining*, Plastics Design Laboratory, 1999.
- *Designing Plastic Parts for Assembly,* by Paul A. Tres, Hanser/Gardner, 1998.
- *Decorating and Assembly of Plastic Parts*, by E. A. Muccio, ASM Publications, 1999.
- *Handbook of Adhesives and Sealants*, by Edward M. Petrie, McGraw-Hill, 1999.

References

1. Engineer's Guide to Plastics, *Materials Engineering,* May 1972.
2. Trauenicht, J. O., Bonding and Joining, Weigh the Alternatives, Part 1: Solvent Cement, Thermal Welding, *Plastics Technology,* August 1970.
3. Gentle, D. F., Bonding Systems for Plastics, *Aspects of Adhesion,* vol. 5, D. J. Almer, Ed., University of London Press, London, 1969.
4. Mark, H. F., Gaylord, N. G., and Bihales, N. M., Eds., *Encyclopedia of Polymer Science and Technology,* vol. 1, John Wiley & Sons, New York, 1964, p. 536.
5. *All about Welding of Plastics,* Seelyte Inc., Minneapolis, MN.
6. How to Fasten and Join Plastics, *Materials Engineering,* Mar. 1971.
7. Spooner, S. A., Designing for Electron Beam and Laser Welding, *Design News,* Sept. 23, 1985.
8. Troughton, M., Lasers and Other New Processes Promise Future Welding Benefits, *Modern Plastics,* Mid-November, 1997.
9. Laser Welding, Chap. 13, *Handbook of Plastics Joining,* Plastics Design Library, Norwich, NY, 1997.
10. EMABOND® Thermoplastic Assembly Systems, Specialty Polymers and Adhesives Division, Ashland Specialty Chemical Company, Norwood, NJ.
11. Leatherman, A., Induction Bonding Finds a Niche in an Evolving Plastics Industry, *Plastics Engineering,* April 1981.
12. EMABOND® Thermoplastic Assembly Systems, Specialty Polymers and Adhesives Division, Ashland Specialty Chemical Company, Norwood, NJ.
13. Spin Welding, Chap. 4, *Handbook of Plastics Joining,* Plastics Design Library, Norwich, NY, 1997.
14. LaBounty, T. J., *Spin Welding Up-Dating and Old Technique,* SPE ANTEC, 1985, pp. 855–856.
15. Grimm, R. A., Welding Process for Plastics, *Advanced Materials and Processes,* March 1995.
16. Ultrasonic Welding, Chap. 5, *Handbook of Plastics Joining,* Plastics Design Library, Norwich, NY, 1997.
17. Branson Sonic Power Company.
18. Ultrasonic Joining Gains Favor with Better Equipment and Knowhow, *Product Engineer,* Jan. 1977.
19. Mainolfi, S. J., Designing Component Parts for Ultrasonic Assembly, *Plastics Engineering,* Dec. 1977.
20. Hauser, R. L., Ultra Adhesives for Ultrasonic Bonding, *Adhesives Age,* 1969.
21. Scherer, R., Vibration Welding Could Make the Impossible Design Possible, *Plastics World,* Sept. 1976.
22. Kagan, V. A., et al., Optimizing the Vibration Welding of Glass Reinforced Nylon Joints, *Plastics Engineering,* Sept. 1996.
23. Raia, D. C., Adhesives—the King of Fasteners, *Plastics World,* June 17, 1975.

24. Engineering Plastics, *Engineered Materials Handbook,* vol. 2, ASM International, Metals Park, OH.
25. *Machine Design,* Nov. 17, 1988.
26. Mechanical Fastening, Chap. 14, *Handbook of Plastics Joining,* Product Design Library, Norwich, NY, 1997.
27. McMaster, W., and Lee, C., New Equations Make Fastening Plastic Components a Snap, *Machine Design,* Sept. 10, 1998.

Chapter

9

Design and Processing of Plastic Parts

John L. Hull

Hull Corporation
Warminster, Pennsylvania

9.1 Introduction

The creation of a plastic part requires a series of conscious decisions regarding type of plastic, method of production, design of mold or tooling, and selection of machine or process. Reaching these decisions requires information as to the intended usage of the part and the conditions of environment (temperature, moisture, exposure to harsh atmospheres, physical and electrical requirements). Furthermore, in most cases, the cost to manufacture the part is a major consideration and is dependent on the choice of materials, the manufacturing process, and, of course, the quantities to be produced per shift, month, or year.

In this chapter, several production processes are discussed broadly, outlining the basics of each process, the equipment involved, the plastic materials for which the process is feasible, and some typical products made by each particular process. In addition, the chapter describes important aspects of plastic part design and considers the closely related topic of mold design and construction.

9.2 Design Procedure

The first step requires a preliminary engineering drawing of the intended part, with approximate overall dimensions, section thicknesses desired, probable location of holes, ribs, bosses, and the like, as well as approximate radii of curves, corners, and so on. Tolerances are not necessary at this stage but will be incorporated before final design.

Next is a list detailing the intended requirements of the final part. Physical considerations regarding stiffness; dimensions and tolerances; impact resistance; surface hardness; compressive, tensile, and torque loads; light transmittance; and so on; all need to be quantified. If the part is intended for electrical applications, determine voltage to be encountered, current-induced temperature extremes, arc resistance and surface tracking resistance, dielectric strength, electrical frequency, and the like. List possible environmental

extremes such as humidity, salt or corrosive spray, ultraviolet exposure, moisture-absorption limitations, service-temperature lows and highs, extreme gravities from acceleration or shock, and possible abrasive exposure. Note also cosmetic requirements such as color or transparency, high gloss, or special tactile surface.

At this point, select several possible materials that appear (from the data sheets) to meet all the intended requirements and also appear to be appropriate for the size of part. If you know of plastic parts similar to your intended size and application requirements, determine the materials from which they are made, and add these to your list of material candidates.

The next step is to consider the several processes that might be appropriate. Later in this chapter, you will find descriptions of many processes, some of which will be worth considering. But you must realize that finding a process that will be compatible to your part requirements as well as to the materials you have tentatively selected is not always easy. Compromises are often made at this stage—compromises in your design, your requirements, and your materials. (In some cases, you may decide not to use plastics as your material of construction!)

The selection of process involves size of part. For example, you can't injection mold 50-foot lengths of garden hose, but you can readily extrude them. Ultrahigh molecular weight polyethylene (UHMWPE) has fantastic resistance to abrasion, but if your part must experience temperatures of 100°C or higher, UHMWPE will lose its physical properties rapidly.

Selecting the process requires consideration of size (injection molding a boat hull is not feasible, but hand lay-up is very practical for such an application), production volume (to build a compression mold for only 100 parts would be extremely expensive, but liquid casting for the same quantity might be ideal), and economics (using a material that costs $10 per pound would be out of the question if your two-ounce part had to sell for $1). Often, choice of process may be dictated by available process equipment already in your manufacturing facility or in your favorite custom molding house. (The part may be appropriate for either transfer molding or injection molding; so if you have transfer molding equipment in your plant but not injection presses, possibly transfer molding will be the "best" process.)

If a practical solution still eludes you, it would be wise to get suggestions from plastic material suppliers, processing-machinery suppliers, or custom molders, all of whose advice will probably be given at no charge.

And if those sources fail to lead to a practical solution, try one or more of the many plastics design consultants for some guidance. The consulting fee will generally be well spent.

9.3 Prototyping

Once you have reached an apparently viable decision as to material and process (having taken manufacturing economics into the formula), you should give strong consideration to starting with an appropriate prototype. Such a prototype may be machined or cast using your selected material, with final dimensions sufficiently close to your design specifications to enable good mechanical and electrical testing. If the prototype can be produced using the selected material and process (as, for example, using a single-cavity injection mold for a part that will ultimately be produced in a multicavity injection mold), you will not only be able to perform various mechanical and electrical tests on the prototype, you may gain valuable molding parameters and cycle rates during the prototyping. You may also discover the need to make changes in wall thickness, corner radii, hole diameters or location, gate and vent location and size, and ejector pin location and size—compromises or improvements that lead to optimal production with minimal costs.

Rapid prototyping is becoming increasingly important for bringing a new product to market in shortest possible time. To date, the newest and fastest process for producing prototype molds utilizes stereo lithography (SLA) in a special computer-operated machine that guides laser beams to create a series of flat styrene layers—two-dimensional *x-y* configured layers (about 0.004 in thick) at progressive increments along the *z* axis of the CAD part design. These layers create a three-dimensional solid plastic part to serve as a pattern. This pattern is then placed in a container filled with liquid silicone rubber. When the rubber has cured, the pattern is removed. The resulting elastomeric cavity may be used to cast rigid parts using epoxies, polyesters, acrylics, and other plastics for further part design analysis and evaluation. Such rigid prototypes can often be produced within a week from start to finish.

If the prototype needs to be molded in an injection, compression, transfer, or blow-molding process, it is possible to create a hard cavity shell by first coating the silicone rubber cavity with a release agent and then metal spray plating the cavity surface. When the plated coating is sufficiently thick, it is removed from the elastomeric cavity and backed up with liquid epoxy and appropriate reinforcing stiffeners to create a prototype mold that, with care, can be used to mold parts in the plastic intended for final part. Time from "art to part" may be merely a few weeks, as compared with the several months generally required for conventional machined steel production molds.

A variation on the above SLA process enables the generation of a powdered metal mold for the CAD designed part. Molds made by this variation have been used as production molds and have withstood millions of cycles of injection molding.

9.4 Processes for Producing Plastic Parts

9.4.1 Liquid-Plastics Processing

For prototyping or for limited production runs, liquid plastics casting offers simplicity of process, relatively low investment in equipment, and fast results. The casting materials may be thermoplastics, such as acrylics, or thermosetting plastics, such as epoxy resins. The plastics may harden by simple cooling, by evaporation of solvents, or by a polymerization or cross-linking reaction. Such plastics are often poured into open molds or cavities. Because pouring is done at atmospheric pressure, molds are simple, often made of soft metals, plaster, or other plastics.

An example of liquid casting is the fabrication of design details such as scrolls and floral or leaf patterns for furniture decoration. Such parts are often made using filled polyester resins. After curing, the parts are simply glued to the wooden bureau drawer or mirror frame. The parts can readily be finished to look like wood. Molds for such parts are often made by casting an elastomeric material over a wood or plastic model of the part. When the elastomer is removed, it generally yields a cavity in which can be cast a very faithful reproduction of the original pattern. The elastomer mold may be used over and over many times.

Another widely used industrial application is casting with liquid plastics to embed objects such as electronic components or circuits in plastic cups, cases, or shells, giving the components mechanical protection, electrical insulation, and a uniform package size. When such applications require fairly high-volume production, machines for mixing and dispensing the liquid plastics may be used for convenience and higher production rates, and curing ovens, conveyors, and other auxiliary capital equipment may be incorporated. In short, the liquid-casting operation may be a low-cost manual one, or it may be highly automated, depending on the nature of the product and the quantities required.

If we cast liquid acrylic into an already-molded acrylic shape in which a coin or emblem has been placed, the result, when the acrylic has hardened, is an attractive paper-

weight. If we cast liquid epoxy into a previously molded container in which has been assembled a hybrid electronic circuit with leads protruding out of the container, the result, when the liquid epoxy has cured, is an attractively packaged and protected circuit. This process is called *potting* when the container ("pot") remains as a permanent element of the finished part. If bubbles or voids are a concern, liquid casting and potting may be performed in a vacuum chamber.

9.4.2 Rotational Molding (Rotomolding or Rotational Casting)

This relatively low-cost process utilizes a closed mold in which is placed an appropriate charge of liquid or granular plastic, generally thermoplastic. The mold is then mounted in a carousel, which enables it to be rotated simultaneously in at least two axes vertical to each other. As the mold spins, the plastic inside coats the inner surface of the mold cavity to a reasonably uniform thickness. The mold is brought to a controlled melt temperature, usually by hot air, which enables the plastic to flow and fuse. Subsequently, as the mold is cooled with cold air or water spray, the plastic hardens. The mold is then opened, and the hollow part is removed. Often the hollow part is cut in half to yield two usable parts, as in thin-walled covers. Applications include storage and feed tanks (as large as a 22,500-gal tank 12 ft wide and 30 ft high!), shipping containers, automotive instrument panels, gearshift covers, door liners, playground equipment, recreational boats, and portable toilets.

Most common thermoplastics, such as polyethylene, polyvinyl chloride, nylon, polycarbonate, acetate butyrate, and polypropylene, are suitable for the rotational molding process. In recent years, some thermosetting formulations have been found successful.

Inexpensive molds are made of cast or machined aluminum, or sheet metal. For finer details, more expensive molds may be made using electroformed or vapor-formed nickel. Release agents of nonstick coatings are generally used for ease of demolding.

9.4.3 Hand Lay-Up (Composites)

When larger plastic parts are required, and often when such parts must be rigid and robust, a process referred to as *hand lay-up* is used. Hand lay-up is closely related to liquid-plastic casting. A reinforcing fabric or mat, frequently fiberglass, is placed into an open mold or over a form, and a fairly viscous liquid resin is poured over the fabric to wet it thoroughly and to penetrate into the weave, ideally with little or no air entrapment. When the plastic hardens, the object is removed from the mold or form, trimmed as necessary, and is then ready for use. Many boats are produced using the hand lay-up process, from small sailing dinghies and bass boats, canoes and kayaks, to large commercial fishing boats and even military landing craft. Unsaturated polyesters are most often used in this process, but epoxies and polyurethanes are also used. Although glass fibers, in mat form or woven, dominate as the reinforcing material, carbon fiber or Kevlar (DuPont's aramid filament) is used where extreme stiffness or strength may be required.

This basic process can be automated as required, with proportioning, mixing, and dispensing machines for liquid resin preparation, using matched molds (that is, two mold halves that are closed after the reinforcing material has been impregnated with the liquid to produce a smooth uniform surface on both top and bottom of the part), conveyors, ovens, and so on.

9.4.4 Resin Transfer Molding

The resin transfer molding (RTM) process is used principally for manufacturing fiber-reinforced composites in moderate to high volumes. The process combines the techniques of hand lay-up, liquid resin casting, and transfer molding. In practice, the fiber mat or pre-

form (a precut and preshaped insert of the reinforcing material) is placed into the open mold, and the heated mold is closed. A liquid catalyzed resin, often epoxy or polyester, is then injected into the cavity at a modest positive pressure (5 to 100 lb/in^2) until all the interstices between the fibers are completely filled. The resulting formulation may have as much as 60 percent glass by weight. Subsequently, the resin mix reacts and hardens. Often vacuum is applied to the mold cavity to remove the air prior to and during cavity fill to minimize the possibility of air entrapment and voids. Cycle times may run several hours, particularly for large aircraft and missile components. During cure, mold temperatures are often ramped up and down to achieve optimum properties of the plastic and therefore of the finished product. Fibers often used include glass, carbon, Kevlar®, or combinations of these. Very high strength-to-weight ratios are achieved in such parts, exceeding those of most metals, and complex configurations are achieved more easily than by machining and forming such high strength-to-weight ratio metals as tantalum or aluminum.

To make such parts, the press and resin transfer equipment for the RTM process is generally less expensive than that for an injection press for comparable-size components, partly because RTM clamping and injection pressures are often only a few hundred pounds per square inch. Control systems have become highly sophisticated, however, raising the cost of equipment, and the long cycle times add to the manufacturing cost of each item produced. Advanced metering and mixing machines for preparing the liquid resin are capable of controlling resin-catalyst ratios over the range of 1:1 to 200:1, and in flow rates from several ounces per minute to 100 lb/min (Fig. 9.1).

9.4.5 Filament Winding

For structural tubes up to several inches in diameter, and for tubes and closed tanks that may hold fluids at high pressure, a continuous filament of glass or other strong polymeric material is drawn through a liquid polyester or epoxy bath and then fed to a rotating mandrel, allowing the wetted filament to closely wind onto the mandrel, often with layering in different directions. When the predetermined number of turns has resulted in the desired thickness, the resin cures, and the mandrel is removed. The resulting tube or container has extremely high hoop strength, ideal for storing or transporting high-pressure liquids or gases.

9.4.6 Thermoforming

Because thermoplastics soften with heat and harden when cooled, many items are produced using extruded thermoplastic sheet as the starting material. The sheet is heated to a temperature that softens the sheet but does not melt it. Then the sheet is shaped or formed while soft and subsequently cooled while being held in the new configuration. Often the heated sheet is forced into an open cavity by vacuum or air pressure. Most blister packages and many low-cost plastic drinking cups and containers are produced by such a thermoforming process. Acrylic (Plexiglas™) rounded canopies for small aircraft and sailplanes are frequently formed by the process, as are many advertising signs on truck doors or retail establishments. One highly successful application is the 60-in long ABS dash panel for a low-volume street sweeper. Bike helmets and snowboards are also produced with this process.

During the process, the material often stretches, making the final wall thickness in some areas less than the original sheet thickness. With proper design and operation, the thermoforming process permits very inexpensive production of three-dimensional thin-wall parts for short or long runs. Many process variations have been developed to accomplish desired results. Such variations include drape forming, matched-mold forming, plug-assist forming, vacuum snap-back, and trapped-sheet forming. Materials suitable for the thermoform-

Figure 9.1 Liquid-silicone-molding machine complete with proportioning, mixing, and dispensing system for two-part resin systems. Machines similar to this are used for resin transfer molding (RTM). *(Photo provided by WABASH/MPI/Carver, Inc.)*

ing processes include ABS, thermoplastic polyesters, polypropylene, polystyrene, acrylics, and polyvinyl chloride.

Equipment costs vary widely but are considerably lower than for comparable injection-molded parts because of the relatively low pressures and temperatures required. Highly automated machines use roll-fed stock. Residual material from trimming can be ground, blended with virgin material, and re-extruded into sheet for subsequent thermoforming use.

9.4.7 Extrusion, Coextrusion, and Pultrusion

The extrusion process starts somewhat like the thermoplastic injection molding process, with a long barrel and rotating augur-type screw advancing thermoplastic crystals from a room-temperature hopper at one end, bringing the material up to melt temperature by the electrically heated barrel and also by the frictional and shearing action of the screw on the material, finally forcing the melt out the nozzle end through a precision-machined die to create the profile or shape of the continuously extruded product and cool it. Then, to cool the extrudate completely, it is generally passed through a water bath kept at a sufficiently low temperature that the material hardens in the shape of the profile. Single-screw extrusion barrels range from 1/2 to 18 in inner diameter, with throughput rates depending on the

size and wall thickness of the profile and the material characteristics but ranging from several pounds per hour to more than five tons per hour. Twin-screw extruder throughputs can be as high as 30 tons per hour (Fig. 9.2).

In recent years, extruder screw architecture has become highly sophisticated, with special configurations for each type of extrudate. Venting in the barrel contributes much to product quality and throughput. As with most high-speed polymer processing systems, closed-loop control systems monitor and regulate process parameters for optimum production.

Profiles may be of closed cross section, as for pipe or conduit, or they may be open sections, as for door jambs and window frames for the construction industry. Extruders can also coat wire on a continuous basis, such as vinyl-insulated wire used for appliance wiring.

Coextrusion is a process in which two different materials are extruded from two separate extruder barrels and then brought together in a complex die to achieve a laminated sheet or profile. Not many materials lend themselves to this process, because temperature and chemical differences often preclude good bonding.

Successful extrusion of thermosetting materials has been achieved and put into commercial practice, but extruded thermosets represent only a tiny fraction of the amount of thermoplastics extruded annually.

Pultrusion is a process somewhat similar to extrusion in concept in that it produces a continuous profile by forcing the material through a precision die configured to the desired profile. The differences between the two processes, however, start with the fact that pultrusion generally uses thermoset plastics rather than thermoplastics, and the process *pulls* the resin and reinforcing fibers or web through the die instead of *pushing* them through. Because the profiles are heavily reinforced, the pull is on the continuous reinforcing material. Thermoplastic materials suited to the pultrusion process include PE and PEEK (Fig. 9.3).

When thermosetting materials are used, the steps in the process involve pulling the reinforcing matrix through a tank filled with catalyzed liquid resin, often polyester, where the fibers become totally saturated. The wet matrix is then passed through a stripper to squeeze out excess liquid and sometimes to start shaping the material into a profile. The material is then pulled into a 20- to 30-in-long heated die at a rate controlled to ensure complete curing of the resin as it passes continuously from one end of the die to the other.

Figure 9.2 Twin-screw extruder, counterrotating, nonintermeshing, for compounding, model HTR2800-511-512-E1. Unit shown has panel open for access to barrel and heater elements. Set of extruder screws shown in front, and seven-zone temperature control panel with loss-in-weight feeder controls and main drive motor controls, stands at right of machine. *(Photo provided by Welding Engineers, Inc.)*

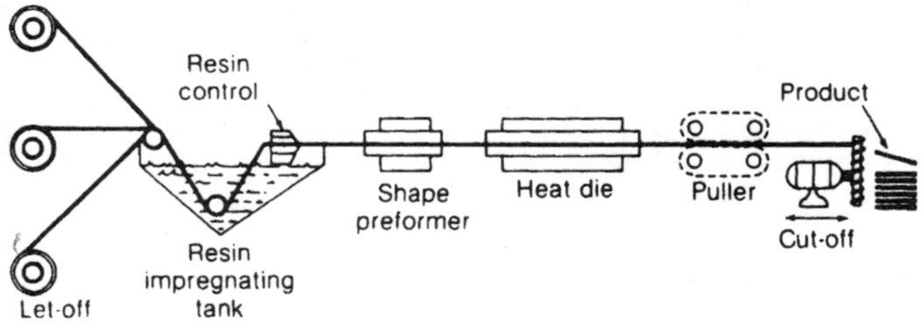

Figure 9.3 Schematic illustration of the pultrusion process.

The material may be partially brought up to temperature prior to entering the curing die by use of high-frequency heating between the stripper and the die, thereby speeding up the curing and enabling faster pulling rates, shorter curing times, or both. Recent process variations inject the catalyzed resin directly into the die, thereby eliminating the above-described matrix saturation and stripping steps.

Perhaps the most crucial element in a pultrusion machine is the gripper mechanism, which grips the cured profile downstream of the curing die and continuously pulls at a carefully regulated rate to ensure total curing in the die, taking into account the degree of advanced cure of the catalyzed resin as it enters the die.

Applications include electric-bus duct, side rails for safety ladders, third-rail covers, walkways, structural supports in harsh chemical environments, and resilient items such as fishing rods, bicycle flagpoles, and tent poles. Common profile sections generally fall between 4×6 in and 8×24 in.

9.4.8 Blow Molding

Ideal for producing plastic beverage bottles and other closed shapes, the blow molding process combines elements of the extrusion process and the thermoforming process in complex, fully automated machines and mold systems to produce thin-walled holloware at fantastically high rates (Fig. 9.4).

In extrusion blow molding, a hollow tube called a *parison* is extruded vertically downward. Immediately after it is extruded, two halves of a mold clamp firmly over the parison, pinching the lower end closed and forming the upper end firmly around a blowing nozzle. Compressed air or some other gas (sometimes chilled) is blown at about 100 lb/in^2 pressure into the upper end of the parison, forcing the parison to expand rapidly in the cavity, where it cools and hardens in the shape of the cavity. The mold at the end in which air is introduced also forms the threads and neck of the hollow bottle or part. The mold is then opened, and the finished part is blown off or extracted onto a conveyor line. In many plants, blow molding is a captive operation, and the molded container is immediately transported to the labeling and then the filling station, then capped and packaged for shipment.

A variation on extrusion blow molding is injection blow molding, a two-stage process. In the first stage, parisons are injection molded into a tubular shape complete with the threaded and formed top. These parisons may then be stored until time for the second step.

The second step involves feeding the premolded parisons into an automatic blow-molding machine where the parison is heated to the softening point and then clamped between die halves and blown as in the extrusion blow molding process. In this two-stage process, the bottom of the parison is not pinched, because it is not open.

Common thermoplastic materials suited to the blow-molding process include high-density polyethylene, polyvinyl chloride, polypropylene, and polyethylene terephthalate.

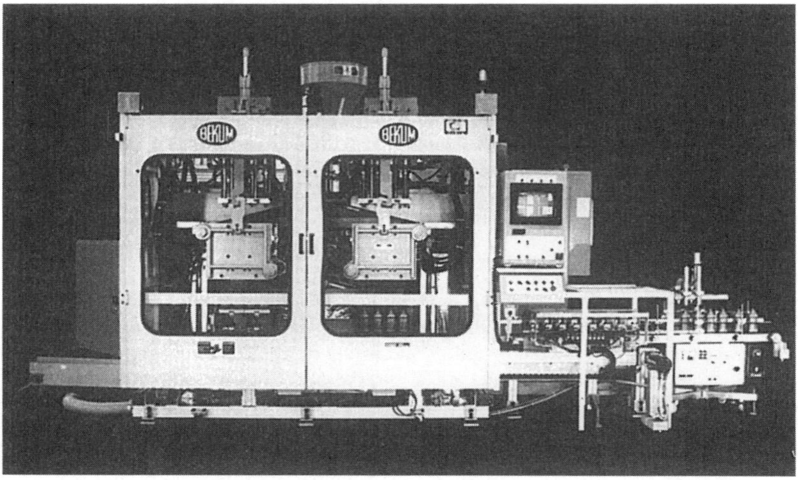

Figure 9.4 Bekum model H-155 double-station extrusion-blow-molding machine with oriented bottle take-out. *(Photo provided by Bekum America Corporation)*

Although the injection blow molding process is not suited to "handleware" bottles, it is rapidly gaining favor over extrusion blow molding for bottles up to 1.5 liters and more. Advantages include the practicality of molding strong, closely toleranced necks and threads suited to childproof caps, and for molding wide-mouthed bottles.

In addition to bottles, a rapidly growing application for blow molding is the production of "technical" parts, such as automotive components—bumpers, ducts, and other fluid containers. For fuel tanks, a coextruded parison has polyethylene as the structural material and special barrier layers to prevent escape of polluting fumes.

Machines and molds may cost between \$200,000 and \$1,000,000, depending on the throughput rates and accessories.

Molds for blow molding are made from beryllium copper and aluminum because of the excellent thermal conductivity of such materials. Stainless steel and hard-chrome plated tool steels are also common.

9.4.9 Compression and Transfer Molding

Compression molding is a process very similar to making waffles. The molding compound, generally a thermosetting material such as phenolic, melamine, or urea, is placed in granular form into the lower half of a hot mold, and the heated upper half is then placed on top and squeezed down until the mold halves come essentially together, forcing the molding compound to flow into all parts of the cavity, where it finally cures, or hardens, under continued heat and pressure. The cure, or polymerization, is an irreversible chemical reaction. When the mold is opened, the part is removed and the cycle repeated. The process can be manual, semiautomatic, or fully automatic (unattended operation), depending on the equipment. Molds are generally made of hardened steel, highly polished and hard-chrome-plated, and the two mold halves, with integral electric, steam, or circulating hot oil heating provisions, are mounted against upper and lower platens in a hydraulic press capable of moving the molds open and closed with adequate pressure to make the plastic flow. Molds may be single cavity or multiple cavity, and the pressure must be adequate to pro-

vide about four tons per square inch of projected area of the molded part or parts at the mold parting surfaces. Overall cycles depend on molding material, part thickness, and mold temperature, and may be about 1 min for parts of 1⁄8-in thickness to 5 or 6 min for parts of 1-in thickness or larger (Fig. 9.5).

The process is generally used for high-volume production, because the cost of a modern semiautomatic press of modest capacity, say 50–75-ton clamping force, may be as much as $50,000, and a moderately sophisticated self-contained multicavity mold may also cost $50,000. Typical applications include melamine dinnerware, toaster legs and pot handles, and electrical outlets, wall plates, and switches—parts that require the rigidity, dimensional stability, heat resistance, and electrical insulating properties typical of thermosetting compounds.

To simplify feeding material into the mold, the molding compound is often precompacted without heat into preforms, or "pills," in a specially designed automatic preformer, which compacts the granular molding compound into cylindrical or rectangular blocks of uniform weight. To reduce the molding cycle time, the preform is often heated with high-frequency electrical energy in a self-contained unit called a *preheater*, arranged beside the press. The preform is manually placed between the electrodes of the preheater before each molding cycle and heated throughout in as little as 10–15 s to about 200°F, at which temperature the plastic coheres but is slightly mushy. It is then placed manually in the bottom mold cavity, and the molding cycle is initiated. Through the use of preheating, the cure

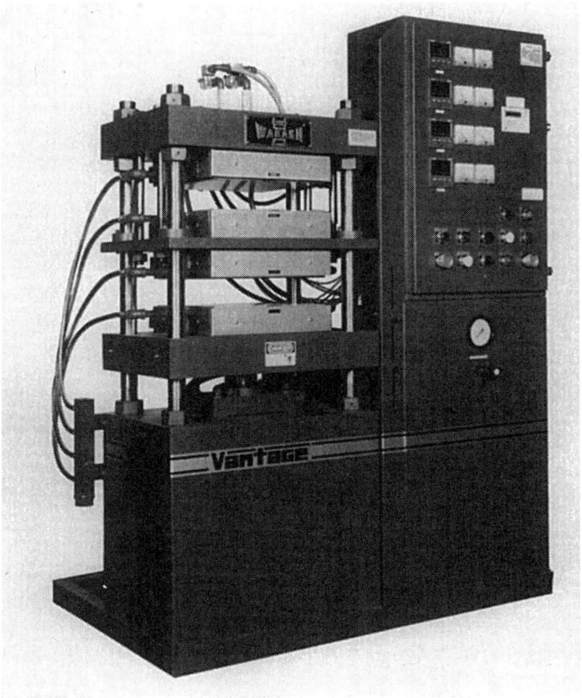

Figure 9.5 Vantage model V502H-18 compression molding press equipped with two heated openings for increased production. *(Photo provided by WABASH/MPI/Carver, Inc.)*

time may be cut in half, mold wear is reduced considerably, and the part quality is often improved.

The process of compression molding is often utilized with sheet molding compounds (SMCs) to produce heavily reinforced composite parts. SMC is generally fiberglass mat or woven cloth, impregnated with B-staged (partially catalyzed) thermosetting epoxy or polyester. Following cavity fill and a suitable cure time, the press opens, and the part is removed.

9.4.10 Transfer Molding

A related process for high-volume molding with thermosetting materials is transfer molding, so called because the material, instead of being compressed between the two halves of a closing mold to make it flow and fill the cavity, is placed into a separate chamber of the mold, called a *transfer pot.* This transfer pot, which generally is cylindrical, is connected by small runners and smaller openings (called *gates*) to the cavity or cavities. In operation, the mold is first closed and held under pressure. The preheated preform is then dropped into the pot and pressed by a plunger, where the material liquefies from the heat of the mold and the pressure of the plunger and flows (or is transferred) through the runners and gates into the cavity or cavities. The plunger maintains pressure on the molding compound until the cavities are full and the material cures. At that point, the mold is opened, the plunger is retracted, and the part or parts, runners, and cull (the material remaining in the pot, generally about 1⁄8 in thick and having the diameter of the pot and plunger) are removed. Because the gate is small, and the cured plastic is relatively rigid, the runners and cull are readily separated from the molded parts at the part surface, leaving a small and generally unobtrusive but visible “gate scar” (Fig. 9.6).

Transfer molding is often used when inserts are to be molded into the finished part as, for example, contacts in an automotive distributor cap or rotor or solenoid coils and protruding terminals for washing machines. Whereas in compression molding, such inserts might be displaced during the flow of the viscous plastic, in transfer molding, the inserts are gently surrounded by a liquid flowing into the cavity at controlled rates and pressures and generally at a relatively low viscosity. Inserts are also rigidly supported by being firmly clamped at the parting line or fitted into close-toleranced holes of the cavity. When dimensions perpendicular to the parting line or parting surfaces of the mold must be held to close tolerances, transfer molding is used, because the mold is fully closed prior to molding. With compression molding, parting-line flash generally prevents metal-to-metal closing of the mold halves, making dimensions perpendicular to the parting line greater by the flash thickness, perhaps by as much as 0.005–0.010 in (Fig. 9.7).

Transfer presses and molds generally cost 5–10 percent more than compression presses and molds, but preheaters and preformers are the same as used in compression molding. Transfer cycle times are often slightly shorter than cycle times for compression molding, because the motion of the compound through the small runners and gates prior to its entering the cavity raises the compound temperature by frictional heat and mechanical shear, therefore accelerating the cure.

One highly significant application of transfer molding is the direct encapsulation of electronic components and semiconductor devices. Adaptation of the basic transfer molding process to successfully mold around the incredibly fragile devices and whisker wires of such items, required first the development of very soft flowing materials, generally epoxies and silicones; then modifications to conventional transfer presses to enable sensitive low-pressure control and accurate speed control (both often programmed through several steps during transfer); and finally new mold design and construction techniques to ensure close-tolerance positioning of the components in the cavities prior to material entry. It can be fairly stated that the successful development of the transfer molding encapsulation process was a large factor in the manufacture of low-cost transistors and integrated circuits.

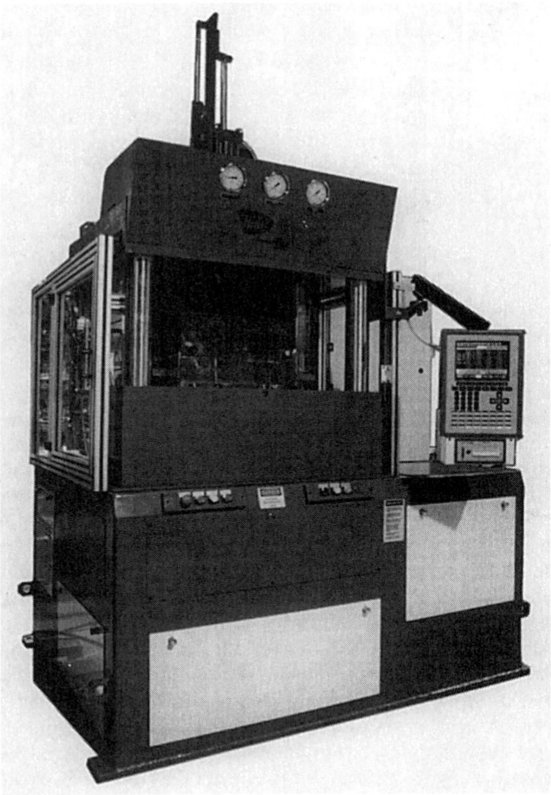

Figure 9.6 Transfer/compression-molding press, model T-200, for high-precision applications. Press shown features a Smart Mold™ software-based control system. *(Photo provided by Hull Corporation)*

In the automotive field, a growing dependence on electronic sensors for highly sophisticated engine controls and for the increasing number of safety features leads to predictions of $2,000 average cost per car for plastics-encapsulated electronic components.

Although compression and transfer molding are used principally with thermosetting compounds, the processes are occasionally used with thermoplastic materials, often thermoplastic composites. Thermoplastic toilet seats, for example, which often have fairly thick cross sections, have been successfully compression-molded at acceptably short cure times, about 4–5 min. Molding thermoplastic requires mold cooling rather than mold heating. In compression molding such materials, therefore, the material is put in the cavities when the mold is hot enough to melt the material, enabling the melt to flow adequately to fill the cavity. Cooling water is then circulated through the mold-cooling channels to cool the plastic below its melt index. When the thermoplastic material in the cavities hardens, the mold is opened and parts removed.

As mentioned in Sec. 9.4.3, liquid transfer molding (LTM) is being increasingly used for reinforced plastics, or composites. Conventional transfer presses may be used in such

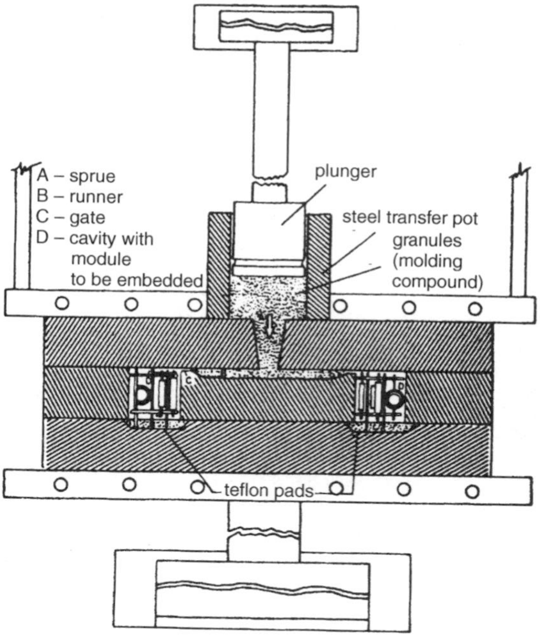

Figure 9.7 Schematic illustration of transfer or plunger mold.

applications, with the plastic being introduced into the transfer pot as a liquid catalyzed thermosetting material such as epoxy, polyester, or silicone, and caused to flow into the cavities by movement of the transfer plunger pushing against the liquid. Following cavity fill, the liquid material cures much the same way as do conventional transfer molding compounds.

9.4.11 Vacuum-Assisted Venting

Inasmuch as transfer molding requires closing the mold before the plastic molding compound is introduced into the cavity or cavities, very thin "vents" are machined into the mold parting surface at one or more locations on each cavity to enable air or gases to flow from the cavity to atmosphere as the molding compound flows into the mold. Such venting eliminates or minimizes air entrapments in the molded part, which could result in voids and/or incomplete cavity fill (Fig. 9.8). Complex part shape may preclude complete evacuation of air prior to cavity fill, because some of the inflowing plastic may close off a vent, rendering the vent ineffective.

Vacuum-assisted venting is a simple method to remove essentially all of the air from the runners and cavities before the plastic enters the cavity. It requires that the small vents in the cavity or cavities in the lower mold half allow air and gases to flow from the cavity to a manifold that accesses all vents. A hose connects the manifold to a vacuum reservoir adjacent to the press. The reservoir is a tank with a capacity of 20 or more gallons, connected to a mechanical vacuum pump that maintains a vacuum of 25 or more inches of mercury in the tank. In operation, a high-temperature rubber "O-ring" gasket surrounds the lower

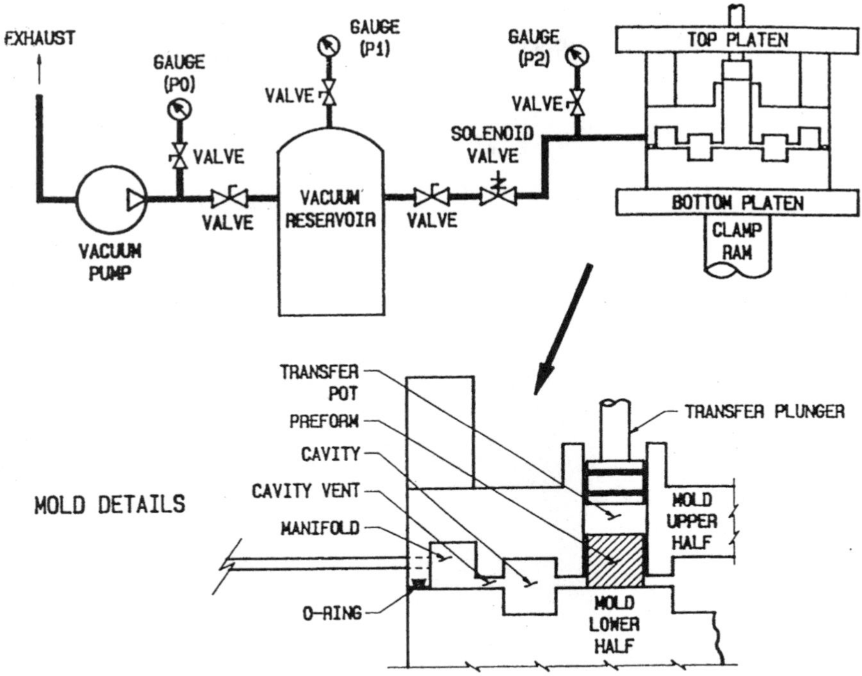

Figure 9.8 Schematic of vacuum-assisted transfer molding system.

mold parting surface, including manifold and cavities and runners, to ensure that no air will enter the mold when the cavities are under vacuum.

In a normal transfer-molding process, as soon as the transfer plunger enters the transfer pot, the solenoid valve in the vacuum line between the mold and the vacuum reservoir opens. Atmospheric pressure air in the runner and cavities flows out through the cavity and vents into the vacuum manifold in the mold, and out through the vacuum line leading to the reservoir. As the plunger drives the molding compound from the pot through the partially evacuated runners and into the cavities, residual air in the mold is pushed ahead of the inflowing plastic through the vents and ultimately into the reservoir. As a result, air leaves the cavities as they are filled with plastic. If a small amount of residual air should be trapped in the cavity when the molding compound seals off the cavity vents, it is readily absorbed into solution in the molding compound and does not create bubbles or voids in the molded part. As soon as the transfer plunger stops its advance, meaning that the cavities are full, the solenoid valve in the vacuum line is automatically closed, allowing the vacuum pump to re-evacuate the reservoir prior to the next cycle.

In much the same manner that vacuum-assisted venting is used with transfer molding, it may be utilized equally successfully in injection molding of thermosets and of thermoplastics.

9.4.12 Injection Molding—Thermosets

Similar in many respects to transfer molding is injection molding of thermosets. The process is also a closed-mold process, and the mold uses runners and gates leading to cavities

in much the same way as does a transfer mold. But instead of a pot and plunger, the injection process generally uses an auger-type screw, rotating inside a long cylindrical tube called a *barrel*. The barrel temperature is closely controlled, usually by circulating hot water in jackets surrounding the barrel. The front of the barrel narrows down to a small opening or nozzle, which is held firmly against a mating opening in the center of one of the mold halves, called a *sprue hole*, which leads into the runner system at the parting surfaces of the mold. The screw and barrel are generally positioned horizontally, and the press opens left and right (as compared to the up-and-down movements traditional with compression and transfer presses), so the mold parting surface is in a vertical plane rather than a horizontal one (Fig. 9.9).

In operation, after the molded parts and runners have been removed from the open mold, the press closes the mold in preparation for the next cycle. By this time, the screw has been rotating in the barrel, conveying granular material forward from the hopper at the back end of the barrel through the screw flights. As the material is conveyed forward, it is heated by the jacketed barrel and also by the mechanical shear caused by the screw rotation in the barrel and the constant motion of the material. The material becomes a viscous, paste-like fluid by the time it reaches the nozzle end. There, it does not have enough pressure to flow through the small nozzle opening, so it exerts a pressure against the front end of the screw, forcing the screw to reciprocate back into the barrel against a controlled hydraulic pressure in a cylinder at the back end of the screw. As the charge of plasticized material accumulates at the nozzle end of the receding screw, it finally reaches the set charge weight or volume for the mold. The screw's backward motion is automatically detected by a limit switch or linear potentiometer, which then stops further backward motion and rotation. The screw is then positioned in the barrel with the correct measured charge of plasticized material between the screw tip and the nozzle end of the barrel. This plasticizing step occurs automatically in the press cycle such that it is completed by the time the mold is closed, ready for another cycle.

When the injection molding machine senses that the mold is closed and is being held closed under full pressure, the screw advances forward rapidly ("reciprocates"), during

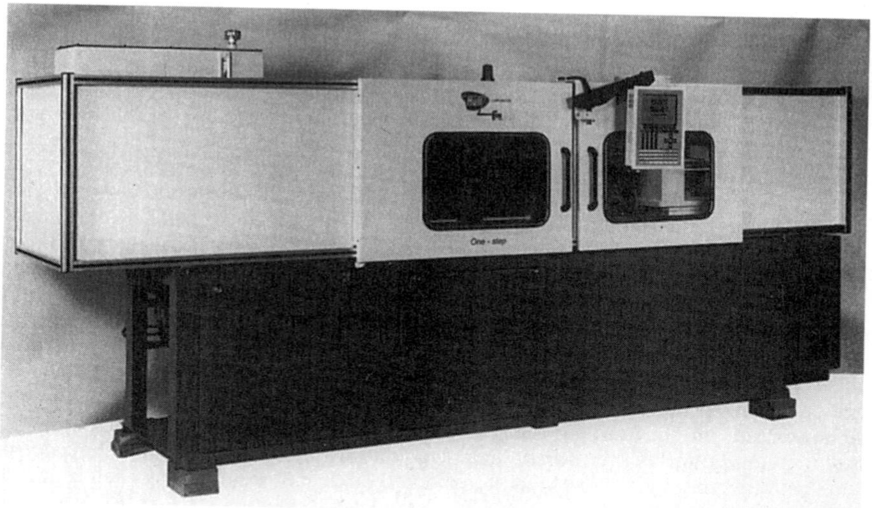

Figure 9.9 110-ton thermoset injection molding machine incorporating integral deflashing system to provide one-step processing of molded parts. *(Photo provided by Hull Corporation)*

which stroke it acts as a piston, driving the plasticized charge of material through the nozzle, sprue, runners, and gates to fill the mold cavities. Fill time is generally from 1–3 s, depending on the charge mass, as compared to 10–30 s in a transfer-molding operation. Frictional heat from the high-velocity flow raises the molding compound temperature rapidly such that the material time-temperature experience assures a rapid cure in the cavity. Overall cycles of thermoset-injection processes are often half those for comparable parts produced by the transfer-molding process.

Modern thermoset injection molding presses are usually fully automatic and produce parts at a high rate. They are ideal for applications requiring high volume of parts at minimal cost. Machines cost about twice as much as comparable-capacity machines for transfer and compression molding. Mold costs are about the same as for transfer operations. No preforming or preheating is required, and the labor content of automatic injection molding is significantly lower than that of semiautomatic transfer and compression molding.

To achieve maximum-strength parts, a high concentration of glass or other reinforcing fibers may be mixed with the molding compound in this process. Bulk molding compounds (BMCs) are often used, in which the formulation, generally polyester and glass fibers up to 1/2 inches in length, is putty-like in consistency. To minimize fiber breakage in BMC injection molding, the screw is often replaced with a plunger, and a special stuffing system is used to load the BMC into the barrel. Many electric switchgear components are produced with BMC injection molding.

Gas-assisted injection molding is a relatively new process that enables certain types of products to be molded as hollow parts without the use of cores. The hollow core is produced by injecting an inert gas, generally nitrogen, into the cavity after it is partially filled with the molding compound. In some respects, the process is like blow molding but without the parison. Parts for this process are generally of relatively simple configuration: a tube or elbow closed on one or both ends. A vital requisite to the process is that the molding compound must adhere to the cavity walls as it flows into the cavity, resulting in a relatively thick contiguous "skin" on cavity walls during initial injection. At the point when cavity walls are well coated, further material injection is halted, and the inert gas is blown into the open space under high pressure. The molding compound is then forced against cavity walls, held there under high pressure until the material is hard, and then ejected (after gas pressure has been reduced to atmospheric).

Most of the common thermoset materials exhibit the mandatory (for gas-assisted injection molding) laminar flow properties—phenolics, urea, melamines, thermosetting polyesters, and epoxies. Inorganic fillers and reinforcing fibers prove more successful than organic reinforcements. Parts molded with the gas-assisted process are lighter than their solid equivalents by 20 percent or more. They also enjoy faster cure and, therefore, shorter cycles, and they generally require less molding pressure in the cavity.

9.4.13 Injection Molding—Thermoplastics

Thermoplastic injection molding came into general practice well before thermoset injection molding, and it is the principal method for volume production of thermoplastic parts. Because thermoplastics are liquefied by heating and hardened by cooling, there are differences between thermoset and thermoplastic injection presses and molds (Figs. 9.10 and Fig. 9.11).

Molds for thermoplastics are often cooled to a temperature well above ambient. The mold temperature is controlled such that it will cool the material in the cavity at the fastest rate possible without causing it to harden before it totally fills the cavity. Molds are often made of metals more easily machined than steel, such as beryllium copper, that will afford optimal heat transfer to the molding material. Cavity surfaces are generally not chrome plated, because thermoplastics are less abrasive than thermosets. Overall cycle times, especially for thin-walled products, are often below 10 s.

Figure 9.10 Elektra 725-ton all-electric injection molding machine. *(Photo provided by Cincinnati Milacron, Inc.)*

Figure 9.11 Model VT550 550-ton toggle-type injection molding machine. *(Photo provided by Cincinnati Milacron, Inc.)*

Screw length-to-diameter ratios are often different from those for thermosets because of the critical need for very thorough mixing to achieve color and temperature uniformity. Because barrel temperatures are generally higher in thermoplastics machines to ensure total melt and low-viscosity flow, barrel heating is usually electric and may be as high as 600°F or more for some of the newer engineering plastics such as nylons and liquid crystalline polymers. Also, with the significantly lower apparent viscosity of the thermoplastics at the time of injection, a check valve is often required at the nozzle end of the barrel to prevent "drooling" or leakage through the nozzle prior to the injection stroke. An economic advantage not possible with thermoset injection molding is that thermoplastic scrap—sprues, runner, and rejects—may usually be reground and mixed with virgin material for reprocessing.

Applications of injection molding of thermoplastics include practically every molded thermoplastic item—picnic ware (including eating utensils); fashion buttons; refrigerator food containers; grocery-store containers for butter, yogurt, and the like; towel racks; and fan and power-tool housings are some examples. Since the thermoplastic materials are

considered less dimensionally stable than thermosetting plastics (because the former exhibit a property called creep, or distortion due to prolonged stress), thermoplastic selection must take into consideration the nature of any loads to be borne by the molded product. And because the heat stability (closely related to the melt index) of thermoplastics extends from about 200°F for some vinyls and styrenes to 600°F and higher for the newer engineering plastics, material selection often depends on the temperature exposure anticipated for the finished product. Where color and superior surface finish are desired, thermoplastics may offer better characteristics than do thermosets.

The costs of injection presses and molds for thermoplastics are comparable to those for thermosets, running from about $30,000 for a 25-ton press to $400,000 for a 500-ton press. Injection presses for thermoplastic molding have been built in sizes of over 5,000-ton clamp force for molds measuring 15 × 15 ft and more.

As with thermoset injection molding, heavily fiber-reinforced parts are also molded from thermoplastics. Such parts still may exhibit creep during prolonged mechanical stress, but they are nevertheless considerably stronger and tougher than the nonreinforced products. Glass-reinforced nylon, for example, is used with the lost-core process for a complex air-intake manifold on a current automotive engine.

Injection-molded thermoset parts requiring hollow sections that are not suited to conventional mold configurations or to mechanical side cores may be produced using the above-mentioned lost-core process, a process borrowed from a similar metal-casting technique. In this molding process, a precisely shaped rigid metal insert is positioned in the mold cavity prior to mold closing. The injection process proceeds in the normal manner. Following ejection, the rigid insert is removed by melting the core. Obviously, a relatively low melting temperature for the core, higher than the molding temperature of the plastic but lower than the heat distortion temperature of the molded part, is mandatory for success in this process. This process is presently used for both thermosetting and thermoplastic injection-molded parts.

Reaction injection molding (RIM) is quite similar to injection molding of thermoplastics, but it is designed to process materials that combine in the press to create a chemical reaction—sometimes to produce a foamed product (such as inexpensive picnic coolers, foamed drinking cups, structural foamed containers, or sporting equipment) and sometimes to process a hybrid plastic such as polyurethane for, say, automotive bumpers. The injection process in RIM machines includes bringing two or more chemicals together at the injection nozzle as they are being injected into the mold, where the mixed chemicals rapidly blend, react, and harden.

Injection-compression molding is a process utilizing a uniquely designed mold in an injection press. The process (used with either thermoset or thermoplastic materials) requires the mold to be closed partially and then held in that position (perhaps 1⁄4 to 1⁄2 in short of full closed) while the charge is delivered into the cavity. Then the mold closes fully. The special mold design prevents injected plastic from escaping at the parting line. The process minimizes the oriented flow lines and ensures a part that is less subject to warpage or shrinkage. Reinforcing fibers are less oriented and may enable a stronger part.

Structural RIM (SRIM) uses the RIM process with a glass-reinforcing mat placed in the mold before closing and injecting the plastic. SRIM parts offer far greater strength than nonreinforced parts.

Expandable polystyrene (EPS) is an extremely low density thermoplastic material (0.2 –10.0 lb per ft^3) that is produced using the RIM process plus live steam injected into the mold cavity as soon as the styrene and blowing agent have been introduced into the cavity. The steam enters the cavity through a number of small holes in the cavity wall to cause rapid expansion of the material. Gases generated escape through the perforated cavity walls. The cavity is chilled prior to mold opening.

Principal EPS applications are in packaging of consumer electronic appliances needing protection against shock and vibration during handling and shipment. Other applications

include disposable food containers, drinking cups, flotation systems for small boats and canoes, and energy-absorbing barriers in automobiles.

One thermoplastic used in RIM is a rubber-toughened nylon that has proven exceptionally ideal for a fender on a Caterpillar™ tractor. The rugged fender weighs only 26 lb, is 101 in long and 33 in wide, and possesses high impact resistance.

9.4.14 Micromolding or Nanomolding

As a subset of thermoplastic injection molding, extremely small part molding processes, requiring highly specialized injection presses, are currently categorized as *micromolding* and *nanomolding*. Micromolding refers to parts as small as 0.5–1.5 cm^3 in volume, whereas nanomolding refers to parts in the 0.5–10 mm^3 in size!

A recently published article describes a nanomolding system (developed by Murray, Inc., Buffalo Grove, IL) capable of insert molding catheter tips directly onto catheters. Used in surgical and medical applications, such catheters are required to deliver miniscule devices into the human anatomy. Murray's fully automatic machine utilizes a one-pellet-at-a-time feeder, a plasticating plunger, and injection via an injection pin, a needle driven by a high-speed linear servomotor. The Sesame molding machine is said to deliver injection pressures up to 50,000 lb/in^2, achieve injection times as short as 0.02 s, and inject shots as small as 2.5 mm^3 to within ± 0.012-mm^3 accuracy.

9.5 Assembly and Machining Guidelines

Many molded plastic parts are complete and usable by themselves, such as melamine dinnerware, closet hooks, and the like. But plastic parts often need to be joined to other plastic or metal parts to become functional.

9.5.1 Adhesives and Welding

Bonding plastics to plastics has become a well developed art through use of adhesives and "welding." For optimal strength in such bonding applications, the design of the joint where the two pieces are to be bonded must provide physical means for the plastic parts to take the loads rather than relying on the bonding alone. The use of tongue and groove, or a molded boss fitting into a molded hole; or a stepped joint, with bonding on all the mating surfaces, ensures that, so long as the two pieces remain together, shear or tension or torque loads will generally be transmitted from one piece to the other without stressing the bond itself (Fig. 9.12).

In addition to the many polymeric adhesives available today for most families of plastics, actual welding of two similar plastics is possible. Many thermoplastics materials can be welded through use of ultrasonic energy. In principle, the process is similar to metal spot welding in that energy is transferred from a "horn," vibrating ultrasonically, through a finite thickness of one of the pieces to the contacting interface of the other piece to which it is temporarily clamped. The contact area is generally established at a properly designed interfacing point. The concentration of energy at the interface melts the contacting plastic surfaces of both pieces in a few seconds. When the joint cools, the pieces are effectively welded (Fig. 9.13).

Spin welding of thermoplastics uses the mechanical friction of one part spinning against another part to generate the heat needed to melt the surfaces at point of contact. As soon as localized melting has occurred, spinning stops, the bond area is cooled, and the parts become fused. Vibration welding creates similar frictional heat for melting and fusing (Fig. 9.14).

Figure 9.12 Structural bonding application for a reinforced plastic bumper assembly for a truck. The equipment dispenses a two-component structural adhesive onto the bumper assembly before it is integrated into the truck. In production, this is typically done robotically with a manual backup system (shown). The system being used is a See-Flo 488 meter/mix system; the dispense gun in the picture is model 2200-250-000 dual-spool Snuf-Bak™ dispense valve that utilizes a No-Flush™ disposable static mixer nozzle. *(Photo provided by Sealant Equipment and Engineering, Inc.)*

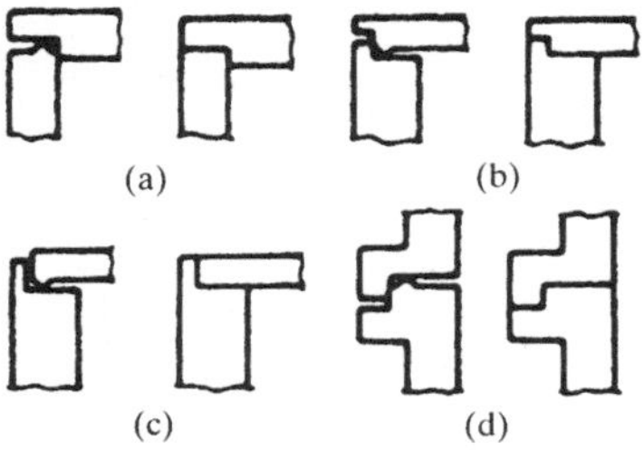

Figure 9.13 Basic joint variations suitable for ultrasonic welding. *(Courtesy of Branson Sonic Power Co.)*

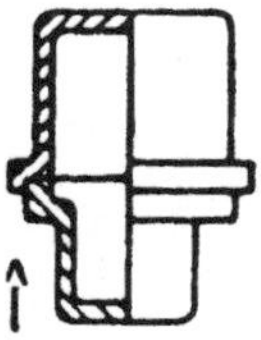

Figure 9.14 Spin welding.

Hot-air welding uses a concentrated blast of hot air directed toward the desired joint area at a temperature above the melt index of the plastic to achieve melting and bonding.

Solvent cementing uses a solvent to temporarily create a solution of softened plastic at the joint. The parts are then clamped together until the solvent has evaporated, thereby effecting fusion. This is a very simple method of bonding acrylics and styrenes for picture frames or desktop decorative items (Fig. 9.15).

A current automobile engine utilizes valve covers molded from glass and mineral-filled nylon 6/6 thermoplastic material, weighing half as much as the formerly used aluminum cover. Ultrasonic welding is used to weld the cover to the oil seal, and vibration welding of the baffle plate to the cover enables reduction of assembly time from three minutes for the former aluminum cover to 10 s for the three-piece plastic cover!

Electromagnetic welding of thermoplastics is a process using induction heating with ferromagnetically filled thermoplastics. Frequencies used range from 3–8 MHz to heat the joint area for melting and fusing.

9.5.2 Threaded Connections

When two pieces of plastic must be joined together in service but must also be suited to simple disassembly, threaded joints are possible. Such joints may be plastic to plastic, as in toothpaste tubes and caps, where male and female threads are molded. If metal screws are to be used, female threads can be molded in a plastic part to accept the metal screw. If such a threaded application does not have to withstand much tension in the screw, a self-tapping screw may be turned into a molded hole of appropriate diameter. Self-tapping screws designed for plastics applications are better suited than those for metal applications. Metal female-threaded inserts may be molded into plastic parts to accept metal screws in a subsequent assembly operation. Metal female-threaded inserts may also be staked into a molded hole of appropriate size, sometimes mechanically, often with heat ("heat staking") to soften the thermoplastics at the interface so as to achieve more positive bonding (Figs. 9.16 and Fig. 9.17).

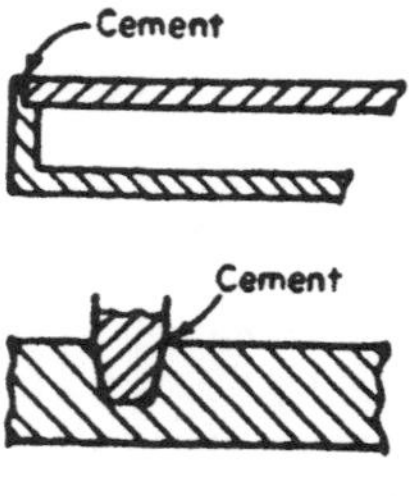

Figure 9.15 Assembly by solvent cement.

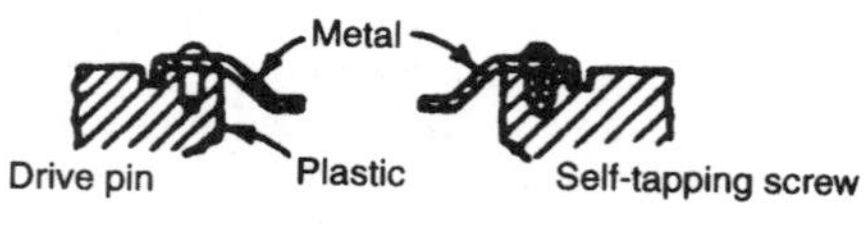

Figure 9.16 Assembly by screws.

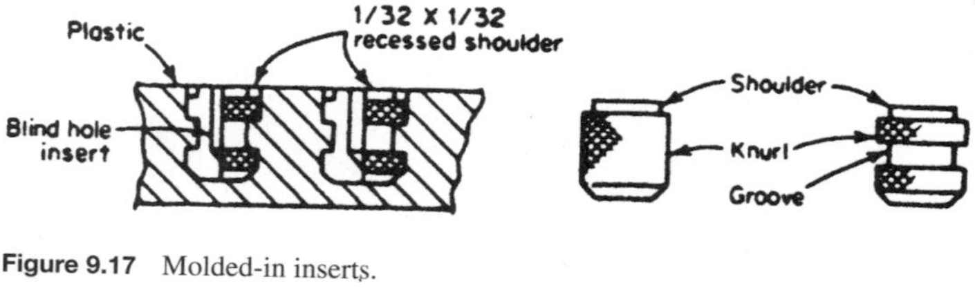

Figure 9.17 Molded-in inserts.

It is important to locate holes for such threaded connections such that there is ample wall thickness to accommodate the concentrated stress in the immediate vicinity of the screw. Screwing a relatively thin plastic part to a more robust part could result in tearing of the thin material at the screw head if separation forces are severe. Designing a thicker section or using an ample washer or oversized screw head will yield more satisfactory results.

Screw joints are especially difficult with brittle plastics such as styrenes, acrylics, polycarbonates, and most thermosetting plastics. In such applications, slight overtightening of the screw may result in cracking of a part or stripping of threads.

9.5.3 Machining

Machining of plastics is often necessary as, for example, in cutting and finishing Formica® (melamine laminate) counter tops, cutting polycarbonate sheet for glazing applications, or drilling structural shapes for attaching to supports. While such operations are commonplace when working in metals or wood, special tools and practices are necessary when machining plastics.

Brittle plastics tend to break or chip under concentrated loads such as those imposed by a saw tooth or a cutting edge of a drill bit. Softer plastics, like many thermoplastics, tend to tear when local load concentrations occur. Furthermore, thermosetting plastics and thermoplastics are good thermal insulators, which means that the high energy imparted by cutting tools turns into frictional heat that, because it does not dissipate easily, quickly reaches the melting point of many plastics and the burning point of others. Essentially all plastics have heat-distortion limits where the plastics lose rigidity and strength.

When conventional woodworking and metalworking cutting and grinding tools are used on plastics, the plastics often become gummy and sticky in the cutting area, binding the cutting tool and distorting the plastic. Therefore, for machining operations on plastics, special cutting tools and special cutting techniques are necessary to achieve desired results.

Cutting tools that are suited to epoxies will not be appropriate for polyethylene and the like. When contemplating such machining operations, therefore, it is critical to contact the plastic supplier for specific recommendations as to cutting-tool configurations and speeds, and procedures often involve cooling.

9.5.3.1 Machining of thermoplastics. Carbide tools should be used. If a mirror-like finish is expected, the cutters should be diamond tipped. Diamond tools must be fed uniformly along the traverse of the cut. Sharp tools carefully ground with minimum cutting edge should be used. Removal of the molded surface layer during machining operations will affect the physical properties of the plastic material. Adding water-soluble agents to the cooling air jet improves the cooling system but requires a subsequent cleaning operation. The coolant should be selected with care, as some liquids may craze, crack, or dissolve the plastic. Glass-reinforced materials present the problem of abrasiveness, which in turn reduces cutter life.

The spiral plastic cuttings that leave the cutting edge create problems of entanglement. These problems usually can be resolved by air jets and vacuum attachments directing the chips away from the cutter. However, the type of thermoplastic to be machined will vary the techniques used.

9.5.3.2 Taps for thermoplastics. For tapping thermoplastics, taps with a slightly negative rake and with two or three flutes are preferred. Some plastics, such as nylon, may

require slightly oversized taps, because the resiliency of the plastic may cause it to compress during cutting, leading to swelling after the tap is removed.

Solid carbide taps and standard taps of high-speed steel with flash-chrome-plated or nitrided surfaces are necessary. During tapping, the tap should be backed out of the hole periodically to clear the threads of chips.

9.5.3.3 Reaming for thermoplastics. High-speed or carbide-steel machine reamers will ream accurately sized holes in thermoplastic. It is advisable to use a reamer 0.001–0.002 in larger than the desired hole size to allow for the resiliency of the plastic. Tolerances as close as ± 0.0005 in can be held in through holes 1⁄4 inch in diameter. Fluted reamers are best for obtaining a good finished surface. Reamer speeds should approximate those used for drilling. The amount of material removed per cut will vary with the hardness of the plastic. Reaming can be done dry, but water-soluble coolants will produce better finishes.

9.5.3.4 Turning and milling. Use tungsten carbide or diamond-tipped tools with negative back rake and front clearance. Milling cutters and end mills will remove undesired material, such as protruding gate scars, on a molded article.

9.5.3.5 Mechanical finishing of thermosets. The machining of thermoset plastics must consider the abrasive interaction with tools but rarely involves the problems of melting from high-speed frictional heat. Although high-speed steel tools may be used, carbide and diamond tools will perform much better with longer tool lives. Higher cutting speeds improve machined finishes, but high-speed abrasion reduces tool life. Since the machining of thermosets produces cuttings in powder form, vacuum hoses and air jets adequately remove the abrasive chips. To prevent grabbing, tools should have an O-rake, which is similar to the rake of tools for machining brass. Adding a water-soluble coolant to the air jet will necessitate a secondary cleaning operation. The type of plastic will vary the machining technique.

9.5.3.6 Drilling thermosets. Drills that are not made of high-speed steel or solid carbide should have carbide or diamond tips. Also, drills should have highly polished flutes and chrome-plated or nitrided surfaces. The drill design should have the conventional land, the spiral with regular or slower helix angle (16–30°), the rake with positive angle (0 to +5°), the point angle conventional (90–118°), the end angle with conventional values (120 to 135°), and the lip clearance angle with conventional values (12—18°). Because of the abrasive material, drills should be slightly oversize by 0.001–0.002 in.

9.5.3.7 Taps for thermosets. Solid carbide taps and standard taps of high-speed steel with flash-chrome-plated or nitrided surfaces are necessary. Taps should be oversize by 0.002–0.003 in and have two or three flutes. Water-soluble lubricants and coolants are preferred.

Machining operations will remove the luster from molded samples. Turning and machining tools should be high-speed steel, carbide or diamond-tipped. Polishing, buffing, waxing, or oiling will return the luster to the machined part, where required.

9.6 Postmolding Operations

9.6.1 Plastic Part Deflashing

Most thermoset molding operations result in some excess material, called *flash,* at the parting line and on molded-in inserts. It is generally necessary to remove this flash, either for cosmetic reasons or, in the case of contacts and leads extending from an encapsulated electric or electronic device, to ensure good electrical contact to the leads.

Robust parts may be tumbled randomly in a wire container to remove the flash. Slow rotation of the containers for 10–15 min, with parts gently falling against one another, will usually suffice. For more thorough flash removal, the tumbling action may be augmented by a blast of moderately abrasive material, or media, of either organic type (such as ground walnut shells or apricot pits) or polymeric type (such as small pellets of nylon or polycarbonate) directed against the tumbling parts. For more delicate parts, a deflashing system passing such components on a conveyor that holds the parts captive as they are conveyed beneath one or more directed blast nozzles, generally proves practical. Such systems have been perfected for transfer-molded electronic components, holding the lead frames captive, often temporarily masking the molded body as the devices pass through the blast area, and using as many as 24 individually positioned blast nozzles to ensure total removal of flash on a continuous basis. Such devices are magazine-fed and collected in magazines to maintain batch separation (Fig. 9.18).

Modern deflashers recycle the blast media and utilize dust collectors to minimize air pollution. The blasting chamber is effectively sealed with entry and exit ports designed to avoid dust and media escape. Chemical deflashing using solvents to remove the flash from such components is also used. In addition, water-honing deflashing has been found successful for some types of devices.

Although very simple tumbling deflashers may be built by the processor, sophisticated applications are best handled by commercial specialists who manufacture a wide variety of special and custom systems.

Wheel deflashers are also used for thermoset parts. A high speed wheel "throws" the deflashing media, by centrifugal force, to the flash area. While the wheel delivery is not as finely directed as the pressure blast through nozzles, the wheel deflasher is often selected, because it is more economical than compressed air deflashers.

9.6.2 Thermoplastic Part Deflashing

Because most thermoplastic materials are more resilient than thermoset plastic parts, conventional wheel or air-blast deflashers are generally ineffective on such molded parts. Many such parts require manual trimming with sharp knives. Cryogenic deflashers, however, designed for rubber and thermoplastic parts deflashing, are often very successful. With cryogenic deflashers, the parts themselves must be essentially frozen to a temperature where the flash becomes brittle and can be removed by tumbling or by blasting with very low-temperature air and media.

9.6.3 Lead Trimming and Forming

Most semiconductor devices encapsulated by transfer molding and a host of other high-volume small and fragile electronic components utilize lead frames as carriers during assembly and molding. These lead frames need to be trimmed off prior to testing, marking, and packing the devices.

Progressive trimming and forming presses and dies have been developed for this application, available both as manually fed and actuated systems and also as fully automated magazine-to-tube carrier systems.

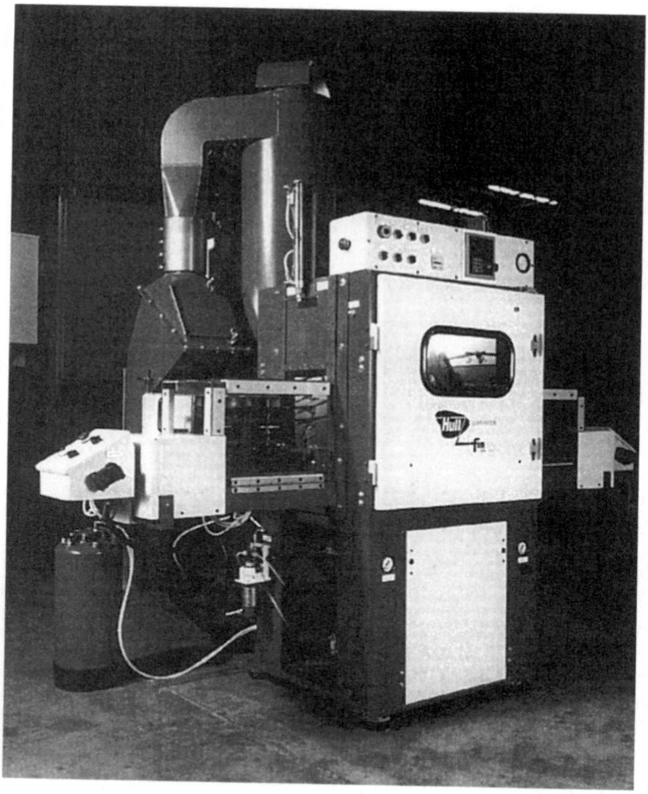

Figure 9.18 Side-delivery deflashing system with conveyor. It feeds, cleans, and deflashes thermoset plastic parts. Parts can be mounted on rotating spindles or indexing fixtures for maximum exposure to blast from nozzles. *(Photo provided by Hull Corporation)*

9.6.4 Cooling Fixtures

Molded parts, both thermoplastic and thermoset, are ejected from the mold while still warm. As cooling to room temperature takes place, parts may warp or deform, due in part to internal stresses or to stresses created because of uneven cooling. Such changes in shape may be minimized by placing the parts in a restraining fixture, which holds them to tolerance during final stages of cooling (Fig. 9.19).

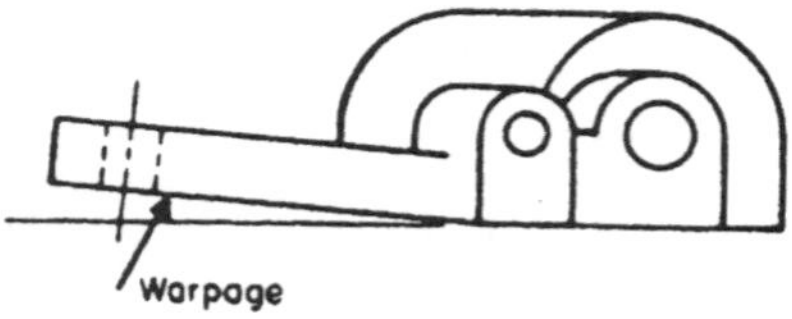

Figure 9.19 Product warpage.

The part design is often at fault for such deformation. Thick sections in combination with thin sections, for example, experience faster cooling of the thinner sections, resulting in localized shrinkage, which produces distortion, while the thick sections are still relatively soft. Designing parts with thin reinforcing ribs rather than thick sections often reduces or eliminates such distortion.

Another cause of internal stresses is the flow pattern as cavities are filled. In general, shrinkage is greater in a direction transverse to flow line than in the direction of flow. Part designers and mold designers need to agree on gate locations and vent locations to minimize such distortion due to flow direction. Other molding parameters, such as temperature and cavity fill pressures and rates, can be adjusted to minimize such distortion. But when all else fails, cooling fixtures may prove to be the final solution.

9.6.5 Postcure

Thermosetting plastics harden by cross-linking under heat and pressure. But at the time of ejection from cavities, especially in relatively short cycles for high production, parts have not fully cross-linked and are to some extent "rubbery" at that stage of cure. They may be distorted from stresses created when ejector pins force parts from cavities or from parts piling on top of each other as they fall into a container or conveyor on being ejected. Longer cure times in the cavities may be necessary to lessen distortion from such conditions immediately following ejection.

Cross-linking, or polymerization, of thermosetting plastics is generally about 90 percent complete at time of ejection, with the irreversible reaction continuing for minutes, possibly hours, and sometimes for days or months, with certain formulations. Mechanical properties of such materials may be improved by a programmed cooling following ejection, a program generally providing staged cooling in ovens instead of conventional cooling in room-temperature air. Molding compound formulators can recommend postcuring cycles where appropriate.

9.7 Process-Related Design Considerations

9.7.1 Flow

As thermoplastics cool and harden into solid shapes following amorphous flow, and as thermosetting materials polymerize, changing from a viscous liquid to a shaped solid, the plastics often carry some history of their flow conditions preceding hardening. This history often adversely affects mechanical and electrical properties, dimensions, cosmetic appearance, and even density of the finished parts.

Flow inside a mold cavity in injection, compression, or transfer molding should ideally be such that the cavity is completely filled while the material is still fluid. If such is the case, the cooling (for thermoplastics) or curing under sustained cavity heat (for thermosetting plastics) will proceed uniformly until the part is sufficiently rigid to withstand the rigors of ejection. But such is rarely the case.

In the case of thermoplastics, as soon as the melted material flows through the gate, it encounters a relatively cool environment and stiffens as its temperature drops to mold temperature. The material pressed against a cavity wall or flowing through restricted passage into a rib or boss has dropped in temperature and increased in stiffness, while relatively hotter fluid continues to enter the cavity, possibly displacing some of the stiffer material on the way. The partially cooled or fully hardened material against the cool cavity walls has begun to shrink, while the later-arriving hotter material continues to enter, starting its hardening and shrinking moments later. The final amount of material passing through the

gate is slowed in its flow, because it is pushing the earlier-arriving material as it packs the mold during final stages of fill. The dynamic temperature change of the melt is nonuniform during cavity fill, and the final part reflects those changes, to a greater or lesser extent, depending on rate of fill and the various cavity obstructions through which or around which the flow was forced—even when cavity fill-time amounts to only a few seconds.

In extreme cases, flow lines showing earlier-hardened material adjacent to later-hardened material will be visible. Material flowing around two sides of a boss or insert in the cavity may have partially hardened before coming together on the other side, showing weld marks or lines where the material actually failed to weld to the material coming from the other side. Material flowing through a thinner section may cool so rapidly that it hardens before it reaches the far side of this section (Figs. 9.20 and 9.21).

A slow cavity fill, or a fill into a mold too chilled, will cause the plastic to stiffen before completely filling the cavity. Corners may not fill out, or detailed configurations in the cavity will not fill completely.

If the material is reinforced, such as a glass-fiber-filled nylon, relatively thin passages in the cavity may cause the low-viscosity plastic to flow into the passage but may "strain" out the glass fiber particles, resulting in a resin-rich but nonreinforced (and therefore weaker) area in the final part.

In thermosetting plastics molding, similar dynamic flow dichotomies occur because of cavity fill rate, obstructions in the cavity configuration, possible separation of resin from filler, and nonuniform cross linking during flow and cavity fill due to slightly differing time-temperature relationships, causing precure or delayed cure in various locations of the cavity during fill. Weld lines, incomplete fill, and resin-rich sections can result in reduced quality of parts and possibly lower yields of acceptable parts.

Cavity venting—machined-in passages at one or more locations along the parting line or along ejector pins—is vital to obtaining complete cavity fill in shortest possible time. Air in the cavity must have a clear and rapid escape route. If it doesn't get out of the way in the few seconds of cavity fill, because it can't find a vent or because the vent is too restricted, a void will result, or possibly a burn mark on the finished part where the highly compressed unvented air overheats according to Boyle's law—the compression of gases

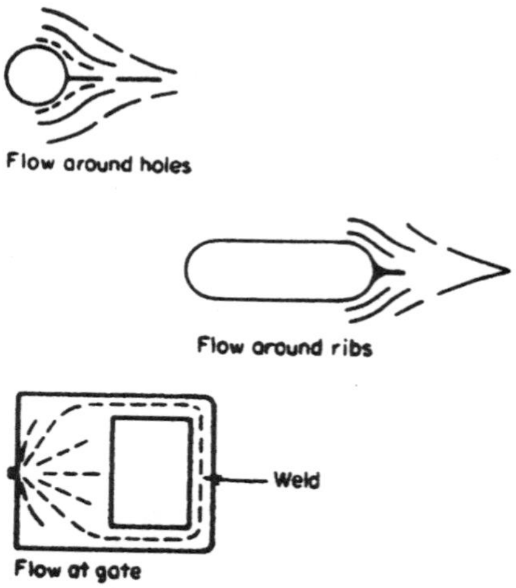

Figure 9.20 Flow patterns around holes, bosses, ribs, and windows.

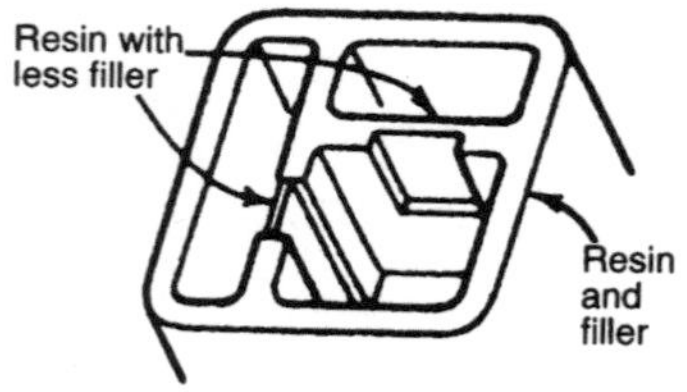

Figure 9.21 Flow patterns in and around thin, delicate sections.

creating very high temperatures, enabling diesel engines to ignite fuel and enabling air in inadequately vented mold cavities to overheat and burn plastic (Figs. 9.22 and Fig. 9.23).

Experienced part designers and mold designers need to understand the flow phenomena of various plastics during the molding process so as to minimize the occurrence of unwanted defects in the final product. But even with the "perfect" design of part and the "perfect" design of mold, the processing parameters of temperatures (of melt and of mold), fill rate, fill pressures, and in-mold dwell are equally critical in achieving quality parts.

Die design for thermoplastic extruding requires extensive understanding of the melt properties of the specific material to be used as well as a full grasp of the parameters of the process. Upstream of the die, parameters of barrel temperature, screw rotation speed, and barrel venting need to be sensitively regulated for the particular plastic and die being used. Additionally, coolant flowing through the die must hold die temperature constant. Because extruding is a continuous process, these parameters must be controlled continuously to ensure a uniform product.

9.7.2 Stretching

In thermoforming processes, including the blowing stage of blow molding, the plastic is not in a true fluid state but more in a relaxed elastomeric state, able to be stretched and formed and mostly stress relieved before finally cooling to a rigid state. Parameters of temperature, pressure, and time must be closely regulated to ensure that the plastic retains ad-

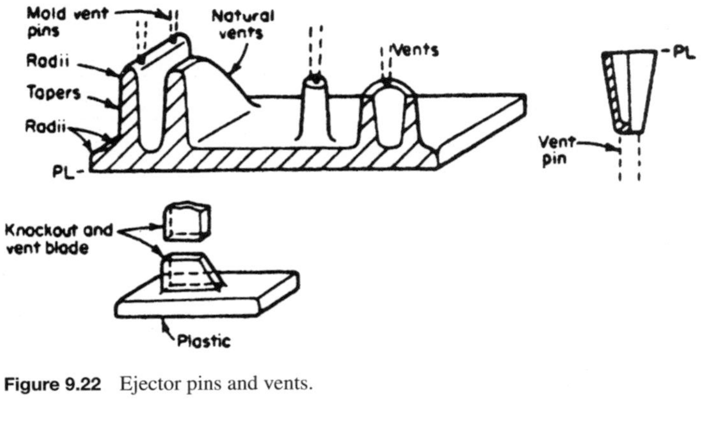

Figure 9.22 Ejector pins and vents.

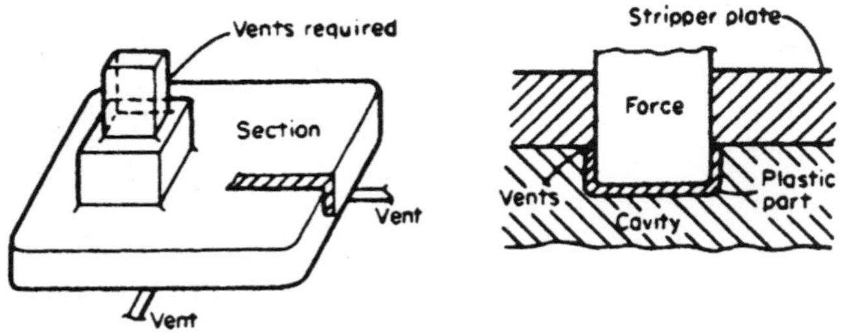

Figure 9.23 Parting-line vents.

equate elasticity to stretch and shape itself to the mold contours without tearing or rupturing. Thickness of material in the sheet (or in the expanded blow-molded part) must be adequate after stretching to meet the mechanical requirements of the finished part.

9.7.3 Final Part Dimensions

Physical dimensions of many processed parts must be held to fairly close tolerances to ensure proper assembly of parts into a complete structure—as, for example, molded fender panels assembled on cars, plastic screw caps for glass jars, and so on. In general, the final dimensions of the processed part will differ from the dimensions of the mold cavity or the extrusion die. Such differences are somewhat predictable but are usually unique to the specific material and to the specific process. The dimensions of a mold cavity for a polycarbonate part requiring close tolerances will often be different from dimensions of a cavity for an identical methyl methacrylate part. Similarly, the dimensions of an extrusion die for a close-toleranced vinyl profile will differ from those of a die for an identical close-toleranced polyethylene part. Both the part designer and the mold or die designer must have a full understanding of the factors affecting final dimensions of the finished product and often need to make compromises in tolerances of both part and cavity dimensions (or even in plastic-material selection) to achieve satisfactory results with the finished product.

The following paragraphs will address the plastic-behavior characteristics affecting dimensional tolerances.

Shrinkage of a plastic as it cools (thermoplastic) or polymerizes (thermoset) is a fact of life, often specified as parts per thousand or, for example, as six thousandths of an inch per inch of mold shrinkage. Such mold shrinkage is reasonably easy to compensate for by making the cavity proportionately larger in all dimensions.

But shrinkage in many materials is different when measured transverse to the material flow, as when measured longitudinal to the flow. In reinforced or heavily filled materials, this difference is significant. Gate location and size, and multiple gates in some instances, must be considered for cavity and part design to minimize effects of such mold shrinkage.

Sink marks in a molded part often occur in relatively thick sections, usually reflecting progressive hardening of the molded or extruded or formed part from cavity wall to inside area. The outside wall hardens, while the mass of plastic in the thick section is still somewhat fluid. As this inside mass subsequently hardens (and shrinks, as most plastics will), the cured outer wall is distorted inward, resulting in a sink mark. The best way to avoid such deformation is to avoid thick sections wherever possible. Often one or more judiciously designed thin ribs in select locations will give a part adequate strength and thickness without the need for thick sections (Fig. 9.24).

Nonuniform hardening of the material during residence in the mold for cooling or curing generally produces internal stresses in a molded part that, after removal from the mold, may distort the part widely from the intended dimensions. Flat panels become concave, straight parts may curve, round holes may elongate, and worse. Part design and cavity design can generally accept some necessary compromises to accommodate such deformations yet still yield a part meeting its functional requirements.

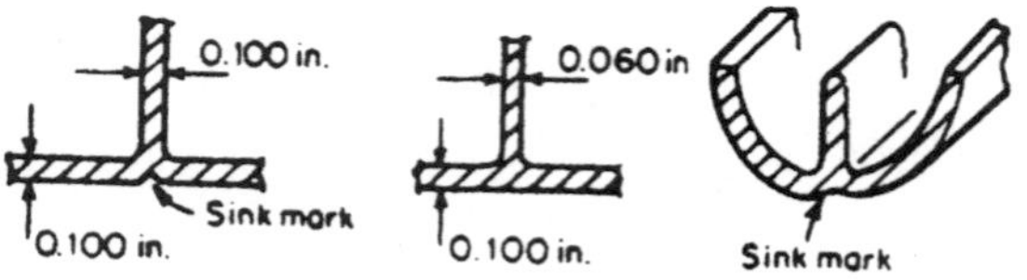

Figure 9.24 Sink marks.

Deep molded parts may require design considerations to ensure minimal stresses during the ejection phase of the molding process. Imagine a straight-walled plastic tumbler of internal and external diameters unchanging from top to bottom. As the part hardens in the cavity, it will tend to shrink around the *force,* or male part of the cavity. When the mold opens, the part will stay with the force. To remove it from the force will require considerable pressure, either from ejector pins or air pressure coming out the end of the force against the bottom of the tumbler or from an ejection ring moving longitudinally from the inner end of the force. The pressure exerted by either of these ejection methods will be considerable until the open end of the tumbler finally slips off the end of the force.

To minimize such ejection stresses, forces for deep molded parts are designed with an appropriate "draft" or taper, up to 5° in some cases, such that very slight movement of the molded part with respect to the force will suddenly free the part from its strong grip on the force, and the remainder of the ejection stroke exerts almost no stress on the part. Such draft is advisable on all plastic parts, even those with depths of only 1⁄4 in, to minimize ejection pressures and to prevent possible localized damage where the knock-out pins push against the plastic (Fig. 9.25).

Parting lines on molded parts require special consideration in part and mold design, especially where two molded parts must come together as, for example, on each half of a molded box with hinged opening (Fig. 9.26 and Fig. 9.27).

In compression molding or injection-compression of thermoset or thermoplastic parts, the mold is fully closed only after the material has been placed into the partially closed cavity. More often than not, some material is forced out of the cavity onto the land area before the mold is fully closed, metal to metal. In effect, then, the mold closing is halted

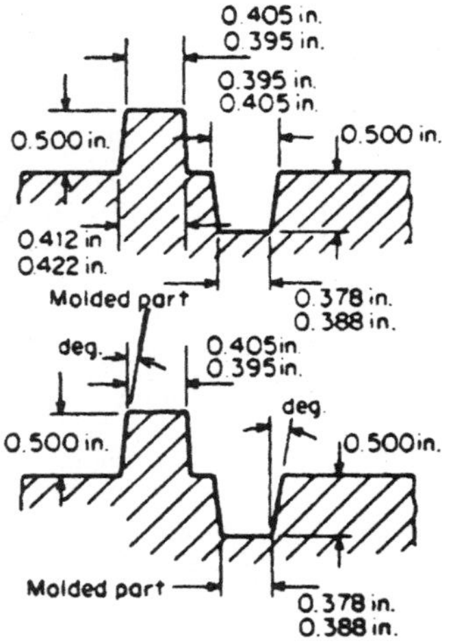

Figure 9.25 Dimensioning tapers.

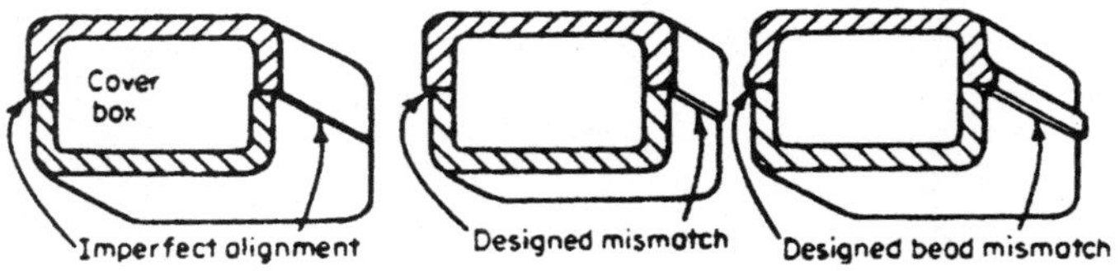

Figure 9.26 Mismatch parting lines.

Figure 9.27 Designed mismatch.

short of full close, perhaps by as much as 0.005 in or more. Such overflow hardens, leaving flash on the molded part. Under these circumstances, the molded part dimension perpendicular to the mold parting surface will be at least 0.005 in greater than intended. The tolerance of such compression-molded parts is kept very wide in deference to the inherent characteristics of the process.

When an assembly of two molded parts is ultimately required, even if the materials and the molding processes are the same, it will be virtually impossible to achieve a perfect match where the parts come together. Slight variations in shrinkage or warpage will yield an easily noticeable or "feelable" mismatch. Intentionally designing mating surfaces with an overlap or a ridge enables ingenious camouflaging of the nonuniformity of mating areas in the final assembly of the two contacting parts.

In designing molded parts, and molds to produce them, it is necessary to consciously determine how the part will be removed from the mold cavity or force and to maintain positive control of the part during mold opening, such that it is ejected as intended. This positive control is especially vital in automatic molding.

Assuming that the decision has been made that the molded part must be ejected from the moving half of the mold (as opposed to the fixed half), then it is necessary to make provisions that the part will not remain in the fixed half of the mold during mold opening but will invariably remain with the moving half.

One common way to accomplish such positive part control is to provide undercuts in the cavity or force of the moving half. These undercuts will enable plastic to flow into them and harden there before the mold is opened. On opening, the hardened plastic in the correctly designed and sized undercuts will hold the molded part in the moving mold-half during opening stroke. After mold opening, the ejector pins or mechanism in the moving half of the mold will then have to push hard enough to allow the molded part to distend sufficiently to be pushed off the undercuts and away from the moving half of the mold.

If undercuts are not practical, the fixed mold-half may be provided with spring-loaded or mechanically actuated hold-down pins, which are ejector pins that assist the molded part to leave the fixed mold-half and to follow the moving mold-half during "breakaway" and initial travel, perhaps 1⁄4 in or more.

The part designer and the mold designer need to consider this aspect of the molding process and agree on how to ensure proper travel of the part to guarantee controlled ejection.

9.8 Mold Construction and Fabrication

This chapter is not intended to cover the broad field of mold and die design and construction, but the plastic-part designer needs to be aware of the several aspects of mold design that can affect the cost of mold construction. With such knowledge, the part designer may be able to achieve the desired finished product with lower mold costs, faster deliveries, lower processing costs (shorter cycles and few, if any, postmolding costs), less stringent processing parameters, and minimal mold maintenance, as well as longer mold life.

9.8.1 Mold Types

For the basic compression, transfer, and injection molding processes, a wide variety of mold types may be considered. Decisions as to the optimal type will often be based on the production volume anticipated and the allowable final part cost, including mold maintenance and amortization costs as well as hourly cost rates for molding machine and labor.

9.8.2 Single-Cavity Molds

If production quantities are as low as a few hundred parts, single-cavity molds may be feasible, even recognizing the longer time period to produce parts one at a time and the increased labor and machine time to produce the required number of parts. Single-cavity molds can be of the hand-molding type, having no mechanical ejection mechanism, no heating or cooling provisions, but requiring a set of universal heating/cooling plates bolted into the press, between which the hand mold is placed and removed each cycle. Such hand molds may be of two-plate or three-plate construction, depending on configuration of parts to be molded. If the part is relatively small, hand molds can be multicavity yet still be light enough that an operator can manually place them into and out of the press each cycle without physical strain.

Although hand molds may be made of soft metals such as aluminum or brass, and the cavities may produce acceptable parts, such metals quickly develop rough surfaces and rapidly lose their practicality after a few dozen cycles. It is best to use conventional metals as used for production molds for the respective process. Molds then may be used not only for prototyping but also for modest production while waiting for full-size production tooling.

9.8.3 Multicavity Molds

Production molds are generally multicavity, and they have integral heating or cooling provisions and ejection systems. And when they follow single-cavity hand molds, cavity dimensions may be fine tuned, and vents and gates can be repositioned, based on experience with the single-cavity molds.

Family molds are multicavity molds that mold one or more sets of a group of parts that are required to make up a complete assembly of the finished product. A base, a cover, and a switch, for example, may be needed for a limit switch assembly. A family mold of, say, 36 cavities may be constructed that will yield 12 assemblies each cycle.

Production molds may also be made using a standard mold base that will accept, say, 12 identically sized cavity inserts. If the component to be produced is a small box with lid and attachable cover, in, say, 12 different sizes, each cavity insert could contain the cavities for one size box and cover. The complete mold would produce 12 boxes and covers each cycle. If sales of any one size require greater quantities than are required for one or more of the other sizes, a second cavity insert could be made to fit into the mold base, enabling twice as many of the faster-selling size each cycle.

For family molds to be successful, all parts should have approximately the same wall thickness so that molding cycles may be optimal for all sizes.

9.8.4 Molds with Removable Cavities

Some molded parts have configurations that require portions of the cavity to be removed before the part can be ejected. Solenoid coil bobbins, as an example, consist of a cylindrical body around which wire will be wound, with two large flat flanges at each end of the cylindrical body to keep the wire contained within the length of the body. Additionally,

there is a hole through the length of the cylindrical body to accommodate the plunger of the solenoid assembly.

Such bobbins may be molded with the cylindrical body axis parallel to the parting surface of the mold. The cavity of the bottom half has the half-round shape to mold half of the body of the bobbin as well as two thin slots to mold half of each flange. The upper mold-half is almost identical to produce the other half of the bobbin. To mold the hole through the body requires a cylindrical metal mandrel, with its outside diameter equal to the inside diameter of the coil body. This removable mandrel is manually placed in the matching half-round shape on the parting line. When the mold is closed, it effectively seals around the mandrel, leaving an open cavity into which the plastic will be injected in an injection or transfer-molding machine. Following hardening of the plastic, the molded part with the mandrel is removed from the mold, the mandrel is pushed out of the bobbin with a simple manually actuated fixture, leaving the finished thin-walled bobbin intact. The mandrel is replaced in the mold for the next cycle.

For fully automatic cycles, molds for such a part may be constructed with cam-actuated or hydraulically actuated side cores that serve as the above-described mandrel. In each cycle, after the mold is closed and prior to injection of material, the side core is automatically actuated into place. Following the cycle, the side core is retracted automatically prior to mold opening and part ejection.

9.8.5 Molds with Inserts

Many plastic parts are produced with molded-in inserts, such as a screwdriver with plastic handle. Molds for such items are designed to accept and hold in the correct location the steel shaft with the flat blade or Phillips head end away from the cavity, and the other end, often knurled or with flats (to ensure that the handle, when rotated in use, won't slip around the shaft) protruding into the handle cavity. After the mold is closed with the insert in place, plastic is injected or transferred into the cavity, where it surrounds the shaft end and fills out the cavity to achieve its shape as a handle. Following hardening of the plastic, the mold opens, and the finished part is removed.

In many insert-molding operations, fully automatic molding becomes possible when mechanisms are installed to put the insert into place before each cycle and to remove the insert with molded part following each cycle.

When inserts are to be molded into plastic parts, close coordination is obviously needed between part designer, mold designer, and manufacturer.

9.9 Summary

Although this chapter has touched on many aspects of part design and processing, it has not covered a myriad of special considerations that may arise in the real world. The less-experienced part designer is advised to consult with others in the field as he develops the part design, selects the material, and chooses the optimal process for production requirements.

Chapter

10

Automotive Plastics and Elastomer Applications

James M. Margolis

Margolis Polymers
Montreal, Quebec, Canada

10.1 Introduction

This chapter primarily covers current and developmental exterior, interior, and under-hood plastics and elastomer applications. The chapter covers drivetrain/powertrain and chassis applications. Components for systems such as brake, fuel delivery, steering, and drivetrain are located in more than one sector. Brake system components extend from the foot pedal in the passenger compartment to the disk and drum brakes in the wheels; fuel delivery components begin with the fuel filler pipe and end with the engine cylinders. The steering system includes the steering wheel, energy absorbing column, steering shaft and arm with steering gears, tie bars, linkages, ball joints, and wheel spindle assembly; power steering adds hose, pump, reservoir, and more gears. The drivetrain transmits power generated by the engine through the transmission to the wheels, which can be front-, rear-, or all-wheel drive. Although designs for assemblies such as air intake manifolds, air and fuel delivery systems, instrument panels, transmissions, and drives differ for specific platforms, typical internal combustion engine vehicles have in common fundamental vehicle concepts.

The chapter describes plastics and elastomer applications that are injection molded, extruded, blow molded, thermoformed, rotational molded, reaction injection molded (RIM), resin transfer molded (RTM), compression molded, mat molded, pultruded, filament wound, and produced via hand lay-up and spray-up. It describes new developments for traditional applications and products new to the twenty-first century such as "cool glass" for fixed and retractable windows, digital/wireless communications, and fuel cell plates and membranes. Opportunities in the drivetrain include filament winding and pultruding for high torque strength, high load-bearing strength, corrosion resistance, and fire retardant applications. Not included in the chapter are fibers/fabrics such as polyester fabrics for seating, nylon fabrics for airbags, polypropylene for carpet, and painting/coating except for important advancements with in-mold coating.

More than one polymer is often specified for any given application, and polymers currently used can be replaced by other polymers. This is the dynamics of automotive design,

and it is an effective proving ground for technology transfer to applications in other industries. This chapter is only a window on plastics and elastomers for automotive applications. More applications and more resin options for any given application would be covered in a longer text. The following list of thermoplastics, thermosetting plastics and elastomers, their copolymers, and blends is a source for automotive polymer selection.[2]

Thermoplastics	Thermosetting plastics	Thermoplastic elastomers	Thermosetting elastomers synthetic rubbers
ABS	Epoxies	Etheresters, esteresters (copolyesters)	Acrylonitrile butadiene copolymers (nitriles)
Acetals	Melamines, ureas	Polyamides	Butyls
Acrylics	Polyimides	Polyolefins	Chlorinated polyethylenes
Fluoroplastics	Polyurethanes	Polyurethanes	Chlorosulfonated polyethylenes
Inonomers	Phenolics	Styrenics	Ethylene propylene copolymers
LCPs	Unsaturated polyesters, vinyl esters	Vulcanizates	Ethylene propylene diene terpolymers
Polyamides (nylons)		Single-phase TPEs	Fluoroelastomers
Poly(amide-imide)s			Polyacrylates
Polycarbonates			Polybutadienes
Polyethylenes			Polychloroprenes (neoprenes)
Poly(either imide)s			Polyepichlorohydrins
Polyimides			Polyisoprenes
Polyketones			Polysulfides
Polyphenylene ethers			Silicones
Polyphenylene sulfides			Styrene butadiene copolymers
Polypropylenes			
Polystyrenes			
Polysulfones			
Polyurethanes, polyureas			
Polyvinylchlorides			
Thermoplastic polyesters PBT, PET			

True properties are determined by the polymer and compound formulation and, for synthetic rubbers, additionally by type of curing agents such as sulfur or peroxides and the state of vulcanization. Typical properties for four polymers often used in automotive applications [acetals, nylons, polycarbonates, and unsaturated polyester (vinyl esters)] are shown in Tables 10.1 through 10.4. Typical property data is readily available from resin producers and compounders. Property data for elastomers is found in the chapter on elastomers in this handbook.

Acetals are produced by Asahi Chemical (homopolymer "Tenac," copolymer "Tenac-C," and conductive copolymer "Lynex-T"), BASF (copolymer "Ultraform"), DuPont (homopolymer "Delrin"), Mitsubishi Gas Chemical (copolymer "Iupital"), Polyplastics (copolymer "Duracon" in Asia) and a joint venture between Ticona and Daicel Chemical Industries, Ticona/Celanese AG (copolymer "Celcon" and "Hostaform" in Europe). Mitsubishi Gas Chemical, Polyplastics, and Ticona are (jointly, directly, or through affiliates) increasing acetal production capacity in the Asia/Pacific area, with a new 60,000 metric ton/year plant to be completed in 2004 in Nantong, Jiangsu province, north of Shanghai.[2]

Typical acetal automotive applications are cams and gears for uses such as starting and accelerating engines, coat hooks, control cables, cup holders, door handles/locks, engine

TABLE 10.1 Typical Properties for "Tenac" SH–4010 Acetal Homopolymer

Property	Value*
Density (g/cc)	1.42
Water absorption percent	0.2
Tensile strength, ultimate lb/in^2 (MPa)	10,008 (69)
Flexural strength, yield lb/in^2 (MPa)	14,939 (103)
Flexural modulus ksi (GPa)	421 (2.9)
Impact strength, Izod ft-lb/in (J/m)	1.5 (80.1)
DTUL/264 lb/in^2 (1.82 MPa) °F (°C)	177 (80.5)
Processing temperature °F (°C) cylinder temperature range	374–410 (190–210)
Mold temperature °F (°C)	140–248 (60–120)

*Room temperature (unless specified otherwise), ATSM test methods (unless specified otherwise).

cooling fan blades and shrouds, exterior trim, fuel delivery system components (e.g., fuel pumps), headrest guides, heater/air conditioner control system components, heater plates, luggage carrier hardware, radiator and gas tank caps, rollover gas shutoff valves, seat belt buckles, retractor covers, release buttons, seat levers, speaker grills, sun visor brackets, trim clips, window cranks, window flexible guide strips, window support brackets, windshield wiper blade holders, wiper bezels, and wiper pivots.[2]

Acetal properties that contribute to automotive applications include fuel and chemical resistance, tensile strength, toughness, wear and abrasion resistance, long-term dimensional stability under stress, creep resistance, UV resistance, and color matching. Acetal homopolymer is composed of alternating –(CH_2O)– oxymethylene groups that polymerize to polyoxymethylene (POM); acetal copolymer is composed of oxymethylene + oxyethylene repeating units.

Table 10.1 shows typical properties for impact-grade "Tenac" SH 4010 acetal homopolymer for seat belt components.[3]

Nylons are produced by Bayer ("Durethan"), DSM Engineering Plastics ("Stanyl"), DuPont ("Zytel" and "Minlon"), Honeywell ("Capron"), Rhodia Engineering Plastics ("Technyl"), Solutia ("Vydyne") marketed through Dow Chemical, and TotalFinaElf ("Rilsan").[2]

Typical automotive applications of nylons are air filter housings, air intake manifolds, alternator components, bearing cages, body panels, brake pedals, cable fasteners, chain tensioners, clips, clutch rings, connectors, door handles, emission control system components, engine covers, fasteners, fuel pump support brackets, fuel delivery system components, license plate frames, oil filter housings, pump assembly components, rearview mirror housings and brackets, rocker covers, seat frames, sensors/switches, shift forks, shrouds, steering wheels, thrust washers, tubing, turbocharger components.

Nylon properties that contribute to automotive applications include compressive, flexural, torque strength, toughness; resistance to creep, fatigue, wear, oils, grease, and underhood chemicals; limitation of PV; and NVH abatement.

Table 10.2 shows typical properties for 40 percent glass fiber reinforced, impact modified nylon 6 "Capron" for injection molding.[4]

TABLE 10.2 Typical Properties for "Capron" 8334G. HS Nylon 6

Property	Value* DAM
Specific gravity (g/cc)	1.47
Water absorption percent 24 hr, ISO test method	0.8
Tensile strength, yield lb/in^2 (MPa)	27,000 (186)
Flexural strength, yield lb/in^2 (MPa)	36,000 (248)
Flexural modulus ksi (GPa)	1,500 (10.3)
Impact strength, Izod ft-lb/in (J/m)	2.8 (150)
DTUL/264 lb/in^2 (1.82 MPa) °F (°C)	410 (210)

*Room temperature (unless specified otherwise), ATSM test methods (unless specified otherwise).

Polycarbonates are produced by Bayer ("Makrolon" PC, "Apec" high-temperature PC, "Bayblend" PC/ABS, "Makroblend" PC blend), Dow Chemical ("Calibre" PC, "Pulse" PC/ABS), and GE Plastics ("Lexan" PC, "Cycoloy" PC/ABS).[2]

Typical automotive applications for polycarbonates and blends are bumper beams, cowl vent grills, door handles, ducts, glazing, lenses for exterior lighting such as fog lights, glove boxes, headlights, parking lights, side lights, turn signal lights, and sunroofs; instrument panel components such as support frames for other components such as air conditioners and steering wheels; interior trim, mirror frames for lighted vanity visors, rearview mirror frames, and reflectors for headlights.

Polycarbonate properties that contribute to automotive applications include resistance to brittle fracture, clarity and optical precision, dimensional stability, impact strength, scratch resistance, and temperature resistance.

Table 10.3 shows typical properties for "Calibre" IM 400 18, impact modified, used for large complex automotive applications such as instrument panel components and retainers. It is produced in MFR 11 and 18 g/10 min.[5]

Thermosetting polyester and vinylester resins are produced by AOC (a joint venture of Owens Corning and Alpha Corporation resin businesses) ("Vipel"), Ashland Chemical/Composite Polymer Division ("Phase Epsilon," "SMC-LITE"), DSM Composite Resins ("Atlac"), Dow Chemical ("Derakane"), and Interplastics ("CoREZYN").

Typical vinyl ester automotive applications are body panels, firewalls, fuel cell plates, lighting components for headlamps, pickup truck cargo boxes, radiator supports, running boards for SUVs, trailer truck cab/roofs and wind deflectors, rocker covers, wheel covers, and windshield surrounds.

Thermosetting polyester and vinyl ester properties that contribute to automotive applications include dimensional stability, strength, toughness, and high service temperatures.

"Derakane" 790 epoxy vinyl esters are based on high-temperature novolac chemistry. Similar to other "Derakane" thermosetting vinyl esters, "Derakane" 790 is diluted with styrene, and the resin can be cured at elevated SMC compression molding temperatures with peroxide initiators.[6,7] A typical catalyst is TBPB (tert-butyl perbenzoate). The resin is chemically thickened with magnesium oxide or other metallic oxide slurries so as to be used as a sheet molding compound (SMC). It is a low-profile resin, containing a shrinkage control additive that produces dimensionally accurate products.[6–8] The resin combines

TABLE 10.3 Typical Properties for "Calibre" IM 400 18 Polycarbonate

Property	Value*
Specific gravity (density) (g/cc)	1.18
MFR (g/10) min	18
Tensile strength, yield lb/in^2 (MPa)	8,400 (58)
Flexural strength lb/in^2 (MPa)	13,000 (89)
Elongation @ break percent	110
Elongation @ yield (ISO) percent	6
Modulus of elasticity in tension (ISO) ksi (GPa)	334 (2.3)
Impact strength, Izod ft-lb/in (J·m)	12 (641)
DTUL/264 lb/in^2 (1.82 MPa), annealed °F (°C)	277 (136)
Vicat softening point °F (°C)	295 (146)

*Room temperature (unless specified otherwise), ATSM test methods (unless specified otherwise).

toughness with dimensional stability, and retains strength properties at elevated temperatures [heat distortion temperature is >550°F (>288°C)]. Vinyl ester and unsaturated polyester resins are produced as SMC, BMC, and TMC and processed by compression molding, filament winding, pultrusion, resin transfer molding, spray-up, and hand lay-up.[6,7]

Table 10.4 shows typical properties for "Derakane" 790 R-65 epoxy vinyl ester for chassis/power train applications such as oil sumps, valve covers, and timing chain covers.[8]

10.2 Plastics

10.2.1 Thermoplastic Resins

10.2.1.1 Typical exterior applications

- Body panels: aerodynamic add-ons (air dams, spoilers), bumper beams, doors, fenders, hoods, filler panels, front grill panels, tailgates, trunk lids, cladding, appliques
- Cowl screens
- Door handles, locks, lock actuators and gears, door lock latch plates, door latch stop plates
- Fascia, front, rear
- Front ends, hybrids (plastics and metals)
- Fuel filler caps

TABLE 10.4 Typical Properties for "Derakane" 790 R–65 Epoxy Vinyl Ester

Property	Value*
Glass fiber content percent (wt)	65†
Specific gravity (g/cc)	1.787
Tensile strength, ultimate lb/in^2 (MPa)	27,400 (188)
Tensile modulus ksi (GPa)	1,850 (12.71)
Tensile elongation percent	1.88
Flexural strength, yield lb/in^2 (MPa)	55,400 (382)
Flexural modulus ksi (GPa)	1,850 (12.71)
Compressive strength lb/in^2 (MPa)	37,600 (258.4)‡
Compressive modulus ksi (GPa)	2,120 (14.6)
Shear strength (in-plane) lb/in^2 (MPa)	16,400 (113)
Shear modulus (in-plane) ksi (GPa)	1,000 (6.87)
Shear strength (interlaminar) lb/in^2 (MPa)	4,510 (31.0)
Izod impact strength, notched ft-lb/in (J/m)	18.3 (977)
Heat distortion temperature °F (°C)	>550 (>288)

*Room temperature (unless specified otherwise), ATSM test methods (unless specified otherwise).
†By glass burnout
‡IITRI compressive test

- Guide rails (for sliding roofs)
- Lighting components: bezels, brackets, mountings, fog lamps, headlamps, lamp sockets and bases, lenses, reflectors, retainers
- Liners for pickup truck cargo boxes
- Luggage carrier rails and hardware
- Power mechanisms (for doors)
- Rearview mirror housings
- Reinforcements and attachments
- Rocker panels
- Running boards
- Trim, trim clips
- Trunk latches, locks
- Sunroofs
- Wheel arch liners

- Wheel nuts
- Wheel covers
- Window guide strips
- Window support brackets
- Windows (glazing): fixed and retractable; windshield, side, rear
- Windshield cowl plenums
- Windshield wiper bezels, wiper blade holders, wiper pivots

Plastic body panels show cost advantages over steel panels, largely due to in-mold painting, in-mold assembly integrating components, and thin-wall resins (higher MFR without sacrificing properties). All three developments significantly lower finished panel costs. Plastic panels have well known advantages on a performance basis—light weight, dent resistance, corrosion resistance, and design versatility. Painting body panels using in-mold colored films (IMCF) with a post-mold clear coating for luster and abrasion resistance is increasingly popular as IMCF injection molding technology is further developed.[9] Several formulations have been developed for in-mold coating such as "Xenoy" PC/PBT, "SollX," "Noryl" GTX, "Surlyn" Reflection series, "Magnum ABS, "Pulse PC/ABS, and "HiFax" propylene polymers.[9,10] These polymers include in-mold paintable grades for instrument panels, consoles and trim. Fascia, radiator grills (front grill panels), body side moldings, trim, and cladding are injection molded from color compounded "Surlyn." Reflection supergloss ionomer/polyamide.[11] The large selection of resins for exterior body panels includes ABS, ASA, PC/ABS, PC/PBT, PC/PET, TPU(E), PC/TPU(E), TPO, polyamides, and thermoplastic polyester blends of PBT/PET. Mineral reinforced blends of PC and polyesters are gaining attention in North America and Europe, providing lower temperature impact resistance, reduced CLTE and automotive class A surface quality for vertical panels, tailgates, and bumpers."Celanex" UV stabilized PBT, "Ultradur" and "Valox," provide stiffness, flexibility, dimensional stability, tight tolerance, and weather resistance for injection molding windshield wiper covers.[12–14]

In-mold assembly (IMA) to integrate functional components has been gaining industry use as a significant cost-saving process that eliminates post-mold assembly costs. More injection molders are using IMA technology, overmolding with multicavity rotating molds, and more self-bonding resins/compounds are available for encapsulation. IMA technology can be used for applications in all automotive segments (e.g., exterior, interior, underhood, and chassis) and for non-automotive applications.

"Azdel" glass fiber mat thermoplastic (GMT), with a polypropylene matrix and other thermoplastics, uses chopped glass fiber strand GMT grades for exterior applications such as bumper beams and spare wheel wells.[15] SuperLite "Azdel" grades are available in 2002 for interior applications.

Long fiber glass (LGF) composites such as "Compel" and "StaMax" P are gaining more recognition for bumper beams, door structures, instrument panel carriers, integrated front end modules, seat bases and splash shields.[16,17] "Stamylan" P glass fiber and mineral reinforced homopolymer and copolymer PP grades are also specified for battery casings, bumper beams, exterior/interior cladding, HV/AC components, instrument panel/dashboard retainers, interior door structures and underhood containers.[16] "Ketan" TP is injection molded into bumper beams for a number of Audi, Mercedes Benz, BMW, and Ford models; and Mazda 323 bumpers have been injection molded from reactor grade elastomer-modified "Kelburn" PP.

"Vetron" composites of polypropylene reinforced with up to 75 percent "Twintex" long glass fibers (LGF) are produced by LNP Engineering, a business unit of G.E. Plastics, for exterior and underhood applications, and RTP compounds a line of LGF reinforced com-

posites for automotive and other applications.[17,18] Husky Injection Molding Systems and Krupp Werner & Pfeidler co-developed in-line compounding-LGF composites by feeding continuous rovings into a compounding extruder with polypropylene and other ingredients.[19] Injection molding is a separate step, but it could be added to the in-line process. "Fiberim" TPU(E) composites are reinforced reaction injection molded (RRIM) into bumper beams and panels, seat pans and sun visors.[20] Processes developed with machinery companies Cannon ("Interwet"), Engel ("Fibermelt"), Hennecke ("FiberTec" Plus) Johnson Controls Interior ("Fibropress"), and Krauss-Maffei are used to produce LGF composites for automotive applications such as in-mold painted and paintable, bumper beams, exterior panels, front ends, instrument panel (i.p.) carriers, interior door panels, battery trays, seat pans, and tailgate inner panels.

Reinforcing fibers can be "Cratec" chopped glass strands produced from continuous glass filaments chopped into specific lengths according to the application.[21]

Modular front end systems, hybrid assemblies of plastics and metal, significantly improve overall functionality and lower installed cost. Hybrid front end assemblies can be produced by injection molding encapsulating plastics with metal pressings. Reinforced nylon and sheet steel front ends replace all-metal units, providing improved torsional, flexural and compressive strength, and help to stabilize the front end section of the car during driving conditions.[22,23] Design can provide uniform distribution of forces such as torque transmitted from the engine. Other advantages of the plastic/metal front end include 40 percent lighter weight than all-steel construction, eliminating the need for tight tolerance joining operations. One process to make these front ends places a perforated deep-drawn sheet steel structure in a mold cavity. Glass fiber reinforced nylon 6 "Durethan" injection molded into the cavity flows through the perforations and around the sheet steel, forming the nylon/steel front end. The process replaces 17 steps with 1 step.

A number of components can be integrated with the front end. A typical front end consists of nylon 66 or nylon 6 and steel, integrating the bumper, fog lamps, headlights, parking lights, radiator grill, and attachments, and even more components can be integrated. The technology can be used for other automotive applications such as front seat construction, and non-automotive applications. DaimlerChrysler Mercedes Benz plastics/metal hybrid constructions are uniquely used in Mercedes-Benz Division CL-class cars with plastics bumpers, front wings, and trunk lid assemblies; aluminum large surface areas such as hood, roof, rear panel, and rear fender; magnesium inner-door panels; and steel for areas that are high stressed in a crash, cross members, side barriers and roof pillars.[24] The company's A-class hybrid construction has a "well balanced diet" of plastics, aluminum, magnesium, and steel which includes plastic front fenders.[25] State-of-the art plastics and elastomeric polymers can offer advantages over aluminum and magnesium applications.

Bumper beams, fascia, and cladding are molded with thermoplastic polyesters, TPU(E) and TPO(E). Bayer developed "Durethan" nylon 6 hybrid assemblies for front ends and "Bayblend" PC/ABS for hybrid assembly of functional components for exterior panels including tailgate panels and interior rear seat back rests.[26]

Thermoplastic and thermosetting resins, including acetals, LCP, nylons, PC, PBT, PEI, PET, BMC, and SMC, replace metals for lighting components—adjusters, attachments, bases, brackets, bezels, fog lamps, hardware, headlamps, lenses, mountings, parking and backup lights, reflectors, retainers, and sockets. The plastics have good creep resistance and dimensional stability; high-temperature properties such as stiffness at elevated temperatures and resistance to hot spots; electrical properties such as dielectric strength and arc tracking resistance; and they are low-cost alternatives to metals and ceramics. Lamp bases and sockets are molded from 30 percent (wt) glass fiber reinforced LCP with high toughness and HDT of 500°F (260°C),[27] 30 percent glass fiber reinforced PET, and heat stabilized 30 percent glass fiber reinforced nylon 6 and nylon 66. Headlamp retainers can be molded from 33 percent glass fiber reinforced nylon 66 such as "Zytel" with good

green strength, and 30 percent and 35 percent glass fiber reinforced PET such as "Rynite" using DuPont's mounting technology. Reflectors are made with metallizable BMC, heat stabilized 40 percent glass fiber/mineral reinforced "Zytel" HTN nylon 66, 40 percent glass fiber/mineral reinforced "Rynite" PET, and "Ultem" PBT for premium priced cars.[28] GE Plastics 3D "Thermal Prediction" software implements headlamp component design, eliminating or reducing the need for prototyping by predicting distortions as a function of temperature increase.

BMC, acetal homopolymers and copolymers, nylon 6 and nylon 66, PBT, PC/PEI blends, and PEI copolymers are used for lighting hardware. Heat-stabilized, dimensionally stable, clear (90 percent light transmittance) PC provides a good balance of property to cost for lenses.[29]

BMC shapeability and low cost often make it the preferred plastics for headlamp and fog lamp reflectors. BMC reflectors provide a good fit with other components in the lighting assembly, good metallizing, and heat resistance to withstand headlamp bulb temperatures of 392°F (200°C).

Nanocomposite-polypropylene with 2.5 percent (wt) reinforcement, compared with about 25 percent talc in conventional mineral reinforced PP, is a harbinger of greater use of nanocomposites for automotive as well as non-automotive applications. Nanocomposites are co-developed for applications such as in-mold painted and paintable vertical panels by GM Technical Center/Materials and Processing Laboratory, Southern Clay Products and Basell/Montell. Vertical body panels show 325,000 lb/in^2 (2.23 GPa) modulus, cladding shows 150,000 lb/in^2 (1.03 GPa) modulus, and lower weight than conventional glass fiber and particulate reinforced composites, without sacrificing properties. Nanocomposites with PP and other engineering thermoplastics include center console, interior door panels, interior trim. knee bolster, pillars.[30]

Luggage carrier rail support (roof rail support) assemblies for the ChryslerDaimler Jeep Wrangler have been molded from 40 percent "Cratec" long (0.5 in, 12.5 mm) chopped glass fiber reinforced polypropylene homopolymer.[31,32] The composite is compounded with UV stabilizer and carbon black for enhanced weather resistance. The composite has 1.20 g/cc specific gravity, flexural modulus 994,000 lb/in^2 (6.85 GPa), Izod impact strength notched 7.22 ft-lb/in (383 J/m), DTUL/264 lb/in^2 (1.82 MPa) 315°F (157°C), and CLTE 113 10^{-5} in/in/°F (205×10^{-5} m/m/°C @ 75–300°F (23–150°C). The rail assembly, integrating several components, snap-fits onto the removable roof. Design and process technologies involve a number of innovations including an insert-molded steel pivot pin, which allows the support rail parts to be folded for easy storage, and patented compression tooling technology.[32]

Dramatic changes are taking place with retractable side windows, fixed windshields, and side and rear windows. New glazing greatly improves driver and passenger safety, security, and the sound damping of road and wind noise. Ultraviolet (UV) and infrared (IR) filtering lead to major changes in plastics selection for instrument panel components, seating, and steering wheel covers, and reduced fuel consumption due to reduced air conditioning. Not all new glazing composites are "cool glass" that filters out or screens the sun's IR. Two window products that have been introduced are

1. Glass laminated with "Butacite" polyvinylbutyral (PVB) film interlayer (used for more than 55 years in shatterproof windshields), and "Mylar" polyester film, developed by DuPont
2. Polycarbonate sheet with an organic or ceramic coating, co-developed by GE Plastics and Bayer

Enhanced Protective Glass (EPG), "SentryGlas," and SentryGlas" Plus for side, rear, and roof windows were tested for intrusion resistance and compared to conventional tempered glass and polycarbonate sheet.[33] DuPont's three types of safety glass laminates are

1. "Butacite" film interlayer in windshields and side windows
2. "SentryGlas" intrusion resistant composite
3. "SentryGlas" Plus, with an ionomeric polymer interlayer for higher impact resistance[33,34]

Automotive EPG is constructed of PVB film laminated between two sheets of glass.[36] EPG blocks up to 95 percent of sunlight UV and up to 55 percent of the sun's IR. This can initially reduce passenger compartment temperature up to 72°F (22°C) in a parked vehicle exposed to direct sunlight. Road and wind noise are damped up to 4 dB. In addition to these attributes, EPG is also intrusion resistant, deterring thieves from breaking into a car, and protecting occupants from being thrown through the window during an accident.[37–39] "SentryGlas" is used in OEM and aftermarkets for windshields, windows, and sidelights. Laminated safety glass (in addition to windshields) is expected to be installed on 1.5 million vehicles in 2002, compared with 400,000 vehicles in 2000.

The Enhanced Protective Glass Automotive Association (EPGAA) was established in September 1999 by Solutia and DuPont for membership by automotive OEMs and glass industry suppliers to "share information and provide overall education on the development of high impact resistant glass."[35]

Solutia produces "Saflex" PVB and Vanceva™ PVB films to complement "Saflex" for security for automotive, architectural, and residential windows. Wedge-shaped "Butacite" PVB film laminated between windshield glass for head-up displays (HUDs) allows displays to show on the windshield.[106]

DaimlerChrysler S-class cars, Mercedes Benz coupe models, have laminated safety glass that significantly reduces UV, and a metal coated plastic film liner on the CL-class cars reduces IR. EPG is installed in Audi and Volvo models.

"Exatec" Plus injection molded polycarbonate glazing produced by Exatec, a joint venture between Bayer and GE Plastics, was developed initially primarily for side, rear, and sunroof windows. Its construction includes a UV barrier organic intermediate layer, a primer to bond this layer to the PC sheet, and an exterior abrasion-resistant surface coating using GE Plastics' plasma deposition of silicone coating for abrasion and weather resistance or Bayer's ceramic-type coating using nanomer material technology. The product is lightweight safety glazing with UV and IR barrier properties, break resistance, and sound damping, as well as abrasion resistance. It is intrusion resistant to deter would-be thieves, and passenger ejection resistant during a crash. PC glazing permits style variations with contoured shapes, and the PC sheet can be injection molded encapsulated to produce integrated glazing-panels for automotive and building products. A 4 mm thick glazing, same thickness as a typical automotive glass window, weighs 21.7 kg, compared with 43.5 kg for the glass window. Large area seamless PC sheet using process technology developed by machinery manufacturers Battenfeld, Engel and Krauss-Maffei, are encapsulated during injection molding to produce integrated windows for automotives and buildings. PC windows for automotives will initially be installed in 2003, and produced in high volume by 2007. The International Organization of Standards (ISO) is establishing global standards for automotive glazing using test methods for mechanical properties, clarity, flammability, and other properties.

10.2.1.2 Typical interior applications (passenger compartment)

- Airbags and components
- Armrests
- Columns (pillar posts)

- Coat hooks
- Consoles
- Control knobs
- Cup holders
- Digital display housings
- Door handles
- Door panels
- Foot pedal brackets
- Gear shifts and handle knobs
- Glove boxes
- Grills: air conditioner, defroster, heater, speaker
- Guide rails for sliding roofs
- Headliner assemblies: skin, foam, substrate
- Headrest guides
- HV/AC components, e.g., vents
- Instrument panel: dashboard components: skin, foam core, substrate, attachments, carriers, frames
- Knee bolster components
- Light covers
- Parking brake fittings
- Rear shelves
- Rearview mirror housings
- Reinforcements and attachments
- Seat belt components: buckles, retractor covers, release buttons
- Seat components, e.g., back rests, frames, levers (e.g., seat adjustor levers), shells
- Steering components, e.g., back rests, frames, levers (e.g., seat adjustor levers), shells
- Steering wheels, columns, switching assemblies
- Sun visors, sun visor brackets
- Trim, trim clips
- Truck trailer liners

"SuperPlug"™ inner door panel assemblies, constructed from 30 percent "Cratec" chopped strand glass fiber reinforced PBT/PC "Xenoy," integrate the window mechanism housing and control, door handle support, armrest supports, speaker mounting, and other system components.[40] Components such as the window regulator, latch, and speakers are assembled off line, according to individual customer specifications. The first application of the inner door assembly had more than 20 variations to meet customer specifications.[40] Chopped strand glass fiber contributes strength, stiffness, dimensional stability, and heat stability; PBT contributes resistance to gasoline, automotive waxes, and cleaners; and PC increases impact resistance over a wide temperature range.[40] Gas assist injection molding is used to produce a hollow channel in the long flow path, increase stiffness, eliminate warpage, and reduce weight.[40] The technology was co-developed by Owens Corning and Delphi Interior Trim and Lighting Systems. Consoles, door panels, pillar trim, and instru-

ment panel component are in-mold painted and paintable and/or low-gloss, high-heat, high-flow, antistat ABS such as "Magnum" and PC/ABS "Pulse."

A foot pedal bracket using glass fiber reinforced "Zytel" nylon 66 replaced steel as it exhibits one-third the weight of steel, lower cost, design flexibility, component integration, and noise and vibration abatement. Plastic accelerator and clutch pedals can be used with the nylon foot pedal bracket.[41]

HV/AC control panel knobs on 2001 Chrysler and Dodge minivans are produced by two-shot insert/overmolding injection molding two acetal copolymer "Celcon" grades. The first shot produces white raised lettering; the second shot is overmolded, retaining the white lettering. Melt viscosity and flow properties are key to the flow around the raised lettering.

Breakthroughs for passenger compartment applications are found with in-mold assembly (IMA), hybridization (in-mold assembly of plastics and metal components), component integration and in-mold painting. Instrument panel (i.p.) substrate + foam core + skin are sequentially in-mold assembled during injection molding. Substrates are in-mold painted with TPU, polyester gels, and colored film inserts, and by spray coating the mold cavity. The i.p. is a proving ground for a selection of resins and processing methods. ABS, ABS/PC, PC, PVC, TPO, and TPU are injection molded, calendered, thermoformed, blow molded, and rotational molded into i.p. components. Instrument panel/dashboard designs greatly differ for different vehicle platforms, but they are increasingly favoring IMA, in-mold painting, and component integration including hybrid construction.

Calendered TPO skin laminated to polyolefin foam core for lower instrument panels are produced by PolyOne Engineered Film for assembly to the i.p. substrate. The instrument panels are assembled by Delphi Automotive Interior Systems from "Mytex" TPO thermoformed skin/foam laminates.[42] TPO provides resistance to heat, UV, scratch, ductility, low gloss grain, soft touch, and minimal fogging.

"Azdel" SuperLite, with improved weight consistency, available in 2002, has improved weight consistency and surface quality for interior applications such as integrated instrument panel components and headliners.[43] Components such as brackets, frames, glove boxes, grills, and knee bolsters are integrated into i.p. assemblies. In-mold paintable injection-molded high heat resistant, impact resistant, low gloss glass fiber reinforced nylon, ABS, ABS/PC, and PC are used for brackets, consoles, frames, inner door panels, knee bolsters, retainers, speaker and defroster grills, and steering wheel covers. High-MFR grades are used for long-flow, thin-wall, complex components. The bar is raised when recyclability is designed into resin selection.

10.2.1.3 Typical underhood (engine compartment, under-the-bonnet) applications

- Air cleaner housings
- Air injector components
- Air intake ducts
- Air intake manifolds
- Air pump components
- Baffle plates
- Battery housings, trays
- Belts, e.g., drive belts, fan belts
- Blower wheels
- Bottles, containers, e.g., window washers

- Cable connectors and covering, e.g., control cables
- Chain guides, chain tensioners
- Clips/fasteners, e.g., brake clips, fuel line clips
- Clutch components: clutch cage, clutch cage cap, clutch ring
- Coil bobbins and spools, e.g., cruise control module
- Coolant system components
- Ducting
- Electronics—connectors, control/sensor housings, throttle control housings, mountings
- Electrical relay components, e.g., relay bases, cases, coil forms
- Engine cooling fan blades, shrouds
- Engine components, e.g., belt guides, engine covers, injector units, rocker covers
- Valve timing chain covers
- Emission control system components
- Fuel cell plates
- Fuel delivery system components: clips (e.g., to fasten hose in place), fuel dam reservoirs, fuel filler necks (filler pipes), fuel filters, fuel hose and lines, fuel pump parts, fuel tanks and canisters
- Hose
- Ignition coil cases, other ignition system components
- Oil filter housings and components
- Oil pans
- Pipes (water and oil pipes in engine systems)
- Power distribution boxes
- Radiator components: caps, support assemblies
- Rocker covers
- Sensors/solenoid coils/switches
- Shrouds
- Splash shields
- Support panels, e.g., firewall, radiator
- Timing chain guides
- Tubing, e.g., convoluted tubing, vacuum tubing
- Turbocharger components
- Valve covers
- Valve lifter guides
- Wiper motor housings

Upper air intake manifolds are part of a ductwork assembly that directs the flow of air or air and fuel into the engine cylinders. The upper manifold distributes air/fuel mixture through the intake ports into the engine cylinders. Air flow begins with the intake of air from under the hood, initiated by a downstroke of an engine cylinder that causes a vacuum, drawing in air through the air filter. The filter removes dirt and other particulates from the ambient air. The cleaned air passes through the air intake manifold and enters a cylinder, typically injected in a stoichiometric air/fuel ratio. Air flow and air/fuel ratio are

essential factors in fuel efficiency. Air that is not combusted exits the cylinder with the exhaust stroke, eventually returning to the atmosphere through the tailpipe. The exhaust manifold collects hot gases from the cylinders and directs the gases into the exhaust system which is basically composed of the following:

1. The exhaust manifold and gasket
2. Connectors and intermediate pipes
3. Muffler, catalytic converter, heat shield, resonator
4. Gaskets, seals, mechanical clamps, hangers
5. Tailpipe and exhaust pipe

Air intake upper manifolds are either lost-core injection molded as a single unit or injection molded in two parts that are subsequently vibration welded together. The latter method, vibration welding two molded parts, is rapidly gaining favor. Twenty five percent to 35 percent (wt) glass fiber reinforced, heat-stabilized, lubricated nylon 6 and nylon 66 are typically used with both processes, although other reinforced engineering thermoplastics can be used.[44–46] Bayer high-temperature stabilized, 30 percent glass fiber reinforced nylon 6 "Durethan" and the company's branched polyamide 6 "Durethan" are two of several nylons used for vibration welded air intake manifolds. The benefits of plastics over aluminum, which it replaces, are as follows:

1. Lighter weight
2. Cost savings through parts integration, elimination of secondary finishing, and subassembly
3. Noise and vibration reduction
4. Design versatility (taking advantage of CAE and FEA)
5. Performance benefits such as smooth air flow through the intake manifold for improved air/fuel combustion efficiency

Air ducts are produced by sequential extrusion or blow molding alternating rigid PBT sections and soft elastomeric polyester accordion sections, which reduces weight, cost, number of components, and NVH.

Underhood electrical connectors are molded from a range of resins from thermoplastic polyesters PBT and PET to LCP, depending on property requirements, particularly temperature stability, stiffness/strength ratio, dimensional stability, and chemical resistance. Recent grades have improved toughness and higher MFR for thin-wall, complex connectors.

Fragile electrical connectors and other delicate electrical/electronic parts are produced by "MuCell" low-pressure injection foam molding and encapsulation injection molding as an alternative to solid injection molding and extrusion.[47] Advantages according to Trexel are as follows:

1. Closer tolerances
2. Improved dimensional stability
3. Low density
4. Low injection pressure, which avoids damage to fragile electronic components
5. Faster cycles

6. Lower clamp force
7. Lower power consumption

Electronic sensor and control housings are getting closer to the engine, and attached to the engine's inner surface. DuPont iTechnology Microcircuit Materials' "Green Tape" and thick films are used for mounting on and inside the engine for engine control circuits, hybrid circuits and high density interconnects in 3-dimensions. "Green Tape" is applied to the integration of passive components in 3-D structures, antilock brake system controls, electroluminescent backlights for instrumentation on instrument panels, transmission control circuits and rear window heaters. The technologies are used at high operating temperatures for high density circuitry, direct chip attachments. The technology's mixed analog/digital/RF and high frequency properties are used for wireless applications, which are being installed on OEM cars and aftermarket retrofits. "Green Tape" and "Fodel" are extensions of iTechnologies' thick film and thin film flexible "Luxiprint" used for electroluminescent backlights and liquid polyimides used in automotive electronic ignition systems.

Electronic throttle control (drive-by-wire) molded from 33 percent nylon 66 "Zytel" replacing mechanical throttle controls is constructed of a control housing that encapsulates a wire lead frame, plus an integrated outer housing/mounting bracket.

Electrical relay components, relay bases, cases, and coil forms make use of a range of engineering thermoplastics, thermoplastic polyesters, nylons, polyphenylenes, LCPs, and thermoplastic polyimide "Vespel" TP can be used for underhood electrical applications due to its high temperature properties, impact resistance, and automotive chemical resistance.[48]

The engine is the end point of fuel delivery and air delivery systems, and the starting point for the drivetrain/powertrain system. The fuel delivery system is basically composed of the fuel cap, filler neck (pipe), fuel lines, fuel tank, fuel pump and pump components, filter, upper air intake manifold, and fuel injection unit. Fuel flow through the filler neck and filter can generate an electrostatic surface on the filler neck or filter.[2] For these applications, electrostatic dissipation (ESD) resins are used, such as glass fiber reinforced ESD nylon, "Finathene" and "Marlex" HDPE, which forms a strong weld bond with the multilayer blow molded HDPE fuel tanks. "Marlex" HDPE, with chemical resistance, creep resistance (apparent modulus), ESCR, long-term impact strength, and MFR 5 g/10 min, is blow molded at 379–430°F (193 to 221°C).[49,50] LCPs, high-temperature resins with high MFR such as "Zenite," are used for the vapor barrier layer in fuel delivery components.

Fuel delivery systems must meet California Air Resources Board (CARB) regulations, including Low Emissions Vehicle LEV II, which goes into effect in 2004 and limits evaporative hydrocarbon emissions to 0.5 g/day maximum—a 75 percent reduction from the regulation in 2000.[51] Other state regulations for evaporative hydrocarbon emissions have been developed. Test methods to measure evaporative hydrocarbon emission as hydrocarbon loss in materials for components and the entire fuel delivery system are standardized by a joint effort between automotive and materials companies. The methods include

1. Gravimetric screening by the cup method for hose and tubing
2. Specialized permeation conditions
3. Mini-shed[52]

Fuel and vapor tubes are typically extruded from fluoroplastics such as "Tefzel" EPTF inner layer and a nylon outer layer. The tubes are temperature resistant above 225°F (107°C) to fuels, fuel vapor permeation, and automotive chemicals, and the retain flexibility above 300°F (149°C). A fuel delivery tube can be constructed with "Viton" fluoroelastomer inner layer, "Vamac" ethylene acrylic synthetic rubber tie layer, and "Kevlar" fiber reinforced "Teflon" FEP middle layer. Fluoroelastomers have high-temperature, fuel-resistant properties that make them ideal for underhood and chassis fuel delivery components.[53]

Kiefel Technologies' plug-assisted twin-sheet thermoformed fuel tank design, aimed at producing under 100,000–300,000 units/year, was initially developed for mid-size trucks. The fuel tank consists of two HDPE outer layers, two tie layers, and an EVOH intermediary barrier layer.[54]

Fuel tank systems are complex, irregular assemblies that are undergoing great changes in resins and designs to meet evaporative hydrocarbon emissions limits. Fortunately, these changes are successfully taking place with a number of viable options. Steel tanks still have a large share of the automotive fuel tank market. The conventional plastic fuel tank is an extrusion blow-molded multilayer unit composed of HDPE with an EVOH barrier layer. Its benefits and disadvantages from tooling to performance are widely documented. Visteon's Partial Zero Emissions Vehicle (PZEV) departs from the conventional, introducing twin-sheet, vacuum-formed, six-layer fuel tanks. Its design minimizes the number of escape paths for evaporative emissions, eliminating most external fuel connections.[55,56] The fuel tank consists of six elements.

1. Carbon canisters
2. Fuel delivery system
3. Fuel filters
4. Fuel/vapor lines
5. On-board recovery
6. The storage tank

Rocker cover construction is rapidly converting from cast aluminum and magnesium to injection-molded/vibration-welded 30 percent to 45 percent (wt) glass fiber reinforced nylon 6 and nylon 66 such as "Zytel," "Vydyne," and "Capron," or mineral reinforced nylon 6 "Minlon," which are heat stabilized, oil resistant above 302°F (150°C), and dimensionally stable.[57,58] The rocker cover and gasket over the engine's cam shaft prevents oil from leaking into the engine compartment. The cover contributes to engine efficiency by sealing off the cylinder head and valves from contamination as well as oil. Rocker performance requirements include a wide operating temperature range, dimensional stability under dynamic stress, stiffness, fuel and oil resistance, and NVH abatement. It can be thin-wall injection molded with ribs for additional stiffening and dimensional stability, and vibration welded. Like the air intake manifold, the rocker cover is in close proximity to the engine. The nylon rocker cover has many of the same benefits as the nylon air intake manifold: lighter weight, cost savings, elimination of secondary finishing and subassembling, design versatility (through CAE and FEA), and improved service performance. Air make-up nipples, heat-staked baffles [separates oil from positive crankcase ventilation (PCV) air supply], cable clips, PCV valves, oil separators, support brackets, and heat-staked inserts for mounting the ignition coil are integrated with the rocker cover, and more components continue to be integrated.[59,60] The sealing gasket is integrated with the rocker cover on certain DaimlerChrysler Neon and Rover models to minimize the possibility of oil leakage.[61]

Windshield wiper pump housings molded from "Celcon" acetal copolymer by Delco Chassis for General Motors eliminate three assembly steps and are snap-fit assembled.

10.2.1.4 Other applications

- Bowden cable guide sleeve
- Cams and gears, e.g., for starting and accelerating engines

- Clutch master cylinder, clutch rings
- Containers, e.g., steering fluid containers
- Crossbeams
- Electrical switches, e.g., light control switches, turn signal switches
- Gears, e.g., helical gears for window lifts
- Gear shift forks
- Heater plates
- Pulleys, e.g, cable lines, seats
- Rollover gas shutoff valves
- Speedometer gear wheels
- Thrust washers
- Transmission levers, components

10.2.2 Thermosetting Resins

Thermosetting resins are usually processed by compression molding, filament winding, hand lay-up, pultrusion, reaction injection molding (RIM, RRIM), resin transfer molding (RTM) and vacuum bagging.

10.2.2.1 Typical exterior applications

- Body panels: front end components, hood panels, RV panels, spoilers, trunk lids
- Lighting components for headlamps, fog lamps, e.g., bezels, baffles, reflectors
- Pickup truck cargo boxes (beds)
- Truck cab roofs and wind deflectors
- Wheel covers
- Windshield surrounds
- Windshield wiper cowl plenum

SMC exterior body panels have been used on Chevrolet Corvette for decades, and thermosetting plastics exterior panels were promoted by the classic picture of Henry Ford striking a hammer on the composite trunk of an early Ford model. Today, horizontal panels, hoods (bonnets), and trunk lids are molded from high-modulus SMC, and RV panels with high contoured, styled bodies are compression molded from SMC. Three characteristics distinguish SMC exterior body panels at this time.

1. Stiffness
2. Versatile stylability
3. In-place cost

SMC is used for the trunk lid on DaimlerChrysler Mercedes class A luxury cars, BMW spoilers, and Plymouth Prowler integrated windshield surrounds.[62]

SMC powertrain applications are developed by DSM Composite Resins and Budd Plastics. The Lonza Group has developed structural grade SMC specifically for the growing use of SMC automotive applications such as pickup truck cargo boxes.[63] The General Mo-

tors pickup truck one-piece cargo box produced by SRIM from polyurethane exhibits dimensional stability down to –40°F (–40°C).[48] Glass fiber reinforced low-profile unsaturated polyesters for trunk lids and rear light bezels are resin transfer molded (RTM).

The inner frame of Alfa Romeo and GTV Coupe hoods are injection-compression molded from BMC produced with low-profile unsaturated polyester and 25 percent (wt) "Cratec" glass fiber, and the outer skins are compression molded from SMC produced with low-profile unsaturated polyester and 28 percent "Cratec" glass fiber.[63] The benefits compared with steel are lower cost, lighter weight, stiffness and dimensional stability, design versatility, and styled contours.[63]

Ninety percent of headlamp reflectors in North America are made from metallized BMC, and 10 percent from "Ultem" PEI for higher-priced cars such as the Ford Lincoln Town Car and BMW Series 3. BMC has dimensional stability, strength properties, and temperature resistance for this application, and it provides a good assembly fit. Temperature resistance is higher than 392°F (200°C). Improvements are made for BMC fog lamps and headlamps with base coatings and metallization.[64]

Two-component SMC windshield surrounds replace 14 steel parts, using "Vetrotex" glass fiber roving reinforced SMC.[62]

A single-part windshield wiper cowl plenum compression molded from 32 percent "Crastin" glass fiber reinforced vinyl ester SMC, replaces 11 steel stampings welded together for DaimlerChrysler minivans.[65] All critical system components are directly mounted to the cowl plenum, which was not possible with the steel assembly, because it did not have the necessary precision fit.[65]

10.2.2.2 Typical interior (passenger compartment) applications

- Airbag/knee bolster supports
- Cross-vehicle beam supports

Low-profile 40 percent glass fiber roving reinforced unsaturated polyester SMC is compression molded into airbag/knee bolster supports on multifunctional cross-vehicle beam supports. The system ties together the "A" pillar and cowl, reducing instrument panel sag and improving steering column stabilization during crash. It integrates HV/AC ducts, simplifies wire harness installation, and reduces NVH.[66]

10.2.2.3 Typical underhood (engine compartment, under-the-bonnet) applications

- Air intake system components
- Electric motor: e.g., components: brush holders, commutators
- Firewalls
- Fuel cell components
- Gaskets for electronic housings
- Oil sumps
- Pulleys
- Radiator support panels
- Seals, e.g., electrical connector assemblies
- Thermostat housings
- Timing belt guides

- Valve covers
- Valve timing chain covers, timing belt covers
- Water pump housings, inlet/outlet tubing

Injection molding of air intake system components with thermosets is on the rise, as indicated with the expansion of Kendrion Backhaus molding facility to produce these components.[62]

Electric motor components, commutators, and brush holders molded from phenolics, are used more due to the increased use of motors for power seats, windows, and sunroofs.[48] The phenolics comeback is further indicated by their use for thermostat housings, water pump housings, and inlet/outlet tubing for engine coolants.[48] Phenolics continue to replace die cast aluminum for stators and other automatic transmission torque converter components.[48] Phenolics have long-term dimensional stability at underhood elevated temperatures, strength, stiffness, chemical resistance to fluids (hydraulics), and resistance to current engine coolants.[48] Timing belt guides made with fiber-mineral reinforced phenolics are used for the resin's abrasion resistance and surface hardness, along with long-term high-temperature dimensional stability, rigidity, and resistance to underhood chemicals.[48] SMC is used for firewalls and radiator support panels on GM models such as the 2001 Oldsmobile Aurora, Buick LaSabre, and Pontiac Bonneville.

Fuel cells will soon be a viable alternative to the gasoline-powered internal combustion engine, and, beyond 2010, they can be expected to be the primary automotive power source. Buses are fuel cell-powered, and "cars are next."[67] A motivation driving fuel cell development for automotives is tougher emission standards required in several eastern states and California.[68] Another motivation is to enhance national security with energy independence. U. S. automobile manufacturers plan to have fuel cell-powered cars on the market by 2004.[69] Thermoplastic elastomers coated with a platinum catalyst compose the fuel cell membrane electrode, referred to in Sec. 10.3.1.3.

Commercially viable fuel cells are developed by DaimlerChrysler, Ford, and Ballard (Burnaby British Columbia); by ExxonMobil and General Motors; and by Ford Motor THINK Technologies (Focus FCV).[68] Fuel cell technologies for vehicles are also being developed by the National Fuel Cell Research Center, Princeton University Center for Energy & Environmental Studies ("Modeling of Fuel Cell Vehicles and Production of H_2 and H_2 Rich Fuels For Fuel Cell Vehicles"); the University of California, Riverside/Advance Vehicle Engineering Group, with several projects including hydrogen fuel cell powered vehicles based on a Ford Ranger; and University of California, Davis, Institute for Transportation Studies.[70]

Electrically conductive BMC with a proprietary conductive filler is developed for proton exchange membrane (PEM) fuel cell conductive plates for passenger cars.[71] Bulk Molding Compounds Inc. vinyl ester compounds are molded into PEM fuel cell plates. Premix subsidiary Quantum Composites "Pemtex" BMC formulations are co-developed for fuel cell plates with Ferromatik Milacron and Apex Plastics technology.[72] The conductive, corrosion-resistant composite is injection molded into bipolar plates for PEM fuel cells.[72]

The core of fuel cell power generating system is the fuel cell stack, which contains a number of fuel cells. Each cell is composed of membrane electrode assembly and enclosed by separator plates.[69] The key element of the proton exchange membrane (PEM*) is an elastomeric thermoplastic membrane coated with a platinum catalyst.[68]

**Author's note:* PEM refers to both *proton exchange membrane* and *polymer electrolyte membrane. In Fuel Cell Technology For Vehicles,*[73] PEM refers to *polymer electrolyte membrane* fuel cells (p. 23) and to *proton exchange membrane* (PEM) technology (p.137).

Unsaturated polyester and vinyl ester SMC molded into radiator support assemblies are integrated with radiator front grill reinforcements and headlamp mountings.

10.2.2.4 Other applications

- Brakes disk brake caliper pistons, brake valves
- Bushings
- Electric motor (DC) brush holders, commutators
- Pulleys for alternator, crankshaft, power-steering
- Run-flat tire support rings, integrated tire-wheel systems
- Seals
- Stators for automatic transmissions
- Suspension links
- Thrust washers in automatic transmissions

Several applications for thermosetting resins are finding increased use. A few include

1. Disc brake caliper pistons, for their dimensional stability, temperature resistance, and lower weight and cost as compared with steel.[48]
2. Pulleys, including crankshaft pulleys and alternator pulleys, molded from phenolics such as "Durez" for the resin's rotary fatigue resistance and durability
3. Bushings, seals, and thrust washers fabricated from thermosetting polyimide "Vespel," which has low coefficient of friction and wear, creep resistance, chemical resistance, and high continuous-use temperature[48]

High glass fiber content (50–75 percent) vinyl ester composite pultruded into suspension links is an example of frontier applications for automotive applications.[73] The suspension links are an assembly of cast aluminum end fittings encapsulated over a composite rod using metal-over-composite "Litecast" process technology. The composite/metal suspension link eliminates fasteners and adhesives, allows parts integration, and provides stiffness and temperature resistance, light weight, and lower tooling costs than the metal link that it replaces.[73] The technology can be transferred to composite/metal cross members, floor pans, and other chassis applications, and it is used with thermoplastics.[73]

10.3 Elastomers

10.3.1 Thermoplastic Elastomers

10.3.1.1 Typical Exterior Applications

- Body panels, doors, fender flairs, liners
- Bumpers
- Front ends
- Glass run channels

- Seals, e.g., fuel delivery caps, panels, windows
- Splash guards
- Tire components
- Trim
- Weather seals
- Weatherstripping
- Windshield, side window lace
- Window lace gaskets
- Windshield wiper blade covers

TPO(E) nanocomposite exterior door panels and rear quarter panels, co-developed by Basell/Montell North America and General Motors, are a sign of greater uses for nanocomposites in automotive applications. These high-aspect-ratio, fine particles with relatively large surface area provide good modulus without sacrificing toughness, dimensional stability, or impact resistance.

Thin-wall TPO(E)s with improved properties account for the increased use of olefinic elastomers for automotive applications. Wall thickness for exterior panels is near 2.0, as illustrated with low-density (0.857 g/cc) "Dexflex" TPO(E) injection molded fascia for a number of models, including fascia for Dodge, Ford, and Mazda. TPU(E) with improved abrasion resistance and tear resistance, such as "Desmopan," "Elastollan" and "Pellethane," are used for fascia, exterior panels, and side moldings. Automotive-grade "Pellethane" is used for single-sided window encapsulation.[2]

Incorporating modifiers "Engage" polyolefin elastomers (POEs), "Affinity" polyolefin plastomers (POPs), and "Exact" polyolefin plastomers into compounds such as TPO(E) significantly improves impact resistance, low-temperature properties, toughness down to -40°F (–40°C), ductility, UV resistance, ozone resistance, resistance to aging, and appearance. Functionalized "Exact" plastomers enhance impact resistance and ESCR in PBT, nylon and PC used in automotive applications.[74–79] "Exxelor" modifiers based on functionalized elastomer and polyolefins are added to technical polymers such as engineering thermoplastics, specialty polyolefins, and blends. "Exxelor" serves as compatibilizers, increasing adhesion between polyolefins and polar polymers; coupling agents, promoting chemical bonding between polymer and reinforcing agents; and as adhesion promoters when added to nonpolar polymers so as to increase adhesion to polar polymers, thermosetting SR, and NR.[80,81] Modifiers are used in TPO(E) compounds for front and rear fascia, fender liners, fender flairs, i.p. skins, NVH applications, and many other applications. Thin-wall, complex, lightweight TPO(E) products, with better knit line strength, are produced when modifiers are incorporated into the compounds. Polyolefin elastomers and plastomers, in addition to being modifiers, are injection molded and extruded.

TPO(E) applications include ducting, front and rear fascia, interior skins, and NVH components.[77,78]

TPV is often used for dynamic seals for doors, seals in glass run channels, O-ring seals, and CVJ boots. Self-bonding TPV grades eliminate the need for adhesive primers for overmolding to thermoplastic substrates including nylon. High-flow "Santoprene" TPV seals are overmolded to encapsulate rear quarter windows.

TPO(E) thermoplastic elastomer and EPDM synthetic rubber are often the elastomers of choice for exterior trim. TPO(E) modified with POE, POP, or EPDM for exterior applications such as trim, fascia, and rocker panels have improved ductility, toughness, impact resistance, and weather resistance.[82,83]

Star-branched butyl rubber is used for inner tubes, nonstaining sidewalls, coverstrips for white sidewalls, body mounts, and curing bladders.[84] "Duradene" solution-SBR grades

for tire tread and sidewalls are based on bound styrene and vinyl content, which determine resilience, abrasion resistance, and hysteresis.[85] Solution-SBR (S-SBR) has a more linear polymer backbone, narrower MWD, and higher Mooney viscosity than emulsion-SBR (E-SBR); consequently, S-SBR and E-SBR require certain different equipment and processing conditions. For example, S-SBR has less dwell time, requiring different types of plates for extruders, and the space between mill rolls for S-SBR is up to 25 percent less than for E-SBR.

TPV illustrated with high-flow "Santoprene" weatherstripping is encapsulated and water-foamed coprocessed for automotive windows such as rear quarter window glass and tailgates.[86,87] TPV has good resistance to compression/relaxation cycles for use as a dynamic seal for doors.[88] EPDM such as "Vistalon" and "Nordel" IP are often the synthetic rubbers of choice for weather seals for hoods, trunk lids, roofs, and body seals.[89,90] They are used for sponge weatherstripping and seals for glass run channels, sunroofs, and trunk panels, and for solid profiles. EPDM provides effective seals to keep out water, noise, odors, and pollutants "Vistalon" EPDM grades are classified as very low diene, low diene, medium diene, high diene, and very high diene content.

Single-phase TPE, "Alcryn," is gaining recognition in weatherstripping applications, windshield lace, and window lace trim.[91] TPEs can be processed in conventional injection molding machines and extruders, but equipment design and processing parameters are different, primarily due to shear sensitivity. Extrusion blow molding single-phase TPE requires *higher* temperature settings at the barrel feed zone, and *lower* temperature settings at the front zone; the *reverse* of conventional temperature settings.

10.3.1.2 Typical interior applications (passenger compartment)

- Airbag covers
- Horn pads
- Instrument panels/dashboard components
- Soft-touch knobs
- Seat belt duct
- Trim

New instrument panel (i.p.) designs comply with safety and recycling requirements, and lower costs are obtained with insert molding, encapsulating, overmolding, and in-mold painting. These processes significantly reduce assembly and decorating costs. Three-component i.p. assembly of skin/foam/core/substrate are produced by a range of designs and processes, including extruded TPO(E) skin laminated to lightly cross-linked polyolefin foam core, calendered TPO(E) skins laminated to polypropylene foam core, in-mold assembly by placing a skin/foam core laminate into a mold cavity, low-pressure injection molding reinforced polypropylene melt into the mold cavity, and vacuum forming a skin/foam core laminate with a polypropylene substrate. "Espolex" TPO(E) is rotational molded (slush molded) into instrument panel skin and inner door liners.[92] TPO and TPO(E) i.p. developments are described both here and in Sec. 10.2.1.2 of this chapter, because these thermoplastic olefins are viewed as either thermoplastic resins and TPEs. Instrument panel skins with resistance to long-term heat aging, chemicals, oil, and scratch, are calendered, extruded, and rotational molded with fine particle resins with high melt strength such as vulcanizate composed of TPU(E) + PP + SBC (styrene block copolymer). "Surlyn" ionomer and high melt strength TPO(E) produce a functionalized copolymer that is deep-draw thermoformed into skins with good grain definition. The range of resin options is further illustrated by the use of Bayer "Makrofol" films and "Lustran" PS for in-

strument panels. Designs and processes for instrument panels can be transferred to inner door panels, other automotive applications, and non-automotive applications.

TPV compounds meet automotive safety requirements for reduced windshield fogging; "Santorpene" TPV fogging tests show 0.7 mg, and "Sarlink" TPV, composed of "Stamylan" P polypropylene and "Kelltan" TP EPDM-modified polypropylene, shows less than 1.0 mg. Fully vulcanized "Santoprene" is used as a modifier in TPO formulations for soft skins, improved grain definition retention, and cold flow for complex shapes.

10.3.1.3 Typical underhood applications (engine compartment)

- Air ducts
- Chain tensioners
- Emission control system components
- Electrical/electronic components
- Engine covers
- Fuel cells
- Gaskets, e.g., engine intake manifolds
- Hose, e.g., air delivery, fuel delivery
- Oil filter components
- O-ring seals
- Rocker panels
- Seals, e.g., driveshaft, interior
- Wire, primary covering—low and medium voltage insulation

Thermoplastic elastomer TPV, such as "Forprene," "Santoprene," and "Uniprene," are increasingly used for air intake ducts; for turbo air intake ducts, "Vamac" ethylene acrylic elastomers represents thermosetting synthetic rubbers with higher temperature properties for underhood applications.

A key element of the automotive fuel cell membrane electrode assembly is the proton exchange membrane (PEM), also referred to as the polymer electrolyte membrane (PEM), which is composed of a thermoplastic elastomer coated with a platinum catalyst. U.S. carmakers expect to have fuel cell-powered cars on the market by 2004.[69] Polymer selection depends on, among other criteria, fuel selection such as Direct Methanol Fuel Cell (DMFC) or Direct Hydrogen Fuel Cell (DHFC).[69] One prototype fuel cell vehicle is the product of the Partnership for a New Generation of Vehicles (PNGV), comprising U.S. automotive companies and the U.S. Department of Energy (DOE).[69]

Rocker Panels can be blow molded from a number of TPEs, including TEEE and TPU(E), but primarily nylons.

10.3.1.4 Other applications

- Boots, e.g., CVJ boots, shock absorber boots, steering gear boots
- Bearing cages
- Bellows
- Body plugs

- Clips
- Clutch rings
- Convoluted tubing
- Fuel delivery primer bulbs
- NVH applications
- O-rings, other seals
- Shift forks

Boots are typically blow molded from thermoplastic elastomers, CVJ boots from TEEE (COPE) and TPV, and shock absorber boots from TPU(E) such as "Desmopan." Thermoplastic etherester elastomers and thermoplastic esterester elastomers TEEE (COPE), "Arnitel," and "Hytrel" are blow molded into transmission and wheel CVJ boots for front-wheel drive (FWD) and rack-and-pinion boots. CVJ boots keep grease inside the joints and keep out dirt, grease, salt, and water. The TEEE CVJ drive axle boots have a balance of flexibility and stiffness, fatigue resistance to compression and expansion strains, chemical resistance, and meet accordion design requirements between –40 to 212°F (–40 to 100°C).[93–96]

10.3.2 Thermosetting Elastomers Synthetic Rubbers

10.3.2.1 Typical exterior applications

- Fascia
- Rocker panels
- Rub strips
- Tires: sidewalls, sidewall coverstrips, inner tubes, inner liners, tread
- Weatherstripping

10.3.2.2 Typical interior applications (passenger compartment)

- Door liners
- Instrument panel skins
- Trim

Tire treads are made from oil extended BIMS, butyls, EPDMs, polybutylenes, SBR, and their blends. Butyls contribute skid resistance, brominated butyls BIIR + SBR can increase wet traction, and chlorinated butyl (CIIR) increases skid resistance and rebound. BIMS also provides good flex crack resistance.[97]

"Nordel" IP is used as a modifier in ETP/TPO skins for dashboard/instrument panels and interior trim skins.

10.3.2.3 Typical underhood applications (engine compartment)

- Belts, e.g., valve timing belts (for low-torque engines)
- Boots, e.g., spark plug boots

- Cables, e.g., ignition cables
- Ducts, e.g., air intake ducts, engine ductwork
- Electronics: connectors
- Fuel delivery system, e.g., fuel filler neck hose, fuel pump diaphragm seals, turbo hose
- Gaskets, e.g., engine gaskets
- Hose, e.g., engine coolant hose, oil cooler hose, return power steering hose, transmission coolant hose, vent hose and tubing
- Seals
- Mounts
- Timing belts
- Tubing

Elastomers for underhood applications cover a complete range from extruded silicone synthetic rubber ignition cable covers and gaskets to "Nordel" IP EPDM, "Tyrin" chlorinated polyethylene, and TPE "Engage" polyolefin ethylene air intake ductwork. Process technology for molded-in gaskets for electronic housings and seals using self-bonding liquid silicones developed by GE Silicones is an example of the transfer of technology, finding uses for non-automotive soft-touch applications.[48]

Fuel filler neck hose is co-injection molded with a 0.6 mm layer of fluoroelastomer "Viton," which reduces hydrocarbon emission to less than 0.01 g/day.[98] "Viton" is used for fuel sender seals and the inner layer of turbo hose for transmitting hot fuel vapors.[98] Fluoroelastomers have good barrier resistance to aromatic hydrocarbons and oxygenated hydrocarbons, which is a reason for their enduring use for fuel delivery applications.

Fluoroelastomers are increasingly in demand to meet California Air Resources Board (CARB) requirements by 2004, which limit evaporative hydrocarbon emissions. "Dyneon" fluoroelastomers, with low permeation number in alcohols and reformulated gasolines, are used in the fuel delivery system, fuel lines, vapor lines, O-rings, and custom seals.[99] "Dyneon" BRE base-resistant elastomers are formulated for resistance to amine additives in automatic transmission fluids (ATF), engine oils, and gear lube oils— in addition to properties in other "Dyneon" fluoroelastomers including high temperature and chemical resistance.[100,101]

Gaskets are a type of seal at the joint between two components, and the engine gasket is critical for engine performance. A range of synthetic rubbers are used for gaskets and hose, largely depending on their different temperature and chemical resistance, including fluoroelastomers such as "Dyneon" and poly(ethylene acrylic) "Vamac." Engine and transmission gaskets are injection molded from high-temperature polyacrylate synthetic rubber, such as "HyTemp" and "Vamac," which have excellent resistance to oil, oxygen, and ozone above 302°F (150°C), and compression set and fuel vapor barrier properties.[102,103]

Epichlorohydrins keep their resiliency and hardness over a wide temperature range and also serve as an impact-resistant enhancer for nylon 6 and nylon 66. Fluoroelastomers and polyacrylates are often used for similar applications, the choice largely depending on high-temperature and chemical resistance requirements. Epichlorohydrin synthetic rubber such as "Hydrin" have low swell in aromatic fuels, oils, aliphatic solvents, low temperature flexibility resistance to ozone, impermeability to gases, and low damping.[104] Polyepichlorohydrin properties are customized by co-, ter-polymerization, blends, reinforcements, and compounding. In addition, synthetic rubber properties are modified by the type of curing agent and milling. Epichlorohydrins are sulfur and peroxide cured. They are used for gaskets, seals, hose, engine mounts, fuel pump diaphragms, and electrostatic dissipation (ESD) terpolymers for air inlet ducts and hose. Engine mounts and belting are made from polyisoprene.

Polychloroprene (neoprene) is used for hose for brake and steering systems, hose for coolants, seals for vibration damping mounts, high flex accessory belts, valve timing belts for low-torque engines, power transmission belts, CVJ boots and liners, air springs, and shock absorbers. Overall, applications are based on polychloroprene's toughness, resistance to damage caused by flexing and twisting (high dynamic snap), low compression set, ignition resistant, NVH damping, resistance to oil, grease, harsh chemicals, and service temperature from –31 to 257°F (–30 to 125°C). Polychloroprene is not resistant to fuels, and it can tear easily once it is punctured. "Hypalon" chlorosulfonated polyethylene is an option for power steering hose, high-temperature timing belts, and spark plug boots.[105] "Tyrin" chlorinated polyethyelene is an option for engine cooling hose, fuel hose, transmission coolant hose, and air intake ducting. Transmission gaskets and lip seals use "Viton" fluoroelastomers for their high-temperature properties, chemical resistance, resistance to fuels and oils (low oil swell), and low compression set—important properties for frontier drivetrain and chassis applications.

10.4 Disclaimer

The chapter author, editors, publisher, and companies referred to are not responsible for the use or accuracy of information in this chapter, such as applications and properties.

References

1. *Modern Plastics World Encyclopedia 2001,* Chemical Week Associates.
2. *Engineering Thermoplastics Global Markets to 2010,* Margolis Polymers.
3. Tenac Acetal Copolymer technical literature, Asahi Chemical.
4. Capron Nylon Property Guide, Honeywell.
5. Calibre Polycarbonate Resins Product Information Guide, Dow Plastics.
6. Preparing a Masterbatch of Derakane Resin, Dow Chemical.
7. Derakane Fabricating Techniques Filament Winding, Resin Transfer Molding, Dow Chemical.
8. Derakane 790 Vinyl Ester Resin for Sheet Molding Compound, Dow Chemical.
9. Resin Producers Answer the Call for Improved Molded-In Color, *Modern Plastics,* April 1999, p. 70.
10. Weather-Resistant ETP Brings Paintless Body Panels a Step Closer, *Modern Plastics,* December 2000, p. 71.
11. Surlyn Reflection Series—New Surlyn SG2001U.
12. Celanex product application sheet, Ticona.
13. Ultradur product applications, BASF.
14. The Reasons to Use GE Plastics Keep Piling Up, Valox PBT, GE Plastics.
15. Azdel Expanding Global Composite Capabilities, G.E. Plastics news item, September 4, 2001.
16. StaMax—Long Glass Reinforced Polypropylene, DSM Automotive Polymers.
17. Long-Fiber Thermoplastics Soar in Auto Applications, *Modern Plastics,* September 2000, p. 34.
18. Twintex Products, Saint Gobain/Vetrotex.
19. In-Line Compounding Cuts Costs of Large Parts, *Modern Plastics,* June 2000, p. 28.
20. Long Fiber Technology Gains Greater Acceptance in RRIM Processing, *Modern Plastics,* April 2001, p. 24.
21. Reinforcements for Multi-Polymer Compatibility, Owens Corning.
22. Modular Front End Systems to Reshape Auto Production, *Modern Plastics*, April 2000, p. 24.

23. Automotive Front End Systems, Dow Plastics.
24. Technological Standard Bearer in the Field Of Body Construction, Mercedes-Benz.
25. Aluminum Instead Of Steel: The Advantages of Lightweight Materials, Mercedes-Benz.
26. Costs Push More Plastics into Cars, *Modern Plastics*, February 2001, p. 58.
27. Zenite LCP Liquid Crystal Polymer Resin Product Information, DuPont.
28. Automotive Lighting Product Information, DuPont.
29. Front and Rear Systems, Dow Plastics.
30. Tiny Mineral Fillers Bring Big Benefits in Compounding, *Modern Plastics*, November 2000, p. 72.
31. Long Fiber Gains Greater Acceptance in RRIM Processing, *Modern Plastics,* April 2001, p. 24.
32. Unique Composite Technologies for Popular 4 × 4, Owens Corning.
33. System Design Parameters in Determining Automotive Sidelite Glazing Performance., Presentation on Test Methodology for Enhanced Protective Glass Material in Automotive Side Windows, 2000 International Body Engineering Conference, October 3–5, 2000, DuPont Automotive.
34. Interlayer for High Penetration Resistance Windshields, DuPont Automotive.
35. Automotive Association to Form to Promote Glass Technology Breakthrough—Enhanced Protective Glass Provides Design Opportunities with Security, Safety, Comfort Benefits, DuPont Automotive Press Release, September 28, 1999.
36. What is Enhanced Protective Glass?, DuPont Automotive.
37. SentryGlas Protects against Crime as Well as Collision, DuPont Automotive.
38. Advantages of Using DuPont SentryGlas Composite, DuPont Automotive.
39. *Ejection Mitigation Using Advanced Glazing*, Status Report II NHTSA Report.
40. The SuperPlug Inner Door: A New Way Of Thinking About Assembly, Owens Corning.
41. Rover Introduces Pedal Bracket of DuPont Zytel, DuPont Automotive.
42. Bonneville's Award Winning Mytex TPO Instrument Panel Skin Is North American OEM FIRST, ExxonMobil Chemical news item, January 25, 2000.
43. New Lightweight Solution For Car Manufacturers of Azdel SuperLite Composites, Azdel news item June, 2000.
44. BMW Rover Active Manifold Wins SPE Award., DuPont Automotive news item, February 1, 2000.
45. Ford 2000 F-150s Expeditions, Econolines Get Improved Manifold., DuPont Automotive news item, March 7, 2000.
46. North America's Largest Nylon Manifold Rolls out on Ford V-8., DuPont Automotive news item, March 7, 2000.
47. New Process Improves Encapsulation, Cuts Molding Costs, DuPont Automotive.
48. Thermosets: More Tailored Durability., *Plastics Engineering,* April 2001, p. 28.
49. TotalFinaElf Product Data Sheets.
50. Chevron Phillips Chemical Marlex C579 HDPE Data Sheet.
51. From Fuel Tanks to Filler Necks—DuPont Fuel System Materials Satisfy Tough Permeation Requirements, DuPont Automotive.
52. DuPont Hosts SAE Forum to Discuss New CARB Regulations, Fuel Trends and Technology, DuPont Automotive.
53. Vamac Product Data Sheet, DuPont; Viton Product Data Sheet, DuPont Dow Elastomers.
54. Thermoformed Fuel Tanks Show Promise, *Modern Plastics*, April 2000, p. 10.
55. Capabilities: Thermoforming Helps Meet PZEV Standards, Visteon news item, June 25, 2001.
56. Visteon Plastic PZEV Fuel Tank System.

57. Ford's First Thermoplastic Rocker Cover Launches on Global Platform, DuPont Automotive news item, March 7, 2000.
58. Rhodia Engineering Plastics Technyl Nylon 66 Technical Information.
59. Nylon Rocker Covers Emerge in North America, *Modern Plastics*, April 2000, p. 25.
60. Key Development Steps Outlined for Successful Nylon Rocker Arm, *Modern Plastics,* June 2000, p. 75.
61. DaimlerChrysler Launches First Thermoplastic Rocker Cover For North American Vehicle, news item, March 7, 2000.
62. FRPs Continue to Prosper in Broad Range of Auto Applications, *Modern Plastics,* June 2000, p. 29.
63. 152-A Cratec Chopped Strands—Reinforcements For Bulk Molding Compounds (BMCs), Owens Corning.
64. BMC Fends Off Challenge In Headlamps, *Modern Plastics,* January 2001, p. 45.
65. Under The Hood Windshield Wiper Cowl Plenum, Owens Corning.
66. Ford Ranger/Cross-Vehicle Beam Structure Support for a Dual Bag System, Owens Corning.
67. A Fuel Cell in Your Phone, *Technology Review,* November, 2001, p. 68, MIT Press.
68. Fill 'er Up With Hydrogen, *Technology Review,* November-December 2000, p. 53 MIT Press.
69. *Fuel Cell technology For Vehicles,* Richard Stobard, Ed. Challenges For Fuel Cells In Transport Applications, Chapter Society of Automotive Engineers (SAE) 2001.
70. Fuel Cell Information: Manufacturers/Developers, National Fuel Cell Research Center.
71. Fuel Cells Present Massive Market Potential For BMC, *Modern Plastics,* August 2000, p. 28.
72. Quantum Composites Develops Innovative Injection Molding Process that Uses Pemtex Conductive Composite to Mold Fuel Cell Plates, news item, June 8, 2000.
73. A Whole New Way to Join Composites and Metals, Owens Corning.
74. Chemical Polymer Modification Applications, ExxonMobil Chemical.
75. Exact Plastomers Envision A Future Of Opportunities, ExxonMobil Chemical.
76. Polymers For Automotive Applications—Exxelor Polyolefinic Modifiers, ExxonMobil Chemical.
77. Engage—The Possibilities Begin Here, DuPont Dow Elastomers.
78. Engage Polyolefin Elastomers–A New Class Of Elastomers, DuPont Dow Elastomers.
79. Exact Plastomer Polymer Modification Applications, ExxonMobil Chemical.
80. Exxelor Modifiers For Technical Polymers—Applying Technology where It Counts, ExxonMobil Chemical.
81. Elastomeric Polymer Global Markets To 2010, Margolis Polymers.
82. Automotive Exterior Trim, DuPont Dow Elastomers.
83. Automotive Exterior Trim–Nordel IP, DuPont Dow Elastomers.
84. Butyl Products—Markets and Products, ExxonMobil Chemical.
85. Firestone Polymers—Diene, Duradene, Stereon.
86. Mercedes-Benz Chooses Santoprene Rubber for Glass Encapsulation on New Sports Utility Vehicle, Advanced Elastomer Systems.
87. Mitsubishi Selects Santoprene Rubber for First All-Type Tailgate Weatherseal, Advanced Elastomer Systems.
88. Ligon Brothers and Lear Develop Quad-Extruded Dynamic Seal with Santoprene TPV (Door Seal), Advanced Elastomer Systems.
89. Automotive Body Sealing in a More Challenging Environment, Vistalon 8700 ExxonMobil Chemical.

90. Stretching the Possibilities—Weatherstrip (such as door and body seal, and trunk, quarter and window seals), Nordel IP. DuPont Dow Elastomers.
91. The Alcryn Advantage–Automotives, Advanced Polymer Alloys.
92. Can We Expect All-Olefinic Interiors in Cars of the Future?, *Plastics Engineering,* September 2000, p. 51.
93. Arnitel General Information on Applications and Properties, DSM Product Portfolio DSM Engineering Plastics.
94. Arnitel—CAE Analysis on CVJ Boots Using Rubber Elastic Modules, DSM Engineering Plastics.
95. Hytrel—A Thermoplastic and an Elastomer All Wrapped in One, DuPont Engineering Polymers.
96. Hytrel Polyester Elastomers Product and Properties Guide, DuPont Engineering Polymers.
97. *Elastomeric Polymer Global Markets,* Margolis Polymers.
98. DuPont Dow Elastomers Meeting Fuel System Emission Regulations, Viton Provides The Answers–Fuel Sender Seals, Turbo Hose, Fuel Filler Neck Hose.
99. Dyneon Fluoroelastomers Product Information, Dyneon.
100. Dyneon Fluoroelastomer Chemical Resistance, Dyneon.
101. Dyneon Base Resistant Elastomers—BRE. Excellent Resistance to Amine Packages In Automotive Lubricating Fluids, Dyneon.
102. Zeon Chemicals Announces New Polyacrylate Elastomer, news item, August 14, 2000.
103. Vamac Ethylene/Acrylate Automotive Applications, DuPont.
104. Zeon Chemicals Hydrin ECO Elastomers Properties and Products.
105. Hypalon Leads in Power Steering Hose, DuPont Dow Elastomers.
106. Wedge-Shaped PVB Gets the Heads Up, *Plastics Engineering,* August 2000, p. 6.

Trademarks

"Atlac"	Registered trademark of DSM Composite Resins
"Affinity"	Registered trademark of Dow Chemical
Aflas™	Trademark of Asahi Glass
"Akulon"	Registered trademark of DSM Engineering Plastics (DSM Engineering Plastics *Products* was acquired by Quadrant Holding)
"Alcryn"	Registered trademark of Advanced Polymer Alloys
"Amplify"	Registered trademark of Dow Chemical
"Apec"	Registered trademark of Bayer
"Arnitel"	Registered trademark of DSM Engineering Plastics
"Ascium"	Registered trademark of DuPont Dow Elastomers
"Atlac"	Registered trademark of DSM Composite Resins
"Azdel"™	Trademark of Azdel/General Electric
"Bayblend"	Registered trademark of Bayer
"Bayflex"	Registered trademark of Bayer
"Bexloy"	Registered trademark of DuPont
"Butacite"	Registered trademark of DuPont
"Butaclor"	Registered trademark of EniChem
"Calibre"	Registered trademark of Dow Chemical
"Capron"	Registered trademark of Honeywell
"Celcon"	Registered trademark of Ticona
"Centrex"	Registered trademark of Bayer
"Compel"	Registered trademark of Ticona/Celanese

"Cratec"	Registered trademark of Owens Corning
"Crastin"	Registered trademark of DuPont
"Cycoloy"	Registered trademark of General Electric
"Delrin"	Registered trademark of DuPont
"Derakane"	Registered trademark of Dow Chemical
"Derakane" Momentum	Trademark of Dow Chemical
"Desmopan"	Registered trademark of Bayer
"Dexflex"	Registered trademark of Solvay Engineered Polymers
"Diene"	Registered trademark of Firestone Polymers
"DuoMod"	Registered trademark of Zeon Chemicals
"Duracon"	Registered trademark of Polyplastics
"Duradene"	Registered trademark of Firestone Polymers
"Durez"	Registered trademark of Durez, Sumitomo Bakelite
Dyneon™	Trademark of Dyneon/3M
"ElastoGrip"	Registered trademark of Goodrich
"Elastollan"	Registered trademark of BASF
"Elvaloy"	Registered trademark of DuPont
"Emerge"	Registered trademark of Dow Chemical
"Engage"	Registered trademark of DuPont Dow Elastomers
"Estane"	Registered trademark of Goodrich
"Exact"	Registered trademark of ExxonMobil
"Exatec"	Registered trademark of Exatec, a joint venture of GE Plastics and Bayer
"Exatec" Plus	Registered trademark of Exatec, a Joint Venture of GE Plastics and Bayer
"Exxelor"	Trademark of Exxon Mobil
"Fiberfil"	Registered trademark of DSM Engineering Plastics
"Fiberim"	Registered trademark of Huntsman
"Finathene"	Registered trademark of TotalFinaElf
"Fluoro-Comp"	Registered trademark of DuPont
"Fodel"	Registered trademark of DuPont
"Forprene"	Registered trademark of PolyOne
Green Tape™	Trademark of DuPont
"HiFax"	Registered trademark of Basell
"Hostaform"	Registered trademark of Ticona/Celanese
"Hivalloy"	Registered trademark of Basell
"Hydrin"	Registered trademark of Zeon Chemicals
"Hypalon"	Registered trademark of DuPont Dow Elastomers
Hytemp™	Trademark of Zeon Chemicals
"Hytrel"	Registered trademark of DuPont
"Inspire"	Registered trademark of Dow Chemical
"Integral"	Registered trademark of Dow Chemical
"Iupac"	Registered trademark of Mitsubishi Gas Chemical
"Kapton"	Registered trademark of DuPont
"Kelburon"	Registered trademark of DSM Elastomers
"Keltan"	Registered trademark of DSM Elastomers
"Kevlar"	Registered trademark of DuPont
"Lexan"	Registered trademark of General Electric
"Lustran"	Registered trademark of Bayer
Luxiprint™	Trademark of DuPont
"Lynex"	Registered trademark of Asahi Chemical
"Makrofil"	Registered trademark of Bayer

"Makroblend"	Registered trademark of Bayer
"Magnum"	Registered trademark of Dow Chemical
"Makroblend"	Registered trademark of Bayer
"Makrolon"	Registered trademark of Bayer
"Margard"	Registered trademark of General Electric
"Marlex"	Registered trademark of Phillips Petroleum
"Minlon"	Registered trademark of DuPont
"MuCell"	Registered trademark of Trexel
"Mytex"	Trademark of Mytex Polymers, a joint venture of ExxonMobil Chemical affiliates and Mitsubishi Petrochemical
"Nipol"	Registered trademark of Zeon Chemicals
"Nordel" IP	Registered trademark of DuPont Dow Elastomers
"Nylatron"	Registered trademark of DSM Engineering Plastics
"Nysyn"	Registered trademark of Zeon Chemicals
"Nysynblak"	Registered trademark of Zeon Chemicals
"Natsyn"	Registered trademark of Goodyear
"Pebax"	Registered trademark of TotalFinaElf
"Pemtex"	Trademark of Premix
"Phase Epsilon"	Registered trademark of Ashland Chemical
"Plaslube"	Registered trademark of DSM Engineering Plastics
"Plexar"	Registered trademark of Equistar Chemicals
"Plexiglas"	Registered trademark of TotalFinaElf
"ProFax	Registered trademark of Basell
"Pulse"	Registered trademark of Dow Chemical
"Pyralux"	Registered trademark of DuPOnt
"Questra"	Registered trademark of Dow Chemical
"Rilsan"	Registered trademark of TotalFinaElf
"Riteflex"	Registered trademark of Ticona/Celanese
"Rynite"	Registered trademark of DuPont
"Ryton"	Registered trademark of Phillips Petroleum
"Santoprene"	Registered trademark of Advanced Elastomer System
"Sarlink"	Registered trademark of DSM Thermoplastic Elastomers
"SentryGlas"	Registered trademark of DuPont
"SentryGlas" Plus	Trademark of DuPont
"SollX"	Registered trademark of General Electric
"SMC-LITE™	Trademark of Ashland Chemical
"Spectrim"	Registered trademark of Dow Chemical
"StaMax"	Registered trademark of StaMax, a joint venture between DSM Automotive Polymers and Owens Corning
"Stamylan"	Registered trademark of DSM Engineering Plastics
"Stamytec"	Registered trademark of DSM Engineering Plastics
"Stanyl"	Registered trademark of DSM Engineering Plastics
SuperPlug™	Trademark of Delphi Automotive Systems
"Surlyn"	Registered trademark of DuPont
"Surlyn" Reflection™	Trademark of DuPont
"Technyl"	Registered trademark of Rhodia Engineering Plastics
"TechnylStar"	Registered trademark of Rhodia Engineering Plastics
"Teflon"	Registered trademark of DuPont
"Tefzel"	Registered trademark of DuPont
"Tenac"	Registered trademark of Asahi Chemical
"Texin"	Registered trademark of Bayer
"Thermoclear"	Registered trademark of General Electric

"Twintex"	Registered trademark of Vetrotex International/Saint Gobain
"Tyril"	Registered trademark of Dow Chemical
"Tyrin"	Registered trademark of DuPont Dow Elastomers
"Ubesta"	Registered trademark of PolyOne Developed with Ube Industries
"Ultem"	Registered trademark of General Electric
"Ultradur"	Registered trademark of BASF
"Ultraform"	Registered trademark of BASF
"Valox"	Registered trademark of General Electric
"Vamac"	Registered trademark of DuPont
Vanceva™	Trademark of Solutia
"Vector"	Registered trademark of Dexco, a Dow Chemical/ExxonMobil Chemical Partnership
"Verton"	Registered trademark of G.E. Plastics (from acquisition of LNP Engineering Plastics)
"Vespel"	Registered trademark of DuPont
"Vipel"	Registered trademark of Owens Corning
"Vistalon"	Registered trademark of Exxon Mobil
"Viton"	Registered trademark of DuPont Dow Elastomers
"Xenoy"	Registered trademark of General Electric
"Zealloy"	Registered trademark of Zeon Chemicals
"Zenite"	Registered trademark of DuPont
"Zetpole"	Registered trademark of Zeon Chemicals
"Zytel"	Registered trademark of DuPont

Chapter

11

Plastics in Packaging

Ruben J. Hernandez, PhD

School of Packaging
Michigan State University
East Lansing, Michigan

Plastics are used extensively in packaging due to their outstanding physical, mechanical, and chemical properties. Plastic are readily available; versatile; easy to process either singly, as multi-plastic, or in combination with other materials such as glass, metal, or cellulose-base; and relatively low in cost. The use of plastics has grown faster than any other group of materials in the packaging industry. The vigorous growth has been driven by a material substitution trend, the flexibility in design, and wide range in product protection. In the last two decades, there has been a strong expansion of the use of plastics, especially in the food and pharmaceutical industries. Market specialty niches like snack foods, baby foods, aseptic food packages, beverages, microwavable foods, and modified atmosphere packaging are examples of popular applications for plastics. Packaging is the largest single application for plastics. As illustrated in Fig. 11.1, in 1997, packaging accounted for about 26 percent of all plastics used in the U.S.A.

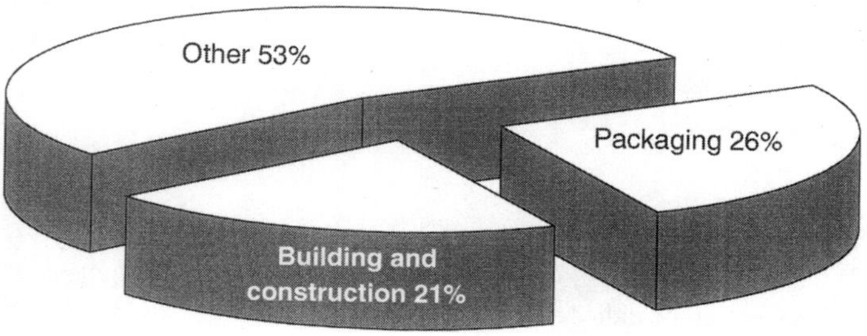

Figure 11.1 U.S. plastic markets, 1997. *(Source: R.J. Hernandez, S.E. Selke, and J.D. Culter,* Plastic Packaging, Properties, Processing, Applications and Regulations, *Hanser, 2000)*

Plastics are used in packaging in diverse forms: single films, sheets, multilayer coextruded structures, sheets, coatings, adhesives, foams, laminations, and rigid and semi-rigid containers. A great variety of products are packaged, delivered, and distributed around the nation and world using plastics. Examples of these products include solids, liquids, chemicals, alkalis, acids, electronics, hardware, foods, beverages, and health-care products. In the U.S.A. alone, more than 14 million tons per year of plastics are used in the packaging industry, with about half of this amount going into films and coatings.

The most widely used plastics in flexible packaging are low-density polyethylene (LDPE), linear low-density polyethylene (LLDPE), polypropylene (PP), and high-density polyethylene (HDPE). HDPE is widely used for rigid containers also, fulfilling almost 50 percent of the total demand. However, the fastest-growing resin for rigid containers continues to be polyethylene terephthalate (PET) and its copolymers. Other resins, such as ethylene vinyl alcohol (EVOH), vinylidene chloride (VDC) copolymers, and nylons, have specialized applications as high barrier materials.

New technologies are continuously improving resin properties as well as generating new resins like metallocenes and polyethylene naphthalate. These technologies widen the processing capability of single and composite structures in the production of flexible and rigid containers, at the same time increasing the number of applications and satisfying very specialized needs. The increasing sophistication in the use of plastics requires a better quantitative treatment of the barrier characteristics of these materials as well as the interactions that occur between the plastic package and the product. The interest in package/product interactions has increased with the use of recycled plastics—most noticeably, recycled PET for food applications. The high percentage of PET recycling (about 30 percent in the U.S.A.), and new processes for cleaning the recycled resin and making bottles available, have prompted the Food and Drug Administration to clear the use of conventionally recycled post-consumer PET in direct-contact food packaging. For example, the European Erema vacuum system for "bottle-to-bottle" recycling has the FDA approval for food contact use.

Plastics however, have some limitations as packaging materials. These limitations are associated with (1) relatively low temperature of use and properties sensitivity to temperature; (2) mass transfer characteristics that allow molecular exchange between a plastic container, its product, and the external environment; (3) tendency to environmental stress cracking; and (4) viscoelastic behavior that may produce physical distortion under external forces and heat. The correct evaluation of the mass transfer phenomena in a package, in conjunction with proper strength and design, is crucial for the optimization of product protection and quality during shelf life and distribution cycles. These aspects determine the design and selection of plastic for food and pharmaceutical packages. This chapter describes the plastics commonly used in the packaging industry; examines their principle properties, characteristics, and evaluations; and studies the fundamental aspects of mass transfer in plastic package systems.

11.1 Packaging Plastics

11.1.1 Polyethylene

Polyethylene (PE) is a group of polymers resulting from the polymerization of ethylene by chain reaction. Polyethylene can be either a homopolymer or a copolymer, and either linear or branched. Homopolymer PE is based mostly on ethylene monomer. In PE copolymers, on the other hand, ethylene can be copolymerized with short alkenes or with compounds having polar functional groups, such as vinyl acetate (VA), acrylic acid (AA), ethyl acrylate (EA), methyl acrylate (MA), or vinyl alcohol (VOH); see Fig. 11.2. When

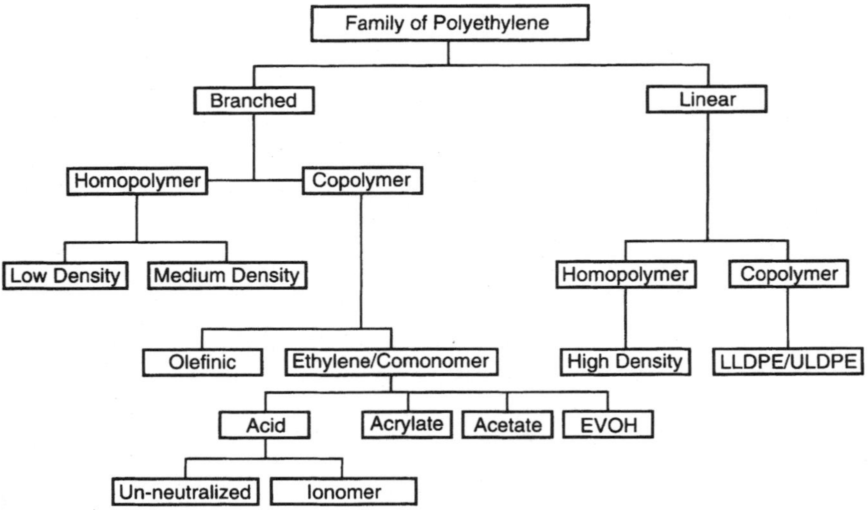

Figure 11.2 Family of polyethylene resins.

the molar percent of the comonomer is less than 10 percent, the polymer can be classified as either copolymer or homopolymer.

Branched polyethylenes are nonlinear, thermoplastic, partially crystalline homopolymers or copolymers of ethylene. They are fabricated at high pressure and temperature conditions by a free radical polymerization process that produces ramification in the main backbone chain. The polymerization of ethylene under these conditions produces a polymer made of a combination of large molecules with different backbone lengths, variable side-chain lengths, and many degrees of side-chain branching. This molecular architecture is characteristic of low-density polyethylene (LDPE), preventing the polymer from reaching high percentage of crystallinity. Nevertheless, branched polyethylene may show crystallinity between 30 and 50 percent and the density ranging from 0.910 to 0.950 g/cm^3. The low crystallinity gives a distinctive clear and transparent appearance to LDPE. The molecular weight distribution of LDPE is controlled by the addition of a comonomer such as propylene or hexene during the polymerization process. Several types of branched polyethylene are commercially available, depending on the reaction conditions and the type and amount of comonomer.

Linear polyethylene is made of long chains, without major branching, which tend to crystallize because the polyethylene molecule stereoregularity is not hindered by branches. Linear PE has, indeed, a high percent of crystallinity ranging from 70 to 90 percent. High-density PE has a density between 0.950 and 0.970 g/cm^3. Therefore, linear PE is normally a highly crystalline polymer having high density, while branched PE is a polymer of relatively low crystallinity with density ranging from 0.89 to 0.94 g/cm^3.

11.1.1.1 Branched homopolymer polyethylene

Low-density polyethylene. Low-density polyethylene (LDPE) is a branched thermoplastic polyethylene with density values ranging from 0.915 to 0.942 g/cm^3, and melt flow index (MFI) between 0.2 and 20 g/10 min. Chain-branching yields desirable characteristics such

as clarity, flexibility, sealability, and ease of processing. The actual values of these properties depend on the balance of the molecular weight, molecular weight distribution, and degree of branching.

Due to its rheological characteristics, LDPE can be processed by blown film extrusion, cast film extrusion, extrusion coating, extrusion molding, or blow molding. Film is the single largest production form of LDPE; in the U.S.A., 55 percent of the total LDPE produced is made into films with thicknesses under 12 mils (300 microns). LDPE has a low melting temperature range (98 to 115°C) and, therefore, is an easily sealable material—a property of great value in flexible packaging. When compared with other plastics, LDPE provides an excellent barrier to water. On the other hand, it shows one of the highest permeability values for oxygen, carbon dioxide, and organic vapors.

Containers, bags for food and clothing, industrial liners, vapor barriers, agricultural films, household products, and shrink- and stretch-wrap films are commonly fabricated from LDPE. Other packaging applications for LDPE are bakery items, snacks, produce, durable consumer goods, textiles, and industrial items.

Medium-density polyethylene. MDPE with density between 0.930 and 0.945 g/cm^3 and melt flow index ranging from 0.02 to 20 g/10 min, is somewhat stronger, stiffer, and less permeable than LDPE. MDPE is processed analogously to LDPE, though usually at slightly higher temperatures. A major competitor material of both MDPE and LDPE is LLDPE. LLDPE provides superior strength at any given density. Nevertheless, LDPE is preferred for high-clarity films and coating substrates.

11.1.1.2 Branched copolymers of polyethylene.

Ethylene monomer can be copolymerized with either alkenes of four to eight carbons or monomers containing polar functional groups like vinyl acetate, acrylic acid, and vinyl alcohol. Branched ethylene/alkene copolymers are equivalent to LDPE, since, in commercial practice, a certain amount of propylene and hexene are always added to help control the average molecular weight. Inserting polar comonomers in branched ethylene copolymers increases their flexibility, widens the range of heat sealing temperatures, improves the barrier properties, and lowers their crystallinities below those of homopolymers.

Ethylene vinyl acetate. The properties of ethylene vinyl acetate (EVA), a random copolymer of ethylene and vinyl acetate, depend on the content of vinyl acetate. EVA resins show better flexibility, toughness, and heat sealability than LDPE. The content of vinyl acetate (VA) in the copolymer ranges from 5 to 50 percent. For optimal food packaging applications, VA content should range from 5 to 20 percent.

As the percent of VA increases, it happens that

1. The crystallinity of EVA decreases, and, in contrast to PE, the density increases.
2. EVA becomes clearer, more flexible at low temperature, and more resistant to impacts.
3. At a 50 percent of VA and above, EVA is totally amorphous and transparent.

The presence of VA enhances the intermolecular bonds between chains, therefore increasing the adhesion strength and tackiness of EVA with respect to LDPE. As the molecular weight increases, the viscosity, toughness, heat-seal strength, hot tack, and flexibility increase. Because of its excellent adhesion and ease of processing, EVA is used in extrusion coating, and in heat sealing layers with PET, cellophane, and biaxially oriented PP films for cheese wrap and medical packages. However, EVA has limited thermal stability and low melting temperature and, therefore, must be processed at relatively low temperatures. Nevertheless, this aspect can be advantageous if toughness is required at low tem-

perature. For this reason, EVA is a good choice for ice bags and stretch wrap for refrigerated meat and poultry.[1]

Ethylene acrylic acid. Copolymerizing ethylene with acrylic acid (AA) produces copolymers containing carboxyl groups along the main and side chains of the molecule. Ethylene acrylic acid (EAA) copolymers are flexible thermoplastics with chemical resistance and barrier properties similar to those of LDPE. EAA is superior to LDPE in strength, toughness, hot tack, and adhesion. Clarity and strength adhesion increase with AA content, while the heat seal temperature decreases by a few degrees. Two major uses of EAA are in blister packaging and as extrusion coating tie layer between aluminum foil and other polymers.

Films of EAA are used in skin packaging, adhesive lamination, and flexible packaging of meat, cheese, snack foods, and medical products. Extrusion coating applications of EAA include coated paperboard, aseptic cartons, composite cans, toothpaste tubes, and food packages. The resin is compatible with LD, LLD, and HDPE.[2] FDA regulations permit the use of ethylene acid copolymers containing up to 25 percent acrylic acid, and 20 percent methylacrylic acid in direct food applications.

Ionomers. Ionomers are unique plastics in that they combine both covalent and ionic bonds in the polymer chain. They are produced by the neutralization of EAA (or similar copolymers), with cations such as Na^+, Zn^{++}, or Li^+. Ionomers show better transparency, toughness, and higher melt strength than the unneutralized copolymer. In general, sodium ion types are better in optical, hot tack, and oil resistance. Zinc ionomers are more inert to water and have better adhesion properties in coextrusion and foil extrusion coating.

The ionic bonds produce random cross-linking between the ionomers chains, yielding solid-state properties characteristics of very high-molecular-weight polymers. But, contrary to covalent cross-linking bond, the ionic bond of ionomers, though strong at low temperatures, weakens as the temperature increases. Therefore, on heating, ionomers behave like normal thermoplastic materials, with normal processing temperatures range from 175 to 290°C. But, at room temperature, they are strong and tough materials. For this reason, ionomers show great pinhole resistance and have low barrier properties. Various grades provide high impact and puncture resistance, have good to excellent abrasion resistance, and retain impact strength in sub-zero temperature[3] as low as –90°C (lower than LDPE).

In packaging, ionomers are commonly used as heat seal layers and applied as coating for nylon, PET, LDPE, PVDC, and composite structures. They are used in coextrusion lamination and extrusion coating. In addition, ionomers adhere very well to aluminum foil. Ionomers are used in packaging where formability, toughness, adhesion, and visual appearance are important. Because of their ionic nature, ionomers are highly resistant to oils and aggressive products while providing reliable seals over a broad range of temperatures. Application of ionomers in food packaging include frozen food (fish and poultry), cheese, snack foods, fruit juice, wine, water, oil, margarine, nuts, and pharmaceutical products. Heavy-gauge ionomer films serve as skin packaging for hardware and electronic products because of their great adhesion to paperboard.

11.1.1.3 Linear polyethylene. Linear polyethylene comprises the following group of polymers:

- Ultralow density PE (ULDPE)
- Linear low-density PE (LLDPE)
- High-density PE (HDPE)
- High-molecular-weight HDPE (HMW-HDPE)

Physical properties of commercially available linear PE are markedly dependent on the average molecular weight, molecular weight distribution, and resin density; see Table 11.1. Common applications of linear polyethylene are illustrated in Fig. 11.3. Linear PEs are produced using gas-phase low-pressure processes and in the presence of transition catalysts such as Zigler-Natta (titanium and aluminum) or Phillips (chromium oxide). The Union Carbide Unipol process produces linear polyethylene based on these transition catalysts. More recent technologies employ single-site catalysts (metallocenes, or constrained-geometry catalysts) to produce polymers with more controlled molecular configurations and very narrower molecular weight distributions than other polyethylenes. Metallocenes, in addition to having improved mechanical properties, show low levels of extractable compounds (an important characteristic in packaging for flavor-sensitive products) due to the lower amount of unreacted monomers and oligomers.

TABLE 11.1 Effects of Density, MW, and MWD on Linear PE Properties

Property	Density	Average molecular weight	Molecular weight distribution
Chemical resistance	I	I	0
Permeability	D	d	0
ESCR	D	I	0
Tensile	I	I	0
Stiffness	I	i	d
Toughness	D	I	D
Melt strength	–	I	I

As variable increases, property undergoes a/n: I = increase; D = decrease; i = slight increase; d = slight decrease; 0 = no significant effect.

Source: J. W. Taylor, *Modern Plastics Encyclopedia Handbook,* McGraw-Hill, New York, 1994, p. 29.

Linear low-density polyethylene. Linear low-density polyethylene (LLDPE) is characterized by narrow molecular weight distributions, a linear structure with very short branches, and low density values (0.916 to 0.940 g/cm^3). Due to its molecular linearity, LLDPE is more crystalline and therefore stiffer than LDPE. This results in melting points 10 to 15°C higher than that of LDPE. LLDPE has better tensile strength, puncture resistance, tear properties, and elongation than LDPE. Because of its greater crystallinity, LLDPE is more hazy and glossy than LDPE. Packaging uses of LLDPE include stretch/cling films, grocery sacks, shopping bags, and heavy-duty shipping sacks.

High-density polyethylene. High-density polyethylene (HDPE) is a linear thermoplastic made of long chains with little branching, with 65 to 90 percent crystallinity, and density ranging from 0.940 to 0.965 g/cc. The high percentage of crystallinity gives HDPE higher moisture and oxygen barriers (Figs. 11.4 and 11.5), better chemical resistance, and more opacity than LDPE. Blow-molding applications of HDPE include industrial chemical drums and containers for milk, detergent, bleach, juice, and water. Thin-walled dairy containers and closures are made by injection blow molding. Cosmetic containers, pharmaceutical bottles, shampoo, and deodorant containers are made by injection blow molding. Both blown and cast films are frequently utilized in flexible packaging applications. HDPE replaces glossine for breakfast cereal, cracker, and snack food packaging. It is also used for wrapping produce bags and wrap for delicatessen products. HDPE strongly resists a wide range of chemical compounds including water-based products, medium-mo-

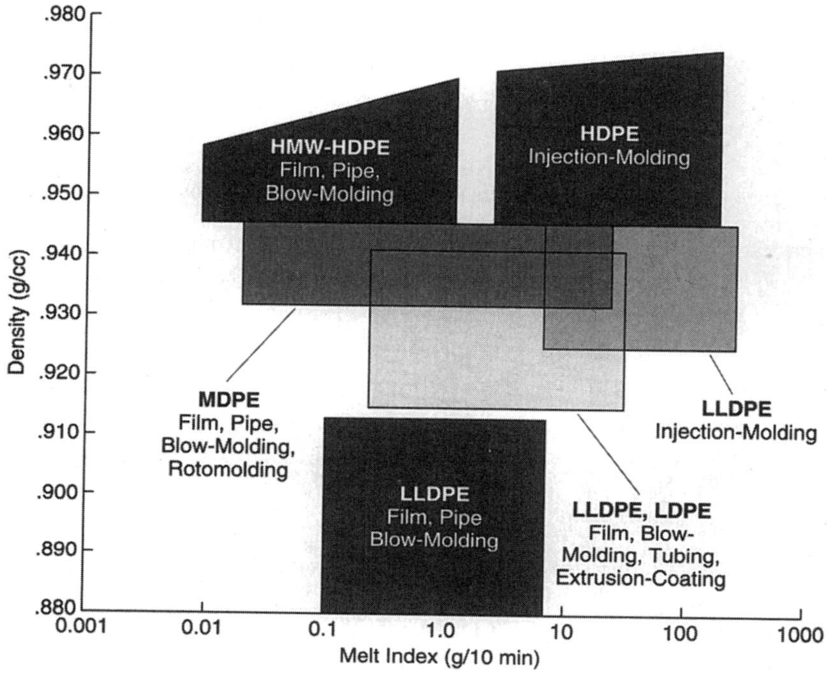

Figure 11.3 Uses of polyethylene as a function of density and melt index. *(Reprinted with permission from P. W. Manders,* Modern Plastics Encyclopedia '95, *McGraw-Hill Inc., 1994)*

lecular-weight aliphatic hydrocarbons, alcohols, ketones, and dilute acids and bases. However, HDPE should not be used with aromatic hydrocarbons such as benzene, which aggressively penetrate the polymer. HDPE is an excellent moisture barrier but shows high permeation coefficient to gases like oxygen and CO_2 as well as organic vapors.

11.1.2 Polypropylene

Polypropylene (PP) is a family of thermoplastic polymers based on the polymerization of the propylene monomer. They are commercially available as PP homopolymers and PP random copolymers. The latter are produced by the addition of small amount of ethylene (2 to 5 percent) during the polymerization process. Thermoplastic PP polymers are characterized by their low density as compared with the rest of polymers (0.89 to 0.92 g/cm^3), by their resistance to chemicals, and by their endurance to mechanical fatigue. PP resins are frequently employed in films and rigid containers.

11.1.2.1 PP homopolymer. Depending on the catalyst and polymerization conditions, the molecular structure of the resulting polymer consists of the three different types of stereo-configurations for vinyl polymers: isotactic, syndiotactic, and atactic. Isotactic PP is synthesized using Ziegler-Natta catalysts under controlled conditions of temperature and pressure. Industrial processes are designed to minimize the production of the atactic PP, a lower-value, noncrystalline, tacking by-product used mainly in adhesives.

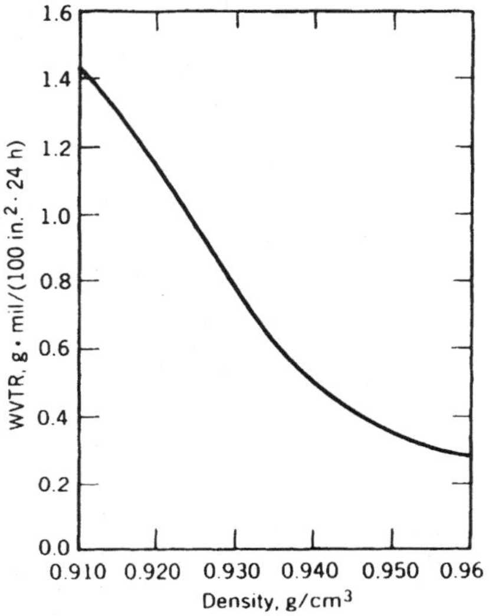

Figure 11.4 Effect of density on water-vapor transmission rate in HDPE. *(Reprinted with permission from M. A. Smith,* The Wiley Encyclopedia of Packaging Technology, *John Wiley and Sons, Inc., New York, 1986)*

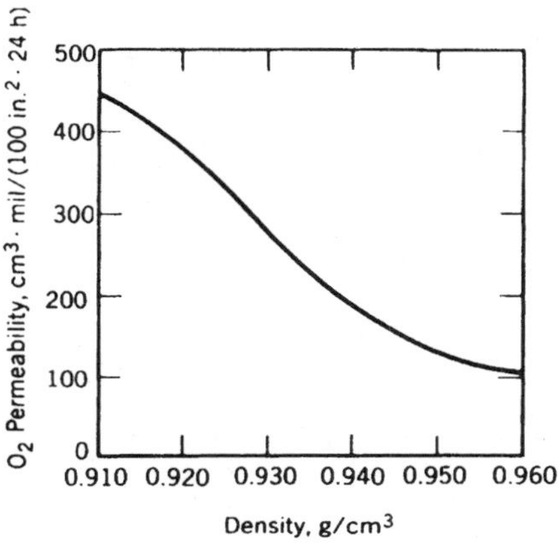

Figure 11.5 Effect of density on oxygen permeability in HDPE. *(Reprinted with permission from M. A. Smith,* The Wiley Encyclopedia of Packaging Technology, *John Wiley and Sons, Inc., New York, 1986)*

Isostatic PP (iso-PP). This is the most common form of PP. The placement of all the methyl groups on the same side of polymer backbone produces a stereo-regular chains and, consequently, highly crystalline materials. Because of its highly crystalline nature, isostatic PP is fairly opaque, has good solvent resistance, and has a higher melting temperature than PE. Nevertheless, clarity in iso-PP can be increased by the addition of chemical agents that prevent the growth of large crystals that give the natural opaqueness to PP. Compared with both LDPE and HDPE, PP has lower density, higher melting point temperature, and higher stiffness (higher modulus of elasticity). These properties determine unique applications for the PP homopolymer. For example, a higher value of stiffness and ease of stretching make PP homopolymers very suitable for wrapping and stretching applications, while the higher heat resistance allows a container made of this material to be autoclavable. On the other hand, iso-PP is more sensitive to heat and light oxidative degradation than PE. An oxidative degradation process during melt processing of PP produces chain scission that reduces the molecular weight, which reduces the properties of PP. For this reason, antioxidant compounds are added to the resin during processing to control oxidation. Antistatic agents are also commonly incorporated for packaging applications to dissipate static charge.

Iso-PP has desirable rheological properties, a wide range of flow rate, and good processing behavior. With molecular weight averages commonly in the range of 200,000 to 600,000 daltons, the melt flow index ranges between 0.5 and 50 g/10 min. Broad MWD materials are easy to process by injection molding. The density of PP, ranging from 0.89 to 0.91 g/cm^3, is one of the lowest among plastics. This low density produces films with high values of yield (or area factor). Yield is calculated as inverse of density times film thickness. PP shows excellent moisture barrier properties like other polyolefins. Films are pro-

duced by both blown and casting methods. Biorientation improves the film's optical appearance and strength (Table 11.2). Oriented PP films are used to wrap compact-disk boxes, toys, games, hardware items, frozen foods, and cigarette packaging.

TABLE 11.2 Effect of Orientation on PP Properties

	Nonoriented PP	Oriented PP
WVTR g mil/m^2 day at 90% RH and 100°F	15	6
Stiffness	Very low	High, similar to cellophane
Propagated tear strength	High	Very low CD, very high MD
Heat sealability	Yes, 350–450°F	No, film distorts
Density	0.902	No change
Optics	Very good	Excellent
Surface adhesivity to inks, etc.	Low	Low
Oxygen permeability, cc mil/m^2 day atm	3700	2500

PP films are commercially available with different coatings (acrylic and PVDC) to improve heat sealing, barrier, and optical characteristics, as well as to lower the coefficient of friction. Also, metallized PP films are available to provide extremely low values of permeability to gases and vapor.[4] PP films are well suited for bag-in-box applications in cereals, crackers, soup mixes, and stand-up pouches. PP is also an excellent material to fabricate injection-molded closures for HDPE, PET, and glass bottles.

PP random copolymer. Random copolymer PP typically contains 1.5 to 7 percent ethylene by weight as comonomer. The random placement of ethylene in the molecular chain prevents the stereo-regularity of the chains and high values of crystallinity as seen in iso-PP. Therefore, polymers with low crystallinity, resulting in better clarity and flexibility, and lower melting points are obtained. Random copolymers with densities ranging 0.89 to 0.90 g/cm^3 are slightly lighter and tougher, and they have lower temperature impact than homopolymer PP. Random copolymers show good chemical resistance to acids, alkalies, alcohols, and low-boiling hydrocarbons (non-aromatic hydrocarbons). Oriented films can be used as shrink wrap for toys and audio products. Moisture-barrier properties are also good. For example, at 100°F and 90 percent RH, permeance is 9.0 g·mil/m^2·d. PP random copolymers are processed as film and by blow and injection molding. Other packaging applications include medical, food and bakery products, produce, and clothing. The 7 percent ethylene copolymer is used as a heat-seal layer in food packaging.

11.1.3 Polyvinyl Chloride

Polyvinyl chloride (PVC) is a homopolymer of vinyl chloride. Eighty percent of commercial PVC in packaging is produced by chain-reaction polymerization using a suspension method. Other methods are emulsion and solution polymerization. Chain-reaction polymerization requires initiators to produce free radicals, then the reaction proceeds until the chain is terminated. The predominant configuration of the monomer in the polymer chain follows a head-to-tail alignment to yield a syndiotactic polymer.

PVC resins start decomposing at temperatures as low as 100°C (212°F) with the generation of HCl corrosive vapors. PVC without plasticizers (rigid PVC) has a glass transition temperature (T_g) of about 82°C. The high T_g value and the tendency to decompose make PVC difficult to process. The addition of plasticizers decreases the glass transition temperature of PVC and its processing temperature, making PVC easier to process. Liquid plasti-

cizers permit the production of a flexible film of PVC with moderate values of oxygen permeability. Plasticizers also permit the production of blow-molded bottles, blown films, and sheets. DOA, or di(2-ethyl hexyl) adipate, is one of the most common plasticizers used with PVC up to 30 to 40 percent. Stabilizers such as Ca/Zn salts, to avoid the decomposition of PVC and the corresponding product of HCl, are also employed during compounding and processing. When used for food applications, these stabilizers must have FDA clearance.

PVC films show good clarity, fair barrier properties, high puncture resistance, and great heat sealability. PVC films and sheets are also tough and resilient. For these reasons, PVC films are extensively used in food packaging, particularly for red and fresh meats. The oxygen permeability coefficient of PVC films is well suited to maintaining the oxygen requirements of the meat to keeping the red color of the meat and its appearance of freshness. In the U.S.A., chilled poultry and tray-packed poultry parts are packaged with PVC stretch films. PVC is also used to wrapping fresh fruits and vegetables. Other food applications include bottles for milk, dairy products, and edible oil, as well as blister packages for fish and produce. Non-food packaging uses of PVC include vacuum-formed blisters and bottles for toiletries, cosmetics, and detergents. Medical packaging applications include tubing and bags for parenteral products as well as for blood and intravenous-solutions.

In the polymerization process of PVC, less than 100 percent of vinyl chloride monomer (VCM) is converted to polymer. This means that relatively high values of VCM may remain unreacted and trapped in the resin. The resin must subsequently be submitted to a process for removing VCM by repeated applications of vacuum to less than 1 ppm. Currently, the industry produces PVC with extremely low levels of VCM in the resin, and the amount of VCM that might migrate to food is well below the sensitivity of common analytical methods. Due to its chlorine content, PVC was the center of a environmental controversy in the 1990s. Several European countries have banned the use of all PVC packaging because of the fear that, during incineration of solid wastes, HCl gas and chlorinated organic compounds (in which dioxines can be found) are generated and can be released into the environment. Although there is available technology that eliminates the emission of such unwanted compounds (Japan incinerates a large portion of its solid waste), European countries maintain the ban on PVC.

11.1.4 Vinylidene Chloride Copolymers

Vinylidene chloride homopolymer and copolymers are known as Saran®, a registered trademark of Dow Chemical. The company developed them during the 1930s. Saran is commonly, though wrongly, referred as PVDC. The polymers produced are all based on vinylidene chloride (or 1,1-dichloroethylene) and comonomers such as vinyl chloride (VC), acrylates (methyl acrylate), and vinyl nitriles. VDC homopolymer has a melting range of 388°C to 401°C, but it decomposes at 205°C. These conditions, similarly to PVC, make vinylidene chloride (VDC) homopolymer difficult to process. By copolymerization, the melting point of the resin is decreased to a range of 140 to 175°C, making melt-processing feasible. Saran polymer contains 2 to 10 percent plasticizer (for example, dibutyl sebacate or diisobutyl adipate) and heat stabilizers. Molecular weight ranges from 65,000 to 150,000 daltons. The most notable attribute of Saran copolymers are their extremely low permeability coefficient values to gases and vapors, which are comparable to dry EVOH resins. Oxygen permeability coefficient values of Saran range from 0.5 to 15.0 $cc \cdot mil/m^2 \cdot day \cdot atm$. Some Saran-base films show permeability values that increases with water humidity.

Saran is available in the following forms:

- F-resins (with acrylonitrile as copolymer) used as solvent-soluble polymer for barrier coating

- Aqueous emulsion latexes for barrier coatings
- Extrusion resins, which are melt-processable in rigid multilayer coextruded containers, extrusion of films, and sheets

F-resins include F-239 and F-278 types. These resins are used for coating plastic films such as cellophane and polyester. Resin F-310 is used for paper coating. Heat sealing temperature of F-resins are in the 100 percent-130°C range (Dow Chemical Co. Form 190-305-1084). Latexes are applied to coat paper, paperboard, and plastics films such as PP and PE. Also, PET, PVC, PS, and PE rigid containers can be coated with latexes. Barrier properties of latexes are similar to those of F-resins (Dow Form 190-309-1084). Extrusion resins are used for flexible packaging in mono- and multilayer (coextruded or laminated) structures for meat and other food applications. For nonrefrigerated foods in rigid containers, extrusion resins can be coextruded with PP or PS resins. Extrusion resins provide poorer barriers than do F-Resins or latexes (Dow Form 190-320-1084). Saran HB Films (with vinyl chloride as comonomer) are better barriers than F-resins (Dow Form 500-1083-586).

Saran resins can be processed by conventional methods: extrusion, coextrusion, laminating resin, latex coating, injection molding, blown extrusion film, and cast film. The main applications of Saran resins are in food packaging as barrier materials to moisture, gases, flavors, and odors. Monolayer films are widely used in household wrap. Multilayer films, generally coextrusions with polyolefins, are used to package meat, cheese, and other moisture- or gas-sensitive foods. The structures usually contain 10 to 20 percent of VDC as copolymer and are commonly used as shrinkable films to provide a tight barrier around the food product. VDC copolymers are used as barrier layers in semi-rigid thermoformed containers. PVDC coatings are used to improve barrier properties of paper and paperboard, cellophane, plastic films, and even semi-rigid containers such as PET bottles. Other industrial applications of monolayer films include laminations unit dose packaging and pack liners for moisture, oxygen, and solvent-sensitive products in pharmaceutical and cosmetic packaging.[5]

11.1.5 Polystyrene

New copolymerization methods, additives, rubber modification, and blending have made of polystyrene polymer and copolymers versatile packaging materials. Developed in 1930 by BASF, polystyrene (PS) is commonly produced by the continuous bulk polymerization of styrene in the presence of ethylbezene that control product viscosity and heat transfer. PS is hydrophobic, nonhygroscopic, and easily processed by extrusion and thermoforming. Three types of PS are available: general-purpose, impact PS, and foams.

11.1.5.1 General-purpose polystyrene (GPPS). Although referred to as "crystal" PS, this is an amorphous, and therefore highly transparent, material with T_g values ranging from 74 to 105°C and no crystalline melting temperature. Amorphous PS is a linear polymer that is brittle and stiff (high modulus) at room temperature, with mold shrinkage of 0.6 percent or less (lower than semicrystalline PE, PP, and PET).

There are three grades of GPPS: high-heat, medium-flow, and high-flow (or easy flow). High-heat resins have high molecular weight, contain few or no additives, and are brittle. They are used as extruded foams and thermoformed materials for electronic packaging, injection molded jewel boxes, high-quality cosmetic containers, and boxes for compact disks. High-flow resins have low molecular weight and usually contain 3 to 4 percent mineral oil as an additive. This makes crystal PS more flexible (less brittle), with lower a distortion temperature. Typical applications include disposable medical ware,

dinnerware, and coextruded sheets for thermoformed packaging. Medium-flow resins have intermediate molecular weight and contain 1 to 2 percent of mineral oil. These resins are used in blow-molded bottles and coextruded materials for food and pharmaceutical packaging.

11.1.5.2 High-impact polystyrene (HIPS). HIPS contains particles of rubber that are added to enhance impact resistance. This produces an opaque material that is easy to process and thermoformable. Typical food-packaging applications for HIPS are tubs for refrigerated dairy products, serving-size cups, lids, plates, and bowls. Limiting factors for HIPS are low heat resistance, high oxygen permeability, low UV light stability, and low resistance to oil and chemicals. According to Toebe et al.,[6] containers made of HIPS have intense flavor-scalping action on foods.

11.1.5.3 Expandable PS (EPS). Crystal PS is also supplied as partially expanded bead to fabricate foams. EPS foam has good shock absorbing and heat insulation characteristics. Applications in food packaging include egg cartons and meat trays.

11.1.5.4 Commercial copolymers. Styrene-butadiene block copolymer (SBS), acrylonitrile-butadiene-styrene (ABS), styrene-acrylonitrile (SAN), styrene-methylmethacrylate (SMMA), and styrene-maleic anhydride (SMA) are important styrene copolymers.

Recent developments show that SBS resins, when blended with GPPS, produce attractive materials for food packaging. SBS resins have chemical resistance similar to that of GPPS: resistant to alkalis, dilute organic acids, and aqueous solution of most salts. However, they are highly swollen by most organic solvents but resist methanol and ethanol. SBS films have superior values of stiffness and impact resistance. SBS resins can be applied as a film for modified-atmospheric packaging—perishable-food packaging like ready to eat salads and vegetables. Coextrusion with metallocene polyethylene SBS resin provides balanced oxygen, water vapor, and carbon dioxide permeability values to provide breathability value to the package.[7] Other applications of SBS/GPPS include packaging material for medicinal and health-care products.

11.1.6 Ethylene Vinyl Alcohol (EVOH)

Introduced in 1970 in Japan, ethylene vinyl alcohol is produced by a controlled hydrolysis of ethylene vinyl acetate copolymer. The hydrolytic process transforms the vinyl acetate group in vinyl alcohol, VOH. The presence of the alcohol function in the backbone chain, equivalent to substituting a certain number of H atoms in a polyethylene chain, imparts several noticeable properties to EVOH resins: First, the OH group, which is highly polar, increases the intermolecular forces between polymer chains and make EVOH more hydrophilic than PE. Second, the OH group is small enough to yield a polymer with high degree of crystallinity, even if it is randomly distributed in the chain. This yields an outstanding barrier to permeating molecules in the absence of water. When the percent of vinyl alcohol is equivalent to 100 percent, polyvinyl alcohol (PVOH) is obtained.

Contrary to PE, PVOH has exceptional gas and odor barrier properties (the lowest permeability of any polymer available at dry conditions). PVOH is difficult to process, although it is water soluble. When the percent of VOH in EVOH ranges from 52 to 70 percent, the ethylene-vinyl alcohol copolymers obtained combine the processability and

water resistance of polyethylene and the gas- and odor-barrier characteristics of PVOH. EVOH copolymers are highly crystalline, and their processing and barrier properties vary with respect to the equivalent percent of ethylene. When the ethylene percent is near 30 percent, the gas and organic vapor barrier is exceptionally high, but processing conditions become more difficult. As the ethylene content increases, water barrier characteristics and processability improve.

The most important characteristic of EVOH is its outstanding O_2 and odor barrier properties. A layer of EVOH in multilayer structures provides high retention of flavors and protection to oxidation of the food product packaged. EVOH also provides a very high resistance to carbon dioxide, oils, and organic vapors. This resistance decreases somewhat as the polarity of the penetrating compound increases. For example, the barrier resistance to linear and aromatic hydrocarbons is outstanding, but, for ethanol and methanol, it is low (it may absorb up to about 12 percent of ethanol). As indicated, the hydroxyl group OH makes the polymer hydrophilic, attracting water molecules. EVOH films in equilibrium with a humid environment, show higher oxygen permeability values (Fig. 11.6). This

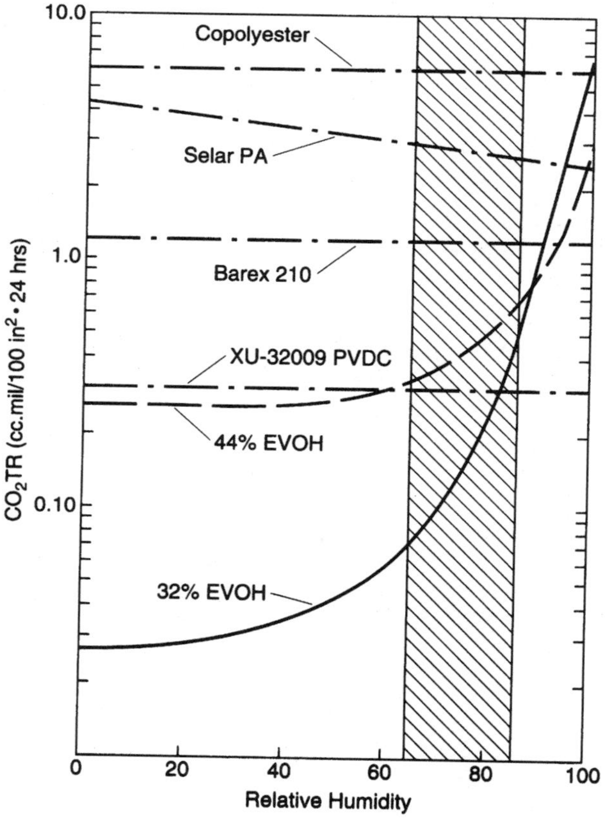

Figure 11.6 Oxygen permeability of selected resins as a function of relative humidity. *(Reprinted with permission from EVAL Company of America)*

poses an interesting challenge in the design of high-barrier packages, since external non-hydrophilic layers are necessary to protect the EVOH oxygen barrier characteristics.

EVOH can be coextruded in numerous combinations with PE or PP; laminated; or coated to several substrates including PET, PE, nylons, and the like. EVOH can also be extruded into films and processed by blow molding, injection molding, and coextrusion blow molding. Due to its great thermal stability, scrap EVOH can be reused with virgin resin. Multilayer containers may have up to 15 percent regrind content of EVOH.[8] The FDA has cleared the use of EVOH resins for direct food contact up to 80 percent of VOH. Applications in packaging include flexible and rigid containers. Typical application are ketchup and barbecue sauce bottles, jelly preserves, vegetable juice, mayonnaise containers, and meat packages. Non-food applications include packaging of solvents and chemicals.

11.1.7 Nylons

Nylons are linear thermoplastic polyamides produced by multiple condensation reactions between acid and amide functions. This yields the amide group (–CONH–), the characteristic recurring unit of nylons. In general, nylons are clear, thermoformable, strong, and tough over a broad range of temperatures. Nylons show excellent chemical resistance and are good barriers to gas, oil, and aromas. They are, however, moisture-sensitive materials like EVOH. When left in normal environmental conditions, nylons can easily absorb 6 to 8 percent of their weight in water. The amount of water sorbed by a nylon sample as a function of relative humidity is described by a sorption isotherm curve.[9] For most packaging applications, nylons are used in film form, as a single component or in multilayer structures.

Nylons are polymers with strong H-bonding between the C=0 and HN groups of different chains. These high intermolecular forces, combined with high crystallinity, yield tough, high-melting thermoplastic materials—Nylon 6,6 has a melting point of 269°C (516°F). Nylons have good puncture resistance, impact strength, and temperature stability. In addition, the flexibility of the aliphatic portion in the chain permits film orientation that enhance strength. Amorphous nylons can be produced by copolymerization of two different acids. For example, isophthalic and terephthalic acids are used to produce the amorphous nylon Selar PA, which is commercialized by Dupont. This copolymer first show a decrease and then an increase in the oxygen permeability coefficient with increasing moisture content.

Nylons are melt-processable using conventional extrusion equipments. Films can be produced by either cast-film processing or blown-film processing. During film production, diverse degrees of crystallinity are obtained, depending on the temperature quenching rate. When the cooling rate is increased, a less crystalline nylon is obtained, since the polymer was not given sufficient time to form crystals. The increase in amorphousness produces a more transparent and more easily thermoformable film. Nylons are used in coextrusion with other plastics, providing strength and toughness to the structure. Polyolefins are commonly used in nylon coextrusions to provide heat sealability, moisture, and low cost. Nylons are used in extrusion coat paperboard to obtain heavy-duty paperboard.

The blow-molding process is used with nylon resins to produce industrial containers, moped fuel tanks, and oil reservoirs. Thermoformed nylons are employed for disposable medical devises, meat and cheese packaging, and in thermoform/fill/seal packaging. For most applications, nylons are combined with other materials, such as LDPE, ionomer, and EVA, to add moisture barrier and heat sealability. Multilayer films containing a nylon layer are used principally in vacuum-packing bacon, cheese, bologna, hot dogs, and other processed meats. Polyvinylidene chloride copolymer coating on nylons are available to improve oxygen, moisture vapor, and grease barrier properties. Biaxial orientation of nylon (BON) films provide increased crack resistance and better mechanical and barrier properties.

11.1.8 Polyethylene Terephthalate

Polyethylene terephthalate (PET), a linear, semicrystalline, thermoplastic homopolymer, has become a dominant packaging material for both carbonated and noncarbonated beverages. Although the largest use of PET is in textile fibers, packaging is the fastest-growing use of PET. The great acceptance of PET as a carbonated beverage packaging material is due to its toughness, clarity, capability of being oriented, reasonable cost, and the development of high-speed bottle processing technology. In 2002, it is expected that PET will hold 42 percent share of the soft drinks packaging market and 5 percent or more of the beer market. Compared with glass, PET containers offer light weight, shatter resistance, acceptable barrier, and recyclability. PET is produced by the condensation of terephthalic acid and ethylene glycol. Its T_g is around 78°C, and its melting temperature is 270°C. Crystallinity of PET ranges between 20 and 40 percent and density between 1.30 and 1.40 g/cm^3.

The presence of moisture during the extrusion of PET reverts the condensation reaction and produces some degree of depolymerization. Prior to processing, moisture content should be less than 0.005 percent to minimize hydrolytic breakdown and loss of properties.

Films can be cast using chill roll, while injection stretch blow molding is used to produce bottles for carbonated beverages. PET is used for packaging food, distilled spirits, carbonated soft drinks, noncarbonated beverages, and toiletries. Typical food products include mustard, picked foods, peanut butter, spices, edible oil, syrups, and cocktail mixers. PET is extensively used for extrusion coating and extrusion into film and sheet. Its crystalline form (CPET) is the basic material for ovenware containers up to 204°C (400 °F). CPET is formed with the addition of crystallization initiators and nucleating agents to highly increase the percent of crystallinity and its opacity. Stretch-blown oriented PET (OPET) is commonly used for beverage containers like water, soft drinks, juices, and alcoholic beverages. Biaxially oriented PET films are used in meat and cheese packaging. PET does not require the use of any additives, such as plasticizers, during processing. And, because of its low values of extractable compounds and absence of toxic by-products, virgin PET resin has been regulated as a safe material for food packaging.

11.1.8.1 Thermoplastic copolyesters. PET copolymers range in degrees of crystallinity, including being totally amorphous. The name *copolyesters* is applied to those polyesters whose synthesis is carried out by incorporating to the polymerization process an additional glycol (e.g., cyclohexanedimetanol or CHDM) or a diacid (e.g., isophthalic acid). The copolyester chain is less regular than the homopolymer, and the degree of crystallinity is lower. PCTA is a copolymer of CHDM and a blend of terephthalic and isophthalic acids. It is an amorphous polyester designed primarily for film forming and sheeting in food, pharmaceutical, and general blister-packaging applications. On the other hand PETG, also an amorphous material, is the copolyester resulting from the reaction of ethylene glycol and CHDM with terephthalic acid. PETG, having outstanding optical properties, can be sterilized and finds applications in pharmaceutical, food, and electronic packaging.

11.1.9 Polyethylene Naphthalene

Polyethylene naphthalate (PEN), a relatively new polyester, is part of the PET family and has a great potential as a resin for bottles. PEN resin is more opaque, shows five times lower oxygen and carbon dioxide permeability coefficients, has a higher glass transition temperature, is stronger, and is stiffer than PET. This makes it more suitable for hot filling and a excellent material for carbonated beverages. Bottles made of PEN provide the product with additional ultraviolet protection. PEN bottles can be returnable, refillable, and re-

cyclable. PEN resins can be processed by blow molding, injection molding, and extrusion thermoforming. PEN is a polyester produced from the reaction of dimethyl 1-2,6 naphthalenedicarboxilate and ethylene glycol.

11.1.10 Polycarbonate

Another polyester, polycarbonate (PC), is a glassy, amorphous thermoplastic material. It has excellent balance of toughness and clarity. Commercial production of PC is based on bisphenol A. The heat deflection temperature of about 130°C, and its glass-transition temperature is 149°C. Polycarbonate has good potential for packaging applications. Commonly, polycarbonate is produced by the reaction of bisphenol-A and carbonyl chloride.

Toughness is the most relevant property of PC. Being tough and clear makes PC a material well suited for reusable bottles, particularly 19-liter (5-gal) water bottles and 1-gal milk bottles. Systems with washing stations have been developed for reusing PC bottles. Polycarbonate films are odorless, tasteless, and do not become stained through normal contact with natural or synthetic coloring agents.

PC has good resistance to fruit juices, aliphatic hydrocarbons, and aqueous solutions of ethanol. It is, however, attacked by some solvents such as acetone an dimethyl ethyl ketone. Since PC is FDA approved, food-contact applications include microwave ware, ovenware, and food storage containers. Other applications include hot filling, modified atmospheric packaging, rigid packaging to substitute for PVC, high gloss for paper, and barrier for fruit juice cartons. PC finds applications in medical device packaging, and it can be sterilized by commercial sterilization techniques such as ethylene oxide, autoclave sterilization, and gamma sterilization.

PC can be processed by injection molding, extrusion, coextrusion, and blow molding. Coextrusions with EVAL or polyamides are carried out with the help of adhesives. PC can be laminated or coextruded to PP, PE, PET, PVC, and PVDC. PCs are hydrophilic polymers and, at ambient conditions, can reach moisture levels of 0.35 percent.

A list of selected properties of plastics commonly used in packaging is presented in Table 11.3.

TABLE 11.3 Selected Properties of Packaging Plastics

	LDPE	HDPE	LLDPE	≥12% VA EVA	Ionomer
Density, g/cc	0.91 to 0.925	0.945 to 0.967	0.918 to 0.923	0.94	0.94 to 0.96
Yield, $in^2/lb \cdot mil \times 10^{-3}$	30	29.0	30.0	29.5	28.6 to 29.5
Tensile strength, kpsi	1.2 to 2.5	3.0 to 7.5	3.5 to 8.0	3 to 5	3.5 to 5.5
Elongation at break, %	225 to 600	10 to 500	400 to 800	300 to 500	300 to 600
Impact strength, kg·cm	7 to 11	1 to 3	8 to 13	11 to 15	6 to 11
Elmendorf tear strength, g/mil	100 to 400	15 to 300	80 to 800	50 to 100	15 to 150
WVTR, g·mil/100 in^2·day @ 100°F and 90% RH	1.2	0.3 to 0.65	1.2	3.9	1.3 to 2.1
Oxygen transmission rate, cm^3·mil/100 in^2·day·atm @ 77°F and 0% RH	250 to 840	30 to 250	250 to 840	515 to 645	226 to 484
CO_2 permeability, cm^3·mil/100 in^2·day·atm @ 77°F and 0% RH	500 to 5000	250 to 645	500 to 5000	2260 to 2900	626 to 1150
Resistance to grease and oil	varies	good	good	varies	good
Dimensional change at high RH, %	0	0	0	0	0
Haze, %	4 to 10	25 to 50	6 to 20	2 to 10	1 to 15
Light transmission, %	65	N/A	N/A	55 to 75	85
Heat-seal temperature range, °F	150–350	275 to 310	250 to 350	150 to 300	225 to 300
Service temperature range, °F	–70 to 180	–60 to 250	–60 to 180	–60 to 140	–150 to 150
Tensile modulus, 1% secant, kpsi	20 to 40	125	25	8 to 20	10 to 50

11.2 Properties of Packaging Plastics

A description of the most important properties of plastics packaging and their test methods is presented in this section. For easy reference, they are grouped under these headings:

1. Morphology and density
2. Thermophysical
3. Mechanical
4. Barrier
5. Surface and adhesion
6. Optical
7. Electrical

11.2.1 Morphology and Density

11.2.1.1 Morphology. Polymer morphology refers to the presence, shape, arrangement, and physical state of amorphous, crystalline, and chain-orientated regions that are generally found co-existing in a polymer. Polymers used in packaging are either homogenous amorphous solids or, more frequently, heterogeneous phase semicrystalline solids. A semicrystalline polymer can be thought of as small crystalline regions embedded in an amorphous phase. LDPE, HDPE, PET, Nylon 6, and EVOH are examples of semicrystalline polymers, while general-purpose polystyrene, butadiene-styrene copolymers, some PET copolymers (like PETG and PCTA), and nylon 6I/6T (Selar PA) are amorphous polymers.

The actual morphology of a polymer depends primarily on three factors: chemical composition, degree of polymerization, and molecular architecture and chain conformation. Other variables affecting the final physical state of a polymer sample include the thermomechanical history and processing methods.

Polymer molecules tend to seek out an arrangement in the lowest energy state possible (lowest Gibbs free energy). The lowest energy level that a compound can achieve is a crystal form. Although crystallization tends to occur naturally, crystallization of most synthetic polymers take place in small regions (in the order of 1×10^{-9} m), if at all. The ability of a polymer to crystallize is largely determined by the regular placement of atoms in the chain. Polymers made of symmetrical unsaturated monomers, such as polyethylene and polyvinylidene chloride, crystallize easily, yielding high-crystalline materials. Asymmetric polymers such as polypropylene (PP) crystallize only within regular configurations, i.e., isotactic or syndiotactic, while atactic PP is totally amorphous. But, for the most part, asymmetric polymers with an atactic configuration can crystallize if the substituents are small and polar as they are EVOH and PAN. On the other hand, polymers with asymmetric, repeating units and regular configurations do not crystallize if the substituent is too bulky, as is the case in polystyrene. Normally, step-reaction polymers, which are synthesized from bifunctional monomers containing alcohol, acid, or amines, can crystallize. These polymers produce highly ordered chains, since the bifunctionality of the monomers forces the chain to grow in only one isomeric configuration. This is the case, for instance, with nylon 6 and nylon 6,6. When the bifunctional monomers contain aromatic and cyclohexane rings, only polymers with substitutions in 1,4 positions, like PET and PC, are crystallizable. A 1,3 linkage in the ring increases the randomness of the chain and, hence, minimizes crystallinity. Step-reaction copolymers combining 1,4 and

1,3 substitutions will produce amorphous polymers as in the case of nylon 6I/6T and PCTA. When the step-reaction polymerization process includes monomers with three and four functional groups, the resulting polymer will form a tri-dimensional network. This is the case in adhesives such as epoxies and polyurethanes, which are both amorphous and thermoset materials.[10]

Amorphous polymeric materials and inorganic glasses do not show crystalline melting points. But amorphous materials do have glass transition temperature, T_g, which is defined as the freezing in (on cooling) or the unfreezing (on heating) of micro-Brownian motion of chain segments 2–50 carbon atoms in length.[11] At low temperature, an amorphous polymer is glassy, hard, and brittle, but, as the temperature increases, it becomes rubbery, soft, and elastic. There is a smooth transition in the polymer's properties from a solid to a flow melt. At the glass transition temperature, properties like specific volume, enthalpy, shear modulus, and permeability show significant changes at the glass transition temperature (see Fig. 11.7).

Woodward[12] describes seven common crystalline morphologies in polymers: faceted single lamellas, non-faceted lamellas, dendritic structures, sheaf-like lamellar ribbons, spherulite arrays, fibrous structures, and epitaxial lamellar overgrowths on microfibrils. Spherulites are complex ordered aggregations of submicroscopic crystals. PE crystal spherulites, for example, are about 10 nm thick. The spherulites are separated from one another by small amorphous regions called *micelles*. The spherulites are larger than the wavelength of visible light, producing light scattering that makes the polymer opaque. Plastic materials with high crystallinity are opaque, while plastics with low degrees of crystallinity are transparent or clear. Amorphous materials are totally transparent.

Process conditions associated with the cooling rate of polymer melts, blown stretching, and film orientation enhance the polymer's natural anisotropy. Unlike low-molecular-

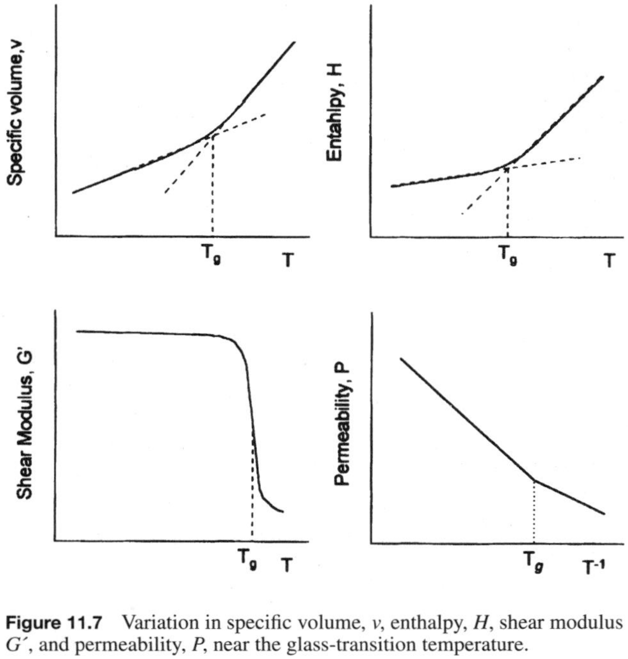

Figure 11.7 Variation in specific volume, v, enthalpy, H, shear modulus G', and permeability, P, near the glass-transition temperature.

weight compounds, polymers show anisotropism in both amorphous and crystalline phases because of the strong covalent bonds in the backbone chain and weak intermolecular forces. This anisotropic behavior increases upon molecular orientation. Unbalanced oriented films, therefore, show different values of tensile and tear strength along machine direction than they do along cross-machine direction. In PVC and PET stretch-blown molded containers, molecular orientation plays an important role in the mechanical, barrier, and optical properties of the container. For a particular molecular weight and stretch ratio, chain orientation increases with strain rate and lower temperatures. Imbalances in orientation can be detected by optical measurements such as birefringence.[13] Several important polymer properties depend on the morphology of the polymer. As crystallinity increases, density, permeability, opacity, strength, and heat sealing temperatures increase, while blocking, clarity, tear strength, impact strength, toughness, ductility, ultimate elongation, and heat sealing temperature ranges decrease.

Degree of crystallinity. Degree of crystallinity is the ratio of crystalline region to amorphous region in a polymer sample. It can be expressed as a volume or mass ratio. Degree of crystallinity is primarily determined by x-ray scattering. In practice, however, this is a tedious and expensive method. The crystallinity of a polymer sample can also be determined by the density gradient method, ASTM D 1505. In this method, two solutions, A and B, are prepared with density in the range of interest. Solution A, with the lowest density, and solution B, with the highest density, are combined in a glass tube to form a vertical column of liquid in which the density varies linearly from the bottom to the top (see Fig. 11.8). The column is calibrated with glass beads of known density. Plastic samples are dropped in the column, and the sample will settle at the corresponding density value, like clouds in the atmosphere. The density of the plastic is calculated from the position of the sample and calibrated beads using the calibration scale. Based on the two-phase model,[14] the degree of crystallinity of a polymer sample is related to its density by

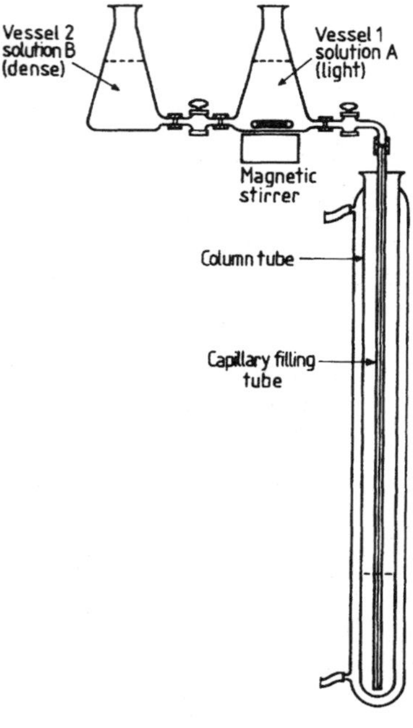

Figure 11.8 Density gradient column. *(Reprinted with permission from R. P. Brown,* Handbook of Plastics Test Methods, *G. Godwin, Ltd., 1981)*

$$a_v = \frac{\rho - \rho_a}{\rho_c - \rho_a}$$

and

$$a_m = a_v \frac{\rho_c}{\rho}$$

where a_v and a_m = volume and mass crystallinity, respectively
ρ, ρ_c, and ρ_a = density of the test sample, the pure crystal, and 100 percent amorphous, respectively

Percent of crystallinity is a_v or a_m multiplied by 100.

Polymer orientation. Molecules in polymer film or sheet are oriented into more orderly morphology in response to external stress above T_g. Orientation tends to increase crystallinity (although not always in the strict sense of the word; e.g., oriented PP is clearer than non-oriented PP, however, the molecules are only more compact) and decrease permeability. In an oriented film, a large fraction of the molecules tend to line up in the direction of stretch. The result of uniaxial molecular orientation is a substantial increase in strength and toughness in the direction of the stretch (but a decrease in strength in the perpendicular direction). Polymers can be unoriented, uniaxially oriented, or biaxially oriented. The films to fabricate bags are generally uniaxially oriented to improve its tensile strength. Films used for pouches are biaxially oriented, because it is expected that the tensile forces may act in both directions, and, if shrinking occurs, the pouch should shrink in both directions. Biaxially orientation can be balanced (stretching the same amount in both directions) or unbalanced.

11.2.2 Thermophysical Properties

11.2.2.1 Relaxation temperatures. A completely amorphous polymer is polymer that does not crystallize even when cooled from the melt at a very slow rate. As the melt cools down, there is a decrease in the degree of thermal agitation of the molecular segments. As the cooling process continues, the rate of segmented movement becomes more and more sluggish and then, upon further decreasing the temperature, the segmental movement finally stops. At this point, the glassy state of the amorphous polymer is reached. The temperature at which this glassy state takes place is called the *glass transition temperature, T_g*. This process is associated with the limitation of segmental chain mobility as defined below. The glassy state consists then of "frozen" entangled chain molecules with a complete absence of stereo-regularity and motion. This absence of stereo-regularity is typical of liquids.

As seen in Fig. 11.7, the change in the specific volume as a function of temperature shows a smooth curve from the melt state to the glassy state. However, there is a change in the slope near T_g. In the glass region, the specific volume is much larger than the value corresponding to the crystal phase. This is as if additional or free volume has been trapped between the entangled and frozen polymeric chain. According to theory, the free volume at the glass transition temperature occupies 2.5 percent of the polymer volume. The reduction in specific volume as the temperature decreases below T_g is associated with molecular movement of the total molecules moving with respect to each other, and it is determined

by intermolecular force. As the free volume increases, density decreases, and properties like permeability increase.

The motions associated with T_g results from the semi-cooperative actions involving torsional oscillation and/or rotations around the backbone bonds in a given chain as well as neighboring chains. Torsional motion of side groups around the axis connecting them to the main chain may also be involved.[11] T_g involves then, both intrachain and interchain segmental motions. In addition to T_g, other relaxation temperatures like T_β, T_{ll}, and $T_{l\rho}$ are important in amorphous polymers.[11] Unlike T_g, T_β is associated mostly with intrachain subgroup motions (2–10 consecutive chain atoms). T_{ll} is a weak transition-relaxation temperature about 1.2 times greater than T_g ($T_{ll} \approx 1.2\ T_g$) associated with the thermal disruption of intermolecular segment-to-segment contacts known as segmental melting. T_{ll} marks the onset of the true liquid state of an amorphous polymer. At a higher temperature, 30 to 50 K above T_{ll}, the molecule chains contain enough energy to break the rotational barrier marking the true liquid nature of an amorphous polymer. This relaxation has been identified as $T_{l\rho}$. According to Boyer,[11] physical aging of polymers occur at all temperatures between T_g and T_β (the higher the temperature, the faster the aging process). The effect of aging can be erased by heating the polymer above T_g. At a given temperature, toughness is associated with the presence of one or more relaxation temperatures below that particular temperature.

An amorphous polymer, whether pure or coexisting with crystalline regions, can be a glassy material (brittle) or rubber-like (soft), depending on the material temperature with respect to T_g. The glass-transition temperature separates a quasi-liquid state (above T_g) behavior from a glassy behavior (below T_g). If an amorphous polymer sample is at a temperature below T_g, it will be brittle and will show aspect of glassy materials. As the temperature of the sample increases to approach T_g, the polymer will exhibit leathery behavior, and there will be a loss in the elastic modulus and a decrease in its barrier properties. T_g marks the lower limit temperature for melt flow, resiliency, and processing conditions for the polymer's molecular orientation. Wetting, tackiness, and adhesion appear to be fully developed at T_{ll}.

For many polymers, it has been determined experimentally that the ratio of the glass transition temperature to the melting temperature, $T_g/T_m \approx 0.6$ (both temperatures in Kelvins).[11] For example, for polypropylene, $T_g = -19°C$ and $T_m = 176°C$, so $T_g/T_m = 0.57$. T_g can be estimated by group-contribution methods.[15]

11.2.2.2 Melting temperature. The melting temperature T_m, is a true transition temperature. This means that, at T_m, both the liquid and solid phase have the same free energy. Most semicrystalline polymers have a melting range, and for an amorphous polymer, do not have T_m at all. Similarly to T_g, T_m can be estimated from contribution groups and an empirical relationship between T_g and T_m. Plastics show T_m as low as 275 K for polyisobutylene, and as high as 728 K in polyethylene terephthalamide.[15,16] ASTM methods D 2117 and D 3418 are methods for measuring T_m.

11.2.2.3 Heat capacity. Heat capacity, or specific heat, is the amount of energy needed to change a unit of mass of a material one degree of temperature. The heat capacity of plastics, which are obtained at constant pressure, is a temperature-dependent parameter, especially near the glass transition temperature. The heat capacity at 25°C ranges from 0.9 to 1.6 J/gK for amorphous polymers, and from 0.96 to 2.3 J/gK for crystalline polymers. For semicrystalline polymer, the heat capacity of the amorphous phase is larger than the heat capacity of the crystalline phase. This implies that the heat capacity values depend on the percent of the polymer's crystallinity. Reliable data regarding the heat capacity of

amorphous and crystalline phases are available for only a limited number of polymers.[15,17] Heat capacity values for polymers can be found in the review by Wunderlich.[18] Usual techniques for measuring specific heat are differential thermal analysis (DTA) and differential scanning calorimetry (DSC).

11.2.2.4 Heat of fusion. The heat of fusion, ΔHm, is the energy involved during the formation and melting of crystalline regions. For semicrystalline polymers, the energy of fusion is proportional to the percent of crystallinity. Amorphous polymers or amorphous polymer regions do not have heat of fusion, since amorphous structures have a smooth transition from the liquid amorphous state to the liquid state. Experimental values of crystalline heat of fusion for common packaging plastics vary from 8.2 kJ/mol for polyethylene, to 43 kJ/mol for Nylon 6,6.[15] ASTM D 3417 describes a method for measuring the heat of fusion and crystallization of a polymer by differential scanning calorimetry (DSC).

11.2.2.5 Thermal conductivity. Thermal conductivity is the parameter in Fourier's law that relates the flow of heat to the temperature gradient. In specific terms, thermal conductivity *k* is a measure of a material's ability to conduct heat. The thermal conductivity of a polymer is the amount of heat conducted through a unit of thickness per unit of area, time, and degree of temperature. Thermal conductivity values control the heat-transfer process in applications such as container-forming, heat-sealing, cooling and heating of plastics, and package sterilization processes. Plastics have values of *k* much lower than metals. Thermal conductivity for plastics ranges from 3×10^{-4} cal/s cm°C for PP to 12×10^{-4} cal/s cm°C for HDPE. For aluminum, k is 0.3 cal/s cm°C, and for steel it is 0.08 cal/s cm°C.[19] For plastic foams, values of *k* are much lower than those of unfoamed. This is due to the presence of air trapped in the cellular structure of the plastic foam. Plastics with low thermal conductivity values do not conduct heat well. Enhanced by low thermal conductivity of air, foams are excellent insulating materials. In addition to being thermal insulator, foams are attractive cushioning, packaging materials. Plastic fillers may increase the thermal conductivity of plastics. Methods for measuring *k* are given in ASTM D4351, C518, and C177.

11.2.2.6 Thermal expansion coefficient. The coefficient of linear (or volume) thermal expansion is the change of length (volume) per unit of length (volume) per degree of temperature change at constant pressure.

$$\beta = \left(\frac{1}{L}\right)\left(\frac{dL}{dT}\right)p$$

and

$$\alpha = \left(\frac{1}{V}\right)\left(\frac{dV}{dT}\right)p$$

Units of α and β are K^{-1} or $°F^{-1}$. Compared with other materials, polymers have high values of thermal expansion coefficients. While metals and glass have values in the range 0.9 to 2.2 K^{-1}, polymers range from 5.0 to 12.4 K^{-1}.[19] Thermal expansion coefficients can be

measured by thermomechanical analysis (TMA). ASTM D 696 describes a method using a quartz dilatometer, while ASTM E 831 describes the determination of the linear thermal expansion of solid materials. Volume contraction of a container from the molding operation temperature down to room temperature is called *shrinkage,* and its measurement is described in the ASTM D 955, D 702, and D 1299.

11.2.3 Mechanical Properties

11.2.3.1 Bursting strength. Bursting strength is the hydrostatic pressure, given in pascals (or lb/in^2) required to produce rupture of a flat material (film or sheet) when the pressure is applied at a controlled increasing rate through a circular rubber diaphragm that is 30.48 mm (1.2 in) in diameter. *Points bursting strength* is the pressure expressed in lb/in^2. ASTM method D 774 describes the measurement of the bursting strength of plastic films.

11.2.3.2 Dimensional stability. Dimensional stability refers to how well a structure maintains its dimensions under changing temperature and humidity conditions. ASTM D 1204 describes a standard method for linear dimensional changes of flexible thermoplastic films and sheets at elevated temperatures. Dimensional stability is a property important in any flexible material converting process. During printing, for example, even a small change in dimensions may lead to serious problems in holding a print pattern. In a flexible structure, dimensional stability may produce different changes in the machine and transverse directions.

11.2.3.3 Folding endurance. This is a measure of the material's resistance to flexure or creasing. Folding endurance is greatly influenced by the polymer's glass transition temperature and the presence of plasticizers. The ASTM D 2176 describes the procedure to determine the number of folds necessary to break the sample film.

11.2.3.4 Impact strength. Impact strength is the material's resistance to breakage under a high-velocity impact. Widely used impact tests are as follows:

- For rigid materials
  - Izod (D 256A)
  - Charpy (D 256)
- For flexible structures
  - Dart drop impact (ASTM D 4272)
  - Pendulum impact resistance (ASTM D 3420)

The free-falling dart method for polyethylene films is described in the ASTM D1709. Unlike low-speed uniaxial tensile tests, the pendulum impact test measures the resistance of film to impact puncture simulating high-speed end-use applications. Dart drop measures the energy lost by a moderate-velocity blunt impact passing through the film. Both pendulum and impact tests measure the toughness of a flexible structure.

11.2.3.5 Pinhole flex test. Pinhole flex resistance is the property of a plastic film to resist the formation of pinholes during repeated folding. A related test is the folding endurance test. Films having a low value of pinhole flex resistance will easily generate pinholes at the fold line under repeated flexing. The test is described by the standard ASTM F 456.

11.2.3.6 Tensile properties. The mechanical behavior of a polymer can be evaluated by its stress-strain tensile characteristics (Fig. 11.9). The stress is measured in force/area expressed in unit of pressure. The strain is the dimensionless fractional length increase.

Modulus of elasticity E is the elastic ratio between the stress applied and the strain produced, giving the material's resistance to elastic deformation. The tensile modulus also gives a measure of the material stiffness; the larger the modulus, the more brittle the material. For example, E of LDPE is 250 mPa, while for crystal PS, it is 2500 mPa. Comparatively, values of tensile modulus in polymers (1.9×10^3 mPa for Nylon) are much lower than for glass (55×10^3 mPa) or mild steel (210×10^3 mPa).[16,20] Elastic elongation is the maximum strain under elastic behavior. Ultimate strength or tensile strength is the maximum tensile stress the material can sustain. Ultimate elongation is the strain at which the sample ruptures.

Toughness is the energy a film can absorb before rupturing, and it is measured by the area under the stress-strain curve. Brittleness is the lack of toughness.[21] Amorphous and semicrystalline polymers become brittle when cooled below their glass transition temperature. Tests for tensile properties are described in the ASTM D 882, for flexural strength in ASTM D 790, and for flexural modulus in ASTM D 790M.

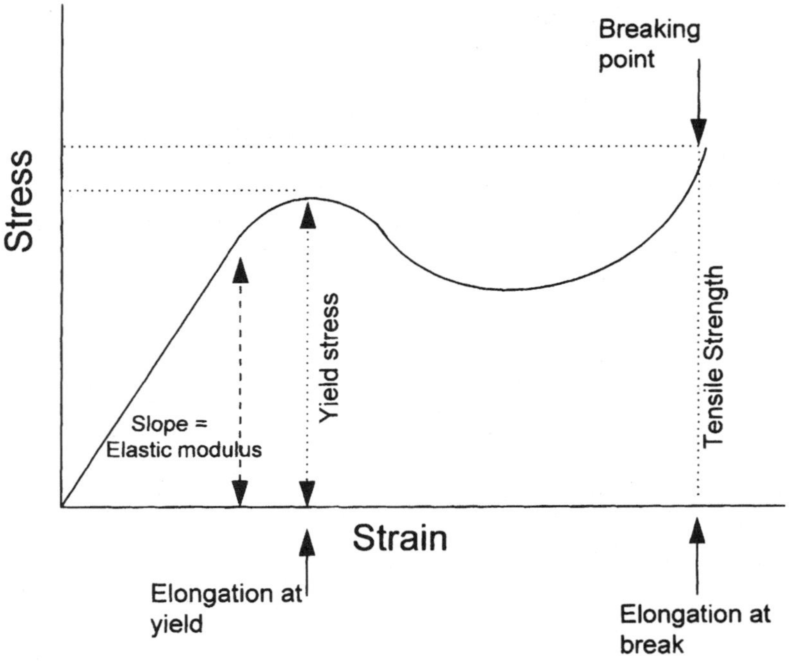

Figure 11.9 Stress-strain curve.

11.2.3.7 Tear strength. The measurement of tear strength evaluates the energy absorbed by a film and sheeting sample during the tear initiation and propagation. Two standard methods are available: ASTM methods D 1004 and D 1922. The former is designed to measure the force to initiate tearing. It covers the determination of the tear resistance of a flexible plastic sample at very low rates of loading, 51 mm/min. ASTM D 1922 refers to propagation of a tear after being initiated by a small hole in the sample. The value of tear strength in a film depends on the orientation stretching ratio and whether the measurement is performed along or across the machine direction.

11.2.3.8 Abrasion resistance. Abrasion is a difficult property to define as well to measure. It is normally accepted that abrasion depends on the polymer's hardness and resilience, frictional forces, load, and actual area of contact. ASTM method D 1044 evaluates the resistance of transparent plastics to one kind of surface abrasion by measuring its effects on the transmission of light. Another test method to evaluate abrasion, ASTM 1242, measures the volume lost by two different types of abrasion machines: loose abrasion and bonded abrasion.

11.2.4 Barrier Properties

Barrier properties of materials indicate their resistance to diffusion and sorption of molecules. A good barrier polymer has low values of both diffusion (D) and solubility (S) coefficients. Since the permeability coefficient P is a derived function of D and S, a high-barrier polymer has low values of P. The diffusion coefficient is a measure of how fast a penetrant will move within the polymer in a particular direction, while the solubility coefficient gives the amount of the penetrant taken (or sorbed) per unit mass of polymer from a contacting phase per unit of sorbate concentration. Both diffusion and solubility can be applied to the reverse process of sorption, that is, the migration of compounds from the polymer to a surrounding media. Several factors influence the effective value of diffusion and solubility coefficients in polymers.

1. Chemical compositions of the polymer and permeant
2. Polymer morphology (diffusion and sorption occur mainly through the amorphous phase and not through crystals)
3. Temperature (as temperature increases, diffusion increases while solubility decreases)
4. Glass transition temperature
5. The presence of plasticizers and fillers

A detailed discussion of barrier properties is presented in Sec. 11.3 of this chapter.

11.2.4.1 Diffusion coefficient. In Fick's law, the diffusion coefficient, D, is a parameter that relates the flux of a penetrant in a medium to its concentration gradient. A diffusion coefficient value is always given for a particular diffusing molecule/polymer pair. For solid polymers, the diffusion coefficient values of a large number low molecular mass substances range from 1×10^{-8} to 1×10^{-14} cm^2/s. The diffusion theory states that diffusion is an activated phenomenon that follows Arrhenius' law. Penetrant concentration and the presence of plasticizers also affect the value of the diffusion coefficient. Methods for the determination of D are discussed in Sec. 11.3.

11.2.4.2 Permeability coefficient. As indicated, the permeability coefficient P combines the effects of the diffusion and solubility coefficients, and the barrier value of polymers is commonly expressed as P. The well known relationship $P = DS$ holds only when D is concentration independent and the S of permeant in the polymer follows Henry's law. Standard methods for measuring the permeability of organic compounds are not yet available for all compounds. ASTM E96 describes a method for measuring the water vapor transmission rate. ASTM D1434 covers a method for the determination of oxygen permeability.

11.2.4.3 Solubility coefficient. The solubility coefficient indicates the sorption capacity of a polymer with respect to a particular sorbate. The simplest solubility coefficient is defined by Henry's law of solubility, which is valid at low concentration values. The solubility of CO_2 in PET at high pressure is described by combining Henry's and Langmuir's laws.

11.2.5 Surface and Adhesion

11.2.5.1 Adhesive bond strength. Adhesive bond strength between an adhesive and a solid substrate is a complex phenomenon. It is controlled (at least in part) by the values of surface tension, solubility parameters, and adhesive viscosity. To obtain good wettability and adhesion between a polymeric substrate and an adhesive, the surface tension of the adhesive must be lower than that of the substrate. Usually, the difference between the two values must be at least 10 dynes/cm. The similarity in solubility parameters between the two phases indicates the similarity of the intermolecular forces between the two phases. For good compatibility, the values of the solubility parameters must be very close. Low viscosity in the adhesive is necessary for good spreadability and wettability of the substrate.[22]

11.2.5.2 Cohesive bond strength. Cohesive bond strength is the force within the adhesive itself when bonding two substrates. The cohesive bond strength depends mainly on the intermolecular forces of the adhesive, molecular mass, and temperature.

11.2.5.3 Blocking. Blocking is the tendency of a polymer film to stick to itself upon physical contact. This effect is controlled by the adhesion characteristic of the polymer. Blocking is enhanced by surface smoothness and by pressure on the films present in stacked sheets or compacted rolls. Blocking can be measured by the perpendicular force needed to separate two sheets, and it can be minimized by incorporating additives such as talc in the polymer film. ASTM D 1893, D 3354, and Packaging Institute Procedure T 3629 present methods to evaluate blocking.

11.2.5.4 Friction. The coefficient of friction (COF) is a measure of the friction forces between two surfaces. It characterizes a film's frictional behavior. The COF of a surface is determined by the surface adhesivity (surface tension and crystallinity), additives (slip, pigment and antiblock agents), and surface finish. Cases in which the material's COF values require careful consideration include film passing over free-running rolls; bag form-

ing; the wrapping of film around a product; and the stacking of bags, pallets, and other containers. In addition to the intrinsic variables affecting a material's COF, environmental factors such as machine speed, temperature, electrostatic buildup, and humidity also have considerable influence on its final value. The static COF is associated with the force needed to start moving an object. It is usually higher than the kinetic COF, which is the force needed to sustain movement. Determination of static COF is described in TAPPI standard T503 and ASTM D1894. Thompson has studied the effect of additives on the COF values of polypropylene.[23]

11.2.5.5 Heat sealing. An important property for wrapping, bag making, or sealing a flexible structure is the heat sealability of the material. At a given thickness, heat-sealing characteristics of flexible webs are determined by the material's composition (which controls strength), average molecular mass (controlling temperature and strength), molecular mass distribution (setting temperature range and molecular entanglement), and the thermal conductivity (controlling dwell time).[24] Tests normally conducted to evaluate the heat sealability of a polymeric material are the cold peel strength (ASTM F 88), and the hot tack strength.[25] Hot tack is the melt strength of a heat seal without mechanical support when the seal interface is still liquid. The hot adhesivity is associated with the molecular entanglement of the polymer chains, viscosity, and intermolecular forces of the material.

11.2.5.6 Surface tension. In both solids and liquids, the forces associated with inside molecules are balanced, because each molecule is surrounded by like molecules. On the other hand, molecules at the surface are not completely surrounded by the same type of molecules, therefore generating unbalanced forces. At the surface, these molecules show additional free surface energy. The intensity of the free energy is proportional to the intermolecular forces of the material. The free surface energy of liquids and solids is called *surface tension*. It can be expressed in mJ/m^2 or dyne/cm. Values of surface tension in polymers range from 20 dyne/cm for Teflon to 46 dyne/cm for nylon 6,6. The experimental determination of the surface tension by contact angle measurement is covered by the ASTM D2578. Several independent methods are available to estimate the surface tension of liquids and solid polymers including the parachor concept.[15]

When two condensed phases are in close contact, the free energy at the interface is called *interfacial energy*. Interfacial energy and surface energy in polymeric materials control adhesion, wetting, printing, surface treatment, and fogging.

11.2.5.7 Wettability. Adhesion and printing operations on a plastic surface depend on the value of the substrate wettability or surface tension. A measure of the wettability of a surface is given by a material's surface tension as described in the ASTM D 2578.

11.2.6 Optical Appearance

Among the most important optical properties of polymers are absorption, reflection, scattering, and refraction. Absorption of light takes place at the molecular level when the electromagnetic energy is absorbed by group of atoms. If visible light is absorbed, a color will appear; however, most polymers show no specific absorption with visible light and are therefore colorless. Reflection is the light that is remitted on the surface. It depends on the refractive indices of air and the polymer. Scattering of light is caused by the polymer surface physical inhomogeneities (like surface roughness), which reflect the light in all direc-

tions. Refraction is the change in direction of light due to the difference between the polymer and air refraction indices. Transparency, opacity, and gloss of a polymer are not directly related to the chemical structure or molecular mass but are mainly determined by the polymer morphology. Optical appearance properties are of two types: optical morphological properties, which correlate with transparency and opacity, and optical surface properties, which produce the specular reflectance and attenuated reflectance.[15,26]

11.2.6.1 Gloss. Gloss is the percentage of incident light that is reflected at an angle equal to the angle of the incident rays (usually 45 percent). It is a measure of the ability of a surface to reflect the incident light. If the specular reflectance is near zero, the surface is said to be matte. A surface with high reflectance has a high gloss which produces a sharp image of any light source and gives a pleasing sparkle. Surface roughness, irregularities, and scratches all decrease gloss. Test method ASTM D 2457 describes the determination of gloss.

11.2.6.2 Haze. Haze is the percentage of transmitted light that, in passing through the sample, deviates by more than 2.5 percent from an incident parallel beam. The appearance of haze is caused by light being scattered by surface imperfections and nonhomogeneity. The measurement of haze is described in the ASTM D 1003.

11.2.6.3 Transparency and opacity. Transmittance is the percent of incident light that passes through the sample. It is determined by the intensity of the absorption and scattering effects. The absorption in polymers is insignificant, so, if the scattering is zero, the sample will be transparent. An opaque material has low transmittance and therefore large scattering power. The scattering power of a polymer results from morphological inhomogeneities and/or the presence of crystals. An amorphous homogeneous polymer such as crystal polystyrene will have little or no scattering power and therefore will be transparent. A highly crystalline polymer such as HDPE will be mostly opaque. Transmittance can be determined according to standard the ASTM D 1003. A transparent material has a transmittance value above 90 percent.

11.2.7 Electrical Properties

Electrical conductivity, dielectric constant, dissipation factor, and triboelectric behavior are electrical properties of polymers subject to low electric field strength. Materials can be classified as a function of their conductivity (κ) in $(\Omega/\text{cm})^{-1}$ as follows:

- Conductors, 0 to 10^{-5}
- Dissipatives, 10^{-5} to 10^{-12}
- Insulators, 10^{-12} or lower

Plastics are considered nonconductive materials (if the newly developed conducting materials are not included). The relative dielectric constant of insulating materials (ε) is the ratio of the capacities of the parallel plate condenser with and without the material between the plates. A correlation between the dielectric constant and the solubility parameter (σ) is given by the relationship $\sigma \approx 7.0\varepsilon$.[15] There is also a relation between resistivity R (inverse of conductivity) and dielectric constant at 298 K: $\log R = 23 - 2\varepsilon$.[15] Values of ε for polymers are presented in Refs. 15 and 17.

When two polymers are rubbed against each other, one becomes positively charged and the other negatively charged. Whether a polymer becomes positive or negative depends on the electron donor-acceptor characteristic of the polymer. A triboelectric series is a listing of polymers according to their charge intensity. The polymers are ordered from more negatively charged (electron acceptors), then neutral polymers, and finally more positively charged polymers (electron donors). The charge of polymer films also takes place by friction during industrial operations such as form-fill-seal. A brief list of the triboelectric series is presented in Table 11.4, where polymers are listed as a function of polymer's dielectric constant. Hydrophilic polymers absorb water and became more conductive as the dielectric constant increases. Standard methods for measuring the triboelectric charge of films and foams are, at the time of this writing, under consideration by the ASTM Committee D10.[27]

TABLE 8.4 Brief Triboelectric Series

Polymer	Dielectric constant
Negative:	
Polypropylene	2.2
Polyethylene	2.3
Polystyrene	2.6
Neutral:	
Polyvinylchloride	2.8
PVDC	2.9
Polyacrylonitrile	3.1
Positive:	
Cellulose	3.7
Nylon 6,6	4.0

Source: Van Kevelen, D. W., *Properties of Polymers,* 3d ed., Elsevier, New York, 1990.

11.3 Mass Transfer in Polymeric Packaging Systems

11.3.1 Packaging Interactions

An ideal packaging material, in addition to containing and enclosing a product, provides an inert separating phase between the product and the environment. With an inert material, there should be no molecular exchange such as oxygen, CO_2, water, ions, product ingredients, and packaging material components between the product and the package. In other words, a truly inert package material will show no interaction with the packaged product. Such a packaging material may not be available, at least at ambient temperature, or may not be of reasonable cost. Although inorganic glasses approach this ideal barrier concept, helium and hydrogen diffuse through glass. In addition to that, inorganic glasses can leach out components like sodium into an aqueous solution. On the other hand, metals are good barrier materials, but metallic ions tend to dissolve and/or react with both the product and the environment through corrosion.

Due to their chemical nature, polymeric materials have the tendency to dissolve molecules of gases, vapors, and other low-molecular-weight substances to a higher degree than inorganic glasses. Dissolved penetrant molecules diffuse through the polymer via an acti-

vated process of random walks promoted by Brownian motion of the polymer chains. The diffusion process is driven by thermodynamic forces tending to equilibrate the penetrant's chemical potential or, more practically, concentration. A molecule diffuses between adjacent phases, because its tendency is to equilibrate its chemical potential in both phases. So, when a polymeric phase is in direct contact with a solid, liquid, or gas phase, diffusing molecules can move between the phases, depending on the values of their respective chemical concentrations. In this respect, plastic materials, then, are not inert packaging materials. When plastics are in direct contact with food product, as in a package, it is certain that molecular exchanges between the product and package will occur. We refer to this phenomenon as *interaction* involving the plastic packaging, product, and environment. The degree of interaction depends on the type of plastic material as well as variables like type of product, compound concentration, and temperature. For a particular molecule, both the diffusivity and solubility in the polymer will depend on the composition of the polymer and the contacting phase (Fig. 11.10). As indicated in Sec. 11.2.4, both the diffusion coefficient and the solubility coefficient are two fundamental parameters controlling the molecular interactions in packaging systems.

11.3.2 Types of Interactions

The study of mass transfer of low-molecular-weight compounds has been of interest to scientists and engineers since 1829, when Thomas Graham performed the first diffusion experiments with natural membranes. In 1855 Adolf Fick derived the fundamental equations of diffusion. Mass-transfer processes (along with physical and mechanical properties) provide the rational for packaging design and modeling the quality and shelf life of packaged products. Molecular interactions in packaging systems start from the moment the package contacts the product during filling and sealing and extend throughout the entire package's shelf life. For easy reference, these interactions can be referred as permeation, migration, and sorption.

11.3.2.1 Permeation. Permeation is the diffusional molecular exchange of gases, vapors, or liquids (permeants) across a plastic packaging material, with the exclusion of per-

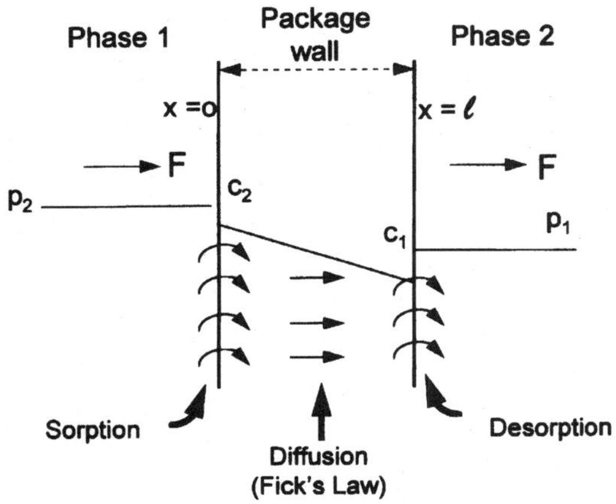

Figure 11.10 Interaction mechanisms in a plastic package.

forations and cracks. A permeation process may significantly impact a product's shelf life, since the product may gain or lose components and/or develop unwanted chemical reactions as a consequence of it. The loss of water or CO_2 from liquid products, the uptake of moisture by dry products, and the oxidation of oxygen-sensitive products affect the product's composition and therefore affect the shelf life. Other consequences of permeation include the transfer of airborne contaminants or volatile components from secondary packages (such as corrugated boxes) or from surrounding packages in a supermarket into the product. These permeants may produce off-flavors in certain flavor-sensitive foods.

While, in many cases, the permeation process may have a negative impact in product quality, there are instances in which a permeation process is not only beneficial but also necessary. Such is the case with modified atmospheric packaging where selected rates of O_2 and CO_2 need to permeate through the package.[28]

11.3.2.2 Migration. Migration is the transfer of substances originally present in the plastic material into a packaged food. These particular substances are called *migrants*. Examples of migrants include residual monomers, solvents, remaining catalysts, and polymer additives. Migration may affect the product's sensorial quality as well as its toxicological characteristics, since it incorporates undesirable components from the package into the food. The presence of potential migrant substances in the packaging material is the subject of legal control by the Food and Drug Administration, (FDA). Vinyl chloride in PVC, styrene in PS, and acrylonitrile in PAN are well known examples of residual monomer migrants. These cases were of object of intense investigations in the 1970s. Resins are commercially produced with residual monomer contents below those allowed by the FDA.

However, even if virgin resins contain very low levels of monomers and additives, migrating substances that may potentially impart off-flavors to products can be developed during converting operations of packaging materials. Excessive temperature-time processing conditions of resins may oxidize and degrade the resin. This may occur during extrusion coating, blown film production, reprocessing of resin in a regrind layer in coextrusion, injection molding, blow molding, or even heat seal operations. Adhesives, inks, pigments, printing solvents, and printing pretreatment ingredients are potential sources of packaging contamination during lamination operations. Residual solvents like toluene, hexane, or pentanol from packages and coupons can produce unwanted odor and taste in food products. The presence of recycled and regrind resins is also a potential source of migrants in the finished package if not sufficiently cleaned. Other sources of migrant substances from plastic containers may arise when contaminated equipment and pipes are used to processing the resin. Similarly, impure air used to blow films can result in a contaminated package.

The sensorial threshold limit may vary widely, depending on the substance and product. It is not unusual that amounts as little as a few parts per billion are sufficient to cause a product to be rejected. The extent of a migration process depends on the initial concentration of the migrant substance in the plastic container and the migrant partition coefficient between the plastic and the food.

11.3.2.3 Sorption. Sorption, sometimes referred as *flavor scalping*, is the uptake of food components [such as flavor, aroma, or color compounds (called sorbates)] by the package material. Many sorption studies of food components have been reported.[29–32,36] Similarity in chemical structures between the sorbate and polymers, as indicated by close values of the solubility parameter,[15] enhance sorption. As the molecular weight distribution (MWD) of the polymer increases, sorption also increases. Metallocenes, or in-site

polyolefins, which are characterized by narrow MWD, are intended to reduce the sorption of volatile compounds as compared with LDPE. Sorption studies of benzaldehyde, citral, and ethyl butyrate showed that sorption levels by ionomers are similar to the sorption of LDPE.[33] As in the case of migration, the extent to which the sorption process occurs depends on the initial concentration of the sorbate in the food and the migrant partition coefficient between the plastic and the food. The dynamic of the diffusion process for permeation, migration, and sorption can be calculated from the corresponding solutions to Fick's laws of diffusion.

11.3.3 Diffusion and Sorption Equations

The fundamental driving force that prompts a molecule to diffuse within the polymer or between a polymer and a surrounding phase is, according to the solution theory, the tendency to equilibrate the specie's activity in both phases. In packaging, polymers are generally contacted by gas (air), liquid phases, and irregular solids. In the case of multilayer structures, where several layers of polymers are in direct contact, a layer is contacted by at least one solid phase. Therefore, any mobile molecular specie that is not in thermodynamic equilibrium within the phases will tend to equilibrate its activity value, as according to Fig. 11.10. As indicated, in a packaging system, mass-transfer sorption and migration processes involves two adjacent phases: the polymer phase and a surrounding liquid or gas. The mobile substance must diffuse in each phase and move across the interface. In a permeation process, the permeant moves across the two film or sheet interphases of the package's wall. The maximum concentration of a substance taken up by the polymer, from a contacting gas or liquid phase, is the solubility. The solubility is controlled by the equilibrium thermodynamics of the system. From solution theory the activity of a specie, a, can be represented by the activity coefficient ξ and concentration c.

$$a = \xi c \tag{11.1}$$

In most packaging situations ξ is approximately 1, and concentration replaces activity.

Fick's laws of diffusion quantitatively describe permeation, migration, and sorption processes in packaging systems.[34] In anisotropic phases, the diffusion theory states that the rate of transfer of a diffusing substance per unit of area, F is given by Fick's first law.

$$F = -D\frac{\partial^2 c}{\partial x^2} \tag{11.2}$$

where c is the penetrant concentration in the polymer, x is the direction of the diffusion, and D is the molecular diffusion coefficient. Quantity in F and c are both expressed in the same quantity unit, e.g., grams or cubic centimeters of gas at standard temperature and pressure. The fundamental equation for unsteady-state, one-dimensional diffusion in an isotropic phase is Fick's second law.

$$\frac{\partial c}{\partial t} = D\frac{\partial^2 c}{\partial x^2} \tag{11.3}$$

where t is time. In systems where the diffusant concentration is relatively low, the diffusion coefficient in Eqs. 11.2 and 11.3 is assumed to be independent of both penetrant concentration and polymer relaxation. Diffusion processes in packaging generally involve low values of diffusant concentration. Also, the diffusion is considered to be unidimensional

and perpendicular to the flat surface of the package, with a negligible amount diffusing through the edges.

The solubility of penetrants in polymers (especially polymers above their glass transition temperature and penetrants at low pressure) is, in many cases, well described by the linear isotherm Henry's law of solubility.

$$c = k\,p \tag{11.4}$$

where c is the concentration of the sorbate in the polymer, p is the partial pressure of penetrant, and k is the Henry's law constant, which is the solubility coefficient and commonly represented by S. It is common to observe that the solubility coefficient S is constant with pressure for gases such as O_2 and CO_2 up to one atmosphere.

For glassy polymers and high-pressure penetrants, as in the cases of CO_2 in PET, a nonlinear Langmuir-Henry's law model is followed.

$$c = kp + \frac{C'_H\,bp}{1 + bp} \tag{11.5}$$

where C'_H is the Langmuir capacity constant, and b is the Langmuir affinity constant.[35]

In closed systems, and when there is an equilibrium in the penetrant activity in the two phases, the partition coefficient K is defined as

$$K = \frac{c_f^*}{c_p^*} \tag{11.6}$$

where c_f^* and c_p^* are the sorbate or migrant equilibrium concentration in the packaged product and polymer, respectively.

For organic vapors, Henry's law is valid, in most cases only at very low penetrant pressure or concentration in air, in the order of milligrams of penetrant per liter of air (ppm). However, the applicability range of Henry's law strongly depends on the particular organic penetrant/polymer system under consideration. Flory-Huggins equation applies to high solvent activity in polymers.[36]

11.3.4 Diffusion across a Flat Single Sheet

11.3.4.1 Steady-state diffusion: permeability. Consider the case of a flat sheet of thickness ℓ that is contacted from both sides with a penetrant at different concentration values as indicated in Fig. 11.10. At the surface $x = 0$, the penetrant concentration $c = c_2$; at $x = 2$, $c = c_1$. Applying these conditions to Eq. 11.2, the penetrant flow rate F across any section of the sheet is given by

$$F = -D\frac{dc}{dx} = D\frac{c_1 - c_2}{\ell} \tag{11.7}$$

In permeability experiments, however, the partial pressure in the gas phase surrounding the sheet is easier to measure than the penetrant concentration c in the polymer. Applying Henry's law, and then replacing DS by P, Eq. 11.7 can be written now as

$$F = P\frac{p_2 - p_1}{\ell} \tag{11.8}$$

where P is the permeability coefficient. Notice that Eq. 11.8 is actually the definition of the permeability coefficient. A material having a low P value for a particular permeant is considered to be a good barrier; i.e., only a small quantity of permeant will be transferred through it. Conversely, a high permeability value indicates a low barrier material. The diffusant flow rate F is given by

$$F = \frac{q}{At} \tag{11.9}$$

where q is the quantity of permeant flowing across the sheet in time t, and A is the sheet area exposed to the permeant. The quantity q can be expressed in mass, volume, or moles.

By combining Eqs. 11.5 and 11.6 we obtain

$$P = \frac{q\ell}{At\Delta p} \tag{11.10}$$

where Δp is $p_2 - p_1$. Equation 11.10 is a simple but very useful equation that can be used to estimate that main the design parameters of plastic packages at steady state.

The water vapor transmission rate, WVTR, relates to P by

$$F = \left(\frac{q}{At}\right)\frac{1}{\Delta p} = \text{WVTR}\frac{\ell}{\Delta p} \tag{11.11}$$

Example The WVTR of a film 25 microns thick measured according ASTM (100°F and 90 percent RH) is 0.1 g/day m^2. Calculate P.

Solution The saturation vapor pressure of water at 100°F is 49.7 mmHg, since

$$p_2 = 0, p = p_1 - p_2 = 49.7 \times 0.9 = 44.73 \text{ mmHg}$$

$$P = \text{WVTR } (\ell/\Delta p) = 0.1 \times 2.5/44.73 = 0.0056 \text{ g·microns/m}^2\text{·d·mmHg}$$

In SI units, P = 0.412 × 10^{-18} kg·m/m^2·s·Pa = 0.412 attosecond (as)

It is common in the industry to report the barrier values as water vapor transmission rate or gas transmission rates rather than as permeability coefficient. This practice, although apparently practical, is in effect confusing and misleading, since it does not allow easy comparison of the barrier characteristics of polymers measured at different conditions or in different countries. This is even more confusing when different units are used. The use of permeability coefficient should be encouraged as common practice in the industry.

As indicated above, the permeability coefficient P combines two more fundamental parameters: the diffusion and solubility coefficients. When the polymer diffusion coefficient D is independent of the permeant concentration, and the penetrant dissolves in the polymer according to Henry's law of solubility, Eqs. 11.8 and 11.7 yield the well known relationship

$$P = DS \tag{11.12}$$

The permeability of a polymer, therefore, depends on the diffusion and solubility coefficients, which can be related by a simple expression like Eq. 812 or alternatively by a

more complex one. For instance, the solubility of CO_2 in PET follows the Langmuir-Henry's law model (Eq. 11.5), and P is given by[37]

$$P = kD_D + \frac{C'_H b D_H}{1 + bp} \tag{11.13}$$

where D_D and D_H are the diffusion coefficients for the Henry's and Langmuir's laws' populations, respectively.

Figure 11.10 illustrates the mechanisms of the permeation process, which involves three steps.

1. The permeant originally in phase 1 dissolves in the polymer interphase polymer/phase 1.
2. The permeant diffuses within the polymer film layer from the side of high concentration toward the low-concentration side.
3. The permeant diffuses out from the opposite polymer interphase.

These steps are always present in any system, regardless of whether D and S follow Fick's and Henry's laws, respectively A good barrier material has a low value of combined diffusion coefficient and solubility coefficient values for a particular penetrant. Preferably, both D and S should be low. For instance, polyethylene is a excellent barrier to water, because water has a very low solubility and diffusion coefficient values.

The plastics industry offers a large number of polymeric structures to cover a wide range of barrier characteristics to satisfy the diverse need within the packaging industry. There are high barrier materials available to protect a product from oxygen, water, or organic vapors, as well as structures with high permeability values for oxygen or carbon dioxide that are needed, for example, in modified atmosphere packaging for produce.

11.3.4.2 Variables affecting permeability

Structure-permeability relationships. The chemical structure of the polymer's constitutional unit ultimately determines the polymer barrier behavior. Chemical composition, polarity, stiffness of the polymer chain, bulkiness of side and backbone-chain groups, and degree of crystallinity significantly impact the sorption and diffusion of penetrants. Salame[38] has developed a comprehensive semi-empirical correlation of polymer structure and gas permeability based on the permachor parameter, provided the permeant does not swell the polymer. The permachor concept is based on the molecular forces holding the polymer together, cohesive energy density, and free-volume fraction of the polymer.

The permeability P is related to permachor π by

$$P = Ae^{-s\pi} \tag{11.14}$$

where A and s are constants based on the permeant gas. Table 11.5 presents values of A, s, and π for selected gases and polymers. From Eq. 11.14 and Table 11.5, the permeability ratio for CO_2 and O_2 for any polymer at 25°C is calculated as

$$\frac{P_{CO_2}}{P_{O_2}} = \frac{Ae^{-s\pi}\big|_{CO_2}}{Ae^{-s\pi}\big|_{O_2}} = \frac{Ae^{-s}\big|_{CO_2}}{Ae^{-s}\big|_{C_2}} \approx 6.2 \tag{11.15}$$

which is very close to experimental values. Figure 11.11 shows a plot of gas permeability and permachor values. Unfortunately, the permachor concept has not been yet extended to organic permeants.

TABLE 11.5 Selected Values of *A* and *s* for Permachor Equation at 25°C

Gas	*A* cc-m/m²d atm	*s*
O_2	3.48	0.112
N_2	1.18	0.121
CO_2	21.69	0.122

These values are good for both glassy and rubber materials only at 25°C.
Source: Salame, M., Prediction of Gas Barrier Properties of High Polymers, *Polymer Eng. Sci.*, 26:22, 1543, 1986.

Temperature. Considering that the diffusion, sorption, and permeability are activated processes, *D, S* and *P* are related to the temperature by an Arrhenius-type equation

$$\Gamma = \Gamma_o e^{-\frac{E_\Gamma}{RT}} \tag{11.16}$$

where Γ represents either *D, S,* or *P;* Γ_o is the respective pre-exponential term, E_Γ is the respective activation energy (sorption enthalpy for *S*), *R* is the gas constant, and *T* is temperature in Kelvins. Equation 11.16 is valid over a relatively small range of temperatures, which should not include the polymer's glass transition temperature. In particular, for *P*, polymers show values of *Ep* larger at temperatures above T_g than below *T.* Equation 11.16 can be used to estimate the permeability coefficient at a desired temperature from a known value.

Example Calculate the permeability coefficient of oxygen through PET at 50°C (P_2) considering that the permeability coefficient at 25°C (P_1) is 220 cc·mil/m²·d·atm.

Solution From Eq. 11.16, P_2 is related to P_1 by the following equation:

$$P_2 = P_1 e^{\frac{Ep}{R}\left(\frac{1}{T_1} - \frac{1}{T_2}\right)} \tag{11.17}$$

which allows calculation the permeability value at a temperature T_2 if the value of *P* is given at T_1 and *Ep* is known.

$$Ep = 43.8 \text{ KJ/mol} = 10{,}500 \text{ cal/g mol}$$[17]

$$\exp[(10{,}500/1.987)\ (298^{-1} - 333^{-1}) = 3.94$$

$$P_2 = P_1 \times 3.94 = 220 \times 3.94 = 887 \text{ cc·mil/m}^2\text{·d·atm}$$

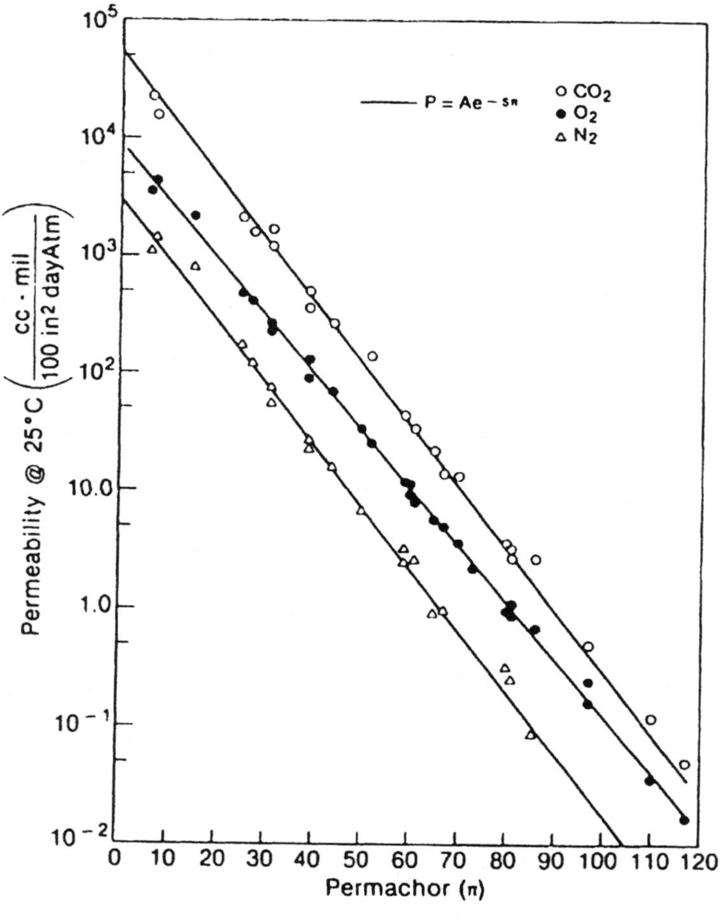

Figure 11.11 Correlation of gas permeability and permachor. *(Reprinted with permission from M. Salame,* Polymer Eng. Sci., *26:22, 1543, 1986)*

Humidity. Hydrophilic polymers such as polyamides and EVOH absorb water from humid air. For example, water sorption isotherms for Nylon have been determined by Hernandez and Gavara.[39] The presence of water in a hydrophilic polymer affects the permeability of, oxygen, carbon dixide, organic vapor, and flavor and aroma compounds. An increase in moisture content increases oxygen permeability in EVOH (Fig. 11.6). Nylon 6, while for amorphous nylon the permeability tends to decrease (Fig. 11.12).

Morphology. In the development of the diffusion sorption, and permeation Eqs. 11.2 through 11.13, it is assumed that the polymer phase is a homogeneous and isotropic phase; that is, an isotropic amorphous polymer. The presence of a crystalline microphase complicates this assumption considerably and makes the diffusion process in semicrystalline polymers a complex phenomenon. Semicrystalline polymers consist of a microcrystalline phase dispersed in an amorphous phase. The dispersed crystalline phase decreases the

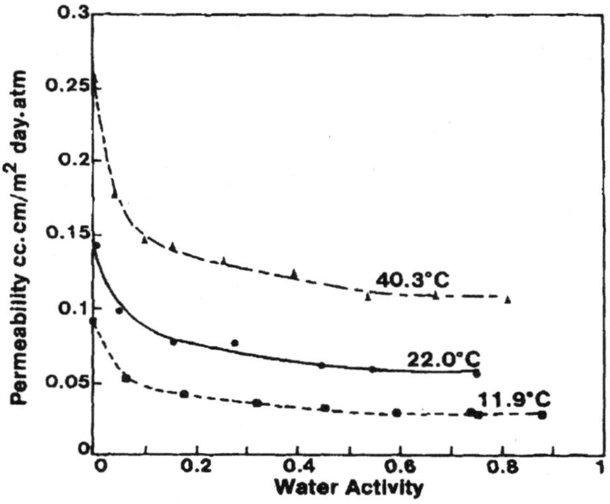

Figure 11.12 Oxygen permeability in nylon 6I/6T. *(Reprinted with permission from R. J. Hernandez,* J. Food Eng., *22, 495, 1994)*

sorption of penetrants whenever the crystal conformations produce regions of higher density than the amorphous polymer. A closer atomic packing tends to exclude relatively large molecules such as CO_2 or O_2. For this reason it is generally accepted that gases and vapors are normally sorbed, and therefore able to diffuse, only in the polymer's rubbery or amorphous phase. The dispersed microcrystals are impermeable to penetrant diffusion and create a more tortuous path for the diffusing molecule. Additionally, the microcrystalline phase also acts as a tridimensional cross linking agent increasing the nonisotropism of the polymer. The combined decrease in sorption and diffusion contributes then to a lower permeability.

The solubility S of a semicrystalline polymer S, having a crystalline volume fraction of α and a 100 percent amorphous phase solubility S_a, is experimentally given by

$$S = (1 - \alpha)S_a \tag{11.18}$$

One of the earliest evidences of Eq. 11.18 was reported by Van Amerongen.[40] The work of Michaels[41] and coworkers on PET and PE supported the model presented in Eq. 11.18. Puleo et al.[42] reported the gas sorption and permeation in a semicrystalline polymer for which the crystal phase has a lower density than the amorphous phase. At 100 percent crystallinity, the sorption of CO_2 and CH_4 was about 25 to 30 percent of the solubility of the amorphous phase. The diffusion coefficient in the amorphous phase, D_a, has been shown to be related to the diffusion coefficient in the semicrystalline phase D by

$$D_a = D\beta\tau \tag{11.19}$$

where β is a "chain immobilization factor," and τ is a "geometric impedance factor." Both β and $\tau > 1$. Michaels et al. applied Eq. 10.19 to semicrystalline PET.[43] Application of Eq. 11.19 does not work well, especially for annealed polymers.[44]

Other factors affecting permeation. Chain orientation, permeant concentration, plasticizers, and fillers affect permeability. Chain orientation in general decreases the permeability to gases (Table 11.2). Permeant concentration of gases below one atmosphere of pressure in general does not effect the permeability coefficient. However, strong effects has been observed in the permeability of organic compounds. For instance, the permeability of organic vapors such as aromas, flavors, and solvents in general are strongly depend on concentration.[45] The addition of plasticizers usually, but not always, increases the permeability. Film thickness *per se* does not affect permeability, diffusion, or solubility of a penetrant, provided the polymer morphology is not affected. However, polymer film produced in different thickness may have different morphology due to different cooling conditions during processing. Molecular weight of a polymer has been found to have little effect on the permeability of polymer except in the very low range of molecular weight. Inorganic mineral fillers such as talc, $CaCO_3$, or TiO_2, used as much as 40 percent, affect the permeability of a film. When coupling agents, such as titanates, are used to improve the interfacial bond between polymer and filler, the permeability to gases and vapors decreases. The absence of bonding agents may increase permeability (Table 11.5).

11.3.5 Diffusion across a Composite Structure

Many packaging materials and containers are produced by the combination of more than one plastic material. Common configurations include multilayer structures, two-phase polymer blends, and plastics with fillers. While multilayer structures offer the lowest permeability for a given composition, the performance of two phase-blends depends on which polymer provides the continuous phase. The effectiveness of a filler to enhance or diminish the barrier value of a polymer depends on the adhesivity and wetting interphase characteristics of the polymer-filler pair.

11.3.5.1 Multilayer configuration. Multilayer structures of two or more polymers are commonly produced by coextrusion, adhesion lamination, coating, extrusion coating, or their combination. In a lamination, the contribution of the tie layer to the barrier of the structure is often neglected, although in some cases it may have some effect.

Plane configuration. Consider a plane multilayer structure whose layers are perpendicular to the permeation flow. This can be referred to as a structure in series.

Figure 11.13 illustrates the case of a three-layer structure during a steady-state permeation process. Applying Fick's first law to each layer and considering that the flow rate F is the same through each layer, the following can be concluded:

$$F = D_1 \frac{c_1^* - c_1}{\ell_1} \tag{11.20}$$

$$F = D_2 \frac{c_2^* - c'_2}{\ell_2} \tag{11.21}$$

$$F = D_3 \frac{c_3^* - c'_3}{\ell_3} \tag{11.22}$$

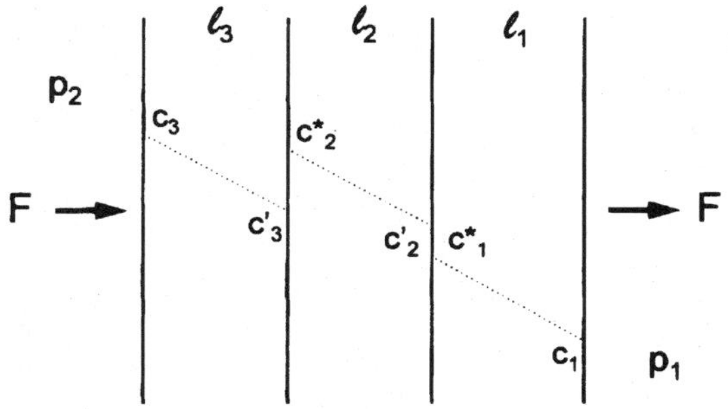

Figure 11.13 Permeability through a plane multilayer structure.

where ℓ_n = thickness of each layer
D_n = diffusion coefficient of the penetrant in each layer
c^* and c' = the penetrant concentrations at the layer's interphases

The partition coefficients between the layers are calculated as

$$K_{12} = \frac{c_2^*}{c'_2} \tag{11.23}$$

and

$$K_{23} = \frac{c_2^*}{c'_3} \tag{11.24}$$

When the concentrations at the interfaces are eliminated from Eqs. 11.20 through 11.24, the following expression is obtained:

$$F_{SS} = \frac{c_3 K_{23} K_{12} - c_1}{\frac{\ell_3}{D_3} K_{23} K_{12} + \frac{\ell_2}{D_2} K_{12} + \frac{\ell_1}{D_1}} \tag{11.25}$$

which gives the steady state flux value F_{SS} as a function of external concentrations c_1 and c_3, partition coefficients, layer thickness, and diffusion coefficients. Equation 11.25 can be extended to n layers[44] as

$$F_{SS} = \frac{c_n K_{(n-1)n} \ldots K_{12} - c_1}{\frac{\ell_n}{D_n} K_{(n-1)n} \ldots K_{12} + \ldots + \frac{\ell_3}{D_3} K_{23} K_{12} + \frac{\ell_2}{D_2} K_{12} + \frac{\ell_1}{D_1}} \tag{11.26}$$

where $K_{(n-1)n}$ = partition coefficient of penetrant between two contacting layers.

Consider two practical cases of a three-layer structure.

Case 1 Layers 1 and 3 are the same polymer, therefore, $D_1 = D_3$, and $K_{12} = 1/K_{23} \equiv K$. Equation 11.25 reduces to

$$F_{SS} = \frac{c_3 - c_1}{\dfrac{\ell_1 + \ell_3}{D_1} + \dfrac{\ell_2}{D_2}K} \tag{11.27}$$

Similarly, in Eqs. 11.20 through 11.22, an apparent diffusion coefficient, D_{eff}, of the whole structure can be defined as

$$F_{SS} = D_{eff}\frac{c_3 - c_1}{\ell_T} \tag{11.28}$$

where ℓ_T is the total thickness of the structure.

From Eqs. 11.27 and 11.28, the D_{eff} is related to D_1 and D_2 and K by

$$D_{eff} = \frac{\ell_T}{\dfrac{\ell_1 + \ell_3}{D_1} + \dfrac{K\ell_2}{D_2}} \tag{11.29}$$

Furthermore, assuming that Henry's law of solubility applies at both external layers interphases in contact with the permeant gas phase, the effective permeability P_{eff} of the multilayer is

$$P_{eff} = SD_{eff} = \frac{S\ell_T}{\dfrac{\ell_1 + \ell_3}{D_1} + \dfrac{\ell_2 K}{D_2}} \tag{11.30}$$

When the partition coefficient K favors the external layer, i.e., the permeant preferentially dissolves much better in the external layers rather than in the middle layer, $K \gg 1$. Additionally, if $K/D_2 \gg 2/D$, then P_{eff} is controlled by the permeant solubility S in the external layer, the diffusion coefficient of the middle layer D_2, and K.

$$P_{eff} \approx \frac{3SD_2}{K} \tag{11.31}$$

For the middle layer to be a high barrier, the permeant diffusion coefficient D_2 must be very small and its solubility must also be low.

Case 2 (a) First consider that the external layers are of equal thickness, $\ell_1 = \ell_3 \equiv \ell$, and $\ell_2 < \ell$. For $K = 1$ (similar preferential sorption of permeant in both polymers) and D_2 being large, Eq. 11.30 simplifies to $P_{eff} \approx SD$. This indicates, as expected, that the middle layer does not contribute at all to the barrier, because the it is similar to the external layers.

(b) If $\ell_2 = \varepsilon\ell$ ($\varepsilon < 1$) and $\varepsilon K \gg 1$, assuming that $D \approx D_2$, then Eq. 11.30 becomes $P_{eff} \approx 2SD_2/\varepsilon K$, and the middle layer is now a high-barrier material. Additionally, if $D_2 < D$, then $P_{eff} \approx (2+\varepsilon)SD_2/\varepsilon k$, which is the best combination for a middle-layer barrier.

When the values of permeability for each layer of a n-layer structure are known instead of the diffusion and partition coefficients, P_{eff}, by the well known relationship, is

$$P_{eff} = \frac{\sum_{i=n} \ell_i}{\sum_{i=n} \frac{\ell_i}{P_i}} \tag{11.32}$$

where P_{eff} is the effective permeability coefficient of the structure, and ℓ_i and P_i are the thickness and permeability coefficient of each layer, respectively.

Cylindrical and spherical multilayer structures. For a cylindrical structure with concentric parallel layers of radius $R_o < R_1 \ldots < R_n$, P_{eff} is[34]

$$\frac{\ln \frac{R_n}{R_o}}{P_{eff}} = \frac{\ln \frac{R_1}{R_o}}{P_1} + \frac{\ln \frac{R_2}{R_1}}{P_2} + \ldots + \frac{\ln \frac{R_n}{R_{n-1}}}{P_n} \tag{11.33}$$

while for a multilayer spherical shell,

$$\frac{\frac{1}{R_o} - \frac{1}{R_n}}{P_{eff}} = \frac{\frac{1}{R_o} - \frac{1}{R_1}}{P_1} + \frac{\frac{1}{R_1} - \frac{1}{R_2}}{P_2} + \ldots + \frac{\frac{1}{R_{n-1}} - \frac{1}{R_n}}{P_n} \tag{11.34}$$

where $\ell_i = R_i - R_{i-1}$, $\Sigma\ell = R_n - R_o$, and $P_i = D_i k_i$. Barrer[46] discussed the case of a laminate in which the diffusion coefficient of each layer is permeant-concentration dependent.

Parallel structures. Although multilayer structures parallel to flow are not common in packaging, the case of two layers, or phases, is illustrated in Fig. 11.14. The effective permeability of two material having P_1 and P_2 permeability coefficients can be estimated by,

$$P_{eff} = \phi_1 P_1 + \phi_2 P_2 \tag{11.35}$$

where ϕ_1 and ϕ_2 are the volume fractions of respective polymers. This is the least efficient way to improve barrier when combining two materials. However, this type of structure can

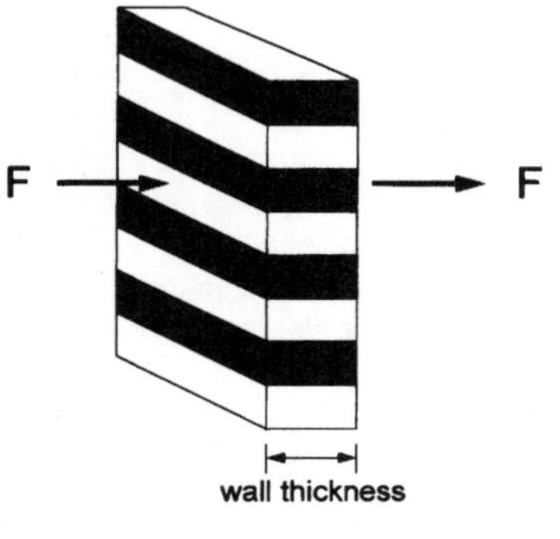

Figure 11.14 Permeability in parallel structures.

be used, in principle, in modified atmospheric packaging applications where high permeability values are often needed.

Immiscible blends. Immiscible polymer blends having good interphase bonding provide a continuous two-phase heterogenous medium to diffusion and sorption. A detailed study of these cases is beyond the scope of this chapter. Let us consider the simple case of a two-polymer immiscible blend in which one phase remains continuous across a composition range, and the other is a discontinuous phase dispersed as spherical particles. The effective permeability is described by the following equation:[47]

$$P_{eff} = \frac{P_c[P_d + 2P_c - 2\phi_d(P_c - P_d)]}{P_d + 2P_c + \phi_d(P_c - P_d)} \tag{11.36}$$

where P_c, P_d, and ϕ_d are the permeability and volume fraction of the continuous and discontinuous phase, respectively. Equation 10.36 is for the continuous phase polymer. In an immiscible two-polymer blend, there is a value of ϕ between 0 and 1 in which a phase inversion occurs. At that point, the continuous phase becomes a discontinuous phase and vice-versa. Figure 11.15 shows the values of P_{eff} calculated by Eq. 11.37 for two polymers having permeability 0.15 and 150 cm^3·mil/100 in^2·day·atm, respectively.

Permeability of polymers with fillers. Permeability of polymers with inorganic fillers can be estimated by Eq. 11.37,[48] provided good adhesion and wettability exist between the polymer and the filler.

$$P_{eff} = P_p\phi_p\left(1 + \frac{L}{2W}\phi_f\right)^{-1} \tag{11.37}$$

where P_p and ϕ_p are the permeability and volume fractions of the unfilled polymer, ϕ_f is the volume fraction of the filler, and L/W is the aspect ratio. Aspect ratio is the average dimension of the dispersed filler particles parallel to the plane of the film, L, divided by the average dimension perpendicular to the film, W. When there is not good adhesion between the filler and polymer, permeability may increase in a less predictable form, by diffusion through interphase microvoids. Equation 11.37 has been applied to correlate the barrier

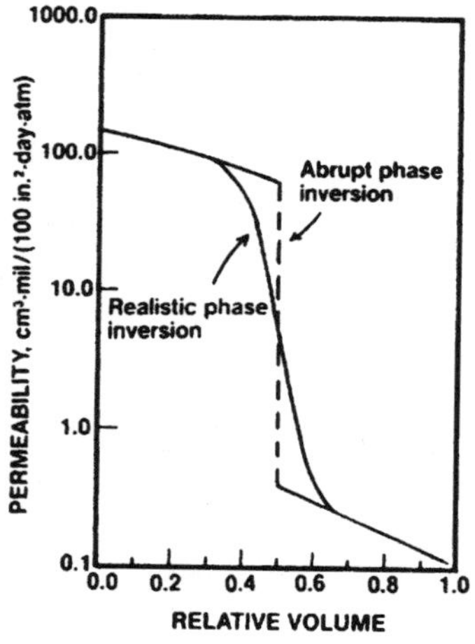

Figure 11.15 Calculated values of permeability in an immiscible blend. *(Reprinted with permission from P. T. DeLassus,* TAPPI J., *3, 126, 1988)*

properties of nanocomposites films. Table 11.6 shows values of oxygen permeability of LDPE filled with calcium carbonate.[49]

TABLE 11.6 Oxygen Permeability of LDPE with Calcium Carbonate

Filler, % by volume	Surface treated?	Oxygen permeability $cc\text{-}m/m^2\cdot d\cdot atm$
0	No	0.189
15	Yes	0.098
25	Yes	0.059
15	No	0.394
25	No	0.787

Source: Steingeser, S., G. Rubb, and M. Salame, *Encyclopedia of Chemical Technology,* vol. 3, 3d ed., 1980, p. 480.

Miscible blends. Miscible blends include polymer blends, random copolymers, and polymer/plasticizer blends.[50] The effective permeability in a miscible blend can be described by the following expression:[51]

$$\ln P_{eff} = \phi_1 \ln P_1 + \phi_2 \ln P_2 \tag{11.38}$$

The permeability of O_2, CO_2, and H_2O in a vinylidene chloride copolymer increased exponentially with increasing plasticizer concentration.[47]

11.3.6 Permeability Measurement

Standard methods for measuring the permeability of oxygen, carbon dioxide gas, and water vapor through flat polymer structures are described in ASTM D3985-81 and E96-80, respectively. The measurement of permeability of organic vapors has been the subject of numerous investigations.[52–56]

Methods for determining permeability are of two types, continuous flow and lag-time or quasi-isostatic. A brief summary of the methods follows. For more details, the reader is referred to the book by Hernandez and Gavara.[57]

11.3.6.1 Continuous flow method. Figure 11.16[58] shows a schematic of a continuous-flow method setup for measuring oxygen permeability in the presence of water vapor. The solution to Fick's second law from a continuous flow permeation experiment with D independent of concentration is given by[59]

$$\frac{F_t}{F_{SS}} = \left(\frac{4}{\sqrt{\pi}}\right)\left(\sqrt{\frac{\ell^2}{4Dt}}\right)\sum_{1,3\ldots}^{\infty} \exp\left(\frac{-n^2\ell_2}{4Dt}\right) \tag{11.39}$$

where F_t is the flow rate of the permeant through the film sample, and F_{ss} is the steady-state flow at large times. A typical curve obtained using the continuous flow method is presented in Fig. 11.17. The experimental permeation flow up to a value of the flow ratio of 0.95 can be virtually described by the first term of Eq. 11.39 as follows:

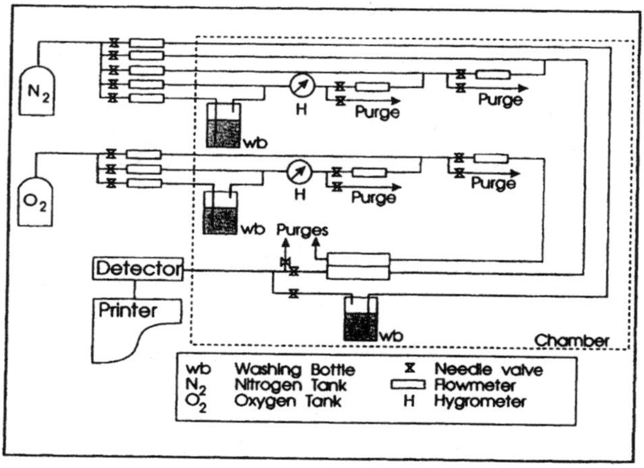

Figure 11.16 Continuous flow permeation method. *(Reprinted with permission from R. Gavara and R. J. Hernandez,* J. Polym. Sci., *Part B, 32, 2375, 1994)*

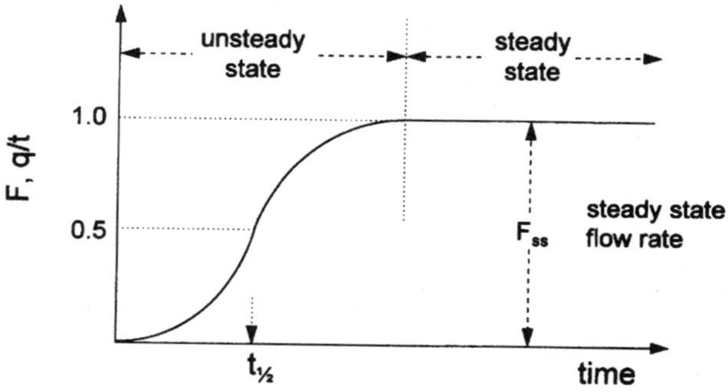

Figure 11.17 Flow profile in a continuous-flow method.

$$\psi = \frac{F_t}{F_{SS}} = \left(\frac{4}{\sqrt{\pi}}\right) X^{\frac{1}{2}} \exp(-X) \qquad (11.40)$$

where $X = \ell^2/4Dt$. The Newton-Rhapson method can be used to evaluate X versus t in Eq. 11.40, and the diffusion coefficient D is obtained from the slope of the linear fitting of X^{-1} versus t for $0.05 < \psi < 0.95$.[54] The permeability coefficient P is calculated from the value of F_{ss}.

$$P = \frac{F_{SS}\ell}{\Delta p} \qquad (11.41)$$

where Δp is the partial pressure across the film during the permeation experiment. For the simple case of constant diffusion coefficient, D can be approximately calculated by

$$D = \frac{\ell^2}{7.20 t_{1/2}} \tag{11.42}$$

where $t_{1/2}$ is the time required to reach $F_t/F_{ss} = 0.5$ from the start of the experiment.

Gavara and Hernandez[60] have presented a simple method to analyze the consistency of experimental Fickian permeability data.

11.3.6.2 Lag-time method. Figure 11.18 shows an apparatus for measuring the permeation and sorption of oxygen in polymers films.[61] The solution of Fick's second law when the initial concentration of permeant in the film is zero is[34]

$$Q(\ell) = \frac{Dc_2}{\ell}\left(t - \frac{\ell^2}{6D}\right) - \frac{2\ell c_2}{\pi^2}\sum_1^{\infty}\frac{(-1)^n}{n^2}\exp\left(\frac{-Dn^2\pi^2 t}{\ell^2}\right) \tag{11.43}$$

where $Q(\ell)$ is the quantity of permeant passing at $x = \ell$ per unit of the membrane area in time t. Profiles of Eq. 11.43 are presented in Fig. 11.19, where the partial pressure of per-

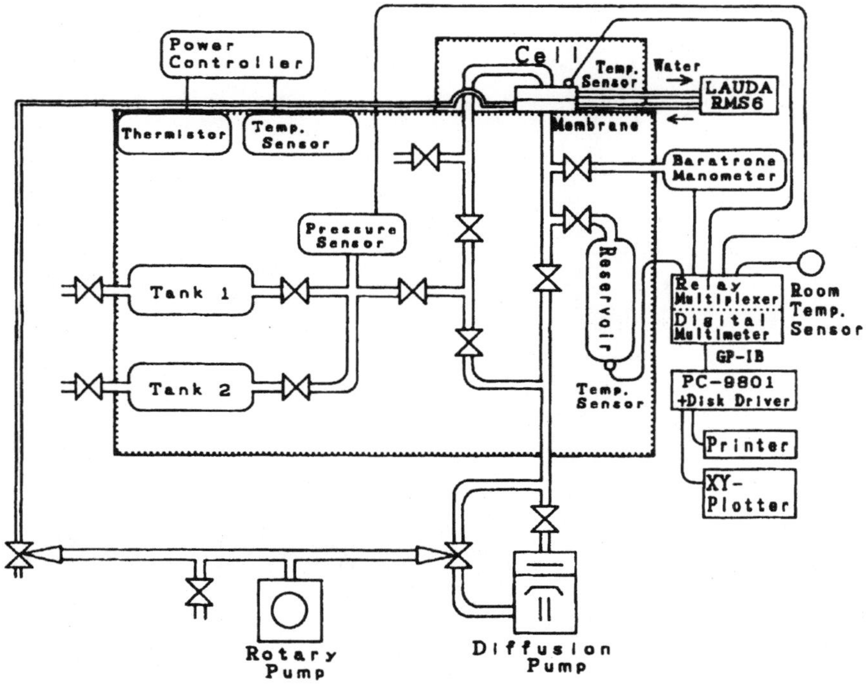

Figure 11.18 Lag-time permeation apparatus. *(Reprinted with permission from K. Toi, H. Suzuki, I. Ikenasto, T. Ito, and T. Kasi,* J. Polym. Sci., *Part B, 33, 777, 1995)*

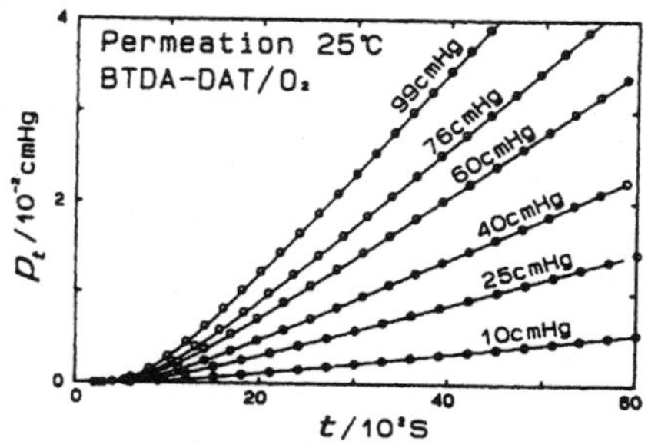

Figure 11.19 Plot of lag-time permeation experiment. *(Reprinted with permission from K. Toi, H. Suzuki, I. Ikenasto, T. Ito, and T. Kasi,* J. Polym. Sci., *Part B, 33, 777, 1995)*

meant, which is proportional to $Q(\ell)$, is plotted versus time t. At the steady state, i.e., at very large t values, Q is given by

$$Q(\ell) = \frac{Dc_2}{\ell}\left(t - \frac{\ell^2}{6D}\right) \tag{11.44}$$

The intercept of the steady state line with t-axis, which occurs at $Q(\ell) = 0$, defines the lag time, θ.

$$\theta = \frac{\ell^2}{6D} \tag{11.45}$$

For a penetrant/polymer system that follows the dual-mode sorption model of Eqs. 11.5 and 11.13, the lag time is given by the following expression:

$$\theta = \frac{\ell^2}{6D}[1 + f(C'_H, k_D, D_H, D_D, b, p)] \tag{11.46}$$

where $f(C'_H, k_D, D_H, D_D, b, p)$ is a nonlinear function of the indicated variables.[62]

11.3.7 Permeant Transport and Packaging Shelf Life

The barrier to the diffusion of small molecules provided by a container is of primary importance when protecting food, cosmetic, and pharmaceutical products.

11.3.7.1 Simplified shelf-life calculation. For many applications, the steady-state permeation equation (Eq. 11.10) is an acceptable approximation to correlate a product's shelf life τ with package area A, wall thickness ℓ, and environmental conditions of pressure and temperature. This approximation is valid for thin films and when the time to reach steady-state flow rate, t_{ss}, is very small compared with the package shelf life. Consider a permeation process through a container wall as illustrated in Fig. 11.20.

Example Estimate the film thickness ℓ for a plastic pouch containing an oxygen-sensitive product, provided that not more than 5 cc of oxygen can allowed to permeate the package during six months. Assume the pouch area $A = 400$ cm^2, and the material permeability $P = 165$ cc micron/m^2·day·atm.

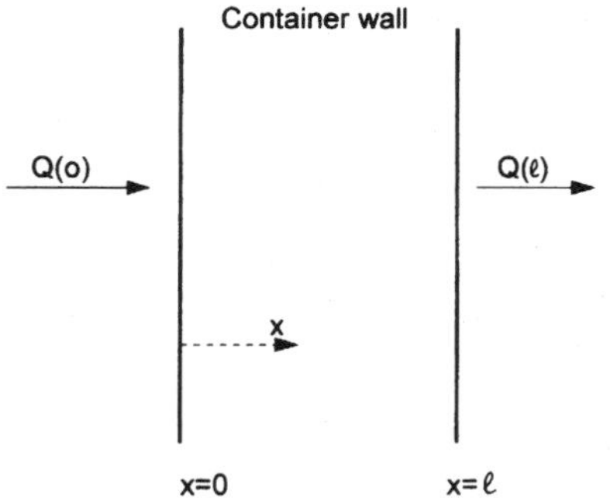

Figure 11.20 Permeation through a container wall of thickness ℓ. At $x = 0$, total quantity entered is $Q(0)$; at $x = \ell$, total quantity permeated is $Q(\ell)$.

Solution Since this is an oxygen-sensitive product, the headspace partial pressure of oxygen is near zero, $p_1 \approx 0$ atm, and Δp across the package can be considered constant during the shelf life, therefore, Eq. 11.10 can be applied.

$$\ell = P\,t\,A\,p_2/q = 165 \times 180 \times 0.04 \times 0.21/5 = 50 \text{ microns}$$

If the diffusion coefficient of the material is 5×10^{-10} cm^2/s, the time to reach the steady-state flow is 3 times the lag time. By Eq. 10.45, $t_{ss} = 6.9$ hr, which is small compared with 180 days. Figure 11.21 shows plots of $t_{ss} = 3\Theta$, calculated for different thickness and diffusion coefficient values.

When p changes during the shelf life as in the case of a moisture sensitive product, Eq. 11.10 can be written as

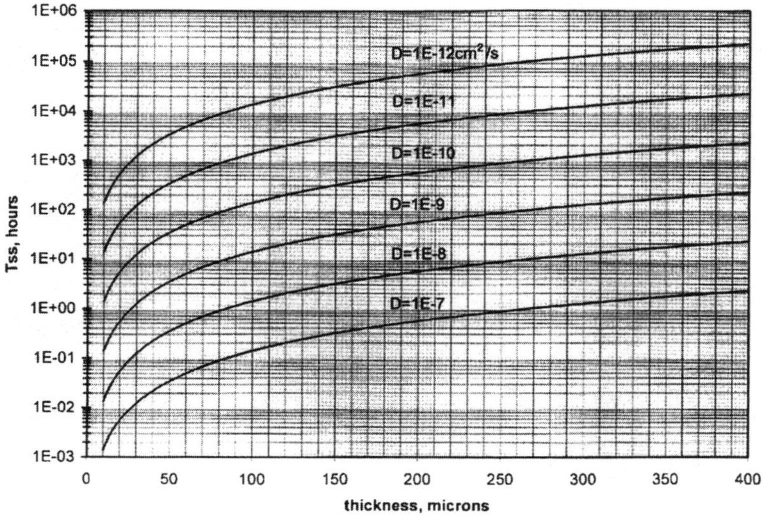

Figure 11.21 Values of t_{ss} calculated for different thicknesses and diffusion coefficient values; $t_{ss} = 3\Theta$ (Eq. 8.45).

$$t = \frac{\ell}{PA}\int \frac{dq}{p_o - p(M)} \tag{11.47}$$

where p_o is the water partial pressure outside the package, and $p(M)$ is the internal partial pressure in equilibrium with the product's moisture content.[63] The finite elements method has been applied by Kim et al.[64] to calculate the uptake of water by solving simultaneously the diffusion equation in both the product and package.

11.3.7.2 Gaining and losing permeant. The unsteady state equation describing the mass transfer in Fig. 11.20 can be written for both sides of the film, i.e., $x = 0$ and $x = \ell$. Both resulting equations have practical importance. A packaged product that loses a component by permeation through the package's wall will start losing it by sorption at the internal side of the package, that is at $x = 0$. If the product uptakes permeant from the environment, the permeant will be incorporated into the product only after the permeant reaches $x = 2$ and has totally penetrated the packaging wall. The distinction of these two cases is especially relevant when the permeant's diffusion coefficient is very small and/or the packaging wall is relatively thick. To evaluate the quantity of permeant at $x = 0$ and $x = \ell$, Fick's second law needs to be solved subject to the following conditions: the initial permeant concentration in the plane sheet is zero, and for $t > 0$, one side, $x = 0$, will be at c_2; the other, $x = 2$, will be at $c_1 = 0$. The quantity permeated at $x = \ell$, $Q(\ell)$ is given by Eq. 11.43, while the quantity entering the plane at $x = 0$, $Q(o)$ is[34]

$$Q(o) = \frac{c_2}{3} + Dc_2\ell t - \frac{2d_2}{\pi^2}\sum_1^{\infty}\frac{(-1)^n}{n^2}\exp\left(\frac{-Dn^2\pi^2 t}{\ell^2}\right) \tag{11.48}$$

Figure 11.22 shows a plot of Eqs. 11.43 and 11.48. After a long time, the steady state is reached and both lines, $Q(o)$ and $Q(\ell)$, become parallel. At steady state, the difference between $Q(o)$ and $Q(\ell)$ is the quantity of permeant sorbed by the polymer; that is, $c_2/2$. The intersection of the asymptotic steady state of $Q(\ell)$ with Dt/ℓ^2 axis is $D\Theta/\ell^2 = 1/6$, where Θ is the lag-time of Eq. 11.45.

The asymptotic lines on both curves indicate the pseudo-steady state solution given by Eq. 11.10 and have the same slope, $F = Q/t$, at both sides of the container wall. As indicated in Fig. 11.22, the pseudo-steady state solution agrees within less than 1 percent of Eq. 11.37 at $t > 2.64\Theta$, or at a value of $Dt/\ell^2 > 0.44$. However, it is generally recommended to consider steady state flow at $Dt/\ell^2 > 0.50$, which is equivalent to 3 times the lag time Θ or 3.6 times $t_{1/2}$ in Eq. 11.42.

$Dt/\ell^2 = 0.5$ can be used to estimate the time needed for a permeant to reach the steady-state permeation rate from the moment the product containing the permeant is put in a package and immediately sealed.

Example If $D = 1 \times 10^{-10}$ cm^2/s, and the packaging thickness $\ell = 25$ microns (1 mil), $t = \ell^2/2D = (25 \times 10^{-4})^2/2 \times 10^{-10} = 31{,}125$ s = 8.68 hr. It will take 8.68 hr to reach the steady-state permeation rate, t_{ss}.

The criterion $Dt/\ell^2 = 0.5$ to reach steady-state permeation is also useful when measuring the permeability coefficient: steady-state data will be reached after an experiment time $t = \ell^2/2D$. Other values of t_{ss} as a function of D and thickness ℓ are presented in Fig. 11.21.

A plot of $Q(o)$ of Eq. 11.48 as a function of thickness is presented in Fig. 11.23 for t in seconds and D in cm^2/s. From this figure, it can be seen that the value of $Q(o)$ at a given time t depends on both the thickness and the diffusion coefficient D. For a given D, the

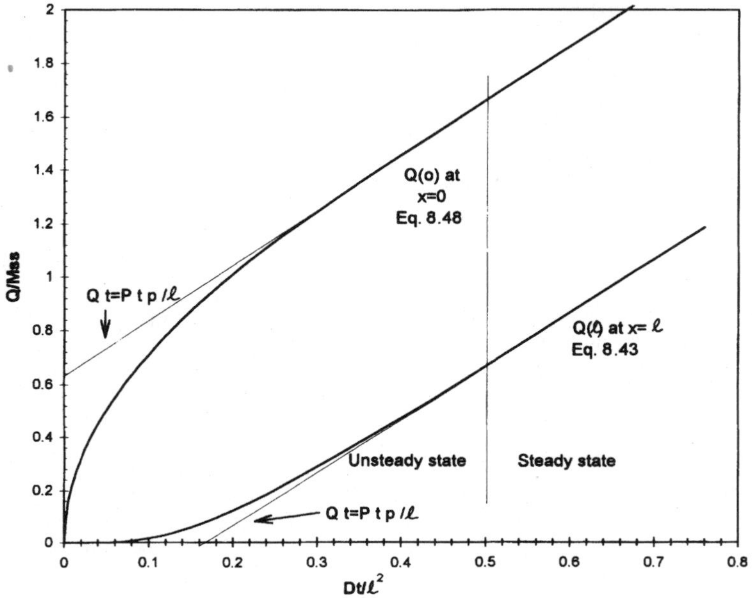

Figure 11.22 Quantity permeated at $x = 0$ and $x = \ell$. *(Adapted from P. Masi and D. Paul,* J. Memb. Sci., *12, 137, 1982)*

loss of a permeant decreases with thickness to a value, ℓ_{min}, beyond which there is no real increase of barrier power. This value is[65]

$$\ell_{min} = (3Dt)^{1/3} \tag{11.49}$$

Another important feature of Fig. 11.23 is that it shows that the pseudo-steady state Eq. 11.10 wrongly indicates that $Q(o)$ continues to decreasing with thickness, even below the minimum value ℓ_{min}. This means that Eq. 11.10 is only valid for $\ell < \ell_{min}$.

11.3.7.3 Loss of pressurized CO_2 through PET. As indicated in Eq. 11.13, the transport of pressurized gases in glassy polymers such as PET and PAN exhibit the dual sorption mode behavior. Masi and Paul[66] carried out simulations based on the dual mobility sorption model of Eq. 11.13 as well as the simple sorption model of Henry's law. Comparison of these two models is shown in Fig. 11.24. As seen, Henry's law model overestimates the loss calculated as $Q(o)$. However, for $Q(\ell)$ the difference between the model is not very great. Also, Henry's law model predicts a larger amount of sorbed CO_2 than the dual mobility model. The total amount of permeant sorbed by the polymer at a given time is $Q(o) - Q(\ell)$.

11.3.8 Migration and Sorption in a Semi-infinite Polymer plane

Consider Fig.11.25, representing packages in which a fluid product of known composition is in direct contact with a plastic material. From the moment the product is put in contact

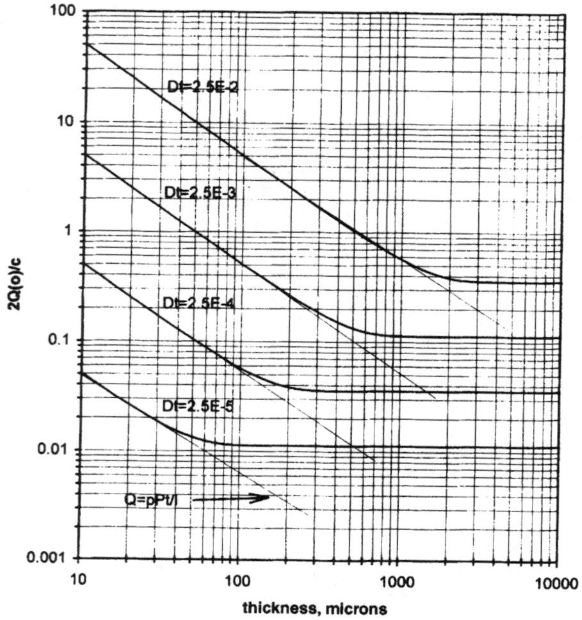

Figure 11.23 Plot of quantity permeated at $Q(o)$ as a function of diffusion coefficient and container thickness. *(Adapted from P. Masi and D. Paul,* J. Memb. Sci., *12, 137, 1982)*

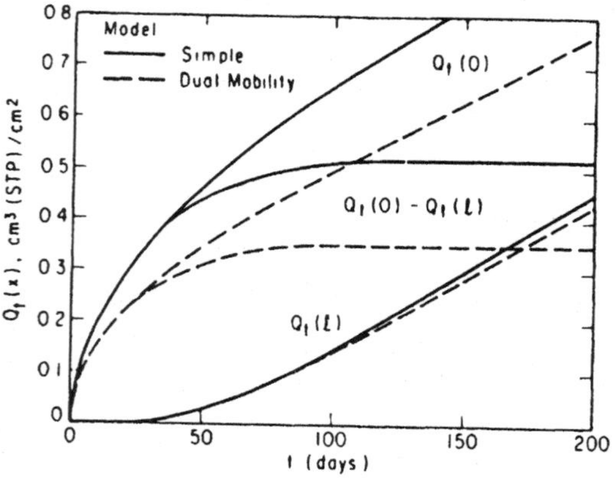

Figure 11.24 Comparison of quantity permeated at $x = 0$ and $x = \ell$ of simple model (Eq. 8.10) and dual mobilized model (Eq. 8.13). *(Reprinted with permission from P. Masi and D. Paul,* J. Memb. Sci., *12, 137, 1982)*

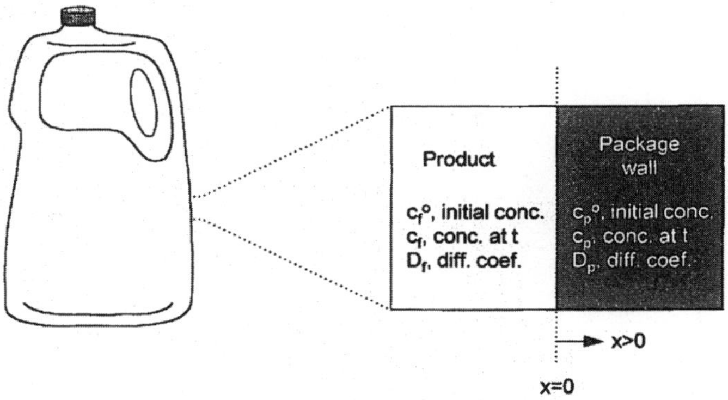

(a)

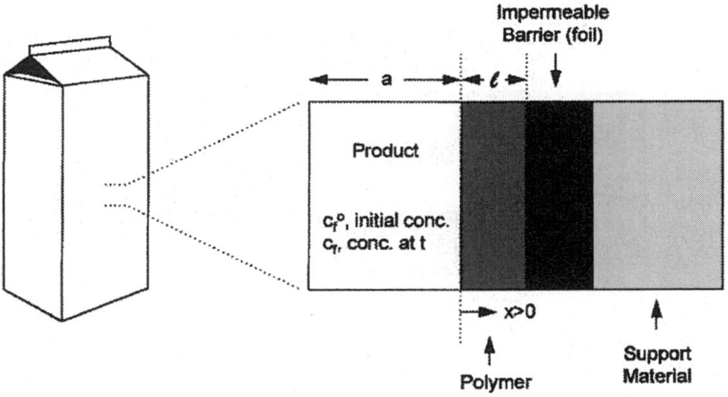

(b)

Figure 11.25 (a) Single-wall plastic package and (b) package with an impermeable foil layer.

with the package, molecules from the product will be sorbed by the polymer and, conversely, low-molecular-weight components from the plastic will migrate into the product. Selected sorption and migration cases are reviewed now.

Since the transfer processes are controlled by diffusion, the following assumptions can be made to simplify the analysis:

1. The time-dependent diffusion process is described by Fick's second law.
2. The migration and sorption are limited to a single component of known concentration that is initially homogeneously distributed in the polymer or in the product, respectively. In addition, the diffusion of one adjuvant is not affected by the presence of other adjuvants.
3. For a given plastic material, the diffusion coefficient of the adjuvant is only a function of temperature.

4. There is no swelling of the polymer by any sorbed component, or, if swelling does occur, the above assumptions still hold.
5. A flat sheet is considered of infinite thickness if less than 40 percent of any migrant originally contained in the sheet is extracted by the contacting media.

D_p is the adjuvant diffusion coefficient in the polymer, and D_f is the adjuvant diffusion coefficient in the contacting phase. In solving Fick's second law, the initial conditions are

$$c_p = c_p^o \text{ at } t = 0, \text{ for } x > 0; \qquad c_f = c_f^o \text{ at } t = 0, \text{ for } x < 0 \tag{11.50}$$

and the boundary condition is

$$F = D_p\left(\frac{\partial c_p}{\partial x}\right) \text{ at } x = 0 \tag{11.51}$$

which gives the rate of adjuvant transferred at the polymer/product interphase.

11.3.8.1 Transient migration

Less than 40 percent of the original migrant is transferred to a contacting well mixed liquid. Consider a single polymer layer with an initial migrant concentration c_p^o, at $t = 0$ for $x > 0$, as in Fig. 11.25*a*. The migrant substance in the polymer can be residual polymerization reactants, processing additives, converting operation residuals, or contaminants in a single layer of post-consumer recycled resin. The contacting product with the polymer is considered to be very large as compared with the potential amount of migrant to be transferred from the plastic. The initial concentration of migrant in the product is 0; that is, $c_f^o = 0$ for $x < 0$ at any time. This case can be simulated by a diffusional process in a semi-infinite/infinite domain[34] with a mass transfer coefficient of infinite value, i.e., a well agitated liquid. The solution to Fick's law gives the amount of migrant transferred from the polymer to the liquid.[34]

$$M = 2c_p^o\left(\frac{D_p t}{\pi}\right) \tag{11.52}$$

where M is the total quantity of migrant lost from the polymer per unit area of containing phase at time t, and D_p is the diffusion coefficient of migrant in the polymer. From Eq. 11.52, it follows that a plot of M/c_p^o versus $(Dt)^{1/2}$ is a straight line. This is illustrated in Fig. 11.26, which presents experimental data on the migration of styrene monomer from a polystyrene sheet in contact with oil at 40°C.[67,68]

Equation 11.52 can be rewritten to give the quantity of migrant transfer per unit of mass of product $W_f = M/\zeta$ as a function of the dimensionless group $D_p t/\ell^2$.

$$\frac{\zeta W_f}{c_p^o} = \frac{2}{\pi^{1/2}}\sqrt{\frac{D_p t}{\ell^2}} \tag{11.53}$$

where ζ is the ratio of the mass of contacting phase/package area. For food products, the U.S. Food and Drug Administration uses $\zeta = 1.55$ g/cm^2. A plot of Eq. 11.53 is shown in Fig. 11.27. Equations 11.52 and 11.53 show that the diffusion coefficient of migrant in the polymer is a key parameter.

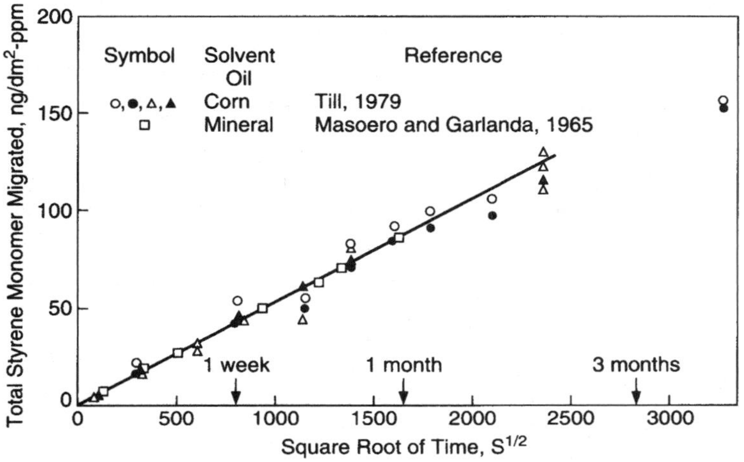

Figure 11.26 Migration of styrene monomer into solvent oil at 40°C. *(Reprinted with permission from R. C. Reid et al.,* Ind. Eng. Chem. Prod. Res. Dev., *19 (4), 580, 1980)*

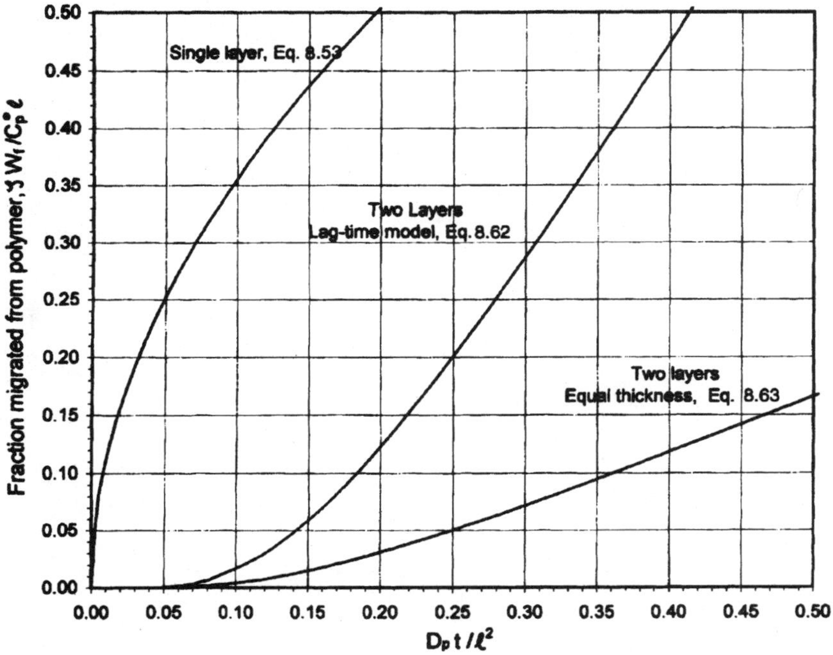

Figure 11.27 Mass fraction migrated from a polymer film or sheet.

Example Estimate the concentration of a migrant in a foodstuff that is in direct contact with a monolayer of recycled polymeric package when $D_p t/\ell^2 = 0.1$, $c_p^o = 23\ \mu g/cm^3$ and thickness $\ell = 50\ \mu m$.

Solution

$$\zeta W_f / c_p^o\ \ell = 2 \times (D_p t/\ell^2)^{1/2}/\pi^{1/2} = 0.3568$$

$W_f = 0.3568 \times c_p^o \times \ell/\zeta = 0.3568 \times 23 \times 0.005/1.55 = 0.0265\ \mu g/kg = 26.5$ ppb. This corresponds to 36 percent of the original amount of migrant in the polymer, for $\zeta = 1.55$ g/cm^2. Therefore, the assumption of semi-infinite polymer is acceptable. Furthermore, if $D_p = 1 \times 10^{-12}$ cm^2/s, calculate how long will it take to transfer the calculated 36 percent of migrant: $t = 0.1 \times \ell^2/D_p = 0.1 \times 0.005^2/1 \times 10^{-12} = 2.6 \times 10^6$ s = 28.9 days.

Equation 11.53 represents the worse possible scenario to estimate migration from the plastic. Since the contacting phase is well mixed, the difference in concentration between the polymer and liquid is the highest possible. Also, Eq. 11.53 does not include a partition coefficient value that would slow the migration process. Therefore, this case gives the shorter possible time for migration, or, for a given time, the larger amount of migrant transferred from the plastic. Calculated values from this case can be used to estimate an upper bound of migrant leaving the polymer in the transient state. The effect of temperature in Eq. 11.52 was studied by Markelov and co-workers for the migration of acrylonitrile in commercial copolymers.[69]

Semi-infinite polymer contacted by a well mixed finite amount of liquid phase of volume V_f. As illustrated in Fig. 11.25*b*, the amount of contacting liquid is given by $V_f = Aa$. Since the liquid is considered to be in agitation (or regularly agitated), there is not any concentration gradient of the migrant into the liquid and the mass transfer coefficient would be very high. To solve this case, the following additional boundary condition applies.

$$a\frac{dc_f}{dt} = D_p\frac{\partial c_p}{\partial x} \text{ at } x = 0 \tag{11.54}$$

which indicates that the flux of migrant is transferred entirely into the contacting phase with no gradient in concentration in the latter. The quantity migrated by unit of area M in time t given in dimensionless form is[67]

$$Y = \frac{M}{aKc_p^o} = [1 - \exp(z^2)erfc\ z] \tag{11.55}$$

where $z = (D_p t)^{1/2}/aK$, $erfc\ z$ is the error function of z, and Y is the ratio of M to the maximum quantity of migrant transferred at equilibrium, aKc_p^o. $Y = 1$ at equilibrium, otherwise $0 < Y < 1$. K is the partition coefficient of the migrant molecule between the product and polymer as defined in Eq. 11.6. K is considered to be independent of migrant concentration, which is a realistic assumption, since concentration of migrant molecules in packaging systems are commonly very low, excepting some plasticizers. A plot of Eq. 11.55 is presented in Fig. 11.28, and values of *erfc z* are given in Table 11.7. For $z < 0.05$ (large liquid volume, small migration time t, or very low D), Eq. 11.54 reduces to Eq. 11.52. The assumption of a infinitely thick polymer is fully acceptable if the quantity of migrant diffused into the liquid is less than 40 percent. Therefore, for $Y < 0.40$, the polymer thickness ℓ is not a significant variable.[67]

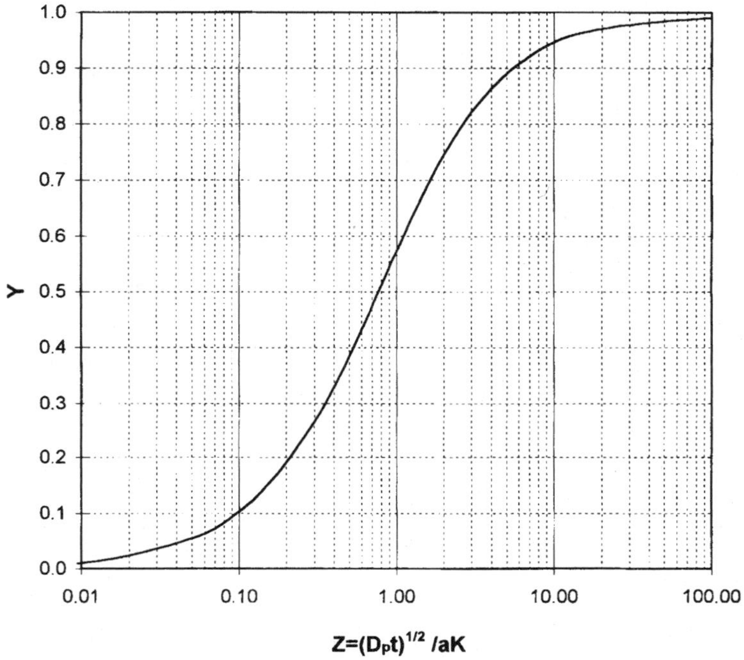

Figure 11.28 Plot of Y versus z in Eq. 8.55.

TABLE 11.7 Error Function

z	$erf\,c\,z$	z	$erf\,c\,z$	z	$erf\,c\,z$
0	1.0	0.7	0.322199	1.8	0.010909
0.05	0.943628	0.75	0.288844	1.9	0.007210
0.1	0.887537	0.8	0.257899	2.0	0.004678
1.15	0.832004	0.85	0.229322	2.1	0.002979
0.2	0.777297	0.9	0.203092	2.2	0.001863
0.25	0.723674	0.95	0.179109	2.3	0.001143
0.3	0.671373	1.0	0.157299	2.4	0.000689
0.35	0.620618	1.1	0.119795	2.5	0.000407
0.4	0.571608	1.2	0.089686	2.6	0.000236
0.45	0.524518	1.3	0.065992	2.7	0.000134
0.5	0.479500	1.4	0.047715	2.8	0.000075
0.55	0.436677	1.5	0.033895	2.9	0.000041
0.6	0.396144	1.6	0.023652	3.0	0.000022
0.65	0.357971	1.7	0.016210		

Example Consider a similar system as in the previous example, but now the contacting phase has a limited volume Aa, with $a = 2.5$ cm. Estimate the time to transfer 25 percent of the total migrant that would potentially migrate at equilibrium (infinite time) if $K = 1 \times 10^{-2}$ and $D_p = 1 \times 10^{-10}$ cm^2/s.

Solution From Eq. 11.55 or Fig. 11.28, at $Y = 0.50$, $z = 0.23$, $t = (z \cdot a \cdot K)^2/D_p = 3.3 \times 10^5$ s = 3.8 days.

Semi-infinite polymer contacted by a semi-infinite immobile phase. In this case, the product is still, and mixing is insignificant or does not exist at all. Since there is no bulk transport due to the absence of mixing, the mass transfer takes place only by molecular diffusion. This can be the case with a viscous liquid such as honey contained in a plastic jar. In addition to the initial conditions of Eq. 11.50 and 11.51, the following boundary condition describes the system:

$$D_p \frac{\partial c_p}{\partial z} = D_f \frac{\partial c_f}{\partial z} \text{ at } x = 0 \tag{11.56}$$

which indicates that every migrant that is diffused out from the polymer is diffused in the product; this is equivalent to saying that the migrant diffusion coefficient D_f in the immobile product cannot be neglected. When the amount of product is very large compared with the amount of migrant, the concentration of migrant can be considered near zero in the product, or $c_f = 0$ at any time, and the solution is[67]

$$M = \frac{2c_p^o}{1+\beta}\left(\frac{D_f t}{\pi}\right)^{1/2} \tag{11.57}$$

where $\beta = K(D_f/D_p)^{1/2}$. If $\beta > 100$, Eq. 11.57 reduces to Eq. 11.52 with an error of less than 1 percent. Large β values might result from large K (the migrant is very soluble in the contacting phase), high values of D_f, or very small D_p. However, if K values are very small (which indicated low solubility of the migrant in of the liquid phase as compared to the polymer), β may be also very small. Equation 11.57 then reduces to

$$M = 2c_p^o K\left(\frac{D_f t}{\pi}\right) \tag{11.58}$$

which indicates that the diffusion coefficient of the diffusing molecule in the polymer, D_p, no longer affects the mass transfer process, and the diffusion process is controlled only by D_f.

Semi-infinite polymer layer contacted by a limited volume of immobile phase. The additional boundary condition for this case is $D_f\, \partial c_f/\partial x = 0$ at $x = a$; that is, no flow at $x = a$. The corresponding solution is[67]

$$Y = \frac{2\beta z}{\pi^{1/2}(1+\beta)}[1 + F(\beta, z)] \tag{11.59}$$

where $F(\beta, z)$ is the following function of β and z:

$$f(\beta, z) = \frac{2}{1+\beta}\sum_{n=1}^{\infty}\left(\frac{\beta-1}{\beta+1}\right)^{n-1}\exp\left(\frac{-n^2}{\beta^2 z^2}\right)\left[1 - E\left(\frac{n}{\beta_z}\right)\right] \tag{11.60}$$

where $E(n/\beta z) = \pi^{1/2}\,(n/\beta z)\exp(n/\beta z)^2\, erf\, c(n/\beta z)$.

Equation 11.59 is plotted in Fig. 11.29. For values of $\beta > 5$, the mixing does not have any effect in the mass transfer, and Eq. 11.59 reduces to Eq. 11.55. The linear portion of Fig. 11.29 is given by

$$Y = \frac{1\beta z}{\pi^{1/2}(1+\beta)} \tag{11.61}$$

Example If $z = 1$ and $\beta = 0.2$, $Y = 0.203$; then, 20.3 percent of the migrant has passed to the contacting phase.

Contaminant transferred from a coextruded recycled polymer and a virgin resin layer. The migration of contaminants from a package made of a single-layered recycled polymer is described by Eq. 11.52, for low migrant concentration levels. Consider now the migration from a layer of recycled polymer coextruded with a layer of virgin resin of the same polymer as illustrated in Fig. 11.30, where r and ℓ are the thickness of the recycled and virgin layers, respectively. The thickness r can be expressed as a function of ℓ: $r = \gamma\ell$, where $\gamma > 0$. The migrating substances must diffuse out from the recycled layer and through the virgin resin layer to reach the food contacting phase. It is further assumed that there is no direct diffusion from the recycled layer into the opposite side. The following two solutions have been proposed:

1. *Lag-time approach.* Here, the virgin polymer layer is considered to be a membrane having a constant contaminant concentration c_p^o at the recycled polymer interphase and zero contacting phase. The corresponding solution is Eq. 11.43, the lag-time equation for permeation,[70] which can be rewritten in dimensionless form as

$$\frac{\zeta W_f}{c_p^o \ell} = \frac{Dt}{\ell^2} - \frac{1}{6} - \frac{2}{\pi^2}\sum_1^{\infty} \frac{(-1)^n}{n^2} \exp\left(\frac{-Dn^2\pi^2 t}{\ell^2}\right) \tag{11.62}$$

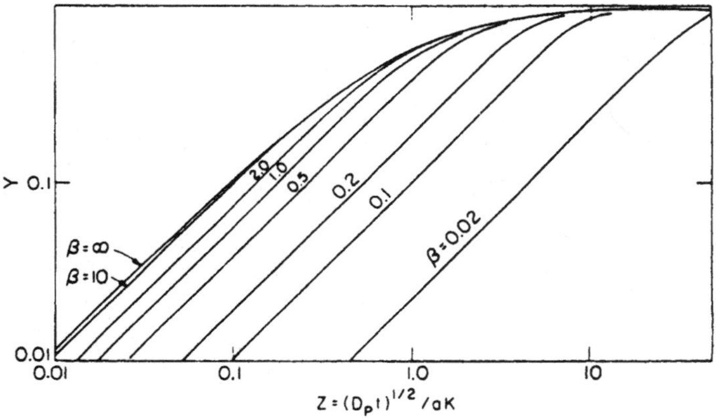

Figure 11.29 Migration from a semi-infinite polymer contacted by a limited volume of immobile phase, Eq. 8.59. *(Reprinted with permission from R. C. Reid et al.,* Ind. Eng. Chem. Prod. Res. Dev., *19, 4, 580, 1980)*

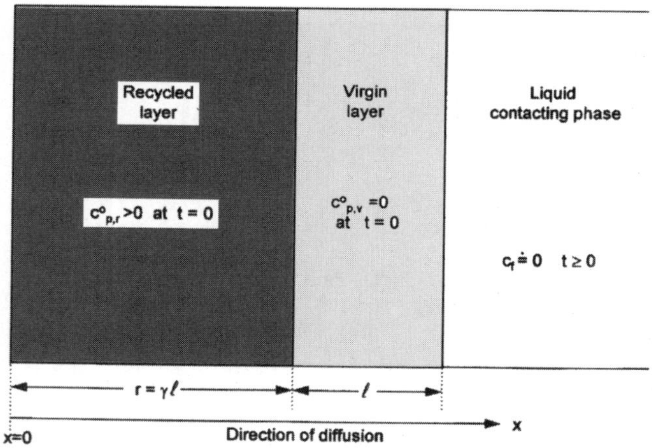

Figure 11.30 Migration from a layer of recycled resin and coextruded with virgin layer of the same resin.

As expected, this equation grossly overestimates the amount of migrant transferred into the foodstuff for $D_p t/\ell^2 > 0.052$, because it assumes that at the recycled/virgin polymer interphase c_p^o always remains constant. For comparison, Eq. 11.62 is plotted along with Eq. 11.53 in Fig. 11.27.

2. *Continued diffusion approach.* This approach considers that the contaminant continuously diffuses through the recycled and virgin resin layers with the corresponding decrease in concentration at the interphase. Figure. 11.31 illustrates the concentration profile within the coextruded structure for $r = \ell$. The migrant diffuses out only at the virgin resin/product interphase, and the contacting phase is well mixed, which implies a mass transfer coefficient of infinite value. The corresponding solution is[71]

$$\frac{\zeta W_f}{c_r^o \ell} = 1 - \frac{8}{\pi^2}\frac{\gamma+1}{\gamma}\sum_0^{\infty}\frac{(-1)^n}{(n+1)^2}\sin\left[\frac{(2n+1)\pi\gamma}{2(\gamma+1)}\right]\exp\left[\frac{(2n+1)^2\pi^2}{4(\gamma+1)^2}\frac{D_p t}{\ell^2}\right] \qquad (11.63)$$

where c_r^o is the initial concentration of the contaminant in the recycled layer of thickness $r = \gamma\ell$. Equation 11.63 is plotted in Fig. 11.27 for $\gamma = 1$; that is, the virgin and recycled layer have both the same thickness ℓ. Equation 11.63 gives the transferred fraction of migrant for a given value of $D_p t/\ell^2$.

Example Estimate the maximum migrant concentration in a mixed foodstuff in direct contact with a coextruded structure made of a recycled and virgin resin of equal thickness $r = \ell = 25$ µm, during a 30-day period. The diffusion coefficient $D_p = 1 \times 10^{-12}$ cm²/s, and $c_r^o = 23$ µg/cm³.

Solution At $D_p t/\ell^2 = 1 \times 10^{-12} \times 30 \times 24 \times 3600/(0.0025)^2 = 0.415$. From Eq. 11.63 or Fig. 11.27 for the line corresponding to Equation 11.63, $zW_f/c_r^o\,\ell = 0.125$; therefore, $W_f = 0.125 \times 23 \times 0.0025/1.55 = 0.0047$ µg/kg = 4.7 ppb. This corresponds to 8 percent of the original amount of migrant in the polymer.

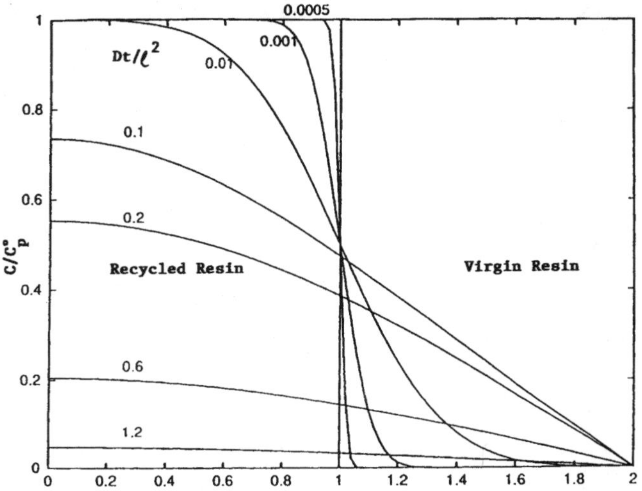

Figure 11.31 Concentration profile of a migrant within a coextruded recycled/virgin structure in contact with a well agitated liquid. *(Reprinted with permission from S. Laoubi et al.,* Pack. Techn. and Sci., *8, 249, 1995)*

Equation 11.63 still overestimates the amount of migrant in the contacting phase, since it is assumed that all migrant diffuses only in the direction of contacting phase, which is perfectly mixed.

11.3.8.2 Transient sorption

Sorption at one side of the film. Foodstuffs packaged in polymeric containers such as the one depicted in Fig. 11.25*b* lose aroma through sorption. The penetration of the sorbate takes place on only one side of the polymer and on the inside of the package. The polymer layer is of thickness ℓ, and its initial concentration $c = c_p^o$. Assuming that the sorbate concentration in the contacting phase remains essentially constant during the sorption process (that is, the amount sorbed is very small compared with the total amount originally in the food), the solution giving the amount of total of sorbed compound after a very large time (to infinity) is[34]

$$\frac{M}{M_{ss}} = 1 - \frac{8}{\pi^2}\sum_{n=1}^{\infty}\frac{1}{(2n+1)^2}\exp\left[-(2n+1)^2\pi^2\frac{D_p t}{4\ell^2}\right] \tag{11.64}$$

The corresponding solution for shorter times is

$$\frac{M}{M_{ss}} = 2\left(\frac{D_p t}{\ell^2}\right)^{1/2}\left[\pi^{-1/2} + 2\sum_{n=1}^{\infty}(-1)^n ierfc\frac{n\ell}{(D_p t)^{1/2}}\right] \tag{11.65}$$

where *iercf* is the inverse error function.

Sorption from both sides. This occurs when the polymer film of thickness ℓ is immersed in a fluid phase (liquid or gas) of constant sorbate concentration. For long times, the amount sorbed is given by

$$\frac{M}{M_{ss}} = 1 - \frac{8}{\pi^2}\sum_{n=0}^{\infty}\frac{1}{(2n+1)^2}\exp\left[-(2n+1)^2\pi\frac{D_p t}{4\ell^2}\right] \tag{11.66}$$

Equation 11.66 is commonly used to determine the diffusion coefficient of a sorbate in a polymer film. The film is exposed on both sides to a flow of sorbate carried in a gas stream, and the increase in mass is continuously monitored.[39,72] The corresponding solution for short times of Eq. 11.66 is

$$\frac{M}{M_{ss}} = 4\left(\frac{D_p t}{\ell^2}\right)^{1/2}\left[\pi^{-1/2} + 2\sum_{n=0}^{\infty}(-1)^n ierfc\frac{n\ell}{2(D_p t)^{1/2}}\right] \tag{11.67}$$

Reduced sorption equations. Because the first term in the above series is dominant, for large times, Eq. 11.66 can be reduced to

$$\frac{M}{M_{ss}} = 1 - \frac{8}{\pi^2}\exp\left(\frac{-\pi^2 D_p t}{\ell^2}\right) \tag{11.68}$$

with a relative error of 0.1 percent, $M/M_{ss} \geq 0.55$.

Similarly, for short times, Eq. 11.67 can be reduced to

$$\frac{M}{M_{ss}} = \frac{4}{\pi^{1/2}}\left(\frac{D_p t}{\ell^2}\right) \tag{11.69}$$

with an error of 1 percent for $0.5 < M/M_{ss} \leq 0.63$, and 0.1 percent for $M/M_{ss} \leq 0.50$.

For $M/M_{ss} = 0.5$, Equation 11.68 can be further simplified to the well known expression

$$D_p = 0.049\frac{\ell^2}{t_{0.5}} \tag{11.70}$$

where $t_{0.5}$ is the time required to reach $M/M_{ss} = 0.5$.

Equations 11.64 through 11.70 also describe a desorption process and the diffusion coefficient can be obtained from a desorption experiment.

Sorption from a stirred solution of limited volume. When a plane sheet or film of thickness 2ℓ is immersed in a limited volume of a solution of width $2a$ per unit area containing the sorbate, the sorbate concentration falls with time due to the sorption process by the polymer. The space occupied by the film is $-\ell \leq x \leq \ell$, and by the liquid $-(\ell + a) \leq x \leq -\ell$, and $\ell \leq x \leq \ell + a$. Initial conditions are $c = 0$ for $-\ell < x < \ell$ at $t = 0$.

The boundary condition contemplates that all sorbate lost by the liquid is taken up by the polymer.

$$aK\frac{\partial c}{\partial t} = \pm D_p\frac{\partial c}{\partial x} \quad \text{at } x = \pm\ell \quad \text{for } t > 0 \tag{11.71}$$

where K is the partition coefficient ($K = c^*_f/c^*_p$) and c^*_f and c^*_p are the equilibrium concentrations of the adjuvant in the contacting phase and polymer, respectively. The solution of Fick's second law subject to these conditions is[34]

$$\frac{M}{M_{ss}} = \sum_{n=1}^{\infty} \frac{2\alpha(\alpha+1)}{1+\alpha+\alpha^2 q_n^2} \exp\left(\frac{-Dq_n^2 t}{\ell^2}\right) \tag{11.72}$$

where M is the quantity of diffusing molecule entering the sheet or film, $q_n s$ are the non-zero positive roots of $\tan q_n = -\alpha q_n$, and $\alpha = aK/\ell$. The fraction uptake, U, is

$$U = \frac{M_{ss}}{2ac_f^o} = \frac{V_p c_f^*}{V_f c_f^o} = \frac{1}{1+\alpha} \tag{11.73}$$

where V_f is the volume of contacting phase and V_p the volume of plastic in the container.

11.3.9 Migration and Sorption at Equilibrium

All diffusion processes involving the transfer of a diffusant *from* or *to* a polymeric material asymptotically reaches a steady condition at long times. In any closed system (like a Tetra Pak carton) that has reached this condition, there is no effective mass transfer between the polymer and the surrounding phases; the adjuvant concentrations are at equilibrium, and the transferred material is at the maximum possible amount. In reference to Fig. 11.25 b, consider a sorption or migration process that has reached equilibrium. V_f is the volume of contacting phase, and V_p is the volume of plastic in the container, and c^*_f and c^*_p are the equilibrium concentrations of the adjuvant in the contacting phase and polymer, respectively. For the case of sorption of a substance from the food phase by the polymer, the mass balance at equilibrium is

$${}^s c_p^* V_p + c_f^* V_f = c_f^o V_f \tag{11.74}$$

and, for a migration process,

$${}^m c_p^* V_p + c_f^* V_f = c_p^o V_p \tag{11.75}$$

The equilibrium concentration, ${}^s c_f^*$ for a sorption process and ${}^m c_f^*$ for a migration process, are expressed as a function only of adjuvant initial concentration (in the polymer of foodstuff), the partition coefficient K, and the ratio of polymer and food volume.

$${}^s c_f^* = \frac{c_f^o}{K^{-1}\dfrac{V_p}{V_f} + 1} \tag{11.76}$$

$${}^m c_f^* = \frac{c_p^o}{K^{-1} + \dfrac{V_f}{V_p}} \tag{11.77}$$

where the s and m superscripts at the right of c refer to a sorption and migration process, respectively. Figures 11.32 and 11.33 plot ${}^{m}c^{*}_{f}$ and ${}^{s}c^{*}_{f}$, respectively, for a range of K and V_f/V_p values.

References

1. *Modern Plastics World Encyclopedia Handbook*, McGraw-Hill Co., Inc., New York, A-193, 2000.
2. *Modern Plastics World Encyclopedia Handbook*, McGraw-Hill Co., Inc., New York, A-192, 2000.

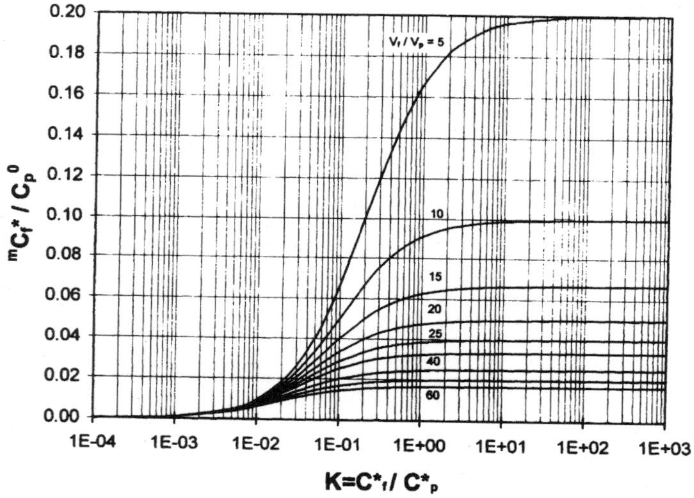

Figure 11.32 Equilibrium concentration in a migration process.

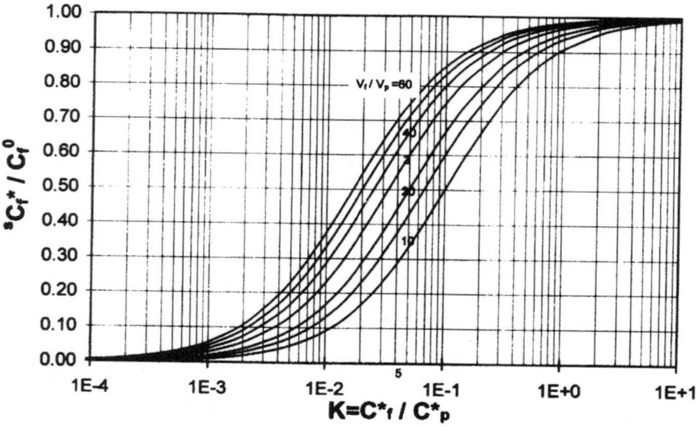

Figure 11.33 Equilibrium concentration in a sorption process.

3. *Modern Plastics World Encyclopedia Handbook*, McGraw-Hill Co., Inc., New York, A-194, 2000.
4. Anonymous, *Mobil OPP Films Product Characteristics*, 1994.
5. *Modern Plastics World Encyclopedia Handbook*, McGraw-Hill Co., Inc., New York, A-270, 2000.
6. Toebe, J., H. Hoojjat, R.J. Hernandez, J. Giacin, and B. Harte, Interaction of Flavour Components from a Onion/Garlic Flavored Sour Cream with HIPS, *Packaging Technology and Science*, 3, 133, 1990.
7. Viola, J.P. *Modern Plastics World Encyclopedia Handbook*, McGraw-Hill Co., Inc., New York, A-253, 2000.
8. *Modern Plastics World Encyclopedia Handbook*, McGraw-Hill Co., Inc., New York, A-32, 2000.
9. Hernandez, R.J., Effect of Water Vapor on the Transport Properties of Oxygen through Polyamides Packaging Materials, *J. Food Eng.*, 22, 495, 1994.
10. Eisenberg, A., The Glassy State, in *The physical Properties of Polymer*, 2d ed., American Chemical Society, Washington DC, 1993.
11. Boyer, R.F., Transitions and Relaxations in Amorphous and Semicrystalline Organic Polymers and Copolymers, reprint from *Encyclopedia of Polymer Science and Technology Supplement* vol. II, John Wiley & Sons, Inc., New York, 745–839, 1977.
12. Woodward, A. E., *Understanding Polymer Morphology*, Hanser Publishers, New York, 1995.
13. Birley, A.W., B. Haworth, and J. Batchelor, *Physics of Plastics Processing Properties and Materials Engineering*, Hanser Publishers, New York, 1992.
14. Peterlin, A., Polymer Morphology, in *Encyclopedia of Polymer Science and Engineering*, vol. 10, 1987.
15. Van Krevelen, D.W., *Properties of Polymers*, 3d ed., Elsevier, New York, 1990.
16. *Guide to Plastics Properties and Specification Charts*, Modern Plastics, McGraw-Hill Inc., New York, (1987).
17. Brandrup, J., E. Immergut, and E. Grulke, Eds., *Polymer Handbook*, Fourth Ed., Wiley-Interscience, New York, 1999.
18. Wunderlich, B., S. Cheng, and K. Loufakis, *Encyclopedia of Polymer Science and Engineering*, Edited by Mark, Bikales, and Overberger, vol. 16, 787–807, New York, 1989.
19. Rosato, D., *Rosato's Plastic Encyclopedia and Dictionary*, Hanser Publisher, New York, 1993.
20. Brady, G.S. and H.R. Clauser, *Materials Handbook*, thirteen ed., McGraw Hill Inc., New York, 1991.
21. Nielsen, L.A. and R.F. Landel, *Mechanical Properties of Polymers and Composite*, 2d ed., Marcel Dekker, New York, 1993.
22. Packham, D.E., *Handbook of Adhesives*, Wiley and Sons, New York, 1993.
23. Thompson, K.I., *TAPPI Journal*, 9, 157, 1988.
24. Spink, J.W., R.J. Hernandez, and J.R. Giacin, Correlation of Heat Sealing and Hot Tack Parameters wit molecular properties of EVA Heat Sealant Copolymers, *TAPPI Proceedings Polymer Laminations and Coatings Conference*, 597, 1991.
25. Theller, H., Heat sealability of Flexible Web Materials in Hot-Bar Sealing Applications, *J. Plastic Film and Sheeting*, 5:1, 66, 1989.
26. Bor, M. and E. Wolf, *Principle of Optics*, 6th ed., Pergamon Press, Oxford, 1980.
27. Singh, S.P. and E. El-Khateeb, Evaluation of Proposed Test Method to Measure Surface and Volume Resistance of Static Dissipative Packaging Materials. *Packaging Technology and Science*, **7**, 283–289, 1994.

28. Cameron, A.C., R.M. Beaudry, N. Banks, and M. Yelanich, Modified-Atmosphere Packaging of Blue Berry Fruit: Modeling Respiration and Package Oxygen Partial Pressure as Function of Temperature. *J. Amer. Soc. Hort. Sci.*, 119:3, 534, 1994.
29. Ikegami, T. Nagashima, M. Shimoda, Y. Tanaka, and Y. Osajima, Sorption of Volatile Compounds in Aqueous Solutions by EVA Films, *J. Food Sci.*, 56, (2), 500, 1991.
30. Koszinoswski, J., Diffusion and Solubility of n-Alkanes in Polyolefins, *J. Applied Polym. Sci.*, 31, 1805, 1986.
31. Tseng, D.T., R. Matthews, J. Gregory, C. Wei, and R. Littell, Sorption of Ethyl Butyrate and Octanal Constituents of Oranges Essences by Polymeric Absorbents, *J. Food Sci.*, 59, (4), 801, 1993.
32. Nielsen, T.J., Limonene and Mircene Sorption into Refillable Polyethylene Terephthalate, and Washing Effects on Removal of Sorbed Compounds, *J. Food Sci.*, 59, (1), 227, 1994.
33. Kwapong, O.Y. and J.H. Hotchkiss, Comparative Sorption of Aroma Compounds by Polyethylene and Ionomer Food-Contact Plastics, *J. Food Sci.*, 52, (3) 761, 1987.
34. Crank, J., *The Mathematics of Diffusion*, second ed., Clarendon Press, Oxford, 1975.
35. Michaels A.S., Wieth W.R., and J.A. Barrie, Diffusion of gases in Poly(ethylene Terephthalate). *J. Appl. Phys.*, 34:13, 1963)
36. Berens, A.R., The Solubility of Vinyl Chloride in PVC, *Angew. Makromol. Chem.*, 47, 85, 1985.
37. Frisch, H.C., Sorption and Transport in Glassy Polymers—A Review, *Polym. Eng. Sci.*, 20 (1), 2, 1980.
38. Salame, M., Prediction of Gas Barrier Properties of High Polymers, *Polym. Eng. Sci.*, 26: 22, 1543–6, 1986.
39. Hernandez, R.J. and R. Gavara, Sorption and Transport of Water in Nylon-6 Films, *J. Polym. Sci.: Part B: Polym. Phys*, 32, 2367, 1994.
40. Van Amerongen, G.J., Solubility, Diffusion, and Permeation in Gutta-percha, *J. Polym. Sci.*, 2, 318, 1947.
41. Michaels, A.S., W.R. Vieth and J.A. Barrier, Solution of Gases in Polyethylene Terephthalate, *J. Appl. Phys.*, 34, (1), 1–12, 1963.
42. Puleo, A.C., D.R. Paul and P.K. Wong, Gas Sorption and Transport in semicrystalline poly(4-methyl-1-pentene), *Polymer*, 30, 1357–66, 1989.
43. Michaels, A.S., W.R. Vieth and J.A. Barrier, Diffusion of Gases in Polyethylene Terephthalate, *J. Appl. Phys.*, 34, (1), 13–30, 1963.
44. Veith W.R., *Diffusion In and Through Polymers*, Hanser Publishers, New York, New York, 1991.
45. Meares, P., Transient Permeation of Organic Vapors Through Polymers Membranes, *J. Appl. Polym. Sci.*, 9, 917, 1965.
46. Barrer, R.M., Diffusion and Permeation in Heterogeneous Media, in *Diffusion in Polymers*, Crank and Park (eds.), Academic Press, New York, 165–217, 1968.
47. DeLassus, P.T., Barrier Expectations for Polymer Combinations, *TAPPI Journal*, March, 216, 1988.
48. Nielsen, L.E., Models for the Permeability of Filled Polymer Systems, *J. Macromol. Sci. (Chem.)*, A1(5):929, 1967.
49. Steingiser, S., G.J. Rubb, and M. Salame, *Encyclopedia of Chemical Technology*, vol. 3, 3d ed., 480, 1980.
50. Bernabeo, A.E., W.S. Creasy, and L.M. Robeson, *J. Polym. Sci.: Polym. Chem. Ed.*, 13: 1979, 1975.
51. Paul, D.R., *J. Mem. Sci.* 18: 75, 1984.
52. Pye, D.G., M.M. Moher, and M. Panar, Measurement of Gas Permeability of Polymers, II. Apparatus for determination of the Permeability of Mixed gases and Vapors, *J. Appl. Polym. Sci.*, 20, 1921, 1976.

53. Niebergall, W., A. Humeid, and W. Blochl, The Aroma permeability of Packaging Films and Its Determination by Means of a Newly Developed Measuring Apparatus, *Lebesn. Wiss U. Technol.*, 11 (1) 1, 1978.
54. Hernandez, R.J., J.R. Giacin, and A.L. Baner, The Evaluation of the Aroma barrier Properties of Polymer Films, *J. Plastic Film and Sheeting*, 2, 187, 1986.
55. Zobel, M.G., The Odour Permeability of Polypropylene Packaging Films, *Polymer Testing*, 5 (2) 153, 1985.
56. Hilton B. W., and S.Y. Nee, Permeability of Organic Vapors Through Packaging Films, *Ind. Eng. Chem. Prod. Res. Dev.*, 17 (1) 80, 1978.
57. Hernandez, R.J. and R. Gavara, Plastic Packaging. *Methods for Studying Mass Transfer Interactions*. Pira International Reviews, Leatherhead, Surrey, United Kingdom, 1999.
58. Gavara, R. and R.J. Hernandez, The Effect of Water on the Transport of Oxygen Through Nylon-6 Films, *J. Polym. Sci. Part B*, 32, 2375, 1994.
59. Pasternak R.A., J.F. Scimscheimer, and J. Heller, A Dynamic approach to Diffusion and permeation experiments, *J. Polym. Sci. Part A-2*, 8:467, 1978.
60. Gavara, R. and R.J. Hernandez, Consistency Test for Continuous Flow Permeability Experimental Data, *J. Plastic and Film Sheeting*, 9, 126, 1993.
61. Toi, K., H. Suzuki, I. Ikemoto, T. Ito, and T. Kasai, Permeation and Sorption for Oxygen and Nitrogen into Polyimide Membrane, *J. Polym. Sci: Part B: Polym. Phys.*, 33, 777, 1995.
62. Paul, D.R. and W.J. Koros, Effect of Partially Immobilized Sorption on Permeability and Diffusion Lag-Time *J. Polym. Sci.:Polym. Phys. Ed.*, 14, 675, 1976.
63. Mizrahi S. and M. Karel, Moisture Transfer in a Packaged Product in Isothermal Storage, *J. Food Processing and Preservation,* 1, 225, 1977.
64. Kim, J, R.J. Hernandez, and G. Burgess, Application of the Finite Difference Method to Estimate the Moisture Content of a Pharmaceutical Tablet in a Blister Package, *J. Packag. Technol. and Sci.*, submitted.
65. Lee, L.M., *Modern Plastics*, 53 (12) 66, 1976.
66. Masi, P. and D.R. Paul, Modeling Gas Transport in Packaging Applications, *J. Mem. Sci.* 12, 137, (1982).
67. Reid R.C., K.R Didman, A.D. Schwope, and D.E. Till, Loss of Adjuvants from Polymer Films to Food or food Simulants, *Ind. Chem.Eng. Prod. Res. Dev.*, 19 (4), 580, 1980.
68. Goydan R., A.D. Schwope, R.C. Rear, and G. Cramer, High Temperature Migration of Antioxidants from Polyolefins, *Food Addit. Contam.*, 7, 323, 1990.
69. Markelov, M., M.M. Alger, T.D. Lickly, and E.M. Rosen, Migration Studies of Acrylonitrile from Commercial Copolymers, *Ind. Eng. Chem. Res.*, 31, 2140, 1992.
70. Begley T.H., and H.C Hollifield, Recycled Polymers in Food Packaging: Migration Considerations, *Food Technology,* 11, 109, 1993.
71. Laoubi S., A. Feigenbaum, and J.M. Vergnaud, Effect of Functional Barrier Thickness for a Food Package with Recycled Polymer, *Packaging Technol. and Sci.*, 8, 249, 1995.
72. Berens A.R., Diffusion of Gases and vapors in Rigid PVC, *J. Vinyl Technol.*, 1:1, 8, 1979.

Chapter

12

Plastics Recycling

Susan E. Selke, Ph.D.

School of Packaging
Michigan State University
East Lansing, Michigan

12.1 Introduction

The amount of plastics being recycled continues to increase around the world. As the use of plastics continues to increase as well, plastics recycling rates have declined in some areas. Public pressure for plastics recycling waned in the last half of the 1990s in the U.S.A., although there are some signs that it is increasing in the early 2000s. The early focus of plastics recycling efforts was packaging materials. More recently, recycling of plastics in consumer products such as carpet, automobiles, and appliances is drawing attention.

This chapter deals almost exclusively with recycling of post-consumer plastics. While recycling of process scrap is important for the economic viability of many operations using plastics, recovery and reuse of clean process scrap is regarded, by and large, as a normal part of operations and is not generally classified as recycling. Recovery of contaminated scrap, while more difficult and less routine, is still often regarded as not "real" recycling. Such materials are sometimes processed along with post-consumer plastics, and some processors and users of recycled plastics depend on the known composition of such materials to help control product properties. Nevertheless, when most people think of plastics recycling, they have in mind the processing for reuse of materials that have served their intended purpose. Post-consumer materials, by this definition, include those generated by business and industry as well as those generated by individuals. The key to the definition is that the products have been used for the purpose for which they were manufactured. For example, a spool that once held cable wire is a post-consumer material, while an in-mold labeled laundry detergent bottle that was defective in some way and therefore was never filled is not.

The discussion in this chapter will also focus on the United States, although it is not limited to it.

12.1.1 Municipal Solid Waste

In the U.S.A., substantial public pressure to develop plastics recycling first emerged in the late 1980s, when a number of municipalities experienced problems with adequate disposal

capacity for municipal solid waste (MSW). Municipal solid waste, as defined by EPA, includes residential, institutional, office, and commercial waste.[1] It does not include industrial process waste, construction and demolition waste, non-household hazardous waste, and a variety of other waste streams. It should be noted that definitions of MSW differ, which can make comparison from country to country, or even state to state, problematic.

In the mid to late 1980s, many U.S. landfills were closing due to reaching capacity or because they had been identified as sources of groundwater pollution, but few new landfills were opening because of public resistance to siting such undesirable facilities. New regulatory requirements made new facilities more costly, and lack of capacity further increased tipping fees (the fee charged to leave the waste at the disposal facility). This adversely affected municipal as well as individual budgets, and people began talking about the "landfill crisis."

In response, some communities increased reliance on incineration, but many found their citizens even more opposed to incineration than to landfill, and the cost of incineration was usually higher than landfill. Other communities relied on shipping their MSW to places with adequate disposal facilities—a practice that continues to draw criticism from the "host" locations unless they feel adequately compensated. While both of these solutions tended to be unpopular, many communities found that, in contrast, there was substantial public support for instituting recycling programs.

As time went by, a number of new, often very large, landfills were opened. While the total number of landfills has continued to decline in the U.S.A., the amount of available capacity has increased. This increase, coupled with a decrease in the amount of waste requiring disposal as recycling grew, resulted in a decrease in tipping fees in many areas. The crisis atmosphere disappeared. Some predicted that recycling would also disappear, especially when the location or materials make it more costly than disposal. However, the public has proven to be much more supportive of recycling than most experts anticipated. Individuals across the U.S.A. readily took up the habit of recycling and are resistant to giving it up.

While the total amount of MSW in the United States has continued to increase, the rate of increase has slowed since 1990. The amount and proportion of MSW going to landfill have fallen, while the amounts recycled and composted have increased substantially, as shown in Fig. 12.1. Most of the increase in the total amount of MSW is due to population increase, as the per capita rate of generation of MSW has been fairly stable since 1990 (Fig. 12.2).

EPA categorizes MSW by product type and by material type. Product categories are durable goods, nondurable goods, containers and packaging, food wastes, yard trimmings, and miscellaneous inorganic wastes. Material categories are paper and paperboard, glass, metals (subdivided into ferrous, aluminum, and other nonferrous), plastics, rubber and leather, textiles, wood, food, yard trimmings, miscellaneous inorganic wastes, and other. Figures 12.3 and 12.4 show the relative contribution, by weight, of these product and material categories to MSW in 1998.

In Europe, concerns about solid waste disposal capacity were and remain more real than in the U.S.A. Governments in the European Union have taken a very active role in requiring recycling, as will be discussed in Sec. 12.18. Canada has actively pushed for reduction in disposal of municipal solid waste, packaging in particular, with impressive success. Japan, Korea, and other Asian countries are rapidly increasing recycling, including plastics recycling. Recycling is also growing in Australia and New Zealand, in some South American countries, and in other parts of the world.

12.1.2 Plastics in Municipal Solid Waste

As can be seen in Fig. 12.4, plastics make up only slightly more than 10 percent of U.S. MSW, by weight.[1] However, both the total amount and the proportion of plastics in MSW

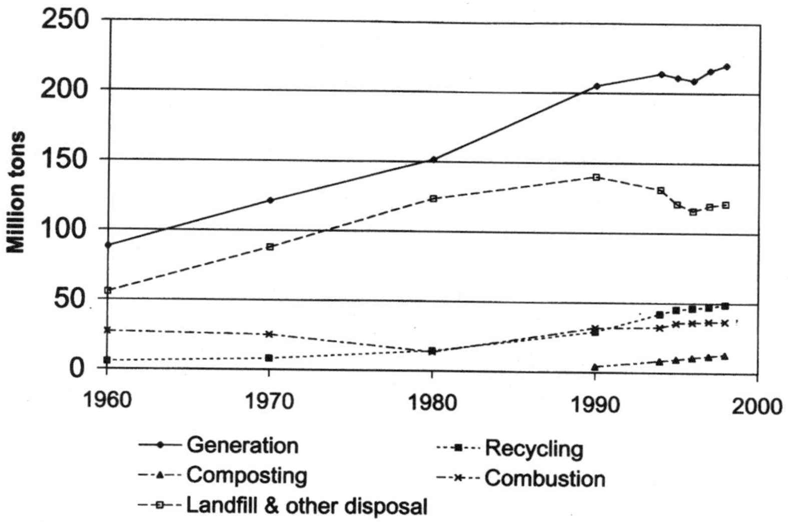

Figure 12.1 Generation and disposal of municipal solid waste in the U.S.A.[1]

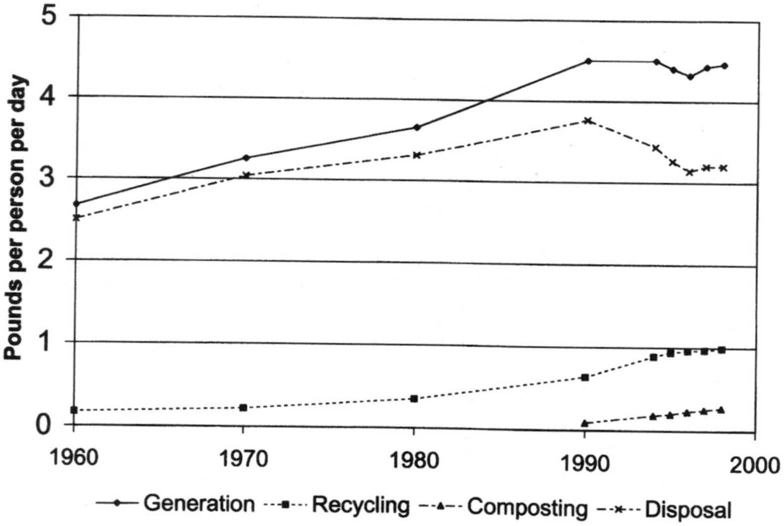

Figure 12.2 Generation and disposal of municipal solid waste in the U.S.A., per capita.[1]

have increased dramatically over the last several decades, as shown in Fig. 12.5. Furthermore, when landfill disposal is involved, contribution by volume is obviously much more important than contribution by weight. However, it is very difficult to determine accurately. While EPA's series of reports on U.S. MSW have occasionally attempted to quantify contributions by volume, these values are much less reliable than the weight estimates.

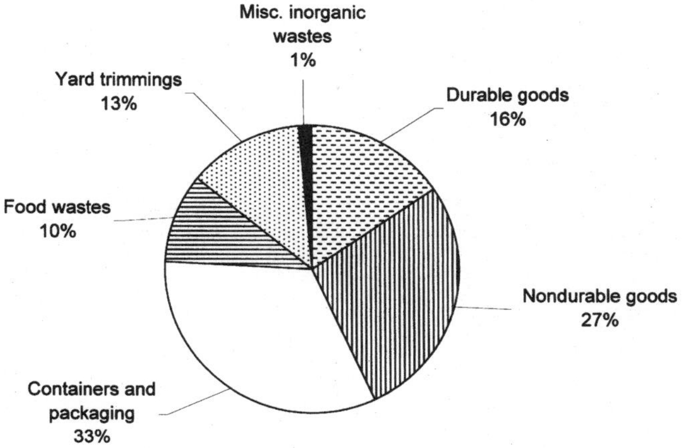

Figure 12.3 Products in the U.S. municipal solid waste stream, 1998.[1]

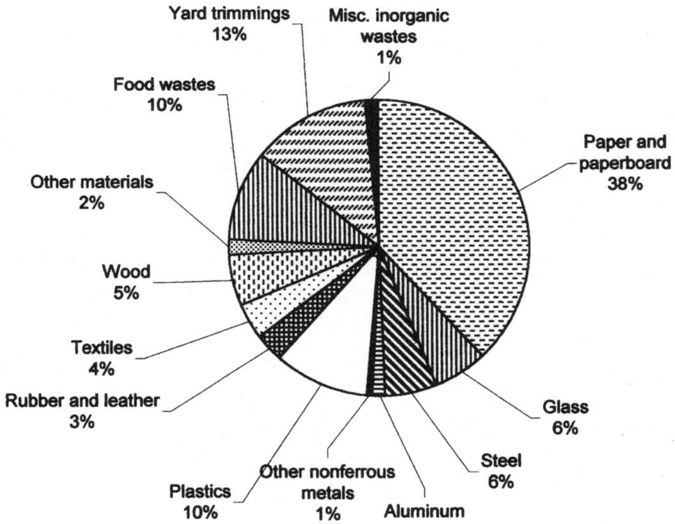

Figure 12.4 Materials in the U.S. municipal solid waste stream, 1998.[1]

What can be said with certainty is that the contribution of plastics to land-filled MSW measured by volume is considerably larger than measured by weight. EPA's most recent attempt to characterize volume contributions evaluated the 1996 waste stream and estimated that while plastics made up only 9.4 percent of all MSW generated (before recycling) by weight, they made up 25.1 percent of MSW going to disposal (after recycling) by volume.[2] While these estimates must be regarded with caution, it is clear that plastics do make up a sizeable component of MSW going to disposal facilities.

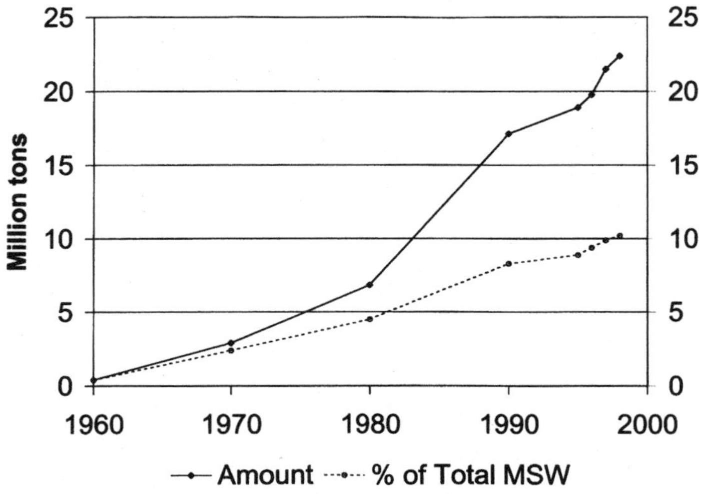

Figure 12.5 Plastics in the U.S. municipal solid waste stream.[1]

12.1.3 Plastics Recycling Rates

The recycling rates for plastics in the U.S.A. are, in general, considerably lower than for many other materials. The rates differ for plastics in different categories of goods, and for different plastic resins within the same category, as will be discussed in more detail. As can be seen in Fig. 12.6, the *containers and packaging* category has both the highest overall recycling rate, 40 percent, and the highest recycling rate for plastics, 9.6 percent. Plastics in durable goods are recycled at a rate of 3.8 percent, compared with 16.7 percent for the category as a whole. Nondurable goods have an overall recycling rate of 23.7 percent, but there is only negligible recycling of plastics in this category. The containers and packaging category also has the largest share of plastics in MSW (Fig. 12.7).

The amount of plastics recycled in 1998 was 260,000 tons from durable goods, and 950,000 tons from containers and packaging, for a total of 1.2 million tons.[1] *Plastics News* reported sales of North American plastics recyclers and brokers totaled \$1.74 billion in 2000, with 29 percent of that post-consumer plastics and the remainder post-industrial and brokered plastics. The total amount of plastics reprocessed was 5.54 billion pounds.[3]

Historical trends in plastics recycling amounts and rates in the U.S.A. and in Europe are shown in Figs. 12.8 and 12.9. These values include feedstock recycling but do not include energy recovery. While both are limited to recovery from post-consumer materials only, there are some differences in the types of materials included, so care should be taken in comparing the values. For example, recovery of agricultural and automotive plastics is included in the European figures but not, for the most part, in those for the U.S.A. Figure 12.10 shows the breakdown by category for plastics recycled in Europe. As can be seen, only 43 percent of the total amount of recycled plastics comes from municipal solid waste.[4] However, straightforward comparisons between amounts of plastic recycled from "municipal solid waste" cannot be made either, since definitions differ. Some, but not all, of the material identified in Europe as belonging to the distribution and industry and to the electrical and electronics sectors, for example, would be classified as part of municipal solid waste in the U.S.A.

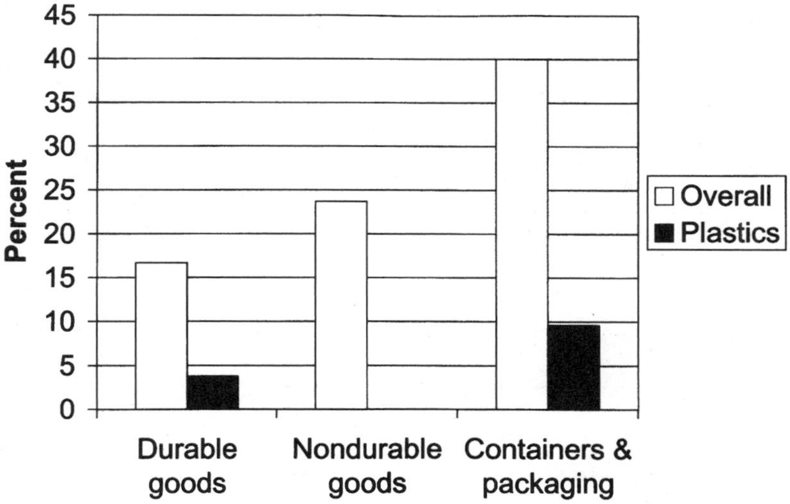

Figure 12.6 Plastics recycling rates in the U.S.A., 1998, by product category.[1]

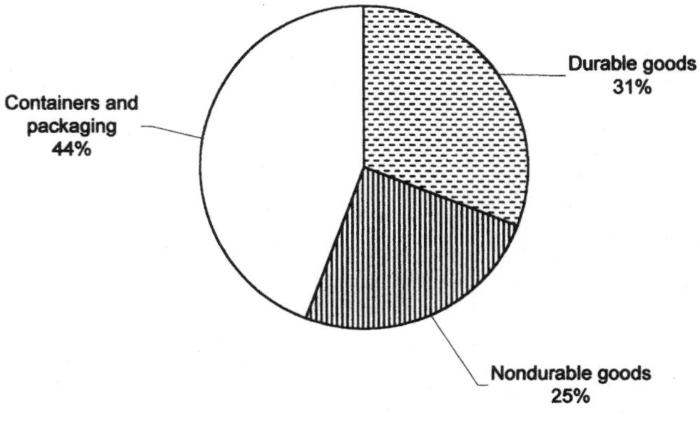

Figure 12.7 Plastics in U.S. MSW, 1998, by product category.[1]

Countries within western Europe differ considerably in recycling rates for plastics. The Association of Plastics Manufacturers in Europe (APME) reports that Germany has the highest rate at nearly 30 percent, while the rate in Greece was under 2 percent (Fig. 12.11). Recycling rates for packaging plastics are higher than the rates for plastics as a whole, ranging from 51 percent in Germany to 2 percent in Greece (Fig. 12.12).[4]

In Australia, a total of 167, 673 tonnes of plastics were recovered for recycling. Of this, 74 percent was processed domestically, and the remainder was exported, mostly to Asia. Plastics recycling has increased significantly in the last decade, with the 2000 amount 180 percent higher than that of 1992 and 80 percent higher than in 1997. While some of this may be due to better accounting, much of it is due to increased recycling. This amount includes both pre-consumer and post-consumer industrial and domestic waste materials.

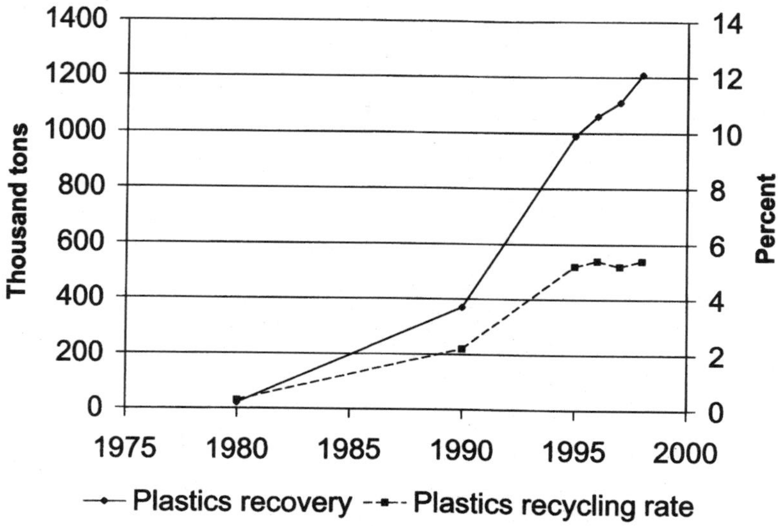

Figure 12.8 Changes in plastics recycling amounts and rates, U.S.A.[1]

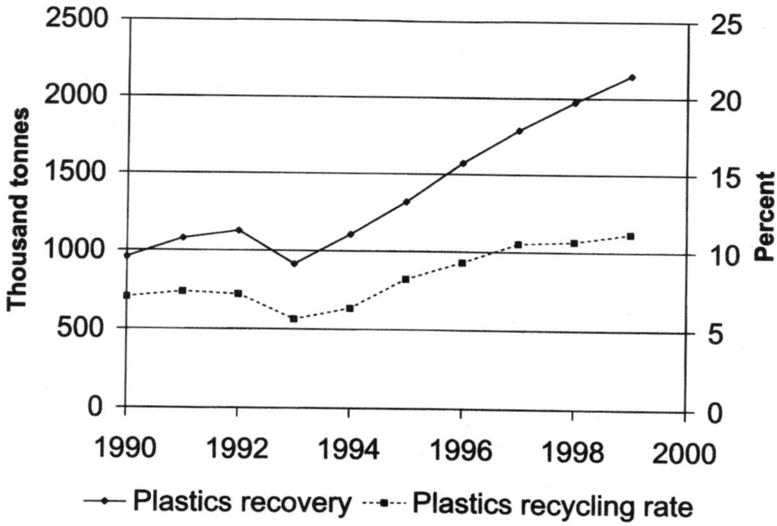

Figure 12.9 Changes in plastics recycling amounts and rates, Europe.[4]

Post-consumer domestic plastics represent only 30.4 percent of the plastics recycled within Australia but are believed to represent a major proportion of the exported material. Post-industrial materials made up 19.4 of the locally reprocessed plastics. Packaging represented 42.1 percent of the locally reprocessed materials, and manufacturing scrap represented 49.5 percent.[5]

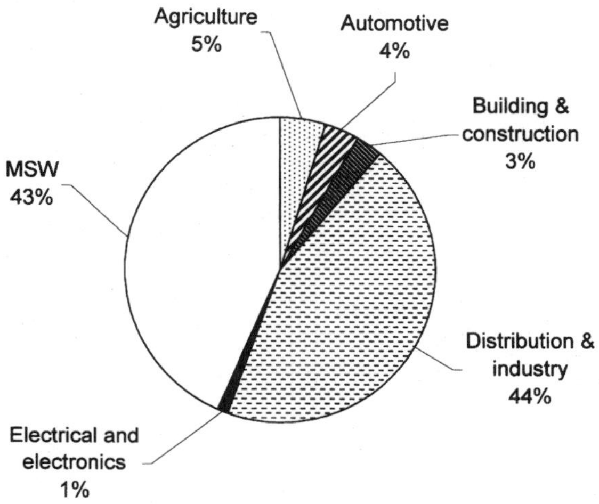

Figure 12.10 Sources of recycled plastics, Europe.[4]

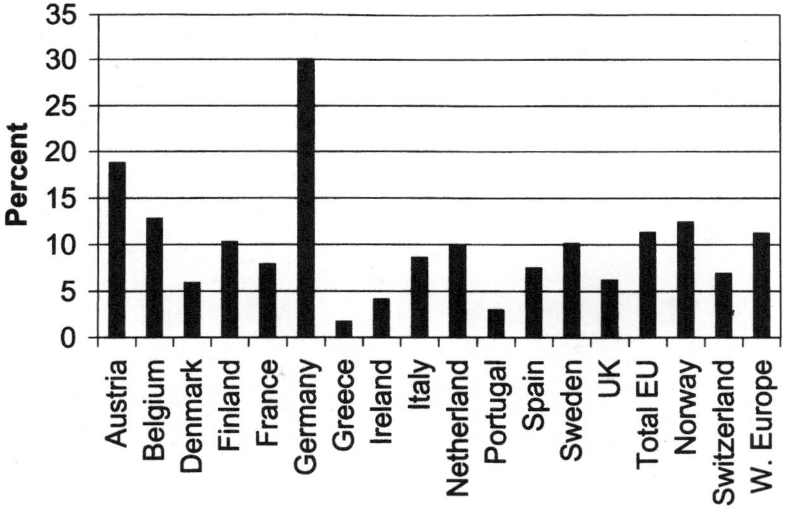

Figure 12.11 Plastics recycling rates in Western Europe, 1999.[4]

Polyethylene terephthalate (PET) has the highest recycling rate of all plastics in U.S. MSW, followed by polystyrene (PS) and high-density polyethylene (HDPE), as can be seen in Fig. 12.13. The most prevalent plastic in MSW is low- and linear low-density polyethylene (LDPE/LLDPE), followed by HDPE. Table 12.1 shows the amounts of the major plastic resins in U.S. MSW in 1997 and 1998 and the amounts recovered for recycling.

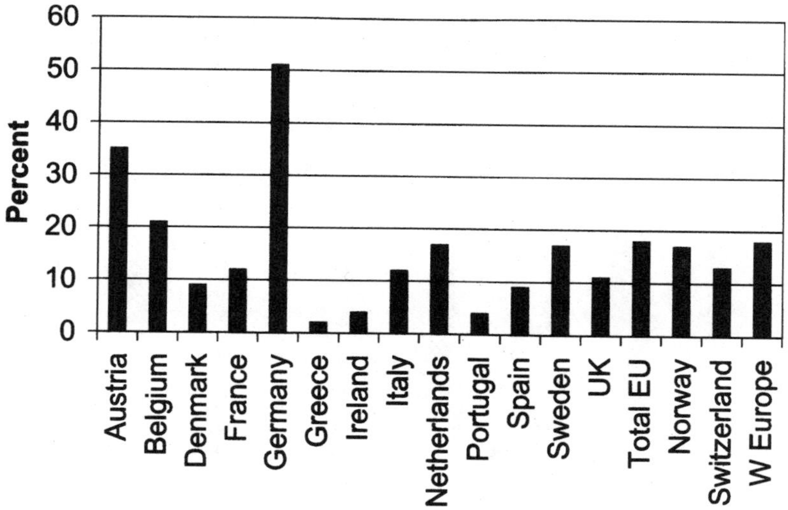

Figure 12.12 Recycling rates for plastics packaging in Western Europe, 1999.[4]

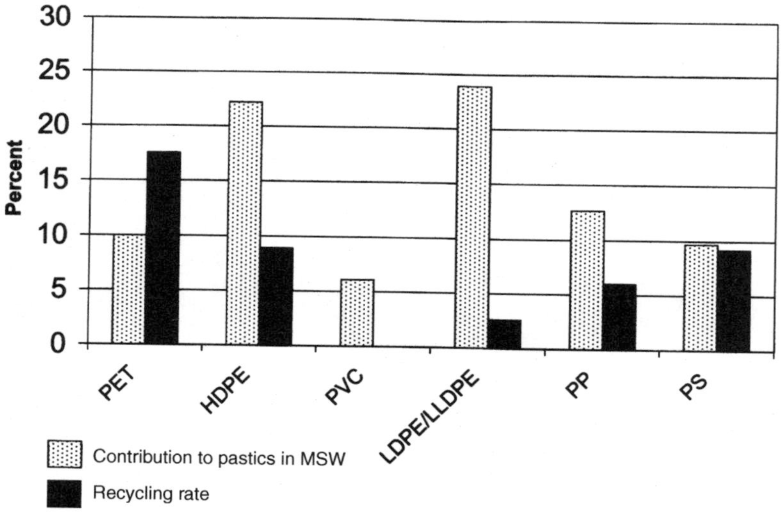

Figure 12.13 Recycling rates for plastics in MSW, by resin type, U.S.A., 1998.[1]

In Australia, where reported recycling rates include both pre- and post-consumer materials, PET also had the highest recovery rate, 32.0 percent, followed by HDPE at 14.7 percent and LDPE/LLDPE at 11.0 percent (Fig 12.14).[5]

Calculation of recycling rates is controversial. There have been charges in the past that surveys that ask recyclers for data receive inflated figures, and thus such surveys inflate re-

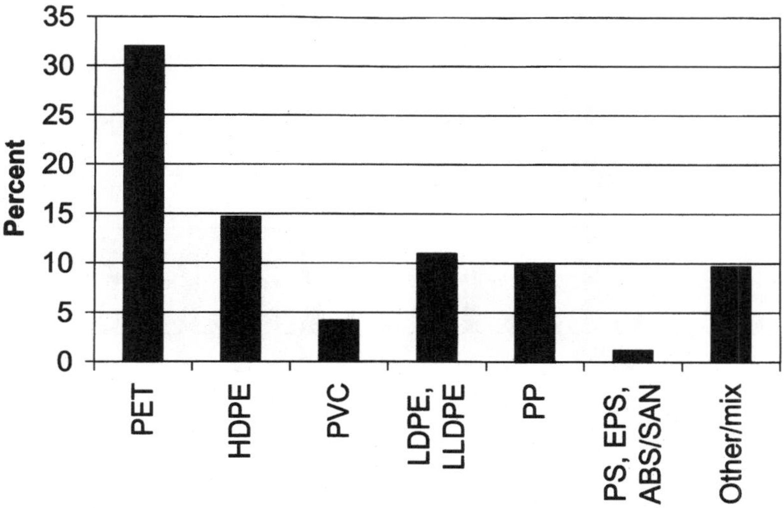

Figure 12.14 Plastics recycling rates by resin type (pre- and post-consumer), Australia, 2000.[5]

Table 12.1 Major Plastic Resins in U.S. MSW (Thousands of Tons)[1,6]

	1997		1998	
Resin	In MSW	Recovered	In MSW	Recovered
HDPE	4630	420	4960	440
LDPE/LLDPE	5380	100	5340	140
PET	1900	360	2230	390
PP	2790	120	2840	170
PS	2100	10	2170	20
PVC	1320	Neg.	1370	Neg.

cycling rates. Surveying organizations take steps to minimize this problem but cannot totally eliminate it. The reverse problem is neglecting organizations that do recycling, thus underestimating recycling rates.

A more fundamental problem than data accuracy is the matter of definition—what should count as *recycled?* The two most common options are determining the amount of material collected for recycling and determining the amount of material delivered for reuse. Since 5 to 15 percent of collected material typically is lost during processing (mostly because it is some type of contaminant such as a paper label, product residue, unwanted variety of plastic, etc.), recycling rates calculated using these two methods can differ substantially.

In the U.S.A., the American Plastics Council (APC) is a major source of information about plastics recycling rates. In 1997, APC switched, in determining recycling rates, from using the amount of cleaned material ready for use to using the amount of material collected for processing. They justified this decision by claiming it is more in keeping with

the way recycling rates are calculated for other materials—a claim that is true for some materials such as paper, but not true for others such as aluminum. This change brought considerable criticism, exacerbated by the fact that it occurred at a time when recycling rates, calculated in the same fashion, were declining. The APC was accused of trying to mask the extent of the decline by the change in methodology. For example, the PET bottle recycling rate in 1997 was 27.1 percent based on material collected, but only 22.7 percent based on clean material ready for reuse.[7] APC drew additional criticism by deleting polystyrene food service items from the definition of plastic packaging, beginning in 1995, which also had the effect of increasing the reported recycling rate for packaging plastics.

12.1.4 Benefits of Recycling

An obvious benefit of recycling is that it reduces the requirement for disposal of waste materials. Items that are recycled are, by definition, diverted from the waste stream. In many, although not all, cases, another benefit of recycling is cost reduction. Use of regrind, for example, became routine because of the monetary savings it provided. Similarly, certain plastics industries for years have relied on a combination of off-spec and recycled plastics because of their lower price. The desire to benefit from consumer preferences for recycled material, coupled, in some cases, with legislative pressures, has led, on occasion, to the anomalous situation of recycled plastic being worth more per pound than virgin resin, but these situations are usually short lived.

Additional benefits from recycling of plastics result from the fact that use of recycled resin displaces use of virgin materials and thus reduces depletion of natural resources. Recycling processes generally produce fewer environmental effluents than do processes that produce virgin resin, so use of recycled plastics usually results in a decrease in air and water pollution.

A factor that is likely to become increasingly important in the next decade is that the use of recycled plastics often results in significant energy savings, compared with use of virgin resin. For example, Fenton[8] calculated the total energy requirement for a low-density polyethylene grocery bag to be 1400 kJ, while a bag with 50 percent recycled content required only 1164 kJ, for a savings of nearly 17 percent. A DOE report concluded that recycling PET products such as soft drink and ketchup bottles requires only about a third of the energy needed to produce the PET from virgin materials.[9] Electricity shortages in California and large increases in the cost of natural gas have focused attention once again on energy. Efforts to reduce emissions of greenhouse gases, which are here or on their way in much of the world, if not in the U.S.A., will fuel energy conservation efforts. Plastics as a whole are likely to benefit, and recycled plastics may benefit even more.

12.2 Recycling Processes

For a material to get recycled, it must be collected, processed into a usable form, and then supplied to an end market and purchased by a subsequent consumer.

12.2.1 Collection of Materials

For post-consumer materials, including plastics, the most difficult part of the recycling process may be getting the material collected in the first place. Industrial scrap is "owned" by the industrial entity that produced it. If the owners cannot get the scrap recycled, they will either have to dispose of it or pay some other business to do so. For much consumer scrap, there is little or no monetary incentive for its owner, the individual consumer, to direct it into a recycling system. Furthermore, industrial scrap tends to be concentrated, with

substantial amounts of material in relatively few locations, making it relatively easy to collect. Post-consumer materials are typically very diffuse. It is estimated that the typical consumer generates only about 35 lb of plastic bottles per year.[10] Thus, a fairly elaborate collection infrastructure is needed to get this material gathered together in quantities that make its processing economically viable.

A number of systems have been developed to accomplish the task of collection. Among the most well developed in the United States are bottle-deposit systems.

12.2.1.1 Bottle deposit systems. Nine states (ten if California is counted) have a deposit system on certain types of beverage containers (Table 12.2). The consumer pays a deposit, usually 5 cents, when buying the container and then receives a refund of that fee when the bottle is returned to a designated collection point. In most cases, any retailer that sells beverages is obligated to accept the returns and refund the deposit. The majority of states provide for a handling fee for the retailer to at least partially offset the costs of managing the system. In most states, the deposit was originally restricted to carbonated soft drinks and beer, although it has since been expanded in several states. In all cases except California, the primary original motivation for the deposit was to reduce litter. Deposits have proved to be a powerful incentive for consumers to return the covered containers, with redemption rates usually exceeding 90 percent. Thus, deposit legislation resulted in the collection of large numbers of PET soft drink bottles. This, in turn, spurred the development of effective reprocessing systems for these bottles and end markets for the recovered resin. The existence of bottle-deposit legislation is in large part responsible for the successful development of PET recycling and for the fact that, in the United States, PET for a number of years had a much higher recycling rate than any other plastic. Even with the growth in curbside recycling in non-deposit states, deposit systems accounted for over 55 percent of the PET soft drink bottles recycled in 1995, while including less than 30 percent of the U.S. population.[13]

Several years ago, Maine extended its early deposit law in an explicit attempt to increase recycling. The state now has deposits in place on most beverages, with the exception of milk.

California has a system that is sometimes counted with deposit system but differs in significant ways. Customers pay a refund value rather than a deposit, and until the law was changed effective January 2000, this was buried in the product price rather than charged as a separate item. Containers can be returned only to designated redemption centers, so return of containers is significantly less convenient than it is in most deposit states, where containers can be returned to any retailer selling the covered beverage. The size of the refund value (currently 2.5 cents for most containers) is only half the typical deposit value (5 cents). Recycling rates in California for covered containers are higher than the national average but lower than the rates in deposit states.

California recently extended its refund value system to a wide variety of beverages, again in an explicit attempt to increase recycling of plastic bottles. Effective in January 2000, water and fruit juice containers are included, along with several other beverages (see Table 12.2). This has produced an increase in the containers covered of about 20 percent and a decrease, at least temporarily, in recycling rates. Redemption rates for the first half of 2000 fell to 70 percent from 80 percent in the first half of 1999. The decline was especially steep for PET, which dropped from 83 percent to 40 percent. The overall California redemption rate in 1999 for containers covered by the refund value system was 76 percent. The rate in the second half of the year was lower than in the first half, as has been the case for the last several years.[12]

A few other countries also have deposit laws. For example, the Netherlands requires deposits on PET and glass containers for soft drinks and waters. Sweden has voluntary de-

TABLE 12.2 States with Bottle Deposit Legislation[11,13]

State	Containers Covered	Characteristics
Connecticut	Beer, malt beverages, carbonated soft drinks, soda water, mineral water	Five-cent deposit
California	Beer, malt beverages, soft drinks, wine coolers, water, fruit drinks & juice, sports drinks, coffee, tea, vegetable juice—sold in aluminum, glass, plastic, and bimetal* containers	Refund system, 2.5 cents on most containers less than 24 oz, 5 cents on larger containers; vegetable juice > 24 oz exempt
Delaware	Non-alcoholic carbonated beverages, beer, and other malt beverages	Five-cent deposit, aluminum cans exempt
Iowa	Beer, soft drinks, wine, liquor	Five-cent deposit
Maine	Beer, soft drinks, distilled spirits, wine, juice, water, and other non-carbonated beverages	Five-cent deposit, 15 cents on wine and liquor, no deposit on milk
Massachusetts	Carbonated soft drinks, mineral water, beer, and other malt beverages	Five-cent deposit; containers 2 gal or larger exempt
Michigan	Beer, soda, wine coolers, carbonated water, mineral water, wine coolers	Ten-cent deposit; 5 cents on some refillable glass bottles
New York	Beer, soda, wine coolers, carbonated mineral water, soda water	Five-cent deposit
Oregon	Beer, malt beverages, soft drinks, carbonated and mineral water	Five-cent deposit; 3 cents on standard refillable bottles
Vermont	Beer and soft drinks, liquor	Five-cent deposit; 15 cents on liquor bottles; all glass bottles must be refillable

*Steel-only cans are defined as bimetal in California's refund value program.

posits on PET beverage bottles. Several provinces in Canada have deposit systems.[14] South Australia has a deposit system that was modeled on that of Oregon.[15]

From a recycling perspective, the most significant aspect of bottle deposit legislation is that, in most cases, the financial incentive provided does an excellent job of getting people to return their empty plastic bottles to appropriate places. The most negative aspect is that the per-container cost of managing these systems, as they are currently designed, is higher than the cost of alternative collection systems. There are also very real sanitary concerns, especially when deposits are expanded to non-carbonated beverages.

12.2.1.2 Other deposit systems. The idea behind beverage bottle deposits has also been applied to other products. Automobile batteries are subject to deposits in many states. While the primary motivation is to avoid the introduction of lead into landfills and incinerators, these systems have been very successful at facilitating the recycling of the polypropylene (PP) battery cases.

12.2.1.3 Buy-back systems. Another way to put a monetary incentive into the return for recycling of plastic materials is similar to the California bottle refund system. Recycling centers have been set up in a number of places that pay consumers for the materials that they bring to the center, generally on a price-per-pound basis. These centers have been very successful for collection of aluminum beverage cans and have also been used to facilitate the recycling of plastic beverage bottles. They typically consist of only one, or at most a few, locations within a metropolitan area. A variation that has been successful in some areas is reverse-vending machines, in which consumers insert the bottles and receive a redeemable receipt or coupon for the refund value. For plastic bottles, these have been used primarily in deposit states. In the last several years, a number of buy-back centers have closed down, as implementation of curbside collection has cut into their success in collecting materials.

12.2.1.4 Drop-off systems. Drop-off systems function much like buy-back systems, consisting of centralized locations for people to bring in their recyclable materials. The difference is that no monetary compensation is offered. This difference also facilitates unattended operation of the centers. Drop-off systems encompass a wide range of designs, including barrels in supermarkets for people to place their plastic grocery sacks, roving multi-material collections centers coming to a location once a month, permanent multi-material centers in a centralized location in a community (or in an out-of-the-way location), collection bins in apartment-building laundry rooms, and even sophisticated garbage and recyclables chutes in high-rise apartment buildings.

Drop-off systems, especially in their simplest designs, can offer the lowest-cost collection option. Their major drawbacks are high rates of contamination and low rates of collection. In many communities, a combination of curbside recycling for single-family residences, drop-off collection for apartment complexes, and an additional set of drop-off sites for the general public (sometimes including materials that are not collected at other sites) provides a range of collection options that function very well together.

12.2.1.5 Collection systems. A general rule in recycling is that the easier you make it for the consumer, the higher the rate of participation—and the higher the rate of diversion of material from the waste stream—you will achieve. The major method in the United States for collecting plastics for recycling from single-family dwellings, outside of deposit systems, is curbside recycling. By going to the consumer to get the materials instead of asking the consumer to go to the recycling point, significantly higher rates of collection can be achieved. Collection systems differ in design but fall into three general categories. *Collection of commingled recyclables* refers to systems in which the participant places all the recyclables together, usually in a container provided by the operator of the system. Other systems require consumers to *separate the recyclables by type* and thus use multiple containers, usually provided by the individual consumer, for set-out of the materials. Many systems are *hybrids*, with most materials collected in a commingled form and others collected separately. For example, a common design is a bin for commingled bottles and cans, with newspapers bundled separately. Virtually all collection systems accept multiple materials.

Systems in which consumers set out commingled recyclables at the curb can be further divided into three categories, depending on how the materials are handled in the collection vehicle.

In a few communities, recycled materials are placed into bags (usually blue in color) and collected in the same vehicles (standard compactor trucks) at the same time as the garbage is collected. When the load is dumped, the blue bags, and sometimes other readily identifiable recyclable materials, are sorted out. While some of these systems seem to work reasonably well, others have experienced significant contamination problems. Even

without losses due to contamination, the yield of recyclables in general is lower than in systems that provide separate collection, simply because not all the bags are recovered intact. One frequently encountered problem is contamination of newspapers with broken glass. Some systems, therefore, require that newspaper be bagged separately from the other recyclables. The largest city to try this type of collection system is Chicago, where operation began in December 1995.

The second category includes systems that use a separate truck, or at least a separate compartment, for commingled recyclables. The recyclables are then delivered to a sorting facility called a materials recovery facility (MRF), where they are separated by material type (and for plastics, sometimes by resin type as well, although this may take place at a separate facility dedicated to plastics only). While the first-generation MRFs relied almost exclusively on hand sorting, modern MRFs are becoming increasingly mechanized. The major advantages of this system are efficiency in the time on route and in the filling of the vehicle. Disadvantages include the need for a separate vehicle and crew, a dedicated sorting facility, and sometimes high residual levels of unwanted materials. A variation of this system uses ordinary garbage trucks for collection of recyclables rather than a specialized recycling vehicle, but, unlike in the first category, garbage and recyclables are not collected together.

The third category includes systems in which the commingled recyclables are sorted at truckside into several categories. The separated streams may or may not require additional processing at a MRF before sale, depending on the materials included. The major advantage of this system is the quality control that can be practiced by the driver, coupled with ongoing education of consumers. If householders put unacceptable items into the bin, they will find the materials left there, ideally with an explanatory flyer, so they can learn from their errors. Another advantage is that a dedicated processing facility may not be required. The major disadvantages are increased time per stop and the potential for the truck filling one compartment and therefore having to leave the route and off-load, even though other compartments are not full. The general recommendation is that truckside sorting works well for moderate- to small-sized communities, and commingled collection and a MRF work best for large communities.

General rules of thumb for effective design of curbside recycling systems are

- Collect the recyclables in commingled form and require little if any preparation beyond cleaning.
- Provide a readily identifiable container for use by the householder in putting out the recyclables.
- Collect recyclables weekly on the same day as garbage collection.
- Put considerable effort into ongoing education and publicity efforts.

Providing a container is particularly important. The container serves several functions. First, it is a convenience for the householder, providing a useful place to deposit the recyclables. It is a visible reminder of the importance of the recycling program. This is key when a program is initiated—a mailing telling people about the program can easily get lost in the junk mail, but it is hard to overlook a large plastic bin. Furthermore, the bins serve as a potent source of peer pressure. When everyone else has a recycling bin at the curb and you do not, you are likely to feel like a bad citizen, which hopefully will be incentive for you to participate in recycling the next time. According to some reports, participation rates average 70 to 80 percent for curbside programs that provide containers, and only 30 to 40 percent for programs that do not.[16]

Another difference between recycling programs is whether they are voluntary or mandatory. The majority of curbside programs are voluntary, but several states and a large number of municipalities have instituted mandatory programs. There seems to be general

agreement that mandatory programs increase participation if enforcement efforts are included. If no enforcement takes place, results are not as clear. Typically, enforcement activities involve a series of warnings, ending in refusal to pick up the garbage for a set period of time. While fines and even jail terms may be permitted by the mandating ordinances, they are seldom employed.

More than 90 percent of the approximately 9,300 curbside collection programs in the United States include plastic bottles among the materials collected.[13] A significant problem in including plastics in curbside programs is the space the containers take up in the truck relative to their value. Educational programs urging consumers to compact the plastic bottles by stepping on them before placing them in the recycling bin can help. The use of on-truck compacting equipment is more effective in reducing volume. Disadvantages of on-truck compacting include the space consumed on the truck by the compactor itself, as well as issues associated with more difficult sorting of compacted containers. On-truck compacting of commingled recyclables must also contend with problems caused by broken glass. In this regard, presence of plastic bottles is an asset, since they reduce glass breakage. Shredding or chipping the plastic on the truck is generally not seen as viable, in large part because of the lack of reliable and efficient methods for separating chipped plastics by resin type.

Most collection systems, including curbside collection, accept only PET and HDPE bottles, which together represent about 95 percent of all plastic bottles (48 percent PET, 47 percent HDPE). Approximately 11 million U.S. households, nearly 70 percent of the total, have access to collection programs for PET and HDPE bottles, 7 million through curbside and 4 million through drop-off programs.[17] However, evidence that collection of these desired containers can be increased substantially by collecting all plastic bottles and the support of the plastics industry for such programs have led to an increasing number of communities accepting all types of plastic bottles.

A study carried out by the American Plastics Council found that programs targeting only PET and HDPE bottles received the same number of bottles of other resins and three times as many non-bottle containers as did programs targeting all plastic bottles. On average, PET and HDPE bottles made up 93 percent of the all-bottle program plastic stream but only 89 percent of the PET and HDPE-only plastic stream. When the city of Mesa, AZ, switched from HDPE and PET only to all plastic bottle collection in 1999, recovery of PET and HDPE bottles increased by 12.1 percent, and recovery rates for pigmented HDPE and custom PET bottles grew even more—36 percent and 18 percent, respectively. The percentage of non-PET and HDPE bottles collected actually declined. Windham County, VT, had a similar experience, with recovery of PET and HDPE bottles increasing from 61 percent to 63 percent, no increase in bottles other than HDPE and PET, and contamination from non-bottle containers cut in half, from 4.2 percent to 2.0 percent of the collected plastic in the curbside program. Drop-off sites in the county, serving the rural population, had a 24 percent increase in PET and HDPE bottles and a 72 percent decrease in non-bottle plastic containers. The net result was a 61 percent increase in the PET recovery rate and a 9 percent increase in HDPE recovery. Results in a variety of other communities, in both bottle-bill and non-bottle-bill states, showed substantial increases in plastic bottle collection on switching from HDPE and PET only to all plastic bottles. However, most communities have been unable to find markets for the non-PET and HDPE bottles, disposing of them as residue.[17]

Costs of curbside collection programs are generally intermediate between bottle deposit programs, which are the most costly, and drop-off programs, which cost the least.

12.2.1.6 Mixed waste processing. Another approach to recycling plastics and other materials is not to ask consumers to do any special sorting or preparation but instead

to recover recyclables from the garbage stream. The advantage of these systems is that, since they do not require any particular cooperation by consumers, they have the potential to recover the largest amount of recyclables.

The major disadvantages are the high cost of such systems and the low quality of the collected materials. The U.S. Bureau of Mines began experimenting with mixed-waste sorting facilities in the 1970s. Techniques employed drew from the mineral processing industries and included size reduction and various types of size- and density-based sortation methods. For plastics, the result of such processing is a mixed stream of plastics, with the resultant problems. Residual contamination is also a major concern. While research on this type of recovery continues, the vast majority of plastics is recovered through source-separation-based programs, where the "free" labor of the individuals who keep the recyclable plastics separate from the garbage is crucial, both in terms of overall economics and in quality of the recovered materials.

12.2.1.7 Types of recycling processes. Recycling processes for plastics can be classified in a variety of ways. One useful categorization differentiates between primary, secondary, tertiary, and sometimes quaternary recycling.

Primary recycling is sometimes defined as applications producing the same or similar products, while secondary recycling generates products with less demanding specifications. An alternative definition considers use of in-plant scrap as primary recycling and use of post-consumer material as secondary recycling. Tertiary recycling uses the recycled plastic as a chemical raw material. Quaternary recycling uses the plastic as a source of energy. This last category is often not considered to be true recycling.

An alternative categorization that is gaining in popularity is mechanical and feedstock recycling. Mechanical recycling, as the name indicates, uses mechanical processes to convert the plastic to a usable form, thus encompassing the primary and secondary processes outlined above. Feedstock recycling is essentially equivalent to tertiary recycling, using the recycled plastic as a chemical raw material, generally (but not always) for the production of new plastics. In Europe, the term *recovery* is often used to encompass mechanical and feedstock recycling plus incineration with energy recovery.

Plastic resins differ in which recycling technologies are appropriate. Thermoplastics are more amenable to mechanical recycling than thermosets, which cannot be melted and reshaped. Typically, condensation polymers such as PET, nylon, and polyurethane are more amenable to feedstock recycling than addition polymers such as polyolefins, polystyrene, and PVC. Most addition polymers produce a complex mixture of products that is difficult to use economically as a chemical feedstock, while condensation polymers usually produce relatively pure one- or two-component streams.

Germany and Austria do a substantial amount of feedstock recycling of plastics packaging, a total of 346,000 tonnes in 1999.[4] In the U.S.A., feedstock recycling is mostly limited to nylon 6, polyurethane, and PET.

12.2.2 Separation and Contamination Issues

When plastics are collected for recycling, they are not pure. They contain product residues, dirt, labels, and other materials, and they often contain more than one type of plastic resin, resins with different colors, additive packages, and so on. This contamination is one of the major stumbling blocks in increasing the recycling of plastic materials. Usefulness of the recovered plastic is greatly enhanced if it can be cleaned and purified. Therefore, technologies for cleaning and separating the materials are an important part of most plastics recycling systems.

12.2.2.1 Separation of non-plastic contaminants. Inclusion of non-plastic contaminants in recycled resins can affect both processing of the material and performance of the products manufactured from these materials. The presence of remnants of paper labels, for example, can result in black specks in plastic bottles, detracting from their appearance and rendering them unsuitable for some applications. These paper fragments can also build up in the screens in the extruder during processing, resulting in greater operating pressures (and energy use) and requiring more frequent screen changes. The presence of solid inclusions in the polymer can adversely affect the physical performance of the molded parts, resulting in premature failure. Mechanical properties can be decreased to the extent that thicker sections are required to obtain the desired performance.

Separation of plastics from non-plastic contaminants typically relies on a variety of fairly conventional processing techniques. Typically, the plastic is granulated, sent through an air classifier to remove light fractions such as labels, washed with hot water and detergent to remove product residues and dirt and to remove or soften adhesives, and screened to remove small heavy contaminants such as dirt and metal. Magnetic separation is used to remove ferrous metals, and techniques such as eddy current separators or electrostatic separators are also often used to remove metals. Many of these techniques were originally developed for mineral processing and have been adapted to plastics recycling.

12.2.2.2 Separation by resin type. Contamination of one resin with another can also result in diminished performance. One of the most fundamental problems is that most polymers are mutually insoluble. Thus, a blend of resins is likely to consist, on a microscopic scale, of domains of one resin embedded in a matrix of the other resin. While this sometimes results in desirable properties, more often it does not. To further complicate matters, the actual morphology, and thus the performance, will be strongly dependent not only on the composition of the material but also on the processing conditions. Therefore, for most high-value applications, it is essential to separate plastics by resin type.

Another problem arises from differences in melting temperatures. When PET is contaminated with polyvinyl chloride (PVC), for example, the PVC decomposes at the PET melt temperatures, resulting in black flecks in the clear PET. A very small amount of PVC contamination can render useless a large quantity of recovered PET. On the other hand, at PVC processing temperatures, PET flakes fail to melt, resulting in solid inclusions in the PVC articles that can cause them to fail. Again, a small amount of PET contamination can render recovered PVC unusable.

More subtle problems can arise, even from very similar resins. When injection-molded HDPE base cups from soft drink bottles were contaminated by newly developed blow-molded HDPE base cups, the recovered HDPE consisted of a blend of a high-melt-flow resin with a low-melt-flow resin that neither blow molders nor injection molders found usable. This caused serious difficulty for some recyclers. While this problem disappeared with the discontinuation of base cups, the difficulty in separating injection-molded HDPE bottles from blow-molded HDPE bottles is an ongoing concern.

Mixing resins of different colors can also be a problem. Laundry detergent bottle producers were able to fairly easily incorporate unpigmented milk bottle HDPE in a buried inner layer in detergent bottles but found it much more difficult to use recycled laundry detergent bottles. The color tended to show through the thin pigmented layer, especially in lighter colored bottles. Motor oil bottlers who used black bottles had no such problem. As a general rule, the lighter the color of the plastic article, the more difficult it is to incorporate recycled content; conversely, the lighter the color of the plastic article, the easier it is to find a use for it when it is recycled (and hence the higher its value).

Separation of plastics by resin type, or by a combination of resin type and color, is, therefore, an important part of most plastics recycling systems. Such separation can be classified as macrosorting, microsorting, or molecular sorting. Macrosorting refers to the

sorting of whole or nearly whole objects. Microsorting refers to sorting of chipped or granulated plastics. Molecular sorting refers to sorting of materials whose physical form has been totally disrupted, such as by dissolution.

Much macrosorting, such as separating PET bottles from HDPE bottles, nylon carpet from polyester carpet, and so forth, is still done by hand, often by workers picking materials off conveyor belts and placing them in the appropriate receptacle. However, mechanized means of sorting have been developed that make the process more economical and reliable. The various devices commercially available to separate plastics by resin type typically rely on differences in the absorption or transmission of certain wavelengths of electromagnetic radiation, on machine vision systems that recognize materials by shape or color, or on some combination of these. Many of these systems can separate plastics by color as well as by resin type.

Magnetic Separation Systems (MSS) of Nashville, TN, developed a system that sorts two to three plastic bottles per second, separating by resin type and color, using four sensors and seven computers. X-ray transmission is used to detect PVC, an infrared light high-density array separates clear from translucent or opaque plastics, a machine vision color sensor identifies bottle color (even ignoring the label), and a near-infrared spectrum detector identifies resin type.[18] A more recent MSS high-capacity plastic bottle separator uses a single sensor for both color and resin identification.[19]

Frankel Industries, of Edison, NJ, has a system that combines manual sorting with differences in optical dispersion and refraction for separating PET and PVC from each other as well as from PETG and polystyrene. Workers wear special goggles that give the different resins a distinctive appearance when a special light shines on them.[20]

Several companies have developed equipment that uses infrared scanning to identify resin type for plastics from items such as appliances, carpet, and automobile parts. The device uses a computer to compare the spectrum to known types of plastics, providing an identification. Some of these systems are portable, while others are designed to be used on a processing line. Some identify a wide range of plastics, while others are specialized to differentiate among a select few.[21–23]

Until recently, sorting of black plastics posed a difficult problem, as standard spectrometers could not be used. Their high carbon content causes black plastics to absorb light to such a degree that, when intense light sources such as lasers are used to analyze them, they heat up and can emit light or even ignite. While Raman spectroscopy can be used to identify dark and intensely pigmented plastics, it requires low laser power, resulting in long measurement times on the order of 10 s. In early 2001, SpectraCode, Inc., a manufacturer of spectrographic plastic identification devices, announced that it had developed a new device that provides for instantaneous identification of post-consumer black plastics. It contains a modified probe that uses a sampling technique to test black samples at full laser power with no burning, allowing identification in half a second or less.[24]

Microsorting of plastics is commonly practiced for separation of lighter-than-water from heavier-than-water plastics. While such separation can be done in a simple float-sink tank, hydrocyclones are often used because of their advantages in size and throughput. Application of density-based separation for mixtures of plastics that are all heavier or all lighter than water is relatively rare. The Multi-Products Recycling Facility operated by wTe Corp. is designed to recover metals and engineering plastics from durable goods. It uses air classifiers to remove light materials and a series of sink/float classifiers operating with water solutions at different specific gravities to separate chipped plastics by density. The system also employs infrared technology to identify plastics before grinding.[25] KHD Humboldt Wedag AG, in Cologne, Germany, has a system that uses water and various salt solutions in centrifuges to separate plastics by density.[26]

Systems that use differences in triboelectric behavior have been developing rapidly. These differ from devices for plastic/metal separation, which have been available for some

time. They operate by charging the materials to be separated. The metal quickly loses its charge, while the plastic remains charged and therefore sticks to a charged drum. For plastics separation, a slight positive or negative static charge is applied to flakes of different resins as they collide with each other and with the walls of a charging chamber. The charge a flake takes on depends on the relative triboelectric characteristics of the two colliding materials, which differs for different combinations. Since the separation is between positively and negatively charged particles, it works best for separating mixtures of only two different resins.[27]

Carpco has several electrostatic plastic separation systems in place in the U.S.A. and Europe. One system used at Nationwide Recycling Division of Crown Cork and Seal in Polkton, NC, separates PVC from PET bottle flake at a rate of 1500 lb/hr.[27] Electrostatic separation can even be used to separate epoxy-painted thermoplastic olefin automobile bumper flakes from non-painted flakes.[27]

Hamos GmbH, of Penzberg, Germany, has developed several types of electrostatic separator systems, including one designed especially for separating scrap streams from auto parts. The company has a number of systems in operation in Europe and at least one in the U.S.A. They include systems designed to separate PP and PE bottle flake.[19,27]

Plas-Sep Ltd. of Canada also sells electrostatic separators for mixed plastics.[27]

There has also been interest in developing microsorting systems that use electromagnetic radiation in much the same manner as for macrosorting. SRC Vision, Inc., of Medford, Oregon, has an optical sorting technology that is used by some large processors for color-sorting single resins, such as in separating green from clear PET flake. Union Carbide has used the system to separate colored HDPE flake into red, yellow, blue, green, black, and white product streams. The full system uses a combination of x-rays, ultraviolet light, visible light, infrared light, reflectance, and both monochromatic and color cameras. ESM International, Inc., of Houston, TX, has also developed an optical sorting system.[28,29]

Other attributes that have been used for microsorting include differences in melting point, surface characteristics, and differences in particle size and shape after either conventional or cryogenic grinding.[28] Recovery Processes International of Salt Lake City, Utah, designed a froth flotation system to separate PET from PVC.[28] A novel European process is designed to separate plastics from durable goods, including laminated materials, by blowing them apart at supersonic speeds. The different deformation behaviors of various plastics permits their separation by use of sieving along with classification based on differences in size, geometry, specific gravity, and ballistic behavior in systems using fluid bed separators and other equipment.[30]

The Salyp N.V. Company in Eiper, Belgium, is using an infrared sorting technology for separating various types of thermoplastics recovered from a system for recycling automotive shredder residue. The system uses infrared energy to heat and dry cleaned shredded plastics and to soften, but not melt, the targeted plastic. The mixed plastic stream is then fed through a set of rollers, and the softened plastic sticks to the roller and is removed. The remaining material is again heated, softening the next desired plastic, and the process repeated until all the plastics have been separated. Since infrared radiators emit at different wavelengths, IR energy can provide selective heating of different thermoplastics, by choosing appropriate emission wavelengths. Therefore, the system does not rely simply on differences in melting point for separating the plastics. Full-scale operation of the facility is scheduled for 2001.[31,32]

There has been little success in using molecular sorting for separation of plastics. Such systems typically use dissolution in one or more organic solvents, and the need to control emissions and to recover the solvents results in high costs. Residual solvent in the recovered plastics is also a concern.

There has been some effort to facilitate plastics separation by incorporating chemical tracers into the plastics to facilitate their identification and separation. Such approaches re-

main rare. Molded-in resin identification symbols on certain types of plastic articles will be discussed in Sec. 12.18.

In total, more than 300 sorting modules have been installed for separating plastic containers or flake. In the U.S.A., most of these are at processing facilities dedicated to plastics, rather than at MRFs, which handle multiple materials and generally rely on manual sorting. In contrast, about 150 of the modules are being used in European MRFs, with Germany having the highest level of automation.[33]

12.2.2.3 Safety concerns. Even when plastics are sorted by type, the performance of recycled plastics may differ from virgin plastic because of the effects of the use cycle. These changes may be due to chemical changes within the polymer, sorption of materials into the polymer, or other factors. If materials are sorbed, there is potential for later release of these substances. For some critical applications, such possible or actual change in behavior of recycled plastics poses unacceptable risks. For example, it is probably safe to conclude that recycled plastics will not be used for implantable medical devices. It is highly unlikely that recycled plastics will be used for the packaging of sensitive drugs. Other examples, of course, could also be cited where the small but real risk of unacceptable performance, or of release of some damaging substance, coupled with the critical nature of the application, is likely to rule out the use of recycled plastics.

For less critical applications, such as the use of recycled packaging for food products, the conventional wisdom used to be that recycled plastics should not even be considered. This has changed dramatically within the last decade. In the U.S.A., one of the earliest applications of recycled plastic for packaging of food products was recycled PET in egg cartons. The physical barrier of the egg shell provided a degree of added protection to the food that FDA agreed was sufficient to allow ordinary recycled PET to be used. Next came the use of recycled plastic in buried inner layers of packaging, such as a recycled PS clamshell used for hamburgers, in which the contact between the food and the recycled plastic was mediated by a layer of virgin plastic that acted as at least a partial barrier. Next, repolymerized PET was used in direct contact with food (blended with virgin material). The repolymerization process, with its crystallization steps, provided assurance that any impurities present would be removed. Next came FDA approval of specific systems for intensive cleaning of physically reprocessed PET, coupled with limitation of incoming material to relatively pure streams of soft drink bottles returned for deposits. Then, production of 100 percent recycled content PET bottles using a process for physically reprocessing bottles collected from curbside was approved. Systems for processing recycled HDPE have been approved for limited direct food contact applications as well.

The concern over use of recycled plastics in food contact falls in two general areas. First is concern about biological contaminants. In most cases, the processing steps for production of plastic packaging materials provide a sufficient heat history to destroy disease-producing organisms. Therefore, this is not a major concern.

A more significant concern is the possible presence of hazardous substances in the recycled feedstock. FDA regulations require food packagers to ensure that the materials they use are safe for food contact and that they do not contain substances that might migrate into the food and cause deleterious effects on human health. Recycled resins, by their very nature, often have a somewhat unknown history. What if, for example, someone put some insecticide—or some gasoline, or weed killer, or any of a myriad of toxic substances—into a soft drink bottle and later turned that bottle in for recycling? How can we prevent that container from contaminating new plastic packages? What we have seen in this area, as in others, is movement at the FDA away from absolute prohibitions and toward a more reasonable evaluation of risk. In particular, FDA has laid out guidelines for challenging recycling processes with known model contaminants and evaluating the abil-

ity of the process to remove those contaminants, thus providing some assurance that unacceptable levels of migration will not occur.

12.2.3 Quality Issues

As discussed, the use history of a recycled plastic can affect its properties and performance. It is well known that plastics undergo chemical changes during processing and use that ultimately lead to deterioration in properties. In fact, much of the history of plastics is related to the development of appropriate stabilizing agents to prevent this degradation. We routinely stabilize plastics against thermo-oxidative degradation that would otherwise occur during processing. We know that some resins are much more sensitive than others. Depending on the amount of stabilizer initially present, the history of the resin, and the type of resin, a recycled plastic resin may or may not require additional stabilizer to be successfully utilized.[34]

Similarly, plastics that are designed to be used outdoors must, in general, be stabilized against photodegradation. Recycled materials are likely to need additional stabilizer to retain adequate performance.

When regrind began to be a common ingredient in plastics processing in the late 1970s, much effort was devoted to studying the effects of multiple processing cycles on polymer performance. For many polymers, three major types of chemical reaction occur. First is oxidation. Reaction of the polymer structure with oxygen results in the incorporation of oxygen-containing groups in the polymer, with concomitant changes in properties and increased potential for further reactions. Either with or without oxidation, chain cleavage can also occur. This results in a decrease in molecular weight, with a consequent decrease in many performance properties. Chain cleavage can be followed by cross-linking, the forming of new molecular bonds that increases molecular weight and also changes properties. In some polymers, one or the other of these reactions predominates. In others, such as polyethylene, the effects of one tend to be balanced by the effects of the other. Some molecular structures are much more reactive than others. Polypropylene, for example, is significantly more susceptible to photo-oxidations than is polyethylene. Furthermore, for some materials, it is feasible to upgrade the material during reprocessing (such as in solid-stating of recycled PET), while for others it is not.

In summary, the general rule is that recycled polymers will have somewhat different properties from those of virgin polymers. These changes are usually detrimental and range in nature from virtually unnoticeable to major. Just as not all polymers are equally sensitive, not all properties are equally sensitive. It is not unusual, for example, for a recycled HDPE to have virtually the same tensile strength as virgin but significantly decreased Izod impact strength.

12.3 Polyethylene Terephthalate Recycling

The largest source of PET in the MSW stream in the U.S.A. is packaging, as shown in Fig. 12.15. PET has long been the most recycled plastic, largely due to recycling of soft-drink bottles. In the U.S.A. in 2000, 20 plants owned by 19 companies produced recycled PET flake from post-consumer bottles. By the end of the year, four of these plants had gone out of operation, and one plant that closed in 1998 reopened, leaving a total of 16 facilities, with a combined processing capacity of 836 million pounds. The aggregate amount processed was 732 million pounds, for a capacity utilization rate of 84.5 percent.[35] In 1999, two companies, Mohawk, of Summerville, GA, and Wellman, Inc., of Johnsonville, SC, dominated PET bottle recycling, with capacities of 200 and 190 million pounds per year, respectively.[36]

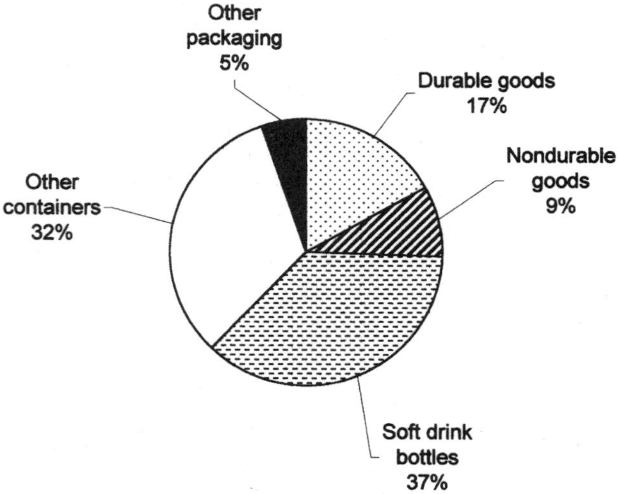

Figure 12.15 Sources of PET in U.S. municipal solid waste, 1998.[1]

Soft drink bottles remain the largest single use of PET in packaging, but non-beverage bottle use continues to grow at a faster rate, so the aggregate use of PET in "custom" (non-soft-drink) bottles now exceeds its use in soft drink bottles. Use of small size soft drink bottles is growing, while use of liter and larger bottles is declining. The most popular individual size is 20 oz, which now represents 47 percent of all PET soft drink bottles used in the U.S.A. Fruit juice and water represent a sizeable segment of custom bottle use. NAPCOR estimates a total of 3.445 billion pounds of PET bottles and jars were available for recycling in the U.S.A. in 2000.[35]

EPA reported an overall recycling rate for PET containers of 21.7 percent in 1998, a total of 720 million pounds.[1] NAPCOR reported a PET bottle recycling rate of 24.8 percent in 1998, for a total of 745 million pounds. The rate fell to 23.7 percent, 771 million pounds, in 1999 and fell again to 22.3 percent, 769 million pounds, in 2000.[35] Recycling rates for PET have been falling since 1994 (Fig. 12.16). The U.S.A. is both an importer and exporter of post-consumer PET, in baled bottle or dirty flake form. Exports totaled 170 million pounds in 2000, all but 19.5 million of these going to China. Imports totaled 69 million pounds, with bottles coming from Canada, Mexico, and Europe.[35]

EPA reported a 1998 recycling rate of 7.9 percent for PET in durable goods, and negligible recycling of nondurables. In the non-container segment, the recycling rate for PET packaging was also negligible. The overall PET packaging recycling rate was 21.7 percent in 1998, and the overall recycling rate for PET in municipal solid waste was 8.9 percent.[1]

In contrast to the declining rates for PET bottle recycling in the U.S.A., rates continue to increase in much of the world. For example, substantial recycling of PET bottles in Japan began only in 1993. The Japanese Council for PET Bottle Recycling reports that the aggregate recycling rate for PET bottles designated as second class (soft drinks, soy sauce, and liquors) reached 31.8 percent in 2000, after starting at only 0.4 percent in 1993 (Fig. 12.17). The "second class" bottles accounted for about 90 percent of all PET bottles produced in Japan in 2000.[37] A draft federal plan in Japan would require a 50 percent recycling rate by 2004.[19]

In Europe, PETCORE reports that 270,000 tonnes of PET bottles were recycled in 2000, a 23 percent increase compared with 1999 (Fig. 12.18), and forecasts an average

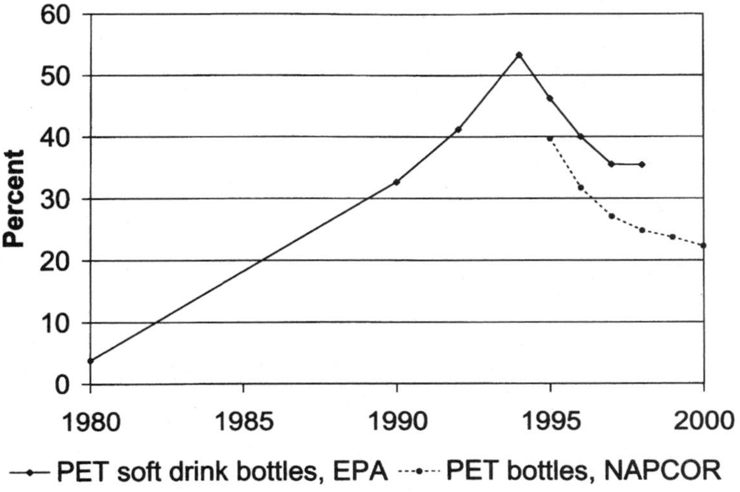

Figure 12.16 Recycling rates for PET in U.S. municipal solid waste.[1,2,35]

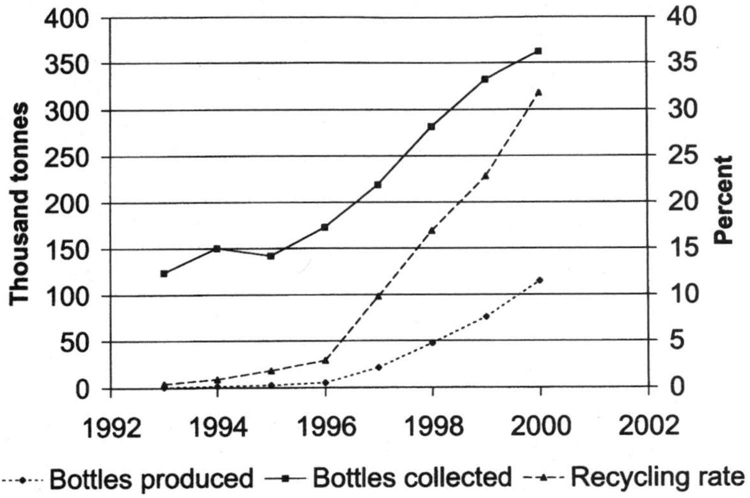

Figure 12.17 Production and recycling of PET soft drink, soy sauce, and liquor bottles, Japan.[37]

growth rate of 20 percent per year for the next five years.[38] The overall PET bottle recycling rate in Europe was reported to be 14 percent in 1999. There are substantial differences in recycling rates between countries, with collection rates for PET bottles of 70 to 80 percent in Sweden and Switzerland, but less than 5 percent in the U.K. Italy, France, Belgium, and Switzerland together accounted for over 75 percent of all PET bottles recycled in Europe in 1999.[39]

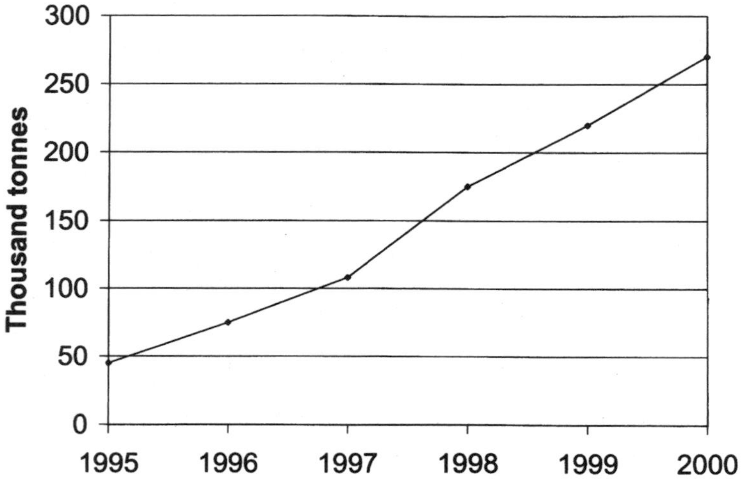

Figure 12.18 Recycling of PET bottles in Europe.[38]

Switzerland claimed the highest recovery rate for PET bottles in the world in 1999, 83 percent, for a total of 21,300 tonnes, collected through a network of nearly 30,000 collection bins supplemented with return through the original distribution channels. The system is operated totally by the Swiss PET recycling organization, PET-Recycling Schweiz (PRS).[40]

In Brazil, the tonnage of PET recycled has continued to grow, with a substantial growth in the recycling rate between 1997, when it stood at 16.2 percent, and 2000, when it reached 26.3 percent (Fig. 12.19), after declining rates in 1996 and 1997.[41]

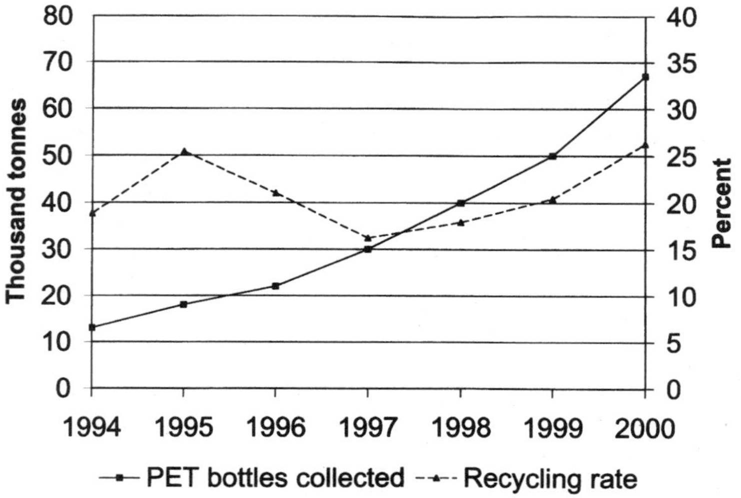

Figure 12.19 Recycling of PET bottles in Brazil.[41]

Global recycling of PET bottles in 1999 was reported to be about 900,000 tonnes, for a recycling rate of about 17 percent.[39]

12.3.1 Soft Drink Bottle Recycling

As mentioned above, soft drink bottle recycling got its start with the introduction of deposit legislation, which resulted in collection of significant volumes of material and recognition of the economic value embedded in them. In the U.S.A., recycling rates for PET bottles grew until 1994 and then began to decline, as did recycling rates for PET soft drink bottles (Fig. 12.16). The decline in recycling rate for PET soft drink bottles is attributed in large part to a substantial increase in the tonnage of PET used in single-serving bottles, which are more likely to be consumed away from home and thus less likely to reach curbside recycling collection. Between 1994 and 1998, the total tonnage and number of soft drink bottles recycled increased, even though the recycling rate fell. However, even the tonnage of PET bottles (all types) collected for recycling declined in 1996, 1997, and again in 2000, although the number of bottles recycled continued to increase (Fig. 12.20).[35,42] The overall recycling rate for PET soft drink bottles in 1998 was 35.4 percent, for a total of 580 million pounds.[1]

The Container Recycling Institute reports that the recycling rate for beer and soft drink containers (all types) in deposit states (including California) averages 80 percent while, in non-deposit states, it averages only 40 percent.[13] Michigan, where the deposit is 10 cents on nearly all containers, rather than the 5 cents charged in most states, reports a 99 percent redemption rate for covered containers,[43] although this may be an overestimate. Maine reports a 96 percent recycling rate for its expanded bottle bill, which covers all beverage containers except dairy products and cider.[44] The difference in recycling rates between deposit and non-deposit states results in nearly two-thirds of all soft drink bottles recycled in the U.S.A. originating in the 10 deposit states, even though they have only about 28 percent of the population.[13]

The decline in recycling rates for soft drink containers has resulted in pressure on major soft drink manufacturers, first Coca-Cola and then PepsiCo, to increase PET soft drink

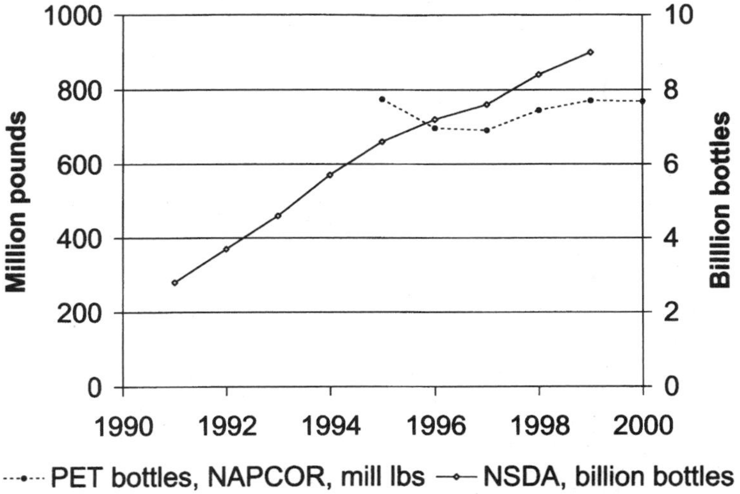

Figure 12.20 Recycling of PET bottles, U.S.A.[35,42]

bottle recycling and, in particular, to use recycled content in bottle manufacture. The GrassRoots Recycling Network (GRRN) and other organizations even took out full-page ads in the *New York Times* and *Wall Street Journal* blasting Coca-Cola for abandoning promises it made in 1990 to use recycled content in soft drink bottles. In 2001, at stockholder meetings for both companies, resolutions were introduced to use recycled content and to work toward increasing recycling rates. In both cases, the resolutions failed, but supporters hailed the larger than expected votes in favor of the resolutions as a successful first step and vowed to bring the issue back in 2002.

At least partially in response to consumer pressure, Coca-Cola began using recycled content in a portion of its U.S. soft drink bottles in 1998, although the company did not make any public announcement that it was doing so until 2000.[45] The company had been using recycled content in Australia and a few other countries for a number of years. After the 2001 stockholder meeting, Coca Cola committed to work toward incorporation of 10 percent recycled content in its soft drink bottles by 2005 and also agreed to cooperate with Businesses and Environmentalists Allied for Recycling (BEAR) in efforts to increase PET bottle recycling.[46]

12.3.2 Recycling of Custom PET Bottles

Recycling rates for custom PET bottles are considerably lower than rates for soft drink bottles. Some custom bottles for beverages such as fruit juice and drinking water are covered by the expanded deposit legislation in Maine and in California. PET beer bottles are covered by deposits in all deposit states. However, non-beverage containers such as peanut butter jars and shampoo bottles are not covered by deposits anywhere in the U.S.A.

The introduction of PET bottles in colors other than green and clear has introduced new complications into PET recycling. For example, the blue color of some bottled water has sparked criticism. It appears that as long as the blue color remains a small fraction of collected materials, it can be incorporated into the green PET stream with no problem. However, larger amounts would create problems. Many of the other colors of PET are handled by diverting the containers into disposal rather than recycling.

The recent introduction of PET bottles for beer has sparked a new round of concern. The amber color of some of these containers means they must be separated from the clear and green PET. Furthermore, these bottles contain additional materials to provide the required level of protection against oxygen permeation. When ethylene vinyl alcohol (EVOH) is used as the barrier, the resulting structure is very similar to bottles for ketchup and some other foods. When the PET/EVOH ketchup bottle structure was introduced, it was shown that the bottle, which is designed to delaminate, is compatible with normal PET recycling systems, at least at current use and recovery levels. Most of the EVOH is removed during the washing and rinsing stages, and the small fraction that remains does not cause performance problems. For PET beer bottles containing a nylon barrier layer, estimates are that 30 to 40 percent of the barrier layer will be removed during processing, leaving the recycled PET containing about 3 percent nylon by weight. Manufacturers of some other bottle variants, such as those using an activated carbon coating, also claim either that the barrier material will be removed during the recycling operation or that the tiny amount of material used in a bottle will not be enough to interfere with end uses of the recovered material.[47]

12.3.3 Mechanical Recycling of PET Bottles

Most PET recycling processes use mechanical processing to convert the collected bottles into a usable form. While precise designs are proprietary, most operate similarly to the pilot recycling facility developed at Rutgers University under the sponsorship of the Plastics

Recycling Foundation. This process begins with color separation, followed by shredding the whole bottles, usually in two steps with initial shredding, followed by a finer granulation step. Next, the shredded material is sent through an air classifier to blow off the light particles, which consist primarily of fines and label fragments. The material is then washed in hot detergent to remove product residues and soften and remove adhesive. Washing is followed by screening and rinsing. Next, a density-based separation using hydrocyclones separates the heavier-than-water PET from the lighter-than-water polyolefins, which consist predominantly of polypropylene from caps and, to a lesser extent, from labels. The PET is dried and then sent through a metal removal process, often using an electrostatic separator.

The Rutgers process was originally developed to handle beverage bottles at a time when nearly all bottles had paper labels, HDPE base cups, and aluminum closures. The disappearance of the base cups and change from aluminum to PP closures has greatly facilitated the recycling process. Aluminum, in particular, was difficult to remove and caused serious performance problems in the recycled material. Metal caps are still used on a few PET containers, but nearly all now use plastic caps and lids. A remaining source of aluminum in recycled PET is fragments of inner seal materials containing a foil layer, which are sealed to the container during the packaging process and may not be removed completely when the consumer opens the container. Voluntary design guidelines discourage the use of aluminum inner seals that are not readily removed when the container is opened, but not all manufacturers adhere to these guidelines.[48] Change of the labels from paper to plastic has also facilitated recycling. In current processes, the PP label fragments that are not removed during previous process steps will be removed with the PP caps in the hydrocyclone.

12.3.4 Recycling of Non-bottle PET

While most PET recycling processes in the U.S.A. handle only bottles, facilities in Germany handle mixed PET packaging, including bottles, tubs, dishes, and film, from yellow bag or bin collection of plastics through the Duales System Deutschland (DSD). Most PET recycling in Germany has utilized feedstock recycling (see Sec. 12.2.6), but the plan is to significantly increase mechanical recycling beginning in 2001.[49]

PET x-ray film represents another source of recycled material. Since these materials generally are coated with silver, there has long been a potent economic incentive for their recovery, and silver from x-ray film has been recovered since the early 1900s.[50] In such processes, recovered PET can be obtained as a by-product of silver recovery. Its recycling is complicated by the fact that it is generally coated with PVDC. Gemark is reported to have a proprietary process to remove the PVDC.[51] In the U.S.A., one recycler of x-ray film is United Resource Recovery Corp. (URRC) of Spartanburg, SC.[52]

DuPont operated a feedstock recycling facility, using its "Petretec" process to recover PET materials such as x-ray film, from 1995 to 1998 but discontinued the operation due to poor market conditions.[53]

12.3.5 Feedstock Recycling of PET

Recovered PET can be chemically broken down into small molecular species, purified, and then repolymerized. The two major processes for tertiary recycling of PET are glycolysis and methanolysis. Both result in PET that is essentially chemically identical to virgin resin and has been approved by the U.S. Food and Drug Administration for food contact applications. However, PET produced by these processes is more costly than virgin resin, which significantly limits its use.

In 1991, Goodyear obtained a letter of no objection from the U.S. FDA for the use of its "Repete" tertiary recycled PET in food contact applications. The process, later sold to

Shell Chemical Co., used glycolysis to partially break down PET, followed by purification and repolymerization. In tests using model contaminants, the contaminants were removed down to a 50- to 100-ppm level. That same year, both Eastman Chemical Co. and Hoechst-Celanese Corp. received letters of no objection from FDA for their methanolysis-based PET depolymerization processes.[54] Eastman's "Superclean" process reportedly can handle PET bottles of any color, and including multilayer bottles such as those containing oxygen barriers, producing new PET equivalent to virgin.[55] Methanolysis processes provide full depolymerization and can remove colorants and certain impurities that cannot be removed by glycolysis.

Petrecycle Pty. Ltd., of Melbourne, Australia, announced in February 2001 that it would install a chemical recycling system capable of processing more than 22 million pounds of post-consumer PET a year in the M&G Finanziaria Industriale SpA virgin PET production facility in Point Pleasant, WV. Petrecycle's "Renew" technology will enable M&G to produce a blend of virgin and recycled PET, reportedly for lower costs than those associated with other recycling technologies.[56]

Other tertiary recycling processes include a Freeman Chemical Corporation process to convert PET bottles and film to aromatic polyols used for manufacture of urethane and isocyanurates.[54] Glycolized PET, preferably from film, since it is often lower in cost than bottles, can be reacted with unsaturated dibasic acids or anhydrides to form unsaturated polyesters. These can then be used in applications such as glass-fiber-reinforced bathtubs, shower stalls, and boat hulls. United States companies that have been involved include Ashland Chemical, Alpha Corporation, Ruco Polymer Corporation, and Plexmar.[51] Unsaturated polyesters have also been used in polymer concrete, where the very fast cure times facilitate repair of concrete structures. Basing polymer concrete materials, for repair or precast applications, on recycled PET reportedly leads to 5 to 10 percent cost savings and comparable properties to polymer concrete based on virgin materials. However, they are still approximately 10 times the cost of portland cement concrete.[57]

12.3.6 Food-Grade Mechanically Recycled PET

A variety of processes have received official non-objection from the U.S. FDA for use of mechanically recycled PET in food packaging. The earliest processes relied on insensitive uses or on imposition of a physical barrier between the food and the recycled plastic. For example, the first approval was in 1989 for use of recycled PET in egg cartons.[51] Continental PET Technologies received approval in 1993 for a coinjected multilayer PET bottle with a 1-mil (0.001-in) layer of virgin PET between the core layer of recycled PET and the container contents. The approach was used initially for soft drink bottles in Australia, New Zealand, and Switzerland.[58–60] In 1994 and 1995, Wellman, Inc., obtained approval for use of mechanically recycled PET in multilayer packaging for a variety of food products.[61]

The first U.S. approval for use of mechanically recycled PET in direct contact with food came in 1994, for Johnson Controls' Supercycle recycled PET. In 2000, 6 of the 20 operating PET recycling plants used technologies that have received letters of non-objection from FDA for direct contact with foods and beverages.[35] These processes rely on intensive cleaning, often in combination with control over the source material, and have been validated by challenge with known amounts of model contaminants.

In 1999, Phoenix Technologies LP became the first company to receive FDA approval for use of 100 percent curbside recycled PET in food containers. In 2001, the company gained approval for use of this material in hot-filled bottles. Commercial production of food-grade curbside recycled PET began in December, 2000. The process is in use in Australia to make 25 percent recycled-content bottles for Coca-Cola.[62]

United Resource Recovery Corporation (URRC) of Spartanburg, SC, became the second U.S. company to get approval for food-grade recycled PET from curbside collection

in 2000, for its "Hybrid-UnPET" technology. While Phoenix Technologies was expected to focus on the soft drink market, URRC was targeting the bottled water market, particularly in Europe. A plant in Frauenfeld, Switzerland, is licensed to use the technology.[52] The process reportedly involves mechanical recycling without hot water followed by a thermal treatment using sodium hydroxide and a final stage for removal of residual contaminants.[63] The treatment with caustic soda results in a solid-phase reaction in which the outer surface of the PET chips is stripped off, and the resulting ethylene glycol and terephthalic acid are recovered as by-products. Any contamination adhering to the outer surface is removed during this stage. Residual contaminants are removed using a combination of air blowing and controlled temperature. The resultant mixture of salt and clean PET granules is separated by mechanical filtration followed by washing and then removal of any small metal particles by a metal separator.[40] This process, therefore, could be regarded as a mix of mechanical and feedstock recycling.

Another process producing food-grade recycled PET is the "Stehning BtoB Process," developed by OHL Apparatebau & Verfahrenstechnik GmbH, of Limburg, Germany, which received U.S. FDA approval in 1999. The first production unit began operation in October 1999 at PET Kunststoffrecycling GmbH (PKR) in Beselich, Germany. Two other facilities are also in operation, worldwide, using this technology. In this process, the clean PET bottle flakes, without preliminary drying, are fed into a modified twin screw extruder where it is dried and degassed and then melt-filtered and pelletized. The amorphous PET chips are fed into a discontinuous solid-stating process for crystallization and condensation/decontamination under vacuum. Reportedly, the sensory characteristics of the recycled material are superior to those of virgin PET. In particular, acetaldehyde and ethylene glycol levels are lower. The German facility has a production capacity of 7,500 tonnes per year.[64]

12.3.7 Properties of Recycled PET

Mechanically recycled PET in general retains very favorable properties. Some reduction in intrinsic viscosity is usual, but it can be reversed by solid-stating. Residual adhesives from attachment of labels are a common contaminant concern. Some of the adhesive residue can become trapped in the PET granules and is not removed by washing. Since these adhesives often contain rosin acids and ethylene vinyl acetate, the rosin acids plus acetic acid from ethylene vinyl acetate hydrolysis can catalyze hydrolysis of the PET during processing. A similar problem can be caused by residues of caustic soda or alkaline detergents from the wash step. Considerable loss of molecular weight can result, and darkening of the adhesive residues can cause discoloration.

PET is very susceptible to damage from PVC contamination, and vice versa. Contamination in the range of 4 to 10 ppm can cause serious adverse effects.[65] Because the densities of PET and PVC overlap, density-based separation methods are ineffective. Technologies have been developed for very effective sorting of whole-bottle PVC and PET. PVC contamination from materials such as coatings, closure liners, labels, etc., is more difficult to handle. Appropriate package design to avoid the use of PVC or PVDC with PET containers is the most effective strategy. The Association of Post-Consumer Plastic Recyclers issued a report detailing the effect of PVC contamination of PET, in which they estimated that the cost to the domestic PET recycling industry of addressing PVC contamination in 1998 totaled \$6.5 million. Sorting accounted for 37 percent of the cost, with depreciated equipment, laboratory labor, and maintenance also representing major costs. The average cost was 1.67 cents per pound of PET produced. Because costs were lower in larger reclaimers, as well as in those specializing in deposit containers, the weighted average was 0.86 cents per pound.[66]

Repolymerized PET is essentially identical in performance to virgin PET.

12.3.8 Markets for Recycled PET

One of the earliest large-volume uses for recycled PET was as polyester fiberfill for applications such as ski jackets and sleeping bags. The range of applications has grown enormously and now includes items as diverse as carpet, automobile distributor caps, produce trays, and soft drink bottles. Fiber applications remain the largest market and continue to grow (see Table 12.3). In fact, Aoki International, of Tokyo, introduced an all-recycled men's suit made from recycled soft drink bottles. The company also promised to recycle used suits returned to their stores into buttons and linings for new suits.[67]

TABLE 12.3 Uses of Recycled PET in the U.S.A. (Millions of Pounds)[35,54]

	1990	2000
Fiber	165	452
Strapping	12	101
Sheet and film	2	65
Food and beverage bottles	0	54
Non-food bottles	1	40
Other	21	32

The first 100 percent recycled PET container in the U.S.A. was introduced in 1988 by Proctor & Gamble for household cleanser. Bottles, including those for food and beverages, are now a significant market for recycled PET, and pressure for bottle-to-bottle recycling continues to grow.

Other markets for recycled PET include strapping, sheet, and film. A number of high-performance engineering alloys and compounds utilizing recycled PET have been developed, especially for the automotive industry.[51] Recycled PET is being used in manufacture of a drainage filtration mesh for roadways by Viy Plastics, an Australian joint venture.[68] A German company, Remaplan Anlagenbau BmbH, has developed a plastic pallet made from 75 percent post-consumer PET, 20 percent post-consumer polyolefins, and 5 percent additives. It claims the pallets will sell for $10 to $12 each, about the same as wood pallets. The company can also produce boxes, crates, trays, and similar items.[69]

As is generally the case for recycled materials, the market situation is strongly affected by the supply and demand situation for virgin resin. When virgin PET supply is low and prices are up, demand for recycled resin is strong. During the last half of the 1990s, there was a significant downturn in recycled PET demand caused by a large increase in production capacity for virgin resin that drove down price. The situation was exacerbated by a temporarily plentiful supply of off-spec resin from new facilities entering production. Some PET recyclers did not survive the lean years. Markets recovered somewhat as demand grew but have not returned to the 1995 situation where baled PET bottles sold for $0.27 to $0.35 per pound.[70] Prices in 2000 ranged from $0.07 to $0.20 per pound.[35]

12.4 High-Density Polyethylene Recycling

The sources of high-density polyethylene (HDPE) in the U.S. MSW stream are shown in Fig. 12.21. HDPE is the second most recycled plastic but has been steadily closing the gap

Figure 12.21 Sources of HDPE in U.S. municipal solid waste, 1998.[1]

on PET. In the U.S.A., the recycling rate for HDPE milk and water bottles reached 31.4 percent in 1998, for a total of 440 million pounds. Other HDPE containers were recycled at a rate of 12.3 percent, for an overall bottle and container recycling rate of 19.3 percent. The recycling rate for HDPE packaging as a whole was 10.3 percent. The recycling rate for HDPE in durable goods was 8.2 percent, and in nondurables was negligible, for an overall HDPE recycling rate of 8.9 percent.[1] While the recycling rates for HDPE are lower than for PET in some cases, the total amount of HDPE recycled exceeds that of PET in all relevant major categories: durables, bottles & containers, all packaging, and all MSW (Fig. 12.22).

The American Plastics Council (APC) reported a 1998 recycling rate for HDPE bottles of 25.2 percent, for a total of 734 million pounds, higher than the 24.4 percent reported for PET bottles.[71] APC reported rates of 24.4 percent in 1996 and 24.7 percent in 1996.[71,72]

Recovery rates for HDPE fell somewhat in 1999, with the rate for natural HDPE bottles (milk and water, primarily) falling to 30 percent from 31 percent in 1998.[73]

HDPE bottle recycling in the U.S.A. is dominated by KW Plastics of Troy, AL, with a capacity of 600 million pounds per year. The next largest companies are USPLC, with facilities in Los Angeles, Chicago, and Auburn, MA, for a combined capacity of 125 million pounds per year, and Clean Tech of Dundee, MI, with 90 million pounds per year.[36]

12.4.1 Recycling of HDPE Bottles

Unpigmented HDPE milk and water bottles are the most valuable type of HDPE for recycling. They are made from a high-quality fractional melt index homopolymer HDPE, usually unpigmented, that is suitable, when recycled, for a wide variety of uses. Considering that these bottles are generally not covered by deposits, their 31.4 percent recycling rate in the U.S.A. in 1998 compared favorably with the 35.4 percent rate for PET soft drink bottles.[1] Extrusion blow molded HDPE bottles, both pigmented and unpigmented, are accepted for recycling in most community recycling programs. Injection blow molded bottles, which are made from a high-melt-flow HDPE, are undesirable contaminants. Other containers that are unacceptable in many programs include motor oil bottles and

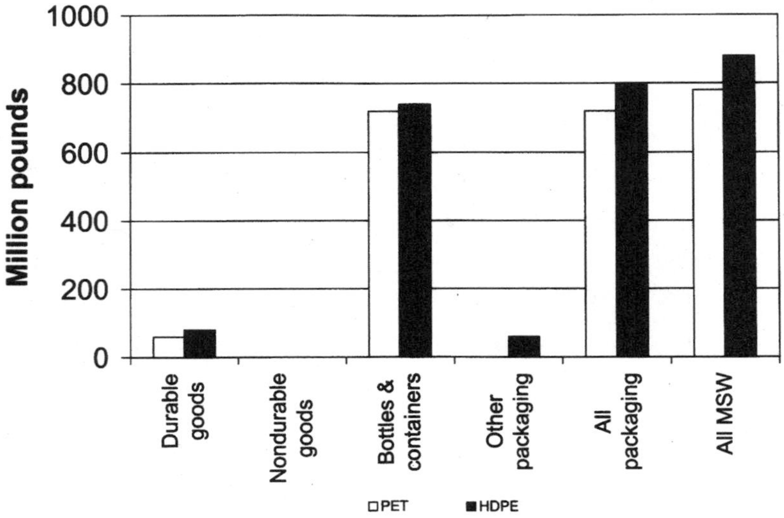

Figure 12.22 Amounts of recycled HDPE and PET from U.S. MSW, by product category, 1998.[1]

those that contained caustic cleansers, insecticides, or other materials where product residues could pose a risk.

Ontario, Canada, has had a deposit system for HDPE milk bottles for a number of years, charging 25 cents per bottle, which is one of the reasons milk sold in flexible pouches is popular. When the government considered discontinuing the system in 1999, Ontario milk producers, retailers, and packaging suppliers objected.[74]

The Alberta (Canada) Dairy Council launched a voluntary recycling program in 1999 in an effort to increase the 35 percent recycling rate for HDPE milk containers to 70 percent. Municipalities and recycling programs receive a payment for collected and densified HDPE, subsidized by the Dairy Council, in the form of a guaranteed price of $400 (Canadian) per tonne. The three Alberta milk processors voluntarily pay two cents for each four-liter milk bottle and one cent for each two-liter bottle into a Container Recovery Fund to support the program. In its first year of operation, July 1, 1999, to June 30, 2000, the program collected 1,197 tonnes of material, a 32 percent increase over the previous year, bringing the province-wide recovery rate to 40 percent. Sixteen communities and recycling authorities achieved recovery rates of 70 percent or more.[75]

Motor oil bottle recycling is a significant issue, not only because of the volume of bottles involved but because of the potential adverse environmental consequences and the value of the oil remaining in the bottles. Honeywell Federal Manufacturing & Technologies, of Kansas City, has developed a system for recycling motor oil bottles and recovering both plastic and oil. The system is licensed to ITec International Technologies, Inc., a subsidiary of Beechport Capital Inc., which is marketing the systems worldwide. ITec estimates that about two billion plastic motor oil containers are discarded each year in the U.S.A., each containing, on average, an ounce of oil, for a total of 250 million pounds of plastic and more than 15 million gallons of motor oil. ITec announced in 2001 that it had received an order for two of its systems from OPT Srl in Italy.[76]

Some programs for used oil recovery also recover oil bottles. For example, in Canada, the Alberta Used Oil Management Association (AUOMA) has developed a program for

used oil recycling that also includes recycling of oil filters and motor oil containers under 30 liters in size. Participating companies pay an Environmental Handling Charge (EHC) of $0.05 (Canadian) per liter of container size, at the wholesale level. The collectors of the used materials receive payment through the program for the returned materials. Saskatchewan and Manitoba are implementing similar programs, and British Columbia is considering doing so.[77]

12.4.2 Recycling of Other HDPE

While most recycled HDPE comes from bottles, limited recycling of other HDPE materials also occurs. Some recovery of HDPE film occurs along with LDPE when retail bags are collected for recycling. This collection has declined substantially in the last several years. DuPont operates a recycling program for its Tyvek envelopes. As mentioned above, some recycling of HDPE in durable goods also occurs, according to the U.S. EPA.

12.4.3 Recycling Processes for HDPE

Recycling processes for HDPE bottles are similar to those for PET. First, the collected HDPE is typically sorted to separate the higher-value unpigmented containers. In some cases, the pigmented HDPE is further separated into color families. Sorting of the unpigmented bottles is often done prior to initial baling but may be done at a later stage.

At the plastics processor, the baled HDPE containers are typically shredded, washed, and sent through either a float-sink tank or a hydrocyclone to separate heavy contaminants. Air classification may be done prior to washing. The clean materials are dried and then usually pelletized in an extruder equipped with a melt filter to remove residual nonplastic contaminants. If mixed colors are processed, the result from typical curbside or drop-off programs is a grayish-green color, which is most often combined with a black color concentrate to produce black products.

Several types of contamination are a concern in HDPE recycling. The first is contaminants that add undesired color to natural HDPE. A prime culprit is caps on bottles. While consumers are generally told to remove caps before turning the bottles in for recycling, a significant number arrive with the caps still in place, and the caps are generally brightly colored. Most of these are polypropylene, with the next largest fraction polyethylene. Neither of these materials are separable in the usual recycling systems, and hence they usually remain with the HDPE, where they result in discoloration of the resin. Typically, the amounts are low enough that mechanical properties are not affected, but they do impart a grayish color to the material. The introduction of pigmented HDPE milk bottles, which seem to periodically pop up in various places, is a concern to recyclers since, if they were widely adopted, they could significantly cut into the use of the more valuable natural bottles. Pigmented HDPE recycled resin typically sells for only 60 percent of the price of natural HDPE.[78]

A second type of contamination is mixing of high melt flow injection-molding grades of HDPE with low melt flow blow-molding grades. The result can be a resin with intermediate flow properties that is not desired by either injection or blow molders. Since the coding system for plastic bottles (see Sec. 12.18) does not differentiate between the two, it is difficult to convey to consumers in any simple fashion which bottles are desired in the recycling system and which are not. The recycling process does not separate the two grades, since their densities and most other properties are equivalent. Some programs simply accept the resulting contamination, while others try to get the message to consumers, sometimes by specifying bottles "with a seam." Fortunately, the vast majority of HDPE bottles, especially in larger sizes, are extrusion blow molded.

Mixing of polypropylene into the HDPE stream is also a concern. As discussed above, much of this arises from bottle caps left in place. Some also arises from fitments on detergent bottles and from inclusion of PP bottles in the recycling stream. Since both PP and HDPE are lighter than water, the density-based separation systems commonly employed will not separate the two resins. Fortunately, in most applications, a certain level of PP contamination can be tolerated. However, particularly in the pigmented HDPE stream, levels of PP contamination are often high enough to limit the amount of the recycled material that can be used, forcing manufacturers to blend the post-consumer materials with other scrap that is free of PP, or with virgin (often off-grade) HDPE. While triboelectric systems for separating chipped PP from HDPE are now available, most recycling facilities do not have such systems.

Finally, contamination of the HDPE with chemical substances that may later migrate from a container with recycled content to the product can present problems. This is a more serious issue with HDPE than with PET for two primary reasons. First, the solubility of foreign substances of many types is greater in HDPE than in PET. Thus, the level of contamination that may be present is higher. Second, the diffusion of most substances is faster in HDPE than in PET. Combined, these factors create a significantly greater potential for migration of possibly hazardous contaminants out of recycled HDPE into container contents. Nevertheless, use of some types of recycled HDPE for limited food contact applications has been approved by the U.S. FDA.

The strategies for dealing with potential migration of hazardous substances from recycled HDPE are essentially the same as with PET. First, a combination of selection of starting materials and processing steps can be used to minimize the contamination levels that are present. The first company to obtain a letter of non-objection from FDA for recycled HDPE in direct food contact was Union Carbide, which later sold the technology to Eco-plast, who also received a letter of non-objection.[79]

Recycled HDPE can also be used in a multilayer structure that provides a layer of virgin polymer as the product contact phase. This is the standard approach for laundry products, where FDA approval is not an issue, but where consumer acceptability issues associated with objectionable odors in the product surfaced early in the development stage. In this case, the multilayer bottles used have a layer of virgin polymer on the outside of the bottle as well as on the inside. This not only solved appearance problems that were also associated with the use of recycled plastic, it permitted a significant savings in the amount of fairly expensive colorants that are required. The middle layer in such structures is composed of a blend of recycled HDPE with process regrind. An additional benefit from the inner layer of virgin polymer is the better environmental stress crack resistance of the copolymer virgin HDPE as compared with the regrind/recycle mix.

12.4.4 Markets for Recycled HDPE

A major early market for recycled HDPE was agricultural drainage pipe. Pipe continues to be a significant market, but a number of additional markets have developed as well. In particular, coextruded bottles containing an inner layer of recycled HDPE have developed into a major market. Nearly all laundry products sold in plastic bottles in the U.S.A. use this structure, typically incorporating about 25 percent recycled content. Motor oil is often sold in single-layer bottles made from a blend of virgin and recycled HDPE. Figure 12.23 shows the proportion of recycled HDPE going into various market categories in the U.S.A. in 1996.

As can be seen, another significant market for recycled HDPE is plastic lumber, 25 percent of total HDPE markets in 1996.[2] Use of plastic lumber is increasing, as its benefits of long life compared to treated wood, freedom from the hazardous chemicals often used in outdoor grades of lumber, and maintenance of color without painting, are recognized.

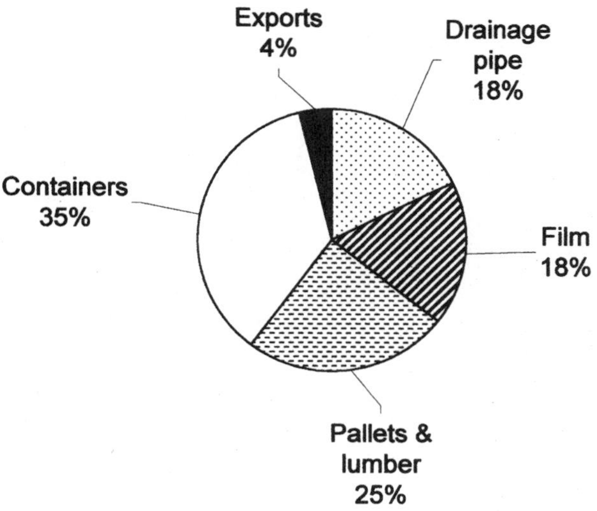

Figure 12.23 Uses of recycled HDPE bottles in the U.S.A., 1996.[2]

Plastic lumber does carry a higher initial purchase price than wood equivalents, but life cycle costing generally shows substantial benefits for plastic. Composites of wood fiber and plastic, either HDPE or LDPE, are also growing in use. Plastics lumber is discussed in more detail in Sec. 12.18.

Recycled HDPE is also used in manufacture of film, especially for merchandise bags. Often, the material used is recycled milk bottles. Recycled milk bottles are being used in manufacture of milk crates in Australia.[80] They are also often used in production of curbside recycling bins.

United States companies that are major users of recycled HDPE include Procter & Gamble, which uses 25 to 100 percent recycled HDPE in most of their household products, Clorox, and DowBrands.[81] DuPont uses 25 percent recycled HDPE in its Tyvek envelopes.[82] Hancor, which manufactures drainage pipe, was one of the first large-scale users of recycled HDPE and continues to be a major user.

12.5 Recycling of Low-Density Polyethylene and Linear Low-Density Polyethylene

Because of the similarity in properties and uses of low-density polyethylene (LDPE) and linear low-density polyethylene (LLDPE), and because they are often blended in a variety of applications, use and recycling of LDPE and LLDPE are often reported and carried out together. Therefore, we will use the term *LDPE* to refer to both LDPE and LLDPE.

Approximately half of the LDPE found in municipal solid waste originates in packaging, as shown in Fig. 12.24.[1] Another sizable fraction comes from nondurable goods, especially trash bags. The two main sources of recycled LDPE are both in the bags, sacks, and wraps category: stretch wrap and merchandise bags. In contrast to PET and HDPE, curbside recycling is not a significant factor in recycling of LDPE in the U.S.A. Stretch wrap is collected primarily from establishments such as warehouses and retailers, where large quantities of goods arrive in pallet loads unitized with the wrap. Sending such material to

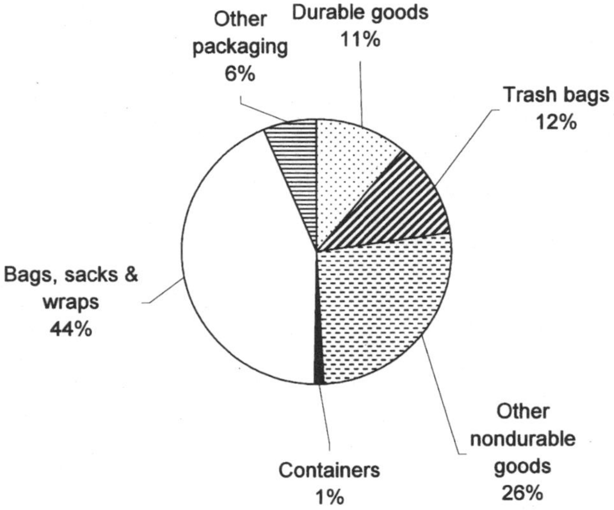

Figure 12.24 Sources of LDPE/LLDPE in U.S. municipal solid waste, 1998.[1]

be recycled, rather than paying to dispose of it, often makes economic sense for the companies involved.

Recycling of grocery and other merchandise sacks is generally carried out at drop-off locations. At one time, there was a wide network of such sites across the U.S.A., but many merchants discontinued the program due to contamination and other concerns. Bag recycling through schools is also available in a significant number of locations.

In 1997, New Jersey had a pilot recycling program targeted at nursery and greenhouse film. In the three month project period, it collected nearly 450,000 lb of film, about 45 percent of the total used by growers in the state. The biggest problem encountered was the dirt in the film. The program was successful enough to expand in 1998.[83] Chevron operated a pilot project for agricultural film in Vermont in 1997 and 1998.[84]

Toro Ag Irrigation, based in El Cajon, CA, began a program in 1999 to recycle its Aqua-Traxx thin-walled drip irrigation hose, made from linear low-density polyethylene. The used irrigation tape could be delivered to Toro dealers and was then sent for recycling. Toro estimated that more than 1 billion feet of drip irrigation tape is sold each year in California.[85]

The U.S. EPA calculated that 120 thousand tons of LDPE/LLDPE bags, sacks, and wraps were recovered in 1998, for a recycling rate of 5.2 percent. About 20 thousand tons of LDPE and LLDPE were recovered from durable goods, a rate of 3.3 percent. The overall recycling rate for LDPE in MSW was 2.6 percent, up from 1.9 percent in 1997.[1]

In Canada, the Plastic Film Manufacturers Association of Canada and the Environment and Plastics Institute of Canada (EPIC) have sponsored curbside recycling programs for plastic film of all types. By 1996, the program had grown to 146 communities in Ontario and 19 in the Montreal, Quebec, area. In 1998, EPIC published *The Best Practices Guide for the Collection and Handling of Polyethylene Plastic Bags and Film in Municipal Curbside Recycling Programs*.[86]

Processing of film plastics is more difficult than processing of containers. The lower bulk density of the film leads to difficulty in handling the material, and it is more difficult

to remove contaminants. Historically, a large fraction of merchandise bags collected in the U.S.A. were shipped to the Far East, where low labor costs permitted hand sorting. Stretch wrap is less contaminated, especially if paper labels are not used, and is often handled domestically.

The Environment and Plastics Industry Council, of Mississauga, Ontario, has devoted considerable effort to identifying processing systems capable of handling post-consumer polyethylene film, as well as to collection of the material, as mentioned above. They chose to focus on dry processing as key to keeping the cost of pelletizing the materials down. EPIC found that pneumatic separation of polyethylene film can remove 98 percent of contaminants from film that has first been chopped, shredded, or granulated. For agglomerating the shredded material, continuous-feed agglomeration was found to be generally superior to pellet mill processing and batch-style agglomeration. While the equipment was more costly than batch agglomeration, labor costs were lower, and product quality was higher.[87]

A major market for recycled plastic film and bags is manufacture of trash bags, typically in a blend with virgin resin. Recycled plastic has also been used in manufacture of new bags, bubble wrap, housewares, and other applications. A rapidly growing application is use of recycled LDPE in plastic lumber, as a composite with wood fibers. The leading company in this area is Trex Co., of Winchester, VA, which has production facilities in both Nevada and Virginia. The company was cited by *Industry Week* in 2000 as one of the most successful small manufacturers in the U.S.[19] Boise-Cascade of Boise, Idaho, and Louisiana-Pacific of Portland, Oregon, also manufacture wood/plastic composites from wood fiber and recovered LDPE, primarily from pallet wrap. Trex also recycles contaminated agricultural film and some material recovered from MRFs.[73]

12.6 Recycling of Polypropylene

Sources of polypropylene (PP) in MSW in the U.S.A. are shown in Fig. 12.25. Packaging represents 33 percent of the total, and durable goods represent 40 percent. There is little recycling of PP from packaging, a rate of only 3.2 percent in 1998, nearly all from the "other plastics packaging" category. Most recycled PP comes from durable goods, where the recycling rate was 12.4 percent in 1998. A significant amount of this material comes from recycling of PP automotive battery cases. The overall recycling rate for PP in U.S. MSW in 1998 was 6 percent.[1]

Lead-acid automotive batteries are prohibited from MSW disposal facilities in 37 U.S. states, primarily due to concern about the effects of lead. Several states impose deposits on batteries. Effective recycling programs for these batteries have existed for a number of years. The Battery Council International reports a 1996 recycling rate of 96.5 percent for lead-acid batteries. PP makes up about 7 percent of the battery, by weight, and is recovered along with the lead. The primary market for the recovered PP is new battery cases. A typical battery contains 60 to 80 percent recycled PP.[88]

PP spools and wheel counters in disposable cameras are recovered for reuse in programs for recycling these cameras, as will be discussed further in Sec. 12.7. Polypropylene hangers from department stores are also sometimes recycled, and some PP is recycled from appliances.

12.7 Recycling of Polystyrene

Nondurable goods represent by far the largest category of polystyrene (PS) in U.S. MSW, 63 percent, with plastic plates and cups alone representing 39 percent (Fig. 12.26). Durable goods account for about 27 percent, with packaging amounting to about 10 percent.[1] A

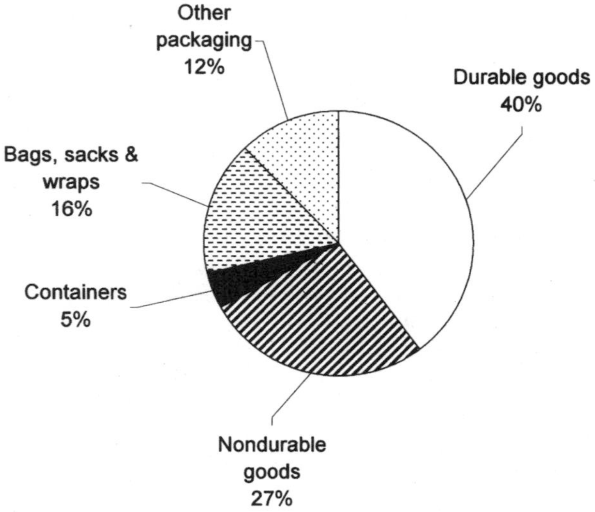

Figure 12.25 Sources of PP in U.S. municipal solid waste, 1998.[1]

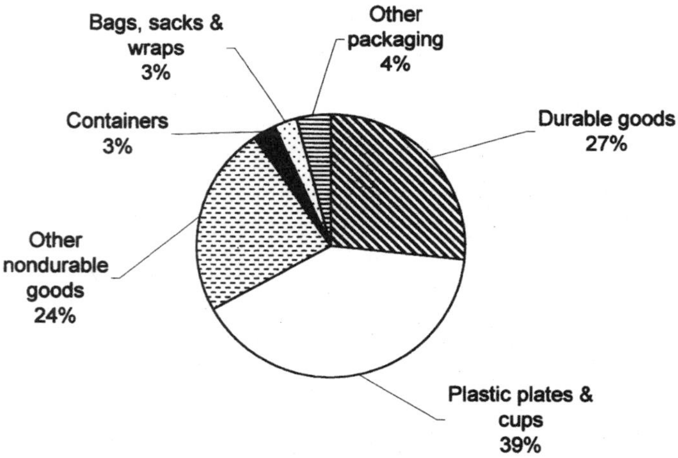

Figure 12.26 Sources of PS in U.S. municipal solid waste, 1998.[1]

substantial amount of PS is used in the building and construction industry, mostly for insulation materials, but these wastes are not considered part of the U.S. MSW stream.

The EPA reports recovery of about 10 thousand tons of PS from durable goods in 1998, a recycling rate of only 1.7 percent. No significant recovery of nondurable goods was reported. Another 10 thousand tons were recovered from the packaging category, for a 4.8 percent recycling rate, resulting in an overall recycling rate for polystyrene of only 0.9 percent.[1]

There is very little recycling of food-service PS in the U.S.A. In 1989, the National Polystyrene Recycling Company (NPRC) was formed by several PS producers, with a goal of achieving a 25 percent recycling rate for food-service PS by 1995. It focused on institutional generators of such wastes, primarily schools, and other cafeterias. However, the operation was plagued by high levels of contamination with food wastes and was unable to operate profitably. Recycling rates and amounts for PS food-service items, including packaging, declined after an initial period of growth. In 1999, NPRC and its remaining two recycling facilities were sold to Elm Packaging Company and its name changed to Polystyrene Recycling Company of America.[89] By late 2000, the company evidently stopped accepting PS food packaging for recycling.[90]

Nearly all recycling of PS from packaging in the U.S.A. now comes from recycling of foam cushioning materials. Such recycling has been much more successful than recycling of food-service PS. Recycling rates have, for the most part, been at the 9 to 10 percent level for the last several years (Fig. 12.27). The American Plastics Council reported a recycling rate for EPS packaging of 9.6 percent in 1999, up from 9.5 percent in 1998. The amount recycled increased by 1 million pounds, reaching 20.2 million pounds in 1999.[91,92] The Alliance of Foam Packaging Recyclers reports availability of more than 200 collection sites in the U.S.A. and Canada for post-consumer EPS protective packaging. The organization web site includes information for consumers about collection locations available in their area.[91] Much of the collected material is used in manufacture of new EPS packaging, often in a blend with virgin resin. Cushioning materials made with 100 percent recycled PS are also available.

In addition to recycling, there is considerable reuse of expanded polystyrene loose fill. The Plastic Loose Fill Council operates a toll-free "Peanut Hotline" to provide information to consumers about where to take EPS loose fill for reuse. The organization reports that over 30 percent of all EPS loose fill is reused, and more than 1,500 collection sites for the material are available in the U.S.A. In addition, post-consumer recycled-content loose fill is widely available.[93]

One of the leaders in EPS recycling in the U.S.A. is FR International, which recycled 10.9 million pounds of expanded polystyrene in 2000, an increase of nearly 2 million

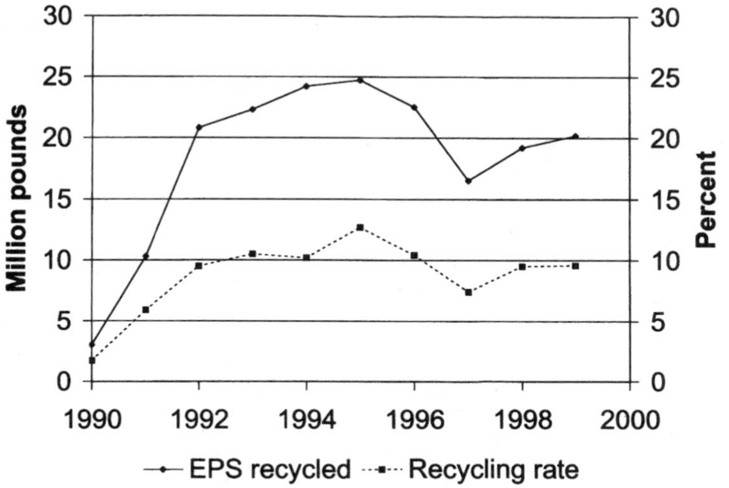

Figure 12.27 Recycling of EPS cushioning, U.S.A.[91,92]

pounds over 1999. The company, which began PS recycling in 1990, now has five recycling operations in the U.S.A. and also has a UK subsidiary that recycled over 2.5 million pounds of polystyrene in 2000.[94]

One of the problems in recycling EPS is the very low bulk density of the material, which makes shipping it over long distances uneconomical. International Foam Solutions, Inc. (IFS), of Delray, FL, has developed a process that dissolves EPS in a citrus-based solvent, producing a gel and eliminating 90 percent of the volume. The "Polygel" is stored in drums and shipped to IFS for processing. IFS further dilutes the gel, filters out contaminants, and produces new PS products. Contaminant levels are reportedly reduced to less than 1 ppm.[95] In addition to cushioning materials, the system can successfully recycle PS from food service use. The company sells or leases its "IFS Solution Machine" to customers and has several food service operations as customers.[96] Sony Corporation Research Center, in Yokohama, Japan, has also developed a solvent-based PS recycling system that uses *d*-limonene.[97]

Another company attempting to use food-service PS is Rastra Technologies, Inc. It manufactures wall panels made of recycled PS and cement. In 1999, it participated in a pilot with a Florida school district to recycle used PS foam lunch trays. In the early 1990s, five McDonalds restaurants were constructed using the company's insulated concrete form (ICF) panels containing recycled PS hamburger boxes.[98]

Kodak operates a recycling program for PS in disposable camera bodies, and also recycles film containers. Recovered camera bodies are ground, mixed with virgin resin, and used in the production of new disposable cameras. The PS internal frame and chassis of the cameras are recovered intact and reused in new cameras. The cameras are collected from photofinishers, who are reimbursed for the cameras they return. The company's program is active in over 20 countries, and achieves a recycling return rate of over 70 percent in the U.S.A. and approximately 60 percent worldwide. Kodak reports that by weight, an average of 86 percent of Kodak one-time-use cameras are recycled or reused.[99,100]

While there appears to be no current commercial use of the system, some years ago, the Toyo Dynam company, in Japan, developed a prototype system for feedstock recycling of PS in which foam PS was ground and sprayed with styrene monomer to dissolve it and separate it from contaminants. The solution was then cracked and vaporized in a heated reflux vessel.[101]

Some PS is also recycled from appliances.

In Japan, one of the goals of the Japan Expanded Polystyrene Recycling Association (JEPSRA) is to increase recycling of expanded PS. The organization reports that 53.6 percent of EPS in Japan is used in containers, 32.5 percent in transport packaging, and 13.9 percent in building materials and civil engineering applications. With a network of more than 1000 recycling sites, Japan achieved a 35 percent recycling rate for EPS in 2000 (Fig. 12.28). JEPSRA has a goal of a 40 percent rate by 2005.[102]

Korea reports a recycling rate of 48.8 percent for EPS in 1999, excluding building insulation and food containers, for a total of 24,371 tonnes. The growth in EPS recycling is shown in Fig. 12.29.[103]

The Canadian Polystyrene Recycling Association (CPRA) reports that more than a million households in Ontario and Manitoba can recycle polystyrene at curbside. The CPRA recycling plant in Mississauga, Ontario, has a capacity of up to 3,500 tonnes of PS a year. It accepts both food service PS and cushioning materials.[104]

Europe also has some recycling of PS food service items as well as cushioning. In the UK, the Expanded Polystyrene Packaging Group (EPS) reports that 4,500 tonnes of expanded polystyrene packaging were recycled in 2000, for a recycling rate of nearly 15 percent. This was a significant increase from the 3,000 tonnes recycled in 1999, a rate of 10 percent. The growth was attributed to interest in wood replacement products and an increase in the price for virgin PS resin as well as efforts by the group to develop recycling processes and raise awareness about the potential benefits of EPS recycling.[105]

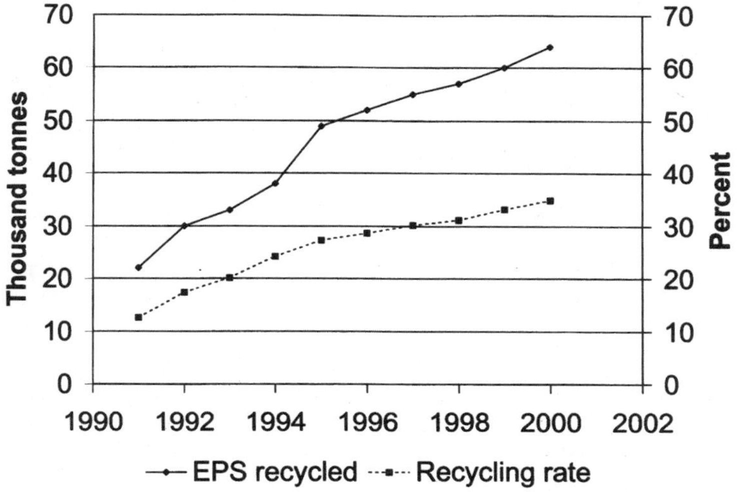

Figure 12.28 Recycling of EPS, Japan.[102]

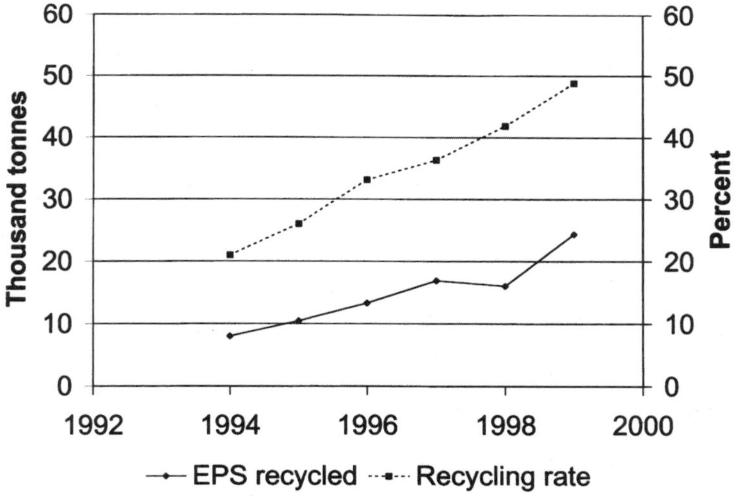

Figure 12.29 Recycling of EPS, Korea.[103]

In October 2000, producers of expanded polystyrene from Asia, Europe, and North America, representing 28 countries, met in Malaysia to form the International EPS Alliance (INEPSA) with a major goal of enhancing EPS recycling.[106]

Products manufactured from recycled polystyrene include cushioning materials, horticultural trays, video and audio cassette housings, rulers and other desktop accessories, clothes hangers, and other items.

12.8 Recycling of Polyvinyl Chloride

Polyvinyl chloride (PVC) in U.S. MSW comes mostly from nondurable goods, followed closely by packaging and durable goods (Fig. 12.30). The U.S. EPA reports no significant recycling of PVC from municipal solid waste in the U.S.[1] The PVC bottle recycling rate in 1997 was estimated to be only 0.1 percent in 1997, down from 2 percent in 1996 after discontinuation of a subsidized program by Occidental Chemical Corporation for buyback of PVC bottles.[107] There is limited recycling of PVC from other waste streams, especially construction and demolition material, which is not classified as MSW. Since non-MSW waste streams account for considerably more PVC and tend to be more uniform than MSW PVC waste streams, most efforts have focused on these materials.

In the U.S.A., the Vinyl Institute publishes a directory of North American companies involved in vinyl recycling, which is also available in searchable on-line form.[108] The 2000 edition lists about 270 U.S. and Canadian companies, with the type of scrap they accept. Nearly 100 are listed as accepting post-consumer materials. The directory also lists companies manufacturing products from recycled PVC. A study commissioned by the Institute reported that about 18 million pounds of post-consumer PVC were recycled in 1997 from sources such as carpet backing, medical products, windows and siding, and packaging. Recycling of post-industrial materials was a much larger amount, representing about 78 percent of all recycled vinyl. Since nearly 70 percent of PVC production goes into products expected to last 10 years or more, recycling of post-consumer materials is expected to be more important in the future.[109] The report found that demand for rigid post-consumer vinyl was smaller than the supply, despite the low recovery rates. The situation was somewhat better for flexible materials, but these were valued mostly for their plasticizer content, not the PVC itself.[110]

PVC has been criticized by Greenpeace and also by the Association of Postconsumer Plastic Recyclers, which in 1998 labeled PVC a recycling contaminant, pointing to the failure of a year-long effort to find markets for recycled PVC and the fact that many of their members had to landfill recovered PVC bottles because they could find no markets for them.[111] After that report, the Vinyl Institute funded a pilot project aimed to jump-start

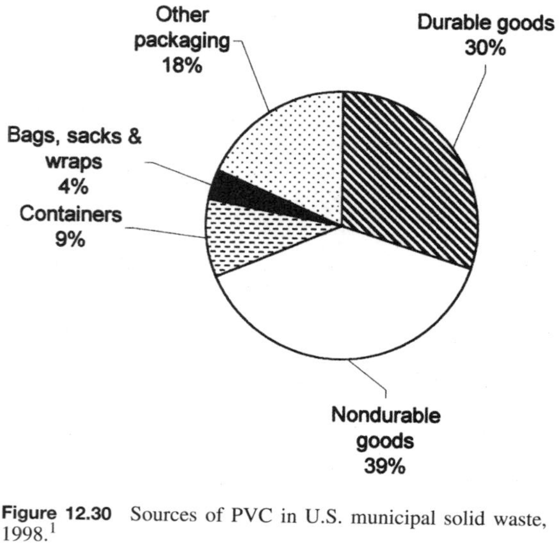

Figure 12.30 Sources of PVC in U.S. municipal solid waste, 1998.[1]

the PVC recycling market,[107] but recycling of PVC bottles in the U.S.A. remains extremely rare.

In 1999, the National Electronic Distributors Association developed an industry position paper in which they recommended recycling of PVC tubes used for packaging semiconductor devices.[112] In 2001, a consortium of New Hampshire circuit board assemblers reported success in doing so. One company reported saving $1000 per year while diverting over 25,000 lb of integrated circuit chip tubes at just one of the company facilities. The recycled tubes are made into a variety of products including new circuit chip tubes and drainage pipe.[113]

There is some recycling of PVC intravenous bags from hospitals. Beth Israel Medical Center, in New York City, was one of the pioneers in these efforts.[114] Baxter Healthcare Corporation, of Round Lake, IL, participated in a pilot collection program for hospital IV bags. Products from the recycled material included floor tiles manufactured by Turtle Plastics of Cleveland, OH.[109] Conigliaro Industries, of Massachusetts, recycled PVC from more than 60 hospitals and biotech firms, accepting IV bags, sterile packaging, employee ID cards, and other medical plastics. It also handles roofing membrane, siding, and industrial scrap, for a total of 500,000 lb per year.[109]

A number of projects have focused on recycling of vinyl siding, usually on scrap originating during building construction or remodeling, and often with financial support from the Vinyl Institute. As a result of one such pilot project in Grand Rapids, MI, recycling of vinyl siding waste is included in the EPA-funded publication *Residential Construction Waste Management: A Builders' Field Guide.*[115] Among the most high-profile pilot projects are those involving Habitat for Humanity, which builds housing for low-income families. The Vinyl Institute and other PVC-related industry organizations have donated money and materials to some of these projects in addition to supporting recycling efforts for the PVC scrap generated during construction. In 1997, Polymer Reclaim and Exchange, in Burlington, NC, was recycling about 300,000 lb per month of vinyl siding scrap through drop-off sites located at landfills and near manufacturers of mobile and manufactured homes.[116] A common market for recycled siding is skirting for mobile homes, containing between 30 and 100 percent recycled content. Other markets include gutters and pipe.[109]

Other recycling efforts targeted building-related PVC wastes have focused on window profiles, carpet backing, and pipe. Often, these, along with vinyl siding scrap, are pre-consumer rather than post-consumer wastes. For example, Mikron Industries, of Kent, WA, and Richmond, KY, buys back its customers' window scrap and uses the material, along with their own process scrap, in new vinyl window extrusions containing 15 to 25 percent recycled content. Excess scrap is used in the manufacture of vinyl fencing, drain channels, and window packaging.[109]

Several firms recycle PVC wire and cable insulation, primarily from the telecommunications industry. The recovered material typically contains small amounts of metal, so it is suitable only for applications that do not require high purity, such as truck mud flaps, flower pots, traffic stops, and reflective bibs for construction workers. The source of most of the material is phone and business equipment wiring that is being replaced by fiber optic cable.[117] One such company is Philip Environmental, in Hamilton, Ontario, which recycles about 125 million pounds per year of the approximately 500 million pounds of wire and cable scrap generated in North America. It purifies the material and resells it for use in products such as car floor mats, truck mud flaps, and sound-deadening panels for cars.[109]

Some post-consumer recycled PVC comes from carpet recycling programs, and carpet backing is also a market for recycled PVC. Collins & Aikman Floorcoverings manufactures a vinyl carpet backing made of 100 percent recycled PVC. While the material can accommodate 100 percent post-consumer PVC, lack of supply means current levels are only 20 to 25 percent post-consumer, with the remainder being industrial scrap, mostly from the manufacture of automotive products.[118]

There is also considerable interest in recycling of both pre-consumer and post-consumer automotive scrap, especially in Europe, with legislative requirements for automobile recycling. For example, in France, the Autovinyle recycling program for PVC automotive scrap recycled 1,740 tonnes of PVC in its first year of operation, 1997–1998, and reported a goal of 5000 by 2000.[119] It appears this goal was not met, since its most current report is that 2200 tonnes were recycled in 1999. It focuses on plasticized PVC used in dashboard, seat, door panel, and carpet coatings.[120]

Delphi Automotive Systems, working with the University of Wuppertal, in Germany, developed a recycling method for PVC insulation on automotive wiring that avoids contamination of plastic by wiring by not shredding the wiring. Instead, a physical/chemical process including a solvent is used to soften the insulation on the cable so that it falls off the wires. The solvent is removed and the PVC regranulated and used for new cable insulation. The copper wire is also recovered. As of 2000, Delphi was producing approximately 400 metric tons of recycled PVC annually.[121]

Recycling of post-consumer PVC is more common in Europe than in the U.S.A. For example, the Vinyl Institute estimates that about half of the 4 billion PVC mineral water bottles used in France each year are recycled.[109]

European PVC producers, through the European Council of Vinyl Manufacturers (ECVM), have committed to recycling of PVC. The organization has initiated recycling programs for products that are easy to identify and separate from the waste stream or which can be kept relatively clean, so that a high quality recycled resin can be produced. Products included are bottles, flooring, pipes, roof covering membranes, and window profiles. Target markets for bottles are non-food bottles, pipes, profiles, fittings, sweaters, and shoe soles. Flooring, pipes, and window profiles can be recycled into new products of the same type. The target market for roof covering membranes is waterproofing liners. Some take-back recycling programs have also been initiated for blister packs, ID cards, and stationery items. Some programs for PVC from cable and leather cloth have been initiated, where markets for these materials, which cannot be separated into pure PVC, have been identified. Cable is made into industrial flooring, and leather cloth is made into mats and carpet backing.[122]

The European PVC Flooring group is investigating mechanical recycling of PVC flooring. Currently, only about 1,500 tonnes of such materials are recycled each year, due to quality issues. The European PVC Window Profiles and related building Products Association (EPPA) has initiated a project to examine and implement collection and recycling systems for PVC window frames.[123] VEKA AG has operated a PVC recycling facility in Behringer, Germany, since 1994, which recycles scrap from manufacture of vinyl windows, roller blinds, sheet products, and profiles.[109]

ECVM also proposes recycling of PVC through feedstock recycling processes, primarily those treating mixed plastics packaging with up to 10 percent PVC. Pretreatment of the wastes is required to get halogen content down to the required ppm range. Thermal dehalogenation can be used, either in a liquid or fluidized bed hydrolysis, to produce hydrochloric acid, which is either neutralized or recovered. For materials in which the PVC content is 30 percent or more, processes need to be designed primarily for the recovery of hydrochloric acid. ECVM reports that there are a number of such processes that have been proposed, but all are at early stages of development.[122]

ECVM is funding a pilot plant at Solvay, in Tavaux, France, for feedstock recycling of PVC. The system uses a molten slag bath containing silicates to decompose PVC to hydrogen chloride, hydrogen, and carbon monoxide. The initial feedstock is cable scrap and packaging, with other potential sources pharmaceutical blister packs, flooring, and car dashboards.[121] The products will be used in production of new PVC. Capacity of the pilot plant is 2,000 tonnes per year. Another feedstock recycling trial is taking place in Schkopau, Germany, using a rotary kiln incineration process to recover HCl and energy. A facil-

ity at Stigsnaes, in Denmark, is investigating recycling of mixed rigid and flexible PVC in a two-step process involving thermal hydrolysis followed by pyrolysis. The aqueous salt solution produced in the first step will be purified and then discharged into the Baltic Sea. The hydrocarbons, fillers, and heavy metals generated in the second stage will be recovered for reuse.[123]

Solvay is also involved in a consortium building a plant for mechanically recycling PVC in Ferrara, Italy, that is designed for PVC from cable insulation and packaging. The facility will use a process, called Vinyloop, in which the PVC is first separated from other plastics using biodegradable solvents and then filtered and precipitated, and the solvent is recovered. The recycled PVC is reported to be equal in quality to virgin material.[125] The facility, expected to begin production in November 2001, is designed to handle 10,000 tonnes of post-consumer PVC waste per year, mostly composite materials that also contain non-PVC substances.[123]

IBM reports that, in the UK, it has successfully recycled more than 500,000 lb of PVC since 1993, using recycled PVC monitor housings to make 100 percent recycled content computer keyboard backs.[109]

In Japan, the Vinyl Environmental Council, Japan PVC Environmental Affairs Council, and the Plastic Waste Management Institute are involved in developing PVC recycling technologies. Many efforts are directed toward removing PVC from mixed plastics from municipal solid waste so that the remaining materials can be used in energy generation, such as in blast furnaces and cement kilns, without the problems presented by chlorine content. There are also efforts to mechanically recycle construction scrap, producing drain pipes and also coextruded window sashes with about 70 percent recycled PVC by weight in a buried inner layer.[126]

In fall 2000, Australia announced initiation of nationwide plastic bottle recycling, which is expected to capture more than 20 million PVC bottles per year when fully implemented. In 2000, recycling rates for PVC bottles were only about 5 percent. Recycling rates for building materials were about 50 percent.[127]

A variety of products can be made from recycled vinyl. Some have already been mentioned. Re-New Wood, Inc., of Wagoner, OK, uses recycled PVC and wood to make roofing shingles that are claimed to be fire, wind, and hail resistant. The Eco-Shake shingles are made from 100 percent recycled post-industrial PVC from manufacture of IV bags, along with sawdust. The company claims the shingles themselves can be recycled at the end of their 50-year expected life.[128,129]

Rhoyl, a French clothing manufacturer, is producing sweaters, scarves, and socks from 70 percent recycled PVC mineral water bottles blended with 30 percent wool. About 27 bottles are needed to produce one sweater.[109]

12.9 Recycling of Nylon and Carpet

Nylon recycling has been growing rapidly, with most efforts focused on recovery of carpet. Estimates are that the U.S.A. produces 450,000 tonnes per year of nylon 6 carpet waste, about 25 percent of all waste carpet.[130]

BASF Corp. claims to have initiated the first comprehensive recycling program for used commercial carpet in the U.S.A. and Canada. Its 6ix Again program, which began in 1994, provides recycling of BASF nylon 6 carpet manufactured after Feb. 1, 1994, and containing 100 percent Zeftron nylon 6 yarn. Carpets must not contain halogenated SBR latex; polyester, jute, PVC, or polyurethane backing, nor can it contain arsenic-containing treatments, and it must not be contaminated with any hazardous substances. Carpets are sent to collection centers located throughout the U.S.A. and Canada. BASF recycles the carpet using chemical depolymerization into caprolactam and repolymerization and uses the material in new carpet, resulting in closed-loop recycling. In 1997, BASF expanded carpet re-

cycling to customers replacing ineligible carpets with qualifying carpet. This program accepts all carpet yarn types and backing systems, and it recycles them if a viable method is available or incinerates the old carpet if recycling is not available. The expanded program, unlike the 6ix Again program, assesses a fee to the customers to cover the costs of recycling or disposal of the old carpet. The program was recently expanded again to include products containing Ultramid nylon 6 plastic with recycled content.[131,132]

DSM and Honeywell (formerly Allied Signal) formed a joint venture, Evergreen Nylon Recycling (ENR), to recycle nylon 6 carpet. Like the BASF process, the nylon 6 is depolymerized to caprolactam. The $85 million ENR plant, in Augusta, GA, began operation in November 1999 with a capacity of more than 200 million pounds of carpet waste per year, producing over 100 million pounds of caprolactam. About 140 million pounds of the 200 million pounds processed each year is post-consumer carpet. The recovered caprolactam is used partly for production of new nylon 6 for carpet and other products and partly as a feedstock for production of other engineering plastics. Honeywell calculates that the process saves about 700,000 barrels of oil each year. The program accepts nylon 6 carpet made by any manufacturer.[133,134] In 2000, DSM and Honeywell announced plans to set up carpet recycling ventures in Europe and Asia, but that first an economically feasible way to collect the old carpet needed to be established.[135] The Evergreen process involves feeding entire carpets into the reactor. Polypropylene and latex backing and calcium carbonate filler from the carpets exit as a sticky brown substance that is sent to a cement kiln for incineration. The calcium carbonate in the mix becomes part of the cement.[132] One customer for repolymerized nylon produced by Honeywell is Ford Motor Co, which is using the Infinity nylon in throttle-body adaptors for more than 250,000 of its vehicles.[136]

DuPont has also developed a patented process for recycling nylon, with the focus on nylon 6,6 rather than nylon 6. The ammonolysis process produces hexamethyldiamine and adiponitrile from nylon 6,6 and caprolactam from nylon 6, which then are used to produce new polymer, for a closed-loop recycling system. In 2000, DuPont built a demonstration plant in Maitland, Ontario, a scale-up of the pilot-scale facility in Kingston, Ontario, that it had been operating since 1993. DuPont has been mechanically recycling old carpet since 1991, with collection through a network of about 80 carpet installation facilities. A facility in Chattanooga, TN, strips the carpet fiber from the backing and grinds it into nylon fluff, which has found use compounded with virgin nylon in air cleaner housings, as carpet underpadding, and in soundproofing materials. DuPont intends to begin sending this material to the Maitland plant for feedstock recycling.[132]

Other companies involved in nylon recycling, primarily for carpet, include Rhodia and Polyamid 2000 A.G. The Polyamid 2000 plant in Premnitz, Germany, mechanically recycles nylon 6,6 and depolymerizes nylon 6.[132]

Collins & Aikman Floorcoverings, of Dalton, GA, recycles vinyl-backed carpet, including nylon, from its customers, making mixed plastic products, including recycled nylon-reinforced carpet backing for its modular tile carpet products. These products can themselves be recycled. The company prides itself on recycling 100 percent of the recovered carpet rather than just some of the components.[137,138]

Wellman, Inc., of Johnsonville, SC, recycles nylon 6 and nylon 66 carpet, using a proprietary process and formulates the product into moldable engineered nylon resins, with both filled and unfilled grades. Product purity is about 98 percent, but the company is working on improving that by reducing levels of dirt, calcium carbonate, and PP.[19]

The Midwestern Workgroup on Carpet Recycling was established in the U.S.A. to address concerns about the growing amounts of carpet entering the disposal system. The final outcome was a Memorandum of Understanding signed January 8, 2001. Governments and the carpet industry agreed to establish reuse and recycling rates for discarded carpet, and the carpet industry agreed to create, fund, and manage a third-party organization to be responsible for achieving the negotiated targets as well as to establish model procurement guidelines for adoption by public entities.[139]

12.10 Recycling of Polyurethane

Recycling of polyurethane is more difficult than recycling of thermoplastics, since the thermoset polyurethane cannot be melted and reformed. Some mechanical recycling of polyurethane, using ground-up material as a filler in new materials, has been attempted. Most efforts, however, are devoted at chemical treatment to break down the material into polyol, which can then be used in production of new plastics.

BASF teamed with Philip Services to open the first facility in North America for recycling rigid and semi-rigid polyurethane, in 1997. This plant used a BASF-patented glycolysis process and was designed first for automotive waste, with the intention to expand to other waste streams.[140] However, the facility appears to be no longer in operation.[141,142] Similarly, United Technologies Automotive, of Dearborn, MI, announced in 1998 that it planned to regrind and reuse the polyurethane core in the automotive headliners it manufactures.[143] The company does not appear to have continued this practice.[144]

In 1999, NV Salyp, of Ieper, Belgium, began building the Salyp ELV Center in Belgium, which uses technology licensed from the Argonne National Laboratory of Illinois to recover polyurethane foam and other plastics from auto shredder residue, along with technology licensed from a German firm, KUTEC, for separating different types of thermoplastics from the Argonne technology reject stream. About five percent of automotive shredder residue, the material remaining after metals are recovered, is polyurethane. The Argonne technology separates the fluff into three streams: fines, foam, and a thermoplastic-rich stream. The foam stream is cleaned and sold for markets such as rebond foam in carpet underlay and for padding in automobiles. The thermoplastic stream is sorted further for recovery of a variety of pure resins. Even the fines stream may be recovered for applications such as cement, substituting for iron or mill scales that are now used to provide the needed iron.[31,32,144] Salyp reports that the process can also recover PUR foam from shredded mattresses.[27]

Dow Chemical Company announced in April 2001 that it has entered into an agreement with Mobius Technologies to promote polyurethane foam recycling technology developed by Mobius. This system pulverizes scrap foam into powder of less than 50 microns, which can be mixed into polyols and used as a filler in production of new foam, replacing up to 10 to 12 percent of the new polyol and thus lowering costs. The scrap can be from either flexible or molded sources and can be production scrap or scrap from post-consumer applications such as old automobile seats.[146]

ISOLA claims to be the largest rigid polyurethane recycling company in Europe, recycling 5,000 tonnes of rigid polyurethane waste each year, including production waste from polyurethane panel and block industries and polyurethane recovered from end-of-life refrigerators in Germany, the Netherlands, Switzerland, Austria, and Scandinavia.[147]

The St. Vincent de Paul Society, of Lane County, in Eugene, OR, in collaboration with the International Sleep Products Association, opened a mattress components recycling facility in Alameda County, CA, designed to recycle about 500,000 lb of polyurethane foam a year. The facility can handle 250 to 500 mattresses a day. It is designed to recover steel, wood, and cotton from the mattresses, as well as polyurethane. The system first slices the incoming mattresses so that layers of polyurethane foam and cotton fiber can be removed from the steel framework. In the future, whole mattresses will be shredded.[148]

In August 2001, Conigliaro Industries, of Framingham, MA, received a grant from the Massachusetts Dept. of Environmental Protection for development and start-up of a mattress shredding and recycling facility.[149]

12.11 Recycling of Polycarbonate

Some recycling of polycarbonate (PC) from products such as automobile bumpers, compact discs, computer housings, and telephones has been carried out.

General Electric carried out pilot programs for recycling five-gallon polycarbonate water bottles, parts from automobiles, and other materials made from GE PC resins.[150]

Bayer built Europe's first polycarbonate CD recycling facility in Leverkusen, Germany, in 1995. The PC was separated from aluminum coatings, protective layers, and imprinting. The product was blended with virgin PC for use in a variety of products.[151] In 1998, Bayer reported the facility had recycled 350 million unused compact discs since 1994.[152]

Nisim Corp., in Ohio, reprocesses metallized CDs using a batch washing system to recover PC. It uses a hot caustic solution to remove and recover metallized coatings, such as gold, from the CDs. The company recently added a batch solution process to recover PC from laminated and printed signs.[27]

Polycarbonate winding levers and outer covers on disposable cameras are recovered for reuse.[99]

12.12 Recycling of Acrylonitrile/Butadiene/Styrene Copolymers

Most acrylonitrile/butadiene/styrene (ABS) recycling is focused on appliances. British Telecommunications plc recovers and recycles telephones, using the recovered ABS from the housing in molded products such as printer ribbon cassettes and car wheel trims.[153] AT&T Bell Laboratories recycles ABS phone housings into mounting panels for business telephone systems.[154] GE Specialty Chemicals recycles ABS from refrigerator liners.[155]

Hewlett-Packard recycles computer equipment and uses recovered ABS in manufacture of printers. The recycled content in the printer cases is reported to be 25 percent or more.[156]

HM Gesellschaft für Wertstoff-Recycling recovers ABS from vacuum cleaners.[157]

Siemens Nixdorf Informationsysteme (SNI), of Munich, Germany, recycles a variety of its equipment, recovering blends of ABS and PC along with other materials. The recovered resins are used in manufacture of equipment housings.[157]

12.13 Recycling of Other Plastics

A few other types of plastics are recycled in small amounts. For example, Arco Chemical Company developed a process for recycling glass-reinforced styrene maleic anhydride from industrial scrap.[158]

The University of Nottingham has worked on developing recycling for thermoset materials, including glass and carbon-fiber reinforced materials.[159] The Fraunhofer Institute in Teltow has also worked on thermoset recycling.[160]

Imperial Chemical Industries plc and Mitsubishi Rayon Co. Ltd. have investigated recycling of acrylics through depolymerization.[161] Nisim Corporation in Ohio recovers acrylic from laminated and printed signs.[27]

AECI Ltd. of South Africa built a pilot plant for recycling of polymethyl methacrylate (PMMA) and using it in production of acrylic sheet. The process uses microwaves to depolymerize PMMA through thermal decomposition, and replaces existing molten-bath recycling reactors.[162]

Additional examples could also be cited.

12.14 Commingled Plastics and Plastic Lumber

When mixed plastics streams are collected, separation by resin type may not be feasible or economical. Similarly, heavily contaminated plastics streams may not be suitable for high-

value end markets. Use of plastics in a commingled form, typically in applications where plastics replaces wood or concrete, offers relatively low value end uses that are much more tolerant of contamination than traditional plastic resin markets. In addition, the plastic substitutes for lumber and concrete tend to be considerably more durable than the materials they replace. This is especially true for plastics lumber, where the plastics also have the benefit of not employing some of the toxic chemical systems used to protect wood from degradation in moist outdoor environments. Many companies have been into and out of the commingled plastics business. The products typically sell for more than the wood or concrete items they replace, and convincing people to buy the products, even when they will last considerably longer than the cheaper alternatives, has been difficult.

As mentioned, one application for recycled commingled plastics is in plastics lumber. The plastics lumber business has grown rapidly in the last five years. Not all plastics lumber is made using recycled plastics, but much of it is. Some companies using recycled plastics for plastics lumber use mixed plastics, and others use relatively pure streams of material. Some use primarily post-consumer materials, while others use mostly industrial scrap. Many manufacturers make composite plastic lumber out of wood fiber, with HDPE, LDPE, or PVC serving as the plastic component.

From 1992 to 1997, the U.S. Army Construction Engineering Research Laboratory (CERL) and Rutgers University, along with several RPL manufacturers, carried out a project to optimize the production process for RPL and develop demonstration projects.[163] In 1998 to 2000, the Plastic Lumber Trade Association and Battelle, with four RPL manufacturers, several government agencies, and others, completed a three-year cooperative effort to develop the use of RPL in structural applications in decking, marine waterfront, and material handling.[164] As a result of these and other projects, the usefulness of plastic lumber in several types of applications is now accepted.

In 1999, the Plastic Lumber Trade Association reported that recycled plastic lumber (RPL) was growing at a 30 percent annual rate. Growth of plastic lumber has been fueled by growing knowledge about performance properties of the material and how they relate to composition, as well as by the development of performance standards such as the standard test methods developed by ASTM.[163] By the end of 2000, eight ASTM test methods for RPL had been established, and several others were at various stages of the development process. The Association has also participated in a number of demonstration projects.[164]

Residential decking is a large component of plastic lumber sales. Major manufacturers using recycled plastics include Trex and Advanced Environmental Recycling Technologies (AERT) with plastic/wood fiber composites, and U.S. Plastic Lumber Corp. with both structurally foamed HDPE and plastic/wood composites.[165] Some major forest product companies are beginning to market RPL products or to introduce their own.[164] A 1999 estimate was that plastic/wood composites had 2 percent of the residential decking market, and plastic-only had 1 percent.[165] In 2000, the combined market share for plastic and wood/plastic composite decking was estimated at 5 percent, with a forecast that it could grow to half the market. Plastic decks cost about 15 percent more than standard wood decks.[166]

Recycled plastic railroad ties have very large scale potential. A number of test projects with a few ties each were carried out between 1996 and 1997. In 1998, the Chicago Transit Authority became the first commercial purchaser, buying 250 ties for a test on its elevated train line.[167] U.S. Plastic Lumber has been one of the major producers of these materials, and it claims that they have twice the life span of wood ties. Polywood produces RPL for railroad ties as well as for bridges and boardwalks from a blend of polyethylene and polystyrene.[164,168] TieTek, Inc., a subsidiary of North American Technologies Group, Inc., produces railroad ties from a mixture of recycled plastics and recycled rubber, with a total recycled content of 75 percent. The company's plant facility in Houston, TX, began pro-

duction in July 2000. In 2001, TieTek announced plans to build a production facility in New Zealand.[168,169]

Plastics lumber has been used in constructing bridges, walkways, and piers. One of the first major uses was a bridge at Fort Leonard Wood, in St. Robert, Mo., that used 13,000 lb of commingled recycled plastics from a number of different suppliers. The bridge is primarily for pedestrian use but can support light trucks. It is expected to last 50 years without maintenance, significantly longer than the 15-year life span of a treated wood equivalent.[170] Manufacturers include Polywood, as well as U.S. Plastic Lumber, which manufactures fiberglass reinforced RPL as well as foamed HDPE lumber.[164]

As mentioned, several companies have combined recycled plastic with wood fiber, sawdust, or recovered paper fiber. For example, Advanced Environmental Recycling Technologies Inc. (AERT), of Springdale, AR, has been producing window frames from recycled LDPE film (mostly pallet wrap) and wood fiber for a number of years.[171] Trex Company of Winchester, VA, which also has production facilities in Nevada, manufactures recycled plastic/wood composites, primarily for decking, from LDPE and HDPE, originating in pallet wrap and grocery bags, and waste wood fiber. Trex was selected by *Industry Week* as one of the 25 most successful small manufacturers in the U.S.[19,172] Atma Plastics Pvt. Ltd., of Chandigarh, India, produces a recycled PE/PP/wood composite for use as a wood substitute, or as a filler in cast polyester for furniture.[173] Natural Fiber Composites Inc., of Baraboo, WI, commercialized a pelletized wood fiber-filled plastic using recovered paper or wood fibers with PP, HDPE, or PS.[174] Mikron Industries manufactures a wood-plastic composite for window frames using an undisclosed mixture of plastics.[175]

Conigliaro Industries, in addition to recycling mixed plastics from computers and electronics and using them in cold-patch adhesive, as discussed in Sec. 12.15, has developed concrete blocks using mixed plastics as a substitute for some of the aggregate. The blocks weigh only half as much as conventional concrete blocks and nest into each other when the wall is erected, providing additional strength.[149]

12.15 Recycling Plastics from Computers and Electronics

Consumer electronics is getting increasing attention as a waste problem, as the boom in personal computers and their short lives before obsolescence result in a burgeoning number entering the waste stream. The hazardous heavy metals, such as lead and cadmium, present in some computer components make their disposal problematic, so there is a growing movement toward instituting recycling programs for these materials in particular. In Europe, manufacturers will be held responsible for the ultimate recycling of these items, so they have an added financial incentive to improve the products' recyclability and to develop recycling systems. In the U.S.A., several computer manufacturers have instituted voluntary take-back programs for computers. However, most charge customers for taking back the items, which significantly limits recovery.

In 2001, a group of manufacturers, recyclers, and government representatives in the U.S.A. formed the National Electronics Product Stewardship Initiative, with the major goal of creating a framework for product stewardship for electronics in the United States. Other product stewardship initiatives were previously initiated in the Northeast and the Western regions of the U.S.[176] The Electronic Industries Alliance operates a web site that directs consumers to electronics recycling and reuse opportunities.[177]

Because computers and electronics contain 15 to 20 different types of plastics, in turn containing a variety of plasticizers, colorants, flame retardants, and fillers, recovery of plastics from these materials is complex.[178] Much of the research has been directed toward development of some of the microsorting techniques discussed in Sec. 12.2.3. Other efforts have been devoted toward design of computers and electronics to facilitate recycling,

incorporating guidelines such as minimizing the use of different types of plastic resins, providing for identification of the types of resins used in computer parts, and designing items to be easily disassembled. Electronics manufacturers have also increased use of recycled materials in their products.

In 1992, IBM became the first major computer maker to code plastic parts for ease in identification for recycling. In 1999, IBM reported that 675,000 lb of PVC, ABS, and PC/ABS resins recovered from old computers were reused in its products in 1998.[179] Nearly all the internal plastic parts in IBM PCs now have recycled content.[19]

Sony Electronics Incorporated in cooperation with the State of Minnesota and Waste Management, Inc., initiated a five-year program in 2000 to take back all Sony electronic products in Minnesota, at no cost to consumers.[180]

Butler-MacDonald Inc., of Indianapolis, has recovered plastics from computer electronics for many years, initially concentrating on ABS and other plastics from telephones. They have relied primarily on density-based separation systems but are adding new separation technologies.[181]

MBA Polymers, of Richmond, CA, with funding from the American Plastics Council, developed automated sortation techniques for plastics from computers and electronics.[178] The company processes several million pounds of plastic each month.[182]

In Japan, Kobe Steel Ltd. developed a sandwich technique to apply virgin resin to a core of recovered plastic, permitting increased use of recycled plastics in computers and other products.[183]

3M developed recycling-compatible label materials to enhance the recyclability of durable goods, including computers and electronics, by permitting the label to remain in place when the equipment housing is recycled, without detracting from resin performance.[184]

Mixed recycled plastics from computers and other electronics equipment are being used by Conigliaro Industries of New England in an asphalt road-paving mix.[185] The company also developed a system for recycling the plastic equipment housings from computer and electronic products, and a unique cold patch asphalt/recycled plastic mix for filling potholes.[186]

Dell announced in 1997 that the computers in its OptiPlex PC line, sold to business, government, and education customers, would be designed to be recyclable by using an ABS chassis without fillers or coatings and would also be designed to make the computer easier to maintain and upgrade.[187]

Several computer companies have minimized the number of types of plastics used in computers. Panasonic once used 20 different polystyrene resin grades but has cut that to 4. Hewlett-Packard (HP) tries to use a single polymer in products, if possible, preferably HIPS or ABS. HP also codes any piece of plastic that weighs more than 25 g with a resin identification number.[188]

Another design change is to facilitate disassembly by using standard screws, and quick-release connections, and by minimizing the number of fasteners. Molded-in ornamental components or information has replaced some painted decorations and adhesive labels. Many manufacturers are identifying the hazardous materials used in products and working toward reducing or eliminating them. On the other hand, some of the new, vibrant colors being used in electronics will make recycling more difficult. Also, smaller and less expensive products, while achieving source reduction, can make repair, reuse, and upgrade more difficult.[188]

In 2000, the American Plastics Council published a report on recycling of plastics from residential electronics equipment. Plastics accounted for about one-third of materials collected in two pilot residential electronics recycling programs, which operated during 1997 and 1998 in Somerville, MA, and Binghamton, NY. Only about 25 percent of the plastics fraction was "clean plastics," homogeneous and free of contaminants. The report identifies

16 different generic plastic resins sold into the electrical and electronics sector (excluding wire and cable) in significant amounts in 1995. The six most common were polystyrene, acrylonitrile/butadiene/styrene (ABS), PP, PU, PC, and phenol formaldehyde (PF) (Fig. 12.31). Of plastic materials collected during a two-week period in Hennepin County, MN, through its curbside and drop-off collection programs for electronics, only 35 percent were found acceptable for further processing for recycling. Of these materials, high-impact polystyrene (HIPS) was the most common material, followed by ABS and polyphenylene oxide (PPO), as shown in Fig. 12.32. There was considerable variation between categories of consumer electronics. Televisions had 75 percent HIPS, while computers contained mostly ABS and PPO. A similar project in San Francisco showed that recovered fans contained mostly polypropylene. APC concluded that, while televisions accounted for nearly half the electronics recovered, they were not a rich source of high-value engineer-

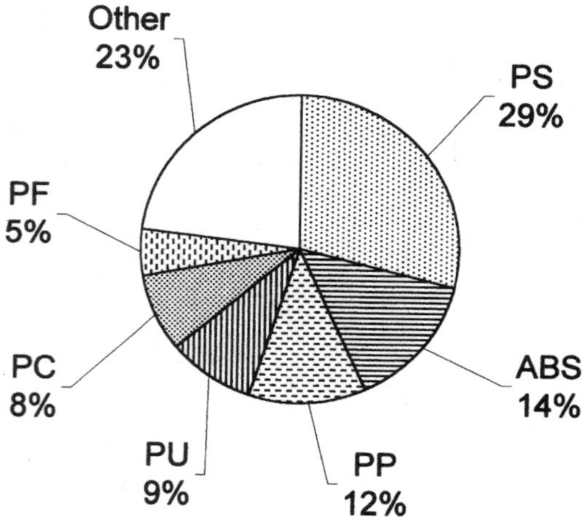

Figure 12.31 Plastic resins in electronics, 1995.[188]

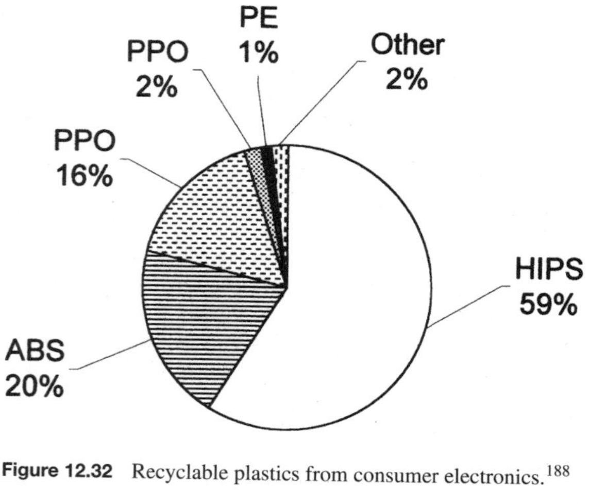

Figure 12.32 Recyclable plastics from consumer electronics.[188]

ing plastics. Computers, however, were and hence had a better chance of being effectively recycled.[189]

A recent report by researchers from Stockholm University drew attention to the problem of release of potentially toxic flame retardants during dismantling of consumer electronics. Airborne levels of flame-retardant additives were found to be two to three orders of magnitude higher in such facilities than in other work environments, and flame retardants were found in the blood of plant workers. Some of these compounds tend to bioaccumulate. Sweden has proposed banning polybrominated diphenyl ethers (PBDEs), which evidence suggests may act as endocrine disruptors.[190] This may to lead to additional design changes in the future.

12.16 Recycling Automotive Plastics

As is the case for electronics, regulations have spurred interest in recycling plastics from automobiles as well as in using recycled plastics and designing to facilitate recycling. European regulations are a major factor, as manufacturers will be responsible for achieving 95 percent recovery and 85 percent recycling of end-of-life vehicles by 2015.[27] All major automobile manufacturers are working on design changes to improve recyclability and on incorporating more recycled materials in automobiles. For example, in late 1998, DaimlerChrysler set standards for recycled content that including asking suppliers of plastics parts to provide at least 20 percent recycled content in 2000, increasing to 30 percent beginning in 2002.[191] In 1999, it demonstrated two Dodge Stratus sedans developed with 26 supplier companies to show the potential for use of recycled materials. More than 500 parts were modified to increase recycled content, and up to 40 percent of the plastic materials used were from recycled materials. DaimlerChrysler is working toward making its vehicles 95 percent recyclable within the next few years.[192]

One of the problems with recycling automotive plastics is paint removal. This applies not only to end-of-life vehicles but also to thermoplastic olefin (TPO) painted parts, which have a high scrap rate. Polymer Sciences Inc. announced plans in late 1999 to build recycling facilities in Duncan, SC, and somewhere in Europe, using the company's mechanical process for stripping paint from TPO parts before they are ground and pelletized. The South Carolina plant was to be completed by mid 2000, but the company's web site, as of mid 2001, still referred to the facility as "under construction."[193,194]

A mechanical paint-removal process is also used by American Commodities Inc., based in Flint, MI, and it is reported to be the largest recoverer of post-consumer automotive plastics in North America. The company has capacity to recycle 40.5 million pounds of TPO a year. One user of the American Commodities process is Visteon, which recycles 1 million pounds of painted TPO parts each year, using about 15 percent recycled old TPO bumpers in manufacture of new bumpers. Even higher amounts of scrap can be used when available, as the company is approved to use 100 percent scrap. Visteon's Milan plant increased scrap use to 60,000 lb per week of recycled TPO in 2000.[193,195,196]

Visteon's Berlin plant reports that it converts 99.3 percent of all purchased plastic into usable products: radiator grills, consoles, instrument panels, and door linings. It deals with the relatively high rate of production of scrap in auto manufacturing by carefully collecting the scrap segregated by type and color, which permits it to be reused. The facility reports that it avoids production of almost 2,000 tons of waste plastic each year.[197]

In early 2000, American Commodities, in a joint venture with Wipag Polymer Technique of Neuberg, Germany, began recycling scrap instrument panels, which contain styrene maleic anhydride (SMA), polyurethane foam, and PVC. The process separates the three materials so that each can be recovered. It can also be employed on door panels, where it also separates the textile inserts. Wipag has been using the process on instrument panels in Europe, where it has facilities in Neuberg, Germany, and Kent, England. Materi-

als are returned to 99.98 percent purity, at a cost savings of about 30 percent as compared with virgin material. Recovered plastic is used in production of new automotive parts.[196,198]

While many automotive recycling processes target disassembled components, including scrap parts produced during the manufacturing process, others are directed toward the more complex problem of recovering usable plastics from the residue from automobile shredder operations designed primarily for metals recovery. Auto shredder residue typically contains 15 to 27 percent plastics. The plastics fraction contains thermosets and thermoplastics. The thermosets, in addition to rubber are mostly rigid polyurethane and fiberglass/polyester composites. The thermoplastics fraction includes nearly 50 different resins, but about 80 percent is PP and ABS.[27] One of the systems that has been developed for recovery of plastics from ASR was discussed in Sec. 12.10, since its major recovery target is polyurethane foam.

In fall 2000, DaimlerChrysler announced that it was working with two other companies in testing recycling technology developed by Recovery Plastics International (RPI), of Salt Lake City, for recovery of plastics from auto shredder residue (ASR) using an automated skin flotation process. In this process, fiber and lint (about 50 percent of ASR by volume) are removed in an aspirator tower. The remaining materials go to a float/sink system maintained at a density of 1.4 g/cc using a suspension of fine iron ore in water. The rubber and thermoset plastics, along with wire and metal, sink, and the thermoplastics float. The thermoplastic fraction is ground to 0.5-in flake and washed in a strong hot sodium hydroxide solution. A surfactant plus aeration causes the PE and PP, including talc-filled PP, to float to the surface, and ABS, PC, and PUR foam sink to the bottom. The PE and PP are not further separated but are used as a blend. In a subsequent washing step, the foam is floated off. Another froth flotation step, using a low concentration of plasticizer, causes ABS and ABS/PC flakes to float so they can be recovered. The remaining heavy materials are a complex mixture of thermoplastic rubber, wood, and trace amounts of many other thermoplastics.[199] RPI's technology is expected to reach full-scale commercial status by 2002.[200]

Another company working on recovery of plastics from ASR is Galloo Plastics in France. Galloo crushes the ASR into 1-in granules at the shredder and then shreds then to less than 0.5-in size at the recycling facility. Ferrous and nonferrous metals and small glass particles are removed, and then the ASR is aspirated to remove PUR foam. A drum tumbler removes wire and metal waste, leaving a plastics wood mix. A proprietary float/sink process recovers filled and unfilled PP. The PUR foam fluff has been tested for sound-deadening applications but was not found to be economically viable. The PP is sold, and the company plans to add PE and ABS as products.[199]

12.17 Design Issues

As has been mentioned, the design of items made of plastics can affect their recyclability. Obviously, design can also affect other environmental impacts associated with the item.

12.17.1 Life Cycle Assessment

There is general agreement that, if one seeks to minimize the environmental consequences of a product, it is necessary to examine the entire life cycle of the product, since decisions based on a single factor, such as recyclability, may be erroneous. For example, a plastic product that is more easily recycled may or may not have a net overall environmental benefit as compared with an alternative plastic product that cannot be recycled effectively but uses significantly less material. The technique of life cycle assessment has been developed in part to aid in such tough decisions. There is considerable interest in life cycle assessment (LCA), but applying it effectively remains very difficult. The fundamental problem is

that using it to make decisions often requires comparing very different types of impacts, and impacts that may be location dependent. For example, what amount of benzene emitted into the air is better/worse than 1 g of lead emitted into the water? Does it matter where the air and water are located? While LCA is continuing to develop and standardize, very difficult questions remain. In general, the more alike the two alternatives are, the easier it is to use life cycle assessment in decision making; the more dissimilar they are, the more difficult LCA is to use. A thorough discussion of LCA is beyond the bounds of this chapter. The Society of Environmental Toxicology and Chemistry (SETAC) has a number of publications in this area.[201] The U.S. EPA has sponsored research and published a number of reports.[202] The Association of Plastics Manufacturers in Europe (APME) has published a series of eco-profiles of plastics that are now available online.[203] A number of other resources are also available.

12.17.2 Design for Recycling

As mentioned, design features can either improve or reduce the recyclability of products when they reach the end of their useful lives. Several guides have been produced to aid manufacturers in building in recyclability to products at the design stage.

One of these guides was produced through the Plastic Redesign Project, a coalition of government agencies and regional associations, funded by the U.S. EPA and the states of Wisconsin, New York, and California. The Project presented a set of general guidelines for design of plastic bottles.[204] The Association of Postconsumer Plastic Recyclers has published a set of specific guidelines for PET, HDPE, PP, and PVC bottles.[205] The American Plastics Council with the Society of the Plastics Industry published a set of design guidelines for information and technology equipment.[206] The EPA is addressing design issues as part of their product stewardship initiatives.[207] These are only some examples of guidance that is available.

12.17.3 Environmental Certification Programs

Another influence on design is the existence of programs to certify, in some fashion, the "environmental goodness" of a product or packaging. While recycling is not the only criterion, inclusion of recycled content is sometimes the basis for such certification.

Environmental certification programs include Green Cross and Green Seal in the United States, the Blue Angel program in Germany, the EcoMark in Japan, and the Environmental Choice label in Canada.

In all these programs, manufacturer participation is voluntary, with a fee generally charged to cover the cost of the program. The impetus for manufacturer participation is the marketing benefit derived from the eco-label.

12.17.4 Green Marketing

Significant marketing advantages can be derived from presenting products as having some positive environmentally related attributes, including recyclability or use of recycled content. When this phenomenon started, there was a great deal of misleading advertising. Such claims are subject to requirements under general legislation related to deceptive advertising, but, in a number of places, specific laws have been enacted to regulate environmental marketing claims. Companies considering making environmental claims about products need to be aware of the legal requirements. In the U.S.A., laws differ somewhat from state to state. General guidance has been provided by the Federal Trade Commission with its document *Complying with the Environmental Marketing Guides*.[208] One of the requirements is that a claim of recyclability cannot be made about a product or package un-

less a substantial majority of consumers in the locations where the claim is made have access to recycling programs that accept the item for recycling.

12.18 Legislation

One way to increase recycling of end-of-life plastics, and thus reduce their burden on disposal systems, is to institute legal requirements in the form of specific laws designed to aid recycling in some manner. Deposits on plastic bottles, discussed in Sec. 12.12.1, fall in this category. Several U.S. states and a large number of municipalities have instituted mandatory recycling for certain products, and plastic bottles are sometimes included. Many states have target levels for recycling or for reduction of waste going to disposal.

Another way to increase recycling is to require target levels of recycling and mandate that the producers of the items take responsibility for ensuring that this occurs, and that they bear the costs for doing so. This version of the "polluter pays" principle is known as *extended producer responsibility (EPR).* A variant, also denoted EPR, is *extended product responsibility.* The difference between the two is that extended product responsibility spreads the burden past the producer to government entities and to users of the products involved, stating that they all must bear some of the responsibility. Most often, EPR is interpreted to mean extended producer responsibility in Europe and extended product responsibility in the U.S.A. A term that is being used increasingly in the U.S.A. and Canada is *product stewardship,* which is essentially equivalent to extended product responsibility. It has the benefit of avoiding the confusion of the two different meanings that have been associated with EPR.

12.18.1 Resin Coding

The majority of U.S. states have requirements for coding of plastic bottles and other containers so that the resin they are made from can be identified. The coding system used was developed by the Society of the Plastics Industry (SPI) and assigns number codes 1 to 6 for PET, HDPE, PVC, LDPE, PP, and PS, respectively, with other plastics coded 7. A number of other countries have adopted this or similar systems. The U.S. Federal Trade Commission has held that conspicuous display of the resin codes constitutes a claim of recyclability, while inconspicuous use does not.[208,209]

12.18.2 Recycling Rates and Recycled Content

Wisconsin, Oregon, and California have laws related to recyclability and recycled content of rigid plastic containers. In Wisconsin, plastic containers, except those for food, beverages, drugs and cosmetics, are required to contain 10 percent recycled content. The exemptions do not apply if the U.S. FDA has approved the use of recycled content. However, the law allows "remanufactured" material (regrind, etc.) to be counted as recycled, and it has had little effect.

Oregon requires that rigid plastic containers contain at least 25 percent recycled content, be recycled at an aggregate rate of at least 25 percent, be made of plastic that is recycled in Oregon at a rate of at least 25 percent, or be a reusable container that is made to be reused at least 5 times. Containers for medical foods, drugs, and devices, and containers of food other than beverages, are exempt, as are packages that have been reduced in weight by at least 10 percent as compared with packaging for the same product used 5 years earlier. Since the law went into effect, the recycling rate for covered containers has exceeded the 25 percent requirement, so no action by companies has been required. However, in determining the projected rate for 2000, the state warned that the recycling rate had been de-

clining and might fall below the target 25 percent level as early as 2002. In the projection for 2001, the Oregon Department of Environmental Quality reported a continued decline and reiterated the warning that the rate might be below 25 percent in 2002, thus putting manufacturers of non-exempt products on warning.[210]

California has a law that is very similar to that in Oregon, except the law sets a higher recycling target for PET—55 percent. However, for the last several years, California's overall rigid plastic packaging container (RPPC) recycling rate has been below the 25 percent required. Therefore, companies manufacturing, distributing, or importing non-exempt products in RPPCs have had to comply with the law in some other way. Another difference in California is that the state does not yet forecast rates, so manufacturers may not know ahead of time whether they will have to comply through something other than the recycling rate provisions. This was especially critical when the rate first fell below 25 percent, since manufacturers were caught unaware and found themselves in the position of having to certify compliance with the law two years after the fact. Since most manufacturers had assumed they would comply automatically under the 25 percent provision, most had not taken other action to come into compliance. Noncompliance with the law remains a significant issue. Between April and August, 2001, the California Integrated Waste Management Board (CIWMB) reached compliance agreements with at least 59 companies that had not been in compliance in 1997, 1998, and/or 1999, in which the businesses committed to take action to come into compliance.[211]

12.18.3 Extended Producer Responsibility and Product Stewardship

Extended producer responsibility got its start in Germany with take-back and recycling requirements for packaging. The government instituted requirements that the producers of products be responsible for recovery and recycling of product packages. The government set targets that must be met but left it up to industry to decide how to do so. What emerged, for consumer packages, was the Duales System Deutschland (DSD) or Green Dot System. It is an industry-funded system to collect and recycle packaging for consumer products.

This system, while widely criticized at first, became policy for the whole European Union and is also in place in several countries that are working toward becoming EU members. The EPR approach has also spread far beyond the borders of Europe. It has also spread beyond packaging. The EU has adopted EPR for automobiles and is in the process of finalizing EPR requirements for electronics.

Canada adopted a National Packaging Protocol that required a 50 percent reduction in packaging waste going to disposal by 2000. The target was actually achieved ahead of schedule. Canadian provinces have taken different tactics to achieve this goal, and voluntary efforts by industry have also been a significant factor.

In the U.S.A., extended producer responsibility has not yet been a major factor. Product stewardship, however, has been officially adopted as state policy in Minnesota, and other states are considering the approach. In Minnesota, product stewardship is currently being implemented for electronics, paint, and carpet. The state has set up voluntary programs to test collection strategies and recycling for household electronics waste and for paint. It has gone a step farther for carpet, negotiating to set national recovery and recycling goals, and has been instrumental in the carpet industry, setting up a third-party organization to achieve those goals. In all cases, cooperation and voluntary agreements with industry have been involved.[212] While the product stewardship approach is still relatively uncommon, it seems likely that this approach to waste management and prevention will increase in the future.

EPR and product stewardship are being adopted in many other countries around the world. Examples include Japan, Korea, Brazil, Australia, and many more. In many cases, the systems are imposed by regulation. In others, governments and businesses are forging

voluntary agreements to head off mandatory requirements. It seems certain that industry will be called on increasingly to assume some of the consequences of handling the wastes from the products they manufacture, and this means, of course, that industry will increasingly be paying attention to the disposal-related consequences of their product and package design decisions.

References

1. U.S. Environmental Protection Agency, *Environmental Fact Sheet: Municipal Solid Waste Generation, Recycling and Disposal in the United States: Facts and Figures for 1998,* EPA530-F-00-024, 2000.
2. U.S. Environmental Protection Agency, *Characterization of Municipal Solid Waste in The United States: 1997 Update,* EPA530-R-98-007, 1998
3. 2001 Survey of N. American Plastics Recyclers & Brokers, *Plastics News*, May 21, 2001, p. 9.
4. Association of Plastics Manufacturers in Europe (APME), *An Analysis of Plastics Consumption and Recovery in Western Europe 1999,* Spring 2001.
5. Plastics and Chemicals Industries Association (PACIA), *Plastics Recycling Survey 2001,* Richmond, Vic., Australia, July 2001.
6. U.S. Environmental Protection Agency, *Characterization of Municipal Solid Waste in The United States: 1998 Update,* EPA530-R-99-021, 1999.
7. Toloken, S., Contaminants Seep into Rates for Recycling, *Plastics News*, Aug. 31, 1998, p. 1.
8. Fenton, R., Reuse Versus Recycling: A Look at Grocery Bags, *Resource Recycling,* vol. 11, no. 3, March 1992, pp. 105–110.
9. Miller, C., DOE Report: Recycle Plastic and Metal, Refill Glass, Burn Paper, *Waste Age's Recycling Times,* March 7, 1995, p. 10.
10. Siegler, T., Collecting Plastic Bottles More Efficiently, *Resource Recycling,* Sept. 1994, pp. 27–42.
11. Thompson Publishing Group, *Environmental Packaging: U.S. Guide to Green Labeling, Packaging and Recycling,* Washington, D.C., 2001.
12. California Dept. of Conservation, www.consrv.ca.gov.
13. Container Recycling Institute, www.container-recycling.org.
14. Container Recycling Institute, www.bottlebill.org.
15. Environment Protection Agency, Govt. of South Australia, www.wastecom.sa.gov.au.
16. Moore, W., Collection and Separation, in R. Ehrig (ed.), *Plastics Recycling: Products and Processes,* Hanser Pub., Munich, 1992, pp. 45–72.
17. Perkins, R. and B. Halpin, Breaking Bottlenecks in Plastic Bottle Recovery, *Resource Recycling,* June 2000, pp. 28–33.
18. Powell, J., The PRFect Solution to Plastic Bottle Recycling, *Resource Recycling*, Feb. 1995, pp. 25–27.
19. Powell, J. (Plas'tik): in a Flexible or Changing State, *Resource Recycling,* Jan. 2001, pp. 32–37.
20. Plastics Recyclers Stay on the Cutting Edge, *BioCycle*, May 1996, pp. 42–46.
21. Colvin, R., Sorting Mixed Polymers Eased by Hand-held Unit, *Modern Plastics*, April, 1995.
22. Stambler, I., Plastic Identifiers Groomed to Cut Recycling Roadblocks, *R&D Magazine,* Oct. 1996, pp. 29–30.
23. Smith, S., PolyAna System Identifies Array of Plastics, *Plastics News*, Dec. 15, 1997, p. 14.

24. SpectraCode Breaks Black Plastics Recycling Barrier, *Purdue News,* Feb. 2001, www.purdue.edu.
25. Maten, A., Recovering Plastics from Durable Goods: Improving the Technology, Resource Recycling, Sept. 1995, pp. 38–43.
26. Schut, J., Centrifugal Force Puts New Spin on Separation, *Plastics World,* Jan. 1996, p. 15.
27. Schut, J., Commingled Plastic Waste: New Gold Mine for Automotive Processors, *Plastics Technology Online,* May 2001, www.plasticstechnology.com.
28. Apotheker, S., Flake and Shake, Then Separate, *Resource Recycling,* June 1996, pp. 20–27.
29. Smith, S., Fry-Sorting Process Promising in Post-consumer Plastics Use, *Plastics News,* Nov. 11, 1996, p. 27.
30. Schut, J., Process for Reclaiming Durables Takes Off in U.S., *Modern Plastics*, March 1998, pp. 56–57.
31. Salyp ELV, www.salyp.com.
32. DeGaspari, J., Infrared Recycler, Mechanical Engineering, 2000, www.memagazine.org/contents/current/feature/infrared/infrared.html.
33. Hottenstein, F. and G. Kenny, Sensing Changes When Sorting Plastics, *Resource Recycling,* May 2000, pp. 20–23.
34. Sitek, F., Restabilization Upgrades Post-consumer Recyclate, *Modern Plastics*, Oct. 1993, pp. 64–68.
35. NAPCOR, 2000 Report on Post Consumer PET Container Recycling Activity, www.napcor.com.
36. Leading Reclaimers and Processors of Plastic Containers, *Resource Recycling*, Oct. 2000, pp. 30–31.
37. The Council for PET Bottle Recycling, www.petbottle-rec.gr.jp/english/.
38. PET Container Recycling Europe, PET Container Recycling—Strong as Ever, *PETCORE*, 3(1), April 2001.
39. PET Container Recycling Europe, PET Bottle Recycling in the UK, *PETCORE*, 2(3), Dec. 2000.
40. Swiss Recycling Plant Inaugurated, *PETplanet Insider*, August 2000.
41. Associação Brasileira dos Fabricantes de Embalagens de PET (ABEPET), www.abepet.com.br.
42. National Soft Drink Association, www.nsda.org.
43. Stutz, J. and C. Gilbert, *Michigan Bottle Bill, A Final Report to: Michigan Great Lakes Protection Fund,* July 10, 2000, www.deq.state.mi.us.
44. Environmental Preservation Initiatives, Ltd., www.waste-tips.org.uk.
45. Coca-Cola press release, Consumer Advisory: Coca-Cola System Using Recycled Plastic, March 1, 2000.
46. Grass Roots Recycling Network press release, "Coca-Cola Shareholder Vote on Recycling—Better Than Expected," April 18, 2001.
47. The Plastic Redesign Project, The Potential Impacts of Plastic Beer Bottles on Plastics Recycling, working paper, Sept. 1999, www.plasticredesign.org/files/plasticbeerbottles.html.
48. Association of Postconsumer Plastic Recyclers, www.plasticsrecycling.org.
49. Green Light for PET Recycling in Germany, *PETplanet Insider*, Feb. 1, 2001, www.petpla.net.
50. DeWinter, W., Poly(ethylene terephthalate) Film Recycling, in F. LaMantia (ed.) *Recycling of Plastic Materials,* ChemTec Pub., Ontario, 1993, pp. 1–15.
51. Milgrom, J., Polyethylene Terephthalate (PET), in R. Ehrig, (ed.) *Plastics Recycling: Products and Processes,* Hanser Pub., Munich, 1992, pp. 17–44.

52. Doba, J., FDA Gives Go-ahead for Recycled-PET Use, *Plastics News,* Feb. 21, 2000, p. 4.
53. DuPont Ends Recycling Experiment, *Plastics News,* Jan. 25, 1999.
54. Bisio, A. and M. Xanthos (eds.) *How to Manage Plastics Waste: Technology and Market Opportunities,* Hanser Pub., Munich, 1994.
55. PET Container Recycling Europe, Eastman Announces Breakthrough in PET Packaging Recycling, *PETCORE,* 1(2), Dec. 1999.
56. Tilley, K., PET Resin Giant Investing in U.S. Recycling, *Plastics News,* Feb. 12, 2001, p. 4.
57. Rebeiz, K., D. Fowler, and D. Paul, Recycling Plastics in Construction Applications, *J. of Resource Management and Tech.,* vol. 21, no. 2, 1993, pp. 76–81.
58. Bakker, M., Using Recycled Plastics in Food Bottles: The technical Barriers, *Resource Recycling,* May 1994, pp. 59–64.
59. Myers, J., Coca-Cola Closes the Loop with Multi-layer PET Bottle, *Modern Plastics,* April 1994, pp. 28–30.
60. Rabasca, L., Coca-Cola Introduces Recycled PET Bottle in Switzerland, *Waste Age's Recycling Times,* Feb. 21, 1995, p. 1.
61. Ford, T., FDA Clears Wellman's Recycled Sheet, *Plastics News,* Aug. 28, 1995, p. 6.
62. Toloken, S., FDA Approves Phoenix Process, *Plastics News,* April 30, 2001, p. 3.
63. PET on Everybody's Lips, *PETplanet Insider,* April 2000.
64. PET Bottle-to-Bottle with the Stehning Process, *PETplanet Insider,* August 2000.
65. Babinchak, S., Current Problems in PET Recycling Need to be Resolved, *Resource Recycling,* Oct. 1997, pp. 29–31.
66. PVC Costs in PET Bales, *Resource Recycling,* April 2000, p. 56.
67. Polyester Man is Here, *Resource Recycling,* Feb. 1998, p. 56.
68. Recycled PET Mesh Filtering Roadways, *Plastics News,* May 22, 2000, p. 26.
69. Doba, J., PET Pallets Made Cheap, *Plastics News,* July 24, 2000, p. 21.
70. Apotheker, S., The Bottle Is the Bottleneck, *Resource Recycling*, Sept. 1994, pp. 27–42.
71. American Plastics Council, 1998 Recycling Rate Study, www.plasticsresource.com.
72. Toloken, S., Plastic Bottle Recycling Rate Keeps Sliding, *Plastics News,* Aug. 24, 1998, p. 1.
73. Powell, J., Plastics Recycling: Changing Markets are the News, *Resource Recycling*, Oct. 2000, pp. 19–23.
74. Milk Run, *Solid Waste & Recycling*, Feb./Mar. 1999.
75. Alberta Dairy Council, *Plastic Milk Jug Recycling Program, Annual Report 1999–2000,* www.milkjugrecycling.com.
76. ITec Environmental Systems, Inc., news release, June 11, 2001
77. Alberta Used Oil Management Association, www.usedoilrecycling.com/auoma.
78. Toloken, S., Recyclers Worried over Opaque Milk Bottles, *Plastics News,* Oct. 27, 1997, p. 7.
79. Ecoplast Gets FDA Nod, *Plastics News,* April 6, 1998.
80. Firms Collaborate on Recycled Crates, *Plastics News,* April 23, 2001, p. 3.
81. Newcorn, D., Plastic's Broken Loop, *Packaging World,* June 1997, pp. 22–24.
82. DuPont, www.dupont.com/tyvek.
83. Statewide Program to Recycle Nursery Greenhouse Film Expands, BioCycle, Sept. 1998, p. 19.
84. Pilot Project Encourages Farmers to Recycle, Chevron Solid Waste Issues Newsletter, 9(1), 1998, p. 5.
85. E. Jackson, Toro Recycling its Hoses, *Plastics News,* April 26, 1999.
86. Environment and Plastics Industry Council, EPIC Moving Ahead on Several Initiatives, www.plastics.ca.

87. Cirko, C., Identifying Processing Efficiencies for PE Film, *Resource Recycling,* May 1999, pp. 28–32.
88. Battery Council International, www.batterycouncil.org.
89. Toloken, S., Thermoformer Elm Packaging buys NPRC, *Plastics News,* July 5, 1999, p. 1.
90. Back, B., Portland Schools' Foam Trays Could End up in Wilsonville, *The Business Journal of Portland,* Feb. 2, 2001
91. Alliance of Foam Packaging Recyclers, www.epspackaging.org.
92. Buwalda, T. and C. McLendon, Taking Out the Trash, *Molding the Future,* 7(2), Summer 2000.
93. The Plastic Loose Fill Council, www.loosefillpackaging.com.
94. FP International, press release, Jan. 31, 2001.
95. Alliance of Foam Packaging Recyclers, EPS Recycling—What's Next? *Molding the Future* 5(2), Oct. 1998, pp. 1, 4.
96. International Foam Solutions, www.internationalfoamsolutions.com.
97. Solution Based Recycling of EPS (Expanded Polystyrene), *Food Packaging Bulletin* 7(7, 8), pp. 13–14, July/Aug. 1998.
98. Acohido, B., Firm Hopes to Turn School Waste into Walls, *Plastics News*, Sept. 13, 1999, p. 18.
99. Eastman Kodak Recycles 50 Million Cameras, *Plastics News,* Aug. 14, 1995, p. 10.
100. Kodak, www.kodak.com.
101. A Noncatalytic Process Reverts Polystyrene to its Monomer, *Chemical Engineering,* Feb. 1993, p. 19.
102. Japan Expanded Polystyrene Recycling Association, www.jepsra.gr.jp/en.
103. Korea Foam-Styrene Recycling Association, www.eps.or.k.
104. Canadian Polystyrene Recycling Association, www.cpra-canada.com.
105. Expanded Polystyrene Packaging Group, www.eps.co.uk.
106. EPS Recycling International, www.epsrecycling.org.
107. Toloken, S., Recycling Program in the Works for PVC, *Plastics News,* Aug. 31, 1998, p. 5.
108. The Vinyl Institute, *Directory of U.S. and Canadian Companies Involved in the Recycling of Vinyl Plastics and Manufacturing Products from Recycled Vinyl,* August 2000.
109. The Vinyl Institute, www.vinylinfo.org.
110. Principia Partners, *Post-industrial and Post-consumer Vinyl Reclaim: Material Flow and Uses in North America,* July 1999, www.principiaconsulting.com/reports_vinyl.pdf.
111. Toloken, S., Recyclers Tag PVC as Contaminant, *Plastics News,* April 20, 1998.
112. National Electronic Distributors Association, www.nedassoc.org.
113. American Plastics Council Technical Assistance Program, http://tap.plastics.org.
114. Kiser, J., Hospital Recycling Moves Ahead, *BioCycle,* Nov. 1994, pp. 30–33.
115. Wisner, D., Recycling Post-consumer Durable Vinyl Products, presented at The World Vinyl Forum, Sept. 7–9, 1997, Akron, Ohio.
116. Alaisa, C., Giving a Second Life to Plastic Scrap, *Resource Recycling*, Feb. 1997, pp. 29–31.
117. Ford, T., Nortel May Add to its Recycling Process, *Plastics News,* Dec. 18, 1995.
118. The Vinyl Institute, *EnVIronmental Briefs,* May 2000.
119. The Vinyl Institute, *EnVIronmental Briefs,* Aug. 1998.
120. Autovinyle, www.autovinyle.com.
121. Delphi Automotive Systems, press release, April 20, 2000.
122. European Council of Vinyl Manufacturers (ECVM), www.ecvm.org.

123. ECVM, EVPC, ESPA, *Voluntary Commitment of the PVC Industry*, Progress Report 2001, March 23, 2001
124. *Plastics Technology*, Tavaux—PVC Recycling Project, France, www.plastics-technology.com/projects/tavaux.
125. *Plastics Technology*, Ferrara—PVC Recycling Plant, Italy, www.plastics-technology.com/projects/ferrara.
126. WTEC Workshop on Environmentally Benign Manufacturing (EBM) Technologies, July 13, 2000, http://itri.loyola.edu/ebm.
127. Australia Commits to Nationwide Plastic Bottle Recycling, Environment News Service, Oct. 19, 2000.
128. The Vinyl Institute, *EnVIronmental Briefs*, May 1999.
129. Ledson, S., Shingle Maker Weathers the Storm, *Plastics News*, April 15, 1999, p. 1.
130. Nooter Corp., press release, Jan. 5, 2000.
131. BASF Corp., www.basf.com.
132. Tullo, A., DuPont, Evergreen to Recycle Carpet Forever, *Chemical & Engineering News*, Jan. 24, 2000, pp. 23–24.
133. Evergreen Nylon Recycling, Inc., press release, Nov. 15, 1999.
134. Honeywell Nylon Products, press release, March 5, 2001.
135. Steady Progress Seen for Nylon Chemical Recycling, *Modern Plastics*, June 2000, p. 17.
136. Miel, R., Ford Using Honeywell's Repolymerized Nylon, *Plastics News*, Nov. 6, 2000, p. 4.
137. J. Powell, Magic Carpet Ride: The Coming of a new Recyclable. *Resource Recycling*, April 1997, pp. 42–46.
138. Collins & Aikman Floorcoverings, press release, June 13, 2000.
139. Midwestern Workgroup on Carpet Recycling, www.moea.state.mn.us/policy/carpet/index.cfm.
140. BASF Corp., Philip Services Corp., press release, Sept. 16, 1997.
141. BASF, www.basf.com.
142. Philip Services, www.contactpsc.com.
143. J. Pryweller, Two Suppliers Step Up Recycling, *Automotive News*, July 20, 1998.
144. United Technologies Corp., www.utc.com.
145. D. Reed, Turning Trash into Cash, Urethanes Technology, 16(5), Oct. 1999.
146. Dow Chemical Co., press release, April 17, 2001.
147. Isola N.V., http://intersight.org/isola/eng/recycl.htm.
148. Alliance for the Polyurethanes Industry, press release, Oct. 18, 2000.
149. Conigliaro Industries, www.conigliaro.com.
150. White, K., GE Plastics Begins Buyback Program Aimed at Plastic Auto Scrap, *Waste Age's Recycling Times*, March 7, 1995, p. 8.
151. CD Recycling Plant is Europe's First, *Modern Plastics*, Sept. 1995, p. 13.
152. Bayer Puts New Spin on Reject Compact Discs, *Modern Plastics*, April 1998, p. 16.
153. British Telecommunications plc, www.bt.co.uk.
154. Texas Society of Professional Engineers, www.tspe.org/recycle-4.HTM.
155. Graff, G., Additive Makers Vie for Reclaimed Resin Markets, *Modern Plastics*, Feb. 1996, pp. 51–53.
156. Ford, T., Hewlett-Packard Printers Use Recycled ABS, *Plastics News*, Aug. 7, 1995, p. 40.
157. Myers, S., Recyclers of Appliances, Durables Looking to Germany's Proposals, *Modern Plastics*, March 1995, pp. 14–15.
158. Arco Chemical Devises SMA Recycling Process, *Plastics News*, March 11, 1996.
159. British Project for Recycling Polymer Composites Launched, *Chemical & Engineering News*, June 19, 1995, p. 13.

160. Thermoset Reclaim Bets on Chemical Process, *Modern Plastics,* April 1995, p. 13.
161. Higgs, R., ICI, Mitsubishi Research Acrylic Recycling, *Plastics News,* May 26, 1997.
162. Methacrylates Recycling Process Makes Debut, *Chemical & Engineering News,* Oct. 12, 1998, p. 21.
163. Standards Boost an Industry, *ASTM Standardization News*, July 1999, pp. 22–26.
164. Robbins, A., 2000 State of the Recycled Plastic Lumber Industry, Plastic Lumber Trade Association, www.plasticlumber.org.
165. Sparks, K. Stiff Competition: Plastic Lumber Makers Shape the Market, *Resource Recycling,* May 1999, pp. 33–38.
166. Carlton, J., Going Green: Plastic Lumber Builds a Market, *Wall Street Journal,* March 31, 2000, p. B1.
167. Bregar, B., Plastic Rail Ties Gaining Favor, *Plastics News,* May 11, 1998.
168. Black, M., Crossties: New Technology and Old Standbys that Still Get the Job Done, *Railway Track & Structures,* Sept. 1999, pp. 17–23.
169. North American Technologies Group, Inc., www.natk.com.
170. Urey, C., Uncle Sam Recruits Recycled Plastic Lumber, *Plastics News,* July 13, 1998.
171. Bregar, B., AERT Wins Major Contract for Its Recycled Material, *Plastics News,* May 20, 1991, p. 1.
172. Trex Company, www.trex.com.
173. Moore, S., Plastics Recycling Profit Soars in India, *Modern Plastics,* June 1995, pp. 19– 21.
174. Wood-fiber Composite is Now Commercial, *Modern Plastics,* May 1997, p. 16.
175. Urey, C., Wood Composites Make Show at Meeting, *Plastics News,* Aug. 17, 1998, p. 8.
176. Duff, S., Group Addresses Electronics Recycling, *Plastics News,* May 7, 2001, p. 16.
177. Electronic Industries Alliance, www.eiae.org.
178. J. Porter, Computers & Electronics Recycling: Challenges and Opportunities, *Resource Recycling,* April 1998, pp. 19–22.
179. Durables Recycling Is Surprisingly Healthy, *Modern Plastics,* March 1999, p. 14.
180. Hansen, B., Minnesota Offers Unique Electronics Recycling Plan, *Environment News Service,* October 18, 2000.
181. Toloken, S. and J. Doba, Industry Wrestles with Electronics Recycling, *Plastics News,* Nov. 27, 2000.
182. MBA Polymers, www.mbapolymers.com.
183. Japan Firms Reusing Plastics in Electronics, *Plastics News*, Oct. 5, 1998.
184. 3M, press release, November 1998.
185. Old Computers Turn up in Road-paving Product, *Modern Plastics,* June 1999, p. 12.
186. Dunbar, J. and G. Conigliaro, Plastics from Recovered Electronics Pave the New Information Highway, *Resource Recycling,* May 1999, p.
187. Dell Computer Corp., www.dell.com.
188. T. Krause, Design for Environment: A Last Will and Testament for Scrap Electronics, *Resource Recycling,* March 2001, pp. 14–23.
189. American Plastics Council, *Plastics from Residential Electronics Recycling, Report 2000,* Arlington, VA, 2000.
190. K. Betts, Could Flame Retardants Deter Electronics Recycling? *Environmental Science & Technology,* 35(3), pp. 58A–59A, 2001.
191. Pryweller, J., DaimlerChrysler Sets Recycling Standards, *Plastics News,* Jan. 4, 1999, p. 9.
192. DaimlerChrysler Corporation Working with Recyclers to Increase Recovery, Reuse of Plastics from Automobiles, *Environment News Service,* Sept. 19, 2000.

193. J. Pryweller, New Plants Will Recycle Painted TPO Parts, *Plastics News,* Nov. 15, 1999, p. 12.
194. Polymer Sciences, Inc., www.polymersciences.com.
195. Plastics Technology, Milan Injection Moulding Plant—USA, www.plastics-technology.com.
196. Brooke, L., Plastics Recycling Goes Global, *Automotive Industries,* Feb., 2000.
197. *Plastics Technology,* Berlin Automotive Plant Upgrade—Germany, www.plastics-technology.com.
198. Miel, R., ACI "Liberates" Instrument-Panel Plastics, *Plastics News,* April 24, 2000, p. 20.
199. DaimlerChrysler Tests Recycling System, *American Metal Market,* Sept. 25, 2000.
200. Doba, J. and R. Miel, The Funk over Junk, *Plastics News,* May 21, 2001.
201. Society of Environmental Toxicology and Chemistry, www.setac.org.
202. U.S. Environmental Protection Agency, www.epa.gov.
203. Association of Plastics Manufacturers in Europe, www.apme.org.
204. The Plastic Redesign Project, Design for Recyclability: Recommendations for the Design of Plastic Bottles, www.plasticredesign.org/files/DesignforRecycling.html.
205. Association of Postconsumer Plastic Recyclers, www.plasticsrecycling.org.
206. American Plastics Council, *A Design Guide for Information and Technology Equipment,* www.plasticsresource.com/downloads/design-guide.pdf.
207. U.S. Environmental Protection Agency, www.epa.gov/epr.
208. U.S. Federal Trade Commission, *Complying with the Environmental Marketing Guides,* www.ftc.gov/bcp/conline/pubs/buspubs/greenguides.pdf.
209. Toloken, S., FTC Cracks down on Resin Code Placement, *Plastics News,* May 4, 1998, pp. 5, 24.
210. Oregon Dept. of Environmental Quality, www.deq.state.or.us.
211. California Integrated Waste Management Board, www.ciwmb.ca.gov.
212. Minnesota Office of Environmental Assistance, www.moea.state.mn.us.

Appendix

A

Glossary of Terms and Definitions

A-stage Stage in which thermosetting reactants are mixed, but at which the polymerization reaction has not yet begun.

abhesive Film or coating, such as a mold release, applied to a surface to prevent adhesion or sticking.

ablative plastics Plastics or resins, the surface layers of which decompose when surface is heated, leaving a heat-resistant layer of charred material. Successive layers break away, exposing a new surface. These plastics are especially useful in applications such as outer skins of spacecraft, which heat up to high temperatures on reentry into the Earth's atmosphere.

abrasion resistance Capability of a material to withstand mechanical forces, such as scraping, rubbing, or erosion, that remove material from the surface.

ABS plastics Abbreviated phrase referring to acrylonitrile-butadiene-styrene copolymers; elastomer-modified styrene.

absorption Penetration of one substance into the mass of another, such as moisture or water absorption of plastics.

accelerated test Test in which conditions are intensified to obtain critical data in shorter time periods, such as accelerated life testing.

accelerator Chemical used to speed up a reaction or cure. Term is often used interchangeably with *promoter*. For example, cobalt naphthanate is used to accelerate the reaction of certain polyester resins. An accelerator is often used along with a catalyst, hardener, or curing agent.

activation Process, usually chemical, of modifying a surface so that coatings will more readily bond to that surface.

activator Chemical material used in the activation process. (*See* **activation**.)

addition reaction or polymerization Chemical reaction in which simple molecules (monomers) are added to each other to form long-chain molecules without forming byproducts.

additive Substance added to materials, usually to improve their properties. Prime examples are plasticizers, flame retardants, and fillers added to plastic resins.

adhere To cause two surfaces to be held together by adhesion.

adherend Body held to another body by an adhesive.

adhesion State in which two surfaces are held together by interfacial forces, which may consist of valence forces, interlocking action, or both.

adhesive Substance capable of holding materials together by surface attachment.

adhesive, anaerobic Adhesive that sets only in the absence of air; for instance, one that is confined between plates or sheets.

adhesive, contact Adhesive that is apparently dry to the touch and will adhere to itself instantaneously on contact; also called *contact-bond adhesive* or *dry-bond adhesive.*

adhesive, heat-activated Dry adhesive film that is rendered tacky or fluid by application of heat or heat and pressure to the assembly.

adhesive, pressure-sensitive Viscoelastic material that in solvent-free form remains permanently tacky. Such material will adhere instantaneously to most solid surfaces with the application of very slight pressure.

adhesive, room-temperature-setting Adhesive that sets in the temperature range of 20 to 30°C (68 to 86°F), in accordance with the limits for standard room temperature as specified in ASTM Methods D 618, Conditioning Plastics and Electrical Insulating Materials for Testing.

adhesive, solvent Adhesive having a volatile organic liquid as a vehicle.

adhesive, solvent-activated Dry adhesive film that is rendered tacky just prior to use by application of a solvent.

aging Change in properties of a material with time under specific conditions.

air vent Small gap in a mold to avoid gases being entrapped in the plastic part during the molding process.

airless spraying High-pressure spraying process in which pressure is sufficiently high to atomize liquid coating particles without air.

alcohols Characterized by the fact that they contain the hydroxyl (–OH) group, alcohols are valuable starting points for the manufacture of synthetic resins, synthetic rubbers, and plasticizers.

aldehydes In general, volatile liquids with sharp, penetrating odors that are slightly less soluble in water than the corresponding alcohols. The group (–CHO), which characterizes all aldehydes, contains the most active form of the carbonyl radical and makes the aldehydes important as organic synthetic agents. They are widely used in industry as chemical building blocks in organic synthesis.

aliphatic hydrocarbon *See* **hydrocarbon.**

allowables Statistically derived estimate of a mechanical property based on repeated tests. Value above which at least 99 percent of the population of values is expected to fall with a confidence of 95 percent.

alloy Blend of polymers, copolymers, or elastomers under controlled conditions.

alpha particle Heavy particle emitted during radioactive decay, consisting of two protons and two neutrons bound together. It has the lowest penetration of the various emitted particles and will be stopped after traversing through only a few centimeters of air or a very thin solid film.

ambient temperature Temperature of the surrounding cooling medium, such as gas or liquid, that comes into contact with the heated parts of the apparatus.

amine adduct Products of the reaction of an amine with a deficiency of a substance containing epoxy groups.

amorphous plastic Plastic that is not crystalline, has no sharp melting point, and exhibits no known order or pattern of molecule distribution.

anhydride Organic compound from which water has been removed. Epoxy resins cured with anhydride-curing agents are generally characterized by long pot life, low exotherm during cure, good heat stability, and good electrical properties.

antioxidant Chemical used in the formulation of plastics to prevent or slow down the oxidation of material exposed to the air.

antistatic agents Agents that, when added to the molding material or applied onto the surface of the molded object, make it less conducting (thus hindering the fixation of dust).

aramid Generic name for highly oriented organic material derived from a polyamide but incorporating an aromatic ring structure.

arc resistance Time required for an arc to establish a conductive path on the surface of an organic material.

areal weight Weight of fabric or prepreg per unit width.

aromatic amine Synthetic amine, derived from the reaction of urea, thiourea, melamine, or allied compounds with aldehydes, that contains a significant amount of aromatic subgroups.

aromatic hydrocarbon *See* **hydrocarbon.**

aspect ratio Ratio of length to width for a flat form, or of length to diameter for a round form such as a fiber.

assembly Group of materials or parts, including adhesive, that has been placed together for bonding or that has been bonded together.

atactic State when the radical groups are arranged heterogeneously around the carbon chain. (*See also* **isotactic.**)

atomic oxygen resistance Ability of a material to withstand atomic oxygen exposure. This is related to the use of plastic and elastomers in space applications.

autoclave Closed vessel for conducting a chemical reaction or other operation under pressure and heat.

autoclave molding After lay-up, the entire assembly is placed in a steam autoclave at 50 to 100 lb/in^2. Additional pressure achieves higher reinforcement loadings and improved removal of air.

average molecular weight Molecular weight of most typical chain in a given plastic. There will always be a distribution of chain sizes and, hence, molecular weights in any polymer.

B-allowable Statistically derived estimate of a mechanical property based on numerous tests; value above which at least 90 percent of the population of values is expected to fall with a confidence of 95 percent.

B-stage Intermediate stage in the curing of a thermosetting resin. In this stage, resins can be heated and caused to flow, thereby allowing final curing in the desired shape. The term *A-stage* is used to describe an earlier stage in the curing reaction, and the term *C-stage* is sometimes used to describe the cured resin. Most molding materials are in the B stage when supplied for compression or transfer molding.

bag molding Method of applying pressure during bonding or molding in which a flexible cover, usually in connection with a rigid die or mold, exerts pressure on the material being molded, through the application of air pressure or drawing of a vacuum.

bagging Applying an impermeable layer of film over an uncured part and sealing the edges so that a vacuum can be drawn.

Barcol hardness Hardness value obtained using a Barcol hardness tester, which gages hardness of soft materials by indentation of a sharp steel point under a spring load.

basket weave Weave in which two or more warp threads cross alternately with two or more filling threads. The basket weave is less stable than the plain weave but produces a flatter and stronger fabric. It is also a more pliable fabric than the plain weave, and a certain degree of porosity is maintained without too much sleaziness, but not as much as with the plain weave.

benzene ring Basic structure of benzene, which is a hexagonal, six-carbon-atom structure with three double bonds; also basic aromatic structure in organic chemistry. Aromatic structures usually yield more thermally stable plastics than do aliphatic structures. *(See also* **hydrocarbon**.)

binder Resin or plastic constituent of a composite material, especially a fabric-reinforced composite.

blister Raised area on the incompletely hardened surface of a molding caused by the pressure of gases inside it.

bismaleimide Type of polyimide that cures by addition rather than a condensation reaction; generally of higher temperature resistance than epoxy.

blind fastener Fastener designed for holding two rigid materials, with access limited to one side.

blow molding Method of fabrication of thermoplastic materials in which a parison (hollow tube) is forced into the shape of the mold cavity by internal air pressure.

blowing agent Chemical that can be added to plastics and generates inert gases upon heating. This blowing, or expansion, causes the plastic to expand, thus forming a foam; also known as *foaming agent.*

bond Union of materials by adhesives; to unite materials by means of an adhesive.

bond strength Unit load, applied in tension, compression, flexure, peel, impact, cleavage, or shear, required to break an adhesive assembly, with failure occurring in or near the plane of the bond.

boron fibers High-modulus fibers produced by vapor deposition of elemental boron onto tungsten or carbon cores. Supplied as single strands or tapes.

boss Projection on a plastic part designed to add strength, to facilitate alignment during assembly, to provide for fastenings, and so on.

bridging Suspension of tensioned fiber between high points on a surface, resulting in uncompacted laminate.

bulk density **Density of a** molding material in loose form (granular, nodular, and the like), expressed as a ratio of weight to volume (for instance, g/cm^3 or lb/ft^3).

bulk rope molding compound Molding compound made with thickened polyester resin and fibers less than 0.5 in. Supplied as rope, it molds with excellent flow and surface appearance.

capacitance (capacity) That property of a system of conductors and dielectrics that permits the storage of electricity when potential difference exists between the conductors. Its value is expressed as the ratio of the quantity of electricity to a potential difference. A capacitance value is always positive.

carbon, fibers Fiber produced by the pyrolysis of organic precursor fibers such as rayon, polyacrylonitrile (PAN), or pitch, in an inert atmosphere.

cast To embed a component or assembly in a liquid resin, using molds that separate from the part for reuse after the resin is cured. Curing or polymerization takes place without external pressure. (*See* **embed, pot**.)

catalyst Chemical that causes or speeds up the cure of a resin but does not become a chemical part of the final product. Catalysts are normally added in small quantities. The peroxides used with polyesters are typical catalysts.

catalytic curing Curing by an agent that changes the rate of the chemical reaction without entering into the reaction.

caul plate Rigid plate contained within vacuum bag to impart a surface texture or configuration to the laminate during cure.

cavity Depression in a mold that usually forms the outer surface of the molded part; depending on the number of such depressions, molds are designated as single-cavity or multicavity.

centrifugal casting Fabrication process in which the catalyzed resin is introduced into a rapidly rotating mold where it forms a layer on the mold surfaces and hardens.

charge Amount of material used to load a mold for one cycle.

chlorinated hydrocarbon Organic compound having hydrogen atoms and, more important, chlorine atoms in its chemical structure. Trichloroethylene, methyl chloroform, and methylene chloride are chlorinated hydrocarbons.

circuit board Sheet of copper-clad laminate material on which copper has been etched to form a circuit pattern. The board may have copper on one (single-sided) or both (double-sided) surfaces. Also called *printed-circuit board* or *printed-wiring board.*

cleavage Imposition of transverse or "opening" forces at the edge of adhesive bond.

coat To cover with a finishing, protecting, or enclosing layer of any compound (such as varnish).

coefficient of thermal expansion (CTE) Change in unit length or volume resulting from a unit change in temperature. Commonly used units are 10^{-6} cm/cm/°C.

cohesion State in which the particles of a single substance are held together by primary or secondary valence forces. As used in the adhesive field, the state in which the particles of the adhesive (or the adherend) are held together.

cold flow (creep) Continuing dimensional change that follows initial instantaneous deformation in a nonrigid material under static load.

cold-press molding Molding process wherein inexpensive plastic male and female molds are used with room-temperature-curing resins to produce accurate parts. Limited runs are possible.

cold pressing Bonding operation in which an assembly is subjected to pressure without the application of heat.

compaction In reinforced plastics and composites, application of a temporary press bump cycle, vacuum, or tensioned layer to remove trapped air and compact the lay-up.

composite Homogeneous material created by the synthetic assembly of two or more materials (a selected filler or reinforcing elements and compatible matrix binder) to obtain specific characteristics and properties.

compound Some combination of elements in a stable molecular arrangement.

compression molding Technique of thermoset molding in which the molding compound (generally preheated) is placed in the heated open mold cavity, and the mold is closed under pressure (usually in a hydraulic press), causing the material to flow and completely fill the cavity, with pressure being held until the material has cured.

compressive strength Maximum compressive stress a material is capable of sustaining. For materials that do not fail by a shattering fracture, the value is arbitrary, depending on the distortion allowed.

condensation polymerization Chemical reaction in which two or more molecules combine with the separation of water or other simple substance.

condensation resins Any of the alkyd, phenol-aldehyde, and urea-formaldehyde resins.

conductivity Reciprocal of volume resistivity.

conformal coating Insulating coating applied to printed-circuit-board wiring assemblies that covers all of the components and provides protection against moisture, dust, and dirt.

copolymer *See* **polymer.**

corona resistance Resistance of insulating materials, especially plastics, to failure under the high-voltage state known as *partial discharge.* Failure can be erosion of the plastic material, decomposition of the polymer, thermal degradation, or a combination of these three failure mechanisms.

coupling agent Chemical or material that promotes improved adhesion between fiber and matrix resin in a reinforced composite, such as an epoxy-glass laminate or other resin-fiber laminate.

crazing Fine cracks that may extend in a network on or under the surface or through a layer of a plastic material.

creep Dimensional change with time of a material under load following the initial instantaneous elastic deformation; time-dependent part of strain resulting from force. Creep at room temperature is sometimes called *cold flow.* See ASTM D 674, Recommended Practices for Testing Long-Time Creep and Stress-Relaxation of Plastics under Tension or Compression Loads at Various Temperatures.

cross-linking Process whereby chemical links set up between molecular chains of a plastic. In thermosets, cross-linking makes one infusible supermolecule of all the chains, contributing to strength, rigidity, and high-temperature resistance. Thermoplastics (like polyethylene) can also be cross-linked (by irradiation or chemically through formulation) to produce three-dimensional structures that are thermoset in nature and offer improved tensile strength and stress-crack resistance.

crowfoot satin Type of weave having a 3-by-1 interlacing; that is, a filling thread floats over the three warp threads and then under one. One side of this type of fabric looks different from the other. Such fabrics are more pliable than either the plain or the basket weave and consequently are easier to form around curves. (*See* **four-harness satin**.)

crystalline melting point Temperature at which the crystalline structure in a material is broken down.

crystallinity State of molecular structure referring to uniformity and compactness of the molecular chains forming the polymer and resulting from the formation of solid crystals with a definite geometric pattern. In some resins, such as polyethylene, the degree of crystallinity indicates the degree of stiffness, hardness, environmental stress-crack resistance, and heat resistance.

cull Material remaining in a transfer chamber after the mold has been filled. Unless there is a slight excess in the charge, the operator cannot be sure the cavity is filled. The charge is generally regulated to control the thickness of the cull.

cure To change the physical properties of a material (usually from a liquid to a solid) by chemical reaction, by the action of heat and catalysts, alone or in combination, with or without pressure.

curing agent *See* **hardener.**

curing temperature Temperature at which a material is subjected to curing.

curing time In the molding of thermosetting plastics, the time it takes for the material to be properly cured.

cycle One complete operation of a molding press from closing time to closing time.

damping In a material, the ability to absorb energy to reduce vibration.

decorative laminates High-pressure laminates consisting of a phenolic-paper core and a melamine-paper top sheet with a decorative pattern.

deflashing Any finishing technique used to remove the flash (excess unwanted material) from a plastic part; examples are filing, sanding, milling, tumbling, and wheelabrating. *(See* **flash**.)

deflection temperature Formerly called heat-distortion temperature (HDT).

degas To remove gases, usually air, from liquid resin mixture, usually achieved by placing mixture in a vacuum. Entrapped gases or voids in a cured plastic can lead to premature failure, either electrically or mechanically.

delamination Separation of the layers of material in a laminate, either locally or in a large area. Can occur during cure or later during operational life.

denier Numbering system for fibers or filaments equal to the weight in grams of a 9000-m long fiber or filament. The lower the denier, the finer the yarn.

design allowables Tested, statistically defined material properties used for design. Usually refers to stress or strain.

dessicant Substance that will remove moisture from materials, usually due to absorption of the moisture onto the surface of the substance; also known as a *drying agent.*

diallyl phythalate Ester polymer resulting from reaction of allyl alcohol and phthalic anhydride.

dielectric constant (permittivity, specific inductive capacity) That property of a dielectric that determines the electrostatic energy stored per unit volume for unit potential gradient.

dielectric loss Time rate at which electric energy is transformed into heat in a dielectric when it is subjected to a changing electric field.

dielectric-loss angle (dielectric-phase difference) Difference between 90° and the dielectric-phase angle.

dielectric-loss factor (dielectric-loss index) Product of the dielectric constant and the tangent of the dielectric-loss angle for a material.

dielectric-phase angle Angular difference in phase between the sinusoidal alternating potential difference applied to a dielectric and the component of the resulting alternating current having the same period as the potential difference.

dielectric-power factor Cosine of the dielectric-phase angle (or sine of the dielectric-loss angle).

dielectric sensors Sensors that use electrical techniques to measure the change in loss factor (dissipation) and in capacitance during cure of the resin in a laminate. This is an accurate measure of the degree of resin cure or polymerization.

dielectric strength Voltage that an insulating material can withstand before breakdown occurs, usually expressed as a voltage gradient (such as volts per mil).

differential scanning Calorimetry measurement of the energy absorbed (endotherm) or produced (exotherm) as a resin system is cured.

diluent Ingredient usually added to a formulation to reduce the concentration of the resin.

diphenyl oxide resins Thermoplastic resins based on diphenyl oxide and possessing excellent handling properties and heat resistance.

dissipation factor (loss tangent, loss angle, tan δ, approximate power factor) Tangent of the loss angle of the insulating material.

doctor blade Straight piece of material used to spread and control the amount of resin applied to roving, tow, tape, or fabric.

domes In a cylindrical container, that portion that forms the integral ends of the container.

dry To change the physical state of an adhesive on an adherend through the loss of solvent constituents by Evaporation or absorption, or both.

drying agent *See* **dessicant**.

ductility Ability of a material to deform plastically before fracturing.

E-glass Family of glasses with low alkali content, usually under 2.0 percent, most suitable for use in electrical-grade laminates and glasses. Electrical properties remain more stable with these glasses due to the low alkali content. Also called *electrical-grade glasses.*

eight-harness satin Type of weave having a 7-by-1 interlacing; that is, a filling thread floats over seven warp threads and then under one. Like the crowfoot weave, its appearance on one side is different from that of the other side. This weave is more pliable than any of the others and is especially adaptable where it is necessary to form around compound curves, such as on radomes.

elastic limit Greatest stress a material is capable of sustaining without any permanent strain remaining when the stress is released.

elasticity Property of a material by virtue of which it tends to recover its original size and shape after deformation. If the strain is proportional to the applied stress, the material is said to exhibit Hookean or ideal elasticity.

elastomer Material that, at room temperature, can be stretched repeatedly to at least twice its original length and, upon release of the stress, will return with force to its approximate original length. Plastics with such or similar characteristics are known as *elastomeric plastics*. The expression is also used when referring to a rubber (natural or synthetic) reinforced plastic, as in elastomer-modified resins.

electric strength (dielectric strength, disruptive gradient) Maximum potential gradient that a material can withstand without rupture. The value obtained for the electric strength will depend on the thickness of the material and on the method and conditions of test.

electrode Conductor of the metallic class through which a current enters or leaves an electrolytic cell, at which there is a change from conduction by electrons to conduction by charged particles of matter, or vice versa.

elongation Increase in gage length of a tension specimen, usually expressed as a percentage of the original gage length. (*See also* **gage length**.)

embed To encase completely a component or assembly in some material—a plastic, for instance. (*See* **cast, pot**.)

encapsulate To coat a component or assembly in a conformal or thixotropic coating by dipping, brushing, or spraying.

engineering plastics Plastics, the properties of which are suitable for engineered products. These plastics are usually suitable for application up to 125°C, well above the thermal stability of many commercial plastics. The next higher grade of plastics, called *high-performance plastics,* is usually suitable for product designs requiring stability of plastics above 175°C.

environmental-stress cracking Susceptibility of a thermoplastic article to crack or craze under the influence of certain chemicals, aging, weather, or other stress. Standard ASTM test methods that include requirements for environmental stress cracking are indexed in Index of ASTM Standards.

epoxy Thermosetting polymers containing the oxirane group; mostly made by reacting epichlorohydrin with a polyol such as bisphenol A. Resins may be either liquid or solid.

eutectic Mixture, the melting point of which is lower than that of any other mixture of the same ingredients.

exotherm Characteristic curve of resin during its cure that shows heat of reaction (temperature) versus time. Peak exotherm is the maximum temperature on this curve.

exothermic Chemical reaction in which heat is given off.

extender Inert ingredient added to a resin formulation chiefly to increase its volume.

extrusion Compacting of a plastic material and forcing of it through an orifice.

fabric Planar structure produced by interlacing yarns, fibers, or filaments.

failure, adhesive Rupture of an adhesive bond such that the separation appears to be at the adhesive–adherend interface.

fiber washout Movement of fiber during cure because of large hydrostatic forces generated in low-viscosity resin systems.

fiberglass Individual filament made by attenuating molten glass. A continuous filament is a glass fiber of great or indefinite length; a staple fiber is a glass fiber of relatively short length (generally less than 17 in).

filament winding Process for fabricating a composite structure in which continuous reinforcements (filament, wire, yarn, tape, or other), either previously impregnated with a matrix material or impregnated during the winding, are placed over a rotating and removable form or mandrel in a previously prescribed way to meet certain stress conditions. Generally, the shape is a surface of revolution and may or may not include end closures. When the right number of layers is applied, the wound form is cured and the mandrel removed.

fill *See* **weft**.

filler Material, usually inert, that is added to plastics to reduce cost or modify physical properties.

film adhesive Thin layer of dried adhesive; also class of adhesives provided in dry-film form with or without reinforcing fabric, that are cured by heat and pressure.

finish, fiber Mixture of materials for treating glass or other fibers to reduce damage during processing or to promote adhesion to matrix resins.

fish paper Electrical-insulation grade of vulcanized fiber in thin cross section.

flame retardants Materials added to plastics to improve their resistance to fire.

flash Extra plastic attached to a molding along the parting line. Under most conditions, it would be objectionable and must be removed before the parts are acceptable.

flexibilizer Material that is added to rigid plastics to make them resilient or flexible. It can be either inert or a reactive part of the chemical reaction. Also called a *plasticizer* in some cases.

flexural modulus Ratio, within the elastic limit, of stress to the corresponding strain. It is calculated by drawing a tangent to the steepest initial straight-line portion of the load-deformation curve and calculating by the following equation:

$$E_B = \frac{L^3 m}{4bd^3}$$

where E_B = modulus
b = width of beam tested
d = depth of beam
m = slope of tangent
L = span, inches

flexural strength Strength of a material in bending expressed as the tensile stress of the outermost fibers of a bent test sample at the instant of failure.

fluorocarbon Organic compound having fluorine atoms in its chemical structure. This structure usually lends chemical and thermal stability to plastics.

four-harness satin Fabric, also named *crowfoot satin* because the weaving pattern when laid out on cloth design paper resembles the imprint of a crow's foot. In this type of weave, there is a 3-by-1 interlacing; that is, a filling thread floats over the three warp threads and then under one. One side of this type of fabric looks different from the other side. Fabrics with this weave are more pliable than either the plain or basket weave and, consequently, are easier to form around curves. (*See* **crowfoot satin**.)

fracture toughness Measure of the damage tolerance of a matrix containing initial flaws or cracks. G_{1c} and G_{2c} are the critical strain energy release rates in the 1 and 2 directions.

gage length Original of that portion of the specimen over which strain is measured.

gate Orifice through which liquid resin enters mold in plastic molding processes.

gel Soft, rubbery mass that is formed as a thermosetting resin goes from a fluid to an infusible solid. This is an intermediate state in a curing reaction and a stage in which the resin is mechanically very weak.

gel coat Resin applied to the surface of a mold and gelled prior to lay-up. The gel coat becomes an integral part of the finished laminate and is usually used to improve surface appearance and so on.

gel point The point in a curing reaction at which gelatin begins. (*See* **gel.**)

gelation Point in resin cure at which the viscosity has increased such that the resin barely moves when probed with a sharp point.

glass-transition point Temperature at which a material loses its glass-like properties and becomes a semiliquid. (*See also* **glass-transition temperature.**)

glass-transition temperature Temperature at which a plastic changes from a rigid state to a softened state. Both mechanical and electrical properties degrade significantly at this point, which is usually a narrow temperature range rather than a sharp point, as in freezing or boiling.

glue line (bond line) Layer of adhesive that attaches two adherends.

glue-line thickness Thickness of the fully dried adhesive layer.

glycol Alcohol containing two hydroxyl (–OH) groups.

graphite fibers High-strength, high-modulus fibers made by controlled carbonization and graphitization of organic fibers, usually rayon, acrylonitrile, or pitch.

hand lay-up Process of placing in position (and working) successive plies of reinforcing material or resin-impregnated reinforcement on a mold by hand.

hardener Chemical added to a thermosetting resin for the purpose of causing curing or hardening. Amines and acid anhydrides are hardeners for epoxy resins. Such hardeners are a part of the chemical reaction and a part of the chemical composition of the cured resin. The terms *hardener* and *curing agent* are used interchangeably. Note that these can differ from catalysts, promoters, and accelerators. (*See* **catalyst, promoter, accelerator.**)

hardness *See* **indentation hardness.**

heat-deflection temperature *See* **heat-distortion point.**

heat-distortion point Temperature at which a standard test bar deflects 0.010 in under a stated load of either 66 or 264 lb/in^2. See ASTM D 648, Standard Method of Test for Deflection Temperature of Plastics under Load.

heat sealing Method of joining plastic films by simultaneous application of heat and pressure to areas in contact. Heat may be supplied conductively or dielectrically.

helical pattern Pattern generated when a filament band advances along a helical path, not necessarily at a constant angle, except in the case of a cylinder, in a filament-wound object.

high-frequency preheating Plastic to be heated forms the dielectric of a condenser to which a high-frequency (20 to 80 MHz) voltage is applied. Dielectric loss in the material is the basis. The process is used for sealing vinyl films and preheating thermoset molding compounds.

high-performance plastics In general, plastics that are suitable for use above 175°C. (*See also* engineering **plastics.**)

homopolymer Polymer resulting from polymerization of a single monomer. (*See also* **monomer.**)

honeycomb Manufactured product of resin-impregnated sheet material or metal foil, formed into hexagonal cells. Skins are bonded to top and bottom surfaces to achieve strength.

hot-melt adhesive Thermoplastic adhesive compound, usually solid at room temperature, that is heated to a fluid state for application.

hydrocarbon Organic compound having hydrogen and carbon atoms in its chemical structure. Most organic compounds are hydrocarbons. Aliphatic hydrocarbons are straight-chained hydrocarbons, and aromatic hydrocarbons are ringed structures based on the benzene ring. Methyl alcohol, trichloroethylene, and the like are aliphatic; benzene, xylene, toluene, and the like are aromatic.

hydrolysis Chemical decomposition of a substance involving the addition of water.

hydrophilic Materials having a tendency to absorb water or to be wetted by water.

hydrophobic Materials having a tendency to repel water; usually materials exhibiting a low surface energy, measured by wetting angle.

hydroxyl group Chemical group consisting of one hydrogen atom plus one oxygen atom.

hygroscopic Tending to absorb moisture.

immiscible Two fluids that will not mix to form a homogeneous mixture or that are mutually insoluble.

impact strength Strength of a material when subjected to impact forces or loads.

impregnate To force resin into every interstice of a part. Cloths are impregnated for laminating, and tightly wound coils are impregnated in liquid resin using air pressure or vacuum as the impregnating force.

in-situ joint Joint between a composite and another surface that is formed during cure of the composite.

indentation hardness Hardness evaluated from measurements of area or indentation depth caused by pressing a specified indentation into the material surface with a specified force.

inhibitor Chemical added to resins to slow down the curing reaction. Inhibitors are normally added to prolong the storage life of thermosetting resins.

injection molding Molding procedure whereby a heat-softened plastic material is forced from a cylinder into a cavity that gives the article the desired shape. Used with all thermoplastic and some thermosetting materials.

inorganic chemicals Chemicals, the structures of which are based on atoms other than the carbon atom.

insulation, electrical Protection against electrical failure in an electrical product.

insulation resistance Ratio of applied voltage to total current between two electrodes in contact with a specified insulator.

insulator Material that provides electrical insulation in an electrical product.

interpenetrating network (IPN) Two or more polymers that have been formed together so that they penetrate each other in the final polymer form.

isotactic Molecules that are polymerized in parallel arrangements of radicals on one side of the carbon chain. (*See also* **atactic**.)

Izod impact strength Measure of the toughness of a material under impact as measured by the Izod impact test.

Izod impact test One of the most common ASTM tests for testing the impact strength of plastic materials.

joint Location at which two adherends are held together with a layer of adhesive.

joint, lap Joint made by placing one adherend partly over another and bonding together the overlapped portions.

joint, scarf Joint made by cutting away similar angular segments of two adherends and bonding the adherends with the cut areas fitted together.

joint, starved Joint that has an insufficient amount of adhesive to produce a satisfactory bond.

Kevlar Trademark for a group of DuPont aromatic polyimides that are frequently used as fibers in reinforced plastics and composites. Major characteristics are low thermal expansion, light weight, and good electrical properties, coupled with stiffness in laminated form. One special application area is in high-performance circuit boards requiring low x-y axis thermal expansion.

laminae Set of single plies or layers of a laminate (plural of *lamina*).

laminate To unite sheets of material by a bonding material, usually with pressure and heat (normally used with reference to flat sheets); product made by so bonding.

latent curing agent Curing agent that produces long-time stability at room temperature but rapid cure at elevated temperature.

lay-up As used in reinforced plastics, the reinforcing material placed in position in the mold; resin-impregnated reinforcement; process of placing the reinforcing material in position in the mold.

leno weave Locking-type weave in which two or more warp threads cross over each other and interlace with one or more filling threads. It is used primarily to prevent shifting of fibers in open fabrics.

limited-coordination specification (or standard) Specification (or standard) that has not been fully coordinated and accepted by all the interested activities. Limited-coordination specifications and standards are issued to cover the need for requirements unique to one particular department. This applies primarily to military agency documents.

liquid-crystal polymer (LCP) Polymers that spontaneously order themselves in the melt, allowing relatively easy processing at relatively high temperatures. They are characterized as rigid rods. Kevlar and Nomex are examples, as is Xydar thermoplastic.

liquid injection molding Fabrication process in which catalyzed resin is metered into closed molds.

loss angle See **dissipation factor**.

loss tangent See **dissipation factor**.

macerate To chop or shred fabric for use as a filler for a molding resin; molding compound obtained when so filled.

mandrel Form around which resin-impregnated fiber is wound to make pipes, tubes, or vessels by the filament-winding process.

mat Reinforcing material composed of randomly oriented, short, chopped fibers. Manufactured in sheet or blanket form and commonly used as alternative to woven fabric (especially glass) in fabrication of laminated plastic forms.

matched metal molding Method of molding reinforced plastics between two close-fitting metal molds mounted in a hydraulic press.

matrix Essentially homogeneous material in which the fiber system of a composite resides.

matrix manipulations Mathematical method of relating stresses and strains.

mechanical properties Material properties associated with elastic and inelastic reactions to an applied force.

melamines Thermosetting resins made from melamine and formaldehyde and possessing excellent hardness, clarity, and electrical properties.

melt Molten plastic, in the melted phase of material during a molding cycle.

microcracks Cracks formed in composites when thermal stresses locally exceed strength of matrix. These cracks generally do not penetrate or cross fibers.

micrometer (micron) Unit of length equal to 10,000 Å, 0.0001 cm, or approximately 0.000039 in.

mock leno weave Open-type weave that resembles a leno and is accomplished by a system of interlacings that draws a group of threads together and leaves a space between the next group. The warp threads do not actually cross each other as in a real leno, and therefore no special attachments are required for the loom. This type of weave is generally used when a high thread count is required for strength and, at the same time, the fabric must remain porous.

modifier Chemically inert ingredient added to a resin formulation that changes its properties.

modulus of elasticity Ratio of unidirectional stress to the corresponding strain (slope of the line) in the linear stress-strain region below the proportional limit. For materials with no linear range, a secant line from the origin to a specified point on the stress-strain curve or a line tangent to the curve at a specified point may be used.

moisture absorption Amount of water pickup by a material when that material is exposed to water vapor. Expressed as percent of original weight of dry material.

moisture resistance Ability of a material to resist absorbing moisture, either from the air or when immersed in water.

moisture vapor transmission Rate at which moisture vapor passes through a material at specified temperature and humidity levels. Expressed as grams per mil of material thickness per 24 hr per 100 in^2.

mold Medium or tool designed to form desired shapes and sizes; to process a plastics material using a mold.

mold release Lubricant used to coat a mold cavity to prevent the molded piece from sticking to it, and thus to facilitate its removal from the mold. Also called release agent.

mold shrinkage Difference in dimensions, expressed in inches per inch, between a molding and the mold cavity in which it was molded, both the mold and the molding being at room temperature when measured.

molecular weight Sum of the atomic masses of the elements forming the molecule.

monofilament Single fiber or filament.

monomer Small molecule that is capable of reacting with similar or other molecules to form large chain-like molecules called polymers.

multilayer printed circuits Electric circuits made on thin copper-clad laminates, stacked together with intermediate prepreg sheets and bonded together with heat and pressure. Subsequent drilling and electroplating through the layers result in a three-dimensional circuit.

necking Localized reduction of the cross-sectional area of a tensile specimen that may occur during loading.

NEMA standards Property values adopted as standard by the National Electrical Manufacturers Association.

notch sensitivity Extent to which the sensitivity of a material to fracture is increased by the presence of a disrupted surface such as a notch, a sudden change in section, a crack, or a scratch. Low notch sensitivity is usually associated with ductile materials, and high notch sensitivity with brittle materials.

nuclear radiation resistance Ability of a material to withstand nuclear radiation and still perform its designated function.

nylon Generic name for all synthetic polyamides. These are thermoplastic polymers with a wide range of properties.

olefin Family of unsaturated hydrocarbons with the formula C_nH_n, named after the corresponding paraffins by adding "ene" or "ylene" to the stem; for instance, ethylene. Paraffins are aliphatic hydrocarbons. (*See* **hydrocarbon**.)

oligomer Polymer containing only a few monomer units, such as a dimer or trimer.

orange peel Undesirably uneven or rough surface on a molded part, resembling the surface on an orange.

organic Composed of matter originating in plant or animal life, or composed of chemicals of hydrocarbon origin, either natural or synthetic. Used in referring to chemical structures based on the carbon atom.

orthotropic Having three mutually perpendicular planes of elastic symmetry.

parting agent *See* **mold release**.

paste Adhesive composition having a characteristic plastic-type consistency; that is, a high order or yield value such as that of a paste prepared by heating a mixture of starch and water and subsequently cooling the hydrolyzed product.

peel Imposition of a tensile stress in a direction perpendicular to the adhesive bond line, to a flexible adherend.

peel strength Strength of an adhesive in peel; expressed in pounds per inch of width.

penetration Entering of one part or material into another.

permanence Resistance of a given property to deteriorating influences.

permeability Ability of a material to allow liquid or gaseous molecules to pass through a film.

permittivity *See* **dielectric constant**.

phenolic Synthetic resin produced by the condensation of an aromatic alcohol with an aldehyde, particularly of phenol with formaldehyde.

phenylsilane Thermosetting copolymer of silicone and phenolic resins; furnished in solution form.

pitch fibers Fibers made from high-molecular-weight residue from the destructive distillation of coal or petroleum products.

plain weave The simplest and most commonly used weave, in which the warp and filling threads cross alternately. Plain woven fabrics are generally the least pliable, but they are also the most stable. This stability permits the fabrics to be woven with a fair degree of porosity without too much sleaziness.

plastic Material containing an organic substance of large molecular weight that is solid in its final condition and that, at some earlier time, was shaped by flow. (*See* **resin, polymer**.)

plastic deformation Change in dimensions of an object under load that is not recovered when the load is removed; opposed to elastic deformation.

plasticity Property of plastics that allows the material to be deformed continuously and permanently without rupture upon the application of a force that exceeds the yield value of the material.

plasticize To soften a material and make it plastic or moldable by means of either a plasticizer or the application of heat.

plasticizer Material incorporated in a resin formulation to increase its flexibility, workability, or distensibility. The addition of a plasticizer may cause a reduction in melt viscosity, lower the temperature of second-order transition, or lower the elastic modulus of the solidified resin.

plastisols Mixtures of vinyl resins and plasticizers that can be molded, cast, or converted to continuous films by the application of heat. If the mixtures contain volatile thinners, they are also known as *organosols*.

Poisson's ratio Absolute value of the ratio of transverse strain to axial strain resulting from a uniformly applied axial stress below the proportional limit of the material.

polyacrylonitrile (PAN) Synthetic fiber used as base material or precursor in manufacture of certain carbon fibers.

polyesters Thermosetting resins produced by reacting unsaturated, generally linear, alkyd resins with a vinyl-type active monomer such as styrene, methyl styrene, or diallyl

phthalate. Cure is effected through vinyl polymerization using peroxide catalysts and promoters or heat to accelerate the reaction. The resins are usually furnished in liquid form.

polyimide High-temperature resins made by reacting aromatic dianhydrides with aromatic diamines.

polymer Compound formed by the reaction of simple molecules having functional groups that permit their combination to proceed to high molecular weights under suitable conditions. Polymers may be formed by polymerization (addition polymer) or polycondensation (condensation polymer). When two or more monomers are involved, the product is called a *copolymer*. Also, any high-molecular-weight organic compound, the structure of which consists of a repeating small unit. Polymers can be plastics, elastomers, liquids, or gums and are formed by chemical addition or condensation of monomers. (*See also* **addition reaction or polymerization, condensation polymerization**.)

polymerize To unite chemically two or more monomers or polymers of the same kind to form a molecule with higher molecular weight.

pot To embed a component or assembly in a liquid resin, using a shell, can, or case that remains as an integral part of the product after the resin is cured. (*See* **embed, cast**.)

pot life Time during which a liquid resin remains workable as a liquid after catalysts, curing agents, promoters, and the like are added; roughly equivalent to gel time; sometimes also called *working life*.

power factor Cosine of the angle between the voltage applied and the current resulting.

precursor PAN or pitch fibers from which carbon graphite fiber is made.

preform Pill, tablet, or biscuit used in thermoset molding. Material is measured by volume, and the bulk factor of powder is reduced by pressure to achieve efficiency and accuracy.

preheating Heating of a compound prior to molding or casting to facilitate operation, reduce cycle, and improve product.

premix Molding compound prepared prior to and apart from the molding operations and containing all components required for molding: resin, reinforcement fillers, catalysts, release agents, and other compounds.

prepolymer Polymer in some stage between that of the monomers and the final polymer. The molecular weight is therefore also intermediate. As used in polyurethane production, reaction product of a polyol with excess of an isocyanate.

prepreg Ready-to-mold sheet that may be cloth, mat, or paper impregnated with resin and stored for use. The resin is partially cured to a B-stage and supplied to the fabricator, who lays up the finished shape and completes the cure with heat and pressure. (*See also* **B-stage**.)

pressure-bag molding Process for molding reinforced plastics in which a tailored flexible bag is placed over the contact lay-up on the mold, sealed, and clamped in place. Fluid pressure, usually compressed air, is placed against the bag, and the part is cured.

primer Coating applied to a surface prior to the application of an adhesive to improve the performance of the bond.

printed-circuit board *See* **circuit board.**

printed-circuit laminates Laminates, either fabric- or paper-based, covered with a thin layer of copper foil and used in the photofabrication process to make circuit boards.

printed-wiring board *See* **circuit board.**

promoter Chemical, itself a weak catalyst, that greatly increases the activity of a given catalyst.

proportional limit Greatest stress a material can sustain without deviating from the linear proportionality of stress to strain (Hooke's law).

pultrusion Reversed "extrusion" of resin-impregnated roving in the manufacture of rods, tubes, and structural shapes of a permanent cross section. The roving, after passing through the resin dip tank, is drawn through a die to form the desired cross section.

qualified products list (QPL) List of commercial products that have been pretested and found to meet the requirements of a specification, especially government specifications.

reactive diluent As used in epoxy formulations, a compound containing one or more epoxy groups that functions mainly to reduce the viscosity of the mixture.

refractive index Ratio of the velocity of light in a vacuum to its velocity in a substance; also ratio of the sine of the angle of incidence to the sine of the angle of refraction.

regrind Excess or waste material in a thermoplastic molding process that can be reground and mixed with virgin raw material, within limits, for molding future parts.

reinforced molding compound Plastic to which fibrous materials such as glass or cotton have been added to improve certain physical properties such as flexural strength.

reinforced plastic Plastic with strength properties greatly superior to those of the base resin, resulting from the presence of reinforcements in the composition.

reinforced thermoplastics Reinforced molding compounds in which the plastic is thermoplastic.

relative humidity Ratio of the quantity of water vapor present in the air to the quantity which would saturate it at any given temperature.

release agent *See* **mold release**.

release paper Impermeable paper film or sheet that is coated with a material to prevent adhering to prepreg.

resin High-molecular-weight organic material with no sharp melting point. For general purposes, the terms *resin, polymer,* and *plastic* can be used interchangeably. (*See* **polymer**.)

resistivity Ability of a material to resist passage of electric current, either through its bulk or on a surface. The unit of volume resistivity is the ohm-centimeter (Ω-cm), and the unit of surface resistivity is ohms per square (Ω/❑).

rheology Study of the change in form and flow of matter, embracing elasticity, viscosity, and plasticity.

rigidsol Plastisol having a high elastic modulus, usually produced with a cross-linking plasticizer.

Rockwell hardness number Number derived from the net increase in depth of impression as the load on a penetrator is increased from a fixed minimum load to a higher load and then returned to minimum load. Penetrators include steel balls of several specified diameters and a diamond-cone penetrator.

rotational casting (or molding) Method used to make hollow articles from thermoplastic materials. Material is charged into a hollow mold capable of being rotated in one or two planes. The hot mold fuses the material into a gel after the rotation has caused it to cover all surfaces. The mold is then chilled and the product stripped out.

roving Collection of bundles of continuous filaments, either as untwisted strands or as twisted yarns. Rovings may be lightly twisted, but for filament winding they are generally wound as bands or tapes with as little twist as possible.

rubber Elastomer capable of rapid elastic recovery; usually natural rubber, Hevea. (*See* **elastomer**.)

runners All channels in the mold through which molten or liquid plastic raw materials flow into mold.

S-glass Glass fabric made with very high tensile strength fibers for high-performance-strength requirements.

sandwich construction Panel consisting of some lightweight core material bonded to strong, stiff skins on both faces. (*See also* **honeycomb**.)

separator ply *See* **shear ply**.

set, mechanical Strain remaining after complete release of the load producing the deformation.

set, polymerization To convert an adhesive into a fixed or hardened state by chemical or physical action, such as condensation, polymerization, oxidation, vulcanization, gelation, hydration, or evaporation of volatile constituents.

shape factor For an elastomeric slab loaded in compression, ratio of loaded area to force-free area.

shear Action or stress resulting from applied forces that causes two contiguous parts of a body or two bodies to slide relative to each other in a direction parallel to their plane of contact.

shear ply Low-modulus layer, rubber, or adhesive interposed between metal and composite to reduce differential shear stresses.

shear strength Maximum shear stress a material is capable of sustaining. In testing, the shear stress is caused by a shear or torsion load and is based on the original specimen dimensions.

sheet molding compound Compression-molding material consisting of glass fibers longer than 0.5 in and thickened polyester resin. Possessing excellent flow, it results in parts with good surfaces.

shelf life Time a molding compound can be stored without losing any of its original physical or molding properties.

Shore hardness Procedure for determining the indentation hardness of a material by means of a durometer. Shore designation is given to tests made with a specified durometer instrument.

silicones Resinous materials derived from organosiloxane polymers, furnished in different molecular weights, including liquids and solid resins.

sink mark Depression or dimple on the surface of an injection-molded part due to collapsing of the surface following local internal shrinkage after the gate seals. May also be an incipient short shot.

sizing agent Chemical treatment containing starches, waxes, and the like that is applied to fibers, making them more resistant to breakage during the weaving process. The sizing agent must be removed after weaving, as its presence would cause delamination and moisture pickup problems if it remained in the final laminate made from the woven fiber.

slipping Lateral movement of tensioned fiber on a surface to a new unanticipated fiber angle.

slush molding Method for casting thermoplastics in which the resin in liquid form is poured into a hot mold where a viscous skin forms. The excess slush is drained off, the mold is cooled, and the molding is stripped out.

solvent Any substance, usually a liquid, that dissolves other substances.

specific heat Ratio of a material's thermal capacity to that of water at 15°C.

specific modulus Young's modulus divided by material density.

specific strength Ultimate tensile strength divided by material density.

spiral flow test Test method for measuring flow properties of a resin wherein the resin flows along a spiral path in a mold in the molding press. The variation in flow for different resins or different molding conditions can be compared. Flow is expressed in inches of flow in a standard spiral flow mold.

spray-up Process in which fiber reinforcement is wetted with resin applied from a spray gun. The fiber is fed through a chopper and into a stream of resin that is sprayed onto a form or into a mold.

stabilizers Chemicals used in plastics formulation to assist in maintaining physical and chemical properties during processing and service life. A specific type of stabilizer, known as an *ultraviolet stabilizer,* is designed to absorb ultraviolet rays and prevent them from attacking the plastic.

stacking sequence Sequence of laying plies into mold. Different stacking sequences have a great effect on off-axis mechanical properties.

starved area Part of a laminate or reinforced plastic structure in which resin has not completely wetted the fabric.

storage life *See* **shelf life**.

strain Deformation resulting from a stress, measured by the ratio of the change to the total value of the dimension in which the change occurred; unit change, due to force, in the size or shape of a body referred to its original size or shape. Strain is nondimensional but frequently expressed in inches per inch or centimeters per centimeter.

strength, dry Strength of an adhesive joint determined immediately after drying under specified conditions or after a period of conditioning in the standard laboratory atmosphere.

strength, wet Strength of an adhesive joint determined immediately after removal from a liquid in which it has been immersed under specified conditions of time, temperature, and pressure.

stress Unit force or component of force at a point in a body acting on a plane through the point. Stress is usually expressed in pounds per square inch.

stress relaxation Time-dependent decrease in stress for a specimen constrained in a constant strain condition.

substrate Material upon the surface of which an adhesive-containing substance is spread for any purpose, such as bonding or coating; broader term than adherend. Material upon the surface of which a circuit is formed.

surface preparation Physical and/or chemical preparation of an adherend to render it suitable for adhesive joining.

surface resistivity Resistance of a material between two opposite sides of a unit square of its surface. Surface resistivity may vary widely with the conditions of measurement.

syntactic foams Lightweight systems obtained by the incorporation of prefoamed or low-density fillers in the systems.

tack Property of an adhesive that enables it to form a bond of measurable strength immediately after adhesive and adherend are brought into contact under low pressure.

tan δ *See* **dissipation factor**.

tensile strength Maximum tensile stress a material is capable of sustaining. Tensile strength is calculated from the maximum load during a tension test carried to rupture and the original cross-sectional area of the specimen.

thermal conductivity Ability of a material to conduct heat; physical constant for the quantity of heat that passes through a unit cube of a material in a unit of time when the difference in temperatures of two faces is 1°C.

thermal stress cracking Crazing and cracking of some thermoplastic resins resulting from overexposure to elevated temperatures.

thermoforming Process of creating a form from a flat sheet by combinations of heat and pressure, which first soften the sheet and then form the sheet into some three-dimensional shape. This is one of the simplest, most economical plastic forming processes. There are numerous variations of this process.

thermoplastic Plastics capable of being repeatedly softened or melted by increases in temperature and hardened by decreases in temperature. These changes are physical rather than chemical.

thermoset Material that will undergo, or has undergone, a chemical reaction by the action of heat, catalysts, ultraviolet light, and the like, leading to a relatively infusible state that will not remelt after setting.

thermosetting Classification of resin that cures by chemical reaction when heated and, when cured, cannot be remelted by heating.

thinner Volatile liquid added to an adhesive to modify the consistency or other properties.

thixotropic Material that is gel-like at rest but fluid when agitated.

time, assembly Time interval between spreading of the adhesive on the adherend and application of pressure or heat, or both, to the assembly.

time, curing Period of time during which an assembly is subjected to heat, pressure, or both to cure the adhesive.

tow Untwisted bundle of continuous fibers. Commonly used in reference to synthetic fibers, particularly carbon and graphite, but also glass and aramid. A tow designated as 12K has 12,000 filaments.

tracking Conductive carbon path formed on surface of a plastic during electrical arcing. (*See also* **arc resistance**.)

transfer molding Method of molding thermosetting materials in which the plastic is first softened by heat and pressure in a transfer chamber and then forced or transferred by high pressure through suitable sprues, runners, and gates into a closed mold for final curing.

transformation Mathematical (tensor analysis) method of obtaining stress from strain values or vice versa, or to find angular properties of a laminate.

transverse isotropy Having essentially identical mechanical properties in two directions but not the third.

transverse, properties Properties perpendicular to the axial (x,1 or 1,1) direction. May be designated as *Y* or *Z*, 2 or 3 directions.

twill weave Basic weave characterized by a diagonal rib or twill line. Each end floats over at least two consecutive picks, permitting a greater number of yarns per unit area than a plain weave while not losing a great deal of fabric stability.

twist Spiral turns about its axis per unit length for a textile strand; expressed as turns per inch.

ultrasonic bonding Bonding of plastics by vibratory mechanical pressure at ultrasonic frequencies due to melting of plastics being joined, heat being generated as frictional heat.

ultraviolet Shorter wavelengths of invisible radiation that are more damaging than visible light to most plastics.

ultraviolet stabilizers Additives mixed into plastic formulations for the purpose of improving resistance of plastic to ultraviolet radiation.

undercured State of a molded article that has not been adequately polymerized or hardened in molding process, usually due to inadequate temperature-time-pressure control in molding process.

vacuum bag Impermeable film applied to outside of lay-up to facilitate conformability to mold form and air removal during cure.

vacuum-bag molding Process for molding reinforced plastics in which a sheet of flexible transparent material is placed over the lay-up on the mold and sealed. A vacuum is applied between sheet and lay-up. The entrapped air is mechanically worked out of the lay-up and removed by the vacuum, and the part is cured.

vacuum-injection molding Molding process in which, using a male and a female mold, reinforcements are placed in the mold, a vacuum is applied, and a room-temperature-curing liquid resin is introduced that saturates the reinforcement.

vent Small opening placed in a mold for allowing air to exit mold as molding material enters. This eliminates air holes, voids, or bubbles in finally molded part.

Vicat softening point One standard test for measuring temperature at which a thermoplastic will soften, involving the penetration of a flat-ended needle into the plastic under controlled conditions.

viscoelastic Characteristic mechanical behavior of some materials that is a combination of viscous and elastic behaviors.

viscosity Measure of the resistance of a fluid to flow (usually through a specific orifice).

void Air bubble that has been entrapped in a plastic part during molding process. (*See also* **vent**.)

volume resistivity (specific insulation resistance) Electrical resistance between opposite faces of a 1-cm cube of insulating material, commonly expressed in ohm-centimeters. The recommended test is ASTM D 257-54T.

vulcanization Chemical reaction in which the physical properties of an elastomer are changed by causing it to react with sulfur or other cross-linking agents.

vulcanized fiber Cellulosic material that has been partially gelatinized by action of a chemical (usually zinc chloride) and then heavily compressed or rolled to required thickness, leached free from the gelatinizing agent, and dried.

warp Fibers that run lengthwise in a woven fabric. (*See also* **weft**.)

warpage Dimensional distortion in a plastic object after molding.

water absorption Ratio of the weight of water absorbed by a material to the weight of the dry material.

water-extended polyester Casting formulation in which water is suspended in the polyester resin.

weave Pattern in which a fabric is woven. There are standard patterns, usually designated by a style number.

weft Fibers that run perpendicular to warp fibers; sometimes also called *fill* or *woof*. (*See also* **warp**.)

wet lay-up Reinforced plastic structure made by applying a liquid resin to an in-place woven or mat fabric.

wetting Ability to adhere to a surface immediately upon contact.

woof *See* **weft**.

working life *See* **pot life**.

woven fabric Flat sheet formed by interwinding yarns, fibers, or filaments. Some standard fabric patterns are plain, satin, and leno.

woven roving Heavy glass-fiber fabric made by the weaving of roving.

x-y axis Directions parallel to fibers in a woven-fiber-reinforced laminate. Thermal expansion is much lower in the *x-y* axis, since this expansion is more controlled by the fabric in the laminate. (*See also* **z axis**.)

yield value (yield strength) Lowest stress at which a material undergoes plastic deformation. Below this stress, the material is elastic; above it, the material is viscous. Also, stress at which a material exhibits a specified limiting deviation from the proportionality of stress to strain.

Young's modulus Ratio of normal stress to corresponding strain for tensile or compressive stresses at less than the proportional limit of the material.

z axis Direction perpendicular to fibers in a woven-fiber-reinforced laminate; that is, through the thickness of the laminate. Thermal expansion is much higher in the z axis, since this expansion is more controlled by the resin in the laminate. (*See also* **x-y axis**.)

Appendix

B

Some Common Abbreviations Used in the Plastics Industry*

ABS	acrylonitrile butadiene styrene
AI	artificial intelligence
AMC	alkyd molding compound
AMS	alpha methyl styrene
ANSI	American National Standards Institute
ASTM	American Society for Testing and Materials
ATH	aluminum trihydrate
BMC	bulk molding compound
CA	cellulose acetate
CAB	cellulose acetate butyrate
CAD	computer-aided design
CAM	computer-aided manufacturing
CAN	cellulose acetate nitrate
CAP	cellulose acetate propionate
CIM	computer-integrated manufacturing
CNC	computer numerical control
CPE	chlorinated polyethylene
CTE	coefficient of thermal expansion
CTFE	chlorotrifluoroethylene
CVD	chemical vapor deposition
DAIP	diallyl isophthalate

* Several of these abbreviations are also used for elastomers.

DAP	diallyl phthalate
DGEBA	diglycidyl ether of bisphenol A
DODISS	Department of Defense Index of Specifications and Standards
DOP	dioctyl phthalate
EB	ethyl benzene
EC	ethyl cellulose
ECTFE	ethylenechlorotrifluoroethylene
EDM	electrical-discharge machine
EMA	ethylene methyl acrylate
EMI	electromagnetic interference
EP	epoxy
EPR	ethylene propylene rubber
EPS	expanded polystyrene
ETFE	ethylenetetrafluoroethylene
EVA	ethylene vinyl acetate
EVOH	ethylene vinyl alcohol
FDA	Food and Drug Administration
FEA	finite-element analysis
FEP	fluoroethylene propylene copolymer
FR	fiber-reinforced; flame-retardant
FRP	fiber-reinforced plastic
HDPE	high-density polyethylene
HDT	heat-deflection temperature; heat-distortion temperature
HIPS	high-impact polystyrene
IPN	interpenetrating polymer network
JIT	just in time
LCP	liquid-crystal polymer
LDPE	low-density polyethylene
LED	light-emitting diode
LIM	liquid injection molding
LLDPE	linear low-density polyethylene
LMC	low-pressure molding compound
LOI	limiting oxygen index
MA	maleic anhydride
MDPE	medium-density polyethylene
MEK	methyl ethyl ketone
MEKP	methyl ethyl ketone peroxide
MF	melamine formaldehyde
MVT	moisture vapor transmission

MW	molecular weight
MWD	molecular-weight distribution
NEAT	nothing else added to it
NR	natural rubber
OSHA	Occupational Safety and Health Act
PA	polyamide (nylon)
PAI	polyamide imide
PAN	polyacrylonitrile
PAS	polyarylsulfone
PB	polybutylene
PBI	polybenzimidazol
PBT	polybutylene terephthalate
PC	polycarbonate
PCTFE	polychlorotrifluoroethylene
PE	polyethylene
PEEK	poiyetheretherketone
PEI	polyetherimide
PEKK	poiyetherketoneketone
PES	polyethersulfone
PET	polyethylene terephthalate
PF	phenol formaldehyde
PFA	perfluoroalkoxy resin
PHR	parts per hundred resin
PI	polyimide
PMDA	pyromellitic dianhydride
PMMA	polymethyl methacrylate (acrylic)
PMP	polymethyl pentene
PP	polypropylene
PPE	polyphenylene ether
PPO	polyphenylene oxide
PPS	polyphenylene sulfide
PPSS	polyphenylene sulfide sulfone
PS	polystyrene
PTFE	polytetrafluoroethylene
PUR	polyurethane
PVA	polyvinyl alcohol
PVDC	polyvinylidene chloride
PVDF	polyvinylidene fluoride
PVF	polyvinyl fluoride

RFI	radio-frequency interference
RIM	reaction injection molding
RLP	reactive liquid polymer
RP	reinforced plastics
RRIM	reinforced reaction molding
RTM	resin transfer molding
RTV	room-temperature vulcanizing
SAN	styrene acrylonitrile
SBR	synthetic butyl rubber
SBS	styrene butadiene styrene
SI	silicone
SMA	styrene maleic anhydride
SMC	sheet molding compound
SPC	statistical process control
SPE	Society of Plastics Engineers
SPI	Society of the Plastics Industry
SQC	statistical quality control
SRIM	structural reaction injection molding
TCE	trichloroethylene
TDI	toluene diisocyanate
TFE	polytetrafluoroethylene
TPE	thermoplastic elastomer
UF	urea formaldehyde
UHMWPE	ultrahigh-molecular-weight polyethylene
UL	Underwriters Laboratories
UV	ultraviolet
VA	vinyl acetate
WVT	water vapor transmission

Appendix

C

Important Properties for Designing with Plastics*

For most product design engineers, designing with plastics presents difficult problems, often resulting in missed opportunities or less than optimal plastic or plastic-containing products. The primary reason is that most design engineers are not experienced with plastics, since their education and training have prepared them for designing with metals. Furthermore, the chemical nature of plastics coupled with the extremely large number of plastics makes it often difficult for designers without specific training to properly understand and differentiate between kinds of plastics. The myriad grades and formulations only complicate the problem.

Based on this fundamental industry problem, the requirement exists for a well-categorized listing of plastic resins and compounds and their important properties for use by design engineers. This table, adapted from *Modern Plastics Encyclopedia '98*, well meets this need. Coupled with the discussions on these materials in the text of this handbook, this comprehensive data table will provide the product design engineer with useful guidelines ranging from processing pressures and temperatures used for molding plastic materials to part shrinkage during molding and on to all of the important mechanical, thermal, and physical properties needed for part design. Also listed are the ASTM test methods used in determining these properties, for those who need further understanding of these properties and their limitations. Lastly, major suppliers are provided for each of the materials, along with a listing of addresses. This will enable source selection and help in obtaining more detailed data for each of the plastic materials listed.

* The following table is reprinted from the *Modern Plastics Encyclopedia '98,* pp. B-152 to B-210, Resins and Compounds section.

Materials	Properties	ASTM test method	ABS						
				Flame-retarded grades, molding and extrusion					Injection molding grades
			Extrusion grade	ABS	ABS/PVC	ABS/PC	ABS/Nylon	ABS/PC injection molding and extrusion	Heat-resistant
Processing	1a. Melt flow (gm./10 min.)	D1238	0.4-6.86	1.2-1.7; 6	1.9				1.1-1.8
	1. Melting temperature, °C. T_m (crystalline)								
	T_g (amorphous)		88-120	110-125					110-125
	2. Processing temperature range, °F. (C = compression; T = transfer; I = injection; E = extrusion)		E: 350-500	C: 350-500 I: 380-500	370-410	I: 425-520	I: 460-520	I: 460-540 E: 450-500	C: 325-500 I: 475-550
	3. Molding pressure range, 10^3 p.s.i.			8-25		10-20	8-25	10-20	8-25
	4. Compression ratio		2.5-2.7	1.1-2.0	2.0-2.5	1.1-2.5	1.1-2.0	1.1-2.5	1.1-2.0
	5. Mold (linear) shrinkage, in./in.	D955	0.004-0.007	0.004-0.008	0.003-0.006	0.004-0.007	0.003-0.010	0.005-0.008	0.004-0.009
Mechanical	6. Tensile strength at break, p.s.i.	D638[b]	2500-8000	3300-8000	5800-6500	5800-9300	4000-6000	5800-7400	4800-7500
	7. Elongation at break, %	D638[b]	2.9-100	1.5-80		20-70	40-300	50-125	3-45
	8. Tensile yield strength, p.s.i.	D638[b]	4300-7250	4000-7400	4300-6600	7700-9000	4300-6300	3500-8500	4300-7000
	9. Compressive strength (rupture or yield), p.s.i.	D695	5200-10,000	6500-7500		11,000-11,300			7200-10,000
	10. Flexural strength (rupture or yield), p.s.i.	D790	4000-14,000	6200-14,000	7900-10,000	12,000-14,500	8800-10,900	8700-13,000	9000-13,000
	11. Tensile modulus, 10^3 p.s.i.	D638[b]	130-420	270-400	325-380	350-455	260-320	350-380	285-360
	12. Compressive modulus, 10^3 p.s.i.	D695	150-390	130-310		230			190-440
	13. Flexural modulus, 10^3 p.s.i. 73° F.	D790	130-440	300-600	320-400	350-400	250-310	290-375	300-400
	200° F.	D790							
	250° F.	D790							
	300° F.	D790							
	14. Izod impact, ft.-lb./in. of notch (1/8-in. thick specimen)	D256A	1.5-12	1.4-12	3.0-18.0	4.1-14.0	15-20	6.4-12.0	2.0-6.5
	15. Hardness Rockwell	D785	R75-115	R100-120	R100-106	R115-119	R93-105	R95-120	R100-115
	Shore/Barcol	D2240/ D2583			Shore D-73				
Thermal	16. Coef. of linear thermal expansion, 10^{-6} in./in./°C.	D696	60-130	65-95	46-84	67	90-110	62-72	60-93
	17. Deflection temperature under flexural load, °F. 264 p.s.i.	D648	170-220	158[g]; 181; 190-225 annealed	169-200 annealed	180-220	130-150	210-240	220-240 annealed 181-193[g]
	66 p.s.i.	D648	170-235	210-245 annealed		195-244	180-195	220-265	230-245 annealed
	18. Thermal conductivity, 10^{-4} cal.-cm./sec.-cm.^{2}-°C.	C177							4.5-8.0
Physical	19. Specific gravity	D792	1.02-1.08	1.16-1.21	1.13-1.25	1.17-1.23	1.06-1.07	1.07-1.15	1.05-1.08
	20. Water absorption (1/8-in. thick specimen), % 24 hr.	D570	0.20-0.45	0.2-0.6		0.24		0.15-0.24	0.20-0.45
	Saturation	D570							
	21. Dielectric strength (1/8-in. thick specimen), short time, v./mil	D149	350-500	350-500	500	450-760		430	350-500
		SUPPLIERS[a]	Albis; American Polymers; Ashley Polymers; Bamberger Polymers; BASF; Bayer Corp; Diamond Polymers; Dow Plastics; Federal Plastics; GE Plastics; LG Chemical; RSG Polymers; A. Schulman; Shuman	Albis; Ashley Polymers; BASF; Bayer Corp.; ComAlloy; Diamond Polymers; DSM; Federal Plastics; GE Plastics; LG Chemical; M.A. Polymers; Polymer Resources; RSG Polymers; RTP; A. Schulman; Shuman	ComAlloy; CONDEA Vista; Novatec; A. Schulman; Shuman	American Polymers; Ashley Polymers; Bayer Corp.; ComAlloy; Diamond Polymers; Dow Plastics; GE Plastics; LG Chemical; M.A. Polymers; Polymer Resources; RTP	Bayer Corp.	Ashley Polymers; Bayer Corp.; ComAlloy; Diamond Polymers; Dow Plastics; GE Plastics; LG Chemical; M.A. Polymers; Polymer Resources; RTP	American Polymers; Ashley Polymers; BASF; Bayer Corp.; Diamond Polymers; Dow Plastics; Federal Plastics; GE Plastics; LG Chemical; Polymer Resources; RSG Polymers; RTP; A. Schulman; Shuman

	ABS (Cont'd)										
	Injection molding grades (Cont'd)								EMI shielding (conductive)		
	Medium-impact	High-impact	Platable grade	Transparent	20% glass fiber-reinforced	30% glass fiber-reinforced	20% long glass fiber-reinforced	40% long glass fiber-reinforced	6% stainless steel fiber	7% stainless steel fiber	10% stainless steel fiber
1a.	1.1-34.3	1.1-18	1.1								
1.											
	102-115	91-110	100-110	120	100-110	100-110	100-110	100-110	100-110	100-110	100-110
2.	C: 325-350 I: 390-525	C: 325-350 I: 380-525	C: 325-400 I: 350-500	455-500	C: 350-500 I: 350-500	I: 400-460	I: 400-460	I: 400-460	I: 400-460	I: 400-460	I: 400-460
3.	8-25	8-25	8-25		15-30						
4.	1.1-2.0	1.1-2.0	1.1-2.0								
5.	0.004-0.009	0.004-0.009	0.005-0.008	0.009-0.067	0.001-0.002	0.002-0.003	0.001-0.002	0.001	0.004-0.006	0.004	0.004-0.006
6.	5500-7500	4400-6300	5200-6400	5000	10,500-13,000	13,000-16,000	13,000	16,000	6300-9200	6000	7100
7.	2.2-60	3.5-75		20	2-3	1.5-1.8	2.0	1.5	3.8	3.8	2.5
8.	5000-7200	2600-6300	6700	7000							
9.	1800-12,500	4500-8000			13,000-14,000	15,000-17,000	14,000	17,000			
10.	7100-13,000	5400-11,000	10,500-11,500	10,000	14,000-17,500	17,000-19,000	20,000	25,000	8700-11,000	10,000-12,000	12,100
11.	300-400	150-350	320-380	290	740-880	1000-1200	900	1000	300-400		400
12.	200-450	140-300			800						
13.	310-400	179-375	340-390		650-800	1000	850	1100	280-410	430	500
14.	3.0-9.6	6.0-10.5	4.0-8.3	1.5-2.0	1.1-1.4	1.2-1.3	2.0	2.5	1.2-1.4	1.0-1.1	1.4
15.	R102-115	R85-106	R103-109	R94	M85-98, R107	M75-85	M85-95	M90-100			
16.	80-100	95-110	47-53	60-130	20-21						
17.	174-220 annealed	205-215 annealed; 192 unannealed	190-222 annealed	194	210-220	215-230	210	215	190	185-200	190
	192-225 annealed	210-225 annealed	215-222 annealed	207	220-230	230-240	225	225			
18.					4.8						
19.	1.03-1.06	1.01-1.05	1.04-1.07	1.06	1.18-1.22	1.29	1.23	1.36	1.10-1.28	1.12	1.14
20.	0.20-0.45	0.20-0.45		0.35	0.18-0.20	0.3	0.2	0.2	0.4	0.4	
21.	350-500	350-500	420-550		450-460						
	Albis; American Polymers; Ashley Polymers; Bamberger Polymers; BASF; Bayer Corp.; Diamond Polymers; Dow Plastics; Federal Plastics; GE Plastics; LG Chemical; Polymer Resources; RSG Polymers; RTP; A. Schulman; Shuman	Albis; American Polymers; Ashley Polymers; Bamberger Polymers; BASF; Bayer Corp.; Diamond Polymers; Dow Plastics; Federal Plastics; GE Plastics; LG Chemical; Polymer Resources; RSG Polymers; RTP; A. Schulman; Shuman	American Polymers; Ashley Polymers; Bamberger Polymers; Bayer Corp.; Diamond Polymers; Dow Plastics; Federal Plastics; GE Plastics; LG Chemical; RSG Polymers; A. Schulman	BASF; Bayer Corp.; Diamond Polymers; GE Plastics; LG Chemical; A. Schulman	Albis; American Polymers; Ashley Polymers; ComAlloy; Diamond Polymers; DSM; Ferro; LG Chemical; LNP; M.A. Polymers; RTP; A. Schulman; Thermofil	Albis; American Polymers; Ashley Polymers; ComAlloy; Diamond Polymers; DSM; Ferro; LG Chemical; LNP; M.A. Polymers; RTP	DSM	DSM	Federal Plastics; Ferro; LNP; RTP; Ticona	DSM; Ferro; LNP; RTP; Ticona	RTP; Ticona

Materials	Properties	ASTM test method	ABS (Cont'd) EMI shielding (conductive) (Cont'd) 20% PAN carbon fiber	ABS EMI shielding 20% graphite fiber	ABS EMI shielding 40% aluminum flake	ABS Rubber-modified Injection molding and extrusion grades	Acetal Homo-polymer	Acetal Copolymer	Acetal Impact-modified homo-polymer
Processing	1a. Melt flow (gm./10 min.)	D1238					1-20	1-90	0.5-7.0
	1. Melting temperature, °C. T_m (crystalline)						172-184	160-175	175
	T_g (amorphous)		100-110						
	2. Processing temperature range, °F. (C = compression; T = transfer; I = injection; E = extrusion)		I: 415-500	I: 420-530	I: 400-550		I: 380-470	C: 340-400 I: 360-450	I: 380-420
	3. Molding pressure range, 10^3 p.s.i.		15-30				10-20	8-20	6-12
	4. Compression ratio						2.0-4.5	3.0-4.5	
	5. Mold (linear) shrinkage, in./in.	D955	0.0005-0.004	0.001	0.001		0.018-0.025	0.020 (Avg.)	0.012-0.019
Mechanical	6. Tensile strength at break, p.s.i.	D638[b]	15,000-16,000	15,200-15,800	3300-4200	5400-7100	9700-10,000		6500-8400
	7. Elongation at break, %	D638[b]	1.0-2.0	2.0-2.2	1.9-5	20-30	10-75	15-75	60-200
	8. Tensile yield strength, p.s.i.	D638[b]				6100	9500-12,000	8300-10,400	5500-7900
	9. Compressive strength (rupture or yield), p.s.i.	D695	17,000	16,000-17,000	6500		15,600-18,000 @ 10%	16,000 @ 10%	7600-11,900 @ 10%
	10. Flexural strength (rupture or yield), p.s.i.	D790	23,000-25,000	23,000	6200	8900-13,000	13,600-16,000	13,000	5800-10,000
	11. Tensile modulus, 10^3 p.s.i.	D638[b]	1800-2000	1660	370	310	400-520	377-464	190-350
	12. Compressive modulus, 10^3 p.s.i.	D695					670	450	
	13. Flexural modulus, 10^3 p.s.i. 73° F.	D790	890-1800	1560	400-600	260-380	380-490	370-450	150-350
	200° F.	D790					120-135		50-100
	250° F.	D790					75-90		33-60
	300° F.	D790							
	14. Izod impact, ft.-lb./in. of notch (1/8-in. thick specimen)	D256A	1.0	1.3	1.4-2.0	3.12-7.34	1.1-2.3	0.8-1.5	2.0-17
	15. Hardness Rockwell	D785	R108		R107	90-105	M92-94, R120	M75-90	M58-79
	Shore/Barcol	D2240/ D2583							
Thermal	16. Coef. of linear thermal expansion, 10^{-6} in./in./°C.	D696	18	20	40		50-112	61-110	92-117
	17. Deflection temperature under flexural load, °F. 264 p.s.i.	D648	215-225	216	190-212	181-212	253-277	185-250	148-185
	66 p.s.i.	D648	225-230	240	220		324-342	311-330	293-336
	18. Thermal conductivity, 10^{-4} cal.-cm./sec.-cm.^{2}-°C.	C177	9.6				5.5	5.5	
Physical	19. Specific gravity	D792	1.13-1.14	1.17	1.54-1.61	103-119	1.42	1.40	1.32-1.39
	20. Water absorption (1/8-in. thick specimen), % 24 hr.	D570	0.17	0.15	0.23		0.25-1	0.20-0.22	0.30-0.44
	Saturation	D570					0.90-1	0.65-0.80	0.75-0.85
	21. Dielectric strength (1/8-in. thick specimen), short time, v./mil	D149					400-500 (90 mil)	500 (90 mil)	400-480 (90 mil)
		SUPPLIERS[a]	DSM; Ferro; LNP; RTP	Albis; LNP; RTP; Thermofil	ComAlloy; Thermofil	Diamond Polymers; RSG Polymers	Ashley Polymers; DuPont; RTP; Shuman	American Polymers; Ashley Polymers; BASF; LG Chemical; M.A. Hanna Eng.; Network Polymers; RTP; Ticona	Ashley Polymers; DuPont

Acetal (Cont'd)

	Impact-modified copolymer	Mineral-filled copolymer	Extrusion and blow molding grade (terpolymer)	Copolymer with 2% silicone, low wear	UV stabilized copolymer	20% glass-reinforced homo-polymer	25% glass-coupled copolymer	40% long glass fiber-reinforced	21% PTFE-filled homo-polymer	2-20% PTFE-filled copolymer	1.5% PTFE-filled homo-polymer
1a.	5-26		1.0	9	2.5-27.0	6.0			1.0-7.0		6
1.	160-170	160-175	160-170	160-170	170	175-181	160-180	190	175-181	160-175	175
2.	I: 360-425	I: 360-450	E: 360-400	I: 360-450	I: 360-450	I: 350-480	I: 365-480	380-430	I: 370-410	I: 350-445 I: 325-500 E: 360-500	I: 400-440
3.	8-15	10-20		10-20	10-20	10-20	8-20	8-12	10-20	8-20	
4.	3.0-4.5	3.0-4.0	3.0-4.0	3.0-4.0	3.0-4.0		3.0-4.5	3.0-4.0		3.0-4.5	
5.	0.018-0.020	0.015-0.019	0.02	0.022	0.022	0.009-0.012	0.004 (flow) 0.018 (trans.)	0.003-0.010	0.020-0.025	0.018-0.029	
6.		6400-11,500				8500-9000	16,000-18,500	17,400	6900-7600	8300	10,000
7.	60-300	5-55	67	60	30-75	6-12	2-3	1.3	10-22	30	13
8.	3000-8000	6400-9800	8700	7400	8800-9280	7500-8250	16,000		6900-7600	8300	10,000
9.			16,000		16,000	18,000 @ 10%	17,000 @ 10%	20,400	13,000 @ 10%	11,000-12,600	
10.	7100	12,500-13,000	12,800	12,000	13,000	10,700-16,000	18,000-28,000	27,000	11,000	11,500	13,900
11.	187-319	522-780			410-910	900-1000	1250-1400	1700	410-420	250-280	450
12.					450						
13.	120-300	430-715	350	350	375-380	600-730	1100	1430	340-380	310-360	430
						300-360			110-120		150
						250-270			80-85		95
14.	1.7-4.7	0.9-1.2	1.7	1.1-1.4	1.12-1.5	0.5-1.0	1.0-1.8	6.9	0.7-1.2	0.5-1.0	1.0
15.	M35-70; R110	M83-90	M84	M75	M80-85	M90	M79-90, R110		M78, M110	M79	M93
16.	130-150	80-90				33-81	17-44		75-113	52-68	
17.	132-200	200-279	205	230	221-230	315	320-325	320	210-244	198-225	277
	308-318	302-325	318	316	316-320	345	327-331		300-334	280-325	342
18.										4.7	
19.	1.29-1.39	1.48-1.64	1.41	1.4	1.41	1.54-1.56	1.58-1.61	1.72	0.15-1.54	1.40	1.42
20.	0.31-0.41	0.20	0.22	0.21	0.22	0.25	0.22-0.29		0.20	0.15-0.26	0.19
	1.0-1.3	0.8-0.9	0.8	0.8	0.8	1.0	0.8-1.0		0.72	0.5	0.90
21.						490 (125 mil)	480-580		400-460 (125 mil)	400-410	450 (90 mils)
	Ashley Polymers; BASF; LG Chemical; M.A. Hanna Eng.; Network Polymers; Ticona	BASF; LG Chemical; M.A. Hanna Eng.; Network Polymers; Ticona	Network Polymers; Ticona	LG Chemical; M.A. Hanna Eng.; Network Polymers; RTP; Ticona	BASF; LG Chemical; M.A. Hanna Eng.; Network Polymers; Ticona	ComAlloy; DSM; DuPont; Ferro; LNP; RTP	BASF; ComAlloy; DSM; Ferro; LNP; M.A. Hanna Eng.; Network Polymers; RTP; Ticona	Ticona	Adell; DSM; DuPont; Ferro; LNP; RTP	ComAlloy; DSM; Ferro; LG Chemical; LNP; M.A. Hanna Eng.; RTP; Ticona	DuPont

Materials	Properties	ASTM test method	Acetal (Cont'd)						Acrylic
							EMI shielding (conductive)		Sheet
	Properties	ASTM test method	Chemically lubricated homo-polymer	UV stabilized homo-polymer	UV stabilized, 20% glass-filled homo-polymer	Extrusion grade homo-polymer	30% carbon fiber	10% PAN carbon fiber, 10% PTFE-filled copolymer	Cast
Processing	1a. Melt flow (gm./10 min.)	D1238	6	1-6	6	1-6			
	1. Melting temperature, °C. T_m (crystalline)		175	175	175	175	166	163-175	
	T_g (amorphous)								90-105
	2. Processing temperature range, °F. (C = compression; T = transfer; I = injection; E = extrusion)		I: 400-440	I: 400-440	I: 400-440	E: 380-420	I: 350-400	I: 350-410	
	3. Molding pressure range, 10^3 p.s.i.						10-20		
	4. Compression ratio						3-4		
	5. Mold (linear) shrinkage, in./in.	D955					0.003-0.005	0.002	1.7
Mechanical	6. Tensile strength at break, p.s.i.	D638[b]	9500	10,000	8500	10,000	7500-11,500	12,000	66-11,000
	7. Elongation at break, %	D638[b]	40	40-75	12	40-75	1.5-2	1.3	2-7
	8. Tensile yield strength, p.s.i.	D638[b]	9500	10,000	8500				
	9. Compressive strength (rupture or yield), p.s.i.	D695							11,000-19,000
	10. Flexural strength (rupture or yield), p.s.i.	D790	13,000	14,100-14,300	10,700	14,100-14,300	12,500-16,500	14,000	12,000-17,000
	11. Tensile modulus, 10^3 p.s.i.	D638[b]	450	400-450	900	400-450	1350	1300	310-3100
	12. Compressive modulus, 10^3 p.s.i.	D695							390-475
	13. Flexural modulus, 10^3 p.s.i. 73° F.	D790	400	420-430	730	420-430	1100-1200	1000	320-3210
	200° F.	D790	130	130-135	360	130-135			
	250° F.	D790	80	90	270	90			
	300° F.	D790							
	14. Izod impact, ft.-lb./in. of notch (1/8-in. thick specimen)	D256A	1.4	1.5-2.3	0.8	1.5-2.3	0.7-0.8	0.7	0.3-0.4
	15. Hardness Rockwell	D785	M90	M94	M90	M94			M80-102
	Shore/Barcol	D2240/ D2583							
Thermal	16. Coef. of linear thermal expansion, 10^{-6} in./in./°C.	D696							50-90
	17. Deflection temperature under flexural load, °F. 264 p.s.i.	D648	257	257-264	316	257-264	320	320	98-215
	66 p.s.i.	D648	329	334-336	345	334-336	325	325	165-235
	18. Thermal conductivity, 10^{-4} cal.-cm./sec.-cm.^{2}-°C.	C177							4.0-6.0
Physical	19. Specific gravity	D792	1.42	1.42	1.56	1.42	1.43-1.53	1.49	1.17-1.20
	20. Water absorption (1/8-in. thick specimen), % 24 hr.	D570	0.27	0.25	0.25	0.25	0.22-0.26	0.25	0.2-0.4
	Saturation	D570	1.00	0.90	1.00	0.90			
	21. Dielectric strength (1/8-in. thick specimen), short time, v./mil	D149	400 (125 mils)	500 (90 mils)	490 (125 mils)	500 (90 mils)			450-550
		SUPPLIERS[a]	DuPont	DuPont	DuPont	DuPont	DSM; LNP; RTP; Ticona	DSM; Ferro; LNP; M.A. Hanna Eng.; RTP	Aristech; AtoHaas; Cyro; DuPont; ICI Acrylics

	Acrylic (Cont'd)							Acrylonitrile			Allyl
	Sheet (Cont'd)	Molding and extrusion compounds									
	Coated	Acrylic/PC alloy	PMMA	MMA-styrene copolymer	Impact-modified	Heat-resistant	Acrylic multi-polymer	Extrusion	High-impact extrusion	Injection	Allyl diglycol carbonate cast sheet
1a.		3-4	1.4-27	1.1-24	1-11	1.6-8.0	2-14	3	3	12	
1.										135	Thermoset
	90-110	140	85-105	100-105	80-103	100-165	80-105			95	
2.		I: 450-510 E: 430-480	C: 300-425 I: 325-500 E: 360-500	C: 300-400 I: 300-500	C: 300-400 I: 400-500 E: 380-480	C: 350-500 I: 400-625 E: 360-550	I: 400-500 E: 380-470	380-420	380-410	C: 320-345 I: 410 E: 380-410	
3.		5-20	5-20	10-30	5-20	5-30	5-20	25	25	20	
4.			1.6-3.0			1.2-2.0		2-2.5	2-2.5	2-2.5	
5.		0.004-0.008	0.001-0.004(flow) 0.002-0.008(trans.)	0.002-0.006	0.002-0.008	0.002-0.008	0.004-0.008	0.002-0.005	0.002-0.005	0.002-0.005	
6.	10,500	8000-9000	7000-10,500	8100-10,100	5000-9000	9300-11,500	5500-8200			9000	5000-6000
7.	3	58	2-5.5	2-5	4.6-70	2-10	5-28	3-4	3-4	3-4	
8.			7800-10,600		5500-8470	10,000		9500	7500	7500-9500	
9.	18,000		10,500-18,000	11,000-15,000	4000-14,000	15,000-17,000	7500-11,500	12,000	11,500	12,000	21,000-23,000
10.	16,000	11,300-12,500	10,500-19,000	14,100-16,000	7000-14,000	12,000-18,000	9000-13,000	14,000	13,700	14,000	6000-13,000
11.	450	320-350	325-470	430-520	200-500	350-650	300-430	500-550	450-500	500-580	300
12.	450		370-460	240-370	240-370	450					300
13.	450	320-350	325-460	450-460	200-430	450-620	290-400	490	390	500-590	250-330
						150-440					
						350-420					
14.	0.3-0.4	26-30	0.2-0.4	0.3-0.4	0.40-2.5	0.2-0.4	1.0-2.5	5.0	9.0	2.5-6.5	0.2-0.4
15.	M105	46-49	M68-105	M80-85	M35-78	M94-100	22-56	M60	M45	M60-M78	M95-100
16.	40	52	50-90	60-80	48-80	40-71	44-50	66	66	66	81-143
17.	205	214	155-212	208-211	165-209	190-310	180-194	156	151	151-164	140-190
	225	253	165-225		180-205	200-315		170	160	166-172	
18.	5.0		4.0-6.0	4.0-5.0	4.0-5.0	2.0-4.5	5.3	6.1	6.1	6.1-6.2	4.8-5.0
19.		1.15	1.17-1.20	1.08-1.13	1.11-1.18	1.16-1.22	1.11-1.12	1.15	1.11	1.11-1.15	1.3-1.4
20.	<0.4	0.3	0.1-0.4	0.11-0.17	0.19-0.8	0.2-0.3	0.3			0.28	0.2
21.	500		400-500	450	380-500	400-500		220-240	220-240	220-240	380
	DuPont	Cyro	American Polymers; AtoHaas; Continental Acrylics; Cyro; DuPont; ICI Acrylics; LG Chemical; Network Polymers; Plaskolite; RTP	Network Polymers; NOVA Chemicals	AtoHaas; Continental Acrylics; Cyro; DuPont; ICI Acrylics; Network Polymers; RTP	AtoHaas; Cyro; ICI Acrylics; Network Polymers; Plaskolite; RTP	Cyro	BP Chemicals	BP Chemicals	BP Chemicals	PPG

Materials			Allyl (Cont'd)		Cellulosic				
			DAP molding compounds			Cellulose acetate		Cellulose acetate butyrate	Cellulose acetate propionate
	Properties	ASTM test method	Glass-filled	Mineral-filled	Ethyl cellulose molding compound and sheet	Sheet	Molding and extrusion compound	Molding and extrusion compound	Molding and extrusion compound
Processing	1a. Melt flow (gm./10 min.)	D1238							
	1. Melting temperature, °C. T_m (crystalline)		Thermoset	Thermoset	135	230	230	140	190
	T_g (amorphous)								
	2. Processing temperature range, °F. (C = compression; T = transfer; I = injection; E = extrusion)		C: 290-360 I: 300-350	C: 270-360	C: 250-390 I: 350-500		C: 260-420 I: 335-490	C: 265-390 I: 335-480	C: 265-400 I: 335-515
	3. Molding pressure range, 10^3 p.s.i.		2000-6000	2500-5000	8-32		8-32	8-32	8-32
	4. Compression ratio		1.9-10.0	1.2-2.3	1.8-2.4		1.8-2.6	1.8-2.4	1.8-3.4
	5. Mold (linear) shrinkage, in./in.	D955	0.0005-0.005	0.002-0.007	0.005-0.009		0.003-0.010	0.003-0.009	0.003-0.009
Mechanical	6. Tensile strength at break, p.s.i.	D638[b]	6000-11,000	5000-8000	2000-8000	4500-8000	1900-9000	2600-8100	2000-7800
	7. Elongation at break, %	D638[b]	3-5	3-5	5-40	20-50	6-70	40-88	29-100
	8. Tensile yield strength, p.s.i.	D638[b]					2500-7600	1600-7200	
	9. Compressive strength (rupture or yield), p.s.i.	D695	25,000-35,000	20,000-32,000			3000-8000	2100-7500	2400-7000
	10. Flexural strength (rupture or yield), p.s.i.	D790	9000-20,000	8500-11,000	4000-12,000	6000-10,000	2000-16,000	1800-10,100	2900-11,400
	11. Tensile modulus, 10^3 p.s.i.	D638[b]	1400-2200	1200-2200				50-200	60-215
	12. Compressive modulus, 10^3 p.s.i.	D695							
	13. Flexural modulus, 10^3 p.s.i. 73° F.	D790	1200-1500	1000-1400			1000-4000	90-300	120-350
	200° F.	D790							
	250° F.	D790							
	300° F.	D790							
	14. Izod impact, ft.-lb./in. of notch (1/8-in. thick specimen)	D256A	0.4-15.0	0.3-0.8	0.4	2.0-8.5	1.0-7.8	1.0-10.9	0.5-No break
	15. Hardness Rockwell	D785	E80-87	E60-80E	R50-115	R85-120	R17-125	R11-116	R10-122
	Shore/Barcol	D2240/ D2583							
Thermal	16. Coef. of linear thermal expansion, 10^{-6} in./in./°C.	D696	10-36	10-42	100-200	100-150	80-180	110-170	110-170
	17. Deflection temperature under flexural load, °F. 264 p.s.i.	D648	330-550+	320-550	115-190		111-195	109-202	111-228
	66 p.s.i.	D648					120-209	130-227	147-250
	18. Thermal conductivity, 10^{-4} cal.-cm./sec.-cm.^{2}-°C.	C177	5.0-15.0	7.0-25	3.8-7.0	4-8	4-8	4-8	4-8
Physical	19. Specific gravity	D792	1.70-1.98	1.65-1.85	1.09-1.17	1.28-1.32	1.22-1.34	1.15-1.22	1.17-1.24
	20. Water absorption (1/8-in. thick specimen), % 24 hr.	D570	0.12-0.35	0.2-0.5	0.8-1.8	2.0-7.0	1.7-6.5	0.9-2.2	1.2-2.8
	Saturation	D570							
	21. Dielectric strength (1/8-in. thick specimen), short time, v./mil	D149	400-450	400-450	350-500	250-600	250-600	250-475	300-475
		SUPPLIERS[a]	Cosmic Plastics; Rogers Corp.	Cosmic Plastics; Rogers Corp.	Dow Chem.; Federal Plastics; Rotuba	Rotuba	Albis; Eastman; Rotuba	Albis; Eastman; Rotuba	Albis; Eastman; Rotuba

	Cellulosic (Cont'd)	Chlorinated PE	Epoxy								
			Bisphenol molding compounds			Sheet molding compound (SMC)		Novolak molding compounds		Casting resins and compounds	
	Cellulose nitrate	30-42% Cl extrusion and molding grades	Glass fiber-reinforced	Mineral-filled	Low density glass sphere-filled	Glass fiber-reinforced	Carbon fiber-reinforced	Mineral- and glass-filled, encapsulation	Mineral- and glass-filled, high temperature	Unfilled	Silica-filled
1a.											
1.		125	Thermoset	Thermoset	Thermoset	Thermoset	Thermoset	Thermoset	Thermoset	Thermoset	Thermoset
								145-155	155-195		
2.	C: 185-250	E: 300-400	C: 300-330 T: 280-380	C: 250-330 T: 250-380	C: 250-300 I: 250-300	C: 250-330 T: 270-330	C: 250-330 T: 270-330	C: 280-360 I: 290-350 T: 250-380	T: 340-400		
3.	2-5		1-5	0.1-3	0.1-2	0.5-2.0	0.5-2.0	0.25-3.0	0.5-2.5		
4.			3.0-7.0	2.0-3.0	3.0-7.0	2.0	2.0		1.5-2.5		
5.			0.001-0.008	0.002-0.010	0.006-0.010	0.001	.001	0.004-0.008	0.004-0.007	0.001-0.010	0.0005-0.008
6.	7000-8000	1400-3000	5000-20,000	4000-10,800	2500-4000	20,000-35,000	40,000-50,000	5000-12,500	6000-15,500	4000-13,000	7000-13,000
7.	40-45	300-900	4			0.5-2.0	0.5-2.0			3-6	1-3
8.											
9.	2100-8000		18,000-40,000	18,000-40,000	10,000-15,000	20,000-30,000	30,000-40,000	24,000-48,000	30,000-48,000	15,000-25,000	15,000-35,000
10.	9000-11,000		8000-30,000	6000-18,000	5000-7000	50,000-70,000	75,000-95,000	10,000-21,800	10,000-21,800	13,000-21,000	8000-14,000
11.	190-220		3000	350		2000-4000	10,000	2100	2300-2400	350	
12.				650					660		
13.			2000-4500	1400-2000	500-750	2000-3000	5000	1400-2400	2300-2400		
						1500-2500					
						1200-1800					
14.	5-7		0.3-10.0	0.3-0.5	0.15-0.25	30-40	15-20	0.3-0.5	0.4-0.45	0.2-1.0	0.3-0.45
15.	R95-115		M100-112	M100-M112				M115		M80-110	M85-120
		Shore A60-76				B: 55-65	B: 55-65	Barcol 70-75	Barcol 78		
16.	80-120		11-50	20-60		12	3	18-43	35	45-65	20-40
17.	140-160		225-500	225-500	200-250	550	550	300-500	500	115-550	160-550
18.	5.5		4.0-10.0	4-35	4.0-6.0	1.7-1.9	1.4-1.5	10-31	17-24	4.5	10-20
19.	1.35-1.40	1.13-1.26	1.6-2.0	1.6-2.1	0.75-1.0	0.10	0.10	1.6-2.05	1.85-1.94	1.11-1.40	1.6-2.0
20.	1.0-2.0		0.04-0.20	0.03-0.20	0.2-1.0	1.4	1.6	0.04-0.29	0.15-0.17	0.08-0.15	0.04-0.1
								0.15-0.3			
21.	300-600		250-400	250-420	380-420			325-450	440-450	300-500	300-550
	P.D. George	Dow Plastics	Amoco Electronic; Cytec Fiberite	Amoco Electronic; Cytec Fiberite	Cytec Fiberite	Quantum Composites	Quantum Composites	Amoco Electronic; Cosmic Plastics; Cytec Fiberite; Rogers	Amoco Electronic; Cosmic Plastics; Cytec Fiberite; Rogers	Ciba Specialty Chemicals; Conap; Dow Plastics; Emerson & Cuming; Epic Resins; ITW Devcon; Shell; United Mineral	Conap; Emerson & Cuming; Epic Resins; ITW Devcon

Materials	Properties	ASTM test method	Epoxy (Cont'd) Casting resins and compounds (Cont'd) Aluminum-filled	Epoxy (Cont'd) Casting resins and compounds (Cont'd) Flexibilized	Epoxy (Cont'd) Casting resins and compounds (Cont'd) Cyclo-aliphatic	Ethylene vinyl alcohol	Fluoroplastics Polychloro-trifluoro-ethylene	Fluoroplastics Polytetrafluoroethylene Granular	Fluoroplastics Polytetrafluoroethylene 25% glass fiber-reinforced
Processing	1a. Melt flow (gm./10 min.)	D1238				0.8-14.0			
	1. Melting temperature, °C. T_m (crystalline)		Thermoset	Thermoset	Thermoset	142-191		327	327
	T_g (amorphous)					49-72	220		
	2. Processing temperature range, °F. (C = compression; T = transfer; I = injection; E = extrusion)					I: 365-480 E: 365-480	C: 460-580 I: 500-600 E: 360-590		
	3. Molding pressure range, 10^3 p.s.i.						1-6	2-5	3-8
	4. Compression ratio					3-4	2.6	2.5-4.5	
	5. Mold (linear) shrinkage, in./in.	D955	0.001-0.005	0.001-0.010			0.010-0.015	0.030-0.060	0.018-0.020
Mechanical	6. Tensile strength at break, p.s.i.	D638[b]	7000-12,000	2000-10,000	8000-12,000	5405-13,655	4500-6000	3000-5000	2000-2700
	7. Elongation at break, %	D638[b]	0.5-3	20-85	2-10	180-330	80-250	200-400	200-300
	8. Tensile yield strength, p.s.i.	D638[b]				7385-10,365	5300		
	9. Compressive strength (rupture or yield), p.s.i.	D695	15,000-33,000	1000-14,000	15,000-20,000		4600-7400	1700	1000-1400 @ 1% strain
	10. Flexural strength (rupture or yield), p.s.i.	D790	8500-24,000	1000-13,000	10,000-13,000	230-285	7400-11,000		2000
	11. Tensile modulus, 10^3 p.s.i.	D638[b]						58-80	200-240
	12. Compressive modulus, 10^3 p.s.i.	D695		1-350	495	300-385	150-300	60	
	13. Flexural modulus, 10^3 p.s.i. 73° F.	D790					170-200	80	190-235
	200° F.	D790					180-260		
	250° F.	D790							
	300° F.	D790							
	14. Izod impact, ft.-lb./in. of notch (1/8-in. thick specimen)	D256A	0.4-1.6	2.3-5.0		1.0-1.7	2.5-5	3	2.7
	15. Hardness Rockwell	D785	M55-85				R75-112		
	Shore/Barcol	D2240/ D2583		Shore D65-89			Shore D75-80	Shore D50-65	Shore D60-70
Thermal	16. Coef. of linear thermal expansion, 10^{-6} in./in./°C.	D696	5.5	20-100			36-70	70-120	77-100
	17. Deflection temperature under flexural load, °F. 264 p.s.i.	D648	190-600	73-250	200-450			115	
	66 p.s.i.	D648					258	160-250	
	18. Thermal conductivity, 10^{-4} cal.-cm./sec.-cm.^{2}-°C.	C177	15-25				4.7-5.3	6.0	8-10
Physical	19. Specific gravity	D792	1.4-1.8	0.96-1.35	1.16-1.21	1.12-1.20	2.08-2.2	2.14-2.20	2.2-2.3
	20. Water absorption (1/8-in. thick specimen), % 24 hr.	D570	0.1-4.0	0.27-0.5		6.7-8.6	0	<0.01	
	Saturation	D570							
	21. Dielectric strength (1/8-in. thick specimen), short time, v./mil	D149		235-400			500-600	480	320
		SUPPLIERS[a]	Conap; Emerson & Cuming; Epic Resins; ITW Devcon	Conap; Dow Plastics; Emerson & Cuming; Epic Resins; ITW Devcon	Ciba Specialty Chemicals; Union Carbide	Eval Co. of America	Ciba Specialty Chemicals; Elf Atochem N.A.	Ausimont; DuPont; Dyneon; ICI Americas	Ausimont; DuPont; Dyneon; ICI Americas; RTP

Fluoroplastics (Cont'd)

	PFA fluoro-plastic	Fluorinated ethylene propylene: Unfilled	Fluorinated ethylene propylene: 20% milled glass fiber	Polyvinyl fluoride film: Unfilled	Polyvinyl fluoride film: Filled	Polyvinylidene fluoride: Molding and extrusion	Polyvinylidene fluoride: Wire and cable jacketing	Polyvinylidene fluoride: EMI shielding (conductive); 30% PAN carbon fiber	Modified PE-TFE: Unfilled	Modified PE-TFE: 25% glass fiber-reinforced	THV-200
1a.											
1.	300-310	275	262	192	192	141-178	168-170		270	270	120
						−60 to −20	−30 to −20				
2.	C: 625-700 I: 680-750	C: 600-750 I: 625-760	I: 600-700			C: 360-550 I: 375-550 E: 375-550	E: 420-525	I: 430-500	C: 575-625 I: 570-650	C: 575-625 I: 570-650	E: 450
3.	3-20	5-20	10-20			2-5	1.10		2-20	2-20	
4.	2.0	2.0				3	3				
5.	0.040	0.030-0.060	0.006-0.010			0.020-0.035	0.020-0.030	0.001	0.030-0.040	0.002-0.030	
6.	4000-4300	2700-3100	2400	6000-16,000	6000-16,000	3500-7250	7100	14,000	6500	12,000	3500
7.	300	250-330	5	100-250	75-250	12-600	300-500	0.8	100-400	8	600
8.	2100					2900-8250	4460				
9.	3500	2200				8000-16,000	6600-7100		7100	10,000	
10.			4000			9700-13,650	7000-8600	19,800	5500	10,700	
11.	70	50		300-380	300-380	200-80,000	145-190	2800	120	1200	
12.						304-420	180				
13.	95-120	80-95	250			170-120,000	145-260	2100	200	950	12
									80	450	
									60	310	
									20	200	
14.	No break	No break	3.2			2.5-80	7	1.5	No break	9.0	
15.						R79-83, 85	R77		R50	R74	
	Shore D64	Shore D60-65				Shore D80, 82 65-70	Shore D75		Shore D75		44
16.	140-210		22			70-142	121-140		59	10-32	
17.			150			183-244	129-165	318	160	410	
	166	158				280-284			220	510	
18.	6.0	6.0		0.0014-0.0017	0.0014-0.0017	2.4-3.1	2.4-3.1		5.7		
19.	2.12-2.17	2.12-2.17		1.38-1.40	1.38-1.72	1.77-1.78	1.76-1.77	1.74	1.7	1.8	1.95
20.	0.03	<0.01	0.01			0.03-0.06	0.03-0.06	0.12	0.03	0.02	
21.	500	500-600		2000-3300 *(D150-81)	2000-3300 *(D150-81)	260-280	260-280		400	425	
	Ausimont; DuPont; Dyneon	DuPont	RTP	DuPont	DuPont	Ausimont; Elf Atochem N.A.; Solvay Polymers	Ausimont; Elf Atochem N.A.; Solvay Polymers	RTP	DuPont	Ausimont; DuPont; RTP	Dyneon

Materials	Properties	ASTM test method	Fluoroplastics (Cont'd)			Ionomer		Ketones	
								Polyaryletherketone	
			THV-400	THV-500	PE-CTFE	Molding and extrusion	Glass- and rubber-modified; molding and extrusion	Unfilled	30% glass fiber-reinforced
Processing	1a. Melt flow (gm./10 min.)	D1238						4-7	15-25
	1. Melting temperature, °C. T_m (crystalline)		150	180	220-245	81-96	81-220	323-381	329-381
	T_g (amorphous)								
	2. Processing temperature range, °F. (C = compression; T = transfer; I = injection; E = extrusion)		E: 470	E: 480	C: 500 I: 525-575 E: 500-550	C: 280-350 E: 300-450	C: 300-400 I: 300-550 E: 350-525	I: 715-805	I: 715-805
	3. Molding pressure range, 10^3 p.s.i.				5-20	2-20	2-20	10-20	10-20
	4. Compression ratio					3	3	2	2
	5. Mold (linear) shrinkage, in./in.	D955			0.020-0.025	0.003-0.010	0.002-0.008	0.008-0.012	0.001-0.009
Mechanical	6. Tensile strength at break, p.s.i.	D638[b]	3400	3300	6000-7000	2500-5400	3500-7900	13,500	23,700-27,550
	7. Elongation at break, %	D638[b]	500	500	200-300	300-700	5-200	50	2.2-3.4
	8. Tensile yield strength, p.s.i.	D638[b]			4500-4900	1200-2300	1200-4500	15,000	
	9. Compressive strength (rupture or yield), p.s.i.	D695						20,000	30,000
	10. Flexural strength (rupture or yield), p.s.i.	D790			7000			18,850-24,500	34,100-36,250
	11. Tensile modulus, 10^3 p.s.i.	D638[b]			240	to 60		520-580	1410-1754
	12. Compressive modulus, 10^3 p.s.i.	D695							
	13. Flexural modulus, 10^3 p.s.i. 73° F.	D790		30	240	3-55	8-700	530	1600
	200° F.	D790						530	1520
	250° F.	D790						520	1460
	300° F.	D790						500	1390
	14. Izod impact, ft.-lb./in. of notch (1/8-in. thick specimen)	D256A			No break	7-No break	2.5-1.8 No break	1.6-2.7	1.8-1.9
	15. Hardness Rockwell	D785			R93-95	R53		M98	M102
	Shore/Barcol	D2240/ D2583	53	54	Shore D75	Shore D25-66	Shore D43-70	Shore D86	Shore D90
Thermal	16. Coef. of linear thermal expansion, 10^{-6} in./in./°C.	D696			80	100-170	50-100	41-44.2	18.5-20
	17. Deflection temperature under flexural load, °F. 264 p.s.i.	D648			170	93-100		323-338	619-662
	66 p.s.i.	D648			240	113-125	131-400	482-582	644-662
	18. Thermal conductivity, 10^{-4} cal.-cm./sec.-cm.^{2}-°C.	C177			3.8	5.7-6.6		7.1	
Physical	19. Specific gravity	D792	1.97	1.78	1.68-1.69	0.93-0.96	0.95-1.2	1.3	1.47-1.53
	20. Water absorption (1/8-in. thick specimen), % 24 hr.	D570			0.01	0.1-0.5	0.1-0.5	0.1	0.07
	Saturation	D570						0.8	0.5
	21. Dielectric strength (1/8-in. thick specimen), short time, v./mil	D149			490-520	400-450		355	370
		SUPPLIERS[a]	Dyneon	Dyneon	Ausimont	DuPont; Exxon; Network Polymers	DuPont; Network Polymers; A. Schulman	Amoco Polymers	RTP

	Ketones (Cont'd)						Liquid Crystal Polymer				
	Polyaryletherketone (Cont'd)			Polyetheretherketone							
	40% glass fiber-reinforced	Modified, 40% glass	30% carbon fiber	Unfilled	30% glass fiber-reinforced	30% carbon fiber-reinforced	45% glass fiber-reinforced, HDT	40% glass fiber-reinforced	30% mineral-filled	Glass fiber-reinforced for SMT	Unfilled medium melting point
1a.	15-25	15-25	15-25								
1.	329	324	329	334	334	334					280-421
2.	I: 715-805	I: 715-765	I: 715-805	I: 660-750 E: 660-720	I: 660-750	I: 660-800	I: 660-730	I: 590-645	I: 660-690	I: 610-680	I: 540-770
3.	10-20	10-20	10-20	10-20	10-20	10-20			4-8		1-16
4.	2	2	2	3	2-3	2	2.5-3	2.5-3	2.5-3	2.5-3	2.5-4
5.	0.001-0.009	0.001-0.009	0.002-0.008	0.011	0.002-0.014	0.0005-0.014					0.001-0.006
6.	25,000	22,500	30,000	10,200-15,000	22,500-28,500	29,800-33,000	18,200	21,240	15,950	19,600	15,900-27,000
7.	2	1.5	1.5	30-150	2-3	1-4	2.1	1.5	4.0	1.6	1.3-4.5
8.				13,200							
9.	32,500	33,000	33,800	18,000	21,300-22,400	25,000-34,400	10,900				6200-19,000
10.	40,000	34,000	40,000	16,000	33,000-42,000	40,000-48,000	22,400	27,000	18,415	25,000	19,000-35,500
11.	1900	2250	2700		1250-1600	1860-3500	2170	2890		2700	1400-2800
12.							491		590		400-900
13.	2100	2100	2850	560	1260-1600	1860-2600	22.4	2280	1400	1950	1770-2700
	2010	1960	2790	435	1400	1820					1500-1700
	1910	1750	2480		1350	1750					1300-1500
	1790	1500	2100	290	1100	1400					1200-1450
14.	2	1.2	1.5	1.6	2.1-2.7	1.5-2.1	2.0	8.3	3.0	1.8	1.7-10
15.	M102	M103									M76; R60-66
16.	18	18.5	7.9	<150°C: 40-47 <150°C: 108	<150°C: 12-22 >150°C: 44	<150°C: 15-22 >150°C: 5-44	12.5-80.2		8-22		5-7
17.	619	586	634	320	550-599	550-610	572	466	455	520	356-671
	644	643	652			615					
18.	10.5	8.8			4.9	4.9	2.02				2
19.	1.55	1.6	1.45	1.30-1.32	1.49-1.54	1.42-1.44	1.75	1.70	1.63	1.6	1.35-1.84
20.	0.05	0.04	<0.2	0.1-0.14	0.06-0.12	0.06-0.12	<.1			<.1	0-<0.1
				0.5	0.11-0.12	0.06					<0.1
21.	420	385					900				800-980
	RTP	Amoco Polymers; RTP	Amoco Polymers; RTP	Victrex USA	DSM; LNP; RTP; Victrex USA	DSM; LNP; RTP; Victrex USA	RTP	Amoco Polymers; RTP	DuPont; RTP	Amoco Polymers; RTP	Ticona

Materials			Liquid Crystal Polymer (Cont'd)						
	Properties	ASTM test method	30% carbon fiber-reinforced	50% mineral-filled	30% glass fiber-reinforced	30% glass fiber-reinforced, high HDT	Unfilled platable grade	PTFE-filled	15% glass fiber-reinforced
Processing	1a. Melt flow (gm./10 min.)	D1238							
	1. Melting temperature, °C. T_m (crystalline)		280	327	280-680	355		281	280
	T_g (amorphous)								
	2. Processing temperature range, °F. (C = compression; T = transfer; I = injection; E = extrusion)		555-600	I: 605-770	I: 555-770	I: 625-730	I: 600-620		
	3. Molding pressure range, 10^3 p.s.i.		1-14	1-14	1-14	4-8			
	4. Compression ratio		2.5-4	2.5-4	2.5-4	2.5-4	3-4	3:1	3:1
	5. Mold (linear) shrinkage, in./in.	D955	0-0.002	0.003	0.001-0.09	-0.01		0-3	0-3
Mechanical	6. Tensile strength at break, p.s.i.	D638[b]	35,000	10,400-16,500	16,900-30,000	18,000-21,025	13,500	24,500-25,000	28,000
	7. Elongation at break, %	D638[b]	1.0	1.1-2.6	1.7-2.7	1.7-2.2	2.9	3.0-5.2	3
	8. Tensile yield strength, p.s.i.	D638[b]			19,600				
	9. Compressive strength (rupture or yield), p.s.i.	D695	34,500	6800-7500	9900-21,000	9600-12,500			
	10. Flexural strength (rupture or yield), p.s.i.	D790	46,000	14,200-23,500	21,700-25,000	24,000-25,230		18,500-29,000	30,000
	11. Tensile modulus, 10^3 p.s.i.	D638[b]	5400	1500-2700	700-3000	2330-2600	1500	1100-1600	1100-2100
	12. Compressive modulus, 10^3 p.s.i.	D695	4800	490-2016	470-1000	447-770			
	13. Flexural modulus, 10^3 p.s.i. 73° F.	D790	4800	1250-2500	1660-2100	1800-2050	1500	1000-1400	1600
	200° F.	D790							
	250° F.	D790			900	1100			
	300° F.	D790			800	1100			
	14. Izod impact, ft.-lb./in. of notch (1/8-in. thick specimen)	D256A	1.4	0.8-1.5	2.0-3.0	2.0-4.2	0.6	2.1-3.8	5.5
	15. Hardness Rockwell	D785	M99	82	77-87, M61	M63			
	Shore/Barcol	D2240/ D2583							
Thermal	16. Coef. of linear thermal expansion, 10^{-6} in./in./°C.	D696	-2.65	9-65	4.9-77.7	14-36			
	17. Deflection temperature under flexural load, °F. 264 p.s.i.	D648	440	429-554	485-655, 271°C	518-568	410	352-435	430
	66 p.s.i.	D648			400-530				
	18. Thermal conductivity, 10^{-4} cal.-cm./sec.-cm.^{2}-°C.	C177		2.57	1.73	1.52			
Physical	19. Specific gravity	D792	1.49	1.84-1.89	1.60-1.67	1.6-1.66		1.50-1.62	1.5
	20. Water absorption (1/8-in. thick specimen), % 24 hr.	D570	<0.1	<0.1	<0.1%, 0.002	<0.1	0.03		
	Saturation	D570			0.05				0.2
	21. Dielectric strength (1/8-in. thick specimen), short time, v./mil	D149		900-955	640-1000	900-1050	600		
		SUPPLIERS[a]	RTP; Ticona	RTP; Ticona	Amoco Polymers; Dupont; RTP; Ticona	Amoco Polymers; Dupont; RTP; Ticona	Ticona	RTP; Ticona	RTP; Ticona

	Liquid Crystal Polymer (Cont'd)	Melamine formaldehyde		Phenolic						
				Molding compounds, phenol-formaldehyde						
									Impact-modified	
	Glass/mineral-filled	Cellulose-filled	Glass fiber-reinforced	Glass	PAN carbon	Woodflour-filled	Woodflour- and mineral-filled	High-strength glass fiber-reinforced	Cotton-filled	Cellulose-filled
1a.										
1.	280	Thermoset	Thermoset	Thermoset	Thermoset	Thermoset	Thermoset	Thermoset	Thermoset	Thermoset
2.		C: 280-370 I: 200-340 T: 300	C: 280-350			C: 290-380 I: 330-400	C: 290-380 I: 330-390 T: 290-350	C: 300-380 I: 330-390 T: 300-350	C: 290-380 I: 330-400	C: 290-380 I: 330-400
3.		8-20	2-8			2-20	2-20	1-20	2-20	2-20
4.	3:1	2.1-3.1	5-10			1.0-1.5		2.0-10.0	1.0-1.5	1.0-1.5
5.	0-3	0.005-0.015	0.001-0.006	0.0012	0.0016	0.004-0.009	0.003-0.008	0.001-0.004	0.004-0.009	0.004-0.009
6.	21,000-25,000	5000-13,000	5000-10,500	13,500	29,750	5000-9000	6500-7500	7000-18,000	6000-10,000	3500-6500
7.	1.4-2.3	0.6-1	0.6	1.6	1.15	0.4-0.8		0.2	1-2	1-2
8.										
9.	18,000-21,000	33,000-45,000	20,000-35,000	55,600	62,700	25,000-31,000	25,000-30,000	16,000-70,000	23,000-31,000	22,000-31,000
10.	31,000-34,000	9000-16,000	14,000-23,000	31,700	49,000	7000-14,000	9000-12,000	12,000-60,000	9000-13,000	5500-11,000
11.	2400-3100	1100-1400	1600-2400	1280	2490	800-1700	1000-1800	1900-3300	1100-1400	
12.	2000-2800							2740-3500		
13.	2200-2900	1100		1570	2800	1000-1200	1200-1300	1150-3300	800-1300	900-1300
14.	1.6-3.8	0.2-0.4	0.6-18	0.3	0.4	0.2-0.6	0.29-0.35	0.5-18.0	0.3-1.9	0.4-1.1
15.	76-79	M115-125	M115			M100-115	M90-110	E54-101	M105-120	M95-115
								Barcol 72		
16.	6-8	40-45	15-28	10-51	5-50	30-45	30-40	8-34	15-22	20-31
17.	437	350-390	375-400			300-370	360-380	350-600	300-400	300-350
18.		6.5-10	10-11.5			4-8	6-10	8-14	8-10	6-9
19.	1.68-1.89	1.47-1.52	1.5-2.0	1.50-1.52	1.37-1.40	1.37-1.46	1.44-1.56	1.69-2.0	1.38-1.42	1.38-1.42
20.		0.1-0.8	0.09-1.3			0.3-1.2	0.2-0.35	0.03-1.2	0.6-0.9	0.5-0.9
	0.2							0.12-1.5		
21.	840	270-400 175-215 @100°C	130-370			260-400	330-375	140-400	200-360	300-380
	RTP; Ticona	Cytec Fiberite; Patent Plastics; Perstorp; Plastics Mfg.	Cytec Fiberite	Cytec Fiberite	Cytec Fiberite	Amoco Electronic; OxyChem; Plaslok; Plastics Eng.; Rogers	Amoco Electronic; OxyChem; Plaslok; Plastics Eng.; Rogers	Cytec Fiberite; OxyChem; Plastics Eng.; Quantum Composites; Resinoid; Rogers	Bayer Corp.; Cytec Fiberite; OxyChem; Plaslok; Plastics Eng.; Resinoid; Rogers	Amoco Electronic; Cytec Fiberite; OxyChem; Plaslok; Plastics Eng.; Resinoid; Rogers

Materials	Properties	ASTM test method	Phenolic (Cont'd)				Polyamide	
			Molding compounds, phenol-formaldehyde (Cont'd)		Casting resins		Nylon alloys	Nylon, Type 6
			Impact-modified (Cont'd)	Heat-resistant				
			Fabric and rag-filled	Mineral- or mineral- and glass-filled	Unfilled	Mineral-filled	Ceramic and glass fiber-reinforced	Molding and extrusion compound
Processing	1a. Melt flow (gm./10 min.)	D1238	0.5-10					0.5-10
	1. Melting temperature, °C. T_m (crystalline)		Thermoset	Thermoset	Thermoset	Thermoset		210-220
	T_g (amorphous)							
	2. Processing temperature range, °F. (C = compression; T = transfer; I = injection; E = extrusion)		C: 290-380 I: 330-400 T: 300-350	C: 270-350 I: 330-380 T: 300-350				I: 440-550 E: 440-525
	3. Molding pressure range, 10^3 p.s.i.		2-20	2-20				1-20
	4. Compression ratio		1.0-1.5	2.1-2.7				3.0-4.0
	5. Mold (linear) shrinkage, in./in.	D955	0.003-0.009	0.002-0.006			0.002	0.003-0.015
Mechanical	6. Tensile strength at break, p.s.i.	D638[b]	6000-8000	6000-10,000	5000-9000	4000-9000	25,000-30,000	6000-24,000
	7. Elongation at break, %	D638[b]	1-4	0.1-0.5	1.5-2.0		3-4	30-100[c]; 300[d]
	8. Tensile yield strength, p.s.i.	D638[b]						13,100[c]; 7400[d]
	9. Compressive strength (rupture or yield), p.s.i.	D695	20,000-28,000	22,500-36,000	12,000-15,000	29,000-34,000		13,000-16,000[c]
	10. Flexural strength (rupture or yield), p.s.i.	D790	10,000-14,000	11,000-14,000	11,000-17,000	9000-12,000	39,000-45,000	15,700[c]; 5800[d]
	11. Tensile modulus, 10^3 p.s.i.	D638[b]	900-1100	2400	400-700			380-464[c]; 100-247[d]
	12. Compressive modulus, 10^3 p.s.i.	D695						250[d]
	13. Flexural modulus, 10^3 p.s.i. 73° F.	D790	700-1300	1000-2000			1800-2150	390-410[c]; 140[d]
	200° F.	D790						
	250° F.	D790						
	300° F.	D790						
	14. Izod impact, ft.-lb./in. of notch (1/8-in. thick specimen)	D256A	0.8-3.5	0.26-0.6	0.24-0.4	0.35-0.5	1.6-2.0	0.6-2.2[c]; 3.0[d]
	15. Hardness Rockwell	D785	M105-115	E88	M93-120	M85-120	R-120	R119[c]; M100-105[c]
	Shore/Barcol	D2240/ D2583		Barcol 70				
Thermal	16. Coef. of linear thermal expansion, 10^{-6} in./in./°C.	D696	18-24	19-38	68	75		80-83
	17. Deflection temperature under flexural load, °F. 264 p.s.i.	D648	325-400	275-475	165-175	150-175	410-495	155-185[c]
	66 p.s.i.	D648					420-515	347-375[c]
	18. Thermal conductivity, 10^{-4} cal.-cm./sec.-cm.^{2}-°C.	C177	9-12	10-24	3.5			5.8
Physical	19. Specific gravity	D792	1.37-1.45	1.42-1.84	1.24-1.32	1.68-1.70	1.59-1.81	1.12-1.14
	20. Water absorption (1/8-in. thick specimen), % 24 hr.	D570	0.6-0.8	0.02-0.3	0.1-0.36		0.35-0.50	1.3-1.9
	Saturation	D570		0.06-0.5				8.5-10.0
	21. Dielectric strength (1/8-in. thick specimen), short time, v./mil	D149	200-370	200-350	250-400	100-250		400[c]
		SUPPLIERS[e]	Cytec Fiberite; OxyChem; Resinoid; Rogers	Amoco Electronic; Cytec Fiberite; OxyChem; Plaslok; Plastics Eng.; Resinoid; Rogers	Ametek; Schenectady; Solutia; Union Carbide	Schenectady; Solutia	Network Polymers; RTP; Thermofil	Adell; Albis; AlliedSignal; ALM; Ashley Polymers; BASF; Bamberger Polymers; Bayer Corp.; ComAlloy; Custom Resins; DuPont; EMS; Federal Plastics; M.A. Hanna Eng.; Network Polymers; Nyltech; Polymer Resources; Polymers Intl.; A. Schulman; Thermofil; Ticona; Wellman

	Polyamide (Cont'd)								
	Nylon, Type 6 (Cont'd)								
								Semiaromatic semi-crystalline copolymer	
	15% glass fiber-reinforced	25% glass fiber-reinforced	30-35% glass fiber-reinforced	50% glass fiber-reinforced	30% long glass fiber-reinforced	40% long glass fiber-reinforced	50% long glass fiber-reinforced	35% glass fiber-reinforced	45% glass fiber-reinforced
1a.									
1.	220	220	210-220	220	210-220	210-271	220	300	300
2.	520-555	520-555	I: 460-550	535-575	I: 460-550	I: 460-550	I: 480-540	610-625	610-625
3.			2-20		10-20	10-20	10-20	5-20	5-20
4.			3.0-4.0		3.0-4.0	3.0-4.0	3-4	3	3
5.	2-3x10^{-3}	2x10^{-3}	0.001-0.005	1x10^{-3}	0.003-0.009	0.002-0.010	0.002-0.008	0.002-0.003	0.001-0.002
6.	18,900^c; 10,200^d	23,200^c; 14,500^d	24-27,600^c; 18,900^d	33,400^c; 23,200^d	25,200-26,000^c	30,400-31,300	35,400-36,200^c	31,000-30,500	35,500-33,500
7.	3.5^c; 6^d	3.5^c; 5^d	2.2-3.6^c	3.0^c; 3.5^d	2.3-2.5^c	2.2-2.3	2.0-2.1^c	2.4-2.2	2.2-2.2
8.									
9.			19,000-24,000^c		24,000-32,200^c	33,800-37,400	39,700-39,900^c	48,600	48,600
10.			34-36,000^c; 21,000^d		38,800-40,000^c	45,700	53,900^c	44,100	47,500
11.	798^c; 508^d	1160^c; 798^d	1250-1600^c; 1090^d	2320^c; 1740^d	1300	1800	2200-2270^c	1,750,000	2,230,000
12.								550,000	560,000
13.	700^d; 420^d	910^c; 650^d	1250-1400^c; 800-950^d	1700^c; 1570^d	1200^c	1600	1930-2000^c	1,500,000	2,000,000
								1,450,000	1,770,000
								740,000	850,000
14.	1.1	2.0	2.1-3.4^c; 3.7-5.5^d		4.2^c	6.2-6.4	8.4-8.6^c	2.1	2.2
15.	M92^c; M74^d	M95^c; M83^d	M93-96^c; M78^d	M104^c; M93^d	M93-96	M93		124	124
16.	52	40	16-80	30	22^c			15-48	15-48
17.	374	410	392-420^c	419	420^c	405	415	500	502
	419	428	420-430^c	428	425^c				
18.			5.8-11.4						
19.	1.23	1.32	1.35-1.42	1.55	1.4	1.45	1.56	1.47	1.58
20.	2.6	2.3	0.90-1.2	1.5	1.3			0.4	0.27
	8.0	7.1	6.4-7.0	4.8				3.5	2.8
21.			400-450^c		400				
	Ashley Polymers; BASF; Bayer Corp.; ComAlloy; EMS; M.A. Hanna Eng.; M.A. Polymers; Network Polymers; Nyltech; Polymers Intl.; RTP; A. Schulman; Wellman	Ashley Polymers; BASF; Bayer Corp.; EMS; M.A. Hanna Eng.; Network Polymers; Nyltech; Polymers Intl.; RTP; A. Schulman	Adell; Albis; AlliedSignal; ALM; Ashley Polymers; BASF; Bamberger Polymers; Bayer Corp.; ComAlloy; DSM; EMS; Ferro; LNP; M.A. Hanna Eng.; M.A. Polymers; Network Polymers; Nyltech; Polymers Intl.; RTP; A. Schulman; Thermofil; Ticona; Wellman	Ashley Polymers; BASF; Bayer Corp.; ComAlloy; EMS; M.A. Hanna Eng.; Network Polymers; Nyltech; Polymers Intl.; RTP	Adell; ALM; DSM; Ferro; LNP; RTP; Ticona	Adell; ALM; DSM; Ferro; LNP; RTP; Ticona	RTP; Ticona	DuPont	DuPont

Materials	Properties	ASTM test method	Polyamide (Cont'd)				
			Nylon, Type 6 (Cont'd)				
			Toughened		Flame-retarded grade		
			Unreinforced	33% glass fiber-reinforced	30% glass fiber-reinforced	40% mineral- and glass fiber-reinforced	40% mineral-reinforced
Processing	1a. Melt flow (gm./10 min.)	D1238					
	1. Melting temperature, °C. T_m (crystalline)		210-220	210-220	210-220	210-220	210-220
	T_g (amorphous)						
	2. Processing temperature range, °F. (C = compression; T = transfer; I = injection; E = extrusion)		I: 520-550	I: 520-550	I: 520-580	I: 450-550	I: 450-550
	3. Molding pressure range, 10^3 p.s.i.				12-25	2-20	2-20
	4. Compression ratio				3.0-4.0	3.0-4.0	3.0-4.0
	5. Mold (linear) shrinkage, in./in.	D955	0.006-0.02	0.001-0.003	0.001	0.003-0.008	0.003-0.008
Mechanical	6. Tensile strength at break, p.s.i.	D638[b]	6500-7900[c]; 5400[d]	17,800[c]	18,800-22,000[c]	17,400[c]; 19,000[d]	11,000-11,300
	7. Elongation at break, %	D638[b]	65.0-150[c]	4.0[c]	1.7-3.0[c]	3[c]; 2-6[d]	3.0
	8. Tensile yield strength, p.s.i.	D638[b]			22,000[c]	19,000-20,000	
	9. Compressive strength (rupture or yield), p.s.i.	D695			23,000[c]	14,000-18,000[c]	
	10. Flexural strength (rupture or yield), p.s.i.	D790	9100[c]	25,800[c]	28,300-31,000[c]	23,000-30,000[c]	
	11. Tensile modulus, 10^3 p.s.i.	D638[b]	290[c]; 102[d]		1200[c]-1700	1160-1400[c]; 725[d]	
	12. Compressive modulus, 10^3 p.s.i.	D695					
	13. Flexural modulus, 10^3 p.s.i. 73° F.	D790	250[c]	1110[c]	1160-1400[c]	900-1300[c]; 650-996[d]	650-700
	200° F.	D790					
	250° F.	D790					
	300° F.	D790					
	14. Izod impact, ft.-lb./in. of notch (1/8-in. thick specimen)	D256A	16.4[c]	3.5[c]	1.5[c]-2.2	0.6-4.2[c]; 5.0[d]	1.8-2.0
	15. Hardness Rockwell	D785				R118-121[c]	
	Shore/Barcol	D2240/ D2583					
Thermal	16. Coef. of linear thermal expansion, 10^{-6} in./in./°C.	D696				11-41	
	17. Deflection temperature under flexural load, °F. 264 p.s.i.	D648	135[c]; 122	400[c]	380-400[c]	390-405[c]	270-285
	66 p.s.i.	D648	158	430[c]	420	410-425[c]	
	18. Thermal conductivity, 10^{-4} cal.-cm./sec.-cm.^{2}-°C.	C177					
Physical	19. Specific gravity	D792	1.07; 1.08	1.33	1.62-1.7	1.45-1.50	1.45
	20. Water absorption (1/8-in. thick specimen), % 24 hr.	D570		0.86	0.5-0.6	0.6-0.9	
	Saturation	D570				4.0-6.0	
	21. Dielectric strength (1/8-in. thick specimen), short time, v./mil	D149				490-550[c]	
		SUPPLIERS[a]	Adell; Albis; AlliedSignal; Ashley Polymers; BASF; Bamberger Polymers; Bayer Corp.; Custom Resins; DSM; EMS; Ferro; M.A. Hanna Eng.; Network Polymers; Nyltech; Polymers Intl.; A. Schulman; Wellman	Adell; AlliedSignal; Ashley Polymers; BASF; Bamberger Polymers; Bayer Corp.; ComAlloy; DSM; EMS; Ferro; LNP; M.A. Hanna Eng.; Network Polymers; Polymers Intl.; RTP; A. Schulman; Wellman	AlliedSignal; ComAlloy; DSM; Ferro; LNP; M.A. Hanna Eng.; Network Polymers; Nyltech; Polymers Intl.; RTP; Ticona	Adell; AlliedSignal; Ashley Polymers; BASF; Bayer Corp.; ComAlloy; DSM; EMS; Ferro; LNP; M.A. Hanna Eng.; M.A. Polymers; Network Polymers; Nyltech; Polymers Intl.; RTP; A. Schulman; Thermofil; Wellman	Albis; AlliedSignal; Ashley Polymers; Bayer Corp.; M.A. Hanna Eng.; Network Polymers; Polymers Intl.; RTP; A. Schulman; Wellman

Polyamide (Cont'd)

	Nylon, Type 6 (Cont'd)									Nylon, Type 66
	High-impact copolymers and rubber-modified compounds	Unfilled with molybdenum disulfide	Impact-modified; 30% glass fiber-reinforced	EMI shielding (conductive); 30% PAN carbon fiber	Cast	Cast, heat-stabilized	Cast, oil-filled	Cast, plasticized	Cast, Type 612 blend	Molding compound
1a.	1.5-5.0									
1.	210-220	215	220	210-220	227-238	227-238	227-238	227-238	227-238	255-265
2.	I: 450-580 E: 450-550	I: 460-500	I: 480-550	I: 520-575						I: 500-620
3.	1-20	5-20	3-20							1-25
4.	3.0-4.0	3.0-4.0	3.0-4.0							3.0-4.0
5.	0.008-0.026	1.1	0.003-0.005	0.001-0.003						0.007-0.018
6.	6300-11,000^c	11,500^c	21,000^c; 14,500^d	30,000-36,000^c	12,000-13,500	12,000-13,500	9,500-11,000	9,700-10,800	8,700-11,500	13,700^c; 11,000^d
7.	150-270^c	60-80	5^c-8^d	2-3^c	20-45	20-30	45-55	25-35	25-80	15-80^c; 150-300^d
8.	9000^c-9500	11,500-12,300								8000-12,000^c; 6500-8500^d
9.	3900^c			29,000^c	16,000-18,000	16,000-18,000	12,000-14,000	22,000-25,000		12,500-15,000^c (yld.)
10.	5000-12,000^c	13,000		46,000-51,000^c	15,500-17,500	15,000-16,000	14,000-16,000	12,000-13,000	15,000-20,000	17,900-1700^c; 6100^d
11.		440	1220^c-754^d	2800-3000^c	485-550	485-550	375-475	375-440	240-330	230-550^c; 230-500^d
12.					300-350	300-310	275-375	250-275		
13.	110-320^c; 130^d	400	1160^c-600^d	2500-2700^c	420-440	420-440	275-375	330-350	285-385	410-470^c; 185^d
	60-130^c									
14.	1.8-No break^c 1.8-No break^d	0.9-1.0	2.2^c-6^d	1.5-2.8^c	0.7-0.9	0.7-0.9	1.4-1.8	0.82-0.91	0.9-1.4	0.55-1.0^c; 0.85-2.1^d
15.	R81-113^c; M50	R119-120		E70^c	115-125	110-115	110-115	110-115	100-115	R120^c; M83^c; M95-M105^d
					78-83	76-78	74-78	76-78	75-81	
16.	72-120		20-25	14.0^c	50	45	35	45	40-45	80
17.	113-140^c	200	410^c	415-490^c	330-400	330-400	330-400	330-400	330-400	158-212^c
	260-367^c	230	428^c	425-505^c	400-430	400-430	400-430	400-430	400-430	425-474^c
18.										5.8
19.	1.07-1.17	1.17-1.18	1.33	1.28	1.15-1.17	1.15-1.17	1.14-1.15	1.14-1.16	1.10-1.13	1.13-1.15
20.	1.3-1.7	1.1-1.4	2.0	0.7-1.0	0.5-0.6	0.5-0.6	0.5-0.6	0.5-0.6	0.5-0.6	1.0-2.8
	8.5		6.2		5-6	5-6	2-2.5	5-6	4-5	8.5
21.	450-470^c				500-600	500-600	500-600	500-600	500-600	600^c
	Adell; AlliedSignal; Ashley Polymers; BASF; Bamberger Polymers; Bayer Corp.; Custom Resins; EMS; M.A. Hanna Eng.; M.A. Polymers; Network Polymers; Nyltech; Polymers Intl.; RTP; A. Schulman; Wellman	Ashley Polymers; DSM; LNP; M.A. Hanna Eng.; Nyltech; RTP; Ticona	Adell; Albis; AlliedSignal; Ashley Polymers; BASF; Bamberger Polymers; Bayer Corp.; ComAlloy; EMS; Ferro; LNP; M.A. Hanna Eng.; M.A. Polymers; Network Polymers; Polymers Intl.; RTP	ComAlloy; DSM; Ferro; LNP; Nyltech; RTP; Thermofil	Cast Nylons Ltd.	Cast Nylons Ltd.	Cast Nylons Ltd.	Cast Nylons Ltd.	Cast Nylons Ltd.	Adell; Albis; ALM; Ashley Polymers; BASF; Bamberger Polymers; Bayer Corp.; ComAlloy; DSM; DuPont; EMS; M.A. Hanna Eng.; MRC; Network Polymers; Nyltech; Polymer Resources; Polymers Intl.; A. Schulman; Solutia; Thermofil; Ticona; Wellman

Materials	Properties	ASTM test method	Polyamide (Cont'd) — Nylon, Type 66 (Cont'd): 13% glass fiber-reinforced, heat-stabilized	15% glass fiber-reinforced	30-33% glass fiber-reinforced	50% glass fiber-reinforced	30% long glass fiber-reinforced
Processing	1a. Melt flow (gm./10 min.)	D1238					
	1. Melting temperature, °C. T_m (crystalline)		257	260	260-265	260	260-265
	T_g (amorphous)						
	2. Processing temperature range, °F. (C = compression; T = transfer; I = injection; E = extrusion)		I: 520-570	535-575	I: 510-580	555-590	I: 530-570
	3. Molding pressure range, 10^3 p.s.i.		7-20		5-20		10-20
	4. Compression ratio		3.0-4.0		3.0-4.0		
	5. Mold (linear) shrinkage, in./in.	D955	0.005-0.009	$4x10^{-3}$	0.002-0.006	$1x10^{-3}$	0.003
Mechanical	6. Tensile strength at break, p.s.i.	D638[b]	15,000-17,000	18,900[c]; 11,600[d]	27,600[c]; 20,300[d]	33,400[c]; 26,100[d]	24,000-28,000[c]
	7. Elongation at break, %	D638[b]	3.0-5	3[c]; 6[d]	2.0-3.4[c]; 3-7[d]	2[c]; 3[d]	2.1-2.5[c]
	8. Tensile yield strength, p.s.i.	D638[b]			25,000[c]		
	9. Compressive strength (rupture or yield), p.s.i.	D695			24,000-40,000[c]		28,000-34,200[c]
	10. Flexural strength (rupture or yield), p.s.i.	D790	27,500[c]; 15,000[d]	25,600[c]; 17,800[d]	40,000[c]; 29,000[d]	46,500[c]; 37,500[d]	40,000-40,300[c]
	11. Tensile modulus, 10^3 p.s.i.	D638[b]		870[c]; 653[d]	1380[c]; 1090[d]	2320[c]; 1890[d]	
	12. Compressive modulus, 10^3 p.s.i.	D695					
	13. Flexural modulus, 10^3 p.s.i. 73° F.	D790	700-750	720[c]; 480[d]	1200-1450[c]; 800[d]; 900	1700[c]; 1460[d]	1200-1300[c]
	200° F.	D790					
	250° F.	D790					
	300° F.	D790					
	14. Izod impact, ft.-lb./in. of notch (1/8-in. thick specimen)	D256A	0.95-1.1	1.1	1.6-4.5[c]; 2.6-3.0[d]	2.5	4.0-5.1[c]
	15. Hardness Rockwell	D785	95M/R120	M97[c]; M87[d]	R101-119[c]; M101-102[c]; M96[d]	M102[c]; M98[d]	E60[c]
	Shore/Barcol	D2240/ D2583					
Thermal	16. Coef. of linear thermal expansion, 10^{-6} in./in./°C.	D696		52	15-54	0.33	23.4
	17. Deflection temperature under flexural load, °F. 264 p.s.i.	D648	450-470	482	252-490[c]	482	485-495
	66 p.s.i.	D648	494	482	260-500[c]	482	505
	18. Thermal conductivity, 10^{-4} cal.-cm./sec.-cm.^{2}-°C.	C177			5.1-11.7		
Physical	19. Specific gravity	D792	1.21-1.23	1.23	1.15-1.40	1.55	1.36-1.4
	20. Water absorption (1/8-in. thick specimen), % 24 hr.	D570	1.1		0.7-1.1		0.9
	Saturation	D570	7.1	7	5.5-6.5	4	
	21. Dielectric strength (1/8-in. thick specimen), short time, v./mil	D149			360-500		500
		SUPPLIERS[a]	Adell; Albis; ALM; Ashley Polymers; ComAlloy; DSM; LNP; M.A. Hanna Eng.; Network Polymers; Polymers Intl.; RTP; A. Schulman; Ticona; Wellman	Ashley Polymers; BASF; EMS; M.A. Hanna Eng.; Network Polymers; Nyltech; Polymers Intl.; RTP; Wellman	Adell; Albis; ALM; Ashley Polymers; BASF; Bamberger Polymers; Bayer Corp.; ComAlloy; DSM; DuPont; EMS; Ferro; LNP; M.A. Hanna Eng.; MRC; Network Polymers; Nyltech; Polymer Resources; Polymers Intl.; RTP; A. Schulman; Solutia; Thermofil; Ticona; Wellman	Ashley Polymers; BASF; ComAlloy; EMS; M.A. Hanna Eng.; Network Polymers; Nyltech; Polymers Intl.; RTP	Ashley Polymers; DSM; Ferro; LNP; RTP; Ticona

Polyamide (Cont'd)

	Nylon, Type 66 (Cont'd)							
				Toughened			Flame-retarded grade	
	40% long glass fiber-reinforced	50% long glass fiber-reinforced	60% long glass fiber-reinforced	Unreinforced	15-33% glass fiber-reinforced	Modified high-impact, 25-30% mineral-filled	Unreinforced	20-25% glass fiber-reinforced
1a.								
1.	257-265	260	260	240-265	256-265	250-260	249-265	200-265
2.	I: 520-570	I: 550-580	I: 560-600	I: 520-580	I: 530-575	I: 510-570	I: 500-560	I: 500-560
3.	10-20	10-20	10-20	1-20				
4.	3.0-4.0	3-4				10-20		
5.	0.002-0.10	0.002-0.007	0.002-0.006	0.012-0.018	0.0025-0.0045[c]	0.01-0.018	0.01-0.016	0.004-0.005
6.	32,800-32,900	37,200-38,000[c]	40,500-41,800[c]	7000[c]-11,000; 5800[d]	10,900-20,300[c]; 14,500[d]	9000-18,900[c]	8500[c]-10,600	20,300[c]-14,500[d]
7.	2.1-2.5	2.0-2.1[c]	2.0-2.1[c]	4-200[c]; 150-300[d]	4.7[c]; 8[d]	5-16[c]	4-10.0[c]	2.3-3[c]
8.				7250-7500[c]; 5500[d]			10,600	
9.	37,700-42,700	42,900-44,900[c]	47,000-47,900[c]		15,000[c]-20,000		25,000[c]	
10.	49,100	54,500-57,000[c]	65,000[c]	8500-14,500[c]; 4000[d]	17,400-29,900[c]	20,000[c]	14,000-15,000[c]	23,000[c]
11.	1700-1790	2270-2300[c]	2900-3170[c]	290[c]-123[d]	1230[c]; 943[d]		420[c]	1230[c]; 870[d]
12.								
13.	1560-1600	1900-1920[c]	2580-2600[c]	225-380[c]; 125-150[d]	479-1100[c]	600[c]-653	400-420[c]	1102
	810	990	1440					
14.	6.2; 6.6[c]; 6.9	8.0-9.2[c]	10.0-11.0[c]	12.0[c]-19.0; 3-N.B.[c]; 1.4-N.B.[d]	>3.2-5.0	1.0-3.0[c]	0.5-1.5[c]	1.1[c]
15.				M60[c]; R100[c]; R107[c]; R113; R114-115[c]; M50[d]	R107[c]; R115; R116; M86[c]; M70[d]	R120; M86[c]	M82[c]; R119	M98[c]; M90[d]
16.				80	43	30		50
17.	490	500	505	140-175[c]	446-470	300-470	170-200	482
	490			385-442[c]	480-495	399-460	410-415	482
18.								
19.	1.45	1.56	1.69	1.06-1.11	1.2-1.34	1.28-1.4	1.25-1.42	1.3-1.51
20.				0.8-2.3	0.7-1.5	0.9-1.1	0.9-1.1	0.7
				7.2	5			
21.							520	430
	DSM; Ferro; LNP; RTP; Ticona	Ashley Polymers; RTP; Ticona	RTP; Ticona	Adell; ALM; Ashley Polymers; BASF; Bamberger Polymers; ComAlloy; DSM; DuPont; Ferro; LNP; M.A. Hanna Eng.; MRC; Network Polymers; Nyltech; Polymers Intl.; RTP; A. Schulman; Solutia; Ticona; Wellman	Adell; ALM; Ashley Polymers; BASF; Bamberger Polymers; ComAlloy; DSM; DuPont; EMS; Ferro; LNP; M.A. Hanna Eng.; Network Polymers; Nyltech; Polymers Intl.; RTP; Solutia; Ticona; Wellman	ALM; Ashley Polymers; ComAlloy; DSM; Ferro; LNP; M.A. Hanna Eng.; Network Polymers; RTP; Wellman	Ashley Polymers; BASF; ComAlloy; DSM; DuPont; EMS; M.A. Hanna Eng.; Nyltech; Polymers Intl.; RTP; Wellman	BASF; ComAlloy; DSM; DuPont; Ferro; LNP; M.A. Hanna Eng.; Nyltech; Polymers Intl.; RTP; Wellman

Materials	Properties	ASTM test method	Polyamide (Cont'd) — Nylon, Type 66 (Cont'd): 40% glass- and mineral-reinforced	40-45% mineral-filled	EMI shielding (conductive): 30% graphite or PAN carbon fiber	40% aluminum flake	5% stainless steel, long fiber	6% stainless steel, long fiber
Processing	1a. Melt flow (gm./10 min.)	D1238						
	1. Melting temperature, °C. T_m (crystalline)		250-260	250-265	258-265	265	260-265	260-265
	T_g (amorphous)							
	2. Processing temperature range, °F. (C = compression; T = transfer; I = injection; E = extrusion)		I: 510-590	I: 520-580	I: 500-590	I: 525-600	I: 530-570	I: 530-570
	3. Molding pressure range, 10^3 p.s.i.		9-20	5-20	10-20	10-20		5-18
	4. Compression ratio		3-4	3.0-4.0				
	5. Mold (linear) shrinkage, in./in.	D955	0.001-0.005	0.012-0.022	0.001-0.003	0.005	0.004	0.004-0.006
Mechanical	6. Tensile strength at break, p.s.i.	D638[b]	15,500-31,000[c]; 13,100[d]	14,000[c]; 11,000[d]	27,600-35,000[c]	6000[c]	10,000[c]	10,000-11,300[c]
	7. Elongation at break, %	D638[b]	2-7[c]	5-10[c]; 16[d]	1-4[c]	4[c]	5.0[c]	2.9; 5.0[c]
	8. Tensile yield strength, p.s.i.	D638[b]		13,900				
	9. Compressive strength (rupture or yield), p.s.i.	D695	18,000-37,000[c]	15,500-22,000[c]	24,000-29,000[c]	7500[c]		
	10. Flexural strength (rupture or yield), p.s.i.	D790	24,000-48,000[c]	22,000[c]; 9000[d]	45,000-51,000[c]	11,700[c]	16,000[c]	16,000-38,700[c]
	11. Tensile modulus, 10^3 p.s.i.	D638[b]		900[c]; 500[d]	3200-3400[c]	720[c]	450[c]	450-1500[c]
	12. Compressive modulus, 10^3 p.s.i.	D695		370[c]				
	13. Flexural modulus, 10^3 p.s.i. 73° F.	D790	985-1750[c]; 600[d]	900-1050[c]; 400[d]	1500-2900[c]	690[c]	450[c]	500-1400
	200° F.	D790	600[c]					
	250° F.	D790						
	300° F.	D790						
	14. Izod impact, ft.-lb./in. of notch (1/8-in. thick specimen)	D256A	0.6-3.8[c]	0.9-1.4[c]; 3.9[d]	1.3-2.5[c]	2.5[c]	1.3[c]	0.7; 2.2
	15. Hardness Rockwell	D785	M95-98[c]	R106-121[c]	R120[c]; M106[c]	R114[c]		
	Shore/Barcol	D2240/ D2583						
Thermal	16. Coef. of linear thermal expansion, 10^{-6} in./in./°C.	D696	20-54	27	11-16	22		
	17. Deflection temperature under flexural load, °F. 264 p.s.i.	D648	432-485[c]	300-438	470-500	380	285	175-480
	66 p.s.i.	D648	480-496[c]	320-480[c]	500-510	400	295	
	18. Thermal conductivity, 10^{-4} cal.-cm./sec.-cm.^{2}-°C.	C177	11	9.6	24.1			
Physical	19. Specific gravity	D792	1.42-1.55	1.39-1.5	1.28-1.43	1.48	1.27	1.19-1.45
	20. Water absorption (1/8-in. thick specimen), % 24 hr.	D570	0.4-0.9	0.6-0.55	0.5-0.8	1.1	0.12	
	Saturation	D570	5.1	6.0-6.5				
	21. Dielectric strength (1/8-in. thick specimen), short time, v./mil	D149	300-525	450[c]				
		SUPPLIERS[a]	Adell; ALM; Ashley Polymers; BASF; ComAlloy; DSM; DuPont; EMS; Ferro; LNP; M.A. Hanna Eng.; MRC; Network Polymers; Nyltech; RTP; Solutia; Thermofil; Ticona; Wellman	Adell; Albis; ALM; Ashley Polymers; ComAlloy; DSM; DuPont; Ferro;LNP; M.A. Hanna Eng.; MRC; Network Polymers; Nyltech; RTP; Solutia; Thermofil; Ticona; Wellman	ComAlloy; DSM; Ferro; LNP; RTP; Thermofil	Thermofil	DSM; RTP	RTP; Ticona

	Polyamide (Cont'd)									
	Nylon, Type 66 (Cont'd)									
	EMI shielding (conductive) (Cont'd)							Lubricated		
	10% stainless steel, long fiber	**50% PAN carbon fiber**	**30% pitch carbon fiber**	**40% pitch carbon fiber**	**15% nickel-coated carbon fiber**	**40% nickel-coated carbon fiber**	**Antifriction molybdenum disulfide-filled**	**5% silicone**	**10% PTFE**	**30% PTFE**
1a.										
1.	257-265	260-265	260-265	260-265	260-265	260-265	249-265	260-265	260-265	260-265
2.	I: 530-570	I: 530-570	I: 530-570	I: 530-570	I: 530-570	I: 530-570	I: 500-600	I: 530-570	I: 530-570	I: 530-570
3.	5-20						5-25			
4.	3.0-4.0									
5.	0.003-0.004	0.0005	0.003	0.002	0.005	0.001	0.007-0.018	0.015	0.01	0.007
6.	11,500	38,000[c]	15,500[c]-15,600	17,500[c]; 19,250	14,000[c]	20,000[c]	10,500-13,700[c]	8500[c]	9500[c]	5500[c]
7.	2.6	1.2[c]	2.0[c]	1.5[c]	1.6[c]	2.5[c]	4.4-40[c]			
8.										
9.			20,500	24,000			12,000-12,500[c]			
10.	18,100	54,000[c]	24,800-26,000[c]	28,000[c]-29,000	21,000[c]	27,000[c]	15,000-20,300[c]	15,000[c]	13,000[c]	8000[c]
11.	600	5000[c]	1400-2000[c]	2250-2600[c]	1100[c]		350-550[c]			
12.										
13.	500	4200[c]	1200-1500[c]	1800-2000[c]	1000[c]	2000[c]	420-495[c]	300[c]	420[c]	460[c]
14.	0.7; 1.9[c]	2.0[c]	0.6-0.7[c]	0.7-0.8[c]	0.7[c]	1.0[c]	0.9-4.5[c]	1.0[c]	0.8[c]	0.5[c]
15.							R119[c]			
16.			16.0-19.0	9.0-14.0			54	63.0	35.0	45.0
17.	175-480	495	465-490	475-490	460	470	190-260[c]	170	190	180
		505	490-500	498-500			395-430			
18.										
19.	1.24	1.38	1.30-1.31	1.36-1.38	1.20	1.46	1.15-1.18	1.16	1.20	1.34
20.		0.5	0.6	0.5	1.0	0.8	0.8-1.1	1.0	0.7	0.55
							8.0			
21.							360[c]			
	RTP; Ticona	DSM; LNP; RTP	ComAlloy; LNP; Polymers Intl.; RTP	ComAlloy; LNP; RTP	DSM; RTP	DSM; RTP	Adell; ALM; Ashley Polymers; ComAlloy; DSM; LNP; M.A. Hanna Eng.; Nyltech; RTP; Thermofil	ComAlloy; DSM; Ferro; LNP; M.A. Hanna Eng.; RTP	ComAlloy; DSM; Ferro; LNP; M.A. Hanna Eng.; RTP	ComAlloy; DSM; Ferro; LNP; M.A. Hanna Eng.; RTP

Materials			Polyamide (Cont'd)					
			Nylon, Type 66 (Cont'd)		Nylon, Type 610			
			Lubricated (Cont'd)					Flame-retarded grade
	Properties	ASTM test method	5% Molybdenum disulfide and 30% PTFE	Copolymer	Molding and extrusion compound	30-40% glass fiber-reinforced	30-40% long glass fiber-reinforced	30% glass fiber-reinforced
Processing	1a. Melt flow (gm./10 min.)	D1238						
	1. Melting temperature, °C. T_m (crystalline)		260-265	200-255	220	220	220	220
	T_g (amorphous)							
	2. Processing temperature range, °F. (C = compression; T = transfer; I = injection; E = extrusion)		I: 530-570	I: 430-500	I: 445-485 E: 480-500	I: 510-550	I: 510-550	I: 500-560
	3. Molding pressure range, 10^3 p.s.i.			1-15	1-19			
	4. Compression ratio				3-4			
	5. Mold (linear) shrinkage, in./in.	D955	0.01	0.006-0.015	0.005-0.015	0.015-0.04	0.013-0.03	0.002
Mechanical	6. Tensile strength at break, p.s.i.	D638[b]	7500[c]	7400-12,400[c]	10,150[c]; 7250[d]	22,000-26,700[c]	22,400-25,400[c]	19,000[c]
	7. Elongation at break, %	D638[b]		40-150[c]; 300[d]	70[c]; 150[d]	4.3-4.7[c]	4.0-4.1[c]	3.5[c]
	8. Tensile yield strength, p.s.i.	D638[b]						
	9. Compressive strength (rupture or yield), p.s.i.	D695				20,400-21,000[c]	20,000[c]	23,000[c]
	10. Flexural strength (rupture or yield), p.s.i.	D790	1200[c]	12,000	350[c]; 217[d]	32,700-38,000[c]	34,000-37,400[c]	28,000[c]
	11. Tensile modulus, 10^3 p.s.i.	D638[b]		150-410[c]		800[c]	950-1600[c]	
	12. Compressive modulus, 10^3 p.s.i.	D695						
	13. Flexural modulus, 10^3 p.s.i. 73° F.	D790	400[c]	150-410[c]		1150-1500[c]	1200-1430[c]	1230[c]
	200° F.	D790						
	250° F.	D790						
	300° F.	D790						
	14. Izod impact, ft.-lb./in. of notch (1/8-in. thick specimen)	D256A	0.6[c]	0.7[c]; No break[d]		1.6-2.4	3.2-4.2[c]	1.5[c]
	15. Hardness Rockwell	D785		R114-119; R83[d]; M75[c]		E43-48[c]	E42-56[c]	M89[c]
	Shore/Barcol	D2240/ D2583						
Thermal	16. Coef. of linear thermal expansion, 10^{-6} in./in./°C.	D696						
	17. Deflection temperature under flexural load, °F. 264 p.s.i.	D648	185	135-170[c]		410-415	425	390
	66 p.s.i.	D648		430[c]-440		430	430	
	18. Thermal conductivity, 10^{-4} cal.-cm./sec.-cm.^{2}-°C.	C177						
Physical	19. Specific gravity	D792	1.37	1.08-1.14		1.3-1.4	1.33-1.39	1.55
	20. Water absorption (1/8-in. thick specimen), % 24 hr.	D570	0.55	1.5-2.0	1.4	0.17-0.19	0.17-0.21	0.16
	Saturation	D570		9.0-10.0	3.3			
	21. Dielectric strength (1/8-in. thick specimen), short time, v./mil	D149		400[c]			500	
		SUPPLIERS[e]	ComAlloy; DSM; M.A. Hanna Eng.; RTP	AlliedSignal; Ashley Polymers; BASF; DuPont; EMS; M.A. Hanna Eng.; Nyltech; Polymers Intl.; Solutia	EMS; M.A. Hanna Eng.	Ashley Polymers; DSM; Ferro; LNP; M.A. Hanna Eng.; RTP	DSM; LNP	DSM; LNP; RTP

Polyamide (Cont'd)

	Nylon, Type 612									Nylon, Type 46	
				Toughened		Flame-retarded grade	Lubricated				
	Molding compound	30-35% glass fiber-reinforced	35-45% long glass fiber-reinforced	Unreinforced	33% glass fiber-reinforced	30% glass fiber-reinforced	10% PTFE	15% PTFE, 30% glass fiber-reinforced	10% PTFE, 30% PAN carbon fiber	Extrusion	Unreinforced
1a.											
1.	195-219	213-217	195-217	195-217	195-217	195-217	195-217	195-217	195-217	295	295
2.	I: 450-550 E: 464-469	I: 450-550	I: 510-550	I: 510-550	I: 510-550	I: 500-560	I: 510-550	I: 510-550	I: 510-550	E: 560-590	I: 570-600
3.	1-15	4-20									5-15
4.										3-4	3-4
5.	0.011	0.002-0.005	0.001-0.002				0.012	0.002-0.003	0.0013		0.018-0.020
6.	6500-8800[c]	22,000[c]; 20,000[d]	26,000-29,000[c]	5500[c]	18,000[c]	18,000-19,000[c]	7000[c]	19,500-20,000[c]	28,000[c]	8,500	14,400
7.	150[c]; 300[d]	4[c]; 5[d]	2.9-3.2[c]	40[c]	5[c]	2.0-3.5[c]		2.5		60	25
8.	5800-8400[c]; 3100[d]										
9.		22,000[c]	23,000[c]			15,000-21,000[c]		19,000			13,000
10.	11,000[c]; 4300[d]	32,000-35,000[c]	39,000-44,000[c]	6500[c]	27,000[c]	28,000[c]		30,500-31,000[c]	42,000[c]	11,500	21,700
11.	218-290[c]; 123-180[d]	1200[c]; 900[d]				1000[c]-1400		1200		250	435
12.											319
13.	240-334[c]; 74-100[d]	1100-1200[c]; 900[d]	1200-1500[c]	195[c]	1050[c]	1200-1250[c]	100[c]	1100-1200[c]	2600[c]	270	460
14.	1.0-1.9[c]; 1.4-No break[d]	1.8-2.6[c]	4.2-6.3[c]	12.5[c]	4.5[c]	1.0-1.5[c]	1.0[c]	2.5-3.0[c]	2.4[c]	17	1.8
15.	M78[c]; M34[d]; R115	M93[c]; E40-50[d]; R116	E40[c]			M89[c]					R113
	D72-80[c]; D63[d]										D85
16.			21.6-25.2					18.0			
17.	136-180[c]	390-425[c]	410-415	135	385	385-390	202[c]	385	390	194	320
	311-330[c]	400-430[c]	420-425			400					
18.	5.2	10.2									
19.	1.06-1.10	1.30-1.38	1.34-1.45	1.03	1.28	1.55-1.60	1.13	1.42-1.45	1.30	1.10	1.18
20.	0.37-1.0	0.2	0.2	0.3	0.2	0.16	0.2	0.13	0.15	1.84	2.3
	2.5-3.0	1.85									
21.	400[c]	520[c]				450					673
	ALM; Ashley Polymers; Creanova; DuPont; EMS; M.A. Hanna Eng.; A. Schulman	ALM; Ashley Polymers; ComAlloy; DSM; DuPont; Ferro; LNP; M.A. Hanna Eng.; RTP	DSM; RTP	DSM; DuPont; M.A. Hanna Eng.; RTP	DSM; DuPont; LNP; M.A. Hanna Eng.; RTP	ComAlloy; DSM; LNP; M.A. Hanna Eng.; RTP	ComAlloy; DSM; Ferro; LNP; M.A. Hanna Eng.; RTP	ComAlloy; DSM; Ferro; LNP; M.A. Hanna Eng.; RTP	ComAlloy; DSM; LNP; RTP	DSM	DSM

Materials	Properties	ASTM test method	Polyamide (Cont'd) Nylon, Type 46 (Cont'd) Super-tough	15% glass-reinforced	15% glass-reinforced, V-0	30% glass-reinforced	30% glass-reinforced, V-0	50% glass-reinforced	50% glass and mineral-filled
Processing	1a. Melt flow (gm./10 min.)	D1238							
	1. Melting temperature, °C. T_m (crystalline)		295	295	295	295	295	295	295
	T_g (amorphous)								
	2. Processing temperature range, °F. (C = compression; T = transfer; I = injection; E = extrusion)		I: 570-600	I: 570-600	I: 570-600	I: 570-600	I: 570-600	I: 570-600	I: 570-600
	3. Molding pressure range, 10^3 p.s.i.		5-15	5-15	5-15	5-15	5-15	5-15	5-15
	4. Compression ratio		3-4	3-4	3-4	3-4	3-4	3-4	3-4
	5. Mold (linear) shrinkage, in./in.	D955	0.018-0.020	0.005-0.009	0.006-0.009	0.004-0.006	0.004-0.006	0.002-0.004	0.003-0.006
Mechanical	6. Tensile strength at break, p.s.i.	D638[b]	8,500	21,500	16,500	30,000	23,000	34,000	20,000
	7. Elongation at break, %	D638[b]	60	3	8	4	3	3	2
	8. Tensile yield strength, p.s.i.	D638[b]							
	9. Compressive strength (rupture or yield), p.s.i.	D695				33,000	34,000		
	10. Flexural strength (rupture or yield), p.s.i.	D790	11,500	31,900	27,000	43,000	34,000	50,750	34,000
	11. Tensile modulus, 10^3 p.s.i.	D638[b]	250	841	1,000	1,300	1,500	2,320	2,100
	12. Compressive modulus, 10^3 p.s.i.	D695				507	688		
	13. Flexural modulus, 10^3 p.s.i. 73° F.	D790	270	798	1125	1,200	1,300	2,030	1700
	200° F.	D790							
	250° F.	D790							
	300° F.	D790							
	14. Izod impact, ft.-lb./in. of notch (1/8-in. thick specimen)	D256A	17	1.6	.5	2.0	1.3	2.2	1.1
	15. Hardness Rockwell	D785				R120	R120		
	Shore/Barcol	D2240/ D2583				D89	D88		
Thermal	16. Coef. of linear thermal expansion, 10^{-6} in./in./°C.	D696							
	17. Deflection temperature under flexural load, °F. 264 p.s.i.	D648	194	480	480	545	545	545	545
	66 p.s.i.	D648							
	18. Thermal conductivity, 10^{-4} cal.-cm./sec.-cm.^{2}-°C.	C177							
Physical	19. Specific gravity	D792	1.10	1.3	1.47	1.41	1.68	1.62	1.6
	20. Water absorption (1/8-in. thick specimen), % 24 hr.	D570	1.84			1.5	0.9	1.15	
	Saturation	D570							
	21. Dielectric strength (1/8-in. thick specimen), short time, v./mil	D149				863	838		
		SUPPLIERS[a]	DSM	DSM	DSM	DSM	DSM	DSM	DSM

	Polyamide (Cont'd)								Polyamide-imide	
	Nylon, Type 11	Nylon, Type 12	Aromatic polyamide		Partially aromatic					
	Molding and extrusion compound	Molding and extrusion compound	Amorphous transparent copolymer	Aramid molded parts, unfilled	35% glass fiber-reinforced impact modified	40% glass fiber-reinforced	50% glass fiber-reinforced	60% glass fiber-reinforced	Unfilled compression and injection molding compound	30% glass fiber-reinforced
1a.										
1.	180-190	160-209		275	260	260	260	260		
		125-155							275	275
2.	I: 390-520 E: 390-475	I: 356-525 E: 350-500	I: 480-610 E: 520-595		270-310	270-310	270-310	270-310	C: 600-650 I: 610-700	C: 630-650 I: 610-700
3.	1-15	1-15	5-20						2-40	15-40
4.	2.7-3.3	2.5-4							1.0-1.5	1.0-1.5
5.	0.012	0.003-0.015	0.004-0.007		0.15/0.95	0.15/0.95	0.15/0.90	0.10/0.80	0.006-0.0085	0.001-0.0025
6.	8000[c]-9,500	5100-10,000[c]; 8000[d]	7600-14,000[c]; 13,000[d]	17,500[c]	26,000	31,500	34,700	36,200	22,000	
7.	300[c]-400	250-390[c]	40-150[c]; 260[d]	5[c]	5	3	3	3	15	7
8.		3000-6100[c]	11,000-14,861[c]; 11,000[d]						27,800	29,700
9.	7300-7800		17,500[c]; 14,000[d]	30,000[c]			28,800		32,100	38,300
10.		1400-8100[c]	10,000-16,400[c]; 14,000[d]	25,800[c]	39,400	48,900	54,400	57,100	34,900	48,300
11.	185[c]	36-180[c]	275-410[c]; 270[d]						700	1560
12.	180[c]		340[c]	290[c]						1150
13.	44-180[c]	27-190[c]	306-400[c]; 350[d]	640[c]	1,270	1,634	2,106	2,680	730	1700
	20									
14.	1.8[c]-N.B.	1.0-N.B.	0.8-3.5[c]; 1.8-2.7[d]	1.4[c]	2.9	2.3	2.3	2.3	2.7	1.5
15.	R108[c]; R80	R70-109[c]; 105[d]	M77-93[c]	E90[c]	86	89	89	91	E86	E94
		D58-75[d]	D83[c]; D85[d]							
16.	100	61-100	28-70	40	18	14	14	11	30.6	16.2
17.	104-126	95-135[c]	170-268	500[c]	440	455	460	460	532	539
	300[c]	158-302[c]	261-330[c]							
18.	8	5.2-7.3	5	5.2					6.2	8.8
19.	1.03-1.05	1.01-1.02	1.0-1.19	1.30	1.39	1.47	1.58	1.72	1.42	1.61
20.	0.4	0.25-0.30	0.4-1.36	0.6	0.59	0.56	0.45	0.36		
	1.9	0.75-1.6	1.3-4.2		5.0	4.6	4.0	3.80	0.33	0.24
21.	650-750	450[c]	350[c]	800[c]					580	840
	Elf Atochem N.A.	ALM; Ashley Polymers; Creanova; Elf Atochem N.A.; EMS	AlliedSignal; Ashley Polymers; Bayer Corp.; Creanova; DuPont; EMS; M.A. Hanna Eng.; Nyltech	DuPont	EMS	EMS	EMS	EMS	Amoco Polymers	Amoco Polymers

Materials			Polyamide-imide (Cont'd)						Poly-buta-diene
	Properties	ASTM test method	Graphite fiber-reinforced	Bearing grade	High compressive strength	Wear resistant for speeds	Stiffness and lubricity	Cost/performance ratio	Casting resin
Processing	1a. Melt flow (gm./10 min.)	D1238							
	1. Melting temperature, °C. T_m (crystalline)								Thermoset
	T_g (amorphous)		275	275					
	2. Processing temperature range, °F. (C = compression; T = transfer; I = injection; E = extrusion)		C: 630-650 I: 610-700	I: 580-700					
	3. Molding pressure range, 10^3 p.s.i.		15-40	15-40					
	4. Compression ratio		1.0-1.5	1.0-1.5					
	5. Mold (linear) shrinkage, in./in.	D955	0.000-0.0015	0.0025-0.0045	3.5-6.0	3.5-6.0	0.0-1.5		
Mechanical	6. Tensile strength at break, p.s.i.	D638[b]							
	7. Elongation at break, %	D638[b]	6	7	7	9	6	7	
	8. Tensile yield strength, p.s.i.	D638[b]	26,000	22,000	23,700	17,800	26,000	31,800	
	9. Compressive strength (rupture or yield), p.s.i.	D695	36,900		24,100	18,300	30,300	46,700	
	10. Flexural strength (rupture or yield), p.s.i.	D790	50,700	27,100-31,200	31,200	27,000	40,100	52,000	8000-14,000
	11. Tensile modulus, 10^3 p.s.i.	D638[b]	3220	870-1130	950	870		2020	560
	12. Compressive modulus, 10^3 p.s.i.	D695	1430						
	13. Flexural modulus, 10^3 p.s.i. 73° F.	D790	2880	910-1060	1000	910	2440	2100	
	200° F.	D790							
	250° F.	D790							
	300° F.	D790							
	14. Izod impact, ft.-lb./in. of notch (1/8-in. thick specimen)	D256A	0.9	1.2-1.6	1.2	1.3	1.0	1.5	
	15. Hardness Rockwell	D785	E94	E66-E72	72	66		107	R40
	Shore/Barcol	D2240/ D2583							
Thermal	16. Coef. of linear thermal expansion, 10^{-6} in./in./°C.	D696	9.0	25.2-27.0	14	15	7	7	
	17. Deflection temperature under flexural load, °F. 264 p.s.i.	D648	540	532-536	534	532	534	536	
	66 p.s.i.	D648							
	18. Thermal conductivity, 10^{-4} cal.-cm./sec.-cm.^{2}-°C.	C177	12.7		3.7				
Physical	19. Specific gravity	D792	1.48	1.46-1.51	1.46	1.50	1.50		0.97
	20. Water absorption (1/8-in. thick specimen), % 24 hr.	D570		0.17-0.33					0.03
	Saturation	D570	0.26	0.33	0.28	0.17		0.21	
	21. Dielectric strength (1/8-in. thick specimen), short time, v./mil	D149					Conductive	490	630
		SUPPLIERS[a]	Amoco Polymers	Amoco Polymers	Amoco Polymers	Amoco Polymers	Amoco Polymers	Amoco Polymers	OxyChem

	Polybutylene			Polycarbonate							
				Unfilled molding and extrusion resins		Glass fiber-reinforced					
	Extrusion compound	Film grades	Adhesive resin	High viscosity	Low viscosity	10% glass	30% glass	20-30% long glass fiber-reinforced	40% long glass fiber-reinforced	50% long glass fiber-reinforced	35% random glass mat
1a.				3-10	10-30	7.0					
1.	126	118	90								
				150	150	150	150	150	150	150	
2.	C: 300-350 I: 290-380 E: 290-380	E: 380-420	300-350	I: 560	I: 520	I: 520-650	I: 550-650	I: 590-650	I: 575-620	570-620	C: 560-600
3.	10-30			10-20	8-15	10-20	10-30	10-30	10-20	10-20	2-3.5
4.	2.5	2.5		1.74-5.5	1.74-5.5					3	
5.	0.003 (unaged) 0.026 (aged)			0.005-0.007	0.005-0.007	0.002-0.005	0.001-0.002	0.001-0.003	0.001-0.003	0.001-0.003	0.002-0.003
6.	3800-4400	4000	3500	9100-10,500	9100-10,500	7000-10,000	19,000-20,000	18,000-23,000	23,100	25,100-26,000	18,000
7.	300-380	350	500	110-120	110-150	4-10	2-5	1.9-3.0	1.4-1.7	1.3-1.5	2.5
8.	1700-2500	1700	600	9000	9000	8500-11,600					18,000
9.				10,000-12,500	10,000-12,500	12,000-14,000	18,000-20,000	18,000-29,800	31,600	32,700	14,800
10.	2000-2300			12,500-13,500	12,000-14,000	13,700-16,000	23,000-25,000	22,000-36,900	36,400	40,800	30,000
11.	30-40	30	10-15	345	345	450-600	1250-1400	1200-1500	1700	2200	1100
12.	31			350	350	520	1300				
13.	45-50		13	330-340	330-340	460-580	1100	800-1500	1500-1700	2100	1000
				275	275	440	960				
				245	245	420	900				
14.	No break	No break	No break	12-18 @ 1/8 in. 2.3 @ 1/4 in.	12-16 @ 1/8 in. 2.0 @ 1/4 in.	2-4	1.7-3.0	3.5-4.7	5.0	6.6	11.8
15.		D45	A90	M70-M75	M70-M75	M62-75; R118-122	M92, R119	M85-95			
	Shore D55-65		D25								
16.	128-150			68	68	32-38	22-23				
17.	130-140			250-270	250-270	280-288	295-300	290-300	305	310	290
	215-235			280-287	273-280	295	300-305	305			
18.	5.2			4.7	4.7	4.6-5.2	5.2-7.6				
19.	0.91-0.925	0.909	0.895	1.2	1.2	1.27-1.28	1.4-1.43	1.34-1.43	1.52	1.63	1.39
20.	0.01-0.02			0.15	0.15	0.12-0.15	0.08-0.14	0.09-0.11			
				0.32-0.35	0.32-0.35	0.25-0.32					
21.	>450			380->400	380->400	470-530	470-475				528
	Shell	Shell	Shell	Albis; American Polymers; Ashley Polymers; Bamberger Polymers; Bayer Corp.; Dow Plastics; Federal Plastics; GE Plastics; MRC; Network Polymers; Polymers Intl.; Polymer Resources; RTP; Shuman	Albis; American Polymers; Ashley Polymers; Bayer Corp.; Dow Plastics; Federal Plastics; GE Plastics; MRC; Network Polymers; Polymers Intl.; Polymer Resources; RTP; Shuman	Albis; American Polymers; Ashley Polymers; Bayer Corp.; ComAlloy; DSM; Dow Plastics; Federal Plastics; Ferro; GE Plastics; LNP; M.A. Polymers; MRC; Network Polymers; Polymers Intl.; Polymer Resources; RTP	Albis; American Polymers; Ashley Polymers; ComAlloy; DSM; Ferro; GE Plastics; LNP; M.A. Polymers; MRC; Network Polymers; Polifil; Polymers Intl.; Polymer Resources; RTP	DSM; RTP; Ticona	RTP; Ticona	RTP; Ticona	Azdel

Materials	Properties	ASTM test method	Polycarbonate (Cont'd)						
			Flame-retarded grade	High-heat			Conductive polycarbonate	EMI shielding (conductive)	
			20-30% glass fiber-reinforced	Polyester copolymer	Polycarbonate copolymer	Impact-modified polycarbonate/polyester blends	6% stainless steel fiber	10% stainless steel fiber	20% PAN carbon fiber
Processing	1a. Melt flow (gm./10 min.)	D1238							
	1. Melting temperature, °C. T_m (crystalline)								
	T_g (amorphous)		149	160-195	160-205		150	150	150
	2. Processing temperature range, °F. (C = compression; T = transfer; I = injection; E = extrusion)		I: 530-590	I: 575-710	I: 580-660	I: 475-560		I: 590-650	I: 590-650
	3. Molding pressure range, 10^3 p.s.i.			8-20	8-20	15-20	10-20		
	4. Compression ratio			1.5-3	2-3	2-2.5	3		
	5. Mold (linear) shrinkage, in./in.	D955	0.002-0.004	0.007-0.010	0.007-0.009	0.006-0.009	0.004-0.006	0.003-0.006	0.001
Mechanical	6. Tensile strength at break, p.s.i.	D638[b]	14,000-20,000	9500-11,300	8300-10,000	7600-8500	9800	10,110-11,000	18,000-20,000
	7. Elongation at break, %	D638[b]	2.0-3.0	50-122	70-90	120-165	4.7	4.0	2.0
	8. Tensile yield strength, p.s.i.	D638[b]		8500-9800	9300-10,500	7400-8300			
	9. Compressive strength (rupture or yield), p.s.i.	D695	18,000-21,000	11,500		7000			18,500
	10. Flexural strength (rupture or yield), p.s.i.	D790	21,000-30,000	10,000-13,800	12,000-14,000	10,900-12,500	14,000	16,300-17,000	27,000-28,000
	11. Tensile modulus, 10^3 p.s.i.	D638[b]	1000-1100	320-340	320-340		410	500	2000
	12. Compressive modulus, 10^3 p.s.i.	D695							
	13. Flexural modulus, 10^3 p.s.i. 73° F.	D790	900-1200	294-340	320-340	280-325	400	500	1500-1800
	200° F.	D790							
	250° F.	D790							
	300° F.	D790							
	14. Izod impact, ft.-lb./in. of notch (1/8-in. thick specimen)	D256A	1.8-2.0	1.5-10	1.5-12	2-18	0.8-1.7	1.1-1.7	1.4-2.0
	15. Hardness Rockwell	D785	M77-85	M74-92	M75-91	R114-122			
	Shore/Barcol	D2240/ D2583							
Thermal	16. Coef. of linear thermal expansion, 10^{-6} in./in./°C.	D696		70-92	70-76	80-95		1410	180
	17. Deflection temperature under flexural load, °F. 264 p.s.i.	D648	288-305	285-335	284-354	190-250	270	270-295	290-295
	66 p.s.i.	D648	295	305-365	306-383	223-265			300
	18. Thermal conductivity, 10^{-4} cal.-cm./sec.-cm.^{2}-°C.	C177		4.7-5.0	4.7-4.8	4.3			
Physical	19. Specific gravity	D792	1.36-1.45	1.15-1.2	1.14-1.18	1.20-1.22	1.28	1.26-1.35	1.28
	20. Water absorption (1/8-in. thick specimen), % 24 hr.	D570	0.15-0.17	0.15-0.2	0.15-0.2	0.12-0.16		0.12	0.2
	Saturation	D570				0.35-0.60			
	21. Dielectric strength (1/8-in. thick specimen), short time, v./mil	D149		509-520	>400	440-500			
		SUPPLIERS[a]	ComAlloy; DSM; Ferro; GE Plastics; LNP; Polymer Resources; RTP	Ferro; GE Plastics	Bayer Corp.	Bayer Corp.; Eastman; GE Plastics; MRC; Polymer Resources	RTP; Ticona	DSM; LNP; RTP; Ticona	ComAlloy; DSM; Ferro; LNP; RTP

	Polycarbonate (Cont'd)						Polydicyclopentadiene	Polyester, thermoplastic		
	EMI shielding (conductive) (Cont'd)		Lubricated					Polybutylene terephthalate		
	30% graphite fiber	40% PAN carbon fiber	10-15% PTFE	30% PTFE	10-15% PTFE, 20% glass fiber-reinforced	2% silicone, 30% glass fiber-reinforced	RIM solid; unfilled	Unfilled	30% glass fiber-reinforced	30% long glass fiber-reinforced
1a.										
1.						Thermoset	Thermoset	220-267	220-267	235
	149-150		150	150	150	150	90-165			
2.	I: 540-650	I: 580-620	I: 590-650	I: 590-650	I: 590-650	I: 590-650	T: <95-<100	I: 435-525	I: 440-530	I: 480-540
3.	15-20	15-20					<0.050	4-10	5-15	10-20
4.										3-4
5.	0.001-0.002	0.0005-0.001	0.007	0.009	0.002	0.002	0.008-0.012	0.009-0.022	0.002-0.008	0.001-0.003
6.	20,000-24,000	23,000-24,000	7500-10,000	6000	12,000-15,000	16,000	5300-6000	8200-8700	14,000-19,500	20,000
7.	1-5	1-2	8-10		2		5-70	50-300	2-4	2.2
8.							5000-6700	8200-8700		
9.	19,000-26,000	22,000	110,000		11,000		8500-9000	8600-14,500	18,000-23,500	26,000
10.	30,000-36,000	34,000-35,000	11,000-11,500	7600	18,000-23,000	22,000	10,000-11,000	12,000-16,700	22,000-29,000	35,000
11.	250-2150	3000-3100	340		1200		240-280	280-435	1300-1450	1400
12.								375	700	
13.	240-1900	2800-2900	250-300	460	850-900	900	260-280	330-400	850-1200	1300
14.	1.8	1.5-2.0	2.5-3.0	1.3	1.8-3.5	3.5	5.0-9.0	0.7-1.0	0.9-2.0	5.7
15.	R118, R119	R119						M68-78	M90	
							D72-D84			
16.	9	11.0-14.4	58.0		21.6-23.4	12.0	46-49 in./in./°F.	60-95	15-25	
17.	280-300	295-300	270-275	260	280-290	290	217-240	122-185	385-437	405
	295	300			290		239	240-375	421-500	
18.	16.9	17.3						4.2-6.9	7.0	
19.	1.32-1.33	1.36-1.38	1.26-1.29	1.39	1.43-1.5	1.46	1.03-1.04	1.30-1.38	1.48-1.54	1.56
20.	0.04-0.08	0.08-0.13	0.13	0.06	0.11	0.12	0.09	0.08-0.09	0.06-0.08	
								0.4-0.5	0.35	
21.								420-550	460-560	
	ComAlloy; DSM; LNP; RTP; Thermofil	ComAlloy; DSM; Ferro; LNP; RTP	ComAlloy; DSM; GE Plastics; LNP; Polymer Resources; RTP	DSM; Polymer Resources; RTP	ComAlloy; DSM; Ferro; GE Plastics; LNP; Polymer Resources; RTP	ComAlloy; DSM; Ferro; LNP; RTP	BFGoodrich; Hercules	Albis; Ashley Polymers; BASF; ComAlloy; Creanova; DuPont; GE Plastics; M.A. Hanna Eng.; RTP; Ticona	Albis; Adell; Ashley Polymers; BASF; ComAlloy; Creanova; DSM; DuPont; Ferro; GE Plastics; LNP; M.A. Hanna Eng.; Polymer Resources; RTP; Ticona	Ticona

Materials	Properties	ASTM test method	Polyester, thermoplastic (Cont'd) — Polybutylene terephthalate (Cont'd): 40% long glass fiber-reinforced	50% long glass fiber-reinforced	60% long glass fiber-reinforced	25% random glass mat	35% random glass mat	40-45% glass fiber- and mineral-reinforced	35% glass fiber- and mica-reinforced
Processing	1a. Melt flow (gm./10 min.)	D1238							
	1. Melting temperature, °C. T_m (crystalline)		235	235	235			220-228	220-224
	T_g (amorphous)								
	2. Processing temperature range, °F. (C = compression; T = transfer; I = injection; E = extrusion)		I: 470-540	I: 480-540	I: 490-530	C: 520-560	C: 520-560	I: 450-520	I: 480-510
	3. Molding pressure range, 10^3 p.s.i.		10-15	10-20	10-20	1.5-3	2-3	10-15	9-15
	4. Compression ratio		3.5-4.0	3-4	3-4			3-4	
	5. Mold (linear) shrinkage, in./in.	D955	0.001-0.008	0.001-0.007	0.001-0.006	0.0035-0.0045	0.003-0.004	0.003-0.010	0.003-0.012
Mechanical	6. Tensile strength at break, p.s.i.	D638[b]	22,900	24,000	20,000	12,000	15,000	12,000-14,800	11,400-13,800
	7. Elongation at break, %	D638[b]	1.4	1.3	1.0	2.8	2.1	2-5	2-3
	8. Tensile yield strength, p.s.i.	D638[b]				12,000	15,000		
	9. Compressive strength (rupture or yield), p.s.i.	D695	24,500	24,600	24,600		14,700	15,000	
	10. Flexural strength (rupture or yield), p.s.i.	D790	35,200	35,500	41,000	28,000	32,000	18,500-23,500	18,000-22,000
	11. Tensile modulus, 10^3 p.s.i.	D638[b]	1900	1900	2100	980	1300	1350-1800	
	12. Compressive modulus, 10^3 p.s.i.	D695						1000	
	13. Flexural modulus, 10^3 p.s.i. 73° F.	D790	1600	2200	2500	900	1200	1250-1600	1200-1600
	200° F.	D790							
	250° F.	D790							
	300° F.	D790							
	14. Izod impact, ft.-lb./in. of notch (1/8-in. thick specimen)	D256A	6.6	8.5	8.0		13.0	0.7-2.0	0.7-1.8
	15. Hardness Rockwell	D785						M75-86	M50-76
	Shore/Barcol	D2240/ D2583							
Thermal	16. Coef. of linear thermal expansion, 10^{-6} in./in./°C.	D696				29	20	1.7	
	17. Deflection temperature under flexural load, °F. 264 p.s.i.	D648	415	420	450	403	425	388-395	330-390
	66 p.s.i.	D648						408-426	410-416
	18. Thermal conductivity, 10^{-4} cal.-cm./sec.-cm.^{2}-°C.	C177	1.72						
Physical	19. Specific gravity	D792	1.65	1.75	1.87	1.45	1.59	1.58-1.74	1.59-1.74
	20. Water absorption (1/8-in. thick specimen), % 24 hr.	D570						0.04-0.07	0.04-0.11
	Saturation	D570							
	21. Dielectric strength (1/8-in. thick specimen), short time, v./mil	D149					440	540-590	450-600
		SUPPLIERS[a]	Ticona	Ticona	Ticona	Azdel	Azdel	Ashley Polymers; ComAlloy; DSM; GE Plastics; LNP; M.A. Hanna Eng.; Polymer Resources; RTP; Ticona	ComAlloy; GE Plastics; LNP; RTP; Ticona

	Polyester, thermoplastic (Cont'd)									
	Polybutylene terephthalate (Cont'd)				EMI shielding (conductive); 30% carbon fiber	Polyester alloy			PCT	
			Flame-retarded grade							
	Impact-modified	50% glass fiber-reinforced	7-15% glass fiber-reinforced	30% glass fiber-reinforced		Unfilled	7.0-30% glass fiber-reinforced	Unreinforced flame retardant	15% glass fiber-reinforced	30% glass fiber-reinforced
1a.										
1.	225	225	220-260	220-260	222					
2.	I: 482-527	I: 482-527	I: 490-560	I: 490-560	I: 430-550	I: 475-540	I: 475-540	I: 470-510	I: 565-590	I: 555-595
3.					5-20	5-17	5-17	5-17		
4.						3.0-4.0	3.0-4.0	3.0-4.0		
5.	20x10-3	4x10-3	0.005-0.01	0.002-0.006	0.001-0.004	0.016-0.018	0.003-0.014		0.001-0.004	0.001-0.004
6.		21,800	11,500-15,000	17,400-20,000	22,000-23,000	4900-6300	7200-12,000	5000	13,700	18,000-19,500
7.		2.5	4.0	2.0-3.0	1-3	150-300	4.5-45	50	2.0	1.9-2.3
8.	6530									
9.			17,000	18,000						
10.			18,000-23,500	30,000	29,000-34,000	7100-8700	12,400-19,000	8500	23,900	24,000-29,800
11.	276	2470	800	1490; 1700	3500					
12.										
13.			580-830	1300-1500	2300-2700	210-280	395-925	260	812	1200-1450
14.	3.3	1.4	0.6-1.1	1.3-1.6	1.2-1.5	NB	2.8-4.1	NB	0.76	1.3-1.8
15.			M79-88	M88; M90	R120	R101-R109; >R115	R104-R111; >R115	R105	M88	>R115
16.	135	25	5	1.5		20	18			20
17.	122	419	300-450	400-450	420-430	105-125	170-374	130	475	500
	248	428	400-490	425-490		180-260	379-417	240	518	>500
18.					15.8	6.9	8.3			6.9
19.	1.20	1.73	1.48-1.53	1.63	1.41-1.42	1.23-1.25	1.30-1.47	1.31	1.33	1.45
20.			0.06	0.06-0.07	0.04-0.45	0.1	0.11	0.10		0.04-0.05
	0.3	0.3								
21.			460	490		460	435		462	440-460
	Ashley Polymers; BASF; M.A. Hanna Eng.	Ashley Polymers; BASF	Albis; Ashley Polymers; ComAlloy; DSM; DuPont; GE Plastics; LNP; M.A. Hanna Eng.; M.A. Polymers; Polymer Resources; RTP; Ticona	Albis; Ashley Polymers; BASF; ComAlloy; DSM; DuPont; Ferro; LNP; M.A. Hanna Eng.; M.A. Polymers; Polymer Resources; RTP; Ticona	ComAlloy; DSM; LNP; RTP	Eastman; GE Plastics; Ticona	Albis; GE Plastics; Ticona	Ticona	Eastman	Eastman; GE Plastics

Materials	Properties	ASTM test method	Polyester, thermoplastic (Cont'd): PCT (Cont'd): 40% glass fiber-reinforced	27-30% glass fiber- and mineral-reinforced	40% glass fiber- and mineral-reinforced	20% glass, flame retarded	30% glass, flame retarded	40% glass, flame retarded	PCTA: 15% glass fiber-reinforced
Processing	1a. Melt flow (gm./10 min.)	D1238							
	1. Melting temperature, °C. T_m (crystalline)								285
	T_g (amorphous)								92
	2. Processing temperature range, °F. (C = compression; T = transfer; I = injection; E = extrusion)		I: 565-590	I: 565-590	I: 565-590	I: 565-590	I: 565-590	I: 565-590	I: 560-590
	3. Molding pressure range, 10^3 p.s.i.								8-16
	4. Compression ratio								2.5-3.5
	5. Mold (linear) shrinkage, in./in.	D955	0.0005-0.003	0.002-0.005	0.002-0.005	0.001-0.004	0.002-0.004	0.001-0.003	0.004-0.006
Mechanical	6. Tensile strength at break, p.s.i.	D638[b]	22,000	17,500	17,000	15,400	19,000	20,600	11,600
	7. Elongation at break, %	D638[b]	2.1	2.4	1.18	1.4-2.0	1.7	1.4	3.4
	8. Tensile yield strength, p.s.i.	D638[b]							
	9. Compressive strength (rupture or yield), p.s.i.	D695							
	10. Flexural strength (rupture or yield), p.s.i.	D790	33,000	26,700	27,400	24,000	29,000	32,000	18,900
	11. Tensile modulus, 10^3 p.s.i.	D638[b]							
	12. Compressive modulus, 10^3 p.s.i.	D695							
	13. Flexural modulus, 10^3 p.s.i. 73° F.	D790	1690	1180	1550	1080	1450	1910	560
	200° F.	D790							
	250° F.	D790							
	300° F.	D790							
	14. Izod impact, ft.-lb./in. of notch (1/8-in. thick specimen)	D256A	1.5	1.0	1.0	0.9	1.0	1.4	1.8
	15. Hardness Rockwell	D785	M88	M96	R119	M96	R122	M94	R119
	Shore/Barcol	D2240/ D2583							
Thermal	16. Coef. of linear thermal expansion, 10^{-6} in./in./°C.	D696							
	17. Deflection temperature under flexural load, °F. 264 p.s.i.	D648	491	482	500	442	477	489	284
	66 p.s.i.	D648	518	523	527	514	527	518	495
	18. Thermal conductivity, 10^{-4} cal.-cm./sec.-cm.^{2}-°C.	C177							
Physical	19. Specific gravity	D792	1.53	1.43	1.55	1.54	1.62	1.70	1.31
	20. Water absorption (1/8-in. thick specimen), % 24 hr.	D570							
	Saturation	D570							
	21. Dielectric strength (1/8-in. thick specimen), short time, v./mil	D149	420	452	462	444	430	440	
		SUPPLIERS[a]	Eastman	Eastman	Eastman	Eastman	Eastman	Eastman	Eastman

Polyester, thermoplastic (Cont'd)

	PCTA (Cont'd)			Polyethylene terephthalate							
	20% glass fiber-reinforced	30% glass fiber-reinforced	Unfilled	Unfilled	15% glass fiber-reinforced	30% glass fiber-reinforced	40-45% glass fiber-reinforced	35% glass, super toughened	15% glass, easy processing	30% glass, flame retarded, V-0 1/32	15% glass, flame retarded, V-0 1/32
1a.											
1.	285	285	285	212-265		245-265	252-255	245-255	245-255	245-255	245-255
	92	92	92	68-80							
2.	I: 560-590	I: 560-590	I: 299-316 E: 299-302	I: 440-660 E: 520-580	I: 540-570	I: 510-590	I: 500-590	I: 525-555	I: 525-555	I: 525-555	I: 525-555
3.	8-16	8-16		2-7		4-20	8-12	8-18	8-18	8-18	8-18
4.	2.5-3.5	2.5-3.5		3.1	2-3	2-3	4	3:1	3:1	3:1	3:1
5.	0.004-0.006	0.003-0.005	0.004	0.002-0.030	0.001-0.004	0.002-0.009	0.002-0.009	0.002-0.009	0.003-0.010	0.002-0.009	0.003-0.010
6.	12,600	14,100		7000-10,500	14,600	20,000-24,000	14,000-27,500	15,000	11,500	22,000	15,500
7.	3.3	3.1	25-250	30-300	2.0	2-7	1.5-3	6.0	6.0	2.3	2.6
8.			5900-9000	8600		23,000					
9.				11,000-15,000		25,000	20,500-24,000	11,700	13,500	25,000	25,000
10.	20,600	22,700		12,000-18,000	20,000	30,000-36,000	21,000-42,400	21,000	13,500	32,000	23,000
11.				400-600		1300-1440	1800-1950				
12.											
13.	690	980	240-285	350-450	830	1200-1590	1400-2190	1000	525	1500	850
						520	489	360	185	620	350
						390	320	275	155	420	220
14.	2.2	2.8	1.5-NB	0.25-0.7	1.9	1.5-2.2	0.9-2.4	4.4	2.2	1.6	1.2
15.	R118	R113	105-122	M94-101; R111	R121	M90; M100	R118; R119	M62; R107	M58, R111	M100, R120	M88, R120
16.			5.8×10^{-5}	65×10^{-6}		18-30	18-21	1.5	1.0	1.1	1.0
17.	381	430	69-95	70-150	400	410-440	412-448	428	405	435	410
	507	514	83-142	167	464	470-480	420-480	475	454	475	471
18.			5×10^{-4} or 5	3.3-3.6		6.0-7.6	10.0				
19.	1.33	1.41	1.195-1.215	1.29-1.40	1.33	1.55-1.70	1.58-1.74	1.51	1.39	1.67	1.53
20.				0.1-0.2		0.05	0.04-0.05	0.25	0.24	0.05	0.07
				0.2-0.3							
21.			422-441	420-550	475	405-650	415-600	530	450	430	490
	Eastman	Eastman	Eastman	DuPont; Eastman; M.A. Polymers; A. Schulman; Ticona; Wellman	Eastman	Albis; AlliedSignal; ComAlloy; DSM; DuPont; EMS; Eastman; Ferro; GE Plastics; M.A. Polymers; MRC; RTP; Ticona; Wellman	Albis; AlliedSignal; ComAlloy; DSM; DuPont; EMS; Eastman; Ferro; GE Plastics; RTP; Ticona	DuPont	DuPont	Albis; DuPont	DuPont

Materials	Properties	ASTM test method	Polyester, thermoplastic (Cont'd) — Polyethylene terephthalate (Cont'd): 15-20% glass, flame retarded	30% glass, flame retarded	40% glass, flame retarded	35-45% glass fiber- and mica-reinforced	30% long glass fiber-reinforced	40% long glass fiber-reinforced	50% long glass fiber-reinforced
Processing	1a. Melt flow (gm./10 min.)	D1238							
	1. Melting temperature, °C. T_m (crystalline)					252-255	275	275	275
	T_g (amorphous)								
	2. Processing temperature range, °F. (C = compression; T = transfer; I = injection; E = extrusion)		I: 540-560	I: 540-560	I: 540-560	I: 500-590	I: 470-530	480-540	I: 470-530
	3. Molding pressure range, 10^3 p.s.i.					5-20	10-20	10-20	10-20
	4. Compression ratio		2-3	2-3	2-3	4	3-4		3-4
	5. Mold (linear) shrinkage, in./in.	D955	0.0015-0.004	0.001-0.004	0.001-0.004	0.002-0.007	0.001-0.008	0.001-0.005	0.001-0.008
Mechanical	6. Tensile strength at break, p.s.i.	D638[b]	13,700-15,700	18,600	19,000	14,000-26,000	20,200	23,200	23,500
	7. Elongation at break, %	D638[b]	2.2	1.8	1.5	1.5-3	1.4	1.4	1.0
	8. Tensile yield strength, p.s.i.	D638[b]							
	9. Compressive strength (rupture or yield), p.s.i.	D695				20,500-24,000	31,000	34,200	35,100
	10. Flexural strength (rupture or yield), p.s.i.	D790	20,000-23,200	27,500	29,300	21,000-40,000	29,300	35,400	36,500
	11. Tensile modulus, 10^3 p.s.i.	D638[b]				1800-1950	1700	2100	2400
	12. Compressive modulus, 10^3 p.s.i.	D695							
	13. Flexural modulus, 10^3 p.s.i. 73° F.	D790	850-1090	1540	2020	1400-2000	1500	1900	2100
	200° F.	D790				489			
	250° F.	D790							
	300° F.	D790				320			
	14. Izod impact, ft.-lb./in. of notch (1/8-in. thick specimen)	D256A	1-1.2	1.4	1.6	0.9-2.4	4.0	5.0	6.2
	15. Hardness Rockwell	D785	M83	M84	M79	R118, R119			
	Shore/Barcol	D2240/ D2583							
Thermal	16. Coef. of linear thermal expansion, 10^{-6} in./in./°C.	D696				18-21			
	17. Deflection temperature under flexural load, °F. 264 p.s.i.	D648	383-409	425	429	396-440	470	475	480
	66 p.s.i.	D648	455-462	459	466	420-480			
	18. Thermal conductivity, 10^{-4} cal.-cm./sec.-cm.^{2}-°C.	C177				10.0			
Physical	19. Specific gravity	D792	1.60-1.63	1.71	1.78	1.58-1.74	1.61	1.70	1.85
	20. Water absorption (1/8-in. thick specimen), % 24 hr.	D570				0.04-0.05			
	Saturation	D570							
	21. Dielectric strength (1/8-in. thick specimen), short time, v./mil	D149	437-460	427	399	550-687			
		SUPPLIERS[a]	Eastman	Eastman	Eastman	Albis; AlliedSignal; Bayer Corp.; ComAlloy; DSM; DuPont; M.A. Polymers; MRC; RTP; Ticona; Wellman	Ticona	Ticona	Ticona

Polyester, thermoplastic (Cont'd)

	Polyethylene terephthalate (Cont'd)					PETG	PCTG	Polyester/polycarbonate blends			Wholly aromatic (liquid crystal)	
	60% long glass fiber-reinforced	EMI shielding (conductive); 30% PAN carbon fiber	Recycled content, 30% glass fiber	Recycled content, 45% glass fiber	Recycled content, 35% glass/mineral-reinforced	Unfilled	Unfilled	High-impact	30% glass fiber-reinforced	Flame retarded	Unfilled medium melting point	Unfilled high melting point
1a.												
1.	270-280										280-421	400
						81-91						
2.	I: 500-530	I: 550-590	I: 530-550	I: 530-550	I: 530-550	I: 480-520 E: 490-550	I: 530-560	I: 460-630	I: 470-560	I: 480-550	I: 540-770	I: 700-850
3.	10-20					1-20	1-20	8-18	10-18	10-20	1-16	5-18
4.	3-4		2-3	2-3	2-3	2.4-3	2.4-3				2.5-4	2.5-3
5.	0.001-0.007	0.001-0.002				0.002-0.005	0.002-0.005	0.0005-0.019	0.003-0.009	0.005-0.007	0.001-0.008	0-0.002
6.	23,500	25,000	24,000	28,500	14,000-15,000	4100	7600	4500-9000	12,000-13,300	8900	15,900-27,000	12,500
7.	0.9	1.4	2.0	2.0	2.1-2.2	110	330	100-175		130	1.3-4.5	2
8.						7300	6500	5000-8100		7400		
9.	35,100							8600-10,000	10,860-11,600		6200-19,000	10,000
10.	40,100	38,000	35,500	45,000	21,500-22,000	10,200	9600	8500-12,500	20,000	14,100	19,000-35,500	19,000
11.	3000	3600						240-325			1400-2800	1750
12.											400-900	309
13.	2600	2700	1400	2100	1400	300	260	310	780-850	380,000	1770-2700[a]	1860
											1500-1700[a]	
											1300-1500[a]	450F: 870
											1200-1450[a]	575F: 450
14.	8.0	1.5	1.5	2.0	1.1	1.9	NB	12-19-No break	3.1-3.2	13	1.7-10	1.2
15.						R106	R105	R112-116	R109-110	R122	M76; R60-66	R97
16.								58-150	25	0.64	5-7	8.9
17.	480	430	435	445	395-420	147	149	140-250	300-330	212	356-671	606
		470				158	165	210-265	400-415	239		
18.								5.2			2	
19.	1.91	1.42	1.58	1.70	1.60	1.27	1.23	1.20-1.26	1.44-1.51	1.3	1.35-1.84	1.79
20.		0.05				0.13	0.13	0.13-0.80	0.09-0.10	0.07	0-<0.1	0
								0.30-0.62		0.22	<0.1	0.02
21.			565	540	450-550			396-500		660	600-980	470
	Ticona	ComAlloy; DSM; Ferro; RTP	Ticona; Wellman	Ticona	Ticona; Wellman	Eastman	Eastman	Bayer Corp.; ComAlloy; Eastman; Ferro; GE Plastics; M.A. Polymers; MRC	ComAlloy; Ferro; GE Plastics; M.A. Polymers; MRC; RTP	Bayer Corp.	Amoco Polymers; Ticona	Amoco Polymers

Materials	Properties	ASTM test method	Polyester, thermoplastic (Cont'd) — Wholly aromatic (liquid crystal) (Cont'd): 30% carbon fiber-reinforced	40% glass fiber-filled	40% glass plus 10% mineral-filled	30-50% mineral-filled	30% glass fiber-reinforced	30% glass-reinforced, high HDT	Unfilled platable grade
Processing	1a. Melt flow (gm./10 min.)	D1238							
	1. Melting temperature, °C. T_m (crystalline)		280			327	280	355	
	T_g (amorphous)								
	2. Processing temperature range, °F. (C = compression; T = transfer; I = injection; E = extrusion)		555-600	I: 660-770	I: 660-770	I: 605-770	I: 555-770	I: 625-730	I: 600-620
	3. Molding pressure range, 10^3 p.s.i.		1-14	5-14	5-14	1-14	1-14	4-8	
	4. Compression ratio		2.5-4	2.5-3	2.5-3	2.5-4	2.5-4	2.5-4	3-4
	5. Mold (linear) shrinkage, in./in.	D955	0-0.002			0.003	0.001-0.09		
Mechanical	6. Tensile strength at break, p.s.i.	D638[b]	35,000	13,600	14,200	10,400-16,500	16,900-30,000	18,000-21,000	13,500
	7. Elongation at break, %	D638[b]	1.0	1.8	2.3	1.1-4.0	1.7-2.7	1.7-2.2	2.9
	8. Tensile yield strength, p.s.i.	D638[b]							
	9. Compressive strength (rupture or yield), p.s.i.	D695	34,500	10,400	9700	6900-7500	9900-21,000	9800-12,500	
	10. Flexural strength (rupture or yield), p.s.i.	D790	46,000	20,500	19,700	14,200-23,500	21,700-24,600	24,000-26,000	
	11. Tensile modulus, 10^3 p.s.i.	D638[b]	5400	1870	1870	1500-2700	700-3000	2330-2600	1500
	12. Compressive modulus, 10^3 p.s.i.	D695	4800	420	473	490-2016	470-1000	447-700	
	13. Flexural modulus, 10^3 p.s.i. 73° F.	D790	4600	1320	1730	1250-2500	1680-2100	1800-2050	1500
	200° F.	D790							
	250° F.	D790					900	1100	
	300° F.	D790					800	1100	
	14. Izod impact, ft.-lb./in. of notch (1/8-in. thick specimen)	D256A	1.4	1.6	1.9	0.8-3.0	2.0-3.0	2.0-2.5	0.6
	15. Hardness Rockwell	D785	M99	79		82	61-87	63	
	Shore/Barcol	D2240/ D2583							
Thermal	16. Coef. of linear thermal expansion, 10^{-6} in./in./°C.	D696	−2.65	14.9	12.9-52.8	9-65	4.9-77.7	14-36	
	17. Deflection temperature under flexural load, °F. 264 p.s.i.	D648	440	606	493	429-554	485-655	518-568	410
	66 p.s.i.	D648					489-530		
	18. Thermal conductivity, 10^{-4} cal.-cm./sec.-cm.^{2}-°C.	C177			2.5	2.57	1.73	1.52	
Physical	19. Specific gravity	D792	1.49	1.70	1.78	1.63-1.89	1.60-1.67	1.6-1.66	
	20. Water absorption (1/8-in. thick specimen), % 24 hr.	D570	<0.1	<0.1	<0.1	<0.1	<0.1	<0.1	0.03
	Saturation	D570							
	21. Dielectric strength (1/8-in. thick specimen), short time, v./mil	D149		510	1145	900-955	640-900	900-1050	600
		SUPPLIERS[a]	RTP; Ticona	Amoco Polymers; RTP	Amoco Polymers; RTP	Amoco Polymers; DuPont; RTP; Ticona	Amoco Polymers; DuPont; Eastman; RTP; Ticona	Amoco Polymers; DuPont; RTP; Ticona	Ticona

	Polyester, thermoset and alkyd									
	Cast		Glass fiber-reinforced							
	Rigid	Flexible	High	Unidirectional Glass	Preformed, chopped roving	Premix, chopped glass	Woven cloth	SMC	SMC, BMC low-density	SMC low-pressure
1a.										
1.	Thermoset	Thermoset			Thermoset	Thermoset	Thermoset		Thermoset	Thermoset
2.			C: 260-320	C: 260-320	C: 170-320	C: 280-350	C: 73-250	C: 270-380 I: 280-310 T: 280-310	C: 270-330 I: 270-350 T: 270-350	C: 270-330
3.			0.2-2.0	0.4-2.0	0.25-2	0.5-2	0.3	0.3-2	0.5-2	0.25-0.8
4.			1.0	1.0	1.0	1.0		1.0	1.0	1.0
5.			0.001	0.000	0.0002-0.002	0.001-0.012	0.0002-0.002	0.0005-0.004	0.0002-0.001	0.0002-0.001
6.	600-13,000	500-3000	50,000	90,000	15,000-30,000	3000-10,000	30,000-50,000	7000-25,000	4000-20,000	7000-25,000
7.	<2.6	40-310	1-2	1-2	1-5	<1	1-2	3	2-5	3
8.										
9.	13,000-30,000		42,000		15,000-30,000	20,000-30,000	25,000-50,000	15,000-30,000	15,000-30,000	15,000-30,000
10.	8500-23,000		90,000	160,000	10,000-40,000	7000-20,000	40,000-80,000	10,000-36,000	10,000-35,000	10,000-36,000
11.	300-640		3500		800-2000	1000-2500	1500-4500	1400-2500	1400-2500	1400-2500
12.			2700							
13.	490-610		3000	6000	1000-3000	1000-2000	1000-3000	1000-2200	1000-2500	1000-22,000
							660			
							430			
							270			
14.	0.2-0.4	>7	35	70	2-20	1.5-16	5-30	7-22	2.5-18	7-24
15.										
	Barcol 35-75	Shore D84-94	60		Barcol 50-80	Barcol 50-80	Barcol 60-80	Barcol 50-70		Barcol 40-70
16.	55-100		15.5	6	20-50	20-33	15-30	13.5-20	6-30	6-30
17.	140-400		>500	>500	>400	>400	>400	375-500	>375	375-500
18.										
19.	1.04-1.46	1.01-1.20	1.9	1.95	1.35-2.30	1.65-2.30	1.50-2.10	1.65-2.6	1.0-1.5	1.65-2.30
20.	0.15-0.6	0.5-2.5	0.08	0.10	0.01-1.0	0.06-0.28	0.05-0.5	0.1-0.25	0.4-0.25	0.1-0.25
21.	380-500	250-400	310		350-500	345-420	350-500	380-500	300-400	380-500
	AOC; Aristech Chem; ICI Americas; Interplastic; Reichhold	AOC; Aristech Chem.; ICI Americas; Interplastic; Reichhold	Quantum Composites	Quantum Composites	Glastic; Haysite; Jet Moulding; Premix; Reichhold; Rostone	Bulk Molding Compounds; Cytec Fiberite; Glastic; Haysite; Jet Moulding; Premix; Reichhold; Rostone	Glastic; Haysite; Jet Moulding; Premix; Reichhold; Rostone	Budd; Haysite; Interplastic; Jet Moulding; Plastics Mfg.; Polyply; Premix; Rostone	Cytec Fiberite; Interplastic; Jet Moulding; Rostone	Interplastic; Jet Moulding; Rostone

Materials	Properties	ASTM test method	Polyester, thermoset and alkyd (Cont'd) — Glass fiber-reinforced (Cont'd): SMC low-shrink	Glass fiber-reinforced (Cont'd): BMC, TMC	EMI shielding (conductive): SMC, TMC	EMI shielding (conductive): BMC	High-strength SMC: 50% glass fiber-reinforced	Alkyd molding compounds: Granular and putty, mineral-filled	Alkyd molding compounds: Glass fiber-reinforced
Processing	1a. Melt flow (gm./10 min.)	D1238							
	1. Melting temperature, °C. T_m (crystalline)		Thermoset	Thermoset	Thermoset	Thermoset	Thermoset	Thermoset	Thermoset
	T_g (amorphous)								
	2. Processing temperature range, °F. (C = compression; T = transfer; I = injection; E = extrusion)		C: 270-330 I: 270-380	C: 280-380 I: 280-370 T: 280-320	C: 270-380 I: 270-370 T: 280-320	C: 310-380 I: 300-370 T: 280-320	C: 270-330	C: 270-350 I: 280-390 T: 320-360	C: 290-350 I: 280-380
	3. Molding pressure range, 10^3 p.s.i.		0.5-2	0.4-1.1	0.5-2		0.5-2	2-20	2-25
	4. Compression ratio		1.0	1.0	1.0		1.0	1.8-2.5	1-11
	5. Mold (linear) shrinkage, in./in.	D955	0.0002-0.001	0.0003-0.007	0.0002-0.001	0.0005-0.004	0.002-0.003	0.003-0.010	0.001-0.010
Mechanical	6. Tensile strength at break, p.s.i.	D638[b]	4500-20,000	3000-13,000	7000-8000	4000-4500		3000-9000	4000-9500
	7. Elongation at break, %	D638[b]	3-5	1-2			3-5		
	8. Tensile yield strength, p.s.i.	D638[b]					27,000-30,000		
	9. Compressive strength (rupture or yield), p.s.i.	D695	15,000-30,000	14,000-30,000	20,000-24,000	18,000		12,000-38,000	15,000-36,000
	10. Flexural strength (rupture or yield), p.s.i.	D790	9000-35,000	11,000-33,000	18,000-20,000	12,000	45,000-50,000	6000-17,000	8500-26,000
	11. Tensile modulus, 10^3 p.s.i.	D638[b]	1000-2500	1500-2500			1200-1600	500-3000	2000-2800
	12. Compressive modulus, 10^3 p.s.i.	D695						2000-3000	
	13. Flexural modulus, 10^3 p.s.i. 73° F.	D790	1000-2500	1500-1800	1400-1500	1400-1500	900-1400	2000	2000
	200° F.	D790							
	250° F.	D790							
	300° F.	D790							
	14. Izod impact, ft.-lb./in. of notch (1/8-in. thick specimen)	D256A	2.5-15	2-18.5	10-12	5-7	22-28	0.3-0.5	0.5-16
	15. Hardness Rockwell	D785						E98	E95
	Shore/Barcol	D2240/ D2583	Barcol 40-70	Barcol 50-65	Barcol 45-50	Barcol 50	50-60		
Thermal	16. Coef. of linear thermal expansion, 10^{-6} in./in./°C.	D696	6-30	20			6-30	20-50	15-33
	17. Deflection temperature under flexural load, °F. 264 p.s.i.	D648	375-500	320-536	395-400+	400+	>425	350-500	400-500
	66 p.s.i.	D648							
	18. Thermal conductivity, 10^{-4} cal.-cm./sec.-cm.^{2}-°C.	C177		18-22				12-25	15-25
Physical	19. Specific gravity	D792	1.6-2.4	1.72-2.10	1.75-1.80	1.80-1.85	1.77-1.83	1.6-2.3	2.0-2.3
	20. Water absorption (1/8-in. thick specimen), % 24 hr.	D570	0.01-0.25	0.1-0.45			0.19-0.25	0.05-0.5	0.03-0.5
	Saturation	D570			0.5	0.5			
	21. Dielectric strength (1/8-in. thick specimen), short time, v./mil	D149	380-450	300-390				350-450	259-530
		SUPPLIERS[a]	Budd; Haysite; Interplastic; Jet Moulding; Polyply; Premix; Rostone	BP Chemicals; Budd; Bulk Molding Compounds; Cytec Fiberite; Epic Resins; Glastic; Haysite; Jet Moulding; Plaslok; Polyply; Premix; Rostone	Jet Moulding; Premix; Rostone	Jet Moulding; Premix; Rostone	Jet Moulding	Cytec Fiberite; Plastics Eng.; Premix	Cosmic Plastics; Cytec Fiberite; Plastics Eng.; Premix; Rogers

	Polyester, thermoset and alkyd (Cont'd)	Polyetherimide			Polyethersulfone					Polyethylene and ethylene copolymers	
	Vinyl Ester BMC									Low and medium density	
	Glass fiber-reinforced	Unfilled	30% glass fiber-reinforced	EMI shielding (conductive); 30% carbon fiber	10% glass fiber-reinforced	20-30% glass fiber-reinforced	20% mineral-filled	30% carbon-filled	Polyether-sulfone unreinforced	Branched homopolymer	Linear copolymer
1a.									12.5-30	0.25-27.0	
1.	Thermoset									98-115	122-124
		215-217	215	215						—25	
2.	C: 290-350 I: 290-350	I: 640-800	I: 620-800	I: 600-780	I: 680-715	I: 660-765	I: 662-716	I: 680-734		I: 300-450 E: 250-450	I: 350-500 E: 450-600
3.	0.5-2	10-20	10-20	10-30						5-15	5-15
4.	1.0	1.5-3	1.5-3	1.5-3						1.8-3.6	3
5.	0-2	0.005-0.007	0.001-0.004	0.0005-0.002	0.005-0.006	0.004-0.006	0.007-0.008		0.007	0.015-0.050	0.020-0.022
6.	9-12	14,000	23,200-28,500	29,000-34,000	16,500	20,000-22,000	7975	15,800	12,000	1200-4550	1900-4000
7.	1-2	60	2-5	1-3	4.3	2.1-2.8	5	1.4		100-650	100-965
8.		15,200	24,500							1300-2100	1400-2800
9.	23,000-28,000	21,900	23,500-30,700	32,000							
10.	20,000-31,000	22,000	33,000	37,000-45,000	24,500	26,500-28,500			16,100		
11.		430	1300-1600	2600-3300	740	1150-1550	522	1740	385	25-41	38-75
12.		480	550-938								
13.		480	1200-1300	2500-2600	650	980-1300			420	35-48	40-105
		370	1100								
		360	1060								
		350	1040								
14.	5-15	1.0-1.2	1.7-2.0	1.2-1.6	1.3	1.4-1.7	1.5		1.6	No break	1.0-No break
15.		M109-110	M114, M125, R123	M127	M94	M96-M97					
	45-65									Shore D44-50	Shore D55-56
16.	20	47-56	20-21		19	12-14	47	12	49	100-220	
17.	500+	387-392	408-420	405-420	414	419	399	433	400		
		405-410	412-415	410-425	419	430	414	440		104-112	
18.	15-25	1.6	6.0-9.3	17.6						8	
19.	1.7-1.9	1.27	1.49-1.51	1.39-1.42	1.45	1.53-1.60	1.52	1.53	1.37	0.917-0.932	0.918-0.940
20.	0.1-0.2	0.25	0.16-0.20	0.18-0.2					1.85	<0.01	
		1.25	0.9		1.9	1.5-1.7	1.7				
21.	300-400	500	495-630						380	450-1000	
	Cytec Fiberite	GE Plastics	ComAlloy; DSM; Ferro; GE Plastics; LNP; Polymer Resources; RTP	ComAlloy; DSM; Ferro; LNP; RTP	Amoco Polymers; BASF; RTP	BASF; RTP	BASF; RTP	BASF; RTP	Amoco Polymers	American Polymers; Bamberger Polymers; Chevron; Dow Plastics; DuPont; Eastman; Equistar; Exxon; Huntsman; Mobil; Network Polymers; NOVA Chemicals; RSG Polymers; A. Schulman; Union Carbide; Wash. Penn; Westlake	Bamberger Polymers; Chevron; Dow Plastics; DuPont; Eastman; Equistar; Exxon; Exxon Chemical Canada; Mobil; Montell NA; Network Polymers; NOVA Chemicals; RSG Polymers; A. Schulman; Solvay Polymers; Union Carbide

Materials	Properties	ASTM test method	Polyethylene and ethylene copolymers (Cont'd)						
			Low and medium density (Cont'd)			High density			
			LDPE copolymers					Copolymers	
			Ethylene-vinyl acetate	Ethylene-ethyl acrylate	Ethylene-methyl acrylate	Polyethylene homopolymer	Rubber-modified	Low and medium molecular weight	High molecular weight
Processing	1a. Melt flow (gm./10 min.)	D1238	1.4-2.0			5-18			5.4-6.8
	1. Melting temperature, °C. T_m (crystalline)		103-110		83	130-137	122-127	125-132	125-135
	T_g (amorphous)								
	2. Processing temperature range, °F. (C = compression; T = transfer; I = injection; E = extrusion)		C: 200-300 I: 350-430 E: 300-380	C: 200-300 I: 250-500	E: 200-620	I: 350-500 E: 350-525	E: 360-450	I: 375-500 E: 300-500	I: 375-500 E: 375-475
	3. Molding pressure range, 10^3 p.s.i.		1-20	1-20		12-15		5-20	
	4. Compression ratio					2		2	
	5. Mold (linear) shrinkage, in./in.	D955	0.007-0.035	0.015-0.035		0.015-0.040		0.012-0.040	0.015-0.040
Mechanical	6. Tensile strength at break, p.s.i.	D638[b]	2200-4000	1600-2100	1650	3200-4500	2300-2900	3000-6500	2500-4300
	7. Elongation at break, %	D638[b]	200-750	700-750	740	10-1200	600-700	10-1300	170-800
	8. Tensile yield strength, p.s.i.	D638[b]	1200-6000		1650	3800-4800	1400-2600	2600-4200	2800-3900
	9. Compressive strength (rupture or yield), p.s.i.	D695		3000-3600		2700-3600		2700-3600	
	10. Flexural strength (rupture or yield), p.s.i.	D790							
	11. Tensile modulus, 10^3 p.s.i.	D638[b]	7-29	4-7.5	12	155-158		90-130	136
	12. Compressive modulus, 10^3 p.s.i.	D695							
	13. Flexural modulus, 10^3 p.s.i. 73° F.	D790	7.7			145-225		120-180	125-175
	200° F.	D790							
	250° F.	D790							
	300° F.	D790							
	14. Izod impact, ft.-lb./in. of notch (1/8-in. thick specimen)	D256A	No break	No break		0.4-4.0		0.35-6.0	3.2-4.5
	15. Hardness Rockwell	D785							
	Shore/Barcol	D2240/ D2583	Shore D17-45	Shore D27-38		Shore D66-73	Shore D55-60	Shore D58-70	Shore D63-65
Thermal	16. Coef. of linear thermal expansion, 10^{-6} in./in./°C.	D696	160-200	160-250		59-110		70-110	70-110
	17. Deflection temperature under flexural load, °F. 264 p.s.i.	D648							
	66 p.s.i.	D648				175-196		149-176	154-158
	18. Thermal conductivity, 10^{-4} cal.-cm./sec.-cm.^{2}-°C.	C177				11-12		10	
Physical	19. Specific gravity	D792	0.922-0.943	0.93	0.942-0.945	0.952-0.965	0.932-0.939	0.939-0.960	0.947-0.955
	20. Water absorption (1/8-in. thick specimen), % 24 hr.	D570	0.005-0.13	0.04	0.0	<0.01		<0.01	
	Saturation	D570							
	21. Dielectric strength (1/8-in. thick specimen), short time, v./mil	D149	620-760	450-550		450-500		450-500	
		SUPPLIERS[a]	AT Plastics; Chevron; DuPont; Equistar; Exxon; Federal Plastics; Huntsman; Mobil; Network Polymers; A. Schulman; Union Carbide; Westlake	Network Polymers; Union Carbide	Chevron; Exxon; Network Polymers; A. Schulman	American Polymers; Bamberger Polymers; Chevron; Dow Plastics; Eastman; Equistar; Exxon; Exxon Chemical Canada; Federal Plastics; Fina; M.A. Polymers; Mobil; Network Polymers; NOVA Chemicals; Phillips; RSG Polymers; A. Schulman; Shuman; Solvay Polymers; Ticona; Union Carbide	Exxon; M.A. Polymers; Network Polymers	AlphaGary; American Polymers; Bamberger Polymers; Chevron; Dow Plastics; Eastman; Equistar; Exxon; Exxon Chemical Canada; Fina; Mobil; Network Polymers; NOVA Chemicals; Phillips; RSG Polymers; A. Schulman; Shuman; Ticona; Union Carbide	AlphaGary; Amoco Chemical; Bamberger Polymers; BASF; Chevron; Dow Plastics; Equistar; Exxon; Fina; Mobil; Network Polymers; NOVA Chemicals; Phillips; Solvay Polymers; Ticona; Union Carbide

	Polyethylene and ethylene copolymers (Cont'd)				Polyimide					
	High density (Cont'd)		Crosslinked		Thermoplastic					
	Ultra high molecular weight	30% glass fiber-reinforced	Molding grade	Wire and cable grade	Unfilled	30% glass fiber-reinforced	30% carbon fiber-reinforced	30% carbon fiber-reinforced, crystallized	15% graphite-filled	40% graphite-filled
1a.					4.5-7.5					
1.	125-138	120-140			388	388	388		388	
					250-365	250	250		250	365
2.	C: 400-500	I: 350-600	C: 240-450 I: 250-300	E: 250-400	C: 625-690 I: 734-740 E: 734-740	I: 734-788	I: 734-788	I: 734-788	I: 734-788	C: 690
3.	1-2	10-20			3-20	10-30	10-30	10-30	10-30	3-5
4.					1.7-4	1.7-2.3	1.7-2.3	1.7-2.3	1.7-2.3	
5.	0.040	0.002-0.006	0.007-0.090	0.020-0.050	0.0083	0.0044	0.0021		0.006	
6.	5800-7000	7500-9000	1600-4600	1500-3100	10,500-17,100	24,000	33,400	31,700	8000-8400	7600
7.	350-525	1.5-2.5	10-440	180-600	7.5-90	3	2	1	3.5	3
8.	3100-4000			1200-2000	12,500-13,000					
9.		6000-7000	2000-5500		17,500-40,000	27,500	30,200		25,000	
10.		11,000-12,000	2000-6500		10,000-28,800	35,200	46,700	43,600	11,000-14,100	18,000
11.		700-900	50-500		300-400	1720	3000			14,000
12.			50-150		315-350	458	573		330	
13.	130-140	700-800	70-350	8-14	360-500	1390	2780	3210	460-500	
										700
					210	1175	2450		260	
14.	No break	1.1-1.5	1-20		1.5-1.7	2.2	2.0	2.4	1.1	0.7
15.	R50	R75-90			E52-99,R129,M95	R128, M104	R128, M105			E27
	Shore D61-63		Shore D55-80	Shore D30-65						
16.	130-200	48	100	100	45-56	17-53	6-47		41	38
17.	110-120	250	105-145	100-173	460-680	469	478	>572	680	680
	155-180	260-265	130-225							
18.		8.6-11			2.3-4.2	8.9	11.7			41.4
19.	0.94	1.18-1,28	0.95-1.45	0.91-1.40	1.33-1.43	1.56	1.43	1.47	1.41	1.65
20.	<0.01	0.02-0.06	0.01-0.06	0.01-0.06	0.24-0.34	0.23	0.23		0.19	0.14
					1.2					0.6
21.	710	500-550	230-550	620-760	415-560	528			250	
	Montell NA; Network Polymers; Ticona	ComAlloy; DSM; Ferro; LNP; M.A. Polymers; RTP; A. Schulman; Thermofil	Equistar; Mobil; Phillips; A. Schulman	AlphaGary; AT Plastics; Equistar; A. Schulman; Union Carbide	Ciba Specialty Chemicals; DuPont; Mitsui Chemicals America; Solutia	Mitsui Chemicals America; RTP	Mitsui Chemicals America; RTP	Mitsui Chemicals America; RTP	DuPont; Mitsui Chemicals America	DuPont

Materials			Poly-imide (Cont'd)	Polyketone			Poly-methylpentene		Poly-phenylene oxide,
			Thermoset						Alloy with polystyrene
	Properties	ASTM test method	Unfilled	Unfilled	30% carbon	Polyketone	Unfilled	Filled	Low glass transition
Processing	1a. Melt flow (gm./10 min.)	D1238				6.0	26	30	
	1. Melting temperature, °C. T_m (crystalline)		Thermoset			428°F	230-240	240	
	T_g (amorphous)								100-112
	2. Processing temperature range, °F. (C = compression; T = transfer; I = injection; E = extrusion)		480-485	715-805	715-805	I: 480-510	I: 510-610 E: 510-650	I: 510-610	I: 400-600 E: 420-500
	3. Molding pressure range, 10^3 p.s.i.		7-29				1-10	1-10	12-20
	4. Compression ratio		1-1.2	gp screw	gp screw	2.5-3.0:1	2.0-3.5	2.0-3.5	1.3-3
	5. Mold (linear) shrinkage, in./in.	D955	0.001-0.01	0.008-0.012	0.002-0.008	0.028	0.016-0.021	0.014-0.017	0.005-0.008
Mechanical	6. Tensile strength at break, p.s.i.	D638[b]	4300-22,900	13,500	30,000	8000	2300-2500	2400	6800-7800
	7. Elongation at break, %	D638[b]	1	50	1.5	7300	20-120	25	48-50
	8. Tensile yield strength, p.s.i.	D638[b]	4300-22,900			8700	2200-3400	3400	6500-7800
	9. Compressive strength (rupture or yield), p.s.i.	D695	19,300-32,900	17,200	33,800				12,000-16,400
	10. Flexural strength (rupture or yield), p.s.i.	D790	6500-50,000	24,500	40,000	8000	6300-8300		8300-12,800
	11. Tensile modulus, 10^3 p.s.i.	D638[b]	460-4650	520	2700	230	160-280		310-380
	12. Compressive modulus, 10^3 p.s.i.	D695	421				114-171		
	13. Flexural modulus, 10^3 p.s.i. 73° F.	D790	422-3000	2000	10,000	230	70-190	270	325-400
	200° F.	D790					36		260
	250° F.	D790					26		
	300° F.	D790	1030-2690				17		
	14. Izod impact, ft.-lb./in. of notch (1/8-in. thick specimen)	D256A	0.65-15	1.6	1.5		2-3		3-6
	15. Hardness Rockwell	D785	110M-120M			105	R35-85	90	R115-116
	Shore/Barcol	D2240/ D2583				75D			
Thermal	16. Coef. of linear thermal expansion, 10^{-6} in./in./°C.	D696	15-50			1X10-4	65		38-70
	17. Deflection temperature under flexural load, °F. 264 p.s.i.	D648	572->575	323	634	221	120-130		176-215
	66 p.s.i.	D648		582	652	410	180-190	230	230
	18. Thermal conductivity, 10^{-4} cal.-cm./sec.-cm.^{2}-°C.	C177	5.5-12				4.0		3.8
Physical	19. Specific gravity	D792	1.41-1.9	1.30	1.45	1.24	0.833-0.835	1.08	1.04-1.10
	20. Water absorption (1/8-in. thick specimen), % 24 hr.	D570	0.45-1.25	0.10	0.20	0.40	0.01	0.11	0.06-0.1
	Saturation	D570				2.1			
	21. Dielectric strength (1/8-in. thick specimen), short time, v./mil	D149	480-506	355	Conductive	320	1096-1098		400-665
		SUPPLIERS[a]	Ciba Specialty Chemicals	Amoco Polymers	Amoco Polymers	Shell	Mitsui Petrochemical	Mitsui Petrochemical	Ashley Polymers; GE Plastics; Polymer Resources; Shuman

	Polyphenylene oxide, modified (Cont'd)									Polyphenylene sulfide	
	Alloy with polystyrene (Cont'd)								Alloy with nylon		
							EMI shielding (conductive)				
	High glass transition	**Impact-modified**	**15% glass fiber-reinforced**	**20% glass fiber-reinforced**	**30% glass fiber-reinforced**	**Mineral-filled**	**30% graphite fiber**	**40% aluminum flake**	**Un-reinforced**	**Unfilled**	**10-20% glass fiber-reinforced**
1a.											
1.										285-290	275-285
	117-190	135	100-125	100-125	100-125	110-135				88	
2.	I: 425-670 E: 460-525	I: 425-550	I: 400-630 E: 460-525	I: 400-630 E: 460-525	I: 400-630 E: 460-525	I: 540-590 E: 470-530	I: 500-600	I: 500-600		I: 590-640	I: 600-675
3.	12-20	10-15	10-40	10-40	10-40	12-20	10-20	10-20		5-15	
4.	1.3-3					2-3				2-3	
5.	0.006-0.008	0.006	0.002-0.004	0.002-0.004	0.001-0.004	0.005-0.007	0.001	0.001	0.013-0.016	0.006-0.014	0.002-0.005
6.	9600	7000-8000	10,000-12,000	13,000-15,000	15,000-18,500		18,700	6500	8500	7000-12,500	7500-14,000
7.	60	35	5-8	8.0	2-5	25	2.5	3.0	60	1-6	1.0-1.5
8.	7000-9000				14,500	9500-11,000					
9.	16,400	10,000			17,900		20,000	6000		16,000	17,000-20,000
10.	9500-14,000	8200-11,000			20,000-23,000		24,000	9500	315	14,000-21,000	9500-20,000
11.	355-380	345-360			1000-1300		1150	750		480	850-1200
12.											
13.	330-400	325-345	450-500	760	1100-1150	425-500	1100	850	290-315	550-600	900-1200
	305				1000						
								35			
14.	5	6.8	1.1-1.3	1.5	1.7-2.3	3-4	1.3	0.6	3.8	<0.5	0.7-1.2
15.	R118-120	L108, M93	R106-110	R115	R115-116	R121	R111	R110		R123-125	R121
16.	33-77				14-25		11	11		27-49	16-20 Trans. dir.: 15.36-45
17.	225-300	190-275	252-260	262-275	275-317	190-230	240	230	250	212-275	440-480
	279	205-245	273-280	280-290	285-320		265	250	350	390	500-520
18.	5.2				3.8-4.1					2.0-6.9	
19.	1.04-1.09	1.27-1.36			1.27-1.36	1.24-1.25	1.25	1.45	1.10	1.35	1.39-1.47
20.	0.06-0.12	0.1-0.07			0.06	0.07	0.04	0.03	0.3	0.01-0.07	0.05
									1.0		
21.	500-700	530			550-630	490				380-450	
	Creanova; GE Plastics; Polymer Resources	GE Plastics; Polymer Resources	Ashley Polymers; Polymer Resources	Ashley Polymers; Polymer Resources	Ashley Polymers; ComAlloy; Ferro; GE Plastics; LNP; Polymer Resources; RTP	ComAlloy; GE Plastics; Polymer Resources; RTP	ComAlloy; LNP; RTP	ComAlloy	Ashley Polymers; GE Plastics	Phillips; Ticona	Akzo; LNP; RTP

Materials	Properties	ASTM test method	Polyphenylene sulfide (Cont'd): 30% glass fiber-reinforced	40% glass fiber-reinforced	53% glass/mineral-reinforced, high elong. and impact	30% glass fiber, 15% PTFE	30% long glass fiber-reinforced	40% long glass fiber-reinforced	50% long glass fiber-reinforced
Processing	1a. Melt flow (gm./10 min.)	D1238		30-41	50	30			
	1. Melting temperature, °C. T_m (crystalline)		275-285	275-290	354	275-290	310	299	315
	T_g (amorphous)		90	88-90					
	2. Processing temperature range, °F. (C = compression; T = transfer; I = injection; E = extrusion)		I: 590-640	I: 600-675	I: 590-620	640-660	580-620	590-620	I: 580-620
	3. Molding pressure range, 10^3 p.s.i.		8-12	5-20	7-15		8-10	7-20	8-10
	4. Compression ratio		3	3	3-4		3	3-4	3
	5. Mold (linear) shrinkage, in./in.	D955	0.003-0.005	0.002-0.005	0.001-0.003	0.002-0.004	0.001-0.007	0.001-0.003	0.001-0.005
Mechanical	6. Tensile strength at break, p.s.i.	D638[b]	22,000	17,500-29,100	24,000	21,000	21,000	23,000	23,200
	7. Elongation at break, %	D638[b]	1.5	0.9-4	1.5	1.8	1.2	1.1	1.0
	8. Tensile yield strength, p.s.i.	D638[b]							
	9. Compressive strength (rupture or yield), p.s.i.	D695		21,000-31,200	30,000		32,400	32,000-34,000	34,200-35,000
	10. Flexural strength (rupture or yield), p.s.i.	D790	28,000	22,700-43,600	35,000	29,000	32,900	35,000-36,400	37,300-38,000
	11. Tensile modulus, 10^3 p.s.i.	D638[b]		1100-2100		1340	1800	2200	2600
	12. Compressive modulus, 10^3 p.s.i.	D695							
	13. Flexural modulus, 10^3 p.s.i. 73° F.	D790	1700	1700-2160	2300	1360	1700	2300	2400
	200° F.	D790							
	250° F.	D790		1000					
	300° F.	D790		730					
	14. Izod impact, ft.-lb./in. of notch (1/8-in. thick specimen)	D256A	1.3	1.1-2.0	1.5	1.6	4.6	4.8-5.3	5.0-5.5
	15. Hardness Rockwell	D785	M102.7-M103	R123, M100-104	100M	R116			
	Shore/Barcol	D2240/ D2583							
Thermal	16. Coef. of linear thermal expansion, 10^{-6} in./in./°C.	D696		Flow dir: 12.1-22 Trans. dir.: 14.4-45	Flow: 19-32 Trans.: 32-80				
	17. Deflection temperature under flexural load, °F. 264 p.s.i.	D648	507	485-515	510	480	490	500	505
	66 p.s.i.	D648	534	536					
	18. Thermal conductivity, 10^{-4} cal.-cm./sec.-cm.^{2}-°C.	C177		6.9-10.7					
Physical	19. Specific gravity	D792	1.38-1.58	1.60-1.67	1.80	1.60	1.52-1.62	1.62	1.72
	20. Water absorption (1/8-in. thick specimen), % 24 hr.	D570	<0.03	0.004-0.05	0.02	0.005			
	Saturation	D570							
	21. Dielectric strength (1/8-in. thick specimen), short time, v./mil	D149	10^6; 380	347-450	300				
		SUPPLIERS[a]	Ferro; GE Plastics; LNP; Phillips; RTP; Ticona	Albis; DSM; Ferro; GE Plastics; LNP; Phillips; RTP; Ticona	Ticona	Ferro; GE Plastics; LNP; RTP	RTP; Ticona	RTP; Ticona	RTP; Ticona

	Polyphenylene sulfide (Cont'd)										Polyphthalamide
							EMI shielding (conductive)				
	60% long glass fiber-reinforced	65% mineral- and glass-filled	50% glass and mineral-reinforced	55% glass and mineral-reinforced	60% glass and mineral-reinforced	Encapsulation grades	60% stainless steel	30% carbon fiber	40% PAN carbon fiber	30% PAN carbon fiber, 15% PTFE	Extra-tough
1a.		20-35	97	85	63	287					
1.	315	285-290	275-285	275-285	275-285	275-285	299	275-285	275-285	275-285	310
		88	90								
2.	I: 580-620	I: 600-675	I: 600-675	I: 600-675	I: 600-675	I: 600-675	I: 590-620	I: 500-675	I: 600-675	I: 600-675	I: 610-660
3.	8-10	5-20	4-12				7-20	5-20			5-15
4.	3	3	3				3.0-4.0				2.5-3
5.	0.001-0.005	0.001-0.004	0.002-0.004	0.003-0.005	0.007	0.006-0.007	0.005-0.007	0.005-0.003	0.0005	0.0008	0.015-0.020
6.	22,500	13,000-23,100	18,000-22,000	12,700	8100	9800	8400	20,000-27,000	26,000-29,000	25,500	11,000
7.	1.0	1.3; <1.4	1.0-3.5	1.6	2.0	1.0		0.5-3	1.0-1.5	2.6	30
8.		11,000									10,500
9.	30,800	11,000-32,300						26,000	27,000		
10.	38,300	17,500-33,900	28,000-33,700	20,300	11,200	15,100	17,800	26,000-36,000	40,000	34,000	17,000
11.	3000							2500-3700	4400-4800	3600	350
12.											
13.	3000	1800-2400	2240-2400	1750	1441	1574	590	2450-3300	3900-4100	3000	368
		2000									
		1200									
14.	5.5	0.5-1.37	0.8-1.3	0.7	0.5	0.4	0.2	0.8-1.2	1.2-1.5	1.2	18
15.		R121, M102	R120	M66	M85	M98		R123	R123		120
16.		12.9-20 14.3	12.2-14.6		14.3-14.8	18.0-18.2		6-16	8		
17.	510	500-510	500-510	491	340	328	450	500-505	505	500	243
		534						>505	505		
18.								8.6-17.9			
19.	1.84	1.78-2.03	1.78-1.8	1.82	1.90	1.86	1.37	1.42-1.47	1.46-1.49	1.47	1.15
20.		0.02-0.07	0.02-0.07	0.07	0.08	0.03		0.01-0.02	0.02	0.06	0.68
21.		280-450	280-343			338					
	Ticona	Ferro; LNP; Phillips; RTP; Ticona	Albis; LNP; RTP; Ticona	LNP; Phillips; RTP	Albis; LNP; Phillips; RTP	Phillips	Ticona	DSM; Ferro; LNP; RTP	DSM; Ferro; LNP; RTP	DSM; Ferro; LNP; RTP	Amoco Polymers

Materials	Properties	ASTM test method	Polyphthalamide (Cont'd)						
			45% glass-reinforced	33% glass-reinforced, V-0	40% mineral-reinforced	40% glass/mineral-reinforced, heat-stabilized	50% glass/mineral reinforced, V0	51% glass/mineral reinforced	15% glass-reinforced
Processing	1a. Melt flow (gm./10 min.)	D1238							
	1. Melting temperature, °C. T_m (crystalline)		310	310	310	310	310	310	
	T_g (amorphous)								
	2. Processing temperature range, °F. (C = compression; T = transfer; I = injection; E = extrusion)		I: 610-660	I: 610-660	I: 610-660	I: 620-650	I: 610-640	I: 610-640	I: 610-640
	3. Molding pressure range, 10^3 p.s.i.		5-15	5-15	5-15				
	4. Compression ratio		2.5-3	2.5-3	2.5-3				
	5. Mold (linear) shrinkage, in./in.	D955	0.002-0.006	0.002-0.004	0.008-0.010	0.004	0.003-0.3		0.006-0.007
Mechanical	6. Tensile strength at break, p.s.i.	D638[b]	38,000-39,400	26,000-27,000	15,800-17,000	23,700	22,200	17,200	
	7. Elongation at break, %	D638[b]	2.0-2.6	1.3	1.6	2.2-2.5	1.3	1.9-2	2.0
	8. Tensile yield strength, p.s.i.	D638[b]				23,500	21,000	16,200	19,000
	9. Compressive strength (rupture or yield), p.s.i.	D695	45,500		26,000	33,000			30,000
	10. Flexural strength (rupture or yield), p.s.i.	D790	54,000-55,100	35,900-37,300	28,600-30,000	31,500-34,300	31,000	24,000-28,400	23,700
	11. Tensile modulus, 10^3 p.s.i.	D638[b]	2500	2000	1300	1600	2400	1530	
	12. Compressive modulus, 10^3 p.s.i.	D695							
	13. Flexural modulus, 10^3 p.s.i. 73° F.	D790	2100-2160	1900-1930	1080-1300	1200-1410	2040	1300	1090
	200° F.	D790							
	250° F.	D790							
	300° F.	D790							
	14. Izod impact, ft.-lb./in. of notch (1/8-in. thick specimen)	D256A	2.1-2.5	1.1-1.5	0.8	0.8-0.9	1.2	1.3	1.1
	15. Hardness Rockwell	D785	125	125	125	125			127
	Shore/Barcol	D2240/ D2583							
Thermal	16. Coef. of linear thermal expansion, 10^{-6} in./in./°C.	D696	8		19	2.9-4.6			3.5
	17. Deflection temperature under flexural load, °F. 264 p.s.i.	D648	527-549	511-523	316-361	478-527	490-505	482	531
	66 p.s.i.	D648							
	18. Thermal conductivity, 10^{-4} cal.-cm./sec.-cm.^{2}-°C.	C177	2.6		2.6				
Physical	19. Specific gravity	D792	1.56-1.60	1.71	1.53-1.54	1.54	1.82	1.66-1.68	1.26
	20. Water absorption (1/8-in. thick specimen), % 24 hr.	D570	0.12	0.18	0.14	0.16	0.12		0.30
	Saturation	D570							
	21. Dielectric strength (1/8-in. thick specimen), short time, v./mil	D149	560	458	>560	505	635		480
		SUPPLIERS[a]	Amoco Polymers; LNP; RTP	Amoco Polymers; RTP	Amoco Polymers; RTP	Amoco Polymers; RTP	Amoco Polymers	Amoco Polymers; RTP	RTP

	Polyphthalamide (Cont'd)				Polypropylene					
					Homopolymer					
	15% glass-reinforced, V0	45% glass-reinforced, V0	40% glass/mineral reinforced	40% mineral-reinforced, heat-stabilized	Unfilled	10-40% talc-filled	10-40% calcium carbonate-filled	10-30% glass fiber-reinforced	10-50% mica-filled	40% glass fiber-reinforced
1a.					0.4-100	0.1-30.0	0.1-30.0	1-20	4-10	1-20
1.	310	310		310	160-175	158-168	168	168	168	168
					—20					
2.	I: 610-640	I: 610-640	I: 610-650	I: 610-650	I: 375-550 E: 400-500	I: 350-550	I: 375-525	I: 425-475		I: 450-550
3.			5-15		10-20	10-20	8-20			10-25
4.			2.5-3		2.0-2.4					
5.	0.005	0.002	0.004-0.007		0.010-0.025	0.008-0.022	0.007-0.018	0.002-0.008	0.002-0.015	0.003-0.005
6.	18,600	28,100		16,500	4500-6000	3545-5000	3400-4500	6500-13,000		8400-15,000
7.	2.0		2.5	1.6	100-600	3-60	10-245	1.8-7	3-10	1.5-4
8.	18,400	29,500	23,500	17,000	4500-5400	3500-5000	3000-4600	7000-10,000	4700-6500	
9.			33,000	24,000	5500-8000	7500	3000-7200	6500-8400		8900-9800
10.	24,400-26,600	39,000-41,100	31,500	28,000-29,800	6000-8000	7000-9200	5500-7000	7000-20,000		10,500-22,000
11.	1400			1300	165-225	450-575	375-500	700-1000		1100-1500
12.					150-300					
13.	1000-1700	2500-2600	1200	1150-1200	170-250	210-670	230-450	310-780	420-1150	950-1000
					50	400	320			
					35					
14.	0.6-0.8	1.24-1.7	0.9	0.7-0.9	0.4-1.4	0.4-1.4	0.5-1.0	1.0-2.2	0.50-0.85	1.4-2.0
15.			125	125	R80-102	R85-110	R78-99	R92-115	R82-100	R102-111
									066-78	
16.			2.9-4.6	3.8-4.1	81-100	42-80	28-50	21-62		27-32
17.	466-504	522-527	527	315-325	120-140	132-180	135-170	253-288		300-330
					225-250	210-290	200-270	290-320		330
18.					2.8	7.6	6.9	5.5-6.2		8.4-8.8
19.	1.58	1.78-1.79	1.54	1.54-1.57	0.900-0.910	0.97-1.27	0.97-1.25	0.97-1.14	0.99-1.35	1.22-1.23
20.	0.28	0.17	0.16	0.14	0.01-0.03	0.01-0.03	0.02-0.05	0.01-0.05	0.01-0.06	0.05-0.06
										0.09-0.10
21.			505	455	600	500	410-500			500-510
	Amoco Polymers; RTP	Amoco Polymers; RTP	Amoco Polymers; RTP	Amoco Polymers; RTP	American Polymers; Amoco Chemical; Aristech Chem.; Bamberger Polymers; ComAlloy; Dow Plastics; Epsilon; Equistar; Exxon; Federal Plastics; Ferro; Fina; Huntsman; M.A. Polymers; Montell NA; Network Polymers; Phillips; RSG Polymers; A. Schulman; Shuman; Solvay Polymers; Union Carbide; Wash. Penn	Adell; Albis; Amoco; Bamberger Polymers; ComAlloy; DSM; Exxon; Federal Plastics; Ferro; M.A. Polymers; Montell NA; MRC; Network Polymers; Polifil; RSG Polymers; RTP; A. Schulman; Spartech Polycom; Thermofil; Wash. Penn	Adell; Albis; Bamberger Polymers; ComAlloy; DSM; Federal Plastics; Ferro; M.A. Polymers; Montell NA; Network Polymers; Polifil; RSG Polymers; RTP; A. Schulman; Spartech Polycom; Thermofil; Wash. Penn	Adell; Albis; Amoco Polymers; Bamberger Polymers; ComAlloy; DSM; Federal Plastics; Ferro; LNP; M.A. Polymers; Montell NA; MRC; Network Polymers; Polifil; RSG Polymers; RTP; A. Schulman; Spartech Polycom; Thermofil; Wash. Penn	Network Polymers; Polifil; RTP; A. Schulman; Spartech Polycom; Washington Penn	Adell; Albis; Amoco Polymers; ComAlloy; DSM; Ferro; LNP; M.A. Polymers; Montell NA; MRC; Network Polymers; Polifil; RTP; A. Schulman; Thermofil

Materials			Polypropylene (Cont'd)					
			Homopolymer (Cont'd)					Copolymer
	Properties	ASTM test method	20-30% long glass fiber-reinforced	40% long glass fiber-reinforced	30% random glass mat	Impact-modified, 40% mica-filled	EMI shielding (conductive); 30% PAN carbon fiber	Unfilled
Processing	1a. Melt flow (gm./10 min.)	D1238						0.6-100
	1. Melting temperature, °C. T_m (crystalline)		168	163	168	168	168	150-175
	T_g (amorphous)							-20
	2. Processing temperature range, °F. (C = compression; T = transfer; I = injection; E = extrusion)		I: 360-440	I: 370-410	C: 420-440	I: 350-470	I: 360-470	I: 375-550 E: 400-500
	3. Molding pressure range, 10^3 p.s.i.			6-12	1-2			10-20
	4. Compression ratio			3-4				2-2.4
	5. Mold (linear) shrinkage, in./in.	D955	0.0025-0.004	0.001-0.003	0.002-0.003	0.007-0.008	0.001-0.003	0.010-0.025
Mechanical	6. Tensile strength at break, p.s.i.	D638[b]	7500-14,100	10,500-15,600	12,000	4500	6800	4000-5500
	7. Elongation at break, %	D638[b]	2.1-2.2	1.7	3	4	0.5	200-500
	8. Tensile yield strength, p.s.i.	D638[b]			12,000			3000-4300
	9. Compressive strength (rupture or yield), p.s.i.	D695	6500-14,400	10,400-15,900				3500-8000
	10. Flexural strength (rupture or yield), p.s.i.	D790	10,000-22,800	20,800-25,200	20,000	7000	9000	5000-7000
	11. Tensile modulus, 10^3 p.s.i.	D638[b]	750-900	970-1120	670	700	1750	130-180
	12. Compressive modulus, 10^3 p.s.i.	D695						
	13. Flexural modulus, 10^3 p.s.i. 73° F.	D790	550-800	920-1000	620	600	1650	130-200
	200° F.	D790						40
	250° F.	D790						30
	300° F.	D790						
	14. Izod impact, ft.-lb./in. of notch (1/8-in. thick specimen)	D256A	3.5-7.8	8.0-10.04	12.2	0.7	1.1	1.1-14.0
	15. Hardness Rockwell	D785	R105-117					R65-96
	Shore/Barcol	D2240/ D2583						Shore D70-73
Thermal	16. Coef. of linear thermal expansion, 10^{-6} in./in./°C.	D696			15			68-95
	17. Deflection temperature under flexural load, °F. 264 p.s.i.	D648	250-295	300	310	205	245	130-140
	66 p.s.i.	D648	305					185-220
	18. Thermal conductivity, 10^{-4} cal.-cm./sec.-cm.^{2}-°C.	C177	2.35					3.5-4.0
Physical	19. Specific gravity	D792	1.04-1.17	1.21	1.1	1.23	1.04	0.890-0.905
	20. Water absorption (1/8-in. thick specimen), % 24 hr.	D570	0.05				0.12	0.03
	Saturation	D570						
	21. Dielectric strength (1/8-in. thick specimen), short time, v./mil	D149						600
		SUPPLIERS[a]	DSM; LNP; Montell NA; RTP; Ticona	LNP; Montell NA; RTP; Ticona	Azdel	Albis; ComAlloy; DSM; Federal Plastics; Ferro; M.A. Polymers; Polifil; A. Schulman; Spartech Polycom	ComAlloy; DSM; LNP; RTP	American Polymers; Amoco Chemical; Aristech Chem.; Bamberger Polymers; ComAlloy; Dow Plastics; Epsilon; Equistar; Exxon; Federal Plastics; Ferro; Fina; Huntsman; M.A. Polymers; Montell NA; Network Polymers; NOVA Chemicals; Phillips; RSG Polymers; A. Schulman; Shuman; Solvay Polymers; Spartech Polycom; Union Carbide; Wash. Penn

	Polypropylene (Cont'd)			Polystyrene and styrene copolymers						
	Copolymer (Cont'd)			Polystyrene homopolymers			Rubber-modified		Styrene copolymers	
									Styrene-acrylonitrile (SAN)	
	Unfilled, impact-modified	10-20% glass fiber-reinforced	10-40% talc-filled	High and medium flow	Heat-resistant	20% long and short glass fiber-reinforced	Flame-retarded, UL-V0	High-impact	Molding and extrusion	Olefin rubber-modified
1a.		0.1-20	0.1-30					5.8	1.4	
1.	150-168	160-168								
	–20			74-105	100-110	115		9.3-105	100-200	–55/110
2.	I: 390-500 E: 400-500	I: 350-480	I: 350-470 E: 425-475	C: 300-400 I: 350-500 E: 350-500	C: 300-400 I: 350-500 E: 350-500	I: 400-550	I: 400-450 E: 375-425	I: 350-525 E: 375-500	C: 300-400 I: 360-550 E: 360-450	I: 480-530 E: 435-460
3.	10-20		15-20	5-20	5-20	10-20	6-15	10-20	5-20	1-2
4.	2-2.4		2-2.5	3	3-5		3	4	3	2.7-3.2
5.	0.010-0.025	0.003-0.01	0.009-0.017	0.004-0.007	0.004-0.007	0.001-0.003	0.003-0.006	0.004-0.007	0.003-0.005	0.005-0.007
6.	3500-5000	5000-8000	3000-3775	5200-7500	6440-8200	10,000-12,000	2650-4100	1900-6200	10,000-11,900	5100
7.	200-700	3.0-4.0	20-50	1.2-2.5	2.0-3.6	1.0-1.3	30-50	20-65	2-3	15-30
8.	1600-4000		2800-4100		6440-8150		3100-4400	2100-6000	9920-12,000	5000-6000
9.	3500-6000	5500-5600		12,000-13,000	13,000-14,000	16,000-17,000			14,000-15,000	
10.	4000-6000	7000-11,000	4500-5100	10,000-14,600	13,000-14,000	14,000-18,000	4500-7500	3300-10,000	11,000-19,000	7700-8900
11.	50-150			330-475	450-485	900-1200	240-300	160-370	475-560	300
12.				480-490	495-500				530-580	
13.	60-160	355-510	160-400	380-490	450-500	950-1100	280-330	160-390	500-610	280-300
14.	2.2-No break	0.95-2.7	0.6-4.0	0.35-0.45	0.4-0.45	0.9-2.5	1.9-3.3	0.95-7.0	0.4-0.63	13-15
15.	R50-60	R100-103	R83-88	M60-75	M75-84	M80-95, R119	R38-65	R50-82; L-60	M80, R83, 75	R100-102
	Shore D45-55									
16.	68-95			50-83	68-85	39.6-40	45	44.2	65-68	80
17.	115-135	260-280	100-165	169-202	194-217	200-220	180-205	170-205	203-220	197-200
	167-192	305	195-260	155-204	200-224	220-230	176-181	165-200	220-224	
18.	3.5-4.0			3.0	3.0	5.9			3.0	
19.	0.880-0.905	0.98-1.04	0.97-1.24	1.04-1.05	1.04-1.05	1.20	1.15-1.17	1.03-1.06	1.06-1.08	1.02
20.	0.03	0.01	0.02	0.01-0.03	0.01	0.07-0.01	0.0	0.05-0.07	0.15-0.25	0.09
				0.01-0.03	0.01	0.3			0.5	
21.	500			500-575	500-525	425	550		425	420
	American Polymers; Amoco Polymers; ComAlloy; Dow Plastics; Epsilon; Equistar; Exxon; Federal Plastics; Huntsman; LG Chemical; M.A. Polymers; Montell NA; Network Polymers; Phillips; A. Schulman; Solvay Engineered Polymers; Solvay Polymers; Spartech Polycom; Wash. Penn	Adell; Albis; ComAlloy; DSM; Federal Plastics; Ferro; LG Chemical; LNP; M.A. Polymers; Montell NA; Polifil; RTP; A. Schulman; Solvay Engineered Polymers; Spartech Polycom; Thermofil; Wash. Penn	Adell; Albis; Bamberger Polymers; ComAlloy; DSM; Federal Plastics; Ferro; LG Chemical; M.A. Polymers; Montell NA; Polifil; RTP; A. Schulman; Solvay Engineered Polymers; Spartech Polycom; Wash. Penn	American Polymers; Bamberger Polymers; BASF; Chevron; Deltech Polymers; Dow Plastics; Federal Plastics; Fina; Huntsman; LG Chemical; Mobil; Network Polymers; RSG Polymers; A. Schulman	American Polymers; BASF; Chevron; Deltech Polymers; Dow Plastics; Federal Plastics; Fina; Huntsman; LG Chemical; Mobil; Network Polymers; A. Schulman	DSM; Ferro; LG Chemical; LNP; M.A. Polymers; Mobil; RTP	BASF; Dow Plastics; Huntsman; LG Chemical; Mobil; Network Polymers; RTP; A. Schulman	American Polymers; Bamberger Polymers; BASF; Bayer Corp.; Chevron; Dow Plastics; Federal Plastics; Fina; Huntsman; LG Chemical; Mobil; NOVA Chemicals; RSG Polymers; RTP; A. Schulman; Shuman	Albis; American Polymers; BASF; Bamberger Polymers; Bayer Corp.; Dow Plastics; Federal Plastics; Huntsman; LG Chemical; Network Polymers; RSG Polymers; A. Schulman	Dow Plastics; Huntsman; Network Polymers

Materials	Properties	ASTM test method	Polystyrene and styrene copolymers (Cont'd)					
			Styrene copolymers (Cont'd)					
			Styrene-acrylonitrile (SAN) (Cont'd)	Clear styrene-butadiene copolymers (SBC)	Acrylate-styrene-acrylonitrile (ASA)		Styrene-maleic anhydride (SMA)	
			20% glass fiber-reinforced		ASA extrusion, blow molding, injection molding grades	Clear styrene-butadiene copolymers	Molding and extrusion	Impact-modified
Processing	1a. Melt flow (gm./10 min.)	D1238		6-18		7-15		
	1. Melting temperature, °C. T_m (crystalline)							
	T_g (amorphous)		120			108	114	
	2. Processing temperature range, °F. (C = compression; T = transfer; I = injection; E = extrusion)		I: 400-550	I: 375-425 E: 375-427	E: 380-450 I: 400-470	I: 380-450 E: 380-440	I: 430-510 E: 400-500	I: 450-550 E: 425-525
	3. Molding pressure range, 10^3 p.s.i.		10-20		9-15		12-17	12-17
	4. Compression ratio				2.8-3.0			2.5-2.7
	5. Mold (linear) shrinkage, in./in.	D955	0.001-0.003	0.002-0.008	0.004-0.006	0.004-0.010	0.004-0.006	0.004-0.006
Mechanical	6. Tensile strength at break, p.s.i.	D638[b]	15,500-18,000		4000-7500		8100	4500-5500
	7. Elongation at break, %	D638[b]	1.2-1.8	20-180	25-40	20-180	1.8-30	10-35
	8. Tensile yield strength, p.s.i.	D638[b]		1900-4500	5200-5600	1900-4400	5200-8100	4500-6400
	9. Compressive strength (rupture or yield), p.s.i.	D695	17,000-21,000					
	10. Flexural strength (rupture or yield), p.s.i.	D790	20,000-22,700	2700-6400	6000-8000	2700-6400	8000-14,200	7600-13,000
	11. Tensile modulus, 10^3 p.s.i.	D638[b]	1200-1710				340-390	270-360
	12. Compressive modulus, 10^3 p.s.i.	D695						
	13. Flexural modulus, 10^3 p.s.i. 73° F.	D790	1000-1280	150-240	220-341	153-215	320-470	280-490
	200° F.	D790						
	250° F.	D790						
	300° F.	D790						
	14. Izod impact, ft.-lb./in. of notch (1/8-in. thick specimen)	D256A	1.0-3.0	0.3-1.5 NB	9-11	0.4-1.4 NB	0.4-2.0	2.5-6
	15. Hardness Rockwell	D785	M89-100, R122		R85-90		R106-109	R75-109
	Shore/Barcol	D2240/ D2583		Shore D 60-65				
Thermal	16. Coef. of linear thermal expansion, 10^{-6} in./in./°C.	D696	23.4-41.4	~65	59		80	47-88
	17. Deflection temperature under flexural load, °F. 264 p.s.i.	D648	210-230	140-170	185-190	143-170	~214-245	198-235
	66 p.s.i.	D648	220		200-210			
	18. Thermal conductivity, 10^{-4} cal.-cm./sec.-cm.^{2}-°C.	C177	6.6					
Physical	19. Specific gravity	D792	1.22-1.40	1.00-1.01	1.05-1.06	1.01	1.05-1.08	1.05-1.09
	20. Water absorption (1/8-in. thick specimen), % 24 hr.	D570	0.1-0.2	0.08	0.2-0.3	0.08	0.1	0.1-0.5
	Saturation	D570	0.7					
	21. Dielectric strength (1/8-in. thick specimen), short time, v./mil	D149	500		490	300		415-480
		SUPPLIERS[a]	ComAlloy; DSM; Ferro; LG Chemical; LNP; Network Polymers; RTP; Thermofil	Phillips	BASF; Bayer Corp.; Diamond Polymers; GE Plastics; LG Chemical; Network Polymers	Network Polymers	Bayer Corp.; DSM; NOVA Chemicals	Bayer Corp.; NOVA Chemicals

	Polystyrene and styrene copolymers (Cont'd)						Polyurethane			
	Styrene copolymers (Cont'd)						Thermoset		Thermoplastic	
	SMA (Cont'd)	High heat-resistant copolymers					Casting resins			
	20% glass fiber-reinforced	Injection molding	Impact-modified	20% glass fiber-reinforced	Styrene methyl methacrylate	EMI shielding (conductive); 20% PAN carbon fiber	Liquid	Unsaturated	Unreinforced molding	30% long glass fiber-reinforced
1a.			1.0-1.8							
1.							Thermoset	Thermoset	75-137	240
					100-105					
2.	I: 400-550	I: 450-550	I: 425-540 E: 400-500	I: 425-550	I: 375-475	I: 430-500	C: 43-250		I: 370-500 E: 370-510	I: 440-500
3.	10-20				5-20		0.1-5		6-15	10-20
4.					2.5-3.5				3	3
5.	0.002-0.003	0.005	0.003-0.006	0.003-0.004	0.002-0.006	0.0005-0.003	0.020		0.004-0.010	0.002-0.006
6.		7100-8100	4600-5800	10,000-14,000	8100-10,100	14,000	175-10,000	10,000-11,000	4500-9000	24,200
7.	2-3	1.7-1.9	10-20	1.4-3.5	2.1-5.0	1	100-1000	3-6	60-550	3.0
8.	8100-11,000								7,800-11,000	
9.		11,400-14,200					20,000			24,600-25,700
10.	16,300-17,000		8500-10,500	16,300-22,000	14,100-16,000	20,700	700-4500	19,000	10,200-15,000	35,300
11.	750-880	440-490	280-330	850-900	440-520	2000	10-100		190-300	1100
12.					440-480		10-100			
13.	720-800	450-490	320-370	800-1050	450-460	1900	10-100	610	4-310	1100
14.	2.1-2.7	0.4-0.6	1.5-4.0	2.1-2.6	0.3-0.4	0.7	25 to flexible	0.4	1.5-1.8-No break	6.2-7.5
15.	R73		R75-95		M80-85				R >100; M48	
							Shore A10-13,D90	Barcol 30-35	Shore 75A-70D	
16.		65-67	67-79	20	40-72		100-200		0.5-0.8	
17.	229-245	226-249	230-260	231-247	205-210	220	Vanes over wide range	190-200	158-260	185
						230			115-275	
18.							5			
19.	1.21-1.22	1.07-1.10	1.05-1.08	1.20-1.22	1.08-1.13	1.14	1.03-1.5	1.05	1.12-1.24	1.43
20.	0.1	0.1	0.1	0.1	0.11-0.17	0.1	0.2-1.5	0.1-0.2	0.15-0.19	
									0.5-0.6	
21.							300-500		400	
	Bayer Corp.; DSM; NOVA Chemicals; RTP	NOVA Chemicals	Bayer Corp.; NOVA Chemicals	ComAlloy; DSM; LNP; M.A. Polymers; NOVA Chemicals; RTP	Network Polymers; NOVA Chemicals	DSM; LNP; RTP	Bayer Corp.; Cabot; Conap; Dow Plastics; Emerson & Cuming; ITW Devcon; Polyurethane Specialties; Union Carbide	Dow Plastics; Emerson & Cuming; Polyurethane Specialties	Bayer Corp.; BF Goodrich; Dow Plastics	RTP; Ticona

Materials	Properties	ASTM test method	Polyurethane (Cont'd) Thermoplastic (Cont'd) Long glass-reinforced molding compound	Polyurethane (Cont'd) Thermoplastic (Cont'd) EMI shielding (conductive); 30% PAN carbon fiber	Polyvinylidene chloride copolymers Injection molding	Polyvinylidene chloride copolymers Barrier film resins Un-plasticized	Polyvinylidene chloride copolymers Barrier film resins Plasticized	Silicone Casting resins Flexible (including RTV)	Silicone Liquid injection molding Liquid silicone rubber
Processing	1a. Melt flow (gm./10 min.)	D1238							
	1. Melting temperature, °C. T_m (crystalline)		75		172	160	172	Thermoset	Thermoset
	T_g (amorphous)				–15	0.2	-15		
	2. Processing temperature range, °F. (C = compression; T = transfer; I = injection; E = extrusion)		I: 450-500	I: 360-450	C: 260-350 I: 300-400 E: 300-400	E: 320-390	E: 340-400		I: 360-420
	3. Molding pressure range, 10^3 p.s.i.				5-30	3-30	5-30		1-2
	4. Compression ratio				2.5	2-2.5	2-2.5		
	5. Mold (linear) shrinkage, in./in.	D955	0.001	0.001-0.002	0.005-0.025	0.005-0.025	0.005-0.025	0.0-0.006	0.0-0.005
Mechanical	6. Tensile strength at break, p.s.i.	D638[b]	27,000-33,000	13,000	3500-5000	2800	3500	350-1000	725-1305
	7. Elongation at break, %	D638[b]	2	20	160-240	350-400	250-300	20-700	300-1000
	8. Tensile yield strength, p.s.i.	D638[b]	27,000-33,000		2800-3800		4900		
	9. Compressive strength (rupture or yield), p.s.i.	D695			2000-2700				
	10. Flexural strength (rupture or yield), p.s.i.	D790	45,000-57,000	9000	4200-6200				
	11. Tensile modulus, 10^3 p.s.i.	D638[b]	1700-2500	500	50-80	50-80	50-80		
	12. Compressive modulus, 10^3 p.s.i.	D695			55-95				
	13. Flexural modulus, 10^3 p.s.i. 73° F.	D790	1500-2200	500	55-95				
	200° F.	D790							
	250° F.	D790							
	300° F.	D790							
	14. Izod impact, ft.-lb./in. of notch (1/8-in. thick specimen)	D256A	8-16	10	0.4-1.0	0.3-1.0	0.3-1.0		
	15. Hardness Rockwell	D785			M60-65	R98-106	R98-106		
	Shore/Barcol	D2240/ D2583						Shore A10-70	Shore A20-70
Thermal	16. Coef. of linear thermal expansion, 10^{-6} in./in./°C.	D696	0.6-0.8		190	190	190	10-19	10-20
	17. Deflection temperature under flexural load, °F. 264 p.s.i.	D648	200-260	180	130-150	130-150	130-150		
	66 p.s.i.	D648	230-370						
	18. Thermal conductivity, 10^{-4} cal.-cm./sec.-cm.^{2}-°C.	C177			3	3	3	3.5-7.5	
Physical	19. Specific gravity	D792		1.33	1.65-1.72	1.65-1.70	1.68-1.72	0.97-2.5	1.08-1.14
	20. Water absorption (1/8-in. thick specimen), % 24 hr.	D570			0.1	0.1	0.1	0.1	
	Saturation	D570							
	21. Dielectric strength (1/8-in. thick specimen), short time, v./mil	D149			400-600		400-600	400-550	
		SUPPLIERS[a]	Dow Plastics; RTP	LNP; RTP	Dow Plastics	Dow Plastics	Dow Plastics	Bayer Corp.; Dow Corning; Emerson & Cuming; GE Silicones	Bayer Corp.

	Silicone (Cont'd)			Sulfone polymers						
	Molding and encapsulating compounds	Silicone/poly-amide pseudo-interpenetrating networks[f]		Polysulfone						
	Mineral-and/or glass-filled	Silicone/nylon 66	Silicone/nylon 12	Injection molding, flame-retarded, extrusion	Mineral-filled	Extrusion/injection molding grade	20% glass fiber-reinforced	EMI shielding (conductive); 30% carbon fiber	Injection molding platable grade	Polyaryl-sulfone
1a.				3.5-9	7-8.5	3-20			12-18	10-30
1.	Thermoset	Interpenetrating network	Interpenetrating network							
				187-190	190	190	190	190		220
2.	C: 280-360 I: 330-370 T: 330-370	I: 460-525	I: 360-410	I: 625-750 E: 600-700	I: 675-775	I: 610-735 E: 600-680	I: 660-715	I: 550-700	I: 690-750	I: 630-800 E: 620-750
3.	0.3-6	8-15	7-12	5-20	10-20			10-20	5-100	5-20
4.	2.0-8.0			2.5-3.5	2.5-3.5			2.5-3.5	2	
5.	0.0-0.005	0.004-0.007	0.005-0.007	0.0058-0.007	0.004-0.005	0.005-0.007	2.8×10^{-3}	0.0005-0.001	0.0025	0.007-0.008
6.	500-1500	10,100-12,500	5200-7400		9500-9800		16,700	23,000-23,500	1100	9000
7.	80-800	5-20	10	50-100	2-5	40-100	2.4	1.5-2	2.3	40-60
8.				10,200-11,600		11,500				10,400-12,000
9.				13,900-40,000				25,000		
10.		14,000-15,900	8000-8300	15,400-17,500	14,300-15,400	17,500	21,000	31,000-35,000	7700	12,400-16,100
11.				360-390	550-650	390	1020	2150-2800		310-385
12.				374						
13.		360-410	200-216	390	600-750	370	850	1900-2300	881	330-420
				370	570-710					
				350	550-690					
				310	510-650					
14.		0.8-0.9	0.6-0.7	1.0-1.3	0.65-1.0	1.0-1.2	1.5	1.2-1.8		1.6-12
15.				M69	M70-74	M69	M83	M80		
	Shore A10-80									
16.	20-50			56	34-39	31	25	6		31-49
17.	>500			345	345-354	340	363	360-365	410	400
				358		360	369	365-380		
18.	7.18			6.2						
19.	1.80-2.05	1.12-1.13	1.01-1.02	1.24-1.25	1.48-1.61	1.24	1.40	1.36-1.7	1.68	1.29-1.37
20.	0.15	0.6-0.8	0.12-0.15	0.3		0.3		0.15-0.25		
	0.15-0.40			0.8	0.5-0.6	0.8	0.6			1.1-1.85
21.	200-550			425	450				380	370-380
	Cytec Fiberite; Dow Corning; GE Silicones	LNP	LNP	Amoco Polymers; BASF	Amoco Polymers; RTP	Amoco Polymers; BASF	Amoco Polymers; BASF; ComAlloy; RTP	DSM; Ferro; LNP; RTP	Amoco Polymers	LNP; RTP

Materials			Sulfone polymers						
			Polyethersulfone				Modified polysulfone		
	Properties	ASTM test method	Unfilled	10% glass fiber-reinforced	20% glass fiber-reinforced	EMI shielding (conductive); 30% carbon fiber	Poly-phenyl sulfone	Poly-phenyl sulfone, unrein-forced	Mineral-filled
Processing	1a. Melt flow (gm./10 min.)	D1238		12	10		14-20	11.5-17.0	
	1. Melting temperature, °C. T_m (crystalline)								
	T_g (amorphous)		220-230	225	220-225	225	225	220	
	2. Processing temperature range, °F. (C = compression; T = transfer; I = injection; E = extrusion)		C: 645-715 I: 590-750 E: 625-720	I: 660-715	C: 610-750 I: 630-735 E: 570-650	I: 600-750	I: 680-735	680-735	I: 575-650
	3. Molding pressure range, 10^3 p.s.i.		6-20	50-100	6-100	10-20	10-20		5-20
	4. Compression ratio		2-2.5	2:1	2.0-3.5	2.5-3.5	2.2		2.5-3.5
	5. Mold (linear) shrinkage, in./in.	D955	0.006-0.007	0.5	0.002-0.005	0.0005-0.002	0.007	0.007	0.006-0.007
Mechanical	6. Tensile strength at break, p.s.i.	D638[b]	9800-13,800	16,000	15,200-20,000	26,000-30,000	10,100	10,100	
	7. Elongation at break, %	D638[b]	6-80	4.1	2-3.5	1.3-2.5	60-120	60-120	50-100
	8. Tensile yield strength, p.s.i.	D638[b]	12,200-13,000		18,000-18,800				10,500
	9. Compressive strength (rupture or yield), p.s.i.	D695	11,800-15,600		19,500-24,000	22,000			
	10. Flexural strength (rupture or yield), p.s.i.	D790	17,000-18,700	21,000; 24,500	23,500-27,600	36,000-38,000	13,200	13,200	16,500
	11. Tensile modulus, 10^3 p.s.i.	D638[b]	350-410	555; 740	825-1130	2120-2900	340	340	400
	12. Compressive modulus, 10^3 p.s.i.	D695							
	13. Flexural modulus, 10^3 p.s.i. 73° F.	D790	348-380	590; 650	750-980	2000-2600	350	350	480
	200° F.	D790			812-850				370
	250° F.	D790	330		840				340
	300° F.	D790	280		580-842				320
	14. Izod impact, ft.-lb./in. of notch (1/8-in. thick specimen)	D256A	1.4-No break	0.9; 1.3	1.1-1.7	1.2-1.6	13	13.0	1.1
	15. Hardness Rockwell	D785	M85-88	M94	M96-99	R123			M74
	Shore/Barcol	D2240/ D2583							
Thermal	16. Coef. of linear thermal expansion, 10^{-6} in./in./°C.	D696	55	34	23-32	10	17	5.6	53
	17. Deflection temperature under flexural load, °F. 264 p.s.i.	D648	383-397	437	408-426; 437	415-420	405	405	325
	66 p.s.i.	D648	410	423	410-430	420-430			
	18. Thermal conductivity, 10^{-4} cal.-cm./sec.-cm.^{2}-°C.	C177	3.2-4.4						
Physical	19. Specific gravity	D792	1.37-1.46	1.43; 1.45	1.51-1.53	1.47-1.48	1.29-1.3		1.30
	20. Water absorption (1/8-in. thick specimen), % 24 hr.	D570	0.12-1.7		0.15-0.4	0.29-0.35	0.37	0.37	
	Saturation	D570	1.8-2.5	1.9	1.65-2.1; 1.7		1.1		0.8
	21. Dielectric strength (1/8-in. thick specimen), short time, v./mil	D149	400	440	375-500		360	380	460
		SUPPLIERS[a]	Amoco Polymers; BASF	Amoco Polymers; BASF; RTP	Amoco Polymers; BASF; DSM; LNP; RTP	DSM; LNP; RTP	Amoco Polymers	Amoco Polymers	RTP

Thermoplastic elastomers

	Polyolefin							Silicone-based, pseudo-interpenetrating networks[f]	
	Low and medium hardness	High hardness	Copolyester ether	Polyester	Polyether/amide block copolymers	Block copolymers of styrene and butadiene or styrene and isoprene	Block copolymers of styrene and ethylene and/or butylene	Silicone/polyamide	Silicone/polyester
1a.	0.4-20.0	0.4-10.0	0.5-20 @190-240	3-12 @190-240		0.5-20			
1.			150-223	148-230	148-209			Interpenetrating network	Interpenetrating network
	165	163-165	–3						
2.	C: 350-450 I: 360-475 E: 380-450	C: 370-450 I: 350-480 E: 380-475	E: 400-500 I: 435-500	I: 340-500 E: 340-500	C: 365-480 I: 340-482 E: 340-460	C: 250-325 I: 300-425 E: 370-400	C: 300-380 I: 350-480 E: 330-380	I: 360-410 I: 355-580 E: 320-365	I: 360-450 E: 380-450
3.	4-19	6-10	1-15	1-15	8-12	0.3-3	1.5-20	5-25	6-14
4.	2-3.5	2-3.5	2-4	3-3.5		2.0-4.0	2.5-5.0	3-5	2.5-4
5.	0.015-0.020	0.007-0.021	0.004-0.02	0.003-0.018		0.001-0.022	0.003-0.022	0.004-0.007	0.004-0.015
6.	650-2500	950-4000	1000-10,000	1000-6800	2000-7000	100-4350	600-3000	5200-12,500	5000
7.	150-780	20-600	200-1000	170-900	350-680	20-1350	600-940	5-275	950-1050
8.		1600-4000	1370-1750	1350-3900	3000-3500	3700-4400			
9.		22-39							
10.					1100-1900	5100-6400	0.1-100	2200-15,900	1700
11.	1.1-16.4	0.34-34.1	16-18	1.1-130	2-60	0.8-235			18
12.				7.5-48		3.6-120			
13.	1..5-30	2.7-300	18	5-175	2.9-66	4-215	0.1-100	35-410	25
14.	No break	5.0-16.0 No break	NB (–30C)	2.5-No break	4.3-No break	No break	No break	No break; 0.6-0.9	No break
15.				104					
	Shore A64-92	Shore D40-70	55/95 (D/A); 30-85	Shore D35-72	Shore A75-D72	Shore A20-95 D63-75	Shore A5-95	Shore D60	Shore D49-51
16.		36-110	150	85-190	210-230	67-140			
17.					45-135	<0-170			
				111-284	158-257	<0-190			
18.	4.5-5.0		5-6	3.6-4.5		3.6			
19.	0.88-0.98	0.90-1.15	1-1.3	1.10-1.28	1.0-1.03	0.90-1.2	0.9-1.2	1.01-1.21	1.21-1.16
20.	0.01	0-0.28	0.4-50	0.2-3.6	1.01-1.36	0.009-0.39	0.1-0.42	0.5-0.15	0.3-0.8
21.	410-445	390-465	350	350-600		300-520	450-800		
	AlphaGary; DuPont; Exxon; M.A. Polymers; Montell NA; Network Polymers; A. Schulman; Solutia; Solvay Engineered Polymers; Teknor Apex; Union Carbide; Vi-Chem	AlphaGary; Equistar; M.A. Polymers; Montell NA; Network Polymers; A. Schulman; Solutia; Solvay Engineered Polymers; Teknor Apex; Vi-Chem; Wash. Penn	DuPont; Eastman	DuPont; Ticona	Creanova; Elf Atochem N.A.; EMS	Fina; Firestone Synthetic Rubber; Network Polymers; A. Schulman; Shell	AlphaGary; Dow Plastics; Network Polymers; Shell; Teknor Apex	Creanova; LNP	Creanova; LNP

Materials	Properties	ASTM test method	Thermoplastic elastomers (Cont'd): Silicone-based, pseudo-interpenetrating networks[f] (Cont'd): Silicone/polyolefin	Silicone/polystyrene ethylene butadiene-styrene	Silicone/polyurethane: Aromatic and aliphatic polyether and polyester urethane, unfilled	Silicone/polyurethane: Aromatic polyether urethane, 15% carbon fiber-reinforced	Polyurethane: Molding and extrusion compounds: Polyester: Low and medium hardness
Processing	1a. Melt flow (gm./10 min.)	D1238					
	1. Melting temperature, °C. T_m (crystalline)		Interpenetrating network	Interpenetrating network	Interpenetrating network	Interpenetrating network	
	T_g (amorphous)				150-225		120-160
	2. Processing temperature range, °F. (C = compression; T = transfer; I = injection; E = extrusion)		I: 340-425	I: 390-490 E: 325-460	I: 325-450 E: 325-475	I: 410-475	I: 340-435 E: 340-410
	3. Molding pressure range, 10^3 p.s.i.		6-11	6-20	5-15	6-15	0.8-1.4
	4. Compression ratio		2.5-3.5	2-3.5	2.5-3.5	2.5-3.5	
	5. Mold (linear) shrinkage, in./in.	D955	0.010-0.020	0.005-0.040	0.010-0.020	0.011-0.014	0.008-0.015
Mechanical	6. Tensile strength at break, p.s.i.	D638[b]	475-1000	1000-3300	2100-6500	18,500	3300-8400
	7. Elongation at break, %	D638[b]	120-1000	500-1000	400-1300	5	410-620
	8. Tensile yield strength, p.s.i.	D638[b]					
	9. Compressive strength (rupture or yield), p.s.i.	D695					
	10. Flexural strength (rupture or yield), p.s.i.	D790			7000-8000	18,000	
	11. Tensile modulus, 10^3 p.s.i.	D638[b]			9-19		
	12. Compressive modulus, 10^3 p.s.i.	D695					
	13. Flexural modulus, 10^3 p.s.i. 73° F.	D790			2-27	800	4000-6990
	200° F.	D790					
	250° F.	D790					
	300° F.	D790					
	14. Izod impact, ft.-lb./in. of notch (1/8-in. thick specimen)	D256A	No break	No break	No break	1.6	
	15. Hardness Rockwell	D785					
	Shore/Barcol	D2240/ D2583	Shore A57-65	Shore A50-84	Shore A55-87; D55-60	Shore D70	Shore A55-94
Thermal	16. Coef. of linear thermal expansion, 10^{-6} in./in./°C.	D696					
	17. Deflection temperature under flexural load, °F. 264 p.s.i.	D648					
	66 p.s.i.	D648					
	18. Thermal conductivity, 10^{-4} cal.-cm./sec.-cm.^{2}-°C.	C177					
Physical	19. Specific gravity	D792	0.96-0.97	0.90-0.97	1.04-1.19	1.24	1.17-1.25
	20. Water absorption (1/8-in. thick specimen), % 24 hr.	D570	0.01	0.05-0.3	0.3-0.6	0.4	
	Saturation	D570					
	21. Dielectric strength (1/8-in. thick specimen), short time, v./mil	D149					
		SUPPLIERS[a]	LNP	Creanova; LNP	Creanova	LNP	BASF; Bayer Corp.; Dow Plastics; BFGoodrich; A. Schulman

	Thermoplastic elastomers (Cont'd)					Vinyl polymers and copolymers			
	Polyurethane (Cont'd)			Elastomeric alloys		PVC and PVC-acetate MC, sheets, rods, and tubes			
	Molding and extrusion compounds (Cont'd)								
	Polyester (Cont'd)	Polyether							
	High hardness	Low and medium hardness	High hardness	Low and medium hardness	High hardness	Rigid	Rigid, tin-stabilized	Flexible, unfilled	Flexible, filled
1a.		Thermoset							
1.				165	165				
	120-160	120-160	120-160			75-105		75-105	75-105
2.	I: 400-440 E: 370-410	I: 350-430 E: 340-410 C: 72-120	I: 400-435 E: 380-440	I: 350-450	I: 350-450	C: 285-400 I: 300-415	I: 380-400	C: 285-350 I: 320-385	C: 285-350 I: 320-385
3.	0.8-1.4	0.6-1.2	1-1.4	6-10	6-10	10-40		8-25	1-2
4.				2.5-3.5	2.5-3.5	2.0-2.3	2.0	2.0-2.3	2.0-2.3
5.	0.005-0.015	0.008-0.015	0.008-0.012	1.5-2.5	1.5-2.5	0.002-0.006		0.010-0.050	0.008-0.035 0.002-0.008
6.	4000-11,000	158-6750	6000-7240	650-3200	2300-4000	5900-7500		1500-3500	1000-3500
7.	110-550	475-1000	340-600	300-750	250-600	40-80		200-450	200-400
8.					1600	5900-6500	5140-7560		
9.						8000-13,000		900-1700	1000-1800
10.						10,000-16,000	6300-13,600		
11.				0.7-5.0	16-100	350-600	310-435		
12.									
13.	9150-64,000	2480-6190	4000-44,900	1.5-6.6	15-50	300-500	305-420		
14.				No fracture	No fracture	0.4-22	2.1-20.0	Varies over wide range	Varies over wide range
15.									
	Shore D46-78	Shore A13-92	Shore D55-75	Shore 55-95A	Shore 32-72D	Shore D65-85	Shore D69-78	Shore A50-100	Shore A50-100
16.				82	82	50-100	1.0-18.1	70-250	
17.						140-170	69-163		
						135-180	161-166		
18.						3.5-5.0		3-4	3-4
19.	1.15-1.28	1.02	1.14-1.21	0.90-1.39	0.90-1.31	1.30-1.58	1.32-1.45	1.16-1.35	1.3-1.7
20.	0.3					0.04-0.4		0.15-0.75	0.5-1.0
21.		470	470	400-600	400-600	350-500		300-400	250-300
	BASF; Bayer Corp.; Dow Plastics; BFGoodrich; A. Schulman	BASF; Bayer Corp.; Cabot; Dow Plastics; BFGoodrich; A. Schulman	BASF; Bayer Corp.; Dow Plastics; BFGoodrich; A. Schulman	AlphaGary; Goodyear; M.A. Hanna; Montell NA; Network Polymers; Novatec; A. Schulman; Solutia; Vi-Chem	AlphaGary; M.A. Hanna; Montell NA; Network Polymers; A. Schulman; Solutia	AlphaGary; Colorite; CONDEA Vista; Creanova; Formosa; Georgia Gulf; Keysor-Century; LG Chemical; Novatec; OxyChem; Rimtec; RSG Polymers; Shintech; Synergistic; Union Carbide; Vi-Chem	AlphaGary; Colorite; CONDEA Vista; Georgia Gulf; Keysor-Century; Oxychem; Rimtec; Shintech; Synergistic	AlphaGary; Colorite; CONDEA Vista; Creanova; Keysor-Century; LG Chemical; Novatec; Rimtec; A. Schulman; Shintech; Shuman; Synergistic; Teknor Apex; Union Carbide; Vi-Chem	AlphaGary; Colorite; CONDEA Vista; Creanova; Keysor-Century; Novatec; Rimtec; Shintech; Synergistic; Teknor Apex; Union Carbide; Vi-Chem

Materials	Properties	ASTM test method	Vinyl polymers and copolymers (Cont'd) — Molding and extrusion compounds: Chlorinated polyvinyl chloride	Vinyl butyral, flexible	PVC/acrylic blends
Processing	1a. Melt flow (gm./10 min.)	D1238			
	1. Melting temperature, °C. T_m (crystalline)				
	T_g (amorphous)		110-127	49	
	2. Processing temperature range, °F. (C = compression; T = transfer; I = injection; E = extrusion)		C: 350-400 I: 395-440 E: 360-430	C: 280-320 I: 250-340	I: 360-390 E: 390-410
	3. Molding pressure range, 10^3 p.s.i.		15-40	0.5-3	2-3
	4. Compression ratio		1.5-2.8		2-2.5
	5. Mold (linear) shrinkage, in./in.	D955	0.003-0.007		0.003
Mechanical	6. Tensile strength at break, p.s.i.	D638[b]	6500-9000	500-3000	6400-7000
	7. Elongation at break, %	D638[b]	4-100	150-450	35-100
	8. Tensile yield strength, p.s.i.	D638[b]	6000-8200		
	9. Compressive strength (rupture or yield), p.s.i.	D695	9000-22,000		6800-8500
	10. Flexural strength (rupture or yield), p.s.i.	D790	11,500-17,000		10,300-11,000
	11. Tensile modulus, 10^3 p.s.i.	D638[b]	326-475		340-370
	12. Compressive modulus, 10^3 p.s.i.	D695	212-600		
	13. Flexural modulus, 10^3 p.s.i. 73° F.	D790	334-450		350-380
	200° F.	D790			
	250° F.	D790			
	300° F.	D790			
	14. Izod impact, ft.-lb./in. of notch (1/8-in. thick specimen)	D256A	1.0-10.0	Varies over wide range	1-12
	15. Hardness Rockwell	D785	R109-117	A10-100	R106-110
	Shore/Barcol	D2240/ D2583	82-88		
Thermal	16. Coef. of linear thermal expansion, 10^{-6} in./in./°C.	D696	52-78		44-79
	17. Deflection temperature under flexural load, °F. 264 p.s.i.	D648	194-234		167-185
	66 p.s.i.	D648	215-247		172-189
	18. Thermal conductivity, 10^{-4} cal.-cm./sec.-cm.^{2}-°C.	C177	3.3		
Physical	19. Specific gravity	D792	1.39-1.58	1.05	1.26-1.35
	20. Water absorption (1/8-in. thick specimen), % 24 hr.	D570	0.02-0.16	1.0-2.0	0.09-.016
	Saturation	D570			
	21. Dielectric strength (1/8-in. thick specimen), short time, v./mil	D149	600-625	350	480
		SUPPLIERS[a]	BF Goodrich; Elf Atochem N.A.; Georgia Gulf	Solutia; Union Carbide	AlphaGary

Notes for Appendix C

a. See list below for addresses of suppliers.

b. Tensile test method varies with material; D638 is standard for thermoplastics; D651 for rigid thermosetting plastics; D412 for elastomeric plastics; D882 for thin plastics sheeting.

c. Dry, as molded (approximately 0.2 percent moisture content).

d. As conditioned to equilibrium with 50 percent relative humidity.

e. Test method in ASTM D4092.

f. *Pseudo* indicates that the thermosetting and thermoplastic components were in the form of pellets or powder prior to fabrication.

g. Dow Plastics samples are unannealed.

Names and Addresses of Suppliers Listed in Appendix C

Adell Plastics, Inc.
4530 Annapolis Rd.
Baltimore, MD 21227
800-638-5218, 410-789-7780
Fax: 410-789-2804

Ain Plastics, Inc.
249 E. Sandford Blvd.
P.O. Box 151-M
Mt. Vernon, NY 10550
800-431-2451, 914-668-6800
Fax: 914-668-8820

Albis Corp.
445 Hwy. 36 N.
RO. Box 711ß
Rosenberg, TX 77471
800-231-5911, 713-342-3311
Fax: 713-342-3058
Telex: 166-181

Alliedsignal Inc.
Alliedsignal Engineered Plastics
P.O. Box 2332, 101 Columbia Rd.
Morristown, NJ 07962-2332
201-455-5010
Fax: 201-455-3507

ALM Corp.
55 Haul Rd.
Wayne, NJ 07470
201-694-4141
Fax: 201-831-8327

Alpha/Owens-Corning
P.O. Box 610
Collierville, TN 38027-0610
901-854-2800
Fax: 901-854-1183

AlphaGary Corp.
170 Pioneer Dr.
P.O. Box 808
Leominster, MA 01453
800-232-9741, 508-537-8071
Fax: 508-534-3021

American Polymers
P.O. Box 366
53 Milbrook St.
Worcester, MA 01606
508-756-1010
Fax: 508-756-3611

Ametek, Inc.
Haveg Div.
900 Greenbank Rd.
Wilmington, DE 19808
302-995-0400
Fax: 302-995-0491

Amoco Chemical Co.
200 E. Randolph Dr.
Mail Code 7802
Chicago, IL 60601-7125
800-621-4590, 312-856-3200
Fax: 312-856-4151

Applied Composites Corp.
333 N. Sixth St.
St. Charles, IL 60174
708-584-3130
Fax: 708-584-0659

Applied Polymer Systems, Inc.
P.O. Box 56404
Flushing, NY 11356-4040
718-539-4425
Fax: 718-460-4159

Arco Chemical Co.
3801 West Chester Pike
Newtown Square, PA 19073
800-345-0252 (PA Only),
215-359-2000

Aristech Chemical Corp.
Acrylic Sheet Unit
7350 Empire Dr.
Florence, KY 41042
800-354-9858
Fax: 606-283-6492

Ashley Polymers
5114 Ft. Hamilton Pkwy.
Brooklyn, NY 11219
718-851-8111
Fax: 718-972-3256
Telex: 42-7884

AtoHaas North America Inc.
100 Independence Mall W
Philadelphia, PA 19106
215-592-3000
Fax: 215-592-2445

Ausimont USA, Inc.
Crown Point Rd. & Leonards Lane
P.O. Box 26
Thorofare, NJ 08086
800-323-2874, 609-853-8119
Fax: 609-853-6405

Bamberger, Claude P., Molding
Compounds Corp.
111 Paterson Plank Rd.
P.O. Box 67
Carlstadt, NJ 07072
201-933-6262
Fax: 201-933-8129

Bamberger Polymers, Inc.
1983 Marcus Ave.
Lake Success, NY 11042
800-888-8959, 516-328-2772
Fax: 516-326-1005
Telex: 6711357

BASF Corp., Plastic Materials
3000 Continental Dr. N.
Mount Olive, NJ 07828-1234
201-426-2600

Bayer Corp.
100 Bayer Rd.
Pittsburgh, PA 15205-9741
800-662-2927, 412-777-2000

BFGoodrich Adhesive Systems Div.
123 W Bartges St.
Akron, OH 44311-1081
216-374-2900
Fax: 216-374-2860

BFGoodrich Specialty Chemicals
9911 Breckville Rd.
Cleveland, OH 44141-3247
800-331-1144, 216-447-5000
Fax: 216-447-5750

Boonton Plastic Molding Co.
30 Plain St.
Boonton, NJ 07005-0030
201-334-4400
Fax: 201-335-0620

Borden Packaging Div.
Borden Inc.
One Clark St.
North Andover, MA 01845
508-686-9591

BP Chemicals (Hitco) Fibers and Materials
700 E. Dyer Rd.
Santa Ana, CA 92705
714-549-1101

BP Chemicals, Inc.
4440 Warrensville Center Rd.
Cleveland, OH 44128
800-272-4367, 216-586-5847
Fax: 216-586-5839

BP Performance Polymers Inc.
Phenolic Business
60 Walnut Ave.
Suite 100
Clark, NJ 07066
908-815-7843
Fax: 908-815-7844

Budd Chemical Co.
Pennsville-Auburn Rd.
Carneys Point, NJ 08069
609-299-1708
Fax: 609-299-2998

Budd Co.
Plastics Div.
32055 Edward Ave.
Madison Heights, MI 48071
810-588-3200
Fax: 810-588-0798

Bulk Molding Compounds Inc.
3N497 N. 17th St.
St. Charles, IL 60174
708-377-1065
Fax: 708-377-7395

Cadillac Plastic & Chemical Co.
143 Indusco Ct.
P.O. Box 7035
Troy, MI 48007-7035
800-488-1200, 810-583-1200
Fax: 810-583-4715

Cast Nylons Ltd.
4300 Hamann Pkwy
Willoughby, OH 44092
800-543-3619, 216-269-2300
Fax: 216-269-2323

Chevron Chemical Co.
Olefin & Derivatives
P.O. Box 3766
Houston, TX 77253
800-231-3828, 713-754-2000

Ciba-Geigy Corp., Ciba Additives
540 White Plains Rd.
P.O. Box 20005
Tarrytown, NY 10591-9005
800-431-2360, 914-785-2000
Fax: 914-785-4244

Color-Art Plastics, Inc.
317 Cortlandt St.
Belleville, NJ 07109-3293
201-759-2400

CornAlloy International Co.
481 Allied Dr.
Nashville, TN 37211
615-333-3453
Fax: 615-834-9941

Conap, Inc.
1405 Buffalo St.
Olean, NY 14760
716-372-9650
Fax: 716-372-1594

Consolidated Polymer Technologies, Inc.
11811 62nd St. N.
Largo, FL 34643
800-541-6880, 813-531-4191
Fax: 813-530-5603

Cook Composites & Polymers
P.O. Box 419389
Kansas City, MO 64141-6389
800-821-3590, 816-391-6000
Fax: 816-391-6215

Cosmic Plastics, Inc.
27939 Beale Ct.
Valencia, CA 91355
800-423-5613, 805-257-3274
Fax: 805-257-3345

Custom Manufacturers
858 S. M-18
Gladwin, MI 48624
800-860-4594, 517-426-4591
Fax: 517-426-4049

Custom Molders Corp.
2470 Plainfield Ave.
Scotch Plains, NJ 07076
908-233-5880
Fax: 908-233-5949

Custom Plastic Injection
Molding Co. Inc.
3 Spielman Rd.
Fairfield, NJ 07004
201-227-1155

Cyro Industries
100 Enterprise Dr., 7th fi.
Rockaway, NJ 07866
201-442-6000

CYTEC Industries Inc.
5 Garret Mt. Plaza
West Paterson, NJ 07424
800-438-5615, 201-357-3100
Fax: 201-357-3065

CYTEC Industries Inc.
12600 Eckel Rd.
P.O. Box 148
Perrysburg, OH 43551
800-537-3360, 419-874-7941
Fax: 419-874-0951

Diamond Polymers
1353 Exeter Rd.
Akron, OH 44306
216-773-2700
Fax: 216-773-2799

Dow Chemical Co.
Polyurethanes
2040 Willard H. Dow Center
Midland, MI 48674
800-441-4369

Dow Corning Corp.
P.O. Box 0994
Midland, MI 48686-0994
517-496-4000
Fax: 517-496-4586

Dow Corning STI
47799 Halyard Dr.
Suite 99
Plymouth, MI 48170
313-459-7792
Fax: 313-459-0204

Dow Plastics
P.O. Box 1206
Midland, MI 48641-1206
800-441-4369

DSM Copolymer, Inc.
P.O. Box 2591
Baton Rouge, LA 70821
800-535-9960, 504-355-5655
Fax: 504-357-9574

DSM Engineering Plastics
2267 W Mill Rd.
P.O. Box 3333
Evansville, IN 47732
800-333-4237, 812-435-7500
Fax: 812-435-7702

DSM Thermoplastic Elastomers, Inc.
29 Fuller St.
Leominster, MA 01453-4451
800-524-0120, 508-534-1010
Fax: 508-534-1005

DuPont Engineering Polymers
1007 Market St.
Wilmington, DE 19898
800-441-7515, 302-999-4592
Fax: 302-999-4358

Eagle-Picher Industries, Inc.
C & Porter Sts.
Joplin, MO 64802
417-623-8000
Fax: 417-782-1923

Eastman Chemical Co.
P.O. Box 511
Kingsport, TN 37662
800-327-8626

Elf Atochem North America, Inc.
Fluoropolymers
2000 Market St.
Philadelphia, PA 19103
800-225-7788, 215-419-7000
Fax: 215-419-7497

Elf Atochem North America, Inc.,
Organic Peroxides
2000 Market St.
Philadelphia, PA 19103
800-558-5575, 215-419-7000
Fax: 215-419-7591

Emerson & Cuming, Inc./Grace
Specialty Polymers
77 Dragon Ct.
Woburn, MA 01888
800-832-4929, 617-938-8630
Fax: 617-935-0125

Epic, Inc.
654 Madison Ave.
New York, NY 10021
212-308-7039
Fax: 212-308-7266

Epsilon Products Co.
Post Rd. and Blueball Ave.
P.O. Box 432
Marcus Hook, PA 19061
610-497-8850
Fax: 610-497-4694

EVAL Co. of America
1001 Warrenville Rd.
Suite 201
Lisle, IL 60532-1359
800-423-9762, 708-719-4610
Fax: 708-719-4622

Exxon Chemical Co.
13501 Katy Freeway
Houston, TX 77079-1398
800-231-6633, 713-870-6000
Fax: 713-870-6970

Federal Plastics Corp.
715 South Ave. E.
Cranford, NJ 07016
800-541-4424, 908-272-5800
Fax: 908-272-9021

Ferro Corp.
Filled & Reinforced Plastics Div.
5001 O'Hara Dr.
Evansville, IN 47711
812-423-5218
Fax: 812-435-2113

Ferro Corp., World Headquarters
1000 Lakeside Ave.
P.O. Box 147000
Cleveland, OH 44114-7000
216-641-8580
Fax: 216-696-6958
Telex: 98-0165

Fina Oil & Chemical Co.
Chemical Div.
P.O. Box 2159
Dallas, TX 75221
800-344-3462, 214-750-2806
Fax: 214-821-1433

Firestone Canada
P.O. Box 486
Woodstock, Ontario, Canada N45 7Y9
800-999-6231, 519-421-5649 Fax: 519-537-6235

Firestone Synthetic Rubber & Latex Co.
P.O. Box 26611
Akron, OH 44319-0006
800-282-0222
Fax: 216-379-7875
Telex: 67-16415

Formosa Plastics Corp. USA
9 Peach Tree Hill Rd.
Livingston, NJ 07039
201-992-2090
Fax: 201-716-7208

Franklin Polymers, Inc.
P.O. Box 481
521 Yale Ave.
Pitman, NJ 08071-0481
800-238-7659, 609-582-6115
Fax: 609-582-0525

FRP Supply Div.
Ashland Chemical Co.
P.O. Box 2219
Columbus, OH 43216
614-790-4272
Fax: 614-790-4012
Telex: 24-5385 ASHCHEM

GE Plastics
One Plastics Ave.
Pittsfield, MA 01201
800-845-0600, 413-448-7110

GE Silicones
260 Hudson River Rd.
Waterford, NY 12188
800-255-8886
Fax: 518-233-3931

General Polymers Div., Ashland Chemical Co.
P.O. Box 2219
Columbus, OH 43216
800-828-7659
Fax: 614-889-3195

George, P. D., Co.
5200 N. Second St.
St. Louis, MO 63147
314-621-5700

Georgia Gulf Corp.
PVC Div.
P.O. Box 629
Plaquemine, LA 70765-0629
504-685-1200

Glastic Corp.
4321 Glenridge Rd.
Cleveland, OH 44121
216-486-0100
Fax: 216-486-1091

GLS Corp.
Thermoplastic Elastomers Div.
740B Industrial Dr.
Cary, IL 60013
800-457-8777 (not IL), 708-516-8300
Fax: 708-516-8361

Goodyear Tire & Rubber Co.
Chemical Div.
1485 E. Archwood Ave.
Akron, OH 44316-0001
800-522-7659, 216-796-6253
Fax: 216-796-2617

Grace Specialty Polymers
77 Dragon Ct.
Woburn, MA 01888
617-938-8630

Haysite Reinforced Plastics
5599 New Perry Hwy
Erie, PA 16509
814-868-3691
Fax: 814-864-7803

Hercules Inc.
Hercules Plaza
Wilmington, DE 19894
800-235-0543, 302-594-5000
Fax: 412-384-4291

Hercules Moulded Products
R.R. 3
Maidstone, Ontario, Canada NOR 1K0
519-737-6693
Fax: 519-737-1747

Hoechst Celanese Corp.
Advanced Materials Group
90 Morris Ave.
Summit, NJ 07901
800-526-4960, 908-598-4000
Fax: 908-598-4330
Telex: 13-6346

Hoechst Celanese Corp.
Hostalen GUR Business Unit
2520 S. Shore Blvd.
Suite 110
League City, TX 77573
713-334-8500

Huls America Inc.
80 Centennial Ave.
Piscataway, NJ 08855-0456
908-980-6800
Fax: 908-980-6970

Huntsman Chemical Corp.
2000 Eagle Gate Tower
Salt Lake City, UT 84111
800-421-2411, 801-536-1500
Fax: 801-536-1581

Huntsman Corp.
P.O. Box 27707
Houston, TX 77227-7707
713-961-3711
Fax: 713-235-6437
Telex: 227031 TEX UR

Hysol Engineering Adhesives
Dexter Distrib. Programs
One Dexter Dr.
Seabrook, NH 03874-4018
800-767-8786, 603-474-5541
Fax: 603-474-5545

ICI Acrylics Canada Inc.
7521 Tranmere Dr.
Mississauga, Ontario, Canada LSS 1L4
800-387-4880, 905-673-3345 Fax: 905-673-1459

ICI Acrylics Inc.
10091 Manchester Rd.
St. Louis, MO 63122
800-325-9577, 314-966-3111
Fax: 314-966-3117

ICI Americas Inc.
Tatnall Bldg.
P.O. Box 15391
3411 Silverside Rd.
Wilmington, DE 19850-5391
800-822-8215, 302-887-5536
Fax: 302-887-2089

ICI Fiberite
Molding Materials
501 W. Third St.
Winona, MN 55987-5468
507-454-3611
Fax: 507-452-8195
Telex: 507-454-3646

ICI Polyurethanes Group
286 Mantua Grove Rd.
West Deptford, NJ 08066-1732
800-257-5547, 609-423-8300
Fax: 609-423-8580

ITW Adhesives
37722 Enterprise Ct.
Farmington Hills, MI 48331
800-323-0451, 313-489-9344
Fax: 313-489-1545

Jet Composites Inc.
405 Fairall St.
Ajax, Ontario, Canada L1S 1R8
416-686-1707
Fax: 416-427-9403

Jet Plastics
941 N. Eastern Ave.
Los Angeles, CA 90063
213-268-6706
Fax: 213-268-8262

Keysor-Century Corp.
26000 Springbrook Rd.
P.O. Box 924
Santa Clarita, CA 91380-9024
805-259-2360
Fax: 805-259-7937

Kleerdex Co.
100 Gaither Dr.
Suite B
Mt. Laurel, NJ 08054
800-541-7232, 609-866-1700
Fax: 609-866-9728

Laird Plastics
1400 Centrepark
Suite 500
West Palm Beach, FL 33401
800-610-1016, 407-684-7000
Fax: 407-684-7088

LNP Engineering Plastics Inc.
475 Creamery Way
Exton, PA 19341
800-854-8774, 610-363-4500
Fax: 610-363-4749

Lockport Thermosets Inc.
157 Front St.
Lockport, LA 70374
800-259-8662, 504-532-2541
Fax: 504-532-6806

M.A. Industries Inc.
Polymer Div.
303 Dividend Dr.
P.O. Box 2322
Peachtree City, GA 30269
800-241-8250, 404-487-7761

Mitsui Petrochemical Industries Ltd.
3-2-5 Kasumigasekl, Chiyoda-ku
Tokyo, Japan 100
03-3593-1630
Fax: 03-3593-0979

Mitsui Plastics, Inc.
11 Martine Ave.
White Plains, NY 10606
914-287-6800
Fax: 914-287-6850

Mitsui Toatsu Chemicals, Inc.
2500 Westchester Ave.
Suite 110
Purchase, NY 10577
914-253-0777
Fax: 914-253-0790

Mobil Chemical Co.
1150 Pittsford-Victor Rd.
Pittsford, NY 14534
716-248-1193
Fax: 716-248-1075

Mobil Polymers
P.O. Box 5445
800 Connecticut Ave.
Norwalk, CT 06856
203-854-3808
Fax: 203-854-3840

Monmouth Plastics Co.
800 W Main St.
Freehold, NJ 07728
800-526-2820, 908-866-0200
Fax: 908-866-0274

Monsanto Co.
800 N. Lindbergh Blvd.
St. Louis, MO 63167
314-694-1000
Fax: 314-694-7625
Telex: 650-397-7820

Montell Polyolefins
Three Little Falls Centre
2801 Centerville Rd.
Wilmington, DE 19850-5439
800-666-8355, 302-996-6000
Fax: 302-996-5587

Morton International Inc.
Morton Plastics Additives
150 Andover St.
Danvers, MA 01923
508-774-3100
Fax: 508-750-9511

Network Polymers, Inc.
1353 Exeter Rd.
Akron, OH 44306
216-773-2700
Fax: 216-773-2799

Nova Polymers, Inc.
P.O. Box 8466
Evansville, IN 47716-8466
812-476-0339
Fax: 812-476-0592

Novacor Chemicals Inc.
690 Mechanic St.
Leominster, MA 01453
800-225-8063, 508-537-1111
Fax: 508-537-5685

Novacor Chemicals Inc.
Clear Performance Plastics
690 Mechanic St.
Leominster, MA 01453
800-243-4750, 508-537-1111
Fax: 508-537-6410

Novatec Plastics & Chemicals Co. Inc.
P.O. Box 597
275 Industrial Way W
Eatontown, NJ 07724
800-782-6682, 908-542-6600

Nylon Engineering
12800 University Dr.
Suite 275
Ft. Myers, FL 33907
813-482-1100
Fax: 813-482-4202

Occidental Chemical Corp.
5005 LBJ Freeway
Dallas, TX 75244
214-404-3800

Patent Plastics Inc.
638 Maryville Pike S.W.
P.O. Box 9246
Knoxville, TN 37920
800-340-7523, 615-573-5411

Paxon Polymer Co.
P.O. Box 53006
Baton Rouge, LA 70892
504-775-4330

Performance Polymers Inc.
803 Lancaster St.
Leominster, MA 01453
800-874-2992, 508-534-8000
Fax: 508-534-8590

Perstorp Compounds Inc.
238 Nonotuck St.
Florence, MA 01060
413-584-2472
Fax: 413-586-4089

Perstorp Xytec, Inc.
9350 47th Ave. S.W
P.O. Box 99057
Tacoma, WA 98499
206-582-0644
Fax: 206-588-5539

Phillips Chemical Co.
101 ARB Plastics Technical Center
Bartlesville, OK 74004
918-661-9845
Fax: 918-662-2929

Plaskolite, Inc.
P.O. Box 1497
Columbus, OH 43216
800-848-9124, 614-294-3281
Fax: 614-297-7287

Plaskon Electronic Materials, Inc.
100 Independence Mall West
Philadelphia, PA 19106
800-537-3350, 215-592-2081
Fax: 215-592-2295
Telex: 845-247

Plaslok Corp.
3155 Broadway
Buffalo, NY 14227
800-828-7913, 716-681-7755
Fax: 716-681-9142

Plastic Engineering & Technical Services, Inc.
2961 Bond
Rochester Hills, MI 48309
313-299-8200
Fax: 313-299-8206

Plastics Mfg. Co.
2700 S. Westmoreland St.
Dallas, TX 75223
214-330-8671
Fax: 214-337-7428

Polymer Resources Ltd.
656 New Britain Ave.
Farmington, CT 06032
800-423-5176, 203-678-9088
Fax: 203-678-9299

Polymers International Inc.
P.O. Box 18367
Spartanburg, SC 29318
803-579-2729
Fax: 803-579-4476

Polyply Composites Inc.
1540 Marion
Grand Haven, MI 49417
616-842-6330
Fax: 616-842-5320

PPG Industries, Inc.
Chemicals Group
One PPG Place
Pittsburgh, PA 15272
412-434-3131
Fax: 412-434-2891

Premix, Inc.
Rte. 20 & Harmon Rd.
P.O. Box 281
North Kingsville, OH 44068
216-224-2181
Fax: 216-224-2766

Prime Alliance, Inc.
1803 Hull Ave.
Des Moines, IA 50302
800-247-8038, 515-264-4110
Fax: 515-264-4100

Prime Plastics, Inc.
2950 S. First St.
Clinton, OH 44216
216-825-3451

Progressive Polymers, Inc.
P.O. Box 280
4545 N. Jackson
Jacksonville, TX 75766
800-426-4009, 903-586-0583
Fax: 903-586-4063

Quantum Composites, Inc.
4702 James Savage Rd.
Midland, MI 48642
800-462-9318, 517-496-2884
Fax: 517-496-2333

Reichhold Chemicals, Inc.
P.O. Box 13582
Research Triangle Park, NC 27709
800-448-3482, 919-990-7500 Fax: 919-990-7711

Reichhold Chemicals, Inc.
Emulsion Polymer Div.
2400 Ellis Rd.
Durham, NC 27703-5543
919-990-7500
Fax: 919-990-7711

Resinoid Engineering Corp.
P.O. Box 2264
Newark, OH 43055
614-928-6115
Fax: 614-929-3165

Resinoid Engineering Corp.
Materials Div.
7557 N. St. Louis Ave.
Skokie, IL 60076
708-673-1050
Fax: 708-673-2160

Rhone-Poulenc
Rte. 8
Rouseville Rd.
P.O. Box98
Oil City, PA 16301-0098
814-677-2028
Fax: 814-677-2936

Rimtec Corp.
1702 Beverly Rd.
Burlington, NJ 08016
800-272-0069, 609-387-0011
Fax: 609-387-0282

Rogers Corp.
One Technology Dr.
Rogers, CT 06263
203-774-9605
Fax: 203-774-9630

Rogers Corp.
Molding Materials Div.
Mill and Oakland Sts.
P.O. Box 550
Manchester, CT 06045
800-243-7158, 203-646-5500
Fax: 203-646-5503

Ronald Mark Associates, Inc.
P.O. Box 776
Hillside, NJ 07205
908-558-0011
Fax: 908-558-9366

Rostone Corp.
P.O. Box 7497
Lafayette, IN 47903
317-474-2421
Fax: 317-474-5870

Rotuba Extruders, Inc.
1401 Park Ave. S.
Linden, NJ 07036
908-486-1000
Fax: 908-486-0874

R.S.G. Polymers Corp.
P.O. Box 1677
Valrico, FL 33594
813-689-7558
Fax: 813-685-6685

RTP Co.
580 E. Front St.
P.O. Box 5439
Winona, MN 55987-0439
800-433-4787, 507-454-6900
Fax: 507-454-2041

Schulman, A., Inc.
3550 W Market St.
Akron, OH 44333
800-547-3746, 216-668-3751
Fax: 216-668-7204
Telex: SCHN 6874 22

Shell Chemical Co.
One Shell Plaza
Rm. 1671
Houston, TX 77002
713-241-6161

Shell Chemical Co.
Polyester Div.
4040 Embassy Pkwy
Suite 220
Akron, OH 44333
216-798-6400
Fax: 216-798-6400

Shintech Inc.
Weslayan Tower
24 Greenway Plaza
Suite 811
Houston, TX 77046
713-965-0713
Fax: 713-965-0629

Shuman Co.
3232 South Blvd.
Charlotte, NC 28209
704-525-9980
Fax: 704-525-0622

Solvay Polymers, Inc.
P.O. Box 27328
Houston, TX 77227-7328
800-231-6313, 713-525-4000
Fax: 713-522-2435

Sumitomo Plastics America, Inc.
900 Lafayette St.
Suite 510
Santa Clara, CA 95050-4967
408-243-8402
Fax: 408-243-8405

Synergistics Industries (NJ) Inc.
10 Ruckle Ave.
Farmingdale, NJ 07727
908-938-5980
Fax: 908-938-6933

Syracuse Plastics, Inc.
400 Clinton St.
Fayetteville, NY 13066
315-637-9881
Fax: 315-637-9260

Teknor Apex Co.
505 Central Ave.
Pawtucket, RI 02861
800-554-9892, 401-725-8000
Fax: 401-724-6250

Tetrafluor, Inc.
2051 E. Maple Ave.
El Segundo, CA 90245
310-322-8030
Fax: 310-640-0312

Texapol Corp.
177 Mikron Rd.
Lower Nazareth Comm. Park
Bethlehem, PA 18017
800-523-9242, 610-759-8222
Fax: 610-759-9460

Thermofil, Inc.
6150 Whitmore Lake Rd.
Brighton, MI 48116-1990
800-444-4408, 810-227-3500
Fax: 810-227-3824

Union Carbide Corp.
39 Old Ridgebury Rd.
Danbury, CT 06817-0001
800-335-8550, 203-794-5300

Vi-Chem Corp.
55 Cottage Grove St. S.W
Grand Rapids, MI 49507
800-477-8501, 616-247-8501
Fax: 616-247-8703

Vista Chemical Co.
900 Threadneedle
P.O. Box 19029
Houston, TX 77079
713-588-3000

Washington Penn Plastic Co.
2080 N. Main St.
Washington, PA 15301
412-228-1260
Fax: 412-228-0962

Wellman Extrusion
P.O. Box 130
Ripon, WI 54971-0130
800-398-7876, 414-748-7421
Fax: 414-748-6093

Westlake Polymers Corp.
2801 Post Oak Blvd.
Suite 600
Houston, TX 77056
800-545-9477, 713-960-9111
Fax: 713-960-8761

Appendix

D

Electrical Properties of Resins and Compounds

	Volume resistivity, Ω · cm	Dielectric strength, V/mil	Dielectric constant at 1 MHz	Dissipation factor at 1 MHz
ABS	10^{16}	425	2.6	0.007
Acetal copolymer	10^{15}	450	3.7	0.005
Acetal homopolymer	10^{15}	450	3.7	0.005
Acrylic	10^{18}	500	2.2	0.3
Alkyds				
Glass-filled	10^{15}	375	4.6	0.02
Mineral-filled	10^{14}	400	4.7	0.02
Allyls, glass-filled (DAP/DAIP)	10^{15}	400	3.5	0.01
Cellulose acetate	10^{13}	425	5.1	0.05
Chlorinated polyethylene	10^{16}	450	4.3	0.10
Elastomers				
COX	10^{15}	500	10.0	0.05
CR	10^{11}	700	8.0	0.03
CSM	10^{14}	700	8.0	0.07
EPDM	10^{16}	800	3.5	0.007
FPM	10^{13}	700	18.0	0.04
IIR	10^{17}	600	2.4	0.003
NR	10^{16}	800	3.0	0.003
SBR	10^{15}	800	3.5	0.003
SI	10^{15}	700	3.6	0.001
T	10^{12}	700	9.5	0.005
U	10^{12}	500	5.0	0.03
Epoxies				
Glass-filled	10^{16}	360	4.6	0.01
Mineral-filled	10^{16}	400	5.0	0.01
Fluoroplastics				
TFE	10^{18}	480	2.2	0.002
FEP	10^{18}	500	2.2	0.003
ETFE	10^{16}	490	1.7	0.007
ECTFE	10^{16}	490	1.7	0.007
Imides				
Polyamide imide	10^{17}	600	3.5	0.02
Polyetherimide	10^{17}	800	3.2	0.002
Polyimide	10^{16}	575	3.4	0.02

	Volume resistivity, Ω · cm	Dielectric strength, V/mil	Dielectric constant at 1 MHz	Dissipation factor at 1 MHz
Liquid-crystal polymer	10^{14}	510	3.9	0.006
Melamine	10^{11}	300	7.9	0.13
Nylon	10^{14}	385	3.6	0.04
Phenolics				
General-purpose	10^{13}	400	6.0	0.7
Glass-filled	10^{13}	350	6.0	0.8
Mineral-filled	10^{14}	400	6.0	0.10
Polyarylate	10^{16}	400	2.6	0.01
Polybutadiene	10^{16}	500	2.7	0.02
Polycarbonate	10^{15}	425	3.1	0.01
Polyesters				
PBT	10^{15}	425	3.1	0.03
PET	10^{15}	650	3.6	0.002
Thermoset	10^{14}	325	4.7	0.02
Polyetheretherketone	10^{16}	450	3.2	0.1
Polyethylene	10^{19}	475	2.3	0.0005
Polymethyl methacrylate	10^{17}	500	2.8	0.03
Polymethyl pentene	10^{16}	650	2.1	0.0001
Polyphenylene ether	10^{16}	500	2.9	0.001
Polyphenylene oxide	10^{17}	600	2.8	0.001
Polyphenylene sulfide	10^{16}	525	3.1	0.0007
Polypropylene	10^{17}	650	2.3	0.0004
Polystyrene	10^{16}	425	2.4	0.0004
Polyurethane	10^{13}	400	3.4	0.03
Polyvinyl chloride	10^{11}	400	6.0	0.1
Silicone	10^{16}	550	2.7	0.001
Sulfones				
Polyaryl	10^{16}	400	3.5	0.006
Polyether	10^{16}	400	3.5	0.006
Polyphenyl	10^{16}	375	3.4	0.007
Polysulfone	10^{16}	425	3.6	0.004

1. Since there are so many variations in possible plastic resin and compound formulations, the values given in this table are averages for each resin and compound type. *Each material supplier can provide the specific values for its individual products.*

2. These values are representative of values measured at room temperature, atmospheric pressure, and up to an electrical frequency of 1 MHz. Values for volume resistivity and dielectric strength decrease with increasing temperature, while values for dielectric constant and dissipation factor increase with increasing temperature. These changes are gradual (but increasingly significant) up to the glass-transition temperature of each material, at which temperature major changes occur. Increases in humidity and electrical frequency also have significant effects on electrical properties. Electrode configuration and material purity and homogeneity have significant effects on dielectric strength. (See Chap. 2 of the second edition of this handbook for changes in values for selected materials as a function of temperature, frequency, humidity, and other system operational parameters.)

3. Voltage breakdown and the associated dielectric strength decrease at decreasing pressures (and hence at increasing altitudes) according to Paschen's law (see Chap. 12 of the second edition of C. A. Harper, *Electronic Packaging and Interconnection Handbook*, McGraw-Hill, New York, 1991).

4. These property values are representative of values obtained in standard tests of the American Society for Testing and Materials (ASTM). The ASTM test methods for measuring these values are D-257 for volume resistivity, D-149 for dielectric strength, and D-150 for dielectric constant and for dissipation factor. Dissipation factor is also sometimes called power factor, loss tangent, or tan δ. Dielectric constant is also sometimes called permittivity. (See also Chap. 2 of the second edition of this handbook for further discussions.)

5. Due to the broad range of possible formulations for each class of materials, small differences in values shown in this appendix may not be significant. *For final design, supplier values for specific formulations and product application conditions should be used.*

Appendix E

Sources of Specifications and Standards for Plastics and Composites

As in other material categories, the reliable use of plastics and composites requires specifications to support procurement of these materials and standards to establish engineering and technical requirements for processes, procedures, practices, and methods for testing and using these materials. These specifications and standards may be general industry-wide, or they may be specific to one industry, such as the spaceborne industry, wherein plastics and composites must exhibit special stability in the harsh environments of temperature and vacuum. Specifications and standards for this industry, for instance, would be controlled by documents from the National Aeronautics and Space Administration (NASA). Increasingly, the development and use of standards is becoming international. Since the total sum of these specifications and standards is voluminous, no attempt is made herein to itemize all of them. However, the names and addresses of organizations that are sources of these documents, and information concerning them, are listed below. Also, one comprehensive reference book that can be recommended to readers with interest in this area is given as Ref. 1 at the end of this listing. Another excellent reference book dealing with the major issues of flammability standards of plastics and composites is listed as Ref. 2.

E.1 Names and Addresses of Organizational Sources of Specifications and Standards for Plastics and Composites

E.1.1 Industry Standards

American National Standards Institute (ANSI)
1819 L Street
Suite 600
Washington DC 20036

American Society for Testing and Materials (ASTM)
100 Barr Harbor Drive
West Conshohocken, PA 19428-2959

E.1.2 Federal Standards and Specifications

General Services Administration (GSA)
Seventh and D Streets, S.W.
Washington, DC 20407

Global Engineering Documents
(See E.1.1, Industry Standards)

American National Standards Institute
(See E.1.1, Industry Standards)

Document Engineering Company, Inc.
15210 Stagg St.
Van Nuys, CA 91405

E.1.3 Military Specifications and Standards

DODSSP—Customer Service
Standardization Document Order Desk
Building 4D
700 Robbins Ave.
Philadelphia, PA 19111-5090

Global Engineering Documents
(See E.1.1, Industry Standards)

American National Standards Institute
(See E.1.1, Industry Standards)

E.1.4 Spaceborne Plastics and Composites Specifications and Standards

For plastics

Individual NASA Space Centers

For composites

National Aviation and Space Administration (NASA)
ACEE Composites Project Office
Langley Research Center
Hampton, VA 23665-5225

E.1.5 Foreign or International Standards

American National Standards Institute
(See E.1.1, Industry Standards)

International Organization for Standardization (ISO)
1 Rue de Varembe
Case Postale 56
CH-1211 Geneva 20
Switzerland

International Standards and Law Information (ISLI)
P.O. Box 230
Accord, MA 02018

E.1.6 National and International Electrical Standards

National Electrical Manufacturers Association (NEMA)
1300 N. 17th St.
Suite 1847
Rosslyn, VA 22209

International Electrotechnical Commission (IEC)
3 Rue de Varembe
Case Postal 131
CH-1211 Geneva 20
Switzerland

American National Standards Institute
(See E.1.1, Industry Standards)

E.1.7 International Standards

ISO Central Secretariat
International Organization for Standardization (ISO)
1 Rue de Varembe
Case Postale 56
CH-1211 Geneva 20
Switzerland

E.1.8 Japanese Standards and Specifications

Japanese Standards Associations (JSA)
1–24, Akasaka 4, Minatoku
Tokyo 107-8440
Japan

E.1.9 German Standards

German National Standards Organization
Deutsche Industrie Norman (DIN)
10772 Berlin
Germany

E.1.10 European Aerospace Specifications and Standards

AECMA
Gulledelle 94-b.5
B-1200 Brussels
Belgium

Generally available in the U.S.A. from

American National Standards Institute
(See E.1.1, Industry Standards)

E.1.11 European Plastics Manufacturing Standards

Association of Plastics Manufacturers in Europe (APME)
Avenue E. van Nieuwenhuyse 4
Box 3
B-1160 Brussels
Belgium

E.1.12 Plastics Standards in United Kingdom

BSI Group Headquarters
389 Chiswick High Road
London W4 4AL, U.K.

BSI, Inc.
12110 Sunset Hills Road
Reston, VA 20190, USA

E.1.13 European Plastics

Association of Plastics Manufacturers in Europe
Avenue E. van Nieuwenhuyse 4
Box 3
B-1160 Brussels, Belgium

E.1.14 Rubber and Plastic Standards in United Kingdom

RAPRA Technology, Ltd.
Shawburuy, Shrewsbury
Shropshire SY4 4NR, U.K.

E.1.15 Plastics Standards in France

French Association for Standardization (AFNOR)
11, Avenue Francis se Pressenre
93571 Saint-Denis La Plaine Cedex
Paris, France

E.1.16 Specifications and Standards for Aerospace Products

Aerospace Industries Association of America, Inc. (AIA)
1250 I Street, N.W., Suite 1200
Washington, DC 20005-3924

E.1.17 National Standards for U.S. Standards Developing Bodies

American National Standards Institute (ANSI)
1819 L Street
Suite 600
Washington DC 20036

E.1.18 Standards for Testing

American Society for Testing and Materials (ASTM)
100 Barr Harbor Drive
West Conshohocken, PA 19428-2959

E.1.19 Specifications and Standards for the Plastics Industry

Society of the Plastics Industry (SPI)
1801 K Street N.W., Suite 600
Washington, DC 20006

E.1.20 Specifications and Standards for Composites Materials

Characterization

Composite Materials Characterization, Inc.
Attn: Cecil Schneider
Lockheed Martin Aeronautical Systems Company
Advanced Structures and Materials Division
Marietta, GA 30063

E.1.21 Standards in Measurements

National Institute of Standards and Technology (NIST)
Standard Reference Materials Program
100 Bureau Drive, Stop 2322
Gaithersburg, MD 20899-2322

E.1.22 Specifications, Standards, and Industry Approvals for Plastics and Testing of Plastics

Underwriters Laboratories, Inc., (UL)
Corporate Headquarters
333 Pfingston Road
Northbrook, IL 60062-2096

E.1.23 Standardization of Composite Materials Test Methods

Composite Fabricators Association
1655 N. Fort Myer Drive, Suite 510
Arlington, VA 22209

References

1. Traceski, F.T., *Specifications and Standards for Plastics and Composites,* ASM International, Materials Park, OH.
2. Hilado, C.J., *Flammability Handbook for Plastics,* 4th ed., Technomic Publishing Co., Inc., Lancaster, PA.

Index

B

C